# Introductory Chemistry

FOURTH EDITION

## An Active Learning Approach

# Introductory Chemistry

FOURTH EDITION

## An Active Learning Approach

**Mark S. Cracolice**
*The University of Montana*

**Edward I. Peters**

BROOKS/COLE
CENGAGE Learning™

Australia • Brazil • Japan • Korea • Mexico • Singapore • Spain • United Kingdom • United States

**Introductory Chemistry: An Active Learning Approach, Fourth Edition**
**Mark S. Cracolice, Edward I. Peters**

Publisher: Mary Finch
Senior Acquisitions Editor: Lisa Lockwood
Development Editor: Alyssa White
Assistant Editor: Ashley Summers
Editorial Assistant: Elizabeth Woods
Senior Media Manager: Lisa Weber
Marketing Manager: Nicole Hamm
Marketing Assistant: Kevin Caroll
Marketing Communications Manager: Linda Yip
Project Manager, Editorial Production: Teresa Trego
Creative Director: Rob Hugel
Art Director: John Walker
Print Buyer: Karen Hunt
Permissions Editor: Scott Bragg
Production Service: Lachina Publishing Services
Text Designer: Jeanne Calabrese
Photo Researcher: Sue Howard
Copy Editor: Lachina Publishing Services
Illustrator: Lachina Publishing Services
Cover Designer: Yvo Riezebos
Cover Image: Getty Images/Erik Isakson—soccer player, Getty Images/Steven Puetzer—ball in net
Compositor: Lachina Publishing Services



Library of Congress Control Number: 2008943179
ISBN-13: 978-0-495-55853-8
ISBN-10: 0-495-55853-2

Advantage Edition:
ISBN-13: 978-0-495-55854-5
ISBN-10: 0-495-55854-0

**Brooks/Cole**
10 Davis Drive
Belmont, CA 94001-3098
USA

Cengage Learning is a leading provider of customized learning solutions with office locations around the globe, including Singapore, the United Kingdom, Australia, Mexico, Brazil, and Japan. Locate your local office at: **international.cengage.com/region.**

Cengage Learning products are represented in Canada by Nelson Education, Ltd.

For your course and learning solutions, visit **www.cengage.com.**

Purchase any of our products at your local college store or at our preferred online store **www.ichapters.com.**

Printed in the United States of America
1 2 3 4 5 6 7 12 11 10 09

# Dedication

This book is dedicated to the memory of Edward I. Peters, who passed away in 2006. Although Ed was not physically a participant in the writing of this edition, I was still able to benefit from his guidance in spirit. Ed was my mentor, teacher, friend, and respected colleague. May his wisdom continue to live on and teach students through this textbook.

# CONTENTS

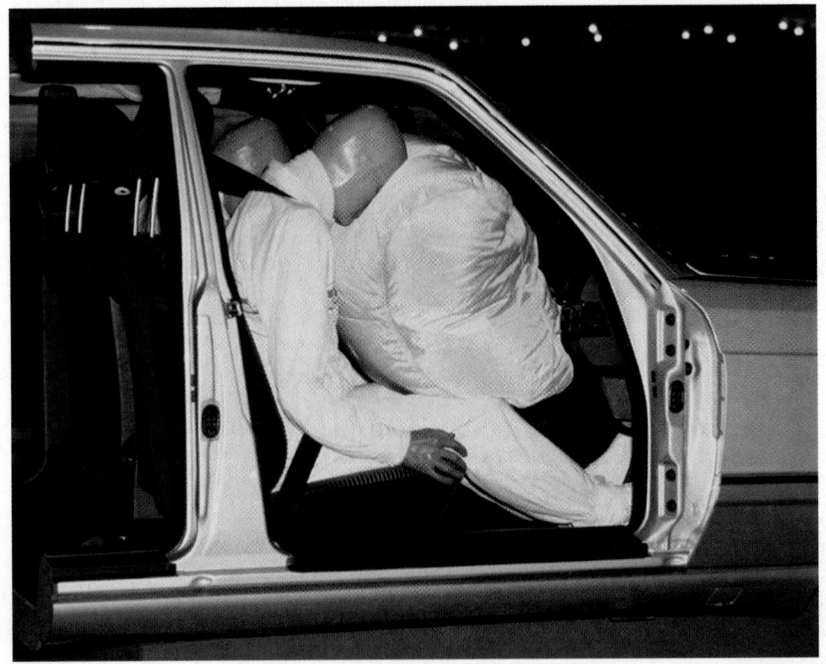
© Trip/Alamy

NASA

© Cengage Learning/Larry Cameron

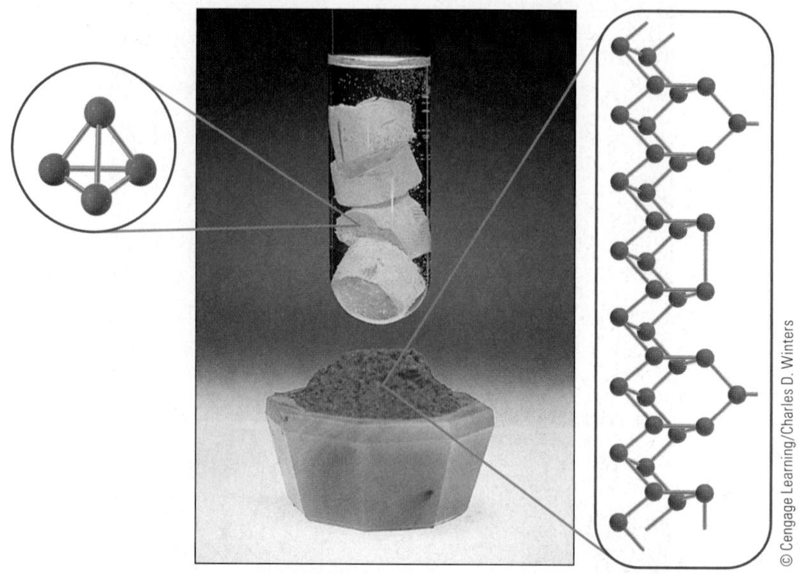

© Hans Pfletschinger/Peter Arnold, Inc.

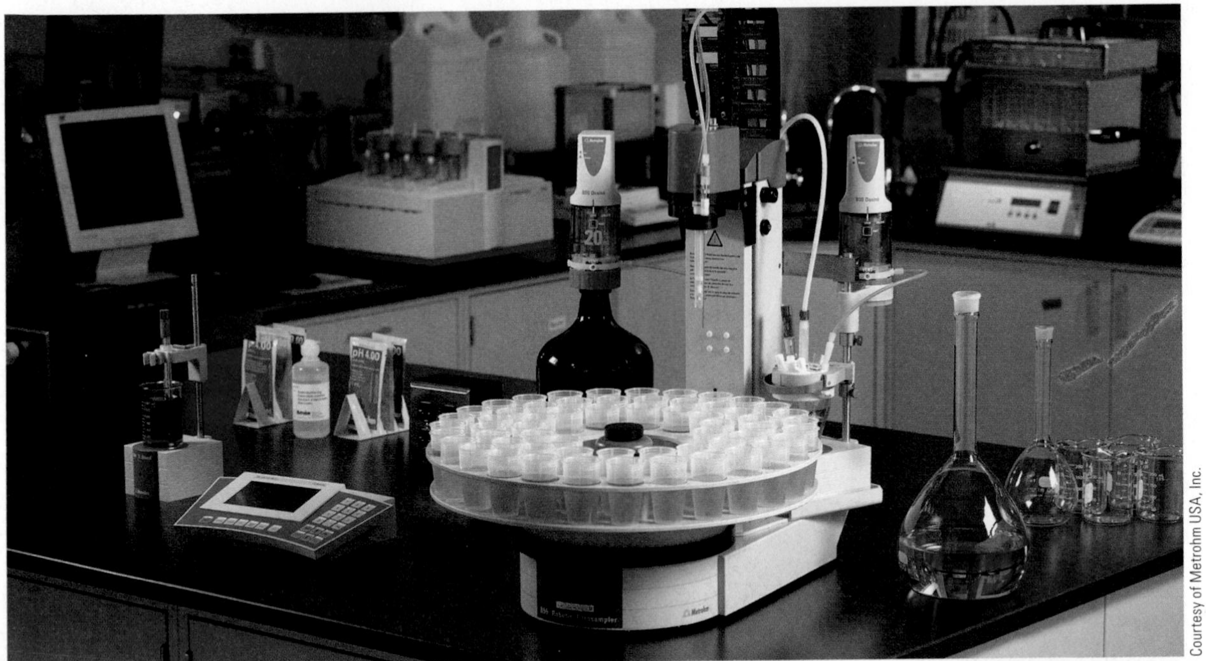
Courtesy of Metrohm USA, Inc.

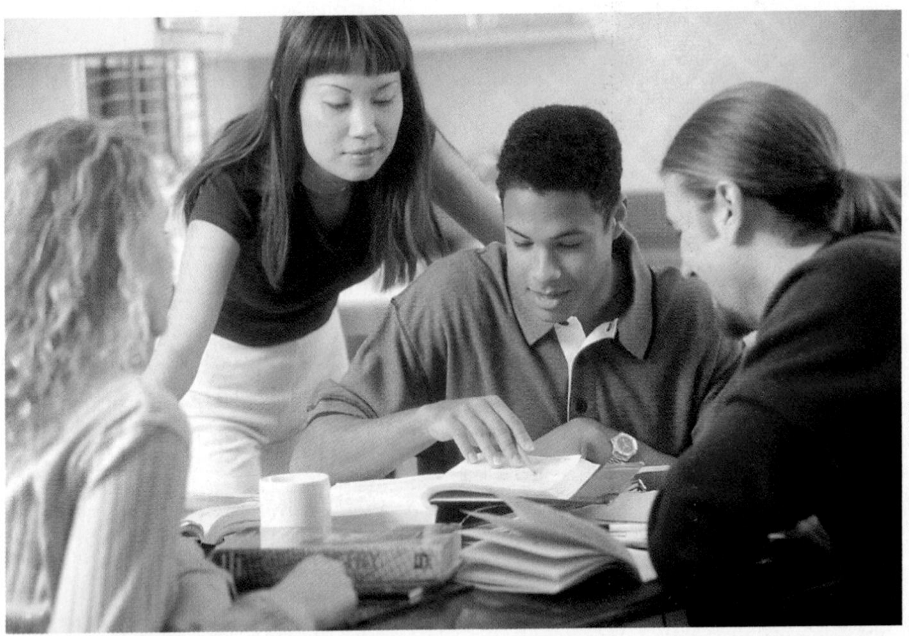
© John Henley Photography/CORBIS

# Preface

The fourth edition of *Introductory Chemistry: An Active Learning Approach* is a textbook for a college-level introductory or preparatory chemistry course for students who later will take a full-fledged general chemistry course and for the first-term general portion of a two-term general, organic, and biological chemistry course. It assumes that this is the student's first chemistry course or, if there has been a prior chemistry course that it has not adequately prepared the student for general chemistry.

The *Active Learning Approach* subtitle of this book refers in part to a question-and-answer presentation in which the student actively *learns* chemistry while studying an assignment, rather than studying now with the intent to learn later. A typical example leads students through a series of steps where they "listen" to the authors guide them to the solution, step by step. As the students solve the problem, they actively write each step, covering the answer with the shield provided in the book. This feature turns the common passive "read the authors' solution" approach to examples into an active "work the problem" approach while guided by the authors' methodology.

### Active Example 3.26

Calculate the mass of 7.04 $cm^3$ of silver.

At first glance, it might seem that you don't have enough information to solve this problem. When this situation occurs, start by assessing what you know from the problem statement. This will often help you discover what else you need to know to solve the problem. What are the GIVEN and WANTED for this problem?

GIVEN: 7.04 $cm^3$ WANTED: mass (assume g)

We assume that grams are the appropriate unit for mass in this problem because the gram is the basic metric unit for mass.

We also provide Target Check questions for students to answer while studying the qualitative material. Many sections have a few fundamental questions for students to complete while studying. This helps them to monitor their progress as they work instead of waiting for the end-of-chapter questions to discover incomplete understandings or misunderstandings.

√ **Target Check 5.4**

*Identify the true statements, and rewrite the false statements to make them true.*

a) The atomic number of an element is the number of protons in its nucleus.

b) All atoms of a specific element have the same number of protons.

c) The difference between isotopes of an element is a difference in the number of neutrons in the nucleus.

d) The mass number of an atom is always equal to or larger than the atomic number.

Active Figures provide presentations of topics that can be enhanced by such an interactive approach. Chapter 9 contains five Active Figures. As an example, Active Figure 9.12 illustrates the reaction of sodium chloride and silver nitrate solutions. Students go to the Introductory Chemistry Web site, where they will first see a photograph of the precipitate forming in solution. They then receive interactive feedback as they are guided through a series of six questions. As they reach the third question, they see particulate-level illustrations of the products of the chemical change. The final two questions link the macroscopic and particulate levels by having students focus on the symbolic representation of the reaction.

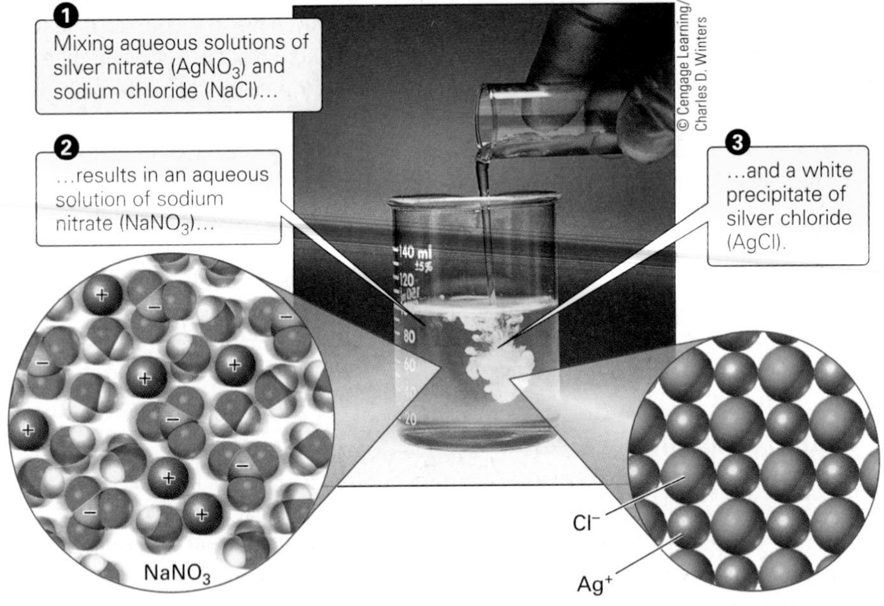
1
Mixing aqueous solutions of silver nitrate ($AgNO_3$) and sodium chloride (NaCl)...
2
...results in an aqueous solution of sodium nitrate ($NaNO_3$)...
3
...and a white precipitate of silver chloride (AgCl).
$NaNO_3$
$Cl^-$
$Ag^+$
© Cengage Learning/Charles D. Winters

**Active Figure 9.12** Sodium chloride and silver nitrate solutions are mixed, producing a white precipitate of silver chloride. Sodium ions and nitrate ions remain in solution. **Watch this active figure at** ***http://www.cengage.com/cracolice.***

## Major New Features in This Edition

**PORTABLE CONTENT CARDS** The most noticeable new feature in this edition is the inclusion of what we call Portable Content Cards. These heavy-stock, detachable cards are essentially the equivalent of a study guide for each chapter, condensed onto one card per chapter (some of the longer chapters are split into two cards). The card for each chapter has a summary list of the chapter goals in one column; these are correlated to a summary of the key concepts associated with each goal in the adjacent column. We encourage students to use these summaries as a preview to help organize their learning as they subsequently study the textbook. When students finish a chapter, the Portable

Content Card provides a chapter test, with answers, to allow students to assess progress toward learning the goals. Because the cards are detachable, students can use them to preview and review when it is inconvenient to carry the whole textbook.

**SMALL-GROUP DISCUSSION QUESTIONS** A growing number of courses feature some sort of groupwork formally integrated within the curriculum. We believe that the end-of-chapter questions typically used as homework are best for individual study, so we've written a new set of questions for each chapter that were designed with groupwork in mind. These questions are typically more conceptual, more challenging, and, potentially, more lengthy than the average end-of-chapter questions. We have tried to make the number of small-group discussion questions approximately proportional to the length of each chapter. We have not provided solutions to these questions in the hope of removing the temptation for students to give up too quickly and look at the solution as a method of learning how to answer the questions. Furthermore, omitting solutions to these questions more closely parallels the true nature of science, where the answer to a question about the natural world comes from interpreting high-quality investigations, not from checking the back of a textbook. We also think that there are plentiful answered questions at the end of each chapter for those instructors and students who want that type of feedback.

GOAL 11 **Distinguish between an empirical formula and a molecular formula.**

An **empirical formula** gives the simplest whole-number ratio of atoms of the elements in a compound. Empirical formulas are calculated from percentage composition data. They are also found from the mass of each element in a sample of a compound. Empirical formulas may or may not be the actual molecular formulas of compounds. The molar mass of the compound is needed to determine molecular formulas from simplest formulas.

GOAL 12 **Given data from which the mass of each element in a sample of a compound can be determined, find the empirical formula of the compound.**

To **find an empirical formula:**

1) Find the masses of different elements in a sample of the compound.
2) Convert the masses into moles of atoms of the different elements.
3) Determine the ratio of moles of atoms.
4) Express the moles of atoms as the smallest possible ratio of integers.
5) Write the empirical formula, using the number for each atom in the integer ratio as the subscript in the formula.

GOAL 13 **Given the molar mass and empirical formula of a compound, or information from which they can be found, determine the molecular formula of the compound.**

## Chapter 7 Test Yourself

Instructions: You may use a "clean" period

1) How many nitrogen atoms and how m
2) Write the formula of the compound co compound.
3) Identify the incorrect statement amon
   (a) A mole is that quantity of a substar
   (b) One mole of any substance contain
   (c) One mole of any substance contain of carbon-12.
   (d) A mole is a quantity of a substance
4) Identify the correct statement among t
   a) The molar mass of atoms of an ele
   (b) The molar mass is the mass in u of
   (c) Molar mass and atomic mass are a
   (d) Molar masses of compounds are al
5) Calculate the molar mass of potassium
6) What is the mass of $1.06 \times 10^{24}$ molecu
7) What mass of sodium nitrate contains
8) How many moles of potassium sulfate
9) Calculate the percentage composition
10) From the following, pick those that are
11) A compound has the percentage comp pound. The molar mass of this compou

ical formulas because they are not in "lowest

## Chapter in Review 7

GOAL 1 **Given the formula of a chemical compound (or a name from which the formula may be written), state the number of atoms of each element in the formula unit.**

A **chemical formula** tells how many atoms of each element are present in the formula unit of a substance. The number of each atom is given by a subscript following the symbol of that atom or a group of atoms. If the number is one, it is omitted in the formula.

GOAL 2 **Distinguish among atomic mass, molecular mass, and formula mass.**

**Atomic mass** is the average mass of all atoms of an element as they occur in nature. It is measured relative to the assignment of a mass of 12 u to an atom of carbon-12. **Molecular (or formula) mass** is the average mass of molecules (or formula units) compared with the mass of an atom of carbon-12, which is 12 atomic mass units.

GOAL 3 **Calculate the formula (molecular) mass of any compound whose formula is given (or known).**

The formula mass of a compound is equal to the sum of all of the atomic masses in the formula unit: **Formula mass = Σ atomic masses in the formula unit.**

GOAL 4 **Define the term *mole*. Identify the number of objects that corresponds to one mole.**

One **mole** of anything contains the same number of objects as the number of atoms in exactly 12 grams of carbon-12. This experimentally determined value is **Avogadro's number, $N_A$, $6.02 \times 10^{23}$.**

GOAL 5 **Given the number of moles (or units) in any sample, calculate the number of units (or moles) in the sample.**

Use dimensional analysis to **convert between moles and number of units:**

$$\text{\# of moles} \times \frac{6.02 \times 10^{23}\ \text{units}}{\text{mol}} = \text{\# of units}$$

$$\text{\# of units} \times \frac{1\ \text{mol}}{6.02 \times 10^{23}\ \text{units}} = \text{\# of moles}$$

GOAL 6 **Define *molar mass* or interpret statements in which the term molar mass is used.**

**Molar mass** is the mass in grams of one mole of a substance.

GOAL 7 **Calculate the molar mass of any substance whose chemical formula is given (or known).**

The molar mass of any substance in grams per mole is numerically equal to the atomic, molecular, or formula mass of that substance in atomic mass units.

GOAL 8 **Given any one of the following for a substance whose formula is given (or known), calculate the other two: (a) mass, (b) number of moles, (c) number of formula units, molecules, or atoms.**

*Molar mass* is the connecting link between the macroscopic world, in which we measure quantities in grams, and the particulate world, in which we count the number of units, usually grouped in moles. Using $N_A$ for Avogadro's number and MM for molar mass,

$$\text{units} \xrightarrow{N_A} \text{mol} \xrightarrow{MM} \text{g} \quad or \quad \text{g} \xrightarrow{MM} \text{mol} \xrightarrow{N_A} \text{units}$$

**Changing from formula units to mass or vice versa is a two-step dimensional analysis conversion.**

GOAL 9 **Calculate the percentage composition of any compound whose formula is given (or known).**

The **percentage composition** of a compound is the percentage by mass of each element in the compound. **Percent** is the amount of one part of a mixture per 100 total parts in the mixture. To calculate the percentage of each element,

$$\%\ \text{Element} = \frac{\text{total molar mass of element in compound}}{\text{Molar mass of compound}} \times 100$$

If you calculate the percentage composition of a compound correctly, the sum of all percents must be 100%.

GOAL 10 **Given the mass of a sample of any compound whose formula is given (or known), calculate the mass of any element in the sample; or, given the mass of any element in the sample, calculate the mass of the sample or the mass of any other element in the sample.**

To find the amount of any element in a known amount of compound, use percentage as a conversion factor, grams of the element PER 100 grams of the compound.

Portable Content Card for Cracolice's *Introductory Chemistry: An Active Learning Approach*


14. Consider the titration of 0.100 M sodium hydroxide into 10.00 mL of 0.100 M sulfuric acid. Sketch a buret and a flask. Illustrate the titration process by illustrating the buret and flask (a) before the titration starts, (b) half way to neutralizing the acid, and (c) at the completion of the titration. Next, calculate the molar concentration of the acid at points (b) and (c). Finally, construct particulate-level sketches of a tiny portion of the contents of the flask at each of the three points.

**ENHANCED ART AND PHOTOGRAPHY PROGRAM** In this edition, we continue to enhance the art and photography. A number of photographs of essentially the same image were reshot to provide better quality, and some new photographs were added. Art was redrawn for both improved technical accuracy and enhanced visible appeal.

**REVISED NARRATIVE** We revised the entire textbook with an emphasis on improving the writing quality, enhancing and tightening up the definitions, improving the clarity, and adding many other subtle changes that are aimed at improving readability. We continue to strive toward an improved, student-friendly narrative with an emphasis on using an active voice, and we continue to work to streamline the text to be as efficient as possible.

## Order of Coverage: A Flexible Format

Topics in a preparatory course or the general portion of a general–organic–biological chemistry course may be presented in several logical sequences, one of which is the order in which they appear in this textbook. However, it is common for individual instructors to prefer a different organization. *Introductory Chemistry* has been written to accommodate these different preferences by carefully writing each topic so that, regardless of when it is assigned, it never assumes knowledge of any concept that an instructor might reasonably choose to assign later in the course. If some prior information is needed at a given point, it may be woven into the text as a preview to the extent necessary to ensure continuity for students who have not seen it before, while affording a brief review for those who have. (See P/Review below.) At other times margin notes are used to supply the needed information. Occasionally, digressions in small print are inserted for the same purpose. There is also an Option feature that actually identifies the alternatives in topic sequence. In essence, we have made a conscious effort to be sure that all students have all the background they need for any topic whenever they reach it.

P/REVIEW This is what a P/Review looks like in this textbook. Information and a section reference are provided in the narrative or as a note in the margin.

*Introductory Chemistry* also offers choices in how some topics are presented. The most noticeable example of this is the coverage of gases, which is spread over two chapters. Chapter 4 introduces the topic through the P-V-T combined gas laws. This allows application of the problem-solving principles from Chapter 3 immediately after they are taught. Then the topic is picked up again in Chapter 14, which uses the Ideal Gas Law. An instructor is free to move Chapter 4 to immediately precede Chapter 14, should a single "chapter" on gases be preferred.

We have a two-chapter treatment of chemical reactivity with a qualitative emphasis, preceding the quantitative chapter on stoichiometry. Chapter 8 provides an introduction to chemical reactivity, with an emphasis on writing and balancing chemical equations and recognizing reaction types based on the nature of the equation. After students have become confident with the fundamentals, we then increase the level of sophistication of our presentation on chemical change by introducing solutions of ionic compounds and net ionic equations. Chapter 9 on chemical change in solution may be postponed to any point after Chapter 8. Chapter 8 alone provides a sufficient background in chemical equation writing and balancing to allow students to successfully understand stoichiometry, the topic of Chapter 10. You may wish to combine Chapter 9 with Chapter 16 on solutions.

On a smaller scale, there are alternative sections, where, again, instructors assign only the one they prefer. Finally, there are minor concepts that are commonly taught in different ways. These may be identified specifically in the book, or mentioned only briefly, but always with the same advice to the student: *Learn the method that is presented in lecture. If your instructor's method is different from anything in the book, learn it the way your instructor teaches it.* Our aim is to have the book support the classroom presentation, whatever it may be.

## The Book and Its Contents

*Introductory Chemistry* is written with the following broad-based goals for its users. On completing the course for which it is the text, the student will be able to do the following:

1. Read, write, and talk about chemistry, using a basic chemistry vocabulary;
2. Write routine chemical formulas and equations;
3. Set up and solve chemistry problems;
4. Think about chemistry on an atomic or molecular level in fundamental theoretical areas and visualize what happens in a chemical change.

To reach these goals, *Introductory Chemistry* helps students deal with three common problems: developing good learning skills, overcoming a weak background in mathematics, and overcoming difficulties in reading scientific material. The first problem is broached in Sections 1.4 through 1.6, which together make up an "introduction to active learning." These sections describe the pedagogical features of the text and how to use them effectively to learn chemistry in the least amount of time—that is, *efficiently*.

*Introductory Chemistry* deals with a weak background in mathematics in Chapter 3, "Measurement and Chemical Calculations." Dimensional analysis and algebra are presented as problem-solving methods that, between them, can be used for nearly all of the problems in the book. The thought processes introduced in Chapter 3 are used in examples throughout the text, constantly reinforcing the student's ability to solve chemistry problems. These thought processes are featured in the examples found in Chapter 3, as well as in the main body of the text.

We aim to help students overcome difficulties in reading scientific material by discussing chemistry in language that is simple, direct, and user-friendly. Maintaining the book's readability continues to be a primary focus in this edition. The book features relatively short sections and chapters to facilitate learning and to provide flexibility in ordering topics.

## Features

THINKING ABOUT YOUR THINKING Throughout *Introductory Chemistry* you will find inserts under the heading *Thinking About Your Thinking*. These features are comments to help students achieve a metacognitive level of understanding the thinking that is associated with the discipline of chemistry. Metacognition is the process of thinking about your thinking. In other words, the feature helps students think about more than just the content of the chemical concepts; it gives them a broader view of the thinking skills used in chemistry. These are written in an effort to encourage students to think as chemists commonly think. By focusing on the thinking skill itself, students can not only learn the context in which it is presented but also improve their competence with the more general skill. These broad thinking skills can then be applied to new contexts in their future chemistry courses, in other academic disciplines, and throughout their lives.

### Thinking About Your Thinking

Proportional Reasoning

**The idea of counting by weighing was introduced in the photograph and caption on the first page of this chapter. Many grocery-store items are sold by weight instead of number because of the inconvenience of counting small objects. Cookies, candies, nuts, and fruits, for example, could be sold by the piece, but instead they are usually sold by the pound in the United States.**

**Let's say that you decide to eat two dozen cherries a day for the next week. When you go to the grocery store, you will not find a price per cherry or per dozen cherries, but rather a price per pound of cherries. Even though you are thinking about the number of cherries, grouped in dozens, your grocer thinks about the number of pounds of cherries. You need to understand both methods for expressing the amount of cherries.**

**Similarly, in chemistry, you have to learn to think about both the number of particles, grouped in moles, and the mass of those particles. Considering moles of particles is useful when you are thinking about models of the particles and their reactivity. On the other hand, when you want a certain number of particles for a chemistry experiment, you will often weigh a specified amount of the macroscopic solid. So thinking about mass is important, too. You have to understand both methods for expressing amount of a substance.**

## Within the Chapter

**GOALS** Learning objectives, identified simply as Goals, appear at the beginning of the section in which each topic is introduced. They focus attention on what students are expected to learn or the skill they are expected to develop while studying the section.

### 6.6 Names and Formulas of Acid Anions

**Goal 6** Given the name (or formula) of an ion formed by the step-by-step ionization of a polyprotic acid from a Group 4A/14, 5A/15, or 6A/16 element, write its formula (or name).

**P/REVIEW** The flexible format of this book is designed so that any common sequence of topics will be supported. A cross reference is called a P/Review because it may refer to a topic already studied or one that is yet to be studied. P/Reviews feature concepts that the student will learn more about later in the course, or they may review concepts that have already been studied. Our aim is to provide a textbook that will work for your curriculum, as opposed to a book that dictates the curriculum design. You—the instructor—should be able to design your syllabus and then find the appropriate book, rather than vice versa. We therefore assume that the chapters will not necessarily be assigned in numerical order. The P/Reviews allow flexibility in chapter order.

**P/REVIEW:** The relationship between charges on monatomic ions and position in the periodic table is explained in Section 11.4. These charges arise when an atom loses or gains one, two, or three electrons to become isoelectronic with a noble-gas atom, that is, when the number and arrangement of electrons is the same as that found in a noble-gas (Group 8A/18) atom.

**SUMMARIES AND PROCEDURES** Clear in-chapter summaries and procedures appear throughout the text. These allow students to reflect on what they've just studied and help them learn how to learn chemistry.

**summary**

**The Nuclear Model of the Atom**

1. Every atom contains an extremely small, extremely dense nucleus.
2. All of the positive charge and nearly all of the mass of an atom are concentrated in the nucleus.
3. The nucleus is surrounded by a much larger volume of nearly empty space that makes up the rest of the atom.
4. The space outside the nucleus is very thinly populated by electrons, the total charge of which exactly balances the positive charge of the nucleus.

**TARGET CHECK** Target Check questions at the end of a section enable students to check their understanding of a topic immediately after studying the section. Although used primarily for concepts rather than problems, where immediate feedback usually appears in an example, some Target Checks are problems that vary slightly from the examples they follow.

## √ Target Check 2.9

*Identify the net electrostatic force (attraction, A; repulsion, R; or none, N) between the following pairs.*

a) two positively charged table tennis balls ______
b) a negatively charged piece of dust and a positively charged dust particle ______
c) a positively charged sodium ion and a positively charged potassium ion (ions are charged particles similar to atoms) ______

**EVERYDAY CHEMISTRY** All chapters have an Everyday Chemistry section that moves chemistry out of the textbook and classroom and into the daily experience of students. This feature gives students a concrete application of a principle within each chapter. We bring the excitement of applications of chemical principles to students in these essays, while simultaneously paying careful attention to the nature of a preparatory course.

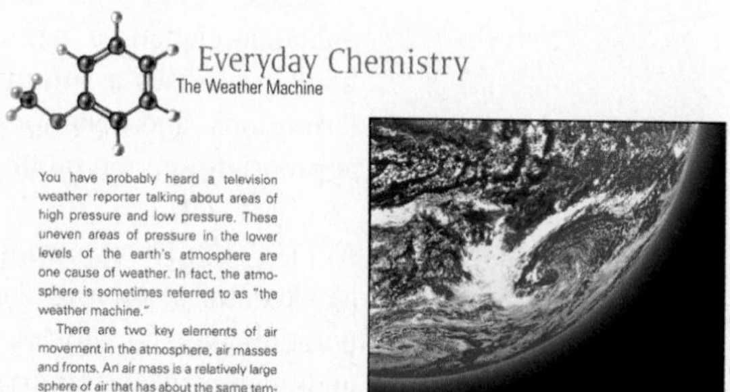


You have probably heard a television weather reporter talking about areas of high pressure and low pressure. These uneven areas of pressure in the lower levels of the earth's atmosphere are one cause of weather. In fact, the atmosphere is sometimes referred to as "the weather machine."

There are two key elements of air movement in the atmosphere, air masses and fronts. An air mass is a relatively large sphere of air that has about the same temperature and water content throughout for a given altitude. A typical cold air mass in the winter may stretch from Montana to Minnesota and from the Canadian border to the Mexican border. In contrast, a front is a small section of the lower atmosphere, usually 1/10 to 1/100 of the size of an air mass. Fronts are where the action is for changes in the weather. The temperature and humidity can vary greatly within a front, and the relatively stable air masses on either side of a front are often very different from one another.

When people refer to an area of high pressure, they are talking about an air mass that has a relatively high pressure when compared to surrounding air. This air, initially cold, descends toward the surface of the earth, compressing and warming. This inhibits cloud formation. High-pressure areas are therefore generally associated with clear weather. Conversely, air in a low-pressure mass rises and cools, which often leads to clouds, rain, or snow.

Weather is a very complex phenomenon, affected by numerous variables, such as the sun, the earth's tilt, the oceans and other bodies of water, and the variations in the land surface. Ultimately, however, each complex variable can be described in terms of simpler concepts such as temperature, pressure, and gas density. Weather experts are known as meteorologists, and their college preparation to understand the science of the atmosphere includes courses in chemistry, physics, geology, and mathematics, as well as advanced courses in meteorology itself. So the next time you hear a weather forecaster discussing high- and low-pressure systems, remember that the functioning of the weather machine is ultimately governed by the basic principles of the gaseous state of matter.

The earth's atmosphere, as seen from space.

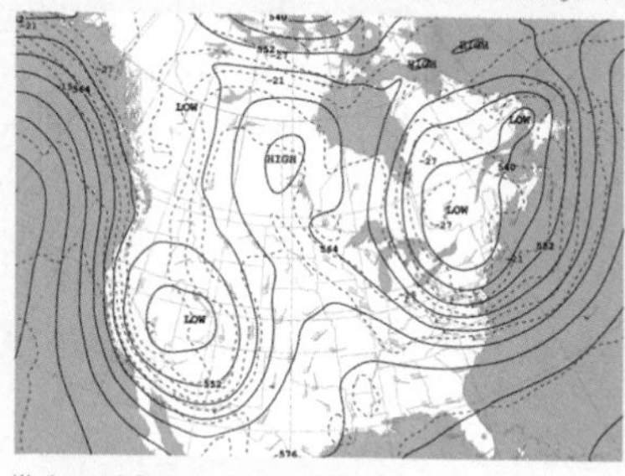


Weather maps indicate areas of high (H) and low (L) pressure.

NOAA Central Library Data Imaging Project

## At the End of the Chapter

**CHAPTER IN REVIEW** A second appearance of the Goals is in the Chapter in Review. Here the goals are restated as an all-in-one-place review of everything the students are to have learned or the skills they are to have acquired while studying the chapter. This review is a focal point for students while they study for a test.

The Chapter in Review includes a list of key terms and concepts. This list of words, names, and ideas introduced and/or used in each chapter is grouped by the section in which the term appears. This is another example of helping students learn how to learn. These lists can be used as a checklist of concepts to review for a quiz or exam.

**STUDY HINTS AND PITFALLS TO AVOID** This end-of-chapter feature has two main purposes: (1) to identify particularly important ideas and offer suggestions on how they can be mastered and (2) to alert students to some common mistakes so they can avoid making them.

**CONCEPT-LINKING EXERCISES** An isolated concept in chemistry often lacks meaning to students until they understand how that concept is related to other concepts. Concept-Linking Exercises ask students to write a brief description of the relationships among a small group of terms of phrases. If they can express those relationships correctly in their own words, they understand the concepts.

**SMALL GROUP DISCUSSION QUESTIONS** This set of questions is specifically designed for groupwork. Answers are not provided so that students must treat the questions like a real scientific problem.

**QUESTIONS, EXERCISES, AND PROBLEMS** Each chapter except Chapter 1 includes an abundant supply of questions, exercises, and problems arranged in three categories. In the first category, the questions are grouped according to sections in the chapter. The category called General Questions includes questions from any section in the chapter, and the final category is More Challenging Problems. Answers for all blue-numbered questions appear in Appendix III. Interactive versions of questions marked with a blue square are available in OWL (Online Web-Based Learning).

## Reference and Resource Materials

THE REFERENCE PAGES *Introductory Chemistry* includes two heavy-stock pages and information on the inside book covers that we refer to as the Reference Pages. One page is made up of tear-apart cards that may be used as shields to cover step-by-step answers while solving examples. One side of each card has a periodic table that gives the student ready access to all the information that table provides. The reverse side of each card contains instructions, taken from Chapter 3, on how to use it in solving examples.

One of the other Reference Pages includes a larger version of the Periodic Table and an alphabetical listing of the elements. The information on the inside covers of the book comprises a summary of nomenclature rules, selected numbers and constants, definitions, and equations, and a mini-index of important text topics, all keyed to the appropriate section number in the text.

APPENDIX The Appendix of *Introductory Chemistry* includes a section on how to use a calculator in solving chemistry problems; a general review of arithmetic, exponential notation, algebra, and logarithms as they are used in this book; and a section on SI units and the metric system. Appendix III provides complete solutions to the odd-numbered questions and many of the general questions and more challenging problems.

GLOSSARY An important feature for a preparatory chemistry course is a glossary. With each end-of-chapter summary of Key Terms and Concepts, we remind students to use their glossary regularly. The glossary provides definitions of many of the terms used in the textbook, and it is a convenient reference source to use to review vocabulary from past chapters.

## For the Instructor

### OWL (Online Web-Based Learning)

OWL Instant Access (1 Semester)
ISBN-10: 0-495-05100-4; ISBN-13: 978-0-495-05100-8

e-Book in OWL Instant Access (1 Semester)
ISBN-10: 0-495-55927-X; ISBN-13: 978-0-495-55927-6

Authored by Roberta Day, Beatrice Botch, and David Gross of the University of Massachusetts, Amherst; William Vining of the State University of New York at Oneonta; and Susan Young of Hartwick College. Developed at the University of Massachusetts, Amherst, and class tested by more than hundreds of thousands of students, OWL is a fully customizable and flexible Web-based learning system. OWL supports mastery learning and offers numerical, chemical, and contextual parameterization to produce thousands of problems correlated to this text. The OWL system also features a database of simulations, tutorials, and exercises, as well as end-of-chapter problems from the text. With OWL, you get the most widely used online learning system available for chemistry with unsurpassed reliability and dedicated training and support. For Cracolice's fourth edition, OWL includes parameterized end-of-chapter questions from the text (marked in the text with ■).

The optional **e-Book in OWL** includes the complete electronic version of the text, fully integrated and linked to OWL homework problems. Most e-Books in OWL are interactive and offer highlighting, notetaking, and bookmarking features that can all be saved.

A fee-based access code is required for OWL. OWL is only available to North American adopters. To view an OWL demo and for more information, visit **www.cengage.com/owl** or contact your Cengage Learning Brooks/Cole representative.

## PowerLecture with ExamView and JoinIn™ Instructor's CD-ROM

ISBN-10: 0-495-55852-4; ISBN-13: 978-0-49555852-1

PowerLecture is a dual platform, one-stop digital library and presentation tool that includes:

- Prepared Microsoft® PowerPoint® Lecture Slides, authored for this text by Mark Cracolice, covering all key points from the text in a convenient format that you can enhance with your own materials or with additional interactive video and animations from the CD-ROM for personalized, media-enhanced lectures.
- Image Libraries in PowerPoint and in JPEG format that provide electronic files for all text art, most photographs, and all numbered tables in the text. These files can be used to print transparencies or to create your own PowerPoint lectures.
- Electronic files for the Instructor's Manual and Test Bank.
- ExamView testing software, with all test items from the printed Test Bank in electronic format, enables you to create customized tests of up to 250 items in print or online.
- JoinIn™ clicker questions, written for this text by Walter K. Dean of Lawrence Technological University, for use with the classroom response system of your choice. Assess student progress with instant quizzes and polls, and display student answers seamlessly within the Microsoft PowerPoint slides of your own lectures and questions. Please consult your Brooks/Cole representative for more details.

## Online Instructor's Manual

ISBN-10: 0-495-55848-6; ISBN-13: 978-0-495-55848-4

The instructor's manual by Mark S. Cracolice and Edward I. Peters includes detailed information about the flexible format and suggestions for its use. It is available for download from the Faculty Companion Web Site at **www.cengage.com/chemistry/cracolice**.

## Test Bank

ISBN-10: 0-495-55849-4; ISBN-13: 978-0-495-55849-1

The Test Bank, revised by Mark Cracolice of The University of Montana, features more than 950 multiple-choice questions for instructors to use for tests, quizzes, or homework assignments. The Test Bank is available on the PowerLecture CD-ROM. BlackBoard and WebCT formatted files for the Test Bank are also available on the Faculty Companion Web Site at **www.cengage.com/chemistry/cracolice**.

## *Introductory Chemistry* Companion Web Site

The Faculty Companion Web Site at **www.cengage.com/chemistry/cracolice** contains the Online Instructor's Manual as well as WebCT and Blackboard versions of the Test Bank.

# For the Student

## OWL (Online Web-Based Learning)

See the above description in the "For the Instructor" section.

## *Introductory Chemistry* Companion Web Site

The Student Companion Web Site contains interactive versions of the Active Figures with questions to test your comprehension of the concepts presented in the media. A Molecular Modeling Database and interactive Periodic Table are also included.

### Laboratory Manual by Susan A. Weiner and Blaine Harrison

ISBN-10: 0-495-11479-0; ISBN-13: 978-0495-11479-6

This lab manual provides laboratory experiments suitable for an introductory chemistry course. The laboratory manual is accompanied by its own instructor's manual.

## Acknowledgments

The development of this textbook was an effort of many more people than the two names that appear on the cover. David Shinn of the University of Hawaii at Manoa checked for accuracy and suggested improvements. A team of reviewers analyzed the third edition via their responses to an extensive questionnaire, and they provided wide-ranging feedback that led to many of the improvements in this edition. We are sincerely appreciative to each member of this team:

K. Kenneth Caswell, University of South Florida

Claire Cohen-Schmidt, The University of Toledo

Mapi Cuevas, Santa Fe Community College

Coretta Fernandes, Lansing Community College

Carol J. Grimes, Golden West College

Rebecca Krystyniak, St. Cloud State University

James C. Morris, The University of Vermont

Felix N. Ngassa, Grand Valley State University

Linda Stevens, Grand Valley State University

Alyssa White was our developmental editor, and she had essential input into the vision for the fourth edition. In addition to commissioning the reviews, Alyssa is primarily responsible for the sticky tabs and the cover design, and she made significant contributions to the art and photography revisions. Teresa Trego was in charge of production, and her book design is outstanding. Katherine Wilson of Lachina Publishing Services coordinated the transition from the many separate pieces that make up a textbook to the beautiful final product you see before you. Our copy editor, Sara Black, continues to help us improve the quality of our writing, and we deeply appreciate her meticulous and detailed work. We would also like to thank Lisa Weber, our Media Editor, and Ashley Summers, Assistant Editor, for their work on the ancillary program as well as Liz Woods, Editorial Assistant, and Nicole Hamm, our Marketing Manager. Finally, we thank Lisa Lockwood, who coordinated the entire project from its inception to the final product, and beyond. We are very appreciative of Lisa's support of our passion for understanding research on human learning and bringing it into a pragmatic form as learning tools for chemistry students.

We are also grateful to the faculty and student users of the first, second, and third editions of *Introductory Chemistry*. Their comments and suggestions over the past twelve years have led to significant improvements in this book. We thank Melvin T. Arnold, Adams State College; Joe Asire, Cuesta College; Caroline Ayers, East Carolina University; Bob Blake, Texas Tech University; Juliette A. Bryson, Las Positas College; Sharmaine Cady, East Stroudsburg State College; Bill Cleaver, University of Vermont; Pam Coffin, University of Michigan–Flint; Jan Dekker, Reedley College; Michelle Driessen, University of Minnesota; Jerry A. Driscoll, University of Utah; Jeffrey Evans, University of Southern Mississippi; Donna G. Friedman, St. Louis Community College at Florissant Valley; Galen C. George, Santa Rosa Junior College; Alton Hassel, Baylor University; Randall W. Hicks, Michigan State University; Ling Huang, Sacramento City College; William Hunter, Illinois State University; Jeffrey A. Hurlburt, Metropolitan State College; C. Fredrick Jury, Collin County Community College; Jane V. Z. Krevor, California State University, San Francisco; Joseph Ledbetter, Contra

Costa College; Jerome Maas, Oakton Community College; Kenneth Miller, Milwaukee Area Technical College; Bobette D. Nourse, Chattanooga State Technical Community College; Brian J. Pankuch, Union County College; Erin W. Richter, University of Northern Iowa; Jan Simek, California Polytechnic State University, San Luis Obispo; John W. Singer, Alpena Community College; David A. Stanislawski, Chattanooga State Tech Community College; David Tanis, Grand Valley State University; Amy Waldman, El Camino College; Andrew Wells, Chabot College; Linda Wilson, Middle Tennessee State University; and David L. Zellmer, California State University, Fresno.

We continue to be very much interested in your opinions, comments, critiques, and suggestions about any feature or content in this book. Please feel free to write us directly or through Cengage, or contact us via e-mail.

Mark S. Cracolice
Department of Chemistry and Biochemistry
The University of Montana
Missoula, MT 59812
mark.cracolice@umontana.edu

# Introductory Chemistry

FOURTH EDITION

## An Active Learning Approach

© John Henley Photography/Corbis

1

CONTENTS



How many of these students are chemistry majors? Probably none of them. How many need chemistry for their major? All of them; that is why they are studying chemistry together. In fact, all educated members of society need to know the fundamentals of chemistry to understand the natural world. In this chapter, we will introduce you to the science and study of chemistry and all of the learning tools available to you, including this textbook.

# Introduction to Chemistry; Introduction to Active Learning

Welcome to your first college chemistry course! Chemistry is the gateway to careers in scientific research and human and animal health. You may be wondering why you, as a biology, pre-medicine, pharmacy, nursing, or engineering major—or as someone with any major other than chemistry—are required to take this course. The answer is that all matter is made up of molecules, and chemistry is the science that studies how molecules behave. If you need to understand matter, you need to know chemistry.

Key terms and definitions are indicated with **boldface print** throughout the textbook.

What lies before you is a fascinating new perspective on nature. You will learn to see the universe through the eyes of a chemist, as a place where you can think of all things large or small as being made up of extremely tiny **molecules**. Let's start by taking a brief tour of some of the amazing variety of molecules in our world.

First consider the simple hydrogen molecules in Figure 1.1(a). This shows you what you would see if you could take a molecular-level look at a cross section from a cylinder filled with pure hydrogen. The molecules are moving incredibly fast—more than 4,000 miles per hour when the gas is at room temperature! The individual molecule is two hydrogen atoms attached by the interaction between minute, oppositely charged particles within the molecule. Even though the hydrogen molecule is simple, it is the high-energy fuel that powers the sun and other stars. It is the ultimate source of most of the energy on earth. Hydrogen is found everywhere in the universe. It is part of many molecules in your body. Hydrogen is also the favorite molecule of theoretical chemists, who take advantage of its simplicity and use it to investigate the nature of molecules at the most fundamental level.

Now look at the DNA molecule (Fig. 1.1[b]). DNA is nature's way of storing instructions for the molecular makeup of living beings. At first glance, it seems complex, but on closer inspection, you can see a simple pattern that repeats to make up the larger

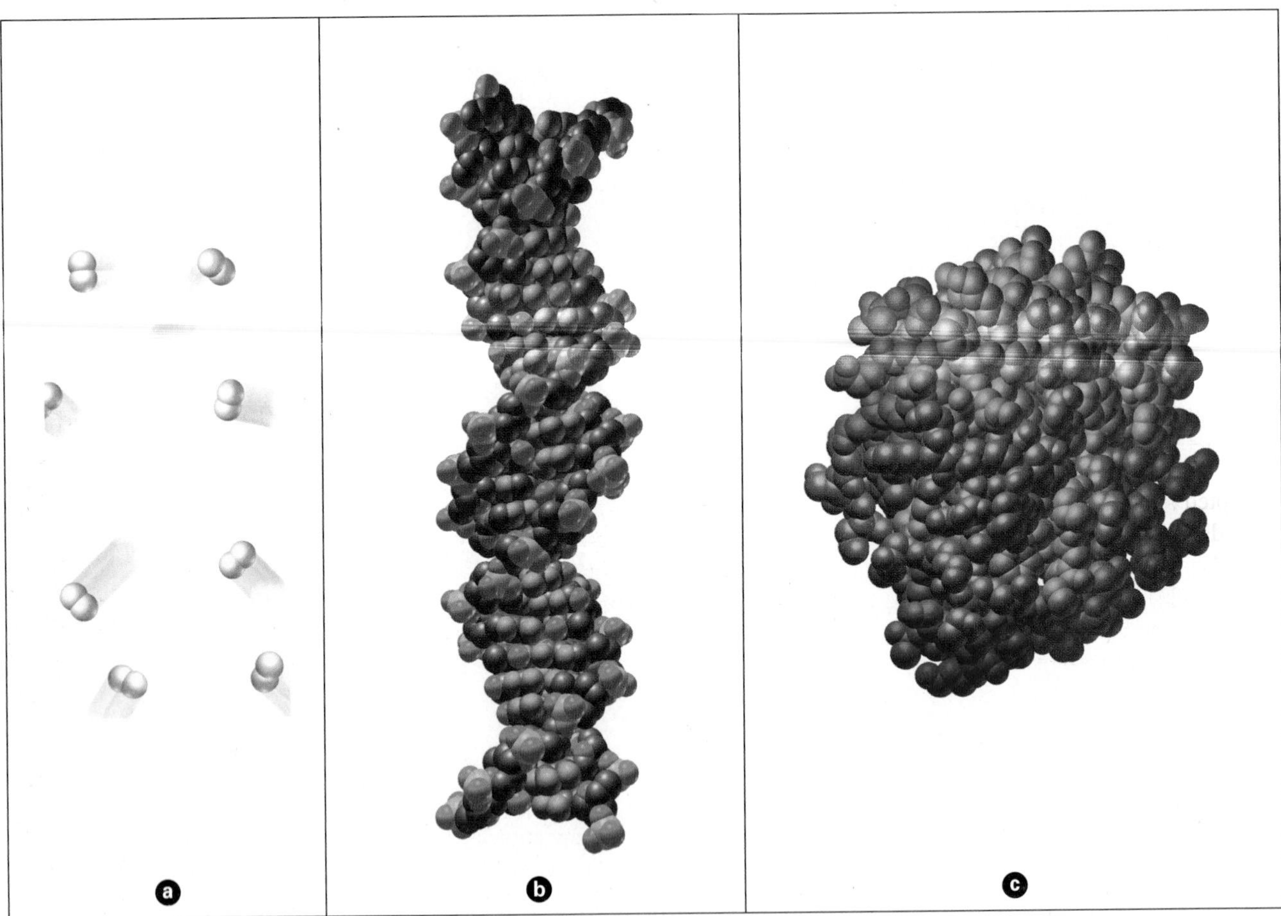


Figure 1.1 A survey of the amazing variety of molecules. (a) A molecular-level view of a tiny sample of pure hydrogen. Each hydrogen molecule is made up of two hydrogen atoms. Hydrogen is a gas (unless pressurized and cooled to a very low temperature), so the molecules are independent of one another and traveling at very high speeds. (b) A molecule of deoxyribonucleic acid, more commonly known as DNA. Notice how the molecule twists around a central axis. Also observe the repeating units of the pattern within the molecule. (c) The protein chymotrypsin, which is one of approximately 100,000 different types of protein molecules in the human body. The function of this molecule is to speed up chemical reactions.

molecule. This illustrates one of the mechanisms by which nature works—a simple pattern repeats many times to make up a larger structure. DNA stands for deoxyribonucleic acid, a compound name that identifies the simpler patterns within the molecule. Even this relatively large molecule is very, very tiny on the human scale. Five million DNA molecules can fit side-by-side across your smallest fingernail. (By the way, if you are a health or life sciences major, we think you'll agree that understanding the DNA molecule is a critical part of your education!)

Speaking of fingernails, they are made of the protein keratin. The human body contains about 100,000 different kinds of protein molecules. Some protein molecules in living organisms act to speed up chemical reactions. Figure 1.1(c) shows one such molecule, known as chymotrypsin. Proteins have many other essential biological functions, including being the primary components of skin, hair, and muscles, as well as serving as hormones.

Before you can truly understand the function of complex molecules such as DNA or proteins, you will have to understand and link together many fundamental concepts. This book and course are your first steps on the journey toward understanding the molecular nature of matter.

Now that you've had a look into the future of your chemistry studies, let's step briefly back to the past and consider the time when the science now called chemistry began.

# INTRODUCTION TO CHEMISTRY

## 1.1 Lavoisier and the Beginning of Experimental Chemistry

Antoine Lavoisier (1743–1794) is often referred to as the father of modern chemistry. His book *Traité Élémentaire de Chime,* published in 1789, marks the beginning of chemistry as we know it today, in the same way Darwin's *Origin of Species* forever changed the science of biology.

Antoine Lavoisier and his wife, Marie. They were married in 1771 when he was 28 and she was only 14. Marie was Antoine's laboratory assistant and secretary.



Lavoisier's experiments and theories revolutionized thinking that had been accepted since the time of the early Greeks. Throughout history, a simple observation defied explanation: When you burn a wooden log, all that remains is a small amount of ash. What happens to the rest of the log? Johann Becher (1635–1682) and Georg Stahl (1660–1734) proposed an answer to the question. They accounted for the "missing" weight of the log by saying that *phlogiston* was given off during burning. In essence, wood was made up of two things, phlogiston, which was lost in burning, and ash, which remained after. In general, Becher and Stahl proposed that *all* matter that had the ability to burn was able to do so because it contained phlogiston.

Lavoisier doubted the phlogiston theory. He knew that matter loses weight when it burns. He also knew that when a candle burns inside a sealed jar, the flame eventually goes out. The larger the jar, the longer it takes for the flame to disappear. How does the phlogiston theory account for these observable facts? If phlogiston is given off in burning, the air must absorb the phlogiston. Apparently a given amount of air can absorb only so much phlogiston. When that point is reached, the flame is extinguished. The more air that is available, the longer the flame burns.

So far, so good; no contradictions. Still, Lavoisier doubted. He tested the phlogiston theory with a new experiment. Instead of a piece of wood or a candle, he burned some phosphorus. Moreover, he burned it in a bottle that had a partially inflated balloon over its top (Fig. 1.2[a]). When the phosphorus burned, its ash appeared as smoke. The smoke was a finely divided powder, which Lavoisier collected and weighed. Curiously, the ash

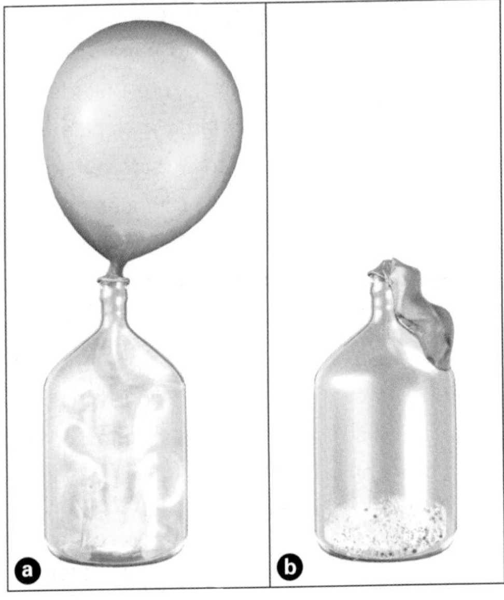

Figure 1.2 Lavoisier's phosphorus-burning experiment. (a) The sample of phosphorus inside the jar is burning. (b) a fine dust of white ash remains after burning. The balloon has collapsed.

weighed more than the original phosphorus. What's more, the balloon collapsed; there was less air in the jar and balloon after burning than before (Fig. 1.2[b]).

What happened to the phlogiston? What was the source of the additional weight? Why did the volume of air go down when it was supposed to be absorbing phlogiston? Is it possible that the phosphorus absorbed something from the air, instead of the air absorbing something (phlogiston) from the phosphorus? Whatever the explanation, something was very wrong with the theory of phlogiston.

Lavoisier needed new answers and new ideas. He sought them in the chemist's workshop, the laboratory. He devised a new experiment in which he burned liquid mercury in air. This formed a solid red substance (Fig. 1.3). The result resembled that of the phosphorus experiment. The red powder formed weighed more that the original mercury. Lavoisier then heated the red powder by itself. It decomposed, re-forming the original mercury and a gas. The gas turned out to be oxygen, which had been discovered and identified just a few years earlier.

These experiments—burning phosphorus and mercury, both in the presence of air and both resulting in an increase in weight—disproved the phlogiston theory. A new hypothesis took its place: When a substance burns, it combines with oxygen in the air. This hypothesis has been confirmed many times. It is now accepted as the correct explanation of the process known as burning.

But wait a moment. What about the ash left after a log burns? It does weigh less than the log. What happened to the lost weight? We'll leave that to you to think about for a while. You probably have a good idea about it already, but, also probably, you aren't really sure. If you were Lavoisier, and you wondered about the same thing, what would you have done? Another experiment, perhaps? We won't ask you to perform an experiment to find out what happens to the lost weight. We'll tell you—but not now. The answer is explained in Chapter 9.

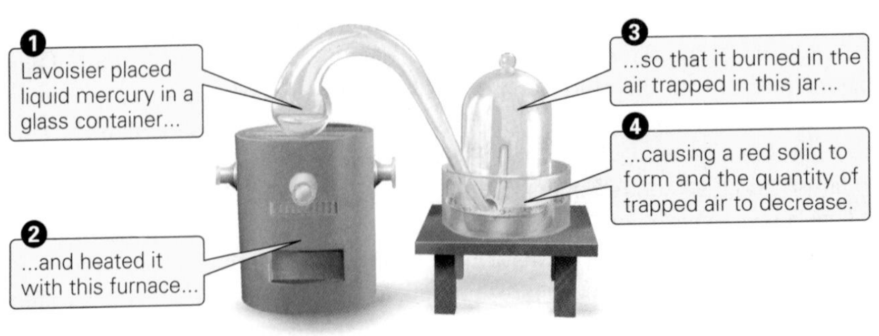
1
Lavoisier placed liquid mercury in a glass container...
2
...and heated it with this furnace...
3
...so that it burned in the air trapped in this jar...
4
...causing a red solid to form and the quantity of trapped air to decrease.
After J. Chem. Educ. Vol. 18, 1941, p. 85.

Figure 1.3 Lavoisier's apparatus for investigating the reaction of mercury and oxygen, as illustrated in his book *Traité Élémentarie de Chime*.

Before leaving Lavoisier, let's briefly visit a spinoff of his phosphorus experiment. Lavoisier was the first chemist to measure the weights of chemicals in a reaction. The concept of measuring weight may seem obvious to you today, but it was revolutionary in the 1700s. We have already noted that the phosphorus gained weight. The weight gained by the phosphorus was "exactly" the same as the weight lost by the air. "Exactly" is in quotation marks because the weighings were only as exact as Lavoisier's scales and balances were able to measure. As you will see in Chapter 3, no measurement can be said to be "exact." In Chapter 2, you will see the modern-day conclusion of Lavoisier's weight observations. It is commonly known as the Law of Conservation of Mass. It says that mass is neither gained nor lost in a chemical change.

## 1.2 Science and the Scientific Method

We have selected a few of Antoine Lavoisier's early experiments to illustrate what has become known as the **scientific method** (Fig. 1.4). Examining the history of physical and biological sciences reveals features that occur repeatedly. They show how science works, develops, and progresses. They include

1. Observing. A wooden log loses weight when it burns.
2. Proposing a hypothesis. A **hypothesis** is a tentative *explanation* for observations. The initial hypothesis posed by scientists before Lavoisier was that wood—and everything else—contains phlogiston. When something burns, it loses phlogiston.
3. Being skeptical. Lavoisier didn't go along with the idea that the loss in weight in a burning log could be explained by the departure of some never-seen and never-described substance called phlogiston.
4. Predicting an outcome that should result if the hypothesis is true. When phosphorus burns, it should lose weight.
5. Testing the prediction by an experiment. Lavoisier burned phosphorus. It gained weight instead of losing it. The new observation required . . .
6. Revising the hypothesis. Lavoisier proposed that burning combines the substance burned and oxygen from the air. (How did Lavoisier know about oxygen?)
7. Testing the new hypothesis and a new predicted outcome by an experiment. The new hypothesis was confirmed when Lavoisier burned mercury and it gained weight.
8. Upgrading the hypothesis to a theory by more experiments. Lavoisier and others performed many more experiments. (How did others get into the process?) All the experiments supported the explanation that burning involves combining with oxygen in the air. When a hypothesis is tested and confirmed by many experiments, without contradiction, it becomes a **theory** or **scientific model.**

The scientific method is not a rigid set of rules or procedures. When scientists get ideas, they most often try to determine if anyone else has had the same idea or perhaps has done some research on it. They do this by reading the many scientific journals in which researchers report the results of their work. Modern scientists communicate with each other through technical literature. Scientific periodicals are also a major source of new ideas, as well as talks and presentations at scientific professional meetings.

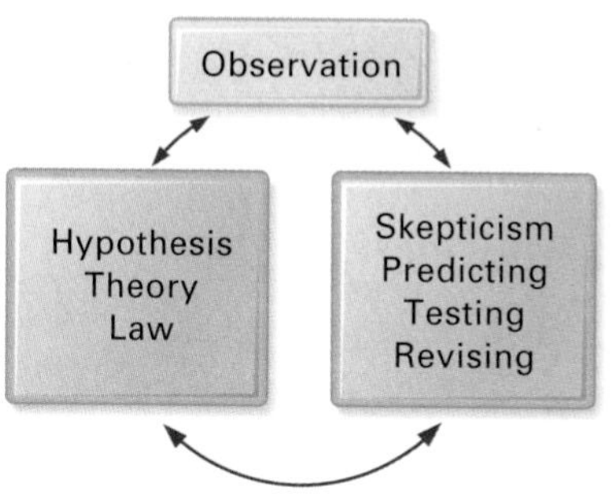


**Figure 1.4** The scientific method.

**Figure 1.5** One volume of the printed version of *Chemical Abstracts*, an index to scientific articles about chemical research and theory (yes, this is just the index!). Most libraries now search *Chemical Abstracts* by computer.

Communication is not usually included in the scientific method, but it should be. Lavoisier knew about oxygen because he read the published reports of Joseph Priestley and Carl Wilhelm Scheele, who discovered oxygen independently in the early 1770s. In turn, other scientists learned of Lavoisier's work and confirmed it with their own experiments. Today, communication is responsible for the explosive growth in scientific knowledge (Fig. 1.5). It is estimated that the total volume of published scientific literature in the world doubles every eight to ten years.

Another term used to describe science in a general way is *law*. In science, a **law** is a summary of repeated observations. Probably the best known is the law of gravity. If you release a rock above the surface of the earth, it will fall to the earth. No rock has ever "fallen" upward. In physics, the law of gravity is expressed precisely in a mathematical relationship.

A scientific law does not try to explain anything, as a hypothesis, theory, or scientific model might. A law simply states what is. Although laws cannot be proved, we do rely on them. The only justification for such faith is that in order for a law to be so classified, it must have no known exceptions. Water never runs uphill.

## 1.3 The Science of Chemistry Today

Chemists study matter and its changes from one substance to another by probing the smallest basic particles of matter to understand how these changes occur. Chemists also investigate energy gained or released in chemical change—heat, electrical, mechanical, and other forms of energy.

Chemistry has a unique, central position among the sciences (Fig. 1.6). It is so central that much research in chemistry today overlaps physics, biology, geology, and other sciences. You will frequently find both chemists and physicists, or chemists and biologists, working on the same research problems. Scientists often refer to themselves with compound words or phrases that include the suffix or word *chemist:* biochemist, geochemist, physical chemist, medicinal chemist, and so on.

Chemistry has five subdivisions: analytical, biological, organic, inorganic, and physical. Analytical chemistry studies what (qualitative analysis) and how much (quantitative analysis) are in a sample of matter. Biological chemistry—biochemistry—is concerned with living systems and is by far the most active area of chemical research today. Organic chemistry is the study of the properties and reactions of compounds that contain carbon. Inorganic chemistry is the study of all substances that are not organic. Physical chemistry examines the physics of chemical change.

You will find the people who practice chemistry—chemists—in many fields. Probably the chemists most familiar to you are those who teach and do chemical research in colleges and universities. Many industries employ chemists for research, product development, quality control, production supervision, sales, and other tasks. The petroleum industry is the largest single employer of chemists, but chemists are also highly visible in medicine, government, chemical manufacturing, the food industry, and mining.

Chemical manufacturers produce many things we buy and take for granted today. However, much of the chemical industry's output rarely reaches the public in raw form. Instead, it is used to make consumer products. Tables 1.1 and 1.2 list the top five inor-

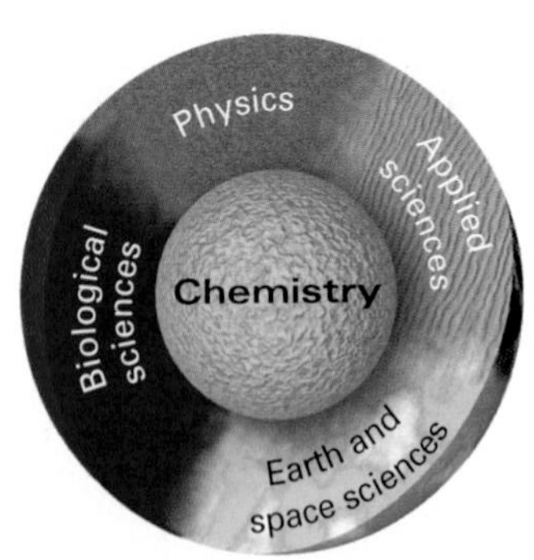


**Figure 1.6** Chemistry is the central science. Imagine all sciences as a sphere. This cross section of the science sphere shows chemistry at the core. If you view the other sciences as surface-to-center samples, each contains a chemistry core.

**Table 1.1** The Top Five Inorganic Chemicals, 2007

| Rank | Name | Production (thousands of tons) |
|---|---|---|
| 1 | Chlorine | 11,958 |
| 2 | Sodium hydroxide | 8,854 |
| 3 | Hydrochloric acid | 4,780 |
| 4 | Aluminum sulfate | 980 |
| 5 | Sodium hydrogen carbonate | 658 |

Source: U.S. Census Bureau

**Table 1.2** The Top Five Fertilizer Materials, 2007

| Rank | Name | Production (thousands of tons) |
|---|---|---|
| 1 | Sulfuric acid | 40,393 |
| 2 | Phosphoric acid | 12,128 |
| 3 | Ammonia | 11,846 |
| 4 | Nitric acid | 8,138 |
| 5 | Ammonium nitrate | 8,061 |

Source: U.S. Census Bureau

**Table 1.3** The Top Ten U.S. Chemical Producers, 2007

| Rank | Name | Sales ($ millions) |
|---|---|---|
| 1 | Dow Chemical | 53,513 |
| 2 | ExxonMobil | 36,826 |
| 3 | DuPont | 29,218 |
| 4 | Lyondell Chemical | 16,165 |
| 5 | Chevron Phillips | 12,534 |
| 6 | PPG Industries | 10,025 |
| 7 | Huntsman Corp. | 9,651 |
| 8 | Praxair | 9,402 |
| 9 | Air Products | 8,820 |
| 10 | Rohm and Haas | 7,837 |

Source: *Chemical and Engineering News* Vol. 86, No. 18, May 5, 2008.

ganic chemical and fertilizer materials produced in 2007. Most of the names may be familiar to you, but you cannot buy any of the pure chemicals at a local store.

You will see many of the chemicals in these tables repeated throughout this course. You will likely find some familiar company names among the top ten makers of industrial chemicals listed in Table 1.3.

# INTRODUCTION TO ACTIVE LEARNING

## 1.4 Learning How to Learn Chemistry

Here is your first chemistry "test" question:

Which of the following is your primary goal in this introductory chemistry course?

A. To learn all the chemistry that I can in the coming term.

B. To spend as little time as possible studying chemistry.

C. To get a good grade in chemistry.

D. All of the above.

If you answered A, you have the ideal motive for studying chemistry—and any other course for which you have the same goal. Nevertheless, this is not the best answer.

If you answered B, we have a simple suggestion: Drop the course. Mission accomplished.

© Jeff Greenberg Rainbow

Scientists at work.

If you answered C, you have acknowledged the greatest short-term motivator of many college students. Fortunately, most students have a more honorable purpose for taking a course, although sometimes they hide it quite well.

If you answered D, you have chosen the best answer.

Let's examine answers A, B, and C in reverse order.

C: There is nothing wrong in striving for a good grade in any course, just as long as it is not your major objective. A student who has developed a high level of skill in cramming for and taking tests can get a good grade even though he or she has not learned much. That helps the grade point average, but it can lead to trouble in the next course of a sequence, not to speak of the trouble it can cause when you graduate and aren't prepared for your career. It is better to regard a good grade as a reward earned for good work.

B: There is nothing wrong with spending "as little time as possible studying chemistry," as long as you *learn* the needed amount of chemistry in the time spent. Soon we'll show why the amount of time required to *learn* (not just study) chemistry depends on *when* you study *and* learn. They should occur simultaneously. Reducing the time required to complete any task satisfactorily is a worthy objective. It even has a name: *efficiency.*

A: There is nothing wrong with learning all the chemistry you can learn in the coming term, as long as it doesn't interfere with the rest of your schoolwork and the rest of your life. The more time you spend studying chemistry, the more you will learn. College is the last period in the lives of most people where the majority of your time can be devoted to intellectual development and the acquisition of knowledge, and you should take advantage of the opportunity. But maintain some balance. Mix some of answer B in your endeavor to learn. Again, the key is efficiency.

To summarize, the best goal for this chemistry course—and for all courses—is to learn as much as you can possibly learn in the smallest *reasonable* amount of time.

The rest of this section identifies choices that you need to make to ensure that you will reach your goal.

## Choice 1: Commit to Sufficient Time Outside of Class

A rule of thumb for college coursework is that an average student in an average course should spend two hours outside of class for every hour in class. Are you ready to *choose* to make this commitment? You may have to spend more time outside of class if your math skills are weak, if you have not recently had a good high school chemistry course, if English is not your native language, or if you have been out of school for some time. To keep your out-of-class time to an efficient minimum, you must study regularly, doing each assignment before the next class meeting. Chemistry builds on itself. If you don't complete today's assignment before the next class meeting, you will not be ready to learn the new material. Many successful students schedule regular study time, just as they would schedule a class. *Failure to commit sufficient time outside of class is the biggest problem when it comes to learning chemistry.*

## Choice 2: Commit to Quality Time When Studying

Efficient learning means learning at the time you are studying. It does not mean just reading your notes or the book and deciding to come back and learn the material later. It takes longer to *learn now* than it does to passively read the textbook, but the payoff comes with all the time you save by not having to learn later. This is so important that we have special *Learn It Now!* reminders throughout the textbook. Are you ready to *choose* to commit to making your study time high quality? If so, you should also commit to studying without distractions—without sounds, sights, people, or thoughts that take your attention away from learning. Every minute your mind wanders while you study must be added to your total study time. Your time is limited, and that wasted minute is lost forever.

## Choice 3: Commit to Utilizing All Learning Resources

College chemistry courses typically have a multitude of learning resources, which may include lecture, this textbook and its accompanying online learning tools, laboratory, discussion sections, help centers, tutors, instructor office hours, and your school library. Are you ready to *choose* to commit to taking advantage of all of the learning tools provided in your course? Let's consider some of these tools in more detail.

**Lecture** Although it is obviously the wrong way to learn, some students choose to skip lecture occasionally. Don't be one of those students. *Attend every lecture.* If you miss just one lecture per month in a semester course, you will probably miss 10% of the material. That is a reduction of one letter grade worth of content in a typical course. You need to learn the role of lecture in your course. If your instructor expects you to listen to his or her discussion and watch presentation slides and/or material written on the board or an overhead projector, you will need to take notes. We recommend that your note-taking procedure follow these general steps: (1) Preview the material by skimming the textbook. Usually, this only needs to be done every few lectures as a new chapter is about to be introduced. Look in particular for new words and the major concepts so that you are not caught unprepared when they are introduced in lecture. (2) Concentrate during lecture and take notes. Don't fool yourself; concentrating over an extended period of time is hard work. Focus on what is being shown and said, and work to transcribe as much material as accurately and quickly as you can. Use a notebook that is exclusively for chemistry lecture. (3) Organize your notes as soon as possible after lecture. Organization is the key. During a classic lecture, you often are mostly working to transcribe the material. True learning occurs when you work to make sense of the material and try to analyze the relationships among the concepts that were discussed. (4) Study the textbook, work the assigned problems, and look for connections between the lecture and the textbook. You will often find that seeing the material presented in a slightly different way is the key to helping you make sense of a concept. Combining your organized lecture notes with the textbook presentation of the same topic is a powerful learning technique.

**Textbook** This book is a central learning resource in your chemistry course. We will help you to become familiar with its structure in the next section.

**OWL** OWL, our **O**nline **W**eb-based **L**earning resource, is a homework system and assessment and feedback tool. Using a question-creation format that varies the amount and type of chemical substance for each online session, OWL can generate more than 100,000 chemistry questions correlated to the book. Instant feedback helps you immediately assess your progress. *(OWL is available for use only within North America.)*

**Student Companion Website** This book's companion Web site at ***http://www.cengage.com/cracolice*** contains Active Figures and other materials to help you learn in ways that a printed textbook cannot.

**Laboratory** If your course includes a laboratory, learn what each experiment is designed to teach. Relate the experiment to the lecture and textbook coverage of the same topic. Seeing something in the laboratory and getting a hands-on experience is often just what you need to fully understand what you read in the textbook and see and hear in the lecture.

**Instructor Office Hours** Many chemistry instructors are available for help outside of class. If your instructor is not, you likely have a teaching assistant with office hours or a tutoring center that you can visit instead. No matter the quality of instructional resources available to you, human help is occasionally needed to accomplish your learning goals. We recommend that you develop a list of questions and/or sample problems that you cannot solve before you attend office hours.

**Active Figure 1.7** How to think like a chemist. You are familiar with the macroscopic view of matter, as seen in this container filled with boiling water. A key characteristic of thinking like a chemist is imagining how the water would appear if you could see it at the particulate level. The particulate circle shows how a chemist views water. To express this viewpoint in writing, chemists use symbols. The symbols in the formula $H_2O$ describe the particulate-level composition of each water molecule. **Watch this active figure at *http://www.cengage.com/cracolice.***

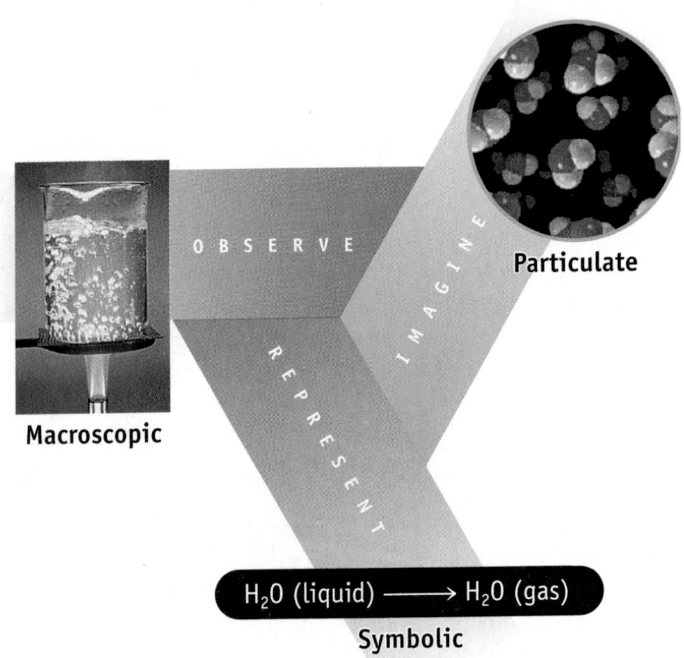


**Library or Learning Center** Many college libraries and learning centers have Internet resources, computer programs, audiotapes, workbooks, and other learning aids that are helpful for practice with using chemical formulas, balancing equations, solving problems, and other routine skills. Find out what is available for your course and use it as needed. Some instructors will also put supplementary materials on reserve. Take advantage of these, if provided.

## Choice 4: Commit to Improvement

By definition, you are changed as a result of learning. You need to be willing to open your mind to new, more powerful ways of thinking about the natural world and the process of personal intellectual development. The purpose of your college education is to make you a better person. Are you willing to *choose* to commit to improving the way you understand nature, becoming a better learner, and developing your intellect? Let's look as some ways to do this within the framework of this chemistry course.

**Think Like a Chemist** The perspective of the chemist is unique, as is the perspective of the philosopher, the mathematician, the geographer, or the linguist. Each course you take in college will expose you to a different way of thinking about the world. In this chemistry course, you should work to understand the distinctive viewpoint of a chemist. In particular, focus on the relationships among the macroscopic, directly observable natural world, the abstract, particulate makeup of those macroscopic materials, and the symbols that chemists use to represent both the macroscopic and particulate world, as illustrated in Active Figure 1.7.

**Think Conceptually** A trap that some students fall into while solving quantitative chemistry problems is to mindlessly crunch numbers without thinking about the underlying concept. Almost certainly, there will be a few routine types of quantitative problem setups that you should master without the need to reinvent the procedure each time you solve such a problem. But many other problems will be more complex. With these more complex problems, it is critical to understand the underlying concept. If you can imagine the particulate-level process described in the problem statement, do so. Remember that it is not the answer that is important when you tackle difficult problems, but rather the process that should be your focus.

**Embrace Multiple Ways of Knowing** This chemistry course will expose you to many ways of obtaining new knowledge. You will likely need to learn, in order of increasing complexity, facts, rules, concepts, and problem solving. Facts are things that you need to memorize, such as the fact that the symbol for hydrogen is H. Rules are connections between things, and they are often expressed as mathematical relationships. For example, the volume of a pure substance is directly proportional to its mass, which can be expressed in symbols as $V \propto m$. Rules also are often expressed in the form of if/then statements. *If* an element forms a monatomic anion, *then* the name of the anion is the name of the element, changed to end in *-ide*. Concepts are mental models of the natural world. We will present relatively simple conceptual models in this introductory course, and as you learn more and more about chemistry in future courses, you will find that you will need to revise and increase the complexity of your conceptual models. Problem solving is a skill that you learn through coaching and practice. Good problem solvers are highly regarded in all aspects of professional life. We will help guide you in developing your problem-solving skills in this textbook, but you will also need to put in a good deal of practice time to become a skilled problem solver. You will likely have your favorite type of learning, and that will probably shape your decision about your major, and ultimately, your career path, but recognize that each mode of learning has its importance in your education. Embrace the opportunity to become a more skilled learner in each kind of way of knowing.

**Think About Your Thinking** It is important not only to learn chemistry content while in this course but also to work to develop the thinking skills that are used by chemists. An example of a thinking skill is proportional reasoning, where you recognize and apply relationships between two variables that are directly proportional to one another. If you learn to see these types of relationships beyond their immediate application, you will be able to utilize these skills in solving problems in many other contexts. We will discuss this further in the next section.

**Utilize Feedback in a Positive Manner** All courses will provide you with feedback on your performance in some way. Typically, courses have exams and/or quizzes that assess your learning. This textbook has many end-of-chapter questions, exercises, and problems that are accompanied by solutions at the end of the book. You can choose to use such feedback as merely a descriptor of your learning history, such as "I earned an 80 on the gases chapter test," or "I got that problem wrong," or you can use the feedback in a positive manner by thinking, "What did I do wrong, and how can I improve?" A critical element of the process of learning is to learn from your mistakes. When you receive a corrected exam or quiz, look at your errors and make a commitment to change your thinking so that you don't repeat the same error. When you solve an end-of-chapter problem incorrectly, assess what you did wrong and restudy the appropriate material so that you can replace the misconception with a more accurate understanding of the concept or procedure.

## 1.5 Your Textbook

The most important tool in most college courses is the textbook. It is worth taking a few minutes to examine the book and look for its unique learning aids. In this section we'll show you the features of the book that are designed specifically to help you learn chemistry.

## Section-by-Section Goals

Goal 1 Read, write, and discuss chemistry using a fundamental and scientifically accurate and precise chemical vocabulary.

Goal 2 Write a fundamental set of inorganic chemical formulas and write names of substances when their formulas are given.

Goal 3 Write, balance, and interpret chemical equations.

Goal 4 Set up and solve elementary chemical problems.

Goal 5 "Think" chemistry in some of the relatively simple theoretical areas and visualize what happens at the particulate level.

Goal 6 Improve your scientific thinking skills, particularly in proportional reasoning and mental modeling.

The goals listed here are not for a section, but for this entire book and the course in which you will use it. They tell you what you will be able to do when you complete the course.

As you approach most sections in this text, you will find one or more goals. They tell you what you should be able to do after you study the section. If you focus your attention on learning what is in the goals, you will learn more in less time. All the goals in a chapter are assembled as a Chapter in Review section at the end of the chapter, which also includes the key terms and concepts associated with those goals.

Few general chemistry textbooks include section-by-section goals, although they sometimes appear in study guides that accompany those books. When you move on to the general chemistry course, it becomes your responsibility to write the goals yourself—to figure out what understanding or ability you are expected to gain in your study. Literally writing your own goals is an excellent way to prepare for an exam.

## *Learn It Now!*

In the previous section, *Choice 2: Commit to Quality Time When Studying* discussed the importance of learning efficiently. We noted that we would provide you with *Learn It Now!* reminders throughout the textbook, printed in red. Most are printed in the margin, but longer entries run the full width of the page.

☞ ***Learn it Now!*** This is what a *Learn It Now!* entry looks like when it is printed in the margin.

When you come to a *Learn It Now!* entry, stop. Do what it says to do. Think about it. Make a conscious effort to understand, learn, and, if necessary, memorize what is being presented. When you are satisfied that this idea is firmly fixed in your mind, then continue on. In short, learn it—*now!* Tomorrow it will take longer. Tomorrow is too late.

## Active Examples

As you study this book, you will acquire certain "chemical skills." These include writing chemical names and formulas, writing and interpreting chemical equations, and solving chemical problems—the things listed previously as Goals 2, 3, and 4. You will develop these skills by studying and actively working the examples in the text.

### Active Example 1.1

This is not an active example, but this sentence and the following paragraph are written in the form of the active examples throughout the book.

All active examples begin with the words "Active Example," followed by the example number, printed in red. The end of the active example is signaled by the end of the green background.

Most active examples in this book take you through a series of questions and answers. Space is provided for you to actually write in a formula or equation or to solve a problem yourself. A small drawing of a pencil appears next to each space in which you should be writing answers. One such example appears in Section 3.2. At that point you will find detailed instructions for working this kind of example problem.

If you are to learn from examples, you need to work through each one as you come to it. Never postpone an example and read ahead. *Learn it now!* Quite often, what you learn in an example is used immediately in the next section. You will not be able to understand that next section without understanding the example.

## Target Checks

Chemical principles, models, and theories are introduced with words and illustrations. Ideally, you will learn and understand these ideas as you study the text and figures. Use a Target Check to find out if you have caught on to the main ideas immediately after they appear in the text. A Target Check is identified by the heading

### √ Target Check

Like Active Examples, Target Checks should be completed as you reach them. Answers to Target Checks can be found in Appendix III. If you answer a Target Check incorrectly, go back and restudy the targeted material before moving on to the next section or example.

The best thing that you can do to maximize your learning from the textbook is to take written notes, work the Active Examples, and complete the Target Checks *while you study*. This means writing in the book and not simply highlighting the textbook. A heavily highlighted textbook is nothing more than a brightly colored list of things you plan to learn later. It is just the opposite of the *Learn It Now!* philosophy. The very act of reading something, thinking about what it means, summarizing it, and writing it down in your own words produces learning—*now!*

## P/Review

Often in the study of chemistry you see some term or concept that was introduced earlier in the course. To understand the idea in its new context, you may wish to review it as presented earlier. At other times a topic is introduced briefly, to meet a present need, even though it may not be necessary to understand it fully. That comes later; the present introduction is a preview.

**P/REVIEW:** This is what a P/Review looks like in the margin.

This book has an optional order of topics, so the same item may be a preview with one instructor and a review with another. We therefore identify this kind of cross-reference as a P/Review. Each P/Review is carefully worded so that as a review it gives you the information you must recall immediately, but not so much that it will be confusing if read as a preview. A P/Review usually appears in the margin, and it always includes a specific chapter or section number that you may refer to if you wish.

## In-Chapter Summaries and Procedures

Throughout this book, you will find summaries and step-by-step procedures that are headed as follows:

**summary**

**Format of Summary and Procedure Boxes**

Each summary or procedure is printed inside a box, as this paragraph is printed. These give you, in relatively few words, the main ideas and methods you should learn from a more general discussion nearby. They should help you clinch your understanding of the topic. Occasionally, summaries are in the form of a table or illustration; some even combine the two. These forms are particularly helpful in reviewing for a test. Not only do they review the topic briefly, they also create a mental image that is easy to recall during an exam.

## Thinking About Your Thinking

You will find passages throughout the text that look like this:

> ### Thinking About Your Thinking
> Name of Skill
>
> **One goal of this textbook is to help you learn the thinking skills that chemists and other scientists commonly use. In these boxes, we discuss thinking skills themselves, somewhat removed from the content of the surrounding text, so that you can clearly see the skill, learn it, and apply it in any context. These boxes will help you to learn to think about chemistry—and many other subjects—far beyond the days you spend in this course.**

When you come to a Thinking About Your Thinking discussion, take a few moments to read it and reflect on the thinking skill it discusses. Ask yourself, "Could I apply this skill in any other context?" Perhaps you've used the skill before in math or physics. Maybe you've used it before in this course. The greater the number of contexts in which you can imagine applying a skill, the more generalizable your thinking skills will become. In this way, you grow intellectually.

## Student Companion Web site

This book's companion Web site at ***www.cengage.com/chemistry/cracolice*** contains Active Figures and other materials to help you learn in ways that a printed textbook cannot.

## Study Hints and Pitfalls to Avoid

Following the Chapter in Review, you will find a brief section suggesting study methods that should make learning easier or more efficient. Some of these are "Remember" statements. Their purpose is to remind you of some word, method, or concept that students often overlook. "Pitfalls" identify the most common errors students make in exams. If you are forewarned of a common mistake, you are less likely to make it.

## Concept-Linking Exercises

After completing your study of a chapter, but before you begin to work the Questions, Exercises, and Problems (described shortly), you will need to have a firm understanding of the key terms and concepts from the chapter and the relationships among them. To help you learn the relationships among the concepts, most chapters include Concept-Linking Exercises. These exercises consist of groups of concepts that you link together with a brief description of how they are related. Answers to the Concept-Linking Exercises can be found after the answers to the Target Checks in Appendix III.

## Small-Group Discussion Questions

After the Concept-Linking Exercises, you will find a group of questions that are written so that they are best solved in collaboration with a small group of classmates. They should be attempted after you have completed your independent work studying the chapter. Whereas the end-of-chapter questions (discussed in the next subsection) are designed for individual work, the Small-Group Discussion Questions are generally more complex and are aimed at helping you to "think chemistry" at a deeper level than you might in the absence of support and assistance from a group of peers in the same course. We do not provide solutions to the Small-Group Discussion Questions. The process of verifying your solutions to the questions without "the" answer being readily provided will help you to better understand how scientists actually work.

## End-of-Chapter Questions, Exercises, and Problems

At the end of all chapters except this one, you will find Questions, Exercises, and Problems. They are grouped by the section in the chapter to which they apply. Some questions are relatively straightforward, similar to the Active Examples and Target Checks. Others are more demanding. You may have to analyze a situation, apply a chemical principle, and then explain or predict some event or calculate some result. General Questions that may be drawn from any section in the chapter follow the section-identified questions. After these are the More Challenging Questions designed to stretch you beyond the goals listed for the chapter.

The Questions, Exercises, and Problems generally are in matched pairs in which the consecutive odd-even numbered combinations involve similar reasoning and, in the case of exercises, similar calculations. Most of the odd-numbered questions out of the matched-pair groups are answered in Appendix III. Answers to exercises and problems include calculation setups. Most General Questions and More Challenging Questions, odd-numbered and even, are answered at the end of the book. Numbers of answered questions are printed in blue; numbers of unanswered questions are printed in **black.** Some of the end-of chapter questions for each chapter (marked in the text with ■) are in OWL as interactive versions.

As you solve problems in the textbook, remember that your *main* objective is to understand the problem, not to get a correct answer. Even when your answer is correct, stop and think about it for a moment. Don't leave the problem until you feel confident that you will recognize any new problem that is worded differently but requires reasoning based on the same principle. Then be confident that you can solve such a problem.

Even more important is what you do when you do *not* get the correct answer to a problem. You may be tempted to return to the Active Examples and Target Checks, find one that matches your problem, and then solve the assigned problem step by step as in the example. *You should resist this temptation.* If you get stuck on a problem, it means that you did not truly *learn* from the earlier examples. Leave the problem. Turn back to the Active Example. Study it again, by itself, until you understand it thoroughly. Then return to the assigned problem with a fresh start and work on it to the end without further reference to the example. Finally, work the remainder of the problems in the group until you are confident you can solve this problem type.

# How to Use Portable Content Chapter-In-Review and Test Yourself Cards

GOAL 1 **Locate and identify the four Active Example shields.**

At the back of the book, you will find a series of heavy-stock detachable cards. One page contains four shields that are to be used with our Active Examples. On one side you will find instructions on how to use the shield, and on the other side, there is a periodic table that you will find to be a handy reference. We will explain the purpose of the shield in more detail in Chapter 3, when you will first need it. For now, you might want to tear out one of the shields and use it as a bookmark.

GOAL 2 **Locate and identify the full-sized periodic table, table of elements, location of useful information table, and the summary of nomenclature system table.**

Another card has a full-sized periodic table on one side and a table of elements on the other. This card will also serve you well as a reference source as you progress through this course and in your future chemistry courses. Although they are not printed on removable cards, you will find a list of the location of useful information on the inside front cover of the book, and a summary of the nomenclature system that will be introduced in Chapter 6 on the inside back cover of the book.

GOAL 3 **Locate and identify the set of Portable Content Cards.**

The remaining cards are what we call Portable Content Cards. They are perforated so they can be detached from the book, allowing you to study anytime and anywhere with the convenience of having to carry just one card rather than the whole book. There is at least one card for each chapter following this introductory chapter. Find the Portable Content Card for Chapter 2 now, and tear it out so that you can follow along as we describe its features and how to use it.

**GOAL 4 Identify how Goals and goal summaries are arranged on a Portable Content Card.**

In the left column, we have listed the Goals for the chapter. Recall that a goal tells you what you should be able to do after you study the relevant section of the textbook. The right column has a brief summary of the content you need to know to satisfy the requirements of each goal. The goals and the summaries are aligned horizontally so that you can clearly see the information that accompanies each goal. Important terms are printed in bold.

**GOAL 5 Identify how the Test Yourself sample test and Answers are arranged on a Portable Content Card.**

At the bottom of the front page of the Portable Content Card for each chapter, you will find the Answers to the Test Yourself sample test. They are printed upside down to help you avoid looking at the answers until you have solved the test problems for yourself. Now look at the back of the card. The goals and summary material continues, and then there is a section we call Test Yourself. This sample test is designed to help you test your knowledge of the chapter so that you can identify any gaps in your learning *before* you take a graded exam in your course.

**GOAL 6 Describe how to use a Portable Content Card to maximize your learning efficiency.**

When used appropriately, the Portable Content Cards are a valuable learning tool. Our recommendations for using the Portable Content Cards are

1. Skim the card for each chapter before that chapter is introduced in lecture. This will allow you to preview the new terminology and concepts that you will be introduced to as you study the chapter.
2. Study the chapter, taking notes and working the Active Examples and Target Checks as you go.
3. Work the end-of-chapter questions assigned by your instructor. If your course does not include directed homework assignments, work, as a minimum, the blue-numbered questions. Check your answers against those in the back of the book, and use the feedback in a positive manner, restudying the textbook, as necessary.
4. Review the Portable Content Card one more time to be sure that you have learned the material.
5. Test Yourself with the sample test on the back of the card. Check your answers against those on the front of the card, and utilize the feedback in a positive manner, filling in any final deficits in your knowledge before you move on to the next chapter.
6. Save the Portable Content Card for each chapter so that you can use them as a tool to review for the final exam in your course.

**GOAL 7 Explain why Portable Content Cards should be used only as a supplement to the textbook.**

Although the Portable Content Cards are an important learning tool, we caution you to use them for what they are intended, as a *supplement* to the textbook. These cards are not a *substitute* for the book. The coverage provided by a two-page summary card is obviously inadequate when compared with the 20 to 40 pages that make up a typical chapter. Use the cards only as a preview and review of each chapter, and to Test Yourself. The purpose of your introductory chemistry course is to provide you with a solid foundational knowledge of the fundamentals of chemistry and to develop your scientific thinking skills, and the only way that can be accomplished is by utilizing the learning tools in your course as they are intended.

## Appendices

The Appendix of this book has four parts.

**1.** *Appendix I: Chemical Calculations.* Here you will find suggestions on how to use a calculator specifically to solve chemistry problems. There is also a general review of the arithmetic and algebraic operations used in this book. You will find these quite helpful if your math skills need dusting off before you can use them.

**2.** *Appendix II: The SI System of Units.* This explains the units in which quantities are measured and expressed in current science textbooks and other scientific publications.

**3.** *Appendix III: Answers.* Here you will find the solutions to the blue-numbered questions that appear at the end of each chapter, plus the answers for the concept-linking exercises that come immediately before the questions and the answers to the Target Checks.

**4.** *Glossary.* Like other fields of study, chemistry has its own special language, in which common words have very specialized and specific meanings. The Glossary lists these words in alphabetical order so you can find them easily and learn to use them correctly. The Glossary contains most of the boldfaced words in the text, as well as many other terms. Use the Glossary regularly; it's a real time saver.

## 1.6 A Choice

*Discipline is the bridge between goals and accomplishments.*

author unknown

You have a choice to make. You can choose to continue learning as before, or you can choose to improve your learning skills. Even if those skills are already good, they can be improved. This chapter gives you some specific suggestions on how to do this. It also helps you upgrade your study habits, beginning here and continuing throughout the book.

If you ever begin to feel that chemistry is a difficult subject, read this chapter again. Then ask yourself, and give an honest answer: "Do I have trouble because the subject is difficult, or is it because I did not choose to improve my learning skills?" Your honest answer will tell you what to do next.

At all stages of our lives we make choices. We then live with the consequences of those choices. Choose wisely—and enjoy learning chemistry.

Understanding the chemistry of living organisms, such as this dolphin, is the goal of many 21st-century chemists. All living organisms—bacteria, plants, and animals—share a set of tiny particles of matter that change from one substance to another by remarkably similar chemical processes.

# Matter and Energy

Chemistry is the study of matter and the energy associated with physical and chemical change. In this chapter, we explain some familiar features of matter and energy and introduce the vocabulary scientists use to talk about their transformations.

## CONTENTS

Online homework for this chapter may be assigned in OWL

## 2.1 Representations of Matter: Models and Symbols

**Goal 1** Identify and explain the differences among observations of matter at the macroscopic, microscopic, and particulate levels.

**Goal 2** Define the term *model* as it is used in chemistry to represent pieces of matter too small to see.

***Learn It Now!*** Performance goals tell you what you should be able to do after you study a section. Always focus your study on the goals. When you complete this section, you should know the differences among observations of matter at the macroscopic, microscopic, and particulate levels and you should be able to define the term *model* as it is used in this text.

**Matter** is anything that has mass (sometimes expressed as *weight*) and takes up space. These two characteristics combine to define matter.

We can describe some forms of matter by observation with the naked eye. These **macroscopic** samples of matter include a huge range of sizes, varying from mountains, rocky cliffs, huge boulders, and all sizes of rocks and stone to gravel and tiny grains of sand. Geologists often study matter at this level.

The prefix *macro-* means large, and the prefix *micro-* means small. The suffix *-scopic* refers to viewing or observing.

Biologists are well known for using microscopes to observe types of matter too small to be seen with the unaided human eye. You've probably used a microscope to observe tiny animals or plants, or perhaps the tiny crystals on the surface of a polished rock. These are examples of **microscopic** samples of matter.

Chemists often think about matter that is too small to be seen even with the most powerful optical microscope. A chemist considers the behavior and transformations of the tiny particles that make up matter. Thinking this way is thinking at the **particulate** level. One of the most valuable skills you will learn in this course is to think about the particulate nature of matter.

Figure 2.1 gives examples of matter at the macroscopic, microscopic, and particulate levels.

Most of the time, chemists work with macroscopic samples in the laboratory, but they imagine what happens at the particulate level while they do so. Biologists frequently work with microscopic samples, but they also think about the particulate-level behavior of their samples to supplement their understanding of what they see through

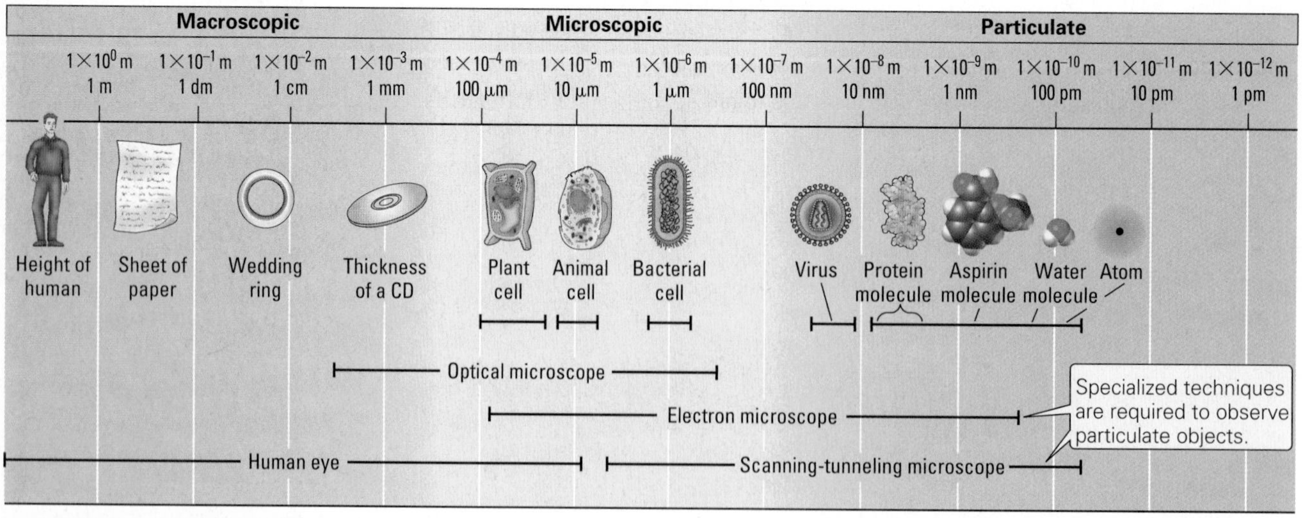

Macroscopic
Microscopic
Particulate
$1\times10^{0}$ m 1 m
$1\times10^{-1}$ m 1 dm
$1\times10^{-2}$ m 1 cm
$1\times10^{-3}$ m 1 mm
$1\times10^{-4}$ m 100 μm
$1\times10^{-5}$ m 10 μm
$1\times10^{-6}$ m 1 μm
$1\times10^{-7}$ m 100 nm
$1\times10^{-8}$ m 10 nm
$1\times10^{-9}$ m 1 nm
$1\times10^{-10}$ m 100 pm
$1\times10^{-11}$ m 10 pm
$1\times10^{-12}$ m 1 pm
Height of human
Sheet of paper
Wedding ring
Thickness of a CD
Plant cell
Animal cell
Bacterial cell
Virus
Protein molecule
Aspirin molecule
Water molecule
Atom
Optical microscope
Electron microscope
Specialized techniques are required to observe particulate objects.
Human eye
Scanning-tunneling microscope

**Figure 2.1** Macroscopic, microscopic, and particulate matter. The particulate-level drawings are *models* of types of matter too small to see with the human eye or an optical microscope. Electron microscopes "shine" a beam of electrons through a sample in much the same way as optical microscopes shine light through the specimen. Scanning-tunneling microscopes depend on electrical properties of matter to create computer-generated images.

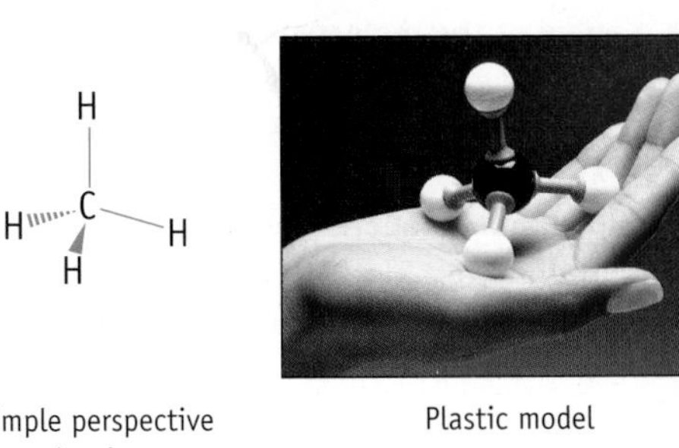




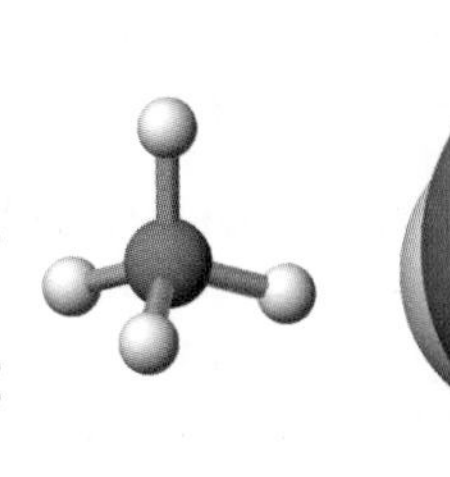

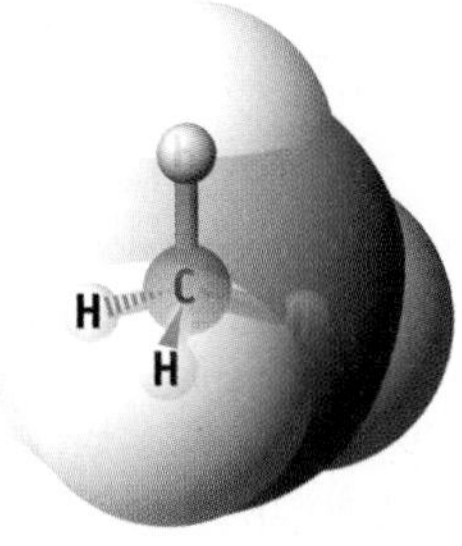


**Active Figure 2.2** Models and symbols used by chemists to represent particulate matter. All of these models and symbols represent a methane molecule. Methane is the primary component of natural gas. A methane molecule consists of a central carbon atom, represented by the symbol C or a black sphere, surrounded by four equidistant hydrogen atoms, represented by the symbol H or a white sphere. The lines in the perspective drawing and the sticks in the ball-and-stick models represent the tiny particles that bond the atoms together. **Watch this Active Figure at *http://www.cengage.com/cracolice.***

the microscope. By understanding and directing the behavior of particles, the chemist and biologist control the macroscopic behavior of matter.

This is, in fact, a distinguishing characteristic of chemistry, biochemistry, and molecular biology. A chemist imagines the nature of the behavior of the tiny particles that make up matter, and then he or she applies this knowledge to carry out changes from one type of macroscopic or microscopic matter to another.

Since matter at the particulate level is too small to see, chemists use models to represent the particles. A **model** is a representation of something. Chemists use models of atoms and molecules, tiny particulate-level entities, that are based on experimental data. We cannot see atoms and molecules directly, so we use data from experiments to infer what they would look like if they were much, much larger. We then construct physical models that match the data.

The two most common models are ball-and-stick models and space-filling models, which are illustrated in Active Figure 2.2. The **ball-and-stick model** shows atoms as balls and linking electrons as sticks connecting the atoms. The **space-filling model** shows the outer boundaries of the particle in three-dimensional space. We use illustrations of both types of models in this book.

The famous scientists James Watson and Francis Crick used models to deduce the structure of DNA, the molecular storehouse of genetic information (Fig 2.3). They used numerical information from experiments, such as the distances between atoms, to construct models of the pieces that make up the DNA molecule. Working with these physical representations of the small pieces, they could see how the pieces fit together to form the double-helix structure of DNA. They completed the scientific research cycle by using their model to predict experimental outcomes, and the actual experimental outcomes were indeed consistent with the model.

**Figure 2.3** James D. Watson (b. 1928) (*left*) and Francis H. C. Crick (1916–2004) (*right*) posing with a model of the DNA molecule. Physical models, such as this one, can be valuable tools in chemistry. More often, however, mental models or computer animations of particulate matter are used to understand how chemical processes work.

## Thinking About Your Thinking

### Mental Models

**A *model* is a depiction of something. You are probably familiar with models of the earth. Globe models are common in geology and earth science classrooms and laboratories. We use these models because the earth is so large. A globe allows us to picture mentally something that otherwise is too big to understand well.**

**Models of molecules are used in chemistry for exactly the opposite reason: Molecules are too small to observe without the aid of models. Chemists need concrete models to understand the behavior of molecules. Figure 2.2 helps you form a mental image of a methane molecule. We will show many molecular models throughout this book.**

P/REVIEW*: Much progress in chemistry has resulted from building a mental model that might explain the observed behavior of a substance (hypothesis). Chemists can then use this model to predict additional behavior and to design experiments to confirm or disprove the prediction (or even the entire model). Recognize these steps in the scientific method described in Section 1.2.

**As you study chemistry, you will develop the skill of visualizing the particulate-level behavior of molecules. Imagine the models of the molecules in three dimensions. Try to think about how they move. When we describe chemical processes in words and two-dimensional illustrations, translate those words and figures into moving three-dimensional "mental movies."**

**We will remind you periodically throughout the book about forming these mental models. You will see the words "Thinking About Your Thinking: *Mental Models*" whenever you should think about this process. Doing so will greatly improve your skill at thinking as a chemist.**

Models are often represented in a simple, easy-to-write form as a **chemical symbol,** or simply as a **symbol.** The symbol H is used to represent a hydrogen atom, and the symbol O represents an oxygen atom. You are probably familiar with how these symbols are combined to represent a water molecule. Its chemical formula is $H_2O$. This formula tells us that a water molecule is composed of two hydrogen atoms and one oxygen atom.

P/REVIEW: Lewis diagrams are formally introduced in Chapter 12, and how to draw them is discussed in Chapter 13.

The information in a formula is limited. It does not tell us anything about the arrangement of the atoms. Another symbolic representation for water is its Lewis diagram, shown in Figure 2.4. This description of a water molecule shows how the atoms are arranged, with the oxygen atom between the two hydrogen atoms. It also gives us some information about the arrangement of the electrons in the molecule. These are represented with dashes and dots.

A chemist can think about matter at many levels, often switching among different representations in one discussion. When talking about a drop of water, for example, a chemist may think about the macroscopic characteristics of the drop, such as its shape, lack of color, and size; make a mental model of the water molecules in the drop; imagine the invisible forces that hold the molecules together; and then use symbols to show how the particles are arranged (Fig. 2.4). One of our major goals in this book is to help you think as a chemist by using models and symbolic representations of the particulate nature of matter.

## √ Target Check 2.1

*Consider the photograph and illustration to the right. Do they include a model? Do they include a depiction of matter at the macroscopic, microscopic, and/or particulate levels? Explain your answers.*

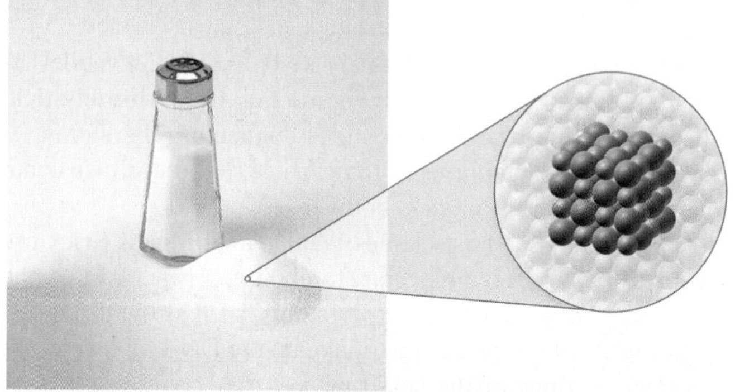

Figure 2.4 Macroscopic, particulate, and symbolic forms and representations of matter. Chemists frequently make mental transformations between visible macroscopic matter and models of the particulate-level molecules that make up the matter. Written symbols serve as simpler representations of the particulate-level models.

*P/Reviews are references to items covered in another chapter of the textbook. For a description of P/Reviews, see Section 1.5.

## 2.2 States of Matter

**Goal** 3 Identify and explain the differences among gases, liquids, and solids in terms of (a) visible properties, (b) distance between particles, and (c) particle movement.

The air you breathe, the water you drink, and the food you eat are examples of the **states of matter** called *gases, liquids,* and *solids*. Water is a common substance that is familiar to us in all three states, as Figure 2.5 depicts. We can explain the differences among gases, liquids, and solids in terms of the **kinetic molecular theory.** According to this theory, all matter consists of extremely tiny particles that are in constant motion. *Kinetic* refers to motion; *molecular* comes from **molecule,** the smallest individual particle in one kind of matter. Water, oxygen, and sugar are examples of three common molecular substances.

Another state of matter is plasma, which consists of positive ions and free electrons in a gaslike state. Examples of plasmas include the substances inside fluorescent lights and neon signs.

According to the kinetic molecular theory, the speed at which particles move is faster at higher temperatures and slower at lower temperatures. As the temperature of a sample rises, the faster-moving particles tend to separate from each other. When that happens, the sample exists as a **gas.** A gas must be held in a closed container to prevent the particles from escaping into the surrounding space. The particles move in a random fashion inside the container. They fill it completely, occupying its full volume.

There are attractions among the particles in any sample of matter, but in a gas the attractions have almost no effect because of the vigorous movement that goes with high temperatures. If the temperature decreases, the particles slow down. The attractions therefore have more of an effect, and the particles clump together to form a **liquid** drop. The drops fall to the bottom of the container, where the particles move freely among themselves, taking on the shape of the container. The volume of a liquid is almost constant, varying only slightly with changes in temperature.

As temperature decreases further, particle movement becomes more and more sluggish. Eventually the particles no longer move among each other. Their movement is reduced to vibrating, or shaking, in fixed positions relative to each other. This is the **solid** state. Like a liquid, a solid effectively has a fixed volume. But unlike a liquid, a solid has its own unique shape that remains the same wherever the sample may be placed.

### Thinking About Your Thinking

#### Mental Models

**Now that you've developed a three-dimensional mental model of the water molecule, you need to refine it so that it's dynamic, or moving. Figure 2.5 shows water molecules in the three states of matter. In the gaseous state, the molecules are far apart and moving very rapidly in straight lines until they collide with one another or the walls of the container. Can you make a "mental movie" of water in the gaseous state?**

**Now imagine the molecules slowing down. They begin to stick together and fall to the bottom of the container. They are now in the liquid state. At the bottom of the container, they take its shape. They move more slowly than in the gaseous state, but still fast enough to move past each other. Collisions are very frequent as the molecules rotate and slip around one another.**

**At even lower temperatures, the molecular movement becomes even more sluggish. The water molecules begin to align in a regularly repeating pattern, remaining in a fixed position relative to each other. This is the solid state. The particles continue to move, but the movement is vibration in a specific location. Each molecule is very close to its neighbors.**

**All other substances have particulate-level behavior similar to that of water with respect to how they behave in and change among the three states of matter. Practice using your mental movie of molecules in different states of matter as you study the rest of this chapter.**

| | Gas | Liquid | Solid |
|---|---|---|---|
| **Water as an example:** | 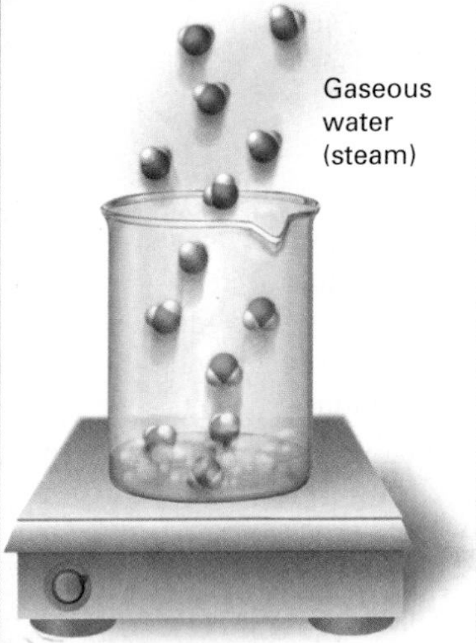 Gaseous water (steam) |  Liquid water |  Solid water (ice) |
| **Shape** | Variable— same as a closed container | Variable— same as the bottom of the container | Constant—rigid, fixed |
| **Volume** | Variable— same as a closed container | Constant | Constant |
| **Particle Movement** | Completely independent (random); each particle may go anyplace in a closed container | Independent beneath the surface, limited to the volume of the liquid and the shape of the bottom of the container | Vibration in fixed position |

**Figure 2.5** Three states of matter illustrated by water.

Figure 2.5 summarizes the properties of gases, liquids, and solids.

***Learn It Now!*** Always complete a Target Check immediately after studying a section. Check the answer at the end of the book. If any answer is wrong, find out why *now*! Reread the text, ask a classmate, and if necessary talk to your instructor at the next office hour or class meeting. Learn it—*now*!

### √ Target Check 2.2

*In the left box, draw a particulate illustration of a substance in the gaseous state. Use single spheres for the particles. Assume that the box represents a tiny, closed container that holds the particles. In the right box, draw a particulate illustration of the same substance after it cools and becomes a liquid.*

## 2.3 Physical and Chemical Properties and Changes

**Goal 4** Distinguish between physical and chemical properties at both the particulate level and the macroscopic level.

**Goal 5** Distinguish between physical and chemical changes at both the particulate level and the macroscopic level.

If you were asked to describe a substance, you would list its **physical properties,** including, perhaps, its color, feel, and smell. Charcoal is black; sulfur is yellow. Glass

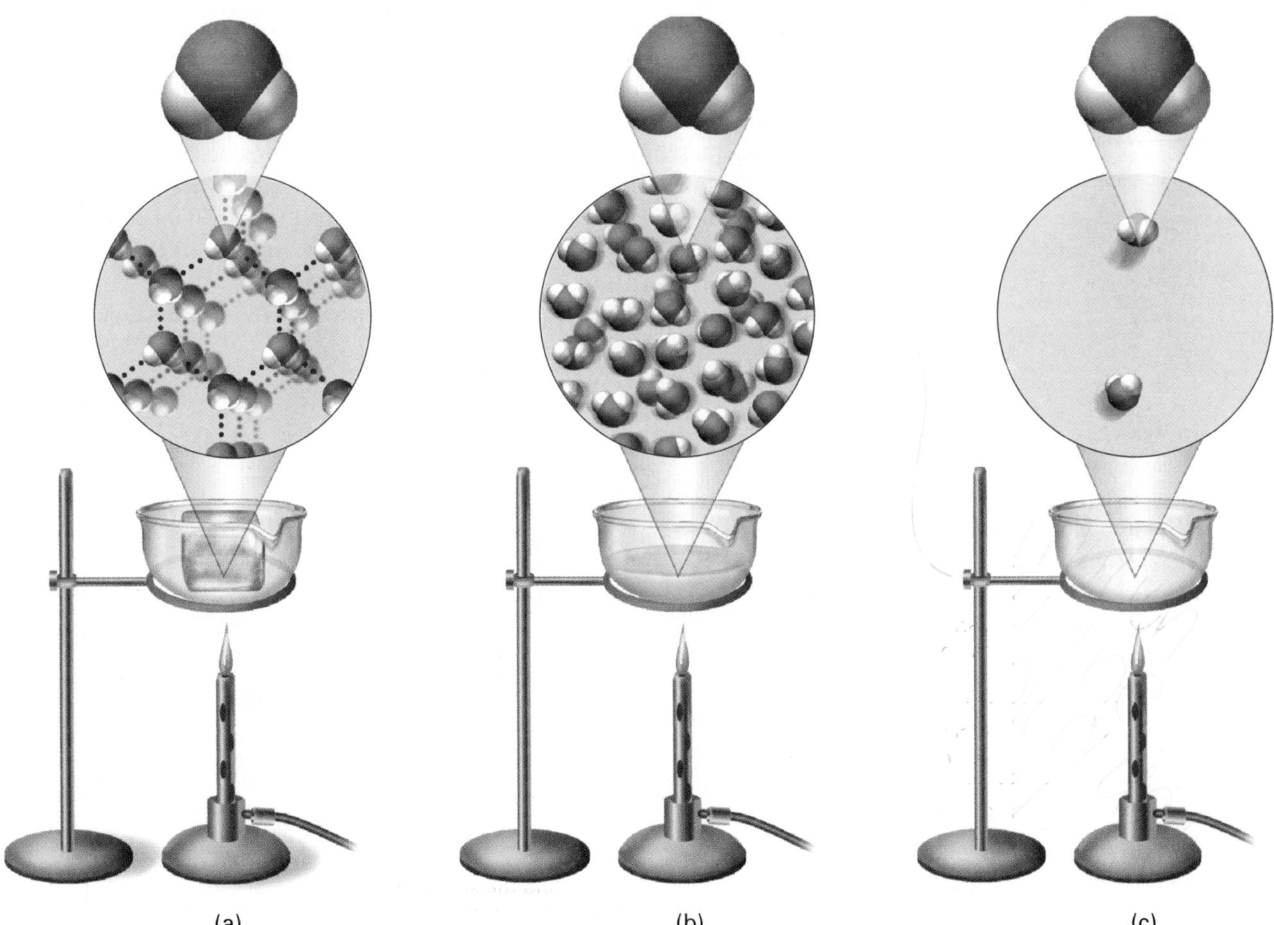


**Figure 2.6** In a physical change, the particles of matter themselves are unchanged. They are water molecules whether in the (a) solid, (b) liquid, or (c) gaseous state. Any individual particle is the same before and after a physical change.

is hard; bread dough is soft. The smell of a rose is pleasant; the odor of ammonia is disagreeable.

You can measure other physical properties in the laboratory. For example, we determine the temperature at which a substance boils or melts, called the boiling point or the melting point. The thickness, or resistance to flow, of liquid substances, such as water versus pancake syrup, compares their viscosities. Boiling point, melting point, and viscosity are all physical properties.

Changes that alter the physical form of matter without changing its chemical identity are called **physical changes** (Fig. 2.6). The melting of ice is a physical change, as is the freezing of liquid water. The substance is still water after it melts and after it refreezes. Dissolving sugar in water is another physical change. The sugar seems to disappear, but if you taste the water, you'll know the sugar is there. You can recover the dissolved sugar by evaporating the water, another physical change.

## Thinking About Your Thinking

### Mental Models

**When you think about a physical change at the particulate level, do not break apart or change the particles in any way. Also note that the physical properties of a macroscopic sample of a substance do not apply at the particulate level. Most physical properties depend on more than just the structure of the individual particle. These properties also depend on the particulate-level arrangement of the particles. Any individual particle does not have color, feel, viscosity, or any other physical property.**

A **chemical change** occurs when the chemical identity of a substance is destroyed and a new substance forms (Fig. 2.7). A chemical change is also called a **chemical reaction.** As a group, all the chemical changes possible for a substance make up its **chemical properties.**

One or more of our five physical senses can usually detect chemical changes. A change of color almost always indicates a chemical change, as when you caramelize sugar, as shown in Figure 2.8. You can feel the heat and see the light given off as a match burns. You can smell and taste milk that becomes sour. Explosions usually give off sound.

Table 2.1 summarizes physical and chemical changes and properties.

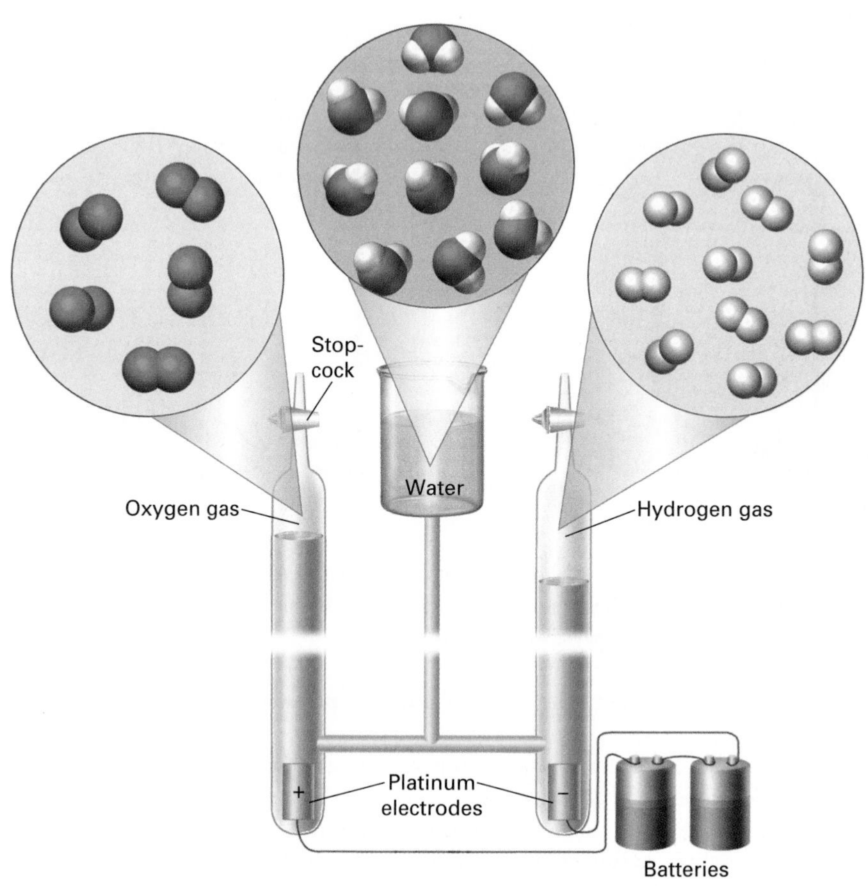
Stop-
cock
Water
Oxygen gas
Hydrogen gas
Platinum
electrodes
+
–
Batteries

Figure 2.7 When electricity is passed through certain water solutions, the water decomposes into its elements, hydrogen and oxygen. This is a chemical change, in which the particles of matter are changed. Water molecules (*center*), made up of hydrogen and oxygen atoms, are destroyed, and oxygen molecules (*left*) and hydrogen molecules (*right*) are created. The number and type of *atoms* that make up the molecules remain the same before and after the chemical change, but the number and type of *molecules* change. The volume of hydrogen on the right is twice the volume of oxygen on the left. This matches the 2:1 ratio in the chemical equation (see Section 2.8).

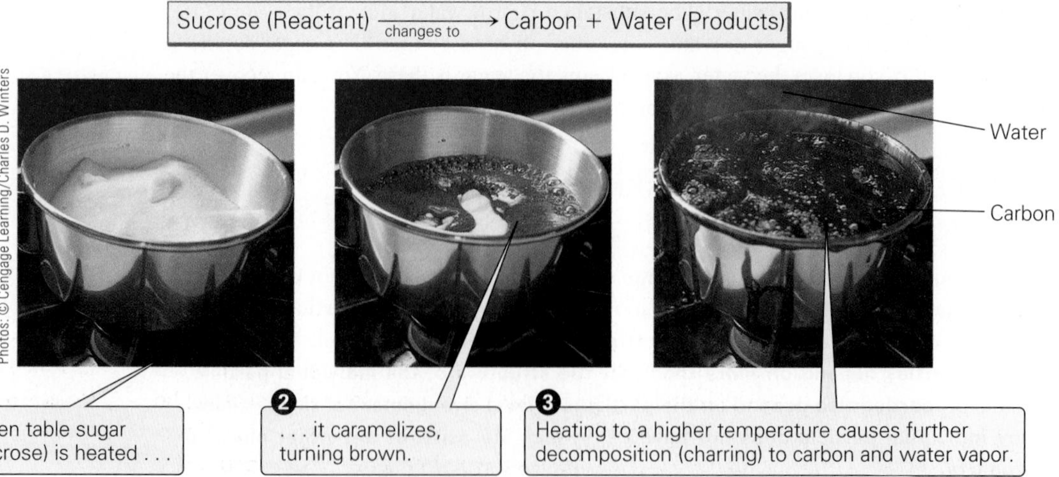
Sucrose (Reactant) changes to → Carbon + Water (Products)
Water
Carbon
1 When table sugar (sucrose) is heated . . .
2 . . . it caramelizes, turning brown.
3 Heating to a higher temperature causes further decomposition (charring) to carbon and water vapor.

Figure 2.8 Chemical change. A change in color is often evidence that a chemical change has occurred.

**Table 2.1** Chemical and Physical Changes and Properties

| | Chemical | Physical |
|---|---|---|
| **Changes** | Old substances destroyed | New form of old substance |
| | New substances formed | No new substances formed |
| **Properties** | Properties defined by types of chemical changes possible | Description by senses such as color, shape, odor<br>Measurable properties such as density, boiling point |

## Thinking About Your Thinking

### Mental Models

**Compare and contrast Figures 2.6 and 2.7 to help form mental models of the difference between a physical change and a chemical change. In a physical change, the particles remain unchanged; in a chemical change, the original particles are destroyed and new particles form.**

### √ Target Check 2.3

*Classify the following changes as chemical (C) or physical (P).*

a) Baking bread

b) Grinding sugar into powder

c)

d)

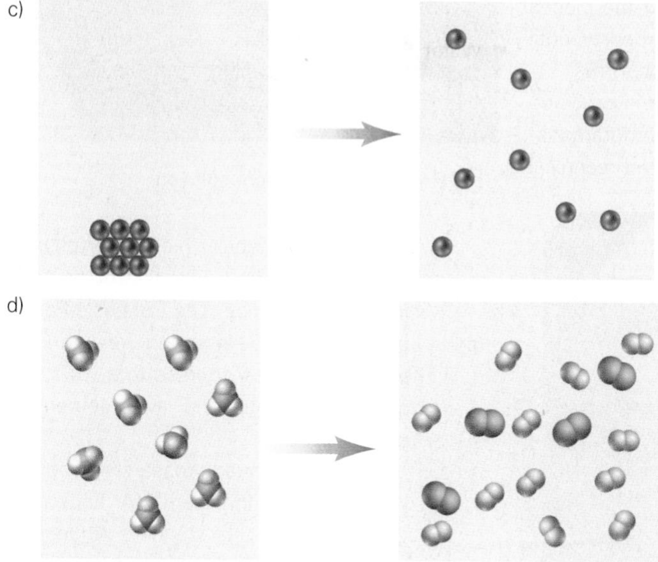

## 2.4 Pure Substances and Mixtures

**Goal 6** Distinguish between a pure substance and a mixture at both the macroscopic level and the particulate level.

**Goal 7** Distinguish between homogeneous and heterogeneous matter.

At normal pressure, pure water boils at 100°C. As boiling continues, the temperature remains at 100°C until all of the liquid has been changed to a gas. Pure water cannot be separated into parts by a physical change. Water from the ocean—salt water—is different. Not only does it boil at a higher temperature, but the boiling temperature continually

When we say that water does not separate into parts as a result of boiling, which is a physical change, we mean that it does *not* become hydrogen and oxygen. Molecules remain the same before and after a physical change. Steam is made up of water molecules in the gaseous state.

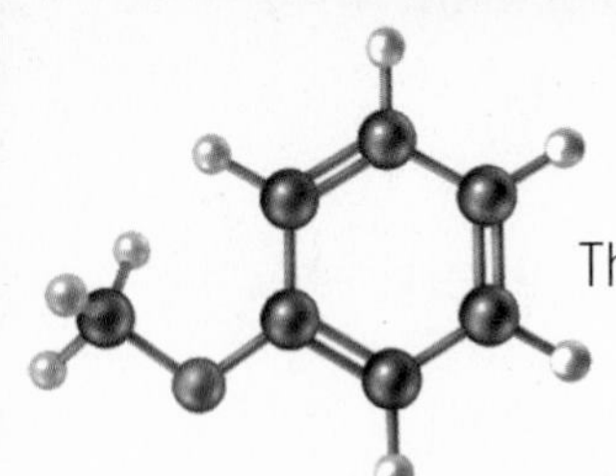

# Everyday Chemistry

## The Ultimate Physical Property?

Chemists engaged in crime analysis are called forensic chemists. Forensic chemists and detectives have a lot in common. They both examine physical evidence in the hope that they can identify some fact, some object, or some person.

The Federal Bureau of Investigation—the FBI—has on file the fingerprints of approximately 79,000,000 people. If two sets of fingerprints share 16 characteristics, they are almost certain to come from the same person. Matching sets of fingerprints by hand is difficult. Computers match fingerprint patterns more quickly than people can match them. In July 1999, the FBI's Integrated Automated Fingerprint Identification System became fully operational, dramatically reducing the time needed to make fingerprint identifications.

What if there were no fingerprints? What if the suspect's fingerprints are not in the automated identification system? Is there another way to reach positive identification?



Fingerprinting is a tool in forensic science.

The genetic information in a person's DNA, deoxyribonucleic acid, governs that individual's physical characteristics. Because every person's DNA is believed to be unique (even identical twins can have tiny differences in their DNA because of mutations in the developing embryos), DNA analysis can, in theory, provide positive identification. As a result, DNA profiling, also called "DNA fingerprinting," has rapidly moved into courtrooms. Unfortunately, DNA profiling cannot prove beyond a doubt a person's guilt. Fortunately for the innocent, however, it can exonerate those falsely accused.

A DNA sample from human tissue is taken at a crime scene and, for comparative purposes, from victims and suspects. At a laboratory, the DNA is extracted from the samples and purified. The pure DNA is then mixed with another substance that, through chemical reactions, fragments the long DNA molecule into smaller pieces. Technicians then treat the small pieces so that they can be visualized, and they sort the pieces by size, a physical property, until an identifying pattern forms.

The FBI has a Combined DNA Index System program that provides software and technical assistance to forensic laboratories across the nation. The system allows all the laboratories to exchange DNA profiles electronically, so that profiles on file can be matched to samples from crime scenes. All states now have laws in place that require certain categories of convicted offenders to submit a DNA sample, which is then placed in the database.

A feature of the Combined DNA Index System is the National DNA Index System, which has operated since October 1998. Over a half million DNA profiles of convicted offenders are now in the system.

Could employment as a forensic chemist be a career option for you? The FBI and the network of forensic laboratories across the country employ scientists who specialize in forensics. An undergraduate degree in chemistry, biochemistry, molecular biology, or physics is usually required, followed by an M.S. in forensic science. The FBI has a Forensic Science Research and Training Center that provides courses to a variety of agents, students, and law-enforcement personnel.

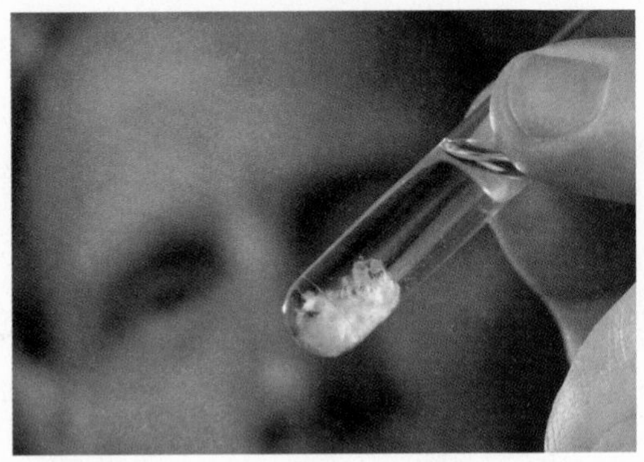



Solid DNA precipitates from solution.

With continued research progress in DNA profiling, the DNA fingerprint could replace the standard fingerprint as the most common forensic identification technique. Currently, people use many different techniques at each step of the identification process, and actively debate the relative merits of each. If repeatable techniques of DNA analysis become widely available, we will have taken a long step toward identifying everyone by the "Ultimate Physical Property."

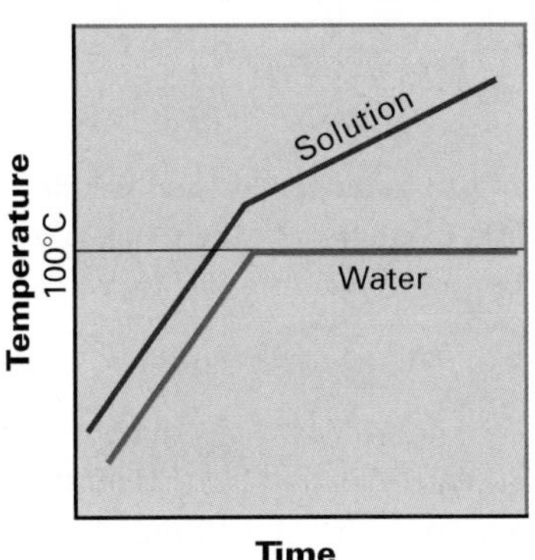


Figure 2.9 Comparison between boiling temperatures of a pure liquid and an impure liquid (solution). As water boils off the solution, what remains becomes more concentrated. This change in concentration causes the solution's boiling temperature to increase.

increases as the boiling proceeds (Fig. 2.9). If boiled long enough, the water boils off as a gas and the salt is left behind as a solid.

The properties of pure water and ocean water illustrate the difference between a pure substance and a mixture. A **pure substance*** is a single chemical, one kind of matter, entirely made up of one type of particle. It has its own set of physical and chemical properties, not exactly the same as the properties of any other pure substance. We may use these properties to identify the substance. A pure substance cannot be separated into parts by physical means.

A **mixture** is a sample of matter that consists of two or more substances. The properties of a mixture depend on the substances in it. These properties vary as the relative amounts of the different substances change. Figure 2.10 shows the differences between pure substances and mixtures.

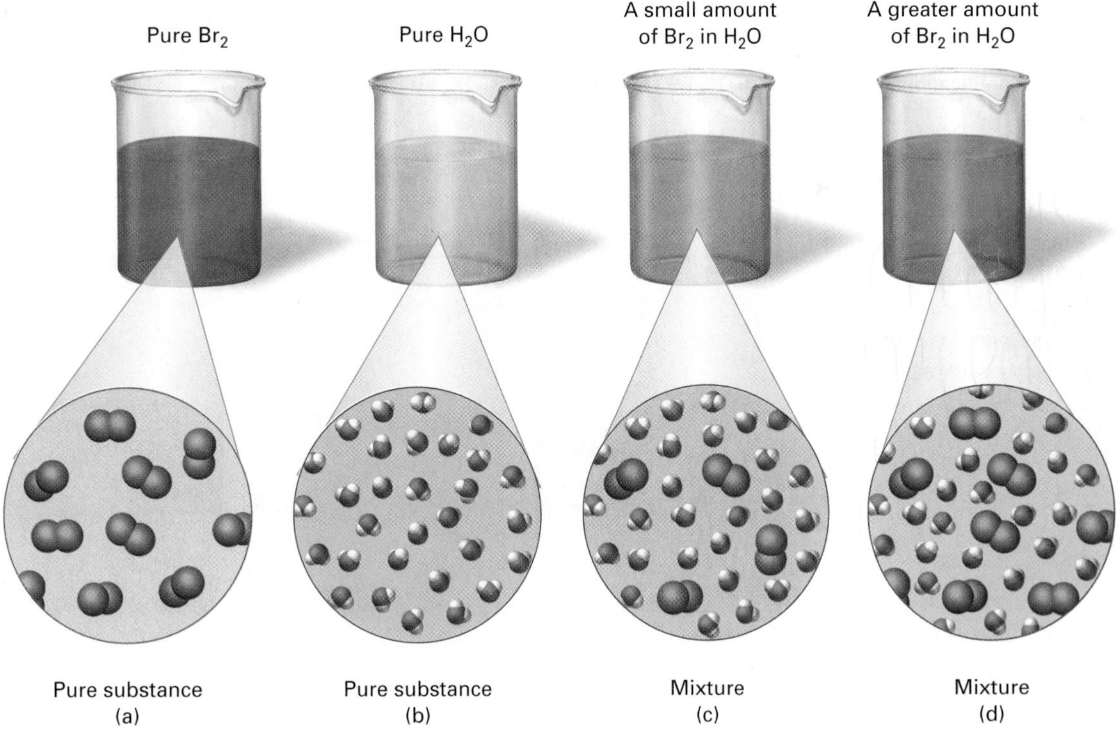


Figure 2.10 You cannot distinguish a pure substance from a mixture of uniform appearance by observation alone at the macroscopic level. On the particulate level, however, the difference is readily apparent. (a) Pure bromine consists of bromine molecules. (b) Pure water consists of water molecules. (c, d) Mixtures of bromine and water consist of bromine and water molecules.

*Technically, a substance is pure by definition. The word is so commonly used for any sample of matter, however, that we include the adjective *pure* when referring to a single kind of matter.

## Thinking About Your Thinking

### Mental Models

**If you can deduce the particulate-level composition of a substance, you can decide whether it is a pure substance or a mixture. All of the particles are identical in a pure substance, and a mixture is a combination of two or more different types of particles.**

When water and alcohol are mixed, they dissolve in each other and form a **solution.** Chemists consider a *homogeneous mixture* and a *solution* to be the same thing. A solution has a uniform appearance, and once properly stirred, it has a uniform composition, too. If you were to take two samples of a given water–alcohol mixture from anyplace in a container, they would have exactly the same composition and properties. This is what is meant by ***homogeneous:*** **If a sample has a uniform appearance and composition throughout, it is said to be homogeneous.** The prefix *homo-* means same.

When cooking oil and water are mixed, they quickly separate into two distinct layers or **phases,** forming a **heterogeneous mixture.** The prefix *hetero-* means different. The different phases in a heterogeneous sample of matter are usually visible to the naked eye. Figure 2.11 summarizes homogeneous pure substances and homogeneous mixtures.

### √ Target Check 2.4

*Specific gravity is a physical property. Beakers A, B, and C hold three clear, colorless liquids. The specific gravities of the liquids are listed in the "before freezing" column. The beakers are placed in a freezer until a solid crust forms across the surface of each. The crusts are removed, and the liquids are warmed to room temperature. Their specific gravities are now given in the "after freezing" column. Which beaker(s) contain(s) a pure substance (P), and which contain(s) a mixture (M)?*

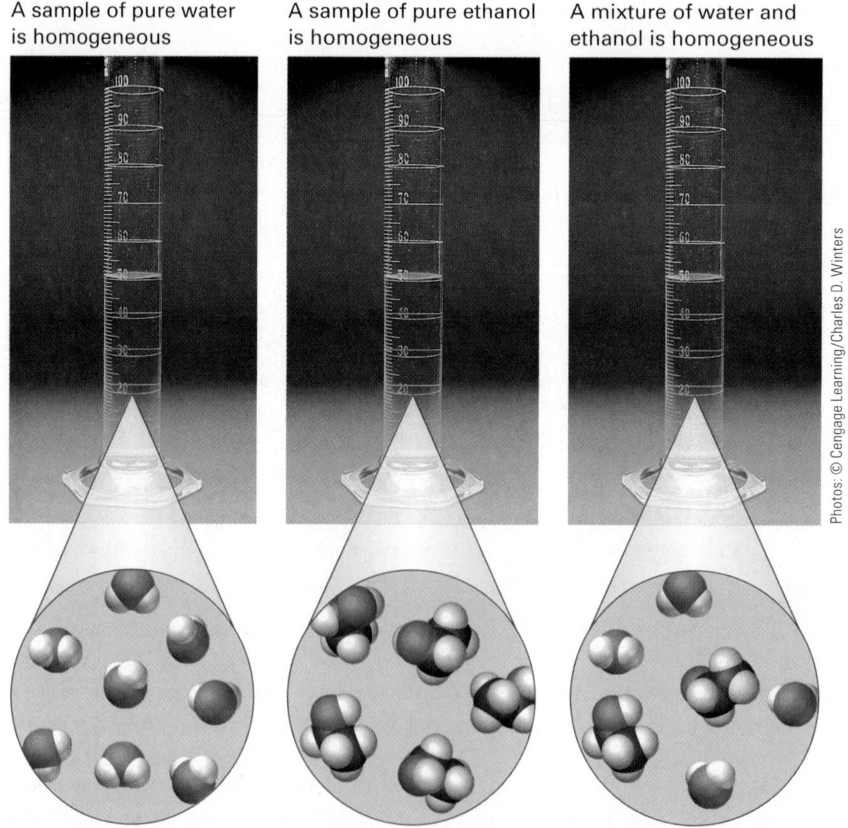
A sample of pure water is homogeneous
A sample of pure ethanol is homogeneous
A mixture of water and ethanol is homogeneous
Photos: © Cengage Learning/Charles D. Winters

Figure 2.11 Homogeneous pure substances and mixtures. The terms *homogeneous* and *heterogeneous* refer to macroscopic samples of matter. They are a reference to the macroscopic *appearance* of the substance. Homogeneous matter may be either a pure substance or a mixture.

| | Before Freezing | After Freezing | P or M |
|---|---|---|---|
| Liquid A | 1.08 | 1.10 | |
| Liquid B | 1.00 | 1.00 | |
| Liquid C | 1.12 | 1.15 | |

### √ Target Check 2.5

*Classify the following as heterogeneous (HET) or homogeneous (HOM).*

a) Contents of beaker on left in photograph
b) Contents of beaker on right in photograph
c) Real lemonade
d) Beach sand

## 2.5 Separation of Mixtures

**Goal 8** Describe how distillation and filtration rely on physical changes and properties to separate components of mixtures.

If you try to think of a pure substance that occurs by itself in nature, you'll begin to appreciate the value of techniques chemists use to separate mixtures into their components. Samples of pure substances in nature are rare. You may initially think of air, but it is a mixture of nitrogen, oxygen, and many other gases. Water in rivers is among the purest form of natural water, but all natural waters contain significant quantities of dissolved substances. Even rainwater contains dissolved gases from the air. Soil is, of course, a mixture. Finding a naturally occurring pure substance is very unusual. Mixtures, on the other hand, abound in nature.

Most methods for separating mixtures into their components depend on differing physical properties among those components. Figure 2.12 shows the removal of iron from a mixture of iron and sulfur. A physical property of iron is that it is attracted to a magnet. Sulfur is not attracted to a magnet. Because of this difference, we can separate components of the mixture.

Another separation technique is **distillation,** as illustrated in Figure 2.13. Water has a lower boiling point than the dissolved solids in a solution of natural water collected

1 Iron and sulfur can be separated by stirring with a magnet.
2 The first time the magnet is removed, much of the iron is removed with it.
3 The sulfur still looks dirty because a small quantity of iron remains.
4 Repeated stirrings eventually leave a bright yellow sample of sulfur that cannot be purified further by this technique.



**Figure 2.12** Separating a mixture of iron and sulfur. Iron and sulfur form a heterogeneous mixture. The physical property of magnetism allows us to remove iron, leaving pure sulfur behind.

**Figure 2.13** Laboratory distillation apparatus. When salt water is heated, the water boils off and is then cooled, condensed (changed back to a liquid), and collected as pure water. Room temperature tap water flows through the outer jacket of the condenser to cool the steam within so that it changes to the liquid state.

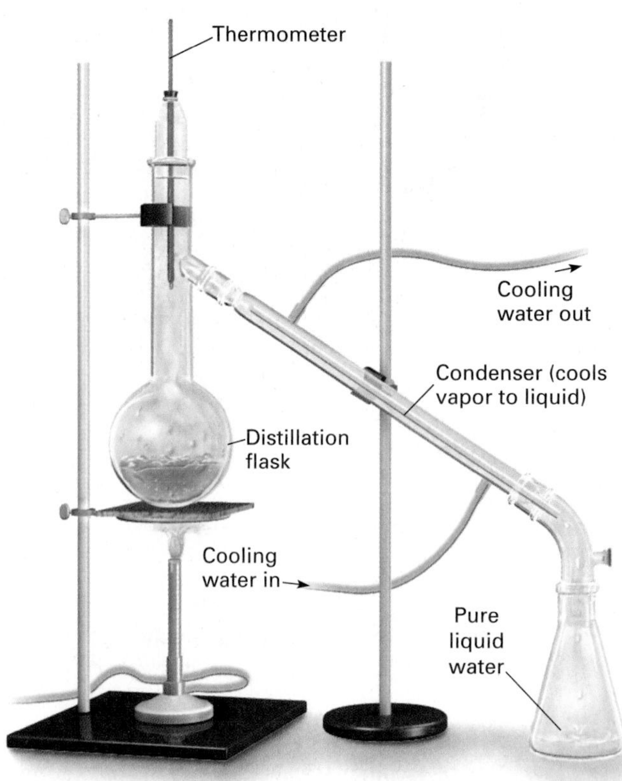


Coffee drinkers are familiar with one of the most common applications of filtration, separating coffee grounds from a coffee solution with filter paper.

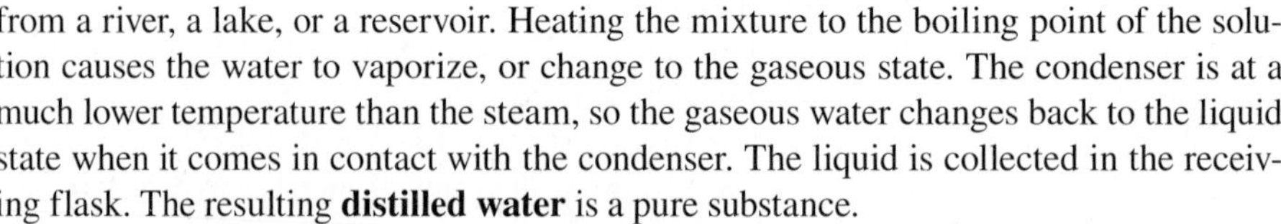

from a river, a lake, or a reservoir. Heating the mixture to the boiling point of the solution causes the water to vaporize, or change to the gaseous state. The condenser is at a much lower temperature than the steam, so the gaseous water changes back to the liquid state when it comes in contact with the condenser. The liquid is collected in the receiving flask. The resulting **distilled water** is a pure substance.

The original natural water is a *homogeneous mixture.* The distillate is a *pure substance,* liquid water. The distillation process separates a component of the mixture—the water—from the other components (dissolved substances) through a *physical change.* The water changes from the liquid state to the gaseous state, and then back to the liquid state. In both states, liquid and gas, water remains water, so it undergoes a physical change in this separation process.

If we collected natural water with some sediment in the sample, our first step would be to separate the sediment from the liquid by **filtration** (Fig. 2.14). A porous medium, such as filter paper, is used to separate the components of the mixture. The pore size in the filtration device must allow one (or more) component(s) of the mixture to pass through while blocking other component(s). The filter paper allows the liquid from the mixture to pass through while preventing passage of the gravel, sand, and dirt in the sediment.

Filtration is based on *physical properties* of a mixture. The particle sizes of the components of the mixture must be significantly larger or smaller than the pore size of the filtration medium. The smaller particles pass through the filter, and the larger particles are left behind.

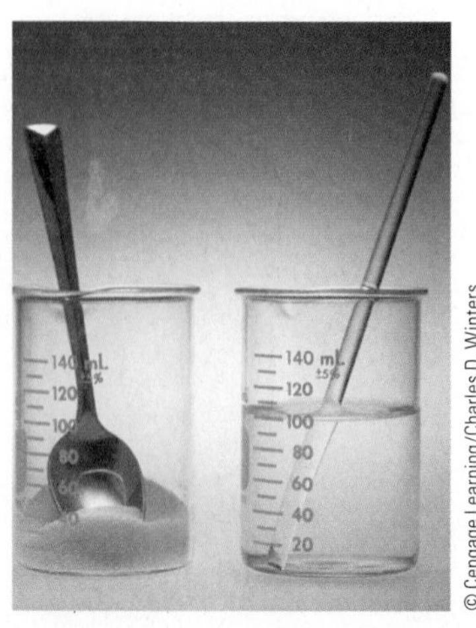



### √ Target Check 2.6

*Table salt from the beaker on the left in the photograph is added to water, forming the solution on the right of the photo. If you want to separate the mixture, would the distillation apparatus in Figure 2.13 or the filtration apparatus in Figure 2.14 be the best choice? Explain.*

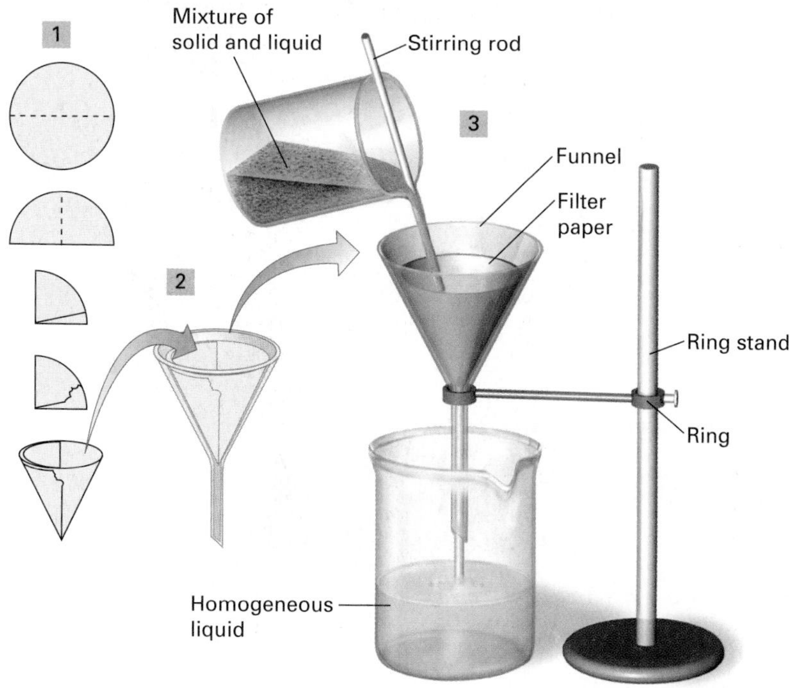


**Figure 2.14** Gravity filtration.

1. A piece of filter paper is selected with a pore size appropriate to the heterogeneous mixture. The filter paper is constructed so that when it is folded in half and then folded again, it opens into a conical shape that fits a filtration funnel. A piece is often torn from the corner to help seal the paper to the funnel.
2. The paper is placed into a filtration funnel.
3. The funnel is placed in a ring on a ring stand. The mixture is poured along a stirring rod, and the filter paper traps the solids as the liquid solution passes into a beaker.

## 2.6 Elements and Compounds

Goal 9 Distinguish between elements and compounds.

Goal 10 Distinguish between elemental symbols and the formulas of chemical compounds.

Goal 11 Distinguish between atoms and molecules.

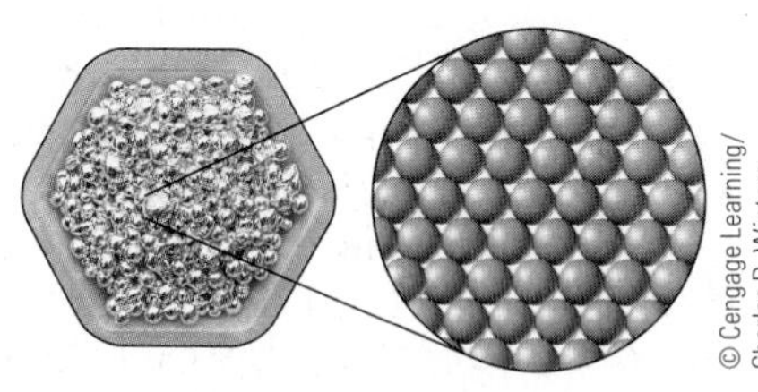


**Figure 2.15** The element silver. Silver cannot be decomposed into other stable pure substances because all atoms are the same element. The magnified view suggests that the silver atoms are arranged in a regularly repeating pattern.

Silver represents one of the two kinds of pure substances. Like all pure substances, it has its own unique set of physical and chemical properties, unlike the properties of any other substance. Among its chemical properties is that silver cannot be decomposed or separated into other stable pure substances. This identifies silver as an **element.** The smallest unit particle of an element is an **atom** (Fig. 2.15).

Water represents the second kind of pure substance. Unlike silver, it can be decomposed into other pure substances. Go back and look at Figure 2.7. Any pure substance that can be decomposed by a chemical change into two or more other pure substances is a **compound.** Be sure to catch the distinction between separating a compound into simpler substances by chemical means and separating a mixture into its components by physical means. This is shown in Figure 2.16.

### Thinking About Your Thinking

#### Mental Models

**Imagine the different forms an element can take at the particulate level. It can be a collection of single atoms. It can be made up of molecules of two atoms, four atoms, or even eight atoms or more. But in any case, all of the atoms of an element are the same. Now compare your mental model of an element to your model of a compound. At least one of the atoms in a molecule or a unit of a compound is different from the other atoms.**

Nature provides us with at least 88 elements. Copper, sulfur, silver, and gold are among the few well-known solid elements that occur uncombined in nature. At common temperatures and pressures, 11 elements occur as gases, 2 (mercury and bromine)

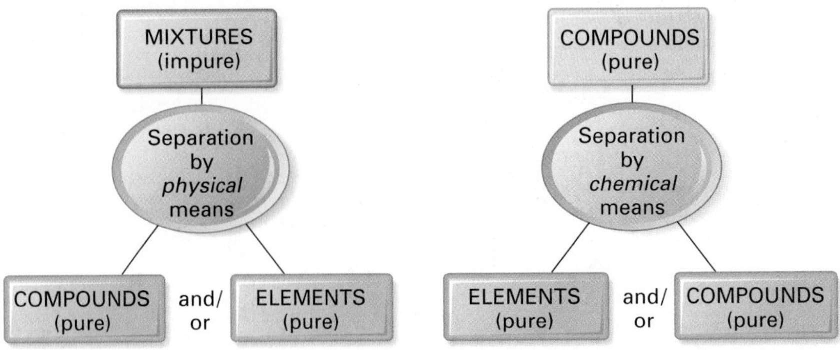


**Figure 2.16** Separations. A mixture—an impure substance—is separated into pure substances by physical means, using physical changes. A compound—a pure substance—is separated into other pure substances by chemical means, using chemical changes.

**Table 2.2** Major Elements of the Human Body

| Element | Percentage Composition by Number of Atoms | |
|---|---|---|
| Hydrogen | 63.0 | These four elements make up 99.3% of the atoms in your body. |
| Oxygen | 25.5 | |
| Carbon | 9.45 | |
| Nitrogen | 1.35 | |
| Calcium | 0.31 | |
| Phosphorus | 0.22 | |
| Chlorine | 0.057 | |
| Sulfur | 0.049 | |
| Sodium | 0.041 | |
| Potassium | 0.026 | |
| Magnesium | 0.013 | |

occur as liquids, and the remainder are solids. Most of the human body is made up of compounds made from just four elements (Table 2.2).

The name of an element is always a single word, such as *oxygen* or *iron*. The chemical names of nearly all compounds have two words, such as *sodium chloride* (table salt) and *calcium carbonate* (limestone). A few familiar compounds have one-word names, such as *water* and *ammonia.* At present, you may use the number of words in the name of a chemical to predict whether it is an element or compound. Figure 2.17 shows some well-known elements and compounds.

Chemists represent the elements with **elemental symbols.** The first letter of the name of the element, written as a capital, is often its symbol. If more than one element begins with the same letter, a second letter, written in lowercase, is added. Thus, the symbol for hydrogen is H, for helium, He, for oxygen, O, for osmium, Os, for carbon, C, and for chlorine, Cl. The symbols of some elements derive from their Latin names, such as Na for sodium (from *natrium*) and Fe for iron (from *ferrum*).

**P/REVIEW** The periodic table is a remarkable source of information that you will use throughout your study of chemistry. Your first use of it will be as an aid to learning the names and symbols of the elements in Section 5.7.

On one side of one of the reference pages described in Section 1.5 is a table that lists the names and symbols of all the elements, plus other information about them. The **periodic table of the elements** also gives the symbols of the elements.

In the next few paragraphs, we describe the particulate character of several pure substances. These are illustrated in Figure 2.18, which also includes, where possible, photographs of macroscopic samples of the substances. We suggest that you refer to Figure 2.18 as you read. It will help you visualize the particulate nature of the substances we name.

## Thinking About Your Thinking

### Mental Models

**Figure 2.18 includes particulate-level illustrations to help you form mental models of atoms, molecules, and crystalline solids.**

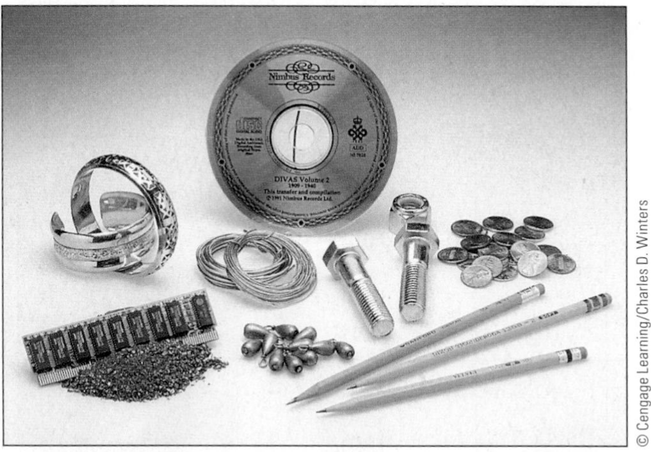


© Cengage Learning/Charles D. Winters

(a) Familiar objects that are nearly pure elements: copper wire and the copper coating on pennies; iron nuts and bolts coated with zinc for corrosion protection; lead sinkers used for fishing; graphite, a form of carbon, which is the "lead" in lead pencils; aluminum in a compact disc; pieces of silicon, used in the computer chips shown next to them; and silver in bracelets.

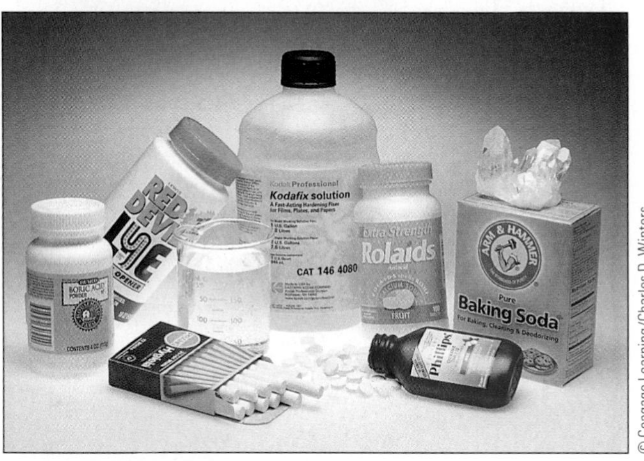


© Cengage Learning/Charles D. Winters

(b) Familiar substances that are compounds: drain cleaner, sodium hydroxide (made up of the elements sodium, hydrogen, and oxygen); photographic fixer, sodium thiosulfate (sodium, sulfur, oxygen); water (hydrogen and oxygen); boric acid (hydrogen, boron, oxygen); milk of magnesia tablets, magnesium hydroxide (magnesium, hydrogen, oxygen); quartz crystal, silicon dioxide (silicon, oxygen); baking soda, sodium hydrogen carbonate (sodium, hydrogen, carbon, oxygen); chalk and an antacid, calcium carbonate (calcium, carbon, oxygen).

**Figure 2.17** Common elements and compounds.

The symbolic representation of the particles of a pure substance is its **chemical formula.** A formula is a combination of the symbols of all the elements in the substance. The formula of most elements is the same as the symbol of the element. This indicates that the element is stable as a single atom. Helium (He), sodium (Na), and barium (Ba) are examples. Other elements exist in nature as stable, distinct, and independent molecules, which are made up of two or more atoms.* Hydrogen, oxygen, and chlorine are three such elements. Their *symbols* are H, O, and Cl, respectively, but their *formulas* are $H_2$, $O_2$, and $Cl_2$. The subscript 2 indicates in each case that a molecule of the element has two atoms. Figure 2.18 shows all of these elements.

The formula of a compound also uses subscript numbers to show the number of atoms of each element in a *formula unit.* Some formula units are molecules. Hydrogen and chlorine, for example, form a compound whose molecules consist of one atom of each element, a ratio of 1:1. Its formula is therefore HCl. A molecule of water contains two atoms of hydrogen and one atom of oxygen, a ratio of 2 hydrogens to 1 oxygen. Its formula is therefore $H_2O$. Notice that if there is only one atom of an element in a formula unit, the subscript for that element is omitted.

Other compounds exist as an orderly, repeating pattern of two or more elements, rather than as independent molecules. The formula expresses the simplest ratio of particles in the solid. Ordinary table salt consists of equal numbers of sodium and chlorine atoms. This 1:1 ratio is expressed in the formula NaCl. Barium chloride has two chlorine atoms for each barium atom. Its formula is $BaCl_2$.

The precise ratio of atoms of different elements in a compound is responsible for the **Law of Definite Composition,** also called the **Law of Constant Composition.**

*The term *molecule* also refers to elemental particles that are stable as individual atoms. In this sense, helium atoms are *monatomic molecules. Mono-* is a prefix that means one.

**Figure 2.18** Particulate and macroscopic views of elements and compounds discussed in Section 2.6. Some elements, such as helium, sodium, and barium, occur in nature as single atoms. Other elements, such as hydrogen, oxygen, and chlorine, occur in nature as multi-atom molecules. Hydrogen chloride, water, sodium chloride, and barium chloride are compounds. The term *crystalline* refers to solids with particles that are arranged in a regularly repeating pattern.

| Substance | Symbol or Formula | Natural Form | Particulate Illustration (not to scale) | Macroscopic Photograph |
|---|---|---|---|---|
| Helium | He | Atom | | A colorless gas |
| Sodium | Na | Atom/ Crystalline solid |  |  |
| Barium | Ba | Atom/ Crystalline solid |  |  |
| Hydrogen | $H_2$ | Molecule |  | A colorless gas |
| Oxygen | $O_2$ | Molecule |  | A colorless gas |
| Chlorine | $Cl_2$ | Molecule | 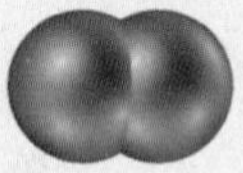 |  |
| Hydrogen chloride | HCl | Molecule | 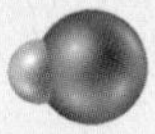 | A colorless gas |
| Water | $H_2O$ | Molecule |  |  |
| Sodium chloride | NaCl | Crystalline solid |  |  |
| Barium chloride | $BaCl_2$ | Crystalline solid | 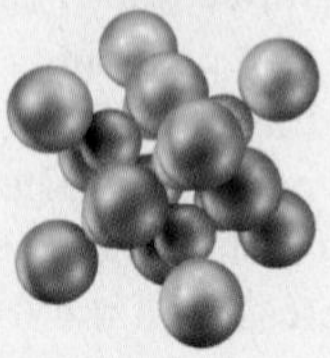 |  |

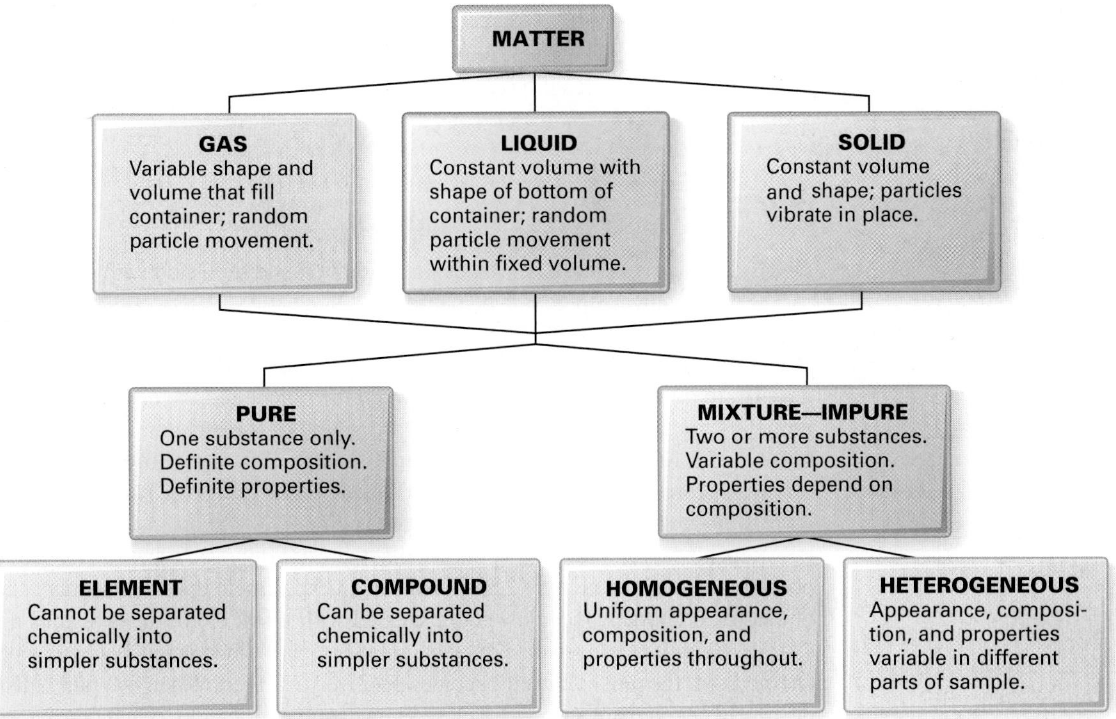


**Figure 2.19** Summary of the classification system for matter.

This law states that **any compound is always made up of elements in the same proportion by mass (weight).** The source of the compound does not matter. For example, 100 grams of pure water always contains 11.1 grams of hydrogen and 88.9 grams of oxygen.* It makes no difference if the water comes from a pond in Kansas, a river in South America, a lake in the Alps, or a comet in the far reaches of the solar system. If 100 grams of the pure material contains 11.1 grams of hydrogen and 88.9 grams of oxygen, it's water.

The properties of compounds are *always different* from the properties of the elements of which they are formed. Sodium is a shiny metal that reacts vigorously when exposed to air or water (see its photograph in Fig. 2.18); chlorine is a yellow-green gas that was the first poison gas used in World War I (also shown in Fig. 2.18). Neither element is pleasant to work with. Yet the compound formed from these two elements, sodium chloride, commonly known as table salt (also shown in Fig. 2.18), is essential in the diets of many animals, including humans.

Figure 2.19 summarizes the classification system for matter.

***Learn It Now!*** Give particular attention to textbook summaries that combine several topics in one place. The top three-box portion of Figure 2.19 summarizes Section 2.2. The middle section and the lower pairs of boxes summarize Sections 2.4 and 2.6.

## √ Target Check 2.7

*Which of the following are compounds, and which are elements?*

a) $Na_2S$
b) $Br_2$
c) potassium hydroxide
d) fluorine

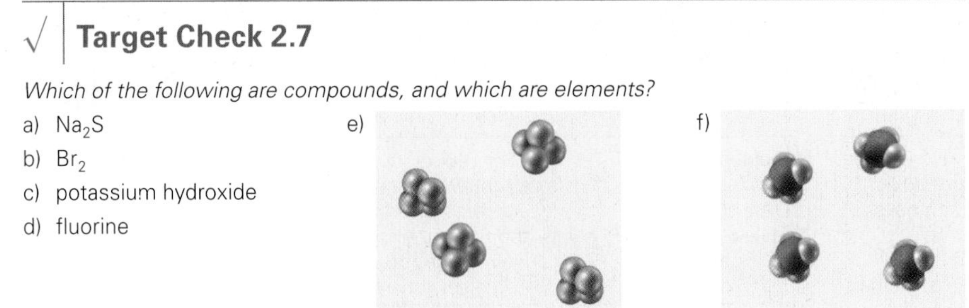


*Chemists treat units of measurement as singular subjects when writing. Thus we say "100 grams . . . contains" instead of "100 grams . . . contain."

√ **Target Check 2.8**

*Is a compound a pure substance or a mixture? On what differences between pure substances and mixtures is your answer based?*

## 2.7 The Electrical Character of Matter

**Goal 12** Match electrostatic forces of attraction and repulsion with combinations of positive and negative charge.

The four fundamental forces in the universe are gravity, the electromagnetic force, the strong force, and the weak force. The strong and weak forces operate within atoms.

If you release an object held above the floor, it falls to the floor. This is the result of gravity, an invisible attractive force between the object and the earth. Gravity is one of four fundamental forces that govern the operation of the universe. Another is the electromagnetic force. Electricity and magnetism are each a part of the electromagnetic force. All forces can be described in terms of **force fields,** or simply **fields.** A force field is the region in space where the force is effective.

Repeated observations have shown that there are only two types of electrical charge: positive and negative. Figure 2.20 illustrates an experiment that demonstrates the nature of electrical charges. If a glass rod is rubbed with a silk cloth, the rod gains a positive charge. If a pith ball, a small spongy ball made of plant fiber, is touched with a positively charged rod, the pith ball itself becomes positively charged. When two pith balls that are positively charged are suspended close to one another, they repel each other.

A hard rubber rod that is rubbed with fur acquires a negative charge. The ebonite rod shown in Figure 2.20 is made from a very hard rubber used to make bowling balls. The positively charged pith balls are attracted to the negatively charged rubber rod.

Electrostatic forces show that matter has electrical properties. These forces are responsible for the energy absorbed or released in chemical changes.

These charges are like those you develop if you scrape your feet across a rug on a dry day. You can discharge yourself by touching another person, each of you receiving a mild shock in the process. In each of these situations, the object acquires an electrical charge that does not move (a *static* electrical charge), so the electrical force is known as **static electricity.** The force is also called an **electrostatic force.**

These experiments show that

- There are only two types of electrical charge, positive and negative.
- Two objects having the *same* charge, both positive or both negative, *repel* each other.
- Two objects having *unlike* charges, one positive and one negative, *attract* each other.

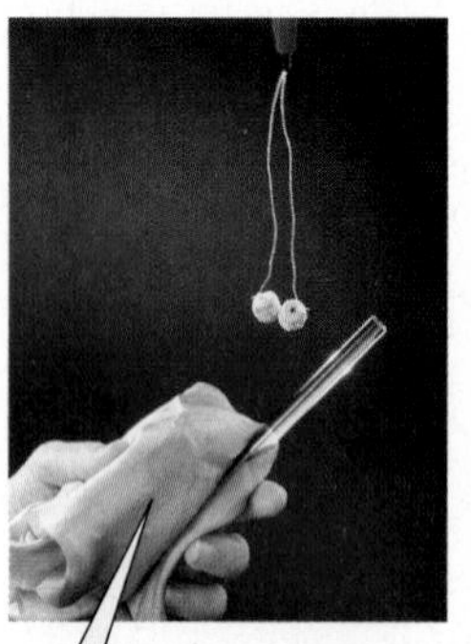

When a silk cloth is rubbed on a glass rod, electrons are transferred from the rod to the cloth, leaving the rod positively charged.

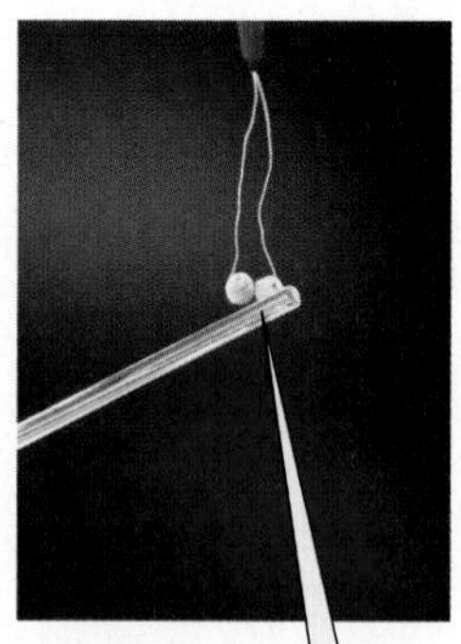

When the positively-charged rod touches two pith balls, it removes some electrons from the pith balls, leaving them with positive charges.

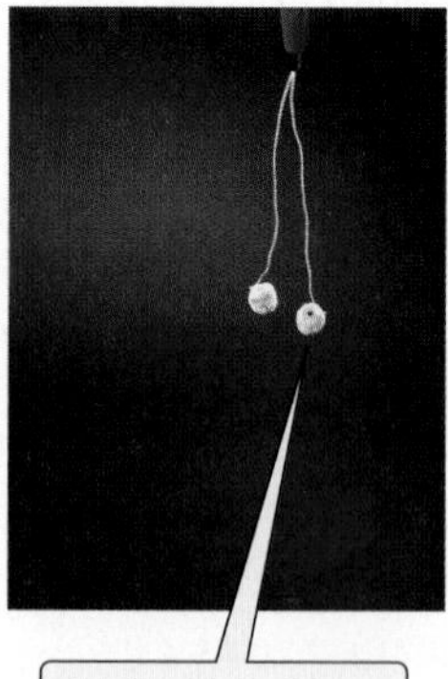

Each pith ball has a positive charge. This shows evidence that like charges repel each other.

When fur is rubbed on a rubber rod, electrons are transferred from the fur to the rod, leaving the rod negatively charged.

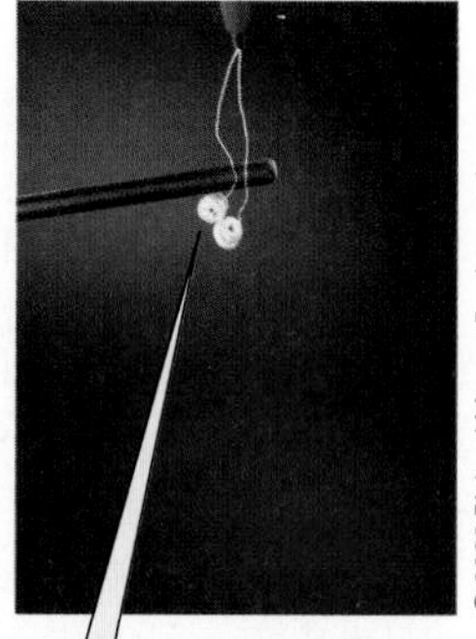

The rod has a negative charge and the pith balls have positive charge. This shows evidence that opposite charges attract each other.



**Figure 2.20** Electrostatic attraction and repulsion. Pith is a spongy tissue extracted from plants. It is lightweight, and it transfers electrical charge very readily.

Electrostatic forces show that matter has electrical properties. These forces are responsible for the energy absorbed or released in chemical changes.

**P/REVIEW:** The quantity of energy transferred in a chemical change is considered in Sections 10.7–9.

### √ Target Check 2.9

*Identify the net electrostatic force (attraction, A; repulsion, R; or none, N) between the following pairs.*

a) two positively charged table tennis balls ______

b) a negatively charged piece of dust and a positively charged dust particle ______

c) a positively charged sodium ion and a positively charged potassium ion (ions are charged particles similar to atoms) ______

## 2.8 Characteristics of a Chemical Change

### Chemical Equations

Goal | **13** Distinguish between reactants and products in a chemical equation.

In Section 2.3 we said that a chemical change occurs when the chemical identity of a starting substance is destroyed and a new substance forms. Chemists describe such a change by writing a **chemical equation.** The formulas of the beginning substances, called **reactants,** are written to the left of an arrow that points to the formulas of the substances formed, called **products.** The equation for the reaction of the element carbon with the element oxygen to form the compound carbon dioxide is

$$C + O_2 \rightarrow CO_2$$

Notice how the equation is a symbolic representation of the essence of a chemical change: The carbon (C) and oxygen ($O_2$) present at the beginning no longer exist as separate pure substances, and a new pure substance, carbon dioxide ($CO_2$), forms.

The decomposition of water (see Fig. 2.7) is another example of a chemical change. The starting compound, water ($H_2O$), decomposes into its elements, hydrogen ($H_2$) and oxygen ($O_2$). The equation is

$$2\ H_2O \rightarrow 2\ H_2 + O_2$$

### Energy in Chemical Change

Goal | **14** Distinguish between exothermic and endothermic changes.

Goal | **15** Distinguish between potential energy and kinetic energy.

If you strike a match and hold your finger in the flame, you learn very quickly that the chemical change in burning wood or paper is releasing energy that we call *heat* in everyday language. Heat is only one form of energy, and it's the one that will concern us the most. A chemical change that releases energy to its surroundings is called an **exothermic reaction.**

Sometimes energy terms are included in chemical equations. The reaction between carbon and oxygen is exothermic, releasing energy to the surroundings. This can be indicated in this way:

$$C + O_2 \rightarrow CO_2 + \text{energy}$$

Sometimes it takes energy to cause a reaction to occur. The chemical change absorbs energy from its surroundings. Electrical energy is needed to decompose water (see Fig. 2.7). This is an **endothermic reaction** whose equation can be written

$$2\ H_2O + \text{energy} \rightarrow 2\ H_2 + O_2$$

Energy is closely associated with the physical concept of work. Work is the application of a force over a distance. If you lift a book from the floor to the table, you do work. You exert the force of raising the book against the attraction of gravity, and you exert this force over the distance from the floor to the table. The energy has been transferred from you to the book. On the table the book has a higher **potential energy** than it had on the floor. The potential energy of an object depends on its position in a field where forces of attraction and/or repulsion are present. In the case of the book, the force is attraction in the earth's gravitational field.

Electrostatic forces exist between charged particles. Just as the book has a higher potential energy as it moves farther from the earth that attracts it, so oppositely charged particles have greater potential energy when they are farther apart. Conversely, the closer the particles are, the lower the potential energy in the system. The relationships are reversed for two particles with the same charge because they repel each other. Their potential energy is greater when they are close together than when they are far apart.

If you push your book off the table, it falls to the floor. Its potential energy is reduced. Physical and chemical systems tend to change in a way that reduces their total energy. "Chemical energy" comes largely from the rearrangement of charged particles in an electrostatic field. Reduction of energy to the smallest amount possible is one of the driving forces that cause chemical reactions to occur. We will mention this from time to time in this text.

A moving automobile, an airplane in flight, and a falling book all possess another kind of energy called **kinetic energy.** We have already noted that the word *kinetic* refers to motion, and motion is the common feature of this automobile, this plane, and this book. Any moving object has kinetic energy. Most of what we call "mechanical energy" is kinetic energy. In another chapter we discuss how the temperature of an object is related to the average kinetic energy of its particles.

Albert Einstein (1879–1955) is possibly the most well-recognized scientist in history. He was awarded the 1921 Nobel Prize in Physics for his contributions to theoretical physics and his explanation of the photoelectric effect.

√ **Target Check 2.10**

a) Is the process of boiling water exothermic or endothermic with respect to the water?

b) A charged object is moved closer to another object that has the same charge. This is an energy change. Is it a change in kinetic energy or potential energy? Is the energy change an increase or a decrease?

## 2.9 Conservation Laws and Chemical Change

### The Conservation Law

Early in the 20th century, Albert Einstein recognized a "sameness" between matter and energy and suggested that it should be possible to convert one into the other. He proposed that matter and energy are related by the equation

The Greek capital letter delta, Δ, is used in the sciences to denote "change in."

$$\Delta E = \Delta m \times c^2$$

Change in energy = (change in mass) × (speed of light)$^2$

The word *mass* refers to quantity of matter. It is closely related to the more familiar term *weight.* Section 3.4 explains the difference between mass and weight. The relationship shown in the equation tells us that matter is an extremely concentrated form of energy! Conversions between matter and energy occur primarily in nuclear reactions.

The conversion of a tiny amount of matter can produce an enormous amount of energy. If it were possible to convert all of a given mass of coal to energy, that energy would be 2,500,000,000—two and one-half billion—times as great as the energy derived from burning that same amount of coal (Fig. 2.21). This is why nuclear bombs are so destructive. It is also why nuclear energy, although troubled by serious questions of safety and cost, is such an attractive alternative to traditional sources of energy.

**P/REVIEW:** Lavoisier's experiments, the disproving of the phlogiston theory, and the proposal of the Law of Conservation of Mass are described more fully in Section 1.1.

### The Law of Conservation of Mass

Goal | **16** State the meaning of, or draw conclusions based on, the Law of Conservation of Mass.

As discussed in Chapter 1, early chemists who studied burning wood concluded that because the ash remaining was so much lighter than the object burned, something called phlogiston was lost in the reaction. However, their reasoning was faulty. Lavoisier realized that oxygen in the air, which could not be seen, was a reactant, and that carbon dioxide and water vapor, also invisible, were products. If we take account of those three gases, we find that

Total mass of reactants = Total mass of products

Mass of (wood + oxygen) = Mass of (ash + carbon dioxide + water vapor)

This equation is the **Law of Conservation of Mass: In a nonnuclear change, mass is conserved. It is neither created nor destroyed.**

The fact that matter can be converted to energy and vice versa does not necessarily repeal the Law of Conservation of Mass. For all nonnuclear changes, the law remains valid within our ability to measure such changes. If we include nuclear changes, the law must be modified by stating that the total of all mass and energy in a change is conserved.

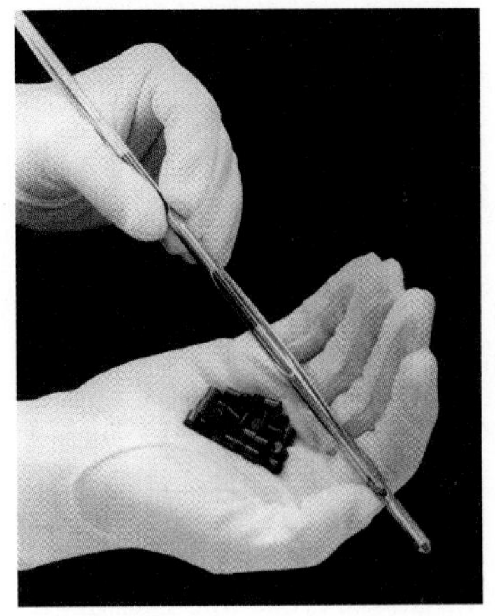


Figure 2.21 Nuclear fuel. A uranium fuel pellet the size of this inert steel cylinder produces energy equal to the energy produced by burning about one ton of coal.

## The Law of Conservation of Energy

Goal | 17 State the meaning of, or draw conclusions based on, the Law of Conservation of Energy.

Energy changes take place all around us—and within us—all the time. Driving an automobile starts with the chemical energy of a battery and fuel. This changes into kinetic, potential, sound, and heat energy as the car moves up and down hills and into light energy when the brake lights go on. The electric alarm clock that moves silently next to your bed converts electrical energy to light energy and then to sound energy when the night is past. Even as you sleep, your body is processing the food you ate into heat energy that maintains body temperature. Figure 2.22 shows other energy conversions.

Careful study of energy conversions shows that the energy lost or used in one form is always exactly equal to the energy gained in another form. This leads to another

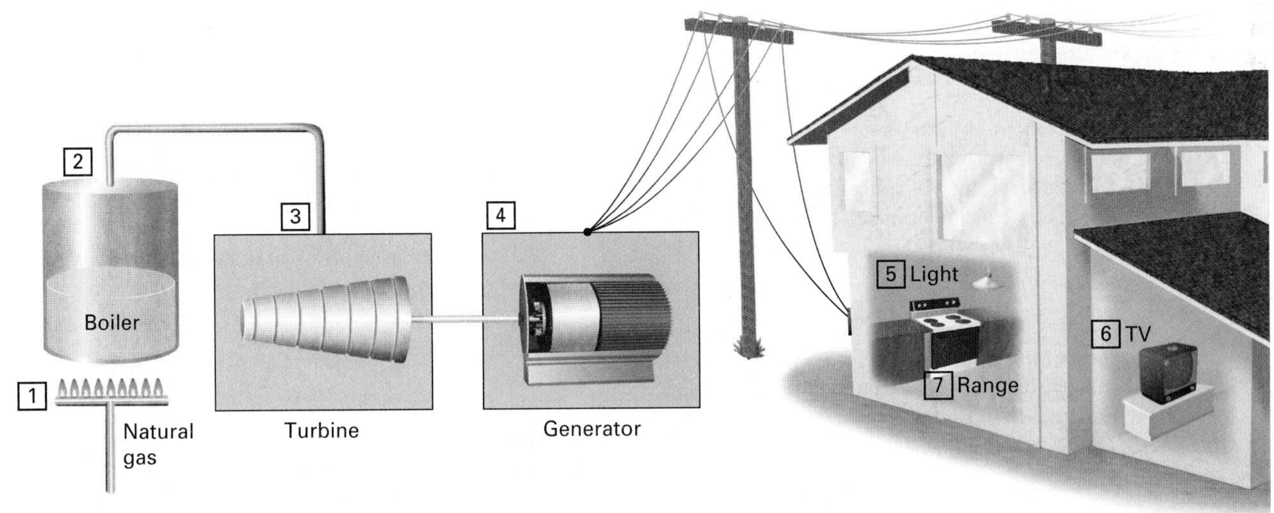


Figure 2.22 Energy changes. Common events in which energy changes from one form to another. (1) Chemical energy of fuel changes to heat energy. (2) Heat energy changes to higher kinetic and potential energy of steam compared to liquid water. (3) Kinetic energy changes to rotating mechanical energy in a turbine. (4) Mechanical energy is transmitted to a generator, where it changes to electrical energy. (5) Electrical energy changes to heat and light energy. (6) Electrical energy is changed to light, sound, and heat energy. (7) Electrical energy changes to heat energy.

conservation law, the **Law of Conservation of Energy: In a nonnuclear change, energy is conserved. It is neither created nor destroyed.**

**Target Check 2.11**

*In everyday language, the term* ***conserved*** *usually refers to protecting something. (She* ***conserved*** *her money.) What does the term* ***conserved*** *mean in scientific language?*

## Chapter 2 in Review

Most of the key terms and concepts and many others appear in the Glossary. Use your Glossary regularly.

**2.1 Representations of Matter: Models and Symbols**

Goal 1 Identify and explain the differences among observations of matter at the macroscopic, microscopic, and particulate levels.

Goal 2 Define the term *model* as it is used in chemistry to represent pieces of matter too small to see.

**Key Terms and Concepts: ball-and-stick model, macroscopic, matter, microscopic, model, particulate, space-filling model, symbol (chemical)**

**2.2 States of Matter**

Goal 3 Identify and explain the differences among gases, liquids, and solids in terms of (a) visible properties, (b) distance between particles, and (c) particle movement.

**Key Terms and Concepts: gas, kinetic molecular theory, liquid, molecule, solid, states of matter**

**2.3 Physical and Chemical Properties and Changes**

Goal 4 Distinguish between physical and chemical properties at both the particulate level and the macroscopic level.

Goal 5 Distinguish between physical and chemical changes at both the particulate level and the macroscopic level.

**Key Terms and Concepts: chemical change, chemical property, chemical reaction, physical change, physical property**

**2.4 Pure Substances and Mixtures**

Goal 6 Distinguish between a pure substance and a mixture at both the macroscopic level and the particulate level.

Goal 7 Distinguish between homogeneous and heterogeneous matter.

**Key Terms and Concepts: heterogeneous mixture, homogeneous, mixture, phase, pure substance, solution**

**2.5 Separation of Mixtures**

Goal 8 Describe how distillation and filtration rely on physical changes and properties to separate components of mixtures.

**Key Terms and Concepts: distillation, distilled water, filtration**

**2.6 Elements and Compounds**

Goal 9 Distinguish between elements and compounds.

Goal 10 Distinguish between elemental symbols and the formulas of chemical compounds.

Goal 11 Distinguish between atoms and molecules.

**Key Terms and Concepts: atom, chemical formula, compound, element, elemental symbol, Law of Definite or Constant Composition, periodic table of the elements**

**2.7 The Electrical Character of Matter**

Goal 12 Match electrostatic forces of attraction and repulsion with combinations of positive and negative charge.

**Key Terms and Concepts: electrostatic force, force field or field, static electricity**

**2.8 Characteristics of a Chemical Change**

Goal 13 Distinguish between reactants and products in a chemical equation.

Goal 14 Distinguish between exothermic and endothermic changes.

Goal 15 Distinguish between potential energy and kinetic energy.

**Key Terms and Concepts: chemical equation, endothermic reaction, exothermic reaction, kinetic energy, potential energy, product, reactant**

**2.9 Conservation Laws and Chemical Change**

Goal 16 State the meaning of, or draw conclusions based on, the Law of Conservation of Mass.

Goal 17 State the meaning of, or draw conclusions based on, the Law of Conservation of Energy.

**Key Terms and Concepts: Law of Conservation of Energy, Law of Conservation of Mass**

## Study Hints and Pitfalls to Avoid

In determining whether a change is physical or chemical, remember that the chemical identities and the starting substances are destroyed in a chemical change and new substances form. If the particles themselves remain the same, the change must be physical.

A common pitfall in this chapter is not recognizing that both elements and compounds are pure substances. A compound is not a mixture. It takes a chemical change to break a compound into its elements (or other compounds). By contrast, a mixture is separated into its components, elements or compounds, by one or more physical changes.

## Concept-Linking Exercises

*Write a brief description of the relationships among each of the following groups of terms or phrases. Answers to the Concept-Linking Exercises are given after answers to the Target Checks in Appendix III.*

*Example:* Natural sciences, physical sciences, biological sciences, chemistry, physics, botany, zoology.

*Solution:* The natural sciences can be divided into two general categories: physical sciences, the study of matter and energy, and biological sciences, the study of living organisms. Botany and zoology are biological sciences. Physics and chemistry are physical sciences, although chemistry overlaps the biological sciences in the fields of biochemistry, biological chemistry, and chemical biology.

1. Matter, state of matter, kinetic molecular theory, gas, liquid, solid
2. Homogeneous, heterogeneous, pure substance, mixture
3. Element, compound, atom, molecule
4. Physical property, physical change, chemical property, chemical change
5. Conservation of mass, conservation of energy, the conservation law
6. Kinetic energy, potential energy, endothermic change, exothermic change

## Small-Group Discussion Questions

*Small-Group Discussion Questions are for group work, either in class or under the guidance of a leader during a discussion section.*

1. A model often is simpler than the natural phenomenon that it represents. How is this an advantage for thinking about matter? How is it a disadvantage?
2. Viscosity is defined as the resistance of a substance to flow. Explain why each state of matter either does or does not have the property of viscosity. How do you suppose viscosity occurs at the particulate level?
3. Describe as many chemical and physical changes and properties as possible that are given in, or that you can deduce from, the following: At the end of a day hiking in the mountains, you set up camp. You gather dry sticks and logs to build a fire, and you light the fire with a match. After roasting marshmallows over the fire, you douse it with cold water from a nearby stream, causing a cloud of steam to form as the fire goes out.
4. If you were to go on a hunt around campus for pure substances, what would you find? List at least as many pure substances as there are members of your group, and for each substance listed, explain why it is pure.
5. Consider the distillation apparatus in Figure 2.13. Explain how you can use the change illustrated as evidence to deduce the composition of the gas that forms in the distillation flask.
6. If you were to go on a hunt around campus for elements, what would you find? List at least as many elements as there are members of your group, and for each element listed, explain why it is an element. How many elements on your list are pure substances? Explain.
7. Compare and contrast the electrical character of matter with the magnetic character of bar magnets. How are they the same? How are they different?
8. Can you have a chemical change without an accompanying physical change? Explain, defining both *chemical change* and *physical change*. Give an example and a counterexample, if possible, to support your explanation.
9. If matter is indeed conserved, how can you explain the fact that a glass of water left alone on a kitchen counter will eventually empty? The outside of a glass of cold water typically becomes wet. What substance wets the glass and where does it come from?

## Questions, Exercises, and Problems

*Questions whose numbers are printed in blue are answered in Appendix III. ■ denotes problems assignable in OWL. In the first section of Questions, Exercises, and Problems, similar exercises and problems are paired in consecutive odd-even number combinations.*

### Section 2.1: Representations of Matter: Models and Symbols

1. Identify the following samples of matter as macroscopic, microscopic, or particulate: (a) a human skin cell; (b) a sugar molecule; (c) a blade of grass; (d) a helium atom; (e) a single-celled plant too small to see with the unaided eye.
2. ■ Classify each of the following as macroscopic, microscopic, or particulate: (a) a cell membrane; (b) a silver atom; (c) iron filings.
3. Suggest a reason for studying matter at the particulate level, given that it is too small to see.
4. How does a chemist think about particles that are so small, they are impossible to see with the naked eye or even the most powerful optical microscope?

### Section 2.2: States of Matter

5. Using spheres to represent individual atoms, sketch particulate illustrations of a substance as it is heated from the solid to the liquid and to the gaseous state.
6. Describe a piece of ice at the particulate level. Then describe what happens to the ice as it is heated until it melts and eventually boils.
7. The word *pour* is commonly used in reference to liquids, but not to solids or gases. Can you pour a solid or a gas? Why or why not? If either answer is yes, can you give an example?

8. The slogan "When it rains, it pours" has been associated with a brand of table salt for decades. How can salt, a solid, be poured? What unique feature—unique at one time, but not today—do you suppose was being emphasized by the slogan? In other words, under what circumstances would one brand of salt "pour" while another brand would not, and why?

9. Which of the three states of matter is most easily compressed? Suggest a reason for this.

10. Compare the volumes occupied by the same sample of matter when in the solid, liquid, and gaseous states.

**Section 2.3: Physical and Chemical Properties and Changes**

11. Classify each of the following properties as chemical or physical: (a) hardness of a diamond; (b) combustibility of gasoline; (c) corrosive character of an acid; (d) elasticity of a rubber band; (e) taste of chocolate.

12. ■ Classify the *italicized* property as chemical or physical: (a) a shiny piece of iron metal *gets rusty* when left outside; (b) the purple crystalline solid potassium permanganate *forms a purple solution* when dissolved in water; (c) the shiny metal mercury *is a liquid at room temperature.*

13. Which among the following are physical changes? (a) blowing glass; (b) fermenting grapes; (c) forming a snowflake; (d) evaporating dry ice; (e) decomposing a substance by heating it.

14. ■ Classify each of the following changes as chemical or physical: (a) grilling a steak; (b) souring of milk; (c) removing nail polish.

15. ■ Is the change illustrated below a physical change or a chemical change? Explain your answer.

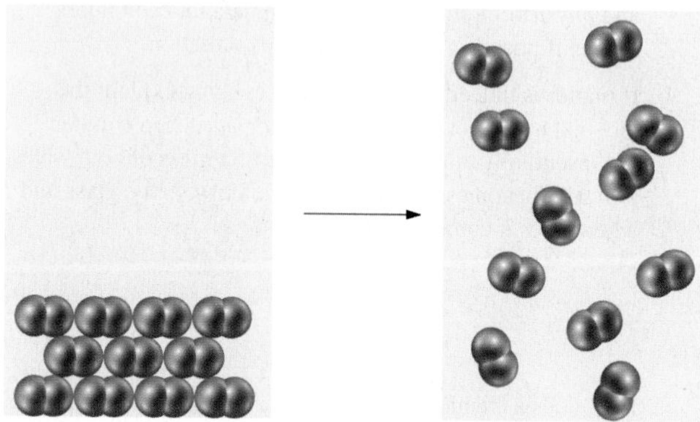

16. Is the change in the illustration below a physical change or a chemical change? Explain your answer.

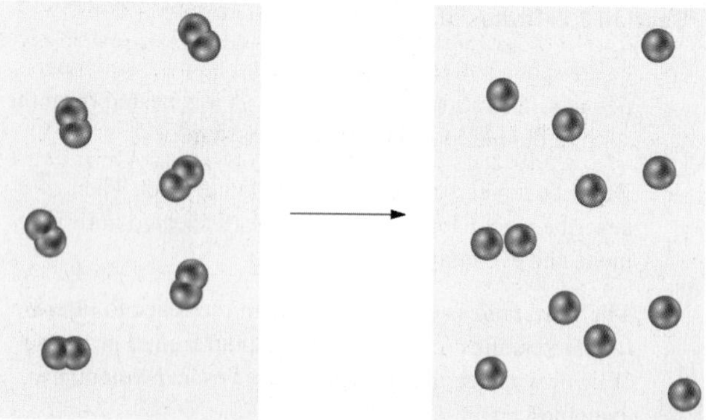

**Section 2.4: Pure Substances and Mixtures**

17. Diamonds and graphite are two forms of carbon. Carbon is an element. Chunks of graphite are sprinkled among the diamonds on a jeweler's display tray. Is the material on the tray a pure substance or a mixture? Is the display homogeneous or heterogeneous? Justify both answers.

18. Aspirin is a pure substance. If you had the choice of buying a widely advertised brand of aspirin whose effectiveness is well known or the generic product of a new manufacturer at half the price, which would you buy? Explain.

19. The substance in the glass below is from a kitchen tap. Is it a pure substance or a mixture? What if it came from a bottle of distilled water?

20. Are the contents of the bottle in the picture below a pure substance or a mixture?

21. ■ Which of the following particulate illustrations represent pure substances and which represent mixtures?

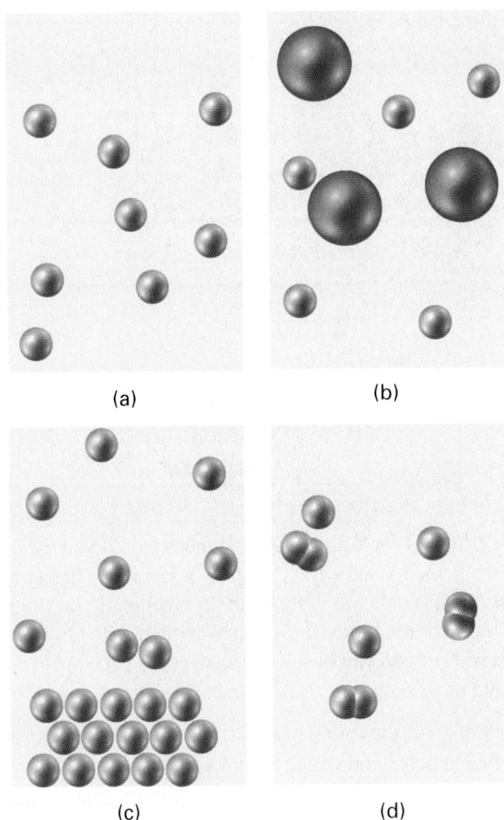

(a) (b) (c) (d)

22. Which of the following particulate illustrations represent pure substances and which represent mixtures?

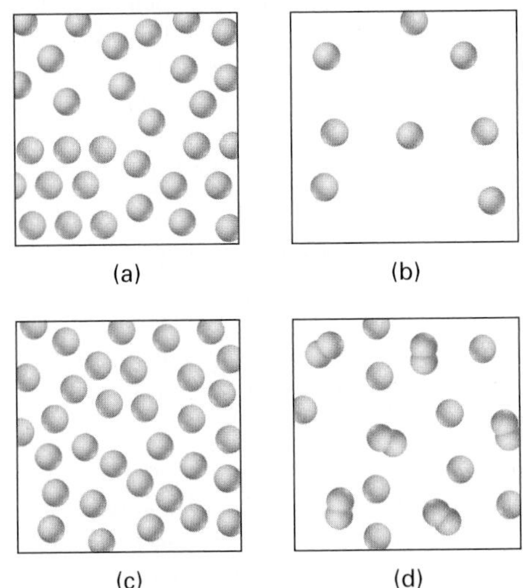

(a) (b) (c) (d)

23. Which of the following are pure substances and which are mixtures? (a) table salt; (b) tap water; (c) clean, dry air; (d) steam.

24. Which of the substances below are pure and which are mixtures? Which could be either? Explain your answers.

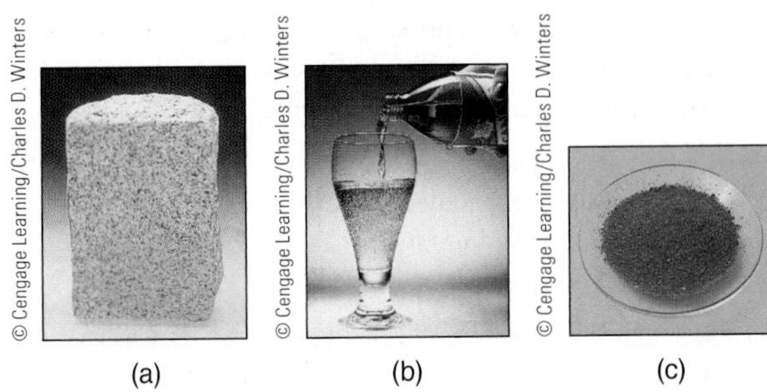


(a) (b) (c)

25. Apart from food, list five things in your home that are homogeneous.

26. Can the terms *homogeneous* and *heterogeneous* be applied to pure substances as well as to mixtures? Explain.

27. Which items in the following list are heterogeneous? (a) sterling silver; (b) freshly opened root beer; (c) popcorn; (d) scrambled eggs; (e) motor oil.

28. ■ Classify each of the following mixtures as either homogeneous or heterogeneous: (a) apple juice; (b) concrete; (c) gin.

29. Some ice cubes are homogeneous and some are heterogeneous. Into which group do ice cubes from your home refrigerator fall? If homogeneous ice cubes are floating on water in a glass, are the contents of the glass homogeneous or heterogeneous? Justify both answers.

30. The brass cylinder in the picture below is a mixture of copper and zinc. Is the cylinder a homogeneous or heterogeneous substance?

31. Draw a particulate-level sketch of a heterogeneous pure substance.

32. Draw a particulate-level sketch of a homogeneous mixture.

**Section 2.5: Separation of Mixtures**

33. Suppose someone emptied ball bearings into a container of salt. Could you separate the ball bearings from the salt? How? Would your method involve no change, be a physical change, or be a chemical change?

34. Suggest at least two ways to separate ball bearings from table-tennis balls. On what property is each method based?

35. A liquid that may be either pure or a mixture is placed in a distillation apparatus (see Fig. 2.13). The liquid is allowed to boil, and some condenses in the receiving flask. The remaining liquid is then removed and frozen, and the freezing point is found to be lower than the freezing point of the original liquid. Is the original liquid pure or a mixture? Explain.

36. You receive a mixture of table salt and sand and have to separate the mixture into pure substances. Explain how you would carry out this task. Is your method based on physical or chemical properties? Explain.

### Section 2.6: Elements and Compounds

37. Classify the following as compounds or elements: (a) silver bromide (used in photography); (b) calcium carbonate (limestone); (c) sodium hydroxide (lye); (d) uranium; (e) tin; (f) titanium.

38. ■ Classify each of the following pure substances as either an element or a compound: (a) silicon dioxide; (b) tungsten; (c) silver.

39. Which of the following are elements, and which are compounds? (a) NaOH; (b) $BaCl_2$; (c) He; (d) Ag; (e) $Fe_2O_3$.

40. ■ Classify each of the following pure substances as either an element or a compound: (a) C; (b) $C_2H_5OH$; (c) $Cl_2$.

41. Classify each substance in the illustrations below as an element or a compound.

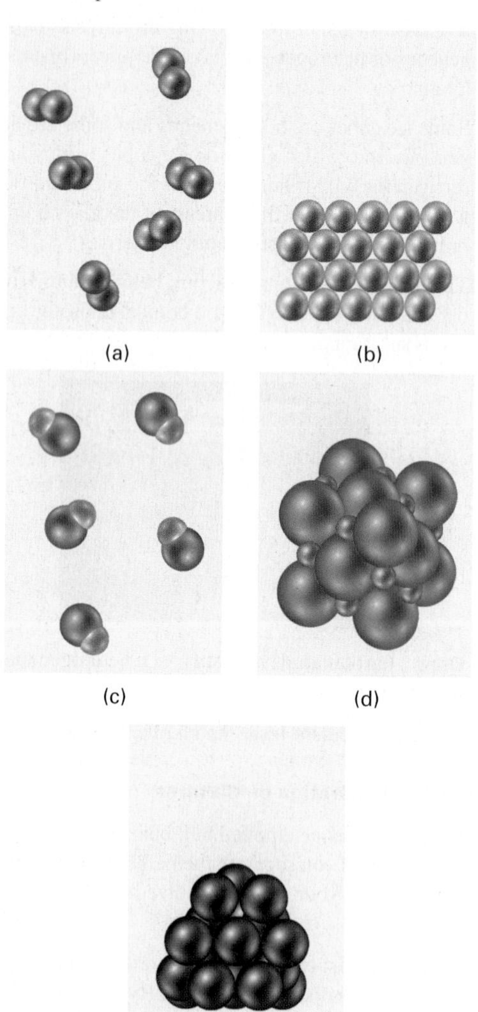


42. ■ Does each of the particulate-level models below depict an element or a compound?

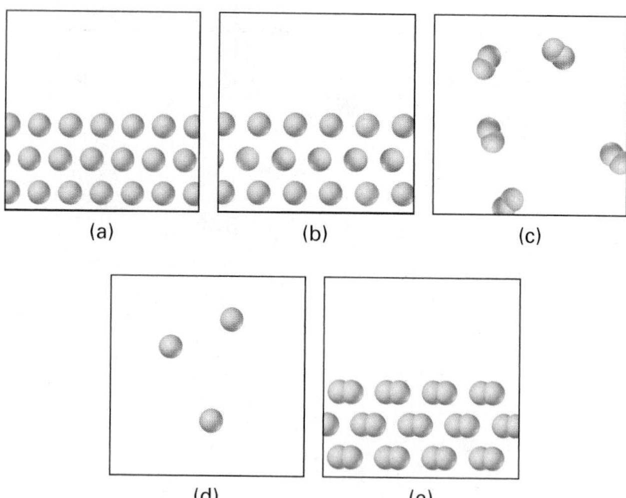


43. (a) Which of the following substances would you expect to be elements and which would you expect to be compounds? (1) calcium carbonate; (2) arsenic; (3) uranium; (4) potassium chloride; (5) chloromethane. (b) On what general rule do you base your answers to part (a)? Can you name any exceptions to this general rule?

44. (a) Which of the following substances would you expect to be elements and which would you expect to be compounds? (1) aluminum sulfate; (2) osmium; (3) radon; (4) lithium carbonate; (5) dimethylhydrazine. (b) On what general rule do you base your answers to part (a)? Can you name any exceptions to this general rule for compounds?

45. Metal A dissolves in nitric acid solution. You can recover the original metal if you place Metal B in the solution. Metal A becomes heavier after prolonged exposure to air. The procedure is faster if the metal is heated. From the evidence given, can you tell if Metal A definitely is or could be an element or a compound? If you cannot, what other information do you need to make that classification?

46. A white, crystalline material that looks like table salt gives off a gas when heated under certain conditions. There is no change in the appearance of the solid that remains, but it does not taste the same as it did originally. Was the beginning material an element or a compound? Explain your answer.

*Questions 47 and 48: Samples of matter may be classified in several ways, including: gas, liquid, or solid (G, L, S); pure substance or mixture (P, M); homogeneous or heterogeneous (Hom, Het); and, for pure substances, element or compound (E, C). For each substance in the left column of the tables shown, place in the other columns the symbol at the top of the column that best describes the substance in its most common state at room temperature and pressure. Assume that the material is clean and uncontaminated. (The first box is filled in as an example.)*

47.

| | G, L, S | P, M | Hom, Het | E, C |
|---|---|---|---|---|
| Factory smokestack emissions | All, but mostly G | | | |
| Concrete (in a sidewalk) | | | | |
| Helium | | | | |
| Hummingbird feeder solution | | | | |
| Table salt | | | | |

48.

| | G, L, S | P, M | Hom, Het | E, C |
|---|---|---|---|---|
| Limestone (calcium carbonate) | | | | |
| Lead | | | | |
| Freshly squeezed orange juice | | | | |
| Oxygen | | | | |
| Butter in the refrigerator | | | | |

**Section 2.7: The Electrical Character of Matter**

49. What is the main difference between electrostatic forces and gravitational forces? Which is more similar to the magnetic force? Can two or all three of these forces be exerted between two objects at the same time?

50. Identify the net electrostatic force (attraction, repulsion, or none) between the following pairs of substances: (a) a small, negatively charged piece of paper and a small, positively charged piece of paper; (b) two positively charged lint balls; (c) a positively charged sodium ion and a negatively charged oxide ion.

**Section 2.8: Characteristics of a Chemical Change**

51. Identify the reactants and products in the equation $AgNO_3 + NaCl \rightarrow AgCl + NaNO_3$.

52. ■ In the following equation for a chemical reaction, the notation (s), ($\ell$), or (g) indicates whether the substance is in the solid, liquid, or gaseous state: $2\ H_2S(g) + 3\ O_2(g) \rightarrow 2\ H_2O(g) + 2\ SO_2(g) + \text{energy}$. Identify each of the following as a product or reactant: (a) $SO_2(g)$; (b) $H_2S(g)$; (c) $O_2(g)$; (d) $H_2O(g)$. When the reaction takes place, is energy released or absorbed? Is the reaction endothermic or exothermic?

53. In the equation $Ni + Cu(NO_3)_2 \rightarrow Ni(NO_3)_2 + Cu$, which of the reactants is/are elements, and which of the products is/are compounds?

54. ■ Write the formulas of the elements that are products and the formulas of the compounds that are reactants in $2\ Na + 2\ H_2O \rightarrow 2\ NaOH + H_2$.

55. Which of the following processes is/are exothermic? (a) water freezing; (b) water vapor in the air changing to liquid water droplets on a windowpane; (c) molten iron solidifying; (d) chocolate candy melting.

56. ■ Classify each of the following changes as endothermic or exothermic with respect to the *italicized* object: (a) cooling a *beer;* (b) burning *leaves;* (c) cooking a *hamburger.*

57. As a child plays on a swing, at what point in her movement is her kinetic energy the greatest? At what point is potential energy at its maximum?

58. A bicycle accelerates from 5 miles per hour to 15 miles per hour. Does its energy increase or decrease? Is the change in potential energy or kinetic energy?

**Section 2.9: Conservation Laws and Chemical Change**

59. After solid limestone is heated, the rock that remains weighs less than the original limestone. What do you conclude has happened?

60. Before electronic flashes were commonly used in photography, a darkened area was lit by a device known as a flashbulb. This one-use device was essentially a glass bulb, filled with oxygen, that encased a metal wire. An electrical discharge from the camera ignited the wire, causing a brief flash of light as the wire quickly burned. How would you expect the mass of a flashbulb before use to compare with its mass after use? Explain.

61. The photograph below shows a beaker of water and a sugar cube, the combined mass of which is balanced by the weights on the right pan. The sugar cube is then placed in the water, and it dissolves completely. Do weights need to be added to or taken away from the right pan to keep the system in balance? Explain.

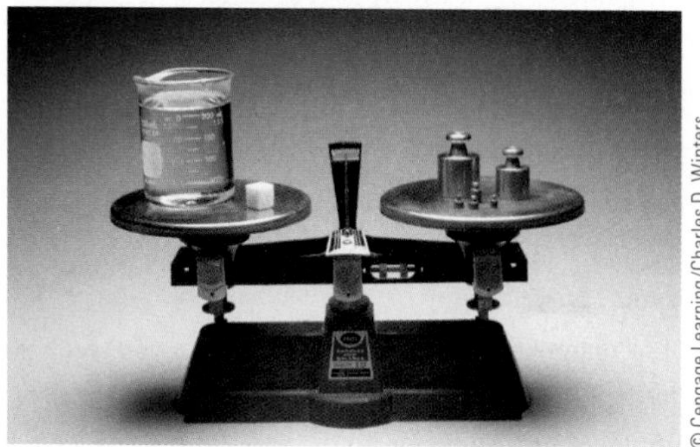


62. Plants manufacture their own food. What is the source of energy for this process? Explain how energy is conserved as a plant makes its food.

63. Identify several energy conversions that occur regularly in your home. State whether each is useful, wasteful, or sometimes useful and sometimes wasteful.

64. List the energy conversions that occur in the process from the time water is about to enter a hydroelectric dam to the burning of an electric lightbulb in your home.

**General Questions**

65. Distinguish precisely and in scientific terms the differences among items in the following groups.
    a) Macroscopic matter, microscopic matter, particulate matter
    b) Physical change, physical property, chemical change, chemical property
    c) Gases, liquids, solids
    d) Element, compound

e) Atom, molecule
f) Pure substance, mixture
g) Homogeneous matter, heterogeneous matter
h) Reactant, product
i) Exothermic change, endothermic change
j) Potential energy, kinetic energy

66. ■ Determine whether each of the following statements is true or false:
a) The fact that paper burns is a physical property.
b) Particles of matter are moving in gases and liquids, but not in solids.
c) A heterogeneous substance has a uniform appearance throughout.
d) Compounds are impure substances.
e) If one sample of sulfur dioxide is 50% sulfur and 50% oxygen, then all samples of sulfur dioxide are 50% sulfur and 50% oxygen.
f) A solution is a homogeneous mixture.
g) Two positively charged objects attract each other, but two negatively charged objects repel each other.
h) Mass is conserved in a nonnuclear endothermic chemical change, but not in a nonnuclear exothermic chemical change.
i) Potential energy can be related to positions in an electric force field.
j) Chemical energy can be converted to kinetic energy.
k) Potential energy is more powerful than kinetic energy.
l) A chemical change always destroys something and always creates something.

67. A natural-food store advertises that no chemicals are present in any food sold in the store. If the ad is true, what do you expect to find in the store?

68. Name some things you have used today that are not the result of human-made chemical change.

69. Name some pure substances you have used today.

70. How many homogeneous substances can you reach without moving from where you are sitting right now?

71. Which of the following can be pure substances: mercury, milk, water, a tree, ink, iced tea, ice, carbon?

72. Can you have a mixture of two elements as well as a compound of the same two elements?

73. Can you have more than one compound made of the same two elements? If yes, try to give an example.

74. Rainwater comes from the oceans. Is rainwater more pure, less pure, or of the same purity as ocean water? Explain.

75. A large box contains a white powder of uniform appearance. One sample is taken from the top of the box and another is taken from the bottom. Analysis reveals that the percentage of oxygen in the sample from the top is 48.2%, whereas in the sample from the bottom it is 45.3%. Answer each question below independently and give a reason that supports your answer.
a) Is the powder an element or a compound?
b) Are the contents of the box homogeneous or heterogeneous?
c) Can you be certain that the contents of the box are either a pure substance or a mixture?

76. If energy cannot be created or destroyed, as the Law of Conservation of Energy states, why are we so concerned about wasting our energy resources?

77. Consider the sample of matter in the illustration below.

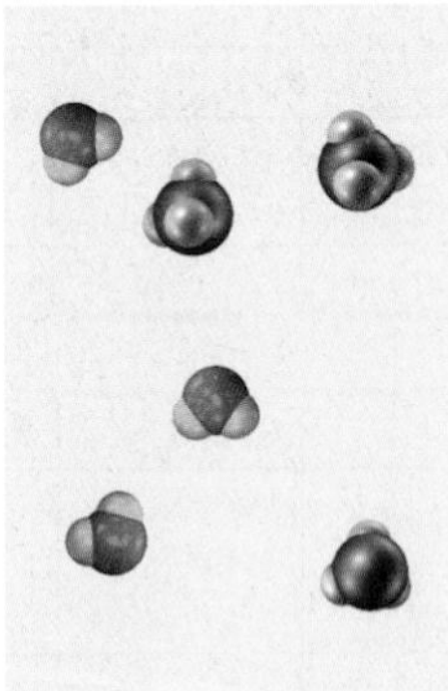

Answer each question independently and explain your answers.
a) Is the sample homogeneous or heterogeneous?
b) Is the sample a pure substance or a mixture?
c) Are the particles elements or compounds?
d) Are the particles atoms or molecules?
e) Is the sample a gas, a liquid, or a solid?

78. A particulate-level illustration of the reaction AB + CD → AD + CB is shown below.

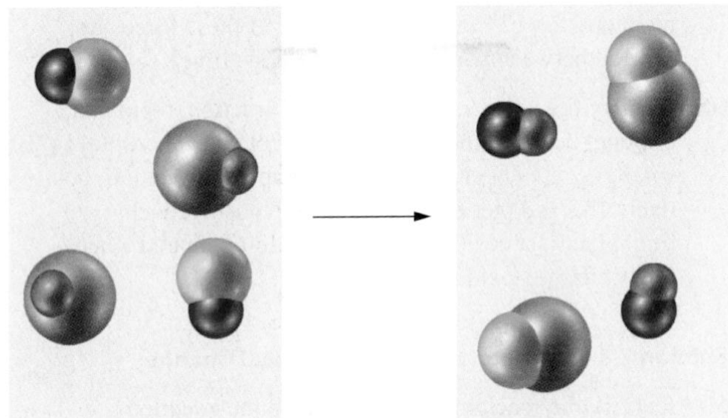

a) Identify the reactants and products in this reaction.
b) Is the change shown chemical or physical?
c) Is the mass of the product particles less than, equal to, or greater than the mass of the reactant particles?
d) If the reaction takes place in a container that allows no energy to enter or to leave, how does the total energy in the container after the reaction compare with the total energy in the container before the reaction?

### More Challenging Problems

79. A clear, colorless liquid is distilled in an apparatus similar to that shown in Figure 2.13. The temperature remains constant throughout the distillation process. The liquid leaving the condenser is also clear and colorless. Both liquids are odorless, and they have the same freezing point. Is the starting liquid a pure substance or a mixture? What single bit of evidence in the preceding description is the most convincing reason for your answer?

80. The density of a liquid is determined in the laboratory. The liquid is left in an open container overnight. The next

morning the density is measured again and found to be greater than it was the day before. Is the liquid a pure substance or a mixture? Explain your answer.

81. There is always an increase in potential energy when an object is raised higher above the surface of the earth, that is, when the distance between the earth and the object increases. Increasing the distance between two electrically charged objects, however, may raise or lower potential energy. How can this be?

82. In the gravitational field of the earth, an object always falls until some physical object prevents it from falling farther. Two electrically charged objects, each of which is made up of unequal numbers of both positive and negative charges, will reach a certain separation distance and stay there without physical support. Can you suggest an explanation for this?

83. Particles in the illustration below undergo a chemical change.

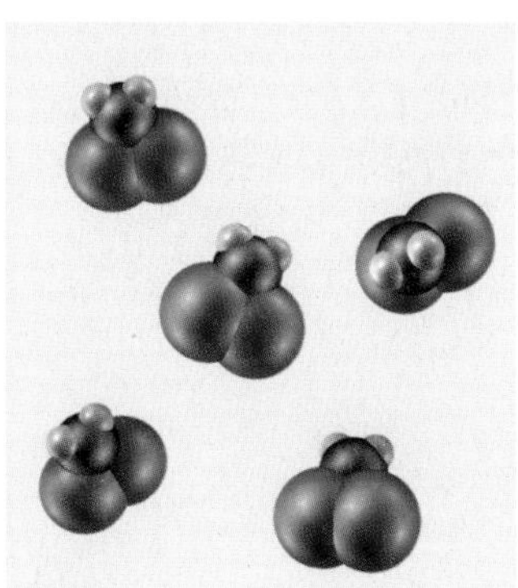

Which among the remaining boxes, (a) through (d), can represent the products of the chemical change? If a box cannot represent the products of the chemical change, explain why.

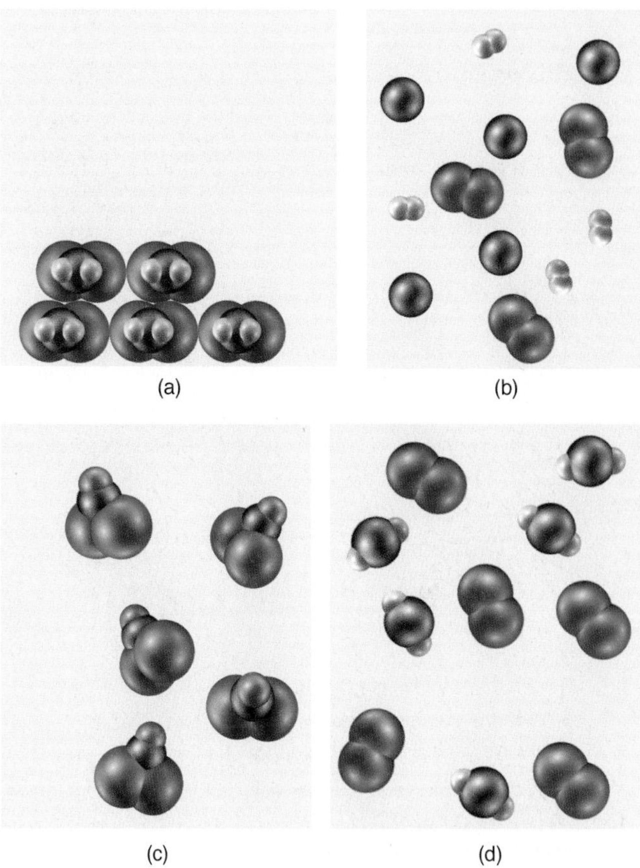

(a) (b) (c) (d)

84. Draw a particulate illustration of five particles in the gas phase in a box. Show the particles at a lower temperature, in the liquid phase, in a new box. Now show the particles at an even lower temperature, in the solid phase, in a new box. Write a description of your illustrations in terms of the kinetic molecular theory.

Why does one full can float while the other sinks? The answer is a difference in density, as discussed in Section 3.8. But why is the liquid in one can more dense than in the other? The answer to this question is based on the differences between the sweeteners. The same volume of one liquid solution, including its dissolved sweetener, weighs more than the other.

# Measurement and Chemical Calculations

CONTENTS



Chemistry is both qualitative and quantitative. In its qualitative role it explains *how* and *why* chemical and physical changes occur. Quantitatively, it considers the amount of a substance measured, used, or produced. Determining the amount involves both measuring and performing calculations, which are the subjects of Chapter 3.

It is likely that this chapter is your introduction to quantitative problem solving in this course. Most of the examples in this book are written in an active-learning, self-teaching style, a series of questions and answers that guide you to understanding a problem. To reach that understanding, you should answer each question before looking at the answer printed on the next line of the page. This requires a shield to cover that answer while you consider the question. Tear-out shields for this purpose are provided in the book.

Find them now and tear one off. On one side you will find instructions on how to use the shield, copied from this section. On the other side is a periodic table that you can use for reference.

Online homework for this chapter may be assigned in OWL

The examples provided in this textbook are designed so that you can actively work them, writing your response to each question. This does not mean that the examples are optional or something to do after you read the text. On the contrary, they are an integral part of the textbook. To maximize learning, you should work each Active Example when you come to it.

We understand the temptation to just read the textbook in the same way you might read a novel. It's certainly much easier than doing the work of writing answers to our questions, comparing your answers to ours, and thinking about your thinking and working to improve. But research on human learning clearly shows that only reading a textbook results in almost no real learning. You have to actively work at learning. This textbook provides that opportunity. A saying often attributed to Confucius is, "I hear and I forget, I see and I remember, I do and I understand." Our textbook follows the "I do and I understand" philosophy.

Active learners learn permanently and grow intellectually. Passive learners remember temporarily and then forget, without mental growth.

This is the point in this course at which you need to ask yourself, "Do I want to learn chemistry as a result of my study? Do I want to learn the thinking skills chemists use?" If the answer is yes, put your wants into action. Make the commitment now to work each example when the textbook presents it. There is no better way to learn the content and process of chemistry than by actively answering questions and solving examples. This is why we've titled the textbook *Introductory Chemistry: An Active Learning Approach.*

procedure

**How to Work an Active Example**

The procedure for solving an Active Example is as follows:

**Step 1:** When you come to an example, locate the point at which the green background changes to white. At this point, a pencil icon indicates that you are to write in the white space:

Use the shield to cover the page below the white space.

**Step 2:** Read the problem statement. Write any answers or calculations needed in the white space above the shield.

**Step 3:** Move the shield down beneath the next white space, if any.

**Step 4:** Compare your answer to the one you can now read in the book. Be sure you understand the example up to that point before going on.

**Step 5:** Repeat the procedure until you finish the example.

***Learn It Now!*** We cannot overemphasize how important it is that you answer the question yourself before you look at the answer in the book. This is the most important of all the *Learn It Now!* statements.

## 3.1 Introduction to Measurement

There is nothing new about measurement. You make measurements every day. How long has that been cooking? How tall are you? How many cups of flour do you need for this recipe? What is the temperature? How many quarts of milk do you want from the store?

What may be new to you in this chapter, if you live in the United States, are the units in which most measurements are made. Scientific measurements and everyday measurements in the rest of the world are made in the **metric system.** Modern scientists often use **SI units,** which are included in the metric system. SI is an abbreviation for the French name for the International System of Units.

The SI system describes seven **base units.** This chapter describes three of these—units of mass, length, and temperature. Other quantities are made up of combined base units; these are called **derived units.** Two of these, units for volume and density, appear in this chapter.

A summary of the SI system appears in Appendix II.

## 3.2 Exponential (Scientific) Notation

**Goal 1** Write in exponential notation a number given in ordinary decimal form; write in ordinary decimal form a number given in exponential notation.

**Goal 2** Using a calculator, add, subtract, multiply, and divide numbers expressed in exponential notation.

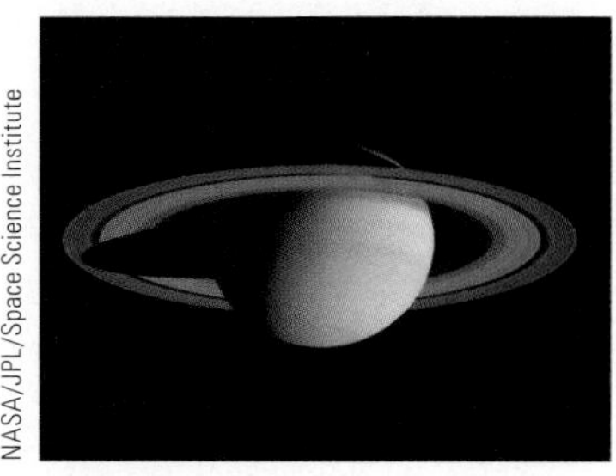

NASA/JPL/Space Science Institute

Scientists deal with large numbers. This picture was taken by the *Cassini* spacecraft on May 7, 2004, when it was $2.82 \times 10^7$ kilometers (17.6 million miles) from Saturn.

Larger and smaller units for the same measurement in the metric system differ by multiples of 10, such as 10 ($10^1$), 100 ($10^2$), 10,000 ($10^4$), and 0.001 ($10^{-3}$). It is often convenient to express these numbers as **exponentials,** as shown.* Furthermore, chemistry calculations and measurements sometimes involve very large or very small numbers. For example, the mass of a helium atom is 0.00000000000000000000000665 gram. In one liter of helium at 0°C and 1 atmosphere of pressure (that's about a quart of the gas at 32°F and the atmospheric pressure found at sea level on a sunny day), there are 26,880,000,000,000,000,000,000 helium atoms. These are two very good reasons to use **exponential notation,** also known as **scientific notation,** for very large and very small numbers. It is also a good reason to devote a section to reviewing calculation methods using these numbers.

A number may be written in **standard exponential notation** as follows:

$$\text{a.bcd} \times 10^{e}$$

Number equal to or greater than 1 and less than 10 — Power of 10

where a.bcd is the **coefficient** and $10^e$ is an exponential. The coefficient a.bcd may have as many digits as necessary after the decimal point or it may have no digits after the decimal point. As examples, 1, 3.4, 8.87, and 4.232990 all are acceptable as coefficients, but 0.23, 12.5, and 200 usually are not. Occasionally, the coefficient may be written in a nonstandard format, outside of the standard range of being equal to or greater than 1 and less than 10. The **exponent,** e, is a whole number (integer); it may be positive or negative.

One way to change an ordinary number to standard exponential notation uses a larger/smaller approach:

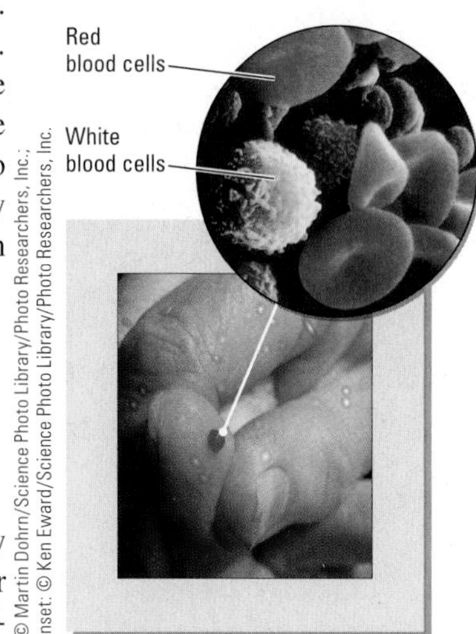



© Martin Dohrn/Science Photo Library/Photo Researchers, Inc.; inset: © Ken Eward/Science Photo Library/Photo Researchers, Inc.

Scientists deal with small numbers. At the microscopic level, blood is a mixture of numerous components. The diameter of a typical human red blood cell is $7 \times 10^{-6}$ meter ($3 \times 10^{-4}$ inch).

**procedure**

**How to Change a Decimal Number to Standard Exponential Notation**

| | | | |
|---|---|---|---|
| | **Sample Problem:** *Write the following numbers in exponential notation:* | *813,000* | *0.000318* |
| **Step 1** | Rewrite the number, placing the decimal after the first nonzero digit. Then write × 10 . | 8.13 × 10 | 3.18 × 10 |
| **Step 2** | Count the number of places the decimal in the original number moved to its new place in the coefficient. Write that number as the exponent of 10. | The decimal moved 5 places. $8.13 \times 10^{5}$ | The decimal moved 4 places. $3.18 \times 10^{4}$ |
| **Step 3** | Compare the original number with the coefficient in Step 1. | | |
| | a) If the coefficient in Step 1 is smaller than the original number, the exponent is larger than 0; it has a positive value. It is not necessary to write the + sign. | $8.13 \times 10^5$ | |
| | b) If the coefficient in Step 1 is larger than the original number, the exponent is smaller than 0; it has a negative value. Insert a minus sign in front of the exponent. | | $3.18 \times 10^{-4}$ |

*An exponential is a number, called the **base,** raised to some power, called the exponent. The mathematics of exponentials is described in Appendix I, Part B. You may wish to review it before studying this section.

It is common, in Step 3, to say that the exponent is positive if the decimal moves left and negative if it moves right. This rule is easy to learn, but just as easy to reverse in one's memory. The larger/smaller approach works no matter which way the decimal moves. You can also use it for relocating the decimal of a number already in exponential notation or for changing a number in exponential notation to ordinary decimal form.

## Thinking About Your Thinking

### Equilibrium

**The larger/smaller rule has the mathematical form a $\times$ b = constant. This type of reasoning is called equilibrium, and it is used frequently in science and in other aspects of everyday life. If a goes up, b must come down; if b goes up, a must come down. The product of the two quantities must always be the same.**

**When applying the equilibrium thinking skill to exponential notation, the product that must stay the same is the value of the number itself. The product of the coefficient and the exponential must equal the number, no matter whether it is in ordinary decimal form or in exponential notation. This leads to the larger/smaller rule. Let's look at the previous example:**

| a | $\times$ | b | = | constant | |
|---|---|---|---|---|---|
| coefficient | $\times$ | exponential | = | constant | |
| 0.000318 | $\times$ | $10^0$ | = | 0.000318 | ($10^0 = 1$) |
| \| | | \| | | \| | |
| *larger* | | *smaller* | | *unchanged* | This is *equilibrium*. |
| ↓ | | ↓ | | ↓ | |
| 3.18 | $\times$ | $10^{-4}$ | = | 0.000318 | |

**As with all *Thinking About Your Thinking* skills, we will revisit equilibrium later in the book.**

Most calculators will convert between exponential notation and ordinary decimal form. You should consult your instruction book to learn how to do this now, if you don't already know.

## Active Example 3.1

Write each of the following in exponential notation: (a) 3,672,199; (b) 0.000098; (c) 0.00461; (d) 198.75.

Locate the white space and the pencil icon below. Place your shield over everything after the white space. Write your answers to the questions in the space provided. When you are finished, move the shield below the next white space, if any.

(a) $3{,}672{,}199 = 3.672199 \times 10^6$ (b) $0.000098 = 9.8 \times 10^{-5}$

(c) $0.00461 = 4.61 \times 10^{-3}$ (d) $198.75 = 1.9875 \times 10^2$

(a) Stepwise thinking: (1) 3,672,199 becomes 3.672199 $\times$ 10 . (2) The decimal moved six places, giving 3.672199 $\times$ $10^6$. (3) The coefficient is *smaller* than the original number. The exponential must then be *larger* than 1. The sign is + and therefore omitted, yielding the final answer $3.672199 \times 10^6$.

(b) Stepwise thinking: (1) 0.000098 becomes $9.8 \times 10$ . (2) The decimal moved five places, giving $9.8 \times 10^{\ 5}$. (3) The coefficient is *larger* than the original number. The exponential must then be *smaller* than 1. The sign is −, yielding the final answer $9.8 \times 10^{-5}$.

(c) (1) $4.61 \times 10$ . (2) $4.61 \times 10^{\ 3}$. (3) 4.61 is larger than 0.00461, $10^{-3}$ is smaller than 1.

(d) (1) $1.9875 \times 10$ . (2) $1.9875 \times 10^{\ 2}$. (3) 1.9875 is smaller than 198.75, $10^2$ is larger than 1.

There are no more white answer spaces, so this is the end of the example. Check your answers against those above. If any answer is different, find out why before proceeding.

If you still have some doubt about whether you understand how to convert from ordinary decimal form to exponential notation, go back and study the textbook some more until you are confident that you have learned this procedure. If (or when) you are satisfied with your learning, continue on with your studies.

Now that you've completed Active Example 3.1, you can see how this textbook promotes active learning. Congratulations! You've just taken the most important step—the first one—toward becoming an active learner.

To change a number written in exponential notation to ordinary decimal form, simply perform the indicated multiplication. The size of the exponent tells you how many places to move the decimal point. A positive exponent indicates a large number, so the coefficient is made larger—the decimal is moved to the right. A negative exponent says the number is small, so the coefficient is made smaller—the decimal is moved to the left. Thus the positive exponent in $7.89 \times 10^5$ says the ordinary decimal number is larger than the coefficient, so the decimal is moved five places to the right: 789,000. The negative exponent in $5.37 \times 10^{-4}$ indicates that the ordinary decimal number is smaller than the coefficient, so the decimal moves four places to the left: 0.000537.

## Active Example 3.2

Write each of the following numbers in ordinary decimal form:

(a) $3.49 \times 10^{-11}$, (b) $3.75 \times 10^{-1}$, (c) $5.16 \times 10^4$, (d) $43.71 \times 10^{-4}$.

(a) $3.49 \times 10^{-11} = 0.0000000000349$ (b) $3.75 \times 10^{-1} = 0.375$
(c) $5.16 \times 10^4 = 51{,}600$ (d) $43.71 \times 10^{-4} = 0.004371$

If two exponentials with the same base are multiplied, the product is the same base raised to a power equal to the sum of the exponents. When exponents are divided, the denominator exponent is subtracted from the numerator exponent:

$$10^a \times 10^b = 10^{a+b} \qquad 10^c \div 10^d = \frac{10^c}{10^d} = 10^{c-d}$$

You will no doubt use your calculator to solve problems involving exponentials. Enter the factors as described in Appendix I, Part A. Be sure to learn how to use your calculator to work with exponential notation as you do the next two examples.

## Active Example 3.3

Perform each of the following calculations. Our answers are rounded off to three digits, beginning with the first nonzero digit. (We will consider *how* to round off in Section 3.5.)

(a) $(3.26 \times 10^4)(1.54 \times 10^6)$ (b) $(8.39 \times 10^{-7})(4.53 \times 10^9)$

(c) $\dfrac{8.94 \times 10^6}{4.35 \times 10^4}$ (d) $\dfrac{(9.28 \times 10^3)(1.13 \times 10^{-5})}{(511)(2.98 \times 10^{-6})}$

(a) $(3.26 \times 10^4)(1.54 \times 10^6) = 5.02 \times 10^{10}$

(b) $(8.39 \times 10^{-7})(4.53 \times 10^9) = 3.80 \times 10^3$

(c) $\dfrac{8.94 \times 10^6}{4.35 \times 10^4} = 2.06 \times 10^2$

(d) $\dfrac{(9.28 \times 10^3)(1.13 \times 10^{-5})}{(511)(2.98 \times 10^{-6})} = 68.9$

To add or subtract exponential numbers without a calculator, you need to align digit values (hundredths, tenths, units, and so on) vertically. This is done by adjusting coefficients and exponents so all exponentials are 10 raised to the same power. The coefficients are then added or subtracted in the usual way. This adjustment is automatic on calculators.

### Active Example 3.4

Add or subtract the following numbers: (a) $3.971 \times 10^7 + 1.98 \times 10^4$, (b) $1.05 \times 10^{-4} - 9.7 \times 10^5$.

(a) $3.971 \times 10^7 + 1.98 \times 10^4 = 3.973 \times 10^7$,

(b) $1.05 \times 10^{-4} - 9.7 \times 10^{-5} = 8 \times 10^{-6}$

You probably did not round off the addition answer as we did, which is fine at this time. The reason for our round-off appears in Section 3.5.

## 3.3 Dimensional Analysis

**Goal 3** In a problem, identify given and wanted quantities that are related by a *PER* expression. Set up and solve the problem by dimensional analysis.

How many days are there in 3 weeks?

You have probably figured out the answer already: 21 days. How did you get it? You probably reasoned that there are 7 days in 1 week, so there must be $3 \times 7$ days in 3 weeks: $3 \times 7 = 21$. In doing this you used the problem-solving method called **dimensional analysis.**

Let's examine this basic days-in-3-weeks problem in detail. The "7 days in each week" relationship between these two units can be stated as "7 days per week." We call this a ***PER* expression.** This *PER* expression can also be written as a fraction or a ratio: 7 days/week, or an equality: 7 days = 1 week. Similarly, there are 24 hours *PER* day (24 hours/day or 24 hours = 1 day) and 60 minutes *PER* hour (60 minutes/hour or 60 minutes = 1 hour). Any *PER* expression can be written as a fraction or an equality. We will identify *PER* expressions in examples by the symbol *PER*.

The mathematical requirement for a PER expression between units for the two quantities is that they are **directly proportional** to each other. What you pay for raw potatoes at the grocery store in units of money, for example, is directly proportional to the amount you buy in units of weight. Two pounds cost twice as much as one pound. If potatoes are priced at 25 cents PER pound (25 cents/pound), three pounds—three times as many pounds—cost 75 cents—three times as many cents.

A PER expression can be used as a **conversion factor.** In a problem setup, a conversion factor is written as a fraction. Each PER expression yields two conversion factors, one the reciprocal of the other. For example, 7 days PER week produces

$$\frac{7 \text{ days}^*}{\text{week}} \quad \textit{and} \quad \frac{1 \text{ week}}{7 \text{ days}}$$

A conversion factor is used to change a quantity of either unit to an equivalent amount of the other unit. The conversion follows a **unit path** (PATH) from the **given quantity** (GIVEN) to the **wanted quantity** (WANTED). A unit path may have any number of steps, but you must know the conversion factor for each step in the path.

In the weeks-to-days example, the one-step unit path is weeks to days, which may be written weeks → days. Mathematically, you multiply the given quantity, 3 weeks, by the conversion factor, 7 days/week, to get the number of days that has the same value as 3 weeks. The calculation setup is

$$3 \cancel{\text{weeks}} \times \frac{7 \text{ days}}{\cancel{\text{week}}} = 21 \text{ days}$$

Notice that the units are always included in the calculation setup. Moreover, *the units are treated in exactly the same way as variables are treated in algebra.* Specifically, they are canceled, as weeks are canceled in the preceding setup. In fact, one way to write a calculation setup is to write the given quantity and then multiply it by the conversion factor that causes the unwanted unit to be canceled out. Eventually, you will probably do this. It is important to understand *why* this method works, just as you understood why you had to multiply 3 by 7 to get the number of days in 3 weeks *before* you saw the calculation setup.

What would have happened if you had selected the wrong conversion factor, the reciprocal of 7 days/week? Let's see:

$$3 \text{ weeks} \times \frac{1 \text{ week}}{7 \text{ days}} = \frac{0.42857\ldots \text{ week}^2}{\text{day}}$$

It wouldn't take long to recognize that this answer is wrong—even if you saw the numerical answer on your calculator! First of all, you know that the number of days in 3 weeks can't be 0.4. . . . The number of days has to be greater than the number of weeks. Second, what is a "$\text{week}^2$/day"? This unit makes no sense; it is what we call a "nonsense unit." In the incorrect setup above, the weeks don't cancel to leave only the wanted days, as they should. Any time your calculation setup yields nonsense units, you can be sure that the answer is wrong in both *numbers* and *units.*

This is one of the valuable features of dimensional analysis. If you get an answer with nonsense units, you *know* you have made a mistake. *Always include units in your calculation setups.*

## Thinking About Your Thinking

### Proportional Reasoning

**This is the first time in this book that you have considered proportional reasoning. This thinking skill has its basis in the mathematics of ratio and proportion. If**

*The conversion factor "7 days/1 week" is usually expressed as "7 days/week." This matches how the factor would be spoken: "seven days per week," not "seven days per one week." In this book the coefficient 1 in a conversion factor is understood to be present but is omitted in print, unless there is some special reason for including it.

a variable, y, is directly proportional to another variable, x, we express this mathematically as

$$y \propto x$$

The symbol $\propto$ represents "is proportional to." This means that any change in either x or y will be accompanied by a corresponding change in the other: Double x, and y is also doubled; reduce y by 1/3, and x will also be reduced by 1/3.

A proportionality can be converted into an equation by inserting a **proportionality constant, m:**

$$y \propto x \xrightarrow{\text{the proportionality changes to an equation}} y = m \times x$$

In the days-in-3-weeks example, y represents days, x represents weeks, and m is the proportionality constant. Proportionality constants sometimes, but not always, express a meaningful physical relationship. This one does. It can be found by solving $y = m \times x$ for the constant, m, and inserting the units of the variables:

$$m = \frac{y}{x} = \frac{y \text{ days}}{x \text{ weeks}} = \frac{7 \text{ days}}{1 \text{ week}} = 7 \text{ days/week}$$

In this form, the proportionality constant becomes a *PER* expression.

Conversion factors can also be derived from **inverse proportionalities,** in which one variable is inversely proportional to another. A familiar example is the relationship between the time (t) it takes to drive a given distance (d) and the speed (s) at which you drive. Greater speed means less time; time is inversely proportional to speed. Another way of saying this is that time is proportional to the inverse of speed. Driving 120 miles at 30 miles per hour takes 4 hours, but at 40 miles per hour the time drops to 3 hours. Algebraically,

$$s \propto \frac{1}{t}$$

Distance is the proportionality constant in this case:

$$s \propto \frac{1}{t} \xrightarrow{\text{the proportionality changes to an equation}} s = d \times \frac{1}{t}$$

Solved for distance, this becomes the familiar

$$d = s \times t$$

in which speed is the *PER* expression that links distance and time, measured in miles per hour, miles/hour, or mph.

## Active Example 3.5

How many weeks are in 35 days?

This is a problem with an easily calculated answer whose purpose is to stress the mechanics of dimensional analysis. It is helpful to plan how to solve a dimensional analysis problem by identifying the *GIVEN* quantity and the *WANTED* units, identifying the *PER* expression you will use, and then writing a unit *PATH*. Begin by identifying the *GIVEN* quantity and its units and the *WANTED* units.

*GIVEN:* 35 days *WANTED:* weeks

The given quantity is almost always a number of some kind of units. The wanted units are whatever it is you are asked to calculate, a number of some other units.

Now write the *PER* expression between days and weeks. Think, ____ days = ____ weeks. Fill in the blanks, and then write it as a *PER* expression.

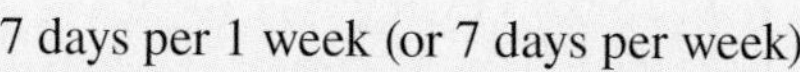

7 days per 1 week (or 7 days per week)

The *PER* expression can be used either as $\frac{7 \text{ days}}{\text{week}}$ or $\frac{1 \text{ week}}{7 \text{ days}}$. All *PER* expressions yield two conversion factors. Now write the *PATH* from the given units to the wanted units. Write the *PER* expression above an arrow linking the units.

*PER:* $\frac{1 \text{ week}}{7 \text{ days}}$

*PATH:* days $\longrightarrow$ weeks

We will write *PER* expressions over the arrows of unit paths in examples throughout this book. We call the combination of these two lines a *PER/PATH*. So, if you are asked to write a *PER/PATH*, you are being asked to write these two lines.

Now write the setup of the problem, using the appropriate conversion factor. Do not calculate the numerical answer yet—just set up the problem.

$$35 \cancel{\text{days}} \times \frac{1 \text{ week}}{7 \cancel{\text{days}}} =$$

Note how the days cancel algebraically. This verifies that the setup is correct.

The remaining units are weeks, the wanted units. Complete the problem by calculating the numerical answer. Put the number *and the associated units* after the equal sign above.

$$35 \cancel{\text{days}} \times \frac{1 \text{ week}}{7 \cancel{\text{days}}} = 5 \text{ weeks}$$

## Thinking About Your Thinking

### Equilibrium

**The equilibrium thinking skill has the form a × b = constant. No matter whether we call it 5 weeks, 35 days, 840 hours, 50,400 minutes, or 3,024,000 seconds, we are describing the exact same amount of time—a constant quantity. Therefore, if we use a relatively large unit, such as the week, it must be associated with a relatively small number of those larger units. If we express the same amount of time in a smaller unit, days in this case, we must have a larger number of those smaller units to describe the same number of "ticks of the clock." A larger *number* of time units compensates for a smaller *unit* in which time is expressed and vice versa.**

The important final step in solving any problem is to be sure the answer makes sense. Let's use Active Example 3.5 to establish a method of checking conversions between units. The two equal quantities, 5 weeks and 35 days, are made up of a smaller number of larger units and a larger number of smaller units:

$$\underset{\text{smaller number of larger units}}{5 \text{ weeks}} = \underset{\text{larger number of smaller units}}{35 \text{ days}}$$

This is the same larger/smaller reasoning we used when moving a decimal point in working with exponential notation in the last section.

In Active Example 3.5 you probably recognized quickly that you could get the answer by dividing 35 by 7. The same reasoning holds for dimensional analysis: You divide 35 days by 7 days per week. The details of this operation are

$$35 \text{ days} \div \frac{7 \text{ days}}{1 \text{ week}} = 35 \text{ days} \times \frac{1 \text{ week}}{7 \text{ days}}$$

Does this ring a memory bell for you? The rule for dividing by a fraction is to invert the divisor and multiply. In different words, to divide by a fraction, multiply by its inverse. This rule works for units just as it works for numbers. Later, you are likely to encounter such a setup as a complex fraction, that is, a fraction in which either the numerator or the denominator, or both, are fractions. In this instance,

$$\frac{35 \; days}{7 \; days/1 \; week} = 35 \; \cancel{\text{days}} \times \frac{1 \text{ week}}{7 \; \cancel{\text{days}}} = 5 \text{ weeks}$$

Most of the time you will be able to see what must be done for the units to cancel, just as we did in the first setup of this problem. At other times, particularly with problems solved by algebra rather than dimensional analysis, you will find "invert and multiply" to be a handy tool for solving the problem.

Some problems have more than one step in their unit paths. To solve a problem by dimensional analysis, you must know the *PER* expression for each step.

In providing space for you to solve example problems, we have assumed you will use the margins. You might find it helpful to write the *GIVEN* quantity and *WANTED* units in the margin next to each example question.

## Active Example 3.6

Calculate the number of weeks in 672 hours.

Most people do not know the number of hours in a week, so they cannot use a one-step unit path from hours to weeks, hours → weeks. But there is an intermediate time unit that breaks the unit path into two steps, each of which has a familiar conversion factor. Can you write the two-step unit path, plus the *PER* expression for each step? Try it as you analyze the problem by listing the *GIVEN*, the *WANTED*, and the *PER*/*PATH*.

*GIVEN:* 672 hours *WANTED:* weeks

*PER:* $\frac{1 \text{ day}}{24 \text{ hours}}$ $\frac{1 \text{ week}}{7 \text{ days}}$

*PATH:* hours ⟶ days ⟶ weeks

Write the dimensional analysis setup for the first step only, the conversion of 672 hours to days, hours → days. Do not calculate the answer, just write the setup.

$$672 \text{ hours} \times \frac{1 \text{ day}}{24 \text{ hours}} \times \frac{\qquad}{\qquad} =$$

We set up this "intermediate answer"—intermediate because it is not the wanted quantity—to show that it does have meaning. If you had calculated it, you would have learned that 672 hours is the same as 28 days. Intermediate quantities are always real quantities with real units if the calculation setup follows the unit path.

You are now ready for the second step of the unit path, the conversion of days to weeks. Insert the conversion factor that completes the setup in the space above and calculate the answer.

$$672 \text{ \sout{hours}} \times \frac{1 \text{ \sout{day}}}{24 \text{ \sout{hours}}} \times \frac{1 \text{ week}}{7 \text{ \sout{days}}} = 4 \text{ weeks}$$

Complete the problem by checking the answer with larger/smaller reasoning.

*CHECK:* 672 hours = 4 weeks

larger number of smaller units = smaller number of larger units

The answer makes sense.

Some conversion factors are established by definition, such as 1 week ≡ 7 days, or 1 minute ≡ 60 seconds. In these examples, the symbol ≡ may be read as "is defined as" or "is identical to": "1 week is defined as 7 days" or "1 minute is identical to 60 seconds." The relationship never changes. The units of measurement are for the same thing—time, in this case. Other conversion relationships are temporary. They may have one value in one problem, and another value in another problem. Speed, measured in kilometers per hour, is an example. If we drive a car at an average of 80 kilometers per hour, we can calculate the distance traveled in 2, 3, or any number of hours. As long as the speed remains the same, the number of kilometers driven is proportional to the number of hours, and 1 hour of driving is said to be "equivalent" to 80 kilometers.* Thus the equivalence between kilometers and hours is 80 kilometers ≈ 1 hour *at 80 kilometers per hour.* But if you slow to 50 km/hr, then 50 kilometers ≈ 1 hour.

## Active Example 3.7

What distance, expressed in miles, will an automobile travel in 3.00 hours at an average speed of 62.0 miles per hour?

Start by writing the *GIVEN*, the *WANTED*, and the *PER/PATH*.

*A relationship between different quantities that are proportional to each other, such as kilometers and hours in 80 kilometers per hour, is sometimes called an *equivalence*. This equivalence would be expressed mathematically as 80 kilometers ≈ 1 hour. The ≈ symbol means "is equivalent to," "is approximately equal to," or "is similar to" (as in similar triangles). A conversion factor derived from an equivalence is valid only while the conditions of the equivalence are maintained.

*Given:* 3.00 hours *Wanted:* miles

*Per:* $\frac{62.0 \text{ miles}}{\text{hour}}$

*Path:* hours $\longrightarrow$ miles

Complete the example with the setup, numerical answer, and check.

$$3.00 \cancel{\text{hours}} \times \frac{62.0 \text{ miles}}{\cancel{\text{hour}}} = 186 \text{ miles}$$

*Check:* We expect a car to travel a greater number of miles than the number of hours it has been traveling, so the answer makes sense.

The decimal-based monetary system in the United States yields calculations that are just like calculations with metric units, which are introduced in the next section.

## Active Example 3.8

If you went into a bank and obtained 325 dollars in dimes, how many dimes would you receive?

Even if you already know the answer to this question, set it up by dimensional analysis.

*Given:* 325 dollars *Wanted:* dimes

*Per:* $\frac{10 \text{ dimes}}{\text{dollar}}$

*Path:* dollars $\longrightarrow$ dimes

$$325 \cancel{\text{dollars}} \times \frac{10 \text{ dimes}}{\cancel{\text{dollar}}} = 3250 \text{ dimes}$$

*Check:* More dimes (smaller unit) than dollars (larger unit). OK.

All you had to do to find the answer was move the decimal point. That's the way it is with a decimal system of units, such as the metric system. But notice how dimensional analysis told you which way to move the decimal point. You didn't need that here with dollars and dimes, but with unfamiliar metrics, you have a different situation.

## Active Example 3.9

A little girl broke open her piggy bank. Inside she found 2608 pennies. How many dollars did she have?

Again, even though you probably know the answer, solve the problem by dimensional analysis. There's a reason; you will understand it soon.

GIVEN: 2608 pennies WANTED: dollars

PER: $\frac{1 \text{ dollar}}{100 \text{ pennies}}$

PATH: pennies ⟶ dollars

$$2608 \text{ ~~pennies~~} \times \frac{1 \text{ dollar}}{100 \text{ ~~pennies~~}} = 26.08 \text{ dollars}$$

CHECK: More pennies (smaller unit) than dollars (larger unit). The answer makes sense.

Notice how the ease of moving a decimal point makes problems like this simple to calculate.

In this section, we have followed a pattern for solving problems by dimensional analysis. This pattern will be used throughout the book. You may wish to adopt it, too. The steps in the pattern are given in the following procedure.

procedure

**How to Solve a Problem by Dimensional Analysis**

**Sample Problem:** *How many nickels are equal to 6 dimes?*

| | |
|---|---|
| **Step 1:** Identify and write down the *GIVEN* quantity. Include units. | *GIVEN:* 6 dimes |
| **Step 2:** Identify and write down the units of the *WANTED* quantity. | *WANTED:* nickels |
| **Step 3:** Write down the *PER/PATH*. | *PER:* $\frac{2 \text{ nickels}}{\text{dime}}$ <br> *PATH:* dimes ⟶ nickels |
| **Step 4:** Write the calculation setup. Include units. | $6 \text{ dimes} \times \frac{2 \text{ nickels}}{\text{dime}} =$ |
| **Step 5:** Calculate the answer. | $6 \text{ ~~dimes~~} \times \frac{2 \text{ nickels}}{\text{~~dime~~}} = 12 \text{ nickels}$ |
| **Step 6:** Check the answer to be sure both the number and the units make sense. | *CHECK:* More nickels (smaller unit) than dimes (larger unit). OK. |

## √ Target Check 3.1

*Which among the following might be GIVEN quantities and, therefore, starting points for a dimensional analysis setup, and which are PER expressions that are used as conversion factors?*
(a) 10 days; (b) 2000 pounds = 1 ton; (c) 2.54 centimeters per inch.

## 3.4 Metric Units

**Goal** 4 Distinguish between mass and weight.

**Goal** 5 Identify the metric units of mass, length, and volume.

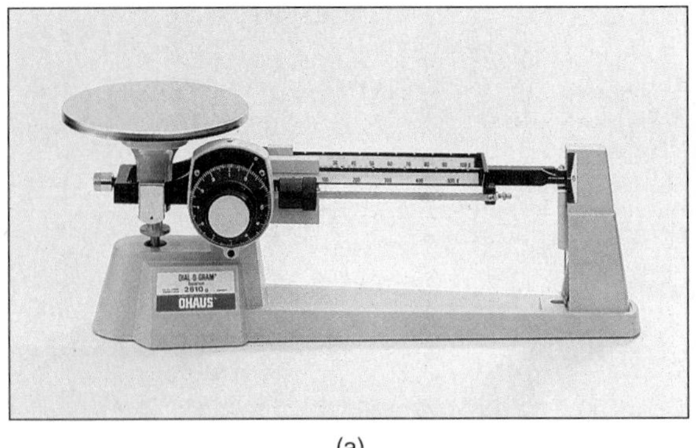


(a)

(b)

(c)

Photos: Ohaus Corporation

**Figure 3.1** Three examples of laboratory balances. (a) A triple-beam balance measures mass with an error of ±0.01 g. It is usually used when high accuracy is not required. (b) This top-loading balance measures mass with an error of ±0.001 g. It has sufficient accuracy for most introductory chemistry applications. (c) An analytical balance measures mass with an error of ±0.0001 g. It is usually used in more advanced courses and in scientific laboratories.

## Mass and Weight

Consider a tool carried to the moon by astronauts. Suppose that tool weighs 6 ounces on the earth. On the surface of the moon it will weigh about 1 ounce. Halfway between the earth and the moon, it will be essentially weightless. Released in midspace, it will remain there, floating, until moved by an astronaut to some other location. Yet in all three locations it would be the same tool, having a constant quantity of matter.

**Mass is a measure of quantity of matter. Weight is a measure of the force of gravitational attraction.** Weight is proportional to mass, but the ratio between them depends on where in the universe you happen to be. Fortunately, this proportionality is essentially constant over the surface of the earth. Therefore, when you weigh something—that is, measure the force of gravity on the object—you can express this weight in terms of mass. In effect, weighing an object is one way of measuring its mass. In the laboratory, mass is measured on a balance (Fig. 3.1).

The SI base unit of mass is the **kilogram, kg.** It is defined as the mass of a platinum–iridium cylinder that is stored in a vault in Sèvres, France. A kilogram weighs about 2.2 pounds, which is too large a unit for most small-scale work in the laboratory. Instead, the **gram, g,** is used. One gram is 1/1000 kilogram, or 0.001 kg. Conversely, we can say that 1 kg is 1000 g.

In the metric system, units that are larger than the basic unit are larger by multiples of 10, that is, 10 times larger, 100 times larger, 1000 times larger, and so on. Similarly, smaller units are 1/10 as large, 1/100 as large, and so forth. This is what makes the metric system so easy to work with. To convert from one unit to another, all you have to do is move the decimal point.

Larger and smaller metric units are identified by metric symbols, or prefixes. The prefix for the unit 1000 times larger than the basic unit is *kilo-,* and its symbol is k. When the *kilo-* symbol, k, is combined with the unit symbol for gram, g, you have the symbol for *kilo*gram, kg. Similarly, *milli-,* symbol m, is the prefix for the unit that is 1/1000 as large as the basic unit. Thus 1/1000 of a gram (0.001 g) is 1 milligram, mg. The unit 1/100 as large as the basic unit is a *centi*unit. The symbol for *centi-* is c. It follows that 1 cg (centigram) is 0.01 g.

It is *essential* that you use upper- and lowercase metric prefixes and symbols appropriately. For example, the symbol for the metric prefix *mega-,* 1,000,000 times larger than the basic unit, is M (uppercase), and the symbol for *milli-,* 1000 times smaller than the basic unit, is m (lowercase).

Table 3.1 lists many metric prefixes and their symbols. Entries for the kilo-, centi-, and milli- units are shown in boldface. Memorize these; you should be able to apply them to any metric unit. We will have fewer occasions to use the prefixes and symbols for other units, but by referring to the table you should be able to work with them, too.

**Table 3.1** Metric Prefixes*

| LARGE UNITS | | | SMALL UNITS | | |
|---|---|---|---|---|---|
| **Metric Prefix** | **Metric Symbol** | **Multiple** | **Metric Prefix** | **Metric Symbol** | **Multiple** |
| tera- | T | $10^{12}$ | Unit (gram, meter, liter) | | $1 = 10^0$ |
| giga- | G | $10^9$ | deci- | d | $0.1 = 10^{-1}$ |
| mega- | M | $1{,}000{,}000 = 10^6$ | **centi-** | **c** | $\mathbf{0.01 = 10^{-2}}$ |
| **kilo-** | **k** | $\mathbf{1{,}000 = 10^3}$ | **milli-** | **m** | $\mathbf{0.001 = 10^{-3}}$ |
| hecto- | h | $100 = 10^2$ | micro- | $\mu$ | $0.000001 = 10^{-6}$ |
| deca- | da | $10 = 10^1$ | nano- | n | $10^{-9}$ |
| Unit (gram, meter, liter) | | $1 = 10^0$ | pico- | p | $10^{-12}$ |

*The most important prefixes are printed in **boldface.**

Your instructor will probably tell you which prefixes from this table to memorize. If not, memorize the three prefixes in **boldface.** You need to recall and use the memorized prefixes, and you need to be able to use the other prefixes, given their value.

## Length

The SI unit of length is the **meter*;** its abbreviation is **m.** The meter has a very precise but awesome definition: the distance light travels in a vacuum in 1/299,792,458 second. Modern technology requires such a precise definition. The meter is 39.37 inches long—about 3 inches longer than a yard.

The common longer length unit, the kilometer (km) (1000 meters), is about 0.6 mile. Both the centimeter and the millimeter are used for small distances. A centimeter (cm) is about the width of a fingernail; a millimeter (mm) is roughly the thickness of a dime. Figure 3.2 compares small metric and U.S. length units. A very tiny unit of length is illustrated in Figure 3.3.

## Volume

The SI volume unit is the cubic meter, $m^3$. This is a derived unit because it consists of three base units, all meters, multiplied by each other. A cubic meter is too large a volume—larger than a cube whose sides are 3 feet long—to use in the laboratory. A more practical unit is the **cubic centimeter, $cm^3$.** It is the volume of a cube with an edge of 1 cm (Fig. 3.4). A teaspoon holds about 5 $cm^3$.

**Figure 3.2** Length measurements: inches, centimeters, and millimeters. This illustration is very close to full scale. The numbers and the long lines on the top scale are inches, and the numbers and the long lines on the bottom scale are centimeters. One inch is defined as 2.54 centimeters, which is the same as 25.4 millimeters (shorter unnumbered lines).

*Outside the United States the length unit is spelled *metre,* and the liter, the volume unit that we will discuss shortly, is spelled *litre.* These spellings and their corresponding pronunciations come from France, where the metric system originated. In this book, we use the U.S. spellings, which match their pronunciations in the English language.

(a) The head of a dead ant is $6 \times 10^{-4}$ m wide. This is 0.6 millimeter or $6 \times 10^{2}$ micrometer. Scientists often shorten the word *micrometer* to micron.

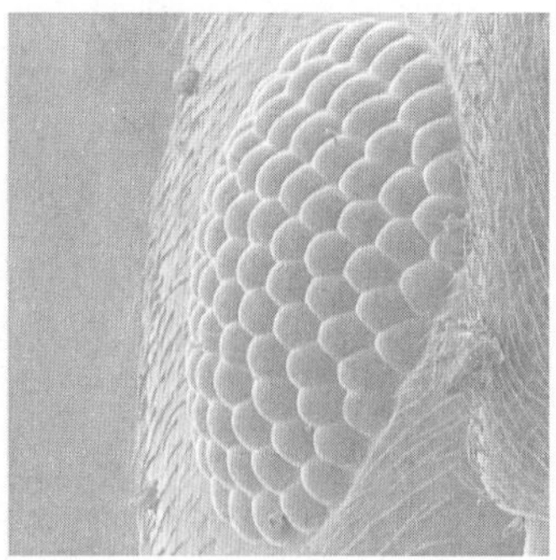

(b) A closeup of the eye of the ant.

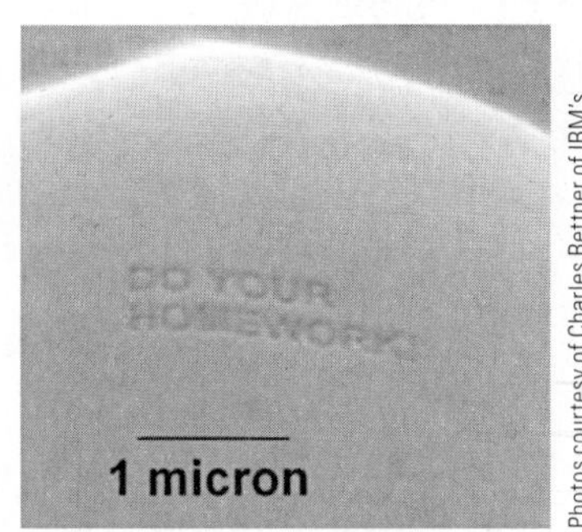


(c) The scientists have inscribed a message to you on one lens of the compound eye. The word *homework!* is just 1.5 micrometer in length.

Photos courtesy of Charles Rettner of IBM's Almaden Research Center

**Figure 3.3** The micron. Scientists at the IBM Almaden Research Center in California used a scanning electron microscope to produce these images.

**Figure 3.4** One cubic centimeter. This full-scale illustration of a cube 1 cm on each side will help you visualize the volume of 1 $cm^3$. One milliliter and one cubic centimeter are the same volume, 1 mL = 1 $cm^3$. The milliliter and the cubic centimeter are the most common volume units you will use in introductory chemistry.

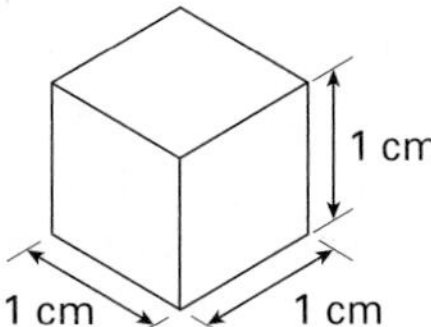


Liquids and gases are not easy to weigh, so we usually measure them in terms of the volumes they occupy. The common unit for expressing their volumes is the **liter, L, which is defined as exactly 1000 cubic centimeters.** Thus there are 1000 $cm^3$/L. This volume is equal to 1.06 U.S. quarts. Smaller volumes are given in **milliliters, mL.** Notice that there are 1000 mL in 1 liter (there are always 1000 milliunits in a unit), and 1 liter is 1000 $cm^3$. This makes 1 mL and 1 $cm^3$ exactly the same volume:

$$1 \text{ mL} = 0.001 \text{ L} = 1 \text{ cm}^3$$

*Learn It Now!* This simple relationship is often missed. There is 1 mL in 1 $cm^3$, not 1000.

Figure 3.5 shows some laboratory devices for measuring volume. Figure 3.6 compares two instruments used to deliver accurately measured volumes in introductory chemistry laboratories.

## Unit Conversions Within the Metric System

**Goal 6** State and write with appropriate metric prefixes the relationship between any metric unit and its corresponding kilounit, centiunit, and milliunit.

**Goal 7** Using Table 3.1, state and write with appropriate metric prefixes the relationship between any metric unit and other larger and smaller metric units.

**Goal 8** Given a mass, length, or volume expressed in metric units, kilounits, centiunits, or milliunits, express that quantity in the other three units.

Conversions from one metric unit to another are applications of dimensional analysis. Goal 8 says you should be able to make these conversions among the unit, kilounit, centiunit, and milliunit. In this context, unit (u) may be gram (g), meter (m), or liter (L).

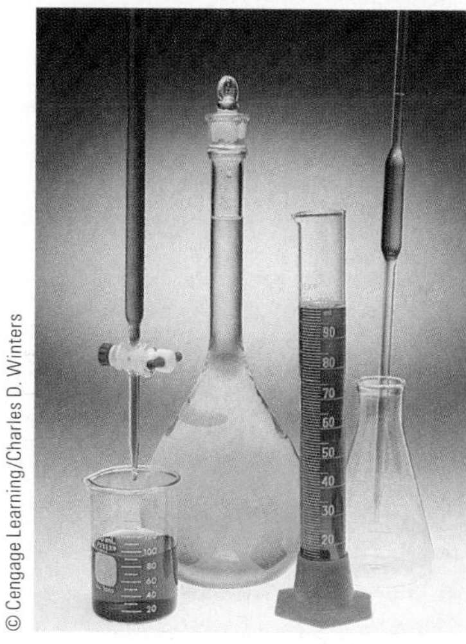


Figure 3.5 Volumetric glassware. The beaker is only for estimating volumes. The tall graduated cylinder is used to measure volume more accurately. The flask with the tall neck (volumetric flask) and the pipet are used to obtain samples of fixed but precisely measured volumes. The buret is used to dispense variable volumes with high precision.

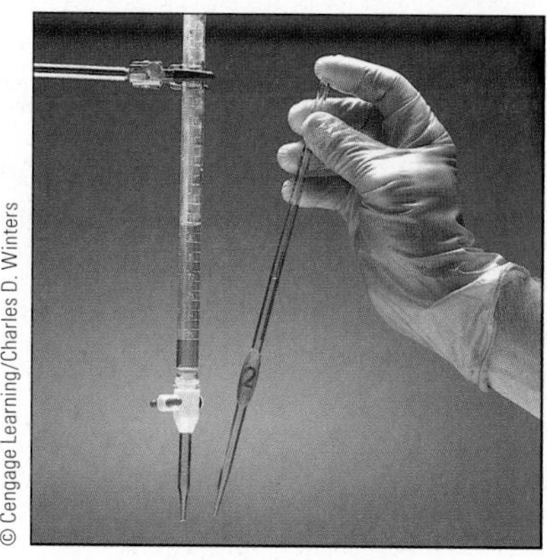


Figure 3.6 A buret and a pipet. The buret on the left is used to deliver variable volumes to an accuracy of ±0.05 mL. The pipet on the right is used to dispense a fixed volume of liquid with an accuracy of ±0.03 mL. These instruments are common in introductory chemistry laboratories, but most research labs use automated titrators and automatic pipets.

These relationships are summarized here as PER expressions and their resulting conversion factors:

1000 units (u) per kilounit (ku) $\quad \frac{1000\ \text{u}}{\text{ku}} \quad \frac{1\ \text{ku}}{1000\ \text{u}}$

100 centiunits (cu) per unit (u) $\quad \frac{100\ \text{cu}}{\text{u}} \quad \frac{1\ \text{u}}{100\ \text{cu}}$

1000 milliunits (mu) per unit (u) $\quad \frac{1000\ \text{mu}}{\text{u}} \quad \frac{1\ \text{u}}{1000\ \text{mu}}$

Your instructor may add other units to those you are required to know from memory. If you can look at Table 3.1, you should be able to convert from any unit in the table to any other unit.

## Thinking About Your Thinking

### Proportional Reasoning

**Conversions between metric units are an example of proportional reasoning. Let's examine the 1000-units-per-kilounit relationship more closely. In any measured quantity, the number of units is directly proportional to the number of kilounits: (# of units) ∝ (# of kilounits). Changing to the form of an equality, (# of units) = m × (# of kilounits). Solving for the proportionality constant, m, gives**

$$m = \frac{\text{\# of units}}{\text{\# of kilounits}}$$

**which, by definition, is $\frac{1000\ \text{units}}{\text{kilounit}}$ (1000 units *PER* kilounit).**

## Active Example 3.10

How many meters are in 2608 centimeters?

You can solve this problem by using dimensional analysis. Identify the *GIVEN* quantity and units, the *WANTED* units, and the *PER/PATH* that you will use.

*GIVEN:* 2608 cm *WANTED:* m *PER:* $\frac{1\text{ m}}{100\text{ cm}}$

*PATH:* cm ⟶ m

Now you have all the information before you. Set up the problem and calculate the answer.

$$2608\ \cancel{\text{cm}} \times \frac{1\text{ m}}{100\ \cancel{\text{cm}}} = 26.08\text{ m}$$

*CHECK:* More centimeters (smaller unit) than meters (larger unit). OK.

When did you first *know* the answer to the preceding example? Was it as soon as you read the question? Look back to Active Example 3.9. You probably knew the answer to that question just as quickly. In essence, that problem was, "How many dollars are in 2608 pennies?" Aside from the units, Active Examples 3.9 and 3.10 are exactly the same problem. In both examples, all you had to do was move the decimal point.

The closest U.S. unit to the centimeter is the inch, and the closest U.S. unit to the meter is the yard. You can compare unit conversions in the metric system with similar conversions in the U.S. system by calculating the number of yards in 2608 inches. Again, it is a dimensional analysis problem:

To avoid confusion with the word *in*, the symbol "in." for inches includes a period. This is the only unit symbol in this book that has a period.

$$2608\ \cancel{\text{in.}} \times \frac{1\text{ yd}}{36\ \cancel{\text{in.}}} = 72.4444\ldots\text{ yd}$$

Which calculation is easier?

The most common error made in metric–metric conversions is moving the decimal the wrong way. The best protection against that mistake is to set up the problem by dimensional analysis, including all units. Always check your result with the larger/smaller rule: If your *GIVENS* and *WANTEDS* have a large number of small units and a small number of large units, and if you've moved the decimal the right number of places, the answer should be correct.

## Active Example 3.11

How many millimeters are in 3.04 centimeters?

If you are sufficiently familiar with the metric system, you can solve this problem in one step. However, at this point we recommend that you convert from the given unit to the basic unit, and then from the basic unit to the wanted unit. Write the *GIVEN*, *WANTED*, and *PER/PATH*, set it up, and calculate the answer. Be sure to check the answer.

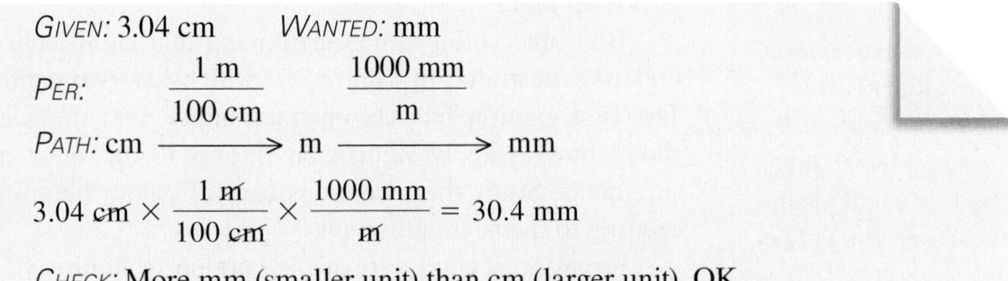

*GIVEN:* 3.04 cm *WANTED:* mm

*PER:* $\frac{1 \text{ m}}{100 \text{ cm}}$ $\frac{1000 \text{ mm}}{\text{m}}$

*PATH:* cm ⟶ m ⟶ mm

$$3.04 \cancel{\text{cm}} \times \frac{1 \cancel{\text{m}}}{100 \cancel{\text{cm}}} \times \frac{1000 \text{ mm}}{\cancel{\text{m}}} = 30.4 \text{ mm}$$

*CHECK:* More mm (smaller unit) than cm (larger unit). OK.

The two conversion factors in the preceding example can be combined: 1000 mm/m and 100 cm/m show that both 1000 mm and 100 cm equal 1 m. Therefore, 1000 mm = 100 cm. This equality can be written as a *PER* expression, 1000 mm/100 cm, which reduces to 10 mm/cm. This relationship is well known to those accustomed to working with metric units. Thus Active Example 3.11 can be solved by either of the following one-step setups:

$$3.04 \cancel{\text{cm}} \times \frac{1000 \text{ mm}}{100 \cancel{\text{cm}}} = 30.4 \text{ mm} \quad or \quad 3.04 \cancel{\text{cm}} \times \frac{10 \text{ mm}}{1 \cancel{\text{cm}}} = 30.4 \text{ mm}$$

As suggested in Active Example 3.11, when you are first learning conversions among metric units, we recommend following the path *GIVEN* unit → basic unit → *WANTED* unit whenever you have to convert between units that are not the basic unit. However, after you have worked a large number of conversion problems, you may find the suggestion in this paragraph to be useful.

### Active Example 3.12

A fruit drink is sold in bottles that contain 1892 mL. Express the volume in cubic centimeters and in liters.

1892 mL = 1892 $\text{cm}^3$. Recall that 1 mL and 1 $\text{cm}^3$ are the same volume.

*GIVEN:* 1892 mL *WANTED:* L *PER:* $\frac{1 \text{ L}}{1000 \text{ mL}}$

*PATH:* mL ⟶ L

$$1892 \cancel{\text{mL}} \times \frac{1 \text{ L}}{1000 \cancel{\text{mL}}} = 1.892 \text{ L}$$

*CHECK:* More mL (smaller unit) than L (larger unit). OK.

## 3.5 Significant Figures

### Counting Significant Figures

**Goal** | **9** State the number of significant figures in a given quantity.

It is important in scientific work to make accurate measurements and to record them correctly. The recorded measurement should indicate the size of its **uncertainty.** One way to do this is to attach a ± value to the recorded number. For example, if a bathroom scale indicates a person's weight correctly to within one pound, and a man reads the scale at 174 pounds, we would record his weight as 174 ± 1 pound. The last digit, 4, is the **uncertain digit.**

The uncertain digit is also called the doubtful digit.

Another way to indicate uncertainty is to use **significant figures.** The number of significant figures in a quantity is **the number of digits that are known accurately plus the uncertain digit.** It follows that, if a quantity is recorded correctly in terms of significant figures, *the uncertain digit is the last digit written.* In 174 pounds, 4 is the uncertain digit.

☞ ***Learn It Now!*** Textbook authors sometimes use illustrations to introduce major concepts. If the text that accompanies the figure is long, it may be written into the caption. This keeps the text and the picture on the same page when the book is printed. When this is done, the caption becomes the text. Figure 3.7 is such an illustration. In fact, it is the only text that addresses Goal 9.

Both uncertainty in measurement and significant figures are illustrated in Figure 3.7, which has two captions. The first caption focuses on uncertainty, and the second shows how to apply significant figures to the same measurements. Study these captions carefully now, before proceeding to the next paragraph.

From the second part of the caption to Figure 3.7, we can state the rule for counting the number of significant figures in any quantity:

**procedure**

**How to Count Significant Figures**

Begin with the first nonzero digit and end with the uncertain digit—the last digit shown.

It should be no surprise that both 0.643 m and 64.3 cm in Figure 3.7(d) have three significant figures. Both quantities came from the same measurement. Therefore, they should have the same uncertainty, the same uncertain digit. Only because the units are different are the decimal points in different places. Therefore, the important conclusion is: *The measurement process, not the unit in which a result is expressed, determines the number of significant figures in a quantity.* **Therefore, the location of the decimal point has nothing to do with significant figures.**

When authors emphasize a point so strongly, there must be a reason. Many errors are made when students relate significant figures to the location of the decimal point. Do not allow yourself to be included among such students!

If 0.643 m is written in kilometers (km) (1000 m = 1 km), it is 0.000643 km. This is also a three-significant-figure number. The first three zeros after the decimal point are not significant, but they are required to locate the decimal point. Counting still begins with the first nonzero digit, 6. *Do not begin counting at the decimal point.*

If the 0.643 m is written in nanometers (nm) (1 m = $1 \times 10^9$ nm), it is 643,000,000 nm. This is still a three-significant-figure number, but the uncertainty is 1,000,000 nm. This time we have six zeros that are not significant, but they are required to locate the decimal point. How do we end the recorded value with the uncertain digit, 3? Write it in exponential notation: $6.43 \times 10^8$ nm. The coefficient shows clearly the number of significant figures in the quantity. Exponential notation works with very small numbers, too: $6.43 \times 10^{-4}$ km.

The last two values show that, in very large and very small numbers, zeros whose only purpose is to locate the decimal point are not significant.

Sometimes—one time in ten, on the average—the uncertain digit is a zero. If so, it still must be the last digit recorded. Suppose, for example, that the length of a board is 75 centimeters plus or minus 0.1 cm. To record this as 75 cm is incorrect; it implies that the uncertainty is ±1 cm. The correct way to write this number is 75.0 cm. If the measurement has been uncertain to 0.01 cm, it would be recorded as 75.00 cm. *The uncertain digit is always the last digit written, even if it is a zero to the right of the decimal point.*

As a matter of fact, if the uncertain digit is a zero, it is best to have it to the right of the decimal point. If 75.0 cm were to be written in the next-smaller decimal unit, it would be 750 mm. The reader of this measurement is faced with the question, "Is the zero significant, or is it a place holder for the decimal point?" With 75.0 cm or $7.50 \times 10^2$ mm, there is no question: The zero is significant.

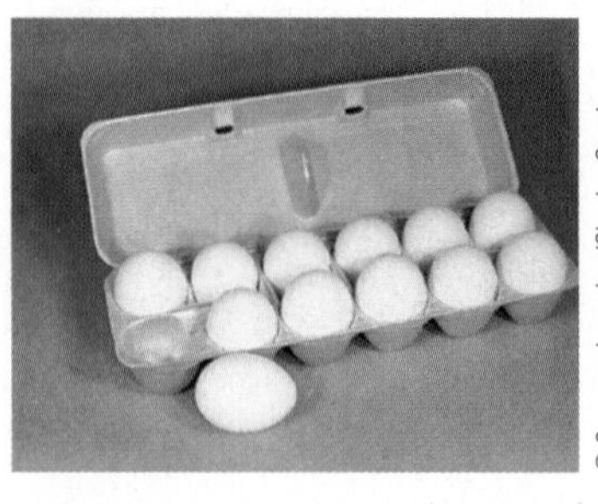



There is no uncertainty in exact numbers, so significant figures do not apply. There are 12 items in a dozen.

Significant figures do not apply to **exact numbers.** An exact number has no uncertainty; it is infinitely significant. Counting numbers are exact. A bicycle has exactly two wheels. Numbers fixed by definition are exact. There are exactly 60 minutes in an hour and exactly 12 eggs in 1 dozen eggs.

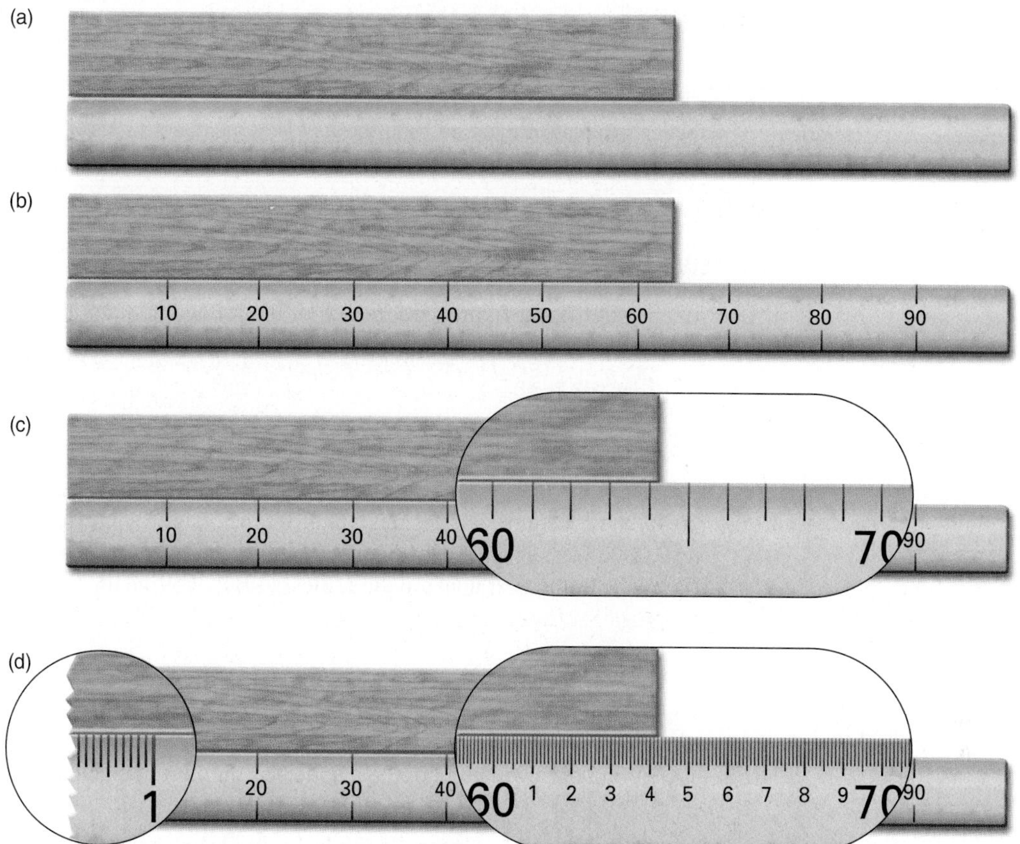


Figure 3.7 Uncertainty in measurement. The length of a board is measured (estimated) by comparing it with meter sticks that have different graduation marks. (a) There are no marks. The board is definitely more than half a meter long, probably close to two-thirds. In decimals this is between 0.6 and 0.7 meter (m). The number of tenths is uncertain. Uncertainty is often added to a measurement as a "plus or minus" (±) value. In this case the length might be recorded as 0.6 ± 0.1 m or 0.7 ± 0.1 m. (b) Graduation marks appear at every 0.1 m but are numbered in centimeters (cm) (100 cm = 1 m). The board is less than halfway between 60 and 70 cm (0.6 and 0.7 m). The length might be closer to 64 cm (0.64 m) than 65 cm (0.65 m), but it is hard to tell. Both 64 ± 1 cm and 0.64 ± 0.01 m are reasonable estimates. (c) Now the centimeter lines are added to the graduations shown in the magnified view. The board's length is clearly closer to 64 cm than 65 cm. It also appears to be about one-fourth to one-third of the way between 64 and 65. Estimating the closest tenth of a centimeter (0.001 m) gives 64.2 ± 0.1 cm (0.642 ± 0.001 m) and 64.3 ± 0.1 cm (0.643 ± 0.001 m) as reasonable estimates. (d) When millimeter (mm) (10 mm = 1 cm) lines are added, the board is clearly closest to, but a little less than, 64.3 cm (0.643 m). Do we estimate to the next decimal? Usually you can estimate between the smallest graduation marks, but in this case wear or roughness at the end of the meter stick can introduce errors as much as several 0.01 cm (0.0001 m). It is best to accept 64.3 ± 0.1 cm or 0.643 ± 0.001 m as the most reliable measurement you can make.

**Counting significant figures**

(a) 0.6 m: The first nonzero digit is 6, so counting starts there. The last digit shown, the same 6, is uncertain, so counting stops there. Therefore 0.6 m has one significant figure.

(b) 0.64 m or 64 cm: The first nonzero digit is 6, so counting starts there. The last digit shown, 4, is uncertain, so counting stops there. Both 0.64 and 64 have two significant figures.

(c) and (d) 0.643 m or 64.3 cm: The first nonzero digit is 6, so counting starts there. The last digit shown, 3, is uncertain, so counting stops there. Both 0.643 and 64.3 have three significant figures.

**summary**

**The Number of Significant Figures in a Measurement**

1. Significant figures are applied to measurements and quantities calculated from measurements. They do not apply to exact numbers.
2. The number of significant figures in a quantity is the number of digits that are known accurately plus the one that is uncertain—the uncertain digit.
3. The measurement process, not the unit in which the result is expressed, determines the number of significant figures in a quantity.

4. The location of the decimal point has nothing to do with significant figures.
5. The uncertain digit is the last digit written. If the uncertain digit is a zero to the right of the decimal point, that zero must be written.
6. Exponential notation must be used for very large numbers to show if final zeros are significant.

### Active Example 3.13

How many significant figures are in each of the following quantities? (a) 45.26 ft, (b) 0.109 in., (c) 0.00025 kg, (d) $2.3659 \times 10^{-8}$ cm

(a) 45.26 ft: 4 significant figures, (b) 0.109 in.: 3 significant figures, (c) 0.00025 kg: 2 significant figures (begin counting significant figures with the first nonzero digit, the 2), (d) $2.3659 \times 10^{-8}$ cm: 5 significant figures

## Rounding Off

Goal | **10** Round off given numbers to a specified number of significant figures.

Sometimes when you add, subtract, multiply, or divide experimentally measured quantities, the result contains digits that are not significant. When this happens, the result must be **rounded off.** Rules for rounding are as follows:

procedure

**How to Round Off a Calculated Number**

**Step 1:** If the first digit to be dropped is less than 5, leave the digit before it unchanged.
**Example:** If we round 5.9936 to three significant figures, we obtain 5.99.

**Step 2:** If the first digit to be dropped is 5 or greater, increase the digit before it by 1.
**Example:** If we round 34.581 to three significant figures, we obtain 34.6.

For any individual round-off by any method, every rule has a 50% chance of being "more correct." Even when "wrong," the rounded-off result is acceptable because only the uncertain digit is affected.

Other rules for rounding vary if the first digit to be dropped is exactly 5. We recommend that you follow your instructor's advice if it differs from ours.

### Active Example 3.14

Round off each of the following quantities to three significant figures. (a) 1.42752 $cm^3$, (b) 45,853 cm, (c) 643.349 $cm^2$, (d) 0.03944498 m

(a) 1.43 $cm^3$, (b) $4.59 \times 10^4$ cm, (c) 643 $cm^2$, (d) 0.0394 m or $3.94 \times 10^{-2}$ m

## Addition and Subtraction

Goal | **11** Add or subtract given quantities and express the result in the proper number of significant figures.

The **significant figure rule for addition and subtraction** can be stated as follows:

**procedure**

**How to Round Off a Sum or Difference**

Round off the answer to the first column that has an uncertain digit. (Active Example 3.15 shows how to apply this rule.)

## Active Example 3.15

A student weighs a beaker and four different chemicals using different balances. The individual masses and their sum are as follows:

| | |
|---|---|
| Beaker | 319.5 g |
| Chemical A | 20.460 g |
| Chemical B | 0.0639 g |
| Chemical C | 45.642 g |
| Chemical D | 4.173 g |
| Total | 389.8389 g |

Express the sum to the proper number of significant figures.

This sum is to be rounded off to the first column that has a doubtful digit. What column is this: hundreds, tens, ones, tenths, hundredths, thousandths, or ten thousandths?

Tenths.

The doubtful digit in 45.6 is in the tenths column. In all other numbers the doubtful digit is in the hundredths column or smaller.

According to the rule, the answer must now be rounded off to the nearest number of tenths. What answer should be reported?

389.8 g

We may use this example to justify the rule for addition and subtraction. A sum or difference digit must be doubtful if any number entering into that sum or difference is uncertain or unknown. In the left addition that follows, all uncertain digits are shown in blue, and all digits to the right of a colored digit are simply unknown:

<table>
<tr><td>319.5</td><td></td><td>319.5</td><td></td><td></td></tr>
<tr><td>20.460</td><td></td><td>20.4</td><td>60</td><td></td></tr>
<tr><td>0.0639</td><td></td><td>0.0</td><td>639</td><td></td></tr>
<tr><td>45.642</td><td></td><td>45.6</td><td>42</td><td></td></tr>
<tr><td>4.173</td><td></td><td>4.1</td><td>73</td><td></td></tr>
<tr><td>389.8389</td><td>= 389.8</td><td>389.8</td><td>389</td><td>= 389.8</td></tr>
</table>

In the left addition, the 8 in the tenths column is the first uncertain digit.

The addition at the right shows a mechanical way to locate the first uncertain digit in a sum. Draw a vertical line after the last column in which every space is occupied.

The uncertain digit in the sum will be just left of that line. The result must be rounded off to the line.

The same rule, procedure, and rationalization hold for subtraction.

## Multiplication and Division

Goal | **12** Multiply or divide given measurements and express the result in the proper number of significant figures.

The **significant figure rule for multiplication and division** is as follows:

procedure

**How to Round Off a Product or Quotient**

Round off the answer to the same number of significant figures as the smallest number of significant figures in any factor. (Active Example 3.16 shows an application of this rule.)

### Active Example 3.16

If the mass of 1.000 L of a gas is 1.436 g, what is the mass of 0.0573 L of the gas?

You can solve this problem with dimensional analysis. If 1.436 grams of the gas occupies 1.000 liter, we can state this relationship in a *PER* expression: $\frac{1.436\text{ g}}{1.000\text{ L}}$. Can you now identify the *GIVEN* quantity, the *WANTED* unit, and the *PER*/*PATH*?

*GIVEN:* 0.0573 L  *WANTED:* g  *PER:* $\frac{1.436\text{ g}}{1.000\text{ L}}$

*PATH:* L $\longrightarrow$ g

The *PER* expression is written $\frac{1.436\text{ g}}{1.000\text{ L}}$ rather than $\frac{1.436\text{ g}}{\text{L}}$ because the purpose of this example is to emphasize significant figures in calculations.

Write a dimensional analysis setup for the problem. Calculate the answer and write it down just as it appears in your calculator.

$$0.0573\text{ }\cancel{\text{L}} \times \frac{1.436\text{ g}}{1.000\text{ }\cancel{\text{L}}} = 0.0822828\text{ g}$$

Three values entered into the multiplication/division problem: 0.0573 L, 1.436 g, and 1.000 L. The answer is to be rounded off to the smallest number of significant figures in any of these measured quantities. What is the smallest number of significant figures?

There are three significant figures in 0.0573 L.

The other numbers have four significant figures.

Round off the calculated 0.0822828 g to three significant figures.

0.0823 g

If you wrote 0.082 g, you forgot that counting significant figures begins at the first nonzero digit.

Active Example 3.16 and its product

$$0.0573 \times 1.436 = 0.0822828 = 0.0823 \text{ in three significant figures}$$

may be used to justify the rule for multiplication and division. Suppose these measurements are the *true* values and they are correctly expressed in terms of significant figures, that is, the final digit is the uncertain digit in each value. Now consider what the product would be if the actual measurements were both 1 higher in the uncertain digit. In that case, the multiplication would be

$$0.0574 \times 1.437 = 0.0824838 = 0.0825 \text{ in three significant figures}$$

If the actual measurements were both 1 lower in the uncertain digit, the multiplication would be

$$0.0572 \times 1.435 = 0.0820820 = 0.0821 \text{ in three significant figures}$$

Compare the three results in three significant figures: 0.0823, 0.0825, and 0.0821. With the third nonzero digit uncertain, these three three-significant-figure products are equal. They might be expressed as 0.0823 ± 0.0002.

Alternately, each product number reached with an uncertain multiplier must itself be uncertain. Blue numbers indicate the uncertain digits in the detailed multiplication:

```
      1 . 4 3 6                      4 significant digits
  × 0 . 0 5 7 3                      3 significant digits
  -------------
          4 3 0 8
      1 0 0 5 2
      7 1 8 0
  -----------------
0 . 0 8 2 2 8 2 8 = 0 . 0 8 2 3      3 significant digits (multiplication/division rule)
```

## Active Example 3.17

How many hours are in 6.924 days? Express the answer in the proper number of significant figures.

This is another problem you can work by applying dimensional analysis. We assume that you already know the PER expression. Set up and solve completely. Round off the answer to the correct number of significant figures.

GIVEN: 6.924 days WANTED: hours

PER: $\frac{24 \text{ hours}}{\text{day}}$

PATH: days ⟶ hours

$$6.924 \text{ days} \times \frac{24 \text{ hours}}{\text{day}} = 166.2 \text{ hours}$$

CHECK: More hours (smaller unit) than days (larger unit). OK.

By definition, there are *exactly* 24 hours in 1 day. Exact numbers are infinitely significant. They never limit the number of significant figures in a calculated result.

## Addition/Subtraction and Multiplication/Division Combined

When a calculation contains both addition/subtraction and multiplication/division, you must apply each individual rule for significant figures separately. If two numbers are to be added and their sum is to be divided by another number, such as in $\frac{(34.49 + 7.3)}{13.80}$, first perform the addition to the correct number of significant figures. Then perform the division, applying the multiplication/division significant-figure rule.

For the addition, we obtain

$$\begin{array}{rl} 34.4|9 & \leftarrow \text{two digits past the decimal point} \\ +\ 7.3| & \leftarrow \text{one digit past the decimal point} \\ \hline 41.7|9 = 41.8 & \leftarrow \text{one digit past the decimal point (addition and subtraction rule)} \end{array}$$

Now apply the multiplication/division rule:

three significant figures ↓

$$\frac{41.8}{13.80} = 3.03 \leftarrow \text{three significant figures (the smallest number of significant figures in any factor)}$$

↑ four significant figures

This calculation can be represented in a "line setup" that shows the key sequence that is used on a typical calculator:

$$(34.49 + 7.3) \div 13.80 = 3.02826087 = 3.03$$

If your calculator uses parentheses, enter each numeral or symbol exactly as shown. If your calculator does not use parentheses, it will be necessary to enter an equal sign (=) after 7.3 to complete the numerator addition before beginning the division.

### Active Example 3.18

The density of a substance is given by

$$\text{density} \equiv \frac{\text{mass}}{\text{volume}}$$

A liquid with a volume of 5.00 mL is placed in a beaker that has a mass of 36.3 g. The mass of the liquid plus the beaker is 42.54 g. What is the density of the liquid?

The problem is solved by first calculating the mass of the liquid, which is the difference between the mass of the liquid plus beaker and the mass of the empty beaker. This difference is then divided by the volume of the liquid. As described above, perform the addition (or subtraction, in this case) first and then do the division. Take it all the way to a calculated answer, expressed in the proper number of significant figures.

1.2 g/mL

If you got the correct answer, good. Before showing the setup that produces that answer, though, let's look at a few wrong answers and the common errors that produce them.

1.248 g/mL comes from ignoring significant figures altogether and taking the answer displayed on the calculator.

1.25 g/mL comes from not recognizing that the numerator has only two significant figures after applying the rule for subtracting and rounding (6.24 ÷ 5.00 = 1.248) to 1.25.

$$\begin{array}{r|l} 42.5 & 4 \\ -\,36.3 & \\ \hline 6.2 & 4 = 6.2 \end{array}$$

1.24 g/mL comes from rounding off the numerator correctly to two significant figures, but not recognizing that those two significant figures limit the final answer to two significant figures: 6.2 ÷ 5.00 = 1.24.

The correct calculation and round-off sequence is

$$(42.54 - 36.3) \div 5.00 = 6.24 \div 5.00 = \underset{\substack{\uparrow \\ \text{Round off subtraction in numerator}}}{6.2} \div 5.00 = 1.24 = \underset{\substack{\uparrow \\ \text{Round off division}}}{1.2\text{ g/mL}}$$

In performing the arithmetic in Active Example 3.18, it is not necessary to round off the subtraction from 6.24 to 6.2. You may never see 6.2 in your calculator display or written on a piece of paper. Both 6.24 ÷ 5.00 = 1.248 and 6.2 ÷ 5.00 = 1.24 round off to 1.2 with two significant figures. What is necessary is to recognize that the numerator in the final division has only two significant figures, and therefore, the answer should be rounded to two significant figures.

## Significant Figures and This Book

Calculators report answers in all the digits they are able to display, usually eight or more. Such answers are unrealistic; never use them. Calculations in this book have generally been made using all digits given, and only the final answer has been rounded off. If, in a problem with several steps, you round off at each step, your answers may differ slightly from those in the book. The difference will be in the uncertain digit; both answers are acceptable.

Be wary of additions and subtractions. They sometimes increase or reduce the number of significant figures in an answer to greater or fewer than the number in any measured quantity.

# 3.6 Metric–USCS Conversions

**Goal 13** Given a metric–USCS conversion factor and a quantity expressed in any unit in Table 3.2, express that quantity in corresponding units in the other system.

Table 3.2 gives common conversion relationships between measurement units in the United States Customary System (USCS) and metric system. All countries of the world use the metric system, except for the United States. The USCS, formerly the British system of units, is the system used in the United States. Table 3.3 gives conversion relationships between USCS measurement units.

**Table 3.2** Metric–USCS Conversion Factors

| | |
|---|---|
| *Length* | 1 in. ≡ 2.54 cm (definition) |
| *Mass* | 1 lb ≡ 453.59237 g (definition) |
| *Volume* | 1 gal ≡ 3.785411784 L (exactly) |

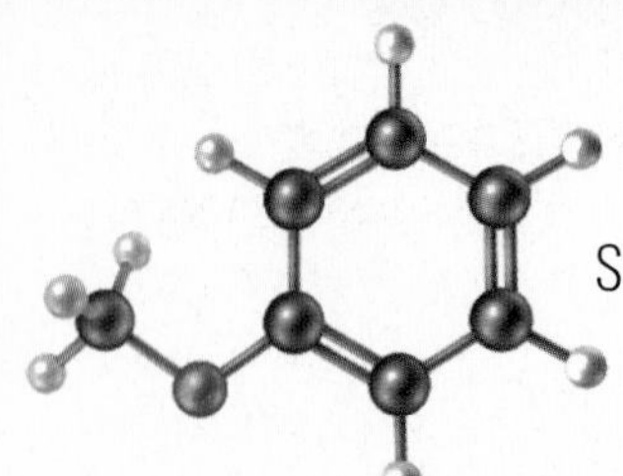

# Everyday Chemistry

## Should the United States Convert to Metric Units? An Editorial

In the 1970s the four major English-speaking nations of the world—the United States, Great Britain, Canada, and Australia—took action to replace what was then called the British system of units with the metric system. Three of these nations have made the change successfully. Anyone living in the United States knows which country did not. In fact, the name of the U.S. measurement system has been changed to the United States Customary System (USCS) because the British no longer use it.

Why does the United States cling to USCS units? Simply because of resistance to change. We've grown up with USCS units. For most Americans, there is no advantage to metrics. One kilogram of potatoes is no more convenient a quantity than 2.2 pounds of potatoes, so why bother to change? Americans can get used to metric quantities, however. We now buy soft drinks in 1- or 2-liter bottles. Is 2 liters less convenient than 2.11 quarts, the equivalent USCS volume? The number is not important once you get used to it, until you get to calculating.

Americans are accustomed to buying soft drinks in containers with metric volumes.

If you must work with measured quantities—calculate as we must in the sciences, or buy, sell, and build as they do in commerce and industry—metrics are so much easier and so much more logical that the choice is obvious. If, in solving the problems in this book, you had to work with quantities expressed in USCS units, your chemistry course would be much more difficult, not because of the chemistry, but because of the USCS units!

The fact is that most people have little day-to-day need to calculate, beyond simple arithmetic at the checkout counter (and even that has been replaced by bar codes and cash registers that figure the sales tax and the customer's change). And the general public in the United States can vote out of office any politician who proposes a mandatory change to metrics.

Changing to metric units is not nearly as difficult as the opponents of change would have us think. The Canadians, British, and Australians did it. Consider all the immigrants to the United States who have changed from their native metric system to the USCS system—a much more difficult change than from USCS to metric. If they can do that, Americans are surely capable of making the easier change in the other direction. So has the conversion to metrics been lost in the United States? Not really. Economic motives for the conversion could be stronger than public opposition to change. If by adopting metrics, and paying the one-time costs that go along with it, a company can (a) make more money, (b) save money, or (c) avoid losing money, you can be quite sure that metrics will be adopted. Here are a few examples.

One major manufacturing firm delivered a large number of appliances to a Middle Eastern country, only to have them rejected because the connecting cord was six feet instead of two meters long.

European countries have told the United States that they may, as a matter of policy, refuse U.S. imports that fail to meet metric standards. In regard to mechanical equipment, their position is, "You may build it with USCS dimensions, but the tools we use to fix it are metric. If we can't fix it, we won't buy it." It's hard to argue against that.

All equipment purchased under U.S. Defense Department contracts must meet metric specifications. The conversion to metrics is at or close to 100% in the U.S. automobile and drug industries.

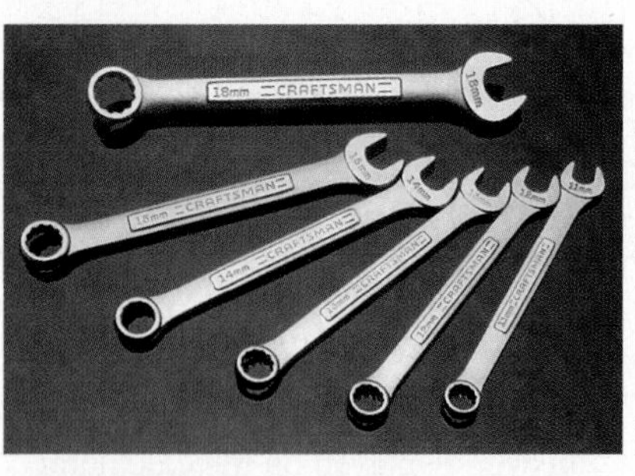




When industrial parts are manufactured in metric sizes, the tools needed to fix them are available everywhere in the world.

One major U.S. manufacturer has reduced the number of screw sizes it holds in inventory from 70 to 15 by using only metric products. Another large exporter reports "saving tens of millions of dollars by avoiding double inventory costs and operating all our 32 domestic and foreign plants on one system."

In the 10 years after one U.S. firm introduced metrics, its number of employees doubled to meet its new export demand. In a two-year slack domestic market, those exports were the only thing that kept the company afloat.

Another American company adopted metrics in order to hold a few Canadian customers. The change opened so many new markets that the company now ships 28% of its output abroad.

Many U.S. companies that have converted to metrics report that the cost of conversion was much lower than expected—less than half in some cases—and that those costs have been recovered quickly and often unexpectedly.

Will the United States ever become a metric nation? As these examples attest, it is already a metric nation where it really counts. Industry and government could no longer wait around for a reluctant populace to do what had to be done. So we are, in effect, a nation of two systems of units. Maybe that's the way it's supposed to be. But metrics are so much easier, so logical, and have so many advantages. . . . Oh, well!

## Thinking About Your Thinking

### Proportional Reasoning

**The relationships between the number of metric units and the number of USCS units are direct proportionalities. You will find it useful to memorize only one conversion in each of three categories: length, mass (weight), and volume, as given in Table 3.2. You can then use familiar metric–metric and/or USCS–USCS conversions to change units within each system of measurement. Although it can add a few steps to a problem, this approach minimizes the amount of memorization necessary.**

**Table 3.3** USCS–USCS Conversion Factors

| |
|---|
| *Length* |
| 1 ft ≡ 12 in. |
| 1 yd ≡ 3 ft |
| 1 mi ≡ 5280 ft |
| *Mass* |
| 1 lb ≡ 16 oz |
| *Volume* |
| 1 qt ≡ 32 fl oz |
| 1 gal ≡ 4 qt |

In Table 3.2, notice two things about the length conversion 1 in. ≡ 2.54 cm: (1) The symbol for the USCS unit is in., including the period. The period distinguishes the unit symbol from the common word *in*. (2) The ≡ symbol indicates a definition, and may be read "one inch is defined as 2.54 centimeters." Thus exactly 1 in. is equal to exactly 2.54 cm, and the numbers are infinitely significant. There are other definitions and exact conversions in Table 3.2 and Table 3.3, also indicated by the ≡ symbol. Since the mass and volume conversion factors in Table 3.2 are lengthy, we will follow the practice of rounding the conversion to one significant digit more than the measured quantity that limits the significant digits in the calculation.

USCS units are defined in terms of metric units.

## Active Example 3.19

What is the length in meters of an object that is 38 inches long?

Write the GIVEN, WANTED, and PER/PATH for the problem. You'll find the metric–USCS conversion factor in Table 3.2. The other conversion factor is a metric–metric conversion you have memorized.

GIVEN: 38 in. WANTED: m

PER: $\frac{2.54 \text{ cm}}{1 \text{ in.}}$ $\frac{1 \text{ m}}{100 \text{ cm}}$

PATH: in. $\longrightarrow$ cm $\longrightarrow$ m

Complete the problem by writing the setup, calculating the answer, and expressing it to the correct number of significant figures.

$$38 \text{ in.} \times \frac{2.54 \text{ cm}}{1 \text{ in.}} \times \frac{1 \text{ m}}{100 \text{ cm}} = 0.97 \text{ m}$$

The number of inches is the only measured quantity, so there are two significant figures in the final answer. If you know that a meter and a yard (36 in.) are about the same length, the answer is reasonable in magnitude.

## Active Example 3.20

The mass of a 250-mL glass beaker is listed as 108,255 milligrams. Express this in ounces for a coworker who needs to calculate a shipping cost based on weight.

Start by listing the *GIVEN* and *WANTED*.

*GIVEN:* 108,255 mg *WANTED:* oz

Notice that the problem gives mass in metric units and asks for weight in USCS units. Therefore you need the appropriate metric–USCS conversion factor from Table 3.2. Look it up and record it in the space below.

1 lb ≡ 453.59237 g

There are two other conversion factors needed for this problem. You need to convert between mg and g and between lb and oz. The USCS–USCS conversion can be found in Table 3.3, and the other is one that you've memorized. List each of these below.

16 oz ≡ 1 lb     1000 mg = 1 g

You are ready to setup the *PER/PATH*.

$$\text{PER:} \quad \frac{1\text{ g}}{1000\text{ mg}} \qquad \frac{1\text{ lb}}{453.59237\text{ g}} \qquad \frac{16\text{ oz}}{\text{lb}}$$

$$\text{PATH: mg} \longrightarrow \text{g} \longrightarrow \text{lb} \longrightarrow \text{oz}$$

Complete the example with the setup and solution.

$$108{,}255\ \cancel{\text{mg}} \times \frac{1\ \cancel{\text{g}}}{1000\ \cancel{\text{mg}}} \times \frac{1\ \cancel{\text{lb}}}{453.59237\ \cancel{\text{g}}} \times \frac{16\text{ oz}}{\cancel{\text{lb}}} = 3.81858\text{ oz}$$

The six-significant-figure measured quantity yields an answer with the same number of significant figures. All conversion factors are definitions or exact.

## Active Example 3.21

A soft drink manufacturer is converting from USCS units to metric units. If the quart bottle initially is to remain at the same volume, what number of milliliters should be printed on the label?

Assume that the dispensing equipment can measure the volume delivered to ±0.01 qt. Go for the entire setup this time.

*GIVEN:* 1.00 qt *WANTED:* mL

*PER:* $\frac{1 \text{ gal}}{4 \text{ qt}}$ $\frac{3.785 \text{ L}}{\text{gal}}$ $\frac{1000 \text{ mL}}{\text{L}}$

*PATH:* qt ⟶ gal ⟶ L ⟶ mL

$$1.00 \text{ qt} \times \frac{1 \text{ gal}}{4 \text{ qt}} \times \frac{3.785 \text{ L}}{\text{gal}} \times \frac{1000 \text{ mL}}{\text{L}} = 946 \text{ mL}$$

We used 3.785 L/gal instead of the full 3.785411784 L/gal given in Table 3.2 because the *GIVEN* quantity has three significant figures, and thus the answer is limited to three significant figures. Using a conversion factor with at least one more significant figure than the *GIVEN* quantity ensures that the uncertain digit in the final answer will not be affected by the number of significant figures in the conversion factor. A conversion factor should never limit the number of significant figures in a calculation.

## 3.7 Temperature

**Goal 14** Given a temperature in either Celsius or Fahrenheit degrees, convert it to the other scale.

**Goal 15** Given a temperature in Celsius degrees or kelvins, convert it to the other scale.

The familiar temperature scale in the United States is the **Fahrenheit scale.** All scientists, including those in the United States, use the **Celsius scale.** Both scales are based on the temperature at which water freezes and boils at standard atmospheric pressure (see Section 4.3). In the Fahrenheit scale, the freezing point of water is 32°F, and boiling occurs at 212°F, a range of 180 degrees. The Celsius scale divides the range into 100 degrees, from 0°C to 100°C. Figure 3.8 compares these scales.

The way Fahrenheit and Celsius temperature scales are defined leads to the relationship:

$$T_{°F} - 32 = 1.8\ T_{°C}$$

$T_{°F}$ is the Fahrenheit temperature and $T_{°C}$ is the Celsius temperature.

### Thinking About Your Thinking

Proportional Reasoning

**Rearranged, the Fahrenheit–Celsius relationship equation is $T_{°F} = 1.8\ T_{°C} + 32$, which corresponds to the slope-intercept form of the equation of a straight line, y = mx + b. However, a y-versus-x plot of that equation does not pass through the origin. When one variable is zero, the other is not. Thus y and x increase and decrease together, but not in the simplest form of a direct proportionality, y = mx. Celsius–Fahrenheit conversions therefore cannot be done by dimensional analysis. They require the use of algebra.**

Archives of the Royal Swedish Academy of Science

Anders Celsius (1701–1744), a Swedish scientist, first proposed setting the freezing point of water at 100° and the boiling point at 0°. The scale was reversed after his death.

The Fahrenheit–Celsius relationship equation is derived by comparing the Celsius and Fahrenheit scales (Fig. 3.8). The freezing point is 0 on the Celsius scale and 32 on the Fahrenheit scale; thus the $T_{°F} - 32$ part of the equation. The range between boiling and freezing points is (100 − 0) = 100 Celsius degrees and (212 − 32) = 180 Fahrenheit degrees. Thus the Celsius degree is 180/100 times as large as the Fahrenheit degree. This ratio is reduced to its decimal equivalent, 1.8, in the equation:

$$\frac{180 \text{ Fahrenheit degrees}}{100 \text{ Celsius degrees}} = \frac{1.8 \text{ Fahrenheit degrees}}{1 \text{ Celsius degree}}$$

**Figure 3.8** A comparison of Fahrenheit, Celsius, and Kelvin temperature scales. The reference or starting point for the Kelvin scale is absolute zero (0 K = −273°C), the lowest temperature theoretically obtainable. Note that the abbreviation K for the kelvin unit is used without the degree sign (°). Also note that 1 Celsius degree = 1 kelvin unit = 180/100 or 9/5 Fahrenheit degree.

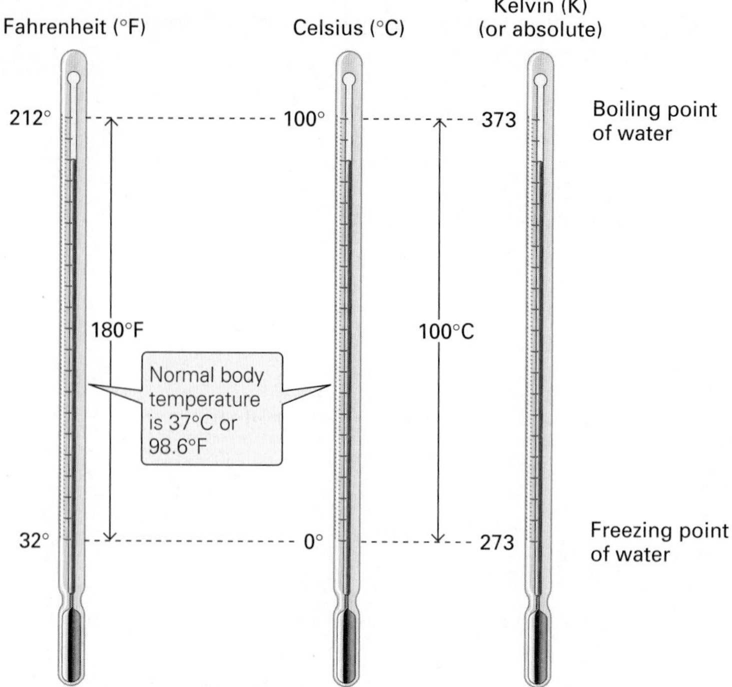


SI units include a third temperature scale known as the **Kelvin** or **absolute temperature scale.** The degree on the Kelvin scale is the same size as a Celsius degree. The origin of the Kelvin scale is discussed in Chapter 4. It is based on zero at the lowest temperature possible, which is 273° below zero on the Celsius scale. The two scales are therefore related by the equation

$$T_K = T_{°C} + 273$$

K in this equation represents kelvins, the actual temperature unit. The degree symbol, °, is not used for Kelvin temperatures.

Converting a temperature from one scale to another is done by algebra, not dimensional analysis. In this book we will always solve algebra problems by first solving the equation *algebraically* for the wanted quantity. When the unknown is the only term on one side of the equation, given values are substituted on the other side, and the result is calculated. The advantage of this procedure is not so apparent in temperature conversions, but it will become clear as you gain more experience in solving algebra-based chemistry problems.

## Active Example 3.22

What is the Celsius temperature on a comfortable 72°F day? What is the Kelvin temperature?

Begin by identifying the *GIVEN* and *WANTED* quantities for the °F-to-°C conversion. Then write the Fahrenheit–Celsius relationship equation, and solve it for the wanted quantity.

*GIVEN:* 72°F  *WANTED:* $T_{°C}$  *EQUATION:* $T_{°F} - 32 = 1.8\,T_{°C}$

$$T_{°F} - 32 = 1.8\,T_{°C} \xrightarrow{\text{Divide both sides by 1.8}} T_{°C} = \frac{T_{°F} - 32}{1.8}$$

To isolate $T_{°C}$, we divided both sides of the equation by 1.8, the coefficient of $T_{°C}$.

Substitute the given value and calculate the temperature in °C.

$$T_{°C} = \frac{T_{°F} - 32}{1.8} = \frac{72 - 32}{1.8} = 22°C$$

Now complete the example by using the Kelvin–Celsius relationship equation to convert $T_{°C}$ to $T_K$.

$$T_K = T_{°C} + 273 = 22 + 273 = 295\text{ K}$$

Scientists usually measure temperature directly in °C with a Celsius thermometer. However, if you ever are asked to convert between $T_{°F}$ and $T_K$, you must determine $T_{°C}$ as an intermediate step, as you did in this example.

An infrared thermometer. When conventional temperature sensors cannot be used, a non-contact device such as this infrared thermometer is used instead. A lens focuses the infrared light emitted from a heated object onto a detector, which is calibrated to convert the wavelength of light to an electronic signal that can be displayed as temperature.

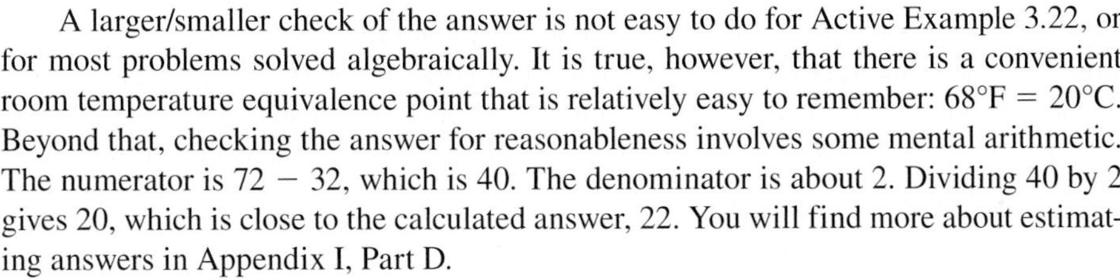

A larger/smaller check of the answer is not easy to do for Active Example 3.22, or for most problems solved algebraically. It is true, however, that there is a convenient room temperature equivalence point that is relatively easy to remember: 68°F = 20°C. Beyond that, checking the answer for reasonableness involves some mental arithmetic. The numerator is 72 − 32, which is 40. The denominator is about 2. Dividing 40 by 2 gives 20, which is close to the calculated answer, 22. You will find more about estimating answers in Appendix I, Part D.

## Active Example 3.23

From each temperature given in the following table, calculate the equivalent temperatures in the other scales.

| °F | °C | K |
|---|---|---|
| | –25 | |
| 105 | | |
| | | 364 |

| °F | °C | K |
|---|---|---|
| –13 | –25 | 248 |
| 105 | 41 | 314 |
| 196 | 91 | 364 |

## 3.8 Proportionality and Density

**Goal 16** Write a mathematical expression indicating that one quantity is directly proportional to another quantity.

**Goal 17** Use a proportionality constant to convert a proportionality to an equation.

**Goal 18** Given the values of two quantities that are directly proportional to each other, calculate the proportionality constant, including its units.

**Goal 19** Write the defining equation for a proportionality constant and identify units in which it might be expressed.

**Goal 20** Given two of the following for a sample of a pure substance, calculate the third: mass, volume, and density.

Consider the following experiment. The mass of a clean, dry container for measuring liquid volume was measured and found to be 26.42 g. Then 5.2 mL of cooking oil was placed into the container. The mass of the oil and the container was 31.06 g. The mass of the 5.2 mL of oil in the container was found by subtracting the mass of the empty container from the mass of the container plus the oil:

| | |
|---|---|
| Mass of 5.2 mL of oil + container | 31.06 g |
| − Mass of empty container | − 26.42 g |
| Mass of 5.2 mL of oil | 4.64 g |

This procedure was repeated several times. The resulting data are shown in the table and summarized in the graph in Figure 3.9.

We can use the results of this experiment to find the relationship between the mass of the cooking oil and its volume. Whenever a graph of two related measurements is a straight line that passes through the origin, the measured quantities are directly proportional to each other. More formally, a direct proportionality exists between two quantities when they increase or decrease at the same rate. "At the same rate" means that if one quantity doubles, the other quantity doubles; if one triples, the other triples; if one is reduced by 20%, the other is reduced by 20%, and so on.

We use direct proportionalities every day. The relationships between defined units of mass, length, and time are proportionalities: Examples include grams per kilogram,

| Volume (mL) | Mass (g) | Mass/Volume (g/mL) |
|---|---|---|
| 5.2 | 4.64 | 0.89 |
| 7.7 | 6.91 | 0.90 |
| 10.4 | 9.28 | 0.892 |
| 13.8 | 12.35 | 0.895 |
| 17.6 | 15.72 | 0.893 |

(a)

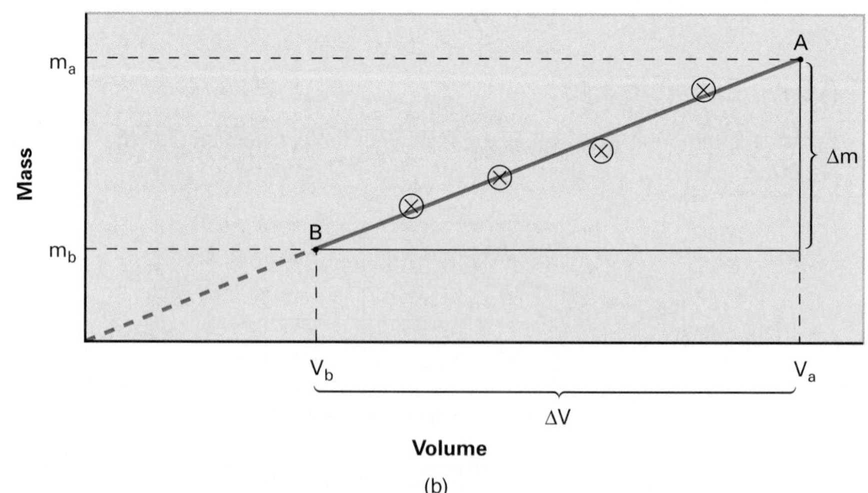


**Figure 3.9** Mass–volume data for a sample of cooking oil. (a) The third column is the result of dividing the measured mass by the measured volume. All quotients are rounded to the number of significant figures justified by the data. (b) The slope of the line between any two points, A and B, on the line, $\Delta m/\Delta V$, is the density of the oil. Notice that the slope of the line of the plot in part (b) is the same as the mass/volume ratio in the third data column in part (a).

inches per foot, and seconds per hour. Speed is a proportionality, whether measured in kilometers per hour, feet per second, or any other distance per time unit. The prices we pay for things are based on proportionalities: dollars per pound or per kilogram, cents per liter or quart, cents per bunch (of carrots), and dollars per six-pack are all money units per mass, volume, or counting unit of some substance.

## Thinking About Your Thinking

### Proportional Reasoning

**The discussion in this part of the text further explains the relationship between direct proportionalities and PER expressions. You will find yourself doing this type of thinking continually throughout this course, in many other science courses, and, as discussed in the body of the text, in everyday situations. Be sure to *Learn It Now!***

From the preceding discussion we can list several conclusions:

**1.** We can describe direct proportionalities between measured quantities with PER expressions.

**2.** Direct proportionalities between measured quantities yield two conversion factors between the quantities.

**3.** Given either quantity in a direct proportionality and the conversion factor between the quantities, we can calculate the other quantity with dimensional analysis.

A direct proportionality between two variables, such as mass (m) and volume (V), is indicated by $m \propto V$, where the symbol $\propto$ means "is proportional to." A proportionality can be changed into an equation by inserting a multiplier called a **proportionality constant.** If we let D be the proportionality constant,

$$m \propto V \xrightarrow{\text{the proportionality changes to an equation}} m = D \times V$$

Solving for the proportionality constant, D, yields the **defining equation** for a physical property of a pure substance called its **density:**

$$m = D \times V \xrightarrow{\text{divide both sides by V}} D \equiv \frac{m}{V}$$

In words, **the density of a substance is its mass per unit volume:**

$$\text{Density} \equiv \frac{\text{mass}}{\text{volume}}$$

The symbol $\equiv$ identifies a definition.

We can think of density as a measure of the relative "heaviness" of a substance, in the sense that a block of iron is heavier than a block of aluminum of the same size. Table 3.4 lists the densities of some common materials.

Some proportionality constants have physical significance. Density, a physical property, is among them. The density of cooking oil in the experiment described at the beginning of this section is equal to the slope of the line in Figure 3.9.

You are familiar with the fact that ice—solid water—floats on liquid water. Why is this? Ice is *less dense* than liquid water. What does this mean on the particulate level? Considering the definition of density, a given volume of ice must have less mass than the same volume of liquid water. In other words, if all water molecules have the same mass and volume, whether liquid or solid, the molecules in liquid water must pack more closely together than molecules in ice. Figure 3.10 shows how solid water forms ice crystals with spaces between the molecules, whereas liquid molecules have fewer and smaller open spaces. So, at the particulate level, the density of a given pure substance in a given state of matter is a measure of how tightly packed the molecules

**Table 3.4** Densities of Some Common Substances ($g/cm^3$ at 20°C and 1 atm)

| Substance | Density |
|---|---|
| Helium | 0.00017 |
| Air | 0.0012 |
| Lumber | |
| Pine | 0.5 |
| Maple | 0.6 |
| Oak | 0.8 |
| Water | 1.0 |
| Glass | 2.5 |
| Aluminum | 2.7 |
| Iron | 7.8 |
| Copper | 9.0 |
| Silver | 10.5 |
| Lead | 11.4 |
| Mercury | 13.6 |
| Gold | 19.3 |

**P/REVIEW:** The reasons for the unusual behavior of water are explained in Section 15.6.*

*P/Reviews are references to items covered in another chapter of the textbook. For a description of P/Reviews, see Section 1.5.

Figure 3.10 Particulate view of solid and liquid water. (a) Water molecules in solid form (ice) are held in a crystal pattern that has open spaces between the molecules. (b) When ice melts, the crystal collapses, the molecules are closer together, and the liquid is therefore denser than the solid. This is why ice floats in water, a solid–liquid property shared by few other substances.

In solid water (ice) each water molecule is close to its neighbors and restricted to vibrating back and forth around a specific location.

In liquid water the molecules are close together, but they can move past each other; each molecule can move only a short distance before bumping into one of its neighbors.

(a)

(b)



are in that state. Water is an unusual substance, and the fact that its solid phase is less dense than its liquid phase is just one of many of its unusual properties. For almost all other substances, the solid phase is more dense than the liquid phase (Fig. 3.11).

The definition of density establishes its units. Mass is commonly measured in grams (g); volume is measured in cubic centimeters ($cm^3$). Therefore, according to the equation $D = m/V$ and the mass-per-unit-volume definition, the units of density are grams per cubic centimeter, $g/cm^3$. Liquid densities are often given in grams per milliliter, g/mL. (Recall that one milliliter is exactly one cubic centimeter by definition.) There are, of course, other units in which density can be expressed, but they all must reflect the definition in terms of mass $\div$ volume. Examples are grams/liter, usually used for gases because their densities are so low, and the USCS pounds/cubic foot.

To find the density of a substance, you must know both the mass and volume of a sample of that substance. Dividing the mass by volume yields density.

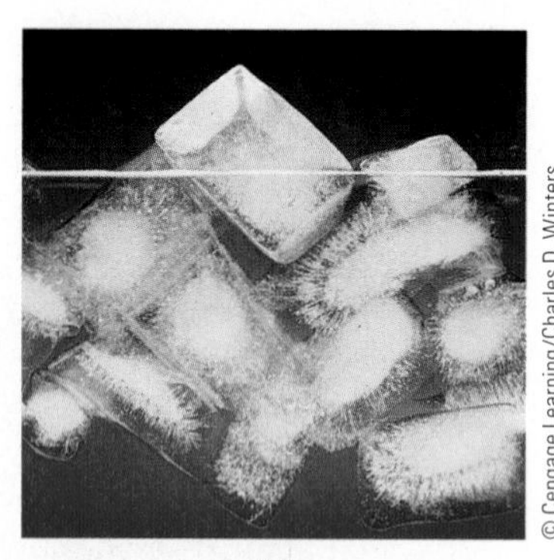

(a)

(b)



Figure 3.11 Densities of solids and liquids. (a) Water is unusual in that its solid phase, ice, will float on its liquid phase. (b) Solid ethanol sinks to the bottom of the liquid. The solid form of almost all substances is more dense than the liquid phase.

## Active Example 3.24

The mass of a 12.0-$cm^3$ piece of magnesium is 20.9 grams. Find the density of magnesium.

This problem simply asks you to apply a definition. One strategy for solving a density problem is to recall and write down the defining equation, rearrange the equation for the wanted variable, if necessary, substitute the given quantities, and solve.

GIVEN: 12.0 $cm^3$; 20.9 g

WANTED: density (mass units divided by volume units, assume $g/cm^3$)

EQUATION: $D \equiv \frac{m}{V} = \frac{20.9\ g}{12.0\ cm^3} = 1.74\ g/cm^3$

CHECK: Units OK; $20/10 = 2$, which is close to 1.74.

Density problems are particularly useful at the beginning of a chemistry course because they illustrate both methods of solving problems, by algebra and by dimensional analysis. Both are used in the next Active Example.

## Active Example 3.25

The density of a certain lubricating oil is 0.862 g/mL. Find the volume occupied by 196 grams of that oil.

Try dimensional analysis first. List the GIVEN, WANTED, and PER/PATH. Set up and solve.

GIVEN: 196 g    WANTED: volume (assume mL)

PER: $\frac{1 \text{ mL}}{0.862 \text{ g}}$

PATH: g ⟶ mL

$$196 \text{ g} \times \frac{1 \text{ mL}}{0.862 \text{ g}} = 227 \text{ mL}$$

Now for algebra. Solve the density equation for volume, substitute the given quantities, and calculate the answer. *Include units when you substitute the given quantities.*

GIVEN: 196 g    WANTED: volume (assume mL)    EQUATION: $D \equiv \frac{m}{V}$

$$D \equiv \frac{m}{V} \xrightarrow{\text{cross-multiply}} V \times D = m \xrightarrow{\text{divide both sides by D}} V = \frac{m}{D}$$

$$V = \frac{m}{D} = \frac{196 \text{ g}}{0.862 \text{ g/mL}} = 196 \text{ g} \times \frac{1 \text{ mL}}{0.862 \text{ g}} = 227 \text{ mL}$$

↑ division is invert and multiply

Notice that the resulting setups are the same, whether you solve the problem by dimensional analysis or by algebra.

Why did we include units in the algebraic setup of the problem? Because about one out of four students solving Active Example 3.25 *without units* would have come up with 169 mL for the answer. Starting with a correct substitution of numbers only into the original equation,

$$0.862 = \frac{196}{V}$$

they would have calculated the answer by multiplying 196 by 0.862 instead of dividing. The difference between the correct answer, 227, and the incorrect 169 is not enough to give an instinctive feeling that 169 is wrong, as is the case in dollar-and-penny conversions, so the numerical answer appears reasonable. With units in the multiplication, however, the result would have been

$$196\ \text{g} \times \frac{0.862\ \text{g}}{\text{mL}} = \frac{169\ \text{g}^2}{\text{mL}}$$

Because $g^2/mL$ are nonsense units, they would have signaled an error in the setup.

## 3.9 A Strategy for Solving Problems

At this point you have solved example problems by the two methods that serve for the majority of the quantitative problems in this text. It is time to organize the methods into a form that will be useful to you. This is done as a flowchart in Figure 3.12. We call this a six-step strategy for solving quantitative chemistry problems. In the first three steps you plan how to solve the problem; in the last three steps you execute your plan by solving the problem. Study Figure 3.12 now, before moving on to the next paragraph.

In examples throughout this textbook, we will ask you to *PLAN* your solution to a problem. *PLAN* is a one-word instruction that means to complete the first three steps given in Figure 3.12.

Step 3 is the heart of problem solving. Once you decide what to do, the rest is usually routine. Deciding is the crucial step. Make your decision by examining the *GIVEN* quantity and units and the *WANTED* units and determining how they are related to each other:

**Learn It Now!** This is a major point. If you can decide which of these two methods works on a problem, you are well on your way to solving the problem.

| The problem can be solved by **dimensional analysis** if the *GIVEN* and *WANTED* can be linked by one or more *PER* expressions and you know or can find the conversion factor for each expression. | The problem can be solved by **algebra** if the *GIVEN* and *WANTED* appear in an algebraic equation in which the *WANTED* is the only unknown. |
|---|---|

The six-step strategy for solving problems is applied in every example in this chapter to which it can be applied. We will use it consistently throughout the book. We omit the last step later, since the check is largely a mental operation. You should never omit the last step. *Always check your answers,* both numbers and units, to be sure they are reasonable. We strongly recommend that you read the section in Appendix I.D on estimating answers for this purpose.

## 3.10 Thoughtful and Reflective Practice

The only way to learn how to solve problems is to solve them yourself. If you have followed the question-and-answer approach to the Active Examples in this chapter and have covered the answers to the steps in the examples until you have figured out the answer for yourself, *you* have solved the problems and not merely looked at how *we* solved them. But you still need more practice.

The end-of-chapter problems with answers give you lots of opportunities to solve problems with immediate feedback on the correctness of your methods. Be sure to keep in mind your reason for solving problems. It is not to get the answer we got, but to *learn how to solve the problem.*

Solving a problem with one finger at its solution in the answer section so you can check your progress is neither solving the problem nor *learning how to solve it.* When you tackle a problem, solve it completely. If you get stuck at any point, put the problem aside and check the part of the chapter that covers that point. Learn there what you

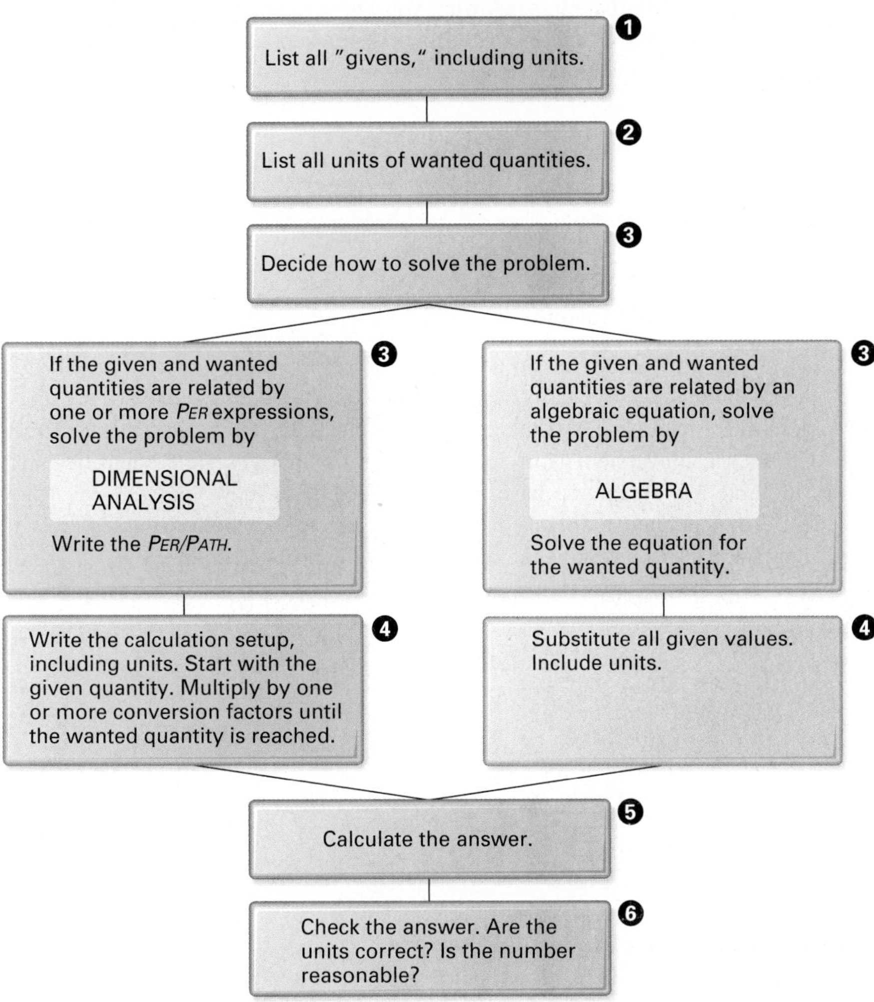
HOW TO SOLVE A QUANTITATIVE CHEMISTRY PROBLEM
Circled numbers correspond to the six steps for solving a problem.
1 List all "givens," including units.
2 List all units of wanted quantities.
3 Decide how to solve the problem.
3 If the given and wanted quantities are related by one or more PER expressions, solve the problem by
DIMENSIONAL ANALYSIS
Write the PER/PATH.
3 If the given and wanted quantities are related by an algebraic equation, solve the problem by
ALGEBRA
Solve the equation for the wanted quantity.
4 Write the calculation setup, including units. Start with the given quantity. Multiply by one or more conversion factors until the wanted quantity is reached.
4 Substitute all given values. Include units.
5 Calculate the answer.
6 Check the answer. Are the units correct? Is the number reasonable?

**Figure 3.12** A six-step strategy for solving quantitative chemistry problems.

need to learn. *Do not check the answer section.* Return to your solution of the problem and complete it. Then compare your solution to the one in the back of the book. If they do not agree, find out why. Be sure you understand the problem before going to the next one.

We close this chapter with a few more examples with which you can practice while we guide you in following the six-step strategy.

## Active Example 3.26

Calculate the mass of 7.04 $cm^3$ of silver.

At first glance, it might seem that you don't have enough information to solve this problem. When this situation occurs, start by assessing what you know from the problem statement. This will often help you discover what else you need to know to solve the problem. What are the GIVEN and WANTED for this problem?

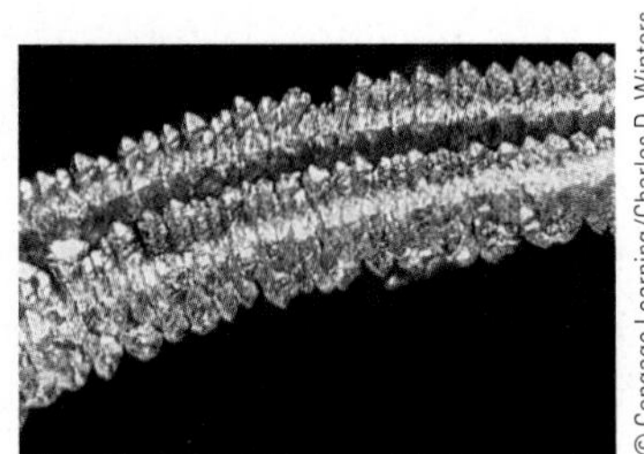

Silver occurs as an uncombined element in nature.

GIVEN: 7.04 $cm^3$ WANTED: mass (assume g)

We assume that grams are the appropriate unit for mass in this problem because the gram is the basic metric unit for mass.

The next step in thinking about this problem is to ask yourself if you know or can find a relationship that connects what you know to what you want. You know volume; you want mass. What physical property of a substance provides a connection between mass and volume?

Density.

Density is mass per unit volume.

Chemistry problems often do not include all of the information necessary to find their solutions. It is typically assumed that you should either know or are able to look up information such as the density of a pure substance, as in this case, or a USCS–metric conversion, as in Section 3.6. In this book, selected densities are given in Table 3.4. When you have what you need, *PLAN* how you will solve the problem. You can choose dimensional analysis, algebra, or both for practice. We will show both solutions. Solve the remainder of the problem completely.

**Dimensional Analysis**

*GIVEN:* 7.04 $cm^3$ *WANTED:* g

*PER:* $\frac{10.5\text{ g}}{\text{cm}^3}$

*PATH:* $\text{cm}^3 \longrightarrow \text{g}$

$$7.04\ \cancel{\text{cm}^3} \times \frac{10.5\text{ g}}{\cancel{\text{cm}^3}} = 73.9\text{ g}$$

**Algebra**

*GIVEN:* 7.04 $cm^3$ *WANTED:* g

*EQUATION:* $D \equiv \frac{m}{V}$

$$D \equiv \frac{m}{V} \xrightarrow{\text{cross-multiply}} V \times D = m$$

$$m = V \times D = 7.04\ \cancel{\text{cm}^3} \times \frac{10.5\text{ g}}{\cancel{\text{cm}^3}} = 73.9\text{ g}$$

## Active Example 3.27

The specific heat of a substance is the amount of heat energy needed to raise the temperature of 1 gram of the substance 1°C. Its units are joules per gram degree, which are written J/g · °C. The raised dot in the denominator is a "times" dot, signifying that the gram unit is multiplied by the °C unit. The equation for finding how much heat, q, is needed to raise the temperature of m grams of the substance by $\Delta T$ degrees is $q = m \times c \times \Delta T$, in which c is the specific heat.

Calculate the amount of heat, expressed in joules, needed to raise the temperature of 43.6 grams of aluminum by 22°C. The specific heat of aluminum is 0.88 J/g · °C.

Believe it or not, this is an easy problem! Don't let the words and unfamiliar terms bother you. By reading the problem, you should be able to decide *how* to solve it. Once you know that, it is routine. *PLAN* your approach by writing the *GIVEN* quantities and the *WANTED* units, and state your decision.

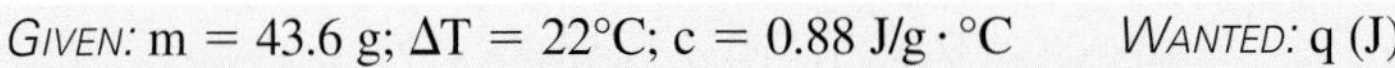

*GIVEN:* m = 43.6 g; $\Delta T$ = 22°C; c = 0.88 J/g · °C *WANTED:* q (J)

Solve by algebra because the *GIVEN* and *WANTED* quantities are related by an equation in which the *WANTED* quantity is the only unknown.

Substitute both numbers and units into the equation and complete the solution.

*EQUATION:* $q = m \times c \times \Delta T = 43.6\ \cancel{g} \times \dfrac{0.88\ J}{\cancel{g} \cdot \cancel{°C}} \times 22\ \cancel{°C} = 8.4 \times 10^2\ J$

The answer is given in two significant figures because two of the factors have only two significant figures. Units check. 43.6 × 0.88 is close to 40 × 1, or 40, and 40 × 20 is 800, which is close to the 840 calculated.

Dimensional analysis can be applied to everyday problems as well as those associated with chemistry. The final example is one you may be able to identify with personally.

## Active Example 3.28

Suppose that you have just landed a part-time job that pays $7.25 an hour. You will work five shifts each week, and the shifts are 4 hours long. You plan to save all of your earnings to pay cash for an MP3 player and headphones that cost $734.26, tax included. You are paid weekly. How many weeks must you work in order to save enough money to buy the sound system? You might also be interested in knowing how much cash you will have left for song downloads or other goodies.

This is a long example, and we ask you to solve it completely and without help. Develop your *PLAN*, set up the problem, and calculate the number of weeks.

*GIVEN:* 734.26 dollars (dol) *WANTED:* weeks

*PER:* $\dfrac{1\ \text{hr}}{7.25\ \text{dol}}$ $\dfrac{1\ \text{shift}}{4\ \text{hr}}$ $\dfrac{1\ \text{week}}{5\ \text{shifts}}$

*PATH:* dol $\longrightarrow$ hr $\longrightarrow$ shifts $\longrightarrow$ weeks

$$734.26\ \cancel{\text{dol}} \times \frac{1\ \cancel{\text{hr}}}{7.25\ \cancel{\text{dol}}} \times \frac{1\ \cancel{\text{shift}}}{4\ \cancel{\text{hr}}} \times \frac{1\ \text{week}}{5\ \cancel{\text{shifts}}} = 5.063862069\ \text{weeks} = 6\ \text{weeks}$$

All of the numbers in the calculation are exact numbers, but it will take your sixth paycheck to get the $734.26 you need.

What will be your total pay at the end of the sixth week? And how much excess cash will there be?

*GIVEN:* 6 weeks *WANTED:* dollars

*PER:* $\frac{5 \text{ shifts}}{\text{week}}$ $\frac{4 \text{ hr}}{\text{shift}}$ $\frac{7.25 \text{ dol}}{\text{hr}}$

*PATH:* weeks ⟶ shifts ⟶ hr ⟶ dol

$$6 \cancel{\text{weeks}} \times \frac{5 \cancel{\text{shifts}}}{\cancel{\text{week}}} \times \frac{4 \cancel{\text{hr}}}{\cancel{\text{shift}}} \times \frac{7.25 \text{ dol}}{\cancel{\text{hr}}} = 870.00 \text{ dollars}$$

$870.00 earned − $734.26 = $135.74 cash remaining

Again, all the numbers are exact numbers, to the penny.

## Chapter 3 in Review

Most of the key terms and concepts and many others appear in the Glossary. Use your Glossary regularly.

### 3.1 Introduction to Measurement

**Key Terms and Concepts: Base unit, derived unit, metric system, SI unit**

### 3.2 Exponential (Scientific) Notation

Goal 1 Write in exponential notation a number given in ordinary decimal form; write in ordinary decimal form a number given in exponential notation.

Goal 2 Using a calculator, add, subtract, multiply, and divide numbers expressed in exponential notation.

**Key Terms and Concepts: Base, coefficient, exponent, exponential, exponential (scientific) notation, standard exponential notation**

### 3.3 Dimensional Analysis

Goal 3 In a problem, identify given and wanted quantities that are related by a *PER* expression. Set up and solve the problem by dimensional analysis.

**Key Terms and Concepts: Conversion factor, dimensional analysis, directly proportional, given quantity (*GIVEN*), inversely proportional, per expression (*PER*), proportional reasoning, proportionality constant, the symbol ≡, unit path (*PATH*), wanted quantity (*WANTED*)**

### 3.4 Metric Units

Goal 4 Distinguish between mass and weight.

Goal 5 Identify the metric units of mass, length, and volume.

Goal 6 State and write with appropriate metric prefixes the relationship between any metric unit and its corresponding kilounit, centiunit, and milliunit.

Goal 7 Using Table 3.1, state and write with appropriate metric prefixes the relationship between any metric unit and other larger and smaller metric units.

Goal 8 Given a mass, length, or volume expressed in metric units, kilounits, centiunits, or milliunits, express that quantity in the other three units.

**Key Terms and Concepts: Cubic centimeter ($cm^3$), kilogram (kg) and gram (g), liter (L) and milliliter (mL), mass, meter (m), weight**

### 3.5 Significant Figures

Goal 9 State the number of significant figures in a given quantity.

Goal 10 Round off given numbers to a specified number of significant figures.

Goal 11 Add or subtract given quantities and express the result in the proper number of significant figures.

Goal 12 Multiply or divide given measurements and express the result in the proper number of significant figures.

**Key Terms and Concepts: Exact numbers, rounding off, significant figures, significant figure rule for addition and subtraction, significant figure rule for multiplication and division, uncertainty (in measurement), uncertain digit**

### 3.6 Metric–USCS Conversions

Goal 13 Given a metric–USCS conversion table and a quantity expressed in any unit in Table 3.2, express that quantity in corresponding units in the other system.

### 3.7 Temperature

Goal 14 Given a temperature in either Celsius or Fahrenheit degrees, convert it to the other scale.

Goal 15 Given a temperature in Celsius degrees or kelvins, convert it to the other scale.

**Key Terms and Concepts: Celsius scale, Fahrenheit scale, Kelvin (absolute) temperature scale**

**3.8 Proportionality and Density**

Goal 16 Write a mathematical expression indicating that one quantity is directly proportional to another quantity.

GOAL 17 Use a proportionality constant to convert a proportionality to an equation.

Goal 18 Given the values of two quantities that are directly proportional to each other, calculate the proportionality constant, including its units.

Goal 19 Write the defining equation for a proportionality constant and identify units in which it might be expressed.

Goal 20 Given two of the following for a sample of a pure substance, calculate the third: mass, volume, and density.

**Key Terms and Concepts: Defining equation, density, proportionality constant**

**3.9 A Strategy for Solving Problems**

**3.10 Thoughtful and Reflective Practice**

## Study Hints and Pitfalls to Avoid

Exponential notation is not usually a problem, except for careless errors when relocating the decimal in the coefficient and adjusting the exponent. These errors will not occur if you make sure the exponent and the coefficient move in opposite directions, one larger and one smaller. It sometimes helps to think about an ordinary decimal number as being written in exponential notation in which the exponential is $10^0$. Thus 0.0024 becomes $0.0024 \times 10^0$, and the larger/smaller changes in the coefficient and exponent are clear when changing to $2.4 \times 10^{-3}$.

A common error in metric-to-metric conversions is moving the decimal point the wrong way. To avoid this, write out fully the dimensional analysis setup. Then challenge your answer. Use the larger/smaller rule. For a given amount of anything, the number of larger units is smaller and the number of smaller units is larger.

*Include units in every problem you solve.* This is the best advice that we can give to someone learning how to solve quantitative chemistry problems.

Challenge every problem answer in both size and units. Many errors would never been seen by a test grader if the test taker had simply checked the reasonableness of an answer. Part D in Appendix I offers some suggestions on how to estimate the numerical result in a problem. Please read this section and put it into practice with every problem you solve.

Significant figures can indeed be troublesome, but they need not be if you learn to follow a few basic rules. There are four common errors to watch for:

1. Starting to count significant figures at the decimal point of a very small number instead of at the first nonzero digit
2. Using the significant figure rule for multiplication/division when rounding off an addition or subtraction result
3. Failing to show an uncertain tail-end zero on the right-hand side of the decimal
4. Failing to use exponential notation when writing larger numbers, thereby causing the last digit shown to be other than the uncertain digit

Most of the arithmetic operations you will perform are multiplications and divisions. Students often learn the rule for those operations well, but then erroneously apply that rule to the occasional addition or subtraction problem that comes along. Products and quotients have the same number of significant figures as the smallest number in any factor. Sums can have more significant figures than the largest number in any number added. Example: $68 + 61 = 129$. Differences can have fewer significant figures than the smallest number in either number in the subtraction. Example: $68 - 61 = 7$.

## Concept-Linking Exercises

*Write a brief description of the relationships among each of the following groups of terms or phrases. Answers to the Concept-Linking Exercises are given after answers to the Target Checks at the end of the book.*

1. Metric system, SI units, derived unit, base unit
2. Dimensional analysis, *PER* expression, conversion factor, unit *PATH*, *GIVEN* quantity, *WANTED* quantity
3. Mass, weight, kilogram, gram, pound
4. Uncertainty in measurement, uncertain (doubtful) digit, significant figures, exact numbers
5. Direct proportionality, inverse proportionality, proportionality constant, the symbol $\equiv$

## Small-Group Discussion Questions

*Small-Group Discussion Questions are for group work, either in class or under the guidance of a leader during a discussion section.*

1. How have you used measurement in your life in the past week? List at least as many measurements as the number of people in your group. What instrument did you use for each measurement?
2. How do you change an ordinary decimal number to scientific notation? Change 3,876,989 to scientific notation, and use this example while explaining the general procedure. How do you change a number in scientific notation to ordinary decimal form? Change $3.99 \times 10^{-5}$ to ordinary decimal form, and use this example while explaining the general procedure.

3. The life expectancy for a person born in the United States in 1990 is 71.8 years for males and 78.8 years for females. How many more times will the heart of an average female beat during her lifespan than an average male? Identify the GIVEN quantity, WANTED units, PER expressions, and the unit PATH, and then setup and solve.

4. A cube measures 1 m × 2 m × 3 m. What is the volume of the cube in liters? If the cube is made of a pure substance that has a density of 2.5 g/mL, what is its mass in kilograms?

5. How many significant figures are usually implied when a person expresses her or his (a) weight, (b) height, (c) age, (d) pulse rate, and (e) number of fingers? Explain each answer. When a three-significant-figure quantity is multiplied by a four-significant-figure quantity, how many significant figures are justified in the product? When a three-significant-figure quantity is subtracted from a four-significant-figure quantity, how many significant figures are justified in the difference?

6. The original definition of the meter was one ten-millionth of the length of the meridian through Paris from pole to the equator. Based on this definition, how many miles is it from the North Pole to the equator? How many milliliters are in a cup? A metric ton is 1,000 kilograms. How many USCS tons (2000 pounds) are equal to a metric ton?

7. Which is larger, a Celsius degree or a Fahrenheit degree? What is the temperature change in Fahrenheit when the temperature increases by 14°C? At what point do the Celsius and Fahrenheit temperatures have the same value? At what point do the kelvin and Celsius temperatures have the same value?

8. The circumference of a circle is directly proportional to its diameter. Write this as a mathematical statement. Change the proportionality to an equation by inserting the appropriate proportionality constant. Write the defining equation for the proportionality constant. What are the units of the proportionality constant?

9. If you know the mass and density of a pure substance, you can calculate its volume. This can be accomplished by either dimensional analysis or algebra. Determine the volume in cubic inches of a gold bar that weighs 1.0 troy ounce by both dimensional analysis and algebra. 12 troy ounces equal 16 ounces. The density of gold is 19.3 g/cm$^3$.

10. A school supplies pencils to its students. The pencils are packaged in boxes of one gross, where 1 gross = 12 dozen. The historical average for the school's pencil needs has been 8.7 pencils per student. If the projected enrollment of the school is 932 students, how many boxes should be ordered for the next school year?

# Questions, Exercises, and Problems

*Blue-numbered questions are answered in Appendix III. ■ denotes problems assignable in OWL. In the first section of Questions, Exercises, and Problems, similar exercises and problems are paired in consecutive odd-even number combinations.*

### Section 3.2: Exponential (Scientific) Notation

1. Write the following numbers in exponential notation: (a) 0.000322, (b) 6,030,000,000, (c) 0.00000000000619

2. ■ Write each of the following numbers in exponential notation: (a) 70,300, (b) 0.0231, (c) 0.000154, (d) 5,040.

3. Write the following exponential numbers in ordinary decimal form: (a) $5.12 \times 10^6$, (b) $8.40 \times 10^{-7}$, (c) $1.92 \times 10^{21}$.

4. ■ Write the ordinary (non-exponential) form of the following numbers: (a) $2.32 \times 10^{-2}$, (b) $9.27 \times 10^4$, (c) $2.54 \times 10^3$, (d) $8.96 \times 10^{-4}$.

5. Complete the following operations:
   a) $(7.87 \times 10^4)(9.26 \times 10^{-8}) =$
   b) $(5.67 \times 10^{-6})(9.05 \times 10^{-7}) =$
   c) $(309)(9.64 \times 10^6) =$
   d) $(4.07 \times 10^3)(8.04 \times 10^{-8})(1.23 \times 10^{-2}) =$

6. ■ Complete the following operations:
   a) $(5.08 \times 10^{-5})(1.83 \times 10^{-7}) =$
   b) $\dfrac{9.42 \times 10^{-4}}{5.98 \times 10^{-4}} =$
   c) $\dfrac{(2.33 \times 10^6)(9.61 \times 10^4)}{(1.83 \times 10^{-7})(8.76 \times 10^4)} =$

7. Complete the following operations:
   a) $\dfrac{6.18 \times 10^4}{817} =$
   b) $\dfrac{4.91 \times 10^6}{5.22 \times 10^5} =$
   c) $\dfrac{4.60 \times 10^7}{1.42 \times 10^3} =$
   d) $\dfrac{9.32 \times 10^4}{6.24 \times 10^7} =$

8. ■ Complete the following operations:
   a) $(2.34 \times 10^6)(4.23 \times 10^5) =$
   b) $\dfrac{8.60 \times 10^{-4}}{1.72 \times 10^{-4} - 9.25 \times 10^{-7}} =$
   c) $\dfrac{(7.54 \times 10^{-5})(1.72 \times 10^{-4})}{8.60 \times 10^{-4}} =$

9. Complete the following operations:
   a) $\dfrac{9.84 \times 10^3}{(6.12 \times 10^3)(4.27 \times 10^7)} =$
   b) $\dfrac{(4.36 \times 10^8)(1.82 \times 10^3)}{0.0856(4.7 \times 10^6)} =$

10. ■ Complete the following operations:
    a) $9.25 \times 10^{-7} + 8.60 \times 10^{-4} =$
    b) $\dfrac{8.60 \times 10^{-4} + 4.23 \times 10^5}{1.72 \times 10^{-4}} =$

c) $\dfrac{7.54 \times 10^{-5}}{(9.25 \times 10^{-7})(8.60 \times 10^{-4})} =$

11. Complete the following operations:
    a) $6.38 \times 10^{7} + 4.01 \times 10^{8} =$
    b) $1.29 \times 10^{-6} - 9.94 \times 10^{-7} =$

12. ■ Complete the following operations:
    a) $8.63 \times 10^{5} + 1.80 \times 10^{-4} =$
    b) $\dfrac{1.80 \times 10^{-4} + 2.90 \times 10^{-7}}{9.53 \times 10^{4}} =$
    c) $\dfrac{1.06 \times 10^{-5}}{(8.63 \times 10^{5})(1.80 \times 10^{-4})} =$

### Section 3.3: Dimensional Analysis

*Use the questions in this section to practice your dimensional analysis skills in an everyday context. Show all setups and unit cancellations. Our answers are rounded off according to the rules given in Section 3.5. Your unrounded answers are acceptable only if you complete these questions before studying Section 3.5.*

13. How long will it take to travel the 406 miles between Los Angeles and San Francisco at an average speed of 48 miles per hour?

14. ■ A student who is driving home for the holidays averages 70.5 miles per hour. How many miles will the student travel if the trip lasts 8.73 hours?

15. How many minutes does it take a car traveling 88 km/hr to cover 4.3 km?

16. ■ How many days are in 89 weeks?

17. What will be the cost in dollars for nails for a fence 62 feet long if you need 9 nails per foot of fence, there are 36 nails in a pound, and they sell for 69 cents per pound?

18. ■ A student working for Stop and Shop is packing eggs into cartons that contain a dozen eggs. How many eggs will the student need in order to pack 72 cartons?

19. An American tourist in Mexico was startled to see $1950 on a menu as the price for a meal. However, that dollar sign refers to Mexican pesos, which on that day had an average rate of 218 pesos per American dollar. How much did the tourist pay for the meal in American funds?

20. ■ How many nickels should you receive in exchange for 89 quarters?

21. How many weeks are in a decade?

22. How many seconds are in the month of January?

### Section 3.4: Metric Units

23. A woman stands on a scale in an elevator in a tall building. The elevator starts going up, rises rapidly at constant speed for half a minute, and then slows to a stop. Compare the woman's weight as recorded by the scale and her mass while the elevator is standing still during the starting period, during the constant rate period, and during the slowing period.

24. A person can pick up a large rock that is submerged in water near the shore of a lake but may not be able to pick up the same rock from the beach. Compare the mass and the weight of the rock when in the lake and when on the beach.

25. What is the metric unit of length?

26. What is the metric unit of mass?

27. *Kilobuck* is a slang expression for a sum of money. How many dollars are in a kilobuck? How about a megabuck (see Table 3.1)?

28. What is the difference between the terms *kilounit* and *kilogram*?

29. One milliliter is equal to how many liters?

30. How many centimeters are in a meter?

31. Which unit, megagrams or grams, would be more suitable for expressing the mass of an automobile? Why?

32. What is the name of the unit whose symbol is nm? Is it a long distance or a short distance? How long or how short?

*Questions 33–40: Make each conversion indicated. Use exponential notation to avoid long integers or decimal fractions. Write your answers without looking at a conversion table.*

33. (a) 5.74 cg to g, (b) 1.41 kg to g, (c) $4.54 \times 10^{8}$ cg to mg

34. ■ (a) 15.3 kg to g and mg, (b) 80.5 g to kg and mg, (c) 58.5 mg to kg and g

35. (a) 21.7 m to cm, (b) 517 m to km, (c) 0.666 km to cm

36. ■ (a) 90.4 mm to m and cm, (b) 11.9 m to mm and cm, (c) 53.6 cm to mm and m

37. (a) 494 $cm^3$ to mL, (b) 1.91 L to mL, (c) 874 $cm^3$ to L

38. ■ (a) 90.8 mL to L and $cm^3$, (b) 16.9 L to mL and $cm^3$, (c) 65.4 $cm^3$ to mL and L

*Questions 39 and 40: Refer to Table 3.1 for less common prefixes in these metric conversions.*

39. (a) 7.11 hg to g, (b) $5.27 \times 10^{-7}$ m to pm, (c) $3.63 \times 10^{6}$ g to dag

40. (a) 0.194 Gg to g, (b) 5.66 nm to m, (c) 0.00481 Mm to cm

### Section 3.5: Significant Figures

*Questions 41 and 42: To how many significant figures is each quantity expressed?*

41. (a) 75.9 g sugar, (b) 89.583 mL weed killer, (c) 0.366 in. diameter glass fiber, (d) 48,000 cm wire, (e) 0.80 ft spaghetti, (f) 0.625 kg silver, (g) $9.6941 \times 10^{6}$ cm thread, (h) $8.010 \times 10^{-3}$ L acid

42. (a) 4.5609 g salt, (b) 0.10 in. diameter wire, (c) $12.3 \times 10^{-3}$ kg fat, (d) 5310 $cm^3$ copper, (e) 0.0231 ft licorice, (f) $6.1240 \times 10^{6}$ L salt brine, (g) 328 mL ginger ale, (h) 1200.0 mg dye

*Questions 43 and 44: Round off each quantity to three significant figures.*

43. (a) $6.398 \times 10^{-3}$ km rope, (b) 0.0178 g silver nitrate, (c) 79,000 m cable, (d) 42,150 tons fertilizer, (e) $649.85

44. (a) 52.20 mL helium, (b) 17.963 g nitrogen, (c) 78.45 mg MSG, (d) 23,642,000 mm wavelength, (e) 0.0041962 kg lead

45. A moving-van crew picks up the following items: a couch that weighs 147 pounds, a chair that weighs 67.7 pounds, a piano at $3.6 \times 10^2$ pounds, and several boxes having a total weight of 135.43 pounds. Calculate and express in the correct number of significant figures the total weight of the load.

46. ■ A solution is prepared by dissolving 2.86 grams of sodium chloride, 3.9 grams of ammonium sulfate, and 0.896 grams of potassium iodide in 246 grams of water. Calculate the total mass of the solution and express the sum in the proper number of significant figures.

47. A buret contains 22.93 mL sodium hydroxide solution. A few minutes later, the volume is down to 19.4 mL because of a small leak. How many milliliters of solution have drained from the buret?

48. An empty beaker has a mass of 94.33 grams. After some chemical has been added, the mass is 101.209 grams. What is the mass of the chemical in the beaker?

49. The mole is the SI unit for the amount of a substance. The mass of one mole of pure table sugar is 342.3 grams. How many grams of sugar are in exactly 1/2 mole? What is the mass of 0.764 mole?

50. Exactly one liter of a solution contains 31.4 grams of a certain chemical. How many grams are in exactly 2 liters? How about 7.37 liters? Express the results in the proper number of significant figures.

51. An empty beaker with a mass of 42.3 g is filled with a liquid, and the resulting mass of the liquid and the beaker is 62.87 g. The volume of this liquid is 19 mL. What is the density of the liquid?

52. ■ Use the definition Density $\equiv \frac{\text{mass}}{\text{volume}}$ to calculate the density of a liquid with a volume of 50.6 mL if that liquid is placed in an empty beaker with a mass of 32.344 g and the mass of the liquid plus the beaker is 84.64 g.

### Section 3.6: Metric–USCS Conversions

*Questions 53–70: You may consult Table 3.2 while answering these questions.*

53. 0.0715 gal = ___ $cm^3$ $\quad$ $2.27 \times 10^4$ mL = ___ gal

54. 19.3 L = ___ gal $\quad$ 0.461 qt = ___ L

55. A popular breakfast cereal comes in a box containing 515 g. How many pounds (lb) of cereal is this?

56. ■ A copy of your chemistry textbook is found to have a mass of $2.60 \times 10^3$ grams. What is the mass of this copy of your chemistry textbook in ounces?

57. The payload of a small pickup truck is 1450 pounds (assume three significant figures). What is this in kilograms?

58. The Hope diamond is the world's largest blue diamond. It weighs 44.4 carats. If 1 carat is defined as 200 mg, calculate the mass of the diamond in grams and in ounces.

59. There is 115 mg of calcium in a 100-g serving of whole milk. How many grams of calcium is this? How many pounds?

60. The largest recorded difference of weight between spouses is 922 lb. The husband weighed 1020 lb, and his wife, 98 lb. Express this difference in kilograms.

61. An Austrian boxer reads 69.1 kg when he steps on a balance (scale) in his gymnasium. Should he be classified as a welterweight (136 to 147 lb) or a middleweight (148 to 160 lb)?

62. A woman gives birth to a 7.5-lb baby. How would a hospital using metric units record this baby's mass?

63. The height of Angel Falls in Venezuela is 399.9 m. How high is this in (a) yards; (b) feet?

64. ■ A penny is found to have a length of 1.97 centimeters. What is the length of this penny in inches?

65. The Sears Tower in Chicago is 1454 feet tall. How high is this in meters?

66. What is the length of the Mississippi River in kilometers if it is 2351 miles long?

67. The summit of Mount Everest is 29,035 ft above sea level. Express this height in kilometers.

68. One of the smallest brilliant-cut diamonds ever crafted had a diameter of about 1/50 in. (0.02 in.). How many millimeters is this?

69. An office building is heated by oil-fired burners that draw fuel from a 619-gal storage tank. Calculate the tank volume in liters.

70. ■ A gas can is found to have a volume of 9.10 liters. What is the volume of this gas can in gallons?

### Section 3.7: Temperature

*Questions 71 and 72: Fill in the spaces in the following tables so that each temperature is expressed in all three scales. Round off answers to the nearest degree.*

71.

| Celsius | Fahrenheit | Kelvin |
|---|---|---|
| 69 | | |
| | −29 | |
| | | 111 |
| | 36 | |
| | | 358 |
| −141 | | |

72.

| Celsius | Fahrenheit | Kelvin |
|---|---|---|
| | 40 | |
| | | 590 |
| −13 | | |
| | | 229 |
| 440 | | |
| | −314 | |

73. "Normal" body temperature is 98.6°F. What is this temperature in Celsius degrees?

74. ■ In the winter, a heated home in the Northeast might be maintained at a temperature of 74°F. What is this temperature on the Celsius and kelvin scales?

75. Energy conservationists suggest that air conditioners should be set so that they do not turn on until the temperature tops 78°F. What is the Celsius equivalent of this temperature?

76. ■ The melting point of an unknown solid is determined to be 49.0°C. What is this temperature on the Fahrenheit and Kelvin scales?

77. The world's highest shade temperature was recorded in Libya at 58.0°C. What is its Fahrenheit equivalent?

78. ■ The boiling point of a liquid is calculated to be 454 K. What is this temperature on the Celsius and Fahrenheit scales?

#### Section 3.8: Proportionality and Density

79. The amount of heat (q) absorbed when a pure substance melts is proportional to the mass of the sample (m). Express this proportionality in mathematical form. Change it into an equation, using the symbol $\Delta H_{fus}$ for the proportionality constant. This constant is the heat of fusion of a pure substance. If heat is measured in calories, what are the units of heat of fusion? Write a word definition of heat of fusion.

80. ■ The distance, d, traveled by an automobile moving at an average speed of s is directly proportional to the time, t, spent traveling. The proportionality constant is the average speed. (a) Express this proportionality in mathematical form. (b) If it takes the automobile 3.88 hours to travel a distance of 239 miles, what are the value and units of the proportionality constant? (c) Assuming the same average speed as in part (b), how long will it take the automobile to travel a distance of 659 miles?

81. It takes 7.39 kilocalories to melt 92 grams of ice. Calculate the heat of fusion of water. (*Hint:* See Question 79. Careful on the units; the answer will be in calories per gram.)

82. ■ If the pressure of a sample of gas is held constant, its volume, V, is directly proportional to the absolute temperature of the gas, T. (a) Write an equation for the proportionality between V and T, in which b is the proportionality constant. (b) For 48.0 grams of $O_2$ gas at a pressure of 0.373 atmosphere, V is observed to be 93.4 L when T is 283 K. What are the value and units of the proportionality constant, b? (c) What volume will this gas sample occupy at a temperature of 375 K?

83. If the temperature and amount of a gas are held constant, the pressure (P) it exerts is inversely proportional to volume (V). This means that pressure is directly proportional to the inverse of volume, or 1/V. Write this as a proportionality, and then as an equation with k′ as the proportionality constant. What are the units of k′ if pressure is in atmospheres and volume is in liters?

84. ■ The mass, m, of a piece of metal is directly proportional to its volume, V, where the proportionality constant is the density, D, of the metal. (a) Write an equation that represents this direct proportion, in which D is the proportionality constant. (b) The density of iron metal is 7.88 g/cm$^3$. What is the mass of a piece of iron that has a volume of 23.5 cm$^3$? (c) What is the volume of a piece of iron metal that has a mass of 172 g?

85. Calculate the density of benzene, a liquid used in chemistry laboratories, if 166 g of benzene fills a graduated cylinder to the 188-mL mark.

86. ■ A general chemistry student found a chunk of metal in the basement of a friend's house. To figure out what it was, she used the ideas just developed in class about density. She measured the mass of the metal to be 175.2 grams. Then she dropped the metal into a measuring cup and found that it displaced 15.3 mL of water. Calculate the density of the metal. What element is this metal most likely to be?

87. Densities of gases are usually measured in grams per liter (g/L). Calculate the density of air if the mass of 15.7 L is 18.6 g.

88. A rectangular block of iron 4.60 cm × 10.3 cm × 13.2 cm has a mass of 4.92 kg. Find its density in g/cm$^3$.

89. Ether, a well-known anesthetic, has a density of 0.736 g/cm$^3$. What is the volume of 471 g of ether?

90. ■ Calculate the mass of 17.0 mL of aluminum, which has a density of 2.72 g/mL.

91. Determine the mass of 2.0 L rubbing alcohol, which has a density of 0.786 g/mL.

92. ■ Calculate the volume occupied by 15.4 grams of nickel, which has a density of 8.91 g/mL.

#### General Questions

93. Distinguish precisely and in scientific terms the differences among items in each of the following groups.
    a) Coefficient, exponent, exponential
    b) PER expression, conversion factor, unit PATH
    c) Mass, weight
    d) Unit, kilounit, centiunit, milliunit
    e) Significant figures, doubtful digit
    f) Uncertainty, exact number
    g) The symbols ≡, =, and ≈
    h) Fahrenheit, Celsius, kelvin
    i) Direct proportionality, proportionality constant

94. Determine whether each statement that follows is true or false:
    a) The SI system includes metric units.
    b) If two quantities are expressed in a PER relationship, they are directly proportional to each other.
    c) The exponential notation form of a number smaller than 1 has a positive exponent.
    d) In changing an exponential notation number whose coefficient is not between 1 and 10 to standard exponential notation, the exponent becomes smaller if the decimal in the coefficient is moved to the right.
    e) There are 1000 kilounits in a unit.
    f) There are 10 milliunits in a centiunit.
    g) There are 1000 mL in a cubic centimeter.
    h) The mass of an object is independent of its location in the universe.
    i) Celsius degrees are smaller than Fahrenheit degrees.
    j) The doubtful digit is the last digit written when a number is expressed properly in significant figures.
    k) The expression 76.2 g means the same as 76.200 g.
    l) The number of significant figures in a sum may be more than the number of significant figures in any of the quantities added.

m) The number of significant figures in a difference may be fewer than the number of significant figures in any of the quantities subtracted.

n) The number of significant figures in a product may be more than the number of significant figures in any of the quantities multiplied.

o) PER means multiply.

p) If the quantity in the answer to a problem is familiar, it is not necessary to check to make sure the answer is reasonable.

q) Dimensional analysis can be used to change from one unit to another only when the quantities are directly proportional.

r) A unit PATH begins with the units of the given quantity and ends with the units in which the answer is to be expressed.

s) There is no advantage to using units in a problem that is solved by algebra.

t) A Fahrenheit temperature can be changed to a Celsius temperature by dimensional analysis.

95. How tall are you in (a) meters; (b) decimeters; (c) centimeters; (d) millimeters? Which of the four metric units do you think would be most useful in expressing people's heights without resorting to decimal fractions?

96. What do you weigh in (a) milligrams; (b) grams; (c) kilograms? Which of these units do you think is best for expressing a person's weight? Why?

97. Standard typewriter paper is the United States is $8\frac{1}{2}$ in. by 11 in. What are these dimensions in centimeters?

98. The density of aluminum is 2.7 g/cm$^3$. An ecology-minded student has gathered 126 empty aluminum cans for recycling. If there are 21 cans per pound, how many cubic centimeters and grams of aluminum does the student have?

**More Challenging Problems**

99. ■ Olga Svenson has just given birth to a bouncing 6 lb, 7 oz baby boy. How should she describe the weight of her child to her sister, who lives in Sweden—in metric units, of course?

100. A student's driver's license lists her height as 5 feet, 5 inches. What is her height in meters?

101. ■ How many grams of milk are in a 12.0-fluid-ounce glass? The density of milk is 64.4 lb/ft$^3$. There are 7.48 gal/ft$^3$; and, by definition, there are 4 qt/gal and 32 fl oz/qt.

102. The fuel tank in an automobile has a capacity of 11.8 gal. If the density of gasoline is 42.0 lb/ft$^3$, what is the mass of fuel in kilograms when the tank is full?

103. A welcome rainfall caused the temperature to drop by 33°F after a sweltering day in Chicago. What is this temperature drop in degrees Celsius?

104. At high noon on the lunar equator the temperature may reach 243°F. At night the temperature may sink to −261°F. Express the temperature difference in degrees Celsius.

105. ■ A recipe calls for a quarter cup of butter. Calculate its mass in grams if its density is 0.86 g/cm$^3$ (1 cup = 0.25 qt).

106. Calculate the mass in pounds of one gallon of water, given that the density of water is 1.0 g/mL.

107. In Active Example 3.28 you calculated that you would have to work six weeks to earn enough money to buy a \$734.26 sound system. You would be working five shifts of four hours each at \$7.25/hr. But, alas, when you received your first paycheck, you found that exactly 23% of your earnings had been withheld for social security, federal and state income taxes, and workers' compensation insurance. Taking these into account, how many weeks will it take to earn the \$734.26?

© Trip/Alamy

# 4

Automobile air bags are an everyday application of the chemical principles you will study in this course. In Chapter 2 you learned that the particles in a solid and a liquid are essentially touching one another, and in a gas, the particles are independent of one another, with large empty spaces between them. The volume a gas occupies is much greater than the volume occupied by the same number of particles in the solid or liquid state. When an automobile air-bag system sensor detects rapid deceleration, as in a collision, a small volume of a solid compound in the steering wheel or dashboard undergoes a reaction to form a relatively large volume of a gas, inflating the air bag.

# Introduction to Gases

CONTENTS



In Chapter 2 we described three common states of matter—gases, liquids, and solids. In this chapter you will study the gaseous state in more detail. Gases were the first state of matter to be understood by early chemists, which is surprising because most gases are invisible. Why, then, were gases the starting point for the science of chemistry? You will explore the answer to this question in this chapter.

Chapter 4 provides an opportunity for you to apply the chemical calculation skills you learned as a result of studying Chapter 3. Scientific notation, dimensional analysis, metric units, significant figures, temperature, proportionality, and density are needed to understand the concepts and work the problems in this introduction to gases. You may find that you occasionally need to review Chapter 3 as you study this chapter. If so, don't be concerned. All successful science students review and refine their understanding of prior material—even content from prior coursework—as they learn new ideas. In fact, we selected the topics of Chapter 4, in part, to give you a chance to apply your calculating skills immediately after you learned them.

Online homework for this chapter may be assigned in OWL

## 4.1 Properties of Gases

**Goal** 1 Describe five macroscopic characteristics of gases.

The air that surrounds us is a sea of mixed gases called the atmosphere. It is not necessary, then, to search very far to find a gas whose properties we may study. Some of the familiar characteristics of air—in fact, of all gases—are the following:

**1.** *Gases may be compressed.* A fixed quantity of air may be made to occupy a smaller volume by applying pressure. Figure 4.1(a) shows a quantity of air in a cylinder having a leakproof piston that can be moved to change the volume occupied by the air. Push the piston down by applying more force, and the volume of air is reduced (Fig. 4.1[b]).

**P/REVIEW:** The distinction between macroscopic and particulate views of matter is discussed in Section 2.1 and illustrated in Figure 2.4.

**2.** *Gases expand to fill their containers uniformly.* If less force were applied to the piston, as shown in Figure 4.1(c), air would respond immediately, pushing the piston upward, expanding to fill the larger volume uniformly. If the piston were pulled up (Fig. 4.1[d]), air would again expand to fill the additional space.

**3.** *All gases have low density.* The density of air is about 0.0012 g/cm$^3$. The density of water is 830 times greater than the density of air, and iron is 6600 times more dense than air, when all are at room temperature (Fig. 4.2).

**4.** *Gases may be mixed.* "There's always room for more" is a phrase that may be applied to gases. You may add the same or a different gas to a gas already occupying a rigid container of fixed volume, provided there is no chemical reaction between them (Fig. 4.3).

**5.** *A confined gas exerts constant pressure on the walls of its container uniformly in all directions.* This pressure, illustrated in Figure 4.4, is a unique property of a gas, independent of external factors such as gravitational forces.

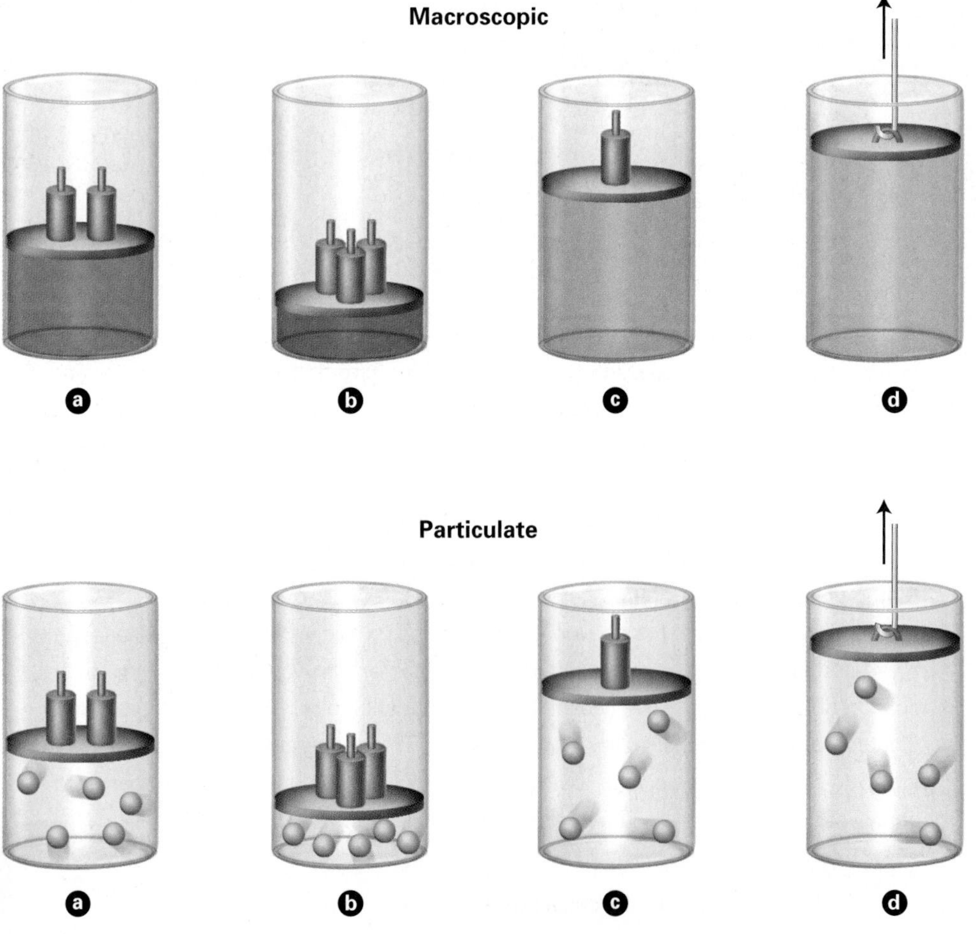


**Figure 4.1** Compression and expansion properties of gases. The piston and cylinder show that gases may be compressed and that they expand to fill the volume available to them.

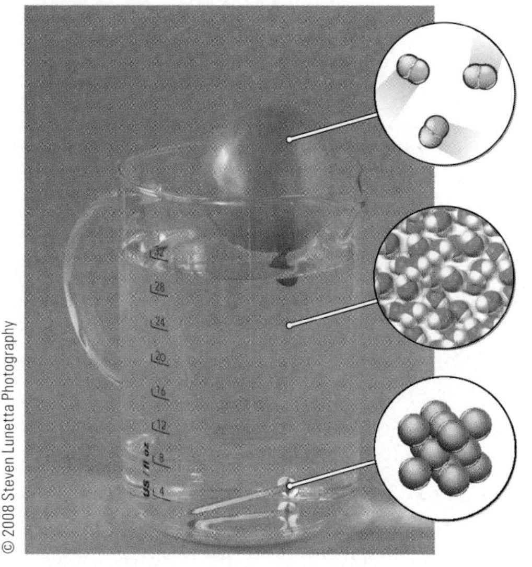


Figure 4.2 Low density of gases. At the macroscopic level, a balloon filled with air floats on water, and an iron nail sinks in water, indicating that the gas is less dense than the liquid and solid. At the particulate level, the low particle density—the relatively small number of particles in a given volume—of a gas is responsible for its low density. The particle density of the solid is somewhat greater than that of the liquid, and the solid particles have a greater mass than the liquid particles.

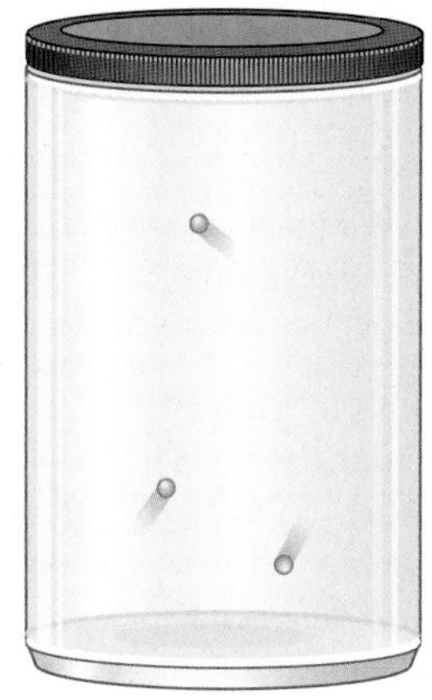

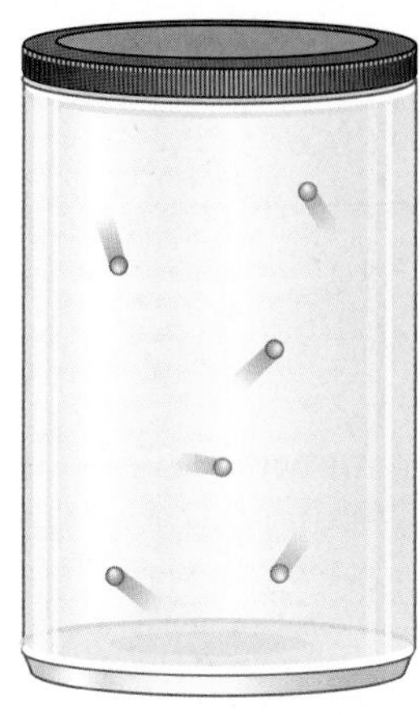

Figure 4.3 Mixing gases. The volume of matter in a container holding a gas is very tiny compared with the volume of the container. If a different gas is added to the container, it occupies the large volume of space that is available.

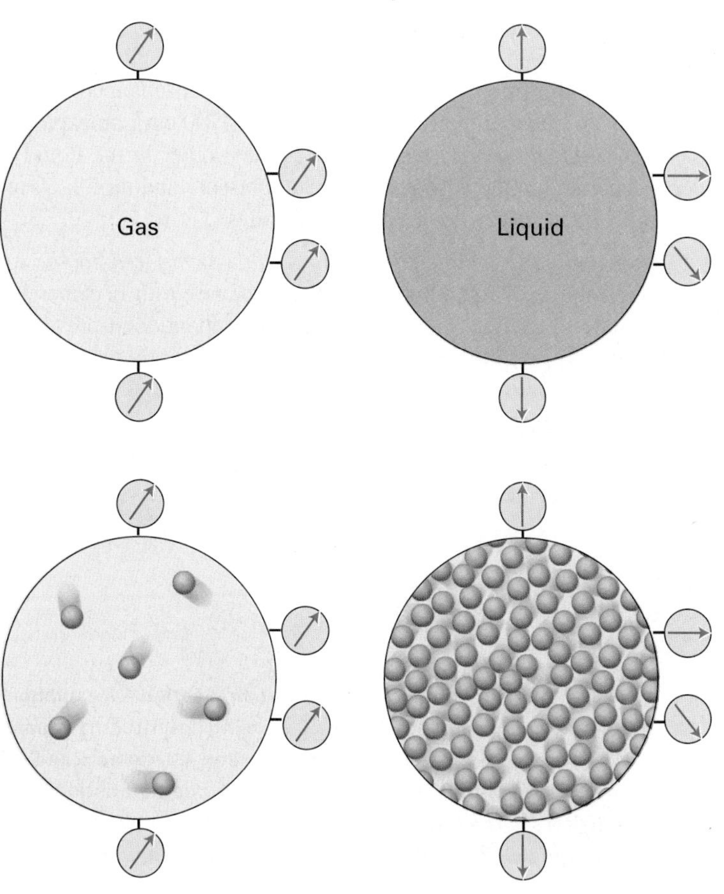


Figure 4.4 Pressure in gases and liquids. Each container has four pressure gauges, one on top, one on bottom, and two on the side. Note how the four gauges on the gas container all read the same pressure, but the liquid gauges show that pressure increases with increasing depth. Gas pressures are exerted uniformly in all directions; liquid pressures depend on the depth of the liquid.

## 4.2 The Kinetic Theory of Gases and the Ideal Gas Model

**Goal** 2 Explain or predict physical phenomena relating to gases in terms of the ideal gas model.

**P/REVIEW:** The kinetic molecular theory proposes that all matter consists of molecules in constant motion. See Section 2.2.

In trying to account for the properties of gases, scientists have devised the **kinetic molecular theory.** The theory describes an **ideal gas model** by which we can visualize the nature of the gas by comparing it with a physical system we can either see or readily imagine. As always, chemists explain observable *macroscopic* phenomena in terms of *particulate* behavior.

The main features of the ideal gas model are as follows:

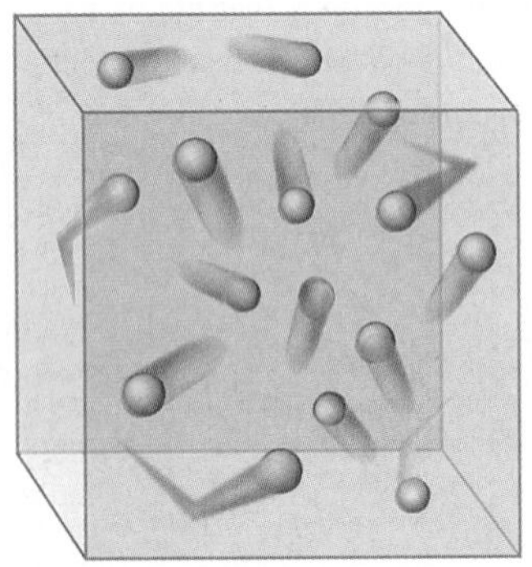

**Figure 4.5** Particle motion in a gas. Particles collide with each other and with the walls of the container, the latter being responsible for the pressure the gas exerts.

1. *Gases consist of particles moving at any given instant in straight lines (Fig. 4.5).* Particle motion explains why gases fill their containers. It also suggests how they exert pressure. When an individual particle strikes a container wall, it exerts a force at the point of collision. When this is added to billions upon billions of similar collisions occurring continuously, the total effect is the steady force that is responsible for gas pressure.

2. *Molecules collide with each other and with the container walls without loss of total kinetic energy (Fig. 4.6).* If gas particles lost energy or slowed down as a result of these collisions, the combined forces would become smaller, and the pressure would gradually decrease. Furthermore, because of the relationship between temperature and average molecular speed, temperature would drop if energy were lost in collisions. Any enclosed gas would eventually become a liquid because of this loss of energy. But these things do not happen, so we conclude that energy is not lost in molecular collisions, either with the walls or between molecules.

$E_1 = E_2 = E_3$

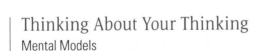

**Figure 4.6** An ideal gas particle collides with the walls of the container without losing energy. The energy of the particle is the same before ($E_1$), during ($E_2$), and after ($E_3$) the collision with the container wall: $E_1 = E_2 = E_3$.

3. *Gas molecules are very widely spaced (Fig. 4.7).* Gas molecules must be widely spaced; otherwise the densities of gases would not be as low as they are. One gram of liquid water at the boiling point occupies 1.04 $cm^3$. When changed to steam at the same temperature, the same number of molecules fills 1700 $cm^3$, an expansion of over 1600 times! If the water molecules were touching each other in the liquid state, they must be widely separated in the gaseous state. Compressing and mixing gases are possible because of the open spaces between gas molecules.

4. *The actual volume of molecules is negligible compared to the space they occupy.* The total volume of the actual molecules in one gram of water is the same regardless of its state of matter. The 1.04-$cm^3$ volume mentioned previously for one gram of water in the liquid state is 0.06% of the 1700-$cm^3$ total volume the molecules occupy as a gas, which qualifies as negligible.

5. *Gas molecules behave as independent particles; attractive forces between them are negligible.* The large distances between gas particles ensure us that attractions between these molecules are negligible.

### Thinking About Your Thinking

Mental Models

**Section 4.1 discussed the familiar characteristics of air. Section 4.2 explains the reasons for those characteristics as a chemist does, by using descriptions of particulate-level behavior. When you complete your study of Section 4.2, go back to Section 4.1 and try to give a particulate-level explanation of each characteristic. Imagine the behavior of gas particles in the form of a mental movie in your mind. Make sure you can use your mental movie to explain each of the macroscopic characteristics of a gas.**

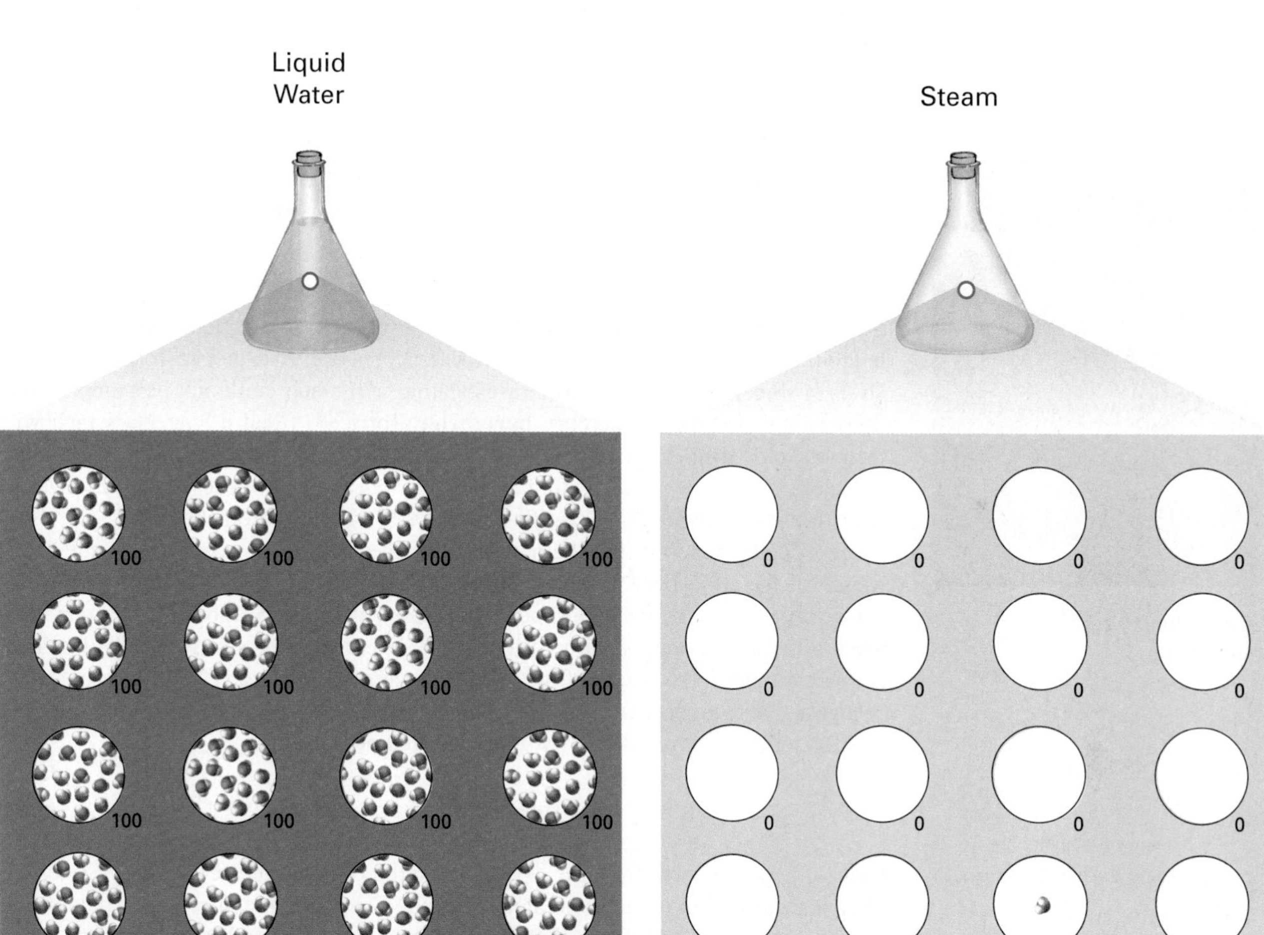


**Figure 4.7** Suppose you were to select a sample of liquid water that contained 100 molecules. If you then selected 15 more samples—total 16—of the same volume, each sample would contain 100 molecules. This would be a total of 1600 molecules in the 16 samples. If you were to then select 16 separate samples of steam, each sample having the same volume as each sample of liquid water, how many water molecules would be in each sample? On average, 15 of the sample volumes of steam would be empty—no molecules—and the 16th sample volume would contain only one water molecule.

In summary, an **ideal gas** is a model of a gas that is constructed of identical particles that occupy no volume and exert no forces on one another. When the particles interact with one another or with the container walls, they do so with no loss of total kinetic energy.

**P/REVIEW:** In Section 2.2 we said that there is an attraction between the particles in any sample of matter. The faster a particle moves, the greater its tendency to overcome these attractive forces will be.

## 4.3 Gas Measurements

Experiments with a gas usually involve measuring or controlling its quantity, volume, pressure, and temperature. Finding a quantity of a gas by weighing it is different from weighing a liquid or a solid, but it is not difficult if the gas is in a closed container. The amount of gas is most often expressed in moles, to which the symbol n is assigned. In this chapter, we consider only a fixed amount of gas; n is constant.

The relationships among pressure, volume, and temperature of a fixed amount of gas were among the earliest quantitative studies in all science. Unlike a liquid or a solid, a gas always fills its container. Consequently, its volume (V) is the same as the volume of the container. The other two variables, pressure (P) and temperature (T), deserve closer examination.

**P/REVIEW:** The mole is sometimes referred to as the "chemist's dozen" because it is commonly used to represent a given number of particles, just as dozen is used to represent 12 units of something. The mole is defined and described fully in Section 7.3.

## Pressure

**Goal 3** Given a gas pressure in atmospheres, torr, millimeters (or centimeters) of mercury, inches of mercury, pascals, kilopascals, bars, or pounds per square inch, express pressure in each of the other units.

**Goal 4** Define *pressure* and interpret statements in which the term *pressure* is used.

Evangelista Torricelli (1608–1647)

Science Source/Photo Researchers, Inc.

Some of the most important investigations in the history of science were those that led to the discovery that the atmosphere exerts pressure. Evangelista Torricelli (1608–1647), a student of Galileo, while investigating why water could not be pumped from deep mines, designed a mercury **barometer.** Torricelli filled a long glass tube with mercury, placed his finger over the end, inverted the tube, and placed it in a dish of mercury. The height of the column of mercury was about 30 inches, although it varied from day to day. Figure 4.8 is an illustration of Torricelli's barometer.

Torricelli proposed that the height of the mercury in the column was due to the weight of the atmosphere pushing the liquid up the tube. Torricelli's hypothesis was confirmed, in part, by the experiments of Blaise Pascal (1623–1662). Pascal found that when a barometer was taken up a mountain, where the atmosphere weighs less than at lower altitudes, the height of the mercury column decreased. Thus the barometer is "weighing" the atmosphere.

By definition, **pressure is the force exerted on a unit area:**

$$\text{Pressure} \equiv \frac{\text{force}}{\text{area}} \quad or \quad P \equiv \frac{F}{A} \qquad \textbf{(4.1)}$$

**P/REVIEW:** Equation 4.1 is a defining equation for pressure, like the defining equation for density in Section 3.8.

Units of pressure come from the definition. In the United States Customary System, if force is measured in pounds and area in square inches, the pressure unit is pounds per square inch (psi). The SI unit of pressure is the **pascal (Pa),** which is one newton per square meter. (The *newton* is the SI unit of force.) Although chemists generally follow SI guidelines, one pascal is a very small pressure; the **kilopascal (kPa)** is a more practical unit. Many other pressure units are commonly used. One is the **bar,** which is $1 \times 10^5$ Pa or 100 kPa. The **millimeter of mercury,** or its equivalent, the **torr,** and the **atmosphere (atm)** are other common units of pressure. The millimeter of mercury is usually abbreviated mm Hg. (Hg is the elemental symbol for mercury.)

Weather bureaus generally report *barometric pressure,* the pressure exerted by the atmosphere at a given weather station, in inches of mercury or kilopascals. Atmospheric pressure is measured by a Torricellian barometer or its mechanical equivalent. On a day when the mercury column in a barometer is 752 mm Hg high, we say that atmospheric pressure is 752 mm Hg.

**One standard atmosphere of pressure is defined as 760 mm Hg.** This is a typical barometric pressure at sea level. Atmospheric pressure on the top of the world's highest mountain is about 270 mm Hg. Atmospheric pressure on the earth's surface therefore varies between 760 and 270 mm Hg, depending on the altitude. The atmosphere unit is particularly useful in referring to very high pressures. To two significant figures, one atmosphere and one bar are equal (1.0 atm = 1.0 bar), and thus the two

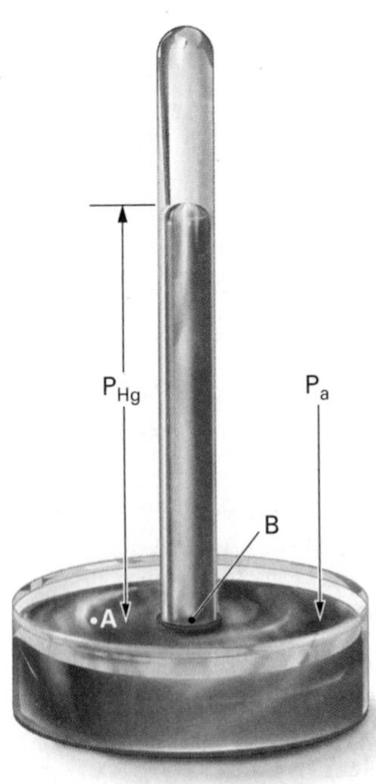


**Figure 4.8** Two operational principles govern the mercury barometer. (1) The total pressure at any point in a liquid system is the sum of the pressures of each gas or liquid phase above that point. (2) The total pressures at any two points at the same level in a liquid system are always equal. Point A at the liquid surface outside the tube is at the same level as Point B inside the tube. The only thing exerting downward pressure at Point A is the atmosphere; $P_a$ represents atmospheric pressure. The only thing exerting downward pressure at Point B is the mercury above that point, designated $P_{Hg}$. A and B being at the same level, the pressures at these points are equal: $P_a = P_{Hg}$.

units can be used as the basis for approximate conversions between barometer-based units (atm, mm Hg) and SI-like units (kPa, bar). Here is a summary of common pressure units and their relationships to one other:

$$1 \text{ atm} \equiv 760 \text{ mm Hg} = 760 \text{ torr} \quad \textbf{(4.2)}$$

$$1 \text{ atm} = 1.013 \times 10^5 \text{ Pa} = 101.3 \text{ kPa} = 1.013 \text{ bar} \quad \textbf{(4.3)}$$

$$1 \text{ atm} = 29.92 \text{ in. Hg} = 14.69 \text{ psi} \quad \textbf{(4.4)}$$

**P/REVIEW:** Numbers fixed by definition are exact; significant figures do not apply to exact numbers (Section 3.5). The relationship 1 atm = 760 mm Hg is such a definition.

The torr, which honors the work of Torricelli, and the millimeter of mercury are essentially identical pressure units. Both terms are widely used; the choice between them is one of personal preference. The advantage of the millimeter of mercury is that it has physical meaning. You can read it by observing an open-end **manometer,** the instrument most commonly used to measure pressure in the laboratory (Fig. 4.9). Torr, however, is easier to say and write. We will use torr hereafter in this text.

Outside the laboratory, mechanical gauges are used to measure gas pressure (Fig. 4.10). A typical tire gauge is probably the most familiar. Tire gauges show the pressure *above* atmospheric pressure, rather than the absolute pressure measured by a manometer. Even a flat tire contains air that exerts pressure. If it did not, the entire tire would collapse, not just the bottom. The pressure of gas remaining in a flat tire is equal to atmospheric pressure. If a tire gauge shows 25 psi, that is the **gauge pressure** of the gas (air) in the tire. The absolute pressure is nearly 40 psi—the 25 psi shown by the gauge plus about 15 psi (1 atm = 14.69 psi) from the atmosphere.

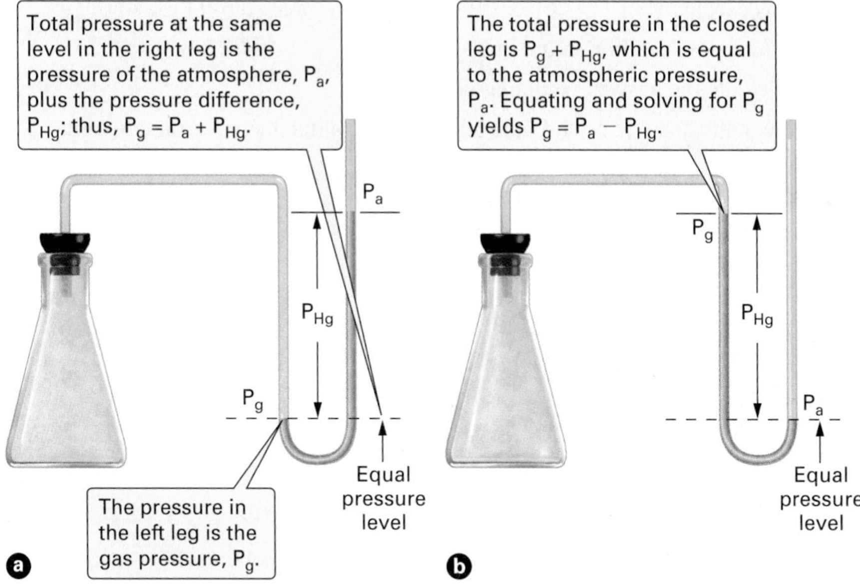
Total pressure at the same level in the right leg is the pressure of the atmosphere, $P_a$, plus the pressure difference, $P_{Hg}$; thus, $P_g = P_a + P_{Hg}$.
$P_a$
$P_{Hg}$
$P_g$
Equal pressure level
The pressure in the left leg is the gas pressure, $P_g$.
a
The total pressure in the closed leg is $P_g + P_{Hg}$, which is equal to the atmospheric pressure, $P_a$. Equating and solving for $P_g$ yields $P_g = P_a - P_{Hg}$.
$P_g$
$P_{Hg}$
$P_a$
Equal pressure level
b

Figure 4.9 Open-end manometers. Open-end manometers are governed by the same principles as mercury barometers (Fig. 4.6). The pressure of the gas, $P_g$, is exerted on the mercury surface in the closed (left) leg of the manometer. Atmospheric pressure, $P_a$, is exerted on the mercury surface in the open (right) leg. With a meter stick, the difference between these two pressures, $P_{Hg}$, may be measured directly in millimeters of mercury (torr). Gas pressure is determined by equating the total pressures at the lower liquid mercury level, indicated with a dashed line.

In (a), the enclosed gas is exerting more pressure than the atmosphere, so the quantities are added to result in a higher pressure. In (b), the enclosed gas exerts less pressure than the atmosphere, and a lower pressure is obtained by subtracting the pressure difference from the atmospheric pressure. In effect, you can determine the pressure of a gas, as measured by a manometer, by adding the pressure difference to, or subtracting the pressure difference from, atmospheric pressure: $P_g = P_a \pm P_{Hg}$.

Change
Rain
Fair
Stormy
Very dry
BAROMETER

Figure 4.10 A mechanical gauge used to measure atmospheric pressure. In general, mechanical gauges work via an air-filled tube that changes shape with changing pressure. Levers and/or gears move in response to the tube, and a pointing needle and scale or an electronic display is calibrated to show the corresponding atmospheric pressure.

*Learn It Now!* Working the Active Examples for yourself, covering our answers until you *write* the answers to the questions, is a critical component of learning. Guard against allowing yourself to fall into the trap of just *reading* the text. Work each step of each Active Example as you study.

## Active Example 4.1

The pressure inside a steam boiler is 1127 psi. Express this pressure in atmospheres.

PLAN the solution. Recall that PLAN, printed as you see it here, means to complete the first three steps in the problem-solving procedure:

(1) Write down what is GIVEN.

(2) Write down what is WANTED.

(3) Decide how to solve the problem. If the given and wanted quantities are related by a PER expression, use dimensional analysis; write the PER/PATH. If the given and wanted quantities are related by an algebraic equation, use algebra by solving the equation for the WANTED quantity. PLAN your solution now.

GIVEN: 1127 psi WANTED: atm PER: $\frac{1 \text{ atm}}{14.69 \text{ psi}}$

PATH: psi $\longrightarrow$ atm

In this case, Equation 4.4 gives you a PER relationship between the GIVEN and WANTED quantities, 1 atm = 14.69 psi. Therefore, the problem is solved by dimensional analysis.

Now execute your PLAN by completing the last three steps of the strategy:

(4) Write the calculation setup, including units.

(5) Calculate the answer, including units.

(6) Check the answer; be sure it is reasonable in numbers and correct in units.

$$1127 \cancel{\text{psi}} \times \frac{1 \text{ atm}}{14.69 \cancel{\text{psi}}} = 76.72 \text{ atm}$$

An atmosphere is a larger unit than a psi, so there should be a smaller number of atm in a given pressure than the number of psi. The number is reasonable, and the unit is what was wanted.

## Temperature

**Goal | 5** Given a temperature in degrees Celsius, convert it to kelvins, and vice versa.

Gas temperatures are ordinarily measured with a thermometer and expressed in Celsius degrees (°C). However, in pressure–volume–temperature gas problems, it is **absolute temperature,** expressed in **kelvins (K),** that enters into proportional relationships with the other two variables. As noted in Section 3.7, the kelvin is the SI unit of temperature. It is related to the Celsius degree by the equation

$$T_K = T_{°C} + 273^* \qquad \textbf{(4.5)}$$

To change Celsius degrees to kelvins, add 273 to the Celsius temperature.

*More precisely, the relationship is $T_K = T_{°C} + 273.15$. For the sake of simplicity, we choose to give up the additional 0.055% precision gained with the more exact relationship.

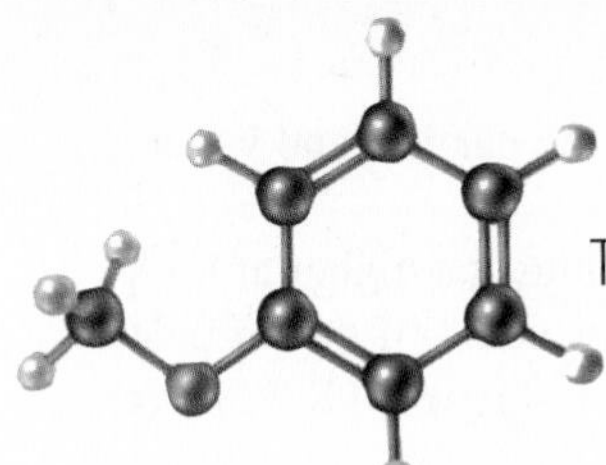

# Everyday Chemistry

## The Weather Machine

Science Photo Library/Photo Researchers, Inc.

The earth's atmosphere, as seen from space.

You have probably heard a television weather reporter talking about areas of high pressure and low pressure. These uneven areas of pressure in the lower levels of the earth's atmosphere are one cause of weather. In fact, the atmosphere is sometimes referred to as "the weather machine."

There are two key elements of air movement in the atmosphere, air masses and fronts. An air mass is a relatively large sphere of air that has about the same temperature and water content throughout for a given altitude. A typical cold air mass in the winter may stretch from Montana to Minnesota and from the Canadian border to the Mexican border. In contrast, a front is a small section of the lower atmosphere, usually 1/10 to 1/100 of the size of an air mass. Fronts are where the action is for changes in the weather. The temperature and humidity can vary greatly within a front, and the relatively stable air masses on either side of a front are often very different from one another.

When people refer to an area of high pressure, they are talking about an air mass that has a relatively high pressure when compared to surrounding air. This air, initially cold, descends toward the surface of the earth, compressing and warming. This inhibits cloud formation. High-pressure areas are therefore generally associated with clear weather. Conversely, air in a low-pressure mass rises and cools, which often leads to clouds, rain, or snow.

Weather is a very complex phenomenon, affected by numerous variables, such as the sun, the earth's tilt, the oceans and other bodies of water, and the variations in the land surface. Ultimately, however, each complex variable can be described in terms of simpler concepts such as temperature, pressure, and gas density. Weather experts are known as meteorologists, and their college preparation to understand the science of the atmosphere includes courses in chemistry, physics, geology, and mathematics, as well as advanced courses in meteorology itself. So the next time you hear a weather forecaster discussing high- and low-pressure systems, remember that the functioning of the weather machine is ultimately governed by the basic principles of the gaseous state of matter.

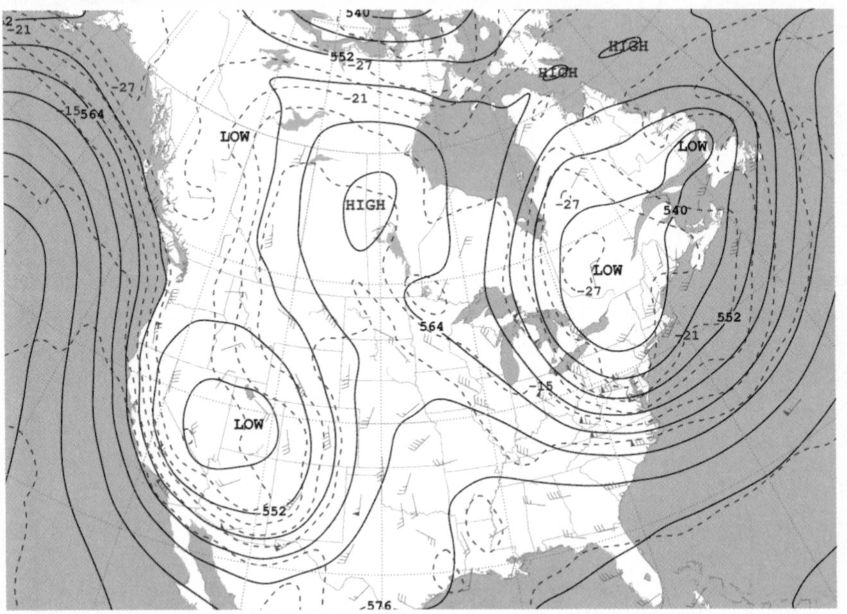


NOAA Central Library Data Imaging Project

Weather maps indicate areas of high (H) and low (L) pressure.

In order to understand gas behavior as it depends on temperature, we need to know just what temperature measures. Experiments indicate that the temperature of a substance is a measure of the average kinetic energy of the particles in the sample. Kinetic energy is the energy of motion as a particle goes from one place to another. It is expressed mathematically as $\frac{1}{2}mv^2$, where m is the mass of the particle and v is its velocity, or speed.

Since the mass of a particle is constant, the particle speed must be higher at high temperatures and lower at low temperatures. If the speed reaches zero—if the particle stops moving through space—the absolute temperature becomes zero, or 0 K. This is referred to as **absolute zero,** the lowest temperature.

Notice the word *average* in the phrase "average kinetic energy." It suggests correctly that not all of the particles in a sample of matter have the same kinetic energy. Some have more, some have less, and as a whole they have an average energy that is proportional to absolute temperature.

## 4.4 Charles's Law

**Goal 6** Describe the relationship between the volume and temperature of a fixed quantity of an ideal gas at constant pressure, and express that relationship as a proportionality, an equation, and a graph.

**Goal 7** Given the initial volume (or temperature) and the initial and final temperatures (or volumes) of a fixed quantity of gas at constant pressure, calculate the final volume (or temperature).

You may have seen Charles's Law in action when a latex rubber or Mylar foil balloon was subjected to a temperature change. A rubber balloon left in a hot car will expand significantly. Either balloon will shrink when brought from a warm home or store to the cold outdoors on a winter day.

The caption for Figure 4.11 describes an experiment performed by a student in a chemistry laboratory. The fact that all such graphs for varying amounts of different gases cross the temperature axis at about −273°C is experimental evidence of the absolute zero discussed in Sections 3.7 and 4.3. By shifting the height (vertical) axis to −273°C and calling it zero, the Kelvin temperature scale appears along the horizontal axis. The kelvin scale and the shifted vertical axis are printed in red.

All graphs of volume versus absolute temperature pass through the origin. This indicates that they are graphs of a direct proportionality:

$$V \propto T \qquad \textbf{(4.6)}$$

where T is absolute temperature. By inserting a proportionality constant, $k_a$, the proportionality can be changed to an equation:

$$V \propto T \xrightarrow{\text{the proportionality changes to an equation}} V = k_aT \qquad \textbf{(4.7)}$$

© Visual Arts Library (London)/Alamy

Jacques Charles (1746–1823)

Equation 4.7 is the mathematical expression of **Charles's Law: The volume of a fixed quantity of gas at constant pressure is directly proportional to absolute temperature.** Charles's Law is named in honor of Jacques Charles, the French scientist who carried out investigations of the volume–temperature relationship. Dividing both sides of the equation by T gives

$$V = k_aT \xrightarrow{\text{divide both sides by T}} \frac{V}{T} = k_a \qquad \textbf{(4.8)}$$

From this it follows that, for a given sample of a gas,

$$\frac{V_1}{T_1} = \frac{V_2}{V_2} \qquad \textbf{(4.9)}$$

where the subscripts 1 and 2 refer to the first and second measurements of the two variables. In words, Equation 4.9 says that *for a fixed amount of a gas at constant pressure,* the ratio of volume to absolute temperature is constant, $\frac{V_1}{T_1} = k_a$. If the volume and/or temperature conditions change, their ratio will still equal the same constant $\frac{V_2}{T_2} = k_a$. Thus the volume-to-temperature ratio at any initial measurement will be the same as at a subsequent measurement, $\frac{V_1}{T_1} = \frac{V_2}{T_2}$.

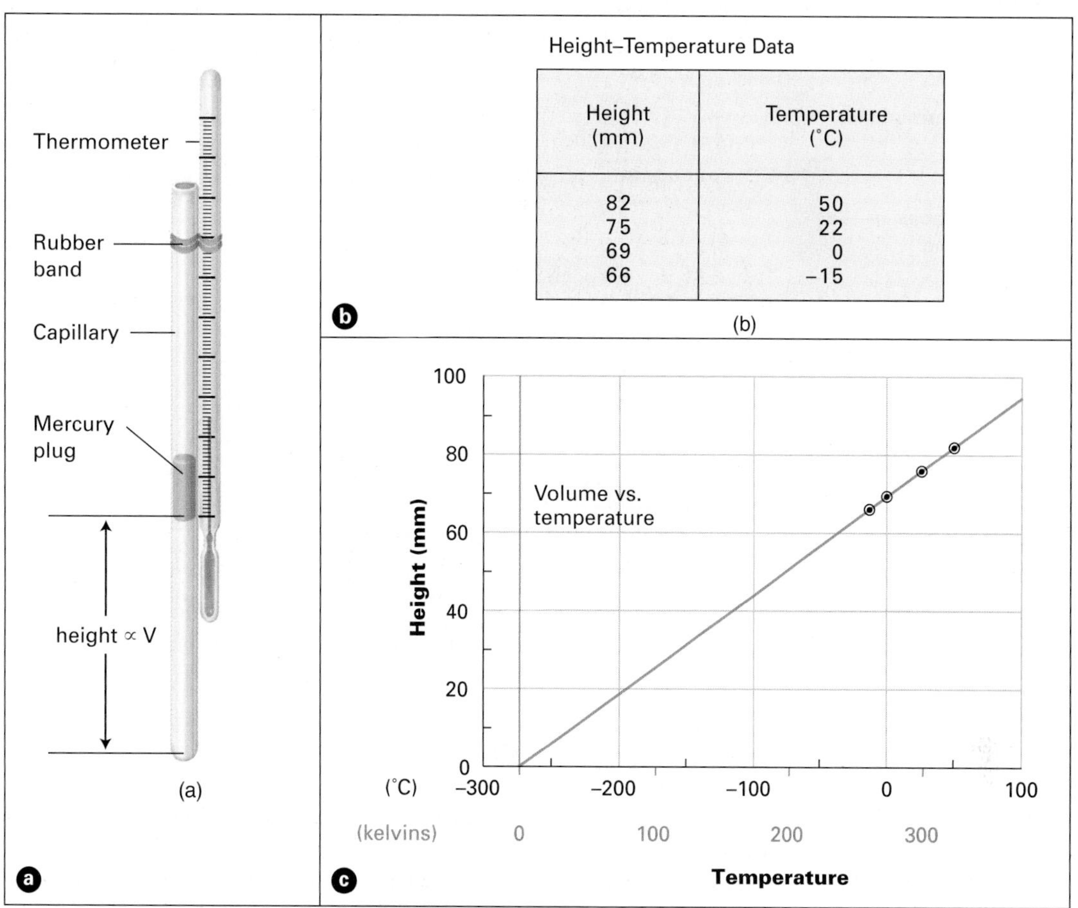


| Height (mm) | Temperature (°C) |
|---|---|
| 82 | 50 |
| 75 | 22 |
| 69 | 0 |
| 66 | −15 |

**Figure 4.11** Volume and temperature. A student performed an experiment to find the relationship between the volume and temperature of a fixed amount of gas at constant pressure. A small plug of liquid mercury was placed in a glass capillary tube sealed at the lower end. The amount of gas (air) trapped between the sealed end and the mercury plug remained fixed, or constant. Since the diameter of the capillary tube was constant, the height of the air column beneath the mercury plug was proportional to the volume of the gas. The capillary tube was attached to a thermometer (a) and submerged in liquid baths at different temperatures.

The student measured the height of the air column at each temperature and recorded the data (b). Part (c) is a graph of the data. Extrapolation to zero pressure suggests that the gas volume drops to zero at −273°C. This result is observed again and again if the experiment is repeated carefully many times. Combined with other experiments, using all kinds and quantities of gases, that all point to the same temperature, we have overwhelming evidence that there is an absolute zero at −273°C. Absolute temperature values are shown on the horizontal axis in red.

Solving Equation 4.9 for $V_2$, we obtain

$$\frac{V_1}{T_1} = \frac{V_2}{T_2} \xrightarrow{\text{cross multiply}} V_1T_2 = V_2T_1 \xrightarrow{\text{divide both sides by } T_1} \frac{V_1T_2}{T_1} = \frac{V_2\cancel{T_1}}{\cancel{T_1}}$$

$$V_2 = \frac{V_1T_2}{T_1} = \frac{V_1T_2}{T_1} = V_1 \times \frac{T_2}{T_1} \qquad \textbf{(4.10)}$$

Equation 4.10 shows three different arrangements of the factors in the $V_1T_2/T_1$ ratio. The first is how you would most likely write it. The second has $T_1$ directly under $T_2$, and the third isolates the $T_2/T_1$ ratio. The arrangements are equivalent; you may use whichever one you wish. In the pages that follow, we will use the third arrangement, with the temperature ratio standing alone. This arrangement best shows that the final volume resulting from a temperature change can be found by multiplying the initial volume by a ratio of temperatures. We use this sort of analysis in deciding if an answer calculated by algebra is reasonable. You will see the process shortly.

You must remember, when working gas law problems involving temperature, that the proportional relationships apply to *absolute* temperatures, not to Celsius

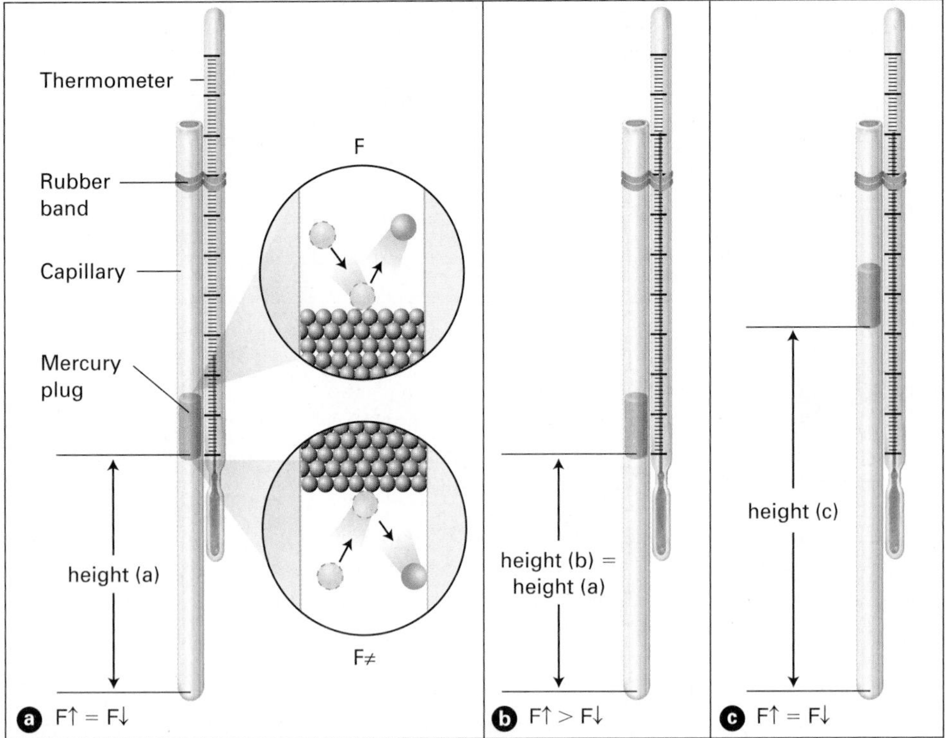


**Figure 4.12** A particulate view of Charles's Law.

(a) When the temperature of the gas trapped between the sealed end of the capillary and the mercury plug is the same as the temperature of the outside air, the average kinetic energy of the air particles and the trapped gas particles is the same. The force exerted upward on the mercury plug by the trapped gas particles (F↑) is equal to the force exerted downward on the mercury plug by the outside air particles (F↓). (This discussion assumes that the downward force exerted by the mercury is negligible.)

(b) As the trapped gas particles are heated, their average kinetic energy increases, so they collide with more force against the mercury plug. More important, at higher temperatures, the gas particles move faster. This increases the frequency with which particles collide with the bottom of the mercury plug, which also adds to F↑ and forces the plug upward.

(c) As the plug rises and the gas particles distribute themselves over the larger volume available to them, there are fewer molecules near the bottom of the mercury plug. The result is that the frequency of collision decreases, which reduces the pressure of the gas on the bottom of the plug. When the pressure finally drops to the point that it again equals the outside pressure, the plug stops moving. Thus the volume of the gas is higher.

temperatures. Before solving a gas law problem, you must convert Celsius temperatures to kelvins.

Figure 4.12 illustrates the Charles's Law experiment on the particulate level.

## Active Example 4.2

A gas with initial volume of 1.67 liters, measured at 32°C, is heated to 55°C at constant pressure. What is the new volume of the gas?

One way to *PLAN* gas law problems is to prepare a table showing the initial (1) and final (2) values of all variables. You then place these values into an equation you already know.

| | Volume | Temperature | Pressure |
|---|---|---|---|
| Initial value (1) | 1.67 L | 32°C; 305 K | Constant |
| Final value (2) | $V_2$ | 55°C; 328 K | Constant |

Notice that the Celsius temperatures have been changed to kelvins in the table. We recommend that you change temperatures to kelvins immediately, before you begin to solve the problem. Then you won't forget to make the change. (This is a very common error in gas law problems.)

Equation 4.10 may be applied directly to this problem, so an algebraic approach is indicated. Set up the problem and calculate the answer.

$$V_2 = V_1 \times \frac{T_2}{T_1} = 1.67\ \text{L} \times \frac{328\ \cancel{\text{K}}}{305\ \cancel{\text{K}}} = 1.80\ \text{L}$$

Arithmetically, the answer appears to be reasonable. The fraction 328/305 is somewhat more than 1, so the value of the number it multiplies should increase accordingly, which it does. But is the fraction itself correct? We'll consider that next.

There are two ways to solve gas law problems: by algebra, as you just did in Active Example 4.2 and by "reasoning." Reasoning is based on the proportionality between the variables. In this case the variables are volume and temperature, which are directly proportional to each other. They move in the same direction; if one goes up, the other goes up (and vice versa), as shown in Figure 4.13, when a sample of gas is heated.

Now notice the nature of Equation 4.10. The initial volume is multiplied by a ratio of temperatures—a temperature fraction. In this example there are two possible temperature ratios,

$$\frac{328\ \text{K}}{305\ \text{K}} \quad \text{and} \quad \frac{305\ \text{K}}{328\ \text{K}}$$

Without thinking of the equation, which temperature ratio is correct? Here's the thought process:

*Temperature increases from 305 K to 328 K.*

*Volume is directly proportional to temperature; they move in the same direction. Therefore, volume must increase.*

*If volume is to increase, the initial volume must be multiplied by a temperature fraction greater than 1.*

*In a fraction greater than 1, the numerator is larger than the denominator. 328 > 305, so 328/305 is the correct fraction.*

The temperature-ratio reasoning approach leads to exactly the same calculation setup as the equation, but the setup is reached by a different thought process.

## Thinking About Your Thinking

### Proportional Reasoning

**The reasoning approach discussed here is the essence of thinking about your thinking on the subject of proportional reasoning. When all other variables are held constant, cooling a gas reduces the volume, so the temperature fraction by which the initial volume will be multiplied will be less than 1. Heating a gas increases the volume, which means that the ratio of temperatures must be greater than 1.**

## Target Check 4.1

*If the final temperature is less than the initial temperature, how will the size of the final volume (pressure and amount constant) compare with the initial volume? Will the ratio of temperatures by which the initial volume is multiplied be greater than, equal to, or less than 1?*

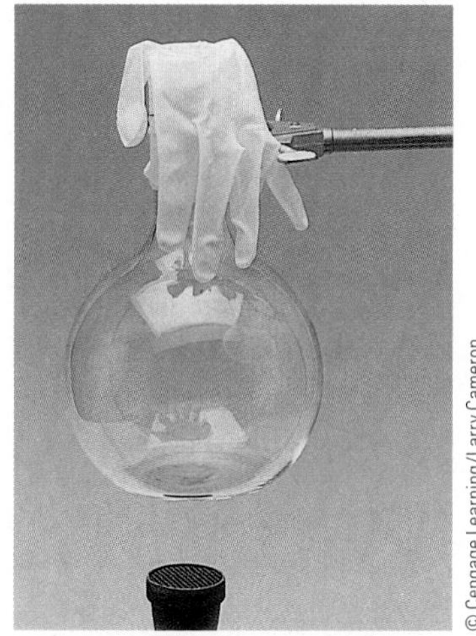

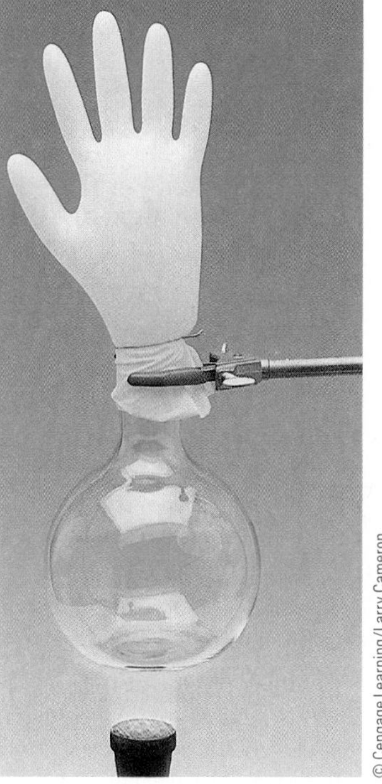





**Figure 4.13** Volume and temperature of a gas. For a fixed amount of a gas at constant pressure, the volume and the absolute temperature of the gas are directly proportional to each other. As the temperature increases, the volume increases.

We suggest that you solve gas law problems by algebra and then check your setup with temperature-ratio reasoning. The check is important. Inverting the fraction by which the beginning quantity is multiplied is the most common error students make when solving gas law problems. Some instructors prefer that their students solve the problems by reasoning rather than by algebra. If yours is among them, we recommend that you follow your instructor's lead rather than ours. If you do it correctly, you will always get the right answer with either method.

Photo by Stock Montage/Getty Images

Robert Boyle (1627–1691)

## 4.5 Boyle's Law

**Goal 8** Describe the relationship between the volume and pressure of a fixed quantity of an ideal gas at constant temperature, and express that relationship as a proportionality, an equation, and a graph.

**Goal 9** Given the initial volume (or pressure) and initial and final pressures (or volumes) of a fixed quantity of gas at constant temperature, calculate the final volume (or pressure).

In the 17th century, Robert Boyle investigated the quantitative relationship between pressure and volume of a fixed amount of gas at constant temperature. A modern laboratory experiment finds this relationship with a mercury-filled manometer such as that shown in Figure 4.14(a). Data are given in the first two columns of the table (Fig. 4.14[b]). A graph (Fig. 4.14[c]) of pressure versus volume measured in the experiment suggests an inverse proportionality between the variables. Expressed mathematically,

$$\text{Pressure} \propto \frac{1}{\text{Volume}} \quad \text{or, in symbols, } P \propto \frac{1}{V} \tag{4.11}$$

Introducing a proportionality constant gives

$$P \propto \frac{1}{V} \xrightarrow{\text{the proportionality changes to an equation}} P = k_b \frac{1}{V} \tag{4.12}$$

(We use the proportionality constant $k_b$ to clearly distinguish it from the $k_a$ we used for Charles's Law.) Multiplying both sides of the equation by V yields

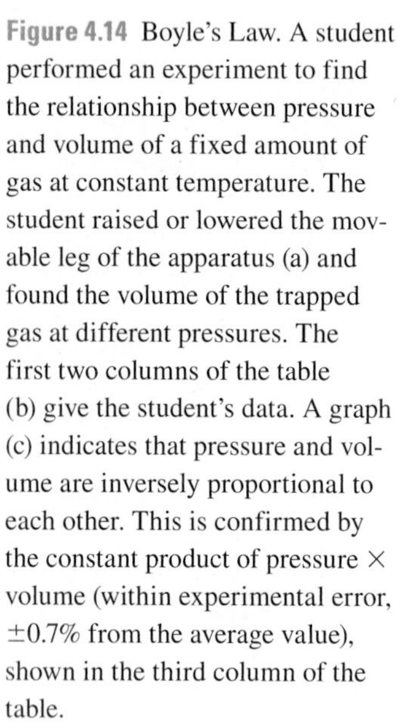

**Figure 4.14** Boyle's Law. A student performed an experiment to find the relationship between pressure and volume of a fixed amount of gas at constant temperature. The student raised or lowered the movable leg of the apparatus (a) and found the volume of the trapped gas at different pressures. The first two columns of the table (b) give the student's data. A graph (c) indicates that pressure and volume are inversely proportional to each other. This is confirmed by the constant product of pressure × volume (within experimental error, ±0.7% from the average value), shown in the third column of the table.

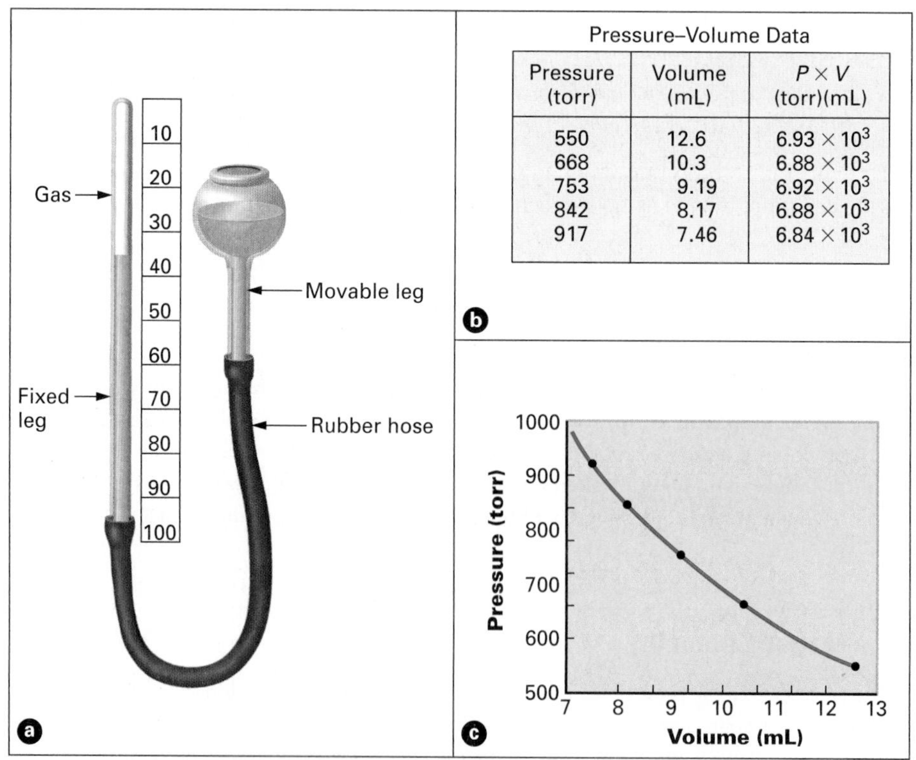


Pressure–Volume Data

| Pressure (torr) | Volume (mL) | $P \times V$ (torr)(mL) |
|---|---|---|
| 550 | 12.6 | $6.93 \times 10^3$ |
| 668 | 10.3 | $6.88 \times 10^3$ |
| 753 | 9.19 | $6.92 \times 10^3$ |
| 842 | 8.17 | $6.88 \times 10^3$ |
| 917 | 7.46 | $6.84 \times 10^3$ |

$$P = k_b \frac{1}{V} \xrightarrow{\text{multiply both sides by V}} PV = k_b \quad \textbf{(4.13)}$$

Within experimental error, PV is indeed a constant (Fig. 4.14[b], third column).

**Boyle's Law,** which this experiment illustrates, **states that for a fixed quantity of gas at constant temperature, pressure is inversely proportional to volume.** Equation 4.13 is the usual mathematical statement of Boyle's Law. Since the product of P and V is constant, when one factor increases the other must decrease. This is what is meant by an inverse proportionality. Notice the difference between an inverse proportionality and the direct proportionality of the previous section, where both variables increase or decrease together.

Also note the shape of the curve in Figure 4.14(c). It is characteristic of an inverse proportionality. Contrast the plot of an inverse proportionality with that of a direct proportionality, Fig 4.11(c).

From Equation 4.13 we see that $P_1V_1 = k_b = P_2V_2$, or

$$P_1V_1 = P_2V_2 \quad \textbf{(4.14)}$$

where subscripts 1 and 2 refer to first and second measurements of pressure and volume at constant temperature. Solving for $V_2$, we have

$$V_2 = V_1 \times \frac{P_1}{P_2} \quad \textbf{(4.15)}$$

Figure 4.15 gives a particulate-level illustration of Boyle's Law.

## √ Target Check 4.2

*Assume constant temperature and amount of gas. (a) If the final pressure is less than the initial pressure, how will the final volume compare with the initial volume? (b) If the final volume is greater than the initial volume, how will the final pressure compare with the initial pressure?*

## Thinking About Your Thinking

### Proportional Reasoning

**An inverse proportionality has the algebraic form $y \propto \frac{1}{x}$. A common example of an inverse proportionality is the relationship between the time needed to travel between two cities and the speed at which you drive. Mathematically, the proportionality is: time $\propto \frac{1}{\text{speed}}$. If you increase your speed, you decrease the value of $\frac{1}{\text{speed}}$ and therefore decrease the time needed for the trip. If your speed doubles, the time is cut in half.**

**To change an inverse proportionality to an equation, a proportionality constant is introduced: $y = k\frac{1}{x}$. The product of the two variables is therefore a constant: $xy = k$. Since the product must always equal a fixed number, when one value goes up, the other must go down.**

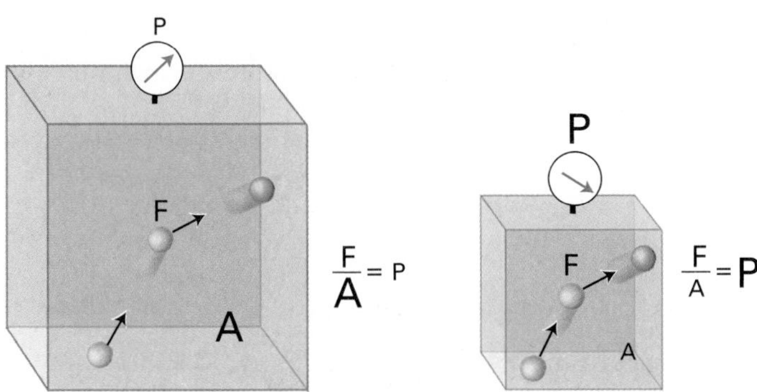


Figure 4.15 Explanation of Boyle's Law. At constant temperature, the particles in a fixed amount of gas have a constant average kinetic energy. Recall that the defining equation for pressure is P ≡ F/A. When the volume of a container is reduced, the particles strike a smaller area. Their force remains the same. Because pressure is inversely proportional to area, when area goes down, pressure goes up. So when volume goes down, pressure goes up, and vice versa.

Boyle's Law works on marshmallows, too! Place some marshmallows in a flask (top), then use a vacuum pump to lower pressure in the flask (bottom). The air trapped inside the marshmallows expands as the pressure decreases. Unfortunately, the process is reversible; when you open the flask to get the giant marshmallows, they shrink back to normal size.

## Active Example 4.3

A gas sample occupies 5.18 liters at 776 torr. Find the volume of the gas if the pressure is changed to 827 torr. Temperature and amount remain constant.

Begin by setting up the table of initial and final values.

| | Volume | Temperature | Pressure |
|---|---|---|---|
| Initial value (1) | | | |
| Final value (2) | | | |

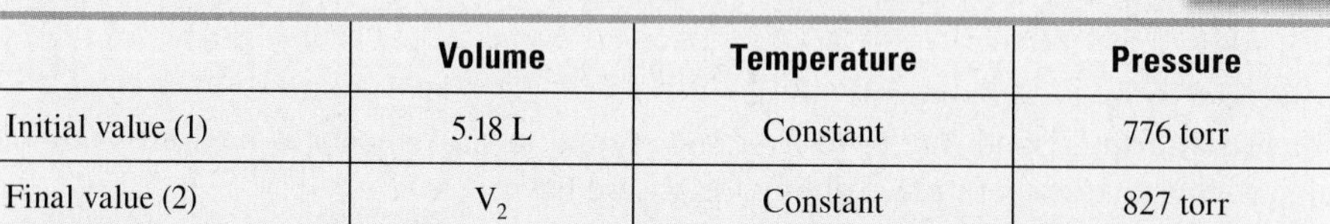

| | Volume | Temperature | Pressure |
|---|---|---|---|
| Initial value (1) | 5.18 L | Constant | 776 torr |
| Final value (2) | $V_2$ | Constant | 827 torr |

Now write the equation, insert values, and solve the problem.

$$V_2 = V_1 \times \frac{P_1}{P_2} = 5.18\text{ L} \times \frac{776\ \cancel{\text{torr}}}{827\ \cancel{\text{torr}}} = 4.86\text{ L}$$

Now check the answer. First, is the pressure increasing or decreasing? From what value to what other value?

Increasing, from 776 torr to 827 torr.

Will the volume increase or decrease as the pressure increases? Explain your reasoning.

Because volume varies inversely with pressure, the volume will decrease. In an inverse proportion the variables go in opposite ways: One goes up and the other goes down.

The multiplier is a ratio of pressures. Should this ratio be greater than 1 or less than 1 if the final pressure is larger than the initial pressure? Why?

The ratio should be less than 1, so the lower pressure is on top.

Is 4.86 a reasonable result when 5.18 is multiplied by 776/827? Explain.

Yes. The fraction is somewhat smaller than 1, and 4.86 is somewhat smaller than 5.18.

## 4.6 The Combined Gas Law

**Goal 10** For a fixed quantity of a confined gas, given the initial volume, pressure, and temperature and the final values of any two variables, calculate the final value of the third variable.

**Goal 11** State the values associated with standard temperature and pressure (STP) for gases.

In the two previous sections we saw that volume and absolute temperature are directly proportional and that volume and pressure are inversely proportional. Mathematically, these may be expressed as $V \propto T$ and $V \propto 1/P$. Whenever the same quantity (V) is proportional to two other quantities (T and 1/P), it is proportional to the product of those quantities:

$$V \propto T \times \frac{1}{P} \quad or \quad V \propto \frac{T}{P} \tag{4.16}$$

This becomes an equation when a proportionality constant, $k_c$, is introduced:

$$V \propto \frac{T}{P} \xrightarrow{\text{the proportionality changes to an equation}} V = k_c \times \frac{T}{P} \tag{4.17}$$

Rearranging, we get

$$V = k_c \times \frac{T}{P} \xrightarrow{\text{multiply both sides by P/T}} \frac{PV}{T} = k_c \tag{4.18}$$

Again, using subscripts 1 and 2 for initial and final values of all variables, we obtain

$$\frac{P_1V_1}{T_1} = k_c = \frac{P_2V_2}{T_2} \quad or \quad \frac{P_1V_1}{T_1} = \frac{P_2V_2}{T_2} \tag{4.19}$$

There is another gas law, **Gay-Lussac's Law,** that states: **The pressure exerted by a fixed quantity of gas at constant volume is directly proportional to absolute temperature ($P \propto T$).** Like Charles's Law, Gay-Lussac's Law predicts an absolute zero at −273°C. Gay-Lussac's Law can also be used to predict how pressure or temperature will change as a result of a change in the other variable.

Problems involving changes in pressure, temperature, and volume of a constant quantity of gas are called **Combined Gas Law** problems. Given five of the six variables in Equation 4.19, the remaining variable can be calculated using algebra. For example, if you know the volume, $V_1$, occupied by a gas at one temperature and pressure, $T_1$ and $P_1$, you can find the volume, $V_2$, that the gas will occupy at another temperature and pressure, $T_2$ and $P_2$, by solving Equation 4.19 for $V_2$:

$$V_2 = V_1 \times \frac{P_1}{P_2} \times \frac{T_2}{T_1} \tag{4.20}$$

**P/REVIEW:** In Section 13.2, we introduce Avogadro's Law, $V \propto n$. The combination of the Combined Gas Law with Avogadro's Law leads to the Ideal Gas Law, $V \propto \frac{nT}{P}$. Introducing R as a proportionality constant gives the equation $V = R\frac{nT}{P}$. The most common form of the Ideal Gas Law rearranges the equation to $PV = nRT$ (Section 14.3).

## Active Example 4.4

A cylinder in an automobile engine has a volume of 352 $cm^3$. This engine takes in air at 21°C and 0.945 atm pressure. The compression stroke squeezes and heats this gas until the pressure is 4.95 atm and the temperature is 95°C. What is the final volume in the cylinder?

Set up a table as before.

| | Volume | Temperature | Pressure |
|---|---|---|---|
| Initial value (1)  | | | |
| Final value (2) | | | |

| | Volume | Temperature | Pressure |
|---|---|---|---|
| Initial value (1) | 352 $cm^3$ | 21°C; 294 K | 0.945 atm |
| Final value (2) | $V_2$ | 95°C; 368 K | 4.95 atm |

You have five of the six variables in the combined gas law, and you are asked to find the sixth. Start by writing the combined gas law, and then solve it algebraically for the unknown in this problem, $V_2$.

$$\frac{P_1V_1}{T_1} = \frac{P_2V_2}{T_2} \xrightarrow{\text{crossmultiply}} P_1V_1T_2 = P_2V_2T_1 \xrightarrow{\text{divide both sides by } P_2T_1}$$

$$\frac{P_1V_1T_2}{P_2T_1} = V_2 \xrightarrow{\text{rearrange}} V_2 = V_1 \times \frac{P_1}{P_2} \times \frac{T_2}{T_1}$$

Now you can substitute values from the table into your equation, solved for $V_2$, and calculate the answer.

$$V_2 = V_1 \times \frac{P_1}{P_2} \times \frac{T_2}{T_1} = 352\text{ cm}^3 \times \frac{0.945\text{ atm}}{4.95\text{ atm}} \times \frac{368\text{ K}}{294\text{ K}} = 84.1\text{ cm}^3$$

Once again, to check our problem setup we'll use a reasoning approach, but this time we'll use it twice. First, does increasing pressure increase or decrease volume? Explain.

Decrease. Volume is inversely proportional to pressure, so volume decreases as pressure increases.

Will the ratio of pressures be greater than 1 or less than 1? Explain.

Less than 1. If the volume is to decrease, it must be multiplied by a ratio smaller than 1.

Now the temperature effect: Will the volume increase or decrease as the temperature increases? Will the temperature ratio be larger than 1 or smaller than 1? Explain your reasoning.

Increase; larger than 1. Volume varies directly with pressure, so it increases as temperature increases. This requires a multiplier greater than 1.

The problem setup looks all right. How about the numerical answer? Is it reasonable or not reasonable? On what logic is your answer based?

Reasonable. We start with about 350. The first fraction is close to 1/5, and one-fifth of 350 is 70 ($35 \div 5 = 7$). Then $368/294 \approx 360/300 = 1.2$, and $1.2 \times 70 = 70\ (1 \times 70) + 14\ (0.2 \times 70) = 84$. The calculated 84.1 looks good.

The volume of a fixed quantity of gas depends on its temperature and pressure. Therefore, it is not possible to state the amount of gas in volume units without also specifying the temperature and pressure. These are often given as 0°C (273 K) and 1 bar, which are known as **standard temperature and pressure (STP).** Many gas law problems require changing volume to or from STP. The procedure is just like the one you used in Active Example 4.4.

## Active Example 4.5

What would be the volume at STP of 3.62 liters of nitrogen gas, measured at 0.843 bar and 16°C?

Solve the problem completely.

| | Volume | Temperature | Pressure |
|---|---|---|---|
| Initial value (1) | | | |
| Final value (2) | | | |

| | Volume | Temperature | Pressure |
|---|---|---|---|
| Initial value (1) | 3.62 L | 16°C; 289 K | 0.843 bar |
| Final value (2) | $V_2$ | 0°C; 273 K | 1 bar |

$$V_2 = V_1 \times \frac{P_1}{P_2} \times \frac{T_2}{T_1} = 3.62\ \text{L} \times \frac{0.843\ \cancel{\text{bar}}}{1\ \cancel{\text{bar}}} \times \frac{273\ \cancel{\text{K}}}{289\ \cancel{\text{K}}} = 2.88\ \text{L}$$

It is difficult to estimate the calculation results, but the 3.62 L initial volume is multiplied by two fractions that are each a little less than 1, so the 2.92 L final volume makes sense.

**☞ *Learn It Now!*** Remembering one equation and understanding the algebraic cancellation of variables is easier than remembering three equations. Be sure that you can derive Boyle's Law and Charles's Law from the Combined Gas Law before you proceed further with your studies.

Now that you have learned the Combined Gas Law, note that Charles's Law and Boyle's Law are simply special cases of the Combined Gas Law. Here's the Combined Gas Law once again:

$$\frac{P_1V_1}{T_1} = \frac{P_2V_2}{T_2}$$

If pressure is constant, $P_1 = P_2$, so we can divide both sides by P:

$$\frac{P_1V_1}{T_1} = \frac{P_2V_2}{T_2} \xrightarrow{\text{divide both sides by P}} \frac{V_1}{T_1} = \frac{V_2}{T_2}\ (\text{constant P})$$

The resulting relationship is Charles's Law. Similarly, Boyle's Law results from the Combined Gas Law when temperature is constant, $T_1 = T_2$:

$$\frac{P_1V_1}{T_1} = \frac{P_2V_2}{T_2} \xrightarrow{\text{multiply both sides by T}} P_1V_1 = P_2V_2\ (\text{constant T})$$

Just remember the Combined Gas Law, and whenever the value of one of the variables remains constant, you can cancel it algebraically.

# Chapter 4 in Review

Most of the key terms and concepts and many others appear in the Glossary. Use your Glossary regularly.

### 4.1 Properties of Gases

Goal 1 Describe five macroscopic characteristics of gases.

### 4.2 The Kinetic Theory of Gases and the Ideal Gas Model

Goal 2 Explain or predict physical phenomena relating to gases in terms of the ideal gas model.

**Key Terms and Concepts: Ideal gas model, kinetic molecular theory**

### 4.3 Gas Measurements

Goal 3 Given a gas pressure in atmospheres, torr, millimeters (or centimeters) of mercury, inches of mercury, pascals, kilopascals, bars, or pounds per square inch, express that pressure in each of the other units.

Goal 4 Define *pressure* and interpret statements in which the term *pressure* is used.

Goal 5 Given a temperature in degrees Celsius, convert it to kelvins, and vice versa.

**Key Terms and Concepts: Absolute temperature, absolute zero, atmosphere (atm) (pressure unit), bar (pressure unit), barometer, gauge pressure, kelvin (K), manometer, millimeter of mercury (mm Hg) (pressure unit), pascal (Pa) and kilopascal (kPa) (pressure units), pressure, torr (pressure unit)**

### 4.4 Charles's Law

Goal 6 Describe the relationship between the volume and temperature of a fixed quantity of an ideal gas at constant pressure, and express that relationship as a proportionality, an equation, and a graph.

Goal 7 Given the initial volume (or temperature) and the initial and final temperatures (or volumes) of a fixed quantity of gas at constant pressure, calculate the final volume (or temperature).

**Key Terms and Concepts: Charles's Law**

### 4.5 Boyle's Law

Goal 8 Describe the relationship between the volume and pressure of a fixed quantity of an ideal gas at constant temperature, and express that relationship as a proportionality, an equation, and a graph.

Goal 9 Given the initial volume (or pressure) and initial and final pressures (or volumes) of a fixed quantity of gas at constant temperature, calculate the final volume (or pressure).

**Key Terms and Concepts: Boyle's Law**

**4.6 The Combined Gas Law**

Goal 10 For a fixed quantity of a confined gas, given the initial volume, pressure, and temperature and the final values of any two variables, calculate the final value of the third variable.

Goal 11 State the values associated with standard temperature and pressure (STP) for gases.

**Key Terms and Concepts: Combined Gas Law, Gay-Lussac's Law, standard temperature and pressure (STP)**

## Study Hints and Pitfalls to Avoid

This chapter introduced two proportional gas laws. Charles's Law ($V \propto T$) is a direct proportionality. If one value increases, the other increases. The other, Boyle's Law ($P \propto 1/V$), is an inverse proportionality. If one value becomes larger, the other becomes smaller.

In all gas calculations, temperature must be expressed in kelvins: K = °C + 273. Failing to make this conversion is a major pitfall in this chapter. If you make this change when you *PLAN* the problem, even before thinking about how to solve it, you will not forget that step.

A second major pitfall is inverting one (or both) of the ratios in a Combined Gas Law setup. Tabulating the initial and final conditions will help you avoid that mistake.

When checking your answers or setting up your problem by reasoning, use the proportional relationship to decide if the change being considered will cause the final value of the wanted quantity to increase or decrease. If it will increase, the ratio must be greater than 1, with the larger number in the numerator. A decrease requires a ratio smaller than 1, with the smaller number in the numerator.

## Concept-Linking Exercises

*Write a brief description of the relationships among each of the following groups of terms or phrases. Answers to the Concept-Linking Exercises are given after answers to the Target Checks at the end of the book.*

1. Gaseous state of matter, compressibility, density, mixability
2. Kinetic molecular theory, ideal gas model, particulate behavior
3. Pressure, barometer, the earth's atmosphere, gauge pressure
4. Temperature, average kinetic energy, absolute zero, kelvin
5. Volume–Temperature Law, Volume–Pressure Law, Combined Gas Law

## Small-Group Discussion Questions

*Small-Group Discussion Questions are for group work, either in class or under the guidance of a leader during a discussion section.*

1. Early scientists did not believe that gases were matter. What apparent properties of gases led to this belief? How do properties of gases contrast with properties of liquids and solids?
2. Use the ideal gas model to formulate a particulate-level explanation for each of the following:
   a) All gases behave approximately the same.
   b) Gases may be compressed.
   c) A tiny quantity of gas will completely fill a large container.
   d) The density of gases is relatively low.
   e) Gases can be mixed.
   f) A confined gas exerts the same pressure on all container walls.
3. Compare Torricelli's mercury barometer with a barometer filled with a liquid exactly half as dense as mercury. Describe how the barometers would differ and how they would be the same.
4. Show how a volume-versus-temperature graph of a fixed quantity of an ideal gas at constant pressure can be used to discover the absolute zero of temperature.
5. Use the data in Figure 4.14 to construct a plot of volume-versus-1/pressure. Explain the nature of the plot in terms of Boyle's Law.
6. Describe how a change in each variable in the Combined Gas Law can be explained by the ideal gas model. For example, if you decrease the pressure of a confined gas at constant temperature, what happens at the particulate level to cause the volume to change?

## Questions, Exercises, and Problems

*Questions whose numbers are printed in blue are answered in Appendix III. ■ denotes problems assignable in OWL. In the first section of Questions, Exercises, and Problems, similar exercises and problems are paired in consecutive odd-even number combinations.*

**Section 4.2: The Kinetic Theory of Gases and the Ideal Gas Model**

1. What properties of gases are the result of the kinetic character of a gas? Explain.
2. What is the meaning of *kinetic* as it is used in describing gases and molecular theory? What is kinetic energy?
3. State how and explain why the pressure exerted by a gas is different from the pressure exerted by a liquid or solid.
4. What causes pressure in a gas?
5. What are the desirable properties of air in an air mattress used by a camper? Show which part of the ideal gas model is related to each property.

6. List some properties of air that make it suitable for use in automobile tires. Explain how each property relates to the ideal gas model.

*Questions 7 through 14: Explain how the physical phenomenon described is related to one or more features of the ideal gas model.*

7. Pressure is exerted on the top of a tank holding a gas, as well as on its sides and bottom.

8. Gases with a distinctive odor can be detected some distance from their source.

9. Balloons expand in all directions when blown up, not just at the bottom as when filled with water.

10. Steam bubbles rise to the top in boiling water.

11. Even though an automobile tire is "filled" with air, more air can always be added without increasing the volume of the tire significantly.

12. The density of liquid oxygen is about 1.4 g/cm$^3$. Vaporized at 0°C and 760 torr, this same 1.4 g occupies 980 cm$^3$, an expansion of nearly 1000 times.

13. Gas bubbles always rise through a liquid and become larger as they move upward.

14. Any container, regardless of size, will be completely filled by one gram of hydrogen.

**Section 4.3: Gas Measurements**

15. What does pressure measure? What does temperature measure?

16. Four properties of gases may be measured. Name them.

17. Explain how a manometer works.

18. Explain how a barometer works.

*Questions 19 and 20: Complete the table by converting the given pressure to each of the other pressures:*

19.

| **atm** | | | | | |
|---|---|---|---|---|---|
| **psi** | | | | | |
| **in. Hg** | | | | | |
| **cm Hg** | | | | | |
| **mm Hg** | 785 | | | | |
| **torr** | | 124 | | | |
| **Pa** | | | $1.18 \times 10^5$ | | |
| **kPa** | | | | 91.4 | |
| **bar** | | | | | 0.977 |

20.

| **atm** | 1.84 | | | |
|---|---|---|---|---|
| **psi** | | 13.9 | | |
| **in. Hg** | | | 28.7 | |
| **cm Hg** | | | | 74.8 |
| **mm Hg** | | | | |
| **torr** | | | | |
| **Pa** | | | | |
| **kPa** | | | | |
| **bar** | | | | |

21. Find the pressure of the gas in the mercury manometer shown if atmospheric pressure is 747 torr.

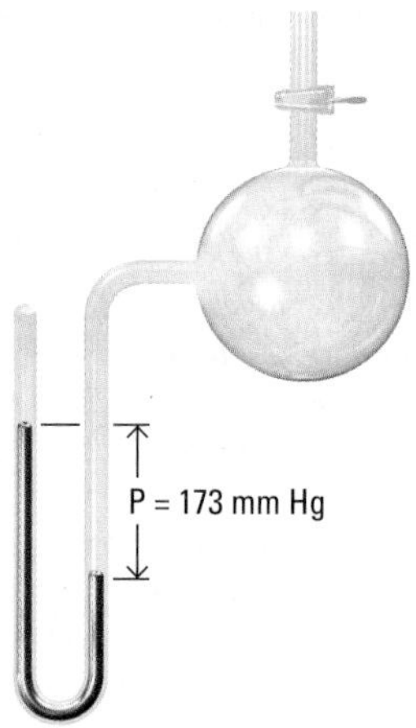


22. ■ A mercury manometer is used to measure pressure in the container illustrated. Calculate the pressure exerted by the gas if atmospheric pressure is 752 torr.

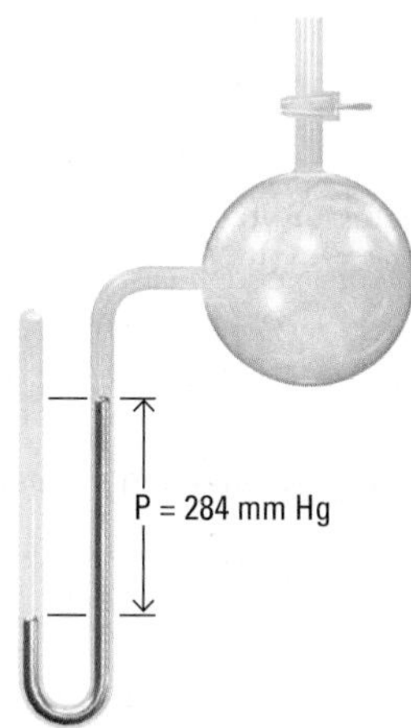


23. A student records a temperature of −18 K in an experiment. What is the nature of things at that temperature? What would you guess the student meant to record? What absolute temperature corresponds to the temperature the student meant to record?

24. What does the term *absolute zero* mean? What physical condition is presumed to exist at absolute zero?

25. If the temperature in the room is 31°C, what is the equivalent absolute temperature?

26. ■ Substances like sulfur dioxide, which is a gas at room temperature and pressure, can often be liquefied or solidified only at very low temperatures. At a pressure of 1 atm, $SO_2$ does not condense to a liquid until −10.1°C and does not freeze until −72.7°C. What are the equivalent absolute temperatures?

27. Hydrogen remains a gas at very low temperatures. It does not condense to a liquid until the temperature is −253°C, and it freezes shortly thereafter, at −259°C. What are the equivalent absolute temperatures?

28. ■ Many common liquids have boiling points that are less than 110°C, whereas most metals are solids at room temperature and have much higher boiling points. The

boiling point of bromoethane, $C_2H_5Br$, is 38.4°C. What is the equivalent absolute temperature? The boiling point of aluminum is 2740.2 K. What is the equivalent temperature on the Celsius scale?

29. Hydrogen cyanide is the deadly gas used in some execution chambers. It melts at 259 K and changes to a gas at 299 K. What are the Celsius temperatures at which hydrogen cyanide changes state?

30. ■ Many common liquids have boiling points that are less than 110°C, whereas most metals are solids at room temperature and have much higher boiling points. The boiling point of propanol, $C_3H_7OH$, is 370.6 K. What is the equivalent Celsius temperature? The boiling point of nickel is 3003.2 K. What is the equivalent temperature on the Celsius scale?

#### Section 4.4: Charles's Law

31. A variable-volume container holds 24.3 L of gas at 55°C. If pressure remains constant, what will the volume be if the temperature falls to 17°C?

32. ■ A sample of argon gas at a pressure of 715 mm Hg and a temperature of 26°C occupies a volume of 8.97 L. If the gas is heated at constant pressure to a temperature of 71°C, what will be the volume of the gas sample?

33. A spring-loaded closure maintains constant pressure on a gas system that holds a fixed quantity of gas, but a bellows allows the volume to adjust for temperature changes. From a starting point of 1.26 L at 19°C, what Celsius temperature will cause the volume to change to 1.34 L?

34. ■ A sample of hydrogen gas at a pressure of 0.520 atm and a temperature of 151°C occupies a volume of 857 mL. If the gas is heated at constant pressure until its volume is $1.02 \times 10^3$ mL, what will be the Celsius temperature of the gas sample?

#### Section 4.5: Boyle's Law

35. If you squeeze the bulb of a dropping pipet (eye dropper) when the tip is below the surface of a liquid, bubbles appear. When you release the bulb, liquid flows into the pipet. Explain why in terms of Boyle's Law.

36. Squeezing a balloon is one way to burst it. Why?

37. The pressure on 648 mL of a gas is changed from 772 torr to 695 torr. What is the volume at the new pressure?

38. ■ A sample of carbon dioxide gas at a pressure of 0.556 atm and a temperature of 21.2°C occupies a volume of 12.9 liters. If the gas is allowed to expand at constant temperature to a volume of 22.3 liters, what will be the pressure of the gas sample (in liters)?

39. A cylindrical gas chamber has a piston at one end that can be used to compress or expand the gas. If the gas is initially at 1.22 atm when the volume is 7.26 L, what will the pressure be if the volume is adjusted to 3.60 L?

40. ■ A sample of krypton gas at a pressure of 905 torr and a temperature of 28.4°C occupies a volume of 631 mL. If the gas is allowed to expand at constant temperature until its pressure is 606 torr, what will be the volume of the gas sample (in liters)?

#### Section 4.6: The Combined Gas Law

41. The gas in a 0.717-L cylinder of a diesel engine exerts a pressure of 744 torr at 27°C. The piston suddenly compresses the gas to 48.6 atm and the temperature rises to 547°C. What is the final volume of the gas?

42. ■ A sample of krypton gas occupies a volume of 6.68 L at 64°C and 361 torr. If the volume of the gas sample is increased to 8.99 L while its temperature is increased to 114°C, what will be the resulting gas pressure (in torr)?

43. A collapsible balloon for carrying meteorological testing instruments aloft is partly filled with 626 liters of helium, measured at 25°C and 756 torr. Assuming the volume of the balloon is free to expand or contract according to changes in pressure and temperature, what will be its volume at an altitude where the temperature is −58°C and the pressure is 0.641 atm?

44. ■ A sample of neon gas occupies a volume of 7.68 L at 59°C and 0.634 atm. If it is desired to increase the volume of the gas sample to 9.92 L while decreasing its pressure to 0.436 atm, what will be the temperature of the gas sample at the new volume and pressure? Answer in °C.

45. Why have the arbitrary conditions of STP been established? Are they realistic?

46. What is the meaning of STP?

47. If one cubic foot—28.3 L—of air at common room conditions of 23°C and 0.985 bar is adjusted to STP, what does the volume become?

48. ■ A sample of helium gas has a volume of 7.06 L at 45°C and 2.12 bar. What would be the volume of this gas sample at STP?

49. An experiment is designed to yield 44.5 mL oxygen, measured at STP. If the actual temperature is 28°C and the actual pressure is 0.894 bar, what volume of oxygen will result?

50. ■ A sample of krypton gas has a volume of 9.92 L at STP. What would be the volume of this gas sample at 40°C and 1.31 bar?

51. What pressure (in bar) will be exerted by a tank of natural gas used for home heating if its volume is 19.6 L at STP and it is compressed to 6.85 L at 24°C?

52. A gas occupies 2.33 L at STP. What pressure (bar) will this gas exert if it is expanded to 6.19 L and warmed to 17°C?

53. A container with a volume of 56.2 L holds helium at STP. The gas is compressed to 23.7 L and 2.09 bar. To what must the temperature change (°C) to satisfy the new volume and pressure?

54. At STP, a sample of neon fills a 4.47-L container. The gas is transferred to a 6.05-L container, and its temperature is adjusted until the pressure is 0.736 bar. What is the new temperature?

#### General Questions

55. Distinguish precisely and in scientific terms the differences among items in each of the following groups.
    a) Kinetic theory of gases, kinetic molecular theory
    b) Pascal, mm Hg, torr, atmosphere, psi, bar

c) Barometer, manometer

d) Pressure, gauge pressure

e) Boyle's Law, Charles's Law, Gay-Lussac's Law, Combined Gas Law

f) Celsius and Kelvin temperature scales

g) Temperature and pressure, standard temperature and pressure, STP, $T_{°C}$, $T_{°F}$, $T_K$

56. Determine whether each of the following statements is true or false:

a) The total kinetic energy of two molecules in the gas phase is the same before and after they collide with each other.

b) Gas molecules are strongly attracted to each other.

c) Gauge pressure is always greater than absolute pressure except in a vacuum.

d) For a fixed amount of gas at constant temperature, if volume increases, pressure decreases.

e) For a fixed amount of gas at constant pressure, if temperature increases, volume decreases.

f) For a fixed amount of gas at constant volume, if temperature increases, pressure increases.

g) At a given temperature, the number of degrees Celsius is larger than the number of kelvins.

h) Both temperature and pressure ratios are larger than 1 when calculating the final gas volume as conditions change from STP to 15°C and 0.834 atm.

### More Challenging Problems

*Questions 57 and 58: Explain how the following physical phenomenon is related to one or more of the features of the ideal gas model.*

57. Very small dust particles, seen in a beam of light passing through a darkened room, appear to be moving about erratically.

58. Properties of gases become less "ideal"—the substance adopts behavior patterns not typical of gases—when subjected to very high pressures so that the individual molecules are close to each other.

59. ■ The volume of the air chamber of a bicycle pump is 0.26 L. The volume of a bicycle tire, including the hose between the pump and the tire, is 1.80 L. If both the tire and the air in the pump chamber begin at 743 torr, what will be the pressure in the tire after a single stroke to the pump?

60. A 1.91-L gas chamber contains air at 959 torr. It is connected through a closed valve to another chamber with a volume of 2.45 L. The larger chamber is evacuated to negligible pressure. What pressure will be reached in both chambers if the valve between them is opened and the air occupies the total volume?

61. A hydrogen cylinder holds gas at 3.67 atm in a laboratory where the temperature is 25°C. To what will the pressure change when the cylinder is placed in a storeroom where the temperature drops to 7°C?

62. Air in a steel cylinder is heated from 19°C to 42°C. If the initial pressure was 4.26 atm, what is the final pressure?

63. ■ A gas storage tank is designed to hold a fixed volume and quantity of gas at 1.74 atm and 27°C. To prevent excessive pressure due to overheating, the tank is fitted with a relief valve that opens at 2.00 atm. To what temperature (°C) must the gas rise in order to open the valve?

64. A gas in a steel cylinder shows a gauge pressure of 355 psi while sitting on a loading dock in the winter when the temperature is −18°C. What pressure (in psi) will the gauge show when the tank is brought inside and its contents warm up to 23°C? The pressure inside the laboratory is 14.7 psi.

65. ■ If 1.62 $m^3$ of air at 12°C and 738 torr is compressed into a 0.140-$m^3$ tank, and the temperature is raised to 28°C while the external pressure is 14.7 psi, what pressure (in psi) will show on the gauge?

66. The pressure gauge reads 125 psi on a 0.140-$m^3$ compressed air tank when the gas is at 33°C. To what volume will the contents of the tank expand if they are released to an atmospheric pressure of 751 torr and the temperature is 13°C?

67. ■ The compression ratio in an automobile engine is the ratio of the gas pressure at the end of the compression stroke to the pressure at the beginning. Assume that compression occurs at constant temperature. The total volume of a cylinder in a compact automobile is 350 $cm^3$, and the displacement (the reduction in volume during the compression stroke) is 309 $cm^3$. What is the compression ratio in that engine?

Photo Researchers

5

CONTENTS



John Dalton's 1808 book *A New System of Chemical Philosophy* set the stage for the modern science of chemistry by reviving the ancient Greek concept that matter was made of elementary particles called atoms. Dalton added to the Greek idea by proposing that atoms of any one element are different from atoms of other elements. This illustration from Dalton's book shows his proposed system of "arbitrary marks or signs," as he wrote, to represent the elements. In this chapter, you will learn about Dalton's atomic theory and the modern system of "arbitrary marks or signs" used by today's scientists to represent the elements.

# Atomic Theory: The Nuclear Model of the Atom

As early as 400 BC, Greek philosophers had proposed that matter consisted of tiny, indivisible particles, which they called **atoms.** In 1808 John Dalton, an English chemist and schoolteacher, revived the concept. We now know that the atom consists of even smaller particles. Today, chemists use some of the most sophisticated research methods ever developed as they continue to seek an understanding of how atoms are put together. But it all started with the vision of John Dalton.

Online homework for this chapter may be assigned in OWL

In this chapter you will begin studying the atom, the smallest particle of any element. You'll learn that the atom is made up of three smaller particles. You will also see that different combinations of these smaller particles account for the different elements. In addition, we introduce the arrangement of elements into groups that have similar properties.

## 5.1 Dalton's Atomic Theory

**Atomic-Theory-Before-Calculations Option** If your instructor has chosen to schedule Chapter 5 before Chapter 3, this may be your first encounter with Active Examples in this book. If so, turn to Chapter 3 and read the introduction. It explains how our self-teaching Active Examples are designed to make you an active learner of chemistry.

Goal 1 Identify the main features of Dalton's atomic theory.

Goal 2 State the meaning of, or draw conclusions based on, the Law of Multiple Proportions.

Dalton knew about the Law of Definite Composition: The percentage by mass of the elements in a compound is always the same (Section 2.6). He was also familiar with the Law of Conservation of Mass: In a chemical change, mass is conserved; it is neither created nor destroyed (Section 2.9). **Dalton's atomic theory** explained these observations. The main features of his theory are as follows (Fig. 5.1).

E. F. Smith Memorial Collection in the History of Chemistry, Department of Special Collections, Van Pelt Dietrich Library, University of Pennsylvania

John Dalton (1766–1844)

summary

**Dalton's Atomic Theory**

1. Each element is made up of tiny, individual particles called atoms.
2. Atoms are indivisible; they cannot be created or destroyed.
3. All atoms of each element are identical in every respect.
4. Atoms of one element are different from atoms of any other element.
5. Atoms of one element may combine with atoms of other elements, usually in the ratio of small whole numbers, to form chemical compounds.

Dalton's theory accounts for chemical reactions in this way: Before the reaction, the reacting substances contain a certain number of atoms of different elements. As the reaction proceeds, the atoms are rearranged to form the products. The atoms are nei-

Figure 5.1 Atoms according to Dalton's atomic theory.

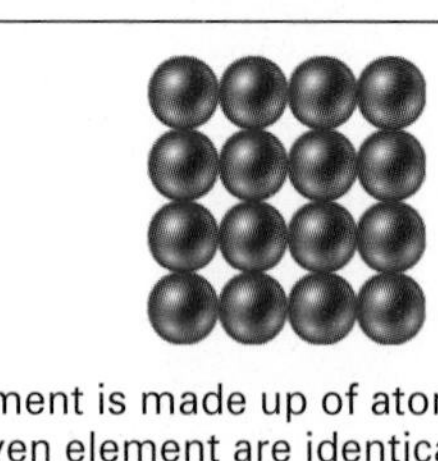

An element is made up of atoms. All atoms of a given element are identical.

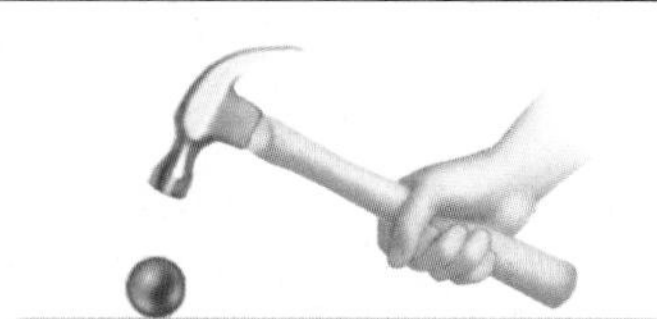

Atoms cannot be destroyed or created.

All atoms of one element have the same mass. Atoms of two different elements have different masses.

Atoms of different elements may combine in the ratio of small, whole numbers to form compounds.

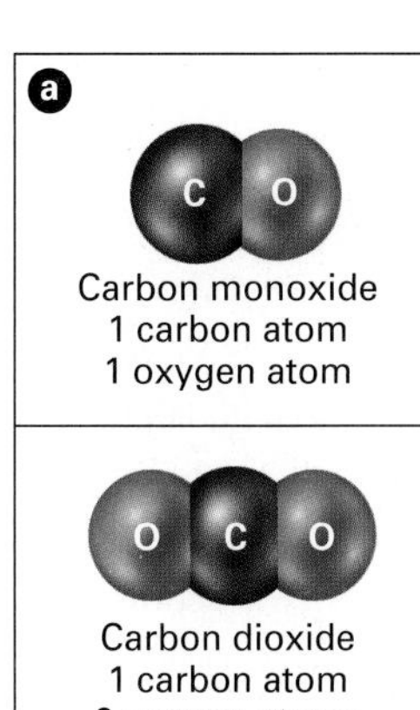


Figure 5.2 Explanation of the Law of Multiple Proportions. Carbon and oxygen combine to form more than one compound. Assume, according to the atomic theory, that carbon and oxygen atoms combine in a 1:1 ratio to form one compound and in a 1-carbon-to-2-oxygen-atom ratio to form another compound. If this assumption at the particulate level is true, the mass of the two oxygen atoms in the 1:2 compound is twice the mass of the one oxygen atom in the 1:1 compound. The oxygen mass ratio is 2:1 in the two compounds with the fixed mass of one carbon atom. This analysis at the particulate level predicts a similar result at the macroscopic level, where masses of combining elements can be measured. Going to the laboratory, we find that (a) 1.0 gram of carbon combines with 1.3 grams of oxygen to form carbon monoxide, CO, and (b) 1.0 gram of carbon combines with 2.6 grams of oxygen to form carbon dioxide, $CO_2$. The macroscopic mass ratio of oxygen that combines with 1.0 gram of carbon in the two compounds is 2.6/1.3, which reduces to 2/1, exactly as predicted by the atomic theory at the particulate level.

ther created nor destroyed, but simply arranged differently. The starting arrangement is destroyed (reactants are destroyed in a chemical change), and a new arrangement is formed (new substances form).

As with many new ideas, Dalton's theory was not immediately accepted. However, it led to a prediction that must be true if the theory is correct. This is now known as the **Law of Multiple Proportions.** It states that when two elements combine to form more than one compound, the different weights of one element that combine with the same weight of the other element are in a simple ratio of whole numbers (Fig. 5.2). This is like threading one, two, or three identical nuts onto the same bolt. The mass of the bolt is constant. The mass of two nuts is twice the mass of one; the mass of three nuts is three times the mass of one. The masses of nuts are in a simple ratio of whole numbers, 1:2:3.

The multiple proportion prediction can be confirmed by experiment. Using a theory to predict something unknown and having the prediction confirmed is convincing evidence that the theory is correct. With supporting evidence such as this, Dalton's atomic theory was accepted.

Michael Faraday (1791–1867)

William Crookes (1832–1919)

√ **Target Check 5.1**

*In an experiment to confirm the Law of Multiple Proportions, a scientist finds that sulfur and oxygen form two different compounds. In each experiment, 1.0 g of sulfur is allowed to react with oxygen. Under one set of conditions, 0.5 g of oxygen reacts, and under another set of conditions, 1.0 g of oxygen reacts. Do these data confirm the Law of Multiple Proportions? Explain.*

## 5.2 Subatomic Particles

Goal 3 Identify the three major subatomic particles by symbol, charge, and approximate atomic mass, expressed in atomic mass units.

Despite the general acceptance of Dalton's atomic theory, it was soon challenged. As early as the 1820s, laboratory experiments suggested that the atom contains even smaller parts, or **subatomic particles.** The brilliant works of Michael Faraday and William Crookes, among others, led to the discovery of the **electron,** but it was not until 1897 that J. J. Thomson described some of its properties. The electrical charge on an electron has been assigned a value of 1−. The mass of an electron is extremely small, $9.109 \times 10^{-28}$ gram ($3.213 \times 10^{-29}$ oz).

The second subatomic particle, the **proton,** was isolated and identified in 1919 by Ernest Rutherford. Its mass is 1836 times greater than the mass of an electron. The proton carries a 1+ charge, equal in size but opposite in sign to the negative charge of the electron. A third particle, the **neutron,** was discovered by James Chadwick in 1932. As

**P/REVIEW:** Some of the electrical properties of matter (Section 2.7) were known in Dalton's day, but there was no explanation for them. Dalton's theory did not account for them. Faraday and Crookes opened the door that led to understanding electricity in terms of parts of atoms.

**Table 5.1** Subatomic Particles

| Subatomic Particle | Symbol | Fundamental Charge | MASS Grams | u* | Location | Discovered |
|---|---|---|---|---|---|---|
| Electron | $e^-$ | 1− | $9.109 \times 10^{-28}$ | $0.000549 \approx 0$ | Outside nucleus | 1897 by Thomson |
| Proton | p or $p^+$ | 1+ | $1.673 \times 10^{-24}$ | $1.00728 \approx 1$ | Inside nucleus | 1919 by Rutherford |
| Neutron | n or $n^0$ | 0 | $1.675 \times 10^{-24}$ | $1.00867 \approx 1$ | Inside nucleus | 1932 by Chadwick |

*An atomic mass unit (u) is a very small unit of mass, $1.66 \times 10^{-24}$ g, used for atomic-sized particles. One atom of carbon-12 has a mass of exactly 12 u.

AIP Emilio Segre Visual Archives, Bainbridge Collection

J. J. Thomson (1856–1940) (*left*) and Ernest Rutherford (1871–1937) (*right*)

its name suggests, it is electrically neutral. The mass of a neutron is slightly more than the mass of a proton. Masses of atoms and masses of parts of atoms are often expressed in atomic mass units (Section 5.5).

Atoms of all elements are made up of electrons, protons, and neutrons. Today we know that protons and neutrons are made up of even smaller particles, but only the three described here are important in this course. Table 5.1 summarizes the properties of subatomic particles.

### √ Target Check 5.2

*Identify the true statements and rewrite the false statements to make them true.*

a) The subatomic particles of an atom are electrons, protons, and neutrons.
b) The mass of an electron is less than the mass of a proton.
c) The mass of a proton is about 1 g.
d) Electrons, protons, and neutrons are electrically charged.

## 5.3 The Nuclear Atom

**Goal** 4 Describe and/or interpret the Rutherford scattering experiments and the nuclear model of the atom.

In 1910 and 1911 Ernest Rutherford, Hans Geiger, and Ernst Marsden performed a series of experiments usually referred to as the **Rutherford experiments.** Figure 5.3 illustrates the experimental apparatus that they used. A narrow beam of positively charged alpha particles (helium atoms stripped of their negatively charged electrons) from a radioactive source was directed at a very thin gold foil. Most of the particles passed right through the foil, striking a fluorescent screen and causing it to glow. Some particles were deflected through moderate angles. The larger deflections were surprises, but the 0.001% of the total that were reflected at acute angles were totally unexpected. Similar results were observed using other metal foils.

The right side of Figure 5.3 illustrates the interpretation of the Rutherford experiments. The atoms are modeled as consisting mostly of open space. At the center is a tiny and extremely dense **nucleus** that contains all of the atom's positive charge and nearly all of its mass. The electrons are thinly distributed throughout the open space. Most of the positively charged alpha particles pass through the open space undeflected, not coming near any positively charged gold nuclei. The few alpha particles that pass fairly close to a nucleus are repelled by positive–positive electrostatic forces and thereby deflected. The very few particles that are on a collision course with gold nuclei are repelled backward at acute angles. Calculations based on the results of the experiments indicated that the diameter of the open-space portion of the atom is about 10,000 times greater than the diameter of the nucleus.

SPL/Photo Researchers, Inc.

James Chadwick (1891–1974)

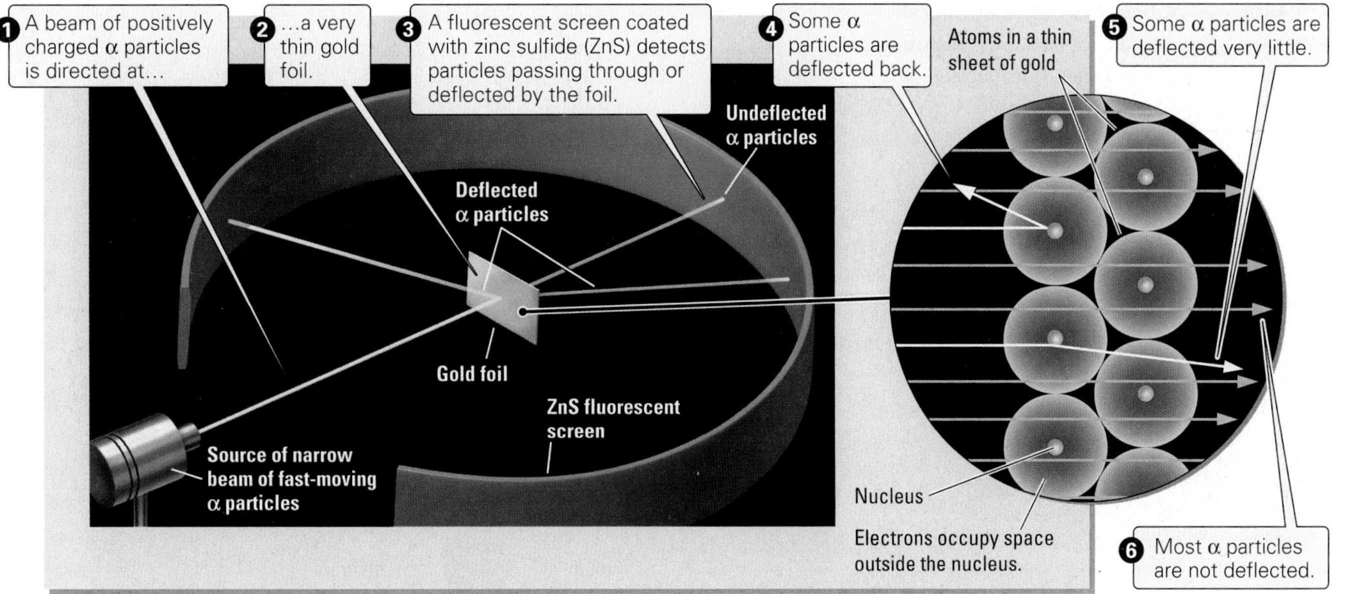


Figure 5.3 Rutherford scattering experiments. A narrow beam of positively charged alpha particles (helium atoms stripped of their negatively charged electrons) from a radioactive source was directed at a very thin gold foil. Most of the particles passed right through the foil, striking a fluorescent screen and causing it to glow (❻). Some particles were deflected through moderate angles (❺). The larger deflections were surprises, but the 0.001% of the total that were reflected at acute angles (❹) were totally unexpected. Similar results were observed using other metal foils.

The data from the Rutherford experiments led to the following conclusions about what is now called the **nuclear model of the atom:**

summary

**The Nuclear Model of the Atom**

1. Every atom contains an extremely small, extremely dense nucleus.
2. All of the positive charge and nearly all of the mass of an atom are concentrated in the nucleus.
3. The nucleus is surrounded by a much larger volume of nearly empty space that makes up the rest of the atom.
4. The space outside the nucleus is very thinly populated by electrons, the total charge of which exactly balances the positive charge of the nucleus.

The emptiness of the atom can be difficult to visualize. Look at the nucleus illustrated in Figure 5.4. If the nucleus of an atom were that size, the diameter of an atom would be the length of a football field. Between this nucleus and its nearest neighboring nucleus would be almost nothing—only a small number of electrons of negligible size and mass. If it were possible to eliminate all of this nearby empty space and pack nothing but nuclei into a sphere the size of a period on this page, that sphere could, for some elements, weigh as much as a million tons!

When protons and neutrons were discovered, scientists concluded that these relatively massive particles make up the nucleus of the atom. In Rutherford's time, it was natural to wonder what electrons did in the vast open space they occupied. The most widely held opinion was that they traveled in circular orbits around the nucleus, much as planets move in orbits around the sun. The atom would then have the character of a miniature solar system. This is called the **planetary model of the atom.**

Figure 5.4 Relative sizes of an atom and its nucleus. The diameter of an atom is approximately 10,000 times the diameter of its nucleus.

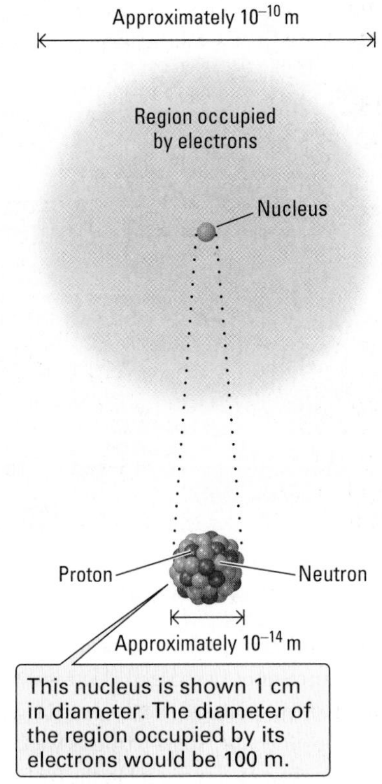


## Thinking About Your Thinking

### Mental Models

**The preceding paragraph about the emptiness of the atom was written to help you form a mental model of the atom. Take a moment to mentally picture the vast amount of empty space that makes up most of an atom's volume.**

**P/REVIEW:** In Chapter 10 you will examine the planetary model of the atom more closely and find out why it is wrong.

### √ Target Check 5.3

*Identify the true statements, and rewrite the false statements to make them true.*

a) Atoms are like small, hard spheres.

b) An atom is electrically neutral.

c) An atom consists mostly of empty space.

## 5.4 Isotopes

**Goal 5** Explain what isotopes of an element are and how they differ from each other.

**Goal 6** For an isotope of any element whose chemical symbol is known, given one of the following, state the other two: (a) nuclear symbol, (b) number of protons and neutrons, (c) atomic number and mass number.

**Goal 7** Identify the features of Dalton's atomic theory that are no longer considered valid, and explain why.

More than 100 years after Dalton's atomic theory was first suggested, another of its features was shown to be incorrect. All atoms of an element are not identical. Some atoms have more mass than other atoms of the same element.

We now know that every atom of a particular element has the same number of protons. This number is called the **atomic number** of the element. It is represented by the symbol **Z.** Atoms are electrically neutral, so the number of electrons must be the same as the number of protons. It follows that the total contribution to the mass of an atom from protons and electrons is the same for every atom of the element. That leaves neutrons. We conclude that the mass differences between atoms of an element must be caused by different numbers of neutrons. **Atoms of the same element that have different masses—different numbers of neutrons—are called isotopes.**

An isotope is identified by its **mass number, A, the total number of protons and neutrons in the nucleus:**

$$\begin{aligned} \text{Mass number} &= \text{number of protons} + \text{number of neutrons} \\ A &= Z + \text{number of neutrons} \end{aligned} \qquad \textbf{(5.1)}$$

The name of an isotope is its elemental name followed by its mass number. Thus an oxygen atom that has 8 protons and 8 neutrons has a mass number of 8 + 8, or 16, and its name is "oxygen sixteen." It is written "oxygen-16."

An isotope is represented by a **nuclear symbol** that has the form

$$^{\text{Number of protons + number of neutrons}}_{\text{number of protons}}\text{Sy} \quad \text{or} \quad ^{\text{mass number}}_{\text{atomic number}}\text{Sy} \quad \text{or} \quad ^{A}_{Z}\text{Sy}$$

Sy is the chemical symbol of the element. The symbol and the mass number are actually all we need to identify an isotope, so the atomic number, Z, is sometimes omitted. The symbol for oxygen-16 is $^{16}_{8}O$ or $^{16}O$.

Two natural isotopes of carbon are $^{12}_{6}C$ and $^{13}_{6}C$, carbon-12 and carbon-13. From the name and symbol of the isotopes and from Equation 5.1, you can find the number of neutrons in each nucleus. In carbon-12, if you subtract the atomic number (protons) from the mass number (protons + neutrons), you get the number of neutrons:

$$\begin{aligned} \text{Mass number} &= \text{protons} + \text{neutrons} = 12 \\ -\ \text{atomic number} &= -\ \text{protons} = -6 \\ \hline \text{neutrons} &= 6 \end{aligned}$$

In carbon-13 there are 7 neutrons: 13 − 6 = 7.

You can find the mass number and nuclear symbol of an isotope from the number of protons and neutrons. A nucleus with 12 protons and 14 neutrons has the atomic number 12, the same as the number of protons. From Equation 5.1, the mass number is 12 + 14 = 26. The symbol of the element may be found by searching for 12 in the atomic number column in the list of elements on the Reference Page. It is easier to find in the periodic table on your shield. The number at the top of each box is the atomic number. The elemental symbol corresponding to Z = 12 is Mg for magnesium. The isotope is therefore magnesium-26, and its nuclear symbol is $^{26}_{12}Mg$.

**P/REVIEW:** The chemical symbols of the elements (Section 2.6) are shown in the alphabetical list of elements on the Reference Page. They also appear on the periodic table on your shield cards.

## Active Example 5.1

Fill in all of the blanks in the following table. Use the table of elements on the Reference Page and Equation 5.1 for needed information. The number at the top of each box in the periodic table is the atomic number of the element whose symbol is in the middle of the box.

We'll help you work through the first line of the table. The nuclear symbol in the first line gives you three pieces of information: the mass number, A, the atomic number, Z, and the elemental symbol. Fill in those three boxes for just the Barium line.

*Learn It Now!* We cannot overemphasize the value of answering the question yourself before you look at the answer in the book. This is the most important of all the *Learn It Now!* Statements.

*continued*

| Name of Element | Elemental Symbol | Atomic Number, Z | Number of Protons | Number of Neutrons | Mass Number, A | Nuclear Symbol | Name of Isotope |
|---|---|---|---|---|---|---|---|
| Barium | | | | | | $^{138}_{56}Ba$ | |
| Oxygen | | | | 10 | | | |
| | | 82 | | | 206 | | |
| | | | | | | | zinc-66 |

| Name of Element | Elemental Symbol | Atomic Number, Z | Number of Protons | Number of Neutrons | Mass Number, A | Nuclear Symbol | Name of Isotope |
|---|---|---|---|---|---|---|---|
| Barium | Ba | 56 | | | 138 | $^{138}_{56}Ba$ | |

The name of the isotope is the elemental name, followed by a hyphen and then the mass number. Also, the atomic number is the number of protons. Fill in those two boxes.

| Name of Element | Elemental Symbol | Atomic Number, Z | Number of Protons | Number of Neutrons | Mass Number, A | Nuclear Symbol | Name of Isotope |
|---|---|---|---|---|---|---|---|
| Barium | Ba | 56 | 56 | | 138 | $^{138}_{56}Ba$ | barium-138 |

One box remains, number of neutrons. The mass number is the total number of protons plus neutrons. You know both the mass number and the number of protons. Complete the final box in the Barium line.

| Name of Element | Elemental Symbol | Atomic Number, Z | Number of Protons | Number of Neutrons | Mass Number, A | Nuclear Symbol | Name of Isotope |
|---|---|---|---|---|---|---|---|
| Barium | Ba | 56 | 56 | 82 | 138 | $^{138}_{56}Ba$ | barium-138 |

$138 - 56 = 82.$

Fill in the remaining blanks; complete the table above.

| Name of Element | Elemental Symbol | Atomic Number, Z | Number of Protons | Number of Neutrons | Mass Number, A | Nuclear Symbol | Name of Isotope |
|---|---|---|---|---|---|---|---|
| Barium | Ba | 56 | 56 | 82 | 138 | $^{138}_{56}Ba$ | barium-138 |
| Oxygen | O | 8 | 8 | 10 | 18 | $^{18}_{8}O$ | oxygen-18 |
| Lead | Pb | 82 | 82 | 124 | 206 | $^{206}_{82}Pb$ | lead-206 |
| Zinc | Zn | 30 | 30 | 36 | 66 | $^{66}_{30}Zn$ | zinc-66 |

The atomic number, Z, and the number of protons are the same in each case, by definition. Also by definition, the mass number, A, is equal to the sum of the number of protons and the number of neutrons. For Z = 82, you must search for 82 in the atomic number column in the table of elements to identify the element as lead.

## √ Target Check 5.4

*Identify the true statements, and rewrite the false statements to make them true.*

a) The atomic number of an element is the number of protons in its nucleus.

b) All atoms of a specific element have the same number of protons.

c) The difference between isotopes of an element is a difference in the number of neutrons in the nucleus.

d) The mass number of an atom is always equal to or larger than the atomic number.

# 5.5 Atomic Mass

Goal 8 Define and use the atomic mass unit (u).

Goal 9 Given the relative abundances of the natural isotopes of an element and the atomic mass of each isotope, calculate the atomic mass of the element.

The mass of an atom is very small—much too small to be measured on a balance. Nevertheless, early chemists did find ways to isolate samples of elements that contained the same number of atoms. These samples were weighed and compared. The ratio of the masses of equal numbers of atoms of different elements is the same as the ratio of the masses of individual atoms. From this, a scale of relative atomic masses was developed. Chemists didn't know about isotopes at that time, so they applied the idea of what was then called "atomic weight" to all of the natural isotopes of an element.

**P/REVIEW:** Equal volumes of gases at the same temperature and pressure have the same number of molecules (Section 14.2).

Today we recognize that samples of most pure elements contain atoms that have different masses. By worldwide agreement, the masses of atoms are expressed in atomic mass units (u). **The atomic mass unit is defined as exactly $\frac{1}{12}$ the mass of a carbon-12 atom:**

$$1\ \text{u} \equiv \frac{1}{12}\ \text{the mass of one carbon-12 atom}$$

**P/REVIEW:** The relationship between the u and the gram is derived from the SI unit for the amount of substance, called the mole. Section 7.3 describes the mole in detail.

$$\frac{12\ \text{g}\ ^{12}\text{C}}{\text{mol}\ ^{12}\text{C}} \times \frac{1\ \text{mol}\ ^{12}\text{C}}{6.02 \times 10^{23}\ ^{12}\text{C atoms}} \times \frac{\frac{1}{12}\ ^{12}\text{C atom}}{1\ \text{u}} = \frac{1}{6.02 \times 10^{23}}\ \text{g/u} = 1.66 \times 10^{-24}\ \text{g/u}$$

Multiplying both sides by 12, it follows that the mass of one carbon-12 atom is 12 u. Since the mass of an electron is nearly zero and a carbon-12 atom has a total of 6 protons plus 6 neutrons, or 12 subatomic particles in the nucleus, both protons and neutrons have atomic masses very close to 1 u (see Table 5.1). We can also think of the u as being simply another mass or weight unit, like the gram, kilogram, or pound. To three significant figures, the relationship between the u and the gram is

$$1\ \text{u} = 1.66 \times 10^{-24}\ \text{g}$$

Samples of most pure elements consist of two or more isotopes, atoms that have different masses because they have varying numbers of neutrons. The u is used to define the **atomic mass of an element: the average mass of all atoms of an element as they occur in nature.** Keep in mind that atomic mass is always measured relative to the mass of an atom of carbon-12, which is defined as 12 u.

To find the (average) atomic mass of an element, you must know the atomic mass of each isotope and the fraction of each isotope in a sample (Fig. 5.5). Fortunately, that fraction is constant for all elements as they occur in nature. Table 5.2 gives the percentage abundance of the natural isotopes of some common elements. The following Active Example shows you how to calculate the atomic mass of an element from these data.

## Active Example 5.2

The natural distribution of the isotopes of boron is 19.9% $^{10}_{5}\text{B}$ at a mass of 10.01294 u and 80.1% $^{11}_{5}\text{B}$ at 11.00931 u. Calculate the atomic mass of boron.

*continued*

**Figure 5.5** Mass spectrum of neon (1+ ions only). The relative abundance plotted on the y-axis is the percentage of each isotope found in a natural sample of the pure element. The mass-to-charge ratio, m/z, is plotted on the x-axis. Since this plot shows 1+ ions only, m/z is the same as the mass of the isotope, expressed in u. Neon contains three isotopes, of which neon-20 is by far the most abundant (90.5%). The mass of that isotope is 19.9924356 u.

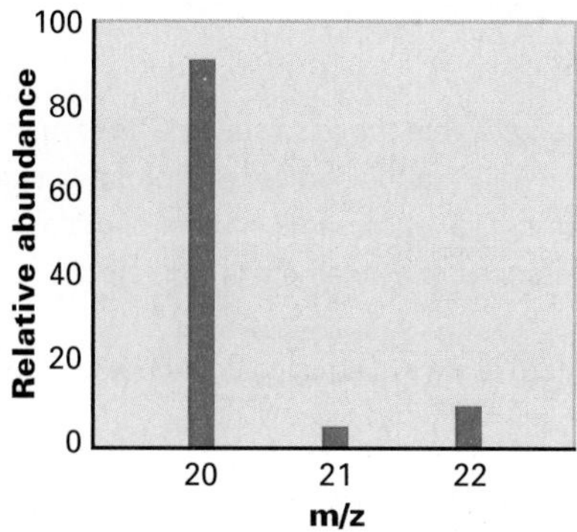


**Table 5.2** Percentage Abundance of Some Natural Isotopes

See notes for set-up for the problems and now to solve →

| Symbol | Mass (u) | Percentage | Symbol | Mass (u) | Percentage |
|---|---|---|---|---|---|
| $^{1}_{1}H$ | 1.007825035 | 99.9885 | $^{19}_{9}F$ | 18.99840322 | 100 |
| $^{2}_{1}H$ | 2.014101779 | 0.0115 | $^{32}_{16}S$ | 31.97207070 | 94.93 |
| $^{3}_{2}He$ | 3.01602931 | 0.000137 | $^{33}_{16}S$ | 32.97145843 | 0.76 |
| $^{4}_{2}He$ | 4.00260324 | 99.999863 | $^{34}_{16}S$ | 33.96786665 | 4.29 |
| $^{12}_{6}C$ | 12 (exactly) | 98.93 | $^{36}_{16}S$ | 35.96708062 | 0.02 |
| $^{13}_{6}C$ | 13.003354826 | 1.07 | $^{35}_{17}Cl$ | 34.968852721 | 75.78 |
| $^{14}_{7}N$ | 14.003074002 | 99.632 | $^{37}_{17}Cl$ | 36.96590262 | 24.22 |
| $^{15}_{7}N$ | 15.00010897 | 0.368 | $^{39}_{19}K$ | 38.9637074 | 93.2581 |
| $^{16}_{8}O$ | 15.99491463 | 99.757 | $^{40}_{19}K$ | 39.9639992 | 0.0117 |
| $^{17}_{8}O$ | 16.9991312 | 0.038 | $^{41}_{19}K$ | 40.9618254 | 6.7302 |
| $^{18}_{8}O$ | 17.9991603 | 0.205 | | | |

19.9% $^{10}_{5}B$ 10.01294 u
80.1% $^{11}_{5}B$ 11.00931 u

The "average" boron atom consists of 19.9% of an atom with mass 10.01294 u. Therefore, it contributes 0.199 × 10.01294 u = 1.99 u to the mass of the average boron atom. Perform a similar calculation for the other isotope.

0.801 × 11.00931 u = 8.82 u

You are multiplying a three significant figure fractional abundance by a seven significant figure mass, so, by following the multiplication and division rule for significant figures, three significant figures are justified in the solution.

To calculate the mass of an average atom of boron, which is its atomic mass, simply add together the contributions from each of the isotopes. Don't forget to consider significant figures when you express the sum.

1.99 u + 8.82 u = 10.81 u

Both masses are known to the hundredths place, so the sum of their masses is expressed to the hundredths place according to the addition and subtraction rule for significant figures. Note that we gained a significant figure in the addition process because we ended with two digits to the left of the decimal point in the sum.

The currently accepted value of the atomic mass of boron is 10.81 u, which matches your calculated value to four significant figures.

One solution of Active Example 5.2 on a calculator gives 10.81103237 u as the answer. Notice that your calculator does not tell you the column to which the result should be rounded off. You must determine that yourself by applying the rules of significant figures to each step in the problem.

**Atomic-Theory-Before-Calculations Option** The rules of significant figures are given in Section 3.5. If you have not yet studied Chapter 3, follow the directions of your instructor regarding significant figures.

Now you try an atomic mass calculation.

### Active Example 5.3

Calculate the atomic mass of potassium (symbol K), using data from Table 5.2.

$$
\begin{aligned}
0.932581 \times 38.9637074 \text{ u} &= 36.3368 \text{ u} \\
0.000117 \times 39.9639992 \text{ u} &= 0.00468 \text{ u} \\
0.067302 \times 40.9618254 \text{ u} &= 2.7568 \text{ u} \\
\hline
& \phantom{=} 39.0983 \text{ u}
\end{aligned}
$$

The accepted value of the atomic mass of potassium is 39.0983 u.

## 5.6 The Periodic Table

**Goal 10** Distinguish between groups and periods in the periodic table and identify them by number.

**Goal 11** Given the atomic number of an element, use a periodic table to find the symbol and atomic mass of that element, and identify the period and group in which it is found.

**Goal 12** Given an elemental symbol or information from which it can be identified, classify the element as either a main group or transition element and either a metal or nonmetal.

During the time of early research on the atom, even before any subatomic particles were identified, some chemists searched for an order among elements. In 1869, two men found an order, independently of each other. Dmitri Mendeleev and Lothar Meyer observed that when elements are arranged according to their atomic masses, certain properties repeat at regular intervals.

Mendeleev and Meyer arranged the elements in tables so that elements with similar properties were in the same column or row. These were the first **periodic tables** of the elements. The arrangements were not perfect. For all elements to fall into the proper groups, it was necessary to switch a few of them in a way that interrupted the orderly increase in atomic masses. Of the two reasons for this, one was anticipated at that time: There were errors in atomic weights (as they were known in 1869). The

Dmitri Mendeleev (1834–1907)

E. F. Smith Memorial Collection in the History of Chemistry, Department of Special Collections, Van Pelt Dietrich Library, University of Pennsylvania

Lothar Meyer (1830–1895)

**Table 5.3** Predicted and Observed Properties of Germanium

| Property | Predicted by Mendeleev | Currently Accepted Values |
|---|---|---|
| Atomic weight | 72 g/mol* | 72.60 g/mol* |
| Density of metal | 5.5 g/cm$^3$ | 5.36 g/cm$^3$ |
| Color of metal | Dark gray | Gray |
| Formula of oxide | $GeO_2$ | $GeO_2$ |
| Density of oxide | 4.7 g/cm$^3$ | 4.703 g/cm$^3$ |
| Formula of chloride | $GeCl_4$ | $GeCl_4$ |
| Density of chloride | 1.9 g/cm$^3$ | 1.887 g/cm$^3$ |
| Boiling point of chloride | Below 100°C | 86°C |
| Formula of ethyl compound | $Ge(C_2H_5)_4$ | $Ge(C_2H_5)_4$ |
| Boiling point of ethyl compound | 160°C | 160°C |
| Density of ethyl compound | 0.96 g/cm$^3$ | Slightly less than 1.0 g/cm$^3$ |

*Mol is the abbreviation for mole, the SI unit for amount of substance. The mole is introduced in Chapter 7.

second reason was more important. About 50 years later, it was found that the correct ordering property is the atomic number, Z, rather than the atomic mass.

We have seen how Dalton used his atomic theory to predict the Law of Multiple Proportions and thus gain acceptance for his theory. Mendeleev did the same with the periodic table. He noticed that there were blank spaces in the table. He reasoned that the blank spaces belonged to elements that were yet to be discovered. By averaging the properties of elements above and below or on each side of the blanks, he predicted the properties of the unknown elements. Germanium is one of the elements about which he made these predictions. Table 5.3 summarizes the predicted properties and their currently accepted values.

## Thinking About Your Thinking

### Classification

**There are many ways to classify a collection of objects. Different criteria satisfy different purposes. Mendeleev and Meyer arranged their periodic tables based on two criteria: The atomic masses of the elements increased across a row and the chemical properties of the elements in a column were similar. Mendeleev is more famous than Meyer as the founder of the periodic table because he subsequently used his classification scheme to make predictions about unknown elements. His predictions were later found to be true.**

**When you formulate a classification scheme, its power is that it allows you to fill in gaps in your knowledge, just as Mendeleev filled in gaps in his periodic table. If you can deduce a classification pattern, you can interpolate and extrapolate to make predictions about unknown things or events in the future. (Interpolation is predicting something within the range of your data. Extrapolation is predicting something beyond the range of your data.)**

**When you practice understanding classification schemes made up by others, such as the periodic table in chemistry, it helps you develop your skill in formulating your own classifications in all aspects of your life.**

**P/REVIEW:** Some atomic masses of elements in the periodic table are in parentheses. These elements are radioactive, and there is no atomic mass in the sense that we have defined it. Instead, parentheses enclose the mass number of the most stable isotope. Radioactivity is discussed in Chapter 20.

The amazing accuracy of Mendeleev's predictions showed that the periodic table made sense, but nobody knew why; that came later. The reason for the shape of the table is explained in Chapter 11.

Figure 5.6 is a modern periodic table. It also appears on your shield. You will find yourself referring to the periodic table throughout your study of chemistry. This is why a periodic table is printed on the shields provided for working Active Examples.

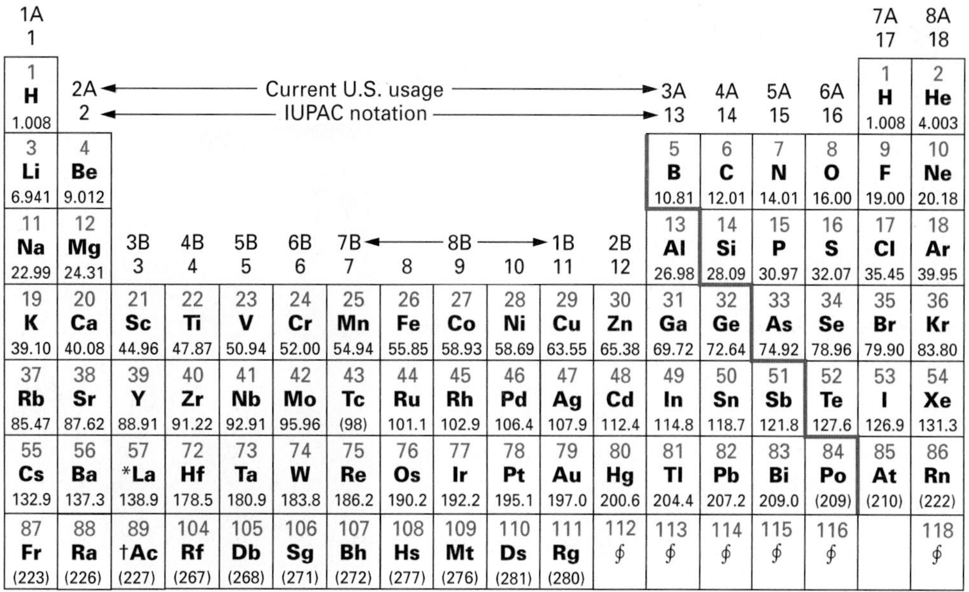

| 1A 1 | | | | | | | | | | | | | | | | 7A 17 | 8A 18 |
|---|---|---|---|---|---|---|---|---|---|---|---|---|---|---|---|---|---|
| 1 **H** 1.008 | 2A ← 2 ← | | | | | Current U.S. usage | | IUPAC notation | | | | → 3A → 13 | 4A 14 | 5A 15 | 6A 16 | 1 **H** 1.008 | 2 **He** 4.003 |
| 3 **Li** 6.941 | 4 **Be** 9.012 | | | | | | | | | | | 5 **B** 10.81 | 6 **C** 12.01 | 7 **N** 14.01 | 8 **O** 16.00 | 9 **F** 19.00 | 10 **Ne** 20.18 |
| 11 **Na** 22.99 | 12 **Mg** 24.31 | 3B 3 | 4B 4 | 5B 5 | 6B 6 | 7B ← 7 | 8B 8 | 8B 9 | 8B → 10 | 1B 11 | 2B 12 | 13 **Al** 26.98 | 14 **Si** 28.09 | 15 **P** 30.97 | 16 **S** 32.07 | 17 **Cl** 35.45 | 18 **Ar** 39.95 |
| 19 **K** 39.10 | 20 **Ca** 40.08 | 21 **Sc** 44.96 | 22 **Ti** 47.87 | 23 **V** 50.94 | 24 **Cr** 52.00 | 25 **Mn** 54.94 | 26 **Fe** 55.85 | 27 **Co** 58.93 | 28 **Ni** 58.69 | 29 **Cu** 63.55 | 30 **Zn** 65.38 | 31 **Ga** 69.72 | 32 **Ge** 72.64 | 33 **As** 74.92 | 34 **Se** 78.96 | 35 **Br** 79.90 | 36 **Kr** 83.80 |
| 37 **Rb** 85.47 | 38 **Sr** 87.62 | 39 **Y** 88.91 | 40 **Zr** 91.22 | 41 **Nb** 92.91 | 42 **Mo** 95.96 | 43 **Tc** (98) | 44 **Ru** 101.1 | 45 **Rh** 102.9 | 46 **Pd** 106.4 | 47 **Ag** 107.9 | 48 **Cd** 112.4 | 49 **In** 114.8 | 50 **Sn** 118.7 | 51 **Sb** 121.8 | 52 **Te** 127.6 | 53 **I** 126.9 | 54 **Xe** 131.3 |
| 55 **Cs** 132.9 | 56 **Ba** 137.3 | 57 ***La** 138.9 | 72 **Hf** 178.5 | 73 **Ta** 180.9 | 74 **W** 183.8 | 75 **Re** 186.2 | 76 **Os** 190.2 | 77 **Ir** 192.2 | 78 **Pt** 195.1 | 79 **Au** 197.0 | 80 **Hg** 200.6 | 81 **Tl** 204.4 | 82 **Pb** 207.2 | 83 **Bi** 209.0 | 84 **Po** (209) | 85 **At** (210) | 86 **Rn** (222) |
| 87 **Fr** (223) | 88 **Ra** (226) | 89 **†Ac** (227) | 104 **Rf** (267) | 105 **Db** (268) | 106 **Sg** (271) | 107 **Bh** (272) | 108 **Hs** (277) | 109 **Mt** (276) | 110 **Ds** (281) | 111 **Rg** (280) | 112 ∮ | 113 ∮ | 114 ∮ | 115 ∮ | 116 ∮ | | 118 ∮ |

| | | | | | | | | | | | | | | |
|---|---|---|---|---|---|---|---|---|---|---|---|---|---|---|
| *Lanthanides | 58 **Ce** 140.1 | 59 **Pr** 140.9 | 60 **Nd** 144.2 | 61 **Pm** (145) | 62 **Sm** 150.4 | 63 **Eu** 152.0 | 64 **Gd** 157.3 | 65 **Tb** 158.9 | 66 **Dy** 162.5 | 67 **Ho** 164.9 | 68 **Er** 167.3 | 69 **Tm** 168.9 | 70 **Yb** 173.1 | 71 **Lu** 175.0 |
| †Actinides | 90 **Th** 232.0 | 91 **Pa** 231.0 | 92 **U** 238.0 | 93 **Np** (237) | 94 **Pu** (244) | 95 **Am** (243) | 96 **Cm** (247) | 97 **Bk** (247) | 98 **Cf** (251) | 99 **Es** (252) | 100 **Fm** (257) | 101 **Md** (258) | 102 **No** (259) | 103 **Lr** (262) |

All atomic masses have been rounded to four significant figures. Atomic masses in parentheses are the mass number of the longest-lived isotope. *Sources:* IUPAC Standard Atomic Weights 2005 and IUPAC Atomic Weights Revised 2007.

∮ The International Union for Pure and Applied Chemistry has not adopted official names or symbols for these elements.

**Figure 5.6** Periodic table of the elements.

The number at the top of each box in our periodic table is the atomic number of the element. The chemical symbol is in the middle, and the atomic mass, rounded to four significant figures, is at the bottom (Fig. 5.7). The boxes are arranged in horizontal rows called **periods.** Periods are numbered from top to bottom, but the numbers are not usually printed. Periods vary in length: The first period has two elements; the second and third have eight elements each; and the fourth and fifth have 18. Period 6 has 32 elements, including atomic numbers 58 to 71, which are printed separately at the bottom to keep the table from becoming too wide. Period 7 also theoretically has 32 elements, but one element in Period 7 was not yet synthesized at the time this was written.

Elements with similar properties are placed in vertical columns called **groups** or **chemical families.** Groups are identified by two rows of numbers across the top of the table. The top row shows the group numbers commonly used in the United States.* European chemists use the same numbers, but a different arrangement of As and Bs. The International Union of Pure and Applied Chemistry (IUPAC) has approved a compromise that simply numbers the columns in order from left to right. This is the second row of numbers at the top of Figure 5.6.

We will use both sets of numbers, leaving it to your instructor to recommend which you should use. When we have occasion to refer to a group number, we will have the U.S. number first, followed by the IUPAC number after a slash. Thus the column headed by carbon, $Z = 6$, is Group 4A/14.

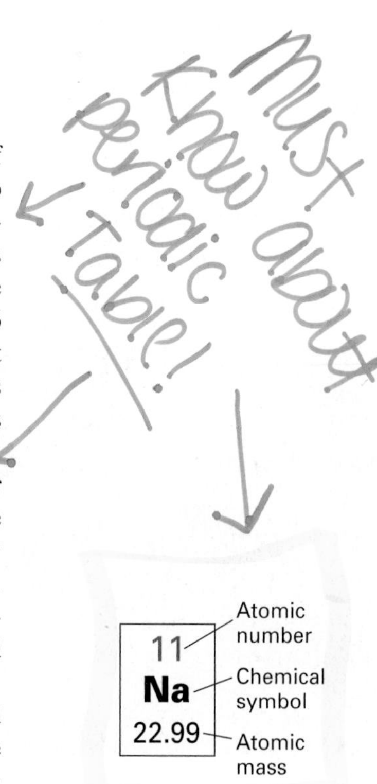


**Figure 5.7** Sample box from the periodic table, representing sodium.

*Roman numerals are sometimes used, for example, IIIA instead of 3A.

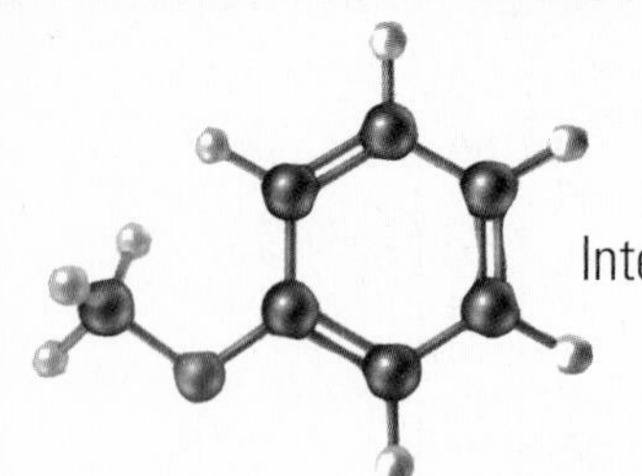

# Everyday Chemistry

## International Relations and the Periodic Table

When a new element is discovered in nature, it has been customary to allow the person who made the discovery to name the element. When an artificial element is newly synthesized, a similar custom has been followed. The person who leads the team of scientists working on the synthesis proposes the name of the new element. That is, this has been the custom until recently.

Because of the expense of the equipment and the highly specialized expertise of research personnel, at the present time only three laboratories in the world work on synthesizing new elements: the Lawrence Berkeley National Laboratory in the United States, the Society for Heavy-Ion Research in Germany, and the Joint Institute for Nuclear Research in Russia. All three laboratories are making similar progress, although they usually use slightly different methods to approach the same problem. You might be able to guess how this leads to conflict. More than once, two different labs have claimed to have synthesized a new element at about the same time. Who gets to choose the name when each lab claims that it was the first? Problems with international relations are not limited to politicians; chemists struggle with issues of national pride, too!

A source of great controversy was the naming of element 106. U.S. chemists endorsed the name seaborgium, in honor of the U.S. chemist Glenn Seaborg, who, over his career, led teams of scientists that synthesized 10 new elements. No person has ever equaled this achievement, so the Americans were confident that their proposal for the name of element 106 would easily gain acceptance from the worldwide scientific community. To their dismay, however, the International Union of Pure and Applied Chemistry (IUPAC) endorsed the name rutherfordium, in honor of Ernest Rutherford (see Section 5.3), for element 106. Moreover, the U.S. chemists were shocked by the IUPAC proposal that element 104 be named dubnium in honor of achievements at the research laboratory in Dubna, Russia. There were serious doubts as to the validity of the Russian chemists' data.

Lawrence Berkeley National Laboratory

Glenn Seaborg (1912–1999) was awarded the Nobel Prize in Chemistry in 1951 for his contributions leading to the discovery of many elements. In this photograph, he points to element 106, named seaborgium in honor of his work.

As with many political debates, the controversy was finally settled with a compromise. Element 106 was named seaborgium. Element 104 is now called rutherfordium and element 105 is dubnium. All parties got "their" name assigned to an element; there was just some shuffling over which element received which name. You can see that the common conception that science is divorced from emotion, governed only by cold logic, is a myth. Scientific disciplines are subject to the frailties and strengths of the human character in the same way as any other human endeavor.

Lawrence Berkeley National Laboratory

The Lawrence Berkeley National Laboratory is located in the hills just east of San Francisco Bay, next to the University of California, Berkeley campus.

Two other regions in the periodic table separate the elements into special classifications. Elements in the A groups (1, 2, and 13 to 18) are called **main group elements.** Main group elements are also known as **representative elements.** Similarly, elements in the B groups (3 to 12) are known as **transition elements,** or **transition metals.** The stair-step line that begins between atomic numbers 4 and 5 in Period 2 and ends between 84 and 85 in Period 6 separates the **metals** on the left from the **nonmetals** on the right. Chemical reasons for these classifications appear in Chapter 11.

The location of an element in the periodic table is given by its period and group numbers.

### Active Example 5.4

List the atomic number, chemical symbol, and atomic mass of the third-period element in Group 6A/16.

Z = 16; symbol, S; atomic mass, 32.07 u

In Group 6A/16, the third column from the right side of the table, you find Z = 16 in Period 3. The element is sulfur.

### √ Target Check 5.5

a) How many Group 3A/13 elements are metals?

b) How many Period 4 elements are transition metals?

## 5.7 Elemental Symbols and the Periodic Table

**Goal 13** Given the name or the symbol of an element in Figure 5.8, write the other.

The periodic table contains a large amount of information, far more than is indicated by the three items in each box. Its usefulness will become apparent gradually as your study progresses. In this section we will use the periodic table to learn the names and symbols of 35 elements. Learning symbols and names is much easier if you learn the location of the elements in the periodic table at the same time. Here's how to do it.

Part (a) of Figure 5.8, the table at the top, gives the name, symbol, and atomic number of the elements whose names and symbols you should learn.* Part (b) is a partial periodic table showing the atomic numbers and symbols of the same elements. Their names are listed in alphabetical order in the caption.

Study part (a) briefly. Try to learn the symbol that goes with each element, but don't spend more than a few minutes doing this. Then cover part (a) and look at the periodic table in part (b). Run through the symbols mentally and see how many elements you can name. If you can't name one, glance through the alphabetical list in the caption and see if it jogs your memory. If you still can't get it, note the atomic number and check part (a) for the elemental name. Do this a few times until you become fairly quick in naming most of the elements from their symbols.

Next, reverse the process with part (a) still covered. Look at the alphabetical list in the caption. For each name, mentally "write"—in other words, think—the symbol.

*Your instructor may require you to learn different names and symbols, or perhaps more or fewer. If so, follow your instructor's directions. If they include elements not among our 35, we recommend that you add them to both parts of Figure 5.8 and also to the caption.

Common Elements

| Atomic Number | Symbol | Element | Atomic Number | Symbol | Element | Atomic Number | Symbol | Element |
|---|---|---|---|---|---|---|---|---|
| 1 | H | Hydrogen | 13 | Al | Aluminum | 28 | Ni | Nickel |
| 2 | He | Helium | 14 | Si | Silicon | 29 | Cu | Copper |
| 3 | Li | Lithium | 15 | P | Phosphorus | 30 | Zn | Zinc |
| 4 | Be | Beryllium | 16 | S | Sulfur | 35 | Br | Bromine |
| 5 | B | Boron | 17 | Cl | Chlorine | 36 | Kr | Krypton |
| 6 | C | Carbon | 18 | Ar | Argon | 47 | Ag | Silver |
| 7 | N | Nitrogen | 19 | K | Potassium | 50 | Sn | Tin |
| 8 | O | Oxygen | 20 | Ca | Calcium | 53 | I | Iodine |
| 9 | F | Fluorine | 24 | Cr | Chromium | 56 | Ba | Barium |
| 10 | Ne | Neon | 25 | Mn | Manganese | 80 | Hg | Mercury |
| 11 | Na | Sodium | 26 | Fe | Iron | 82 | Pb | Lead |
| 12 | Mg | Magnesium | 27 | Co | Cobalt | | | |

(a)

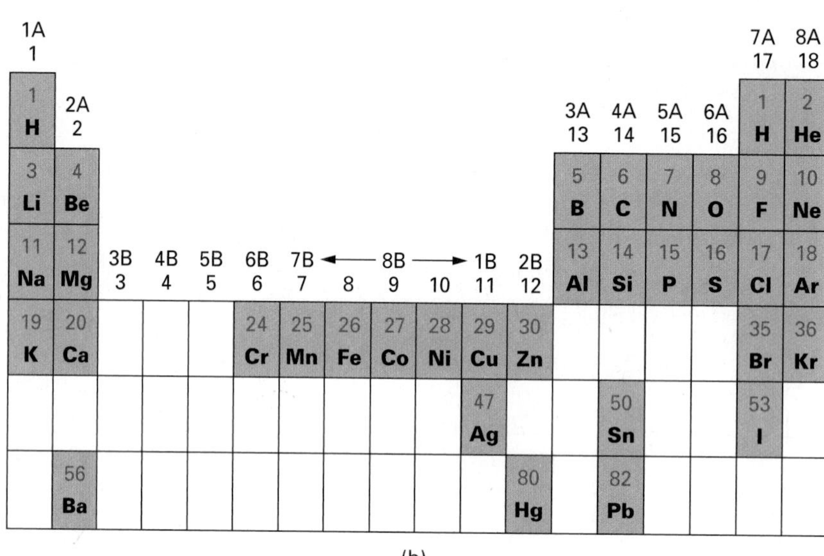


(b)

**Figure 5.8** (a) Table of common elements, with symbols and atomic numbers. (b) Partial periodic table showing the symbols and locations of the more common elements. The table in part (a) and the list below identify the elements that you should be able to recognize or write, referring only to a complete periodic table. Associating the names and symbols with the table makes learning them much easier. The elemental names are

| | | | | | | |
|---|---|---|---|---|---|---|
| aluminum | bromine | cobalt | iodine | magnesium | nitrogen | silver |
| argon | calcium | copper | iron | manganese | oxygen | sodium |
| barium | carbon | fluorine | krypton | mercury | phosphorus | sulfur |
| beryllium | chlorine | helium | lead | neon | potassium | tin |
| boron | chromium | hydrogen | lithium | nickel | silicon | zinc |

Glance up to the periodic table and find the element. Again, use part (a) as a temporary help only if necessary. Repeat the procedure several times, taking the elements in random order. Move in both directions, from name to symbol in the periodic table and from symbol to name.

When you feel reasonably sure of yourself, try the following Active Example. Do not refer to Figure 5.8. Instead, use only the more complete periodic table that is on your shield.

## Thinking About Your Thinking

### Memory

**The greater the number of mental pathways you can form to access memorized information, the more likely you are to recall it. We suggest that you memorize name–symbol pairs and the location of the element in the periodic table at the same time so that you have an additional mental pathway associated with each elemental symbol.**

### Active Example 5.5

For each elemental symbol listed, write the name; for each name, write the symbol: N, F, I, carbon, aluminum, copper. Use no reference other than the periodic table on your shield.

N, nitrogen; F, fluorine; I, iodine; carbon, C; aluminum, Al; copper, Cu.

Look closely at your symbols for aluminum and copper in Active Example 5.5. If you wrote AL or CU, the symbol is wrong. The letters are right, but the symbol is not. Whenever a chemical symbol has two letters, the first letter is always capitalized, but *the second letter is always written in lowercase,* or as a small letter. You can enjoy a long and happy life with a pile of Co in your house, but CO is a potentially serious problem in homes. Co is the metal cobalt, which is sometimes used in steel and pottery, among other things. CO is the deadly gas carbon monoxide, which is present in automobile exhaust and tobacco smoke. Carbon monoxide poisoning is the number one cause of accidental poisoning deaths in the world.

## Chapter 5 in Review

Most of the key terms and concepts and many others appear in the Glossary. Use your Glossary regularly.

### 5.1 Dalton's Atomic Theory

Goal 1 Identify the main features of Dalton's atomic theory.

Goal 2 State the meaning of, or draw conclusions based on, the Law of Multiple Proportions.

**Key Terms and Concepts: Atom, Dalton's atomic theory, Law of Multiple Proportions**

### 5.2 Subatomic Particles

Goal 3 Identify the three major subatomic particles by symbol, charge, and approximate atomic mass, expressed in atomic mass units.

**Key Terms and Concepts: Electron, neutron, proton, subatomic particles**

### 5.3 The Nuclear Atom

Goal 4 Describe and/or interpret the Rutherford scattering experiments and the nuclear model of the atom.

**Key Terms and Concepts: Nuclear model of the atom, nucleus, planetary model of the atom, Rutherford (scattering) experiments**

### 5.4 Isotopes

Goal 5 Explain what isotopes of an element are and how they differ from each other.

Goal 6 For an isotope of any element whose chemical symbol is known, given one of the following, state the other two: (a) nuclear symbol, (b) number of protons and neutrons, (c) atomic number and mass number.

Goal 7 Identify the features of Dalton's atomic theory that are no longer considered valid, and explain why.

**Key Terms and Concepts: Atomic number (Z), isotopes, mass number (A), nuclear symbol**

### 5.5 Atomic Mass

Goal 8 Define and use the atomic mass unit (u).

Goal 9 Given the relative abundances of the natural isotopes of an element and the atomic mass of each isotope, calculate the atomic mass of the element.

**Key Terms and Concepts: Atomic mass, atomic mass unit (u)**

### 5.6 The Periodic Table

Goal 10 Distinguish between groups and periods in the periodic table and identify them by number.

Goal 11 Given the atomic number of an element, use a periodic table to find the symbol and atomic mass of that element, and identify the period and group in which it is found.

Goal 12 Given an elemental symbol or information from which it can be identified, classify the element as either a main group or transition element and either a metal or nonmetal.

**Key Terms and Concepts: Group (in the periodic table) (or chemical family), main group element, metal, nonmetal, period (in the periodic table), periodic table, representative element, transition element (or metal)**

**5.7 Elemental Symbols and the Periodic Table**

Goal 13 Given the name or the symbol of an element in Figure 5.8, write the other.

## Study Hints and Pitfalls to Avoid

The value of the periodic table cannot be overstated. You will probably use it at first in Chapter 6 to learn the system of naming chemical compounds; after that, it will help you in many other ways. For example, you will need atomic masses to solve some of the problems in this course. The periodic table is a readily available source of these values. Other applications of the periodic table will appear as the course progresses. Be sure to locate elements in the periodic table while learning name–symbol pairs.

Students often have no trouble with questions about isotopes while studying Section 5.4. In particular, questions such as those posed in Active Example 5.1 are generally easy to answer at first. However, after studying Section 5.5, students may become confused about the difference between the mass of a specific isotope of an element and the average atomic mass of all of the naturally occurring isotopes of an element. For example, 12.01 u is the atomic mass of carbon, but it is neither the mass number of carbon-12 nor its mass. Additionally, you cannot use the atomic mass of an element to determine the number of neutrons in an isotope of that element. The number of neutrons is the difference between the mass number (A) of a specified isotope and the atomic number (Z) for the element.

The most common error while learning to write symbols and formulas is writing both letters in a two-letter elemental symbol as capitals. The first letter is always a capital letter. If a second letter is present, it is *always* written in lowercase. The language of chemistry is very precise, and correctly written symbols are part of that language. It is also important to learn the correct spelling of elemental names as you come to them. *Flourine* instead of *fluorine* is the most common misspelling of an elemental name.

## Concept-Linking Exercises

*Write a brief description of the relationships among each of the following groups of terms or phrases. Answers to the Concept-Linking Exercises are given after the answers to the Target Checks at the end of the book.*

1. Dalton's atomic theory, nuclear model of the atom, planetary model of the atom, Rutherford's scattering experiment
2. Electron, proton, neutron, subatomic particles
3. Atomic mass unit, carbon-12, gram
4. Isotopes, neutron, proton, mass number, atomic number, atomic mass
5. Periodic table, groups, periods
6. Main group elements, transition elements, transition metals, metals, nonmetals
7. Atomic mass of an element, atomic mass of an isotope, average atomic mass

## Small-Group Discussion Questions

*Small-Group Discussion Questions are for group work, either in class or under the guidance of a leader during a discussion section.*

1. Define the Law of Definite Composition and the Law of Conservation of Mass. Explain how these laws lead to Dalton's atomic theory. What is the Law of Multiple Proportions? How does it provide evidence to support the atomic model proposed by Dalton?
2. Fill in the blanks in the following table. Do not use any references.

| | Symbol | Charge Relative to an Electron at 1− | Mass Relative to a Proton at 1 u | Is it a Nuclear Particle? (Y/N) |
|---|---|---|---|---|
| Electron | | | | |
| Proton | | | | |
| Neutron | | | | |

3. If the nucleus of a single atom was scaled up to a diameter of 1 inch, what would be the diameter of an atom? Answer in USCS units.

4. When expressed to the full, unrounded correct number of significant figures, the atomic mass of zinc is 65.38 u and the atomic mass of nickel is 58.6934 u. Why do the number of significant figures vary among elements? Nickel has five naturally occurring isotopes, nickel-58, -60, -61, -62, and -64. Which is most abundant? How do you know? How many electrons, protons, and neutrons are in each nickel isotope?

5. A balloon is filled with nitrogen gas and placed in a refrigeration unit set at 2°C. The atmospheric pressure is 1 bar. The volume of the balloon is decreased until it is 1 liter. The balloon and its contents are weighted, and the gas is found to have a mass of 1.2 g. The experiment is repeated with oxygen, fluorine, and chlorine gases. What is the mass of the gas in each trial? Equal volumes of gases at the same pressure and temperature have equal numbers of molecules.

6. How many elements fall into each of the following categories? Use the periodic table on your shield when answering this question. Period 4, main group elements, metals, U.S. Group 8B, transition elements, Period 7, nonmetals, IUPAC Group 16, representative elements.

7. Construct a set of flash cards with the symbols of each of the 35 elements to be memorized on one side and their names on the other side. (Make more or fewer cards if your instructor requires you to memorize more or fewer elemental symbols.) Split into pairs and practice giving names for symbols and symbols for names until each person can repeat the entire set rapidly and accurately in both directions.

# Questions, Exercises, and Problems

*Questions whose numbers are printed in* blue *are answered at the end of the book.* ■ *denotes problems assignable in OWL. In the first section of Questions, Exercises, and Problems, similar exercises and problems are paired in consecutive odd-even number combinations.*

### Section 5.1: Dalton's Atomic Theory

1. According to Dalton's atomic theory, can more than one compound be made from atoms of the same two elements?

2. List the major points in Dalton's atomic theory.

3. Show that Dalton's atomic theory explains the Law of Definite Composition.

4. How does Dalton's atomic theory account for the Law of Conservation of Mass?

5. The chemical name for limestone, a compound of calcium, carbon, and oxygen, is calcium carbonate. When heated, limestone decomposes into solid calcium oxide and gaseous carbon dioxide. From the names of the products, tell where you might find the atoms of each element after the reaction. How does Dalton's atomic theory explain this?

6. The brilliance with which magnesium burns makes it ideal for use in flares and flashbulbs. Compare the mass of magnesium that burns with the mass of magnesium in the magnesium oxide ash that forms. Explain this in terms of atomic theory.

7. Sulfur and fluorine form at least two compounds—$SF_4$ and $SF_6$. Explain how these compounds can be used as an example of the Law of Multiple Proportions.

8. When 10.0 g of chlorine reacts with mercury under varying conditions, the reaction consumes either 28.3 g or 56.6 g of mercury. No other combinations occur. Explain these observations in terms of the Law of Multiple Proportions.

### Section 5.2: Subatomic Particles

9. Compare the three major parts of an atom in charge and mass.

10. ■ Which of the following applies to the electron? (a) Charge = 1−; (b) Mass ≈ 1 u; (c) Charge = 0; (d) Mass ≈ 0 u; (e) Charge = 1+.

### Section 5.3: The Nuclear Atom

11. How can we account for the fact that, in the Rutherford scattering experiment, some of the alpha particles were deflected from their paths through the gold foil, and some were even bounced back at various angles?

12. How can we account for the fact that most of the alpha particles in the Rutherford scattering experiment passed directly through a solid sheet of gold?

13. What do we call the central part of an atom?

14. What major conclusions were drawn from the Rutherford scattering experiment?

15. Describe the activity of electrons according to the planetary model of the atom that appeared after the Rutherford scattering experiment.

16. The Rutherford experiment was performed and its conclusions reached before protons and neutrons were discovered. When they were found, why was it believed that they were in the nucleus of the atom?

### Section 5.4: Isotopes

17. Can two different elements have the same atomic number? Explain.

18. Compare the number of protons and electrons in an atom, the number of protons and neutrons, and the number of electrons and neutrons.

19. Explain why isotopes of different elements can have the same mass number, but isotopes of the same element cannot.

20. ■ How many protons, neutrons, and electrons are there in a neutral atom of the isotope represented by $^{8}_{5}B$?

*Questions 21 and 22: From the information given in the following tables, fill in as many blanks as you can without looking at any reference. If there are unfilled spaces, continue by referring to your periodic table. As a last resort, check the table of elements on the Reference Page. All atoms are neutral.*

21.

| Name of Element | Nuclear Symbol | Atomic Number | Mass Number | Protons | Neutrons | Electrons |
|---|---|---|---|---|---|---|
| | | | | | 24 | 21 |
| | $^{76}_{32}Ge$ | | | | | |
| | | | 122 | | 72 | |
| | | | 37 | | | 17 |
| | | 11 | | | 12 | |

22. ■

| Name of Element | Nuclear Symbol | Atomic Number | Mass Number | Protons | Neutrons | Electrons |
|---|---|---|---|---|---|---|
| | $^{138}_{57}La$ | | | | | |
| Phosphorus | | | 28 | | | |
| | | | | 57 | 82 | |
| | | 29 | | | | 36 |
| | $^{63}_{29}Cu$ | | | | | |
| Aluminum | | | 25 | | | |

**Section 5.5: Atomic Mass**

*Although this set of questions is based on material in Section 5.5, some parts of some questions assume that you have also studied Section 5.6 and can use the periodic table as a source of atomic masses.*

23. What advantage does the atomic mass unit have over grams when speaking of the mass of an atom or a subatomic particle?

24. What is an atomic mass unit?

25. ■ The mass of an average atom of a certain element is 6.66 times as great as the mass of an atom of carbon-12. Using either the periodic table or the table of elements, identify the element.

26. The average mass of boron atoms is 10.81 u. How would you explain what this means to a friend who had never taken chemistry?

27. The atomic masses of the natural isotopes of neon are 19.99244 u, 20.99395 u, and 21.99138 u. The average of these three masses is 20.99259 u. The atomic mass of neon is listed as 20.1797 u on the periodic table. Which isotope do you expect is the most abundant in nature? Explain.

28. ■ A certain element consists of two stable isotopes. The first has an atomic mass of 137.9068 u and a percentage natural abundance of 0.09%. The second has an atomic mass of 138.9061 u and a percentage natural abundance of 99.91%. What is the atomic mass of the element?

29. The mass of 60.4% of the atoms of an element is 68.9257 u. There is only one other natural isotope of that element, and its atomic mass is 70.9249 u. Calculate the average atomic mass of the element. Using the periodic table and/or the table of the elements, write its symbol and name.

30. Isotopic data for boron allow the calculation of its atomic mass to the number of significant figures justified by the measurement process. One analysis showed that 19.78% of boron atoms have an atomic mass of 10.0129 u and the remainder have an atomic mass of 11.00931 u. Find the average mass in as many significant figures as those data will allow.

*Questions 31 through 36: Percentage abundances and atomic masses (u) of the natural isotopes of an element are given. (a) Calculate the atomic mass of each element from these data. (b) Using other information that is available to you, identify the element.*

| | Percentage Abundance | Atomic Mass (u) |
|---|---|---|
| 31. | 51.82 | 106.9041 |
| | 48.18 | 108.9047 |
| 32. | 69.09 | 62.9298 |
| | 30.91 | 64.9278 |
| 33. | 57.25 | 120.9038 |
| | 42.75 | 122.9041 |
| 34. | 37.07 | 184.9530 |
| | 62.93 | 186.9560 |
| 35. | 0.193 | 135.907 |
| | 0.250 | 137.9057 |
| | 88.48 | 139.9053 |
| | 11.07 | 141.9090 |
| 36. | 67.88 | 57.9353 |
| | 26.23 | 59.9332 |
| | 1.19 | 60.9310 |
| | 3.66 | 61.9283 |
| | 1.08 | 63.9280 |

**Section 5.6: The Periodic Table**

37. How many elements are in Period 5 of the periodic table? Write the atomic numbers of the elements in Group 3B/3.

38. ■ Write the symbol of the element in each given group and period. (a) Group 1A/1, Period 6; (b) Group 6A/16, Period 3; (c) Group 7B/7, Period 4; (d) Group 1A/1, Period 2.

39. Locate in the periodic table each element whose atomic number is given, and identify first the number of the period it is in and then the number of the group: (a) 20; (b) 14; (c) 43.

40. ■ List the symbols of the elements of each of the following: (a) transition metals in the fourth period; (b) metals in the third period; (c) nonmetals in Group 6A/16 or 7A/17 with $Z < 40$; (d) main group metals in the sixth period.

41. ■ Using only a periodic table for reference, list the atomic masses of the elements whose atomic numbers are 29, 55, and 82.

42. ■ Write the atomic number of the element in each given group and period. (a) Group 4A/14, Period 3; (b) Group 4A/14, Period 2; (c) Group 1B/11, Period 4; (d) Group 2A/2, Period 3.

43. Write the atomic masses of helium and aluminum.

44. ■ Give the atomic mass of the element in each given group and period. (a) Group 8A/18, Period 3; (b) Group 3A/13, Period 4; (c) Group 4B/4, Period 4; (d) Group 2A/2, Period 3.

### Section 5.7: Elemental Symbols and the Periodic Table

45. The names, atomic numbers, or symbols of some of the elements in Figure 5.8 are given in Table 5.4. Fill in the open spaces, referring only to a periodic table for any information that you need.

**Table 5.4** Table of Elements

| Name of Element | Atomic Number | Symbol of Element |
|---|---|---|
| | | Mg |
| | 8 | |
| Phosphorus | | |
| | | Ca |
| Zinc | | |
| | | Li |
| Nitrogen | | |
| | 16 | |
| | 53 | |
| Barium | | |
| | | K |
| | 10 | |
| Helium | | |
| | | Br |
| | | Ni |
| Tin | | |
| | 14 | |

46. ■ The names, atomic numbers, or symbols of some of the elements in Figure 5.8 are given in Table 5.5. Fill in the open spaces, referring only to a periodic table for any information you need.

**Table 5.5** Table of Elements

| Name of Element | Atomic Number | Symbol of Element |
|---|---|---|
| Sodium | | |
| | | Pb |
| Aluminum | | |
| | 26 | |
| | | F |
| Boron | | |
| | 18 | |
| Silver | | |
| | 6 | |
| Copper | | |
| | | Be |
| Krypton | | |
| Chlorine | | |
| | 1 | |
| | | Mn |
| | 24 | |
| Cobalt | | |
| | 80 | |

### General Questions

47. Distinguish precisely and in scientific terms the differences among items in each of the following groups.
    a) Atom, subatomic particle
    b) Electron, proton, neutron
    c) Nuclear model of the atom, planetary model of the atom
    d) Atomic number, mass number
    e) Chemical symbol of an element, nuclear symbol
    f) Atom, isotope
    g) Atomic mass, atomic mass unit
    h) Atomic mass of an element, atomic mass of an isotope
    i) Period, group, or family (in the periodic table)
    j) Main group element, transition element

48. Determine whether each statement that follows is true or false.
    a) Dalton proposed that atoms of different elements always combine on a one-to-one basis.
    b) According to Dalton, all oxygen atoms have the same diameter.
    c) The mass of an electron is about the same as the mass of a proton.
    d) There are subatomic particles in addition to the electron, proton, and neutron.
    e) The mass of an atom is uniformly distributed throughout the atom.
    f) Most of the particles fired into the gold foil in the Rutherford experiment were not deflected.
    g) The masses of the proton and electron are equal but opposite in sign.
    h) Isotopes of an element have different electrical charges.
    i) The atomic number of an element is the number of particles in the nucleus of an atom of that element.
    j) An oxygen-16 atom has the same number of protons as an oxygen-17 atom.
    k) The nuclei of nitrogen atoms have a different number of protons from the nuclei of any other element.
    l) Neutral atoms of sulfur have a different number of electrons from neutral atoms of any other element.
    m) Isotopes of different elements that exhibit the same mass number exhibit similar chemical behavior.
    n) The mass number of a carbon-12 atom is exactly 12 g.
    o) Periods are arranged vertically in the periodic table.
    p) The atomic mass of the second element in the farthest right column of the periodic table is 10 u.
    q) Nb is the symbol of the element for which $Z = 41$.
    r) Elements in the same column of the periodic table have similar properties.
    s) The element for which $Z = 38$ is in both Group 2A/2 and the fifth period.

49. The first experiment to suggest that an atom consisted of smaller particles showed that one particle had a negative charge. From that fact, what could be said about the charge of other particles that might be present?

### More Challenging Problems

50. Sodium oxide and sodium peroxide are two compounds made up of the elements sodium and oxygen. Sixty-two grams of sodium oxide contains 46 g of sodium and 16 g of oxygen; 78 g of sodium peroxide has 46 g of sodium and 32 g of oxygen. Show how these figures confirm the Law of Multiple Proportions.

51. Two compounds of mercury and chlorine are mercury(I) chloride and mercury(II) chloride. The amount of mercury(I) chloride that contains 71 g of chlorine has 402 g of mercury; the amount of mercury(II) chloride that has 71 g of chlorine has 201 g of mercury. Show how the Law of Multiple Proportions is illustrated by these quantities.

52. ■ The *CRC Handbook,* a large reference book of chemical and physical data from which many of the values in this book are taken, lists two isotopes of rubidium ($Z = 37$). The atomic mass of 72.15% of rubidium atoms is 84.9118 u. Through a typographical oversight, the atomic mass of the second isotope is not printed. Calculate that atomic mass.

53. ■ The element lanthanum has two stable isotopes, lanthanum-138 with an atomic mass of 137.9071 u and lanthanum-139 with an atomic mass of 138.9063 u. From the atomic mass of La, 138.9 u, what conclusion can you make about the relative percentage abundance of the isotopes?

54. ■ The atomic mass of lithium on a four-significant-figure periodic table is 6.941 u. Lithium has two natural isotopes with atomic masses of 6.10512 u and 7.01600 u. Calculate the percentage distribution between the two isotopes.

55. When Thomson identified the electron, he found that the ratio of its charge to its mass (the e/m ratio) was the same regardless of the element from which the electron came. This showed that the electron is a unique particle that is found in atoms of all elements. Positively charged particles found at about the same time did not all have the same e/m ratio. (Later it was found that even different atoms of the same element contain positive particles that have different e/m ratios.) What does that suggest about the mass, particle charge, and minimum number of positive particles from different elements?

56. Why were scientists inclined to think of an atom as a miniature solar system in the planetary model of the atom? What are the similarities and differences between electrons in orbit around a nucleus and planets in orbit around the sun?

57. Isotopes were unknown until nearly a century after Dalton proposed the atomic theory. When they were discovered, it was through experiments more closely associated with physics than with chemistry. What does this suggest about the chemical properties of isotopes?

58. ■ A carbon-12 atom contains six electrons, six protons, and six neutrons. Assuming the mass of the atom is the sum of the masses of those parts, as given in Table 5.1, calculate the mass of the atom. Why is it not exactly 12 u, as the definition of the atomic mass unit would suggest?

59. Using the figures in Question 58, calculate the percentage each kind of subatomic particle contributes to the mass of a carbon-12 atom. What percentage of the total mass of the atom is in the nucleus?

60. The element carbon occurs in two crystal forms, diamond and graphite. The density of the diamond form is 3.51 $g/cm^3$, and of graphite, 2.25 $g/cm^3$. The volume of a carbon atom is $1.9 \times 10^{-24}$ $cm^3$. As stated in Section 5.5, one atomic mass unit is $1.66 \times 10^{-24}$ g.
    a) Calculate the average density of a carbon atom.
    b) Suggest a reason for the density of the atom being so much larger than the density of either form of carbon.
    c) The radius of a carbon atom is roughly $1 \times 10^5$ times larger than the radius of the nucleus. What is the volume of that nucleus? (*Hint:* Volume is proportional to the cube of the radius.)
    d) Calculate the average density of the nucleus.
    e) The radius of a period on this page is about 0.02 cm. The volume of a sphere that size is $4 \times 10^{-5}$ $cm^3$. Calculate the mass of that sphere if it were completely filled with carbon nuclei. Express the mass in tons.

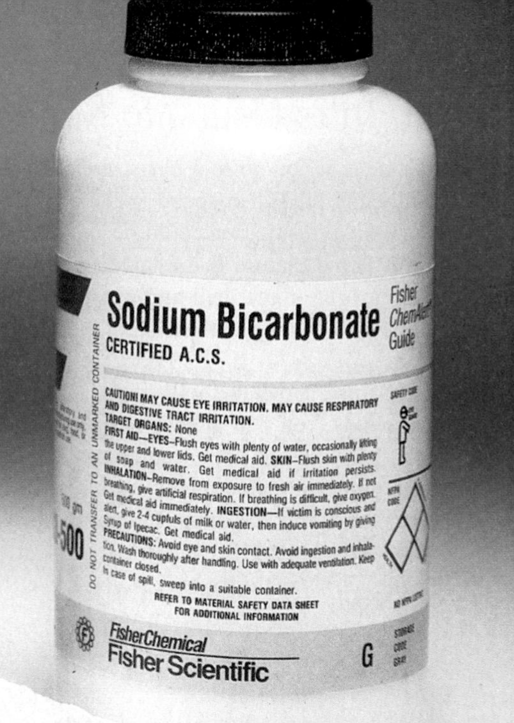


© Cengage Learning/Charles D. Winters

# 6

The same substance may be known by several names, but it can have only one chemical formula. Baking soda is the common name of $NaHCO_3$ because of its widespread use in baking. Its old chemical name, bicarbonate of soda, was replaced by a similar name, sodium bicarbonate. Its official name today is sodium hydrogen carbonate. The current name and formula are part of a system of nomenclature that you will learn in this chapter.

# Chemical Nomenclature

The term ***nomenclature*** comes from a Latin word meaning "calling by name." Thus, "Chemical Nomenclature" is the system of naming chemicals. In this chapter we present the language of chemistry as used by most U.S. chemists today.

Online homework for this chapter may be assigned in OWL

## CONTENTS

## 6.1 Introduction to Nomenclature

Imagine that you go to a party, are introduced to hundreds of people you have never seen before, and then spend the rest of the evening trying to remember their names. This awkward situation could be avoided altogether if you had access to a refined system of nomenclature that enabled you to memorize the names of the people instead of using brute-force drill-and-practice over and over again for each individual. You'd still have to memorize the system, but that's much easier than memorizing hundreds of names.

In this chapter we will introduce nearly 1500 chemical names for elements and compounds. You will memorize a few, but you will know the names of the others simply by learning and applying the system. After you have studied the chapter, if you are given either the name or the chemical formula of an element or compound, you will be able to supply the other. Furthermore, you will be able to do this with hundreds of compounds that aren't even directly covered in the chapter. But you will be able to do these remarkable things only if you LEARN THE SYSTEM.

## 6.2 Formulas of Elements

**Goal** 1 Given a name or formula of an element in Figure 5.8 write the other.

In Chapter 5 you learned the symbols of 35 elements and their location in a partial periodic table that contained only those symbols. That periodic table was Figure 5.8. In this section you will learn the chemical formulas of those elements as they are written in chemical equations.

The formula of most elements is simply the elemental symbol. In a chemical reaction, the macroscopic sample of the element behaves as if it were a collection of individual particles, or atoms. This indicates that the tiniest individual unit in a sample of the element is one atom. We call this individual unit a **formula unit.**

Some elements do not behave as if their smallest particle is a single atom. For example, at common temperatures seven elements form **diatomic molecules.** A molecule is the tiniest independent particle of a pure substance—element or compound—in a sample of that substance. *Di-* is a prefix that means two. Thus, a diatomic molecule consists of two atoms that are chemically bonded to each other. This molecule is a formula unit of the element. Its chemical formula is the elemental symbol followed by a subscript 2. For example, the chemical formula of hydrogen (symbol H) is $H_2$.

**P/REVIEW:** In Section 2.1, atoms and molecules are identified as two of the particles studied in chemistry. They appear again in the introduction to chemical formulas in Section 2.6.

You must be able to recognize the elements that form diatomic molecules and write their formulas correctly. Listed in a way that will help you remember them, they are

| Elements | Formulas |
|---|---|
| Hydrogen, the element that makes up about 90% of the atoms in the universe | $H_2$ |
| Nitrogen and oxygen, the two elements that make up about 98% of the atmosphere | $N_2$, $O_2$ |
| Fluorine, chlorine, bromine, and iodine, the first four elements in Group 7A/17 | $F_2$, $Cl_2$, $Br_2$, $I_2$ |

### Thinking About Your Thinking

Memory

**A mnemonic (pronounced NEE-mahn-ick) is something that helps you remember. Perhaps the first mnemonic device you learned was the alphabet song. By matching the letters of the alphabet with the notes in the song, it became easier for you to recall the letters in their proper order. Even though you must remember more information, it helps to associate one thing with another. There are many other mne-**

monics, such as mental images, pictures, catchy sayings, and jingles. Advertisers often use mnemonics to help you remember the names of their products. You may be able to think of a few advertising jingles right now; notice how the jingle helps you remember the company's slogan.

A mnemonic device to remember the seven diatomic elements is the saying: **Horses Need Oats For Clear Brown I's**. The first letter(s) of each word is the elemental symbol of one of the seven elements. Remembering this phrase is often easier than learning the elemental symbols alone.

As you gain experience in writing the formulas of the elements, your reliance on this mnemonic will gradually decrease, which is something that occurs through repeated practice. When you automatically think "$O_2$" or "$Br_2$" without the mnemonic, you know that you are improving in your command of chemical facts and information.

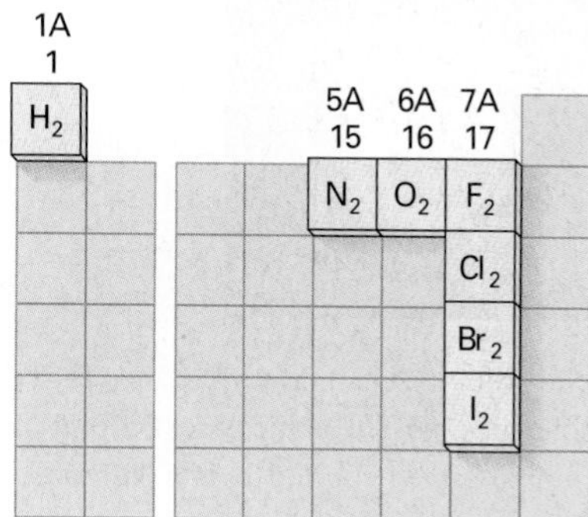


**Figure 6.1** Elements that form stable diatomic molecules and their positions in the periodic table.

The formulas of these elements are shown in their positions in the periodic table in Figure 6.1.

When you write chemical equations, it is *absolutely essential* that you write the formulas of these seven diatomic elements as two-atom molecules with a subscript 2 following the symbol of the element. *Failure to do so is probably the most common mistake made by beginning chemistry students.* The formulas of other elements are the same as their elemental symbols. No subscript is used.

## Active Example 6.1

Write the formulas of the following elements as they would be written in a chemical equation: potassium, fluorine, hydrogen, nitrogen, calcium.

Potassium, K; fluorine, $F_2$; hydrogen, $H_2$; nitrogen, $N_2$; calcium, Ca

At normal temperatures, two other elements, sulfur and phosphorus, form molecules having more than 1 atom. The most common form of solid sulfur is an 8-atom ring. Other combinations of sulfur atoms form sulfur molecules with as few as 2 atoms and as many as 20 atoms. Phosphorus commonly forms 4-atom molecules, but it also forms 2-atom molecules. Carbon can also form molecules consisting of 30 to 600 carbon atoms, and it can exist as huge interconnected networks of uncountable numbers of atoms. Because of the variations in the molecular forms of these elements, we write their formulas simply as S, P, and C. Referring to the polyatomic (many-atom) structure of sulfur, phosphorus, and carbon as monatomic (1-atom) will not affect calculations involving these elements. Figure 6.2 illustrates these molecules.

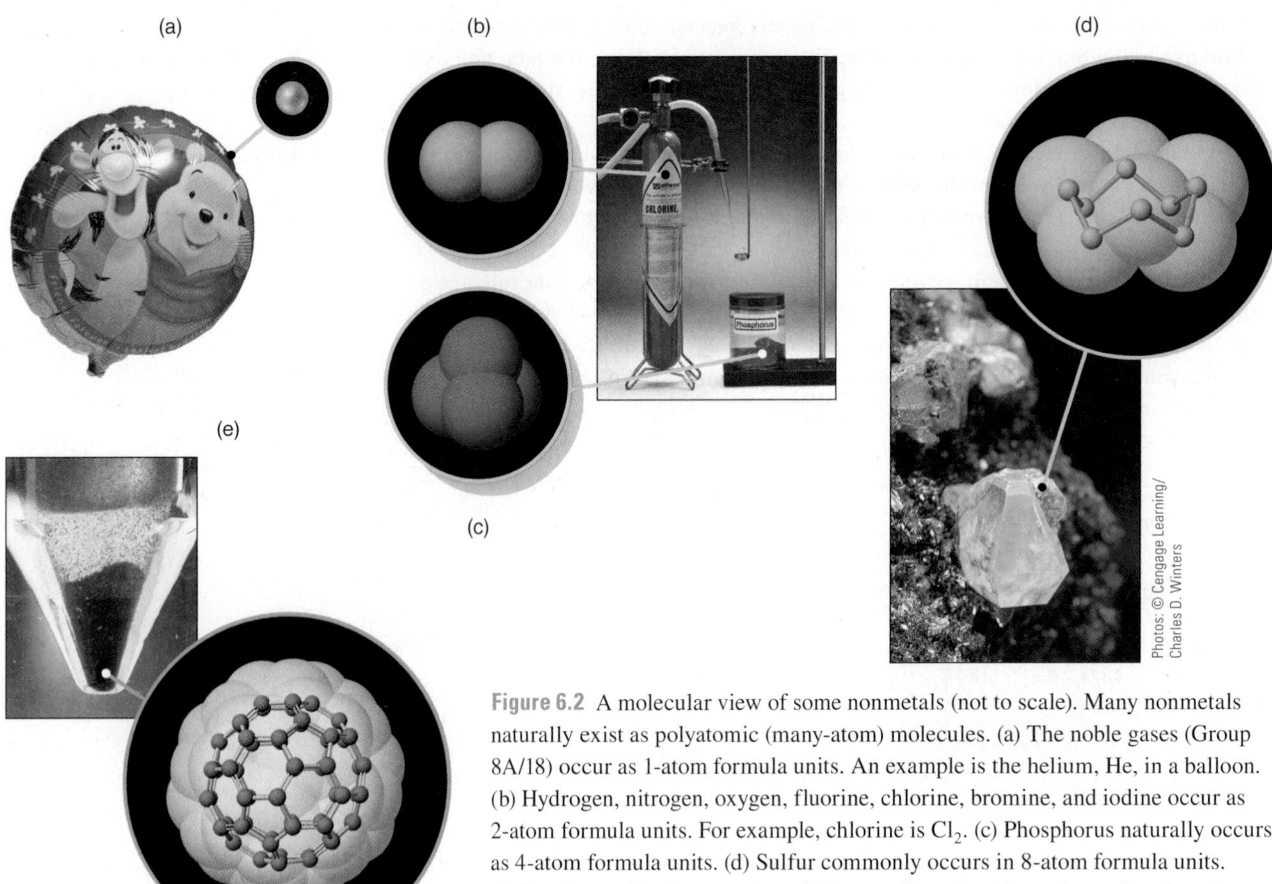


**Figure 6.2** A molecular view of some nonmetals (not to scale). Many nonmetals naturally exist as polyatomic (many-atom) molecules. (a) The noble gases (Group 8A/18) occur as 1-atom formula units. An example is the helium, He, in a balloon. (b) Hydrogen, nitrogen, oxygen, fluorine, chlorine, bromine, and iodine occur as 2-atom formula units. For example, chlorine is $Cl_2$. (c) Phosphorus naturally occurs as 4-atom formula units. (d) Sulfur commonly occurs in 8-atom formula units. (e) One form of carbon consists of 60-atom formula units, commonly referred to as buckyballs. The official name of $C_{60}$ is buckminsterfullerene.

## 6.3 Compounds Made from Two Nonmetals

**Goal** 2 Given the name or formula of a binary molecular compound, write the other.

**Goal** 3 Given the name or the formula of water, write the other; given the name or the formula of ammonia, write the other.

The stair-step line in the periodic table that begins between atomic numbers 4 and 5 and ends between 84 and 85 separates elements on the left that are metals, such as iron, copper, and lead, from elements on the right that are nonmetals, such as hydrogen, oxygen, and nitrogen (Fig. 6.3). Several elements bordering on this stair-step line are called *metalloids* or *semimetals.* Many compounds are formed by two nonmetal elements or by a metalloid and a nonmetal element. These compounds are all molecular; their tiniest individual unit is a molecule. Compounds formed by two nonmetals or a metalloid and a nonmetal are called **binary molecular compounds.** A binary molecular compound contains two elements, both nonmetals (or one can be a metalloid).

**P/REVIEW:** The differences between elements that are metals and those that are nonmetals can be traced to the arrangement of electrons in the atoms. This subject is considered in more detail in Section 11.6.

The name of a molecular compound formed by two nonmetals or a metalloid and a nonmetal has two words:

**summary**

**Names of Compounds Made from Two Nonmetals (or a Metalloid and a Nonmetal)**

1. The first word is the name of the element appearing first in the chemical formula, including a prefix to indicate the number of atoms of that element in the molecule.
2. The second word is the name of the element appearing second in the chemical formula, changed to end in *-ide,* and also including a prefix to indicate the number of atoms in the molecule.

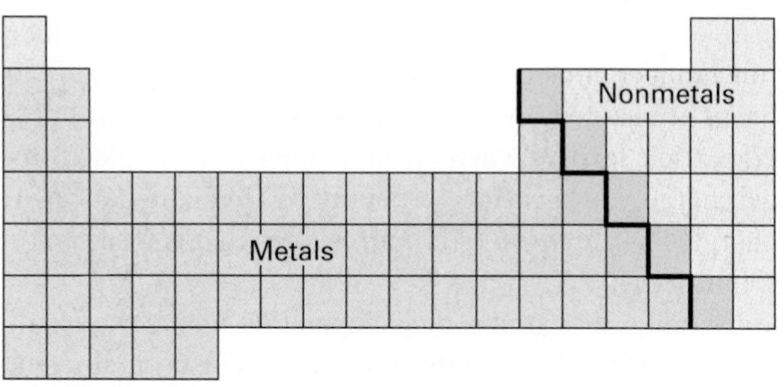


**Figure 6.3** Metals and nonmetals. Green identifies elements that are metalloids (semimetals), which have properties that are intermediate between those of metals and nonmetals.

The same two elements often form more than one binary compound. Their names are distinguished by the prefixes mentioned in the preceding rules. Silicon and chlorine form silicon tetrachloride, $SiCl_4$, and disilicon hexachloride, $Si_2Cl_6$. The prefix *tetra-* identifies four chlorine atoms in a molecule of $SiCl_4$. In $Si_2Cl_6$, *di-* indicates two silicon atoms, and *hexa-* shows six chlorine atoms in the molecule. Technically, $SiCl_4$ should be monosilicon tetrachloride, but the prefix *mono-* for "one" is usually omitted in the first word. If an elemental name has no prefix in a binary molecular compound, you may assume that there is only one atom of that element in the molecule.

Table 6.1 gives the first ten number prefixes. The letter *o* at the end of the prefix *mono-* and the letter *a* in the prefixes for four to ten are omitted if the resulting word sounds better. This usually occurs when the next letter is a vowel. For example, a compound with a formula ending in $O_5$ is a *pentoxide* rather than a *pentaoxide.*

The oxides of nitrogen are ideal for practicing the nomenclature of binary molecular compounds.

**Table 6.1** Number Prefixes Used in Chemical Names

| Number | Prefix |
|---|---|
| 1 | *mono-* |
| 2 | *di-* |
| 3 | *tri-* |
| 4 | *tetra-* |
| 5 | *penta-* |
| 6 | *hexa-* |
| 7 | *hepta-* |
| 8 | *octa-* |
| 9 | *nona-* |
| 10 | *deca-* |

Study Hints and Pitfalls to Avoid at the end of this chapter gives some suggestions to help you memorize the prefixes in Table 6.1.

## Active Example 6.2

For each name below, write the formula; for each formula, write the name.

| | | | |
|---|---|---|---|
| nitrogen monoxide | ________ | $NO_2$ | ________________ |
| dinitrogen monoxide | ________ | $N_2O_3$ | ________________ |
| dinitrogen pentoxide | ________ | $N_2O_4$ | ________________ |

nitrogen monoxide, NO — $NO_2$, nitrogen dioxide
dinitrogen monoxide, $N_2O$ — $N_2O_3$, dinitrogen trioxide
dinitrogen pentoxide, $N_2O_5$ — $N_2O_4$, dinitrogen tetroxide

Nitrogen dioxide could be correctly identified as mononitrogen dioxide.
Two of the preceding compounds continue to be called by their older names: $N_2O$ is nitrous oxide (laughing gas) and NO is nitric oxide.

Two compounds are so common they are always called by their traditional names rather than their chemical names. $H_2O$ is always called water rather than dihydrogen oxide, and $NH_3$ is always called ammonia rather than nitrogen trihydride. These traditional names and formulas are important; memorize them.

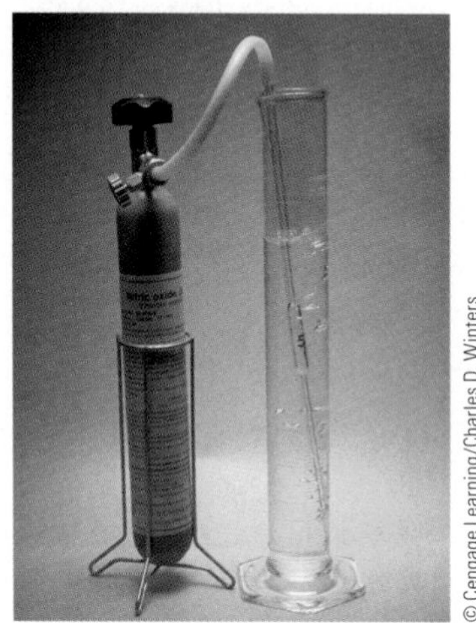



Nitrogen monoxide and nitrogen dioxide. The gas cylinder contains colorless nitrogen monoxide, which can be seen as bubbles in the liquid. When the nitrogen monoxide in the bubbles contacts the oxygen in the air, a reaction occurs, producing red-brown nitrogen dioxide.

## 6.4 Names and Formulas of Ions Formed by One Element

**Goal** 4 Given the name or formula of an ion in Figure 6.4, write the other.

A neutral atom contains the same number of protons (positive charges) and electrons (negative charges). That makes the atom electrically neutral; there is no net charge. But an atom can gain or lose one or more electrons. When it does, the balance between

**P/REVIEW:** In Section 5.6, elements in the A groups (1, 2, and 13 to 18) of the periodic table were identified as main group elements, and elements in the B groups (3 to 12) were called transition elements.

positive and negative charges is upset. Thus, the particle acquires a net electrical charge that is equal to the number of electrons gained or lost. If electrons are lost, the charge is positive; if electrons are gained, the charge is negative.

The charged particle formed when an atom gains or loses electrons is an **ion.** If the ion has a positive charge, it is a **cation** (pronounced CAT-ion, not ca-SHUN). If the ion has a negative charge, it is an **anion** (AN-ion). An ion that is formed from a single atom is a **monatomic ion.**

The rules for naming monatomic ions depend on where the element is located in the periodic table. The rules for naming ions formed from main group elements are different from the rules for most transition elements. We will first consider how to name ions formed from main group elements.

## Main Group Elements

The formation of a monatomic ion from an atom can be expressed in a chemical equation. For example, a magnesium atom forms a magnesium ion by losing two electrons:

Group 2A/2, magnesium ion: $Mg \rightarrow Mg^{2+} + 2\ e^-$

This equation illustrates two important rules about monatomic ions:

1. The formula of a monatomic ion is the symbol of the element followed by its electrical charge, written in superscript. The size of the charge is written in front of the + or − sign. (If the charge is 1+ or 1−, the number 1 is omitted.)
2. The name of a monatomic cation is the name of the element followed by the word *ion*.

Notice the *essential difference* between the formula of an element and the formula of an ion formed by that element. The formula of most elements is the elemental symbol. The formula of the ion is the elemental symbol *followed by the ionic charge, written in superscript.* An elemental symbol, without a charge, is *never* the formula of an ion.

**P/REVIEW:** The relationship between charges on monatomic ions and position in the periodic table is explained in Section 11.4. These charges arise when an atom loses or gains one, two, or three electrons to become isoelectronic with a noble-gas atom, that is, when the number and arrangement of electrons is the same as that found in a noble-gas (Group 8A/18) atom.

The thing that makes the periodic table so helpful in writing chemical formulas is that *all* the elements in Group 2A/2 that form monatomic ions do so by losing two electrons. This loss of two electrons gives the elements of Group 2A/2 their similar chemical properties.

As you might expect, this idea also applies to other groups in the periodic table. Group 1A/1 elements form monatomic ions by losing one electron per atom, and aluminum, a Group 3A/13 element, loses three electrons per atom. The equations for the ions formed by atoms of the Period 3 elements are

Group 1A/1, sodium ion: $Na \rightarrow Na^+ + e^-$

Group 3A/13, aluminum ion: $Al \rightarrow Al^{3+} + 3\ e^-$

Notice that the positive charge of a monatomic cation formed by a main group element is the same as the group number of the element in the periodic table.

Atoms of the nonmetals in Groups 5A/15, 6A/16, and 7A/17 of the periodic table form monatomic ions by gaining electrons. These ions have negative charges; they are anions. The equations for the ions formed by atoms of the Period 3 elements are

Group 5A/15, phosphide ion: $P + 3e^- \rightarrow P^{3-}$

Group 6A/16, sulfide ion: $S + 2e^- \rightarrow S^{2-}$

Group 7A/17, chloride ion: $Cl + e^- \rightarrow Cl^-$

Notice that the name of a monatomic anion comes from the name of the parent element, but is it not exactly the same:

> *The name of a monatomic anion is the name of the element changed to end in -ide, followed by the word ion.*

Thus, when a phosph*orus* atom gains three electrons, it becomes the phosph*ide ion,* sulf*ur* becomes the sulf*ide ion,* and chlor*ine* becomes the chlor*ide ion.**

*As elements, phosphorus, sulfur, and chlorine exist as polyatomic molecules that must separate into atoms before ions can form.

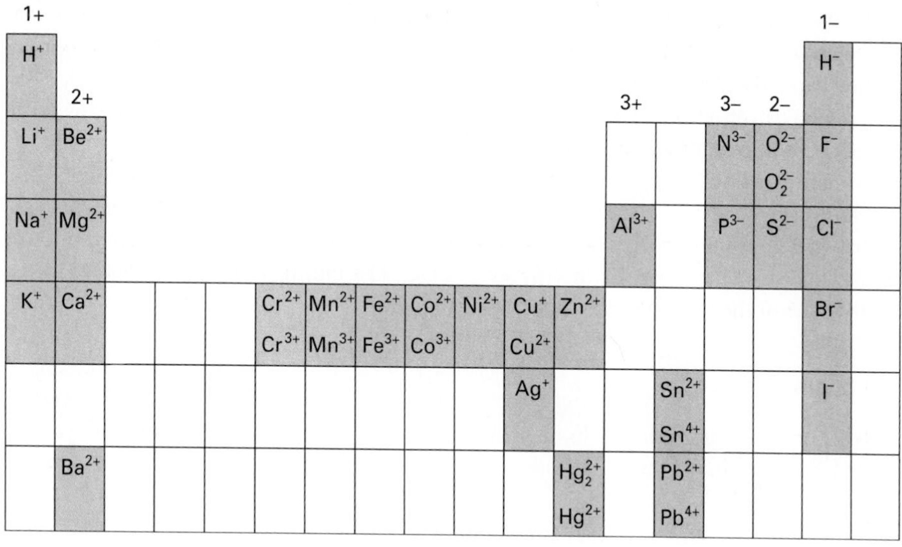

1+
H+
2+
Li+ Be2+
Na+ Mg2+
K+ Ca2+
Ba2+
Cr2+ Mn2+ Fe2+ Co2+ Ni2+ Cu+ Zn2+
Cr3+ Mn3+ Fe3+ Co3+ Cu2+
Ag+
Hg22+
Hg2+
3+
Al3+
Sn2+
Sn4+
Pb2+
Pb4+
3– 2–
N3– O2–
O22–
P3– S2–
1–
H–
F–
Cl–
Br–
I–


**Figure 6.4** Partial periodic table of common ions. $O_2^{2-}$ is the peroxide ion, a diatomic elemental ion. $Hg_2^{2+}$ is also a diatomic elemental ion. Its name is mercury(I) ion, indicating a 1+ charge from each atom in the ion.

## Active Example 6.3

Look only at a complete periodic table as you write (a) the names of $Br^-$ and $Ba^{2+}$ and (b) the formulas of the potassium and fluoride ions.

(a) bromide ion and barium ion; (b) $K^+$ and $F^-$

Notice that the formula of the fluoride ion is $F^-$, not $F_2^-$. Fluorine occurs as a diatomic molecule, $F_2$, when it is uncombined, but it is *never* a diatomic ion.

Figure 6.4 places on a periodic table all the monatomic ions formed by the elements whose symbols you learned in Section 5.7. Thus far we have examined the monatomic ions formed by elements in Groups 1A/1, 2A/2, 3A/3, 5A/15, 6A/16, and 7A/17. Notice that all ions formed by nonmetals are anions (they have negative charges) and all ions formed by metals are cations (positive charges). This is one of the features that distinguish metals from nonmetals.

An important exception to this rule is hydrogen. A hydrogen atom can lose an electron to form a hydrogen ion, $H^+$, or it can gain an electron to form the hydride ion, $H^-$. This is why it appears in both Group 1A/1 and 7A/17 on our periodic table. Even though hydrogen is a nonmetal, it most commonly forms the hydrogen ion, $H^+$, a cation.

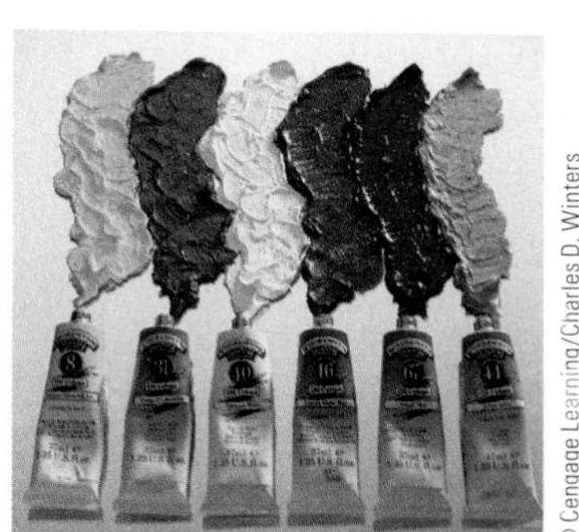


Many highly colored substances contain transition-element ions. The contents of all these paint pigments include ions of elements in the B groups of the periodic table.

## Transition Elements that Form More Than One Ion

Some transition elements—elements in the B groups (Groups 3 to 12) of the periodic table—are able to form two or more different monatomic ions that have different charges. Iron is one example. If a neutral atom loses two electrons, the ion has a 2+ charge, $Fe^{2+}$. A neutral atom can also lose three electrons, resulting in an ion with a 3+ charge: $Fe^{3+}$. To distinguish between the two ions, we include the size of the charge, but not its sign, when naming the ion. Thus, $Fe^{2+}$ is called the "iron two ion," and $Fe^{3+}$ is called the "iron three ion." In writing, the ion charge appears in Roman numerals and is enclosed in parentheses: $Fe^{2+}$ is the iron(II) ion, and $Fe^{3+}$ is the iron(III) ion. Note that there is no space between the last letter of the elemental name and the opening parenthesis.

The charge on a monatomic ion is formally referred to as the **oxidation state,** or **oxidation number,** of the element. Thus the oxidation state of $Na^+$ is +1; of $Fe^{2+}$, +2;

P/REVIEW: Formal rules for assigning oxidation numbers are presented in Section 19.3.

and of $Fe^{3+}$, +3. These oxidation states are part of a broader scheme that assigns oxidation numbers to all elements in a compound or ion, but we do not need these additional features in this chapter. Nevertheless, many instructors choose to include oxidation state as a part of nomenclature. If your instructor is among them, by all means learn the concept of oxidation numbers now.

There is an older, but still widely used, way to distinguish between the $Fe^{2+}$ and $Fe^{3+}$ ions. $Fe^{2+}$ is called the *ferrous* ion and $Fe^{3+}$ is the *ferric* ion. The general rule is that two common charges are distinguished by using an *-ous* ending for the lower charge and an *-ic* ending for the higher charge. The endings are often applied to the Latin name of the element, which for iron is *ferrum*.

### Active Example 6.4

Refer only to a periodic table, if necessary, and write (a) the names of $Mn^{2+}$ and $Mn^{3+}$ and (b) the formulas of copper(I) ion and copper(II) ion.

(a) manganese(II) ion and manganese(III) ion; (b) $Cu^{+}$ and $Cu^{2+}$

## Transition Elements that Form Only One Ion

© Cengage Learning/Charles D. Winters

Water solutions that contain nickel ion are green.

Notice that when we named ions formed by main group elements, the charge was *not* included in the name of the ion. $Na^{+}$ was named "sodium ion." When we named $Fe^{2+}$ and $Fe^{3+}$ the charge was included in the name to distinguish between the iron(II) ion and the iron(III) ion. This leads to an important point:

*The charge is included in the name of an ion only when the ions of an element exhibit more than one common charge.*

Charge is not indicated in the name of the sodium ion because all sodium ions have the same charge, 1+.

Figure 6.4 shows that the positive charge on all main group cations is the same as the group number in the periodic table. Among the transition elements shown, nickel, zinc, and silver form ions of only one charge, $Ni^{2+}$, $Zn^{2+}$, and $Ag^{+}$. Therefore, charge is not included in the names of these ions. Location of these elements in the periodic table gives no hint of their ionic charges; the charges must be memorized. All the other transition elements shown form two ions of different charges. The magnitude of those charges is included in the name to distinguish between those ions.

With any active language, there are differences between everyday verbal usage and more formal written usage. The language of chemistry is no exception. For example, chemists sometimes use "nickelous ion" or "nickel(II) ion" to describe $Ni^{2+}$. As you gain experience in "talking chemistry" with practicing chemists, you will hear many additional examples of informal verbal nomenclature.

### Active Example 6.5

(a) Write the names of $Co^{2+}$ and $Ni^{2+}$. (b) Write the formulas of zinc ion and chromium(III) ion. Refer only to a periodic table.

(a) cobalt(II) ion and nickel ion; (b) $Zn^{2+}$ and $Cr^{3+}$

## Other Monatomic and Diatomic Ions

Figure 6.4 entries for oxygen, mercury, tin, and lead deserve special comment. Oxygen and mercury each form a monatomic ion. These monatomic ions cause no naming problems. However, these two elements also form diatomic ions. $O_2^{2-}$ is called peroxide ion. (Later you will see the prefix *per-* used to indicate an extra oxygen in naming acids and polyatomic ions.) In $Hg_2^{2+}$ the 2+ charge is shared by two mercury atoms, just as if each atom were contributing 1+ to the total charge of the ion. Accordingly, its name is mercury(I) ion.

The tin(II) and lead(II) ions, $Sn^{2+}$ and $Pb^{2+}$, behave like other monatomic ions. Tin and lead in compounds in which the elements appear to have 4+ charges have properties that differ from the properties of other compounds made up of ions. This notwithstanding, the compounds are named as if they were made up of true ions.

Chloride compounds containing tin(II) and lead(II) ions are white solids, and those containing tin(IV) and lead(IV) are colorless liquids.

## 6.5 Acids and the Anions Derived from Their Total Ionization

**Goal 5** Given the name (or formula) of an acid or ion in Table 6.5, write its formula (or name).

**P/REVIEW:** Acids and their "chemical opposites," bases, are discussed in more detail in Chapter 17. Notice how the reaction of an acid involves the *transfer of a proton*, symbolized as $H^+$, from one molecule to another.

### The Hydrogen and Hydronium Ions

The most familiar form of an **acid** is a molecular compound that reacts with water to produce a **hydrated hydrogen ion** and an anion. Examples of acids include HCl, $HNO_3$, and $HC_2H_3O_2$. Notice that the common characteristic among these formulas is that the **ionizable hydrogen** is written first in the formula. If HX represents an acid, its ionization equation can be written as

**P/REVIEW:** An ion is said to be hydrated when it is surrounded by water molecules. In Section 6.10 you will learn about hydrated solid compounds that have water molecules as part of their structure.

$$H_2O + HX \rightarrow H_3O^+ + X^-$$

The **hydronium ion, $H_3O^+$,** is a simplified form of the hydrated hydrogen ion, a hydrogen ion that is bound to a number of water molecules, $H^+ \cdot (H_2O)_x$. Removing a water molecule from both sides of the previous ionization equation leaves

Some chemists disapprove of the term *hydrogen ion*. If your instructor is among them, think and speak of the *hydronium ion* instead.

$$HX \rightarrow H^+ + X^-$$

In this form the acid appears to **ionize,** or separate into ions, one of which is the hydrogen ion. When referring to acids, we understand that the hydrogen ion is hydrated, that is, surrounded by water molecules, even though it may be written as $H^+$.

A hydrogen atom consists of one proton and one electron. To form a hydrogen ion, $H^+$, the neutral atom must lose its electron. That leaves only the proton; a hydrogen ion is simply a proton. Acids are sometimes classified by the number of hydrogen ions, or protons, that a single molecule can release. An acid that fits the general formula HX is a **monoprotic acid.** If an acid has two ionizable hydrogens, $H_2Y$, it is **diprotic,** and a **triprotic acid** has three ionizable hydrogens, $H_3Z$. ***Polyprotic*** is a general term that may be applied to any acid having two or more ionizable hydrogens.

This section discusses only the **total ionization** of an acid, meaning that all of the ionizable hydrogen is removed from the acid molecule. The total ionization equations for diprotic and triprotic acids are therefore

$$H_2Y \rightarrow 2\ H^+ + Y^{2-}$$

$$H_3Z \rightarrow 3\ H^+ + Z^{3-}$$

The intermediate ions coming from the step-by-step ionization of diprotic and triprotic acids are discussed in the next section.

## Acids and Their Corresponding Anions

An acid that contains oxygen in addition to hydrogen and another nonmetal is an **oxyacid.*** When hydrogen ions are removed from an oxyacid, the oxygen stays with the nonmetal as part of an **oxyanion.*** The ionization of nitric acid, $HNO_3$, shows this:

$$HNO_3 \rightarrow H^+ + NO_3^-$$

The name of the anion, $NO_3^-$, is *nitrate ion.*

Nitric acid and the nitrate ion are a perfect example of the nomenclature system that follows:

**summary**

**Formulas of Oxyanions Derived from the Total Ionization of Any Oxyacid with a Name Ending in *-ic***

For the total ionization of any oxyacid with a name ending in *-ic:*

1. The formula of the anion is the formula of the acid without the hydrogen(s).
2. There is a negative charge equal to the number of ionizable hydrogens in the acid.
3. The name of the anion is the name of the central element of the acid changed to end in *-ate.*

To illustrate with a different acid, chloric acid, $HClO_3$, without the hydrogen is $ClO_3$. The difference is one hydrogen, so the charge on the ion is 1−: $ClO_3^-$. To get the name, change chlor*ic* acid to chlor*ate* ion, *-ic* → *-ate.*

The names and formulas of five *-ic* acids and their corresponding *-ate* anions should be memorized. The acids, anions, and ionization equations are in Table 6.2.

Before leaving Table 6.2, we should point out that its only purpose is to show the relationship between the names and formulas of acids and ions. It is not intended to describe the chemical properties of the substances listed. All of the ionizations do take place, some to a large extent, and some only slightly. The acids are not the usual sources of the anions, all of which are abundant in the crust of the earth.

## The Acids of Chlorine and Their Anions

Chlorine forms five acids that furnish quite a complete picture of acids and the anions derived from their total ionization. All five are assembled in Table 6.3. The table illustrates a system of nomenclature that begins with the *-ic* acid, whose name and formula you have memorized from Table 6.2. From there on, it is a system of prefixes (beginnings) and suffixes (endings) based on the number of oxygens in the *-ic* acid. Memorizing these prefixes and suffixes and knowing how to use them is another major step in learning how to figure out names and/or formulas of compounds.

When you memorize the names and formulas of these five acids, you will have taken the first step toward figuring out the names and formulas of more than a thousand chemical compounds.

**Table 6.2** Acids and Anions

| Acid | Ionization Equation | Ion Name |
|---|---|---|
| Chloric acid | $HClO_3 \rightarrow H^+ + ClO_3^-$ | Chlorate ion |
| Nitric acid | $HNO_3 \rightarrow H^+ + NO_3^-$ | Nitrate ion |
| Sulfuric acid | $H_2SO_4 \rightarrow 2\,H^+ + SO_4^{2-}$ | Sulfate ion |
| Carbonic acid* | $H_2CO_3 \rightarrow 2\,H^+ + CO_3^{2-}$ | Carbonate ion |
| Phosphoric acid* | $H_3PO_4 \rightarrow 3\,H^+ + PO_4^{3-}$ | Phosphate ion |

*The carbonic and phosphoric acid ionizations occur only slightly in water solutions. They are used here to illustrate the derivation of the formulas and names of the carbonate and phosphate ions, which are quite abundant from sources other than their parent acids.

**Oxoacid* and *oxoanion* are alternate terms for oxyacid and oxyanion.

**Table 6.3** Prefixes and Suffixes in Acid and Anion Nomenclature (Acids and Anions of Chlorine Given as Examples)

| Number of Oxygen Atoms Compared with *-ic* Acid and *-ate* Anion | Acid Prefix and/or Suffix (Example) | Anion Prefix and/or Suffix (Example) |
|---|---|---|
| One more $HClO_4$ | *per- -ic* (perchloric) | *per- -ate* (perchlorate) |
| Same $HClO_3$ | *-ic* (chloric) | *-ate* (chlorate) |
| One fewer $HClO_2$ | *-ous* (chlorous) | *-ite* (chlorite) |
| Two fewer HClO | *hypo- -ous* (hypochlorous) | *hypo- -ite* (hypochlorite) |
| No oxygen HCl | *hydro- -ic* (hydrochloric) | *-ide* (chloride) |

## Thinking About Your Thinking

Memory

**At the heart of the prefix–suffix system in Table 6.3 lie the four suffixes used for all oxyacids and oxyanions, *-ic, -ate, -ous,* and *-ite.* A mnemonic to remember these is: Ick! I ate a poisonous bite!**

Chloric acid and its chlorate ion have three oxygens. Starting from there and referring to Table 6.3:

summary

**The Acids of Chlorine and Their Anions**

1. If the number of oxygens is one larger than the number in the *-ic* acid, the prefix *per-* is placed before both the acid and anion names: $HClO_4$ is *per*chloric acid, and $ClO_4^-$ is *per*chlorate ion.
2. If the number of oxygens is one smaller than the number in the *-ic* acid, the suffixes *-ic* and *-ate* are replaced with *-ous* and *-ite:* $HClO_2$ is chlor*ous* acid, and $ClO_2^-$ is the chlor*ite* ion.
3. If the number of oxygens is two smaller than the number in the *-ic* acid (one smaller than the number in the *-ous* acid), the prefix *hypo-* is placed before both the acid and anion names, and the *-ous* and *-ite* suffixes are kept: HClO is *hypo*chlor*ous* acid, and $ClO^-$ is the *hypo*chlor*ite* ion.
4. The name of an acid with no oxygen is *hydro-* followed by the name of the nonmetal, changed to end in *-ic:* HCl is *hydro*chlor*ic* acid. The monatomic anion from the acid is named by the rule for monatomic anions, by which the elemental name is changed to end in *-ide:* $Cl^-$ is the chlor*ide* ion.

NASA

Perchlorate ion is one of the components in the solid fuel that propels the booster rockets of the Space Shuttle.

## Acids and Anions of Group 4A/14 to 7A/17 Elements

One of the characteristics of a chemical family is that its members usually form similar compounds. Chlorine is a member of the **halogen family,** which includes three other elements in Group 7A/17: fluorine, bromine, and iodine. With some exceptions, these three halogens form acids similar to the acids of chlorine (Table 6.4). Thus, for example, since chloric acid has the formula $HClO_3$, bromic acid must have the formula $HBrO_3$. Rather than simply supplying the names and formulas of these acids, we will present them through Active Examples and thereby give you the opportunity to *learn the system.*

**PREVIEW:** A chemical family is a group of elements having similar chemical properties because of similar electron arrangements. Elements in a chemical family appear in the same group in the periodic table (Section 10.5).

© Cengage Learning/Charles D. Winters

Hypochlorite ion is the ingredient in bleach responsible for its whitening power.

**Table 6.4** Acids of Chlorine

| Acid | Ionization Equation | Anion Name |
|---|---|---|
| Perchloric acid | $HClO_4 \rightarrow H^+ + ClO_4^-$ | Perchlorate ion |
| Chloric acid | $HClO_3 \rightarrow H^+ + ClO_3^-$ | Chlorate ion |
| Chlorous acid | $HClO_2 \rightarrow H^+ + ClO_2^-$ | Chlorite ion |
| Hypochlorous acid | $HClO \rightarrow H^+ + ClO^-$ | Hypochlorite ion |
| Hydrochloric acid | $HCl \rightarrow H^+ + Cl^-$ | Chloride ion |

## Active Example 6.6

Write the name and formula of the *-ic* acid of selenium (Se, Z = 34) and the name and formula of the anion formed by its total ionization (the loss of all hydrogens in the original acid).

You've memorized the five *-ic* acids found in Table 6.2. (Haven't you? If not, do so now and then return to this Active Example.) Selenium is in the same chemical family as the central atom in which of those *-ic* acids? Write its name and formula.

Sulfuric acid, $H_2SO_4$

Both sulfur, S, and selenium, Se, are found in Group 6A/16 of the periodic table.

The *-ic* acid of selenium will have the same number of hydrogens and oxygens as the *-ic* acid of sulfur. Write its formula and name.

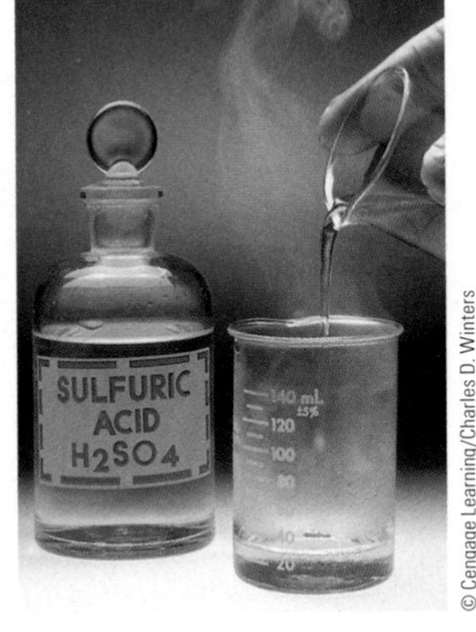


© Cengage Learning/Charles D. Winters

Sulfuric acid is one of the five *-ic* acids whose name and formula must be memorized. A great deal of heat is evolved when sulfuric acid is added to water. Acids are usually purchased in concentrated form and then diluted for laboratory use. An acid is always diluted by adding it to water, to prevent splattering. Never add water to a concentrated acid!

$H_2SeO_4$, selenic acid

$H_2SO_4$ becomes $H_2SeO_4$, and *sulfur*ic acid becomes *selen*ic acid.

Now write the anion formula and name.

$SeO_4^{2-}$, selenate ion

## Active Example 6.7

(a) What is the formula of hydrochloric acid? (b) What is the name of HI?

Fluorine, chlorine, bromine, and iodine are all from the same chemical family. Consider the five memorized *-ic* acids (Table 6.2). Which has a central atom from this family? Write its formula and name.

$HClO_3$, chloric acid

Now apply the system and write the formula of hydrochloric acid to complete part (a).

HCl, hydrochloric acid

The prefix and suffix *hydro-* and *-ic* indicate that all oxygen is removed from the *-ic* acid; thus, the formula is HCl.

You may have noticed that the formulas of hydrochloric acid and hydrogen chloride are the same. When necessary, *state symbols* may be used to distinguish between them. HCl(g) clearly identifies the compound hydrogen chloride as a gas. HCl(aq) refers specifically to the water (*aq*ueous) solution, hydrochloric acid.

(b) Write the name of HI(aq) and, for practice and to clearly understand the distinction, HI(g).

**PREVIEW:** State symbols are commonly used in writing chemical equations, as described in Section 8.1. The two other state symbols are for the solid state, (s), and the liquid state, $(\ell)$.

HI(aq), hydroiodic acid; HI(g), hydrogen iodide

## Active Example 6.8

Complete the name and formula blanks in the table below.

Try to complete this Active Example without referring to the prefixes and suffixes table (Table 6.3), but use it if absolutely necessary. Again, the key is that you may substitute one Group 7A/17 element for another.

| Acid Name | Acid Formula | Anion Formula | Anion Name |
|---|---|---|---|
| Bromic acid | | | |
| | | $IO_3^-$ | |

| Acid Name | Acid Formula | Anion Formula | Anion Name |
|---|---|---|---|
| Bromic acid | $HBrO_3$ | $BrO_3^-$ | Bromate ion |
| Iodic acid | $HIO_3$ | $IO_3^-$ | Iodate ion |

*Thought process:* Bromic acid corresponds with the memorized *-ic* acid; chloric acid, $HClO_3$. The acid and ion formulas come from substituting Br for Cl in the corresponding chlorine formulas. The anion name for an *-ic* acid is the name of the central element changed to end in *-ate*.

$IO_3^-$ corresponds to $ClO_3^-$, the chlorate ion that comes from $HClO_3$, chloric acid. The acid and ion names and formulas are the same, except that iodine

*continued*

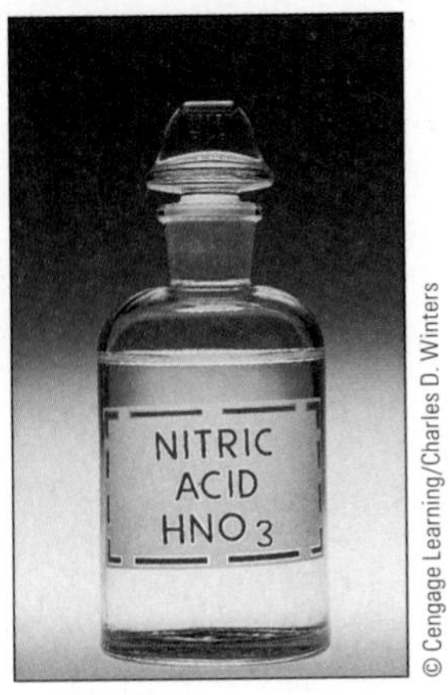


© Cengage Learning/Charles D. Winters

Nitric acid solution, like most acids, is colorless.

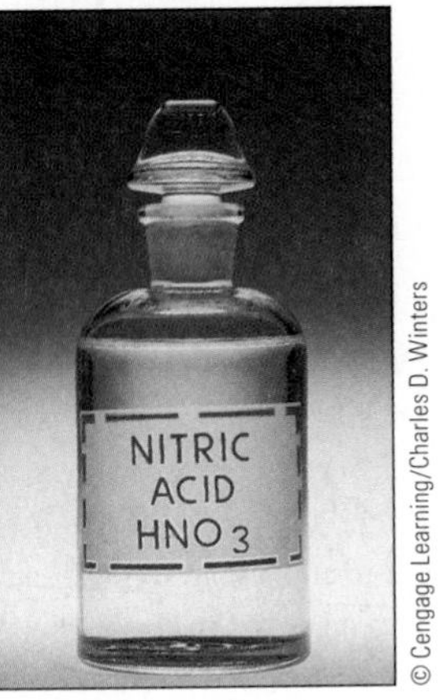


© Cengage Learning/Charles D. Winters

When exposed to sunlight, the solution turns yellow because nitric acid decomposes, forming nitrogen dioxide as one product.

replaces chlorine. Hence, chlorate becomes iodate as the name of $IO_3^-$, chloric becomes iodic as the name of the acid, and $HClO_3$ becomes $HIO_3$ as the acid formula. The acid formula can also be derived directly from the anion formula. The ion has a 1− charge, so the acid must have one hydrogen: $HIO_3$.

Nitric, sulfuric, and phosphoric acids have important variations with different numbers of oxygen atoms. The acid and anion nomenclature system in Table 6.3 remains the same.

## Active Example 6.9

Fill in the name and formula blanks in the following table:

| Acid Name | Acid Formula | Anion Formula | Anion Name |
|---|---|---|---|
| | $HNO_2$ | | |
| | | | Sulfite ion |

| Acid Name | Acid Formula | Anion Formula | Anion Name |
|---|---|---|---|
| Nitrous acid | $HNO_2$ | $NO_2^-$ | Nitrite ion |
| Sulfurous acid | $H_2SO_3$ | $SO_3^{2-}$ | Sulfite ion |

$HNO_2$ has one fewer oxygen atom than nitric acid, $HNO_3$, so its name must be nitr*ous* acid. One hydrogen ion must be removed from the acid formula to produce the ion, $NO_2^-$. The anion from an *-ous* acid has an *-ite* suffix; nitrous acid → nitrite ion.

From the memorized sulfuric acid, $H_2SO_4$, you have the sulfate ion, $SO_4^{2-}$. The sulfite ion has one fewer oxygen, $SO_3^{2-}$. If the anion has a 2− charge, the acid must have two hydrogens: $H_2SO_3$. The name of the acid with one fewer oxygen than sulfuric acid is sulfurous acid.

## Active Example 6.10

Fill in the name and formula blanks in the following table:

| Acid Name | Acid Formula | Anion Formula | Anion Name |
|---|---|---|---|
| Hydrosulfuric acid | | | |
| | | $ClO^-$ | |

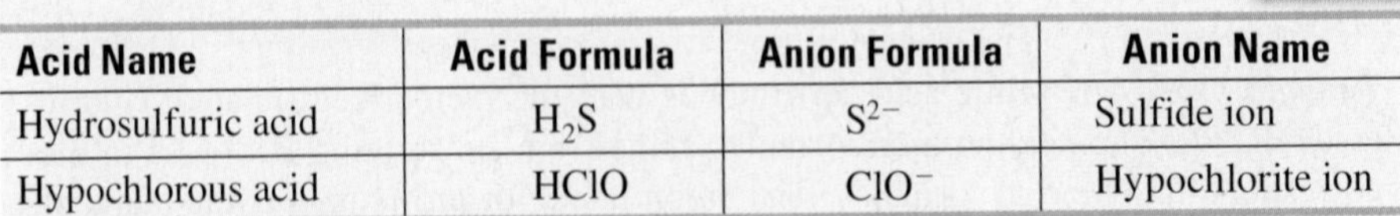

| Acid Name | Acid Formula | Anion Formula | Anion Name |
|---|---|---|---|
| Hydrosulfuric acid | $H_2S$ | $S^{2-}$ | Sulfide ion |
| Hypochlorous acid | $HClO$ | $ClO^-$ | Hypochlorite ion |

We know that hydrosulfuric acid has no oxygen because of the *hydro-* and *-ic* prefix and suffix. Why is its formula $H_2S$ rather than HS? First, sulfur is in Group 6A/16, not 7A/17. Is there another element in Group 6A/16 that forms a compound

**Table 6.5** Acids and Anions Derived from Their Total Ionization

| 4A 14 | 5A 15 | 6A 16 | 7A 17 |
|---|---|---|---|
| $H_2CO_3 \rightarrow CO_3^{2-}$ | $HNO_3 \rightarrow NO_3^-$ | | $HOF \rightarrow OF^-$ |
| | $HNO_2 \rightarrow NO_2^-$ | | $HF \rightarrow F^-$ |
| | $H_3PO_4 \rightarrow PO_4^{3-}$ | $H_2SO_4 \rightarrow SO_4^{2-}$ | $HClO_4 \rightarrow ClO_4^-$ |
| | | $H_2SO_3 \rightarrow SO_3^{2-}$ | $HClO_3 \rightarrow ClO_3^-$ |
| | | $H_2S \rightarrow S^{2-}$ | $HClO_2 \rightarrow ClO_2^-$ |
| | | | $HClO \rightarrow ClO^-$ |
| | | | $HCl \rightarrow Cl^-$ |
| | $H_3AsO_4 \rightarrow AsO_4^{3-}$ | $H_2SeO_4 \rightarrow SeO_4^{2-}$ | $HBrO_4 \rightarrow BrO_4^-$ |
| | | $H_2SeO_3 \rightarrow SeO_3^{2-}$ | $HBrO_3 \rightarrow BrO_3^-$ |
| | | $H_2Se \rightarrow Se^{2-}$ | $HBrO_2 \rightarrow BrO_2^-$ |
| | | | $HBrO \rightarrow BrO^-$ |
| | | | $HBr \rightarrow Br^-$ |
| | | $H_2TeO_4 \rightarrow TeO_4^{2-}$ | $HIO_4 \rightarrow IO_4^-$ |
| | | $H_2TeO_3 \rightarrow TeO_3^{2-}$ | $HIO_3 \rightarrow IO_3^-$ |
| | | $H_2Te \rightarrow Te^{2-}$ | $HIO_2 \rightarrow IO_2^-$ |
| | | | $HIO \rightarrow IO^-$ |
| | | | $HI \rightarrow I^-$ |

The five *-ic* acids from Table 6.2 are highlighted in white. You must memorize these. You can name many other acids and anions by learning the system.

with hydrogen whose formula you know? How about oxygen and its famous hydrogen compound, $H_2O$, also known as water? If the compound of hydrogen and oxygen is $H_2O$, then the compound of hydrogen and sulfur should be $H_2S$. This kind of reasoning shows you how you can use the periodic table to predict names or formulas of compounds.

Table 6.5 summarizes this section. It is in the form of Groups 4A/14 through 7A/17 of the periodic table. Each entry in the table shows the formula of an acid to the left of an arrow. To the right is the formula of the anion that results from the total ionization of the acid. The table includes all the acids and anions whose names and formulas can be figured out by the nomenclature system we have described. The key acids and anions—the ones that are the basis of this part of the nomenclature system—are highlighted in white. If you have memorized the key acids and understand the system, you should be able to figure out any name or formula in this table, given the formula or name.

The following Active Example gives you the opportunity to practice what you have learned about acid and anion nomenclature. If you have memorized what is necessary and know how to apply the rules, you will be able to write the required names and formulas with reference to nothing other than a periodic table. If you have really mastered the system, you will be able to extend it to the last substance in each column.

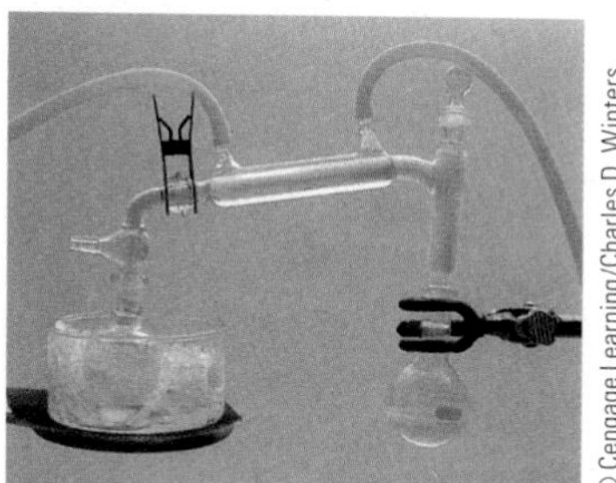


Nitric acid, one of the five memorized *-ic* acids, can be prepared from the reaction of sulfuric acid and sodium nitrate.

## Active Example 6.11

For each of the following names, write the formula; for each formula, write the name.

You may find each name or formula by memorizing the acids found in Table 6.2 and applying the system given in Table 6.3. Review the previous sections of this chapter, as necessary, before proceeding with this Active Example. When you are ready, proceed with no references other than a "clean" periodic table—one that has no information other than what is found on your shield.

*continued*

| | | | |
|---|---|---|---|
| Phosphoric acid | ____________ | $CO_3^{2-}$ | ____________ |
| Sulfate ion | ____________ | HF | ____________ |
| Bromous acid | ____________ | $NO_2^-$ | ____________ |

Phosphoric acid, $H_3PO_4$
Sulfate ion, $SO_4^{2-}$
Bromous acid, $HBrO_2$

$CO_3^{2-}$, carbonate ion
HF, hydrofluoric acid
$NO_2^-$, nitrite ion

## 6.6 Names and Formulas of Acid Anions

**Goal** 6 Given the name (or formula) of an ion formed by the step-by-step ionization of a polyprotic acid from a Group 4A/14, 5A/15, or 6A/16 element, write its formula (or name).

**P/REVIEW:** *Polyprotic acid* is a general term that refers to all acids that have more than one ionizable hydrogen (Section 6.5).

Polyprotic acids do not lose their hydrogens all at once, but rather one at a time. The intermediate anions produced are stable chemical species* that are the negative ions in many ionic compounds. The hydrogen-bearing **acid anion,** as it is called, releases a hydrogen ion when dissolved in water, just like any other acid.

Baking soda, commonly found in kitchen cabinets, contains the acid anion $HCO_3^-$. It can be regarded as the intermediate step in the ionization of carbonic acid:

$$H_2CO_3 \xrightarrow{-H^+} HCO_3^- \xrightarrow{-H^+} CO_3^{2-}$$

$HCO_3^-$ is the hydrogen carbonate ion—a logical name, since the ion is literally a hydrogen ion bonded to a carbonate ion. The ion is also called the bicarbonate ion.

Phosphoric acid, $H_3PO_4$, has three steps in its ionization process:

$$H_3PO_4 \xrightarrow{-H^+} H_2PO_4^- \xrightarrow{-H^+} HPO_4^{2-} \xrightarrow{-H^+} PO_4^{3-}$$

$H_2PO_4^-$ is the dihydrogen phosphate ion, signifying two hydrogen ions attached to a phosphate ion. $HPO_4^{2-}$ is the monohydrogen phosphate ion, or simply the hydrogen phosphate ion. It is essential that the prefix *di-* be used in naming the $H_2PO_4^-$ ion to distinguish it from $HPO_4^{2-}$, but the prefix *mono-* is usually omitted in naming $HPO_4^{2-}$.

A carbonic acid–hydrogen carbonate ion system and a dihydrogen phosphate ion–hydrogen phosphate ion system are two of the three major chemical systems that regulate the acidity of your blood.

If you recognize the logic of this part of the nomenclature system, you will be able to extend it to intermediate ions from the step-by-step ionization of hydrosulfuric, sulfuric, and sulfurous acids. All of these are shown in Table 6.6.

**Table 6.6** Names and Formulas of Anions Derived from the Step-by-Step Ionization of Acids

| | | NAMES OF IONS | |
|---|---|---|---|
| **Acid** | **Ion** | **Preferred** | **Other** |
| $H_2CO_3$ | $HCO_3^-$ | Hydrogen carbonate | Bicarbonate; acid carbonate |
| $H_2S$ | $HS^-$ | Hydrogen sulfide | Bisulfide; acid sulfide |
| $H_2SO_4$ | $HSO_4^-$ | Hydrogen sulfate | Bisulfate; acid sulfate |
| $H_2SO_3$ | $HSO_3^-$ | Hydrogen sulfite | Bisulfite; acid sulfite |
| $H_3PO_4$ | $H_2PO_4^-$ | Dihydrogen phosphate | Monobasic phosphate |
| $H_2PO_4^-$ | $HPO_4^{2-}$ | Hydrogen phosphate | Dibasic phosphate |

*In this book, *species* is used as a generic term for any chemical particle, such as atom, ion, or molecule.

## 6.7 Names and Formulas of Other Acids and Ions

**Goal 7** Given the name (or formula) of the ammonium ion or hydroxide ion, write the corresponding formula (or name).

The U.S. Department of Transportation requires all highway or rail shipments of acids or bases of over 1000 pounds to display the international warning placard for corrosive materials.

A small number of relatively common acids and ions do not fit within our nomenclature system. Follow the advice of your instructor as to whether none, some, or all of these should be committed to memory. We recommend that, at a minimum, you make the ammonium ion and the hydroxide ion part of your chemical vocabulary, as specified in Goal 7.

**The Ammonium Ion and the Hydroxide Ion** The ammonium and hydroxide ions are very common. Both are binary. The ammonium ion, $NH_4^+$, with its 1+ charge, shares many chemical properties with the monatomic cations formed by Group 1A/1 elements. Even its name, with an *-ium* ending, is similar. The *-ide* ending for the hydroxide ion, $OH^-$ and its 1− charge suggest a monatomic anion belonging in Group 7A/17. The chemical behavior of the hydroxide ion is in many ways like that of the monatomic anions formed by Group 7A/17 elements. You will encounter the ammonium and hydroxide ions often. Memorize their names and formulas.

**Acetic Acid and the Acetate Ion** Acetic acid, $HC_2H_3O_2$, is the component of vinegar that is responsible for its odor and taste. Its ionization equation is

$$HC_2H_3O_2 \rightarrow H^+ + C_2H_3O_2^-$$

Notice that only the hydrogen written first in the formula ionizes; the others do not. This is typical of organic acids, which usually produce anions containing carbon, hydrogen, and oxygen. The $C_2H_3O_2^-$ ion is the acet*ate* ion, as might be expected from the *-ic* ending in the name of the acid, acet*ic*.

**Hydrocyanic Acid and the Cyanide Ion** When hydrocyanic acid, HCN, ionizes, it produces the cyanide ion, $CN^-$. Both the acid and the anion are exceptions to the nomenclature rules. The acid name suggests a binary acid, and the anion name suggests a monatomic anion.

**Polyatomic Anions from Transition Elements** Chromium and manganese form some polyatomic anions that theoretically may be traced to acids, but only the ions are important. Their names and formulas are the chromate ion, $CrO_4^{2-}$, the dichromate ion, $Cr_2O_7^{2-}$, and the permanganate ion, $MnO_4^-$.

**Other Ions** The learning objectives identify the ions whose names and formulas you should be able to recognize and write. There are many others, too. Some of these, plus the ions already discussed, are listed in Tables 6.7 and 6.8. We recommend that you use these tables as a reference for the less common ions and only as a last resort if you forget one of the ions you should be able to name by applying the nomenclature system.

## 6.8 Formulas of Ionic Compounds

**Goal 8** Given the name of any ionic compound made up of ions included in Goals 4 through 7, or other ions whose formulas are given, write the formula of that compound.

Ionic compounds are hard and brittle solids at room temperature. This photograph shows (clockwise from the rear center) sodium chloride, calcium fluoride, iron(III) oxide, and copper(II) bromide.

Chemical compounds are electrically neutral. For **ionic compounds** this means that the formula unit must have an equal number of positive and negative charges. A net zero charge is achieved by combining cations and anions in such numbers that positive and negative charges are balanced. This is done in two steps:

procedure

**How to Write the Formula of an Ionic Compound**

**Step 1:** Write the formula of the cation, followed by the formula of the anion, omitting the charges.

**Step 2:** Insert subscripts to show the number of each ion needed in the formula unit to make the sum of the charges equal to zero with the fewest number of ions possible.

a) If only one ion is needed, omit the subscript.

b) If a polyatomic ion is needed more than once, enclose the formula of the ion in parentheses and place the subscript after the closing parenthesis.

**Table 6.7** Cations

| Ionic Charge: 1+ | | Ionic Charge: 2+ | | Ionic Charge: 3+ | |
|---|---|---|---|---|---|
| *Alkali Metals: Group 1A/1* | | *Alkaline Earths: Group 2A/2* | | *Group 3A/3* | |
| $Li^+$ | Lithium | $Be^{2+}$ | Beryllium | $Al^{3+}$ | Aluminum |
| $Na^+$ | Sodium | $Mg^{2+}$ | Magnesium | $Ga^{3+}$ | Gallium |
| $K^+$ | Potassium | $Ca^{2+}$ | Calcium | | |
| $Rb^+$ | Rubidium | $Sr^{2+}$ | Strontium | | |
| $Cs^+$ | Cesium | $Ba^{2+}$ | Barium | | |
| *Transition Elements* | | *Transition Elements* | | *Transition Elements* | |
| $Cu^+$ | Copper(I) | $Cr^{2+}$ | Chromium(II) | $Cr^{3+}$ | Chromium(III) |
| $Ag^+$ | Silver | $Mn^{2+}$ | Manganese(II) | $Mn^{3+}$ | Manganese(III) |
| *Polyatomic Ions* | | $Fe^{2+}$ | Iron(II) | $Fe^{3+}$ | Iron(III) |
| $NH_4^+$ | Ammonium | $Co^{2+}$ | Cobalt(II) | $Co^{3+}$ | Cobalt(III) |
| *Others* | | $Ni^{2+}$ | Nickel | | |
| $H^+$ | Hydrogen | $Cu^{2+}$ | Copper(II) | | |
| *or* | | $Zn^{2+}$ | Zinc | | |
| $H_3O^+$ | Hydronium | $Cd^{2+}$ | Cadmium | | |
| | | $Hg_2^{2+}$ | Mercury(I) | | |
| | | $Hg^{2+}$ | Mercury(II) | | |
| | | *Others* | | | |
| | | $Sn^{2+}$ | Tin(II) | | |
| | | $Pb^{2+}$ | Lead(II) | | |

**Table 6.8** Anions

| Ionic Charge: 1− | | | | Ionic Charge: 2− | | Ionic Charge: 3− | |
|---|---|---|---|---|---|---|---|
| *Halogens: Group 7A/17* | | *Oxyanions* | | *Group 6A/16* | | *Group 5A/15* | |
| $F^-$ | Fluoride | $ClO_4^-$ | Perchlorate | $O^{2-}$ | Oxide | $N^{3-}$ | Nitride |
| $Cl^-$ | Chloride | $ClO_3^-$ | Chlorate | $S^{2-}$ | Sulfide | $P^{3-}$ | Phosphide |
| $Br^-$ | Bromide | $ClO_2^-$ | Chlorite | *Oxyanions* | | *Oxyanion* | |
| $I^-$ | Iodide | $ClO^-$ | Hypochlorite | $CO_3^{2-}$ | Carbonate | $PO_4^{3-}$ | Phosphate |
| *Acid Anions* | | $BrO_3^-$ | Bromate | $SO_4^{2-}$ | Sulfate | | |
| $HCO_3^-$ | Hydrogen carbonate | $BrO_2^-$ | Bromite | $SO_3^{2-}$ | Sulfite | | |
| $HS^-$ | Hydrogen sulfide | $BrO^-$ | Hypobromite | $C_2O_4^{2-}$ | Oxalate | | |
| $HSO_4^-$ | Hydrogen sulfate | $IO_4^-$ | Periodate | $CrO_4^{2-}$ | Chromate | | |
| $HSO_3^-$ | Hydrogen sulfite | $IO_3^-$ | Iodate | $Cr_2O_7^{2-}$ | Dichromate | | |
| $H_2PO_4^-$ | Dihydrogen phosphate | $NO_3^-$ | Nitrate | *Acid Anions* | | | |
| *Other Anions* | | $NO_2^-$ | Nitrite | $HPO_4^{2-}$ | Hydrogen phosphate | | |
| $SCN^-$ | Thiocyanate | $OH^-$ | Hydroxide | *Diatomic Elemental* | | | |
| $CN^-$ | Cyanide | $C_2H_3O_2^-$ | Acetate | $O_2^{2-}$ | Peroxide | | |
| $H^-$ | Hydride | $MnO_4^-$ | Permanganate | | | | |

## Active Example 6.12

Write the formulas of calcium chloride and calcium hydroxide.

Begin by writing the formulas of the three ions in the two compounds.

$Ca^{2+}$ $Cl^-$ $OH^-$

Now decide how many $Ca^{2+}$ ions must combine with how many $Cl^-$ ions to produce a total charge of zero in the formula of calcium chloride. Do the same for the $Ca^{2+}$ and $OH^-$ ions.

1 $Ca^{2+}$ ion + 2 $Cl^-$ ions; 1 $Ca^{2+}$ ion + 2 $OH^-$ ions

This time $(2+) + 2 \times (1-) = 0$ for both compounds.

Now follow the two-step procedure for both compounds. Be careful about where and how you use parentheses.

Calcium chloride, $CaCl_2$; calcium hydroxide, $Ca(OH)_2$

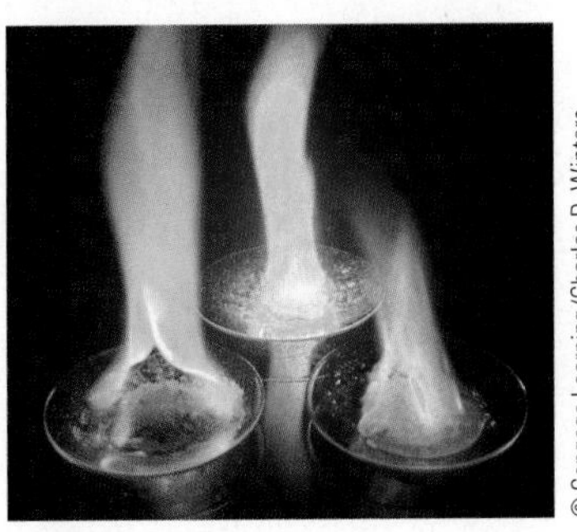



Cations can be identified by the characteristic colors of their flames. Calcium chloride, strontium chloride, and barium chloride were dissolved in methanol, and the solutions were ignited. Calcium ion gives a yellow-orange flame, strontium ion gives a red flame, and the barium ion flame is yellow-green.

Active Example 6.12 illustrates two important points:

1. *Parentheses are used in a chemical formula to enclose a polyatomic ion that is used more than once.* That is the case with calcium hydroxide; two $OH^-$ ions are needed to combine with the one $Ca^{2+}$ ion: $Ca^{2+} + OH^- + OH^-$. There are two tempting but incorrect ways to write the formulas of calcium hydroxide. First, writing $CaO_2H_2$ gives a correct atom count in the formula, 1 Ca, 2 O, and 2 H, but the identity of the hydroxide ion, $OH^-$, has been lost. Second, writing $CaOH_2$ gives an atom count of 1 Ca, 1 O, and 2 H, which is not correct.
2. *Parentheses are used only with* polyatomic *ions.* Notice that calcium chloride is $CaCl_2$, not $Ca(Cl)_2$. Even though its symbol has two letters, $Cl^-$ is a monatomic ion.

## Active Example 6.13

Write the formulas of potassium nitrate and calcium nitrate.

First, write the formulas of the three ions.

$K^+$ $Ca^{2+}$ $NO_3^-$

What will be the numbers of cations and anions in each formula?

*continued*

© Cengage Learning/Charles D. Winters

Potassium nitrate is a white solid at room temperature. Other compounds containing potassium ion include orange potassium dichromate, $K_2Cr_2O_7$, and purple potassium permanganate, $KMnO_4$.

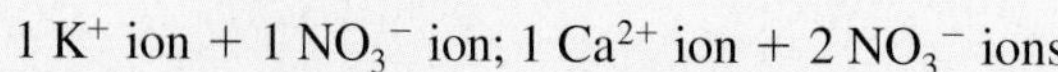

1 $K^+$ ion + 1 $NO_3^-$ ion; 1 $Ca^{2+}$ ion + 2 $NO_3^-$ ions

$(1+) + (1-) = 0$ for potassium nitrate; $(2+) + 2 \times (1-) = 0$ for calcium nitrate.

Now write the two formulas. Again, be careful about how you use parentheses.

Potassium nitrate, $KNO_3$; calcium nitrate, $Ca(NO_3)_2$

The subscript 3 is part of the nitrate ion formula, so it goes inside the parentheses. The two nitrate ions are shown by the subscript 2 after the closing parenthesis.

The most challenging ion combination occurs when one ion has a charge of 3 and the other a charge of 2. The lowest common multiple is 6, so to obtain balance you must have two ions with a charge of 3 and three ions with a charge of 2. The next Active Example illustrates this with both monatomic and polyatomic ions.

### Active Example 6.14

Write the formulas of aluminum oxide and barium phosphate.

Start with aluminum oxide. Remember, you need two 3s and three 2s.

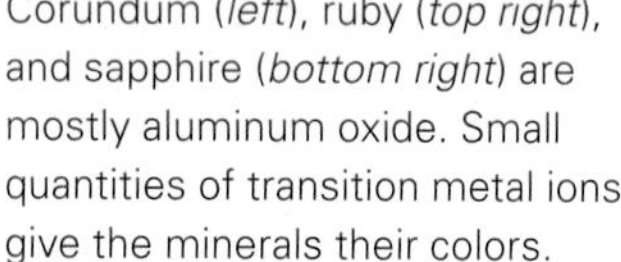

© Cengage Learning/Charles D. Winters

Corundum (*left*), ruby (*top right*), and sapphire (*bottom right*) are mostly aluminum oxide. Small quantities of transition metal ions give the minerals their colors.

$Al_2O_3$

The ions are $Al^{3+}$ and $O^{2-}$. Two $Al^{3+}$ ions and three $O^{2-}$ ions give a charge balance of $2 \times (3+) + 3 \times (2-) = 0$ with the least number of each ion. If you wrote $Al_4O_6$, you have the charge balanced, but you did not do it with the fewest number of ions possible. Both 4 and 6 are divisible by 2 to give $Al_2O_3$.

Now try it with barium phosphate.

John Luke/Index Open

Even though airplanes are frequently exposed to rain and snow, they never rust. A thin layer of aluminum oxide forms on the surface of the aluminum metal used for the exterior skin of the aircraft. The aluminum oxide serves as a chemically inert, protective rust-resistant shield.

$Ba_3(PO_4)_2$

$Ba^{2+}$ and $PO_4^{3-}$ are the ions. Three $Ba^{2+}$ + two $PO_4^{3-}$ give a charge balance of $3 \times (2+) + 2 \times (3-) = 0$.

## 6.9 Names of Ionic Compounds

**Goal** 9 Given the formula of an ionic compound made up of identifiable ions, write the name of the compound.

procedure

**How to Write the Name of an Ionic Compound**

**Step 1:** Write the name of the cation.

**Step 2:** Write the name of the anion.

If you recognize the names of the two ions, you have the name of the compound. Note that prefixes are *not* included in the names of ionic compounds.

There is one case where you might be uncertain about the name of the cation. For example, what is the name of $FeCl_3$? Iron chloride is not an adequate answer. It fails to distinguish between the two possible charges on the iron ion. Is $FeCl_3$ iron(II) chloride or iron(III) chloride? To decide, you must reason from the known charge on the chloride ion, 1−, and the fact that the total charge on the compound is zero. The formula has three chloride ions, so the total negative charge is 3−. This must be balanced by three positive charges from the iron ion, so it must be an iron(III) ion. The compound is iron(III) chloride. If the formula had been $FeCl_2$, you would have reached the name iron(II) chloride by recognizing that the 2− of the two chloride ions is balanced by the 2+ of a single iron(II) ion.

In writing or speaking the name of an ionic compound containing a metal that commonly is *capable of* having more than one ionic charge, *the compound name includes the charge of that metal.*

## Active Example 6.15

Write the name of each of the following compounds:

LiBr ______

$Mg(IO_4)_2$ ______

$AgNO_3$ ______

$MnCl_3$ ______

$Hg_2Br_2$ ______

LiBr, lithium bromide
$Mg(IO_4)_2$, magnesium periodate
$AgNO_3$, silver nitrate
$MnCl_3$, manganese(III) chloride
$Hg_2Br_2$, mercury(I) bromide

In $MnCl_3$, three 1− charges from three $Cl^-$ ions require 3+ from the manganese ion, so it is the manganese(III) ion. In $Hg_2Br_2$, the two 1− charges from two $Br^-$ ions are balanced by the 2+ charge from the diatomic mercury(I) ion.

## 6.10 Hydrates

Goal 10 Given the formula of a hydrate, state the number of water molecules associated with each formula unit of the anhydrous compound.

Goal 11 Given the name (or formula) of a hydrate, write its formula (or name). (This goal is limited to hydrates of ionic compounds for which a name and formula can be written based on the rules of nomenclature presented in this book.)

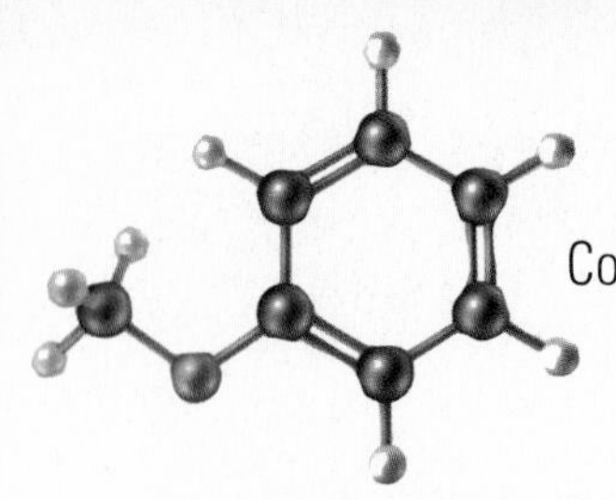

# Everyday Chemistry

## Common Names of Chemicals

Although chemists agree that the system of nomenclature presented in this chapter gives the official names of compounds, many practicing scientists still use *un*official, or *trivial,* names when talking about many substances. Just as there are variations in the English language that depend on whether you are in Australia, Great Britain, Canada, or the United States, or even in a region within the United States, there are variations in the language of chemistry that depend on an individual's specialization within chemistry, or on whether the person is a biologist, another type of scientist, or a nonscientist.

Methanol, also known as methyl alcohol or wood alcohol, is used as a chemical building block in the manufacture of wood paneling, paints, adhesives, and fuels. It is poisonous when consumed by humans.

*Alcohol* is an interesting example of a term that has many meanings. In everyday language, alcohol refers to what is officially called *ethanol,* $CH_3CH_2OH$. However, many chemists use the similar term *ethyl alcohol* for this compound. Additionally, a chemist would *never* refer to this substance simply as alcohol. In chemistry, the term *alcohol* refers to an entire class of compounds that have a certain carbon–oxygen–hydrogen arrangement within a molecule. Prohibition-era bootleggers were careful to distinguish between the very poisonous but legal *wood alcohol* (*methanol* or *methyl alcohol* to a chemist) and *grain alcohol,* which is ethanol. Wood alcohol can be produced from wood smoke, and grain alcohol can be made by fermenting grain, and thus they are aptly named. There's yet another common alcohol, *rubbing alcohol,* which is used to cool the skin rapidly. A chemist would probably call this substance *isopropyl alcohol,* although its official name is *isopropanol.*

*Sugar* is similar to *alcohol* in that it refers to a specific substance in everyday language, but to an entire class of compounds in scientific language. Everyday usage of the term *sugar* refers to the substance *sucrose,* $C_{12}H_{22}O_{11}$, which has the official name *α-D-glucopyranosyl-(1 → 2)-β-D-fructofuranoside.* We doubt that you'll be surprised to learn that most chemists avoid the official name! Ordinary table sugar is also called *cane sugar,* when it is processed from sugar cane, or *beet sugar,* when it comes from sugar beets. Whatever its origin, pure table sugar is pure sucrose. To a chemist, the term *sugar* refers to a class of compounds that have the general formula $C_nH_{2n}O_n$, where n varies from three to nine. There are literally dozens of common sugars, such as lactose (milk sugar), glucose (blood sugar), and fructose (fruit sugar).

People joke about alternate names for *water,* which in itself appears to be a trivial name, but since it is so entrenched in our language, it is also the official chemical name for $H_2O$. Since it is a binary molecular compound, it could be called dihydrogen monoxide. Using the letter abbreviations for the prefixes and stem words, water can be abbreviated DHMO. You may have seen some of the many "warnings" about DHMO circulated by email or at Web sites on the Internet. These are just poking fun at those who are unaware of this way of (improperly) naming water.

Many of these trivial names have been used for a century or more, and many will continue to be used well into the future in spite of attempts to standardize the system for naming chemical compounds. Table 6.10 has many more common names of chemicals.

Invert sugar is produced by the "inversion" (chemical breakdown) of table sugar, resulting in a mixture of glucose and fructose.

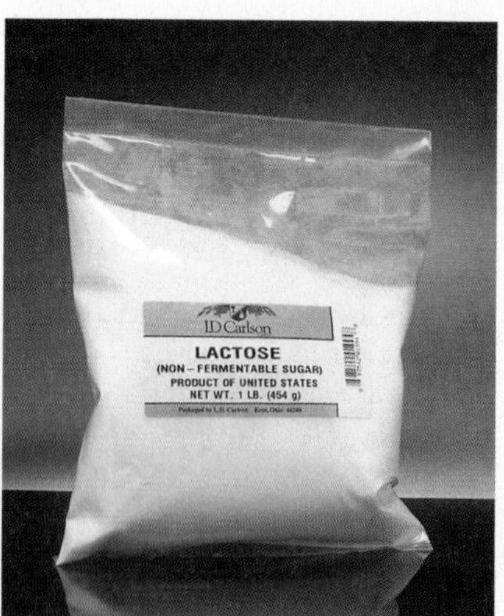




Lactose is the predominant sugar in milk.

Some compounds, when crystallized from water solutions, form solids that include water molecules as part of the crystal structure. Such water is referred to as **water of crystallization** or **water of hydration.** The compound is said to be **hydrated** and is called a **hydrate.** Hydration water can usually be driven from a compound by heating, leaving the **anhydrous compound.**

Copper(II) sulfate is an example of a hydrate. The anhydrous compound, $CuSO_4$, is a nearly white powder. Each formula unit of $CuSO_4$ combines with five water molecules in the hydrate, which is a blue crystal (Fig. 6.5). Its formula is $CuSO_4 \cdot 5\ H_2O$. The number of water molecules that crystallize with each formula unit of the anhydrous compound is shown after the anhydrous formula, separated by a dot. This number is 5 for the hydrate of copper(II) sulfate. The equation for the dehydration of this compound is

$$CuSO_4 \cdot 5\ H_2O \rightarrow CuSO_4 + 5\ H_2O$$

Just as in binary molecular compounds, prefixes are used to indicate the number of water molecules in a formula unit of a hydrate (see Table 6.1). By this system, $CuSO_4 \cdot 5\ H_2O$ is called copper(II) sulfate pentahydrate, since *penta-* is the prefix for 5.

**Figure 6.5** A sample of hydrated and anhydrous copper(II) sulfate pentahydrate. The blue compound in the center of the crucible is $CuSO_4 \cdot 5\ H_2O$; the compound near the edges is $CuSO_4$.



### Active Example 6.16

a) How many water molecules are associated with each formula unit of anhydrous sodium carbonate in $Na_2CO_3 \cdot 10\ H_2O$? Name the hydrate.
b) Write the formula of nickel chloride hexahydrate.

a) Ten; sodium carbonate decahydrate
b) $NiCl_2 \cdot 6\ H_2O$

Gypsum is the common name for $CaSO_4 \cdot 2\ H_2O$, which has the chemical name calcium sulfate dihydrate.

## 6.11 Summary of the Nomenclature System

Throughout this chapter we have emphasized memorizing certain names and formulas and some prefixes and suffixes. They are the basis for the system of chemical nomenclature. Once you have memorized them, you simply apply the system to the different names and formulas you encounter. Table 6.9 summarizes all the ideas that have been presented. It should help you learn this nomenclature system.

## 6.12 Common Names of Chemicals

Every day we encounter and use a large number of chemicals. Many of these are listed in Table 6.10, along with their common names, chemical names, and chemical formulas. How many of these do you recognize? How many did you learn to name in this chapter?

**Table 6.9** Summary of Nomenclature System

| Substance | Name | Formula |
|---|---|---|
| Element | Name of element | Symbol of element; exceptions: $H_2$, $N_2$, $O_2$, $F_2$, $Cl_2$, $Br_2$, $I_2$ |
| Compounds made up of two nonmetals | First element in formula followed by second, changed to end in *-ide,* each element preceded by prefix to show the number of atoms in the molecule | Symbol of first element in name followed by symbol of second element, with subscripts to show number of atoms in molecule |
| Acid | Most common: middle element changed to end in *-ic*<br>One more oxygen than *-ic* acid: add prefix *per-* to name of *-ic* acid<br>One fewer oxygen than *-ic* acid: change ending of *-ic* acid to *-ous*<br>Two fewer oxygens than *-ic* acid: add prefix *hypo-* to name of *-ous* acid<br>No oxygen: Prefix *hydro-* followed by name of second element changed to end in *-ic* | H followed by symbol of nonmetal followed by O (if necessary), each with appropriate subscript.<br><br>*Memorize the following:*<br>Chloric acid $HClO_3$<br>Nitric acid $HNO_3$<br>Sulfuric acid $H_2SO_4$<br>Carbonic acid $H_2CO_3$<br>Phosphoric acid $H_3PO_4$ |
| Monatomic cation | Name of element followed by ion; if element forms more than one monatomic cation, elemental name is followed by ion charge in Roman numerals and in parentheses | Symbol of element followed by superscript to indicate charge |
| Monatomic anion | Name of element changed to end in *-ide* | Symbol of element followed by superscript to indicate charge |
| Polyatomic anion from total ionization of oxyacid | Replace *-ic* in acid name with *-ate,* or replace *-ous* in acid name with *-ite,* followed by ion | Acid formula without hydrogen plus superscript showing negative charge equal to number of hydrogens removed from acid formula |
| Polyatomic anion from step-by-step ionization of oxyacid | Hydrogen followed by name of ion from total ionization of acid (dihydrogen in the case of $H_2PO_4^-$) | Acid formula minus one (or two for $H_3PO_4$) hydrogen(s), plus superscript showing negative charge equal to number of hydrogen removed from acid formula |
| Other polyatomic ions | Ammonium ion<br>Hydroxide ion | $NH_4^+$<br>$OH^-$ |
| Ionic compound | Name of cation followed by name of anion | Formula of cation followed by formula of anion, each taken as many times as necessary to yield a net charge of zero (polyatomic ion formulas enclosed in parentheses if taken more than once) |
| Hydrate | Name of anhydrous compound followed by (number prefix)hydrate, where (number prefix) indicates the number of water molecules associated with one formula unit of anhydrous compound | Formula of anhydrous compound followed by "$\cdot$ n $H_2O$" where n is number of water molecules associated with one formula unit of anhydrous compound |

**Table 6.10** Common Names of Chemicals

| Common Name | Chemical Name | Formula |
|---|---|---|
| Alumina | Aluminum oxide | $Al_2O_3$ |
| Baking soda | Sodium hydrogen carbonate | $NaHCO_3$ |
| Bleach (liquid) | Hydrogen peroxide or sodium hypochlorite | $H_2O_2$ $NaClO$ |
| Bleach (solid) | Sodium perborate | $NaBO_2 \cdot H_2O_2 \cdot 3\,H_2O$ |
| Bluestone | Copper(II) sulfate pentahydrate | $CuSO_4 \cdot 5\,H_2O$ |
| Borax | Sodium tetraborate decahydrate | $Na_2B_4O_7 \cdot 10\,H_2O$ |
| Brimstone | Sulfur | S |
| Carbon tetrachloride | Tetrachloromethane | $CCl_4$ |
| Chile saltpeter | Sodium nitrate | $NaNO_3$ |
| Chloroform | Trichloromethane | $CHCl_3$ |
| Cream of tartar | Potassium hydrogen tartrate | $KHC_4H_4O_6$ |
| Diamond | Carbon | C |
| Dolomite | Calcium magnesium carbonate | $CaCO_3 \cdot MgCO_3$ |
| Epsom salt | Magnesium sulfate heptahydrate | $MgSO_4 \cdot 7\,H_2O$ |
| Freon (refrigerant) | Dichlorodifluoromethane | $CCl_2F_2$ |
| Galena | Lead(II) sulfide | PbS |
| Grain alcohol | Ethyl alcohol; ethanol | $C_2H_5OH$ |
| Graphite | Carbon | C |
| Gypsum | Calcium sulfate dihydrate | $CaSO_4 \cdot 2\,H_2O$ |
| Hypo | Sodium thiosulfate | $Na_2S_2O_3$ |
| Laughing gas | Dinitrogen monoxide | $N_2O$ |
| Lime | Calcium oxide | CaO |
| Limestone | Calcium carbonate | $CaCO_3$ |
| Lye | Sodium hydroxide | NaOH |
| Marble | Calcium carbonate | $CaCO_3$ |
| MEK | Methyl ethyl ketone | $CH_3COC_2H_5$ |
| Milk of magnesia | Magnesium hydroxide | $Mg(OH)_2$ |
| Muriatic acid | Hydrochloric acid | HCl |
| Oil of vitriol | Sulfuric acid (conc.) | $H_2SO_4$ |
| Plaster of Paris | Calcium sulfate-hydrate | $CaSO_4 \cdot \frac{1}{2}\,H_2O$ |
| Potash | Potassium carbonate | $K_2CO_3$ |
| Pyrite (fool's gold) | Iron(II) disulfide | $FeS_2$ |
| Quartz | Silicon dioxide | $SiO_2$ |
| Quicksilver | Mercury | Hg |
| Rubbing alcohol | Isopropyl alcohol; isopropanol | $(CH_3)_2CHOH$ |
| Sal ammoniac | Ammonium chloride | $NH_4Cl$ |
| Salt | Sodium chloride | NaCl |
| Salt substitute | Potassium chloride | KCl |
| Saltpeter | Potassium nitrate | $KNO_3$ |
| Slaked lime | Calcium hydroxide | $Ca(OH)_2$ |
| Sugar | Sucrose | $C_{12}H_{22}O_{11}$ |
| TSP (trisodium phosphate) | Sodium phosphate | $Na_3PO_4$ |
| Washing soda | Sodium carbonate decahydrate | $Na_2CO_3 \cdot 10\,H_2O$ |
| Wood alcohol | Methyl alcohol; methanol | $CH_3OH$ |

# Chapter 6 in Review

*Most of the key terms and concepts and many others appear in the Glossary. Use your Glossary regularly.*

**6.1 Introduction to Nomenclature**

**Key Terms and Concepts: Chemical nomenclature**

**6.2 Formulas of Elements**

Goal 1 Given a name or formula of an element in Figure 5.8, write the other.

**Key Terms and Concepts: Diatomic molecule, formula unit**

**6.3 Compounds Made from Two Nonmetals**

Goal 2 Given the name or formula of a binary molecular compound, write the other.

Goal 3 Given the name or the formula of water, write the other; given the name or the formula of ammonia, write the other.

**Key Terms and Concepts: Binary molecular compound**

**6.4 Names and Formulas of Ions Formed by One Element**

Goal 4 Given the name or formula of an ion in Figure 6.3, write the other.

**Key Terms and Concepts: Anion, cation, ion, monatomic ion, oxidation state (number)**

**6.5 Acids and the Anions Derived from Their Total Ionization**

Goal 5 Given the name (or formula) of an acid or ion in Table 6.5, write its formula (or name).

**Key Terms and Concepts: Acid, halogen family, hydrated hydrogen ion, hydronium ion ($H_3O^+$), ionizable hydrogen, ionize, mono-, di-, tri-, and polyprotic acids, oxyacid, oxyanion, total ionization**

**6.6 Names and Formulas of Acid Anions**

Goal 6 Given the name (or formula) of an ion formed by the step-by-step ionization of a polyprotic acid from a Group 4A/14, 5A/15, or 6A/16 element, write its formula (or name).

**Key Terms and Concepts: Acid anion**

**6.7 Names and Formulas of Other Acids and Ions**

Goal 7 Given the name (or formula) of the ammonium ion or hydroxide ion, write the corresponding formula (or name).

**6.8 Formulas of Ionic Compounds**

Goal 8 Given the name of any ionic compound made up of ions included in Goals 4 through 7, or other ions whose formulas are given, write the formula of that compound.

**Key Terms and Concepts: Ionic compound**

**6.9 Names of Ionic Compounds**

Goal 9 Given the formula of an ionic compound made up of identifiable ions, write the name of the compound.

**6.10 Hydrates**

Goal 10 Given the formula of a hydrate, state the number of water molecules associated with each formula unit of the anhydrous compound.

Goal 11 Given the name (or formula) of a hydrate, write its formula (or name). (This goal is limited to hydrates of ionic compounds for which a name and formula can be written based on the rules of nomenclature presented in this book.)

**Key Terms and Concepts: Anhydrous compound, hydrate, hydrated, water of crystallization (hydration)**

**6.11 Summary of the Nomenclature System**

**6.12 Common Names of Chemicals**

# Study Hints and Pitfalls to Avoid

As noted in the introduction to this chapter, the most important thing you can do to learn nomenclature is to *learn the system.* The system is based on some rules, prefixes, and suffixes that you must memorize. You can then apply the rules in writing the names and formulas of hundreds of chemical substances. This is by far the easiest and quickest way to learn how to write chemical names and formulas.

Remember that the *elements* hydrogen, nitrogen, oxygen, fluorine, chlorine, bromine, and iodine exist as diatomic molecules at common temperatures. Note the limitation; it refers to the uncombined *elements,* not to compounds—either molecules or ions—in which the elements may be present.

Here are a few memory aids that may help you learn the number prefixes in Table 6.1: A *mono*poly is when *one* company controls an economic product or service. A *two*-wheel cycle is a *bi*cycle, but a chemist might call it a *di*cycle. No problem with *three* wheels: it's a *tri*cycle. No help on *tetra-* for four, unless you happen to remember that a *four*-sided solid is called a *tetra*hedron. The *Penta*gon is the *five*-sided building in Washington that serves as headquarters for U.S. military operations. *Six* and *hex-* are the only number/prefix combination that has the letter *x*. If you change the *s* to an *h* in September, the *Hept*ember, *Oct*ober, *Nov*ember, and *Dec*ember list the beginnings of what were once the *seven*th, *eigh*th, *nin(e)*th, and *ten*th months of the year.

Notice that number prefixes are *almost never* used in naming ionic compounds. The *di*hydrogen phosphate ion is the only exception in this chapter, and the *di*chromate ion is in Table 6.8. Number prefixes were used in the past, but today exam graders will find use for their red pens if you write about aluminum trichloride.

Be sure you use parentheses correctly in writing formulas of ionic compounds. They enclose *polyatomic* ions used more than once, but never a monatomic ion. Examples are $BaCl_2$, not $Ba(Cl)_2$; $Ba(OH)_2$, not $BaOH_2$; *but* NaOH, not Na(OH).

A charge, written as a superscript, is included in the formula of *every* ion. Without a charge, it is not an ion. Do not include ionic charge in the formula of an ionic compound, however: $Na_2S$, not $Na^+{}_2S^{2-}$.

The rules for naming binary molecular compounds are sometimes confused with those for naming polyatomic oxyanions. The critical difference in the formulas of the two species is that ions have charges and molecules do not. $NO_2$ is uncharged, and therefore it is a binary molecular compound. Its name is nitrogen dioxide. $NO_2^-$ has a charge; therefore, it is an oxyanion. Its name is nitrite ion.

An oxyacid should be named as an acid, not as an ionic compound. For example, $HNO_3$ is nitric acid, not hydrogen nitrate.

Be sure to use the charge of a cation when naming an ionic compound if the element forms more than one monatomic ion, as with iron(II) ion, $Fe^{2+}$, and iron(III) ion, $Fe^{3+}$. Do *not* use the charge if the element forms only one cation, as with zinc ion, $Zn^{2+}$.

Learning the nomenclature system correctly is the first of two steps. The second is applying the system correctly. To develop this skill, you must practice, practice, and then practice some more until you write names and formulas almost automatically.

The end-of-chapter questions give ample opportunity for practice. Take full advantage of them. In particular, perfect your skill in writing formulas of ionic compounds by completing Formula-Writing Exercises 1 and 2 (Tables 6.13 and 6.14). Your ultimate self-test lies in the last group of questions, in which different kinds of substances are mixed. You must first identify the kind of substance the compound is, then select the proper rule to apply, and then apply the rule correctly. If you make a mistake, use Table 6.9 to trace the error in your logic. Remember that making mistakes while practicing is normal. Mistakes can even be helpful, as long as you use them as a learning tool.

## Concept-Linking Exercises

*Write a brief description of the relationships among each of the following groups of terms or phrases. Answers to the Concept-Linking Exercises are given in Appendix III.*

1. Atom, molecule, formula unit
2. Ion, cation, anion, monatomic ion, polyatomic ion, acid anion
3. Acid, hydrogen ion, hydronium ion, hydrated hydrogen ion
4. Water of hydration, hydrate, anhydrous compound

## Small-Group Discussion Questions

*Small-Group Discussion Questions are for group work, either in class or under the guidance of a leader during a discussion section.*

1. Construct a set of flash cards with the formulas of a variety of elements, binary molecular compounds, ions, acids, ionic compounds, and hydrates on one side and their names on the other side. Split into pairs and practice giving names for formulas and formulas for names until each person can repeat the entire set rapidly and accurately in both directions.

## Questions, Exercises, and Problems

*General Instructions: Most of the questions in this chapter ask that you write the name of any species if the formula is given, or the formula if the name is given. You will be reminded of this briefly at the beginning of each such block of questions. You should try to follow these instructions without reference to anything except a clean periodic table, one that has nothing written on it. Names and/or atomic numbers are given in questions involving elements not shown in Figure 5.8. An asterisk (*) marks a substance containing an ion you are not expected to recognize. If you cannot predict what it is from the periodic table, refer to Table 6.7 or 6.8. Blue-numbered questions are answered in Appendix III. ■ denotes problems assignable in OWL. In the first section of Questions, Exercises, and Problems, similar exercises and problems are paired in consecutive odd-even number combinations.*

### Section 6.2: Formulas of Elements

1. The elements of Group 8A/18 are stable as monatomic atoms. Write their formulas.
2. ■ Write the formulas of all elements that exist as diatomic molecules under normal conditions.

*Questions 3 through 8: Given names, write formulas; given formulas, write names.*

3. Fluorine, boron, nickel, sulfur
4. ■ Germanium (Z = 32), fluorine, argon, gallium (Z = 31)
5. Cr, $Cl_2$, Be, Fe
6. $O_2$, Ca, Ba, Ag
7. Krypton, copper, manganese, nitrogen
8. ■ Chlorine, magnesium, nitrogen, vanadium (Z = 23)

### Section 6.3: Compounds Made from Two Nonmetals

*Questions 9 and 10: Given names, write formulas; given formulas, write names.*

9. Dichlorine monoxide, tribromine octoxide, HBr(g), $P_2O_3$
10. ■ $CCl_4$, $CBr_4$, NO, dinitrogen monoxide, sulfur dioxide, sulfur hexafluoride

### Section 6.4: Names and Formulas of Ions Formed by One Element

11. Explain how monatomic anions are formed from atoms, in terms of protons and electrons.
12. ■ Write an equation that shows the formation of a rubidium ion from a neutral rubidium atom (Z = 37).

*Questions 13 through 16: Given names, write formulas; given formulas, write names.*

13. $Cu^+$, $I^-$, $K^+$, $Hg_2^{2+}$, $S^{2-}$
14. ■ $Ca^{2+}$, $P^{3-}$, $Mn^{2+}$, $Te^{2-}$ (Z = 52)
15. Iron(III) ion, hydrogen ion, oxide ion, aluminum ion, barium ion
16. ■ Calcium ion, nitride ion, manganese(II) ion, lithium ion

### Section 6.5: Acids and the Anions Derived from Their Total Ionization

17. How do you recognize the formula of an acid?
18. What element is present in all acids discussed in this chapter?
19. How many ionizable hydrogens are in monoprotic, diprotic, and triprotic acids?
20. What is the meaning of the suffix *-protic* when describing an acid? What is a polyprotic acid?
21. ■ Table 6.11 has spaces for half the names and formulas covered by Goal 5. One name or formula appears on each line. Fill in the blanks. The first line is completed as an example. (Selenium, Se, Z = 34; tellurium, Te, Z = 52.)

**Table 6.11**

| Acid Name | Acid Formula | Ion Name | Ion Formula |
|---|---|---|---|
| Sulfuric | $H_2SO_4$ | Sulfate | $SO_4^{2-}$ |
| | | Carbonate | |
| | | | $ClO_3^-$ |
| Hydrofluoric | | | |
| | $HBrO_3$ | | |
| | | Sulfite | |
| | | | $AsO_4^{3-}$ |
| Periodic | | | |
| | $H_2SeO_3$ | | |
| | | Tellurite | |
| | | | $IO^-$ |
| Hypobromous | | | |
| | $H_2TeO_4$ | | |
| | | Perbromate | |
| | | | $Br^-$ |

22. Table 6.12 has spaces for the names and formulas covered by Goal 5 that are not in Question 21. Fill in the blanks. (Selenium, Se, Z = 34; tellurium, Te, Z = 52.)

**Table 6.12**

| Acid Name | Acid Formula | Ion Name | Ion Formula |
|---|---|---|---|
| | HCl | | |
| | | Nitrate | |
| | | | $PO_4^{3-}$ |
| Hydrosulfuric | | | |
| | $HNO_2$ | | |
| | | Iodate | |
| | | | $Te^{2-}$ |
| Hypochlorous | | | |
| | $HIO_2$ | | |
| | | Selenide | |
| | | | $ClO_4^-$ |
| Hydroiodic | | | |
| | | Chlorite | |
| | | | $SeO_4^{2-}$ |
| Bromous | | | |

### Section 6.6: Names and Formulas of Acid Anions

23. Explain how an anion can behave like an acid. Is it possible for a cation to be an acid?
24. Write equations for the first and second step in the step-by-step ionization of selenic acid (selenium, Z = 34).

*Questions 25 through 28: Given names, write formulas; given formulas, write names.*

25. Hydrogen sulfite ion, hydrogen carbonate ion
26. Dihydrogen phosphate ion, hydrogen sulfate ion, hydrogen tellurite ion (tellurium, Z = 52)
27. $HSeO_3^-$, $HTe^-$
28. ■ $HCO_3^-$, $HTeO_4^-$, $HSO_3^-$

### Section 6.7: Names and Formulas of Other Acids and Ions

*Questions 29 through 32: Given names, write formulas; given formulas, write names. Refer to Table 6.7 or 6.8 only if necessary.*

29. $NH_4^+$, $CN^-$
30. ■ $HC_2H_3O_2$, $Ga^{3+}$ (Z = 31), $MnO_4^-$
31. Hydroxide ion, cadmium ion (Z = 48)
32. ■ Acetate ion, rubidium ion (Z = 37), chromate ion

### Section 6.8: Formulas of Ionic Compounds

*Complete the formula-writing exercise in Table 6.13 on page 173. When you have developed your skill in writing formulas of ionic compounds, test it by writing the formulas of the compounds in Questions 33 through 38.*

33. Calcium hydroxide, ammonium bromide, potassium sulfate
34. ■ Barium nitrite, sodium hydrogen carbonate, aluminum nitrate, calcium chromate (chromate ion, $CrO_4^{2-}$)
35. Magnesium oxide, aluminum phosphate, sodium sulfate, calcium sulfide
36. ■ Aluminum bromite, potassium dichromate (dichromate ion, $Cr_2O_7^{2-}$), barium dihydrogen phosphate, rubidium selenate (rubidium, Z = 37; selenium, Z = 34)
37. Barium sulfite, chromium(III) oxide, potassium periodate, calcium hydrogen phosphate
38. ■ Cobalt(II) hydroxide, zinc chloride, lead(II) nitrate, copper(I) nitrate

### Section 6.9: Names of Ionic Compounds

*Complete the formula-writing exercise in Table 6.14 on page 173. Then test your skill by naming the compounds in Questions 39 through 44.*

39. $Li_3PO_4$, $MgCO_3$, $Ba(NO_3)_2$
40. ■ PbS, AgOH, $CoBr_3$
41. KF, NaOH, $CaI_2$, $Al_2(CO_3)_3$
42. ■ $Mg(OH)_2$, $Na_2CO_3$, $AgNO_2$, $Co_3(PO_4)_2$
43. $CuSO_4$, $Cr(OH)_3$, $Hg_2I_2$

**Table 6.13** Formula-Writing Exercise Number 1

*Instructions: In each box, write the chemical formula of the compound formed by the cation at the head of the column and the anion at the left of the row. Refer only to the periodic table when completing this exercise. Correct formulas are listed in the Answers to Questions, Exercises, and Problems in Appendix III.**

| Ions | Potassium | Calcium | Chromium (III) | Zinc | Silver | Iron (III) | Aluminum | Mercury (I) |
|---|---|---|---|---|---|---|---|---|
| Nitrate | | | | | | | | |
| Sulfate | | | | | | | | |
| Hypochlorite | | | | | | | | |
| Nitride | | | | | | | | omit |
| Hydrogen sulfide | | | | | | | | |
| Bromite | | | | | | | | |
| Hydrogen phosphate | | | | | | | | |
| Chloride | | | | | | | | |
| Hydrogen carbonate | | | | | | | | |
| Acetate† | | | | | | | | |
| Selenite‡ | | | | | | | | |

*Some compounds in the table are unknown.
†The acetate ion is derived from the ionization of acetic acid, $HC_2H_3O_2$. The ion formula is listed in Table 6.8.
‡The selenite ion contains selenium, Z = 34.

**Table 6.14** Formula-Writing Exercise Number 2

*Instructions: In each box, write the chemical formula of the compound formed by the cation at the head of the column and the anion at the left of the row. Refer only to the periodic table when completing this exercise. Correct formulas are listed in the Answers to Questions, Exercises, and Problems in Appendix III.**

| Ions | $Na^+$ | $Hg^{2+}$ | $NH_4^+$ | $Pb^{2+}$ | $Mg^{2+}$ | $Ga^{3+}$† | $Fe^{3+}$ | $Cu^{2+}$ |
|---|---|---|---|---|---|---|---|---|
| $OH^-$ | | | | | | | | |
| $BrO^-$ | | | | | | | | |
| $CO_3^{2-}$ | | | | | | | | |
| $ClO_3^-$ | | | | | | | | |
| $HSO_4^-$ | | | | | | | | |
| $Br^-$ | | | | | | | | |
| $PO_4^{3-}$ | | | | | | | | |
| $IO_4^-$ | | | | | | | | |
| $S^{2-}$ | | | | | | | | |
| $MnO_4^-$‡ | | | | | | | | |
| $C_2O_4^{2-}$‡ | | | | | | | | |

*Some compounds in the table are unknown.
†Ga is the symbol for gallium, Z = 31.
‡These ions are listed in Table 6.8.

44. ■ $Ba(BrO_3)_2$, $K_2C_2O_4$ ($C_2O_4^{2-}$, oxalate ion), $Al(HTeO_4)_3$ (Te, tellurium, Z = 52), $K_2SeO_3$ (Se, selenium, Z = 34)

#### Section 6.10: Hydrates

45. Among the following, identify all hydrates and anhydrous compounds: $NiSO_4 \cdot 6\ H_2O$, KCl, $Na_3PO_4 \cdot 12\ H_2O$.
46. ■ It is often possible to change a hydrate into an anhydrous compound by heating it to drive off the water (dehydration). Write an equation that shows the dehydration of potassium fluoride dihydrate.
47. Epsom salt has the formula $MgSO_4 \cdot 7\ H_2O$. How many water molecules are associated with one formula unit of $MgSO_4$? Write the chemical name of Epsom salt.
48. ■ How many water molecules are associated with one formula unit of calcium chloride in $CaCl_2 \cdot 2\ H_2O$? Write the name of this compound.
49. Write the formulas of ammonium phosphate trihydrate and potassium sulfide pentahydrate.
50. ■ The compound barium perchlorate forms a hydrate with three water molecules per formula unit. What is the name of hydrate? What is its formula?

*Tables 6.15 and 6.16 and Questions 51 to 80: Items in the tables and remaining questions are selected at random from various sections of the chapter. Unless marked with an asterisk (*), all names and formulas are included in the Goals and should be found with reference to no more than a periodic table. Ions in compounds marked with an asterisk are included in Tables 6.7 and 6.8, or, if the unfamiliar ion is monatomic, the atomic number of the element is given. Complete the exercises in the tables and then test your skills by completing the questions. In all questions, given a name, write the formula; given a formula, write the name.*

**Table 6.15** Name-and-Formula Writing Exercise Number 1

*Instructions: In each box, when given a formula, write the name; given a name, write the formula. Refer to nothing but the periodic table printed on your shield.*

| Formula | Name | Name | Formula |
|---|---|---|---|
| $OH^-$ | | Sodium bromide | |
| PbO | | Ammonium phosphate | |
| $SO_4^{2-}$ | | Bromine | |
| $Cl_2$ | | Hydrogen phosphate ion | |
| $Cl_2O$ | | Nitrite ion | |
| $Cr(ClO_4)_3$ | | Iron(III) phosphate | |
| $H_2PO_4^-$ | | Nickel sulfide | |
| $Al_2S_3$ | | Nitrogen | |
| $CuSO_4$ | | Copper(II) bromide | |
| HF(aq) | | Oxygen difluoride | |
| C | | Potassium ion | |

**Table 6.16** Name-and-Formula Writing Exercise Number 2

*Instructions: In each box, when given a formula, write the name; given a name, write the formula. Refer to nothing but the periodic table printed on your shield.*

| Formula | Name | Name | Formula |
|---|---|---|---|
| $SeO_4^{2-}$ (Se, selenium, Z = 34) | | Gallium sulfate (Gallium, Z = 31) | |
| $HCO_3^-$ | | Perchloric acid | |
| Ne | | Lithium | |
| $N_2O_5$ | | Cobalt(II) chloride hexahydrate | |
| $HNO_2$ | | Barium dihydrogen phosphate | |
| $CI_4$ | | Hydrosulfuric acid | |
| $BaH_2$ | | Magnesium nitride | |
| $CaTeO_3$ (Te, tellurium, Z = 52) | | Selenic acid (Selenium, Z = 34) | |
| HBrO | | Calcium sulfite | |
| $Fe(NO_3)_2$ | | Sodium hydride | |
| $MgSO_4 \cdot 7\ H_2O$ | | Mercury(I) chloride | |

51. Perchlorate ion, barium carbonate, $NH_4I$, $PCl_3$

52. ■ Calcium fluoride, sodium bromide, aluminum iodide

53. $HS^-$, $BeBr_2$, aluminum nitrate, oxygen difluoride

54. ■ $Na_3PO_4$, $CaCO_3$, $Al(C_2H_3O_2)_3$*

55. Mercury(I) ion, cobalt(II) chloride, $SiO_2$, $LiNO_2$

56. ■ $SO_3$, $Na_2O$, ammonium sulfate, dinitrogen tetroxide

57. $N^{3-}$, $Ca(ClO_3)_2$, iron(III) sulfate, phosphorus pentachloride

58. $HPO_4^{2-}$, CuO, sodium oxalate*, ammonia

59. Tin(II) fluoride, potassium chromate*, LiH, $FeCO_3$

60. Hypochlorous acid, chromium(II) bromide, $KHCO_3$, $Na_2Cr_2O_7$*

61. $HNO_2$, $Zn(HSO_4)_2$, potassium cyanide*, copper(I) fluoride

62. $Co_2O_3$, $Na_2SO_3$, mercury(II) iodide, aluminum hydroxide

63. Magnesium nitride, lithium bromite, $NaHSO_3$, KSCN*

64. Calcium dihydrogen phosphate, potassium permanganate*, $NH_4IO_3$, $H_2SeO_4$ (Se is selenium, Z = 34)

65. $Ni(HCO_3)_2$, CuS, chromium(III) iodate, potassium hydrogen phosphate

66. $Hg_2Cl_2$, $HIO_4$, cobalt(II) sulfate, lead(II) nitrate

67. Selenium dioxide (selenium, Z = 34), magnesium nitrite, $FeBr_2$, $Ag_2O$

68. Tetraphosphorus heptasulfide, barium peroxide*, $MnCl_2$, $NaClO_2$

69. SnO, $(NH_4)_2Cr_2O_7$*, sodium hydride*, oxalic acid*

70. $K_2TeO_4$ (Te is tellurium, Z = 52), $ZnCO_3$, chromium(II) chloride, acetic acid*

71. Cobalt(III) sulfate, iron(III) iodide, $Cu_3(PO_4)_2$, $Mn(OH)_2$

72. ■ Barium chromate*, calcium sulfite, CuCl, $AgNO_3$

73. $Al_2Se_3$ (Se is selenium, Z = 34), $MgHPO_4$, potassium perchlorate, bromous acid

74. ■ $Na_2O_2$*, $NiCO_3$, iron(II) oxide, hydrosulfuric acid

75. Strontium iodate (strontium, Z = 38), sodium hypochlorite, $Rb_2SO_4$, $P_2O_5$

76. Zinc phosphide, cesium nitrate (cesium, Z = 55), $NH_4CN$*, $S_2F_{10}$

77. ICl, $AgC_2H_3O_2$*, lead(II) dihydrogen phosphate, gallium fluoride (gallium, Z = 31)

78. ■ $N_2O_3$, $LiMnO_4$*, indium selenide (indium, Z = 49; selenium, Z = 34), mercury(I) thiocyanate*

79. Magnesium sulfate, mercury(II) bromite, $Na_2C_2O_4$*, $Mn(OH)_3$

80. $CdCl_2$ (Cd is cadmium, Z = 48)*, $Ni(ClO_3)_2$, cobalt(III) phosphate, calcium periodate

# 7

How would you like to count the grains in this sample of copper? It would take a while, wouldn't it? But how about counting all the atoms in this sample of copper? It would be impossible, even if you were assisted by all the people who ever lived. Yet "counting" atoms is exactly what you do if you weigh the sample and express the result in a grouping unit chemists call the *mole*. You will learn how to count atoms by weighing in this chapter.

# Chemical Formula Relationships

CONTENTS



In Chapter 2, we introduced a critical component of a chemist's way of thinking: Chemists study the behavior of the tiny molecules that make up matter, and then they apply what they learn to carry out changes from one type of macroscopic matter to another. In this chapter, we introduce quantitative methods to connect the particulate and macroscopic views of matter.

You may be returning to problem solving as you start this chapter. Recall the six-step approach that is outlined in Section 3.9:

Online homework for this chapter may be assigned in OWL

### procedure

**The Plan**

**Step 1:** List everything that is given. Include units.

**Step 2:** List everything that is wanted. Include units.

**Step 3:** Identify the relationship between the given and wanted. Decide how to solve the problem.

**The Execution**

**Step 4:** Write the calculation setup for the problem. Include units.

**Step 5:** Calculate the answer. Include units.

**Step 6:** Challenge the answer. Be sure that the number is reasonable and that the units are correct (make sense).

Note the emphasis on units. They are an essential part of almost all quantitative chemistry problems.

Recall that we used the word *PLAN* to describe the first three steps in the procedure. The first thing we ask you to do in many of our examples is to *PLAN* the problem. In the third step you decide how to solve the problem. The decision is based on the mathematical relationship between the *GIVEN* and *WANTED*. Most problems fit into one of two categories:

1. If there are one or more *PER* expressions (Section 3.3) between the *GIVEN* and *WANTED*, the problem is solved by dimensional analysis. Multiply the *GIVEN* quantity by one or more conversion factors to reach the *WANTED* quantity.
2. If the *GIVEN* and *WANTED* are related by an algebraic equation, the problem is solved by algebra (Section 3.9). Solve the equation for the *WANTED* variable, substitute the *GIVEN* values *and units*, and calculate the answer.

The whole problem-solving procedure is summarized in a flowchart that shows the six steps (Fig. 3.12).

## 7.1 The Number of Atoms in a Formula

**Goal** 1 Given the formula of a chemical compound (or a name from which the formula may be written), state the number of atoms of each element in the formula unit.

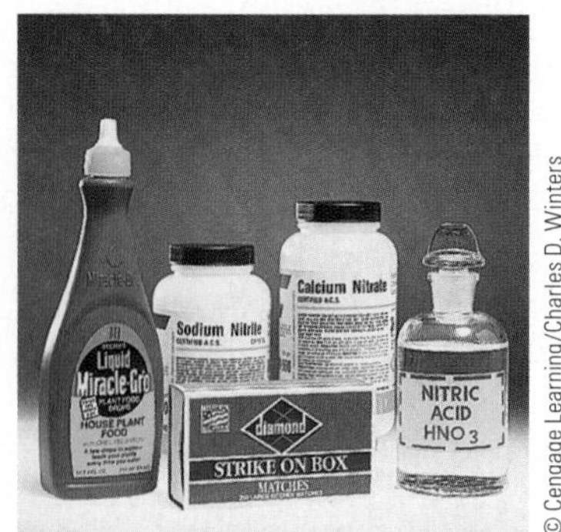


© Cengage Learning/Charles D. Winters

Sodium nitrate is also called Chile saltpeter. It occurs as a natural mineral in Chile; saltpeter means "salt of the rock." It is used in manufacturing sulfuric acid as well as other products. Calcium nitrate is used in explosives and as a corrosion inhibitor in diesel fuels. Both compounds are used in fertilizers and matches and in the manufacture of nitric acid.

In writing the formula of a substance, subscript numbers are used to indicate the number of atoms or groups of atoms of each element in the formula unit of the substance. There is one exception: If that number is one, it is omitted. The formula of sodium nitrate is $NaNO_3$. The formula unit contains one sodium atom, one nitrogen atom, and three oxygen atoms.

The formula of calcium nitrate is $Ca(NO_3)_2$. How many atoms of each element are in this formula? Applying the information from the preceding paragraph, you know that there is one calcium atom. There is one nitrogen atom in *each* of the two nitrate ions, so there are $2 \times 1 = 2$ nitrogen atoms. (This is like asking, "How many seats are on two tricycles?") There are three oxygen atoms in *each* of the two nitrate ions, so there are $2 \times 3 = 6$ oxygen atoms. (How many wheels are on two tricycles?)

In some examples in this chapter, you will be asked to solve quantitative problems that have only formula names in the question. Part of the problem is to write the chemical formula as you learned to do in Chapter 6. This will give you an opportunity to practice your formula-writing skills, even though that is not the purpose in this chapter.

We recommend that you take advantage of this quick review, but don't let it distract you from your immediate goal of learning how to solve problems based on chemical formula relationships. In every case, the correct formula is given in the first step of the example solution so that you may verify it before proceeding with the remainder of the problem.

### Active Example 7.1

How many atoms of each element are in a formula unit of (a) magnesium chloride and (b) barium iodate?

Before you can answer these questions, you need the formulas of magnesium chloride and barium iodate. Use only a periodic table as a guide. (You may skip this step if you are not yet responsible for writing formulas.)

$MgCl_2$ and $Ba(IO_3)_2$

$MgCl_2$ comes from the $Mg^{2+}$ ion giving a 2+ charge and two $Cl^-$ ions giving a total 2− charge. The formula of barium iodate is developed in the same way.

Now you have both formulas. How many atoms of each element are in each formula?

magnesium chloride, $MgCl_2$: 1 magnesium atom, 2 chlorine atoms
barium iodate, $Ba(IO_3)_2$: 1 barium atom, 2 iodine atoms, 6 oxygen atoms

We find the two iodine atoms and six oxygen atoms in $Ba(IO_3)_2$ the same way we found the two nitrogen atoms and six oxygen atoms in $Ca(NO_3)_2$: $2 \times 1$ for iodine, and $2 \times 3$ for oxygen.

## 7.2 Molecular Mass and Formula Mass

Goal 2 Distinguish among atomic mass, molecular mass, and formula mass.

Goal 3 Calculate the formula (molecular) mass of any compound whose formula is given (or known).

In Section 5.5 you learned that the atomic mass of an element is the average mass of its atoms, expressed in atomic mass units, u. But what about compounds? Is there such a thing as a "compound mass"? The answer is yes. It is called the **formula mass,** and it is based on the chemical formula of the compound. For molecular compounds it is called by the more precise term **molecular mass.** These terms are defined exactly the same way as atomic mass: **Molecular (or formula) mass is the average mass of molecules (or formula units) compared with the mass of an atom of carbon-12, which is exactly 12 atomic mass units.**

The formula mass of a compound is equal to the sum of all the atomic masses in the formula unit:

$$\text{Formula mass} = \Sigma \text{ atomic masses in the formula unit}$$

$\Sigma$ is the capital Greek letter sigma. When used as a mathematical symbol, it means "the sum of all values of" whatever follows.

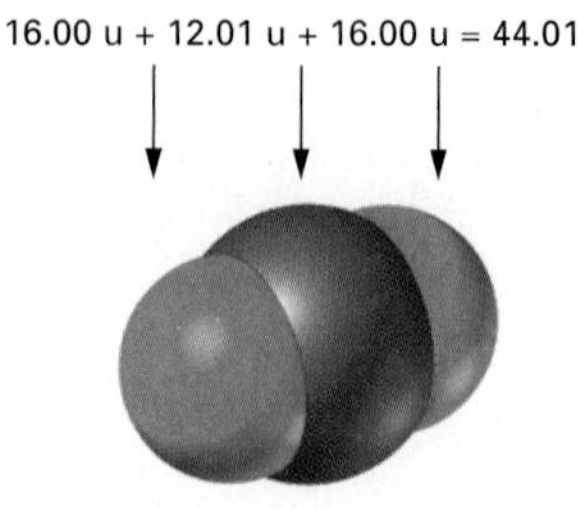


**Figure 7.1** Molecular (or formula) mass. Molecular mass is the sum of the atomic masses of each atom in the molecule. The molecular mass of a carbon dioxide molecule is the sum of the masses of one atom of carbon and two atoms of oxygen.

Here we illustrate the calculation of the molecular mass of carbon dioxide, $CO_2$. There are one carbon atom and two oxygen atoms in the molecule. The molecular mass is the sum of the atomic masses of these three atoms (Fig. 7.1):

| Element | Atoms in Formula | | Atomic Mass | | Mass in Formula |
|---|---|---|---|---|---|
| Carbon | 1 | × | 12.01 u | = | 12.01 u |
| Oxygen | 2 | × | 16.00 u | = | 32.00 u |
| | | | Total molecular mass | = | 44.01 u |

We are accustomed to setting up addition problems in vertical columns, as shown. However, a horizontal setup of the problem is convenient, because when read from left to right it matches the typical calculator sequence:

$$12.01 \text{ u C} + 2(16.00 \text{ u}) \text{ O} = 44.01 \text{ u } CO_2$$

From this point on, our formula mass calculation setups are written horizontally unless we have a special point to illustrate.

**P/REVIEW:** The term *molecular mass* applies to compounds that have covalent bonds, and the term *formula mass* applies to compounds with at least one ionic bond (Sections 12.2 and 12.3).

## Active Example 7.2

Calculate the formula mass of ammonium nitrate.

First, you need the formula. You should be careful, because in writing the formula of an ionic compound the cation always comes first, followed by the anion. Both ions in this compound are polyatomic.

Ammonium nitrate is used in making such common products as anesthetics, matches, fireworks, and fertilizer. It was also one component of the explosive used in the tragic bombing of the Murrah Federal Building in Oklahoma City on April 19, 1995.

$NH_4NO_3$

Both the ammonium ion, $NH_4^+$, and the nitrate ion, $NO_3^-$, contain nitrogen. They are not combined to $N_2H_4O_3$ in writing the formula so that each ion keeps its identity in the compound formula.

Now count up the atoms of each element in $NH_4NO_3$ and calculate its formula mass. Use a vertical setup this time. Be careful about applying the rules for significant figures.

There are two N atoms, four H atoms, and three O atoms in $NH_4NO_3$:

$$\begin{aligned} 2 \times 14.01 \text{ u} &= 28.02 \text{ u} \\ 4 \times 1.008 \text{ u} &= 4.03|2 \text{ u} \\ 3 \times 16.00 \text{ u} &= 48.00 \text{ u} \\ & \quad 80.05|2 \text{ u} = 80.05 \text{ u} \end{aligned}$$

If you had any difficulty getting the correct answer on your calculator, take a few minutes to learn the technique. *Learn it now!* (See Appendix I.A.)

If you had solved Active Example 7.2 with a horizontal setup, you may have missed the fact that the "2" in the calculator answer, 80.052, should be dropped. But a reliable clue can be found in the atomic masses from which the formula mass is calculated. Notice that two of the atomic masses are given to two decimal places. That puts a limit on the number of decimal places that can appear in any addition or subtraction in which those atomic masses appear; it can never be more than two. This leads to a shortcut for expressing formula masses in the correct number of significant figures:

**procedure**

**Rounding Off Formula and Molecular Masses**

**Step 1:** Calculate the formula mass using atomic masses to as many decimal places as they are known.

**Step 2:** Round off the answer to the same number of decimal places as the smallest number of decimal places in any atomic mass used in the calculation.

When considering significant figures in formula and molecular masses, think in terms of the addition/subtraction rule. For example, for $P_4$, think 30.97 u + 30.97 u + 30.97 u + 30.97 u = 123.88 u.

In Active Example 7.2 the atomic masses of both nitrogen and oxygen are given to two decimal places. The formula mass of ammonium nitrate must therefore be rounded off to the second decimal place.

If you use the periodic table on your shield as your source of atomic masses—and we encourage this—most of your molecular or formula masses will be rounded to two decimal places. But there are some exceptions. . . .

## Active Example 7.3

Calculate the formula or molecular mass of (a) ammonium sulfate and (b) hydrogen.

Ammonium sulfate is a familiar fertilizer sold in most garden shops.

Hydrogen is the most abundant element in the universe. Most commercially produced hydrogen is used in manufacturing ammonia.

Begin with the formulas.

$(NH_4)_2SO_4$ and $H_2$

It takes two ammonium ions, each with a 1+ charge, to balance the 2− charge of a single sulfate ion, $SO_4^{2-}$. Hydrogen is one of the seven elements that occur in nature as diatomic molecules.

Now calculate the formula mass of $(NH_4)_2SO_4$.

$$2(14.01\text{ u}) + 8(1.008\text{ u}) + 32.07\text{ u} + 4(16.00\text{ u}) = 132.15\text{ u}$$

In this case, the answer is rounded to two decimal places because the atomic masses of nitrogen, sulfur, and oxygen all are given to two decimal places.

Calculate the molecular mass of $H_2$.

The Bridgeman Art Library/Getty Images

Jacques Charles (Section 4.4) and a passenger flew over Paris in a hydrogen-filled balloon on December 1, 1783.

$$1.008\text{ u} + 1.008\text{ u} = 2.016\text{ u}$$

Did you round your answer to two decimal places? If so, look back at the procedure box. The atomic mass of each hydrogen atom is given to three decimal places on the periodic table on your shield, so the sum is expressed to three decimal places.

© Cengage Learning/Charles D. Winters

Elemental chlorine, $Cl_2$, is a major industrial chemical. It is used to bleach wood pulp in manufacturing paper, to bleach textiles, in manufacturing plastics, and to purify drinking water.

© Cengage Learning/Charles D. Winters

The chemical name of table sugar is sucrose. It is made from the juice of sugar beets or sugar cane.

Atomic masses in our periodic table have been rounded arbitrarily to four significant figures. Most atomic masses are known more precisely than that. The atomic mass of sodium, for example, is known to ten significant figures, eight to the right of the decimal. If more precise information is available, atomic masses from a four-significant-figure source should never limit the number of significant figures in a calculated result. If the data in a problem are known to five or more significant figures, you should calculate formula or molecular masses from other, more precise sources of atomic mass.

## Active Example 7.4

The stable form of elemental chlorine is a diatomic (two-atom) molecule with the formula $Cl_2$. Table sugar, $C_{12}H_{22}O_{11}$, is a molecular compound. Calculate the molecular masses of elemental chlorine and table sugar.

Remember that *molecular* mass is the same as formula mass; it is the term used when speaking of molecular substances. Be sure you find the mass of a $Cl_2$ *molecule.* Calculate both molecular masses.

$Cl_2$: 2(35.45 u) = 70.90 u

$C_{12}H_{22}O_{11}$: 12(12.01 u) + 22(1.008 u) + 11(16.00 u) = 342.30 u

Note that the *atomic* mass of chlorine, Cl, is 35.45 u, but the *molecular* mass of chlorine, $Cl_2$, is two times 35.45 u. This illustrates an important point: *Always calculate formula or molecular masses exactly as the formula is written.* The formula, of course, must be written correctly!

## 7.3 The Mole Concept

**Goal 4** Define the term *mole.* Identify the number of objects that corresponds to one mole.

**Goal 5** Given the number of moles (or units) in any sample, calculate the number of units (or moles) in the sample.

In Section 2.1 we said that you would study chemistry at the particulate level. In other words, chemists are interested in the individual particles that make up a sample of matter. In Section 2.6 we identified atoms and molecules as two of these particles. To understand the amounts of substances in a chemical change, we must know the number of particles of the different substances in the reaction. That's not an easy number to find. Literally counting atoms and molecules is not practical because such numbers are extremely large.

To describe the number of particles, chemists use a quantity called the **mole: One mole is the amount of any substance that contains the same number of units as the number of atoms in exactly 12 grams of carbon-12.** In calculation setups the mole is abbreviated **mol.**

At the time of this writing, scientists have determined Avogadro's number to nine significant figures, $6.02214179 \times 10^{23}$/mol. Our three-significant-figure round-off will be adequate for most of the calculations in this book. However, $N_A$ should never limit the number of significant figures in a calculation.

The definition of a mole refers to a number of particles, but it doesn't say what that number is. By experiment is has been found that, to three significant figures,

$$1 \text{ mol of any substance} = \mathbf{6.02 \times 10^{23}} \text{ units of that substance}$$

This number is called **Avogadro's number, or the Avogadro constant,** $N_A$, in honor of the man whose experiments with gases eventually led to the mole concept.

This is a huge number. To get some appreciation of the size of Avogadro's number, if you were to try to count the atoms in exactly 12 g of carbon-12, and proceeded at the rate of 100 atoms per minute without interruption, 24 hours per day, 365 days per year, the task would require 11,500,000,000,000,000 years. This is about two million times as long as the earth has been in existence!

Technically, the mole is not a number by definition. However, it is convenient to think of the mole as a number, just as we think of a dozen as the number 12. In this sense, as one dozen eggs is 12 eggs, one mole of eggs is $6.02 \times 10^{23}$ eggs. If we wanted to find the number of eggs in three dozen eggs, we would set up the problem this way:

$$3\ \cancel{\text{doz eggs}} \times \frac{12\text{ eggs}}{1\ \cancel{\text{doz eggs}}} = 36\text{ eggs}$$

Similarly, the number of carbon atoms in three moles of carbon atoms is

$$3\ \cancel{\text{mol C atoms}} \times \frac{6.02 \times 10^{23}\text{ C atoms}}{1\ \cancel{\text{mol C atoms}}} = 1.81 \times 10^{24}\text{ C atoms}$$

E. F. Smith Collection/Van Pelt Library/ University of Pennsylvania

Amedeo Avogadro's hypothesis that equal volumes of gases at the same temperature and pressure contain the same number of molecules provided the conceptual foundation on which the mole concept and Avogadro's number were built.

## Thinking About Your Thinking

### Proportional Reasoning

**The number of eggs in a dozen and the number of particles in a mole are both examples of direct proportionalities used in proportional reasoning.**

## Active Example 7.5

How many moles of water are in $1.67 \times 10^{21}$ water molecules?

How would you calculate the number of dozens of eggs in 48 eggs? This problem is solved in exactly the same way. List the *GIVEN*, *WANTED*, and *PER/PATH*, set up the problem, and solve.

*GIVEN:* $1.67 \times 10^{21}$ water molecules *WANTED:* mol water

*PER:* $\dfrac{1\text{ mol water}}{6.02 \times 10^{23}\text{ water molecules}}$

*PATH:* water molecules ⟶ mol water

$$1.67 \times 10^{21}\ \cancel{\text{water molecules}} \times \frac{1\text{ mol water}}{6.02 \times 10^{23}\ \cancel{\text{water molecules}}} = 0.00277\text{ mol water}$$

You may be interested to know that 0.00277 mole of water is about one drop.

## 7.4 Molar Mass

**Goal 6** Define *molar mass* or interpret statements in which the term *molar mass* is used.

**Goal 7** Calculate the molar mass of any substance whose chemical formula is given (or known).

P/REVIEW: The molar mass *PER* relationship yields a defining equation that is a conversion factor for dimensional analysis calculations. See Section 3.3.

It is all very well to calculate the atomic, molecular, and formula masses of atoms, molecules, and other compounds, but since we cannot weigh an individual particle, these masses have a limited usefulness. To make measurements of mass useful, we must express chemical quantities at the macroscopic level. The bridge between the particulate and the macroscopic levels is **molar mass, the mass in grams of one mole of a substance.** The units of molar mass follow from its definition: grams per mole (g/mol). Mathematically, the defining equation of molar mass is

$$\text{molar mass} \equiv \frac{\text{mass}}{\text{mole}} = \frac{\text{g}}{\text{mol}}$$

The definitions of atomic mass, the mole, and molar mass are all directly or indirectly related to carbon-12. This leads to two important facts:

1. The mass of one atom of carbon-12—the atomic mass of carbon-12—is exactly 12 atomic mass units.
2. The mass of one mole of carbon-12 atoms is exactly 12 grams; its molar mass is exactly 12 grams per mole.

Notice that the atomic mass and the molar mass of carbon-12 are numerically equal. They differ only in units; atomic mass is measured in atomic mass units, and molar mass is measured in grams per mole. The same relationships exist between atomic and molar masses of elements, between molecular masses and molar masses of molecular substances, and between formula masses and molar masses of ionic compounds. In other words,

*The molar mass of any substance in grams per mole is numerically equal to the atomic, molecular, or formula mass of that substance in atomic mass units.*

The following table gives examples from sources in this chapter.

| Substance | Atomic/Molecular/ Formula Mass (u) | Molar Mass (g/mol) | Source |
|---|---|---|---|
| O atoms | 16.00 | 16.00 | Section 7.2 |
| $NH_4NO_3$ | 80.05 | 80.05 | Active Example 7.2 |
| $Cl_2$ | 70.90 | 70.90 | Active Example 7.4 |

If you can find the atomic, molecular, or formula mass of a substance, change the units from u to g/mol and you will have its molar mass.

## Active Example 7.6

Calculate the molar mass of elemental fluorine and of calcium fluoride.

Begin with the formulas of fluorine and calcium fluoride.

© Cengage Learning/Charles D. Winters

Calcium fluoride occurs in nature and is known as fluorspar. It is used in making steel.

$F_2$ and $CaF_2$

Now set up and calculate the horizontal additions as if you were calculating formula mass, but use molar mass units.

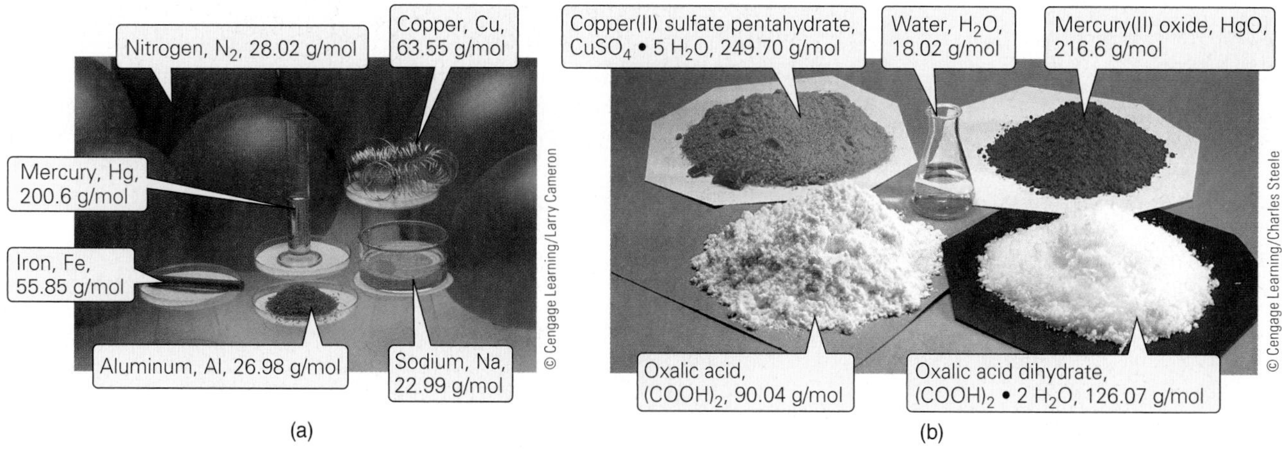


**Figure 7.2** (a) One mole of elements and (b) one mole of molecular and ionic compounds.

$$F_2\text{: } 2(19.00\text{ g/mol F}) = 38.00\text{ g/mol } F_2$$

$$CaF_2\text{: } 40.08\text{ g/mol Ca} + 2(19.00\text{ g/mol F}) = 78.08\text{ g/mol } CaF_2$$

There is one mole of each substance in Figure 7.2. This means that each sample has the same number of atoms, molecules, or formula units. The mass of each sample in grams is numerically the same as the molar mass of each substance.

## 7.5 Conversion Among Mass, Number of Moles, and Number of Units

**Goal** **8** Given any one of the following for a substance whose formula is given (or known), calculate the other two: (a) mass; (b) number of moles; (c) number of formula units, molecules, or atoms.

The *Per* relationship in molar mass, grams per mole, means you can use dimensional analysis to convert from grams to moles or from moles to grams. Molar mass is the conversion factor. This one-step conversion is probably used more often than any other conversion in chemistry because we measure quantities on the macroscopic level in grams. However, chemical reactions take place on the particulate level, and moles are units that express the number of particles.

### Active Example 7.7

You are carrying out a laboratory reaction that requires 0.0360 mole of barium chloride. How many grams of the compound do you measure out?

Barium chloride is used in making paint pigment and in tanning leather.

Before you can use molar mass as a conversion factor, you must calculate its value. This first requires the formula of barium chloride.

$BaCl_2$

Now calculate the molar mass.

*continued*

$$137.3 \text{ g/mol Ba} + 2(35.45 \text{ g/mol Cl}) = 208.2 \text{ g/mol BaCl}_2$$

It is now a one-step conversion from moles to grams, using molar mass as the conversion factor. *PLAN* the problem (see the opening pages of this chapter for a reminder of what we mean by *PLAN*, if necessary) and calculate the answer.

*GIVEN:* 0.0360 mol $BaCl_2$ *WANTED:* g $BaCl_2$

*PER:* $\dfrac{208.2 \text{ g BaCl}_2}{\text{mol BaCl}_2}$

*PATH:* mol $BaCl_2$ ⟶ g $BaCl_2$

$$0.0360 \cancel{\text{mol BaCl}_2} \times \frac{208.2 \text{ g BaCl}_2}{\cancel{\text{mol BaCl}_2}} = 7.50 \text{ g BaCl}_2$$

## Thinking About Your Thinking

### Proportional Reasoning

**The idea of counting by weighing was introduced in the photograph and caption on the first page of this chapter. Many grocery-store items are sold by weight instead of number because of the inconvenience of counting small objects. Cookies, candies, nuts, and fruits, for example, could be sold by the piece, but instead they are usually sold by the pound in the United States.**

**Let's say that you decide to eat two dozen cherries a day for the next week. When you go to the grocery store, you will not find a price per cherry or per dozen cherries, but rather a price per pound of cherries. Even though you are thinking about the number of cherries, grouped in dozens, your grocer thinks about the number of pounds of cherries. You need to understand both methods for expressing the amount of cherries.**

**Similarly, in chemistry, you have to learn to think about both the number of particles, grouped in moles, and the mass of those particles. Considering moles of particles is useful when you are thinking about models of the particles and their reactivity. On the other hand, when you want a certain number of particles for a chemistry experiment, you will often weigh a specified amount of the macroscopic solid. So thinking about mass is important, too. You have to understand both methods for expressing amount of a substance.**

**The link between the number of moles of particles and their mass is molar mass, expressed in grams per mole. The number of grams of a pure substance is directly proportional to the number of moles of that substance. The examples that follow will provide you with practice in thinking as a chemist by counting particles by weighing samples.**

Let's step back and look at some of the dimensional analysis changes we have made:

| Changes We Made | Conversion Factors We Used | Where We Did It |
|---|---|---|
| *mol* ↔ units | $6.02 \times 10^{23}$ units/mol | Section 7.3 |
| *mol* ↔ g | Molar mass | Active Example 7.7 |

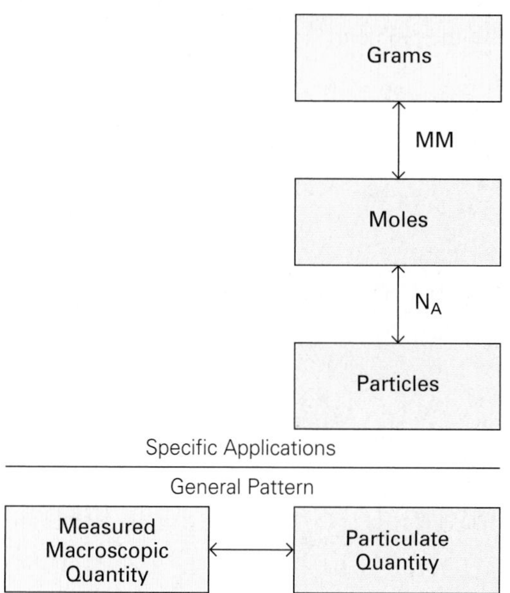


**Figure 7.3** Conversion among grams, moles, and number of particles. Molar mass, MM, is a *PER* expression that links mass in grams, a measurable macroscopic quantity, with moles, the number of particles in the sample. Avogadro's number, $N_A$, links the number of particles grouped in moles to the absolute number of particles. The overarching pattern, illustrated by the two boxes at the bottom of the figure, is that a measured macroscopic quantity in a yellow box (e.g., mass in grams, pounds, etc.) can be converted to a particulate quantity in a blue box.

Notice that the mole is present in both of these changes. In fact, the mole is the connecting link between the macroscopic world, in which we measure quantities in grams, and the particulate world, in which we count the number of units. Using $N_A$ for Avogadro's number, $6.02 \times 10^{23}$/mol, and MM for molar mass in g/mol, we have

*PER* (conversion factor): $\text{units} \xrightarrow{N_A} \text{mol} \xrightarrow{MM} \text{g}$ *or* $\text{g} \xrightarrow{MM} \text{mol} \xrightarrow{N_A} \text{units}$

Changing from units to mass or vice versa is a two-step dimensional analysis conversion: Change the given quantity to moles, then change moles to the wanted quantity. Figure 7.3 summarizes the process of conversion among mass, number of moles, and number of unit particles.

## Active Example 7.8

How many molecules are in 454 g (1 pound) of water?

You need the molar mass of water as one of the conversion factors.

$2(1.008\text{ g/mol H}) + 16.00\text{ g/mol O} = 18.02\text{ g/mol } H_2O$

Now, to see clearly the start, the finish, and the way to solve the problem, *PLAN* it, using the *GIVEN*, the *WANTED*, and the *PER*/*PATH*.

*GIVEN:* 454 g $H_2O$ *WANTED:* molecules $H_2O$

*PER:* $\dfrac{1\text{ mol } H_2O}{18.02\text{ g } H_2O}$ $\dfrac{6.02 \times 10^{23}\text{ molecules } H_2O}{\text{mol } H_2O}$

*PATH:* $\text{g } H_2O \longrightarrow \text{mol } H_2O \longrightarrow \text{molecules } H_2O$

*continued*

Set up and solve the problem.

$$454\ \cancel{g\ H_2O} \times \frac{1\ \cancel{mol\ H_2O}}{18.02\ \cancel{g\ H_2O}} \times \frac{6.02 \times 10^{23}\ \text{molecules}}{\cancel{mol\ H_2O}}$$

$$= 1.52 \times 10^{25}\ \text{molecules}\ H_2O$$

## 7.6 Mass Relationships Among Elements in a Compound: Percentage Composition

**Goal 9** Calculate the percentage composition of any compound whose formula is given (or known).

**Goal 10** Given the mass of a sample of any compound whose formula is given (or known), calculate the mass of any element in the sample; or, given the mass of any element in the sample, calculate the mass of the sample or the mass of any other element in the sample.

The term *cent* refers to 100. For example, there are 100 cents in a dollar and 100 years in a century. **Percent** therefore means "per 100." Thus, **percent is the amount of one part of a mixture per 100 total parts in the mixture.** If the part whose percentage we wish to identify is A, then

$$\%\ A \equiv \frac{\text{parts of A in the mixture}}{100\ \text{total parts in the mixture}} \qquad \textbf{(7.1)}$$

Equation 7.1 is a defining equation for percentage. To calculate percentage, we use a more convenient form that is derived from Equation 7.1:

$$\%\ A = \frac{\text{parts of A}}{\text{total parts}} \times 100 \qquad \textbf{(7.2)}$$

The ratio (parts of A)/(total parts) is the fraction of the sample that is A. Multiplying that fraction by 100 gives the percentage of A. To illustrate, in Active Example 7.6, you calculated the molar mass of calcium fluoride. The calculation setup was

$$40.08\ \text{g/mol Ca} + 2(19.00\ \text{g/mol F}) = 78.08\ \text{g/mol}\ CaF_2$$

The part of a mole that is calcium is 40.08 g. The total mass of a mole is 78.08 g. The fraction of a mole that is calcium is 40.08 g/78.08 g. The percentage of calcium is therefore

$$\%\ \text{Ca} = \frac{\text{parts of A}}{\text{total parts}} \times 100 = \frac{40.08\ \text{g Ca}}{78.08\ \text{g}\ CaF_2} \times 100 = 51.33\%\ \text{Ca}$$

### Active Example 7.9

Calculate the percentage of fluorine in $CaF_2$.

Solve this algebra problem using Equation 7.2.

GIVEN: 2(19.00 g F); 78.08 g $CaF_2$ WANTED: %F

EQUATION: $\% \text{ F} = \dfrac{\text{g F}}{\text{g CaF}_2} \times 100 = \dfrac{2(19.00) \text{ g F}}{78.08 \text{ g CaF}_2} \times 100 = 48.67\% \text{ F}$

Notice that you are finding the percentage of the element fluorine in the compound, which is % F, not % $F_2$.

**The percentage composition of a compound is the percentage by mass of each element in the compound.** The percentage composition of calcium fluoride is 51.33% calcium and 48.67% fluorine. As you have seen with $CaF_2$, percentage composition can be calculated from the same numbers that are used to find the molar mass of a compound.

If you calculate the percentage composition of a compound correctly, the sum of all percents must be 100%. This fact can be used to check your work. With calcium fluoride, 51.33% + 48.67% = 100.00%. When you apply this check, don't be concerned if you are high or low by ±1 or even ±2 in the doubtful digit of the sum of the percents. This can result from legitimate round-offs along the way.

## Active Example 7.10

Calculate the percentage composition of aluminum sulfate.

Start by writing the formula of aluminum sulfate.

Aluminum sulfate is used in manufacturing paper. Unfortunately, moisture in the air reacts with the compound, forming an acid that causes paper to yellow and eventually disintegrate.

$Al_2(SO_4)_3$

The numbers are bigger, but the procedure is just like the one in Active Example 7.9. Calculate the molar mass of $Al_2(SO_4)_3$.

2(26.98 g/mol Al) + 3(32.07 g/mol S) + 12(16.00 g/mol O)
= 342.17 g/mol $Al_2(SO_4)_3$

Now find the percentage composition, the percentage by mass of each element in the compound. Check your result.

*continued*

$$\frac{2(26.98)\,g\,Al}{342.17\,g\,Al_2(SO_4)_3} \times 100 = 15.77\%\,Al \qquad \frac{3(32.07)\,g\,S}{342.17\,g\,Al_2(SO_4)_3} \times 100 = 28.12\%\,S$$

$$\frac{12(16.00)\,g\,O}{342.17\,g\,Al_2(SO_4)_3} \times 100 = 56.11\%\,O \qquad 15.77\% \text{ t}+ 28.12\% + 56.11\% = 100.00\%$$

The result checks.

If you have the percentage composition of a compound, you can find the amount of any element in a known amount of the compound. One way to do this is to use percentage as a conversion factor, grams of the element *PER* 100 grams of the compound. For example, if aluminum sulfate is 15.77% aluminum, the mass of aluminum in 88.9 g $Al_2(SO_4)_3$ is

$$88.9\,g\,\cancel{Al_2(SO_4)_3} \times \frac{15.77\,g\,Al}{100\,g\,\cancel{Al_2(SO_4)_3}} = 14.0\,g\,Al$$

Are there three significant figures in the denominator, 100 g $Al_2(SO_4)_3$? The 100 is a defined quantity, like 12 inches equal 1 foot. The total percentage of anything is defined to be exactly 100. The denominator therefore has an infinite number of significant figures, more than any other measured quantity in the calculation. The significant figures are limited by the measured quantity, 88.9 g $Al_2(SO_4)_3$, which has three significant digits.

## Active Example 7.11

How many grams of fluorine are in 216 g of calcium fluoride?

In Active Example 7.9 you found that calcium fluoride is 48.67% fluorine. Solve the problem.

*GIVEN:* 216 g $CaF_2$ *WANTED:* g F *PER:* $\frac{48.67\,g\,F}{100\,g\,CaF_2}$

*PATH:* g $CaF_2 \longrightarrow$ g F

$$216\,g\,\cancel{CaF_2} \times \frac{48.67\,g\,F}{100\,g\,\cancel{CaF_2}} = 105\,g\,F$$

It is not necessary to know the percentage composition of a compound to change between mass of an element in a compound and mass of the compound. The masses of all elements in a compound and the mass of the compound itself are directly proportional to each other; they are related by *PER* expressions. Once again, the molar mass figures for $CaF_2$ are

$$40.08 \text{ g/mol Ca} + 2(19.00 \text{ g/mol F}) = 78.08 \text{ g/mol CaF}_2$$

From these numbers we conclude that

| | |
|---|---|
| $\text{g Ca} \propto \text{g F}$ | 40.08 g Ca/38.00 g F |
| $\text{g Ca} \propto \text{g CaF}_2$ | 40.08 g Ca/78.08 g $CaF_2$ |
| $\text{g F} \propto \text{g CaF}_2$ | 38.00 g F/78.08 g $CaF_2$ |

Any of these ratios, or their inverses, may be used as a conversion factor from the mass of one species to the mass of the other. For instance, to find the mass of $CaF_2$ that contains 3.55 g Ca from these numbers, we calculate

$$3.55 \text{ g Ca} \times \frac{78.08 \text{ g CaF}_2}{40.08 \text{ g Ca}} = 6.92 \text{ g CaF}_2$$

### Active Example 7.12

Calculate the number of grams of fluorine in a sample of calcium fluoride that contains 2.01 g of calcium.

*PLAN* the solution. Then write the calculation setup and find the answer.

*GIVEN:* 2.01 g Ca *WANTED:* g F *PER:* $\frac{38.00 \text{ g F}}{40.08 \text{ g Ca}}$

*PATH:* g Ca ⟶ g F

$$2.01 \text{ g Ca} \times \frac{38.00 \text{ g F}}{40.08 \text{ g Ca}} = 1.91 \text{ g F}$$

## 7.7 Empirical Formula of a Compound

### Empirical Formulas and Molecular Formulas

**Goal** | **11** Distinguish between an empirical formula and a molecular formula.

Where do chemical formulas come from? They come from the same source as any fundamental chemical information—from experiments, usually performed in the laboratory. Among other things, chemical analysis can give us the percentage composition of a compound. Such data give us the **empirical formula** of the compound. Empirical is a term that means "experimentally determined."

The percentage composition of ethylene is 85.6% carbon and 14.4% hydrogen. Its chemical formula is $C_2H_4$. The percentage composition of propylene, formula $C_3H_6$, is also 85.6% carbon and 14.4% hydrogen. These are, in fact, two of a whole series of compounds having the general formula $C_nH_{2n}$, where n is an integer. In ethylene, n = 2, and in propylene, n = 3. All compounds with the general formula $C_nH_{2n}$ have the same percentage composition.

$C_2H_4$ and $C_3H_6$, ethylene and propylene, are the building blocks from which the plastics polyethylene and polypropylene are made.

$C_2H_4$ and $C_3H_6$ are typical molecular formulas of real chemical substances. If, in the general formula, we let n = 1, the result is $CH_2$. This is the empirical formula for all compounds having the general formula $C_nH_{2n}$. **The empirical formula shows the simplest ratio of atoms of the elements in the compound.** All subscripts are reduced to their lowest terms; they have no common divisor.

Empirical formulas may or may not be molecular formulas of real chemical compounds. There happens to be no stable compound with the formula $CH_2$—and there is good reason to believe that no such compound can exist. On the other hand, the molecular formula of dinitrogen tetroxide is $N_2O_4$. The subscripts have a common divisor, 2. Dividing by 2 gives the empirical formula, $NO_2$. This is also the molecular formula of a real substance, nitrogen dioxide. In other words, $NO_2$ is both the empirical formula and the molecular formula of nitrogen dioxide, as well as the empirical formula of dinitrogen tetroxide.

Nitrogen dioxide, $NO_2$, is responsible for one kind of chemical smog that produces a brown haze in the atmosphere.

## Active Example 7.13

Write EF after each formula that is an empirical formula. Write the empirical formula after each compound whose formula is not already an empirical formula.

$C_4H_{10}$: _____ $C_2H_6O$: _____ $Hg_2Cl_2$: _____ $C_6H_6$: _____

$C_4H_{10}$: $C_2H_5$; $C_2H_6O$: EF; $Hg_2Cl_2$: HgCl; $C_6H_6$: CH

## Determination of an Empirical Formula

**Goal** | **12** Given data from which the mass of each element in a sample of a compound can be determined, find the empirical formula of the compound.

To find the empirical formula of a compound, you must find the whole-number ratio of atoms of the elements in a sample of the compound. When the numbers in the ratio are reduced to their lowest terms—that is, when they have no common divisor—they are the subscripts in the empirical formula. The procedure by which this is done is as follows:

procedure

**How to Find an Empirical Formula**

**Step 1:** Find the masses of different elements in a sample of the compound.

**Step 2:** Convert the masses into moles of atoms of the different elements.

**Step 3:** Determine the ratio of moles of atoms.

**Step 4:** Express the moles of atoms as the smallest possible ratio of integers.

**Step 5:** Write the empirical formula, using the number for each atom in the integer ratio as the subscript in the formula.

It is usually helpful in an empirical formula problem to organize the calculations in a table with the following headings:

| Element | Grams | Moles | Mole Ratio | Formula Ratio | Empirical Formula |
|---|---|---|---|---|---|

We will use ethylene to show how to find the empirical formula of a compound from its percentage composition. As noted, the compound is 85.6% carbon and 14.4% hydrogen. We need masses of elements in Step 1 of the procedure. If we think of percent as the number of grams of one element per 100 g of the compound, then a 100-g sample must contain 85.6 g of carbon and 14.4 g of hydrogen. From this we see that *percentage composition figures represent the grams of each element in a 100-g sample of the compound.* These figures complete Step 1 of the procedure. They are entered into the first two columns of the table.

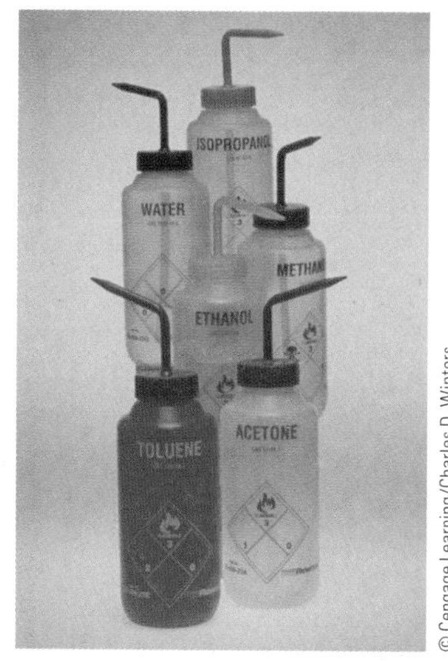

© Cengage Learning/Charles D. Winters

Ethylene is the building block from which the widely used plastic polyethylene is made. Liquids are often packaged or stored in polyethylene bottles.

| Element | Grams | Moles | Mole Ratio | Formula Ratio | Empirical Formula |
|---|---|---|---|---|---|
| C | 85.6 | | | | |
| H | 14.4 | | | | |

We are now ready to find the number of moles of atoms of each element, Step 2 in the procedure. This is a one-step conversion between grams and moles, g ↔ mol, but in the direction opposite to similar conversions in Active Example 7.7.

| Element | Grams | Moles | Mole Ratio | Formula Ratio | Empirical Formula |
|---|---|---|---|---|---|
| C | 85.6 | $\frac{85.6\,g}{12.01\,g/mol} = 7.13\,mol$ | | | |
| H | 14.4 | $\frac{14.4\,g}{1.008\,g/mol} = 14.3\,mol$ | | | |

The ratio of these moles of atoms must now be determined, Step 3 in the procedure. This is most easily done by dividing each number of moles by the smallest number of moles. In this problem the smallest number of moles is 7.13. Thus,

| Element | Grams | Moles | Mole Ratio | Formula Ratio | Empirical Formula |
|---|---|---|---|---|---|
| C | 85.6 | 7.13 | $\frac{7.13}{7.13} = 1.00$ | | |
| H | 14.4 | 14.3 | $\frac{14.3}{7.13} = 2.01$ | | |

The ratio of *atoms* of the elements in a compound is the same as the ratio of *moles* of atoms in the compound. To see this in a more familiar setting, the ratio of seats to wheels in a bicycle is

$$\frac{1\text{ seat}}{2\text{ wheels}}\text{ or, numerically, }\frac{1}{2}$$

In four dozen bicycles there are four dozen seats and eight dozen wheels. This yields a ratio that can be reduced:

$$\frac{4\text{ doz seat}}{8\text{ doz wheels}} = \frac{4 \times \cancel{12}\text{ seats}}{8 \times \cancel{12}\text{ wheels}} = \frac{4\text{ seats}}{8\text{ wheels}} = \frac{1\text{ seat}}{2\text{ wheels}}\text{ or, numerically, }\frac{1}{2}$$

When reduced to lowest terms, the ratio of seats to wheels is the same as the ratio of dozens of seats to dozens of wheels. Similarly, the numbers in the Mole Ratio column are in the same ratio as the subscripts in the empirical formula.

When used in a formula, the numbers must be integers. The numbers represent atoms, and a fraction of an atom does not exist. Accordingly, small round-offs may be necessary to compensate for experimental errors. In this problem, 1.00/2.01 becomes 1/2, and the empirical formula is $CH_2$. Step 4 is now finished. Step 5 is to write the empirical formula. The formula ratio tells us that there is 1 carbon atom for every 2 hydrogen atoms, so the empirical formula is $CH_2$. The entire procedure is now complete and summarized in the following table.

G. Whiteley/Photo Researchers, Inc.

The ratio of seats to wheels on a bicycle is the same as the ratio of dozens of seats to dozens of wheels. In both cases, it is 1:2.

| Element | Grams | Moles | Mole Ratio | Formula Ratio | Empirical Formula |
|---|---|---|---|---|---|
| C | 85.6 | 7.13 | 1.00 | 1 | $CH_2$ |
| H | 14.4 | 14.3 | 2.01 | 2 | |

## Active Example 7.14

A sample of a pure compound is found to be made up of 1.61 g of phosphorus and 2.98 g of fluorine. Find the empirical formula of the compound.

The given information is already placed in the following table. Calculate the number of moles of each element in the space below the table and add the results to the table.

| Element | Grams | Moles | Mole Ratio | Formula Ratio | Empirical Formula |
|---|---|---|---|---|---|
| P | 1.61 | | | | |
| F | 2.98 | | | | |

| Element | Grams | Moles | Mole Ratio | Formula Ratio | Empirical Formula |
|---|---|---|---|---|---|
| P | 1.61 | $\frac{1.61\text{ g}}{30.97\text{ g/mol}} = 0.0520\text{ mol}$ | | | |
| F | 2.98 | $\frac{2.98\text{ g}}{19.00\text{ g/mol}} = 0.157\text{ mol}$ | | | |

Recalling that the mole ratio figures are obtained by dividing each number of moles by the smallest number of moles, calculate those numbers in the space below and place them in the preceeding table.

| Element | Grams | Moles | Mole Ratio | Formula Ratio | Empirical Formula |
|---|---|---|---|---|---|
| P | 1.61 | 0.0520 | $\frac{0.0520}{0.0520} = 1.00$ | | |
| F | 2.98 | 0.157 | $\frac{0.157}{0.0520} = 3.02$ | | |

Now round off the mole ratio numbers to get the whole-number formula ratio. Write the whole numbers in the preceeding table.

| Element | Grams | Moles | Mole Ratio | Formula Ratio | Empirical Formula |
|---|---|---|---|---|---|
| P | 1.61 | 0.0520 | 1.00 | 1 | |
| F | 2.98 | 0.157 | 3.02 | 3 | |

Use the formula ratio numbers, as needed, for subscripts in the empirical formula. Add the empirical formula to complete the table.

| Element | Grams | Moles | Mole Ratio | Formula Ratio | Empirical Formula |
|---|---|---|---|---|---|
| P | 1.61 | 0.0520 | 1.00 | 1 | $PF_3$ |
| F | 2.98 | 0.157 | 3.02 | 3 | |

If either quotient in the Mole Ratio column is not close to a whole number, the Formula Ratio may be found by multiplying both quotients by a small integer. You will be guided to do this in the next Active Example.

## Active Example 7.15

The mass of a piece of iron is 1.62 g. If the iron is exposed to oxygen under conditions in which oxygen combines with all of the iron to form a pure oxide of iron, the final mass increases to 2.31 g. Find the empirical formula of the compound.

As before, you need to know the masses of the elements in the compound. This time you must obtain them from the data. The number of grams of iron in the final compound is the same as the number of grams at the start. The rest is oxygen. How many grams of oxygen combined with 1.62 grams of iron if the iron oxide produced has a mass of 2.31 g?

Iron metal rusts when exposed to the moisture and oxygen in the air. One form of rust is hydrated iron(III) oxide.

g oxygen = ? g iron oxide − ? g iron

2.31 g iron oxide − 1.62 g iron = 0.69 g oxygen

The table is started here, with the symbols and masses of elements. Step 2 is to calculate the number of moles of atoms of each element. Do so in the space beneath the table, and then put the results in the table.

| Element | Grams | Moles | Mole Ratio | Formula Ratio | Empirical Formula |
|---|---|---|---|---|---|
| Fe | 1.62 | | | | |
| O | 0.69 | | | | |

*continued*

| Element | Grams | Moles | Mole Ratio | Formula Ratio | Empirical Formula |
|---|---|---|---|---|---|
| Fe | 1.62 | $\frac{1.62\ g}{55.85\ g/mol} = 0.029\ mol$ | | | |
| O | 0.69 | $\frac{0.69\ g}{16.00\ g/mol} = 0.043\ mol$ | | | |

Now calculate the mole ratios in the space below and place the results in the table.

| Element | Grams | Moles | Mole Ratio | Formula Ratio | Empirical Formula |
|---|---|---|---|---|---|
| Fe | 1.62 | 0.0290 | $\frac{0.0290}{0.0290} = 1.00$ | | |
| O | 0.69 | 0.043 | $\frac{0.043}{0.0290} = 1.5$ | | |

Typical round-offs are of the form 0.5/1, or $^1/_2$/1, which is multiplied by 2/2 to become 1/2; 0.33/1, or $^1/_3$/1, which is multiplied by 3/3 to become 1/3; or 0.25/1, or $^1/_4$/1, which is multiplied by 4/4 to become 1/4. However, don't expect mole ratios to always be exactly 0.50, 0.33, or 0.25 to 1. Slight variances in the uncertain digit are common.

This time the numbers in the mole ratio column are not both integers or very close to integers. But they can be changed to integers and kept in the same ratio by multiplying both of them by the same small integer. Find the smallest whole number that will yield integers when used as a multiplier for 1.00 and 1.5. Use it to obtain the formula-ratio figures. Complete the table and write the empirical formula of the compound.

| Element | Grams | Moles | Mole Ratio | Formula Ratio | Empirical Formula |
|---|---|---|---|---|---|
| Fe | 1.62 | 0.0290 | 1.00 | 2 | $Fe_2O_3$ |
| O | 0.69 | 0.043 | 1.5 | 3 | |

Multiplying the mole ratio numbers by 2 yields $1.00 \times 2 = 2$ and $1.5 \times 2 = 3$, both whole numbers without a common divisor.

$Fe_2O_3$ is a source of iron in an ore called hematite. The compound is used to polish glass, precious metals, and diamonds. Hence another of its names is jeweler's rouge.

Usually a mole ratio that is not a ratio of whole numbers can be made into one by multiplying all ratio numbers by the same integer. In Active Example 7.15, the multiplier was 2. If 2 doesn't work, try 3. If that fails, try 4, and, if necessary, 5. Remember, however, that your final ratio of whole numbers must not have a common divisor.

The procedure is the same for compounds containing more than two elements.

## Active Example 7.16

A compound is found to contain 20.0% carbon, 2.2% hydrogen, and 77.8% chlorine. Determine the empirical formula of the compound.

When percentage composition is given, assume 100 g of the compound. Set up the table and complete the problem.

| Element | Grams | Moles | Mole Ratio | Formula Ratio | Empirical Formula |
|---|---|---|---|---|---|
| C | 20.0 | 1.67 | 1.00 | 3 | |
| H | 2.2 | 2.2 | 1.3 | 4 | $C_3H_4Cl_4$ |
| Cl | 77.8 | 2.19 | 1.31 | 4 | |

To convert from grams to moles:

20.0 g C ÷ 12.01 g/mol C = 1.67 mol C    2.2 g H ÷ 1.008 g/mol H = 2.2 mol H

77.8 g Cl ÷ 35.45 g/mol Cl = 2.19 mol Cl

To convert from moles to mole ratio:

1.67 mol ÷ 1.67 mol = 1.00    2.2 mol ÷ 1.67 mol = 1.3

2.19 mol ÷ 1.67 mol = 1.31

Multiplying the mole ratio figures by 3 yields integers for the formula-ratio column:

1.00 × 3 = 3    1.3 × 3 = 3.9 or 4    1.31 × 3 = 3.93 or 4

## 7.8 Determination of a Molecular Formula

**Goal 13** Given the molar mass and empirical formula of a compound, or information from which they can be found, determine the molecular formula of the compound.

At the beginning of Section 7.7, you learned that there is a series of compounds having the general formula $C_nH_{2n}$, where n is an integer. This can also be written $(CH_2)_n$. $CH_2$ is the empirical formula of all compounds in the series. The molar mass of the empirical formula unit is 12.01 g/mol C + 2(1.008 g/mol H) = 14.03 g/mol $CH_2$. If the actual compound contains two empirical formula units, that is, if n = 2, the molar mass of the real compound is 2 × 14 g/mol = 28 g/mol. If n = 3, the molar mass of the compound is 3 × 14 g/mol = 42 g/mol, and so forth.

Now reverse the process. Suppose an experiment determines that the molar mass of the compound is 70.2 g/mol. To find n, find the number of 14.03 g/mol empirical formula units in one 70.2 g/mol molecular formula unit (that is, find how many 14s there are in 70). That number is five: 70 ÷ 14 = 5. The real compound is $(CH_2)_5$, or $C_5H_{10}$. In general

$$n = \text{empirical formula units in 1 molecule} = \frac{\text{molar mass of compound}}{\text{molar mass of empirical formula}} \qquad \textbf{(7.3)}$$

Note that n *must* be an integer. If any problem yields an n that is not an integer, or very close to an integer, either the empirical formula or the molar mass is incorrect.

Our purpose here is to find n, which is a small integer. An answer expressed in three or four significant figures has no meaning; it must be rounded off to one, or at most, two. Consequently, it is permissible to "bend the rules" of significant figures in making this kind of calculation.

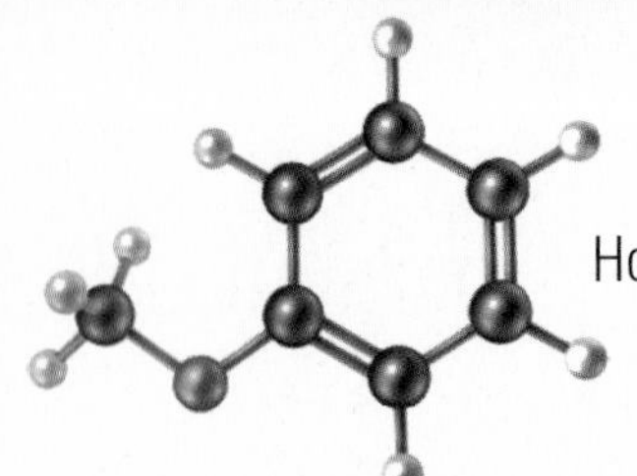

# Everyday Chemistry

## How to Read a Food Label

Hundreds of thousands of food products are available in the United States, courtesy of a multibillion-dollar-a-year industry. The Nutrition Labeling and Education Act of 1990 (NLEA) requires food manufacturers to list amounts of total fat, saturated fat, cholesterol, sodium, total carbohydrates, dietary fiber, sugars, protein, vitamin A, vitamin C, calcium, and iron in each serving of their products. Total Calories and Calories from fat must also be listed, as well as the ingredients. Food labels include the amount per serving of each nutrient (except vitamins and minerals) and the amount of each nutrient as a percentage of a daily value based on a 2000-Calorie diet.

The Food and Drug Administration (FDA) and the American Heart Association recommend a daily diet in which no more than 30% of the Calories come from fat. Fat is listed on a label in grams per serving and as a percentage of daily value. Each gram of fat accounts for 9 Calories. These data can be used to calculate the percentage of fat from any food serving. For example, according to the label on a carton of whole milk, one serving accounts for a total of 160 Calories and has 9 grams of fat. Thus,

$$\frac{9 \text{ g fat}}{\text{serving}} \times \frac{9 \text{ Cal}}{\text{g fat}} = \frac{81 \text{ Cal}}{\text{serving}} \text{ from fat}$$

% Calories from fat

$$= \frac{81 \text{ fat Cal/serving}}{160 \text{ total Cal/serving}} \times 100 = 51\%$$

If you are conscious of the fat content of your diet, would you buy the whole milk? That 51% of Calories from fat seems excessive, doesn't it? But put it into perspective. If that serving of milk is poured over a serving of corn flakes (100 Calories per serving), the 81 fat Calories are now distributed among 260 total Calories. The percentage of Calories from fat drops to $(81 \div 260) \times 100 = 31\%$, which is essentially at the recommended 30%. Add some fruit to the breakfast and you add more total Calories without increasing fat Calories, and the percentage drops some more. Watch out, though, if you eat a couple of pieces of buttered toast with your meal.

Let's say that you want buttered toast, so you decide to reduce the fat from the milk by buying "2% fat" milk. You figure 2% is *much* lower than 51%. Unfortunately, it is not that simple. The fat content of 2% milk is lower, but not by the amount those numbers suggest. The 2% is the percentage by *mass* of the milk that is fat. If you check the label you will see that a serving of 2% milk has 5 grams of fat (5 g × 9 Cal/g = 45 Cal from fat) and 140 Calories. The label also shows that this means 32% of the Calories are from fat (you can check the calculation). That's the number to be compared with 51% for whole milk. Go another step to "1% fat" milk; it has 15% of its Calories from fat. Skim milk has close to 0% Calories from fat.

Figure 7.4 shows a label from a carton of "light" ice cream that complies with the NLEA. It clearly states the percentage of fat by mass and the percentage of Calories from fat.

Isn't there an easier way to arrange a healthful diet than dealing with all these percentages? Yes, there is. The FDA diet suggests a daily fat allowance of 65 grams. This is essentially the same as the recommended 30% of Calories from fat: 65 grams × 9 Calories/gram = 585 Calories, and (585 ÷ 2000) × 100 = 29%. Now you can think of your glass of whole milk simply as 9 of the 65 grams, whether you drink the milk or put it on your corn flakes. Add the grams of fat, regardless of percentages, from everything else you eat. When you have had 65 grams of fat, eat only nonfat foods for the rest of the day and your diet will be acceptable—at least from a fat standpoint.

Perhaps the best advice is still the old saying, "Eat to live; don't live to eat." And read the labels.

**Figure 7.4** Label from a carton of "light" ice cream.

**procedure**

**How to Find the Molecular Formula of a Compound**

**Sample Problem:** *A compound with the empirical formula $C_2H_5$ has a molar mass of 58.12 g/mol. Find the molecular formula of the compound.*

**Step 1:** Determine the empirical formula of the compound. — The empirical formula is given in this example.

| | |
|---|---|
| **Step 2:** Calculate the molar mass of the empirical formula unit. | 2(12.01 g/mol C) + 5(1.008 g/mol H) = 29.06 g/mol |
| **Step 3:** Determine the molar mass of the compound (which will be given in this book). | Given as 58.12 g/mol |
| **Step 4:** Divide the molar mass of the compound by the molar mass of the empirical formula unit to get n, the number of empirical formula units per molecule. | $\frac{58.12\text{ g/mol}}{29.06\text{ g/mol}} = 2$ |
| **Step 5:** Write the molecular formula. | $(C_2H_5)_2 = C_4H_{10}$ |

Fructose is a sugar found in honey, fruit, and some vegetables. It is much sweeter than table sugar.

## Active Example 7.17

Fructose is the sugar found in honey and fruits. It is commonly known as fruit sugar. Its percentage composition is 40.0% carbon, 6.71% hydrogen, and the remainder is oxygen. The molar mass of fructose is 180.16 g/mol. Find the empirical and molecular formulas of the compound.

Start by finding the empirical formula.

| Element | Grams | Moles | Mole Ratio | Formula Ratio | Empirical Formula |
|---|---|---|---|---|---|
| C | 40.0 | 3.33 | 1.00 | 1 | |
| H | 6.71 | 6.66 | 2.00 | 2 | $CH_2O$ |
| O | 53.3 | 3.33 | 1.00 | 1 | |

To use Equation 7.3, you must have the mass of the empirical formula unit. Find the molar mass of $CH_2O$.

12.01 g/mol C + 2(1.008 g/mol H) + 16.00 g/mol O = 30.03 g/mol $CH_2O$

Calculate the number of empirical formula units in the molecule and write the molecular formula.

n = 180.16 g/mol ÷ 30.03 g/mol = 5.999 = 6 $(CH_2O)_6 = C_6H_{12}O_6$

# Chapter 7 in Review

*Most of the key terms and concepts and many others appear in the Glossary. Use your Glossary regularly.*

**7.1 The Number of Atoms in a Formula**

Goal 1 Given the formula of a chemical compound (or a name from which the formula may be written), state the number of atoms of each element in the formula unit.

**7.2 Molecular Mass and Formula Mass**

Goal 2 Distinguish among atomic mass, molecular mass, and formula mass.

Goal 3 Calculate the formula (molecular) mass of any compound whose formula is given (or known).

**Key Terms and Concepts: Formula mass, molecular mass, $\Sigma$**

**7.3 The Mole Concept**

Goal 4 Define the term *mole*. Identify the number of objects that corresponds to one mole.

Goal 5 Given the number of moles (or units) in any sample, calculate the number of units (or moles) in the sample.

**Key Terms and Concepts: Avogadro's number ($N_A$), mole (mol), $6.02 \times 10^{23}$**

**7.4 Molar Mass**

Goal 6 Define *molar mass* or interpret statements in which the term *molar mass* is used.

Goal 7 Calculate the molar mass of any substance whose chemical formula is given (or known).

**Key Terms and Concepts: Molar mass**

**7.5 Conversion Among Mass, Number of Moles, and Number of Units**

Goal 8 Given any one of the following for a substance whose formula is given (or known), calculate the other two: (a) mass; (b) number of moles; (c) number of formula units, molecules, or atoms.

**7.6 Mass Relationships Among Elements in a Compound: Percentage Composition**

Goal 9 Calculate the percentage composition of any compound whose formula is given (or known).

Goal 10 Given the mass of a sample of any compound whose formula is given (or known), calculate the mass of any element in the sample; or, given the mass of any element in the sample, calculate the mass of the sample or the mass of any other element in the sample.

**Key Terms and Concepts: Percent, percentage composition**

**7.7 Empirical Formula of a Compound**

Goal 11 Distinguish between an empirical formula and a molecular formula.

Goal 12 Given data from which the mass of each element in a sample of a compound can be determined, find the empirical formula of the compound.

**Key Terms and Concepts: Empirical formula**

**7.8 Determination of a Molecular Formula**

Goal 13 Given the molar mass and empirical formula of a compound, or information from which they can be found, determine the molecular formula of the compound.

**Key Terms and Concepts: Molecular formula**

# Study Hints and Pitfalls to Avoid

In this chapter you are building a foundation for what is to come. You will never finish with this chapter as long as you use or study chemistry.

Know and understand how to use the mole. The conversion between mass, a macroscopic quantity, and number of particles, grouped in moles, which is a particulate quantity, is probably among the most useful conversions in all of chemistry.

A strong suggestion: As you use dimensional analysis in solving problems, label each entry completely. Specifically, include the chemical formula of each substance in the calculation setup. Always calculate molar masses that correspond to the chemical formula. This is critical in the applications that lie ahead.

Empirical formulas express the ratio of atoms in the formula. The subscripts must therefore be whole numbers, never fractions, since fractions of atoms do not exist. When you calculate a mole ratio in an empirical formula problem that cannot easily be converted to whole numbers, you should carefully check your work in the previous part of the problem.

# Concept-Linking Exercises

*Write a brief description of the relationships among each of the following groups of terms or phrases. Answers to the Concept-Linking Exercises are in Appendix III.*

1. Atomic mass, molecular mass, formula mass, molar mass
2. Mole, $6.02 \times 10^{23}$, carbon-12, Avogadro's number

# Small-Group Discussion Questions

*Small-Group Discussion Questions are for group work, either in class or under the guidance of a leader during a discussion section.*

1. How many sodium ions and how many chloride ions are in a 0.01-g grain of table salt? How is this ion ratio expressed in the formula of the compound? In general, what is the meaning of a formula of an ionic compound?
2. Compare and contrast the terms *atomic mass, molecular mass, molar mass,* and *formula mass.* Write the formulas of five examples that fit into each category. Are there examples that fit two categories simultaneously? Can a single species be described by three or all four types of mass? If yes, give examples.

3. The mole is defined as the number of atoms in 12 g of carbon-12. The Avogadro constant is $6.02214179 \times 10^{23}$/mol. Why do these have different definitions? IUPAC specifies that "when the mole is used, the elementary entities must be specified and may be atoms, molecules, ions, electrons, or other particles, or specified groups of such particles." Why is this needed as part of the definition of the mole?

4. Carbon-12 serves as the basis of a number of interrelated definitions central to chemistry. Atomic mass, the mole, and molar mass are all directly or indirectly related to carbon-12. Write a definition of each. Explain how these definitions lead to the facts: (a) the atomic mass of carbon-12 is exactly 12 u and (b) the molar mass of carbon-12 is exactly 12 g/mol.

5. What experimental procedure would you use to determine the number of atoms in a pure gold coin? Explain, and give an example of the predicted outcome of your experiment for any selected coin weight.

6. What is the definition of the *percentage composition* of a compound? How does this compare with the percentage by number of atoms of each element in a compound? Explain in detail.

7. Explain the relationship between the percentage composition of a compound and its empirical formula. If you know the percentage composition of a compound, can you determine its empirical formula? If you know the empirical formula of a compound, can you determine its percentage composition? If the answer to either or both questions is yes, illustrate with an example.

8. What experimental information is needed to determine the molecular formula of a compound?

# Questions, Exercises, and Problems

*Blue-numbered questions are answered in Appendix III. ■ denotes problems assignable in OWL. In the first section of Questions, Exercises, and Problems, similar exercises and problems are paired in consecutive odd-even number combinations.*

*Many questions in this chapter are written with the assumption that you have studied Chapter 6 and can write the required formulas from their chemical names. If this is not the case, we have placed a list of all chemical formulas needed to answer Chapter 7 questions at the end of the Questions, Exercises, and Problems.*

*If you have studied Chapter 6 and you get stuck while answering a question because you cannot write a formula, we urge you to review the appropriate section of Chapter 6 before continuing. Avoid the temptation to "just peek" at the list of formulas. Developing skill in chemical nomenclature is part of your learning process in this course, and in many cases, you truly learn the material from Chapter 6 as you apply it here in Chapter 7 and throughout your study of chemistry.*

### Section 7.1: The Number of Atoms in a Formula

1. How many atoms of each element are in a formula unit of aluminum nitrate?

2. ■ How many atoms of each element are in a formula unit of ammonium phosphate?

### Section 7.2: Molecular Mass and Formula Mass

3. Why is it proper to speak of the molecular mass of water but not of the molecular mass of sodium nitrate?

4. It may be said that because atomic, molecular, and formula masses are all based on carbon-12, they are conceptually alike. What then are their differences?

5. Which of the three terms, *atomic mass, molecular mass, or formula mass,* is most appropriate for each of the following: ammonia, calcium oxide, barium, chlorine, sodium carbonate?

6. In what units are atomic, molecular, and formula mass expressed? Define those units.

7. Find the formula mass of each of the following substances:
   a) Lithium chloride
   b) Aluminum carbonate
   c) Ammonium sulfate
   d) Butane, $C_4H_{10}$ (molecular mass)
   e) Silver nitrate
   f) Manganese(IV) oxide
   g) Zinc phosphate

8. ■ Determine the formula or molecular mass of each substance in the following list:
   a) Nitrogen trifluoride
   b) Barium chloride
   c) Lead(II) phosphate

### Section 7.3: The Mole Concept

9. What do quantities representing one mole of iron atoms and one mole of ammonia molecules have in common?

10. Explain what the term *mole* means. Why is it used in chemistry?

11. Is the mole a number? Explain.

12. Give the name and value of the number associated with the mole.

13. Determine how many atoms, molecules, or formula units are in each of the following:
    a) 7.75 moles of methane, $CH_4$
    b) 0.0888 mole of carbon monoxide
    c) 57.8 moles of iron
    d) 0.81 mole of magnesium chloride

14. ■ a) How many molecules of boron trifluoride are present in 1.25 moles of this compound?
    b) How many moles of boron trifluoride are present in $7.04 \times 10^{22}$ molecules of this compound?
    c) How many moles of nitrogen dioxide are present in $7.89 \times 10^{22}$ molecules of this compound?
    d) How many molecules of nitrogen dioxide are present in 1.60 moles of this compound?

15. Calculate the number of moles in each of the following:
    a) $2.45 \times 10^{23}$ acetylene molecules, $C_2H_2$
    b) $6.96 \times 10^{24}$ sodium atoms

16. ■ a) How many atoms of hydrogen are present in 2.69 moles of water?

b) How many moles of oxygen are present in 1.39 × $10^{22}$ molecules of water?

**Section 7.4: Molar Mass**

17. In what way are the molar mass of atoms and atomic mass the same?

18. How does molar mass differ from molecular mass?

19. Find the molar mass of all the following substances.

a) $C_3H_8$

b) $C_6Cl_5OH$

c) Nickel phosphate

d) Zinc nitrate

20. ■ Calculate the molar mass of each of the following:

a) Chromium(II) iodide

b) Silicon dioxide

c) Carbon tetrafluoride

**Section 7.5: Conversion Among Mass, Number of Moles, and Number of Units**

*Questions 21 to 24: Find the number of moles for each mass of substance given.*

21. a) 6.79 g oxygen

b) 9.05 g magnesium nitrate

c) 0.770 g aluminum oxide

d) 659 g $C_2H_5OH$

e) 0.394 g ammonium carbonate

f) 34.0 g lithium sulfide

22. a) 53.8 g beryllium

b) 781 g $C_3H_4Cl_4$

c) 0.756 g calcium hydroxide

d) 9.94 g cobalt(III) bromide

e) 8.80 g ammonium dichromate (dichromate ion, $Cr_2O_7^{2-}$)

f) 28.3 g magnesium perchlorate

23. a) 0.797 g potassium iodate

b) 68.6 g beryllium chloride

c) 302 g nickel nitrate

24. a) 91.9 g sodium hypochlorite

b) 881 g aluminum acetate (acetate ion, $C_2H_3O_2^-$)

c) 0.586 g mercury(I) chloride

*Questions 25 to 28: Calculate the mass of each substance from the number of moles given.*

25. a) 0.769 mol lithium chloride

b) 57.1 mol acetic acid, $HC_2H_3O_2$

c) 0.68 mol lithium

d) 0.532 mol iron(III) sulfate

e) 8.26 mol sodium acetate (acetate ion, $C_2H_3O_2^-$)

26. a) 0.542 mol sodium hydrogen carbonate

b) 0.0789 mol silver nitrate

c) 9.61 mol sodium hydrogen phosphate

d) 0.903 mol calcium bromate

e) 1.14 mol ammonium sulfite

27. a) 0.379 mol lithium sulfate

b) 4.82 mol potassium oxalate (oxalate ion, $C_2O_4^{2-}$)

c) 0.132 mol lead(II) nitrate

28. a) 0.819 mol manganese(IV) oxide

b) 8.48 mol aluminum chlorate

c) 0.926 mol chromium(II) chloride

*Questions 29 to 32: Calculate the number of atoms, molecules, or formula units that are in each given mass.*

29. a) 29.6 g lithium nitrate

b) 0.151 g lithium sulfide

c) 457 g iron(III) sulfate

30. a) 85.5 g beryllium nitrate

b) 9.42 g manganese

c) 0.0948 g $C_3H_7OH$

31. a) 0.0023 g iodine molecules

b) 114 g $C_2H_4(OH)_2$

c) 9.81 g chromium(III) sulfate

32. a) 7.70 g iodine atoms

b) 0.447 g $C_9H_{20}$

c) 72.6 g manganese(II) carbonate

*Questions 33 and 34: Calculate the mass of each of the following.*

33. a) 4.30 × $10^{21}$ molecules of $C_{19}H_{37}COOH$

b) 8.67 × $10^{24}$ atoms of fluorine

c) 7.23 × $10^{23}$ formula units of nickel chloride

34. a) 2.58 × $10^{23}$ formula units of iron(II) oxide

b) 8.67 × $10^{24}$ molecules of fluorine

c) 7.36 × $10^{23}$ atoms of gold (Z = 79)

35. On a certain day the financial pages quoted the price of gold at $478 per troy ounce (1 troy ounce = 31.1 g). What is the price of a single atom of gold (Z = 79)?

36. How many carbon atoms has a gentleman given his bride-to-be if the engagement ring has a 0.500-carat diamond? There are 200 mg in a carat. (The price of diamonds doesn't seem so high when figured at dollars per atom.)

37. A person who sweetens coffee with two teaspoons of sugar, $C_{12}H_{22}O_{11}$, uses about 0.65 g. How many sugar molecules is this?

38. The mass of one gallon of gasoline is about 2.7 kg. Assuming the gasoline is entirely octane, $C_8H_{18}$, calculate the number of molecules in the gallon.

*The stable form of certain elements is a two-atom molecule. Fluorine and nitrogen are two of those elements. As you answer Questions 39 and 40, keep in mind that the chemical formula of a molecule identifies precisely what is in the individual molecule.*

39. a) What is the mass of 4.12 × $10^{24}$ N atoms?

b) What is the mass of 4.12 × $10^{24}$ $N_2$ molecules?

c) How many atoms are in 4.12 g N?

d) How many molecules are in 4.12 g $N_2$?

e) How many atoms are in 4.12 g $N_2$?

40. a) How many molecules are in 3.61 g $F_2$?
b) How many atoms are in 3.61 g $F_2$?
c) How many atoms are in 3.61 g F?
d) What is the mass of 3.61 × $10^{23}$ F atoms?
e) What is the mass of 3.61 × $10^{23}$ $F_2$ molecules?

### Section 7.6: Mass Relationships Among Elements in a Compound: Percentage Composition

*Questions 41 and 42: Calculate the percentage composition of each compound.*

41. a) Ammonium nitrate
b) Aluminum sulfate
c) Ammonium carbonate
d) Calcium oxide
e) Manganese(IV) sulfide

42. a) Magnesium nitrate
b) Sodium phosphate
c) Copper(II) chloride
d) Chromium(III) sulfate
e) Silver carbonate

43. Lithium fluoride is used as a flux when welding or soldering aluminum. How many grams of lithium are in 1.00 lb (454 g) of lithium fluoride?

44. Ammonium bromide is a raw material in the manufacture of photographic film. What mass of bromine is found in 7.50 g of the compound?

45. Potassium sulfate is found in some fertilizers as a source of potassium. How many grams of potassium can be obtained from 57.4 g of the compound?

46. Magnesium oxide is used in making bricks to line very-high-temperature furnaces. If a brick contains 1.82 kg of the oxide, what is the mass of magnesium in the brick?

47. Zinc cyanide, $Zn(CN)_2$, is a compound used in zinc electroplating. How many grams of the compound must be dissolved in a test bath in a laboratory to introduce 146 g of zinc into the solution?

48. ■ An experiment requires that enough $C_5H_{12}O$ be used to yield 19.7 g of oxygen. How much $C_5H_{12}O$ must be weighed out?

49. Molybdenum (Z = 42) is an element used in making steel alloys. It comes from an ore called wulfenite, $PbMoO_4$. What mass of pure wulfenite must be treated to obtain 201 kg Mo?

50. ■ How many grams of nitrogen monoxide must be weighed out to get a sample of nitrogen monoxide that contains 14.7 g of oxygen?

51. How many grams of the insecticide calcium chlorate must be measured if a sample is to contain 4.17 g chlorine?

52. ■ If a sample of carbon dioxide contains 16.4 g of oxygen, how many grams of carbon dioxide does it contain?

### Section 7.7: Empirical Formula of a Compound

53. Explain why $C_6H_{10}$ must be a molecular formula, while $C_7H_{10}$ could be a molecular formula, an empirical formula, or both.

54. From the following list, identify each formula that could be an empirical formula. Write the empirical formulas of any compounds that are not already empirical formulas. $C_2H_6O$; $Na_2O_2$; $C_2H_4O_2$; $N_2O_5$.

55. A certain compound is 52.2% carbon, 13.0% hydrogen, and 34.8% oxygen. Find the empirical formula of the compound.

56. ■ A compound is found to contain 15.94% boron and 84.06% fluorine by mass. What is the empirical formula for this compound?

57. A researcher exposes 11.89 g of iron to a stream of oxygen until it reacts to produce 16.99 g of a pure oxide of iron. What is the empirical formula of the product?

58. ■ A compound is found to contain 39.12% carbon, 8.772% hydrogen, and 52.11% oxygen by mass. What is the empirical formula for this compound?

59. A compound is 17.2% C, 1.44% H, and 81.4% F. Find its empirical formula.

60. ■ A compound is found to contain 21.96% sulfur and 78.04% fluorine by mass. What is the empirical formula for this compound?

### Section 7.8: Determination of a Molecular Formula

61. A coolant widely used in automobile engines is 38.7% carbon, 9.7% hydrogen, and 51.6% oxygen. Its molar mass is 62.0 g/mol. What is the molecular formula of the compound?

62. ■ A compound is found to contain 31.42% sulfur, 31.35% oxygen, and 37.23% fluorine by mass. What is the empirical formula for this compound? The molar mass for this compound is 102.1 g/mol. What is the molecular formula for this compound?

63. A compound is 73.1% chlorine, 24.8% carbon, and the balance is hydrogen. If the molar mass of the compound is 97 g/mol, find the molecular formula.

64. ■ A compound is found to contain 25.24% sulfur and 74.76% fluorine by mass. What is the empirical formula for this compound? The molar mass for this compound is 254.1 g/mol. What is the molecular formula for this compound?

### General Questions

65. Distinguish precisely and in scientific terms the differences among items in each of the following groups.
a) Atomic mass, molecular mass, formula mass, molar mass
b) Molecular formula, empirical formula

66. Classify each of the following statements as true or false:
a) The term *molecular mass* applies mostly to ionic compounds.
b) Molar mass is measured in atomic mass units.
c) Grams are larger than atomic mass units; therefore, molar mass is numerically larger than atomic mass.
d) The molar mass of hydrogen is read directly from the periodic table, whether it is monatomic hydrogen, H, or hydrogen gas, $H_2$.
e) An empirical formula is always a molecular formula, although a molecular formula may or may not be an empirical formula.

67. Would you need a truck to transport $10^{25}$ atoms of copper? Explain.

68. ■ The stable form of elemental phosphorus is a tetratomic molecule. Calculate the number of molecules and atoms in 85.0 g $P_4$.

69. Is it reasonable to set a dinner table with one mole of salt, NaCl, in a salt shaker with a capacity of 2 oz? How about one mole of sugar, $C_{12}H_{22}O_{11}$, in a sugar bowl with a capacity of 10 oz?

### More Challenging Problems

70. ■ The quantitative significance of "take a deep breath" varies, of course, with the individual. When one person did so, he found that he inhaled $2.95 \times 10^{22}$ molecules of the mixture of mostly nitrogen and oxygen we call air. Assuming this mixture has an average molar mass of 29 g/mol, what is his apparent lung capacity in grams of air?

71. ■ Assuming gasoline to be pure octane, $C_8H_{18}$ (actually, it is a mixture of many substances), an automobile getting 25.0 miles per gallon would consume $5.62 \times 10^{23}$ molecules per mile. Calculate the mass of this amount of fuel.

72. ■ A researcher took 27.37 g of a certain compound containing only carbon and hydrogen and burned it completely in pure oxygen. All the carbon was changed to 85.9 g of $CO_2$, and all the hydrogen was changed to 35.5 g of $H_2O$. What is the empirical formula of the original compound? (*Hint:* Find the mass in grams of carbon and hydrogen in the original compound.)

73. ■ $Co_aS_bO_c \cdot X\ H_2O$ is the general formula of a certain hydrate. When 43.0 g of the compound is heated to drive off the water, 26.1 g of anhydrous compound is left. Further analysis shows that the percentage composition of the anhydrate is 42.4% Co, 23.0% S, and 34.6% O. Find the empirical formula of (a) the anhydrous compound and (b) the hydrate. (*Hint:* Treat the anhydrous compound and water just as you have treated elements in calculating X in the formula of the hydrate.)

# Formulas

*These formulas are provided in case you are studying Chapter 7 before you study Chapter 6. You should not use this list unless your instructor has not yet assigned Chapter 6 or otherwise indicated that you may use the list.*

1. $Al(NO_3)_3$
2. $(NH_4)_3PO_4$
7. $LiCl$, $Al_2(CO_3)_3$, $(NH_4)_2SO_4$, $AgNO_3$, $MnO_2$, $Zn_3(PO_4)_2$
8. $NF_3$, $BaCl_2$, $Pb_3(PO_4)_2$
13. CO, Fe
14. $BF_3$, $NO_2$
15. Na
16. $H_2O$
19. $Ni_3(PO_4)_2$, $Zn(NO_3)_2$
20. $CrI_2$, $SiO_2$, $CF_4$
21. $O_2$, $Mg(NO_3)_2$, $Al_2O_3$, $(NH_4)_2CO_3$, $Li_2S$
22. Be, $Ca(OH)_2$, $CoBr_3$, $(NH_4)_2Cr_2O_7$, $Mg(ClO_4)_2$
23. $KIO_3$, $BeCl_2$, $Ni(NO_3)_2$
24. NaClO, $Al(C_2H_3O_2)_3$, $Hg_2Cl_2$
25. LiCl, Li, $Fe_2(SO_4)_3$, $NaC_2H_3O_2$
26. $NaHCO_3$, $AgNO_3$, $Na_2HPO_4$, $Ca(BrO_3)_2$, $(NH_4)_2SO_3$
27. $Li_2SO_4$, $K_2C_2O_4$, $Pb(NO_3)_2$
28. $MnO_2$, $Al(ClO_3)_3$, $CrCl_2$
29. $LiNO_3$, $Li_2S$, $Fe_2(SO_4)_3$
30. $Be(NO_3)_2$, Mn
31. $I_2$, $Cr_2(SO_4)_3$
32. I, $MnCO_3$
33. F, $NiCl_2$
34. FeO, $F_2$, Au
35. Au
36. C
41. $NH_4NO_3$, $Al_2(SO_4)_3$, $(NH_4)_2CO_3$, CaO, $MnS_2$
42. $Mg(NO_3)_2$, $Na_3PO_4$, $CuCl_2$, $Cr_2(SO_4)_3$, $Ag_2CO_3$
43. LiF
44. $NH_4Br$
45. $K_2SO_4$
46. MgO
48. $O_2$
50. NO
51. $Ca(ClO_3)_2$
52. $CO_2$
67. Cu

NASA

A huge quantity of energy is necessary to provide enough power for the space shuttle to escape the earth's gravitational pull. The reactants in this exothermic reaction are liquid hydrogen and liquid oxygen; the product, water. Rather than using words, a chemist might write a chemical equation for this combination reaction: $2\ H_2(\ell) + O_2(\ell) \rightarrow 2\ H_2O(g)$. In this chapter you will learn about many kinds of chemical reactions and how to write equations for them.

# Chemical Reactions

CONTENTS



We will now begin to present chemical reactions in detail. You will learn how to write equations for those reactions. Recall from Section 2.8 that an equation shows the formulas of the **reactants**—the starting substances that will be destroyed in the chemical change—written on the left side of an arrow, $\rightarrow$. The formulas of the **products** of the reaction—the new substances formed in the chemical change—are written on the right side of the arrow.

In this chapter you will learn to identify four different kinds of chemical changes. These four types are based on how atoms are rearranged in the chemical equation. They are not a complete description of the chemical change; however, when you combine knowledge of these rearrangements with an understanding of the particulate-level driving forces for reactions (which are introduced in Chapter 9), you will have learned the most common reactions that occur in living organisms, in industry and the laboratory, and in nature. Writing a chemical equation is easier if you can classify the reaction as a certain type. You will be able to look at a particular combination of reactants, recognize what kind of reaction is possible, if any, and predict what products will be formed. The equation follows.

Online homework for this chapter may be assigned in OWL

With these facts in mind, we suggest that you set your sights on the following goals as you begin this chapter:

1. Learn the mechanics of writing an equation.
2. Learn how to identify four different kinds of reactions.
3. Learn how to predict the products of each kind of reaction and write the formulas of those products.
4. Given potential reactants, write the equations for the probable reaction.

## 8.1 Evidence of a Chemical Change

**Goal** 1 Describe five types of evidence detectable by human senses that usually indicate a chemical change.

In Chapter 2 we stated that a chemical change occurs when the chemical identity of the reacting substances is destroyed and new substances form. Particles of matter are literally changed. The number and type of *atoms* that make up molecules are the same before and after a chemical change, but the number and type of *molecules* change. If we could observe matter at the particulate level, we would have a simple method for detecting chemical change. Of course, we cannot directly see what happens at the particulate level, so we must rely on indirect evidence of particulate-level rearrangements.

**P/REVIEW:** Physical and chemical changes were introduced in Section 2.3.

We also stated in Chapter 2 that one or more of our five physical senses can usually detect chemical change. This is true because particulate-level changes are often accompanied by macroscopic changes that we can sense by sight, hearing, touch, smell, or taste.* A **change of color** is almost always evidence that a chemical reaction has occurred. Figure 8.1 shows a violet-colored dye reacting with hydroxide ion to form a colorless product. The original dye molecules react with the hydroxide ions to form new molecules that do not exhibit color in a water solution.

Another visible form of evidence of a chemical change is the **formation of a solid product** when clear solutions are combined (Fig. 8.2). Substances in each of the reacting solutions combine to form the solid product. **Formation of a gas,** as evidenced by bubbles forming in a liquid, is another form of visible evidence of a chemical reaction (Fig 8.3). One product of such a reaction exists in the gaseous state at room conditions, so it bubbles out of the liquid as soon as it forms.

We can often detect the energy changes that accompany a chemical change in the form of feeling **heat** or seeing **light.** Many reactions both emit heat and give off light

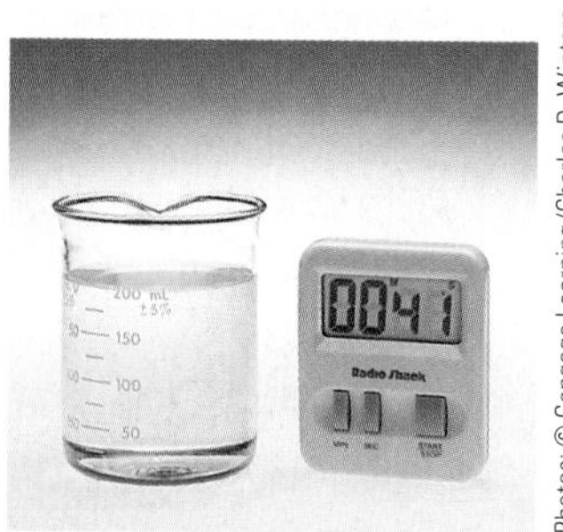




**Figure 8.1** Color change as evidence of a chemical reaction. When a solution containing hydroxide ion is added to crystal violet dye, the intensity of the color decreases with time until it disappears. The product of the reaction is colorless in water solution.

*Tasting is not an acceptable method for detecting chemical change in the laboratory. Many early chemists sacrificed their health and even their lives by using this sense. The other senses must also be used with caution and proper technique.

**Figure 8.2** Formation of a solid as evidence of a chemical reaction. When two clear, colorless solutions are combined, one containing barium ion and the other containing sulfate ion, solid barium sulfate is formed. The white solid will eventually settle to the bottom of the test tube.

**Figure 8.3** Formation of a gas as evidence of a chemical reaction. Alka-Seltzer tablets contain citric acid and sodium hydrogen carbonate. When added to water, these two reactants combine to form carbon dioxide gas as one of the products.

**Figure 8.4** Evolution of heat and light as evidence of a chemical reaction. This photograph shows aluminum reacting with iron(III) oxide to produce molten iron and aluminum oxide. The quantity of heat evolved from this chemical change is large enough to cause the iron produced to be in the liquid (molten) state.

(Fig. 8.4). Burning is a common example. More unusual are reactions such as those that occur in light sticks, which give off light without getting hot (Fig. 8.5). Some reactions absorb heat, such as the reaction of hydrogen and iodine gases to form hydrogen iodide gas.

One must always be cautious when using sensory evidence of a chemical change. Many chemical reactions are not visible, and their heat energy changes are too small to feel. Also, sensory evidence can indicate a physical change, rather than one that is chemical. For example, we feel the coldness when ice is placed on our skin, but there is no chemical change when ice—solid water—changes to the liquid state. Similarly, when water boils, we see bubbles of steam rising through the liquid, but the liquid-to-vapor change is also physical. Water molecules are unchanged, whether in the solid (ice), liquid, or gaseous (steam) state.

Another example of sensory evidence indicating a physical change is the coldness you feel when an instant cold pack is placed on swollen tissue (Fig. 8.6). Most cold packs contain a small sealed container of solid ammonium nitrate that breaks when

**Figure 8.5** Emission of light as evidence of a chemical reaction. The light emitted from a light stick is a form of energy released in a multistep chemical change that involves several reactants. One of the reactants is stored in a fragile glass vial inside of the larger plastic tube. When you bend the plastic tube, you break the glass vial, mixing the reactants.

| **Table 8.1** | Evidence of Chemical Change |
|---|---|
| 1. | Color change |
| 2. | Formation of a solid |
| 3. | Formation of a gas |
| 4. | Absorption or release of heat energy |
| 5. | Emission of light energy |

© Cengage Learning/Charles D. Winters

**Figure 8.6** Heat transfer as evidence of a physical change. When heat is absorbed by a system, as in this instant cold compress, your skin feels cold because you are the source of heat needed to drive the change.

the larger packet is squeezed, allowing the ionic solid to dissolve in water via an endothermic process. This is an example of an endothermic physical change rather than a chemical change because the ammonium ions and the nitrate ions are the same, no matter whether they are in the solid state or the aqueous state.

Table 8.1 summarizes the five types of evidence that indicate the possibility of a chemical change.

### √ Target Check 8.1

*Are any of the following a chemical change? Give evidence for each answer.*

a) A firecracker explodes.

b) A puddle of water evaporates.

c) 

© Cengage Learning/Charles D. Winters

d)  

© YAKOBCHUK VASYL, 2008/Used under license from Shutterstock.com

## 8.2 Evolution of a Chemical Equation

**Goal** 2 Distinguish between an unbalanced and a balanced chemical equation, and explain why a chemical equation needs to be balanced.

If a piece of sodium is dropped into water, a vigorous reaction occurs (Fig. 8.7). A full qualitative description of the chemical change is "solid sodium plus liquid water yields hydrogen gas plus sodium hydroxide solution plus heat." That sentence is translated literally into a chemical equation:

$$Na(s) + H_2O(\ell) \rightarrow H_2(g) + NaOH(aq) + heat \quad \textbf{(8.1)}$$

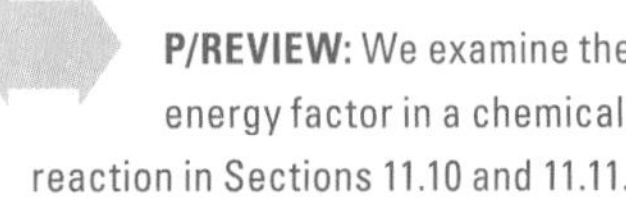

**P/REVIEW:** We examine the energy factor in a chemical reaction in Sections 11.10 and 11.11.

The (s) after the symbol of sodium indicates it is a solid. Similarly, the $(\ell)$ after $H_2O$ and the (g) after $H_2$ show that they are a liquid and a gas, respectively. When a substance is dissolved in water, the mixture is an **aqueous solution** and is identified by

(a)

(b)

(c)

Photos: © Cengage Learning/Charles D. Winters

**Figure 8.7** Sodium reacting with water. (a) A small piece of sodium is dropped into a test tube of water containing phenolphthalein, a substance that turns pink in a solution of a metallic hydroxide. (b) Sodium forms a "ball" that dashes erratically over the water surface, releasing hydrogen as it reacts. Pink color near the sodium indicates that a sodium hydroxide (NaOH) solution is being formed in the region. (c) Dissolved NaOH is now distributed uniformly through the solution, which is hot because of the heat released in the reaction. *Warning:* Do not try this experiment yourself; it is dangerous, potentially splattering hot alkali into eyes and onto skin and clothing.

(aq). (*Aqueous* comes from the Latin *aqua* for "water.") These **state symbols** are sometimes omitted when writing equations, but they are included in most of the equations in this book. They are discussed in more detail in Chapter 9. We suggest that you use or not use them at this time according to the directions of your instructor. State symbols are summarized in Table 8.2.

**Table 8.2** State Symbols and Their Meanings

| Symbol | Meaning |
|---|---|
| (s) | Solid |
| ($\ell$) | Liquid |
| (g) | Gas |
| (aq) | Aqueous (dissolved in water) |

Nearly all chemical reactions involve some transfer of energy, usually in the form of heat. Generally, we omit energy terms from equations unless there is a specific reason for including them.

The Law of Conservation of Mass (Section 2.9) says that the total mass of the products of a reaction is the same as the total mass of the reactants. Atomic theory explains this by saying that atoms involved in a chemical change are neither created nor destroyed, but simply rearranged. Equation 8.1 does not satisfy this condition. There are two hydrogen atoms in $H_2O$ in the left side of the equation, but three atoms of hydrogen on the right—two in $H_2$ and one in NaOH. At this point, the equation is only a *qualitative* description of the reaction. The equation is not **balanced.**

An equation is balanced by placing a coefficient in front of one or more of the formulas, indicating that it is used more than once. Once balanced, an equation is both a *qualitative* and a *quantitative* description of the reaction.

Hydrogen is lacking on the left side of Equation 8.1, so let's try two water molecules:

$$Na(s) + 2\ H_2O(\ell) \rightarrow H_2(g) + NaOH(aq) \quad \textbf{(8.2)}$$

At first glance, this hasn't helped; indeed, it seems to have made matters worse. The hydrogen is still out of balance (four on the left, three on the right), and furthermore, oxygen is now unbalanced (two on the left, one on the right). We are short one oxygen atom and one hydrogen atom on the right side. But look closely: Oxygen and hydrogen are part of the same compound on the right, and there is one atom of each in that compound. If we take two NaOH units

$$Na(s) + 2\ H_2O(\ell) \rightarrow H_2(g) + 2\ NaOH(aq) \quad \textbf{(8.3)}$$

there are four hydrogens and two oxygens on both sides of the equation. These elements are now in balance. Unfortunately, the sodium is now *un*balanced. This condition is corrected by adding a coefficient of 2 to the sodium

$$2\ Na(s) + 2\ H_2O(\ell) \rightarrow H_2(g) + 2\ NaOH(aq) \quad \textbf{(8.4)}$$

The equation is now balanced. Note that, in the absence of a numerical coefficient, as with $H_2$, the coefficient is assumed to be 1.

Balancing an equation involves some important do's and don'ts that are apparent in this example:

*DO: Balance the equation entirely by using coefficients placed before the different chemical formulas.*

*DON'T: Change a correct chemical formula in order to make an element balance.*

*DON'T: Add some real or imaginary chemical species to either side of the equation just to make an element balance.*

A moment's thought shows why the two "don'ts" are improper. The first step in writing an equation is to write formulas of reactants and products. These formulas describe the substances in the reaction, and these do not change. Changing a formula or adding another formula would change the qualitative description of the reaction.

In other words, writing and balancing a chemical equation requires two main steps:

**procedure**

**Writing and Balancing a Chemical Equation**

**Sample Problem:** *An aluminum strip reacts with the oxygen in air to form solid aluminum oxide. Write and balance a chemical equation for the reaction.*

**Step 1:** Write a qualitative description of the reaction. In this step you write the formulas of reactants and products. $Al(s) + O_2(g) \rightarrow Al_2O_3(s)$

**Step 2:** Quantify the description by balancing the equation. Do this by adding coefficients. Do not change the qualitative description of the reaction. $4\ Al(s) + 3\ O_2(g) \rightarrow 2\ Al_2O_3(s)$

### √ Target Check 8.2

*Are the following true or false? Explain your reasoning.*

a) The equation $C_2H_4O + 3\ O_2 \rightarrow 2\ CO_2 + 2\ H_2O$ is balanced.

b) The equation $H_2 + O_2 \rightarrow H_2O$ may be balanced by changing it to $H_2 + O_2 \rightarrow H_2O_2$.

## 8.3 Balancing Chemical Equations

**Goal** **3** Given an unbalanced chemical equation, balance it by inspection.

The balancing procedure in the preceding section is sometimes called "balancing by inspection." It is a trial-and-error method that succeeds in nearly all the reactions you are likely to encounter in a general chemistry course. Most equations can be balanced without following a set series of steps. However, if you prefer a formal written procedure, the following works well, even with equations that are quite complicated.

**procedure**

**Balancing a Chemical Equation: A Formal Approach**

**Step 1:** Place a "1" in front of the formula with the largest number of atoms. If two formulas have the same number of atoms, select the one with the greater number of elements. We will call this formula the **starting formula** in the discussion and Active Examples that follow.

**Step 2:** Insert coefficients that balance the elements *that appear in compounds.* Use fractional coefficients, if necessary. Do not balance element-only formulas, such as Na or $O_2$, at this time. We call these *uncombined elements.* Choosing elements in the following order is usually easiest:

a) Elements in the starting formula that are in only one other compound

b) All other elements from the starting formula

c) All other elements *in compounds*

**Step 3:** Place coefficients in front of formulas of uncombined elements that balance those elements. Use fractional coefficients, if necessary.

**Step 4:** Clear fractions, if any, by multiplying all coefficients by the lowest common denominator. Remove any "1" coefficients that remain.

**Step 5:** Check to be sure the final equation is balanced.

In following this procedure, don't change a coefficient, once written, except to clear fractions. NEVER change or add a formula to balance an equation.

We now apply this procedure to the sodium-plus-water reaction in Section 8.2. Each step is listed, immediately followed by comments or explanations.

**Step 1:** Place the coefficient "1" in front of the starting formula.

NaOH is the starting formula in this example because it has three atoms and three elements. | $Na + H_2O \rightarrow H_2 + 1\ NaOH$

**Step 2:** Balance elements in compounds.

Start by balancing the elements in the starting formula that are in only one other compound formula. Oxygen appears only in $H_2O$. The 1 O in NaOH is balanced by 1 O in $H_2O$. | $Na + 1\ H_2O \rightarrow H_2 + 1\ NaOH$

**Step 3:** Balance uncombined elements.

The 1 Na in NaOH on the right requires 1 Na on the left. | $1\ Na + 1\ H_2O \rightarrow H_2 + 1\ NaOH$

Only 1 H atom in $H_2O$ comes from NaOH, so the other must come from $H_2$. To get 1 atom of H from $H_2$, we need $^1/_2$ of an $H_2$ unit. | $1\ Na + 1\ H_2O \rightarrow {}^1/_2\ H_2 + 1\ NaOH$

**Step 4:** Clear fractions; remove 1s.

Multiply by 2 to clear the $^1/_2$. | $2 \times (1\ Na + 1\ H_2O \rightarrow {}^1/_2\ H_2 + 1\ NaOH)$

$2\ Na + 2\ H_2O \rightarrow H_2 + 2\ NaOH$

**Step 5:** Check your work. | 2 Na, 4 H, 2 O on each side. ✓

Notice that the procedure in this section is not the same as the thought process by which we balanced the same equation in Section 8.2. There is no one "correct" way to balance an equation, but there are techniques that can shorten the process. Look for them in the Active Examples later in this chapter.

Is $Na(s) + H_2O(\ell) \rightarrow {}^1/_2\ H_2(g) + NaOH(aq)$ a legitimate equation? Yes and no. If you think of the equation as "1 Na atom reacts with 1 $H_2O$ molecule to produce $^1/_2$ an $H_2$ molecule and 1 NaOH unit," the equation is not legitimate. There is no such thing as "$^1/_2$ an $H_2$ molecule," any more than a chicken can lay $^1/_2$ an egg. But if you think about "1 mole of Na atoms reacts with 1 mole of $H_2O$ molecules to produce $^1/_2$ mole of hydrogen molecules and 1 mole of NaOH units," the fractional coefficient is reasonable. There can be $^1/_2$ mole of hydrogen molecules just as a chicken can lay $^1/_2$ dozen eggs.

It is sometimes necessary to use fractional coefficients, but few appear in this book. We stay with the standard practice of writing equations with whole-number coefficients. These coefficients should be written in the lowest terms possible; they should not have a common divisor. For example, although $4\ Na(s) + 4\ H_2O(\ell) \rightarrow 2\ H_2(g) + 4\ NaOH(aq)$ is a legitimate equation, it can and should be reduced to Equation 8.4 by dividing all coefficients by 2.

Notice that you can treat chemical equations exactly the same way you treat algebraic equations. Chemical formulas replace x, y, or other variables. You can multiply or divide an equation by some number by multiplying or dividing each term by that number. The order in which you write reactants or products may change; in an equation, $2\ H_2O + 2\ Na$ is the same as $2\ Na + 2\ H_2O$.

**Figure 8.8** Natural gas burning. Methane, $CH_4$, is the principal component of natural gas. When natural gas burns, methane reacts with the oxygen in the air, forming the products carbon dioxide and water vapor.

## Active Example 8.1

Consider the reaction of methane and oxygen, forming carbon dioxide and water, at a temperature at which all reactants and products are in the gas phase (Fig. 8.8). Balance the equation written below and show that the number of atoms of each element remains the same before and after the reaction.

Step 1 in the formal approach is to place a "1" in front of the formula with the greatest number of atoms. Begin there.

$$\_\_\ CH_4(g) + \_\_\ O_2(g) \rightarrow \_\_\ CO_2(g) + \_\_\ H_2O(g)$$

$$1\ CH_4(g) + \_\_\ O_2(g) \rightarrow \_\_\ CO_2(g) + \_\_\ H_2O(g)$$

$CH_4$ has five atoms, more than any other molecule.

The next step is to balance elements in the starting formula that appear in compounds. Carbon is in $CO_2$ and hydrogen is in $H_2O$, so balance both elements by adding the appropriate coefficients to the product compounds.

$$1\ CH_4(g) + \_\_\ O_2(g) \rightarrow \_\_\ CO_2(g) + \_\_\ H_2O(g)$$

$$1\ CH_4(g) + \_\_\ O_2(g) \rightarrow 1\ CO_2(g) + 2\ H_2O(g)$$

Now balance the remaining species, which is an uncombined element.

$$1\ CH_4(g) + \_\_\ O_2(g) \rightarrow 1\ CO_2(g) + 2\ H_2O(g)$$

$$1\ CH_4(g) + 2\ O_2(g) \rightarrow 1\ CO_2(g) + 2\ H_2O(g)$$

Two oxygens in $CO_2$ plus one from each of two water molecules makes a total of four oxygen atoms.

The final balancing step is to remove the "1" coefficients that remain. We've done that for you in the equation below. Draw the number of molecules of each species indicated by the coefficients in the balanced equation in the boxes below the equation. Use our molecular drawings above as a guide.

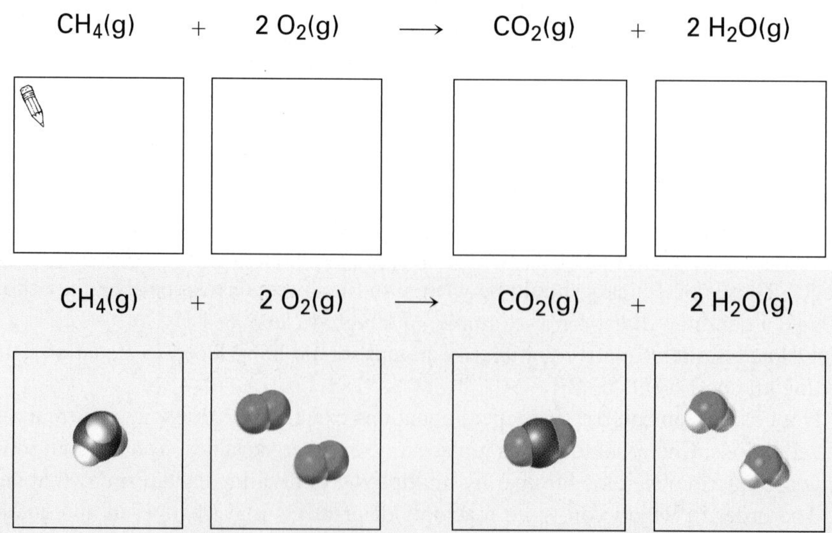

Does the number of atoms balance? To answer this question, count the number of each type of atom in your boxes.

Yes, the atoms balance: 1 C, 4 H, and 4 O on each side.

In a chemical change, the atoms are the same before and after the change, in both number and identity. Their new arrangement in molecules is the essence of the change.

## Active Example 8.2

Balance the equation for the reaction of solid phosphorus pentachloride with liquid water, yielding aqueous solutions of phosphoric acid and hydrochloric acid.

Identify the starting formula—the formula with the greatest number of atoms and/or elements. Give it a coefficient of 1.

$$\_\_ \ PCl_5(s) + \_\_ \ H_2O(\ell) \rightarrow \_\_ \ H_3PO_4(aq) + \_\_ \ HCl(aq)$$

$$\_\_ \ PCl_5(s) + \_\_ \ H_2O(\ell) \rightarrow 1\ H_3PO_4(aq) + \_\_ \ HCl(aq)$$

Now begin to balance the elements in the starting formula that are in only one other formula.

$$\_\_ \ PCl_5(s) + \_\_ \ H_2O(\ell) \rightarrow 1\ H_3PO_4(aq) + \_\_ \ HCl(aq)$$

$$1\ PCl_5(s) + 4\ H_2O(\ell) \rightarrow 1\ H_3PO_4(aq) + \_\_ \ HCl(aq)$$

Both phosphorus and oxygen are in one other formula. One P in $H_3PO_4$ requires 1 $PCl_5$, and 4 O in $H_3PO_4$ is satisfied by 4 $H_2O$. Hydrogen is in two other compounds, so we save that until later.

Are there any elements in only one compound on each side of the equation? Balance them.

$$1\ PCl_5(s) + 4\ H_2O(\ell) \rightarrow 1\ H_3PO_4(aq) + \_\_ \ HCl(aq)$$

$$1\ PCl_5(s) + 4\ H_2O(\ell) \rightarrow 1\ H_3PO_4(aq) + 5\ HCl(aq)$$

Chlorine is the only element appearing in compounds on both sides of the equation. Five Cl in $PCl_5$ are balanced by 5 HCl.

Hydrogen remains, and hydrogen is already balanced at 8 H on each side. When no uncombined elements are present, the last element should already be balanced when you reach it. If not, look for an error on some earlier element.

There are no fractional coefficients, so remove the 1s and make a final check. Write the final balanced equation and show the atom balance.

*continued*

$PCl_5(s) + 4\ H_2O(\ell) \rightarrow H_3PO_4(aq) + 5\ HCl(aq)$

There are 1 P, 5 Cl, 8 H, and 4 O on each side.

### Active Example 8.3

Balance the equation that represents the reaction of liquid propionic acid with gaseous oxygen to form gaseous carbon dioxide and liquid water.

This equation is just a little bit tricky. See if you can avoid the traps. Try to take it all the way to completion.

$\_\_\ C_2H_5COOH(\ell) + \_\_\ O_2(g) \rightarrow \_\_\ CO_2(g) + \_\_\ H_2O(\ell)$

$2\ C_2H_5COOH(\ell) + 7\ O_2(g) \rightarrow 6\ CO_2(g) + 6\ H_2O(\ell)$

| | |
|---|---|
| 1 before starting formula: | $1\ C_2H_5COOH(\ell) + O_2(g) \rightarrow CO_2(g) + H_2O(\ell)$ |
| Balance C: | $1\ C_2H_5COOH(\ell) + O_2(g) \rightarrow 3\ CO_2(g) + H_2O(\ell)$ |
| Balance H: | $1\ C_2H_5COOH(\ell) + O_2(g) \rightarrow 3\ CO_2(g) + 3\ H_2O(\ell)$ |
| Balance O: | $1\ C_2H_5COOH(\ell) + 7/2\ O_2(g) \rightarrow 3\ CO_2(g) + 3\ H_2O(\ell)$ |
| Clear fractions, remove 1s: | $2\ C_2H_5COOH(\ell) + 7\ O_2(g) \rightarrow 6\ CO_2(g) + 6\ H_2O(\ell)$ |
| Final check: | There are 6 C, 12 H, and 18 O on each side. ✓ |

The traps are in the formula $C_2H_5COOH$. The symbols of all three elements appear twice. The formulas of organic compounds are often written this way in order to suggest the arrangement of atoms in the molecule. Students usually count carbon and hydrogen correctly, but they often overlook the oxygen in the original compound when selecting the coefficient for $O_2$.

The reason for balancing uncombined elements last is that you can insert *any* coefficient that is needed without *un*balancing any element that has already been balanced.

## 8.4 Interpreting Chemical Equations

**Goal 4** Given a balanced chemical equation or information from which it can be written, describe its meaning on the particulate, molar, and macroscopic levels.

In the previous two sections, you learned that a chemical equation contains much information. First, it gives a qualitative description of a reaction. The equation $2\ H_2(g) + O_2(g) \rightarrow 2\ H_2O(g)$ says that hydrogen reacts with oxygen to form water and that both reactants and the product are in the gas phase. Second, the equation gives quantitative information about the reaction. You can interpret this quantitative information on a number of different levels.

We stated earlier that standard practice is to write equations with the lowest whole-number coefficients possible. This is because equations are often interpreted on the *particulate level.* Thus $2\ H_2(g) + O_2(g) \rightarrow 2\ H_2O(g)$ means "two molecules of hydrogen react with one oxygen molecule to form two water molecules." The symbols in the equation represent the particles, shown here as space-filling models:

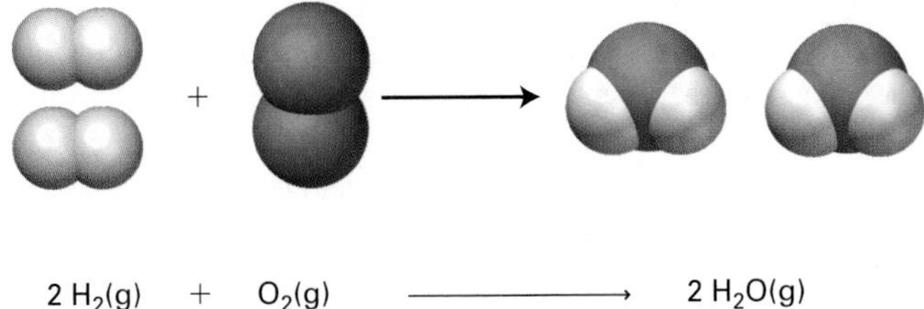


We can scale up our particulate interpretation to any desired multiple of the coefficients in the balanced equation. We can think of the equation in terms of dozens, for example: "two *dozen* molecules of hydrogen react with one *dozen* oxygen molecules to form two *dozen* water molecules":

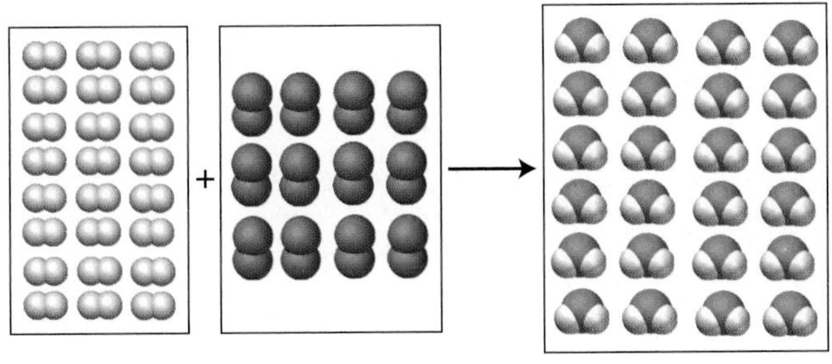


We can also choose the "chemistry dozen," the mole, as our grouping unit: "two *moles* of hydrogen molecules react with one *mole* of oxygen molecules to form two *moles* of water molecules." We simply change the grouping unit from 12 to $6.02 \times 10^{23}$.

The *molar* interpretation of a chemical equation involves reading the coefficients as the number of moles of the reactants and products. This is still a particulate-level explanation, but we are grouping the particles into counting units that make it easier to translate into a macroscopic-level interpretation. On the molar level, fractional coefficients are acceptable. $H_2(g) + \frac{1}{2}\ O_2(g) \rightarrow H_2O(g)$ can be read as "one mole of hydrogen molecules reacts with one-half mole of oxygen molecules to form one mole of water molecules."

## Thinking About Your Thinking

### Mental Models

**The illustrations that accompany the preceding representations of the $2\ H_2(g) + O_2(g) \rightarrow 2\ H_2O(g)$ reaction are designed to help you form mental models of chemical reactivity. If you can imagine what the particles look like as they react—and if you can see in your mind how the various combinations of atoms occur—you are making good progress toward thinking like a chemist. In the following sections, we will introduce four types of chemical reactions. We provide a particulate-level illustration of each reaction type. Use those drawings to form mental models of each reaction type.**

Literally counting particles is impossible. Instead, we measure the mass or some other measurable *macroscopic* property of the sample. But the reaction still occurs as a combination of particles in the ratio given in the chemical equation. The balanced equation does *not* directly give information about masses. We *cannot* say "two grams of hydrogen reacts with one gram of oxygen to form two grams of water." We do know, however, that molar mass provides a link between the number of moles of particles and their masses. Given the molar masses (to the nearest gram)—$H_2$, 2 g/mol, $O_2$,

**P/REVIEW:** Molar mass is the mass in grams of one mole of a substance (Section 7.4).

Photo by Hulton Archive/Getty Images

**Figure 8.9** The end of the dirigible *Hindenburg* in May 1937. The explosion demonstrates the tremendous amount of energy released when hydrogen reacts with oxygen to form water. This disaster ended the use of hydrogen in lighter-than-air craft. Today, the unreactive Group 8A/18 gas helium is used in the airships—blimps—from which aerial views of major sporting events are displayed on television.

32 g/mol, and $H_2O$, 18 g/mol—we can interpret the equation $2\ H_2(g) + O_2(g) \rightarrow 2\ H_2O(g)$ as "4 grams of hydrogen reacts with 32 grams of oxygen to form 36 grams of water." Other masses are acceptable, of course, as long as they reflect the molar relationships in the balanced equation. Notice that the mass of the reactants equals the mass of the product: 4 g + 32 g = 36 g. Figure 8.9 shows a macroscopic demonstration of energy released in this reaction.

You need to be able to interpret a chemical equation on any level. The remainder of this chapter and other, future chapters provide additional practice in this.

### √ Target Check 8.3

*Consider the chemical equation for the decomposition of hydrogen peroxide solution:*

$$2\ H_2O_2(aq) \rightarrow O_2(g) + 2\ H_2O(\ell)$$

*Write a word description of this equation on the (a) particulate, (b) molar, and (c) macroscopic levels.*

## 8.5 Writing Chemical Equations

In the next four sections, you will learn how to write equations for four kinds of chemical reactions for which you are given the names or formulas of the reactants only. It will be up to you to predict the formulas of the products. Your ability to classify a reaction as a certain type will be a big help in predicting what products will form. Thus we modify our two main steps for writing a chemical equation from Section 8.2 by adding a new first step. The overall procedure for writing a chemical equation thus becomes:

**procedure**

**Writing and Balancing a Chemical Equation**

**Step 1:** Classify the reaction type.

**Step 2:** Write a qualitative description of the reaction. In this step you write the formulas of the given reactants to the left of an arrow and the formulas of the given or predicted products to the right.

**Step 3:** Quantify the description by balancing the equation. Do this by adding coefficients. Do not change the qualitative description of the reaction by adding, removing, or altering any formula.

## 8.6 Combination Reactions

**Goal** 5 Write the equation for the reaction in which a compound is formed by the combination of two or more simpler substances.

A reaction in which two or more substances combine to form a single product is a **combination reaction** or **synthesis reaction.** The reactants are often elements but sometimes are compounds or both elements and compounds. The general equation for a combination reaction is

$$A + X \longrightarrow AX$$

The reaction between sodium and chlorine to form sodium chloride is a combination reaction: $2\ Na(s) + Cl_2(g) \rightarrow 2\ NaCl(s)$.

## Active Example 8.4

Write the equation for the formation of solid sodium peroxide, $Na_2O_2$ (Fig. 8.10), by direct combination of solid sodium and oxygen.

This is a combination reaction in which two elements unite to form a compound. Write the formulas of the elements on the left side of an arrow and the formula of the product on the right. In other words, write the qualitative description of the reaction.

$Na(s) + O_2(g) \rightarrow Na_2O_2(s)$

Did you remember to show oxygen as a diatomic molecule?

Balance the equation.

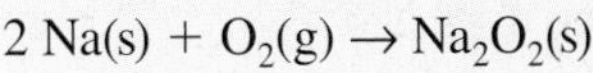

$2\ Na(s) + O_2(g) \rightarrow Na_2O_2(s)$

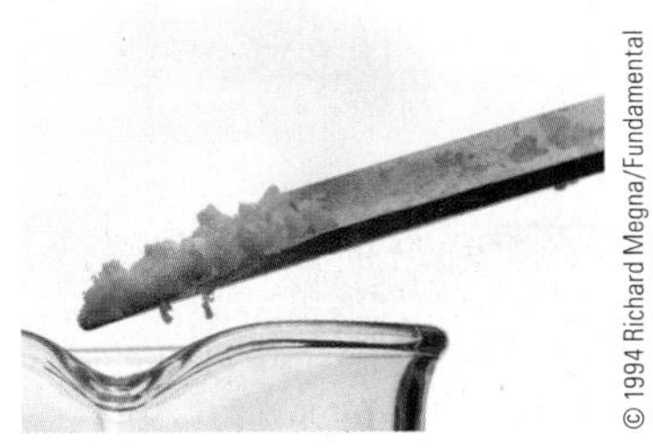



**Figure 8.10** Sodium peroxide is used to bleach animal and vegetable fibers, feathers, bones, and ivory. The textile industry uses it in dyeing. It is used to maintain breathable air in confined areas that develop a high level of carbon dioxide, such as submarines and diving bells, by the reaction $2\ Na_2O_2 + 2\ CO_2 \rightarrow 2\ Na_2CO_3 + O_2$.

We will no longer show the stepwise balancing of equations unless the reaction is complex.

## Active Example 8.5

Carbon dioxide gas is formed when charcoal (carbon) burns in air, as in a backyard barbecue (Fig. 8.11). Write the equation for the reaction.

A word description of a reaction sometimes assumes that you already know something about it. This Active Example assumes you know that when something burns in air, it is reacting chemically with the oxygen in the air. In other words, oxygen is an unidentified reactant, so it should appear on the left side of the equation along with carbon. With that hint, write the qualitative, or unbalanced, equation.

**Figure 8.11** Backyard barbecue. A simple combination reaction occurs in this widely used device.

$C(s) + O_2(g) \rightarrow CO_2(g)$

Now balance the equation.

$C(s) + O_2(g) \rightarrow CO_2(g)$

Sometimes balancing an equation is easy, as when all coefficients are 1!

**summary**

**Combination Reactions**

| | |
|---|---|
| Reactants: | Any combination of elements and/or compounds |
| Reaction type: | Combination |
| Equation type: | A + X ⟶ AX |
| Products: | One compound |

☞ *Learn It Now!* The four points in this summary are components in the thought process you should use to write an equation. Step 1 in the equation-writing procedure is classifying the reaction. After you examine the reactants, you will be able to decide the reaction type and what kind of equation it is. This also enables you to predict the products. This summary and those that follow it are gathered as a chapter summary in Section 8.10.

## 8.7 Decomposition Reactions

**Goal 6** Given a compound that is decomposed into simpler substances, either compounds or elements, write the equation for the reaction.

The decomposition of mercury(II) oxide was a key reaction used by Lavoisier in disproving the phlogiston theory described in Chapter 1.

A **decomposition reaction** is the opposite of a combination reaction, in that a compound breaks down into simpler substances. The products may be any combination of elements and compounds. The general decomposition equation is

$$AX \longrightarrow A + X$$

A typical decomposition reaction occurs when mercury(II) oxide is heated: $2\ HgO(s) \rightarrow 2\ Hg(\ell) + O_2(g)$ (Fig. 8.12).

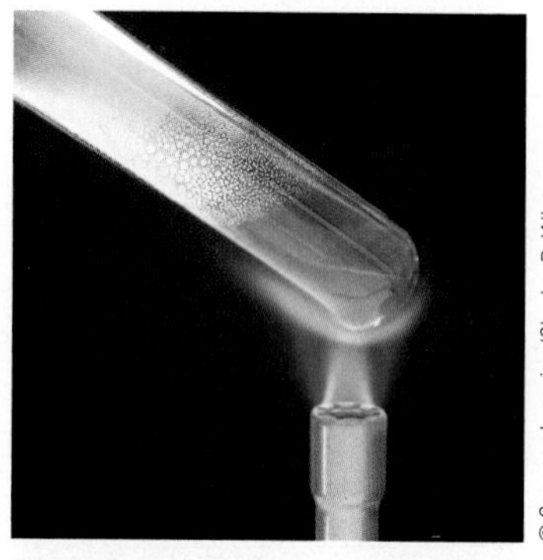



**Figure 8.12** Decomposition of mercury(II) oxide. On heating, the bright orange-red powder undergoes a decomposition reaction to form silver liquid mercury and colorless oxygen gas.

### Active Example 8.6

Water decomposes into its elements when electrical energy is added (Active Fig. 8.13). Write the equation.

Start with the formulas of the reactants and products.

$H_2O(\ell) \rightarrow \quad H_2(g) + \quad O_2(g)$

Now balance the equation.

$2\ H_2O(\ell) \rightarrow 2\ H_2(g) + O_2(g)$

This reaction is literally the reverse of the reaction discussed in Section 8.4.

Most decomposition reactions require an energy input of some sort. The decomposition of water (Active Example 8.6) is accomplished electrolytically, that is, with electrical energy. The reaction in Active Example 8.7 is achieved by heating.

Many chemical changes can be made to go in either direction, as the reversibility of the "hydrogen plus oxygen to form water" reaction suggests. These are called **reversible reactions.** Reversibility is often indicated by a double arrow, $\rightleftharpoons$. Thus

$$2\ H_2O(\ell) \rightleftharpoons 2\ H_2(g) + O_2(g) \text{ and } 2\ H_2(g) + O_2(g) \rightleftharpoons 2\ H_2O(\ell)$$

are equivalent reversible equations.

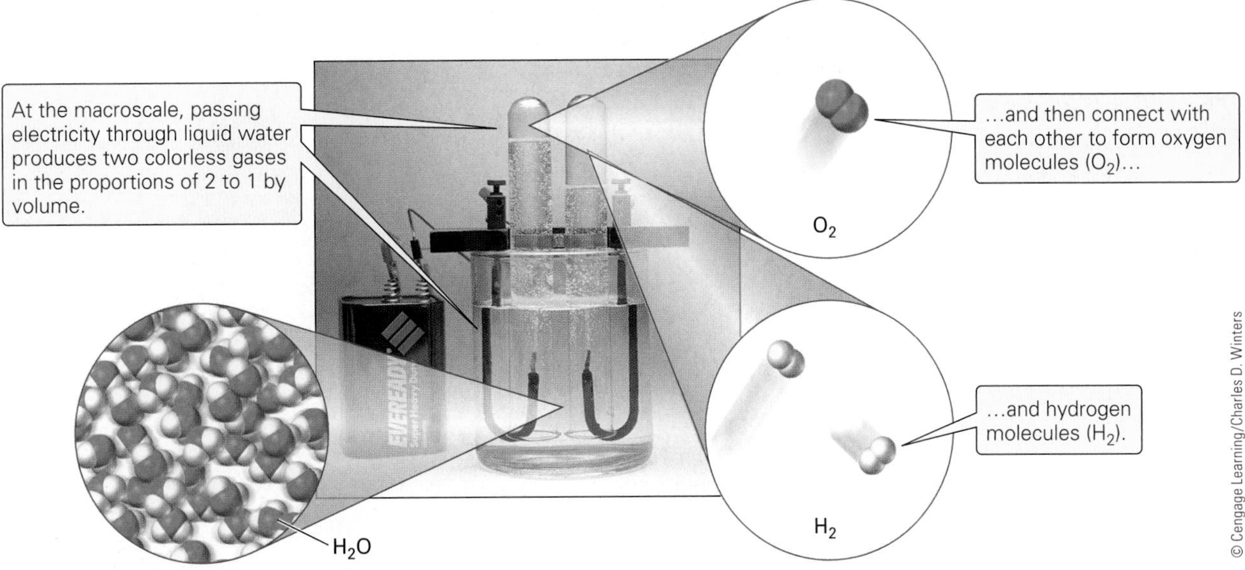
At the particulate level, hydrogen atoms and oxygen atoms originally connected in water molecules ($H_2O$) separate...
At the macroscale, passing electricity through liquid water produces two colorless gases in the proportions of 2 to 1 by volume.
...and then connect with each other to form oxygen molecules ($O_2$)...
$O_2$
...and hydrogen molecules ($H_2$).
$H_2$
$H_2O$
EVEREADY
© Cengage Learning/Charles D. Winters

Active Figure 8.13 Decomposition of water. Electrical energy is used to decompose liquid water to its elements. **Watch this active figure at *http://www.cengage.com/cracolice.***

***Learn It Now!*** Active Figures provide a dynamic animation designed to enhance your understanding of the related concept. When you come to an Active Figure, log on to our Web site at *http://www.cengage.com/cracolice* and watch the Active Figure and take the concept test about the figure to assess your understanding of it. If you don't have immediate access to the Web while you are studying, the textbook figure alone will provide sufficient information for learning. However, you will probably find that viewing the Active Figure and taking the concept test as close in time as possible to when you study the text will maximize the efficiency of your learning.

**P/REVIEW:** Reversible changes, both physical and chemical, are quite common. Changes of state, such as freezing or boiling, are reversible physical changes. So is the process by which a substance dissolves in water. When a reversible change occurs in both directions at the same time and at equal rates, equilibrium is established. Physical equilibria are described in Sections 15.4 and 16.3, and all of Chapter 18 is devoted to chemical equilibrium.

## Active Example 8.7

Lime, CaO(s), and carbon dioxide gas are the products of thermal decomposition of limestone, solid calcium carbonate (Fig. 8.14). (*Thermal* refers to heat. The reaction occurs at high temperature.) Write the equation.

Start with the unbalanced equation.

$CaCO_3(s) \rightarrow \quad CaO(s) + \quad CO_2(g)$

Add coefficients as necessary to complete the equation.

$CaCO_3(s) \rightarrow CaO(s) + CO_2(g)$

Calcium oxide is an industrial chemical used in making construction materials such as bricks, mortar, plaster, and stucco. It is also used in manufacturing sodium carbonate, one of the most widely used of all chemicals, and in manufacturing steel, magnesium, and aluminum.

**Figure 8.14** Thermal decomposition of limestone (calcium carbonate). Calcium oxide, or "quicklime," is prepared by decomposing calcium carbonate in a large kiln at 800°C to 1000°C. Calcium oxide is among the most widely used chemicals in the United States, annual consumption being measured in the millions of tons. Nearly one half the CaO output is used in the steel industry, and much of the remainder is used to make "slaked lime," $Ca(OH)_2$, by reaction with water. (As an exercise, write the equation for the reaction of calcium oxide with water that yields calcium hydroxide.)

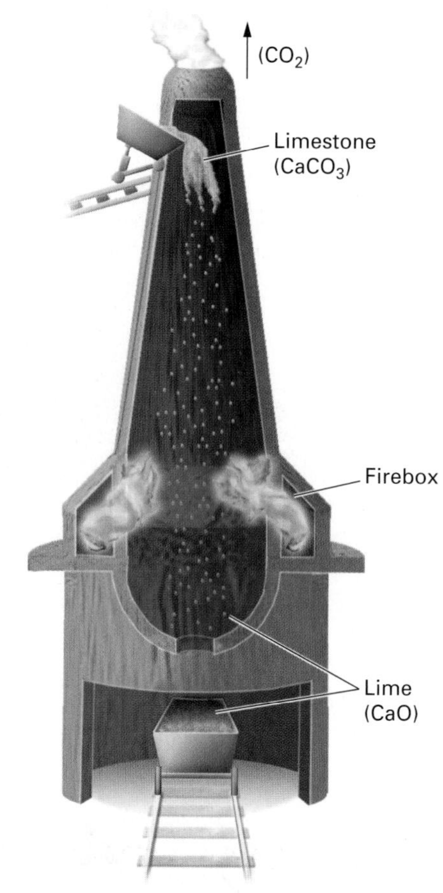


Learn It Now! Compare the Summary for Combination Reactions with that for Decomposition Reactions.

**summary**

**Decomposition Reactions**

| | |
|---|---|
| Reactants: | One compound |
| Reaction type: | Decomposition |
| Equation type: | AX ⟶ A + X |
| Products: | Any combination of elements and compounds |

## 8.8 Single-Replacement Reactions

Goal 7 Given the reactants of a single-replacement reaction, write the equation for the reaction.

P/REVIEW: *Oxidation* and *reduction* are terms that refer to the exchange of electrons in a chemical reaction. Oxidation or burning reactions, and many of the combination and decomposition reactions of the previous sections, are also oxidation–reduction reactions. "Redox" reactions, as oxidation–reduction reactions are called, are discussed in detail in Chapter 19.

Many elements are capable of replacing ions of other elements in aqueous solution. This is one kind of **oxidation–reduction reaction,** or **"redox" reaction.** The equation for such a reaction looks as if one element is replacing another in a compound. It is a **single-replacement equation.** These reactions are sometimes called **single-replacement reactions.** The general equation is

$$A + BX \longrightarrow AX + B$$

Sodium is able to replace hydrogen in water in a single-replacement reaction: $2\ Na(s) + 2\ HOH(\ell) \rightarrow 2\ NaOH(aq) + H_2(g)$ (Fig. 8.15).

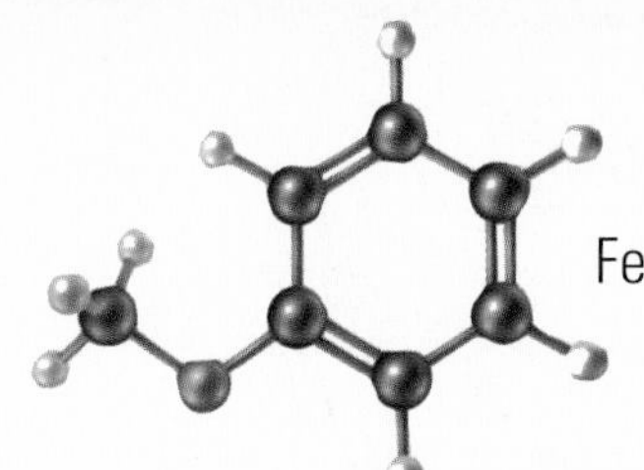

# Everyday Chemistry

## Femtochemistry

At this point in your study of chemistry, you have recognized that chemical reactions are a continual part of your everyday life. Human life itself depends on literally billions of reactions per second. Until recently, though, chemists have had an incomplete idea of how a chemical reaction proceeds. We know a great deal about reactants and products in chemical change, but the nature of the pathway between the two has been a difficult problem to study. This is because chemical change occurs so quickly. Many reactions are complete in as little as $10 \times 10^{-15}$ second.

In the late 1980s, groundbreaking research into the fundamentals of chemical reactivity began. The instrumentation used for this research was, in essence, a high-speed camera designed to take snapshots of a reaction with a "shutter speed" fast enough to capture a chemical change as it occurs. This is accomplished with lasers that emit light flashes on the femtosecond timescale. The first reaction studied with this new technique was a decomposition reaction: $ICN \rightarrow I + CN$. This reaction is complete in 200 femtoseconds, where *femto-* is the metric prefix for $10^{-15}$. The prefix *femto-* was derived from the Scandinavian word for *fifteen*.

A century earlier, in the late 1880s, an unanswered question was: Does a horse always have at least one hoof on the ground while trotting? This question was answered with a relatively new technology, namely, photography. A camera was designed to take a series of photographs at a fixed time interval, every 0.052 second. As you can see in the figure, the photographic evidence shows that all four hooves are off the ground at once.

Photographer Eadweard Muybridge's study of a horse at full gallop in collotype print. In the second photo in the top row, notice that all four hooves are off the ground.

Photo by Eadweard Muybridge/Time Life Pictures/Getty Images

Analogously, femtosecond lasers can be used to "photograph" molecules as they undergo chemical change. However, instead of recording the image every 0.052 second, the image needs to be documented every 0.00000000000001 second! At this miniscule timescale, we can follow the realtime progress of a chemical reaction. The figure starts by showing a complex molecule formed by the association of an iodide molecule with a ring molecule, benzene, $C_6H_6$. In the second frame, the complex molecule is subjected to a flash from a femtosecond laser. An electron jumps from the benzene molecule to the iodine molecule. Even if the electron leaps back to the benzene molecule, the iodine molecule is decomposed into individual iodine atoms in less than $1500 \times 10^{-15}$ second.

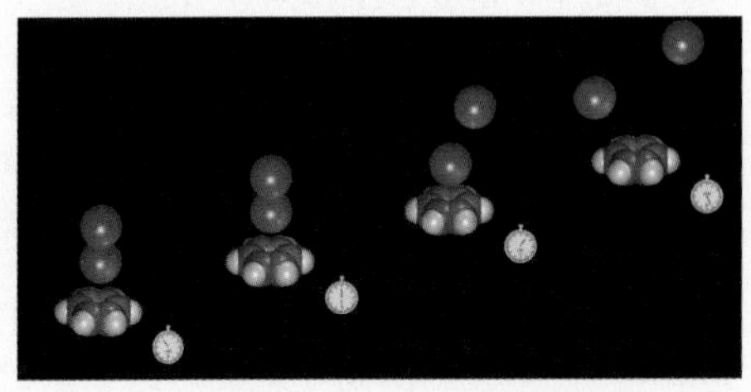

Femtosecond technology reveals the mechanism of how an iodine molecule is split into iodine atoms via the momentary transfer of an electron from the benzene molecule.

California Institute of Technology

Ahmed Zewail is the scientist who pioneered the field now known as *femtochemistry*. He was awarded the 1999 Nobel Prize in Chemistry for his work. Zewail currently directs the Laboratory for Molecular Sciences at Caltech (called *Femtoland!*), where he and his coworkers continue their studies using femtochemistry techniques.

Femtochemistry research holds great promise. Chemists see enormous potential in using it to gain greater control over chemical reactions. Biologists are doing femtochemistry research on photosynthesis. There is a possibility that artificial photosynthesis may be possible, where light energy is converted to energy contained in chemical bonds. Engineers and materials scientists hope to use femtochemistry technology in creating faster and higher-capacity electronics. Medical researchers are investigating the use of drugs activated by light to selectively destroy cancerous tumors, and femtochemistry techniques have the potential to help researchers study the reactions that occur. All of these areas of research have immense promise for having a significant effect on your life in the near future!

Ahmed Zewail (1946–)

AP Photo

**Figure 8.15** Reaction of sodium with water. A single-replacement reaction occurs when sodium, an element, reacts with water, a compound. The metal sodium appears to replace hydrogen in water, 2 Na + 2 HOH (A + BX) $\rightarrow$ 2 NaOH + $H_2$ (AX + B). The products of the reaction are a different compound, sodium hydroxide, and a different element, hydrogen. Sodium hydroxide is a soluble ionic compound, so it exists as sodium ions and hydroxide ions in water solution.

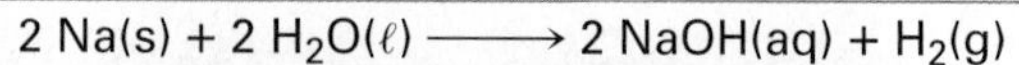


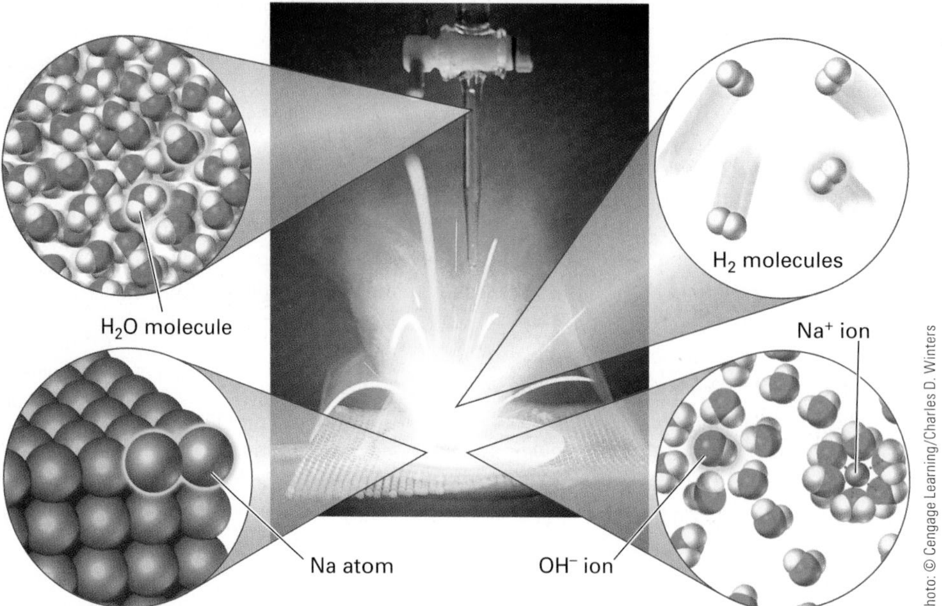


Reactants in a single-replacement equation are always an element and a compound. If the element is a metal, it replaces the metal or hydrogen in the compound. If the element is a nonmetal, it replaces the nonmetal in the compound.

## Active Example 8.8

Write the single-replacement equation for the reaction between solid elemental calcium and an aqueous solution of hydrochloric acid (Fig. 8.16).

Begin by writing the formulas of the reactants to the left of the arrow.

Ca(s) + HCl(aq) $\rightarrow$

Now decide which element in the compound, hydrogen or chlorine, will be replaced by the calcium. Reread the paragraph before this Active Example if you need help. Then write the formulas of the products on the right side of the equation that you started above.

| Ca(s) + | HCl(aq) $\rightarrow$ | $CaCl_2$(aq) + | $H_2$(g) |
|---|---|---|---|
| A + | BX $\rightarrow$ | AX + | B |

The elemental metal reactant, A, becomes the positive ion in the product aqueous solution, AX. The positive ion in the reactant aqueous solution, BX, becomes an uncombined element as a product, B. B is usually a metal, but if the reactant solution is an acid, B is hydrogen. At this time, you have no reason to know that AX ($CaCl_2$, in this case) will be in aqueous solution. However, you do need to recognize that the 2+ calcium ion, $Ca^{2+}$, requires *two* 1− chloride ions, $Cl^-$, for the correct formula for calcium chloride, $CaCl_2$. You also have to recognize that uncombined hydrogen exists in nature as a diatomic molecule, $H_2$.

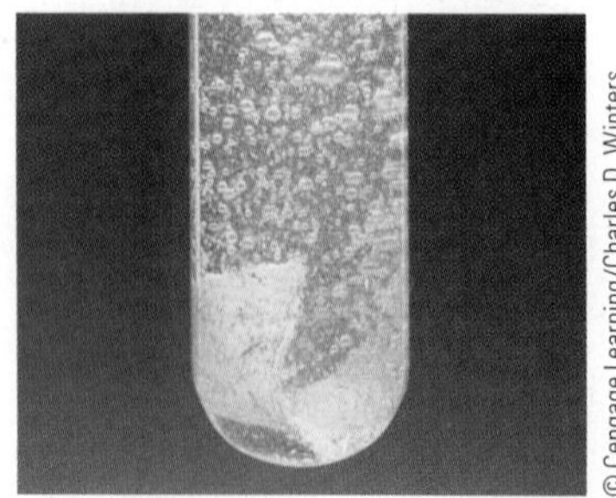


**Figure 8.16** The reaction of calcium metal and hydrochloric acid.

We will discuss how you can decide upon the state symbol of ionic compounds in Section 9.6. For now, we will assume that ionic reactants and products are in aqueous solution unless stated otherwise.

Now balance the equation.

$Ca(s) + 2\ HCl(aq) \rightarrow CaCl_2(aq) + H_2(g)$

## Active Example 8.9

Copper reacts with a solution of silver nitrate (Fig. 8.17). Write the equation for the reaction.

In this case an elemental metal is replacing a metal ion in an aqueous solution. The copper ion that forms is a copper(II) ion. Write the formulas of the reactants.

$Cu(s) + AgNO_3(aq) \rightarrow$

Complete the unbalanced equation.

$Cu(s) + \quad AgNO_3(aq) \rightarrow \quad Cu(NO_3)_2(aq) + \quad Ag(s)$

Now balance the equation.

$Cu(s) + 2\ AgNO_3(aq) \rightarrow Cu(NO_3)_2(aq) + 2\ Ag(s)$

Note: Just because an equation for a reaction can be written does not mean the reaction will occur. For example, if a piece of solid silver and a solution of copper(II) nitrate are given as reactants, you would produce the equation $2\ Ag(s) + Cu(NO_3)_2(aq) \rightarrow 2\ AgNO_3(aq) + Cu(s)$, just the reverse of the equation in Active Example 8.9. This reaction does not occur spontaneously, although it can be forced with some help from a source of electrical current. To find which reaction works, you must try the reactions in the laboratory. In Chapter 9 we show how to use the results of many experiments to predict which reactions will occur and which will not.

(a)

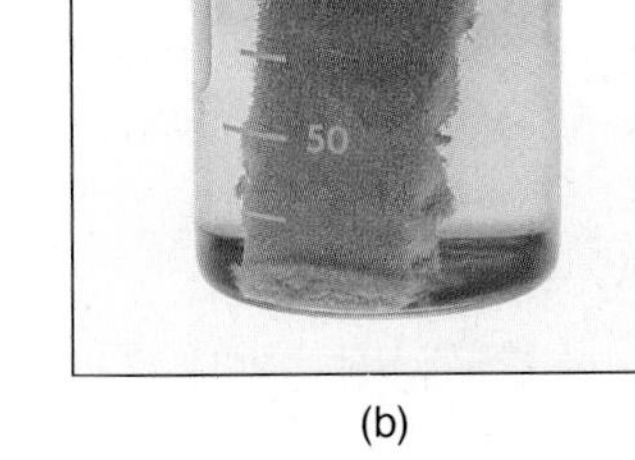


(b)



**Figure 8.17** The reaction between copper and a solution of silver nitrate. (a) A copper strip is placed in a clear, colorless solution of silver nitrate. (b) After about an hour at room temperature, the products of the reaction become visible. The copper strip is covered with solid silver metal. The solution becomes blue because of the presence of hydrated copper(II) ions.

Active Example 8.9 gives us an opportunity to show you a balancing trick that can save you some time. Whenever a polyatomic ion is unchanged in a chemical reaction, the entire ion can be balanced as a unit. The nitrate ion is unchanged in

$$Cu(s) + 2\ AgNO_3(aq) \rightarrow Cu(NO_3)_2(aq) + 2\ Ag(s)$$

Start with $Cu(NO_3)_2$. It has three ions: one copper(II) ion and two nitrate ions. Cu is already balanced. $AgNO_3$ has one nitrate ion, so it takes two $AgNO_3$ to balance the nitrate ions in one $Cu(NO_3)_2$. This, in turn, requires two Ag on the right. When you learn this technique, you will find it quicker and easier than balancing each element in a polyatomic ion separately. But remember the condition: *All of the ions must be unchanged.* This technique will not work, for instance, if there is an $NO_3^-$ compound on one side and an $NO_3^-$ plus an NO or some other nitrogen species on the other side.

**summary**

**Single-Replacement Reactions**

| | |
|---|---|
| Reactants: | Element (A) plus a solution of either an acid or an ionic compound (BX) |
| Reaction type: | Single-replacement |
| Equation type: | A + BX ⟶ AX + B |
| Products: | An ionic compound (usually in solution) (AX) plus an element (B) |

## 8.9 Double-Replacement Reactions

**Goal** 8 Given the reactants in a double-replacement precipitation or neutralization reaction, write the equation.

When solutions of two compounds are mixed, a positive ion from one compound may combine with the negative ion from the other compound to form a solid compound that settles to the bottom. The solid is a **precipitate;** the reaction is a **precipitation reaction.** In the equation for a precipitation reaction, ions of the two reactants appear to change partners. The equation, and sometimes the reaction itself, is a **double-replacement equation** or a **double-replacement reaction.** The general equation, with bridges to show the rearrangement of ions, is

$$AX + BY \longrightarrow AY + BX$$

A typical precipitation reaction occurs between solutions of calcium chloride and sodium fluoride: $CaCl_2(aq) + 2\ NaF(aq) \rightarrow CaF_2(s) + 2\ NaCl(aq)$.

© Cengage Learning/Charles Steele

**Figure 8.18** The precipitation of silver chloride.

### Active Example 8.10

Silver chloride precipitates when solutions of sodium chloride and silver nitrate are combined (Fig. 8.18). Write the equation for the reaction.

Start by writing the formulas of the reactants.

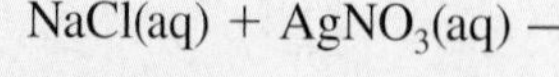

$NaCl(aq) + AgNO_3(aq) \rightarrow$

The statement identifies silver chloride as one of the products. It forms when the silver ion from silver nitrate combines with the chloride ion from sodium chloride. If you can write the formula of the other product, you can write the unbalanced equation. Try it. Remember that positive ions combine with negative ions to form products.

$NaCl(aq) + \quad AgNO_3(aq) \rightarrow \quad NaNO_3(aq) + \quad AgCl(s)$

The $Na^+$ ion from NaCl is paired with the $NO_3^-$ ion from $AgNO_3$ for the second product, $NaNO_3$, which remains in solution. The nitrate ion is unchanged in the reaction, so it may be balanced as a unit, as explained after Active Example 8.9. Note that the solid that forms, AgCl(s), has the state symbol (s) to indicate that it is the precipitate.

Now balance the equation.

$NaCl(aq) + AgNO_3(aq) \rightarrow NaNO_3(aq) + AgCl(s)$

The skeleton equation is balanced, all coefficients being 1. In checking the balance, the nitrate ion again may be thought of as a unit. There is one nitrate on each side of the equation.

An **acid** is a compound that releases hydrogen ions, $H^+$. A substance that contains hydroxide ions, $OH^-$, is a **base.** When an acid is added to an equal amount of base, each hydrogen ion reacts with a hydroxide ion to form a molecule of water. The acid and the base **neutralize** each other in a **neutralization reaction.*** An ionic compound called a **salt** is also formed; it usually remains in solution. The general equation, using M to represent a metal ion, is

$$\underset{\text{acid}}{HX(aq)} + \underset{\text{base}}{MOH(s)} \rightarrow \underset{\text{water}}{H_2O(\ell)} + \underset{\text{salt}}{MX(aq)}$$

**P/REVIEW:** Acids are discussed in more detail in Section 6.5 and in Chapter 17. Some chemists disapprove of the term *hydrogen ion.* If your instructor is among them, think and speak of the hydronium ion instead.

Neutralization reactions are described by double-replacement equations, although it might not seem that way at first glance. The water molecule forms when the hydrogen ion from the acid combines with the hydroxide ion from the base. The double-replacement character of the equation becomes clear if you write the formula of water as HOH rather than $H_2O$. For the neutralization of hydrochloric acid by solid sodium hydroxide, the equation is

$$HCl(aq) + NaOH(s) \rightarrow HOH(\ell) + NaCl(aq)$$

With the formula of water in conventional form, we have

$$HCl(aq) + NaOH(s) \rightarrow H_2O(\ell) + NaCl(aq)$$

*Suggestion:* When balancing a neutralization reaction equation, begin by writing the formula of water as HOH. This will help you see the double-replacement format of the reaction. When you've completely balanced the equation, change the formula of water to $H_2O$.

Most hydroxides do not dissolve in water, but they do appear to "dissolve" in acids. What really happens is that they react with the acid in a double-replacement neutralization reaction. The next Active Example shows one such reaction.

*There are acids that do not contain hydrogen ions and bases that do not contain hydroxide ions. Reactions between them are also called neutralization reactions. The $H^+$-plus-$OH^-$ neutralization is the most common and the only one we will consider here.

## Active Example 8.11

Write the equation for the reaction between sulfuric acid and solid aluminum hydroxide.

What are the formulas of the reactants?

$H_2SO_4(aq) + Al(OH)_3(s) \rightarrow$

Now write the unbalanced equation.

$H_2SO_4(aq) + \quad Al(OH)_3(s) \rightarrow \quad HOH(\ell) + \quad Al_2(SO_4)_3(aq)$

Balance the sulfate ions.

$H_2SO_4(aq) + \quad Al(OH)_3(s) \rightarrow \quad HOH(\ell) + \quad Al_2(SO_4)_3(aq)$

$3\ H_2SO_4(aq) + Al(OH)_3(s) \rightarrow HOH(\ell) + Al_2(SO_4)_3(aq)$

The three sulfate ions in $Al_2(SO_4)_3$ require 3 $H_2SO_4$.

Balance the hydrogen ions. Temporarily treat water as "hydrogen hydroxide" for balancing purposes.

$3\ H_2SO_4(aq) + \quad Al(OH)_3(s) \rightarrow \quad HOH(\ell) + \quad Al_2(SO_4)_3(aq)$

$3\ H_2SO_4(aq) + Al(OH)_3(s) \rightarrow 6\ HOH(\ell) + Al_2(SO_4)_3(aq)$

The six hydrogen ions in 3 $H_2SO_4$ balance with the six in 6 HOH.

Now balance the hydroxide ions.

$3\ H_2SO_4(aq) + \quad Al(OH)_3(s) \rightarrow \quad 6\ HOH(\ell) + \quad Al_2(SO_4)_3(aq)$

$3\ H_2SO_4(aq) + 2\ Al(OH)_3(s) \rightarrow 6\ HOH(\ell) + Al_2(SO_4)_3(aq)$

Six hydroxide ions in 6 HOH require 2 $Al(OH)_3$ to achieve six on the left.

Aluminum ions are the only thing left unchecked. Balance the Al, restore the formula of water to its correct form, and make a final check.

$3\ H_2SO_4(aq) + 2\ Al(OH)_3(s) \rightarrow 6\ HOH(\ell) + \quad Al_2(SO_4)_3(aq)$

$3\ H_2SO_4(aq) + 2\ Al(OH)_3(s) \rightarrow 6\ H_2O(\ell) + Al_2(SO_4)_3(aq)$

6 H, 3 $SO_4$, 2 Al, and 6 OH on each side; or 12 H, 3 S, 18 O, and 2 Al on each side.

summary

**Double-Replacement Reactions**

Reactants: Solutions of two compounds, each with positive and negative ions (AX + BY)

Reaction type: Double-replacement

Equation type:

$$AX + BY \longrightarrow AY + BX$$

Products: Two new compounds (AY + BX), which may be a solid, water, an acid, or an aqueous ionic compound

## 8.10 Summary of Reactions and Equations

All of the information from the equation-writing summaries at the ends of Sections 8.6 through 8.9 has been assembled into Table 8.3. The reactants (first column) are shown for each reaction type (second column). Each reaction type has a certain "equation type" (third column) that yields predictable products (fourth column).

Reading the column heads from left to right follows the "thinking order" by which reactants and products are identified. It will help you organize your approach to writing equations. Given the reactants of a specific chemical change, you can fit them into one of the reactant boxes in the table and thereby determine the reaction type. Once you know what kind of reaction it is, you know the type of equation that describes it. You can then write the formulas of the products on the right. Balance the equation and you are finished.

We indicated earlier that just because an equation can be written, it does not necessarily mean that the reaction will happen. By using the results of experiments performed over many years, we can make reliable predictions. For example, we can predict with confidence that zinc will replace copper in a copper sulfate solution but that silver will not, or that pouring calcium nitrate solution into sodium fluoride solution will yield a precipitate, but that pouring it into sodium bromide solution will not. We have deliberately refrained from making these predictions in this chapter; learning to write equations is enough for now.

**Table 8.3** Summary of Types of Reactions and Equations

| Reactants | Reaction Type | Equation Type | Products |
|---|---|---|---|
| Any combination of elements and compounds that form one product | Combination | $A + X \longrightarrow AX$ | One compound |
| One compound | Decomposition | $AX \longrightarrow A + X$ | Any combination of elements and compounds |
| Element + ionic compound or acid | Single-replacement | $A + BX \longrightarrow AX + B$ | Element + ionic compound |
| Solutions of two compounds, each with positive and negative ions | Double-replacement | $AX + BY \longrightarrow AY + BX$ | Two new compounds, which may be a solid, water, an acid, or an aqueous ionic compound |

## Thinking About Your Thinking

### Classification

**In this chapter, we presented a classification scheme to help organize your knowledge about chemical reactions. You could learn about reactions and equations without an organizing system, but our classification scheme helps you learn more easily and more efficiently than you would without a method of organizing your newly obtained knowledge.**

**There are many ways to classify chemical changes, and our scheme is just one of them. This is the essence of the classification thinking skill: As what you know changes, you need to practice reorganizing your knowledge in different ways. Later in your study of chemistry, you will find that there are three broad classifications of chemical change. One is metathesis reactions, the formation of a solid or molecular substance. Another category is proton-transfer reactions, where a hydrogen ion is transferred. Finally, there are electron-transfer reactions, which are the redox reactions introduced in Section 8.8.**

**Is our four-category classification scheme better than the three-category scheme? No, it's just better for now, given your present understanding of chemistry. Later, you will have a great deal more knowledge about many different types of chemical change, and you will probably find that the three-category scheme is useful when you are considering broad, general differences in reactivity. The four-category scheme you are learning now will be useful in predicting products of reactions. Both schemes will be helpful in organizing your knowledge, and your choice of mode of classification will depend on the situation. You will be forming a hierarchical system, one where different needs lead to different classifications. In doing so, your knowledge will be organized in such a way that the whole is greater than the sum of the parts because of the added value of the classification systems.**

**Always be looking for classification systems to organize your knowledge, and don't hesitate to change your systems when new information yields a better method of classification.**

## Chapter 8 in Review

Most of the key terms and concepts and many others appear in the Glossary. Use your Glossary regularly.

**8.1 Evidence of a Chemical Change**
Goal 1 Describe five types of evidence detectable by human senses that usually indicate a chemical change.

**Key Terms and Concepts: Absorption or release of heat energy, color change, emission of light energy, formation of a gas, formation of a solid product, products, reactants**

**8.2 Evolution of a Chemical Equation**
Goal 2 Distinguish between an unbalanced and a balanced chemical equation, and explain why a chemical equation needs to be balanced.

**Key Terms and Concepts: Aqueous solution, balance (an equation), qualitative and quantitative descriptions of chemical reactions, state symbol**

**8.3 Balancing Chemical Equations**
Goal 3 Given an unbalanced chemical equation, balance it by inspection.

**Key Terms and Concepts: Starting formula**

**8.4 Interpreting Chemical Equations**
Goal 4 Given a balanced chemical equation or information from which it can be written, describe its meaning on the particulate, molar, and macroscopic levels.

**8.5 Writing Chemical Equations**

**8.6 Combination Reactions**
Goal 5 Write the equation for the reaction in which a compound is formed by the combination of two or more simpler substances.

**Key Terms and Concepts: Combination (synthesis) reaction**

**8.7 Decomposition Reactions**
Goal 6 Given a compound that is decomposed into simpler substances, either compounds or elements, write the equation for the reaction.

**Key Terms and Concepts: Decomposition reaction, reversible reaction**

**8.8 Single-Replacement Reactions**
Goal 7 Given the reactants of a single-replacement reaction, write the equation for the reaction.

**Key Terms and Concepts: Oxidation–reduction (redox) reaction, single-replacement equation, single-replacement reaction**

**8.9 Double-Replacement Reactions**

Goal 8 Given the reactants in a double-replacement precipitation or neutralization reaction, write the equation.

**Key Terms and Concepts: Acid, base, double-replacement equation, double-replacement reaction, neutralize and neutralization reaction, precipitate and precipitation reaction, salt**

**8.10 Summary of Reactions and Equations**

## Study Hints and Pitfalls to Avoid

Be careful about the "don'ts" in Section 8.2. It is very tempting to balance an equation quickly by changing a correct formula or adding one that doesn't belong. Another device some creative students invent is slipping what should be a coefficient into the middle of a correct formula, such as changing $NaNO_3$ to $Na2NO_3$ to balance the nitrates in $Ca(NO_3)_2$ on the other side. That doesn't work, either. There is no such thing as $Na2NO_3$. Remember to write the qualitative description of the reaction first and then balance the equation—*using coefficients*—without changing the qualitative description.

Don't forget the "hidden" (unnamed) reactants and products. "Bubbles form . . ." indicates a gaseous product, such as $H_2$, $O_2$, or $CO_2$. In the word description of a reaction, ionic products are often unnamed if their formulas can be derived from the reactants.

An equation-classification exercise and an equation-balancing exercise follow these study hints. We suggest you use them to practice your skills.

Probably the best suggestion for studying this chapter is to focus on Table 8.3. Look for the "big picture." Knowing how things fit together helps in learning the details. Look at the *reactants.* They should tell you the *reaction type.* Each reaction type has a certain *equation type,* and that gives you the *products.* In order from left to right, these are the column heads in the classification exercise table.

After completing the equation-classification and equation-balancing exercises, you might wish to practice further by referring to the unclassified reactions beginning with Question 31. Based on the reactants described in each question, see if you can determine mentally what kind of reaction it is and what its products are. Run through the whole list this way without writing equations until you feel sure of your ability. Then write and balance equations until you are completely confident that you can write any equation without hesitation.

## Equation-Classification Exercise

*Step 1 of our general equation-writing procedure is to "classify the reaction type." In this exercise, we provide the chemical formulas of reactants in a reaction. Without writing the formulas of the products or balancing the equation, classify the following as one of the four reaction types given in this chapter by placing the appropriate abbreviation after each. The first two are completed as examples. Answers to 3–12 appear in Appendix III.*

| Reaction Type | Equation Type | Abbreviation |
|---|---|---|
| Combination | A + X ⟶ AX | Comb |
| Decomposition | AX ⟶ A + X | Decomp |
| Single-replacement | A + BX ⟶ AX + B | SR |
| Double-replacement | AX + BY ⟶ AY + BX | DR |

1. $Mg + N_2 \rightarrow$ Comb
2. $Zn(OH)_2 + H_2SO_4 \rightarrow$ DR
3. $Zn + I_2 \rightarrow$
4. $PbO_2 \rightarrow$
5. $Al + CuSO_4 \rightarrow$
6. $FeCl_2 + Na_3PO_4 \rightarrow$
7. $Al + HCl \rightarrow$
8. $HgO \rightarrow$
9. $Fe + O_2 \rightarrow$
10. $Fe_2(SO_4)_3 + Ba(OH)_2 \rightarrow$
11. $HNO_3 + CsOH \rightarrow$ (Cs, Z = 55)
12. $Cu + AgNO_3 \rightarrow$

## Equation-Balancing Exercise

*Balance the following equations, for which correct chemical formulas are already written. We have provided space for you to insert coefficients directly into the unbalanced equations. Balanced equations are given in the answer section.*

1. $Na + O_2 \rightarrow Na_2O$
2. $H_2 + Cl_2 \rightarrow HCl$
3. $P + O_2 \rightarrow P_2O_3$
4. $KClO_4 \rightarrow KCl + O_2$
5. $Sb_2S_3 + HCl \rightarrow SbCl_3 + H_2S$
6. $NH_3 + H_2SO_4 \rightarrow (NH_4)_2SO_4$
7. $CuO + HCl \rightarrow CuCl_2 + H_2O$
8. $Zn + Pb(NO_3)_2 \rightarrow Zn(NO_3)_2 + Pb$
9. $AgNO_3 + H_2S \rightarrow Ag_2S + HNO_3$
10. $Cu + S \rightarrow Cu_2S$
11. $Al + H_3PO_4 \rightarrow H_2 + AlPO_4$
12. $NaNO_3 \rightarrow NaNO_2 + O_2$
13. $Mg(ClO_3)_2 \rightarrow MgCl_2 + O_2$
14. $H_2O_2 \rightarrow H_2O + O_2$
15. $BaO_2 \rightarrow BaO + O_2$
16. $H_2CO_3 \rightarrow H_2O + CO_2$
17. $Pb(NO_3)_2 + KCl \rightarrow PbCl_2 + KNO_3$
18. $Al + Cl_2 \rightarrow AlCl_3$
19. $C_6H_{14} + O_2 \rightarrow CO_2 + H_2O$
20. ■ $NH_4NO_3 \rightarrow N_2O + H_2O$
21. ■ $H_2 + N_2 \rightarrow NH_3$
22. ■ $Fe + Cl_2 \rightarrow FeCl_3$
23. ■ $H_2S + O_2 \rightarrow H_2O + SO_2$
24. $MgCO_3 + HCl \rightarrow MgCl_2 + CO_2 + H_2O$
25. $P + I_2 \rightarrow PI_3$

## Small-Group Discussion Questions

*Small-Group Discussion Questions are for group work, either in class or under the guidance of a leader during a discussion section.*

1. What is the meaning of the arrow in a balanced chemical equation? Is the meaning different when considering the particulate level versus the macroscopic level? Write a definition of the arrow in a chemical equation that would be appropriate for a chemical dictionary.
2. Chemical equations are usually interpreted at two different levels. Using the reaction of dihydrogen sulfide and oxygen to form sulfur dioxide and water as an example, explain each interpretation of the balanced equation.
3. Write and balance the equation for the decomposition of hydrogen peroxide, $H_2O_2$, to water and oxygen. Is the number of molecules the same before and after the chemical change? Does your answer to the previous question depend on the initial number of reactant molecules? Is the number of atoms the same before and after the chemical change? Justify your answers in terms of the Law of Conservation of Mass.
4. Explain *why* the subscripts in a chemical formula cannot be changed to balance a chemical equation.
5. Write the phrase "chemical equation" on the board. Each student should jot down at least one word or phrase that comes to mind that is related to chemical equations. After the group has come up with a list of related words, separate them into three categories. Each word or phrase should fit only one category. Now give each category a name. Explain the relationship between each category and the phrase "chemical equation." Summarize the overall organization scheme in a few sentences.
6. Write and balance the equation for the reaction of hydrogen and oxygen to form water. If 2 moles of hydrogen react with 1 mole of oxygen, how many moles of water are formed? Explain how you know the answer to this question. How many moles of water will form when 2 moles of hydrogen react with 2 moles of oxygen? Explain. Investigate the number of moles of water that will form from a variety of combinations of numbers of moles of hydrogen and oxygen, such as 14 moles of hydrogen and 20 moles of oxygen, or 21 moles of hydrogen and 8 moles of oxygen. In each case, explain the role of the chemical equation in determining the number of moles of water produced.

# Questions, Exercises, and Problems

*Questions whose numbers are printed in blue are answered in Appendix III. ■ denotes problems assignable in OWL. In the first section of Questions, Exercises, and Problems, similar exercises and problems are paired in consecutive odd-even number combinations.*

*Many questions in this chapter are written with the assumption that you have studied Chapter 6 and can write the required formulas from their chemical names. If this is not the case, we have placed a list of all chemical formulas needed to answer Chapter 8 questions at the end of the Questions, Exercises, and Problems.*

### Section 8.4: Interpreting Chemical Equations

1. Consider the following particulate-level representation of a chemical equation:

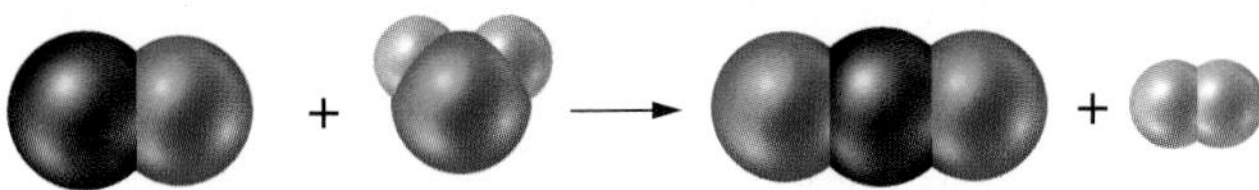

The white spheres represent hydrogen atoms, the black sphere represents a carbon atom, and the red spheres represent oxygen atoms. (a) Write a balanced chemical equation representing this reaction. (b) Write a word description of the reaction on the particulate and molar levels.

2. ■ Give a description, using words and chemical formulas, of the following chemical equation on both the particulate and molar levels: $C_2H_4 + H_2 \rightarrow C_2H_6$.

3. The left box of the diagram below shows the hypothetical elements A (tan atoms) and B (blue diatomic molecules) before they react. Write a balanced chemical equation, using the lowest possible whole-number coefficients, for the reaction that occurs to form the product in the right box.

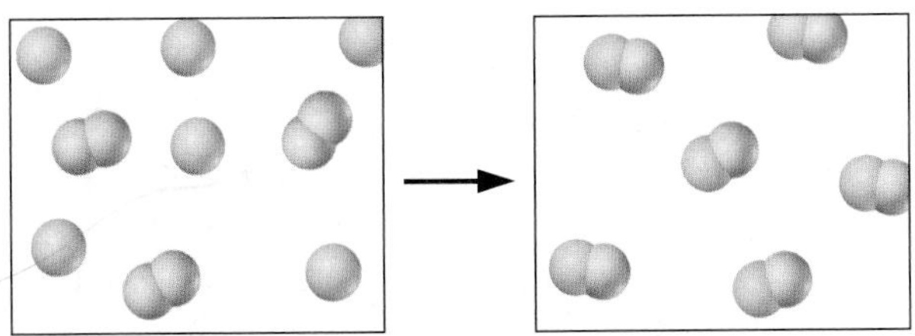

4. Draw a box and then sketch five space-filling models of diatomic molecules within it, similar to those in Question 3. Draw an arrow and draw a product box to the right of your arrow. Add the remaining space-filling models necessary to depict the balanced reaction $X_2 + Y_2 \rightarrow XY_3$ (the reaction is not yet balanced), where your five molecules represent the species $X_2$.

5. A particulate-level sketch of a reaction shows four spheres representing atoms of element A and four spheres depicting atoms of element B before reacting and 4 models of the molecule AB after the reaction. Explain why 4 A + 4 B → 4 AB is not the correct equation for the reaction.

6. Consider the reaction of the elements antimony and chlorine, forming antimony trichloride. Write and balance the chemical equation for this reaction, and then state which of the boxes below best represents the reactants and product of the reaction.

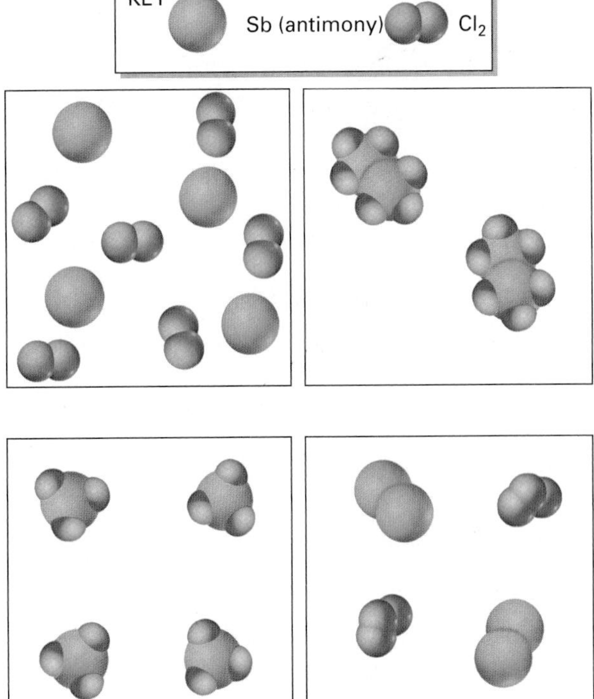


7. Write a balanced chemical equation to represent the reaction illustrated below.

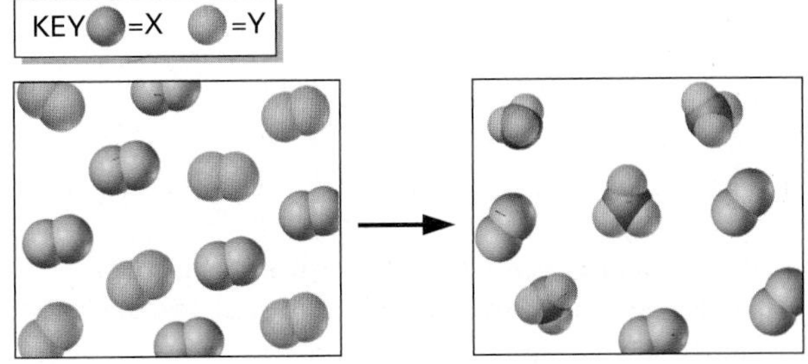


8. Write and balance the equation for the reaction of butane, $C_4H_{10}$, with oxygen to form carbon dioxide and water. Use your equation to help answer the following questions. (a) Write, in words, a description of the reaction on the particulate level. (b) If you were to build physical ball-and-stick models of the reactants and products, what minimum number of balls representing atoms of each element do you need if you show both reactants and products at the same time? (c) What if the models of the reactants from Part (b) were built and then rearranged to form products? How many balls would you need? (d) Use words to interpret the equation on the molar level. (e) Use the molar-level interpretation of the equation from Part (d) and molar masses rounded to the nearest gram to show that mass is indeed conserved in this reaction.

*Questions 9 to 30: Write the equation for each reaction described. Follow your instructor's advice about whether state symbols should be included.*

### Section 8.6: Combination Reactions

9. Lithium combines with oxygen to form lithium oxide.
10. ■ When nitrogen combines with hydrogen, ammonia is formed.
11. Boron combines with oxygen to form diboron trioxide.
12. ■ When nitrogen monoxide combines with oxygen, nitrogen dioxide is formed.
13. Calcium combines with bromine to make calcium bromide.
14. ■ When phosphorus ($P_4$) combines with chlorine, phosphorus trichloride is formed.

### Section 8.7: Decomposition Reactions

15. Pure hydrogen iodide decomposes spontaneously to its elements.
16. ■ When sodium nitrate decomposes, sodium nitrite and oxygen are formed.
17. Barium peroxide, $BaO_2$, breaks down into barium oxide and oxygen.
18. ■ Pure bromine trifluoride decomposes to its elements.

### Section 8.8: Single-Replacement Reactions

19. Calcium reacts with hydrobromic acid.
20. ■ Magnesium is placed into sulfuric acid.
21. Chlorine gas is bubbled through an aqueous solution of potassium iodide.
22. ■ Copper metal combines with aqueous silver nitrate.

### Section 8.9: Double-Replacement Reactions

*It is not necessary for you to identify the precipitates formed in Questions 23 to 26.*

23. Calcium chloride and potassium fluoride solutions react to form a precipitate.
24. ■ A precipitate forms when aqueous solutions of cobalt(III) iodide and lead(II) nitrate are combined.
25. Milk of magnesia is the precipitate that results when sodium hydroxide and magnesium bromide solutions are combined.
26. ■ A precipitate forms when aqueous solutions of chromium(III) nitrate and sodium hydroxide are combined.
27. Sulfuric acid reacts with barium hydroxide solution.
28. ■ A reaction occurs when aqueous solutions of perchloric acid and sodium hydroxide are combined.
29. Sodium hydroxide is added to phosphoric acid.
30. ■ A reaction occurs when aqueous solutions of hydrosulfuric acid and sodium hydroxide are combined.

### Unclassified Reactions

*Questions 31 to 66: Write the equation for the reaction described or for the most likely reaction between given reactants.*

31. Lead(II) nitrate solution reacts with a solution of sodium iodide.
32. ■ A precipitate forms when aqueous solutions of calcium nitrate and potassium carbonate are combined.
33. The fuel butane, $C_4H_{10}$, burns in oxygen, forming carbon dioxide and water.
34. ■ Acetaldehyde, $CH_3CHO$, a raw material used in manufacturing vinegar, perfumes, dyes, plastics, and other organic materials, is oxidized, reacting with oxygen, forming carbon dioxide and water.
35. Sulfurous acid decomposes spontaneously to sulfur dioxide and water.
36. ■ Carbonated beverages contain carbonic acid, an unstable compound that decomposes to carbon dioxide and water.
37. Potassium reacts violently with water. (*Hint:* Think of water as an acid with the formula HOH.)
38. Bubbles form when metallic barium is placed in water. (*Hint:* Think of water as an acid with the formula HOH.)
39. Zinc metal is placed in a silver chlorate solution.
40. ■ Chlorine gas combines with aqueous potassium bromide.
41. Ammonium sulfide is added to a solution of copper(II) nitrate.
42. Silver nitrate is added to a solution of sodium sulfide.
43. Phosphorus tribromide is produced when phosphorus reacts with bromine.
44. ■ When iron reacts with oxygen, iron(III) oxide is formed.
45. When calcium hydroxide (sometimes called slaked lime) is heated, it forms calcium oxide—lime—and water vapor.
46. Hydrogen peroxide, $H_2O_2$, the familiar bleaching compound, decomposes slowly into water and oxygen.
47. Glycerine, $C_3H_8O_3$—used in making soap, cosmetics, and explosives—reacts with oxygen to form carbon dioxide and water.
48. ■ Aqueous solutions of hydrobromic acid and potassium hydroxide are combined.
49. Powdered antimony (Z = 51) ignites when sprinkled into chlorine gas, producing antimony(III) chloride.
50. Fluorine reacts spontaneously with nearly all elements. Oxygen difluoride is produced when fluorine reacts with oxygen.
51. A solution of potassium hydroxide reacts with a solution of zinc chloride.
52. ■ A precipitate forms when aqueous solutions of silver nitrate and magnesium sulfate are combined.
53. Aluminum carbide, $Al_4C_3$, is the product of the reaction of its elements.
54. Magnesium nitride is formed from its elements.
55. A solution of lithium sulfite is mixed with a solution of sodium phosphate.
56. ■ Aqueous solutions of sulfuric acid and zinc hydroxide are combined.
57. A solution of chromium(III) nitrate is one of the products of the reaction between metallic chromium and aqueous tin(II) nitrate.

58. Lithium is added to a solution of manganese(II) chloride.

59. Sulfuric acid is produced when sulfur trioxide reacts with water.

60. ■ Solid zinc and aqueous copper(II) sulfate are combined.

*The remaining reactions are not readily placed into one of the four classifications used in this chapter. Nevertheless, enough information is given for you to write the equations.*

61. ■ A solid oxide of iron, $Fe_3O_4$, and hydrogen are the products of the reaction between iron and steam.

62. Metallic zinc reacts with steam at high temperatures, producing zinc oxide and hydrogen.

63. Aluminum carbide, $Al_4C_3$, reacts with water to form aluminum hydroxide and methane, $CH_4$.

64. ■ When solid barium oxide is placed into water, a solution of barium hydroxide is produced.

65. Magnesium nitride and hydrogen are the products of the reaction between magnesium and ammonia.

66. ■ Solid iron(III) oxide reacts with gaseous carbon monoxide to produce iron and carbon dioxide.

**General Questions**

67. Distinguish precisely and in scientific terms the differences among items in each of the following pairs or groups.
    a) Reactant, product
    b) (g), (ℓ), (s), (aq)
    c) Combination reaction, decomposition reaction
    d) Single replacement, double replacement
    e) Acid, base, salt
    f) Precipitation, neutralization

68. Classify each of the following statements as true or false:
    a) In a chemical reaction, reacting substances are destroyed and new substances are formed.
    b) A chemical equation expresses the quantity relationships between reactants and products in terms of moles.
    c) Elements combine to form compounds in combination reactions.
    d) Compounds decompose into elements in decomposition reactions.
    e) A nonmetal cannot replace a nonmetal in a single-replacement reaction.

69. Each reactant or pair of reactants listed below is *potentially* able to participate in one type of reaction described in this chapter. In each case, name the type of reaction and complete the equation. In Parts (a) and (j), the 2+ ion is formed from the reaction of the corresponding metal. In Part (c), carbon dioxide and water are the products of the reaction (you do not have to name this reaction type).
    a) $Pb + Cu(NO_3)_2 \rightarrow$
    b) $Mg(OH)_2 + HBr \rightarrow$
    c) $C_5H_{10}O + O_2 \rightarrow$
    d) $Na_2CO_3 + CaSO_4 \rightarrow$
    e) $LiBr \rightarrow$
    f) $NH_4Cl + AgNO_3 \rightarrow$
    g) $Ca + Cl_2 \rightarrow$
    h) $F_2 + NaI \rightarrow$
    i) $Zn(NO_3)_2 + Ba(OH)_2 \rightarrow$
    j) $Cu + NiCl_2 \rightarrow$

70. Hydrogen, nitrogen, oxygen, fluorine, chlorine, bromine, and iodine exist as diatomic molecules at the temperatures and pressures at which most reactions occur. Under these normal conditions, when may the formulas of these elements be written without a subscript 2 in a chemical equation?

71. Acid rain is rainfall that contains sulfuric acid originating from organic fuels that contain sulfur. The process occurs in three major steps. The sulfur first burns, forming sulfur dioxide. In sunlight, the sulfur dioxide reacts with oxygen in the air to produce sulfur trioxide. When rainwater falls through the sulfur trioxide, the reaction produces sulfuric acid. Write the equation for each step in the process and tell what kind of reaction it is.

72. One of the harmful effects of acid rain is its reaction with structures made of limestone, which include marble structures, ancient ruins, and many famous statues (Fig. 8.19). Write the equation you would expect for the reaction between acid rain, which contains sulfuric acid, and limestone, solid calcium carbonate. Write a second equation with the same reactants, showing that the expected but unstable carbonic acid decomposes to carbon dioxide and water.

**Figure 8.19** The effect of acid rain on a marble statue.

73. The tarnish that appears on silver is silver sulfide, which is formed when silver is exposed to sulfur-bearing compounds in the presence of oxygen in the air. When hydrogen sulfide is the sulfur-bearing reacting compound, water is the second product. Write the equation for the reaction.

74. Sulfur combines directly with three of the halogens, but not with iodine. The products, however, do not have similar chemical formulas. When sulfur reacts with fluorine, sulfur hexafluoride is the most common product; with chlorine, the product is usually sulfur dichloride; and reaction with bromine usually yields disulfur dibromide. Write the equation for each reaction.

75. One source of the pure tungsten (Z = 74) filament used in light bulbs is tungsten(VI) oxide. It is heated with hydrogen at high temperatures. The hydrogen reacts with the oxygen in the oxide, forming steam. Write the equation for the reaction and classify it as one of the reaction types discussed in this chapter.

*Questions 76 to 79: Write the equation for each reaction described.*

76. ■ Write the equation for the neutralization reaction in which barium nitrate is the salt formed.

77. Lithium sulfate is one of the products of a neutralization reaction.

78. Only the first hydrogen comes off in the reaction between sulfamic acid, $HNH_2SO_3$, and potassium hydroxide. (Figure out the formula of the anion from the acid and use it to write the formula of the salt formed.)

79. The concentration of sodium hydroxide solution can be found from reacting it with oxalic acid, $H_2C_2O_4$. (The formula of oxalic acid should lead you to the formula of the oxalate ion and then to the formula of the salt formed in the reaction.)

## Formulas

*These formulas are provided in case you are studying Chapter 8 before you study Chapter 6. You should not use this list unless your instructor has not yet assigned Chapter 6 or otherwise indicated that you may use the list.*

9. Li, $O_2$, $Li_2O$
10. $N_2$, $H_2$, $NH_3$
11. B, $O_2$, $B_2O_3$
12. NO, $O_2$, $NO_2$
13. Ca, $Br_2$, $CaBr_2$
14. $Cl_2$, $PCl_3$
15. HI
16. $NaNO_3$, $NaNO_2$, $O_2$
17. BaO, $O_2$
18. $BrF_3$
19. Ca, HBr
20. Mg, $H_2SO_4$
21. $Cl_2$, KI
22. Cu, $AgNO_3$
23. $CaCl_2$, KF
24. $CoI_3$, $Pb(NO_3)_2$
25. NaOH, $MgBr_2$
26. $Cr(NO_3)_3$, NaOH
27. $H_2SO_4$, $Ba(OH)_2$
28. $HClO_4$, NaOH
29. NaOH, $H_3PO_4$
30. $H_2S$, NaOH
31. $Pb(NO_3)_2$, NaI
32. $Ca(NO_3)_2$, $K_2CO_3$
33. $O_2$, $CO_2$, $H_2O$
34. $O_2$, $CO_2$, $H_2O$
35. $H_2SO_3$, $SO_2$, $H_2O$
36. $H_2CO_3$, $CO_2$, $H_2O$
37. K, HOH
38. Ba, HOH
39. Zn, $AgClO_3$
40. $Cl_2$, KBr
41. $(NH_4)_2S$, $Cu(NO_3)_2$
42. $AgNO_3$, $Na_2S$
43. $PBr_3$, P, $Br_2$
44. Fe, $O_2$, $Fe_2O_3$
45. $Ca(OH)_2$, CaO, $H_2O$
46. $H_2O$, $O_2$
47. $O_2$, $CO_2$, $H_2O$
48. HBr, KOH
49. Sb, $Cl_2$, $SbCl_3$
50. $F_2$, $OF_2$, $O_2$
51. KOH, $ZnCl_2$
52. $AgNO_3$, $MgSO_4$
54. $Mg_3N_2$
55. $Li_2SO_3$, $Na_3PO_4$
56. $H_2SO_4$, $Zn(OH)_2$
57. $Cr(NO_3)_3$, Cr, $Sn(NO_3)_2$
58. Li, $MnCl_2$
59. $H_2SO_4$, $SO_3$, $H_2O$
60. Zn, $CuSO_4$
61. $H_2$, Fe, $H_2O$
62. Zn, $H_2O$, ZnO, $H_2$
63. $H_2O$, $Al(OH)_3$
64. BaO, $H_2O$, $Ba(OH)_2$
65. $Mg_3N_2$, $H_2$, Mg, $NH_3$
66. $Fe_2O_3$, CO, Fe, $CO_2$
71. S, $SO_2$, $SO_3$, $H_2O$, $H_2SO_4$, $O_2$
72. $H_2SO_4$, $CaCO_3$, $H_2CO_3$, $CO_2$, $H_2O$
73. $Ag_2S$, Ag, $O_2$, $H_2S$, $H_2O$
74. S, $F_2$, $SF_6$, $Cl_2$, $SCl_2$, $Br_2$, $S_2Br_2$
75. W, $WO_3$, $H_2$, $H_2O$
76. $Ba(NO_3)_2$
77. $Li_2SO_4$
78. KOH
79. NaOH

9



This photograph shows the copper coating on a penny reacting with nitric acid. If you were to write the equation for the reaction, as you learned how to do in Chapter 8, you would get this:

$$Cu(s) + 4\ HNO_3(aq) \rightarrow Cu(NO_3)_2(aq) + 2\ NO_2(aq) + 2\ H_2O(\ell)$$

Is this an accurate and real equation? Real, yes—and the simplest equation you can use to calculate quantity relationships among the species in the reaction. Accurate, well, not exactly. In this chapter you will learn why this equation is not altogether correct and how to write an equation that tells exactly what species take part in a chemical change and nothing more.

# Chemical Change

CONTENTS



In Chapter 8 we discussed chemical equations for reactions, some of which occur in water solutions. These equations, however, are not entirely accurate in describing the ions in the solutions and the chemical changes that occur. Net ionic equations identify precisely the species in the solution that experience a change and the species that are produced. In this chapter you will refine and improve your equation-writing skill so that you can more accurately describe aqueous solutions and chemical change.

Our focus in this chapter primarily will be on chemical change that occurs with substances dissolved in water, or aqueous solutions. Chemists define a **solution** as a homogeneous mixture. An aqueous solution, therefore, is a homogeneous mixture of substances dissolved in water. We begin by examining the electrical properties of a solution.

Online homework for this chapter may be assigned in OWL

P/REVIEW: Solutions are discussed in more detail in Chapter 16.

## 9.1 Electrolytes and Solution Conductivity

Goal 1 Distinguish among strong electrolytes, weak electrolytes, and nonelectrolytes.

P/REVIEW: Formulas of ionic compounds were introduced in Section 6.8. Chemical compounds are electrically neutral. A net zero charge is achieved by combining cations and anions in numbers such that positive and negative charges are balanced.

P/REVIEW: Figure 2.5 (Section 2.2) shows that solid particles are held in fixed position relative to each other, whereas liquid particles are free to move. Ions in an ionic solid are arranged in crystals, as shown in Figures 12.2 and 12.3 (Section 12.2). When the solid is dissolved in water, the ions are released from the crystal lattice, and they are free to move among the water molecules, as described in Section 16.3.

Consider the apparatus shown in Figure 9.1. It is made of two metal strips called **electrodes,** an electrical cord attached to a battery, and a lightbulb. When the two electrodes touch, the lightbulb glows. If the electrodes are not in contact, electrons cannot flow through the apparatus, and the bulb does not light.

Suppose we place the apparatus in a solution of a molecular compound dissolved in water, as in Active Figure 9.2(a), which shows drinking alcohol (ethanol) dissolved in water. Nothing happens. But if the liquid is a solution of an ionic compound, the bulb glows brightly, as in Active Figure 9.2(b), which shows copper(II) chloride dissolved in water.

What is it about the solution of ions that allows it to conduct an electric current? Why doesn't the solution of a molecular compound conduct electricity? When the two metal strips in the conductivity apparatus touch, electrons have a path to flow through the filament in the lightbulb and thus make it glow. When the strips are separated, the electrons cannot flow. From this we can reason that when the apparatus is placed in a solution containing ions, there must be a way for electrons to move from one metal strip to the other. The solution of a molecular substance does not have this property.

A solid ionic substance is made up of positively and negatively charged ions that are held in fixed positions. These ions become free to move when they are dissolved in water. It is the movement of ions that makes up an electric current in a solution—a process called **electrolysis.** In fact, the ability of a solution to conduct electricity is regarded as positive evidence that it contains ions (Fig. 9.3).

### Thinking About Your Thinking

Mental Models

**Forming a mental model of the particulate-level process that occurs in a solution is a goal for you to work toward in Section 9.1. You need to incorporate Figures 9.1, 9.2, and 9.3 into this mental model. The key part of forming your model is to make Figure 9.3 into a dynamic, moving, three-dimensional mental image. Ions must be present in solution in order for the solution to conduct electricity. The current is carried by ions. The word *ion* comes from a Greek word meaning "to go." These charged particles move when an external electrical force is applied.**

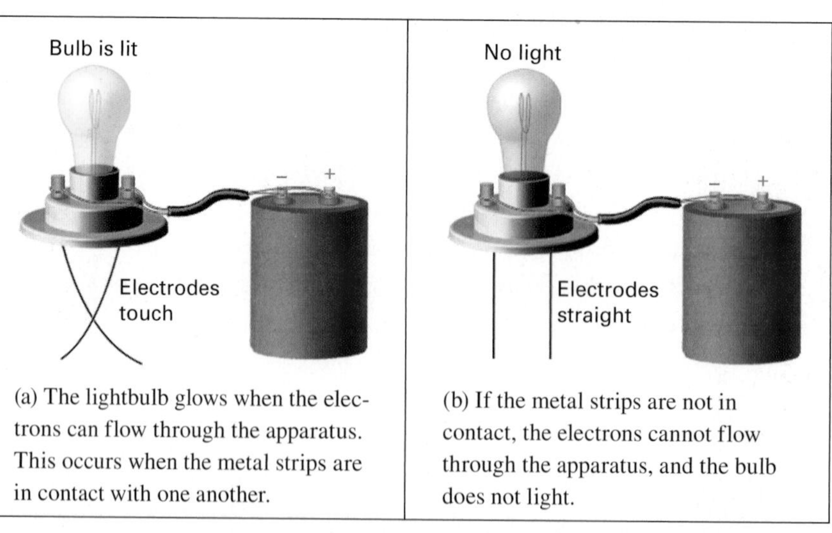


(a) The lightbulb glows when the electrons can flow through the apparatus. This occurs when the metal strips are in contact with one another.

(b) If the metal strips are not in contact, the electrons cannot flow through the apparatus, and the bulb does not light.

**Figure 9.1** A lightbulb conductivity apparatus.

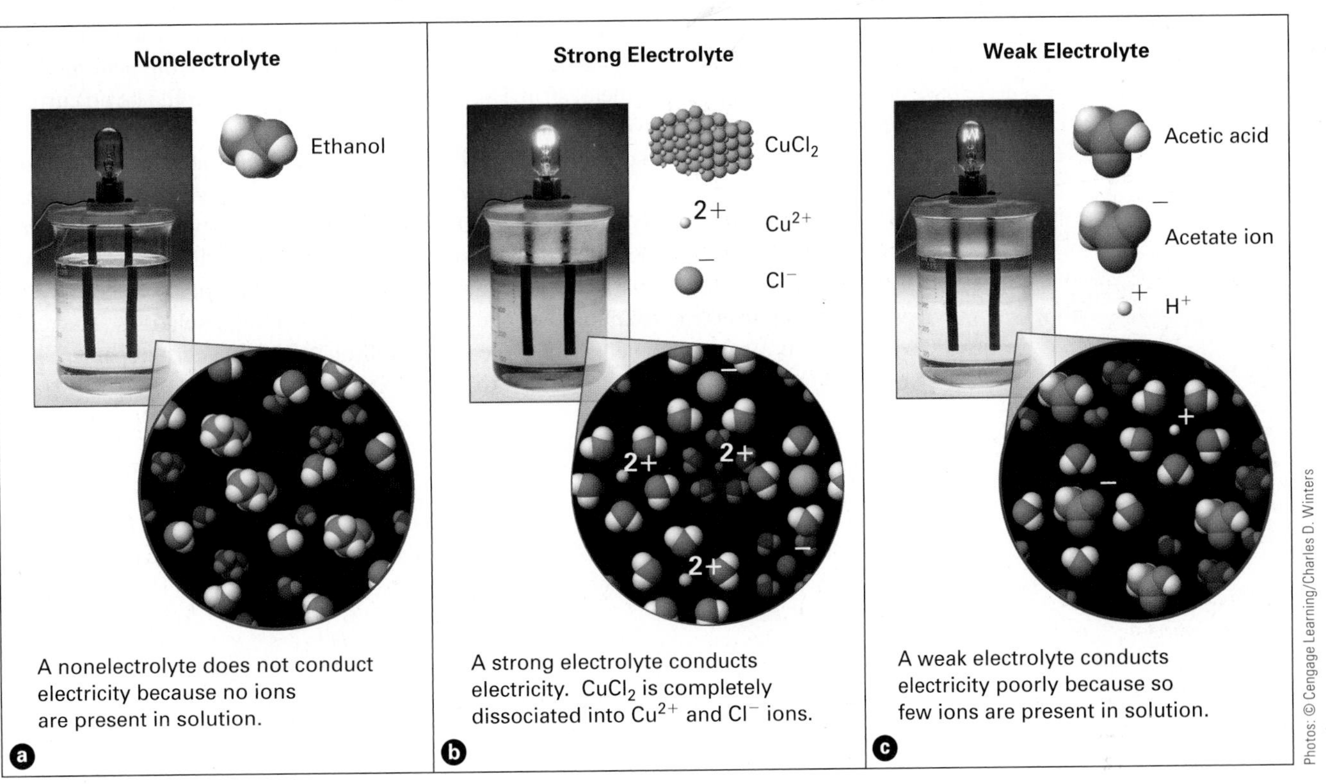

Nonelectrolyte
Ethanol
A nonelectrolyte does not conduct electricity because no ions are present in solution.
a
Strong Electrolyte
$CuCl_2$
$Cu^{2+}$
$Cl^-$
A strong electrolyte conducts electricity. $CuCl_2$ is completely dissociated into $Cu^{2+}$ and $Cl^-$ ions.
b
Weak Electrolyte
Acetic acid
Acetate ion
$H^+$
A weak electrolyte conducts electricity poorly because so few ions are present in solution.
c
Photos: © Cengage Learning/Charles D. Winters


**Active Figure 9.2** Nonelectrolytes, strong electrolytes, and weak electrolytes. The liquid in the beaker is used to "close" the electrical circuit. (a) If the liquid is pure water or a solution of molecular compounds, the bulb does not light. Solutes whose solutions do not conduct electricity are nonelectrolytes. (b) A soluble ionic salt is a strong electrolyte because its solution makes the bulb burn brightly. (c) Some solutes are called weak electrolytes because their solutions conduct poorly, and the bulb glows dimly. **Watch this active figure at *http://www.cengage.com/cracolice.***

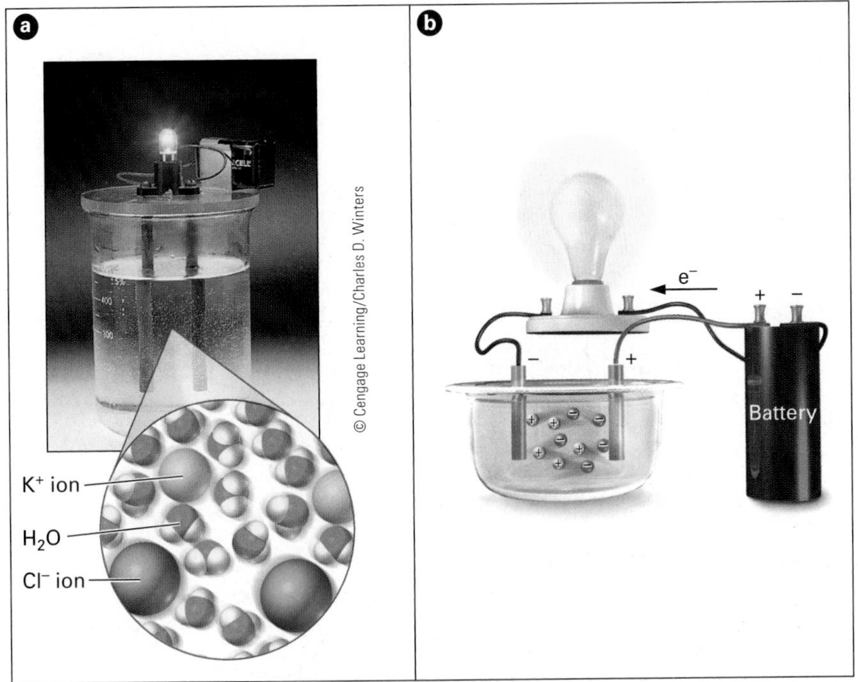


**Figure 9.3** Conductivity in an ionic solution. (a) The illuminated bulb shows that this solution, made by dissolving the solid ionic compound potassium chloride in water, is indeed a conductor. If a solution conducts electricity, it is tangible evidence that mobile ions are present. At the particulate level, we model the solution showing potassium ions and chloride ions free to move among the water molecules. (b) A simplified model of the electrical circuit. Positively charged ions are attracted to the negatively charged metal strip. Similarly, negatively charged ions move to the positively charged metal strip. Note how electrons flow through the system.

Solutions of molecular compounds do not conduct electricity because no ions are present. The ethanol and water molecules are electrically neutral. Being neutral, they do not move toward either metal strip. Even if they did move, there would be no current because the molecules have no charge.

A substance whose solution is a good conductor is called a **strong electrolyte.** Copper(II) chloride is an example. Ethanol, whose solution is a nonconductor, is an example of a **nonelectrolyte.** Some substances are **weak electrolytes.** Their solutions conduct electricity, but poorly, permitting only a dim glow of the lightbulb, as in Active Figure 9.2(c), which shows acetic acid dissolved in water. The poor conductivity of weak electrolytes is the result of low concentrations of ions in their solutions. The term *electrolyte* is also applied generally to the solution through which a current passes. The acid solution in an automobile battery is an electrolyte in this sense.

### √ Target Check 9.1

*Three compounds, P, G, and N, are dissolved in water, and the solutions are tested for electrical conductivity. Solution P is a poor conductor, G is a good conductor, and N does not conduct. Classify P, G, and N as electrolytes (strong, weak, or non-), and state the significance of the conductivities of their solutions.*

## 9.2 Solutions of Ionic Compounds

**Goal** 2 Given the formula of an ionic compound (or its name), write the formulas of the ions present when it is dissolved in water.

When an ionic compound dissolves in water, its solution consists of water molecules and ions. The ions are identified simply by separating the compound into its ions. When sodium chloride dissolves, the solution consists of water molecules, sodium ions, and chloride ions (Fig. 9.4):

$$NaCl(s) \xrightarrow{H_2O} Na^+(aq) + Cl^-(aq)$$

The $H_2O$ above the arrow indicates that the reaction occurs in the presence of water. We will be concerned only with reactions that occur in water in this chapter, so from this point on, we simply assume the presence of water molecules when we discuss solutions.

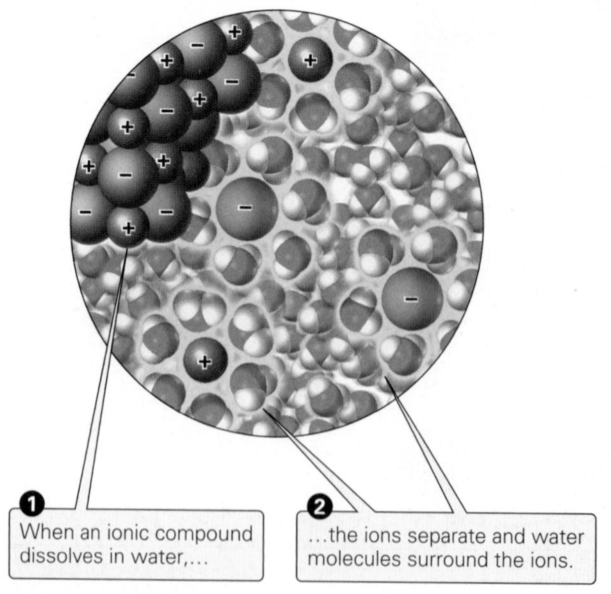
1 When an ionic compound dissolves in water,...
2 ...the ions separate and water molecules surround the ions.

**Figure 9.4** Sodium chloride dissolves in water. The solution that results consists of water molecules, sodium ions, and chloride ions.

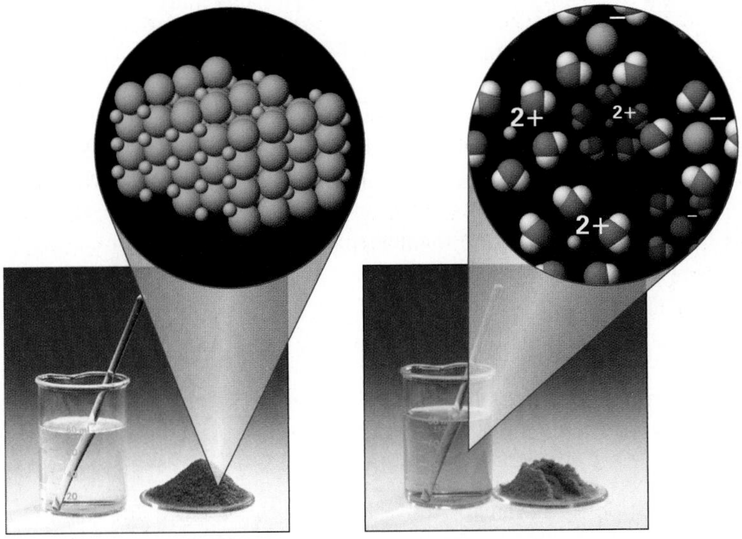

Figure 9.5 Copper(II) chloride is dissolved in water. The solution is made up of water molecules, copper(II) ions, and chloride ions. The copper(II) ions and chloride ions are in a 1:2 ratio.

Photos: © Cengage Learning/ Charles D. Winters

Copper(II) chloride is added to water. Interactions between water and the $Cu^{2+}$ and $Cl^-$ ions allow the solid to dissolve.

The ions are now sheathed in water molecules.

If the dissolved compound is copper(II) chloride, the solution contains copper(II) ions and chloride ions (Fig. 9.5):

$$CuCl_2(s) \rightarrow Cu^{2+}(aq) + 2\ Cl^-(aq)$$

Notice that no matter where a chloride ion comes from, its formula is always $Cl^-$, never $Cl_2^-$ or $Cl_2^{2-}$. The subscript after an ion in a formula, such as the 2 in $CuCl_2$, tells us how many ions are present in the formula unit. This subscript is not part of the ion formula.

## Thinking About Your Thinking

### Mental Models

**We first discussed forming a mental model of the particulate-level process that occurs when a solid ionic solute is dissolved in water in Section 9.1. That model is applied here in Section 9.2. This section asks you to use symbols—chemical formulas of ions—to describe the species in solution when an ionic compound dissolves. These symbols are a written representation of the composition of the solution at the particulate level. Be sure that your mental model matches the symbols you use to describe the particles in solution. Figures 9.4 and 9.5 will help you form your mental models.**

In this chapter you should include state symbols for all species in equations because it helps in writing net ionic equations. Remember also to include the charge every time you write the formula of an ion.

## Active Example 9.1

Write the formulas of all species in solution for the following ionic compounds by writing their dissolving equations:

$NaOH(s)$ $\rightarrow$

$K_2SO_4(s)$ $\rightarrow$

$(NH_4)_2CO_3(s)$ $\rightarrow$

*continued*

| | | |
|---|---|---|
| $NaOH(s)$ | $\rightarrow$ | $Na^+(aq) + OH^-(aq)$ |
| $K_2SO_4(s)$ | $\rightarrow$ | $2\ K^+(aq) + SO_4^{2-}(aq)$ |
| $(NH_4)_2CO_3(s)$ | $\rightarrow$ | $2\ NH_4^+(aq) + CO_3^{2-}(aq)$ |

Polyatomic and monatomic ions are handled in exactly the same way in writing the formulas of ions in solution. In $(NH_4)_2CO_3$, notice that the subscript outside the parentheses tells us how many ammonium ions are present, while the subscript 4 inside the parentheses is part of the polyatomic ion formula.

## Active Example 9.2

Write the formulas of the ions present in solutions of the following ionic compounds *without* writing the dissolving equations: $MgSO_4$; $Ca(NO_3)_2$; $AlBr_3$; $Fe_2(SO_4)_3$.

This question is similar to Active Example 9.1. It asks simply for the products of the reaction without writing an equation. This is how you will write these ions later. Be sure to show the number of each kind of ion released by a formula unit—the coefficient if you were writing the dissolving equation. Also remember state symbols.

$MgSO_4$: 

$AlBr_3$:

$Ca(NO_3)_2$:

$Fe_2(SO_4)_3$:

$MgSO_4$: $Mg^{2+}(aq) + SO_4^{2-}(aq)$

$Ca(NO_3)_2$: $Ca^{2+}(aq) + 2\ NO_3^-(aq)$

$AlBr_3$: $Al^{3+}(aq) + 3\ Br^-(aq)$

$Fe_2(SO_4)_3$: $2\ Fe^{3+}(aq) + 3\ SO_4^{2-}(aq)$

## 9.3 Strong and Weak Acids

**Goal 3** Explain why the solution of an acid may be a good conductor or a poor conductor of electricity.

**Goal 4** Given the formula of a soluble acid (or its name), write the major and minor species present when it is dissolved in water.

**P/REVIEW:** Lone hydrogen ions do not exist by themselves in solution. $H^+$ is always bound to a variable number of water molecules in the form $H^+ \cdot (H_2O)_x$. The hydrated hydrogen ion is sometimes represented by the hydronium ion, $H_3O^+$, which is one hydrogen ion combined with one water molecule. Lewis diagrams (Section 13.1) can be used to show the ionization of an acid and the bonds (Section 12.3) broken and formed in the process:

$$H{:}\ddot{O}{:}H + H{:}\ddot{X}{:} \longrightarrow \left[H{:}\ddot{O}{:}H \atop H\right]^+ + \left[{:}\ddot{X}{:}\right]^-$$

water — acid — hydronium ion — anion

Removing a water molecule from both sides of the equation leaves $HX \rightarrow H^+ + X^-$. See Section 6.5 and Chapter 17.

An **acid** is a hydrogen-bearing molecule or ion that releases a hydrogen ion in water solution. Its formula is HX, $H_2X$, or $H_3X$, where X is any negatively charged ion produced when the acid ionizes:

$$HX \rightarrow H^+ + X^-$$

X may be a monatomic ion, as when HCl ionizes and leaves $Cl^-$:

$$HCl \rightarrow H^+ + Cl^-$$

X may contain oxygen, as the $NO_3^-$ from $HNO_3$:

$$HNO_3 \rightarrow H^+ + NO_3^-$$

X may be an ion that contains hydrogen. Acids that have two or more ionizable hydrogens release them in steps. For sulfuric acid, this is

$$H_2SO_4 \rightarrow H^+ + HSO_4^-$$

$$HSO_4^- \rightarrow H^+ + SO_4^{2-}$$

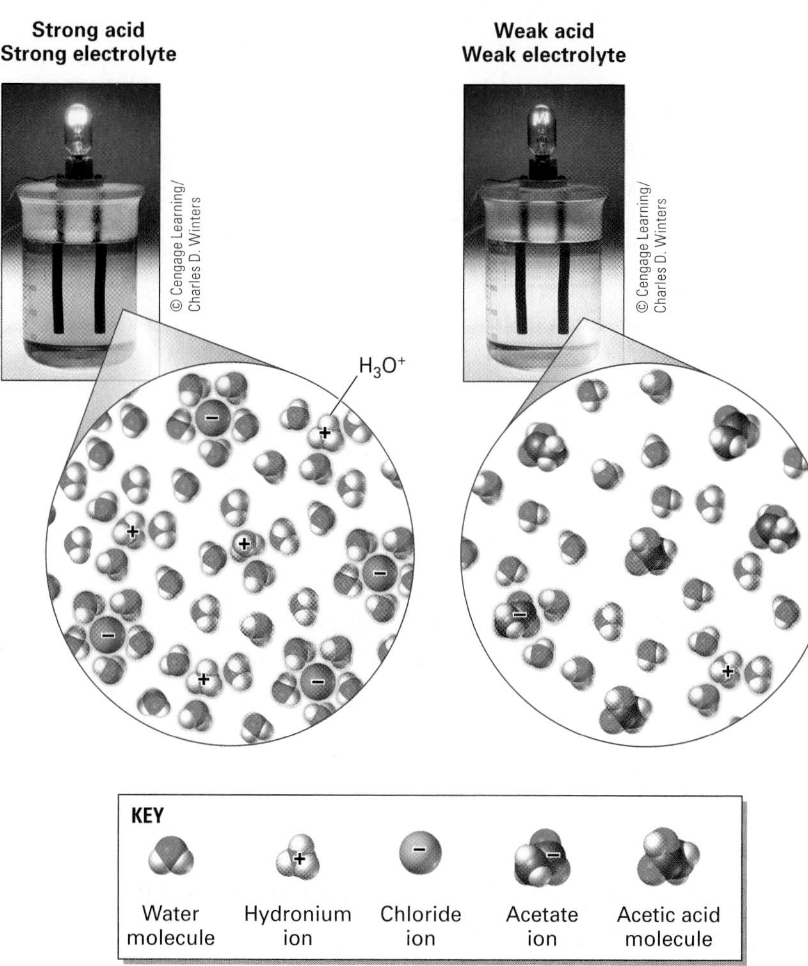


**Figure 9.6** Strong and weak acids at the macroscopic and particulate levels. Strong acids are strong electrolytes because the major species in solution are ions. For example, a hydrochloric acid solution primarily consists of water molecules, hydronium (hydrated hydrogen) ions, and chloride ions. Weak acids are weak electrolytes because the major species in solution are neutral molecules. Some acid molecules ionize, however, providing for the weak conductivity of the solution. Acetic acid is an example of a weak acid. Its solution primarily consists of water molecules and acetic acid molecules.

$HSO_4^-$ is an example of an ion that behaves as an acid. The equations for the step-by-step ionization of an acid are often combined:

$$H_2SO_4 \rightarrow 2\ H^+ + SO_4^{2-}$$

Carbon-containing acids usually contain hydrogen that is not ionizable. Acetic acid, $HC_2H_3O_2$, for example, ionizes to $H^+$ and the acetate ion, $C_2H_3O_2^-$:

$$HC_2H_3O_2 \rightleftharpoons H^+ + C_2H_3O_2^-$$

Do not be concerned if the anion (negatively charged ion) is not familiar. It will behave just like the anion from any other acid.

Acids are classified as strong or weak, depending on the extent to which the original compound ionizes when dissolved in water. In a dilute solution of a **strong acid,** almost all of the molecules of the original compound are converted to ions; very few remain as un-ionized molecules. We say that the **major species** present are ions and the **minor species** are un-ionized molecules. Consequently, strong acids are excellent conductors of electricity (Fig. 9.6[*left*]).

A **weak acid,** on the other hand, is only slightly ionized in solution. The major species in a weak acid are un-ionized molecules. Some ions are present, however; they are the minor species in solution. Because of the low concentration of ions, weak acids are poor conductors of electricity (Fig. 9.6[*right*]).

The differences between strong acids and weak acids are illustrated by comparing the acids formed by two very similar compounds, HCl and HF. Table 9.1 makes this comparison. Notice the four facts listed at the end of the table.

There are **seven common strong acids**. You must memorize their names and formulas. It helps to group them into three classifications:

**P/REVIEW:** A reversible reaction is one in which the products, as an equation is written, change back into the reactants. Reversibility is indicated by a double arrow, one pointing in each direction. Reversible changes are introduced in Section 8.7. Other reversible changes involve liquid–vapor equilibria (Section 15.4) and the formation of solutions (Section 16.3); others are found throughout Chapter 18 on chemical equilibrium.

| Seven Common Strong Acids | | |
|---|---|---|
| **Two Well-Known Acids** | **Three Group 7A/17 Acids** | **Two Chlorine Oxyacids** |
| nitric, $HNO_3$ | hydrochloric, HCl | chloric, $HClO_3$ |
| sulfuric, $H_2SO_4$ | hydrobromic, HBr | perchloric, $HClO_4$ |
| | hydroiodic, HI | |

To decide whether an acid is strong or weak, ask yourself, "Is it one of the seven strong acids?" If it is listed here, it is strong. If it is not one of these, it is weak.

The dividing line between strong and weak acids is arbitrary, and some acids are marginal in their classifications. Sulfuric acid is definitely strong in its first ionization step but marginal in the second. At present, we will avoid questions involving the classification of acids on the borderline between strong and weak.

**Table 9.1** Strong and Weak Acids

| | Hydrochloric Acid | Hydrofluoric Acid |
|---|---|---|
| **Strength of acid**<br>More than half the molecules of a strong acid are ionized. More than half the molecules in a weak acid are in molecular form. | Strong | Weak |
| **Formula as it is written in a conventional equation** | HCl(aq) | HF(aq) |
| **Percentage ionization in 0.1 and 0.01-M solutions**<br>M refers to concentration. (In a 0.1-M solution, 1 liter of solution contains 0.1 mole of acid.) | 0.10 M: 79%<br>0.01 M: 99+% | 0.10 M: 9.6%<br>0.01 M: 22% |
| **Electrical conductivity** | Excellent | Poor |
| **Ionization equation**<br>The relative lengths of the arrows are a visual indication of the extent of ionization. If the longer arrow points to the right, it means more than half of the acid molecules are separated into ions. If the longer arrow points left, more than half of the acid molecules are present as un-ionized molecules. | $HCl(aq) \rightleftharpoons H^+(aq) + Cl^-(aq)$ | $HF(aq) \rightleftharpoons H^+(aq) + F^-(aq)$ |
| **Major species**<br>These are the species present in greater abundance in the reaction; they can be found on the side of the equation pointed to by the longer arrow. Formulas of the major species are written in the net ionic equations you are about to write. | $H^+(aq)$ and $Cl^-(aq)$ | HF(aq) |
| **Minor species**<br>These are the species present in lesser abundance in the reaction; they can be found on the side of the equation pointed to by the shorter arrow. Even though these species are present in a reaction vessel, they do not take part in a reaction in the form shown. Therefore, formulas of the minor species do not appear in the net ionic equation. | HCl(aq) | $H^+(aq)$ and $F^-(aq)$ |

- HCl(aq), which is 79% ionized in a 0.10-M solution (0.10 mole of HCl in 1 liter of solution), is a strong acid, whereas 0.10-M HF, at 9.6% ionized, is a weak acid.
- The double arrows in the ionization equations show that the ionization process is reversible. The longer arrow pointing to the right indicates that it is much more likely for HCl molecules to break into ions than for the ions to combine and form dissolved HCl molecules.
- The major species in the strong acid are the ions, but in the weak acid the major species is the acid molecule. The formulas of the major species appear in the net ionic equation.
- The minor species are present in the reaction vessel, but do not participate in the reaction as shown. Minor species are not included in a net ionic equation.

Note that the terms *strong* and *weak* do *not* refer to the corrosiveness of the acid or its reactivity. Hydrofluoric acid is weak, but it cannot be stored in glass bottles because it will react with the glass.

### Active Example 9.3

Write the major species in solutions of the following acids.

You can, if necessary, write ionization equations. It is better just to identify the ions or molecules without equations, however. Include state symbols.

$HNO_2$: HI:

$HClO_3$: $HC_3H_5O_2$:

$HNO_2$: $HNO_2(aq)$ HI: $H^+(aq) + I^-(aq)$

$HClO_3$: $H^+(aq) + ClO_3^-(aq)$ $HC_3H_5O_2$: $HC_3H_5O_2(aq)$

$HNO_2$ and $HC_3H_5O_2$ are not among the seven strong acids, so the major species in solution are the molecules themselves. HI and $HClO_3$ are strong acids; they break up into ions.

We can now summarize the major species in solutions as they have been described in this section and the last.

summary

**Identifying the Major Species in a Solution**

*Ions* are the major species in the solutions of two kinds of substances:

All soluble ionic compounds

The seven strong acids

*Neutral molecules* are the major species in solutions of everything else, primarily

Weak acids

Weak bases

Water

The major species in solution are the species that appear in net ionic equations, which we examine next.

## 9.4 Net Ionic Equations: What They Are and How to Write Them

Goal 5 Distinguish among conventional, total ionic, and net ionic equations.

In Section 8.9 we showed how to write the double-replacement equation for a precipitation reaction between solutions of two ionic compounds. Such a reaction occurs when a solution of lead(II) nitrate is added to a solution of sodium chloride:

$$Pb(NO_3)_2(aq) + 2\ NaCl(aq) \rightarrow PbCl_2(s) + 2\ NaNO_3(aq) \qquad \textbf{(9.1)}$$

In this chapter we call this kind of equation a **conventional equation.**

A conventional equation serves many useful purposes, including its essential role in understanding quantitative relationships in chemical change. However, it falls short in describing precisely the reaction that occurs in water solution. Usually, it does not describe the reactants or products correctly. Rarely does it describe accurately the chemical changes that occur.

The shortcomings of a conventional equation are illustrated by this fact: The solutions that react in Equation 9.1 contain no substances with the formulas $Pb(NO_3)_2$ or

**P/REVIEW:** The quantitative relationships between substances involved in a chemical reaction are explored in Chapter 10.

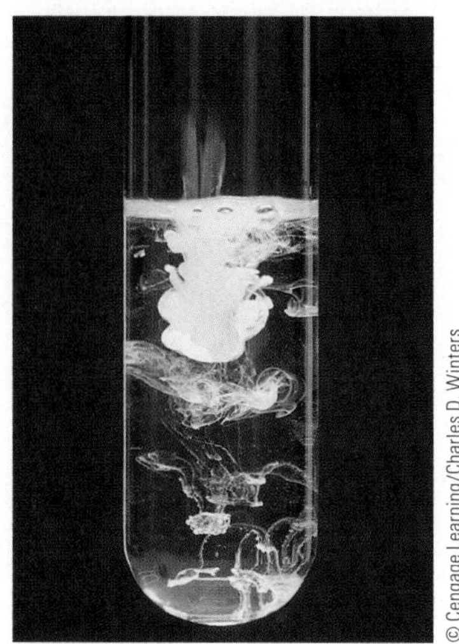
© Cengage Learning/Charles D. Winters

**Figure 9.7** Precipitation of lead(II) chloride. When clear, colorless solutions of lead(II) nitrate and sodium chloride are mixed, white lead(II) chloride can be seen precipitating from the solution. Nitrate ions and sodium ions remain in solution.

NaCl. Actually present are the major species, $Pb^{2+}$ and $NO_3^-$ in one solution and $Na^+$ and $Cl^-$ in the other. The conventional equation doesn't tell you that. Nothing with the formula $NaNO_3$ is formed in the reaction. The $Na^+$ and $NO_3^-$ ions are still there in solution after the reaction. The conventional equation kept that a "secret," too. The only substance in Equation 9.1 that is *really there* is solid lead(II) chloride, $PbCl_2(s)$. If you perform the reaction, you can see the precipitate (Fig. 9.7).

To write an equation that describes the reaction in Equation 9.1 more accurately, we replace the formulas of the dissolved substances with the major species in solution. This produces the **total ionic equation:**

$$Pb^{2+}(aq) + 2\ NO_3^-(aq) + 2\ Na^+(aq) + 2\ Cl^-(aq) \rightarrow PbCl_2(s) + 2\ Na^+(aq) + 2\ NO_3^-(aq) \qquad \textbf{(9.2)}$$

Notice that in order to keep the total ionic equation balanced, we must include the coefficients from the conventional equation. One formula unit of $Pb(NO_3)_2$ gives one $Pb^{2+}$ ion and two $NO_3^-$ ions, two formula units of NaCl give two $Na^+$ ions and two $Cl^-$ ions, and two formula units of $NaNO_3$ give two $Na^+$ ions and two $NO_3^-$ ions.

A total ionic equation tells more than just what species take part in a chemical change. It includes **spectator ions,** or simply **spectators.** A spectator is an ion that is present at the scene of a reaction but experiences no chemical change. It appears on both sides of the total ionic equation. $Na^+(aq)$ and $NO_3^-(aq)$ are spectators in Equation 9.2. To change a total ionic equation into a **net ionic equation,** you remove the spectators:

$$Pb^{2+}(aq) + 2\ Cl^-(aq) \rightarrow PbCl_2(aq) \qquad \textbf{(9.3)}$$

A net ionic equation indicates exactly what chemical change takes place, and nothing else.

In Equations 9.1 to 9.3 you have the three steps to follow in writing a net ionic equation. They are:

**procedure**

**Writing a Net Ionic Equation**

**Step 1:** Write the conventional equation, including state symbols—(g), (ℓ), (s), and (aq). Balance the equation.

$Pb(NO_3)_2(aq) + 2\ NaCl(aq) \rightarrow PbCl_2(s) + 2\ NaNO_3(aq)$

**Step 2:** Write the total ionic equation by replacing each aqueous (aq) substance that is a strong acid or a soluble ionic compound with its major species. *Do not separate a weak acid into ions,* even though its state is aqueous (aq). Also, never change solids (s), liquids (ℓ), or gases (g) into ions. Be sure the equation is balanced in both atoms and charge. (Charge balance is discussed in more detail in the next section.)

$Pb^{2+}(aq) + 2\ NO_3^-(aq) + 2\ Na^+(aq) + 2\ Cl^-(aq) \rightarrow PbCl_2(s) + 2\ Na^+(aq) + 2\ NO_3^-(aq)$

**Step 3:** Write the net ionic equation by removing the spectators from the total ionic equation. Reduce coefficients to lowest terms, if necessary. Be sure the equation is balanced in both atoms and charge.

$Pb^{2+}(aq) + 2\ Cl^-(aq) \rightarrow PbCl_2(s)$

Now try the net ionic equation-writing procedure on two Active Examples.

## Active Example 9.4

When nickel nitrate and sodium carbonate solutions are combined, solid nickel carbonate precipitates, leaving a solution of sodium nitrate. Write the conventional equation, total ionic equation, and net ionic equation for this reaction.

*Step 1* in the procedure is to write the conventional equation, including state symbols. The problem statement gives you all the necessary information to complete this step.

$Ni(NO_3)_2(aq) + Na_2CO_3(aq) \rightarrow NiCO_3(s) + 2\ NaNO_3(aq)$

The next step is to write the total ionic equation. The species that occur as ions in solution are written *as ions* rather than as ionic compounds, which is how they appear in the conventional equation. In other words, you replace each aqueous substance except weak acids with its major species. There are no weak acids in this Active Example, so separate each aqueous substance into its ions. Don't forget to account for the number of ions in the formula unit of each compound as well as the coefficients in the balanced conventional equation.

$$Ni^{2+}(aq) + 2\ NO_3^-(aq) + 2\ Na^+(aq) + CO_3^{2-}(aq) \rightarrow NiCO_3(s) + 2\ Na^+(aq) + 2\ NO_3^-(aq)$$

Note how the two nitrate and two sodium ions on the left come from the compound formula unit and the two on the right result from the coefficients in the balanced equation.

The final step in the procedure is to write the net ionic equation. This is accomplished by identifying and eliminating any spectator ions. Cross out the ions that appear on both sides of the equation, and what remains will be the net ionic equation.

$Ni^{2+}(aq) + CO_3^{2-}(aq) \rightarrow NiCO_3(s)$

Checking to be sure the net ionic equation is balanced, the two plus charges from $Ni^{2+}(aq)$ cancel the two minus from $CO_3^{2-}(aq)$, leaving a net zero charge on the left to balance the zero charge on the right. There are one nickel and one carbonate on each side. The equation remains balanced.

The net ionic equation tells us that nickel ions from one solution combine with the carbonate ions from the other solution to form solid nickel carbonate. The sodium ions and nitrate ions that were in the separate solutions remain as ions in the combined solution.

## Active Example 9.5

When a strip of magnesium metal is placed in an iron(III) chloride solution, a magnesium chloride solution results and solid iron also forms. Develop the net ionic equation for this reaction.

*continued*

Start with the conventional equation. Remember to use state symbols when your goal is to write the net ionic equation.

$3\ Mg(s) + 2\ FeCl_3(aq) \rightarrow 3\ MgCl_2(aq) + 2\ Fe(s)$

The three Cl on the left and two Cl on the right require a total of six Cl on each side to balance the equation. The coefficients on Mg and Fe follow.

Now go for the total ionic equation. Neither of the aqueous species is a weak acid, so they separate into their ions.

$3\ Mg(s) + 2\ Fe^{3+}(aq) + 6\ Cl^-(aq) \rightarrow 3\ Mg^{2+}(aq) + 6\ Cl^-(aq) + 2\ Fe(s)$

The six $Cl^-(aq)$ on each side come from multiplying the coefficient in the balanced equation by the number of ions in the formula unit of the compound.

Remove the spectator ions to arrive at the net ionic equation. Check for charge and atom balance.

$3\ Mg(s) + 2\ Fe^{3+}(aq) \rightarrow 3\ Mg^{2+}(aq) + 2\ Fe(s)$

The six plus charge on the left balances the six plus charge on the right. Three Mg and two Fe on each side give atom balance.

In this case, the net ionic equation tells you that magnesium metal is changed to magnesium ions in solution while the opposite process occurs for iron. The iron ions that were in solution were changed to iron atoms, which collectively form solid iron. The chloride ions in the solution are unchanged while the metals react, and thus they do not appear in the net ionic equation.

## 9.5 Single-Replacement Oxidation–Reduction (Redox) Reactions

**P/REVIEW:** A single-replacement redox reaction is one for which the equation has the form A + BX → AX + B. An uncombined element, A, appears to replace another element, B, in a compound, BX. See Section 8.8 and Table 8.3.

**Goal 6** Given two substances that may engage in a single-replacement redox reaction and an activity series by which the reaction may be predicted, write the conventional, total ionic, and net ionic equations for the reaction that will occur, if any.

An iron nail is dropped into a solution of hydrochloric acid. Hydrogen gas bubbles out (Fig. 9.8). When the reaction ends, the test tube contains a solution of iron(II) chloride. The question is, "What happened?" The answer lies in the net ionic equation.

The conventional equation (*Step 1* in the procedure from Section 9.4) is a single-replacement equation:

$$Fe(s) + 2\ HCl(aq) \rightarrow H_2(g) + FeCl_2(aq) \qquad \textbf{(9.4)}$$

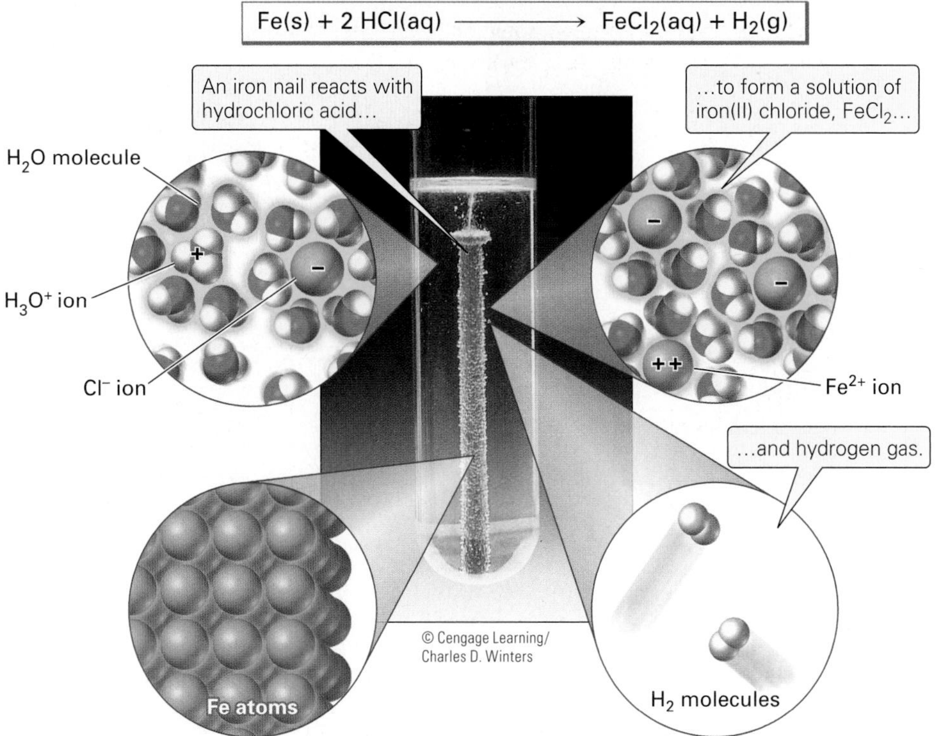


**Figure 9.8** The reaction between iron and hydrochloric acid. This is an example of an oxidation–reduction, or redox, reaction. Redox reactions are electron-transfer reactions.

HCl is a strong acid, and $FeCl_2$ is a soluble ionic compound. Both have ions as the major species in their solutions. These ions replace the compounds in the total ionic equation (*Step 2* in the procedure for writing net ionic equations):

$$Fe(s) + 2\ H^+(aq) + 2\ Cl^-(aq) \rightarrow H_2(g) + Fe^{2+}(aq) + 2\ Cl^-(aq) \qquad \textbf{(9.5)}$$

The total ionic equation remains balanced in both atoms and charge. The net charge is zero on both sides.

The final step (*Step 3*) in writing a net ionic equation is to rid the total ionic equation of spectators. Chloride ions are the only spectators. Taking them away gives

$$Fe(s) + 2\ H^+(aq) \rightarrow H_2(g) + Fe^{2+}(aq) \qquad \textbf{(9.6)}$$

This is the net ionic equation. This is what happened—and no more.

Notice that all equations are balanced. Balancing atoms is discussed in Chapter 8; balancing charge is something new. Neither protons nor electrons are created or destroyed in a chemical change, so the total charge among the products must equal the total charge among the reactants. In Equation 9.5, two plus charges in 2 $H^+(aq)$ added to two negative charges in 2 $Cl^-(aq)$ give a net zero charge on the left. Two plus charges in $Fe^{2+}(aq)$ and two negative charges in 2 $Cl^-(aq)$ on the right also total zero. The equation is balanced in charge.

Notice also that the net charge does not have to be zero on each side for the equation to be balanced. The charges must be *equal*. In Equation 9.6 the net charge is 2+ on each side.

Equation 9.6 illustrates why this is an oxidation–reduction, or redox, reaction. Redox reactions are electron-transfer reactions: Electrons are actually or effectively transferred from one species to another. In Equation 9.6, a neutral iron atom becomes an iron(II) ion with a 2+ charge. It does this by losing two electrons: $Fe(s) \rightarrow Fe^{2+}(aq) + 2\ e^-$. Where do the electrons go? One goes to each of two hydrogen ions so they become neutral hydrogen atoms, and the atoms combine to form a diatomic molecule: $2\ H^+(aq) + 2\ e^- \rightarrow H + H \rightarrow H_2(g)$. Thus, the electrons are literally transferred from one species, iron atoms, to another species, hydrogen ions. The iron atoms, which lost the electrons, are said to have been oxidized, and hydrogen ions, the receivers of electrons, have been reduced.

You will learn more about redox reactions in Chapter 19.

(a)

(b)

Photos: Andrew Lambert Photography/Photo Researchers, Inc.

**Figure 9.9** The reaction between zinc and a solution of copper(II) nitrate. (a) The zinc strip is shiny white before the reaction. (b) After the zinc is dipped into the solution for about two seconds, the strip is covered with tiny, finely divided particles of copper that appear almost black when wet with the solution.

## Active Example 9.6

A reaction occurs when a piece of zinc is dipped into a solution of copper(II) nitrate (Fig. 9.9). Write the conventional, total ionic, and net ionic equations.

Begin by writing the conventional equation (*Step 1*). Do you know what the products are? You know that the equation for this kind of reaction is a single-replacement type (Section 8.8). Something must replace something else in a compound. Don't forget the state designations.

$$Zn(s) + Cu(NO_3)_2(aq) \rightarrow Zn(NO_3)_2(aq) + Cu(s)$$

In the single-replacement equation, zinc appears to replace copper in $Cu(NO_3)_2$.

Now write the total ionic equation (*Step 2*), replacing formulas of the appropriate aqueous compounds in solution with their ions. Only dissolved ionic compounds and strong acids are divided into ions. Be sure your equation is balanced in both atoms and charge.

$$Zn(s) + Cu^{2+}(aq) + 2\ NO_3^-(aq) \rightarrow Zn^{2+}(aq) + 2\ NO_3^-(aq) + Cu(s)$$

The equation is balanced. There is one zinc on each side, an atom on one side and an ion on the other. The same is true for copper. Finally, there are two nitrate ions on each side. The net charge is zero on each side of the equation.

Examine the total ionic equation. Are there any spectators? If so, remove them and write the net ionic equation.

$$Zn(s) + Cu^{2+}(aq) \rightarrow Zn^{2+}(aq) + Cu(s)$$

Zinc and copper are balanced as before. This time the net charge is 2+ on each side of the equation. The equation is balanced in atoms and charge.

If you were to place a strip of copper into a solution of zinc nitrate and go through the identical thought process, the equations would be exactly the reverse of those in Active Example 9.6. Does the reaction occur in both directions? If not, in which way does it occur? How can you tell?

Technically, both reactions occur, or can be made to occur. Only the reaction in Active Example 9.6 takes place without an outside source of energy to drive the reaction. The best way to find out which of two reversible reactions occurs is to try them and see. These experiments have been done, and the results are summarized in the **activity series** in Table 9.2. Under normal conditions, any element in the table will replace the dissolved ions of any element beneath it. Zinc is above copper in the table, so zinc will replace $Cu^{2+}(aq)$ ions in the solution. Copper, being below zinc, will not replace $Zn^{2+}(aq)$ in solution.

Use Table 9.2 to predict whether or not a single-replacement redox reaction will take place. If asked to write the equation for a reaction that does not occur, write NR for "no reaction" on the product side: $Cu(s) + Zn^{2+}(aq) \rightarrow NR$.

**Table 9.2** Activity Series

| | |
|---|---|
| Li<br>K<br>Ba<br>Sr<br>Ca<br>Na | Will replace $H_2$ from liquid water, steam, or acid |
| Mg<br>Al<br>Mn<br>Zn<br>Cr | Will replace $H_2$ from steam or acid |
| Fe<br>Ni<br>Sn<br>Pb | Will replace $H_2$ from acid |
| $H_2$ | |
| Sb<br>Cu<br>Hg<br>Ag<br>Pd<br>Pt<br>Au | Will not replace $H_2$ from liquid water, steam, or acid |

## Active Example 9.7

Write the conventional, total ionic, and net ionic equations for the reaction that occurs, if any, between calcium and hydrobromic acid.

First, write the formula of hydrobromic acid. Remember from Section 9.3 that hydrobromic acid is a strong acid. The formula of hydrobromic acid is not found in the activity series. What dissolved ion will potentially be a reactant in this reaction? What corresponding elemental formula will be in the activity series?

$HBr \rightarrow H^+ + Br^-$, so $H^+$ is the potential reactant, and $H_2$ is the corresponding elemental formula.

Now, will a reaction occur? Check the activity series, then answer and explain.

Yes. Calcium is above hydrogen in the series, so calcium will replace hydrogen ions from solution (Fig. 9.10).

Write the conventional equation. Remember the state designations.

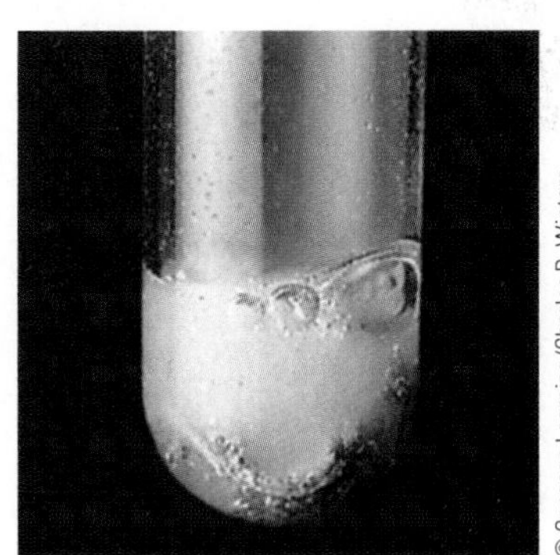

Figure 9.10 Calcium reacts with hydrobromic acid.

$Ca(s) + 2\ HBr(aq) \rightarrow CaBr_2(aq) + H_2(g)$

Now write the total ionic equation.

*continued*

$Ca(s) + 2\ H^+(aq) + 2\ Br^-(aq) \rightarrow Ca^{2+}(aq) + 2\ Br^-(aq) + H_2(g)$

Each HBr(aq) yields one $H^+(aq)$ and one $Br^-(aq)$. The conventional equation has two HBr(aq), so there will be 2 $H^+(aq)$ and 2 $Br^-(aq)$ in the total ionic equation.

Now eliminate the spectators and write the net ionic equation.

$Ca(s) + 2\ H^+(aq) \rightarrow Ca^{2+}(aq) + H_2(g)$

Bromide ion, $Br^-(aq)$, is the only spectator.

## Active Example 9.8

Copper is placed into a solution of silver nitrate (see Active Figure 9.11), forming copper(II) ions as one of the products. Write the three equations.

Start with the conventional equation.

$Cu(s) + 2\ AgNO_3(aq) \rightarrow Cu(NO_3)_2(aq) + 2\ Ag(s)$

Now write the final two equations, please.

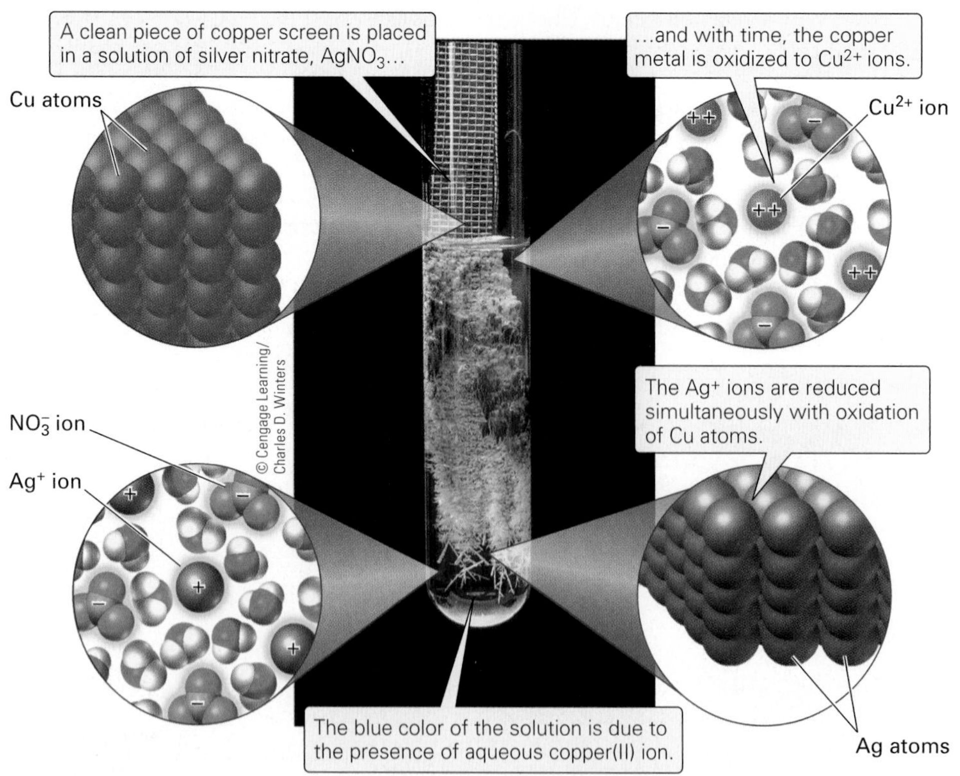
A clean piece of copper screen is placed in a solution of silver nitrate, $AgNO_3$...
...and with time, the copper metal is oxidized to $Cu^{2+}$ ions.
Cu atoms
$Cu^{2+}$ ion
The $Ag^+$ ions are reduced simultaneously with oxidation of Cu atoms.
$NO_3^-$ ion
$Ag^+$ ion
© Cengage Learning/ Charles D. Winters
The blue color of the solution is due to the presence of aqueous copper(II) ion.
Ag atoms

**Active Figure 9.11** The oxidation of copper metal by silver ion at the macroscopic level and particulate level. **Watch this active figure at *http://www.cengage.com/cracolice.***

$$Cu(s) + 2\ Ag^+(aq) + 2\ NO_3^-(aq) \rightarrow Cu^{2+}(aq) + 2\ NO_3^-(aq) + 2\ Ag(s)$$

$$Cu(s) + 2\ Ag^+(aq) \rightarrow Cu^{2+}(aq) + 2\ Ag(s)$$

Does your net ionic equation match the particulate-level illustrations in Active Figure 9.11? Explain.

Yes. The reactants are Cu atoms and $Ag^+$ ions. The products are $Cu^{2+}$ ions and Ag atoms.

$NO_3^-$ ions are spectators and not included in the net ionic equation.

## 9.6 Oxidation–Reduction Reactions of Some Common Organic Compounds

Goal 7 Write the equation for the complete oxidation or burning of any compound containing only carbon and hydrogen or only carbon, hydrogen, and oxygen.

Another type of electron-transfer reaction occurs when a substance burns. A large number of compounds, including fuels, petroleum products, alcohols, some acids, and sugars, consist of only two or three elements: carbon and hydrogen or carbon, hydrogen, and oxygen. When such compounds are burned in air, they react with oxygen in the atmosphere. We say they are oxidized, a term that originally meant reacting with oxygen. The reaction is an oxidation–reduction reaction. Chemists now use the term *oxidized* to mean loss of electrons during an electron-transfer reaction. Electrons are transferred from the carbon–hydrogen–oxygen compound to oxygen.

The products of a complete burning or oxidation of these compounds are always the same: carbon dioxide, $CO_2(g)$, and water, $H_2O(g)$ or $H_2O(\ell)$, depending on the temperature at which the product is examined. The distinction is not important in this chapter, so we will use $H_2O(\ell)$ consistently.

In writing these equations you will be given only the identity of the compound that *burns* or is *oxidized*. These words tell you the compound reacts with oxygen, $O_2(g)$, so you must include it as a second reactant. The formulas of water and carbon dioxide appear on the right side of the equation. The general equation for a complete oxidation (burning) reaction is always

$$C_xH_yO_z + O_2(g) \rightarrow CO_2(g) + H_2O(\ell)\ [\text{or } H_2O(g)]$$

The burning of methane, the principal component of natural gas, is an example:

$$CH_4(g) + 2\ O_2(g) \rightarrow CO_2(g) + 2\ H_2O(\ell)$$

As a rule, these equations are most easily balanced if you take the elements carbon, hydrogen, and oxygen in that order.

Both carbon dioxide gas, $CO_2(g)$, and steam, $H_2O(g)$, are invisible. The white "smoke" commonly seen rising from chimneys and smokestacks is really tiny drops of condensed liquid water, $H_2O(\ell)$. Black smoke comes from carbon that is not completely burned.

### Active Example 9.9

Write the equation for the complete burning of ethane, $C_2H_6(g)$.

The phrase "complete burning" is your cue that the substance reacts with oxygen, forming carbon dioxide and water as products. Remember, balance elements in the order C first, H second, and O last.

*continued*

$$2\,C_2H_6(g) + 7\,O_2(g) \rightarrow 4\,CO_2(g) + 6\,H_2O(\ell)$$

Balancing C: $C_2H_6(g) + O_2(g) \rightarrow 2\,CO_2(g) + H_2O(\ell)$

Balancing H: $C_2H_6(g) + O_2(g) \rightarrow 2\,CO_2(g) + 3\,H_2O(\ell)$

Balancing O: $C_2H_6(g) + 7/2\,O_2(g) \rightarrow 2\,CO_2(g) + 3\,H_2O(\ell)$

Clearing the fraction by multiplying all coefficients by 2 produces the final equation (written above).

## 9.7 Double-Replacement Precipitation Reactions

**Goal 8** Predict whether a precipitate will form when known solutions are combined; if a precipitate forms, write the net ionic equation. (Reference to a solubility table or a solubility guidelines list may or may not be allowed.)

**Goal 9** Given the product of a precipitation reaction, write the net ionic equation.

An **ion-combination reaction** occurs when the cation (positively charged ion) from one reactant combines with the anion (negatively charged ion) from another to form a particular kind of product compound. The conventional equation is a double-replacement type in which the ions appear to "change partners": AX + BY → AY + BX. In this section, the product is an insoluble ionic compound that settles to the bottom of mixed solutions. A solid formed this way is called a **precipitate;** the reaction is a **precipitation reaction.**

**P/REVIEW:** Double-replacement equations are also used to describe precipitation reactions in Section 8.9. See Table 8.3.

### Active Example 9.10

Mixing sodium chloride and silver nitrate solutions produces a white precipitate of silver chloride. Develop the net ionic equation for the reaction.

Start by writing the formulas of the reactants as they will appear in the conventional equation.

$$NaCl(aq) + AgNO_3(aq) \rightarrow$$

The same three steps you used on single-replacement redox reactions are applied to precipitation reactions. In this case, the products can be predicted by "changing partners" of the reacting ions. Complete the equation above and write the other two equations.

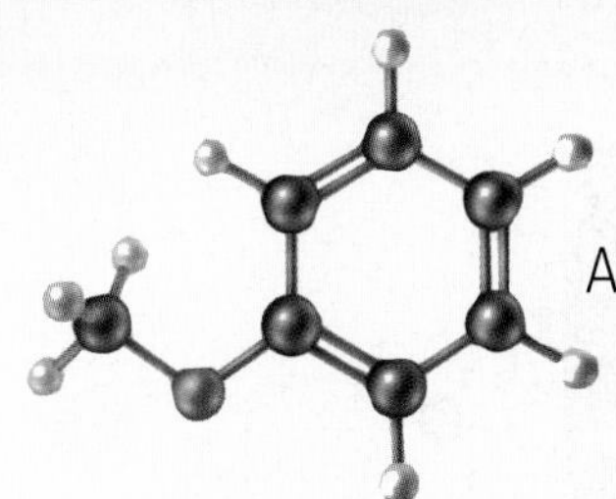

# Everyday Chemistry

## An Every-Moment Type of Chemical Reaction

There is one type of chemical reaction that affects your life each and every day, no matter who you are, no matter where you live, from the day you are born to the day you die. This reaction type is the one found in the preceding section, the complete oxidation, or burning, of organic compounds.

If you live in an urban area, your home is probably heated with natural gas, which is mostly methane. The reaction that occurs in your furnace is

$$CH_4(g) + 2\ O_2(g) \rightarrow CO_2(g) + 2\ H_2O(g) + \text{heat}$$

If you live in a rural area where there are no natural gas pipelines, you probably use liquid propane as a source of energy:

$$C_3H_8(\ell) + 5\ O_2(g) \rightarrow 3\ CO_2(g) + 4\ H_2O(g) + \text{heat}$$

Even in the summer when you don't use your furnace, you probably use one of these reactions to heat the water in your hot-water heater. Both reactions are the oxidation of an organic compound.

You may ask, "What about homes that use electric heat?" Well, it's likely that the power used to generate the electricity was generated either by burning coal, a form of carbon, or by burning natural gas. However, if you live in the western United States, your electricity may be generated by dams, in which case no burning reaction is necessary. Nuclear power is also used to generate electricity without burning organic compounds.

Every time you drive a car, gasoline, diesel fuel, or, less commonly, other forms of carbon–hydrogen or carbon–hydrogen–oxygen compounds are oxidized. A major component in gasoline is octane, $C_8H_{18}$:

$$2\ C_8H_{18}(\ell) + 25\ O_2(g) \rightarrow 16\ CO_2(g) + 18\ H_2O(g) + \text{heat}$$

But what if you are far away from civilization, where there are no home heaters, hot-water heaters, or automobiles? Didn't we say that oxidization is an every-moment chemical reaction? Although there are many steps in what is overall a complex process, the energy you use to sustain your life is released through oxidation reactions. The overall simple form of the reaction by which humans convert blood sugar into energy is

$$C_6H_{12}O_6(aq) + 6\ O_2(g) \rightarrow 6\ CO_2(g) + 6\ H_2O(\ell) + \text{energy}$$

We breathe in—inhale—oxygen from the atmosphere, the source of one of the two reactants. This oxygen then reacts with the blood sugar, and we breathe out—exhale—carbon dioxide and some of the water formed, the products of the reaction.

It may be difficult to imagine that burning gasoline and metabolizing food occur via the same reaction type, but it is true. One major difference is that you can "burn" food in your body at 37°C only with the help of specialized molecules known as *enzymes*. We'll discuss these molecules, which are needed for low-temperature burning, in more detail in Chapter 22, Biochemistry. In the meantime, remember that burning reactions occur in you and around you throughout every moment of your life!

$$NaCl(aq) + AgNO_3(aq) \rightarrow NaNO_3(aq) + AgCl(s)$$

$$Na^+(aq) + Cl^-(aq) + Ag^+(aq) + NO_3^-(aq) \rightarrow Na^+(aq) + NO_3^-(aq) + AgCl(s)$$

$$Cl^-(aq) + Ag^+(aq) \rightarrow AgCl(s)$$

$Na^+(aq)$ and $NO_3^-(aq)$ are spectators in the total ionic equation. The reaction is illustrated in Active Figure 9.12.

Let's try another Active Example that has two interesting features at the end.

### Active Example 9.11

When solutions of silver chlorate and aluminum chloride are combined, silver chloride precipitates. Write the three equations.

Begin with the left-hand side of the conventional equation.

*continued*

Active Figure 9.12 Sodium chloride and silver nitrate solutions are mixed, producing a white precipitate of silver chloride. Sodium ions and nitrate ions remain in solution. **Watch this active figure at *http://www.cengage.com/cracolice.***

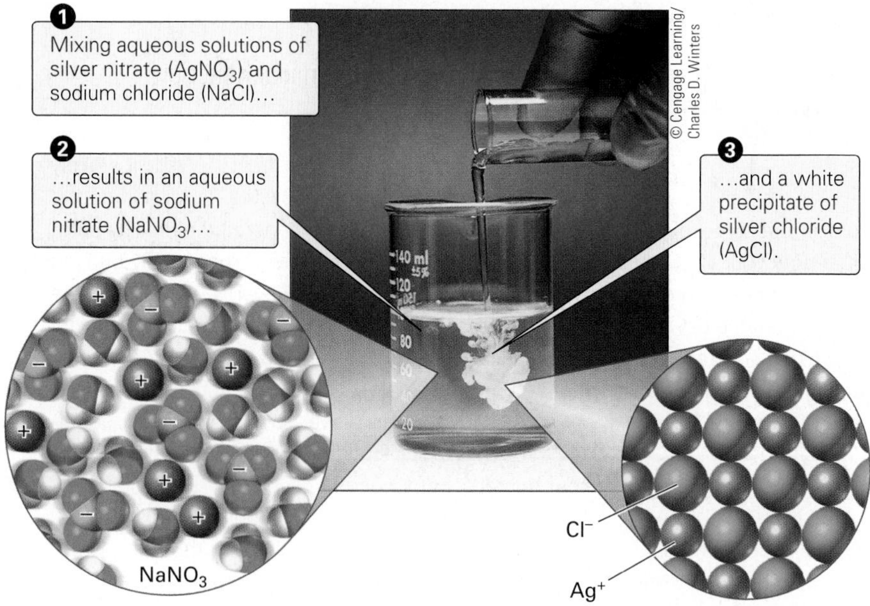

1
Mixing aqueous solutions of silver nitrate ($AgNO_3$) and sodium chloride (NaCl)...
2
...results in an aqueous solution of sodium nitrate ($NaNO_3$)...
3
...and a white precipitate of silver chloride (AgCl).
$NaNO_3$
$Cl^-$
$Ag^+$
© Cengage Learning/ Charles D. Winters

$AgClO_3(aq) + AlCl_3(aq) \rightarrow$

Now proceed as usual. When you get to the net ionic equation, which will have something new, see if you can figure out what to do about it. If necessary, reread *Step 3* of the procedure in Section 9.4.

$$3\ AgClO_3(aq) + AlCl_3(aq) \rightarrow 3\ AgCl(s) + Al(ClO_3)_3(aq)$$

$$3\ Ag^+(aq) + 3\ ClO_3^-(aq) + Al^{3+}(aq) + 3\ Cl^-(aq) \rightarrow 3\ AgCl(s) + Al^{3+}(aq) + 3\ ClO_3^-(aq)$$

$$3\ Ag^+(aq) + 3\ Cl^-(aq) \rightarrow 3\ AgCl(s)$$

$$Ag^+(aq) + Cl^-(aq) \rightarrow AgCl(s)$$

The net ionic equation this time has 3 for the coefficient of all species. *Step 3* of the procedure says to reduce coefficients to lowest terms. The equation may be divided by 3, as shown. This leads to the second interesting feature. . . .

Active Examples 9.10 and 9.11 produced the same net ionic equation, even though the reacting solutions were completely different. Actually, only the spectators were different, but they are not part of the chemical change. Eliminating them shows that both reactions are exactly the same. It will be this way whenever a solution containing $Ag^+(aq)$ ion is added to a solution containing $Cl^-(aq)$ ion. Therefore, if asked to write the net ionic equation for the reaction between such solutions, you can go directly to $Ag^+(aq) + Cl^-(aq) \rightarrow AgCl(s)$. (It is possible that a second reaction may occur between the other pair of ions, yielding a second net ionic equation.)

This simple and direct procedure may be used to write the net ionic equation for the precipitation of any insoluble ionic compound. The compound is the product, and the reactants are the ions in the compound. Try it on the following Active Example.

## Active Example 9.12

Write the net ionic equations for the precipitations of the following from aqueous solutions:

CuS:

$Mg(OH)_2$:

$Li_3PO_4$:

$Cu^{2+}(aq) + S^{2-}(aq) \rightarrow CuS(s)$

$Mg^{2+}(aq) + 2\ OH^{-}(aq) \rightarrow Mg(OH)_2(s)$

$3\ Li^{+}(aq) + PO_4^{3-}(aq) \rightarrow Li_3PO_4(s)$

If we knew in advance which combinations of ions yield insoluble compounds, we could predict precipitation reactions. These compounds have been identified in the laboratory. Table 9.3 shows the results of such experiments for a large number of ionic compounds. Their solubilities have been summarized in a set of "solubility guidelines" that your instructor may ask you to memorize. These guidelines are in Active Figure 9.13.

In the remaining Active Examples in this section and for the end-of-chapter questions, use the table suggested by your instructor—either Table 9.3 or Active Figure 9.13 (or memorize the solubility guidelines in Active Figure 9.13, if required)—to predict precipitation reactions.

**Table 9.3 Solubilities of Ionic Compounds***

| Ions | Acetate | Bromide | Carbonate | Chlorate | Chloride | Fluoride | Hydrogen Carbonate | Hydroxide | Iodide | Nitrate | Nitrite | Phosphate | Sulfate | Sulfide | Sulfite |
|---|---|---|---|---|---|---|---|---|---|---|---|---|---|---|---|
| **Aluminum** | s | aq | | aq | aq | s | | s | — | aq | | s | aq | — | |
| **Ammonium** | aq | aq | aq | aq | aq | aq | aq | — | aq | aq | aq | aq | aq | aq | aq |
| **Barium** | aq | aq | s | aq | aq | s | | aq | aq | aq | aq | s | s | — | s |
| **Calcium** | aq | aq | s | aq | aq | s | | s | aq | aq | aq | s | s | — | s |
| **Cobalt(II)** | aq | aq | s | aq | aq | — | | s | aq | aq | | s | aq | s | s |
| **Copper(II)** | aq | aq | s | aq | aq | aq | | s | | aq | | s | aq | s | |
| **Iron(II)** | aq | aq | s | | aq | s | | s | aq | aq | | s | aq | s | s |
| **Iron(III)** | — | aq | | | aq | s | | s | aq | aq | | s | aq | — | |
| **Lead(II)** | aq | s | s | aq | s | s | | s | s | aq | aq | s | s | s | s |
| **Lithium** | aq | aq | aq | aq | aq | aq | aq | aq | aq | aq | aq | s | aq | aq | aq |
| **Magnesium** | aq | aq | s | aq | aq | s | | s | aq | aq | aq | s | aq | — | aq |
| **Nickel** | aq | aq | s | aq | aq | aq | | s | aq | aq | | s | aq | s | s |
| **Potassium** | aq | aq | aq | aq | aq | aq | aq | aq | aq | aq | aq | aq | aq | aq | aq |
| **Silver** | s | s | s | aq | s | aq | | — | s | aq | s | s | s | s | s |
| **Sodium** | aq | aq | aq | aq | aq | aq | aq | aq | aq | aq | aq | aq | aq | aq | aq |
| **Zinc** | aq | aq | s | aq | aq | aq | | s | aq | aq | | s | aq | s | s |

*Compounds having solubilities of 0.1 mole or more in 1 L of water at 20°C are listed as soluble (aq); if the solubility is less than 0.1 mole per liter of water, the compound is listed as insoluble (s). A dash (—) identifies an unstable species in aqueous solution, and a blank space indicates lack of data. In writing equations for reactions that occur in water solution, insoluble substances, shown by s in this table, have the state symbol of a solid, (s). Dissolved substances, aq in the table, are designated by (aq) in an equation.

**SILVER COMPOUNDS**

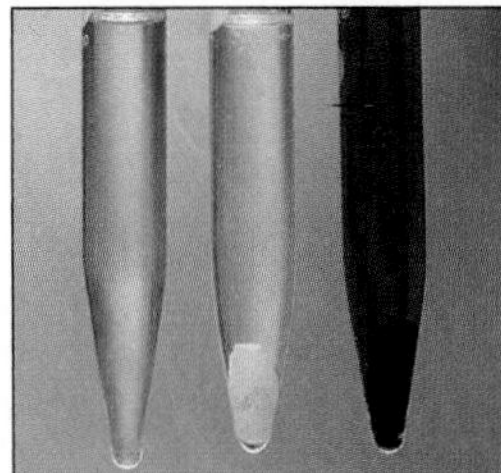


Nitrates are generally soluble, as are chlorides (except AgCl). Hydroxides are generally not soluble.

**SULFIDES**

Sulfides are generally not soluble (exceptions include salts with $NH_4^+$ and $Na^+$).

**HYDROXIDES**

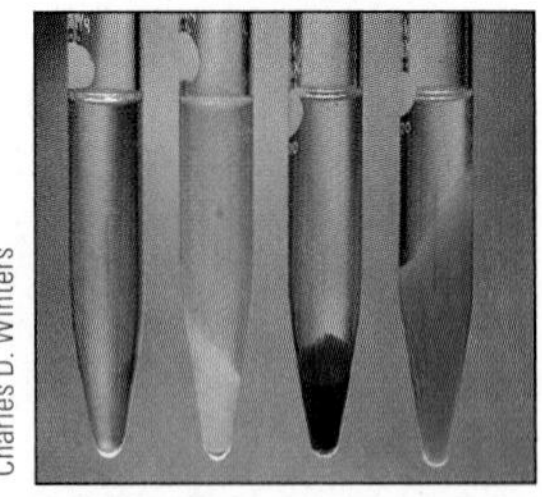


Hydroxides are generally not soluble except when the cation is a Group 1A/1 metal.



| SOLUBLE COMPOUNDS | EXCEPTIONS |
|---|---|
| Almost all salts of $Na^+$, $K^+$, $NH_4^+$ | |
| Salts of nitrate, $NO_3^-$<br>chlorate, $ClO_3^-$<br>perchlorate, $ClO_4^-$<br>acetate, $CH_3CO_2^-$ | Acetates of $Al^{3+}$, $Ag^+$ |
| Almost all salts of $Cl^-$, $Br^-$, $I^-$ | Halides of $Ag^+$, $Hg_2^{2+}$, $Pb^{2+}$ |
| Compounds containing $F^-$ | Fluorides of $Mg^{2+}$, $Ca^{2+}$, $Sr^{2+}$, $Ba^{2+}$, $Pb^{2+}$ |
| Salts of sulfate, $SO_4^{2-}$ | Sulfates of $Ca^{2+}$, $Sr^{2+}$, $Ba^{2+}$, $Pb^{2+}$ |

| INSOLUBLE COMPOUNDS | EXCEPTIONS |
|---|---|
| Most salts of carbonate, $CO_3^{2-}$<br>phosphate, $PO_4^{3-}$<br>oxalate, $C_2O_4^{2-}$<br>chromate, $CrO_4^{2-}$ | Salts of $NH_4^+$ and the alkali metal cations |
| Most metal sulfides, $S^{2-}$ | |
| Most metal hydroxides and oxides | $Ba(OH)_2$ is soluble |

**Active Figure 9.13** Solubility guidelines for ionic compounds. If a compound contains at least one of the ions in the Soluble Compounds list, apart from the exceptions listed, it is likely to be at least moderately soluble in water. Compounds with at least one ion in the Insoluble Compounds list are poorly soluble in water (again, with the exception of the few compounds in the Exceptions list). **Watch this active figure at *http://www.cengage.com/cracolice.***

## Active Example 9.13

Solutions of lead(II) nitrate and sodium bromide are combined. Write the net ionic equation for any precipitation reaction that may occur.

Start with the conventional equation. Leave blank the state symbols of the products for now.

$$Pb(NO_3)_2(aq) + 2\,NaBr(aq) \rightarrow PbBr_2(\quad) + 2\,NaNO_3(\quad)$$

You must determine if the compound is soluble or insoluble. If soluble, the state symbol should be (aq); if insoluble, (s). Fill in the state symbols in your equation above.

$Pb(NO_3)_2(aq) + 2\ NaBr(aq) \rightarrow PbBr_2(s) + 2\ NaNO_3(aq)$

In Table 9.3 the intersection of the lead(II)-ion line and the bromide-ion column shows that lead(II) bromide is insoluble; lead(II) bromide forms a *solid* precipitate, $PbBr_2(s)$. The intersection of the sodium-ion line and the nitrate-ion column shows that sodium nitrate is soluble in water; sodium nitrate remains in *aqueous* solution, $NaNO_3(aq)$. Both conclusions can also be reached from the solubility guidelines: Lead(II) bromide is one of the three insoluble bromides, and all sodium and nitrate compounds are soluble.

Complete the Active Example by writing the total ionic and net ionic equations.

$$Pb^{2+}(aq) + 2\ NO_3^-(aq) + 2\ Na^+(aq) + 2\ Br^-(aq) \rightarrow PbBr_2(s) + 2\ Na^+(aq) + 2\ NO_3^-(aq)$$

$$Pb^{2+}(aq) + 2\ Br^-(aq) \rightarrow PbBr_2(s)$$

You could choose to write the net ionic equation directly in Active Example 9.13. To do this, you would have to decide from the reactants, $Pb(NO_3)_2$ and NaBr, what the products would be in a double-replacement equation. Exchanging the ions gives $PbBr_2$ and $NaNO_3$. The solubility guidelines suggest that $PbBr_2$ is insoluble and $NaNO_3$ is soluble, so $PbBr_2$ will precipitate. Therefore, write the formula of that compound on the right and the formulas of its ions on the left, and you have the net ionic equation, $Pb^{2+}(aq) + 2\ Br^-(aq) \rightarrow PbBr_2(s)$.

## Active Example 9.14

Write the net ionic equation for any reaction that occurs between solutions of aluminum sulfate and calcium acetate. The formula of the acetate ion is $C_2H_3O_2^-$.

This time take it all the way. But be careful!

$$Al_2(SO_4)_3(aq) + 3\ Ca(C_2H_3O_2)_2(aq) \rightarrow 2\ Al(C_2H_3O_2)_3(s) + 3\ CaSO_4(s)$$

$$2\ Al^{3+}(aq) + 3\ SO_4^{2-}(aq) + 3\ Ca^{2+}(aq) + 6\ C_2H_3O_2^-(aq) \rightarrow 2\ Al(C_2H_3O_2)_3(s) + 3\ CaSO_4(s)$$

This time *both* new combinations of ions precipitate. There are no spectators, so the total ionic equation is the net ionic equation. Actually, two separate reactions are taking place at the same time:

$$Al^{3+}(aq) + 3\ C_2H_3O_2^-(aq) \rightarrow Al(C_2H_3O_2)_3(s)$$

$$Ca^{2+}(aq) + SO_4^{2-}(aq) \rightarrow CaSO_4(s)$$

# Everyday Chemistry

## Green Chemistry

*Green chemistry* is a phrase you'll probably hear with increasing frequency in the near future. It refers to carrying out chemical changes in a way that avoids environmentally hazardous substances. This can be accomplished by finding new ways of reacting chemicals, conducting reactions in environmentally friendly solvents, and incorporating more reactant atoms in the final product, eliminating waste.

A central issue in making chemistry "greener" is the efficiency of incorporating the reactant atoms into the product compound. Any atoms that are used in the reaction but are not part of the product are wasted, and this waste must subsequently be disposed of. If chemists can design reactions that do not have wasted atoms, there are no disposal issues and no potential for environmental impact. This type of efficiency can be measured in terms of *percentage atom utilization:*

$$\%\text{ atom utilization} = \frac{\text{molar mass of target product}}{\text{molar mass of all products}} \times 100$$

One well-known case of improving atom economy in the pharmaceutical drug industry occurred in the 1990s, when a new method of synthesizing ibuprofen (sold in generic form and under the brand names Advil, Medipren, Motrin, and Nuprin) was developed. The tried-and-true method was originally developed in the 1960s; it involved six steps and generated a significant quantity of waste. It has a percentage atom utilization of only about 40%! The newer green synthesis requires only three steps and has a 99% atom utilization, partly because one by-product can be recovered. About a quarter of the world's supply of ibuprofen is now produced via the green synthesis.

Another example of green chemistry is the development of hydrogen fuel cells. A typical fuel cell is designed to allow the exothermic reaction of hydrogen and oxygen without explosive combustion, producing water as its only product with 100% atom utilization. At the time this book was written, most automobile manufacturers had developed prototype passenger cars powered with hydrogen fuel cells.

Even though fuel cells are beginning to appear in cars, the primary problem with this technology is producing the hydrogen fuel itself. It is expensive; its production usually involves a pollution-generating process; and the hydrogen requires a large storage volume compared to gasoline. Research is ongoing to find solutions to these problems.

Green chemists continue to work on methods of making product compounds that use all of the reactant atoms. The conventional and net ionic equations for these reactions are usually classified as combination reactions, which were first introduced in Section 8.6. Research on safer chemicals, better reaction conditions, and more efficient reaction pathways continues to make chemical manufacturing more healthful for humans and more friendly to the environment.

## 9.8 Double-Replacement Molecule-Formation Reactions

**Goal 10** Given reactants for a double-replacement reaction that yield a molecular product, write the conventional, total ionic, and net ionic equation.

**P/REVIEW:** The neutralization reactions in Section 8.9 are between acids and hydroxide bases. The products are water and an ionic compound. The general equation is $HX + MOH \rightarrow HOH + MX$.

The reaction of an acid often leads to an ion combination that yields a **molecular product** instead of a precipitate. Except for the difference in the product, the equations are written in exactly the same way. Just as you had to recognize an insoluble product and not break it up in total ionic equations, you must now recognize a molecular product and not break it into ions. Water and weak acids are the two kinds of molecular products you will find.

**Neutralization** reactions are the most common molecular-product reactions.

### Active Example 9.15

Write the conventional, ionic, and net ionic equations for the reaction between solutions of hydrochloric acid and sodium hydroxide.

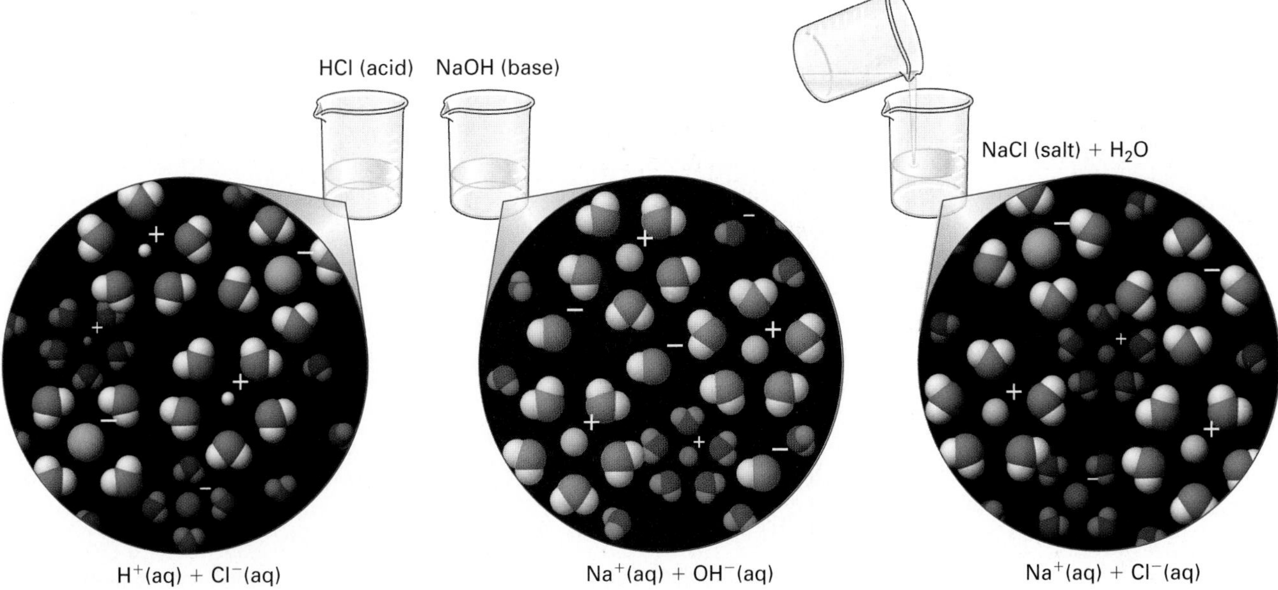


**Active Figure 9.14** The reaction between solutions of hydrochloric acid and sodium hydroxide. Hydrogen ions from one solution combine with hydroxide ions from the other to form water molecules. Sodium ions and chloride ions are spectators. **Watch this active figure at *http://www.cengage.com/cracolice.***

Write the conventional equation.

$HCl(aq) + NaOH(aq) \rightarrow HOH(\ell) + NaCl(aq)$

Now proceed just as you did for precipitation reactions. Watch your state designations.

$H^+(aq) + Cl^-(aq) + Na^+(aq) + OH^-(aq) \rightarrow HOH(\ell) + Na^+(aq) + Cl^-(aq)$

$H^+(aq) + OH^-(aq) \rightarrow H_2O(\ell)$

Water is the molecular product. It is not ionized, and it is in the liquid state. A particulate-level illustration of this reaction is presented in Active Figure 9.14.

The acid in a neutralization may be a weak acid. You must then recall that the major species in a weak acid solution is the weak acid molecule; it is not broken into ions.

## Active Example 9.16

Write the three equations leading to the net ionic equation for the reaction between acetic acid, $HC_2H_3O_2(aq)$, and a solution of sodium hydroxide.

*continued*

The conventional equation, please . . .

$HC_2H_3O_2(aq) + NaOH(aq) \rightarrow H_2O(\ell) + NaC_2H_3O_2(aq)$

Complete the Active Example.

$HC_2H_3O_2(aq) + Na^+(aq) + OH^-(aq) \rightarrow H_2O(\ell) + Na^+(aq) + C_2H_3O_2^-(aq)$

$HC_2H_3O_2(aq) + OH^-(aq) \rightarrow H_2O(\ell) + C_2H_3O_2^-(aq)$

The weak acid molecule appears in its molecular form in the net ionic equation.

When the reactants are a strong acid and the salt of a weak acid, that weak acid is formed as the molecular product. You must recognize it as a weak acid (not one of the seven strong acids) and leave it in molecular form in the total ionic and net ionic equations.

## Active Example 9.17

Develop the net ionic equation for the reaction between hydrochloric acid and a solution of sodium acetate. The formula of the acetate ion is $C_2H_3O_2^-(aq)$.

Write the conventional equation.

$HCl(aq) + NaC_2H_3O_2(aq) \rightarrow HC_2H_3O_2(aq) + NaCl(aq)$

Take it all the way to the net ionic equation.

$H^+(aq) + Cl^-(aq) + Na^+(aq) + C_2H_3O_2^-(aq) \rightarrow HC_2H_3O_2(aq) + Na^+(aq) + Cl^-(aq)$

$H^+(aq) + C_2H_3O_2^-(aq) \rightarrow HC_2H_3O_2(aq)$

Compare the reactions in Active Examples 9.15 and 9.17. The only difference between them is that Active Example 9.15 has the hydroxide ion as a reactant and Active Example 9.17 has the acetate ion as a reactant. In the first case the molecular product is water, formed when the hydrogen ion bonds to the hydroxide ion. In the second case the molecular product is acetic acid, a weak acid, formed when the hydrogen ion bonds to the acetate ion.

Just as you can write the net ionic equation for a precipitation reaction without the conventional and total ionic equations, so you can write the net ionic equation for a reaction that forms a molecule. Again, you must recognize the product from the formulas of the reactants. The acid will contribute a hydrogen ion to the molecular product. It will form a molecule with the anion from the other reactant. If the molecule is water or a weak acid, you have the reactants and product of the net ionic equation.

### Active Example 9.18

Without writing the conventional and total ionic equations, write the net ionic equations for each of the following pairs of reactants.

$H_2SO_4$ and LiOH: 

$KNO_2$ and HBr:

$H_2SO_4$ and LiOH: $H^+(aq) + OH^-(aq) \rightarrow H_2O(\ell)$

$KNO_2$ and HBr: $H^+(aq) + NO_2^-(aq) \rightarrow HNO_2(aq)$

$Li^+(aq)$ and $SO_4^{2-}(aq)$ are spectators in the first equation, a neutralization reaction. $HNO_2$ is recognized as a weak acid in the second reaction because it is not one of the seven strong acids. $K^+(aq)$ and $Br^-(aq)$ are spectators.

There are two points by which you can identify a molecular-product reaction: (1) One reactant is an acid, usually strong, and (2) one product is water or a weak acid. One of the most common mistakes in writing net ionic equations is failing to recognize a weak acid as a molecular product. If one reactant in a double-replacement equation is a strong acid, you can be sure there will be a molecular product. If it isn't water, look for a weak acid.

## 9.9 Double-Replacement Reactions That Form Unstable Products

**Goal** 11 Given reactants that form $H_2CO_3$, $H_2SO_3$, or "$NH_4OH$" by ion combination, write the net ionic equation for the reaction.

Three ion combinations yield molecular products that are not the products you would expect. Two of the expected products are carbonic and sulfurous acids. If hydrogen ions from one reactant react with carbonate ions from another, carbonic acid, $H_2CO_3$, should form:

$$2\ H^+(aq) + CO_3^{2-}(aq) \rightarrow H_2CO_3(aq)$$

But carbonic acid is unstable and decomposes to carbon dioxide gas and water. The correct net ionic equation is therefore

$$2\ H^+(aq) + CO_3^{2-}(aq) \rightarrow CO_2(g) + H_2O(\ell)$$

Sulfurous acid, $H_2SO_3$, decomposes in the same way to sulfur dioxide and water, but the sulfur dioxide remains in solution:

$$2\ H^+(aq) + SO_3^{2-}(aq) \rightarrow SO_2(aq) + H_2O(\ell)$$

The third ion combination that yields unexpected molecular products occurs when ammonium and hydroxide ions react:

$$NH_4^+(aq) + OH^-(aq) \rightarrow \text{“}NH_4OH\text{”}$$

In spite of the existence of printed labels, laboratory bottles with $NH_4OH$ etched on them, and wide use of the name ammonium hydroxide, no substance having the formula "$NH_4OH$" exists at ordinary temperatures. The actual product is a solution of ammonia molecules, $NH_3(aq)$. The proper net ionic equation is therefore

$$NH_4^+(aq) + OH^-(aq) \rightarrow NH_3(aq) + H_2O(\ell)$$

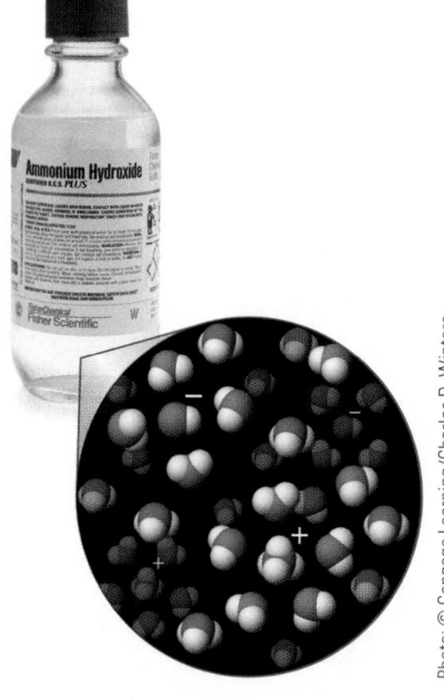

Photo: © Cengage Learning/Charles D. Winters

**Figure 9.15** Aqueous ammonia at the macroscopic and particulate levels. In spite of the fact that bottles are labeled with the name *ammonium hydroxide* ($NH_4OH$), the major species in solution are ammonia molecules and water molecules [$NH_4OH \rightarrow NH_3(aq) + H_2O(\ell)$]. Ammonium ions and hydroxide ions are minor species.

The reaction is reversible and yields a solution in which $NH_3$ is the major species and $NH_4^+$ and $OH^-$ are minor species (Fig. 9.15).

There is no system by which these three "different" molecular product reactions can be recognized. You simply must be alert to them and catch them when they appear. Once again, the predicted but unstable formulas are $H_2CO_3$, $H_2SO_3$, and $NH_4OH$.

## Active Example 9.19

Write the conventional, total ionic, and net ionic equations for the reaction between solutions of sodium carbonate and hydrochloric acid (Fig. 9.16).

Write the conventional equation.

$$Na_2CO_3(aq) + 2\ HCl(aq) \rightarrow 2\ NaCl(aq) + CO_2(g) + H_2O(\ell)$$

The standard double-replacement equation would predict NaCl and $H_2CO_3$ as the products of this reaction, but $H_2CO_3$ is one of the unstable ion combinations. We therefore break it down into its decomposition products, $CO_2$ and $H_2O$.

Complete the Active Example.

$$2\ Na^+(aq) + CO_3^{2-}(aq) + 2\ H^+(aq) + 2\ Cl^-(aq) \rightarrow 2\ Na^+(aq) + 2\ Cl^-(aq) + CO_2(g) + H_2O(\ell)$$

$$CO_3^{2-}(aq) + 2\ H^+(aq) \rightarrow CO_2(g) + H_2O(\ell)$$

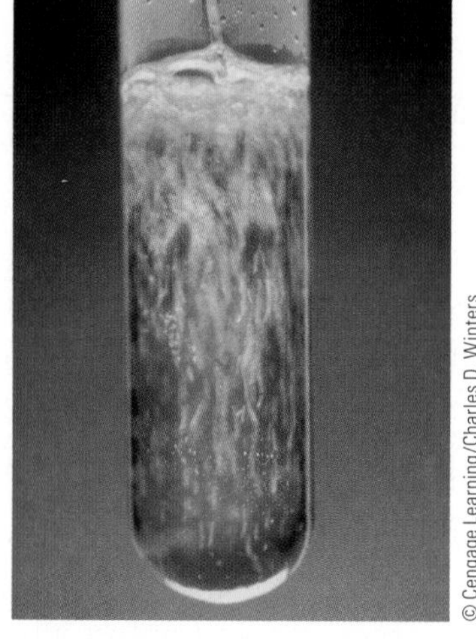

© Cengage Learning/Charles D. Winters

**Figure 9.16** The reaction between hydrochloric acid and a solution of sodium carbonate.

## 9.10 Double-Replacement Reactions with Undissolved Reactants

In every ion-combination reaction considered so far, it has been assumed that both reactants are in solution. This is not always the case. Sometimes the description of the reaction will indicate that a reactant is a solid, liquid, or gas, even though it may be soluble in water. In such a case, write the correct state symbol after the formula in the conventional equation and carry the formula through all three equations unchanged.

A common example of this kind of reaction occurs with a compound that is insoluble in water but reacts with acids. The net ionic equation shows why.

## Active Example 9.20

Write the net ionic equation to describe the reaction when solid aluminum hydroxide "dissolves" in hydrochloric acid and in nitric acid.

This is a neutralization reaction between a strong acid and a solid hydroxide, that is, one that is insoluble in water. Write the conventional equation for the aluminum hydroxide and hydrochloric acid reaction.

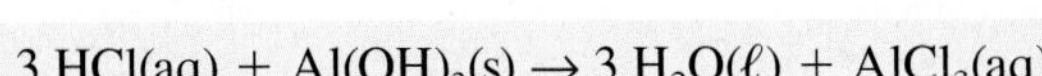

$3\ HCl(aq) + Al(OH)_3(s) \rightarrow 3\ H_2O(\ell) + AlCl_3(aq)$

Now write the total ionic and net ionic equations. Remember that you replace only dissolved substances, designated (aq), with the major species in their solutions.

$3\ H^+(aq) + 3\ Cl^-(aq) + Al(OH)_3(s) \rightarrow 3\ H_2O(\ell) + Al^{3+}(aq) + 3\ Cl^-(aq)$

$3\ H^+(aq) + Al(OH)_3(s) \rightarrow 3\ H_2O(\ell) + Al^{3+}(aq)$

Now think a bit before writing the net ionic equation for the reaction between solid aluminum hydroxide and nitric acid. The question asks only for the net ionic equation. Can you write it directly, without the conventional and net ionic equations? If not, write all three equations.

$3\ H^+(aq) + Al(OH)_3(s) \rightarrow 3\ H_2O(\ell) + Al^{3+}(aq)$

This is the same equation as for the reaction of solid aluminum hydroxide with hydrochloric acid. The chloride and nitrate ions are spectators in the two reactions.

## 9.11 Other Double-Replacement Reactions

### Active Example 9.21

Write the net ionic equation for any reaction that would occur if solutions of sodium chloride and potassium nitrate are combined.

$Na^+(aq) + Cl^-(aq) + K^+(aq) + NO_3^-(aq) \rightarrow NR$

The total ionic equation is

$$Na^+(aq) + Cl^-(aq) + K^+(aq) + NO_3^-(aq) \rightarrow Na^+(aq) + Cl^-(aq) + K^+(aq) + NO_3^-(aq)$$

Nothing says you will have a reaction *every* time you mix solutions of ionic compounds! There is no net ionic equation for this combination because there is no reaction.

**Table 9.4 Summary of Net Ionic Equations**

| Reactants (Conventional) | Reaction Type | Equation Type (Conventional) | Products (Conventional) | Reactants (Net Ionic) | Products (Net Ionic) |
|---|---|---|---|---|---|
| Element + salt *or* Element + strong acid | Oxidation–reduction | Single-replacement | Element + salt | Element + ion | Element + ion |
| Two salts *or* Salt + strong acid *or* Salt + hydroxide base | Precipitation | Double-replacement | Two salts | Two ions | Ionic precipitate |
| Strong acid + hydroxide base | Molecule-formation, ($H_2O$), neutralization | Double-replacement | Salt + $H_2O$ | $H^+ + OH^-$ | $H_2O$ |
| Weak acid + hydroxide base | Molecule-formation, ($H_2O$), neutralization | Double-replacement | Salt + $H_2O$ | Weak acid + $OH^-$ | $H_2O$ + anion from weak acid |
| Strong acid + salt of weak acid | Molecule-formation, (weak acid) | Double-replacement | Salt + weak acid | $H^+$ + anion of weak acid | Weak acid |
| Strong acid + carbonate *or* hydrogen carbonate | Unstable product + decomposition | Double-replacement + decomposition | Salt + $H_2O$ + $CO_2$<br>Salt + $H_2O$ + $CO_2$ | $H^+ + CO_3^{2-}$<br>$H^+ + HCO_3^-$ | $H_2O + CO_2$<br>$H_2O + CO_2$ |
| Strong acid + sulfite *or* hydrogen sulfite | Unstable product + decomposition | Double-replacement + decomposition | Salt + $H_2O$ + $SO_2$<br>Salt + $H_2O$ + $SO_2$ | $H^+ + SO_3^{2-}$<br>$H^+ + HSO_3^-$ | $H_2O + SO_2$<br>$H_2O + SO_2$ |
| Ammonium salt + hydroxide base | "$NH_4OH$" + decomposition | Double-replacement + decomposition | Salt + $NH_3$ + $H_2O$ | $NH_4^+ + OH^-$ | $H_2O + NH_3$ |

## 9.12 Summary of Net Ionic Equations

Table 9.4 summarizes this chapter. The blue area is essentially the same as the last three rows of Table 8.3, Section 8.10, the table that summarized writing conventional equations.

### Thinking About Your Thinking

Classification

**The classification thinking skill calls on your ability to change the way you see relationships among associated concepts when there is a better way of organizing them. Chapter 9 presents a new way of classifying chemical equations that includes the ideas from Chapter 8 but also moves beyond them. Table 9.4 is designed to help you to reorganize your mental arrangement of the relationships among types of equations. Notice, in particular, how the equation types from Chapter 8 fit into the new scheme.**

## Chapter 9 in Review

Most of the key terms and concepts and many others appear in the Glossary. Use your Glossary regularly.

**9.1 Electrolytes and Solution Conductivity**

Goal 1 Distinguish among strong electrolytes, weak electrolytes, and nonelectrolytes.

**Key Terms and Concepts: Conductor and nonconductor, electrodes, electrolysis, solution, strong and weak electrolyte, nonelectrolyte**

**9.2 Solutions of Ionic Compounds**

Goal 2 Given the formula of an ionic compound (or its name), write the formulas of the ions present when it is dissolved in water.

**9.3 Strong and Weak Acids**

Goal 3 Explain why the solution of an acid may be a good conductor or a poor conductor of electricity.

Goal 4 Given the formula of a soluble acid (or its name), write the major and minor species present when it is dissolved in water.

**Key Terms and Concepts: Acid, major species, minor species, seven common strong acids, strong acid, weak acid**

**9.4 Net Ionic Equations: What They Are and How to Write Them**
Goal 5 Distinguish among conventional, total ionic, and net ionic equations.

**Key Terms and Concepts: Conventional equation, net ionic equation, spectator or spectator ion, total ionic equation**

**9.5 Single-Replacement Oxidation–Reduction (Redox) Reactions**
Goal 6 Given two substances that may engage in a single-replacement redox reaction and an activity series by which the reaction may be predicted, write the conventional, total ionic, and net ionic equations for the reaction that will occur, if any.

**Key Terms and Concepts: Activity series**

**9.6 Oxidation–Reduction Reactions of Some Common Organic Compounds**
Goal 7 Write the equation for the complete oxidation or burning of any compound containing only carbon and hydrogen or only carbon, hydrogen, and oxygen.

**9.7 Double-Replacement Precipitation Reactions**
Goal 8 Predict whether a precipitate will form when known solutions are combined; if a precipitate forms, write the net ionic equation. (Reference to a solubility table or a solubility guidelines list may or may not be allowed.)
Goal 9 Given the product of a precipitation reaction, write the net ionic equation.

**Key Terms and Concepts: Ion-combination reaction, precipitate, precipitation reaction**

**9.8 Double-Replacement Molecule-Formation Reactions**
Goal 10 Given reactants for a double-replacement reaction that yield a molecular product, write the conventional, total ionic, and net ionic equation.

**Key Terms and Concepts: Molecular product, neutralization**

**9.9 Double-Replacement Reactions That Form Unstable Products**
Goal 11 Given reactants that form $H_2CO_3$, $H_2SO_3$, or "$NH_4OH$" by ion combination, write the net ionic equation for the reaction.

**9.10 Double-Replacement Reactions with Undissolved Reactants**

**9.11 Other Double-Replacement Reactions**

**9.12 Summary of Net Ionic Equations**

## Study Hints and Pitfalls to Avoid

Table 9.4 summarizes this chapter. It should be a focal point of your study of net ionic equations. The upper left corner of the table is taken from Table 8.3; it describes conventional equations. Be sure to see the connection between these equations and the expanded Table 9.4.

If a conventional equation, *including states,* can be written, it can be converted into a net ionic equation by the three steps discussed in Section 9.4. In order to write these three steps, you must understand how to change a conventional equation into a total ionic equation. This is the critical step in the procedure. The electrolyte-classification exercise that follows will help you master this concept.

You learned to check the activity series, check the solubility table or rules, or look for molecular products (water or a weak acid) to be sure there is a reaction. Look out for double reactions. Recognize unstable products ($H_2CO_3$, $H_2SO_3$, or "$NH_4OH$").

Several pitfalls await you in this chapter. Be careful of these:

1. Incorrect or missing states of reactants and products. Many incorrect formulas are written because students do not recognize the states of some species.
2. Failing to recognize weak acids as molecular products that *do not* separate into ions. This wins the title of "most common error in writing net ionic equations."
3. "Inventing" diatomic ions because there are two atoms of an element in a compound. $H_2^+$ is the most common error.
4. Insufficient practice. Writing net ionic equations is a learning-by-doing skill. Making mistakes and learning from those mistakes is the usual route to success. Students who complete *both* steps *before* the exam are happier students.

## Concept-Linking Exercises

*Write a brief description of the relationships among each of the following groups of terms or phrases. Answers to the Concept-Linking Exercises are given after the answer to the Target Check in Appendix III.*

1. Electrolyte, nonelectrolyte, strong electrolyte, weak electrolyte
2. Conventional equation, total ionic equation, net ionic equation, spectator ion
3. Redox reaction, ion-combination reaction, precipitation reaction, molecule-formation reaction, neutralization reaction

## Electrolyte-Classification Exercise

*A skill you need to write net ionic equations successfully is the ability to identify the major species in an aqueous solution. For each solute formula (in aqueous solution) written below, (a) classify it as a strong electrolyte, a weak electrolyte, or a nonelectrolyte; (b) identify the major species in solution; and (c) identify the minor species in solution (if any). The first two rows are completed as examples.*

| Formula | Electrolyte Classification | Major Species | Minor Species |
|---|---|---|---|
| 1. HI | Strong electrolyte | $H^+(aq)$, $I^-(aq)$ | HI(ag) |
| 2. $Na_2S$ | Strong electrolyte | $Na^+(aq)$, $S^{2-}(aq)$ | none |
| 3. $C_{12}H_{22}O_{11}$* | | | |
| 4. $HNO_2$ | | | |
| 5. HF | | | |
| 6. LiF | | | |
| 7. $HClO_4$ | | | |
| 8. $HCHO_2$ | | | |
| 9. $NH_4NO_3$ | | | |
| 10. $HC_2H_3O_2$ | | | |
| 11. HCl | | | |
| 12. $C_6H_{12}O_6$* | | | |

*This compound is one kind of sugar.

## Small-Group Discussion Questions

*Small-Group Discussion Questions are for group work, either in class or under the guidance of a leader during a discussion section.*

1. Draw particulate-level sketches of a strong electrolyte, a weak electrolyte, and a nonelectrolyte dissolved in water. Explain how each sketch fits the corresponding definition. Describe the conductivity of each solution. Write the formulas of at least three substances that fit each category.

2. Draw a particulate-level sketch of a tiny fraction of a grain of table salt dissolving in water. Start with separate salt and water, then show the salt dissolving, and finally, illustrate the salt water solution.

3. Draw particulate-level sketches of solutions of a strong acid and a weak acid. Based on your sketches, how are the terms *strong* and *weak* used to differentiate between these two acid classifications? How can you use symbols to illustrate the difference between chemical equations illustrating the dissociation of strong and weak acids?

4. A silver-white strip of zinc is placed in a blue solution of copper(II) nitrate. Describe the macroscopic changes that occur as the products of the reaction form. Zinc ion is colorless in aqueous solution and finely divided copper appears black when wet. Draw a particulate-level sketch of this reaction. Explain how your macroscopic description is related to your particulate-level sketch.

5. Gasoline is a mixture of many different compounds. It is made by boiling and cooling crude oil to separate the components of crude oil into fractions, each of which represents a small range of boiling points. Small quantities of additives are mixed into the final product, which contains more than 150 different compounds. Major components include butane ($C_4H_{10}$), pentane ($C_5H_{12}$), hexane ($C_6H_{14}$), heptane ($C_7H_{16}$), toluene ($C_6H_5CH_3$), and xylene ($C_6H_4C_2H_6$). Write the equation for the complete oxidation of each of these components. Then write a procedure for writing and balancing hydrocarbon oxidation equations that could be used by a beginning chemistry student.

6. Cobalt(II) chloride and sodium carbonate solutions are combined, forming a precipitate. Explain how you can determine the identity of the precipitate. Draw particulate-level sketches of the separated solutions, and then illustrate the chemical change that occurs when the solutions are combined.

7. How are double-replacement molecule-formation reactions and double-replacement precipitation reactions similar? How are they different? Draw a particulate-level sketch of the formation of water from the combination solutions of two soluble ionic compounds. Compare this sketch with your sketch of the formation of a precipitate.

8. Copy Table 9.4 onto the board in your classroom. For each of the eight reaction types, give an example conventional, total ionic, and net ionic equation. Explain how each of your example reactions satisfies the description in each of the six columns in the table.

# Questions, Exercises, and Problems

*Blue-numbered questions are answered in Appendix III.* ■ *denotes problems assignable in OWL. In the first section of Questions, Exercises, and Problems, similar exercises and problems are paired in consecutive odd-even number combinations. Many questions in this chapter are written with the assumption that you have studied Chapter 6 and can write the required formulas from their chemical names. If this is not the case, we have placed a list of all chemical formulas needed to answer Chapter 9 questions at the end of the Questions, Exercises, and Problems.*

#### Section 9.1: Electrolytes and Solution Conductivity

1. How does a weak electrolyte differ from a nonelectrolyte?
2. Solid A is a strong electrolyte and solid B is a nonelectrolyte. State how these two compounds differ.
3. How can it be that all soluble ionic compounds are electrolytes but soluble molecular compounds may or may not be electrolytes?
4. Compare the passage of electricity through a wire and through a solution. What conclusion may you draw if a liquid is able to carry an electrical current?

#### Sections 9.2 and 9.3: Solutions of Ionic Compounds and Strong and Weak Acids

*Questions 5 through 12: Write the major species in the water solution of each substance given. All ionic compounds given are soluble.*

5. $(NH_4)_2SO_4$, $MnCl_2$
6. $Mg(NO_3)_2$, $FeCl_3$
7. $NiSO_4$, $K_3PO_4$
8. ■ HCN, $NaNO_3$, $HClO_4$
9. $HNO_3$, HBr
10. ■ HCl, $HCHO_2$, KI
11. $H_2C_4H_4O_4$, HF
12. ■ $NaClO_4$, HI, $HC_2H_3O_2$

#### Section 9.4: Net Ionic Equations: What They Are and How to Write Them

*Questions 13 through 18: For each reaction described, write the net ionic equation.*

13. A zinc chloride solution is mixed with a sodium phosphate solution, forming a precipitate of solid zinc phosphate and a sodium chloride solution.
14. ■ When aqueous solutions of nickel chloride and ammonium phosphate are combined, solid nickel phosphate and a solution of ammonium chloride are formed.
15. Solid iron metal is dropped into a solution of hydrochloric acid. Hydrogen gas bubbles out, leaving a solution of iron(III) chloride.
16. ■ When solid nickel metal is put into an aqueous solution of copper(II) sulfate, solid copper metal and a solution of nickel sulfate result.
17. Aqueous solutions of oxalic acid and sodium chloride form when solid sodium oxalate (oxalate ion, $C_2O_4^{2-}$) is sprinkled into hydrochloric acid.
18. ■ When aqueous solutions of sodium nitrite and hydrobromic acid are mixed, an aqueous solution of sodium bromide and nitrous acid results.

#### Section 9.5: Single-Replacement Oxidation–Reduction (Redox) Reactions

*Questions 19 through 24: For each pair of reactants, write the net ionic equation for any single-replacement redox reaction that may be predicted by Table 9.2 (Section 9.5). If no redox reaction occurs, write NR.*

19. $Cu(s) + Li_2SO_4(aq)$
20. ■ $Sn(s) + Zn(NO_3)_2(aq)$
21. $Ba(s) + HCl(aq)$
22. ■ $Mg(s) + CuSO_4(aq)$
23. $Ni(s) + CaCl_2(aq)$
24. ■ $Zn(s) + HNO_3(aq)$

#### Section 9.6: Oxidation–Reduction Reactions of Some Common Organic Compounds

*Questions 25 through 28: Write the equation for each reaction described.*

25. Propane, $C_3H_8$, a component of "bottled gas," is burned as a fuel in heating homes.
26. ■ Butane, $C_4H_{10}$, is burned as lighter fluid in disposable lighters.
27. Ethanol, $C_2H_5OH$, the alcohol in alcoholic beverages, is oxidized.
28. ■ Acetylene, $C_2H_2$, is burned in welding torches.

#### Section 9.7: Double-Replacement Precipitation Reactions

*Questions 29 through 36: For each pair of reactants given, write the net ionic equation for any precipitation reaction that may be predicted by Table 9.3 or Active Figure 9.13 (Section 9.7). If no precipitation reaction occurs, write NR.*

29. $Pb(NO_3)_2(aq) + KI(aq)$
30. ■ $Co(NO_3)_2(aq) + K_3PO_4(aq)$
31. $KClO_3(aq) + Mg(NO_2)_2(aq)$
32. ■ $Ba(NO_3)_2(aq) + Fe_2(SO_4)_3(aq)$
33. $AgNO_3(aq) + LiBr(aq)$
34. ■ $NiBr_2(aq) + KOH(aq)$
35. $ZnCl_2(aq) + Na_2SO_3(aq)$
36. ■ $NaNO_3(aq) + Ba(OH)_2(aq)$
37. Write the net ionic equations for the precipitation of each of the following insoluble ionic compounds from aqueous solutions: $PbCO_3$; $Ca(OH)_2$.
38. ■ Write the net ionic equations for the precipitation of each of the following insoluble ionic compounds from aqueous solutions: lead(II) sulfide, copper(II) carbonate.

#### Section 9.8: Double-Replacement Molecule-Formation Reactions

*Questions 39 through 44: For each pair of reactants given, write the net ionic equation for the molecule-formation reaction that will occur.*

39. $NaNO_2(aq) + HI(aq)$

40. ■ $KNO_2(aq) + HCl(aq)$

41. $KC_3H_5O_3(aq) + HClO_4(aq)$

42. ■ $NaOH(aq) + HI(aq)$

43. $RbOH(aq) + HF(aq)$

44. ■ $HC_2H_3O_2(aq) + Ba(OH)_2(aq)$

**Section 9.9: Double-Replacement Reactions That Form Unstable Products**

*Questions 45 through 48: For each pair of reactants given, write the net ionic equation for the reaction that will occur.*

45. $MgCO_3(aq) + HCl(aq)$

46. ■ $HCl(aq) + CaSO_3(s)$

47. $Na_2SO_3(aq) + HClO_3(aq)$

48. ■ $(NH_4)_2SO_4(aq) + NaOH(aq)$

**Section 9.12: Summary of Net Ionic Equations**

*The remaining questions include all types of reactions discussed in this chapter. Use the activity series and solubility guidelines to predict whether redox or precipitation reactions will take place. If a reaction will take place, write the net ionic equation; if not, write NR.*

49. Barium chloride and sodium sulfite solutions are combined in an oxygen-free atmosphere. (Sulfites follow the same solubility guidelines as sulfates.)

50. ■ Aqueous solutions of sodium phosphate and silver nitrate are combined.

51. Copper(II) sulfate and sodium hydroxide solutions are combined.

52. ■ Aqueous solutions of iron(III) sulfate and cobalt(II) nitrate are combined.

53. Bubbles appear as hydrochloric acid is poured onto solid magnesium carbonate.

54. ■ A piece of solid nickel metal is put into an aqueous solution of lead(II) nitrate.

55. Nitric acid appears to "dissolve" solid lead(II) hydroxide.

56. ■ Solutions of hydrobromic acid and potassium carbonate are combined.

57. Liquid methanol, $CH_3OH$, is burned.

58. ■ Aqueous solutions of sodium hydroxide and hydrochloric acid are combined.

59. Benzoic acid, $HC_7H_5O_2(s)$, is neutralized by sodium hydroxide solution.

60. ■ Aqueous solutions of sodium hydroxide and potassium nitrate are combined.

61. Nickel is placed into hydrochloric acid.

62. ■ A piece of solid copper metal is put into an aqueous solution of lead(II) nitrate.

63. Hydrochloric acid is poured into a solution of sodium hydrogen sulfite.

64. ■ Aqueous solutions of sodium cyanide (cyanide ion, $CN^-$) and hydrobromic acid are mixed.

65. Solutions of magnesium sulfate and ammonium bromide are combined.

66. ■ Aqueous solutions of sodium hydroxide and ammonium iodide are combined.

67. Magnesium ribbon is placed in hydrochloric acid.

68. ■ Aqueous solutions of barium hydroxide and calcium iodide are combined.

69. Solid nickel hydroxide is apparently readily "dissolved" by hydrobromic acid.

70. ■ A piece of solid lead metal is put into an aqueous solution of nitric acid.

71. Sodium fluoride solution is poured into nitric acid.

72. ■ Nitric acid and solid manganese(II) hydroxide are combined.

73. Silver wire is dropped into hydrochloric acid.

74. ■ Aqueous solutions of silver nitrate and nickel bromide are combined.

75. When solid lithium is added to water, hydrogen is released.

76. ■ Nitric acid and solid calcium sulfite are combined. (Sulfites follow the same solubility guidelines as sulfates.)

77. Aluminum shavings are dropped into a solution of copper(II) nitrate.

78. ■ Aqueous solutions of hydrofluoric acid and potassium hydroxide are combined.

**General Questions**

79. ■ Distinguish precisely, and in scientific terms, the differences among items in each of the following pairs or groups.
    a) Strong electrolyte, weak electrolyte
    b) Electrolyte, nonelectrolyte
    c) Strong acid, weak acid
    d) Conventional, total ionic, and net ionic equations
    e) Burn, oxidize
    f) Ion-combination, precipitation, and molecule-formation reactions
    g) Molecule-formation and neutralization reactions
    h) Acid, base, salt

80. Classify each of the following statements as true or false:
    a) The solution of a weak electrolyte is a poor conductor of electricity.
    b) Ions must be present if a solution conducts electricity.
    c) Ions are the major species in a solution of a soluble ionic compound.
    d) There are no ions present in the solution of a weak acid.
    e) Only seven "important" acids are weak.
    f) Hydrofluoric acid, which is used to etch glass, is a strong acid.
    g) Spectators are included in a net ionic equation.
    h) A net ionic equation for a reaction between an element and an ion is the equation for a single-replacement redox reaction.

i) A compound that is insoluble forms a precipitate when its ions are combined.

j) Precipitation and molecule-formation reactions are both ion-combination reactions having double-replacement conventional equations.

k) Neutralization is a special case of a molecule-formation reaction.

l) One product of a molecule-formation reaction is a strong acid.

m) Ammonium hydroxide is a possible product of a molecule-formation reaction.

n) Carbon is changed to carbon dioxide when a carbon-containing compound burns completely.

## Formulas

*These formulas are provided in case you are studying Chapter 9 before you study Chapter 6. You should not use this list unless your instructor has not yet assigned Chapter 6 or otherwise indicated that you may use the list.*

13. $ZnCl_2$, $Na_3PO_4$, $Zn_3(PO_4)_2$, NaCl
14. $NiCl_2$, $(NH_4)_3PO_4$, $Ni_3(PO_4)_2$, $NH_4Cl$
15. Fe, HCl, $H_2$, $FeCl_3$
16. Ni, $CuSO_4$, Cu, $NiSO_4$
17. $H_2C_2O_4$, NaCl, $Na_2C_2O_4$, HCl
18. $NaNO_2$, HBr, NaBr, $HNO_2$
38. PbS, $CuCO_3$
49. $BaCl_2$, $Na_2SO_3$
50. $Na_3PO_4$, $AgNO_3$
51. $CuSO_4$, NaOH
52. $Fe_2(SO_4)_3$, $Co(NO_3)_2$
53. HCl, $MgCO_3$
54. Ni, $Pb(NO_3)_2$
55. $HNO_3$, $Pb(OH)_2$
56. HBr, $K_2CO_3$
58. NaOH, HCl
59. NaOH
60. NaOH, $KNO_3$
61. Ni, HCl
62. Cu, $Pb(NO_3)_2$
63. HCl, $NaHSO_3$
64. NaCN, HBr
65. $MgSO_4$, $NH_4Br$
66. NaOH, $NH_4I$
67. Mg, HCl
68. $Ba(OH)_2$, $CaI_2$
69. $Ni(OH)_2$, HBr
70. Pb, $HNO_3$
71. NaF, $HNO_3$
72. $HNO_3$, $Mn(OH)_2$
73. Ag, HCl
74. $AgNO_3$, $NiBr_2$
75. Li, $H_2O$, $H_2$
76. $HNO_3$, $CaSO_3$
77. Al, $Cu(NO_3)_2$
78. HF, KOH

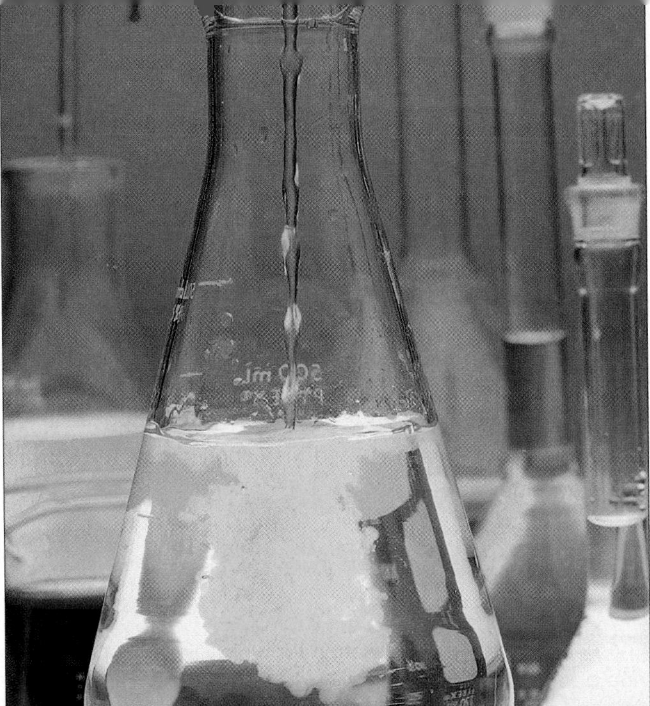

© Cengage Learning/Larry Cameron

This is the precipitation of lead(II) chromate, $PbCrO_4$, that occurs when a solution of sodium chromate, $Na_2CrO_4$, is added to a solution of lead(II) nitrate, $Pb(NO_3)_2$. The equation for the double-replacement precipitation reaction is $Na_2CrO_4(aq) + Pb(NO_3)_2(aq) \rightarrow PbCrO_4(s) + 2\ NaNO_3(aq)$. The question that this chapter asks—and answers—is, "How many grams of lead(II) chromate will precipitate if a specified mass of sodium chromate reacts?"

# 10 Quantity Relationships in Chemical Reactions

## CONTENTS



In this chapter you will learn how to solve **stoichiometry** problems. A stoichiometry problem asks, "How much or how many?" How many tons of sodium chloride must be electrolyzed to produce ten tons of sodium? How many kiloliters of chlorine at a certain temperature and pressure will be produced at the same time? How much energy is needed to do the job? These are only a few examples.

Several introductory comments may help you learn how to solve stoichiometry problems. The problem-solving strategy from Section 3.9 is used repeatedly. You will

Online homework for this chapter may be assigned in OWL

soon see that a series of *PER* relationships link the *GIVEN* and *WANTED* quantities in all problems in this chapter. Therefore, the problems are solved by dimensional analysis.

As usual, our solutions to Active Examples begin with the *PLAN*—identify *GIVEN* and *WANTED* quantities and the *PER/PATH*. We know you will not write out, "Given: so many grams of X," and so forth for every Active Example. But you should *think* the *GIVEN*, *WANTED*, and *PER/PATH*. If you cannot think those steps clearly, then write them. Sometimes when you write thoughts on paper, the missing parts in a thought process will present themselves.

Stoichiometry problems can become long; most unit paths have three steps and some have four or more. But the problems are not difficult if you recognize the unit path, know the conversion factor for each step, and then apply the factors correctly. That's the skill you should develop in this chapter.

## 10.1 Conversion Factors from a Chemical Equation

**Goal** 1 Given a chemical equation, or a reaction for which the equation is known, and the number of moles of one species in the reaction, calculate the number of moles of any other species.

In Section 8.4 we introduced the three levels on which a chemical equation may be interpreted: particulate, molar, and macroscopic. Let's briefly review the particulate and molar interpretations of an equation. Consider the equation

$$4\ NH_3(g) + 5\ O_2(g) \rightarrow 4\ NO(g) + 6\ H_2O(g) \qquad \textbf{(10.1)}$$

On the particulate level, this equation is read, "Four $NH_3$ molecules react with 5 $O_2$ molecules to produce 4 NO molecules and 6 $H_2O$ molecules." If we have 40 $NH_3$ molecules—10 times as many—how many $O_2$ molecules are required for the reaction? The answer is again ten times as many, or 50. But what if we have 308 $NH_3$ molecules? How many $O_2$ molecules are required for the reaction? This answer is less obvious.

The coefficients in a chemical equation give us the conversion factors to get from the number of particles of one substance to the number of particles of another substance in a reaction. For example, Equation 10.1 shows that 5 molecules of $O_2$ are needed to react with every 4 molecules of $NH_3$. In other words, the reaction uses 5 molecules of $O_2$ per 4 molecules of $NH_3$, or $\frac{5\ O_2\ \text{molecules}}{4\ NH_3\ \text{molecules}}$. This *PER* relationship is the conversion factor by which we may convert in either direction between molecules of $O_2$ and molecules of $NH_3$.

### Thinking About Your Thinking

#### Proportional Reasoning

**Each conversion factor from a chemical equation is a direct proportionality. Therefore, in the reaction $4\ NH_3(g) + 5\ O_2(g) \rightarrow 4\ NO(g) + 6\ H_2O(g)$, 5 mol $O_2 \propto$ 4 mol $NH_3$, 5 mol $O_2 \propto$ 4 mol NO, 5 mol $O_2 \propto$ 6 mol $H_2O$, etc.**

Let's return to the question we asked before: How many $O_2$ molecules are required to react with 308 $NH_3$ molecules? The answer to this question comes from applying the *PER* relationship:

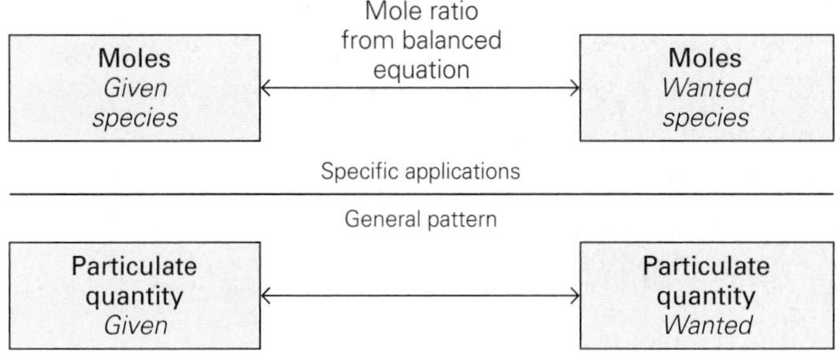


**Figure 10.1** Conversion between moles of one species and moles of another species in a chemical change. The mole ratio between the two species, as given in the balanced chemical equation, is the PER expression that links moles of one species to moles of another. The overarching pattern, illustrated by the two boxes at the bottom of the figure, is that one particulate quantity in a blue box is linked to another.

$$308 \text{ } \cancel{NH_3 \text{ molecules}} \times \frac{5 \text{ } O_2 \text{ molecules}}{4 \text{ } \cancel{NH_3 \text{ molecules}}} = 385 \text{ } O_2 \text{ molecules}$$

Unfortunately, we cannot count particles—at least, not directly. Instead, we measure masses or other macroscopic quantities. These can be converted into numbers of particles, grouped in moles. You already know how to do this conversion with mass, using molar mass as a conversion factor between mass (in grams) and number of particles (in moles) (see Section 7.5). Equation 10.1 can also be interpreted as, "Four *moles* of $NH_3$ molecules react with 5 *moles* of $O_2$ molecules to produce 4 *moles* of NO molecules and 6 *moles* of $H_2O$ molecules."

On the molar level, Equation 10.1 shows that the reaction uses 5 moles of $O_2$ per 4 moles of $NH_3$, or $\frac{5 \text{ mol } O_2}{4 \text{ mol } NH_3}$. Similarly, it takes 5 moles of $O_2$ to produce 4 moles of NO: $\frac{5 \text{ mol } O_2}{4 \text{ mol NO}}$. Five moles of oxygen also yield 6 moles of $H_2O$: $\frac{5 \text{ mol } O_2}{6 \text{ mol } H_2O}$. As always, the inverse of each conversion factor is also valid: $\frac{4 \text{ mol } NH_3}{5 \text{ mol } O_2}$, $\frac{4 \text{ mol NO}}{5 \text{ mol } O_2}$, and $\frac{6 \text{ mol } H_2O}{5 \text{ mol } O_2}$.

If you know the number of moles of any species in a reaction, either reactant or product, there is a one-step conversion to the moles of any other species, illustrated in Figure 10.1. If 3.20 moles of $NH_3$ react according to Equation 10.1, how many moles of $H_2O$ will be produced? The unit path is mol $NH_3 \rightarrow$ mol $H_2O$. Of the several mole relationships that are available in the equation, the one that satisfies the unit path is $\frac{6 \text{ mol } H_2O}{4 \text{ mol } NH_3}$:

$$3.20 \text{ } \cancel{\text{mol } NH_3} \times \frac{6 \text{ mol } H_2O}{4 \text{ } \cancel{\text{mol } NH_3}} = 4.80 \text{ mol } H_2O$$

Note that the 6 and 4 are exact numbers; they do not affect the significant figures in the final answer.

This method may be applied to any equation.

## Active Example 10.1

How many moles of oxygen are required to burn 2.40 moles of ethane, $C_2H_6(g)$?

First you need an equation for this burning reaction. You wrote equations for burning reactions in Section 9.6. Recall the unnamed second reactant and the two products that are always formed.

$2 \, C_2H_6(g) + 7 \, O_2(g) \rightarrow 4 \, CO_2(g) + 6 \, H_2O(\ell)$

Now PLAN the problem. Identify the GIVEN and WANTED quantities and the PER/PATH.

*continued*

GIVEN: 2.40 mol $C_2H_6$ WANTED: mol $O_2$

PER: $\frac{7 \text{ mol } O_2}{2 \text{ mol } C_2H_6}$

PATH: mol $C_2H_6$ ⟶ mol $O_2$

The conversion factor comes from the PER relationship between the GIVEN and WANTED quantities, using their coefficients in the equation.

Complete the problem.

$$2.40 \cancel{\text{mol } C_2H_6} \times \frac{7 \text{ mol } O_2}{2 \cancel{\text{mol } C_2H_6}} = 8.40 \text{ mol } O_2$$

## Active Example 10.2

Ammonia is formed directly from its elements. How many moles of hydrogen are needed to produce 4.20 moles of ammonia?

The procedure is exactly the same as in the previous Active Example. Complete the problem.

$N_2 + 3 H_2 \rightarrow 2 NH_3$

GIVEN: 4.20 mol $NH_3$ WANTED: mol $H_2$

PER: $\frac{3 \text{ mol } H_2}{2 \text{ mol } NH_3}$

PATH: mol $NH_3$ ⟶ mol $H_2$

$$4.20 \cancel{\text{mol } NH_3} \times \frac{3 \text{ mol } H_2}{2 \cancel{\text{mol } NH_3}} = 6.30 \text{ mol } H_2$$

## 10.2 Mass–Mass Stoichiometry

**Goal** 2 Given a chemical equation, or a reaction for which the equation can be written, and the number of grams or moles of one species in the reaction, find the number of grams or moles of any other species.

You are now ready to solve the problem that underlies the manufacture of chemicals and the design of many laboratory experiments: How much product will you get from a certain amount of raw material, or how much raw material will you need to obtain a specific amount of product? In solving this type of problem, you will tie together several skills:

*You will write chemical formulas (Chapter 6)* (if you are required to know nomenclature at this point in your study of chemistry).

*You will calculate molar masses from chemical formulas (Section 7.4).*

*You will use molar masses to change mass to moles and moles to mass (Section 7.5).*

*You will write and balance chemical equations (Chapters 8 and 9).*

*You will use the equation to change from moles of one species to moles of another (Section 10.1).*

And you will do all these things in one problem!

The preceding list should impress upon you how much solving stoichiometry problems depends on other skills. If you have any doubt about these skills, you will find it helpful to review the sections or chapters listed.

Before you can solve any stoichiometry problem, you must have the balanced equation for the reaction and the conversion factors between moles and quantities of *GIVEN* and *WANTED* substances. For convenience, we will use the expression "starting steps" to describe these items. Thus, "complete the starting steps" means to write and balance the equation, if it is not given, and determine the conversion factors. Molar mass is the conversion factor in this section; others will appear later.

After you complete the starting steps, the solution of a stoichiometry problem usually falls into a three-step "mass-to-mass" path. The **mass-to-mass** path is

$$\text{Mass of Given} \longrightarrow \text{Moles of Given} \longrightarrow \text{Moles of Wanted} \longrightarrow \text{Mass of Wanted} \qquad \textbf{(10.2)}$$

$$\text{Mass Given} \times \underbrace{\frac{\text{mol Given}}{\text{mass Given}}}_{\text{Step 1}} \times \underbrace{\frac{\text{mol Wanted}}{\text{mol Given}}}_{\text{Step 2}} \times \underbrace{\frac{\text{mass Wanted}}{\text{mol Wanted}}}_{\text{Step 3}} \qquad \textbf{(10.3)}$$

**P/REVIEW:** The mass-to-mass path expressed in Equation 10.2 will be expanded to include gases in Chapter 14 and again in Chapter 16 to include solutions. The method by which those problems are solved is the same as the method for solving the problems in this section. *Learn it now!* And you will be ready to use it later.

### Thinking About Your Thinking
Proportional Reasoning

**Each step in the mass-to-mass stoichiometry path involves a conversion based on a direct proportionality.**

In words, these three steps are as follows:

procedure

**How to Solve a Stoichiometry Problem: The Stoichiometry Path**

**Step 1:** Change the mass of the given species to moles (Section 7.5).

**Step 2:** Change the moles of the given species to moles of the wanted species (Section 10.1).

**Step 3:** Change the moles of the wanted species to mass (Section 7.5).

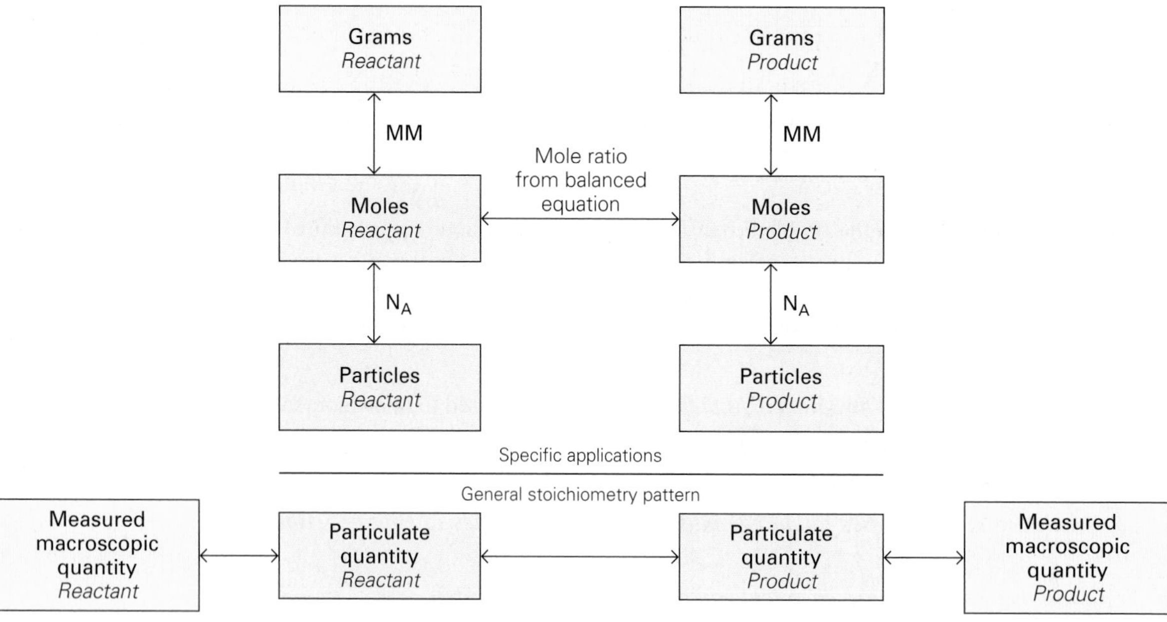


**Figure 10.2** Conversion among grams, moles, and number of particles for two species in a chemical change. This figure combines Figures 7.3 and 10.1. Molar mass, MM, is a PER expression that links mass in grams, a measurable macroscopic quantity, with moles, the number of particles in the sample. Avogadro's number, $N_A$, links the number of particles grouped in moles to the absolute number of particles. The mole ratio between the two species, as given in the balanced chemical equation, is the PER expression that links moles of one species to moles of another. The overarching pattern, illustrated by the four boxes at the bottom of the figure, is that a measured macroscopic quantity in a yellow box (e.g., mass in grams, pounds, etc.) can be converted to a particulate quantity in a blue box, and one particulate quantity can be converted to any other by using the mole ratio from the balanced chemical equation.

Occasionally you may be given the mass of one substance and asked to find the number of moles of a second substance. In this case, *Step 1* and *Step 2* of the mass-to-mass path complete the problem. You may also be given the moles of one substance and asked to find the mass of another. *Step 2* and *Step 3* solve this problem.

Figure 10.2 ties together the grams–moles–number-of-particle conversions from Figure 7.3, the mole–mole conversions from Figure 10.1, and the mass-to-mass stoichiometry path. The top four boxes illustrate the mass-to-mass stoichiometry path. If you tie in what you learned in Chapter 7, you see that you can calculate the number of particles of any species if you can calculate the number of moles of that species. The bottom of Figure 10.2 illustrates the general stoichiometry path. The yellow boxes symbolize macroscopic quantities, which are what we can measure in the laboratory. The blue boxes symbolize particulate quantities, which are the quantities that govern the nature of the chemical change.

The burning of ethane (Active Example 10.1) may be used to illustrate the stoichiometry path.

## Active Example 10.3

Calculate the number of grams of oxygen that are required to burn 155 g of ethane in the reaction $2\ C_2H_6(g) + 7\ O_2(g) \rightarrow 4\ CO_2(g) + 6\ H_2O(\ell)$.

***Solution***

First, we complete the starting steps for all stoichiometry problems. We write the reaction equation and calculate the molar mass or other conversion relationships between moles and the measured quantities of the substances in the problem. In this Active Example the equation is given. We must calculate the molar masses of the given and wanted substances. They are: $C_2H_6$: 2(12.01 g/mol) + 6(1.008 g/mol) = 30.07 g/mol; $O_2$: 2(16.00 g/mol) = 32.00 g/mol. The *PLAN* for the problem is

*GIVEN:* 155 g $C_2H_6$ *WANTED:* g $O_2$

*PER:* $\frac{1 \text{ mol } C_2H_6}{30.07 \text{ g } C_2H_6}$ $\frac{7 \text{ mol } O_2}{2 \text{ mol } C_2H_6}$ $\frac{32.00 \text{ g } O_2}{\text{mol } O_2}$

*PATH:* g $C_2H_6$ ⟶ mol $C_2H_6$ ⟶ mol $O_2$ ⟶ g $O_2$

In *Step 1* we change the mass of the given species to moles. The setup begins

$$155 \cancel{\text{g } C_2H_6} \times \frac{1 \text{ mol } C_2H_6}{30.07 \cancel{\text{g } C_2H_6}} \times \frac{\quad}{\quad} \times \frac{\quad}{\quad} =$$

If we calculated the answer to this point, we would have moles of $C_2H_6$. In *Step 2* the setup is extended to convert the moles of the given species to moles of the wanted species:

$$155 \cancel{\text{g } C_2H_6} \times \frac{1 \cancel{\text{mol } C_2H_6}}{30.07 \cancel{\text{g } C_2H_6}} \times \frac{7 \text{ mol } O_2}{2 \cancel{\text{mol } C_2H_6}} \times \frac{\quad}{\quad} =$$

If we calculated the answer to this point, we would have moles of $O_2$. In *Step 3* the moles of wanted oxygen are converted to mass in grams:

$$155 \cancel{\text{g } C_2H_6} \times \frac{1 \cancel{\text{mol } C_2H_6}}{30.07 \cancel{\text{g } C_2H_6}} \times \frac{7 \cancel{\text{mol } O_2}}{2 \cancel{\text{mol } C_2H_6}} \times \frac{32.00 \text{ g } O_2}{\cancel{\text{mol } O_2}} = 577 \text{ g } O_2$$

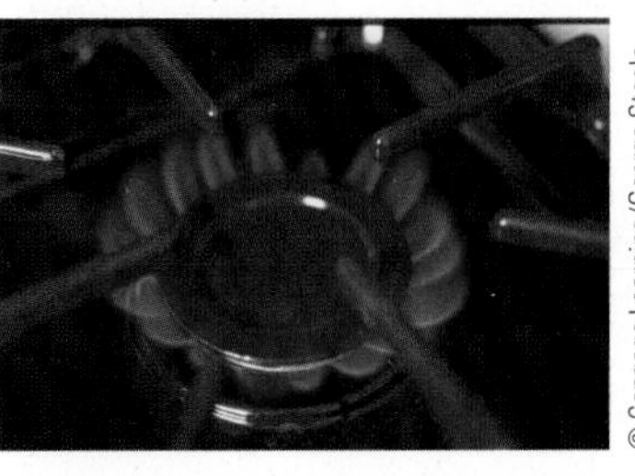

© Cengage Learning/George Steele

Natural gas is about 5 to 9% ethane. Therefore, this reaction occurs in many urban homes every time a gas furnace, a gas range, or a gas water heater is used.

## Active Example 10.4

How many grams of oxygen are required to burn 3.50 moles of liquid heptane, $C_7H_{16}(\ell)$?

Complete the starting steps. Be sure to read the problem carefully.

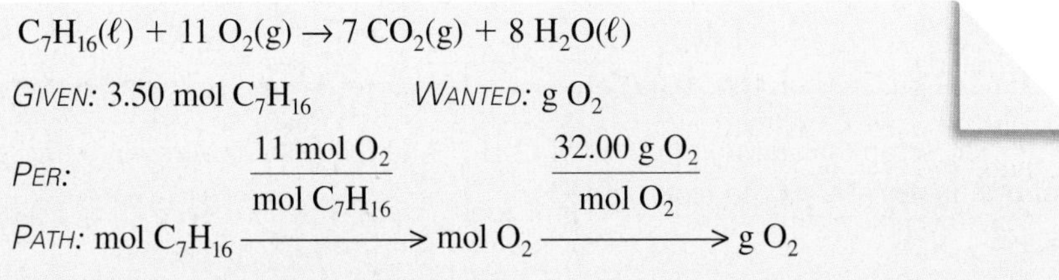

$C_7H_{16}(\ell) + 11\ O_2(g) \rightarrow 7\ CO_2(g) + 8\ H_2O(\ell)$

*GIVEN:* 3.50 mol $C_7H_{16}$ *WANTED:* g $O_2$

*PER:* $\frac{11 \text{ mol } O_2}{\text{mol } C_7H_{16}}$ $\frac{32.00 \text{ g } O_2}{\text{mol } O_2}$

*PATH:* mol $C_7H_{16}$ ⟶ mol $O_2$ ⟶ g $O_2$

The given quantity is *moles* of heptane. In other words, *Step 1* of the stoichiometry path is completed. The wanted quantity of $O_2$ is to be expressed in grams. Therefore, you need its molar mass for the mol → g conversion.

Set up the first step in the problem to change 3.50 moles of $C_7H_{16}$ to moles of $O_2$.

© Lynne Furrer, 2008/Used under license from Shutterstock.com

Heptane is present in gasoline, so the reaction in Active Examples 10.4 and 10.5 occurs in automobile engines.

*continued*

$$3.50 \text{ mol } C_7H_{16} \times \frac{11 \text{ mol } O_2}{\text{mol } C_7H_{16}} \times$$

Now extend the setup above to change moles of $O_2$ to grams and calculate the answer.

$$3.50 \text{ } \cancel{\text{mol } C_7H_{16}} \times \frac{11 \text{ } \cancel{\text{mol } O_2}}{\cancel{\text{mol } C_7H_{16}}} \times \frac{32.00 \text{ g } O_2}{\cancel{\text{mol } O_2}} = 1.23 \times 10^3 \text{ g } O_2$$

Now you will perform the complete three-step mass-to-mass stoichiometry path.

## Active Example 10.5

How many grams of $CO_2$ will be produced by burning 66.0 g $C_7H_{16}$ by the same reaction as in Active Example 10.4, $C_7H_{16}(\ell) + 11\ O_2(g) \rightarrow 7\ CO_2(g) + 8\ H_2O(\ell)$?

Begin as before with the starting steps.

*GIVEN:* 66.0 g $C_7H_{16}$ *WANTED:* g $CO_2$

*PER:* $\frac{1 \text{ mol } C_7H_{16}}{100.20 \text{ g } C_7H_{16}}$ $\frac{7 \text{ mol } CO_2}{\text{mol } C_7H_{16}}$ $\frac{44.01 \text{ g } CO_2}{\text{mol } CO_2}$

*PATH:* g $C_7H_{16}$ $\longrightarrow$ mol $C_7H_{16}$ $\longrightarrow$ mol $CO_2$ $\longrightarrow$ g $CO_2$

Set up, but do not calculate the answer for, *Step 1* in the unit path.

$$66.0 \text{ } \cancel{\text{g } C_7H_{16}} \times \frac{1 \text{ mol } C_7H_{16}}{100.20 \text{ } \cancel{\text{g } C_7H_{16}}} \times \underline{\qquad} \times \underline{\qquad} =$$

The setup thus far gives moles of $C_7H_{16}$. Add the next conversion to the setup above, moles of $C_7H_{16}$ to moles of $CO_2$.

$$66.0 \text{ } \cancel{\text{g } C_7H_{16}} \times \frac{1 \text{ } \cancel{\text{mol } C_7H_{16}}}{100.20 \text{ } \cancel{\text{g } C_7H_{16}}} \times \frac{7 \text{ mol } CO_2}{\cancel{\text{mol } C_7H_{16}}} \times \underline{\qquad} =$$

Now add the final conversion in the space provided above, changing moles of $CO_2$ to grams. Calculate the answer.

$$66.0 \text{ } \cancel{\text{g } C_7H_{16}} \times \frac{1 \text{ } \cancel{\text{mol } C_7H_{16}}}{100.20 \text{ } \cancel{\text{g } C_7H_{16}}} \times \frac{7 \text{ } \cancel{\text{mol } CO_2}}{\cancel{\text{mol } C_7H_{16}}} \times \frac{44.01 \text{ g } CO_2}{\cancel{\text{mol } CO_2}} = 203 \text{ g } CO_2$$

Sometimes it is more convenient to measure mass in larger or smaller units. Most analytical and academic chemists, for example, work in milligrams and millimoles of substance, whereas an industrial chemical engineer is more likely to think in kilograms and kilomoles. The mass-to-mole and mole-to-mass conversions in *Step 1* and *Step 3* are performed in exactly the same way. The conversion factor 44.01 g $CO_2$/mol $CO_2$ is equal to 44.01 kg $CO_2$/kmol $CO_2$:

$$\frac{44.01\ \cancel{g}\ CO_2}{\cancel{mol}\ CO_2} \times \frac{\cancel{1000\ mol}}{kmol} \times \frac{1\ kg}{\cancel{1000\ g}} = \frac{44.01\ kg\ CO_2}{kmol\ CO_2}$$

The factors of 1000 cancel in changing from units to kilounits. The same is true if milliunits are used. 44.01 mg $CO_2$/mmol $CO_2$ is equal to 44.01 g $CO_2$/mol $CO_2$.

## Active Example 10.6

How many milligrams of nickel chloride are in a solution if 503 mg of silver chloride is precipitated in the reaction of silver nitrate and nickel chloride solutions?

Start with the formulas of the reactants and products. Then write the equation for the reaction.

This reaction is used in the laboratory and in electroplating factories to find the nickel ion concentration of a solution.

$$2\ AgNO_3(aq) + NiCl_2(aq) \rightarrow 2\ AgCl(s) + Ni(NO_3)_2(aq)$$

Now set up and solve the problem completely. The setup is exactly the same as it would be with grams and moles, except that the unit g becomes mg and the unit mol becomes mmol.

GIVEN: 503 mg AgCl WANTED: mg $NiCl_2$

PER: $\frac{1\ mmol\ AgCl}{143.4\ mg\ AgCl}$ $\frac{1\ mmol\ NiCl_2}{2\ mmol\ AgCl}$ $\frac{129.59\ mg\ NiCl_2}{mmol\ NiCl_2}$

PATH: mg AgCl ⟶ mmol AgCl ⟶ mmol $NiCl_2$ ⟶ mg $NiCl_2$

$$503\ \cancel{mg\ AgCl} \times \frac{1\ \cancel{mmol\ AgCl}}{143.4\ \cancel{mg\ AgCl}} \times \frac{1\ \cancel{mmol\ NiCl_2}}{2\ \cancel{mmol\ AgCl}} \times \frac{129.59\ mg\ NiCl_2}{\cancel{mmol\ NiCl_2}} = 227\ mg\ NiCl_2$$

**Learn it Now!** You have reached a critical point in your study of chemistry. Students usually have no difficulty changing grams to moles, moles of one substance to moles of another, and moles to grams when those operations are presented as separate problems. When the operations are combined in a single problem, however, trouble sometimes follows. It doesn't help when, as in the following sections, many other ideas are added. In other words, you will not understand the following concepts unless you have already mastered this one.

Just studying the foregoing Active Examples will not give you the skill that you need. That comes only with practice. You should solve some end-of-chapter problems now—before you begin the next section. We strongly recommend that, at the very least, you solve Problems 1, 13, 17, 21, 23, and 27 at this time. If you understand and can solve these problems, you are ready to proceed to the next section.

## 10.3 Percent Yield

**Goal** 3 Given two of the following, or information from which two of the following may be determined, calculate the third: theoretical yield, actual yield, percent yield.

In Active Example 10.5 you found that burning 66.0 grams of $C_7H_{16}$ will produce 203 grams of $CO_2$. This is the **theoretical yield** (theo), the amount of product formed from the *complete* conversion of the given amount of reactant to product. Theoretical yield is always a calculated quantity, calculated by the principles of stoichiometry. In actual practice, factors such as impure reactants, incomplete reactions, and side reactions cause the **actual yield** (act) to be lower than the theoretical yield. The actual yield is a measured quantity, determined by experiment or experience.

If you know the actual yield and the theoretical yield, you can find the **percent yield.** This is the actual yield expressed as a percentage of theoretical yield. As with all percentages, it is the part quantity (actual yield) over the whole quantity (theoretical yield) times 100:

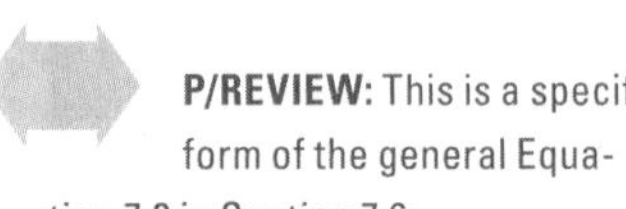

**P/REVIEW:** This is a specific form of the general Equation 7.2 in Section 7.6:

$$\% \text{ of A} = \frac{\text{parts of A}}{\text{total parts}} \times 100$$

When a part quantity and a whole quantity are both given, percentage is calculated by substitution into this equation.

$$\% \text{ yield} = \frac{\text{actual yield}}{\text{theoretical yield}} \times 100 \qquad \textbf{(10.4)}$$

If, in Active Example 10.5, only 181 grams of $CO_2$ had been produced, instead of the calculated 203 g, the percent yield would be

$$\% \text{ yield} = \frac{181 \text{ g}}{203 \text{ g}} \times 100 = 89.2\%$$

### Active Example 10.7

A solution containing 3.18 g of barium chloride is added to a second solution containing excess* sodium sulfate. Barium sulfate precipitates (Active Fig. 10.3). (a) Calculate the theoretical yield of barium sulfate. (b) If the actual yield is 3.37 g, calculate the percent yield.

What is the equation for the reaction?

*The word *excess* as used here means "more than enough," just as there are "more than enough" oxygen molecules to react with a compound burning in air. The atmosphere effectively provides an inexhaustible supply of oxygen molecules. In this case, there is "more than enough" sodium sulfate in the solution to precipitate all the barium in 3.18 grams of barium chloride.

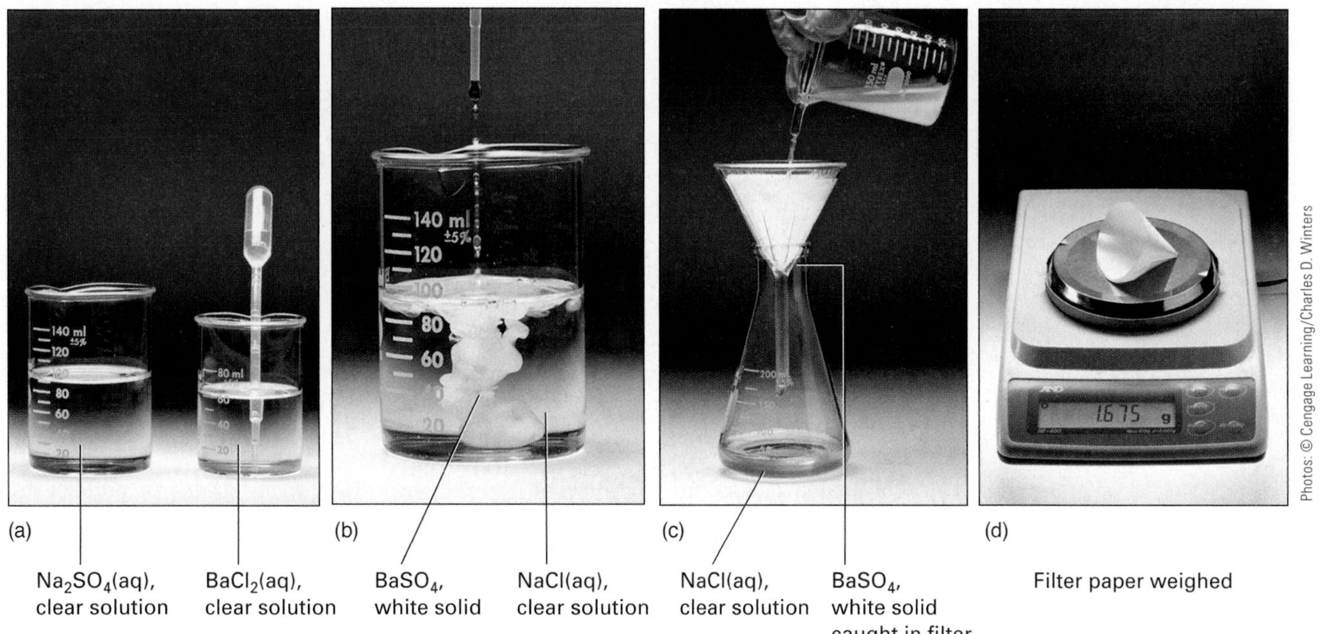


**Active Figure 10.3** A barium chloride solution is added to a sodium sulfate solution. (a) Separate clear, colorless solutions of sodium sulfate and barium chloride before mixing. (b) The barium chloride solution is added to the sodium sulfate solution. A white precipitate of barium sulfate forms. (c) Barium sulfate is separated from the solution by filtration. (d) The actual yield of barium sulfate is determined by drying the product and subtracting the mass of the filter paper from the combined mass of the filter paper and product. **Watch this active figure at *http://www.cengage.com/cracolice.***

$BaCl_2(aq) + Na_2SO_4(aq) \rightarrow BaSO_4(s) + 2\ NaCl(aq)$

Calculating theoretical yield is a typical stoichiometry problem. Solve part (a).

*GIVEN:* 3.18 g $BaCl_2$ *WANTED:* g $BaSO_4$

*PER:* $\frac{1\ mol\ BaCl_2}{208.2\ g\ BaCl_2}$ $\frac{1\ mol\ BaSO_4}{1\ mol\ BaCl_2}$ $\frac{233.4\ g\ BaSO_4}{mol\ BaSO_4}$

*PATH:* g $BaCl_2$ ⟶ mol $BaCl_2$ ⟶ mol $BaSO_4$ ⟶ g $BaSO_4$

$$3.18\ \cancel{g\ BaCl_2} \times \frac{1\ \cancel{mol\ BaCl_2}}{208.2\ \cancel{g\ BaCl_2}} \times \frac{1\ \cancel{mol\ BaSO_4}}{1\ \cancel{mol\ BaCl_2}} \times \frac{233.4\ g\ BaSO_4}{\cancel{mol\ BaSO_4}} = 3.56\ g\ BaSO_4$$

Part (b) is solved by substitution into Equation 10.4, $\%\ yield = \frac{actual\ yield}{theoretical\ yield} \times 100$.

*continued*

© Cengage Learning/Charles D. Winters

Barium sulfate is used in manufacturing photographic papers and linoleum and as a color pigment in wallpaper.

P/REVIEW: Percentage is used this way as a conversion factor in the solution of Active Example 7.11 in Section 7.6. This is a specific example of a conversion factor obtained from a defining equation and *PER* relationship, as introduced in Section 3.8.

*GIVEN:* 3.37 g (act); 3.56 g (theo) *WANTED:* % yield

EQUATION: $\% \text{ yield} = \frac{\text{actual yield}}{\text{theoretical yield}} \times 100 = \frac{3.37 \text{ g}}{3.56 \text{ g}} \times 100 = 94.7\%$

If you know the percent yield, you can use it as a dimensional-analysis conversion factor. Percent yield is grams actual yield per 100 grams theoretical yield. For example, assume that a manufacturer of magnesium hydroxide knows from experience that the percent yield is 81.3% from the production process. This can be used as either of two conversion factors, $\frac{81.3 \text{ g (act)}}{100 \text{ g (theo)}}$ or $\frac{100 \text{ g (theo)}}{81.3 \text{ g (act)}}$. How many grams of product are expected if the theoretical yield is 697 g? Using percent yield as a conversion between g (act) and g (theo), we obtain

$$697 \cancel{\text{g Mg(OH)}_2 \text{ (theo)}} \times \frac{81.3 \text{ g Mg(OH)}_2 \text{ (act)}}{100 \cancel{\text{g Mg(OH)}_2 \text{ (theo)}}} = 567 \text{ g Mg(OH)}_2 \text{ (act)}$$

Note that the g (theo) units cancel, leaving g (act) as the remaining unit.

## Thinking About Your Thinking

### Proportional Reasoning

**Thinking about percentage as a direct proportionality between the number of units in the percentage and 100 total units is a powerful thinking tool that you can use in a variety of problem-solving situations. In this section we translate 81.3% yield into the *PER* expressions $\frac{81.3 \text{ g (act)}}{100 \text{ g (theo)}}$ and $\frac{100 \text{ g (theo)}}{81.3 \text{ g (act)}}$. A number of other *PER* expressions can result from the same percent yield, such as $\frac{81.3 \text{ mg (act)}}{100 \text{ mg (theo)}}$, $\frac{81.3 \text{ kg (act)}}{100 \text{ kg (theo)}}$, or even $\frac{81.3 \text{ lb (act)}}{100 \text{ lb (theo)}}$. Your choice of units in the *PER* expression depends on the other units given in the problem statement.**

**In Chapter 7, we used percentage composition in the same way. For example, sodium chloride is 39.34% sodium. This can be interpreted as $\frac{39.34 \text{ g Na}}{100 \text{ g NaCl}}$, $\frac{100 \text{ g NaCl}}{39.34 \text{ g Na}}$, and through many other equivalent *PER* expressions with different mass units.**

**Remember that when you see a percentage in other contexts, you can always choose to write it as a *PER* expression in terms of units in the percentage per 100 total units.**

A more likely problem for this maker of magnesium hydroxide is finding out how much raw material is required to obtain a certain amount of product.

© Cengage Learning/Charles D. Winters

We began with 20 popcorn kernels and found that only 16 of them popped. The percent yield of popcorn from our "reaction" is (16 ÷ 20) × 100 = 80%. (Since we can count the kernels, this is *exactly* 80.000 . . .%.)

## Active Example 10.8

A manufacturer wants to prepare 800 kg of $Mg(OH)_2$ (assume three significant figures) by the reaction $MgO(s) + H_2O(\ell) \rightarrow Mg(OH)_2(s)$. Previous production experience shows that the process has an 81.3% yield calculated from the initial MgO. How much MgO should the manufacturer use?

***Solution***

We can choose to make this a two-part problem. We are given the actual yield. The stoichiometry problem for finding MgO must be based on theoretical yield. What

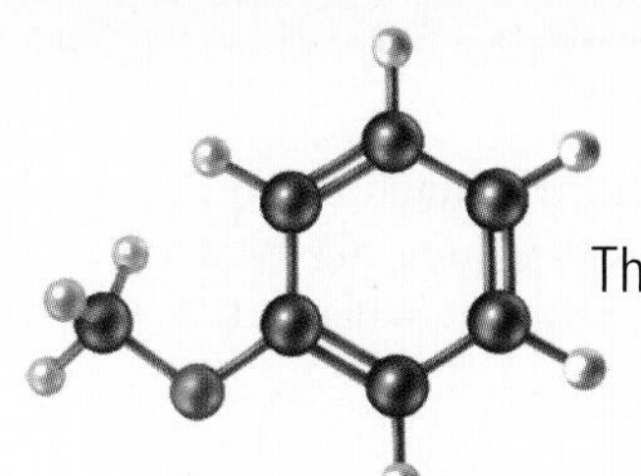

# Everyday Chemistry

## The Stoichiometry of $CO_2$ Emissions in Automobile Exhaust

A common concern of environmentalists, politicians, and most citizens of developed nations is the extent to which automobile exhaust contributes carbon dioxide gas to the atmosphere. There is some evidence that excessive $CO_2$ emissions can contribute to what is known as the greenhouse effect,* a process by which some atmospheric gases effectively store heat energy and increase the temperature of the earth relative to what it would be in the absence of these gases.

The greenhouse effect is necessary to support life on the earth as we know it. Atmospheric gases keep the temperature of the earth much warmer than it would be if the atmosphere was devoid of greenhouse gases. The question presently being debated is, "Are we overwarming the earth because of human-made industrial and automotive emissions of greenhouse gases, the most abundant of which is carbon dioxide?" This is a difficult question to answer, and many scientists are presently studying the effects of the byproducts of modern civilization on the earth's average temperature.

Basic stoichiometry, such as you are now learning, is one part of understanding how automotive emissions contribute to the $CO_2$ quantity in the atmosphere. We'll show you some very approximate calculations so that you can appreciate the amount of $CO_2$ emitted by cars. Let's make some assumptions about an average car: It gets 25 miles per gallon of gasoline (which we'll consider to be pure octane, $C_8H_{18}$, with a density of 0.7 g/mL) and it is driven an average of 50 miles per week. How much carbon dioxide does this car emit in a year? Let's see:

First, we need the reaction equation: $2\ C_8H_{18}(\ell) + 25\ O_2(g) \rightarrow 16\ CO_2(g) + 18\ H_2O(\ell)$. Now we'll determine the mass of $CO_2$ emitted by this single car in a year:

$$\begin{aligned}
&1\ \cancel{\text{year}} \times \frac{52\ \cancel{\text{weeks}}}{\cancel{\text{year}}} \times \frac{50\ \cancel{\text{miles}}}{\cancel{\text{week}}} \\
&\times \frac{1\ \cancel{\text{gal C}_8\text{H}_{18}}}{25\ \cancel{\text{miles}}} \times \frac{3.785\ \cancel{\text{L C}_8\text{H}_{18}}}{\cancel{\text{gal C}_8\text{H}_{18}}} \\
&\times \frac{1000\ \cancel{\text{mL C}_8\text{H}_{18}}}{\cancel{\text{L C}_8\text{H}_{18}}} \times \frac{0.7\ \cancel{\text{g C}_8\text{H}_{18}}}{\cancel{\text{mL C}_8\text{H}_{18}}} \\
&\times \frac{1\ \cancel{\text{mol C}_8\text{H}_{18}}}{114.22\ \cancel{\text{g C}_8\text{H}_{18}}} \times \frac{16\ \cancel{\text{mol CO}_2}}{2\ \cancel{\text{mol C}_8\text{H}_{18}}} \\
&\times \frac{44.01\ \cancel{\text{g CO}_2}}{\cancel{\text{mol CO}_2}} \times \frac{1\ \text{lb CO}_2}{453.6\ \cancel{\text{g CO}_2}} \\
&= 2 \times 10^3\ \text{lb CO}_2\ (\text{or } 8 \times 10^2\ \text{kg CO}_2)
\end{aligned}$$

That's 2000 pounds—one ton—of carbon dioxide emitted into the atmosphere each year by an "average" car. Now, how many cars are there in your city, your state, or in the entire world? How many exceed our 50-miles-per-week estimate? What about all of the trucks, buses, airplanes, ships, and other forms of mass transportation? There's no doubt that industrialization is putting a great deal more $CO_2$ into the atmosphere today than before the internal combustion engine was invented.

On the other hand, plants capture carbon dioxide from the atmosphere and convert it into their biomass and oxygen through photosynthesis. As long as we grow more vegetation and forests, any additional $CO_2$ produced can in turn be consumed. Scientists have calculated that a little more than an acre of trees will consume all of the $CO_2$ emissions one automobile driver can produce in a lifetime. Use this figure with caution, however, because the world's population is over six billion. Even if only one in a hundred people in the world drive a car, 60 million acres of trees—27 times the size of Yellowstone National Park—are needed to consume the $CO_2$. That's a huge number of forested acres that would need to be planted! You are also partly responsible for carbon dioxide emission from power plants and factories that produce the energy you use and the products you buy.

Efforts to understand the effects of $CO_2$ emissions on the atmosphere will no doubt continue throughout your lifetime. It is a complex problem, but our understanding of it continues to progress. The next time you read an article on global climate change or the greenhouse effect, keep in mind that basic stoichiometric concepts are at the heart of all the lengthy calculations in the various sophisticated mathematical models used in scientific studies.

*The term *global warming* refers to the gradual rise in the near-surface temperature of the earth that has been occurring over the past century. Scientists are presently debating whether this is due to natural influences or the greenhouse effect.

is our link between actual and theoretical yield? Percent yield gives us the conversion factor 81.3 g (act)/100 g (theo). First we will find the theoretical yield from the actual yield; then we will calculate the amount of reactant by stoichiometry.

The first part of the problem gives us

$$800\ \cancel{\text{kg Mg(OH)}_2\ \text{(act)}} \times \frac{100\ \text{kg Mg(OH)}_2\ \text{(theo)}}{81.3\ \cancel{\text{kg Mg(OH)}_2\ \text{(act)}}} = 984\ \text{kg Mg(OH)}_2\ \text{(theo)}$$

You may have been thinking about converting from kg to g, and then back to kg again. The setup shown here is shorter.

*continued*

David R. Frazier/Photo Researchers, Inc.

A water mixture of magnesium hydroxide is commonly known as milk of magnesia.

Now we are ready to use the mass-to-mass path to find the amount of MgO that is required to produce 984 kg $Mg(OH)_2$. The starting steps require the molar masses of $Mg(OH)_2$ and MgO as conversion factors. Once these are calculated, the *PLAN* and calculation follow:

*GIVEN:* 984 kg $Mg(OH)_2$ (theo) *WANTED:* kg MgO

*PER:* $\frac{1 \text{ kmol Mg(OH)}_2}{58.33 \text{ kg Mg(OH)}_2}$ $\frac{1 \text{ kmol MgO}}{1 \text{ kmol Mg(OH)}_2}$ $\frac{40.31 \text{ kg MgO}}{\text{mol MgO}}$

*PATH:* kg $Mg(OH)_2$ ⟶ kmol $Mg(OH)_2$ ⟶ kmol MgO ⟶ kg MgO

$$984 \text{ kg Mg(OH)}_2 \times \frac{1 \text{ kmol Mg(OH)}_2}{58.33 \text{ kg Mg(OH)}_2} \times \frac{1 \text{ kmol MgO}}{1 \text{ kmol Mg(OH)}_2} \times \frac{40.31 \text{ kg MgO}}{\text{kmol MgO}} = 6.80 \times 10^2 \text{ kg MgO}$$

This problem can be solved with a single calculation setup by adding the actual → theoretical conversion to the stoichiometry path. The 984 kg of $Mg(OH)_2$ that starts the preceding setup is replaced by the calculation that produced the number—the setup that changed actual yield to theoretical yield. The two unit paths

kg $Mg(OH)_2$ (act) → kg $Mg(OH)_2$ (theo) *and*
kg $Mg(OH)_2$ (theo) → kmol $Mg(OH)_2$ → kmol MgO → kg MgO

are combined to give

kg $Mg(OH)_2$ (act) → kg $Mg(OH)_2$ (theo) → kmol $Mg(OH)_2$ → kmol MgO → kg MgO

The calculation setup becomes

$$\underbrace{800 \text{ kg Mg(OH)}_2 \text{ (act)} \times \frac{100 \text{ kg Mg(OH)}_2 \text{ (theo)}}{81.3 \text{ kg Mg(OH)}_2 \text{ (act)}}}_{= 984 \text{ kg Mg(OH)}_2 \text{ (theo)}} \times \frac{1 \text{ kmol Mg(OH)}_2}{58.33 \text{ kg Mg(OH)}_2}$$

$$\times \frac{1 \text{ kmol MgO}}{1 \text{ kmol Mg(OH)}_2} \times \frac{40.31 \text{ kg MgO}}{\text{kmol MgO}} = 6.80 \times 10^2 \text{ kg MgO}$$

You may solve percent-yield problems such as this as two separate problems or as a single extended dimensional-analysis setup. Either way, note that percent *yield* refers to the *product,* not to a reactant. *Yield always refers to a product.* Accordingly, your percent-yield conversion should always be between actual and theoretical product quantities, not reactant quantities. Sometimes the conversion is at the beginning of the setup, as in Active Example 10.8, and sometimes it is at the end.

Sodium sulfate is used in dyeing textiles and in manufacturing glass and paper pulp.

## Active Example 10.9

A procedure for preparing sodium sulfate is summarized in the equation

$$2\ S(s) + 3\ O_2(g) + 4\ NaOH(aq) \rightarrow 2\ Na_2SO_4(aq) + 2\ H_2O(\ell)$$

The percent yield in the process is 79.8%. Find the number of grams of sodium sulfate that will be recovered from the reaction of 36.9 g NaOH.

The starting steps will help you *PLAN* your strategy for solving this problem. Our *PLAN* will be for a single dimensional-analysis setup from the given quantity to the wanted quantity.

GIVEN: 36.9 g NaOH    WANTED: g $Na_2SO_4$ (act)

PER: $\frac{1 \text{ mol NaOH}}{40.00 \text{ g NaOH}}$  $\frac{2 \text{ mol } Na_2SO_4}{4 \text{ mol NaOH}}$

PATH: g NaOH ⟶ mol NaOH ⟶

$\frac{142.05 \text{ g } Na_2SO_4}{\text{mol } Na_2SO_4}$  $\frac{79.8 \text{ g } Na_2SO_4 \text{ (act)}}{100 \text{ g } Na_2SO_4 \text{ (theo)}}$

mol $Na_2SO_4$ ⟶ g $Na_2SO_4$ (theo) ⟶ g $Na_2SO_4$ (act)

Write the calculation setup to the theoretical yield. If you choose to solve the problem in two steps, calculate the theoretical yield. If you are going to write a single setup for the whole problem, do not calculate that value.

$$36.9 \text{ g NaOH} \times \frac{1 \text{ mol NaOH}}{40.00 \text{ g NaOH}} \times \frac{2 \text{ mol } Na_2SO_4}{4 \text{ mol NaOH}} \times \frac{142.05 \text{ g } Na_2SO_4 \text{ (theo)}}{\text{mol } Na_2SO_4} \times \frac{\qquad}{\qquad} =$$

If you are solving this problem in two steps, your answer to this point—the theoretical yield—is 65.5 g $Na_2SO_4$.

Now multiply by the percent-yield conversion factor that changes theoretical yield to actual yield in the space above.

$$36.9 \text{ g NaOH} \times \frac{1 \text{ mol NaOH}}{40.00 \text{ g NaOH}} \times \frac{2 \text{ mol } Na_2SO_4}{4 \text{ mol NaOH}} \times \frac{142.05 \text{ g } Na_2SO_4 \text{ (theo)}}{\text{mol } Na_2SO_4}$$
$$\times \frac{79.8 \text{ g } Na_2SO_4 \text{ (act)}}{100 \text{ g } Na_2SO_4 \text{ (theo)}} = 52.3 \text{ g } Na_2SO_4$$

## 10.4 Limiting Reactants: The Problem

Goal **4** Identify and describe or explain limiting reactants and excess reactants.

How many pairs of gloves can you assemble from 20 left gloves and 30 right gloves? Picture the solution to this question in your mind. How many unmatched gloves will be left over? Which hand will these unmatched gloves fit? The answers are that 20 pairs of gloves can be assembled and that 10 right gloves will remain.

The same reasoning can be applied to a chemistry question. Carbon reacts with oxygen to form carbon dioxide: $C(g) + O_2(g) \rightarrow CO_2(g)$. Suppose you put three atoms of carbon and two molecules of oxygen in a reaction vessel and cause them to react until either carbon or oxygen is totally used up. How many carbon dioxide molecules will result? How many particles of which element will remain unreacted? Let's draw the answers to these questions:

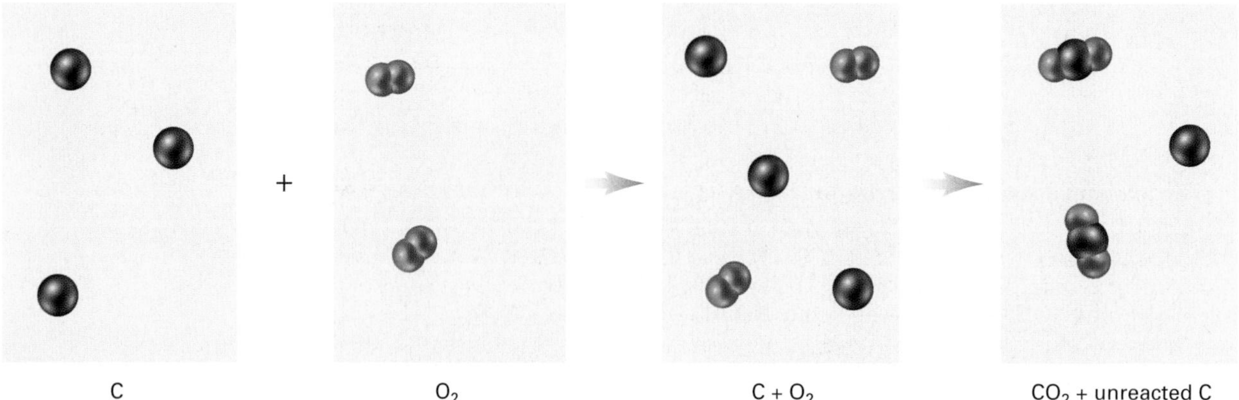


It is very unusual for reactants in a chemical change to be present in the exact quantities that will react completely with each other. This condition is approached, however, in the process of titration, which is considered in Sections 16.12 and 16.13.

The reactants combine in a one-to-one ratio. If you start with three atoms of carbon and two molecules of oxygen, the reaction will stop when the two oxygen molecules are used up. Oxygen, **the reactant that is completely used up by the reaction, is called the limiting reactant.** One atom of carbon, the **excess reactant,** will remain unreacted.

Since we cannot count individual particles, we again must turn to our macroscopic-particulate link, the mole. The amount of product is limited by the number of *moles* of the limiting reactant, and it must be calculated from that number of moles. If we start with 3 moles of carbon and 2 moles of oxygen, we have the same ratio of particles as when we start with three carbon atoms and two oxygen molecules. If 2 moles of the limiting reactant—oxygen—react, 2 moles of carbon dioxide are produced. According to the equation, each mole of carbon dioxide produced requires 1 mole of carbon to react.

The entire analysis on the molar level may be summarized as follows:

| | C | + | $O_2$ | $\rightarrow$ | $CO_2$ |
|---|---|---|---|---|---|
| Moles at start | 3 | | 2 | | 0 |
| Moles used (−) or produced (+) | −2 | | −2 | | +2 |
| Moles after the reaction | 1 | | 0 | | 2 |

## √ Target Check 10.1

*Consider the hypothetical reaction $2\ A + B \rightarrow A_2B$. The illustration below shows a mixture of particles before the reaction, where atoms of A are in gold and atoms of B are in black. Draw a particulate representation after a complete reaction. Explain your reasoning.*

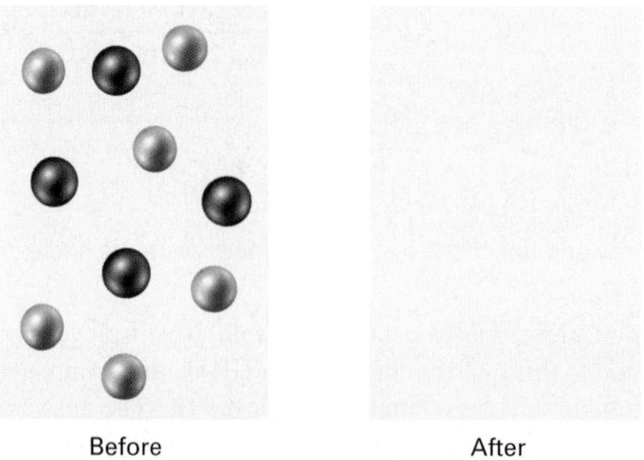

## Limiting Reactants: A Choice

In practice, you must approach analyzing and solving limiting-reactant problems with measurable units, usually mass units. There are two ways to solve these problems. In Section 10.5 we describe the comparison-of-moles method, which is a continuation of the analysis we have been doing. Section 10.6 presents the smaller-amount method, which solves the problem entirely in terms of mass units.

Please note: The two methods are *alternatives*. We recommend that you learn one or the other, not both. Your instructor will probably direct you by assigning either Section 10.5 or Section 10.6. If the choice is left to you, we suggest that you look at both briefly. Both will enable you to achieve the goal that heads each section. Learn the method that looks best to you, and disregard the other.

To help you choose, we will tell you the advantages of each method. The comparison-of-moles method (Section 10.5) is said to yield a better understanding of what happens to all substances present in the reaction. Gaining this understanding now makes it easier to understand more advanced ideas later. On the other hand, the smaller-amount method (Section 10.6) is said to be easier to learn because it requires solving no more than two or three separate stoichiometry problems by methods you already know.

As you proceed from this point, continue with Section 10.5 *or* Section 10.6.

# 10.5 Limiting Reactants: Comparison-of-Moles Method

**Goal** 5 Given a chemical equation, or information from which it may be determined, and initial quantities of two or more reactants, (a) identify the limiting reactant, (b) calculate the theoretical yield of a specified product, assuming complete use of the limiting reactant, and (c) calculate the quantity of the reactant initially in excess that remains unreacted.

The following Active Example shows you what to do when the reactants do not react in a 1:1 mole ratio. Preparing a table, as in the $C(g) + O_2(g) \rightarrow CO_2(g)$ example in the previous section, usually helps.

### Active Example 10.10

When powdered antimony (Z = 51) is sprinkled into iodine gas, antimony triiodide is produced: $2\ Sb + 3\ I_2 \rightarrow 2\ SbI_3$. If 0.167 mol Sb is introduced into a flask that holds 0.267 mol $I_2$, (a) how many moles of $SbI_3$ will be produced if the limiting reactant is completely used up and (b) how many moles of which element will remain unreacted?

***Solution***

| | 2 Sb | + | 3 $I_2$ | → | 2 $SbI_3$ |
|---|---|---|---|---|---|
| Moles at start | 0.167 | | 0.267 | | 0 |
| Moles used (−) or produced (+) | | | | | |
| Moles after the reaction | | | | | |

The limiting reactant is not easy to recognize this time. One way to identify it is to select either reactant and ask, "If Reactant A is the limiting reactant, how many moles of Reactant B are needed to react with all of Reactant A?" If the number of moles of Reactant B needed is more than the number available, Reactant B is the limiting reactant. If the number of moles of Reactant B needed is smaller than the number present, Reactant B is the excess reactant and Reactant A is the limiting reactant. The conclusion is the same regardless of which reactant you select as A and which you select as B. To illustrate:

*continued*

Iodine is a bluish-black lustrous solid at room conditions, as seen on the right. It readily sublimates—changes directly from solid to gas—to a blue-violet gas, as seen on the left.

If antimony is the limiting reactant, how many moles of iodine are required to react with all of the antimony?

$$0.167 \text{ \sout{mol Sb}} \times \frac{3 \text{ mol } I_2}{2 \text{ \sout{mol Sb}}} = 0.251 \text{ mol } I_2$$

The table shows that there are 0.267 mol $I_2$ present, more than enough to react with all of the antimony. Antimony is the limiting reactant.

If iodine is the limiting reactant, how many moles of antimony are required to react with all of the iodine?

$$0.267 \text{ \sout{mol I}}_2 \times \frac{2 \text{ mol Sb}}{3 \text{ \sout{mol I}}_2} = 0.178 \text{ mol Sb}$$

The table shows that there are 0.167 mol Sb present, not enough to react with all of the chlorine. Antimony is therefore the limiting reactant.

Having established 0.167 mole of antimony as the limiting reactant by either of the above assumptions, we can now complete the table:

| | 2 Sb | + | 3 $I_2$ | → | 2 $SbI_3$ |
|---|---|---|---|---|---|
| Moles at start | 0.167 | | 0.267 | | 0 |
| Moles used (−) or produced (+) | −0.167 | | −0.251 | | +0.167 |
| Moles after the reaction | 0 | | 0.016 | | 0.167 |

Do you recognize where the +0.167 mol $SbI_3$ came from? If the reaction uses all of the limiting reactant, 0.167 mol Sb, how many moles of $SbI_3$ will form? The moles of Sb and $SbI_3$ are on a 1:1 ratio—the equation shows 2 mol Sb and 2 mol $SbI_3$—so the moles of $SbI_3$ produced are equal to the moles of Sb used, 0.167 mol.

## Active Example 10.11

How many moles of the fertilizer saltpeter ($KNO_3$) can be made from 7.94 mol KCl and 9.96 mol $HNO_3$ by the reaction 3 KCl(s) + 4 $HNO_3$(aq) → 3 $KNO_3$(s) + $Cl_2$(g) + NOCl(g) + 2 $H_2O$(ℓ)? Also, how many moles of which reactant will be unused?

Potassium nitrate is the major component of tree stump remover. It is also used in making matches and explosives, including fireworks, gunpowder, and blasting powder.

Because $KNO_3$ is the only product asked about, the other products need not appear in your tabulation. Setting up the table effectively *PLANS* this kind of problem. Do that, filling in only the "Moles at start" line.

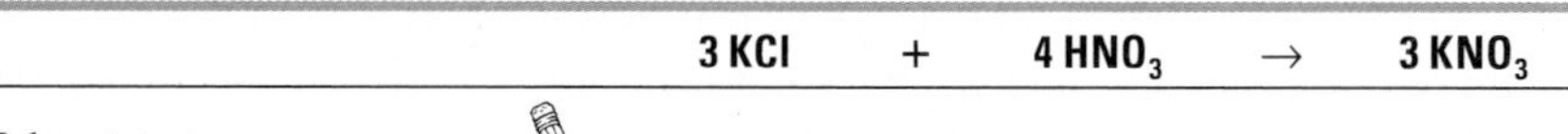

| | 3 KCl | + | 4 $HNO_3$ | → | 3 $KNO_3$ |
|---|---|---|---|---|---|
| Moles at start | | | | | |
| Moles used (−) or produced (+) | | | | | |
| Moles after the reaction | | | | | |

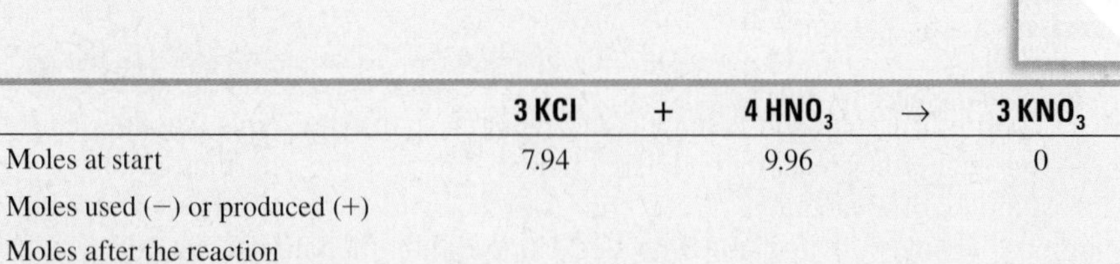

| | 3 KCl | + | 4 $HNO_3$ | → | 3 $KNO_3$ |
|---|---|---|---|---|---|
| Moles at start | 7.94 | | 9.96 | | 0 |
| Moles used (−) or produced (+) | | | | | |
| Moles after the reaction | | | | | |

Now identify the limiting reactant. Guess whether it is KCl or $HNO_3$. Test your guess by using the space below to calculate the amount of the other reactant needed to react with all of what you think is the limiting reactant, as it was done in Active

Example 10.10. When you are sure about the limiting reactant, insert the numbers in the second line of the table above.

This is how your calculations would have gone for either limiting reactant choice:

**Limiting reactant: KCl**

$$7.94 \text{ mol KCl} \times \frac{4 \text{ mol } HNO_3}{3 \text{ mol KCl}} = 10.6 \text{ mol } HNO_3$$

If KCl is the limiting reactant, 10.6 mol of $HNO_3$ is required to react with all of the KCl. Only 9.96 mol of $HNO_3$ is present. This is not enough $HNO_3$, so $HNO_3$ is the limiting reactant.

**Limiting reactant: $HNO_3$**

$$9.96 \text{ mol } HNO_3 \times \frac{3 \text{ mol KCl}}{4 \text{ mol } HNO_3} = 7.47 \text{ mol KCl}$$

If $HNO_3$ is the limiting reactant, 7.47 mol of KCl is required to react with all of the $HNO_3$. There is 7.94 mol of KCl present. This is more than enough KCl, so $HNO_3$ is the limiting reactant.

The second line of the table is therefore:

| | 3 KCl | + | 4 $HNO_3$ | → | 3 $KNO_3$ |
|---|---|---|---|---|---|
| Moles at start | 7.94 | | 9.96 | | 0 |
| Moles used (−) or produced (+) | −7.47 | | −9.96 | | +7.47 |
| Moles after the reaction | | | | | |

The number of moles of each species at the end is simply the algebraic sum of the moles at the start and the moles used or produced. Complete the problem by filling in the third line of the table above.

| | 3 KCl | + | 4 $HNO_3$ | → | 3 $KNO_3$ |
|---|---|---|---|---|---|
| Moles at start | 7.94 | | 9.96 | | 0 |
| Moles used (−) or produced (+) | −7.47 | | −9.96 | | +7.47 |
| Moles after the reaction | 0.47 | | 0 | | 7.47 |

7.47 mol $KNO_3$ is produced and 0.47 mol KCl remains unreacted.

Reactant and product quantities are not usually expressed in moles, but rather in grams. This adds one or more steps before and after the sequence in Active Example 10.11. We return to the antimony/iodine reaction to illustrate the process.

## Active Example 10.12

Calculate the mass of antimony(III) iodide, $SbI_3$, that can be produced by the reaction of 129 g antimony, Sb (Z = 51), and 381 g iodine. Also find the number of grams of the element that will be left.

You must have the equation for every stoichiometry problem. Start there.

*continued*

$2\ Sb + 3\ I_2 \rightarrow 2\ SbI_3$

This time the table begins with the starting *masses* of all species, rather than moles. You must convert the masses to moles. It is convenient to add a line for molar mass, too. The first two lines are completed for you. Fill in the third.

| | 2 Sb | + | 3 $I_2$ | → | 2 $SbI_3$ |
|---|---|---|---|---|---|
| Grams at start | 129 | | 381 | | 0 |
| Molar mass, g/mol | 121.8 | | 253.8 | | 502.5 |
| Moles at start | | | | | |

| | 2 Sb | + | 3 $I_2$ | → | 2 $SbI_3$ |
|---|---|---|---|---|---|
| Grams at start | 129 | | 381 | | 0 |
| Molar mass, g/mol | 121.8 | | 253.8 | | 502.5 |
| Moles at start | 1.06 | | 1.50 | | 0 |

The conversion from mass of each reactant to moles is the usual g → mol division by molar mass.

The work thus far has brought us to what was the starting point of Active Example 10.10, the moles of each reactant before the reaction begins. Extend the table through the next two lines to find the moles of product and excess reactant after the limiting reactant is used up.

| | 2 Sb | + | 3 $I_2$ | → | 2 $SbI_3$ |
|---|---|---|---|---|---|
| Grams at start | 129 | | 381 | | 0 |
| Molar mass, g/mol | 121.8 | | 253.8 | | 502.5 |
| Moles at start | 1.06 | | 1.50 | | 0 |
| Moles used (−), produced (+) | | | | | |
| Moles at end | | | | | |

| | 2 Sb | + | 3 $I_2$ | → | 2 $SbI_3$ |
|---|---|---|---|---|---|
| Grams at start | 129 | | 381 | | 0 |
| Molar mass, g/mol | 121.8 | | 253.8 | | 502.5 |
| Moles at start | 1.06 | | 1.50 | | 0 |
| Moles used (−), produced (+) | −1.00 | | −1.50 | | +1.00 |
| Moles at end | 0.06 | | 0 | | 1.00 |
| Grams at end | | | | | |

The final step is to change moles to grams by means of molar mass. Set up the calculation in the space below (you should be able to fill in two of the three columns without a setup), and then fill in the blanks in the table above to complete the problem.

$$0.06\ \cancel{\text{mol Sb}} \times \frac{121.8\ \text{g Sb}}{\cancel{\text{mol Sb}}} = 7\ \text{g Sb}$$

| | 2 Sb | + | 3 $I_2$ | → | 2 $SbI_3$ |
|---|---|---|---|---|---|
| Grams at start | 129 | | 381 | | 0 |
| Molar mass, g/mol | 121.8 | | 253.8 | | 502.5 |
| Moles at start | 1.06 | | 1.50 | | 0 |
| Moles used (−), produced (+) | −1.00 | | −1.50 | | +1.00 |
| Moles at end | 0.06 | | 0 | | 1.00 |
| Grams at end | 7 | | 0 | | 503 |

We are now ready to summarize the overall procedure for solving a limiting-reactant problem:

procedure

**How to Solve a Limiting-Reactant Problem**

1. Convert the number of grams of each reactant to moles.
2. Identify the limiting reactant.
3. Calculate the number of moles of each species that reacts or is produced.
4. Calculate the number of moles of each species that remains after the reaction.
5. Change the number of moles of each species to grams.

## 10.6 Limiting Reactants: Smaller-Amount Method

Goal 5 Given a chemical equation, or information from which it may be determined, and initial quantities of two or more reactants, (a) identify the limiting reactant, (b) calculate the theoretical yield of a specified product, assuming complete use of the limiting reactant, and (c) calculate the quantity of the reactant initially in excess that remains unreacted.

Limiting-reactant problems are more complicated when the quantities are expressed in measurable mass units. The "3 moles of carbon react with 2 moles of oxygen" question becomes 36.03 grams of carbon (3 moles) combines with 64.00 grams of oxygen (2 moles) until one is totally used up in the reaction $C(g) + O_2(g) \rightarrow CO_2(g)$. How many grams of carbon dioxide, $CO_2$, result? Also, how many grams of which element remain unreacted?

A type of reasoning frequently used in science involves "what if?" thinking. A scientist will *assume* that a particular condition or set of conditions is true and then calculate the answer based on this assumption. If the answer proves to be impossible—if it contradicts known facts—that answer is discarded and another is considered. In limiting-reactant problems, we need to identify the limiting reactant. We can answer this question by assuming that each reactant is limiting.

First, what if carbon is the limiting reactant? How many grams of $CO_2$ will be produced? In other words, if all the carbon is used up in the reaction, how many grams of $CO_2$ will be made? This is a fundamental three-step stoichiometry problem:

$$36.03\ \cancel{\text{g C}} \times \frac{1\ \cancel{\text{mol C}}}{12.01\ \cancel{\text{g C}}} \times \frac{1\ \cancel{\text{mol CO}_2}}{1\ \cancel{\text{mol C}}} \times \frac{44.01\ \text{g CO}_2}{\cancel{\text{mol CO}_2}} = 132.0\ \text{g CO}_2$$

Now, what if oxygen is the limiting reactant? How many grams of $CO_2$ will be produced by 64.00 g of $O_2$?

$$64.00 \text{ g } \cancel{O_2} \times \frac{1 \text{ } \cancel{\text{mol } O_2}}{32.00 \text{ } \cancel{\text{g } O_2}} \times \frac{1 \text{ } \cancel{\text{mol } CO_2}}{1 \text{ } \cancel{\text{mol } O_2}} \times \frac{44.01 \text{ g } CO_2}{\cancel{\text{mol } CO_2}} = 88.02 \text{ g } CO_2$$

We interpret these two results by saying that we have enough carbon to produce 132.0 g of $CO_2$. However, it is impossible to form more than 88.02 g $CO_2$ from only 64.00 g $O_2$. Therefore, oxygen is the limiting reactant, and the reaction stops when the limiting reactant is used up. *The limiting reactant is always the one that yields the smaller amount of product.*

To find the amount of excess reactant that remains, calculate how much of that reactant will be used by the entire amount of limiting reactant:

$$64.00 \text{ g } \cancel{O_2} \times \frac{1 \text{ } \cancel{\text{mol } O_2}}{32.00 \text{ } \cancel{\text{g } O_2}} \times \frac{1 \text{ } \cancel{\text{mol C}}}{1 \text{ } \cancel{\text{mol } O_2}} \times \frac{12.01 \text{ g C}}{\cancel{\text{mol C}}} = 24.02 \text{ g C}$$

We started with 36.03 g C. Reaction with the limiting reactant used up 24.02 g C. The amount that remains is the starting amount minus the amount used:

$$36.03 \text{ g C (initial)} - 24.02 \text{ g C (used)} = 12.01 \text{ g C left}$$

The smaller-amount method can be summarized as follows:

Notice that this procedure speaks of "amount" of reactant. In this section, we are expressing amount in terms of grams. It can be expressed in moles just as well. In fact, that is exactly how amount is described in Section 10.4, where the limiting-reactant concept was introduced.

**How to Solve a Limiting-Reactant Problem**

1. Calculate the amount of product that can be formed by the initial amount of each reactant.
   a) The reactant that yields the smaller amount of product is the limiting reactant.
   b) The smaller amount of product is the amount that will be formed when all of the limiting reactant is used up.
2. Calculate the amount of excess reactant that is used by the total amount of limiting reactant.
3. Subtract from the amount of excess reactant present initially the amount that is used by all of the limiting reactant. The difference is the amount of excess reactant that is left.

© Cengage Learning/Charles D. Winters

Iodine is a bluish-black lustrous solid at room conditions, as seen on the right. It readily sublimates—changes directly from solid to gas—to a blue-violet gas, as seen on the left.

## Active Example 10.13

Calculate the mass of antimony(III) iodide, $SbI_3$, that can be produced by the reaction of 129 g antimony, Sb (Z = 51), and 381 g iodine. Also find the number of grams of the element that will be left.

You must have the equation for every stoichiometry problem. Start there.

$$2 \text{ Sb} + 3 \text{ I}_2 \rightarrow 2 \text{ SbI}_3$$

What if antimony is the limiting reactant? How many grams of $SbI_3$ can be produced with 129 g Sb? Set up and solve the problem.

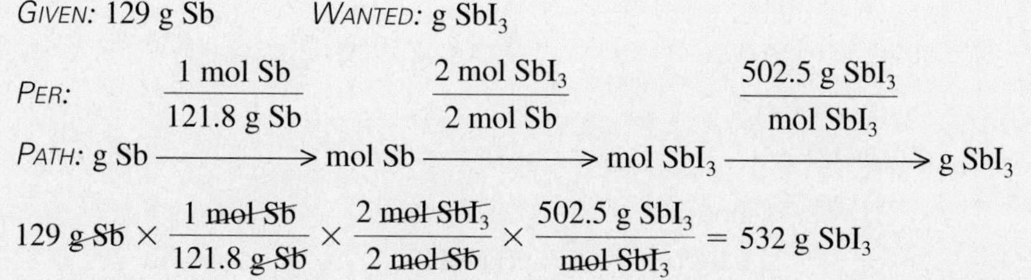

*Given:* 129 g Sb  *Wanted:* g $SbI_3$

*Per:* $\frac{1 \text{ mol Sb}}{121.8 \text{ g Sb}}$  $\frac{2 \text{ mol SbI}_3}{2 \text{ mol Sb}}$  $\frac{502.5 \text{ g SbI}_3}{\text{mol SbI}_3}$

*Path:* g Sb ⟶ mol Sb ⟶ mol $SbI_3$ ⟶ g $SbI_3$

$$129 \text{ g Sb} \times \frac{1 \text{ mol Sb}}{121.8 \text{ g Sb}} \times \frac{2 \text{ mol SbI}_3}{2 \text{ mol Sb}} \times \frac{502.5 \text{ g SbI}_3}{\text{mol SbI}_3} = 532 \text{ g SbI}_3$$

Now, what if the iodine is the limiting reactant? How many grams of $SbI_3$ can be produced with 381 g $I_2$?

*Given:* 381 g $I_2$  *Wanted:* g $SbI_3$

*Per:* $\frac{1 \text{ mol I}_2}{70.90 \text{ g I}_2}$  $\frac{2 \text{ mol SbI}_3}{3 \text{ mol I}_2}$  $\frac{502.5 \text{ g SbI}_3}{\text{mol SbI}_3}$

*Path:* g $I_2$ ⟶ mol $I_2$ ⟶ mol $SbI_3$ ⟶ g $SbI_3$

$$381 \text{ g I}_2 \times \frac{1 \text{ mol I}_2}{253.8 \text{ g I}_2} \times \frac{2 \text{ mol SbI}_3}{3 \text{ mol I}_2} \times \frac{502.5 \text{ g SbI}_3}{\text{mol SbI}_3} = 503 \text{ g SbI}_3$$

You are now able to identify the limiting reactant. What is it?

Iodine, $I_2$

You have determined that 129 g Sb is enough antimony to produce 532 g $SbI_3$, but there is only enough $I_2$ to produce 503 g $SbI_3$. You concluded that iodine is the limiting reactant, and 503 g $SbI_3$ is produced.

Antimony will react until there's no more iodine available. Some antimony will be left. How much? You know how much antimony you started with. If you knew how much you used, you could subtract what was used from what you had at the beginning to find out how much is left. Well, how much antimony *is*

*continued*

used to react with 381 g $I_2$? This is another stoichiometry question. Calculate the answer.

*GIVEN:* 381 g $I_2$ *WANTED:* g Sb

*PER:* $\frac{1 \text{ mol } I_2}{253.8 \text{ g } I_2}$ $\frac{2 \text{ mol Sb}}{3 \text{ mol } I_2}$ $\frac{121.8 \text{ g Sb}}{\text{mol Sb}}$

*PATH:* g $I_2$ ⟶ mol $I_2$ ⟶ mol Sb ⟶ g Sb

$$381 \text{ g } I_2 \times \frac{1 \text{ mol } I_2}{253.8 \text{ g } I_2} \times \frac{2 \text{ mol Sb}}{3 \text{ mol } I_2} \times \frac{121.8 \text{ g Sb}}{\text{mol Sb}} = 122 \text{ g Sb}$$

You now know how many grams of antimony you started with and how many grams were used. How much is left?

129 g Sb (initial) − 122 g Sb (used) = 7 g Sb left

## 10.7 Energy

**Goal** 6 Given energy in one of the following units, calculate the other three: joules, kilojoules, calories, and kilocalories.

In Section 8.2 we pointed out that nearly all chemical changes involve an energy transfer, usually in the form of heat. We now wish to consider the amount of energy that is transferred in a reaction. To do that we must introduce the common units in which energy is measured.

The SI unit for energy is the **joule (J)** (rhymes with *pool*), which is defined in terms of three base units as 1 kg · $m^2/sec^2$. An older energy unit is the **calorie (cal),** originally defined as the amount of energy required to raise the temperature of one gram of water 1°C. The calorie has now been redefined in terms of joules:

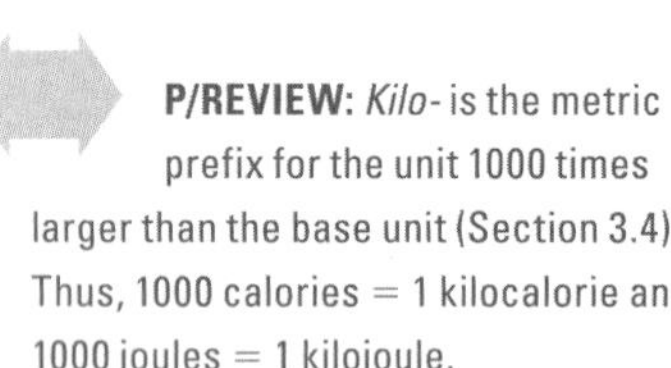

**P/REVIEW:** *Kilo-* is the metric prefix for the unit 1000 times larger than the base unit (Section 3.4). Thus, 1000 calories = 1 kilocalorie and 1000 joules = 1 kilojoule.

$$1 \text{ calorie} \equiv 4.184 \text{ joules} \qquad 4.184 \text{ J/cal}$$

Both the calorie and the joule are small amounts of energy, so the **kilojoule (kJ)** and **kilocalorie (kcal)** are often used instead. It follows that

$$1 \text{ kcal} = 4.184 \text{ kJ} \qquad 4.184 \text{ kJ/kcal}$$

The **Calorie** used when talking about food energy is actually the kilocalorie. It is written with a capital C to distinguish it from the thermochemical calorie, which is written with a lowercase c. Caloric food requirements vary considerably among individuals. A typical diet provides about 2000 Calories (kcal) per day.

## Active Example 10.14

Calculate the number of kilocalories, calories, and joules in 42.5 kJ.

Set up the kJ → kcal conversion first.

GIVEN: 42.5 kJ WANTED: kcal

PER: $\frac{1 \text{ kcal}}{4.184 \text{ kJ}}$

PATH: kJ ⟶ kcal

$$42.5 \cancel{\text{kJ}} \times \frac{1 \text{ kcal}}{4.184 \cancel{\text{kJ}}} = 10.2 \text{ kcal}$$

Now change 42.5 kJ to J and 10.2 kcal to cal. Try to do it by just moving the decimal point.

$42.5 \text{ kJ} = 42{,}500 \text{ J} = 4.25 \times 10^4 \text{ J}$ $10.2 \text{ kcal} = 10{,}200 \text{ cal} = 1.02 \times 10^4 \text{ cal}$

The larger/smaller rule (Section 3.3) states that any quantity may be expressed as a large number of small units or a small number of large units. In this example, this translates into 1000 units in 1 kilounit—1000 units/kilounit. Therefore, the *number* of smaller units (cal and J) is larger than the *number* of larger units (kcal and kJ). The decimal must move right three places. By dimensional analysis,

$$42.5 \cancel{\text{kJ}} \times \frac{1000 \text{ J}}{\cancel{\text{kJ}}} \times 4.25 \times 10^4 \text{ J} \text{ and } 10.2 \cancel{\text{kcal}} \times \frac{1000 \text{ cal}}{\cancel{\text{kcal}}} = 1.02 \times 10^4 \text{ cal}$$

## 10.8 Thermochemical Equations

**Goal** 7 Given a chemical equation, or information from which it may be written, and the heat (enthalpy) of reaction, write the thermochemical equation either (a) with ΔH to the right of the conventional equation or (b) as a reactant or product.

The heat given off or absorbed in a chemical reaction can be measured in the laboratory. We sometimes call this change the **heat of reaction;** more formally, it is the **enthalpy of reaction,** ΔH. H is the symbol for enthalpy and Δ, the Greek uppercase delta, indicates change.* When a system gives off heat to the surroundings—an exothermic change (Fig. 10.4)—the enthalpy of the system goes down, and ΔH has

*Δ always means final value minus initial value.

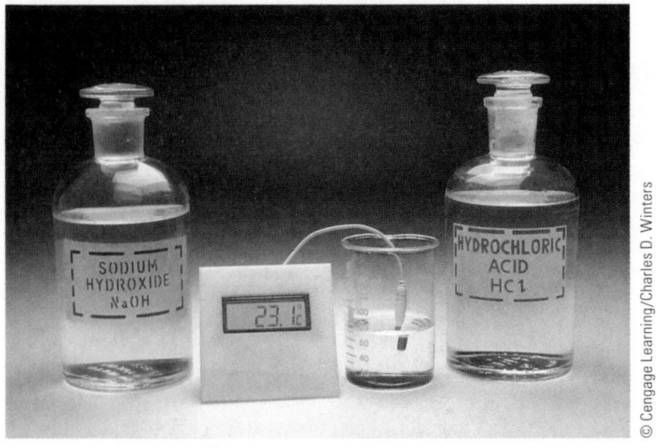


© Cengage Learning/Charles D. Winters

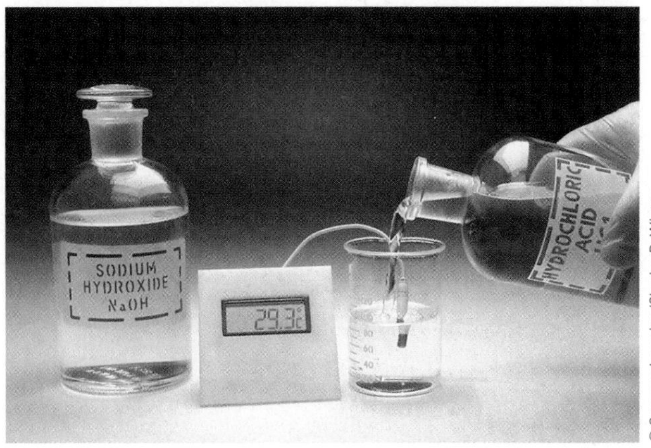


© Cengage Learning/Charles D. Winters

**Figure 10.4** An exothermic reaction. The reaction between solutions of hydrochloric acid and sodium hydroxide is exothermic, releasing heat and causing the temperature of the solution to rise.

a negative value. When a reaction absorbs heat—an endothermic change—enthalpy increases, and $\Delta H$ is positive.*

An equation that includes a change in energy is a **thermochemical equation.** There are two kinds of thermochemical equations. One simply writes the $\Delta H$ of the reaction to the right of the conventional equation. For example, if you burn 2 moles of ethane, $C_2H_6$, 2855 kJ of heat is given off. This is an exothermic reaction, so $\Delta H$ is negative: $\Delta H = -2855$ kJ. The thermochemical equation is

$$2\ C_2H_6(g) + 7\ O_2(g) \rightarrow 4\ CO_2(g) + 6\ H_2O(g) \qquad \Delta H = -2855\ \text{kJ} \qquad \textbf{(10.5)}$$

The second form of thermochemical equation includes energy as if it were a reactant or product. In an exothermic reaction, heat is "produced," so it appears as a positive quantity on the product side of the equation:

$$2\ C_2H_6(g) + 7\ O_2(g) \rightarrow 4\ CO_2(g) + 6\ H_2O(g) + 2855\ \text{kJ} \qquad \textbf{(10.6)}$$

In an endothermic reaction, energy must be added to the reactants to make the reaction happen. Heat is a "reactant." The decomposition of ammonia into its elements is an example:

$$2\ NH_3(g) + 92\ \text{kJ} \rightarrow N_2(g) + 3\ H_2(g)$$

$$2\ NH_3(g) \rightarrow N_2(g) + 3\ H_2(g) \qquad \Delta H = +92\ \text{kJ}$$

NASA

An enormous amount of energy is expended in a space shuttle launch. Would you classify the fuel's reaction as exothermic or endothermic?

When you write thermochemical equations, you *must* use the state symbols (g), ($\ell$), (s), and (aq). The equation is meaningless without them because the size of the enthalpy change depends on the states of the reactants and products. If Equation 10.5 is written with water in the liquid state,

$$2\ C_2H_6(g) + 7\ O_2(g) \rightarrow 4\ CO_2(g) + 6\ H_2O(\ell) \qquad \Delta H = -3119\ \text{kJ} \qquad \textbf{(10.7)}$$

the value of $\Delta H$ is $-3119$ kJ. Unless stated otherwise, we will assume $H_2O(\ell)$ to be the product of burning reactions because most thermochemical processes are tabulated in reference sources at 25°C, and water is a liquid at this temperature.

## Active Example 10.15

The thermal decomposition of limestone, solid calcium carbonate, $CaCO_3(s)$, to lime, solid CaO(s), and gaseous carbon dioxide, $CO_2(g)$, is an endothermic reac-

*This discussion might lead you to think that heat and enthalpy are the same. They are not, but the difference between them is beyond the scope of an introductory text. Under certain common conditions, however, heat flow to or from a system is equal to the enthalpy change. We will limit ourselves to such reactions.

tion requiring 178 kJ per mole of calcium carbonate decomposed. Write the thermochemical equation in two forms.

$CaCO_3(s) \rightarrow CaO(s) + CO_2(g) \qquad \Delta H = +178 \text{ kJ}$

$CaCO_3(s) + 178 \text{ kJ} \rightarrow CaO(s) + CO_2(g)$

## 10.9 Thermochemical Stoichiometry

**Goal** 8 Given a thermochemical equation, or information from which it may be written, calculate the amount of energy released or added for a given amount of reactant or product; alternately, calculate the mass of reactant required to produce a given amount of energy.

If you burn twice as much fuel, it is logical to expect that twice as much energy should be released. The proportional relationships between moles of different substances in a chemical equation, expressed by their coefficients, extend to energy terms, as illustrated in Figure 10.5. Note that the general stoichiometry pattern remains the same.

The equation for burning ethane (Equation 10.7) indicates that for every 2 moles burned, 3119 kJ of energy is released: $\frac{3119 \text{ kJ}}{2 \text{ mol } C_2H_6}$ or $\frac{2 \text{ mol } C_2H_6}{3119 \text{ kJ}}$. Similar conversion factors may be written between kilojoules and any other substance in the equation. These factors are used in solving **thermochemical stoichiometry** problems.

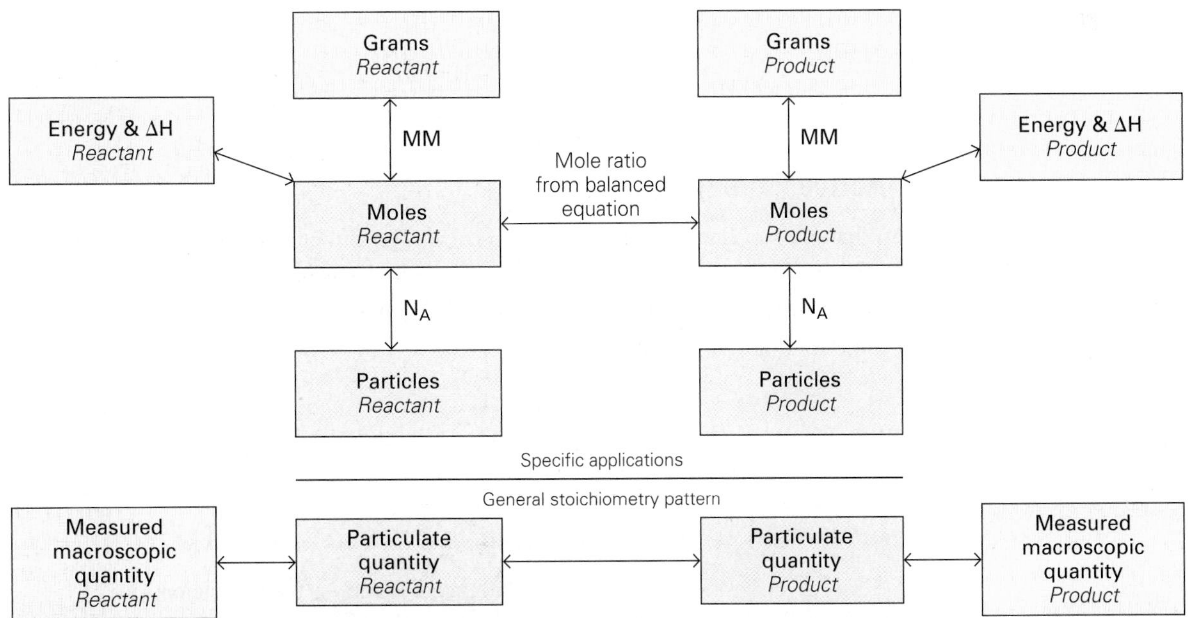


**Figure 10.5** Conversion among grams or energy and ΔH, moles, and number of particles for two species in a chemical change. This figure adds a new measured macroscopic quantity, energy and ΔH, to Figure 10.2. The link between energy and the number of moles is the balanced thermochemical equation, which includes ΔH. The overarching pattern, illustrated by the four boxes at the bottom of the figure, remains the same.

## Thinking About Your Thinking

### Proportional Reasoning

**The direct proportionalities between the moles of the reactants and products in a chemical change and the quantity of energy absorbed or released allow these relationships to be used as *PER* expressions in dimensional-analysis setups.**

### Active Example 10.16

How many kilojoules of energy are released when 73.0 g $C_2H_6$ burns according to $2\ C_2H_6(g) + 7\ O_2(g) \rightarrow 4\ CO_2(g) + 6\ H_2O(\ell) + 3119\ kJ$?

See if you can *PLAN* this new kind of problem without hints.

*GIVEN:* 73.0 g $C_2H_6$ *WANTED:* kJ

$$\text{PER: } \frac{1 \text{ mol } C_2H_6}{30.07 \text{ g } C_2H_6} \qquad \frac{3119 \text{ kJ}}{2 \text{ mol } C_2H_6}$$

$$\text{PATH: g } C_2H_6 \longrightarrow \text{mol } C_2H_6 \longrightarrow \text{kJ}$$

This is a two-step problem. Once you reach moles of any substance, you can go directly to quantity of energy.

Set up and solve the problem.

$$73.0 \text{ g } C_2H_6 \times \frac{1 \text{ mol } C_2H_6}{30.07 \text{ g } C_2H_6} \times \frac{3119 \text{ kJ}}{2 \text{ mol } C_2H_6} = 3.79 \times 10^3 \text{ kJ}$$

### Active Example 10.17

What mass of liquid *n*-octane, $C_8H_{18}(\ell)$, a component of gasoline, must be burned to provide $9.05 \times 10^4$ kJ of heat? $\Delta H = -5471$ kJ/mol $C_8H_{18}$ burned.

Start by writing the thermochemical equation. (Careful! There's a potential pitfall in that $\Delta H$ statement.)

$$2\ C_8H_{18}(\ell) + 25\ O_2(g) \rightarrow 16\ CO_2(g) + 18\ H_2O(\ell) + 10{,}942 \text{ kJ}$$

$\Delta H$ is given in kJ *per mole.* In the balanced equation, the coefficient of $C_8H_{18}$ is 2. The equation represents *2 mol* $C_8H_{18}$. The energy conversion factor is either 10,942 kJ/2 mol $C_8H_{18}$ from the equation or 5471 kJ/mol $C_8H_{18}$ from the statement of the problem. The two energy conversion factors are equivalent. Our setup will use the one with the coefficients from the equation.

*PLAN* the problem, then set up and solve it.

*GIVEN:* $9.05 \times 10^4$ kJ *WANTED:* mass $C_8H_{18}$ (assume g)

*PER:* $\frac{2 \text{ mol } C_8H_{18}}{10{,}942 \text{ kJ}}$ $\frac{114.22 \text{ g } C_8H_{18}}{\text{mol } C_8H_{18}}$

*PATH:* kJ $\longrightarrow$ mol $C_8H_{18}$ $\longrightarrow$ g $C_8H_{18}$

$$9.05 \times 10^4 \cancel{\text{kJ}} \times \frac{2 \cancel{\text{mol } C_8H_{18}}}{10{,}942 \cancel{\text{kJ}}} \times \frac{114.22 \text{ g } C_8H_{18}}{\cancel{\text{mol } C_8H_{18}}} = 1.89 \times 10^3 \text{ g } C_8H_{18}$$

A word about algebraic signs: In these examples we have been able to disregard the sign of $\Delta H$ because of the way the questions were worded. The question, "How much heat . . . ?" is answered simply with a number of kilojoules. The wording of the question tells whether the heat is gained or lost. However, if a question is of the form, "What is the value of $\Delta H$?" the algebraic sign is an essential part of the answer.

## Chapter 10 in Review

Most of the key terms and concepts and many others appear in the Glossary. Use your Glossary regularly.

**10.1 Conversion Factors from a Chemical Equation**

Goal 1 Given a chemical equation, or a reaction for which the equation is known, and the number of moles of one species in the reaction, calculate the number of moles of any other species.

**Key Terms and Concepts: Stoichiometry**

**10.2 Mass–Mass Stoichiometry**

Goal 2 Given a chemical equation, or a reaction for which the equation can be written, and the number of grams or moles of one species in the reaction, find the number of grams or moles of any other species.

**Key Terms and Concepts: Mass-to-mass path**

**10.3 Percent Yield**

Goal 3 Given two of the following, or information from which two of the following may be determined, calculate the third: theoretical yield, actual yield, percent yield.

**Key Terms and Concepts: Actual yield, percent yield, theoretical yield**

**10.4 Limiting Reactants: The Problem**

Goal 4 Identify and describe or explain limiting reactants and excess reactants.

**Key Terms and Concepts: Excess reactant, limiting reactant**

**10.5 Limiting Reactants: Comparison-of-Moles Method**

Goal 5 Given a chemical equation, or information from which it may be determined, and initial quantities of two or more reactants, (a) identify the limiting reactant, (b) calculate the theoretical yield of a specified product, assuming complete use of the limiting reactant, and (c) calculate the quantity of the reactant initially in excess that remains unreacted.

**10.6 Limiting Reactants: Smaller-Amount Method**

Goal 5 Given a chemical equation, or information from which it may be determined, and initial quantities of two or more reactants, (a) identify the limiting reactant, (b) calculate the theoretical yield of a specified product, assuming complete use of the limiting reactant, and (c) calculate the quantity of the reactant initially in excess that remains unreacted.

**10.7 Energy**

Goal 6 Given energy in one of the following units, calculate the other three: joules, kilojoules, calories, and kilocalories.

**Key Terms and Concepts: Calorie (Cal) and calorie (cal), kilocalorie (kcal), joule (J), kilojoule (kJ)**

**10.8 Thermochemical Equations**

Goal 7 Given a chemical equation, or information from which it may be written, and the heat (enthalpy) of reaction, write the thermochemical equation either (a) with $\Delta H$ to the right of the conventional equation or (b) as a reactant or product.

**Key Terms and Concepts: Heat or enthalpy of reaction ($\Delta H$), thermochemical equation**

**10.9 Thermochemical Stoichiometry**

Goal 8 Given a thermochemical equation, or information from which it may be written, calculate the amount of energy released or added for a given amount of reactant or product; alternately, calculate the mass of reactant required to produce a given amount of energy.

**Key Terms and Concepts: Thermochemical stoichiometry**

## Study Hints and Pitfalls to Avoid

The ability to solve stoichiometry problems is probably the most important problem-solving skill you can develop in a beginning chemistry course. Work to understand the steps in the quantity-to-quantity path, not just to be able to "do" the steps from memory or by juggling units. In this chapter, quantity may be measured in moles or grams. You have not finished a problem when you reach an answer, whether it is right or wrong. You must understand each problem. Be sure that you do so before going on.

Take a moment to think through the logic of the calculation sequence in stoichiometry. Look particularly at the mole-to-mole conversion in the middle. If you understand the process, apart from a specific problem, you will be able to recognize and solve other kinds of stoichiometry problems in chapters to come.

There are three kinds of percent-yield problems in this book. You should understand the similarities and differences among them. They are summarized in the following table.

| GIVEN or Calculated from GIVEN | WANTED | Solve by |
|---|---|---|
| Actual and theoretical yields | Percent yield | $\% \text{ yield} = \frac{\text{actual yield}}{\text{theoretical yield}} \times 100$ |
| Reactant quantity and percent yield | Product quantity | Dimensional analysis |
| Product quantity and percent yield | Reactant quantity | Dimensional analysis |

When the product quantity is WANTED, the percent yield is applied to the product at the end of the calculation setup. When the reactant quantity is WANTED, the percent conversion is applied to the product at the beginning of the setup. It will help you keep things straight if you always apply the percent conversion to the product and then distinguish between actual and theoretical product in your setup.

Thermochemical stoichiometry problems have one less step than other stoichiometry problems because they involve only one substance. There is no mole-to-mole conversion, but rather a mole-to-energy change between the single substance and the $\Delta H$ of the reaction. Watch the sign of $\Delta H$ if the wording of the problem is such that it must be taken into account.

## Problem-Classification Exercises

*Five examples from the chapter are repeated here. Each example represents one kind of stoichiometry problem. If you can classify a problem as one of these types, you will find it easier to select the correct procedure for solving it. The problem types are summarized in Table 10.1. Exercises follow the table.*

### Mass Stoichiometry

*Active Example 10.3: Calculate the number of grams of oxygen required to burn 155 g of ethane in the reaction* $2\ C_2H_6(g) + 7\ O_2(g) \rightarrow 4\ CO_2(g) + 6\ H_2O(\ell)$.

*General format: Given one mass, find another mass.*

### Percent Yield

*Active Example 10.9: A procedure for preparing sodium sulfate is summarized in the equation* $2\ S(s) + 3\ O_2(g) + 4\ NaOH(aq) \rightarrow 2\ Na_2SO_4(aq) + 2\ H_2O(\ell)$. *The percent yield in the process is 79.8%. Find the number of grams of sodium sulfate that will be recovered from the reaction of 36.9 g NaOH.*

*General format: (1) Given actual and theoretical yields, find percent yield or (2) given percent yield and either the reactant quantity or the product quantity, find the unknown quantity.*

### Limiting Reactant

*Active Examples 10.12 and 10.13: Calculate the mass of antimony(III) chloride,* $SbCl_3$, *that can be produced by the reaction of 129 g antimony, Sb (Z = 51), and 106 g chlorine. Also find the number of grams of the element that will be left.*

*General format: Given masses of two reactants, find the mass of product and the unreacted mass of the excess reactant.*

**Table 10.1** Summary of Stoichiometry Classifications

| Classification | GIVEN | WANTED |
|---|---|---|
| Mass stoichiometry | Mass of reactant or product | Mass of reactant or product |
| Percent yield | Actual and theoretical yields | Percent yield |
| | Reactant quantity and percent yield | Product quantity |
| | Product quantity and percent yield | Reactant quantity |
| Limiting reactant | Masses of two reactants | Mass of product and mass of unreacted reactant |
| Thermochemical stoichiometry | Mass of reactant or product | Quantity of heat energy |
| | Quantity of heat energy | Mass of reactant or product |

**Thermochemical Stoichiometry**

*Active Example 10.17: What mass of liquid n-octane, $C_8H_{18}(\ell)$, a component of gasoline, must be burned to provide $9.05 \times 10^4$ kJ of heat? $\Delta H = -5471$ kJ/mol $C_8H_{18}$ burned.*

*General format: For a reaction with a known ΔH, given mass, find heat or given heat, find mass.*

*Eight problem-classification examples follow. Test your classification skill by deciding which kind of problem each one represents. It is not necessary to set up and solve the problem now; this exercise is primarily concerned with problem classification.*

1. What mass of magnesium hydroxide will precipitate if 2.09 grams of potassium hydroxide is added to a magnesium nitrate solution?

2. How many grams of octane, a component of gasoline, would you have to burn in your car to liberate $9.48 \times 10^5$ kJ of energy? $\Delta H = -1.09 \times 10^4$ kJ for the reaction $2\ C_8H_{18}(\ell) + 25\ O_2(g) \rightarrow 16\ CO_2(g) + 18\ H_2O(\ell)$.

3. A mixture of tetraphosphorus trisulfide and powdered glass is in the white tip of strike-anywhere matches. The compound is made by the direct combination of the elements: $8\ P_4 + 3\ S_8 \rightarrow 8\ P_4S_3$. If 133 g of phosphorus is mixed with the full contents of a 4-oz (126 g) bottle of sulfur, how many grams of the compound can be formed? How much of which element will be left over?

4. How many grams of sodium hydroxide are needed to neutralize completely 32.6 grams of phosphoric acid?

5. The Haber process for making ammonia from nitrogen in the air is given by the equation $N_2 + 3\ H_2 \rightarrow 2\ NH_3$. Calculate the mass of hydrogen that must be supplied to make $5.00 \times 10^2$ kg ammonia in a system that has an 88.8% yield.

6. Quicklime, the common name for calcium oxide, CaO, is made by heating limestone, $CaCO_3$, in slowly rotating kilns about 2½ meters in diameter and about 60 meters long. The reaction is $CaCO_3(s) + 178\ kJ \rightarrow CaO(s) + CO_2(g)$. How many kilojoules are required to decompose 5.80 kg of limestone?

7. A solution containing 1.63 g barium chloride is added to a solution containing 2.40 g sodium chromate (chromate ion, $CrO_4^{2-}$). Find the number of grams of barium chromate that can precipitate.

8. Ethylacetate, $CH_3COOC_2H_5$, is manufactured by the reaction between acetic acid, $CH_3COOH$, and ethanol, $C_2H_5OH$, by the equation $CH_3COOH + C_2H_5OH \rightarrow CH_3COOC_2H_5 + H_2O$. How much acetic acid must be used to get 62.5 kg of ethylacetate if the percent yield is 69.1%?

# Small-Group Discussion Questions

*Small-Group Discussion Questions are for group work, either in class or under the guidance of a leader during a discussion section.*

1. Natural gas is usually composed of about 80% methane, $CH_4$, 5% ethane, $C_2H_6$, and the balance is other gases. Assume that we have a natural gas sample that is 90% methane and 10% ethane in this problem. (a) Write a balanced chemical equation for the reaction of each gas. (b) What total mass of oxygen is required to burn 1 kg of natural gas? (c) What total mass of products will be formed when 1 kg of natural gas is burned?

2. Titanium tetrachloride, $TiCl_4$, is a colorless liquid at room conditions. It is used in the plastic, electronics, metals, ceramics, and leather industries, and it is sometimes sold in drums that contain 300 kg of titanium tetrachloride. When exposed to water, the liquid reacts to form solid titanium dioxide, $TiO_2$, and hydrogen chloride gas. What volume of water is needed to completely react with a drum of titanium tetrachloride? How many pounds of titanium dioxide will be formed?

3. A mass of 0.500 g of a metal element reacts with oxygen to yield 0.909 g of the metal oxide $MO_2$. What is the metal?

4. An Alka-Seltzer tablet contains 325 mg aspirin, 1916 mg sodium hydrogen carbonate, and citric acid. When dissolved in water, the sodium hydrogen carbonate reacts with citric acid, $H_3C_6H_5O_7$, to form sodium citrate, water, and carbon dioxide. The aspirin does not react, and it serves as a pain reliever, and the sodium citrate serves as an antacid. Assuming that the tablets are manufactured with the correct stoichiometric quantity of citric acid, determine the mass of citric acid per tablet.

5. Chemists consider changes in matter at three levels: the macroscopic, the particulate, and the symbolic, which are symbols that are used to represent matter and its changes. Consider the reaction of charcoal, which is mostly carbon, and oxygen from the air. Carbon dioxide is the product of this reaction. (a) Sketch and describe this reaction at the macroscopic level. Explain why it might lead you to believe that the Law of Conservation of Mass is not correct. (b) Sketch and describe this reaction at the particulate level. (c) Use symbols to describe the reaction, and explain how they are describing the reaction at the macroscopic level and the particulate level.

6. Limestone is rock that is largely composed of calcium carbonate. When limestone is heated in a lime kiln (essentially a large oven), it decomposes to quicklime, which is calcium oxide, and carbon dioxide. A 150.8-kg sample of limestone was heated, yielding 56.1 kg of carbon dioxide. Determine the percent limestone in the original sample.

7. Gold occurs naturally in low amounts in hard rock deposits in some locations in the world. This gold can be extracted from the rock by reacting the gold with sodium cyanide solution (cyanide ion, $CN^-$), oxygen, and water, yielding $NaAu(CN)_2$ and sodium hydroxide in solution. The complex is then reacted with zinc to yield gold metal. One metric ton (1000 kg) of ore that contains 0.02% gold is to be reacted to remove the precious metal. What mass of sodium cyanide is needed for the process?

8. Malonic acid is 34.62% carbon, 3.88% hydrogen, and 61.50% oxygen by mass. Write the balanced equation for its complete combustion, and then determine the masses

of all additional reactants and products consumed and produced when one pound of malonic acid is burned in air.

9. A solution with 24.99 mg of sodium carbonate is combined with a solution containing 16.70 mg of silver nitrate. Determine the mass of each species in the flask (other than water) after the two solutions are combined.

10. One process for producing sodium sulfate is known as the Hargreaves method. Sodium chloride, sulfur dioxide, water, and oxygen react to form hydrogen chloride and sodium sulfate. Water is inexpensive, and oxygen is obtained from the air. The two reactants that control the cost of the manufacturing process are sodium chloride and sulfur dioxide. How many pounds of sodium chloride are needed for each pound of sulfur dioxide consumed in the reaction if excess water and oxygen are available?

# Questions, Exercises, and Problems

*Blue-numbered questions are answered in Appendix III.* ■ *denotes problems assignable in OWL. In the first section of Questions, Exercises, and Problems, similar exercises and problems are paired in consecutive odd-even number combinations. Many questions in this chapter are written with the assumption that you have studied Chapter 6 and can write the required formulas from their chemical names. If this is not the case, we have placed a list of all chemical formulas needed to answer Chapter 10 questions at the end of the Questions, Exercises, and Problems.*

### Section 10.1: Conversion Factors from a Chemical Equation

1. The first step in the Ostwald process for manufacturing nitric acid is the reaction between ammonia and oxygen described by the equation $4\ NH_3 + 5\ O_2 \rightarrow 4\ NO + 6\ H_2O$. Use this equation to answer all parts of this question.
   a) How many moles of ammonia will react with 95.3 moles of oxygen?
   b) How many moles of nitrogen monoxide will result from the reaction of 2.89 moles of ammonia?
   c) If 3.35 moles of water is produced, how many moles of nitrogen monoxide will also be produced?

2. ■ When hydrogen sulfide reacts with oxygen, water and sulfur dioxide are produced. The balanced equation for this reaction is $2\ H_2S(g) + 3\ O_2(g) \rightarrow 2\ H_2O(\ell) + 2\ SO_2(g)$. For all parts of this question, consider what will happen if 4 moles of hydrogen sulfide react.
   a) How many moles of oxygen are consumed?
   b) How many moles of water are produced?
   c) How many moles of sulfur dioxide are produced?

3. Magnesium hydroxide is formed from the reaction of magnesium oxide and water. How many moles of magnesium oxide are needed to form 0.884 mole of magnesium hydroxide, when the oxide is added to excess water?

4. ■ In our bodies, sugar is broken down by reacting with oxygen to produce water and carbon dioxide. How many moles of carbon dioxide will be formed upon the complete reaction of 0.424 moles glucose sugar ($C_6H_{12}O_6$) with excess oxygen gas?

5. When sulfur dioxide reacts with oxygen, sulfur trioxide forms. How many moles of sulfur dioxide are needed to produce 3.99 moles of sulfur trioxide if the reaction is carried out in excess oxygen?

6. ■ Aqueous solutions of potassium hydrogen sulfate and potassium hydroxide react to form aqueous potassium sulfate and liquid water. How many moles of potassium hydroxide are necessary to form 0.636 moles of potassium sulfate?

### Section 10.2: Mass–Mass Stoichiometry

7. The first step in the Ostwald process for manufacturing nitric acid is the reaction between ammonia and oxygen described by the equation $4\ NH_3 + 5\ O_2 \rightarrow 4\ NO + 6\ H_2O$. Use this equation to answer all parts of this question.
   a) How many moles of ammonia can be oxidized by 268 grams of oxygen?
   b) If the reaction consumes 31.7 moles of ammonia, how many grams of water will be produced?
   c) How many grams of ammonia are required to produce 404 grams of nitrogen monoxide?
   d) If 6.41 grams of water result from the reaction, what will be the yield of nitrogen monoxide (in grams)?

8. Butane, $C_4H_{10}$, is a common fuel used for heating homes in areas not served by natural gas. The equation for its combustion is $2\ C_4H_{10} + 13\ O_2 \rightarrow 8\ CO_2 + 10\ H_2O$. All parts of this question are related to this reaction.
   a) How many grams of butane can be burned by 1.42 moles of oxygen?
   b) If 9.43 grams of oxygen is used in burning butane, how many moles of water result?
   c) Calculate the number of grams of carbon dioxide that will be produced by burning 78.4 grams of butane.
   d) How many grams of oxygen are used in a reaction that produces 43.8 grams of water?

9. The explosion of nitroglycerine is described by the equation $4\ C_3H_5(NO_3)_3 \rightarrow 12\ CO_2 + 10\ H_2O + 6\ N_2 + O_2$. How many grams of carbon dioxide are produced by the explosion of 21.0 grams of nitroglycerine?

10. ■ According to the reaction $2\ AgNO_3 + Cu \rightarrow Cu(NO_3)_2 + 2\ Ag$, how many grams of copper(II) nitrate will be formed upon the complete reaction of 26.8 grams of copper with excess silver nitrate?

11. Soaps are produced by the reaction of sodium hydroxide with naturally occurring fats. The equation for one such reaction is $C_3H_5(C_{17}H_{35}COO)_3 + 3\ NaOH \rightarrow C_3H_5(OH)_3 + 3\ C_{17}H_{35}COONa$. The last compound is the soap. Calculate the number of grams of sodium hydroxide required to produce 323 grams of soap by this method.

12. ■ According to the reaction $CH_4 + CCl_4 \rightarrow 2\ CH_2Cl_2$, how many grams of carbon tetrachloride are required for the complete reaction of 24.7 grams of methane, $CH_4$?

13. One way to make sodium thiosulfate, known as "hypo" and used in photographic developing, is described by the equation $Na_2CO_3 + 2\ Na_2S + 4\ SO_2 \rightarrow 3\ Na_2S_2O_3 + CO_2$. How many grams of sodium carbonate are required to produce 681 grams of sodium thiosulfate?

14. One of the methods for manufacturing sodium sulfate, once widely used in making the kraft paper for grocery bags, involves the reaction $4\ NaCl + 2\ SO_2 + 2\ H_2O + O_2 \rightarrow 2\ Na_2SO_4 + 4\ HCl$. Calculate the number of kilograms of sodium chloride required to produce 5.00 kilograms of sodium sulfate.

15. The hard water scum that forms a ring around the bathtub is an insoluble soap, $Ca(C_{18}H_{35}O_2)_2$. It is formed when a soluble soap, $NaC_{18}H_{35}O_2$, reacts with the calcium ion that is responsible for the hardness in water: $2\ NaC_{18}H_{35}O_2 + Ca^{2+} \rightarrow Ca(C_{18}H_{35}O_2)_2 + 2\ Na^+$. How many milligrams of scum can form from 616 milligrams of $NaC_{18}H_{35}O_2$?

16. Trinitrotoluene is the chemical name for the explosive commonly known as TNT. Its formula is $C_7H_5N_3O_6$. TNT is manufactured by the reaction of toluene, $C_7H_8$, with nitric acid: $C_7H_8 + 3\ HNO_3 \rightarrow C_7H_5N_3O_6 + 3\ H_2O$.
    a) How much nitric acid is needed to react completely with 1.90 kilograms of toluene?
    b) How many kilograms of TNT can be produced in the reaction?

17. Pig iron from a blast furnace contains several impurities, one of which is phosphorus. Additional iron ore, $Fe_2O_3$, is included with pig iron in making steel. The oxygen in the ore oxidizes the phosphorus by the reaction $12\ P + 10\ Fe_2O_3 \rightarrow 3\ P_4O_{10} + 20\ Fe$. If a sample of the remains from the furnace contains 802 milligrams of tetraphosphorus decoxide, how many grams of $Fe_2O_3$ was used in making it?

18. ■ According to the reaction $N_2 + O_2 \rightarrow 2\ NO$, how many grams of nitrogen monoxide will be formed upon the complete reaction of 25.4 grams of oxygen gas with excess nitrogen gas?

*The Solvay process is a multi-step industrial method for the manufacture of sodium carbonate, $Na_2CO_3$, which is also known as washing soda. Although most of the industrialized world utilizes the Solvay process for production of sodium carbonate, it is not manufactured in the United States because it can be obtained at lower cost from a large natural deposit in Wyoming. Questions 19 through 22 are based on reactions in that process.*

19. How much NaCl is needed to react completely with 83.0 grams of ammonium hydrogen carbonate in $NaCl + NH_4HCO_3 \rightarrow NaHCO_3 + NH_4Cl$?

20. How many grams of ammonium hydrogen carbonate will be formed by the reaction of 81.2 grams of ammonia in the reaction $NH_3 + H_2O + CO_2 \rightarrow NH_4HCO_3$?

21. By-product ammonia is recovered from the Solvay process by the reaction $Ca(OH)_2 + 2\ NH_4Cl \rightarrow CaCl_2 + 2\ H_2O + 2\ NH_3$. How many grams of calcium chloride can be produced along with 62.0 grams of ammonia?

22. What mass of $NaHCO_3$ must decompose to produce 448 grams of $Na_2CO_3$ in $2\ NaHCO_3 \rightarrow Na_2CO_3 + H_2O + CO_2$?

23. How many grams of sodium hydroxide are needed to neutralize completely 32.6 grams of phosphoric acid?

24. ■ Aqueous ammonium chloride decomposes to form ammonia gas and aqueous hydrochloric acid. How many grams of ammonia will be formed upon the complete reaction of 28.3 grams of ammonium chloride?

25. What mass of magnesium hydroxide will precipitate if 2.09 grams of potassium hydroxide is added to a magnesium nitrate solution?

26. ■ Under specialized conditions, an aqueous solution of hydrobromic acid can decompose to hydrogen gas and liquid bromine. How many grams of hydrobromic acid are needed to form 31.5 grams of bromine?

27. An experimenter recovers 0.521 gram of sodium sulfate from the neutralization of sodium hydroxide by sulfuric acid. How many grams of sulfuric acid reacted?

28. ■ Solid sulfur reacts with carbon monoxide gas to form gaseous sulfur dioxide and solid carbon. How many grams of carbon monoxide are required for the complete reaction of 25.5 grams of sulfur?

29. The reaction of a dry-cell battery may be represented as follows: Zinc reacts with ammonium chloride to form zinc chloride, ammonia, and hydrogen. Calculate the number of grams of zinc consumed during the release of 7.05 grams of ammonia in such a cell.

30. ■ Solid calcium carbonate decomposes to solid calcium oxide and carbon dioxide gas. How many grams of calcium carbonate are needed to form 22.8 grams of carbon dioxide?

### Section 10.3: Percent Yield

31. The function of "hypo" (see Question 13) in photographic developing is to remove excess silver bromide by the reaction $2\ Na_2S_2O_3 + AgBr \rightarrow Na_3Ag(S_2O_3)_2 + NaBr$. What is the percent yield if the reaction of 8.18 grams of sodium thiosulfate produces 2.61 grams of sodium bromide?

32. ■ Butane gas is used as a fuel for camp stoves, producing heat in the reaction $2\ C_4H_{10} + 13\ O_2 \rightarrow 8\ CO_2 + 10\ H_2O$. When 4.14 grams of butane reacts with excess oxygen gas, the reaction yields 11.0 grams of carbon dioxide. What is the percent yield in this reaction?

33. Calcium cyanamide is a common fertilizer. When mixed with water in the soil, it reacts to produce calcium carbonate and ammonia: $CaCN_2 + 3\ H_2O \rightarrow CaCO_3 + 2\ NH_3$. How much ammonia can be obtained from 7.25 grams of calcium cyanamide in a laboratory experiment in which the percent yield will be 92.8%?

34. ■ Hydrochloric acid is known as an enemy of stainless steel, the primary component of which is iron. The reaction is $2\ HCl + Fe \rightarrow FeCl_2 + H_2$. What will be the actual yield of iron(II) chloride when 3.17 grams of iron reacts with excess hydrochloric acid? Assume that the percent yield of iron(II) chloride is 76.6%.

35. The Haber process for making ammonia from nitrogen in the air is given by the equation $N_2 + 3\ H_2 \rightarrow 2\ NH_3$. Calculate the mass of hydrogen that must be supplied to make $5.00 \times 10^2$ kg of ammonia in a system that has an 88.8% yield.

36. ■ Hydrogen iodide is a colorless, nonflammable, and corrosive gas with a penetrating and suffocating odor. It can be prepared by direct reaction of its elements: $H_2 + I_2 \rightarrow 2\ HI$. It is desired to produce 331 grams of hydrogen iodide in a process in which the percent yield is 71.1%. How many grams of hydrogen gas will need to be reacted?

37. Calculate the percent yield in the photosynthesis reaction by which carbon dioxide is converted to sugar if 7.03 grams of carbon dioxide yields 3.92 grams of $C_6H_{12}O_6$. The equation is $6\ CO_2 + 6\ H_2O \rightarrow C_6H_{12}O_6 + 6\ O_2$.

38. ■ The simplest example of the hydrogenation of a carbon–carbon double bond is the reaction between ethene and hydrogen in the presence of nickel: $CH_2{=}CH_2 + H_2 \xrightarrow{Ni} CH_3CH_3$. When 5.20 grams of ethene reacts with excess hydrogen gas, the reaction yields 4.75 grams of ethane. What is the percent yield for this reaction?

39. Ethylacetate, $CH_3COOC_2H_5$, is manufactured by the reaction between acetic acid, $CH_3COOH$, and ethanol, $C_2H_5OH$: $CH_3COOH + C_2H_5OH \rightarrow CH_3COOC_2H_5 + H_2O$. How much acetic acid must be used to get 62.5 kilograms of ethylacetate if the percent yield is 69.1%?

40. ■ Nitrogen monoxide is produced by combustion in an automobile engine. It can then react with oxygen: $2\ NO + O_2 \rightarrow 2\ NO_2$. If 5.74 grams of oxygen gas reacts with excess nitrogen monoxide in a system in which the percent yield of nitrogen dioxide is 80.2%, what is the actual yield of nitrogen dioxide?

### Section 10.4: Limiting Reactants: The Problem

41. Ammonia can be formed from a combination reaction of its elements. A small fraction of an unreacted mixture of elements is illustrated in the following diagram, where white spheres represent hydrogen atoms and blue spheres represent nitrogen atoms. The temperature is such that all species are gases.

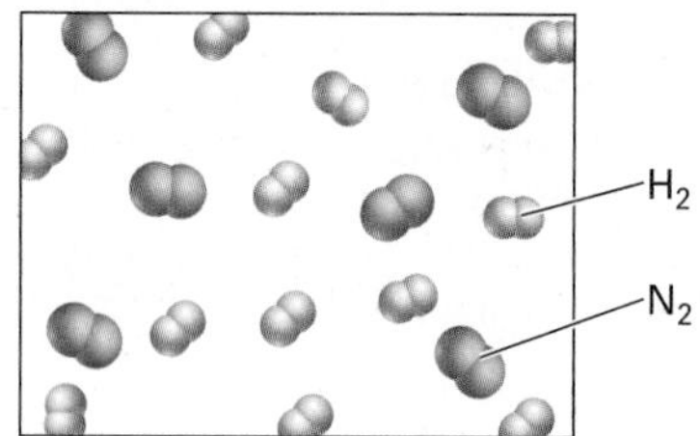


a) Write and balance the equation for the reaction.

b) Which of the following correctly represents the product mixture?

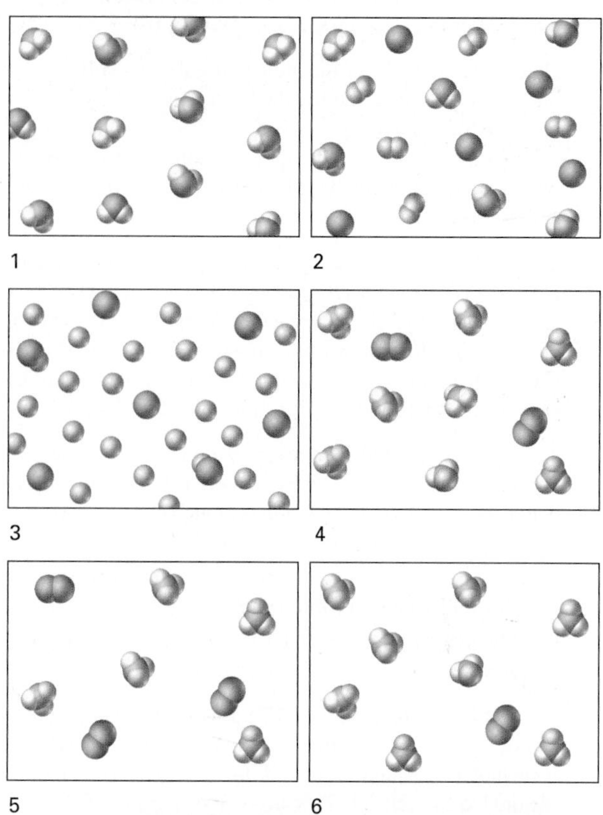


c) Which species is the limiting reactant? Explain.

42. Carbon monoxide reacts with oxygen to form carbon dioxide at a temperature at which all species are in the gas phase. A tiny sample of the pre-reaction mixture is shown in the following diagram, where blue spheres represent oxygen atoms and gold spheres represent carbon atoms. Draw the product mixture. Give a written description of your reasoning.

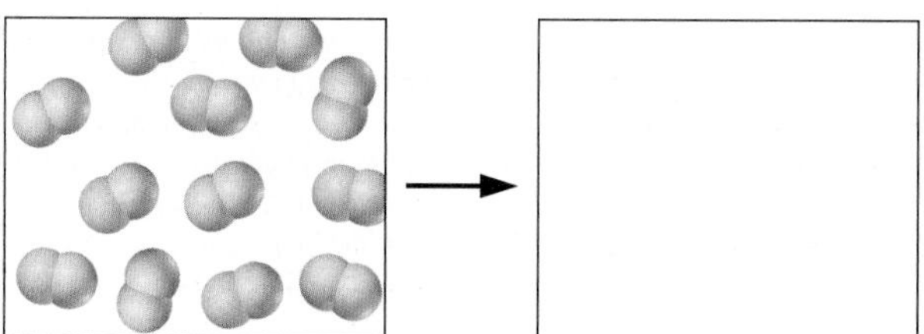

43. An experiment is conducted in which varying amounts of solid iron are added to a fixed volume of liquid bromine. The product of the reaction is a single compound, which can be separated from the product mixture and weighed. The graph shows the relationship between the mass of iron in each trial versus the mass of the product compound. Explain why the graph has a positive slope for low masses of iron and a zero slope when the mass of iron added becomes larger.

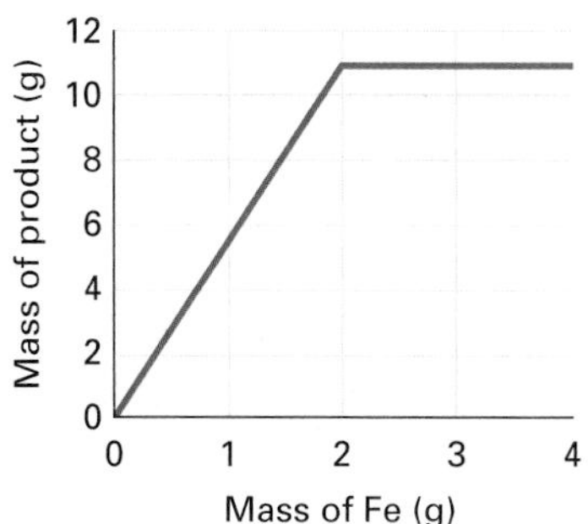


44. The flasks below illustrate three trials of a reaction between varying amounts of zinc and a constant volume of hydrochloric acid solution. The table gives the initial amount of zinc added and the observations at the conclusion of each reaction.

© Cengage Learning/Charles D. Winters

| Mass of Zinc Added (g) | Observations at Conclusion of Reaction |
|---|---|
| 6.10 | Balloon inflated completely<br>Some unreacted zinc remains |
| 2.37 | Balloon inflated completely<br>No zinc remains |
| 0.41 | Balloon not inflated completely<br>No zinc remains |

The reaction produces an aqueous solution of zinc chloride and hydrogen gas. Explain the observed results.

### Sections 10.5 *and* 10.6: Limiting Reactants

45. A solution containing 1.63 grams of barium chloride is added to a solution containing 2.40 grams of sodium chromate (chromate ion, $CrO_4^{2-}$). Find the number of grams of barium chromate that can precipitate. Also determine which reactant was in excess, as well as the number of grams over the amount required by the limiting reactant.

46. ■ Carbon dioxide can be removed from the air by reaction with potassium hydroxide: $CO_2 + 2\ KOH \rightarrow K_2CO_3 + H_2O$. If 15.6 grams of carbon dioxide reacts with 43.0 grams of potassium hydroxide, what mass of potassium carbonate will be formed? What amount of excess reactant remains after the reaction is complete?

47. The equation for one method of preparing iodine is $2\ NaIO_3 + 5\ NaHSO_3 \rightarrow I_2 + 3\ NaHSO_4 + 2\ Na_2SO_4 + H_2O$. If 6.00 kilograms of sodium iodate is reacted with 7.33 kilograms of sodium hydrogen sulfite, how many kilograms of iodine can be produced? Which reactant will be left over? How many kilograms will be left?

48. ■ Carbon monoxide is the chief waste product of the gasoline combustion process. Auto manufacturers use catalytic converters to convert carbon monoxide into carbon dioxide: $2\ CO + O_2 \rightarrow 2\ CO_2$. When 12.6 grams of carbon monoxide is allowed to react with 5.22 grams of oxygen gas, how many grams of carbon dioxide are formed? What amount of the excess reactant remains after the reaction is complete?

49. A mixture of tetraphosphorus trisulfide and powdered glass is in the white tip of strike-anywhere matches. The $P_4S_3$ is made by the direct combination of the elements: $8\ P_4 + 3\ S_8 \rightarrow 8\ P_4S_3$. If 133 grams of phosphorus is mixed with the full contents of a 126-g (4-oz) bottle of sulfur, how many grams of the compound can be formed? How much of which element will be left over?

50. ■ Sodium carbonate can neutralize nitric acid by the reaction $2\ HNO_3 + Na_2CO_3 \rightarrow 2\ NaNO_3 + H_2O + CO_2$. Is 135 grams of sodium carbonate enough to neutralize a solution that contains 188 grams of nitric acid? How many grams of carbon dioxide will be released in the reaction?

### Section 10.7: Energy

51. a) How many kilojoules are equal to 0.731 kcal?
    b) What number of calories is the same as 651 J?
    c) Determine the number of kilocalories that is equivalent to $6.22 \times 10^3$ J.

52. ■ Complete the following:
    a) 504 J = ______ cal = ______ kJ
    b) 192 cal = ______ J = ______ kJ
    c) 0.423 kJ = ______ J = ______ cal

53. When 15 g of carbon is burned, 493 kJ of energy is released. Calculate the number of calories and kilocalories this represents.

54. ■ A list of the calorie content of foods indicates that a piece of lemon meringue pie contains 333 Calories. Express this value in kJ and in J.

55. Each day, $5.8 \times 10^2$ kcal of heat is removed from the body by evaporation. How many kilojoules are removed in a year?

56. Burning sucrose releases $3.94 \times 10^3$ cal/g. Calculate the kilojoules released when 56.7 grams of sucrose is burned.

**Section 10.8: Thermochemical Equations**

*Questions 57 through 62: Thermochemical equations may be written in two ways, either with an energy term as a part of the equation or with $\Delta H$ set apart from the regular equation. In the questions that follow, write both forms of the equations for the reactions described. Recall that state designations are required for all substances in a thermochemical equation.*

57. Energy is absorbed from sunlight in the photosynthesis reaction in which carbon dioxide and water vapor combine to produce sugar, $C_6H_{12}O_6$, and release oxygen. The amount of energy is $2.82 \times 10^3$ kJ per mole of sugar formed.

58. ■ When gaseous nitrogen reacts with gaseous oxygen to form nitrogen dioxide gas, 66.4 kJ of energy is absorbed for each mole of nitrogen that reacts.

59. The electrolysis of water is an endothermic reaction, absorbing 286 kJ for each mole of liquid water decomposed to its elements.

60. ■ When ammonia and oxygen react at a temperature at which both compounds are in the gaseous state, 226 kJ of energy is evolved for each mole of ammonia that reacts. Gaseous nitrogen monoxide and steam are the products of the reaction.

61. The reaction in an oxyacetylene torch is highly exothermic, releasing $1.31 \times 10^3$ kJ of heat for every mole of acetylene, $C_2H_2(g)$, burned. The end products are gaseous carbon dioxide and liquid water.

62. ■ Carbon dioxide gas reacts with hydrogen gas to form carbon monoxide gas and steam, and 41.2 kJ of energy is absorbed for each mole of carbon dioxide that reacts.

**Section 10.9: Thermochemical Stoichiometry**

63. Quicklime, the common name for calcium oxide, CaO, is made by heating limestone, $CaCO_3$, in slowly rotating kilns about 2.5 meters in diameter and about 60 meters long. The reaction is $CaCO_3(s) + 178 \text{ kJ} \rightarrow CaO(s) + CO_2(g)$. How many kilojoules are required to decompose 5.80 kg of limestone?

64. ■ How many grams of hydrogen would have to react to produce 71.9 kJ of energy from the reaction $2\, H_2(g) + O_2(g) \rightarrow 2\, H_2O(g) + 484 \text{ kJ}$?

65. The quicklime produced in Question 63 is frequently converted to calcium hydroxide, sometimes called slaked lime, by an exothermic reaction with water: $CaO(s) + H_2O(\ell) \rightarrow Ca(OH)_2(s) + 65.3 \text{ kJ}$. How many grams of quicklime were processed in a reaction that produced 291 kJ of energy?

66. ■ $\Delta H = -75.8$ kJ for the reaction $S(s) + 2\, CO(g) \rightarrow SO_2(g) + 2\, C(s)$. When 10.1 grams of sulfur react with excess carbon monoxide, how many kilojoules of energy are evolved or absorbed?

67. How many grams of octane, a component of gasoline, would you have to burn in your car to liberate $9.48 \times 10^5$ kJ of energy? $\Delta H = -1.09 \times 10^4$ kJ for the reaction $2\, C_8H_{18}(\ell) + 25\, O_2(g) \rightarrow 16\, CO_2(g) + 18\, H_2O(\ell)$.

68. ■ Calculate the quantity of energy (kJ) evolved or absorbed when 10.4 grams of carbon dioxide reacts with excess hydrogen if $\Delta H = 41.2$ kJ for the reaction $CO_2(g) + H_2(g) \rightarrow CO(g) + H_2O(g)$.

**General Questions**

69. Distinguish precisely and in scientific terms the differences among items in each of the following pairs or groups.
    a) Theoretical, actual, and percent yield
    b) Limiting reactant, excess reactant
    c) Heat of reaction, enthalpy of reaction
    d) Chemical equation, thermochemical equation
    e) Stoichiometry, thermochemical stoichiometry
    f) Joule, calorie

70. Classify each of the following statements as true or false:
    a) Coefficients in a chemical equation express the molar proportions among both reactants and products.
    b) A stoichiometry problem can be solved with an unbalanced equation.
    c) In solving a stoichiometry problem, the change from quantity of given substance to quantity of wanted substance is based on masses.
    d) Percent yield is actual yield expressed as a percent of theoretical yield.
    e) The quantity of product of any reaction can be calculated only through the moles of the limiting reactant.
    f) $\Delta H$ is positive for an endothermic reaction and negative for an exothermic reaction.

71. One of the few ways of "fixing" nitrogen, that is, making a nitrogen compound from the elemental nitrogen in the atmosphere, is by the reaction $Na_2CO_3 + 4\, C + N_2 \rightarrow 2\, NaCN + 3\, CO$. Calculate the mass in grams of $Na_2CO_3$ required to react with 35 g $N_2$.

72. How many grams of calcium phosphate will precipitate if excess calcium nitrate is added to a solution containing 3.98 grams of sodium phosphate?

73. Emergency oxygen masks contain potassium superoxide, $KO_2$, pellets. When exhaled $CO_2$ passes through the $KO_2$, the following reaction occurs: $4\, KO_2(s) + 2\, CO_2(g) \rightarrow 2\, K_2CO_3(s) + 3\, O_2(g)$. The oxygen produced can then be inhaled, so no air from outside the mask is needed. If the mask contains 125 grams of $KO_2$, how many grams of oxygen can be produced?

74. Baking cakes and pastries involves the production of $CO_2$ to make the batter rise. For example, citric acid, $H_3C_6H_5O_7$, in lemon or orange juice can react with baking soda, $NaHCO_3$, to produce carbon dioxide gas: $H_3C_6H_5O_7(aq) + 3\, NaHCO_3(aq) \rightarrow Na_3C_6H_5O_7(aq) + 3\, CO_2(g) + 3\, H_2O(\ell)$.
    a) If 6.00 g $H_3C_6H_5O_7$ reacts with 20.0 g $NaHCO_3$, how many grams of carbon dioxide will be produced?
    b) How many grams of which reactant will remain unreacted?
    c) Can you name $Na_3C_6H_5O_7$? Remember, it comes from citric acid.

75. ■ A laboratory test of 12.8 grams of aluminum ore yields 1.68 grams of aluminum. If the aluminum compound in the ore is $Al_2O_3$ and it is converted to the pure metal by the

reaction $2\ Al_2O_3(s) + 3\ C(s) \rightarrow 4\ Al(s) + 3\ CO_2(g)$, what is the percentage of $Al_2O_3$ in the ore?

76. How much energy is required to decompose 1.42 grams of $KClO_3$ according to the following equation: $2\ KClO_3(s) \rightarrow 2\ KCl(s) + 3\ O_2(g)$? $\Delta H = 89.5$ kJ for the reaction.

### More Challenging Problems

77. A phosphate rock quarry yields rock that is 79.4% calcium phosphate, the raw material used in preparing $Ca(H_2PO_4)_2$, a fertilizer known as *superphosphate*. The rock is made to react with sulfuric acid according to the equation $Ca_3(PO_4)_2 + 2\ H_2SO_4 \rightarrow Ca(H_2PO_4)_2 + 2\ CaSO_4$. What is the smallest number of kilograms of rock that must be processed to yield 0.500 ton of fertilizer?

78. ■ Carborundum, SiC, is widely used as an abrasive in industrial grinding wheels. It is prepared by the reaction of sand, $SiO_2$, with the carbon in coke: $SiO_2 + 3\ C \rightarrow SiC + 2\ CO$. How many kilograms of carborundum can be prepared from 727 kg of coke that is 88.9% carbon?

79. A sludge containing silver chloride is a waste product from making mirrors. The silver may be recovered by dissolving the silver chloride in a sodium cyanide (NaCN) solution, and then reducing the silver with zinc. The overall equation is $2\ AgCl + 4\ NaCN + Zn \rightarrow 2\ NaCl + Na_2Zn(CN)_4 + 2\ Ag$. What minimum amount of sodium cyanide is needed to dissolve all the silver chloride from 40.1 kilograms of sludge that is 23.1% silver chloride?

80. The chemical equation that describes what happens in an automobile storage battery as it generates electrical energy is $PbO_2 + Pb + 2\ H_2SO_4 \rightarrow 2\ PbSO_4 + 2\ H_2O$.
    a) What fraction of the lead in $PbSO_4$ comes from $PbO_2$ and what fraction comes from elemental lead?
    b) If the process uses 29.7 g $PbSO_4$, how many grams of $PbO_2$ must have been consumed?

81. Hydrogen and chlorine are produced simultaneously in commercial quantities by the electrolysis of salt water: $2\ NaCl + 2\ H_2O \rightarrow 2\ NaOH + H_2 + Cl_2$. Yield from the process is 61%. What mass of solution that is 9.6% NaCl must be processed to obtain 105 kilograms of chlorine?

82. A dry mixture of hydrogen chloride and air is passed over a heated catalyst in the Deacon process for manufacturing chlorine. Oxidation occurs by the following reaction: $4\ HCl + O_2 \rightarrow 2\ Cl_2 + 2\ H_2O$. If the conversion is 63% complete, how many tons of chlorine can be recovered from 1.4 tons of HCl? (*Hint:* Whatever you can do with moles, kilomoles, and millimoles, you can also do with ton-moles.)

83. Fluorides retard tooth decay by forming a hard, acid-resisting calcium fluoride layer in the reaction $SnF_2 + Ca(OH)_2 \rightarrow CaF_2 + Sn(OH)_2$. If at the time of a treatment there are 239 mg $Ca(OH)_2$ on the teeth and the dentist uses a mixture that contains 305 mg $SnF_2$, has enough of the mixture been used to convert all of the $Ca(OH)_2$? If no, what minimum additional amount should have been used? If yes, by what number of milligrams was the amount in excess?

84. ■ In the recovery of silver from silver chloride waste (see Question 79) a certain quantity of waste material is estimated to contain 184 g of silver chloride, AgCl. The treatment tanks are charged with 45 grams of zinc and 145 grams of sodium cyanide, NaCN. Is there enough of the two reactants to recover all of the silver from the AgCl? If no, how many grams of silver chloride will remain? If yes, how many more grams of silver chloride could have been treated by the available Zn and NaCN?

85. In 1866 a young chemistry student conceived the electrolytic method of obtaining aluminum from its oxide. This method is still used. $\Delta H = 1.97 \times 10^3$ kJ for the reaction $2\ Al_2O_3(s) + 3\ C(s) \rightarrow 4\ Al(s) + 3\ CO_2(g)$. The large amount of electrical energy required limits the process to areas of relatively inexpensive power. How many kilowatt-hours of energy are needed to produce one pound (454 g) of aluminum by this process, if 1 kw-hr $= 3.60 \times 10^3$ kJ?

86. ■ Nitroglycerine is the explosive ingredient in industrial dynamite. Much of its destructive force comes from the sudden creation of large volumes of gaseous products. A great deal of energy is released, too. $\Delta H = -6.17 \times 10^3$ kJ for the equation $4\ C_3H_5(NO_3)_3(\ell) \rightarrow 12\ CO_2(g) + 10\ H_2O(g) + 6\ N_2(g) + O_2(g)$. Calculate the number of pounds of nitroglycerine that must be used in a blasting operation that requires $5.88 \times 10^4$ kJ of energy.

87. ■ A student was given a 1.6240-g sample of a mixture of sodium nitrate and sodium chloride and was asked to find the percentage of each compound in the mixture. She dissolved the sample and added a solution that contained an excess of silver nitrate, $AgNO_3$. The silver ion precipitated all of the chloride ion in the mixture as AgCl. It was filtered, dried, and weighed. Its mass was 2.056 g. What was the percentage of each compound in the mixture?

88. A researcher dissolved 1.382 grams of impure copper in nitric acid to produce a solution of $Cu(NO_3)_2$. The solution went through a series of steps in which $Cu(NO_3)_2$ was changed to $Cu(OH)_2$, then to CuO, and then to a solution of $CuCl_2$. This was treated with an excess of a soluble phosphate, precipitating all the copper in the original sample as pure $Cu_3(PO_4)_2$. The precipitate was dried and weighed. Its mass was 2.637 g. Find the percent copper in the original sample.

89. How many grams of magnesium nitrate, $Mg(NO_3)_2$, must be used to precipitate as magnesium hydroxide all of the hydroxide ion in 50.0 mL 17.0% NaOH, the density of which is 1.19 g/mL? The precipitation reaction is $2\ NaOH + Mg(NO_3)_2 \rightarrow Mg(OH)_2 + 2\ NaNO_3$.

90. ■ If a solution of silver nitrate, $AgNO_3$, is added to a second solution containing a chloride, bromide, or iodide, the silver ion, $Ag^+$, from the first solution will precipitate the halide as silver chloride, silver bromide, or silver iodide. If excess $AgNO_3(aq)$ is added to a mixture of the above halides, it will precipitate them both, or all, as the case may be. A solution contains 0.230 g NaCl and 0.771 g NaBr. What is the smallest quantity of $AgNO_3$ that is required to precipitate both halides completely?

## Formulas

*These formulas are provided in case you are studying Chapter 10 before you study Chapter 6. You should not use this list unless your instructor has not yet assigned Chapter 6 or otherwise indicated that you may use the list.*

3. $Mg(OH)_2$, $MgO$, $H_2O$
4. $O_2$, $H_2O$, $CO_2$
5. $SO_2$, $O_2$, $SO_3$
6. $KHSO_4$, $KOH$, $K_2SO_4$, $H_2O$
23. $NaOH$, $H_3PO_4$
24. $NH_4Cl$, $NH_3$, $HCl$
25. $Mg(OH)_2$, $KOH$, $Mg(NO_3)_2$
26. $HBr$, $H_2$, $Br_2$
27. $Na_2SO_4$, $NaOH$, $H_2SO_4$
28. $S$, $CO$, $SO_2$, $C$
29. $Zn$, $NH_4Cl$, $ZnCl_2$, $NH_3$, $H_2$
30. $CaCO_3$, $CaO$, $CO_2$
41. $NH_3$, $N_2$, $H_2$
42. $CO$, $O_2$, $CO_2$
45. $BaCl_2$, $Na_2CrO_4$, $BaCrO_4$
57. $CO_2$, $H_2O$, $O_2$
58. $N_2$, $O_2$, $NO_2$
59. $H_2O$, $H_2$, $O_2$
60. $NH_3$, $O_2$, $NO$, $H_2O$
61. $CO_2$, $H_2O$
62. $CO_2$, $H_2$, $CO$, $H_2O$
72. $Ca_3(PO_4)_2$, $Ca(NO_3)_2$, $Na_3PO_4$
87. $NaNO_3$, $NaCl$

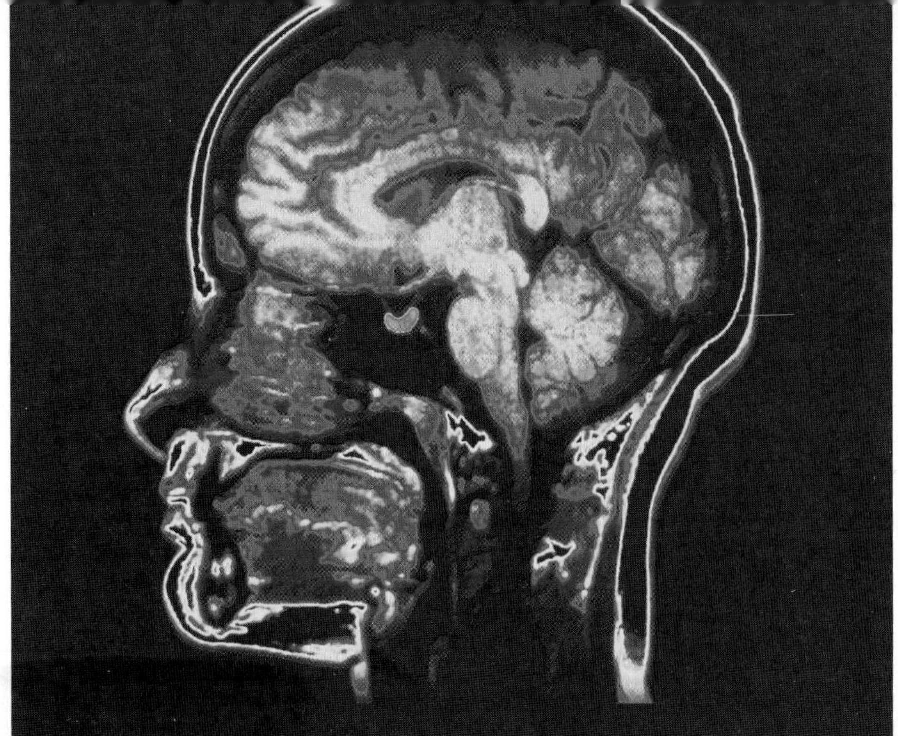

Scott Camazine/Photo Researchers, Inc.

# 11

Magnetic resonance imaging (MRI) scans, such as this computer-enhanced scan of a human head, are possible because of scientists' understanding of the quantum mechanical model of the atom. To make such an image, the MRI patient is placed in the opening of a long, narrow tube that is surrounded by a magnet. The magnetic field then interacts with the hydrogen atom protons in the part of the body being scanned. The protons first absorb energy from the magnet, and then they release the energy in such a way that the characteristics of their signals can be used to form an image. This image shows a normal, healthy brain.

# Atomic Theory: The Quantum Model of the Atom

CONTENTS



The four decades from 1890 to 1930 were a period of rapid progress in learning about the atom. Chapter 5 discusses the first half of this period; in this chapter we cover the second half.

In 1911, Rutherford's alpha-particle scattering experiments were controversial. In the Rutherford model of the atom, all the positive charge was crammed into the dense, tiny nucleus. Like charges repel, so the nucleus of the atoms should not be stable, yet it was. The relationships of classical physics that worked so well in explaining large-scale

Online homework for this chapter may be assigned in OWL

systems did not work on atom-sized systems. Thus, someone had to develop a new approach to understanding the atom. The breakthrough that was needed was the development of the field of study now known as **quantum mechanics.**

## 11.1 Electromagnetic Radiation

Goal 1 Define and describe electromagnetic radiation.

Goal 2 Distinguish between continuous and line spectra.

When you first open your eyes in the morning after a night's rest, the light that strikes your eyes stimulates nerve cells, which send signals to your brain that you interpret as the scene you see. Simultaneously, additional energy in the same form as the visible light—radio and television waves—also strikes your eyes, but that energy does not stimulate your optical nerve cells. Visible light, radio and television waves, and microwaves are some of the variations of the same type of energy, called **electromagnetic radiation.** Our eyes have evolved to detect just a small fraction of the entire **electromagnetic spectrum** (Fig 11.1), in a manner similar to how your radio is engineered to detect only radio waves without responding to television waves, microwaves, and visible light. Other parts of the electromagnetic spectrum include gamma rays, X-rays, ultraviolet radiation, and infrared radiation.

Electromagnetic radiation is a form of energy that consists of both electric and magnetic fields, and although this energy transmission seems instantaneous in our everyday experiences, it actually travels at a finite speed, $3.00 \times 10^8$ meters per second (186,000 miles per hour). This is called the **speed of light** and usually appears in equations with the symbol **c.** This speed is so great that we cannot sense the slight delay between the time light reflects off an object and when our brains interpret it. Light takes a little more than 8 minutes to travel from the sun to the earth and just over one second to travel from the moon to the earth. The timespan for light to travel short distances, such as from this page to your eye, is very, very tiny and beyond the limit of human perception.

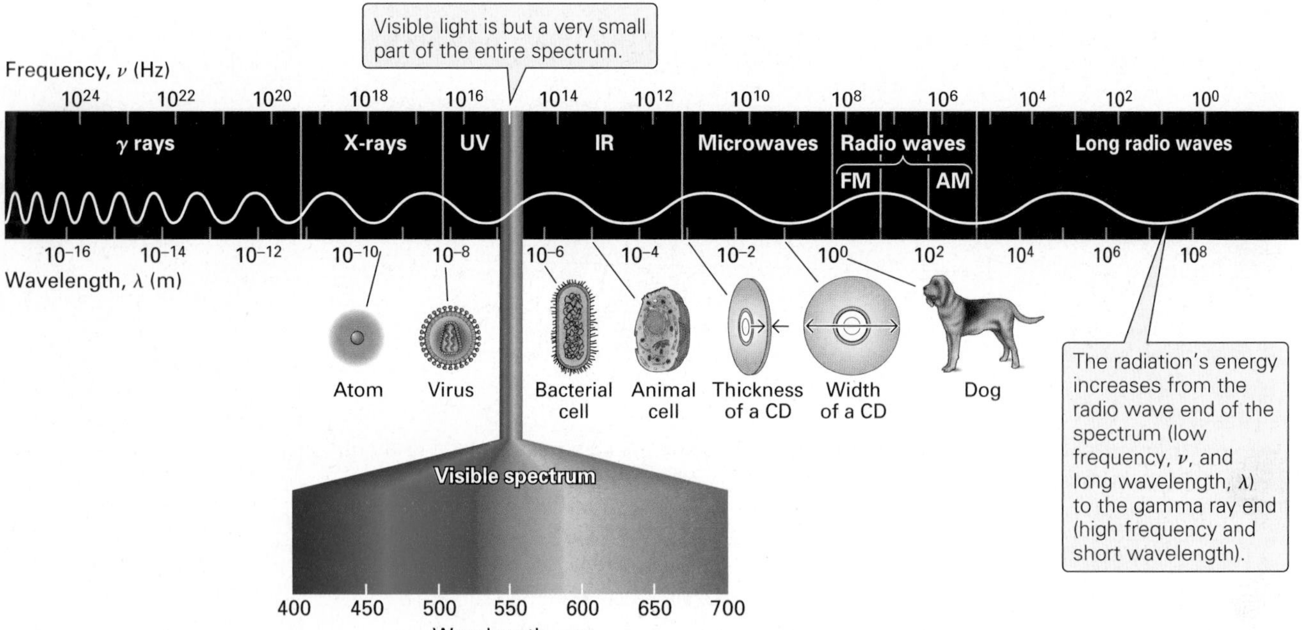


**Figure 11.1** The electromagnetic spectrum. Visible light is only a small portion of the electromagnetic spectrum, covering a wavelength range of about 400 to 700 nm. There seems to be no upper or lower limit to the theoretically possible lengths of electromagnetic waves, although experimentally measured waves vary between about $10^{11}$ m and $10^{-15}$ m.

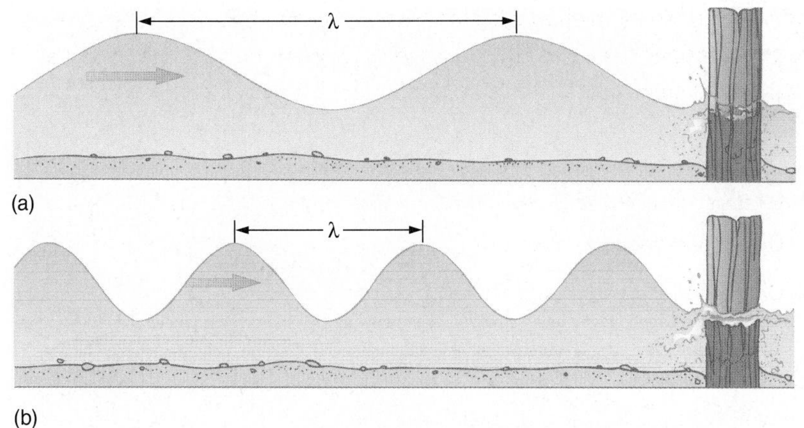


**Figure 11.2** Wave properties. Water waves are similar to electromagnetic waves in that they can be described by properties such as velocity, wavelength, and frequency. Velocity, v, is the linear speed of a point on a wave. In this illustration, the two waves are traveling at the same velocity. Wavelength, $\lambda$, is the distance between corresponding points on a wave. The wavelength in (a) is longer than that in (b). Frequency, $\nu$, is the number of complete waves passing a point per second. In this illustration, the wave crests in (b) will hit the post more often than those in (a), indicating a greater frequency.

Electromagnetic radiation has wavelike properties. Waves can be mathematically described by a **wave equation.** Every wave can be described by its properties, which include velocity (v or, specifically for electromagnetic radiation, c), wavelength ($\lambda$), and frequency ($\nu$) (Fig. 11.2). $\lambda$ and $\nu$ are the Greek lowercase letters lambda (pronounced *lam-duh*) and nu (pronounced *new*), respectively. Wavelength is the distance between identical parts of a wave, and frequency is the number of complete waves passing a point in a given period of time. The velocity of an electromagnetic wave is related to its wavelength and frequency by the equation $c = \lambda\nu$. Since the speed of light is essentially a fixed quantity, the wavelength and frequency of electromagnetic waves are inversely proportional. As the wavelength increases, the frequency decreases, and vice versa.

Light from standard lightbulbs, known as white light, produces a **continuous spectrum** when passed through a prism (Fig. 11.3). However, when an element is placed in a glass container and subjected to an electrical discharge, it glows, or emits light, and when that light is passed through a prism, the light forms a **line spectrum** as in Active Figure 11.4. These separate lines of color are known as **discrete** lines to indicate that they are individually distinct. Each element has a unique line spectrum (Fig. 11.5).

Why does the light from an element produce a discrete line spectrum? We won't be able to answer that question fully until Section 11.2, but an important piece of the answer was provided in 1900 by Max Planck, a German theoretical physicist. When a

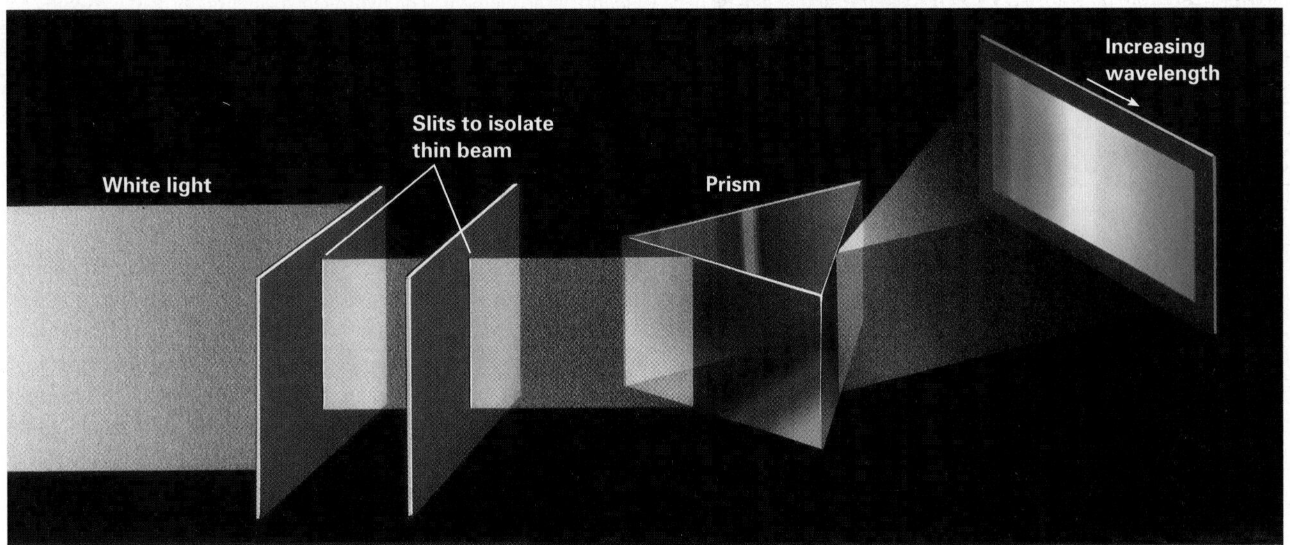


**Figure 11.3** Dispersion of white light by a prism. White light is passed through slits and then through a prism. It is separated into a continuous spectrum of all wavelengths of visible light. This corresponds to the visible spectrum in Figure 11.1.

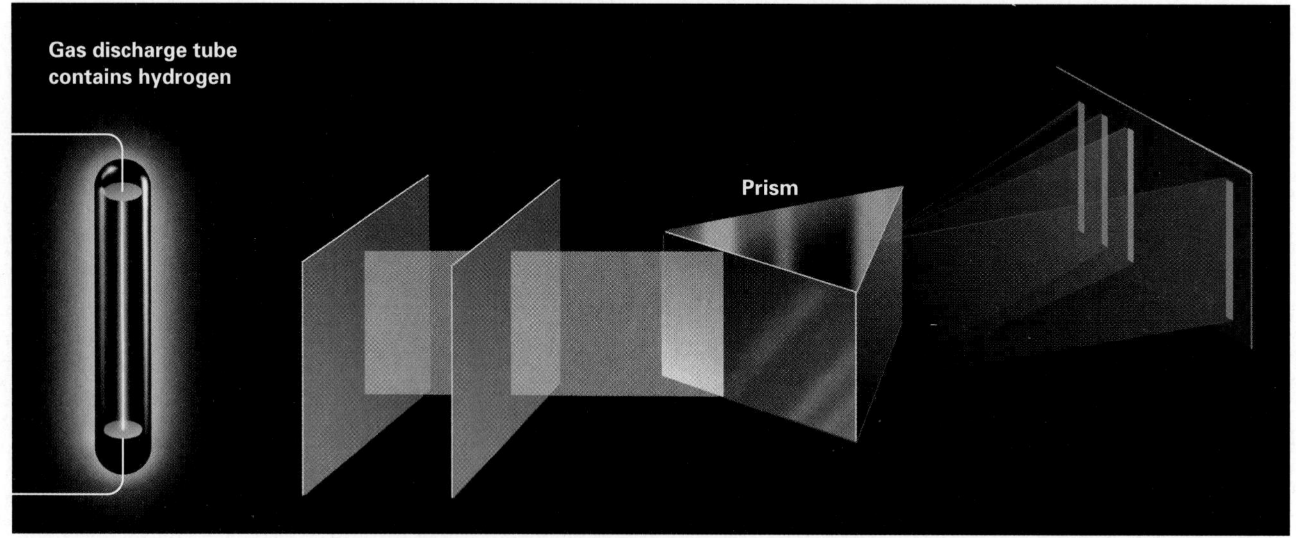


**Active Figure 11.4** Dispersion of light from a gas discharge tube filled with hydrogen. This is like a "neon" light, except that neon gives red light. The magenta light from hydrogen is passed through slits and then through a prism. It is separated into a line spectrum made up of four wavelengths of visible light. **Watch this active figure at *http://www.cengage.com/cracolice.***

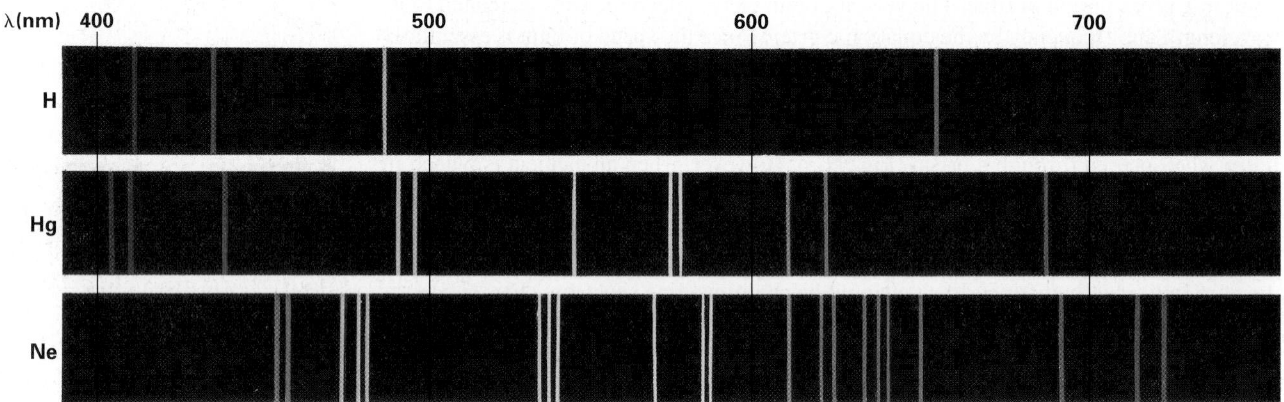


**Figure 11.5** Line spectra of hydrogen, mercury, and neon. Each element produces a unique spectrum that can be used to identify the element. The hydrogen spectrum corresponds with that shown in Active Figure 11.4. The neon spectrum is a combination of the colors we see as the red light of a neon sign.

AIP/Photo Researchers, Inc.

Max Planck (1858–1947). Planck was awarded the 1918 Nobel Prize in Physics for his discovery of the quantization of energy.

substance is heated sufficiently, it emits light. You see this when you turn on the burner of an electric stove. The burner emits red light as it gets hotter. The nature of the relationship between the intensity of the light emitted by a heated object and its frequency was explained by Planck. He stated that the energy of the electrons in the atoms of the material is directly proportional to their frequency: $E \propto \nu$, where E is energy. Albert Einstein improved on Planck's idea by proposing that energy is released by electrons in the form of a massless "packet" of electromagnetic radiation known as a **photon.** A photon is a particle of light.

You may be wondering how we can say that light is a wave and then say that light is a particle. If so, you are in good company because many brilliant scientists of the early 20th century also were intrigued by what is often referred to as the **wave-particle duality.** Today, we recognize that not only do electromagnetic waves have particle-like properties, but particles also have wavelike properties! As you will see later in this chapter, the nature of the universe as we observe it at the macroscopic level differs from what we know of how photons of energy and tiny particles behave. Light has properties that have no analogy at the macroscopic level; thus, we have to combine two different ideas to describe its behavior.

 **Target Check 11.1**

*Figure 11.1 shows that a decrease in frequency of electromagnetic radiation corresponds to an increase in wavelength. Explain why this is true.*

√ **Target Check 11.2**

*How does a continuous spectrum differ from a line spectrum? What is the source of each?*

## 11.2 The Bohr Model of the Hydrogen Atom

Goal 3 Describe the Bohr model of the hydrogen atom.

Goal 4 Explain the meaning of quantized energy levels in an atom and show how these levels relate to the discrete lines in the spectrum of that atom.

Goal 5 Distinguish between ground state and excited state.

**P/REVIEW:** Two objects having opposite charges, one positive and one negative, attract each other. Two objects with the same charge repel each other. See Section 2.7.

Chemical Heritage Foundation

Niels Bohr (1885–1962). Bohr won the 1922 Nobel Prize in Physics for his work in investigating the structure of the atom.

In 1913 Niels Bohr, a Danish scientist, suggested that an atom consists of an extremely dense nucleus that contains all of the atom's positive charge and nearly all of its mass. Negatively charged electrons of very small mass travel in orbits around the nucleus. The orbits are huge compared to the nucleus, which means that most of the atom is empty space. Bohr's model of the atom was based on the results of many scientific investigations that were known at the time, including Rutherford's nuclear atom, Planck's $E \propto \nu$ relationship, and Einstein's particles of light.

Bohr reasoned that the energy possessed by the electron in a hydrogen atom and the radius of its orbit are **quantized.** An amount that is quantized is limited to specific values; it may never be between two of those values. By contrast, an amount is **continuous** if it can have any value; between any two values there is an infinite number of other acceptable values (Fig. 11.6). A line spectrum is quantized, but the spectrum of white light is continuous. Bohr said that the electron in the hydrogen atom has **quantized energy levels.** This means that, at any instant, the electron may have one of several possible energies, but at no time may it have an energy between them.

Bohr calculated the values of the quantized energy levels by using an equation that contains an integer, *n,* that is, 1, 2, 3, . . . , and so forth. The results for the integers 1 to 4 are shown in Figure 11.7.

As noted above, the radius of the electron orbit is also quantized in the Bohr model of the hydrogen atom. This leads to one of the most interesting and strangest results from Bohr's model: The electron can orbit the nucleus at certain specified distances, but *it is never found between them.* This means that the electron simply disappears

**Figure 11.6** The quantum concept. A woman on a ramp can stop at any level above ground. Her elevation is not quantized. A woman on stairs can stop only on a step. Her elevation is quantized at $h_1$, $h_2$, $h_3$, or H.

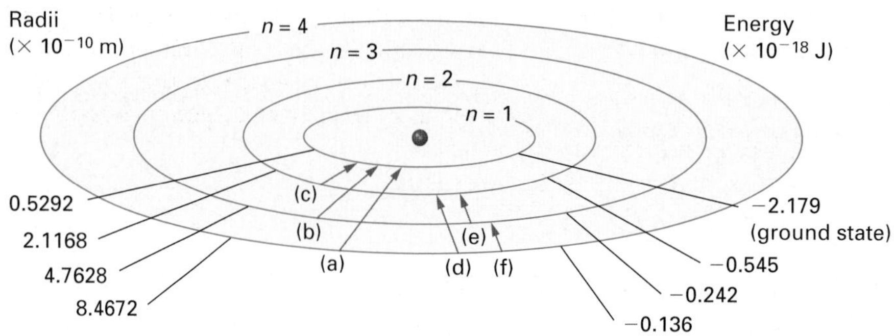


**Figure 11.7** The Bohr model of the hydrogen atom. The electron is allowed to circle the nucleus only at certain radii and with certain energies, the first four of which are shown. An electron in the ground-state level, $n = 1$, can absorb the exact amount of energy to raise it to any other level, such as $n = 2$, 3, or 4. An electron at such an excited state is unstable and drops back to the $n = 1$ level in one or more steps. Electromagnetic energy is radiated with each step dropped. Jumps *a*, *b*, and *c* are in the ultraviolet portion of the spectrum, *d* and *e* are in the visible range, and *f* is in the infrared region.

from one orbit and reappears in another! The process by which an electron moves between orbits is called a **quantum jump** or a **quantum leap.**

The electron is normally found in its **ground state, the condition when all electrons in an atom occupy the lowest possible energy levels.** For the single electron of a hydrogen atom, ground state is when the electron is at the $n = 1$ energy level. If the atom absorbs energy, the electron can be raised to an **excited state,** the condition at which one or more electrons in an atom has an energy level above ground state.

An electron in an excited state is unstable. It falls back to the ground state, sometimes in one quantum jump, sometimes in two or more. In doing this, it releases a photon of light energy with a frequency proportional to the energy difference between the two levels (Active Fig. 11.8). This energy release appears as a line in the spectrum of the element.

Using different values of *n* in his equation, Bohr was able to calculate the energies of all known lines in the spectrum of the hydrogen atom. He also predicted additional lines and their energies. When the lines were found, his predictions were proven to be correct. In fact, all of Bohr's calculations correspond with measured values to within one part per thousand. It certainly seemed that Bohr had discovered the structure of the atom.

There were problems, however. First, hydrogen is the *only* atom that fits the Bohr model. The model fails for any atom with more than one electron. Second, it is a fact that a charged body moving in a circle radiates energy. This means the electron itself should lose energy and promptly—in about 0.00000000001 second!—crash into the nucleus. This suggests that circular orbits violate the Law of Conservation of Energy (Section 2.9). So, for 13 years scientists accepted and used a theory they knew was only partly correct. They worked on the faulty parts and improved them, and finally they replaced the old theory with a new model that better explained the data, which we will introduce in the next section.

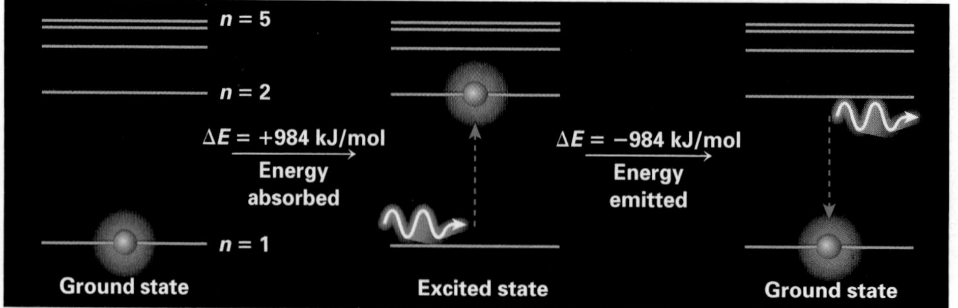


**Active Figure 11.8** Ground and excited states. When an electron in the ground state absorbs energy, it is promoted to an excited state. When the electron returns to the ground state, energy can be released in the form of a photon of electromagnetic radiation that corresponds to one of the lines in that element's line spectrum. $\Delta E$ is the amount of energy absorbed or emitted. **Watch this active figure at *http://www.cengage.com/cracolice.***

Niels Bohr made two huge contributions to the development of modern atomic theory. First, he suggested a reasonable explanation for the atomic line spectra in terms of electron energies. Second, he introduced the idea of quantized electron energy levels in the atom. These levels appear in modern theory as **principal energy levels;** they are identified by the **principal quantum number, *n*.**

### √ Target Check 11.3

*Identify the true statements, and rewrite the false statements to make them true.*

a) The speed of automobiles on a highway is quantized.
b) Paper money in the United States is quantized.
c) The weight of canned soup on a grocery store shelf is quantized.
d) The volume of water coming from a faucet is quantized.
e) A person's height is quantized.
f) Bohr described mathematically the orbits of electrons in a sodium atom.

## 11.3 The Quantum Mechanical Model of the Atom

In 1924 Louis de Broglie, a French scientist, suggested that matter in motion has properties that are normally associated with waves. He also said that these properties are especially applicable to subatomic particles. Between 1925 and 1928, Erwin Schrödinger applied the principles of wave mechanics to atoms and developed the **quantum mechanical model of the atom.** This model has been tested for more than 75 years. It explains more satisfactorily than any other theory all observations to date, and no exceptions have appeared. In fact, attempts to disprove the quantum mechanical model have served only to reinforce it. Today it is the generally accepted model of the atom.

The quantum mechanical model is both mathematical and conceptual. It keeps the quantized energy levels that Bohr introduced. In fact, it uses four *quantum numbers* to describe electron energy. These refer to (1) the principal energy level, (2) the sublevel, (3) the orbital, and (4) the number of electrons in an orbital.* The model is summarized at the end of this section. You might find it helpful to keep a finger at that summary and refer to it as details of the model are developed.

Oesper Collection in the History of Chemistry/University of Cincinnati

Louis de Broglie (1892–1987). De Broglie earned the 1929 Nobel Prize in Physics for his discovery of the wave characteristics of electrons.

### Principal Energy Levels

Goal | 6 Identify the principal energy levels in an atom and state the energy trend among them.

Following the Bohr model, principal energy levels are identified by the principal quantum number, $n$. The first principal energy level is $n = 1$, the second is $n = 2$, and so on. Mathematically, there is no end to the number of principal energy levels, but the seventh level is the highest occupied by ground-state electrons in any element now known.

The energy possessed by an electron depends on the principal energy level it occupies. In general, energies increase as the principal quantum numbers increase:

$$n = 1 < n = 2 < n = 3 \ldots < n = 7$$

Oesper Collection in the History of Chemistry/University of Cincinnati

Erwin Schrödinger (1887–1961). Schrödinger shared the 1933 Nobel Prize in Physics for his contributions to atomic theory.

### Sublevels

Goal | 7 For each principal energy level, state the number of sublevels, identify them, and state the energy trend among them.

*The formal names of these numbers are principal, azimuthal, magnetic, and electron spin. We use the name and number of the principal quantum number, but not the other three. All, however, are described to the extent necessary to specify the distribution of electrons in an atom.

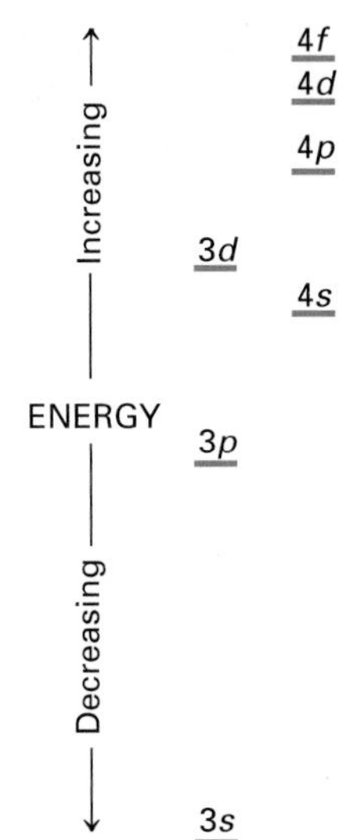


**Figure 11.9** Energies of $n = 3$ and $n = 4$ sublevels.

For each principal energy level there are one or more **sublevels.** They are the ***s, p, d,*** and ***f*** **sublevels,** using initial letters that come from terms formerly used in spectroscopy, the study of the interaction of matter and electromagnetic radiation.* A specific sublevel is identified by both the principal energy level and sublevel. Thus the *p* sublevel in the third principal energy level is the 3*p* sublevel. An electron that is in the 3*p* sublevel may be referred to as a "3*p* electron."

The total number of sublevels within a given principal energy level is equal to *n*, the principal quantum number. For $n = 1$ there is one sublevel, designated 1*s*. At $n = 2$ there are two sublevels, 2*s* and 2*p*. When $n = 3$ there are three sublevels, 3*s*, 3*p*, and 3*d*; $n = 4$ has four sublevels, 4*s*, 4*p*, 4*d*, and 4*f*. Quantum theory describes sublevels beyond *f* when $n = 5$ or more, but these are not needed for the ground state electron configurations of elements known today.

For elements other than hydrogen, the energy of each principal energy level spreads over a range related to the sublevels. These energies increase in the order *s, p, d, f*. Thus,

at $n = 2$, the increasing order of energy is $2s < 2p$

at $n = 3$, the increasing order of energy is $3s < 3p < 3d$

at $n = 4$, the increasing order of energy is $4s < 4p < 4d < 4f$

Beginning with principal quantum numbers 3 and 4, the energy ranges overlap. This is shown in Figure 11.9. When "plotted" vertically, the highest $n = 3$ electrons (3*d*) are at higher energy than the lowest $n = 4$ electrons (4*s*). Note, however, that for the same sublevel, $n = 3$ electrons always have lower energy than $n = 4$ electrons: $3s < 4s$; $3p < 4p$; $3d < 4d$.

## Electron Orbitals

**Goal** | **8** Sketch the shapes of *s* and *p* orbitals.

**Goal** | **9** State the number of orbitals in each sublevel.

According to modern atomic theory, it is not possible to know at the same time both the position of an electron in an atom and its velocity. This means it is not possible to describe the path an electron travels. There are no clearly defined orbits, as in the Bohr atom. However, we can describe mathematically a region in space around a nucleus in which there is a high probability of finding an electron. These regions are called **orbitals.** Notice the uncertainty of the *orbital*, stated in terms of "probability," compared to a Bohr *orbit* that states exactly where the electron is, where it was, and where it is going (Fig. 11.10).

Each sublevel has a certain number of orbitals. There is only one orbital for every *s* sublevel. All *p* sublevels have three orbitals, all *d* sublevels have five, and all *f* sub-

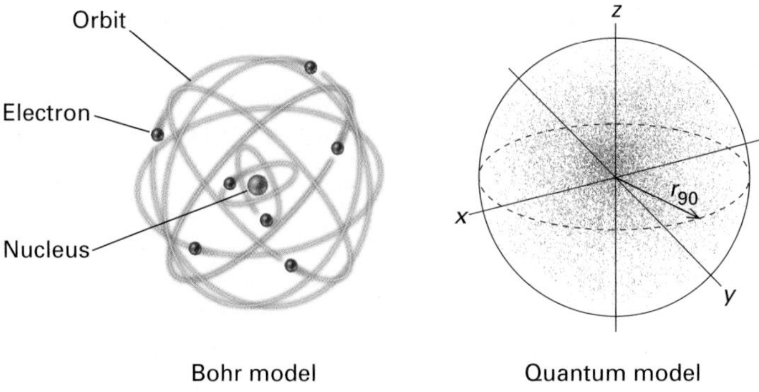


**Figure 11.10** The Bohr model of the atom compared with the quantum model. Bohr described the electron as moving in circular orbits of fixed radii around the nucleus. The quantum model says nothing about the precise location of the electron or the path in which it moves. Instead, each dot represents a possible location for the electron. The higher the density of dots, the higher the probability that an electron is in that region. The $r_{90}$ dimension is a radius; 90% of the time the electron is inside the dashed line, and 10% of the time it is farther from the nucleus.

*The terms are sharp, principal, diffuse, and fundamental. They describe the appearance of the spectral lines in a line spectrum.

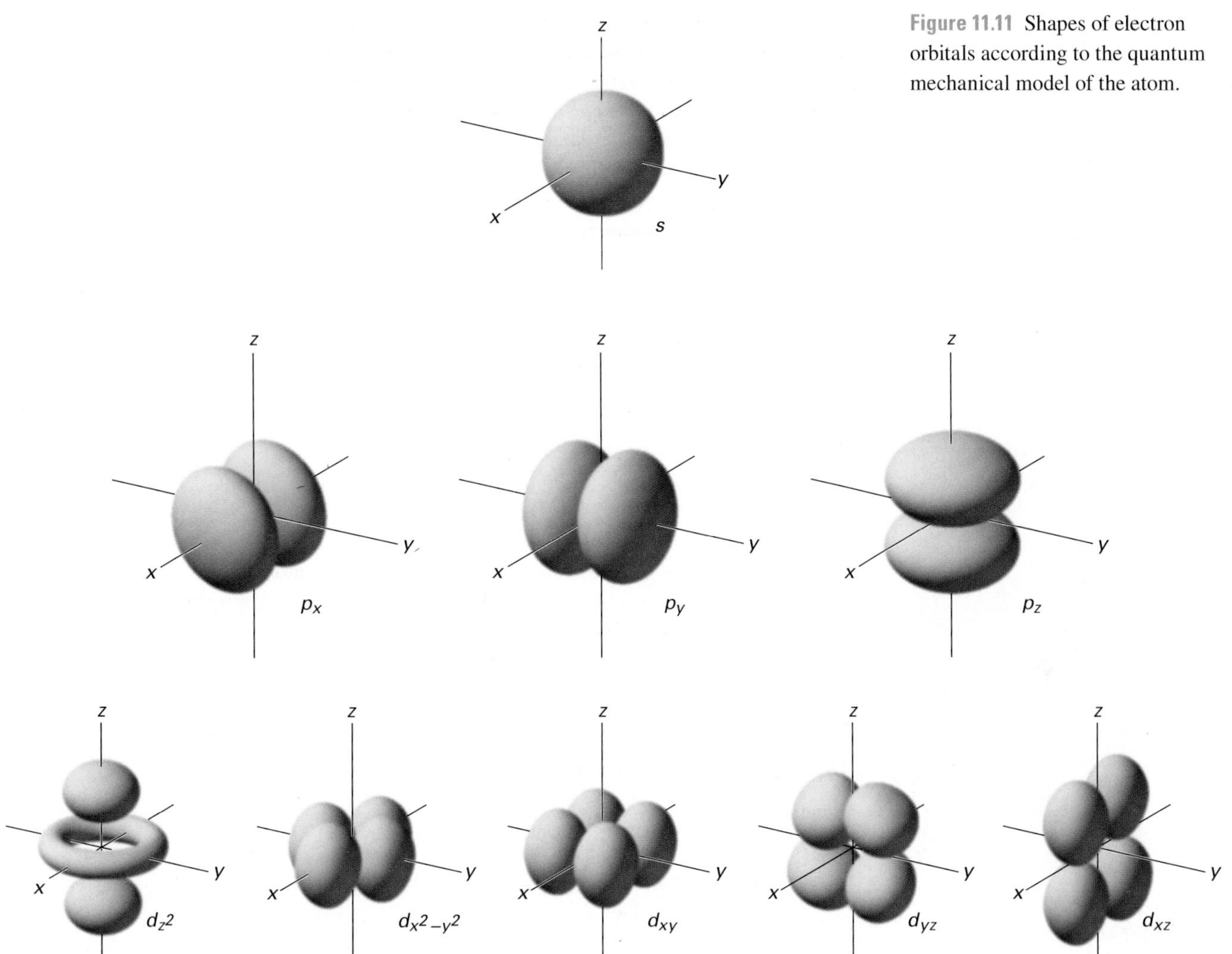


**Figure 11.11** Shapes of electron orbitals according to the quantum mechanical model of the atom.

levels have seven. This 1–3–5–7 sequence of odd numbers continues through higher sublevels.

Figure 11.11 shows the shapes of the *s, p,* and *d* orbitals. The seven *f* orbitals have even more complex shapes. The *x, y,* and *z* axes around which these shapes are drawn are from the mathematics of the quantum theory. We will be concerned with the shapes of only the *s* and *p* orbitals.

All *s* orbitals are spherical. As the principal quantum number increases, the size of the orbital increases. Thus, a 2*s* orbital is larger than a 1*s* orbital, 3*s* is larger than 2*s*, and so forth. Similar increases in size through constant shapes are present with *p, d,* and *f* orbitals at higher principal energy levels.

## Thinking About Your Thinking

### Probability

**The quantum mechanical model is probabilistic in nature, as opposed to the Bohr model, which is deterministic. A probabilistic phenomenon is one for which we cannot predict the future with certainty, even if we know all there is to know about the system at the present moment. A deterministic phenomenon is one for which we can determine the future state of the system if we know all of the system's present conditions. For example, in the Bohr model, if we know the speed of an electron and its orbit path, we can predict exactly where the electron will be at some specified future time. This is a deterministic phenomenon.**

**The quantum mechanical model is an example of a probabilistic phenomenon. Even if we know everything that can be simultaneously measured about an electron at a given moment in time, we cannot predict things such as the future position or speed of the electron. We can only predict *probabilities* about the future. Thus quantum mechanics gives us *orbitals,* regions in space where an electron is likely to be found at a certain probability level, rather than the deterministic orbits of the Bohr model.**

**Many natural phenomena are probabilistic in nature. This is one concept that distinguishes modern science from the thinking accepted at the beginning of the 20th century, when scientists believed that given knowledge of all present conditions, the future conditions could always be predicted. Alas, nature is not that simple. Look for more examples of probabilistic relationships and phenomena in your future studies in science.**

© Bettmann/Corbis

Wolfgang Pauli (1900–1958) discovered the exclusion principle and was awarded the 1945 Nobel Prize in Physics for his work.

## The Pauli Exclusion Principle

Goal | **10** State the restrictions on the electron population of an orbital.

The last detail of the quantum mechanical model of the atom comes from the **Pauli exclusion principle.** Its effect is to limit the population of any orbital to two electrons. At any instant an orbital may be (1) unoccupied, (2) occupied by one electron, or (3) occupied by two electrons. No other occupancy is possible.

### summary

**The Quantum Mechanical Model of the Atom**

***Principal Energy Levels***

Principal energy levels are identified by the principal quantum number, $n$, a series of integers: $n = 1, 2, 3, \ldots, 7$. Generally, energy increases with increasing $n$: $n = 1 < n = 2 < n = 3, \ldots$.

***Sublevels***

Each principal energy level—each value of $n$—has $n$ sublevels. These sublevels are identified by the principal quantum number followed by the letter *s, p, d,* or *f.* Sublevels that are not needed for the ground state electron configurations of elements known today appear in color.

| Energy Trend | $n$ | Number of Sublevels | Identification of Sublevels |
|---|---|---|---|
| Increasing energy ↓ | 1 | 1 | 1*s* |
| | 2 | 2 | 2*s*, 2*p* |
| | 3 | 3 | 3*s*, 3*p*, 3*d* |
| | 4 | 4 | 4*s*, 4*p*, 4*d*, 4*f* |
| | 5 | 5 | 5*s*, 5*p*, 5*d*, 5*f*, 5*g* |
| | 6 | 6 | 6*s*, 6*p*, 6*d*, 6*f*, 6*g*, 6*h* |
| | 7 | 7 | 7*s*, 7*p*, 7*d*, 7*f*, 7*g*, 7*h*, 7*i* |

For any given value of $n$, energy increases through the sublevels in the order of *s, p, d, f*: $2s < 2p$; $3s < 3p < 3d$; $4s < 4p < 4d < 4f$; and so on.

*Note:* The range of energies in consecutive principal energy levels may overlap. Example: $4s < 3d < 4p$. However, for any given sublevel, energy and orbital size increase with increasing $n$: $1s < 2s < 3s \ldots$, $2p < 3p < 4p \ldots$, and so on.

***Orbitals and Orbital Occupancy***

Each kind of sublevel contains a definite number of orbitals that begin with 1 and increase in order with odd numbers: *s*, 1; *p*, 3; *d*, 5; *f*, 7.

An orbital may be occupied by 0, 1, or 2 electrons, but never more than 2. Therefore, the maximum number of electrons in a sublevel is twice the number of orbitals in the sublevel.

| Sublevel | Orbitals | Maximum Electrons per Sublevel |
|---|---|---|
| $s$ | 1 | $1 \times 2 = 2$ |
| $p$ | 3 | $3 \times 2 = 6$ |
| $d$ | 5 | $5 \times 2 = 10$ |
| $f$ | 7 | $7 \times 2 = 14$ |

## Thinking About Your Thinking

### Formal Models

**The quantum mechanical model of the atom is an example of a formal model, one where the concept being modeled has abstract parts that have to be imagined. Figures 11.10 and 11.11 are the keys to building a model in your mind that you can use to think about the quantum mechanical model.**

**Figure 11.10 is designed to help you think about how electrons behave in atoms. The Bohr model is an analogy to the solar system. Electrons orbit the nucleus like planets orbit the sun. This model is incorrect. The quantum model, on the right of Figure 11.10, is the mental model you should form. Each blue dot represents the position of a single electron at one point in time. Imagine that you can take a photograph of the single electron in a hydrogen atom. You center your camera so that the nucleus is at the convergence of the *x, y,* and *z* axes in the figure. When you take a picture, you make an overhead transparency of the single blue dot. Now you overlap a few hundred such transparencies. This will give the quantum model illustration you see. The dashed line that is labeled as the $r_{90}$ distance is the radius that encloses 90% of the dots. In other words, 90% of the time, you will find the electron within this radius.**

**Now look at the *s* orbital in Figure 11.11. It represents the same thing as the quantum model illustration in Figure 11.10. The radius of the sphere in Figure 11.11 corresponds to the $r_{90}$ radius of the circle in Figure 11.10. It illustrates in three dimensions the limit of a region in space inside of which there is a 90% probability of finding an electron at any given instant. The *p* and *d* orbitals of Figure 11.11 also represent the regions in which the *p* and *d* electrons may be found 90% of the time.**

**Your mental models of the *s* and *p* electron orbitals, plus your model of the behavior of an electron in an atom, combine to make the quantum mechanical model you should hold in your mind at this point in your chemistry studies.**

### √ Target Check 11.4

*Identify the true statements, and rewrite the false statements to make them true. If possible, avoid looking at the summary of the quantum mechanical model.*

a) There is one *s* orbital when $n = 1$, two *s* orbitals when $n = 2$, three s orbitals when $n = 3$, and so on.

b) All $n = 3$ orbitals are at lower energy than all $n = 4$ orbitals.

c) There is no *d* sublevel when $n = 2$.

d) There are five *d* orbitals at both the fourth and sixth principal energy levels.

# 11.4 Electron Configuration

## Electron Configurations and the Periodic Table

**Goal | 11** Use a periodic table to list electron sublevels in order of increasing energy.

Many of the chemical properties of an atom or ion depend on its **electron configuration,** the ground-state distribution of electrons among the orbitals of the species. Two rules guide the assignments of electrons to orbitals:

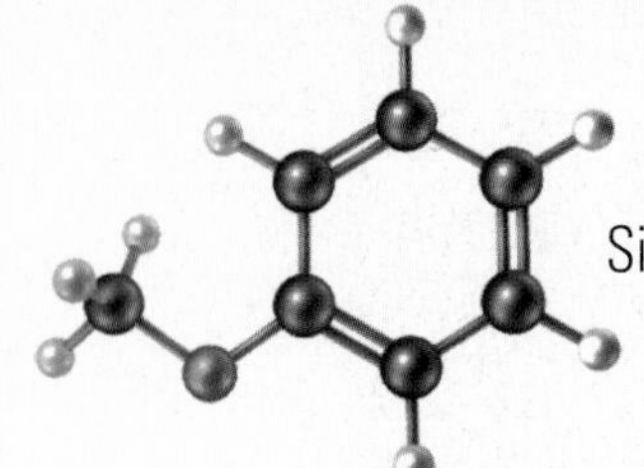

# Everyday Chemistry

## Simply Pure Darn Foolishness?

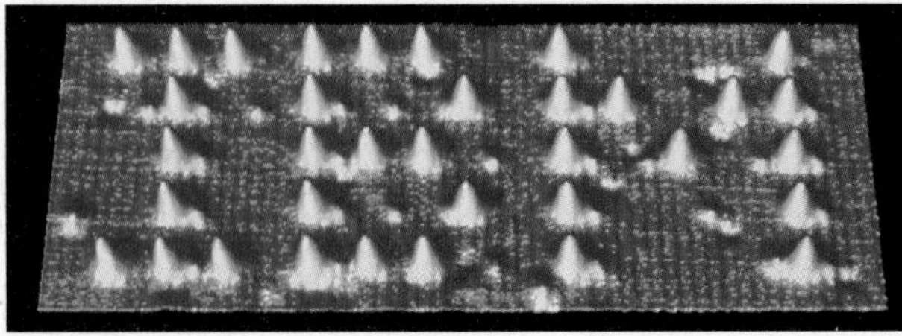

IBM Corporation, Research Division, Almaden Research Center

Chemistry serves marketing. The logo is composed of 35 xenon (Z = 54) atoms, *individually* moved using a scanning tunneling microscope. The space between the atoms is about 1.3 × $10^{-7}$ cm. As printed, the magnification is 4.7 million.

The rules of quantum mechanics have no direct correlation in physical systems much larger than atoms. It's easy to believe that this chapter has nothing to do with "real life" and is just something else to memorize and forget later. Given a choice (with the instructor out of earshot), some students would argue that the *s, p, d,* and *f* classifications for electron orbitals stand for "**S**imply **P**ure **D**arn **F**oolishness."

Nothing could be further from the truth.

Because it's so different, quantum mechanics has been tested by many scientists since Erwin Schrödinger proposed it in 1927. It has passed all the tests. Indeed, Albert Einstein spent portions of the last 30 years of his life futilely searching for an alternative explanation of atomic structure.

The most rigorous test a scientific model can pass is to predict the results of an experiment and then have those results confirmed in the laboratory. The first major prediction of quantum mechanics that came true was the transistor, invented in 1947 at Bell Laboratories. The small size and low power consumption of the transistor make possible the complicated electrical circuits in the personal-computer microprocessor. This textbook was typed on a desktop computer, a tool that has immensely changed writers' lives for the better since the 1980s.

Today, microprocessors are everywhere. Of course, they are in your computer and calculator, but microprocessors are also in many simple things we take for granted. Personal stereos, appliances with digital displays, automobiles with exhaust emission controls, even quartz watches and clocks (jumping second hand) all are controlled by microprocessors. One of the most common uses for microprocessors in the United States today is for controlling traffic lights.

Look back at Figure 11.11. See the *p* orbitals with the two lobes. You know

© sciencephotos/Alamy

© Joe White, 2008. Used under license from Shutterstock.com

The transistor (left, center) replaced the vacuum tube (left) on about a 1-to-1 basis, making battery-powered portable radios possible in the 1950s. A modern integrated circuit (right) contains millions of transistors.

**P/REVIEW:** We are showing both the traditional U.S. A–B numbering scheme for groups in the periodic table and the 1-to-18 system that has been approved by the International Union of Pure and Applied Chemistry (IUPAC) (Section 5.6). Use—and "read"—the one selected by your instructor.

1. At ground state the electrons fill the *lowest*-energy orbitals available.
2. No orbital can have more than two electrons.

Although the first periodic tables were based on properties of the elements, quantum mechanics shows that its true basis lies in the arrangement of electrons as predicted by the quantum model. As the number of electrons in an atom increases, specific sublevels are filled in different regions of the periodic table. This is indicated by color in Active Figure 11.12. In Groups 1A/1 and 2A/2 the *s* sublevels are the highest *occupied* energy sublevels. Other sublevels have higher energy, but they have no electrons in them, and thus they are *not occupied*. The *p* orbitals are filled in order across Groups 3A/13 to 8A/18. The *d* electrons appear in the B groups (3 to 12). Finally, *f* electrons show up in the lanthanide and actinide series.

When you read the periodic table from left to right across the periods in Active Figure 11.12, you get the order of increasing sublevel energy. The first period gives only the 1*s* sublevel. The 1*s* orbital is the lowest-energy orbital in any atom. Period 2

that those orbitals are high-probability regions, but you may not have known that the lobes of a *p* orbital do not touch each other. Between those two lobes is a node, a region in which there is zero probability of finding an electron. Yet we know from experiment that a single 2*p* electron is in *both* lobes at the same time! Quantum mechanics holds that an electron can tunnel from one lobe to the other without going through the node that is between them. This is unbelievable on the macroscopic level, but true at the level of subatomic particles.

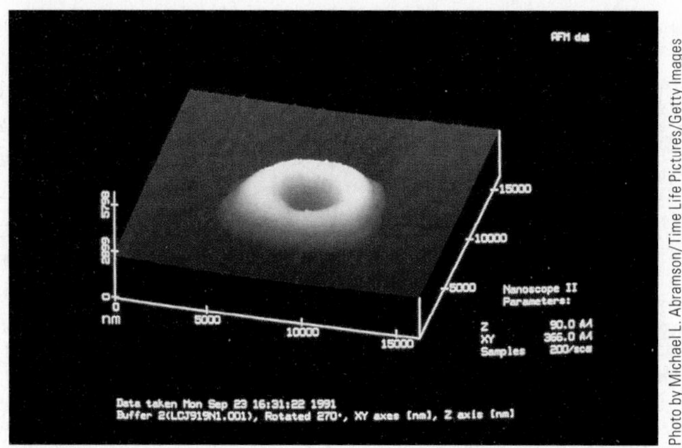


Photo by Michael L. Abramson/Time Life Pictures/Getty Images

The Atomic Force Microscope (AFM), a modification of the scanning tunneling microscope, allows us to see groups of atoms. This is a single red blood cell. The AFM can also slice the cell to reveal individual protein molecules inside.

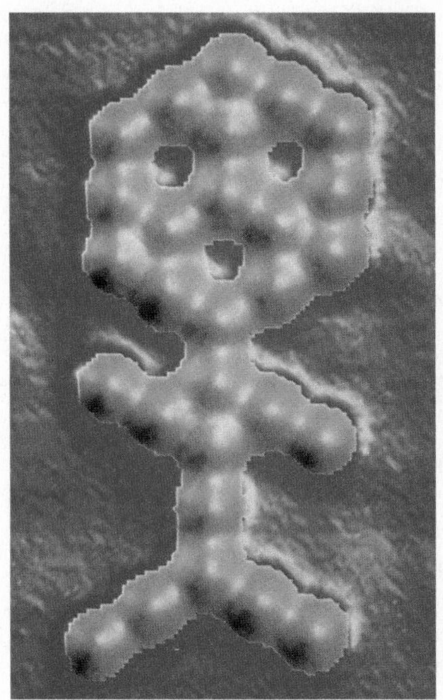

IBM Corporation, Research Division, Almaden Research Center

Molecular Man is made of 28 carbon monoxide molecules on a platinum surface. He is all of 5 X $10^{-7}$ cm tall.

In 1981, Gerd Binning and Heinrich Rohrer, two researchers at IBM in Zurich, Switzerland, invented the scanning tunneling microscope (STM). They shared the 1986 Nobel Prize in Physics (with Ernst Ruska, inventor in 1933 of the electron microscope) for this invention, which is rapidly changing the way chemists think about atoms. The STM probes surfaces with a tungsten needle that is at most a few atoms wide. At very short distances, electrons tunnel from the needle across the gap between the needle and the surface under study. This generates a minute current that can be converted into images of individual atoms.

After mapping individual atoms, scientists using STMs learned they could apply a voltage pulse at the needle to pick up atoms one at a time and move them! Their first effort was an IBM logo. The dots are individual atoms of xenon that were moved approximately 4 × $10^{-7}$ millimeter per second. The process took about 22 hours. Molecular Man was drawn by arranging 28 CO molecules in the stick figure shown.

The implications of moving individual atoms are stunning. For example, as scientists mapping out the full set of human chromosomes learn which human diseases are caused by molecular errors in a chromosome, tools such as the STM could theoretically be used to move individual atoms to fix the error and cure the disease.

Hang on. Your study of chemistry places you in the middle of the most exciting new developments in medicine, science, and technology. You ain't seen nothin' yet!

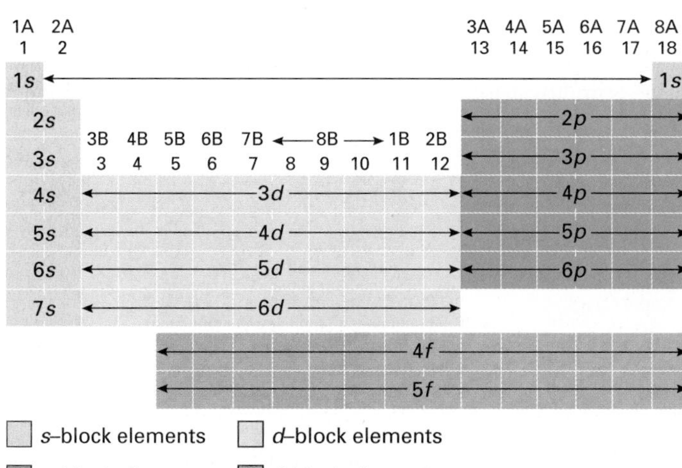


**Active Figure 11.12** Arrangement of periodic table according to atomic sublevels. The highest-energy sublevels occupied at ground state are *s* sublevels in Groups 1A/1 and 2A/2. This region of the periodic table is the *s*-block. Similarly, *p* sublevels are the highest occupied sublevels in Groups 3A/13 to 8A/18, the *p*-block. The *d*-block includes the B Groups (3 to 12) whose highest occupied energy sublevels are *d* sublevels. Finally, the *f*-block is made up of the elements whose *f* sublevels hold the highest-energy electrons. **Watch this active figure at *http://www.cengage.com/cracolice.***

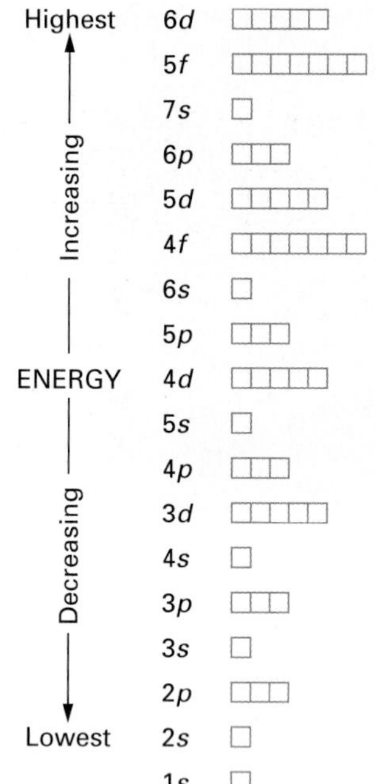


**Figure 11.13** Sublevel energy diagram. All sublevels are positioned vertically according to a general energy-level scale shown at the left. Each box represents an orbital. When orbitals are filled from the lowest energy level, each orbital will generally hold two electrons before any orbital higher in energy accepts an electron.

**P/REVIEW:** Z is the atomic number of an element. It stands for the number of protons in the nucleus of each atom of the element. See Section 5.4.

takes you through 2*s* and 2*p*. Similarly, the third period covers 3*s* and 3*p*. Period 4 starts with 4*s*, follows with 3*d*, and ends up with 4*p*, and so forth. Ignoring minor variations in Periods 6 and 7, the complete list of sublevel energies is given in Figure 11.13, in order of increasing sublevel energy. The periodic table therefore is a guide to the order of increasing sublevel energy.

Think, for a moment, how remarkable this is. Mendeleev and Meyer developed their periodic tables from the physical and chemical properties of the elements. They knew nothing of electrons, protons, nuclei, wave functions, or quantized energy levels. Yet, when these things were found some 60 years later, the match between the first periodic tables and the quantum mechanical model of the atom was nearly perfect.

*If you are ever required to list the sublevels in order of increasing energy without reference to a periodic table, the following diagram taken from the summary of the quantum model may be helpful. Beginning at the upper left, the diagonal lines pass through the sublevels in the sequence required.*

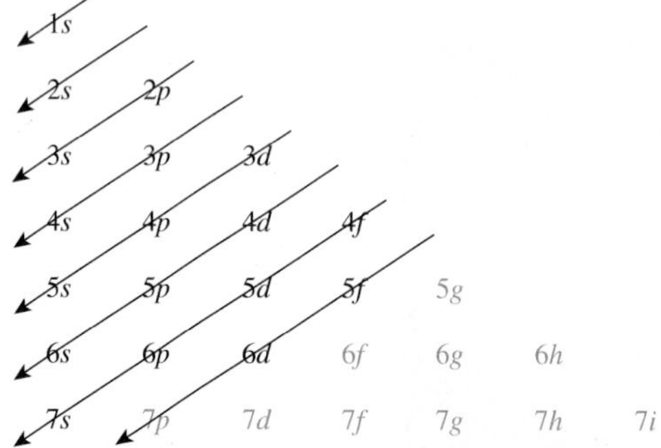


*The sublevels shown in color are not needed for the elements known today, but the mathematics of the quantum mechanical model predict the order of increasing energy indefinitely.*

Let's look at examples of ground-state electron configurations of atoms:

The one electron of a hydrogen atom (H, Z = 1) occupies the lowest-energy orbital in any atom. Both Figures 11.12 and 11.13 show that this is the 1*s* orbital. The total number of electrons in any sublevel is shown by a superscript number. Therefore, the electron configuration of hydrogen is $1s^1$ (Fig. 11.14). Helium (He, Z = 2) has two electrons, and both fit into the 1*s* orbital. The helium configuration is $1s^2$.

These and other electron configurations to be developed appear in the first four periods of the periodic table in Figure 11.15.

Lithium (Li, Z = 3) has three electrons. The first two fill the 1*s* orbital, as before. The third electron goes to the next orbital up the energy scale that has a vacancy. According to the order of increasing energy derived from reading the periodic table as in Active Figure 11.12, this is the 2*s* orbital. The electron configuration for lithium is therefore $1s^22s^1$. Similarly, beryllium (Be, Z = 4) divides its four electrons between the two lowest orbitals, filling both: $1s^22s^2$. These configurations are also in Figure 11.15.

The first four electrons of boron (B, Z = 5) fill 1*s* and 2*s* orbitals. The fifth electron goes to the next highest level, 2*p*, according to the periodic table. The configuration for boron is $1s^22s^22p^1$. Similarly, carbon (C, Z = 6) has a $1s^22s^22p^2$ configuration.*

**Figure 11.14** Electron configuration notation. An electron configuration specifies the number of electrons in each sublevel for any specified atom. The value of *n* that represents the principal energy level is written first, followed by the letter that indicates the sublevel. The number of electrons in each sublevel is written as a superscript.

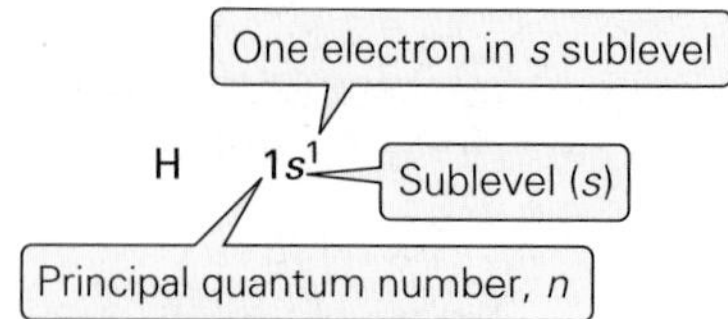


*Although we will not emphasize the point, the two 2*p* electrons occupy different 2*p* orbitals. In general, all orbitals in a sublevel are half-filled before any orbital is completely filled.

| 1 **H** $1s^1$ | | | | | | | | | | | | | | | | | 2 **He** $1s^2$ |
|---|---|---|---|---|---|---|---|---|---|---|---|---|---|---|---|---|---|
| 3 **Li** $1s^2$ $2s^1$ | 4 **Be** $1s^2$ $2s^2$ | | | | | | | | | | | 5 **B** $1s^2$ $2s^2 2p^1$ | 6 **C** $1s^2$ $2s^2 2p^2$ | 7 **N** $1s^2$ $2s^2 2p^3$ | 8 **O** $1s^2$ $2s^2 2p^4$ | 9 **F** $1s^2$ $2s^2 2p^5$ | 10 **Ne** $1s^2$ $2s^2 2p^6$ |
| 11 **Na** [Ne] $3s^1$ | 12 **Mg** [Ne] $3s^2$ | | | | | | | | | | | 13 **Al** [Ne] $3s^2 3p^1$ | 14 **Si** [Ne] $3s^2 3p^2$ | 15 **P** [Ne] $3s^2 3p^3$ | 16 **S** [Ne] $3s^2 3p^4$ | 17 **Cl** [Ne] $3s^2 3p^5$ | 18 **Ar** [Ne] $3s^2 3p^6$ |
| 19 **K** [Ar] $4s^1$ | 20 **Ca** [Ar] $4s^2$ | 21 **Sc** [Ar] $4s^2 3d^1$ | 22 **Ti** [Ar] $4s^2 3d^2$ | 23 **V** [Ar] $4s^2 3d^3$ | 24 **Cr** [Ar] $4s^1 3d^5$ | 25 **Mn** [Ar] $4s^2 3d^5$ | 26 **Fe** [Ar] $4s^2 3d^6$ | 27 **Co** [Ar] $4s^2 3d^7$ | 28 **Ni** [Ar] $4s^2 3d^8$ | 29 **Cu** [Ar] $4s^1 3d^{10}$ | 30 **Zn** [Ar] $4s^2 3d^{10}$ | 31 **Ga** [Ar] $4s^2$ $3d^{10} 4p^1$ | 32 **Ge** [Ar] $4s^2$ $3d^{10} 4p^2$ | 33 **As** [Ar] $4s^2$ $3d^{10} 4p^3$ | 34 **Se** [Ar] $4s^2$ $3d^{10} 4p^4$ | 35 **Br** [Ar] $4s^2$ $3d^{10} 4p^5$ | 36 **Kr** [Ar] $4s^2$ $3d^{10} 4p^6$ |

**Figure 11.15** Ground-state electron configurations of neutral atoms.

The next four elements increase the number of electrons in the three $2p$ orbitals until they are filled with six electrons for neon (Ne, Z = 10). All of these configurations appear in Figure 11.15.

The first ten electrons of sodium (Na, Z = 11) are distributed in the same way as the ten electrons in neon. The eleventh sodium electron is a $3s$ electron: $1s^2 2s^2 2p^6 3s^1$. The configurations for all elements whose atomic numbers are greater than 10 begin with the neon configuration, $1s^2 2s^2 2p^6$. This part of the configuration is often shortened to the **neon core,** represented by [Ne]. For sodium this becomes [Ne]$3s^1$; for magnesium (Mg, Z = 12), [Ne]$3s^2$; for aluminum (Al, Z = 13), [Ne]$3s^2 3p^1$; and so on to argon (Ar, Z = 18), [Ne]$3s^2 3p^6$. The sequence is exactly as it was in Period 2. The neon core is used for Period 3 in Figure 11.15.

Potassium (K, Z = 19) repeats at the $4s$ level the development of sodium at the $3s$ level. Its complete configuration is $1s^2 2s^2 2p^6 3s^2 3p^6 4s^1$. All configurations for atomic numbers greater than 18 distribute their first eighteen electrons in the configuration of argon, $1s^2 2s^2 2p^6 3s^2 3p^6$. This may be shortened to the **argon core,** [Ar]. Accordingly, the configuration for potassium may be written [Ar]$4s^1$, and calcium (Ca, Z = 20) is [Ar]$4s^2$.

The periodic table tells us that five $3d$ orbitals are next available for electron occupancy. The next three elements fill in order, as predicted, to vanadium (V, Z = 23); [Ar]$4s^2 3d^3$.* Chromium (Cr, Z = 24) is the first element to break the orderly sequence in which the lowest-energy orbitals are filled. Its configuration is [Ar]$4s^1 3d^5$, rather than the expected [Ar]$4s^2 3d^4$. This is generally attributed to an extra stability found when all orbitals in a sublevel are half-filled or completely filled. Manganese (Mn, Z = 25) puts us back on the track, only to be derailed again at copper (Cu, Z = 29): [Ar]$4s^1 3d^{10}$. Zinc (Zn, Z = 30) has the expected configuration: [Ar]$4s^2 3d^{10}$. Examine the sequence for atomic numbers 21 to 30 in Figure 11.15 and note the two exceptions.

By now the pattern should be clear. Atomic numbers 31 to 36 fill in sequence in the next orbitals available, which are the $4p$ orbitals. This is shown in Figure 11.15.

Our consideration of electron configuration ends with atomic number 36, krypton. If we were to continue, we would find the higher $s$ and $p$ orbitals fill just as they do in Periods 2 to 4. The $4d$, $4f$, $5d$, and $5f$ orbitals have several variations like those for chromium and copper, so their configurations must be looked up. But you should be able to reproduce the configurations for the first 36 elements—*not from memory or from Figure 11.15, but by referring to a periodic table.*

*Some chemists prefer to write this configuration [Ar]$3d^3 4s^2$, putting the $3d$ before the $4s$. This is equally acceptable. There is, perhaps, some advantage at this time in listing the sublevels in the order in which they fill, which can be "read" from the periodic table, as you will see shortly. This same idea continues as the $f$ sublevels are filled, but with frequent irregularities. We will not be concerned with $f$-block elements.

## Writing Electron Configurations

**Goal** | **12** Referring only to a periodic table, write the ground-state electron configuration of an atom of any element up to atomic number 36.

You can write the electron configuration of an atom with help from the periodic table if you can

1. List the sublevels in order of increasing energy (Goal 11),
2. State the maximum number of electrons that occupy each sublevel, and
3. Establish the number of electrons in the highest occupied energy sublevel of the atom.

The number of electrons in the highest occupied energy sublevel of an atom is related to the position of the element in the periodic table. For atoms of all elements in Group 1A/1, that number is 1. This can be written as $ns^1$, where $n$ is the highest occupied principal energy level. For hydrogen, $n$ is 1; for lithium, $n$ is 2; for sodium, 3; and so forth. In Group 2A/2 the highest occupied sublevel is $ns^2$. In Groups 3A/13 to 8A/18, the number of $p$ electrons is found by counting from the left from 1 to 6. For the main group elements:

| **Group (U.S.):** | 1A | 2A | | 3A | 4A | 5A | 6A | 7A | 8A |
|---|---|---|---|---|---|---|---|---|---|
| **Group (IUPAC):** | 1 | 2 | | 13 | 14 | 15 | 16 | 17 | 18 |
| ***s* electrons:** | 1 | 2 | ***p* electrons:** | 1 | 2 | 3 | 4 | 5 | 6 |
| **Electron configuration:** | $ns^1$ | $ns^2$ | | $np^1$ | $np^2$ | $np^3$ | $np^4$ | $np^5$ | $np^6$ |

A similar count-from-the-left order appears among the transition elements, in which the $d$ sublevels are filled. There are interruptions, however. Among the $3d$ electrons the interruptions appear at chromium (Z = 24) and copper (Z = 29).

| **Group (U.S.):** | 3B | 4B | 5B | 6B | 7B | ← | 8B | → | 1B | 2B |
|---|---|---|---|---|---|---|---|---|---|---|
| **Group (IUPAC):** | 3 | 4 | 5 | 6 | 7 | 8 | 9 | 10 | 11 | 12 |
| ***d* electrons:** | 1 | 2 | 3 | 5 | 5 | 6 | 7 | 8 | 10 | 10 |
| **Electron configuration:** | $3d^1$ | $3d^2$ | $3d^3$ | $3d^5$ | $3d^5$ | $3d^6$ | $3d^7$ | $3d^8$ | $3d^{10}$ | $3d^{10}$ |

**P/REVIEW:** Main group elements are those in the A groups of the periodic table, or Groups 1, 2, and 13 to 18 by the IUPAC system. The B groups (3 to 12) are transition elements (Section 5.6).

Fix these electron populations and their positions in the periodic table firmly in your thought now. Then cover both of these summaries and refer only to a full periodic table as you try the following Active Example.

### Active Example 11.1

Write the electron configuration of the highest occupied energy sublevel for each of the following elements: beryllium, phosphorus, manganese.

beryllium ________ phosphorus ________ manganese ________

beryllium, $2s^2$ phosphorus, $3p^3$ manganese, $3d^5$

Counting from the left, beryllium is in the second box (Group 2A/2) among the $2s$ sublevel elements, so its configuration is $2s^2$. Phosphorus is in the third box (Group 5A/15) among the $3p$ sublevel elements, so its configuration is $3p^3$. Manganese is in the fifth box (Group 7B/7) among the $3d$ sublevel elements, so its configuration is $3d^5$.

You are now ready to write electron configurations. The procedure follows.

procedure

### How to Write Electron Configurations

***Sample Problem:*** *Write the complete electron configuration for chlorine (Cl, Z = 17).*

**Step 1:** Locate the element in the periodic table. From its position in the table, identify and write the electron configuration of its highest occupied energy sublevel. (Leave room for writing lower-energy sublevels to its left.)

From its position in the periodic table (Group 7A/17, Period 3), the electron configuration of the highest occupied energy sublevel of chlorine is $3p^5$.

**Step 2:** To the left of what has already been written, list all lower-energy sublevels in order of increasing energy.

The sublevels having lower energies than $3p$ can be "read" across the periods from left to right in the periodic table, as in Figure 11.15: $1s\ 2s\ 2p\ 3s\ 3p^5$. If the neon core were to be used, these would be represented by [Ne]$3s\ 3p^5$.

**Step 3:** For each filled lower-energy sublevel, write as a superscript the number of electrons that fill that sublevel. (There are two *s* electrons, $ns^2$, six *p* electrons, $np^6$, and ten *d* electrons, $nd^{10}$. Exceptions: For chromium and copper, the $4s$ sublevel has only one electron, $4s^1$.)

A filled *s* sublevel has two electrons, and a filled *p* sublevel has six. Filling in these numbers yields $1s^22s^22p^63s^23p^5$, or [Ne]$3s^23p^5$.

**Step 4:** Confirm that the total number of electrons is the same as the atomic number.

$2 + 2 + 6 + 2 + 5 = 17 = Z$

The last step checks the correctness of your final result. The atomic number is the number of protons in the nucleus of the atom, which, for a neutral atom, is equal to the number of electrons. Therefore, the sum of the superscripts in an electron configuration, which is the total number of electrons in the atom, must be the same as the atomic number. For example, the electron configuration of oxygen (Z = 8) is $1s^22s^22p^4$. The sum of the superscripts is 2 + 2 + 4 = 8, the same as the atomic number.

## Active Example 11.2

Write the complete electron configuration (no Group 8A/18 core) for potassium (K, Z = 19).

First, what is the electron configuration of the highest occupied energy sublevel? (When you write the answer, leave space for the lower-energy sublevels.)

$4s^1$ (Group 1A/1 elements have one *s* electron.)

Now list to the left of $4s^1$ all lower-energy sublevels in order of increasing energy.

$1s\ 2s\ 2p\ 3s\ 3p\ 4s^1$

Finally, add the superscripts that show how many electrons fill the lower-energy sublevels. Check the final result.

$1s^22s^22p^63s^23p^64s^1$ $\quad 2 + 2 + 6 + 2 + 6 + 1 = 19 = Z$

*continued*

Rewrite the configuration with a core from the closest Group 8A/18 element that has a smaller atomic number.

$[Ar]4s^1$

## Active Example 11.3

Develop the electron configuration for cobalt (Co, Z = 27).

This is your first example with *d* electrons. The procedure is the same. Write both a complete configuration and one with a Group 8A/18 core.

$1s^22s^22p^63s^23p^64s^23d^7$ or $[Ar]4s^23d^7$

By steps,

**Step 1:** $3d^7$
**Step 2:** $1s\ 2s\ 2p\ 3s\ 3p\ 4s\ 3d^7$ or $[Ar]4s\ 3d^7$
**Step 3:** $1s^22s^22p^63s^23p^64s^23d^7$ or $[Ar]4s^23d^7$
**Step 4:** $2 + 2 + 6 + 2 + 6 + 2 + 7 = 27 = Z$ or $18 + 2 + 7 = 27$

We show two different ways to calculate Step 4. The first sum is the addition of the superscripts in the full electron configuration. The second sum is based on the configuration with a Group 8A/18 core. Argon, Ar, has Z = 18, so we started with 18 in the second sum. We then added the remaining superscripts to complete the check.

## 11.5 Valence Electrons

Goal 13 Using *n* for the highest occupied energy level, write the configuration of the valence electrons of any main group element.

Goal 14 Write the Lewis (electron-dot) symbol for an atom of any main group element.

**P/REVIEW:** The vertical groups in the periodic table make up families of elements that have similar chemical properties (Section 5.6).

It is now known that many of the similar chemical properties of elements in the same column of the periodic table are related to the total number of *s* and *p* electrons in the highest occupied energy level. These are called **valence electrons.** In sodium, $1s^22s^22p^63s^1$, the highest occupied energy level is three. There is a single *s* electron in that sublevel, and there are no *p* electrons. Thus, sodium has one valence electron. With phosphorus, $1s^22s^22p^63s^23p^3$, the highest occupied energy level is again three. There are two *s* electrons and three *p* electrons, a total of five valence electrons.

**P/REVIEW:** One way that valence electrons act in forming chemical compounds, and thereby determine the chemical properties of an element, is described briefly in the next section. The topic is discussed more fully in Chapter 12.

Using *n* for any principal quantum number, we note that $ns^1$ is the configuration of the highest occupied principal energy level for all Group 1A/1 elements. All members of this family have one valence electron. Similarly, all elements in Group 5A/15 have the general configuration $ns^2np^3$, and they have five valence electrons.

The highest occupied sublevels of all families of main group elements can be written in the form $ns^xnp^y$. These are shown in the second row of Table 11.1. In all cases the number of valence electrons (third row) is the sum of the superscripts, x + y. Notice that for every group the number of valence electrons is the same as the group number in the U.S. system, or the same as the only or last digit in the IUPAC system.

**Table 11.1** Lewis Symbols of the Elements

| | | | | | | | | |
|---|---|---|---|---|---|---|---|---|
| **Group (U.S.)** | 1A | 2A | 3A | 4A | 5A | 6A | 7A | 8A |
| **Group (IUPAC)** | 1 | 2 | 13 | 14 | 15 | 16 | 17 | 18 |
| **Highest-Energy Electron Configuration** | $ns^1$ | $ns^2$ | $ns^2np^1$ | $ns^2np^2$ | $ns^2np^3$ | $ns^2np^4$ | $ns^2np^5$ | $ns^2np^6$ |
| **Number of Valence Electrons** | 1 | 2 | 3 | 4 | 5 | 6 | 7 | 8 |
| **Lewis Symbol of Third-Period Element** | Na· | Mg: | Al: | ·Si: | ·P: | ·S: | :Cl: | :Ar: |

## Active Example 11.4

Try to answer these questions without referring to anything; if you cannot, use only a full periodic table. (a) Write the electron configuration for the highest occupied energy level for Group 6A/16 elements. (b) Identify the group whose electron configuration for the highest occupied energy level is $ns^2np^2$.

(a) $ns^2np^4$ (b) Group 4A/14

(a) A Group 6A/16 element has six valence electrons, as indicated by the group number. The first two must be in the *ns* sublevel, and the remaining four must be in the *np* sublevel. (b) The total number of valence electrons is 2 + 2 = 4. The group is therefore 4A/14.

**P/REVIEW:** Elements in different areas of the periodic table are identified in Section 5.6. Among them are the main group elements that are found in the A groups of the table (1, 2, and 13 to 18) and the transition elements in the B groups (3 to 12).

Another way to show valence electrons uses **Lewis symbols,** which are also called **electron-dot symbols.** The symbol of the element is surrounded by the number of dots that matches the number of valence electrons. Dot symbols for the main group elements in Period 3 are given in Table 11.1. Paired electrons, those that occupy the same orbital, are usually placed on the same side of the symbol, and single occupants of one orbital are by themselves. This is not a fixed rule; exceptions are common if other positions better serve a particular purpose.

Group 8A/18 atoms have a full set of eight valence electrons, two in the *s* orbital and six in the *p* orbitals. This is sometimes called an **octet of electrons.** Elements in Group 8A/18 are particularly unreactive; only a few compounds of these elements are known. The filled octet is responsible for this chemical property.

## Active Example 11.5

Write electron-dot symbols for the elements whose atomic numbers are 38 and 52.

Locate in the periodic table the elements whose atomic numbers are 38 and 52. Write their symbols. Determine the number of valence electrons from the group each element is in, and surround its symbol with the number of dots that matches the number of valence electrons.

*continued*

Sr:   $\cdot\ddot{\dot{Te}}:$

The periodic table gives Sr for the symbol of atomic number 38. The element is strontium, which is in Group 2A/2, indicating two valence electrons. Te (tellurium) is the symbol for Z = 52. It is in Group 6A/6, so there are six electron dots.

## 11.6 Trends in the Periodic Table

Mendeleev and Meyer developed their periodic tables by trying to organize some recurring physical and chemical properties of the elements. Some of these properties are examined in this section.

### Atomic Size

**Goal** | **15** Predict how and explain why atomic size varies with position in the periodic table.

Active Figure 11.16 shows the sizes of atoms of main group elements. With a few exceptions two trends can be identified. Moving across the table from left to right, the atoms become smaller. Moving down the table in any group, atoms ordinarily increase in size. These observations are believed to be primarily the result of two influences:

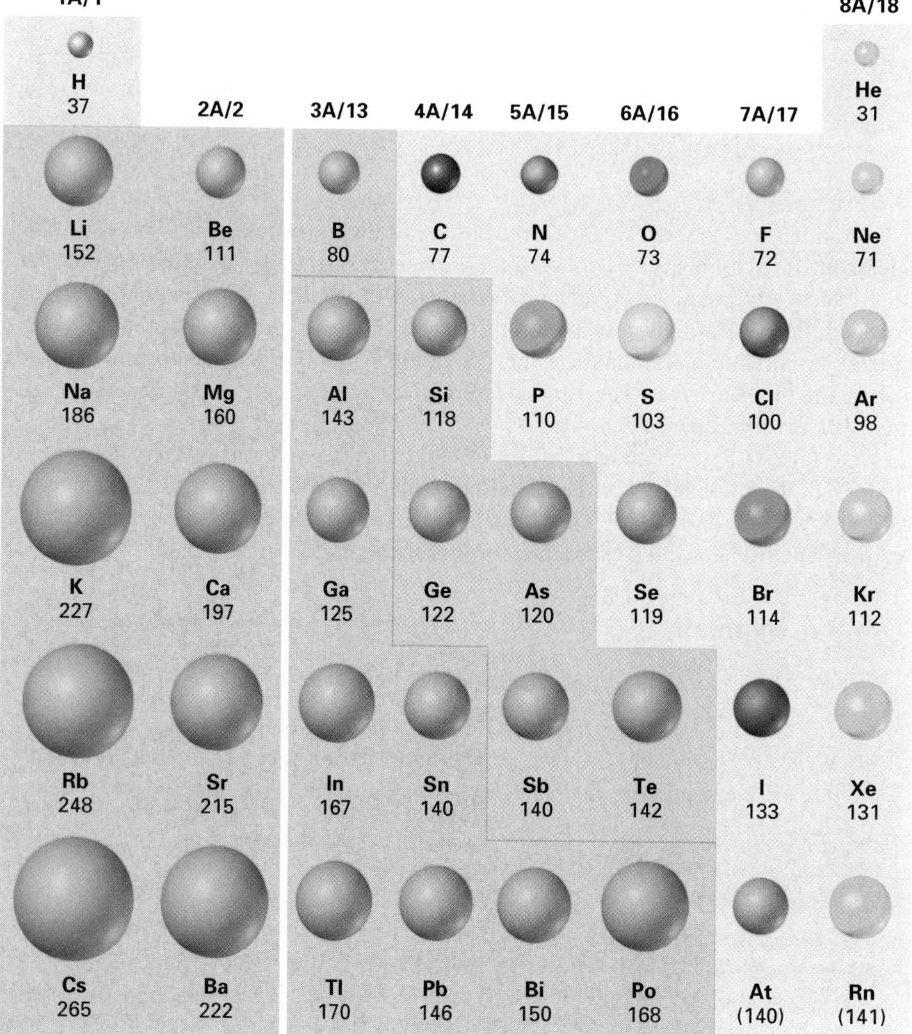


**Active Figure 11.16** Sizes of atoms of main group elements, expressed in picometers (1 pm = $10^{-12}$ m). Trends to observe: (1) Sizes of atoms increase with increasing atomic number in any group, that is, going down any column. (2) Sizes of atoms decrease with increasing atomic number within a given period, that is, going from left to right across any row. **Watch this active figure at *http://www.cengage.com/cracolice.***

1. *Highest occupied principal energy level.* As valence electrons occupy higher and higher principal energy levels, they are generally farther from the nucleus, and the atoms become larger. For example, the valence electron of a lithium atom is a 2*s* electron, whereas the valence electron of a sodium atom is a 3*s* electron. Sodium atoms are therefore larger.
2. *Nuclear charge.* Within any period, the valence electrons are all in the same principal energy level. As the number of protons in an atom increases, the positive charge in the nucleus also increases. This pulls the valence electrons closer to the nucleus, so the atom becomes smaller. For example, the atomic number of sodium is 11 (11 protons in the nucleus), and the atomic number of magnesium is 12 (12 protons). The 12 protons in a magnesium atom attract the 3*s* valence electrons more strongly than the 11 protons of a sodium atom. The magnesium atom is therefore smaller.

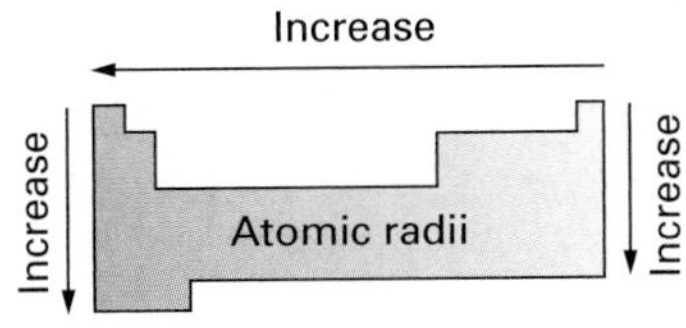


General trends in atomic radii with position in periodic table.

**summary**

**Atomic Size**

Atomic size generally increases from right to left across any row of the periodic table and from top to bottom in any column. The smallest atoms are toward the upper right corner of the table, and the largest are toward the bottom left corner.

## Active Example 11.6

Referring only to a periodic table, list atomic numbers 15, 16, and 33 in order of increasing atomic size.

The preceding summary and accompanying sketch in the margin should guide you into selecting the smallest and the largest of the three atoms.

$16 < 15 < 33$

The smallest atom is toward the upper right (Z = 16) and the largest is toward the bottom left (Z = 33). Specifically, Z = 16 (sulfur) atoms are smaller than Z = 15 (phosphorus) atoms because sulfur atoms have a higher nuclear charge to attract the highest-energy 3*s* and 3*p* electrons. The highest occupied energy level in a phosphorus atom is $n = 3$, but for a Z = 33 (arsenic) atom it is $n = 4$. Therefore, the phosphorus atom is smaller than the arsenic atom.

## Ionization Energy

Goal | **16** Predict how and explain why first ionization energy varies with position in the periodic table.

A sodium atom (Z = 11) has 11 protons and 11 electrons. One of the electrons is a valence electron. Mentally separate the valence electron from the other 10. This is pictured in the larger block of Figure 11.17. The valence electron, with its 1− charge, is still part of the neutral atom. The rest of the atom has 11 protons and 10 electrons (11 plus charges and 10 minus charges), giving it a net charge of 1+. If we take away the valence electron, the particle that is left keeps that 1+ charge. This particle is a sodium ion, $Na^+$. **An ion is an atom or group of atoms that has an electrical charge because of a difference in the number of protons and electrons.**

It takes work to remove an electron from a neutral atom. Energy must be spent to overcome the attraction between the negatively charged electron and the positively

**P/REVIEW:** A monatomic (one-atom) ion is an atom that has gained or lost one, two, or three electrons. If the atom loses electrons, the ion has a positive charge and is a cation; if the atom gains electrons it becomes a negatively charged anion (Section 6.4).

**Figure 11.17** The formation of a sodium ion from a sodium atom.

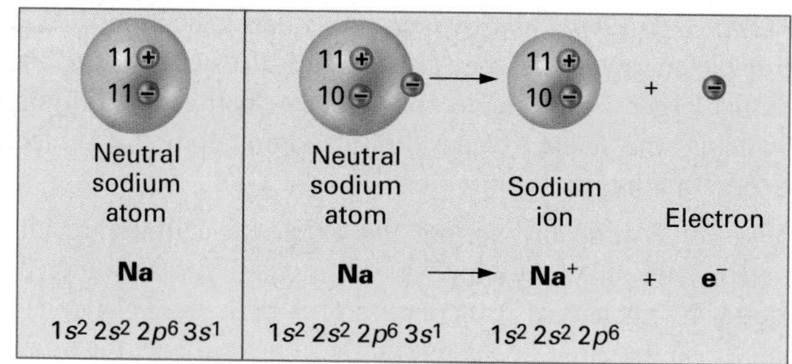


charged ion that is left. **The energy required to remove one electron from a neutral gaseous atom of an element is the ionization energy of that element.**

Ionization energy is one of the more striking examples of a periodic property, particularly when graphed (Fig. 11.18). Notice the similarity of the shape of the graph between Li and Ne (Period 2 in the periodic table), between Na and Ar (Period 3), and between K and Kr (Period 4). Notice also that the peaks are elements in Group 8A/18, and the low points are from Group 1A/1.

Continue observing the trends in Figure 11.18. As atomic number increases within a period, the general trend in ionization energy is an increase. The general trend among elements in the same group in the periodic table is a decrease. You can see this by looking at the elements in Group 1A/1 (Li, Na, and K), 2A/2 (Be, Mg, and Ca), and 8A/18 (He, Ne, Ar, and Kr). Ionization energies are lower as the atomic number increases within the group. If the graph is extended to the right, we find the same general shapes and trends in all periods.

The energy required to remove a second electron from the 1+ ion of an element is its **second ionization energy,** and removing a third electron requires the **third ionization energy.** In all cases the ionization energy displays a big jump in size when the valence electrons are gone and the next electron must be removed from a full octet of electrons. This adds to our belief that valence electrons are largely responsible for the chemical properties of an element. This also provides evidence to support the validity of the quantum model of the atom.

Periodic trends in ionization energies are explained by the same influences that explain atomic size trends:

1. *Highest occupied principal energy level.* Ionization energy decreases down a group because the electron that is removed is farther from the nucleus. The attractive force between the electrons and the nucleus decreases with increasing distance.

**Figure 11.18** First ionization energy plotted as a function of atomic number, to show periodic properties of elements. Ionization energies of elements in the same period generally increase as atomic number increases. Ionization energies of elements in the same group generally decrease as atomic number increases.

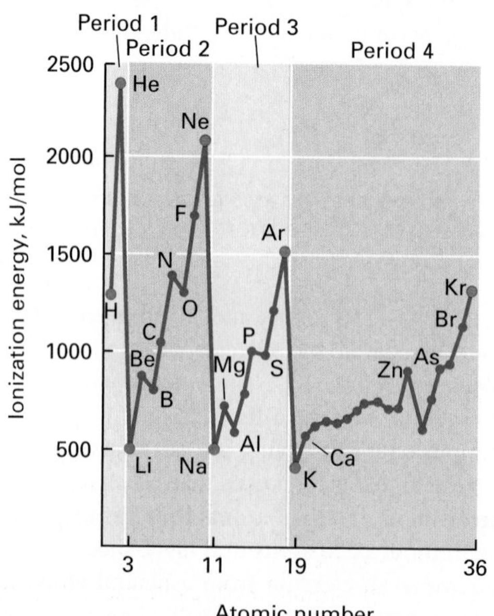

2. *Nuclear charge.* The outermost electron in atoms of main-group elements within a period are in the same principal energy level. Ionization energy increases across a period because the electron that is removed is attracted by an increasing quantity of positive charge. As the attractive force becomes stronger, the energy required to remove the electron will increase.

**Ionization Energy**

Ionization energy generally increases from left to right across any row of the periodic table and from bottom to top in any column. The atoms with the highest ionization energy are toward the upper right corner, and those with the lowest ionization energy are toward the bottom left corner.

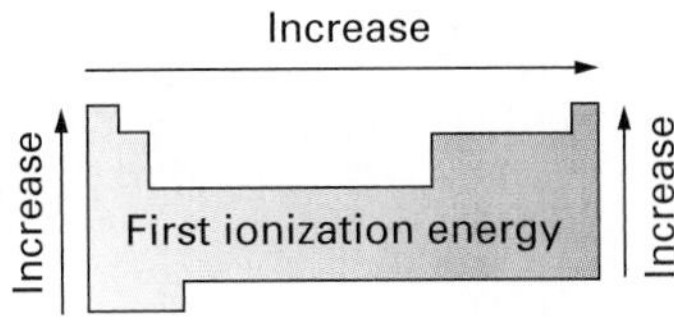


General trends in first ionization energy.

## Active Example 11.7

List each of the following groups of elements in order of increasing first ionization energy: (a) Be, N, F; (b) F, Cl, Br.

You are ready to take this all the way. Review the preceding discussion if you are unsure about how to answer.

(a) Be < N < F (b) Br < Cl < F

In part (a), ionization energy increases across a period because of the increasing nuclear charge holding electrons that are in the same principal energy level. In part (b), ionization energy increases up a group because the highest occupied principal energy level is decreasing and the electrons are closer to the positive charge of the nucleus.

## Chemical Families

Goal | **17** Explain, from the standpoint of electron configuration, why certain groups of elements make up chemical families.

Goal | **18** Identify in the periodic table the following chemical families: alkali metals, alkaline earths, halogens, noble gases.

Elements with similar chemical properties appear in the same group, or vertical column, in the periodic table. Several of these groups form **chemical families.** Family trends are most apparent among the main group elements. Active Figure 11.19 shows family trends in chemical properties for two groups. In this text, we will consider four families: the alkali metals in Group 1A/1, the alkaline earths in Group 2A/2, the halogens in Group 7A/17, and the noble gases in Group 8A/18. As these families are discussed, the symbol X is used to refer to any member of the family.

**Alkali Metals  Valence Electrons: $ns^1$  X·**

With the exception of hydrogen, Group 1A/1 elements are known as **alkali metals** (Figs. 11.19, 11.20, and 11.21). The single valence electron is easily lost, forming an ion with a 1+ charge. All ions with a 1+ charge, such as the $Na^+$ ion described earlier, tend to combine with other elements in the same way. This is why the chemical properties of the elements in a family are similar.

Figure 11.18 shows that ionization energies of the alkali metals decrease as the atomic number increases. The higher the energy level in which the negatively charged

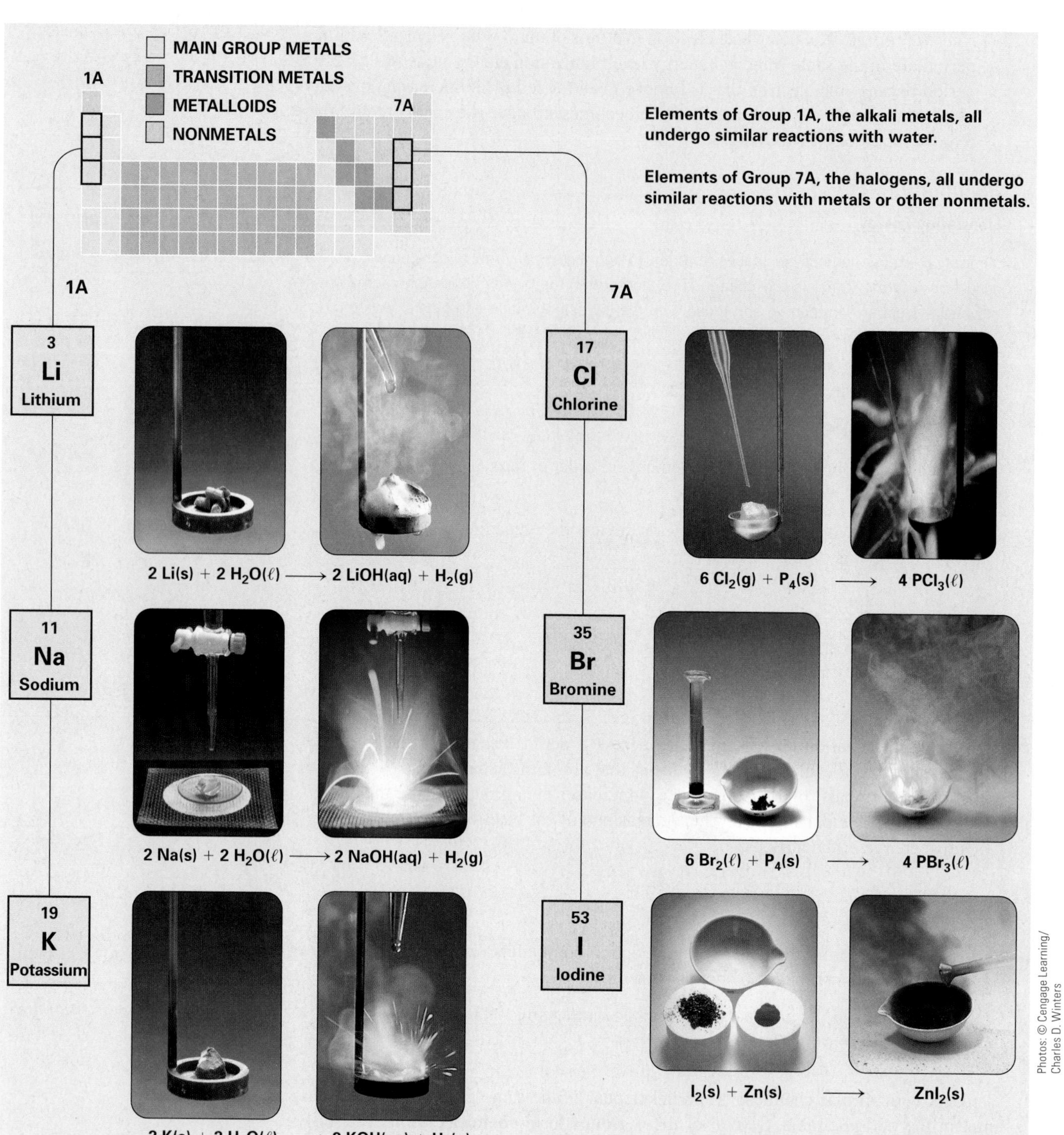



**Active Figure 11.19** Chemical families. Elements with similar chemical properties form chemical families. These elements are in the same group in the periodic table. Lithium, sodium, and potassium are three members of Group 1A/1, and they undergo similar reactions with water. Atoms of each element in the group have one valence electron. Chlorine, bromine, and iodine, members of Group 7A/17, each have atoms with seven valence electrons. This leads to similar reactivity among all members of the chemical family. **Watch this active figure at *http://www.cengage.com/cracolice.***

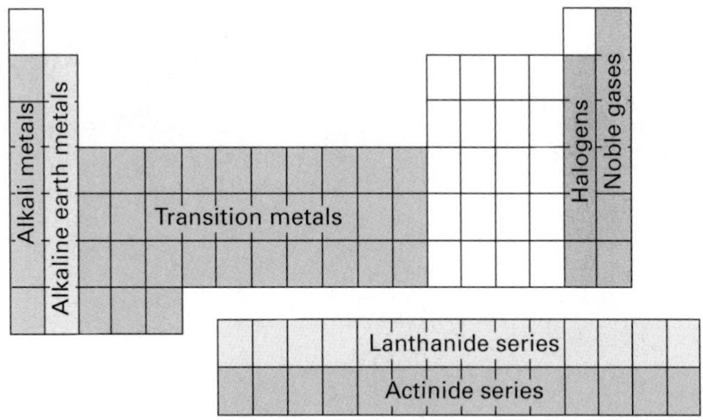


**Figure 11.20** Chemical families and regions in the periodic table.

$ns^1$ electron is located, the farther it is from the positively charged nucleus, and therefore, the more easily it is removed. As a direct result of this, the **reactivity** of the element—that is, its tendency to react with other elements to form compounds—increases as you go down the column.

When an alkali metal atom loses its valence electron, the ion formed is **isoelectronic** with a noble gas atom. (The prefix, *iso-*, means "same," as in isotope.) For example, the electron configuration of sodium is $1s^22s^22p^63s^1$ or $[Ne]3s^1$. If the $3s^1$ electron is removed, $1s^22s^22p^6$ is left. This is the configuration of the noble gas neon. In each case the alkali metal ion reaches the same configuration as the noble gas just before it in the periodic table. Its highest-energy octet is complete; all electron orbitals are filled. This is a highly stable electron distribution. The chemical properties of most main-group elements can be explained in terms of their atoms becoming isoelectronic with a noble gas atom.

Alkali metals do not normally look like common, everyday metals. This is because they are so reactive that they combine with oxygen in the air to form an oxide coating, which hides the bright metallic luster that can be seen in a freshly cut sample. These elements possess other common metallic properties, too. For example, they are good conductors of heat and electricity, and they are easy to form into wires and thin foils.

We can see distinct trends in the physical properties of alkali metals. Their densities increase as atomic number increases. Boiling and melting points generally decrease

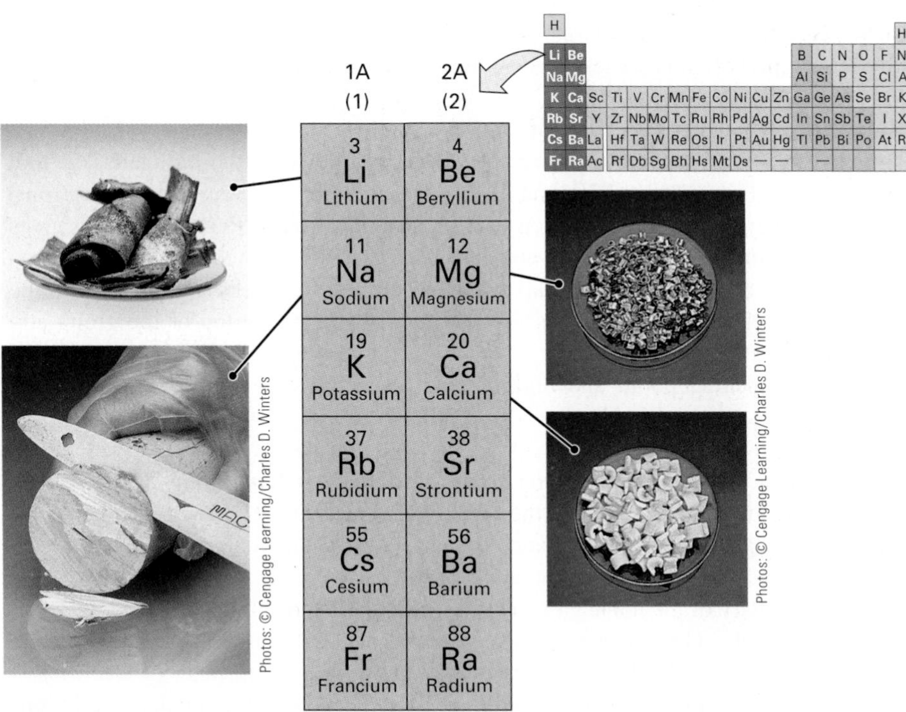


**Figure 11.21** The alkali metal and alkaline earth metal families. Alkali metal atoms (Group 1A/1) have the valence electron configuration $ns^1$. Alkaline earth atoms (Group 2A/2) have the valence electron configuration $ns^2$.

**Figure 11.22** The halogen family. Halogen atoms (Group 7A/17) have the valence electron configuration $ns^2np^5$, for a total of seven valence electrons.

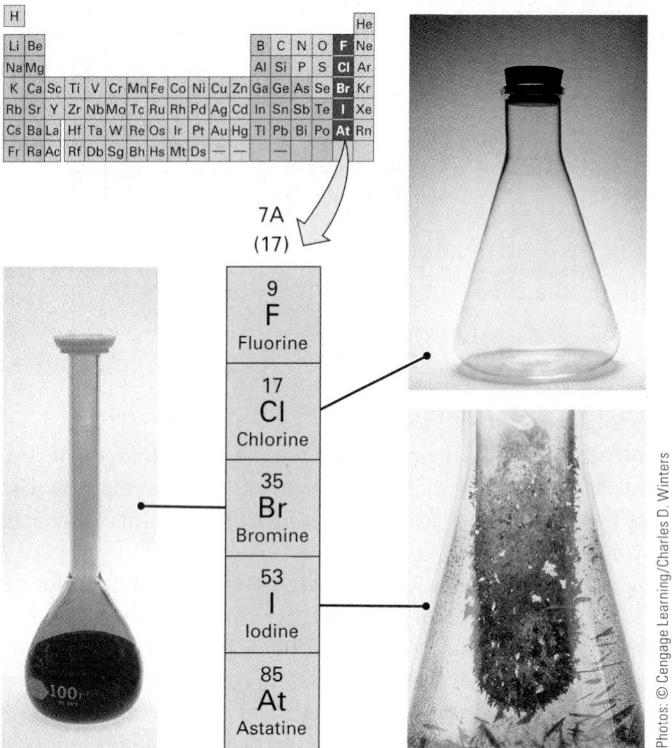


as you go down the periodic table. The single exception is cesium (Z = 55), which boils at a temperature slightly higher than the boiling point of rubidium (Z = 37).

**Alkaline Earths Valence Electrons: $ns^2$ X:**

Group 2A/2 elements are called **alkaline earths** or **alkaline earth metals** (see Figs. 11.20 and 11.21). Both the first and second ionization energies are relatively low, so the two valence electrons are given up readily to form ions with a 2+ charge. Again, the ions have the configuration of a noble gas. If magnesium, [Ne]$3s^2$, loses two electrons, only the [Ne] core is left.

Trends like those noted with the alkali metals are also seen with the alkaline earths. Reactivity again increases as you go down the column in the periodic table. Physical property trends are less evident among the alkaline earths.

**Halogens Valence Electrons: $ns^2np^5$ :X:**

The elements in Group 7A/17 make up the family known as the **halogens,** or "salt formers" (see Figs. 11.19, 11.20, and 11.22). Halogens have seven valence electrons. The easiest way for a halogen to reach a full octet of electrons is to gain one. This gives it the configuration of a noble gas and forms an ion with a 1– charge. The tendency to gain an electron is greater in small atoms, where the added electron is closer to the nucleus. Consequently, reactivity is greatest for fluorine at the top of the group and least for iodine at the bottom.

Density, melting point, and boiling point all increase steadily with increasing atomic number among the halogens.

© David R. Frazier Photolibrary, Inc./Alamy

The "lighter than air" feature of an airship comes from the low-density helium that fills it. Helium is safe for this purpose because it is unreactive, unlike even lower-density hydrogen. See Figure 8.9.

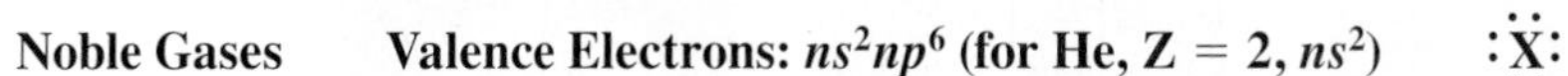

**Noble Gases Valence Electrons: $ns^2np^6$ (for He, Z = 2, $ns^2$) :X:**

The elements of Group 8A/18 are the **noble gases** (see Fig. 11.20). In chemistry the word *noble* means having "a reluctance to react." Only a small number of compounds of the noble gases are known, and none occur naturally.

The inactivity of the noble gases is believed to be the result of their filled valence electron sublevels. The two electrons of helium fill the 1*s* orbital, the only valence orbital helium has. All other elements in the group have a full octet of electrons—completely filled *s* and *p* valence electron orbitals. This configuration apparently represents a "minimization of energy" arrangement of electrons that is very stable. The high ionization energies of a full octet have already been noted, so the noble gases resist forming positively charged ions. Nor do the atoms tend to gain electrons to form negatively charged ions.

The noble gases provide excellent examples of periodic trends in physical properties. Without exception, the densities, melting points, and boiling points increase as you move down the column in the periodic table.

**P/REVIEW:** In Section 2.8 the expression "minimization of energy" was introduced to identify a natural tendency for chemical and physical change to occur spontaneously if the result of that change is a lower total energy within the system. That is why objects fall in a gravitational field and why oppositely charged particles attract each other and similarly charged particles repel each other in an electrical field.

**Hydrogen  Valence Electron: $1s^1$  H·**

You have probably wondered why hydrogen appears twice in our periodic table, at the tops of Groups 1A/1 and 7A/17. Hydrogen is neither an alkali metal nor a halogen, although it shares some properties with both groups. Hydrogen combines with some elements in the same ratio as alkali metals, but the way the compounds are formed is different. Hydrogen atoms can also gain an electron to form an ion with a 1− charge, similar to a halogen. But other properties of hydrogen differ from those of the halogens. The way the periodic table is used makes it handy to have hydrogen in both positions, although it really stands alone as an element.

## Metals and Nonmetals

Goal 19 Identify metals and nonmetals in the periodic table.

Goal 20 Predict how and explain why metallic character varies with position in the periodic table.

Both physically and chemically, the alkali metals, alkaline earths, and transition elements are metals. At the particulate level, an element is a **metal** if it can lose one or more electrons and become a positively charged ion. An element that lacks this quality is a **nonmetal.** The larger the atom, the more easily the outermost electron is removed. Therefore, the **metallic character** of elements in a group increases as you go down a column in the periodic table.

An atom becomes smaller as the nuclear charge increases across a period in the table. The larger number of protons holds the outermost electrons more strongly, making it more difficult for them to be lost. This makes the metallic character of elements decrease as you go from left to right across the period.

Table 11.2 compares the properties of metals and nonmetals. Chemically, the distinction between metals and nonmetals—elements that lose electrons in chemical reactions and those that do not—is not sharp. It can be drawn roughly as a stair-step line beginning between atomic numbers 4 and 5 in Period 2 and ending between 84 and 85 in Period 6 (Fig. 11.23). Elements to the left of the line are metals; those to the right are nonmetals.

**Table 11.2** Some Physical and Chemical Properties of Metals and Nonmetals

| Metals | Nonmetals |
|---|---|
| Lose electrons easily to form cations | Tend to gain electrons to form anions |
| 1, 2, or 3 valence electrons | 4 or more valence electrons |
| Low ionization energies | High ionization energies |
| Form compounds with nonmetals, but not with other metals | Form compounds with metals and with other nonmetals |
| High electrical conductivity | Poor electrical conductivity (carbon in the form of graphite is an exception) |
| High thermal conductivity | Poor thermal conductivity; good insulator |
| Malleable (can be hammered into sheets) | Brittle |
| Ductile (can be drawn into wires) | Nonductile |

**Figure 11.23** Metals and nonmetals. Green identifies elements that are metalloids (semimetals), which have properties that are intermediate between those of metals and nonmetals.

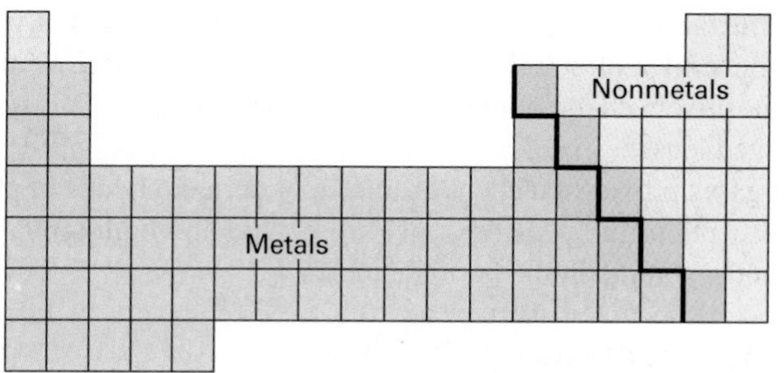


Most of the elements next to the stair-step line have some properties of both metals and nonmetals. They are often called **metalloids** or **semimetals.** Included in the group are silicon and germanium, the semiconductors on which the electronics industry has been built. Indeed, silicon is so important to the industry that the area south of San Francisco where many major electronics manufacturers are located is known as "Silicon Valley."

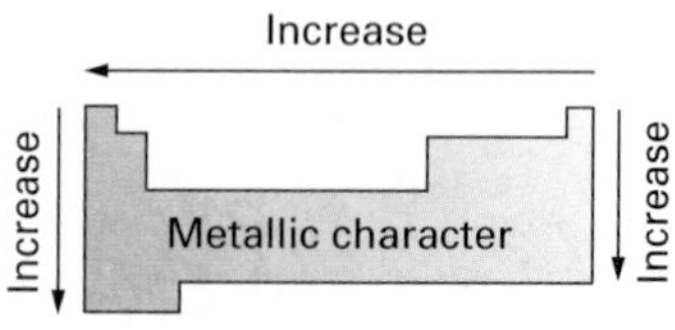


General trends in metallic character with position in periodic table.

summary

**Metallic Character**

Metallic character generally increases from right to left across any row of the periodic table and from top to bottom in any column. The least metallic character is toward the upper right corner of the table, and the most is toward the bottom left corner.

### √ Target Check 11.5

a) List the following in order of increasing ionization energy: C, N, F.
b) Write the symbol of the Period 2 element in each of the following chemical families: halogens, alkali metals, noble gases, alkaline earths.
c) List the following in order of increasing atomic size: N, O, P.
d) Which Period 3 elements are metals? Which are nonmetals? Which are metalloids?

## Chapter 11 in Review

Most of the key terms and concepts and many others appear in the Glossary. Use your Glossary regularly.

**11.1 Electromagnetic Radiation**

Goal 1 Define and describe electromagnetic radiation.
Goal 2 Distinguish between continuous and line spectra.

**Key Terms and Concepts: Discrete, electromagnetic radiation, photon, quantum mechanics, spectrum (electromagnetic, continuous, line), speed of light (c), wave-particle duality, wave equation**

**11.2 The Bohr Model of the Hydrogen Atom**

Goal 3 Describe the Bohr model of the hydrogen atom.
Goal 4 Explain the meaning of quantized energy levels in an atom and show how these levels relate to the discrete lines in the spectrum of that atom.
Goal 5 Distinguish between ground state and excited state.

**Key Terms and Concepts: Continuous, excited state, ground state, principal energy levels, principal quantum number *(n)*, quantized, quantized energy levels, quantum jump (leap)**

**11.3 The Quantum Mechanical Model of the Atom**

Goal 6 Identify the principal energy levels in an atom and state the energy trend among them.
Goal 7 For each principal energy level, state the number of sublevels, identify them, and state the energy trend among them.
Goal 8 Sketch the shapes of *s* and *p* orbitals.
Goal 9 State the number of orbitals in each sublevel.
Goal 10 State the restrictions on the electron population of an orbital.

**Key Terms and Concepts: Orbital, Pauli exclusion principle, quantum mechanical model of the atom, sublevels, *s, p, d, f***

**11.4 Electron Configuration**

Goal 11 Use a periodic table to list electron sublevels in order of increasing energy.
Goal 12 Referring only to a periodic table, write the ground-state electron configuration of an atom of any element up to atomic number 36.

**Key Terms and Concepts: Argon or neon core, electron configuration**

### 11.5 Valence Electrons

Goal 13 Using *n* for the highest occupied energy level, write the configuration of the valence electrons of any main group element.

Goal 14 Write the Lewis (electron-dot) symbol for an atom of any main group element.

**Key Terms and Concepts: Lewis (electron-dot) symbols, octet of electrons, valence electrons**

### 11.6 Trends in the Periodic Table

Goal 15 Predict how and explain why atomic size varies with position in the periodic table.

Goal 16 Predict how and explain why first ionization energy varies with position in the periodic table.

Goal 17 Explain, from the standpoint of electron configuration, why certain groups of elements make up chemical families.

Goal 18 Identify in the periodic table the following chemical families: alkali metals, alkaline earths, halogens, noble gases.

Goal 19 Identify metals and nonmetals in the periodic table.

Goal 20 Predict how and explain why metallic character varies with position in the periodic table.

**Key Terms and Concepts: Alkali metals (family), alkaline earths, alkaline earth metals (family), chemical family, first, second, and third ionization energy, halogens (family), ion, cation, anion, monatomic ion, ionization energy, isoelectronic, metal, nonmetal, metalloid, semimetal, metallic character, noble gases (family), nuclear charge, reactivity**

## Study Hints and Pitfalls to Avoid

It takes many words to describe the quantum model of the atom, even at this introductory level. The words are not easy to remember unless they are organized into some kind of pattern. The summary at the end of Section 11.3 gives you this organization.

Be sure you know what *quantized* means. Also understand the difference between a Bohr *orbit* (a fixed path the electron travels around the nucleus) and the quantum *orbital* (a mathematically defined region in space in which there is a high probability of finding the electron).

In writing electron configurations, we recommend that you use the periodic table to list the sublevels in increasing energy rather than the slanting-line memory device. The periodic table is the greatest organizer of chemical information there is, and every time you use it you strengthen your ability to use it in all other ways.

Understand well the two influences that determine atomic size. The same thinking appears with other properties later.

## Concept-Linking Exercises

*Write a brief description of the relationships among each of the following groups of terms or phrases. Answers to the Concept-Linking Exercises are given after answers to the Target Checks in Appendix III.*

1. Electromagnetic spectrum, continuous spectrum, line spectrum
2. Quantized energy levels, ground state, excited state, continuous values
3. Bohr model of the hydrogen atom, orbit, orbital
4. Quantum mechanical model of the atom, principal energy level, sublevel, electron orbital, Pauli exclusion principle
5. Valence electrons, Lewis symbols, octet of electrons
6. Ionization energy, second ionization energy, third ionization energy
7. Chemical families, alkali metals, alkaline earths, halogens, noble gases, hydrogen
8. Atomic size, highest occupied principal energy level, nuclear charge
9. Metal, nonmetal, metalloid, semimetal

## Small-Group Discussion Questions

*Small-Group Discussion Questions are for group work, either in class or under the guidance of a leader during a discussion section.*

1. Radio stations identify themselves by the frequency at which they broadcast, such as 98.5 FM, which stands for a signal of 98.5 megahertz carried by a FM (frequency modulation) wave. (A hertz is a wave cycle per second.) Could radio stations identify themselves by wavelength? If yes, what wavelength is emitted by 98.5 FM? If no, why not?
2. Explain how the existence of line spectra for elements leads to the Bohr model of the atom. Is a quantum jump a large or small quantity of energy? How does this quantity of energy compare with the everyday usage of the term *quantum leap*? Why does an electron in a Bohr orbit farther from the nucleus have more energy than one that is closer to the nucleus?
3. Carefully define each and distinguish between *orbit* and *orbital*. How many orbitals are within each of the seven ground-state principal energy levels?
4. Using nothing but a periodic table, write the ground-state electron configuration for each of the first 36 elements on the periodic table.
5. Write the electron-dot symbol for each of the first 18 elements in the periodic table. Explain the relationship between each electron configuration from Question 4 and the electron-dot symbol.
6. In a plot of ionization energy versus atomic number, ionization energy tends to increase within a period in the periodic table. However, in Period 2, there are two instances where ionization energy decreases: Be to B and N to O. Write the electron configurations of these elements and use them to propose an explanation for these disruptions in the general trend.

# Questions, Exercises, and Problems

*Blue-numbered questions are answered in Appendix III. ■ denotes problems assignable in OWL. In the first section of Questions, Exercises, and Problems, similar exercises and problems are paired in consecutive odd-even number combinations.*

### Section 11.1: Electromagnetic Radiation

1. The visible spectrum is a small part of the whole electromagnetic spectrum. Name several other parts of the whole spectrum that are included in our everyday vocabulary.
2. What is meant by a discrete line spectrum? What kind of spectra do not have discrete lines? Have you ever seen a spectrum that does not have discrete lines? Have you ever seen one with discrete lines?
3. Identify measurable wave properties that are used in describing light.
4. What do visible light and radio waves have in common?

### Section 11.2: The Bohr Model of the Hydrogen Atom

5. Which among the following are *not* quantized? (a) number of plastic bags in a box; (b) cars passing through a toll plaza in a day; (c) birds in an aviary; (d) flow of a river in $m^3$/hr; (e) percentage of salt in a solution.
6. ■ Which of the following are quantized? (a) canned soup from a grocery store; (b) weight of jelly beans; (c) elevation of a person on a ramp.
7. What kind of light would atoms emit if the electron energy were not quantized? Why?
8. ■ In the Bohr model of the hydrogen atom, the electron occupies distinct energy states. In one transition between energy states, an electron moves from $n = 1$ to $n = 2$. Is energy absorbed or emitted in the process? Does the electron move closer to or further from the nucleus?
9. What must be done to an atom, or what must happen to an atom, before it can emit light?
10. What experimental evidence leads us to believe that electron energy levels are quantized?
11. Which atom is more apt to emit light, one in the ground state or one in an excited state? Why?
12. What is meant when an atom is "in its ground state"?
13. Using a sketch of the Bohr model of an atom, explain the source of the observed lines in the spectrum of hydrogen.
14. Draw a sketch of an atom according to the Bohr model. Describe the atom with reference to the sketch.
15. Identify the major advances that came from the Bohr model of the atom.
16. Identify the shortcomings of the Bohr theory of the atom.

### Section 11.3: The Quantum Mechanical Model of the Atom

17. Compare the relative energies of the principal energy levels within the same atom.
18. What is the meaning of the principal energy levels of an atom?
19. How many sublevels are present in each principal energy level?
20. ■ How many sublevels are there in an atom with $n = 4$?
21. How many orbitals are in the *s* sublevel? The *p* sublevel? The *d* sublevel? The *f* sublevel?
22. What is an orbital? Describe the shapes of *s* and *p* orbitals using words and sketches.
23. Each *p* sublevel contains six orbitals. Is this statement true or false? Comment on this statement.
24. ■ How many orbitals are there in an atom with $n = 3$?
25. The principal energy level with $n = 6$ contains six sublevels, although not all six are occupied for any element now known. Is this statement true or false? Comment on this statement.
26. Although we may draw the 4*s* orbital with the shape of a ball, there is some probability of finding the electron outside the ball we draw. Is this statement true or false? Comment on this statement.
27. An orbital may hold one electron or two electrons, but no other number. Is this statement true or false? Comment on this statement.
28. What is the significance of the Pauli exclusion principle?
29. What is your opinion of the common picture showing one or more electrons whirling around the nucleus of an atom?
30. An electron in an orbital of a *p* sublevel follows a path similar to a figure 8. Comment on this statement.
31. What is the largest number of electrons that can occupy the 4*p* orbitals? The 3*d* orbitals? The 5*f* orbitals?
32. ■ How many 4*f* orbitals are there in an atom?
33. Energies of the principal energy levels of an atom sometimes overlap. Explain how this is possible.
34. What general statement may be made about the energies of the principal energy levels? About the energies of the sublevels?
35. Which of the following statements is true? The quantum mechanical model of the atom includes (a) all of the Bohr model; (b) part of the Bohr model; (c) no part of the Bohr model. Justify your answer.
36. Is the quantum mechanical model of the atom consistent with the Bohr model? Why or why not?

### Section 11.4: Electron Configuration

37. What do you conclude about the symbol $2d^5$?
38. What is the meaning of the symbol $3p^4$?
39. What element has the electron configuration $1s^22s^22p^63s^23p^4$? In which period and group of the periodic table is it located?
40. ■ In what group and period do you find elements with the electron configurations given?
    a) $1s^22s^22p^63s^24s^23d^8$
    b) $1s^22s^22p^63s^2$

41. What is the argon core? What is its symbol, and what do you use it for?

42. What is meant by [Ne] in [Ne]$3s^23p^1$?

*Questions 43 through 46: Identify the elements whose electron configurations are given.*

43. a) $1s^22s^22p^63s^23p^64s^23d^{10}4p^4$
b) $1s^22s^22p^4$ c) $1s^22s^22p^63s^2$

44. ■ a) $1s^22s^22p^63s^24s^23d^2$ b) $1s^22s^22p^6$

45. a) [He]$2s^22p^3$ b) [Ne]$3s^23p^1$ c) [Ar]$4s^23d^7$

46. a) [He]$2s^22p^2$ b) [Ne]$3s^23p^5$ c) [Ar]$4s^23d^{10}4p^3$

*Questions 47 through 50: Write complete ground-state electron configurations of an atom of the elements shown. Do not use a noble gas core.*

47. Magnesium and nickel

48. ■ Bromine and manganese

49. Chromium and selenium (Z = 34)

50. Calcium and copper

51. If it can be done, rewrite the electron configurations in Questions 47 and 49 with a neon or argon core.

52. ■ a) Write the complete ground-state electron configuration for the silicon atom.
b) Use a noble gas core to write the electron configuration of phosphorus.

53. Use a noble gas core to write the electron configuration of vanadium (Z = 23).

54. ■ a) Write the complete ground-state electron configuration for the scandium (Z = 21) atom.
b) Use a noble gas core to write the electron configuration of iron.

**Section 11.5: Valence Electrons**

55. Why are valence electrons important?

56. What are valence electrons?

57. What are the valence electrons of aluminum? Write both ways they may be represented.

58. ■ Write the name of the element with the valence electron configuration given.
a) $2s^2$ b) $4s^24p^4$

59. ■ Using $n$ for the principal quantum number, write the electron configuration for the valence electrons of Group 6A/16 atoms.

60. ■ In what group and period do you find elements with the valence electron configurations given?
a) $2s^22p^2$ b) $4s^24p^5$

**Section 11.6: Trends in the Periodic Table**

61. Identify an atom of a main-group element in Period 4 that is (a) larger than an atom of arsenic (Z = 33) and (b) smaller than an atom of arsenic.

62. ■ Using only the periodic table, arrange the following elements in order of increasing atomic radius: aluminum, boron, indium (Z = 49), gallium (Z = 31).

63. Why does atomic size decrease as you go left to right across a row in the periodic table?

64. ■ Using only the periodic table, arrange the following elements in order of increasing atomic radius: sulfur, magnesium, sodium, chlorine.

65. Even though an atom of germanium (Z = 32) has more than twice the nuclear charge of an atom of silicon (Z = 14), the germanium atom is larger. Explain.

66. The text describes the formation of a sodium ion as the removal of one electron from a sodium atom. Predict the size of a sodium ion compared to the size of a sodium atom. Explain your prediction.

67. What is the general trend in ionization energies of elements in the same chemical family?

68. ■ Using only the periodic table, arrange the following elements in order of increasing ionization energy: radon, helium, neon, xenon.

69. Explain the reason for the general trend of increasing ionization energy across a period.

70. ■ Using only the periodic table, arrange the following elements in order of increasing ionization energy: phosphorus, aluminum, silicon, chlorine.

71. What is it about strontium (Z = 38) and barium that makes them members of the same chemical family?

72. ■ Give the symbol for an element that is: (a) a halogen; (b) an alkali metal; (c) a noble gas; (d) an alkaline earth metal.

73. To what family does the electron configuration $ns^2np^5$ belong?

74. ■ a) What is the name of the alkali metal that is in Period 6?
b) What is the name of the halogen that is in Period 6?
c) What is the name of the alkaline earth metal that is in Period 3?
d) What is the name of the noble gas that is in Period 3?

75. Expressed as $ns^xnp^y$, what electron configuration is isoelectronic with a noble gas?

76. ■ Which of the following describes the element Ba? Choose all that apply: (a) is very reactive as a metal; (b) consists of diatomic molecules in elemental form; (c) forms an ion with a charge of 2+; (d) forms a cation that is isoelectronic with a noble gas; (e) reacts with alkali metals to form salts; (f) is one of the group of the least reactive elements.

77. Identify the chemical families in which (a) krypton (Z = 36) and (b) beryllium are found.

78. ■ Which of the following describes the element Br? Choose all that apply: (a) reacts vigorously as a metal; (b) belongs to a group consisting entirely of gases; (c) consists of diatomic molecules in elemental form; (d) is one of the group of least reactive elements; (e) reacts vigorously with alkali metals to form salts; (e) forms a cation with a charge of 1+.

79. Account for the chemical similarities between chlorine and iodine in terms of their electron configurations.

80. Where in the periodic table do you find the transition elements? Are they metals or nonmetals?

81. What property of atoms of an element determines that the element is a metal rather than a nonmetal?

82. ■ Arrange the following elements in order of increasing metallic character: sodium, aluminum, phosphorus, silicon.

83. What are *metalloids* or *semimetals*? How do their properties compare with those of metals and nonmetals?

84. ■ Arrange the following elements in order of increasing metallic character: nitrogen, antimony (Z = 51), arsenic (Z = 33), phosphorus.

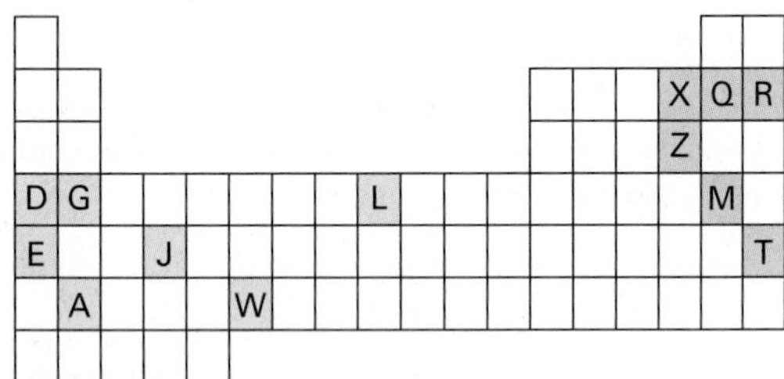


**Figure 11.24** Periodic table for Questions 85 to 90.

*Use the periodic table in Figure 11.24 to answer Questions 85 through 90. Answer with letters (D, E, X, Z, and so on) from the figure.*

85. Give the letters that are in the positions of (a) alkaline earth metals and (b) noble gases.

86. Give the letters that are in the positions of (a) halogens and (b) alkali metals.

87. List the letters that correspond to nonmetals.

88. Give the letters that correspond to transition elements.

89. List the elements Q, X, and Z in order of increasing atomic size (smallest atom first).

90. List the elements D, E, and G in order of decreasing atomic size (largest atom first).

## GENERAL QUESTIONS

91. Distinguish precisely, and in scientific terms, the differences among items in each of the following pairs or groups.
    a) Continuous spectrum, discrete line spectrum
    b) Quantized, continuous
    c) Ground state, excited state
    d) Principal energy level, principal quantum number
    e) Bohr model of the atom, quantum mechanical model of the atom
    f) Principal energy level, sublevel, orbital
    g) *s*, *p*, *d*, *f* (sublevels)
    h) Orbit, orbitals
    i) First, second, third ionization energies
    j) Metal, nonmetal, metalloid, semimetal

92. Determine whether each statement that follows is true or false:
    a) Electron energies are quantized in excited states but not in the ground state.
    b) Line spectra of the elements are experimental evidence of the quantization of electron energies.
    c) Energy is released as an electron passes from ground state to an excited state.
    d) The energy of an electron may be between two quantized energy levels.
    e) The Bohr model explanation of line spectra is still thought to be correct.
    f) The quantum mechanical model of the atom describes orbits in which electrons travel around the nucleus.
    g) Orbitals are regions in which there is a high probability of finding an electron.
    h) All energy sublevels have the same number of orbitals.
    i) The 3*p* orbitals of an atom are larger than its 2*p* orbitals but smaller than its 4*p* orbitals.
    j) At a given sublevel, the maximum number of *d* electrons is 5.
    k) The halogens are found in Group 7A/17 of the periodic table.
    l) The dot structure of the alkaline earths is X·, where X is the symbol of any element in the family.
    m) Stable ions formed by alkaline earth metals are isoelectronic with noble gas atoms.
    n) Atomic numbers 23 and 45 both belong to transition elements.
    o) Atomic numbers 52, 35, and 18 are arranged in order of increasing atomic size.
    p) Atomic numbers 7, 16, and 35 are all nonmetals.

93. Figure 11.5 shows the spectra of hydrogen and neon. Hydrogen has four lines, one of which is sometimes difficult to see. Yet we say there are more lines in the hydrogen spectrum. Why do we not see them?

94. What is meant by the statement that something behaves like a wave?

## MORE CHALLENGING PROBLEMS

95. ■ Write the electron configurations you would expect for barium and technetium (Z = 43).

96. ■ Write the electron configurations you would expect for iodine and tungsten (Z = 74).

97. Why do you suppose the second ionization energy of an element is always greater than the first ionization energy?

98. Why is the definition of atomic number based on the number of protons in an atom rather than the number of electrons?

99. Compare the ionization energies of aluminum and chlorine. Why are these values as they are?

100. Suggest why the ionization energy of magnesium is greater than the ionization energies of sodium, aluminum, and calcium.

101. ■ Do elements become more metallic or less metallic as you (a) go down a group in the periodic table, (b) move left to right across a period in the periodic table? Support your answers with examples.

102. One of the successes of the Bohr model of the atom was its explanation of the lines in atomic spectra. Does the quan-

tum mechanical model also have a satisfactory explanation for these lines? Justify your answer.

103. Although the quantum model of the atom makes predictions for atoms of all elements, most of the quantitative confirmation of these predictions is limited to substances whose formulas are H, $He^+$, and $H_2^+$. From your knowledge of electron configurations and the limited information about chemical formulas given in this chapter, can you identify a single feature that all three substances have that makes them unique and makes them relatively easy to investigate?

104. What do you suppose are the electron configuration and the formula of the monatomic ion formed by scandium (Z = 21)?

105. Carbon does not form a stable monatomic ion. Suggest a reason for this.

106. Consider the block of elements in Periods 2 to 6 and Groups 5A/15 to 7A/17. In Group 7A/17 are the halogens, a distinct chemical family in which the different elements share many properties. Elements in Group 6A/16, with the exception of polonium (Z = 84), have enough similarity to be considered a family; they are called the chalogen family. In Group 5A/15, however, family similarities are weak, and those that exist belong largely to nitrogen, phosphorus, and to some extent, arsenic (Z = 33). Why do you suppose family similarities break down in Group 5A/15?

107. Xenon (Z = 54) was the first noble gas to be chemically combined with another element. Xenon is present in nearly all of the small number of noble gas compounds known today. Note the ionization energy trend begun with the other noble gases in Figure 11.18. What do these facts suggest about the relative reactivities of the noble gases and the character of noble gas compounds?

108. ■ Iron forms two monatomic ions, $Fe^{3+}$ and $Fe^{2+}$. From which sublevels do you expect electrons are lost in forming these ions? (*Hint:* It is possible for electrons other than those in the *s* and *p* sublevels to be involved in forming ions.)

109. Figure 11.18 indicates a general increase in the first ionization energy from left to right across a row of the periodic table. However, there are two sharp breaks in Periods 2 and 3. (a) Suggest why ionization energy should increase as atomic number increases within a period. (b) Can you correlate features of electron configurations with the locations of the breaks?

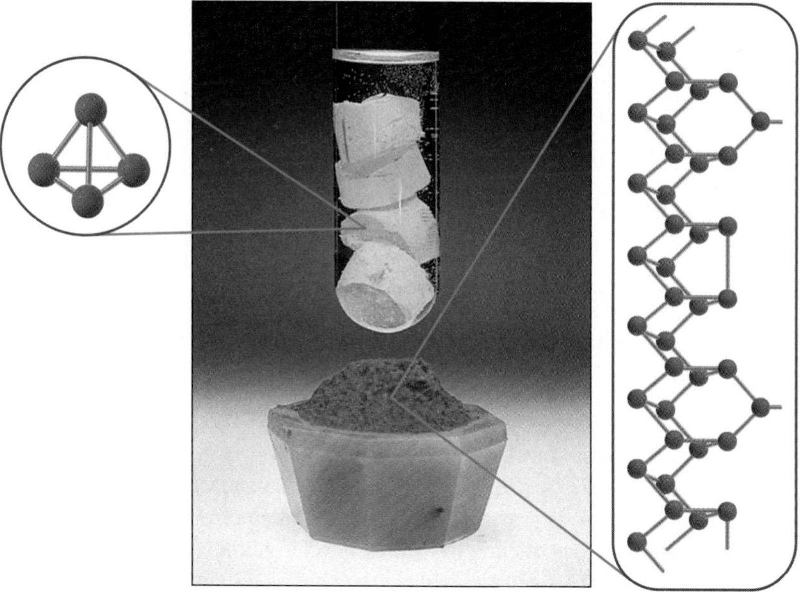

© Cengage Learning/Charles D. Winters

12

CONTENTS



The two substances in the photograph appear very different, but they are the same element, phosphorus. On top is white phosphorus, $P_4$, stored in water because it catches fire when exposed to air. Notice that its particulate-level structure is a collection of discrete four-atom molecules. When white phosphorus is heated, it changes to red phosphorus, shown in the bottom of the photograph. This substance is stable in air, as you can see. At the particulate level, red phosphorus is a network of phosphorus atoms with no individual molecules. How can two forms of the same pure element be so different? The answer lies in understanding the differing arrangements of the chemical bonds, the subject of this chapter.

# Chemical Bonding

In his atomic theory, John Dalton said that atoms of different elements combine to form compounds. He didn't say how they combined, or why. We now believe we understand *how* most chemical compounds form. *Why* atoms combine was touched on in Section 2.8: "Minimization of energy is one of the driving forces that cause chemical reactions to occur." This minimization refers to lowering the potential energy resulting from the attractions and repulsions among charged particles in the structure of the compound.

The term **chemical bond** describes the forces that hold atoms together to form molecules or polyatomic ions, that hold atoms together in metals, or that hold oppositely charged ions together to form ionic compounds. Chemical bonds can break and re-form in new combinations when atoms, molecules, or ions collide. The first contact in such a collision is between the outermost electrons of two particles. These

Online homework for this chapter may be assigned in OWL

P/REVIEW: The valence electrons of an atom (Section 11.5) are the electrons in the *s* and *p* sublevels of the highest occupied principal energy level when an atom is in the ground state. When these sublevels contain a total of eight electrons, two in the *s* sublevel and six in the *p*, we say the octet of electrons is filled. Two particles are isoelectronic if they have identical electron configurations (Sections 11.4 and 11.6).

outermost electrons are the valence electrons of the bonded atoms. Our study of chemical bonds therefore focuses on the role of valence electrons in forming a bond between two atoms.

## 12.1 Monatomic Ions with Noble-Gas Electron Configurations

Goal 1 Define and distinguish between cations and anions.

Goal 2 Identify the monatomic ions that are isoelectronic with a given noble-gas atom and write the electron configuration of those ions.

P/REVIEW: In Section 11.6 an ion is defined as an atom or group of atoms that has an electrical charge because of a difference in the number of protons and electrons. A more detailed discussion of the formation of both cations and anions appears in Section 6.4.

Elements in the same chemical family usually form monatomic ions having the same charge (Section 6.4). They do this by gaining or losing valence electrons until they become isoelectronic with a noble-gas atom and the octet of electrons is complete. If an atom loses one or more electrons, the ion has a positive charge and is called a **cation.** If the atom gains electrons, the ion is negatively charged and is called an **anion.**

Table 12.1 shows how electrons are gained and lost when nitrogen, oxygen, fluorine, sodium, magnesium, and aluminum form monatomic ions that are isoelectronic with a neon atom. The pattern built around neon is duplicated for other noble gases. Thus phosphorus, sulfur, chlorine, potassium, calcium, and scandium (Z = 21) are the elements that form ions that are isoelectronic with argon, and five of the elements on either side of krypton form ions that duplicate the electron configuration of krypton.

The monatomic hydride and lithium ions, $H^-$ and $Li^+$, duplicate the electron configuration of helium with just two electrons: $1s^2$. Unlike other elements in Groups 2A/2 and 3A/13, beryllium and boron tend to form covalent bonds by sharing electrons rather than forming ions. We will look at covalent bonds later in this chapter. Figure 12.1 is a periodic table that summarizes the elements that form monatomic ions that are isoelectronic with noble-gas atoms.

Elements below aluminum on the periodic table—gallium (Ga), indium (In), and thallium (Tl)—are not included in Figure 12.1 because these elements are uncommon and have an increasing tendency to form 1+ ions as you move down the group.

P/REVIEW: The hydrated hydrogen ion is present in acids. It forms when a hydrogen-bearing compound reacts with water (Section 6.5).

You may wonder about the hydrogen ion, $H^+$. This ion does not normally exist by itself, but rather exists as a hydrated hydrogen ion, which means that the hydrogen ion is associated with one or more molecules of water. This can be represented as $H^+ \cdot (H_2O)_x$, where x is some number of water molecules. One form of the hydrated hydrogen ion is

| | | | | | | | | | | | | | | | | | |
|---|---|---|---|---|---|---|---|---|---|---|---|---|---|---|---|---|---|
| | | | | | | | | | | | | | | | | $H^-$ | He |
| $Li^+$ | | | | | | | | | | | | | | $N^{3-}$ | $O^{2-}$ | $F^-$ | Ne |
| $Na^+$ | $Mg^{2+}$ | | | | | | | | | | | $Al^{3+}$ | | $P^{3-}$ | $S^{2-}$ | $Cl^-$ | Ar |
| $K^+$ | $Ca^{2+}$ | $Sc^{3+}$ | | | | | | | | | | | | | $Se^{2-}$ | $Br^-$ | Kr |
| $Rb^+$ | $Sr^{2+}$ | $Y^{3+}$ | | | | | | | | | | | | | $Te^{2-}$ | $I^-$ | Xe |
| $Cs^+$ | $Ba^{2+}$ | $La^{3+}$ | | | | | | | | | | | | | | | |

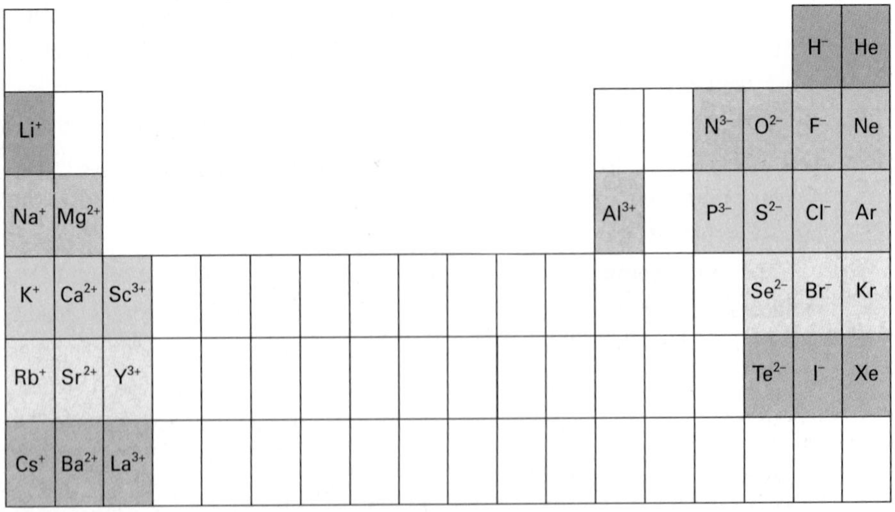

**Figure 12.1** Monatomic ions with noble-gas electron configurations. Each color group includes one noble-gas atom and the monatomic ions that are isoelectronic with that atom. Beryllium and boron are not included because these elements more commonly form covalent bonds (Section 12.3).

**Table 12.1** Formation of Monatomic Ions That Are Isoelectronic with Neon Atoms*

| Element | Atom | | Electron(s) | | Monatomic Ion | Atom/Ion Electron Count | | |
|---|---|---|---|---|---|---|---|---|
| Nitrogen | $7\ p^+$ $7\ e^-$ | + | $e^- + e^- + e^-$ | → | $7\ p^+$ $10\ e^-$ | *Start* | *Change* | *Final* |
| Z = 7 | | | | | | 7 | +3 | 10 |
| Group 5A/15 | $\cdot\ddot{N}:$ | + | $\cdot(e^-) + \cdot(e^-) + \cdot(e^-)$ | → | $[:\ddot{N}:]^{3-}$ $N^{3-}$ | | | |
| | $1s^22s^22p^3$ | + | $3\ e^-$ | → | $1s^22s^22p^6$ | | | |
| Oxygen | $8\ p^+$ $8\ e^-$ | + | $e^- + e^-$ | → | $8\ p^+$ $10\ e^-$ | *Start* | *Change* | *Final* |
| Z = 8 | | | | | | 8 | +2 | 10 |
| Group 6A/16 | $\cdot\ddot{O}:$ | + | $\cdot(e^-) + \cdot(e^-)$ | → | $[:\ddot{O}:]^{2-}$ $O^{2-}$ | | | |
| | $1s^22s^22p^4$ | + | $2\ e^-$ | → | $1s^22s^22p^6$ | | | |
| Fluorine | $9\ p^+$ $9\ e^-$ | + | $e^-$ | → | $9\ p^+$ $10\ e^-$ | *Start* | *Change* | *Final* |
| Z = 9 | | | | | | 9 | +1 | 10 |
| Group 7A/17 | $:\ddot{F}:$ | + | $\cdot(e^-)$ | → | $[:\ddot{F}:]^-$ $F^-$ | | | |
| | $1s^22s^22p^5$ | + | $e^-$ | → | $1s^22s^22p^6$ | | | |
| Neon | $10\ p^+$ $10\ e^-$ | | | | $10\ p^+$ $10\ e^-$ | *Start* | *Change* | *Final* |
| Z = 10 | | | | | | 10 | | 10 |
| Group 8A/18 | $:\ddot{Ne}:$ | | | | Ne | | | |
| | $1s^22s^22p^6$ | | | | $1s^22s^22p^6$ | | | |
| Sodium | $11\ p^+$ $11\ e^-$ | − | $e^-$ | → | $11\ p^+$ $10\ e^-$ | *Start* | *Change* | *Final* |
| Z = 11 | | | | | | 11 | −1 | 10 |
| Group 1A/1 | $Na\cdot$ | − | $\cdot(e^-)$ | → | $Na^+$ | | | |
| | $1s^22s^22p^63s^1$ | − | $e^-$ | → | $1s^22s^22p^6$ | | | |
| Magnesium | $12\ p^+$ $12\ e^-$ | − | $e^- + e^-$ | → | $12\ p^+$ $10\ e^-$ | *Start* | *Change* | *Final* |
| Z = 12 | | | | | | 12 | −2 | 10 |
| Group 2A/2 | $Mg\cdot$ (with dot below) | − | $\cdot(e^-) + \cdot(e^-)$ | → | $Mg^{2+}$ | | | |
| | $1s^22s^22p^63s^2$ | − | $2\ e^-$ | → | $1s^22s^22p^6$ | | | |
| Aluminum | $13\ p^+$ $13\ e^-$ | − | $e^- + e^- + e^-$ | → | $13\ p^+$ $10\ e^-$ | *Start* | *Change* | *Final* |
| Z = 13 | | | | | | 13 | −3 | 10 |
| Group 3A/13 | $\cdot Al\cdot$ (with dot below) | − | $\cdot(e^-) + \cdot(e^-) + \cdot(e^-)$ | → | $Al^{3+}$ | | | |
| | $1s^22s^22p^63s^23p^1$ | − | $3\ e^-$ | → | $1s^22s^22p^6$ | | | |

*This table shows how monatomic anions (pink) and cations (yellow) that are isoelectronic with neon atoms (blue) are formed. A neon atom has ten electrons, including a full octet of valence electrons. Its electron configuration is $1s^22s^22p^6$. Nitrogen, oxygen, and fluorine atoms form anions by gaining enough electrons to reach the same configuration. Sodium, magnesium, and aluminum atoms form cations by losing valence electrons to reach the same configuration. Dots around each elemental symbol represent valence electrons. The notation "$\cdot(e^-)$" represents an electron added to or subtracted from an electron-dot symbol.

commonly called the **hydronium ion** and written $H_3O^+$. The ion is not truly monatomic, and therefore it is not properly included in this section.

## Active Example 12.1

Write the electron configurations for calcium and chloride ions. With what noble gases are these ions isoelectronic?

What are the formulas of the calcium ion and the chloride ion?

P/REVIEW: Lewis symbols use dots distributed around the symbol of an element to represent valence electrons. See Section 11.5.

$Ca^{2+}$ and $Cl^{-}$

Note the location of calcium and chlorine in the periodic table. Write their Lewis symbols and from them state the number of electrons that must be gained or lost to achieve complete octets at the highest energy level.

Ca: must lose two electrons to achieve an octet, and :C̣l: must gain one to achieve an octet.

You are now ready to answer the main questions: What are the electron configurations for $Ca^{2+}$ and $Cl^{-}$, and with which noble gases are they isoelectronic?

Both ions are isoelectronic with argon, $1s^22s^22p^63s^23p^6$.

The calcium atom starts with the configuration $1s^22s^22p^63s^23p^64s^2$. In losing two electrons to yield $Ca^{2+}$, it reaches the electron configuration of argon. Chlorine, with the configuration $1s^22s^22p^63s^23p^5$, must gain one electron to become $Cl^{-}$, which is also isoelectronic with argon.

### √ Target Check 12.1

*Which ions among the following are isoelectronic with noble-gas atoms?*

$Cu^{2+}$ $S^{2-}$ $Fe^{3+}$ $Ag^{+}$ $Ba^{2+}$

## 12.2 Ionic Bonds

**Goal** 3 Use Lewis symbols to illustrate how an ionic bond can form between monatomic cations from Groups 1A, 2A, and 3A (1, 2, 13) and anions from Groups 5A, 6A, and 7A (15–17) of the periodic table.

P/REVIEW: This section explains why the formulas of ionic compounds are what they are. Writing formulas of ionic compounds is discussed in Section 6.8. You must account for all of the electrons in all of the atoms in the formula unit of an ionic compound.

We have been discussing the formation of monatomic ions as neutral atoms that gain or lose electrons. For most elements this is not a common event, but an accomplished fact. The natural occurrence of many elements is in **ionic compounds**—compounds made up of ions—or solutions of ionic compounds. Nowhere in nature, for example, are sodium or chlorine atoms to be found, but there are large natural deposits of sodium chloride (table salt) that are made up of sodium ions and chloride ions. The compound, along with other ionic compounds, may also be obtained by evaporating seawater, which contains the ions in solution.

Sodium and chlorine are both highly reactive elements. If they are brought together, after having been prepared from any natural source, they will react vigorously to form the compound sodium chloride. In that reaction a sodium atom literally

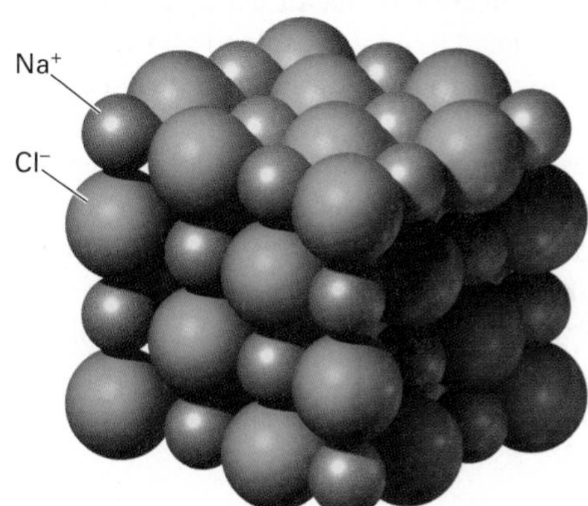


**Figure 12.2** Arrangement of ions in sodium chloride. The gray spheres represent sodium ions and the green spheres are chloride ions. The number of positive charges is the same as the number of negative charges, making the crystal electrically neutral.

loses an electron to become a sodium ion, $Na^+$, and a chloride ion gains an electron to become a chloride ion, $Cl^-$. Lewis symbols show the electron transfer clearly

$$Na\cdot + \cdot\ddot{\underset{..}{Cl}}: \longrightarrow Na^+ + \left[:\ddot{\underset{..}{Cl}}:\right]^- \longrightarrow \text{NaCl crystal}$$

Ions are not always present in a 1:1 ratio, as in sodium chloride. Calcium atoms have two valence electrons to lose to form a calcium ion, $Ca^{2+}$, but a chlorine atom receives only one electron when forming a chloride ion, $Cl^-$. It therefore takes two chlorine atoms to receive the two electrons from a single calcium atom:

$$Ca: + \begin{matrix}\cdot\ddot{\underset{..}{Cl}}: \\ \cdot\ddot{\underset{..}{Cl}}:\end{matrix} \longrightarrow Ca^{2+} + 2\left[:\ddot{\underset{..}{Cl}}:\right]^- \longrightarrow CaCl_2 \text{ crystal}$$

The 1:2 ratio of calcium ions to chloride ions is reflected in the formula of the compound formed, calcium chloride, $CaCl_2$. Several combinations of charges appear in ionic compounds, but they are always in such numbers that the compound is electrically neutral.

Nearly all ionic compounds are solids at normal temperatures and pressures. The solid has a definite geometric structure called a **crystal.** Ions in a crystal are arranged so the potential energy resulting from the attractions and repulsions between them is at a minimum (Section 2.8). The precise form of the crystal depends on the kinds of ions in the compound, their sizes, and the ratio in which they appear. Figure 12.2 shows the structure of a sodium chloride crystal. The strong electrostatic forces that hold the ions in fixed position in the crystal are called **ionic bonds.**

Ionic crystals are not limited to monatomic ions; polyatomic ions—ions consisting of two or more atoms—also form crystal structures. Atoms within a polyatomic ion are held together by covalent bonds (Section 12.3). Figure 12.3 is a model of a

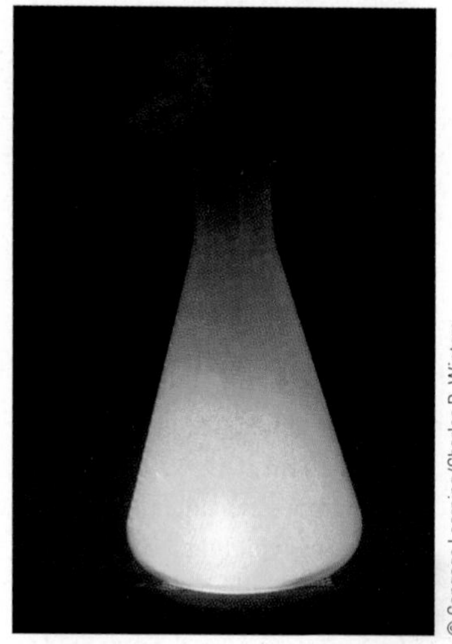


The reaction of sodium and chlorine. Sodium metal reacts violently with chlorine gas to form sodium chloride.

**P/REVIEW:** Electrostatic forces are unmoving electrical forces (Section 2.7).

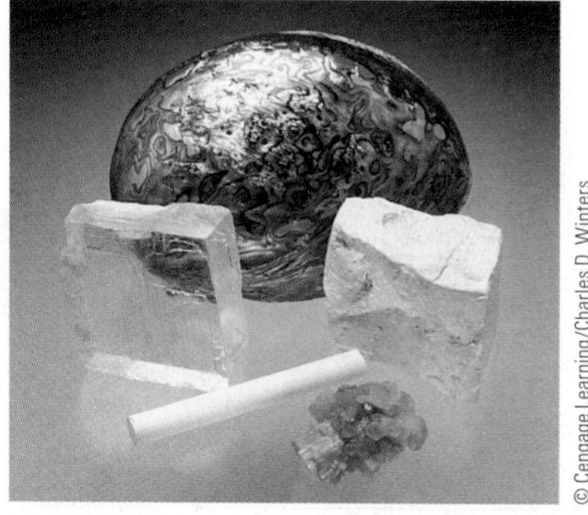


The many forms of calcium carbonate (*described clockwise from top*). An abalone shell is composed of thin overlapping layers of calcium carbonate, interspersed with a protein (a class of macromolecules described in Chapter 22). Limestone is a mixture of calcium carbonate and other compounds found in sedimentary rocks. Aragonite is one of the crystalline solid forms of calcium carbonate, characterized by the arrangement of carbonate ions in two planes that point in opposite directions. Blackboard chalk is often incorrectly believed to be made from the mineral chalk, which is a form of calcium carbonate, but, in fact, it is made from calcium sulfate. Iceland spar is the clear form of the mineral calcite, which is calcium carbonate arranged in a different crystal structure from that of aragonite.

P/REVIEW: The unit-like behavior of polyatomic ions is used when balancing chemical equations (Section 8.8).

P/REVIEW: Solutions that conduct electric current because of the movement of ions are called electrolytes (Section 9.1).

calcium carbonate crystal. The formula of a carbonate ion, one of which is circled in Figure 12.3, is $CO_3^{2-}$. The carbon atom is surrounded by three covalently bonded oxygen atoms. The carbonate ion is a distinct unit in the structure of the crystal, and it also behaves as a unit in many chemical changes.

The bonds in an ionic crystal are very strong, which is why nearly all ionic compounds are solids at room temperature. A high temperature is required to break the many ionic bonds, free the ions from one another, and melt the crystal to become a liquid. Solid ionic compounds are poor conductors of electricity because the ions are locked in place in the crystal (Fig. 12.4[a]). When the substance is melted or dissolved, the crystal is destroyed. The ions are then free to move and able to carry electric current (Fig. 12.4[b]). Liquid ionic compounds and water solutions of ionic compounds are good conductors.

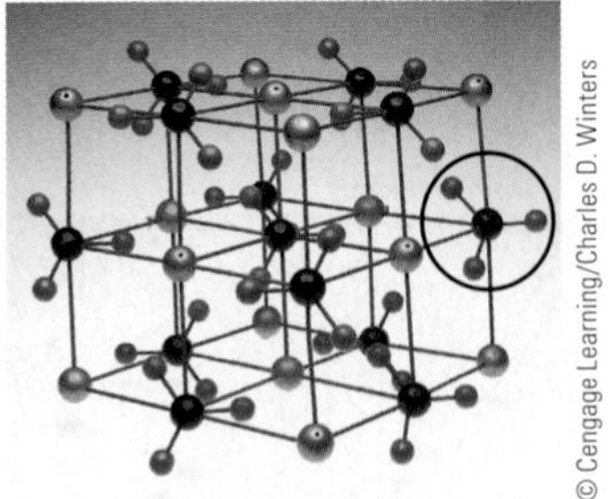


**Figure 12.3** Model of a calcium carbonate crystal. The gray spheres represent calcium ions, $Ca^{2+}$. The black spheres with three red spheres attached (see circle) are carbonate ions, $CO_3^{2-}$. There are equal numbers of calcium and carbonate ions, yielding a compound that is electrically neutral.

## Thinking About Your Thinking

### Mental Models

**Figures 12.2 and 12.3 should help you form a mental model of an ionic compound. A key characteristic of your mental model needs to be an image of the ion itself as an individual particle. There are no such things as "sodium chloride molecules." The ionic compound sodium chloride consists of equal numbers of sodium ions and chloride ions, but there are no discrete sodium chloride units. Each positively charged sodium ion is surrounded by negatively charged chloride ions, and each chloride ion is surrounded by sodium ions.**

**Recall that the process of dissolving sodium chloride in water is a physical change. (Physical and chemical changes are described in Section 2.3.) The sodium ions and chloride ions originally locked in the solid crystal are themselves unchanged as they become sodium ions and chloride ions surrounded by water molecules. If the water evaporates, you have the same sodium ions and chloride ions, but they are in a solid crystalline state once again. No chemical change has occurred; the sodium ions and chloride ions remain unchanged throughout the process. If you can imagine how this occurs at the particulate level, you have a good understanding of the structure of ionic compounds.**

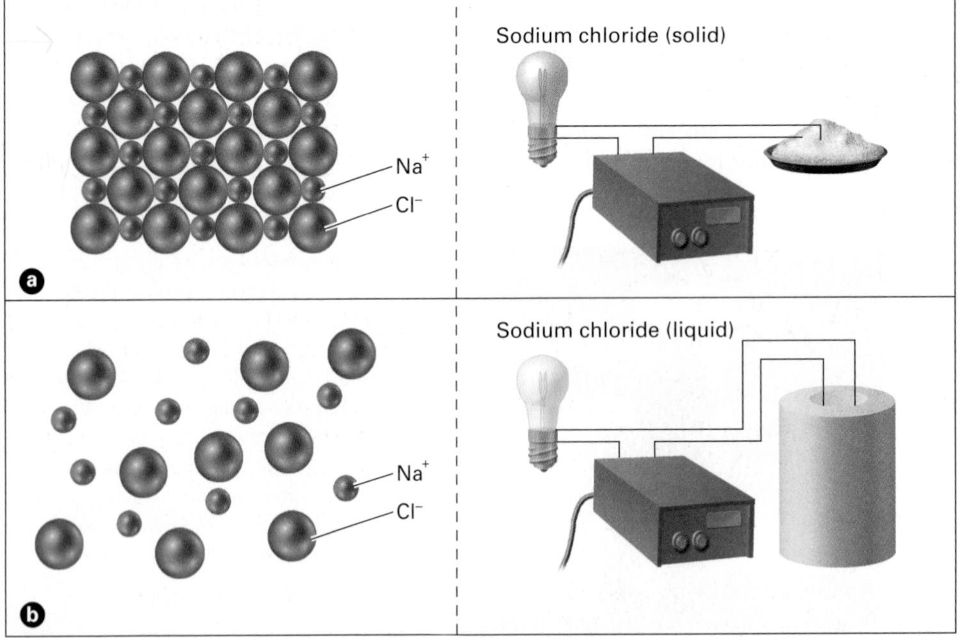


**Figure 12.4** Electrical conductivity. (a) Ionic solids do not conduct electrical current but (b) are good conductors when melted or dissolved in water.

√ **Target Check 12.2**

*Use Lewis diagrams to show the electron transfer that occurs when a potassium atom reacts with a fluorine atom to form a potassium fluoride crystal.*

## 12.3 Covalent Bonds

Goal **4** Describe, use, or explain each of the following with respect to forming a covalent bond: electron cloud, charge cloud, or charge density; valence electrons; half-filled electron orbital; filled electron orbital; electron sharing; orbital overlap; octet rule or rule of eight.

Goal **5** Use Lewis symbols to show how covalent bonds are formed between two nonmetal atoms.

Goal **6** Distinguish between bonding electron pairs and lone pairs.

Let's take a look at two familiar white crystals—salt and sugar. Both dissolve easily in water, but the salt water conducts electricity, while the sugar water does not (Fig. 12.5). Salt melts at 801°C to give a colorless liquid, but sugar doesn't melt at all. Instead, at only 160°C, sugar begins to char, emitting the aroma of caramel. (Indeed, that's how caramel is made.)

We know that the properties of table salt and other ionic compounds are explained by ionic bonding. Other compounds, such as sugar, have properties so different from those of ionic compounds that a different type of bonding must be responsible for holding together atoms in these compounds.

Water, $H_2O$; hydrogen fluoride, HF; and methane, $CH_4$, are compounds of the "other kind." Hydrogen fluoride and methane are gases at room temperature and pressure, and

For electrical current to flow and light the bulb, the solution in which the electrodes are immersed must contain ions, which carry electrical charge.

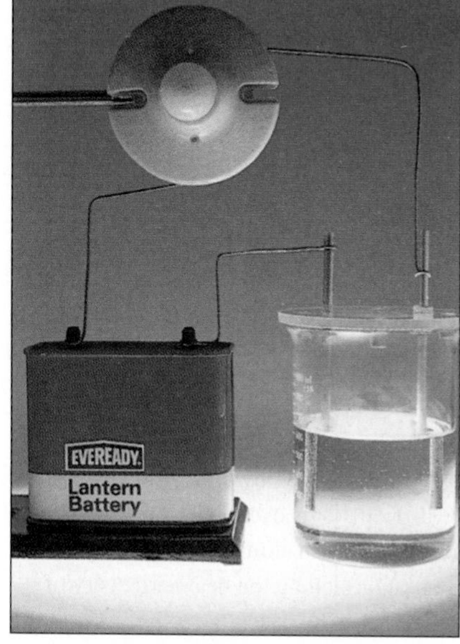


The solution of pure water does not contain ions and thus does not light the bulb.

(a) pure water

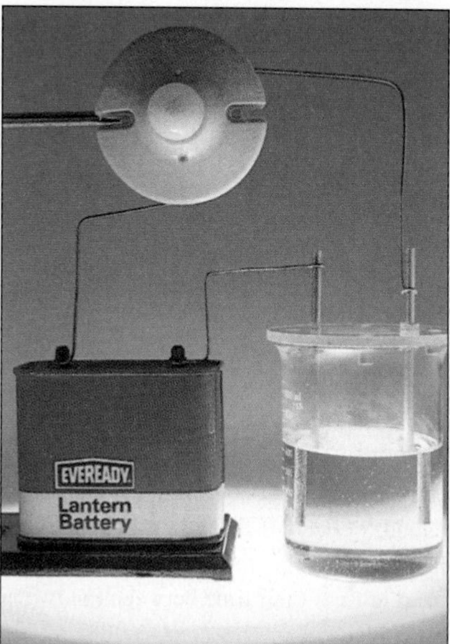


The solution of sucrose (table sugar) and pure water also lacks ions, and fails to light the bulb.

(b) table sugar

The solution of sodium chloride (NaCl) and pure water does contain ions, and thus lights the bulb.

(c) table salt



**Figure 12.5** Conductivity of solutions. Ions are held together by ionic bonds in salt, but another type of bonding must be responsible for the properties of sugar.

Oesper Collection in the History of Chemistry/University of Cincinnati

Gilbert N. Lewis (1875–1946). Electron-dot structures are called Lewis diagrams in his honor.

water is a liquid. When hydrogen fluoride and methane are condensed to liquids, they are nonconductors, like water. These are **molecular compounds,** whose ultimate structural unit is an individual particle known as a **molecule.** The physical and chemical properties of a molecule are different from the properties of the atoms that make up the molecule.

In 1916 Gilbert N. Lewis proposed that two atoms in a molecule are held together by a **covalent bond** in which they share one or more pairs of electrons. The idea is that when the bonding electrons can spend most of their time between two atomic nuclei, they attract *both* positively charged nuclei and "couple" the atoms to each other, much like a car and a trailer are held together by the trailer hitch between them. The result is a bond that is permanent until broken by a chemical change.

The simplest covalent bond appears in hydrogen, $H_2$. Using Lewis symbols, the formation of a molecule of $H_2$ can be represented as

$$\mathrm{H\cdot + \cdot H \longrightarrow H:H} \quad \textit{or} \quad \mathrm{H—H}$$

The two dots or the straight line drawn between the two atoms represent the covalent bond that holds the atoms together. In quantum mechanical terms we say that the **electron cloud** or **charge density** formed by the two electrons is concentrated in the region between the two nuclei. This is where there is the greatest probability of locating the bonding electrons. The atomic orbitals of the hydrogen atoms **overlap,** coupling them together to form the hydrogen molecule.

The formation of a bond between two hydrogen atoms is shown in Figure 12.6. The caption describes and explains in some detail how and why this bond forms. Study both the illustration and the caption carefully. In particular, notice the last sentence. When bonding electrons are *between two nuclei,* both nuclei are attracted to the electrons. The electrons link the nuclei together; the electrons are the "glue" that bonds the atoms to each other.

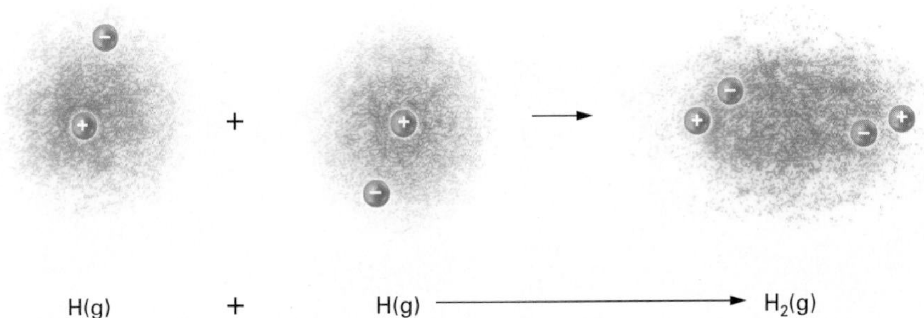


Recall that illustrations and their captions are sometimes used to explain major concepts. That's what is being done here. The caption is the text.

**Figure 12.6** The formation of a hydrogen molecule from two hydrogen atoms. Each location in the electron cloud represents an instantaneous position of the electron in the atom. To the left of the arrow, the circled minus signs represent the instantaneous position of the 1*s* electron in each separate atom *at the same instant,* before the bond is formed. To the right of the arrow the circled minus signs represent the location of those valence electrons *at the same instant* as they make up a bond between the atoms. The electron pair is shared by both atoms. The charge clouds of the half-filled 1*s* orbitals of the two hydrogen atoms overlap in the combined atoms in such a way that the two electrons fill the 1*s* orbitals of both atoms. The bonding electrons spend most of their time between the two nuclei, as suggested by the heavier density of electron-position dots in that area.

To understand why covalent bonds form, consider the attraction and repulsion forces both within each atom and between the two atoms when the atoms are separated and after the bond has been formed. Before bonding (left of the arrow), attractions and repulsions between the separated atoms are negligible. Within the atoms there is an attraction between the electron and the nucleus. After bonding (right of the arrow), the atoms are close to each other, and the electrons are attracted to *both* nuclei. *It is this attraction of the bonding electrons to two nuclei that holds the nuclei together in a covalent bond.*

A similar approach shows the formation of the covalent bond between two fluorine atoms to form a molecule of $F_2$ and between one hydrogen atom and one fluorine atom to form an HF molecule:

$$:\ddot{\underset{..}{F}}\cdot + \cdot\ddot{\underset{..}{F}}: \longrightarrow :\ddot{\underset{..}{F}}:\ddot{\underset{..}{F}}: \quad \textit{or} \quad :\ddot{\underset{..}{F}}\text{—}\ddot{\underset{..}{F}}: \quad \textit{or} \quad F\text{—}F$$

$$H\cdot + \cdot\ddot{\underset{..}{F}}: \longrightarrow H:\ddot{\underset{..}{F}}: \quad \textit{or} \quad H\text{—}\ddot{\underset{..}{F}}: \quad \textit{or} \quad H\text{—}F$$

Fluorine has seven valence electrons. The 2*s* orbital and two of the 2*p* orbitals are filled, but the remaining 2*p* orbital has only one electron. The $F_2$ bond is formed by the overlap of the half-filled 2*p* orbitals of two fluorine atoms. In the HF molecule, the bond forms from the overlap of the half-filled 1*s* orbital of a hydrogen atom with the half-filled 2*p* orbital of a fluorine atom.

When used to show the bonding arrangement between atoms in a molecule, electron-dot symbols are commonly called **Lewis diagrams, Lewis formulas,** or **Lewis structures.** Notice that the unshared electron pairs of fluorine are shown for two of the Lewis diagrams for $F_2$ and HF, but they are omitted in the F—F and H—F diagrams. Technically, they should always be shown, but they are frequently omitted when not absolutely needed. Unshared electron pairs are often called **lone pairs.**

When two atoms share two bonding electrons, the electrons effectively "belong" to both atoms. They count as valence electrons for each bonded atom. Thus each hydrogen atom in $H_2$ and the hydrogen atom in HF has two electrons, the same number as an atom of the noble gas helium. Each fluorine atom in $F_2$ as well as the fluorine atom in HF has eight valence electrons, matching neon and the other noble-gas atoms.

These and many similar observations lead us to believe that the stability of a noble-gas electron configuration is the result of a minimization of energy associated with that configuration. This generalization is known as the **octet rule,** or **rule of eight,** because each atom has "completed its octet." The tendency toward a complete octet of electrons in a bonded atom reflects the natural tendency of a system to move to the lowest energy state possible.

In Section 2.8, minimization of energy was identified as "one of the driving forces that cause chemical reactions to occur." Many laboratory measurements show that the energy of a system is reduced as bonds form that reach the noble-gas electron configuration.

### √ Target Check 12.3

*Identify the true statements and rewrite the false statements to make them true.*

a) Atoms in molecular compounds are held together by covalent bonds.

b) A lone pair of electrons is not shared between two atoms.

c) An octet of valence electrons usually represents a low energy state.

## 12.4 Polar and Nonpolar Covalent Bonds

Goal 7 Distinguish between polar and nonpolar covalent bonds.

Goal 8 Predict which end of a polar bond between identified atoms is positive and which end is negative. (You may refer to a periodic table.)

Goal 9 Rank bonds in order of increasing or decreasing polarity based on periodic trends in electronegativity values or actual values, if given. (You may refer to a periodic table.)

The two electrons joining the atoms in the $H_2$ molecule are shared equally by the two nuclei. **A bond in which bonding electrons are shared equally is a nonpolar covalent bond.** A more formal way of saying this is that the charge density is centered in the region between the bonded atoms. A bond between identical atoms, as in $H_2$ or $F_2$, is always nonpolar.

The fluorine atom in an HF molecule has a stronger attraction for the bonding electron pair than the hydrogen atom has. The bonding electrons therefore spend more

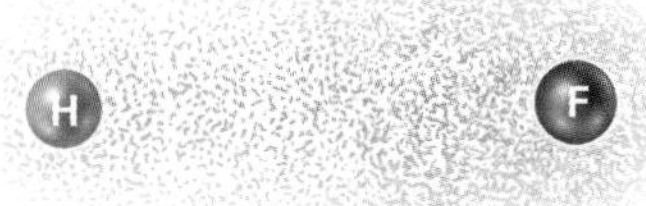


**Figure 12.7** A polar bond. Fluorine in a molecule of HF has a higher electronegativity than hydrogen. The bonding electron pair is therefore shifted toward fluorine. The uneven distribution of charge yields a polar bond.

time nearer the fluorine atom than the hydrogen atom. **A bond in which bonding electrons are shared, but shared unequally, is a polar covalent bond.** The charge density is shifted toward the fluorine atom and away from the hydrogen atom. Because the negative charge density is closer to the fluorine atom, that atom acts as a negative pole, and the hydrogen atom acts as a positive pole (Fig. 12.7). When the charge density shift is extreme, the bonding electrons are effectively transferred from one atom to another, and an ionic bond results (Fig. 12.8).

## Thinking About Your Thinking

### Mental Models

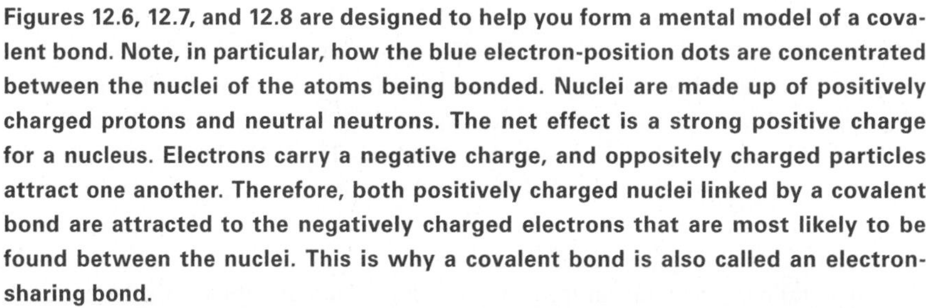

**Figures 12.6, 12.7, and 12.8 are designed to help you form a mental model of a covalent bond. Note, in particular, how the blue electron-position dots are concentrated between the nuclei of the atoms being bonded. Nuclei are made up of positively charged protons and neutral neutrons. The net effect is a strong positive charge for a nucleus. Electrons carry a negative charge, and oppositely charged particles attract one another. Therefore, both positively charged nuclei linked by a covalent bond are attracted to the negatively charged electrons that are most likely to be found between the nuclei. This is why a covalent bond is also called an electron-sharing bond.**

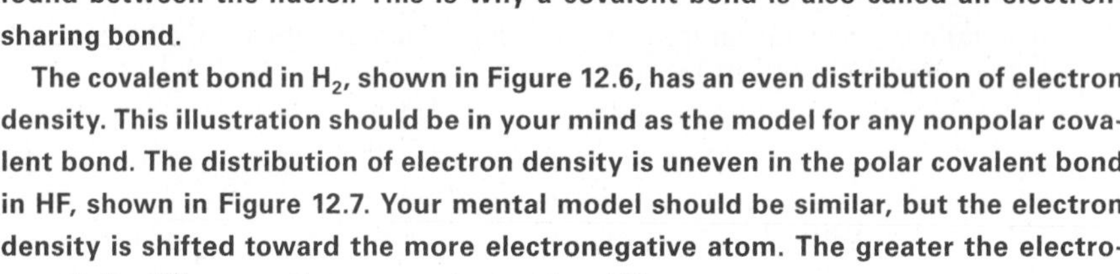

**The covalent bond in $H_2$, shown in Figure 12.6, has an even distribution of electron density. This illustration should be in your mind as the model for any nonpolar covalent bond. The distribution of electron density is uneven in the polar covalent bond in HF, shown in Figure 12.7. Your mental model should be similar, but the electron density is shifted toward the more electronegative atom. The greater the electronegativity difference, the more extreme the shift.**

© Roger Ressmeyer/Corbis

Linus Pauling (1901–1994). The concept of electronegativity was proposed by Pauling, who was awarded the 1954 Nobel Prize in Chemistry and the 1963 Nobel Peace Prize.

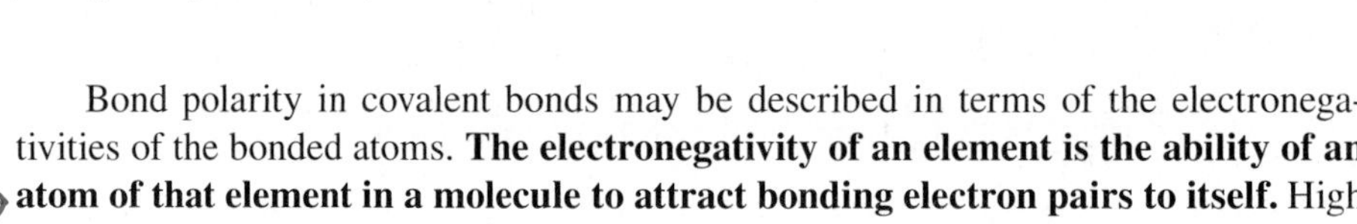

Bond polarity in covalent bonds may be described in terms of the electronegativities of the bonded atoms. **The electronegativity of an element is the ability of an atom of that element in a molecule to attract bonding electron pairs to itself.** High electronegativity identifies an element with a strong attraction for bonding electrons.

Electronegativity values of the elements are shown in Figure 12.9. Notice that electronegativities tend to be greater at the top of any column. In a smaller atom the bonding electrons are closer to the nucleus and therefore are attracted by it more strongly. Electronegativities also increase from left to right across any row of the periodic table. This matches the increase in nuclear charge among atoms whose bonding electrons are in the same principal energy level. In general, electronegativities are highest in the upper right region of the periodic table, and lowest in the lower left region.

**P/REVIEW:** These explanations of electronegativity trends are identical with those given for atomic size (Section 11.6).

You can estimate the polarity of a bond by calculating the difference between the electronegativity values for the bonded elements: The greater the difference, the more polar the bond. In $H_2$ and $F_2$, where two atoms of the same element are bonded, the

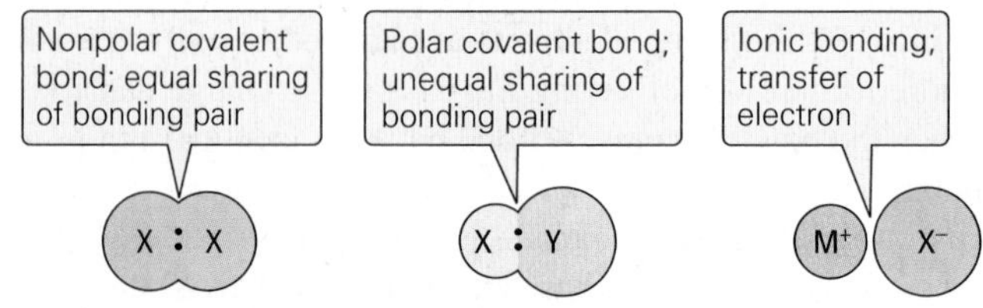


**Figure 12.8** A comparison of nonpolar covalent, polar covalent, and ionic bonding.

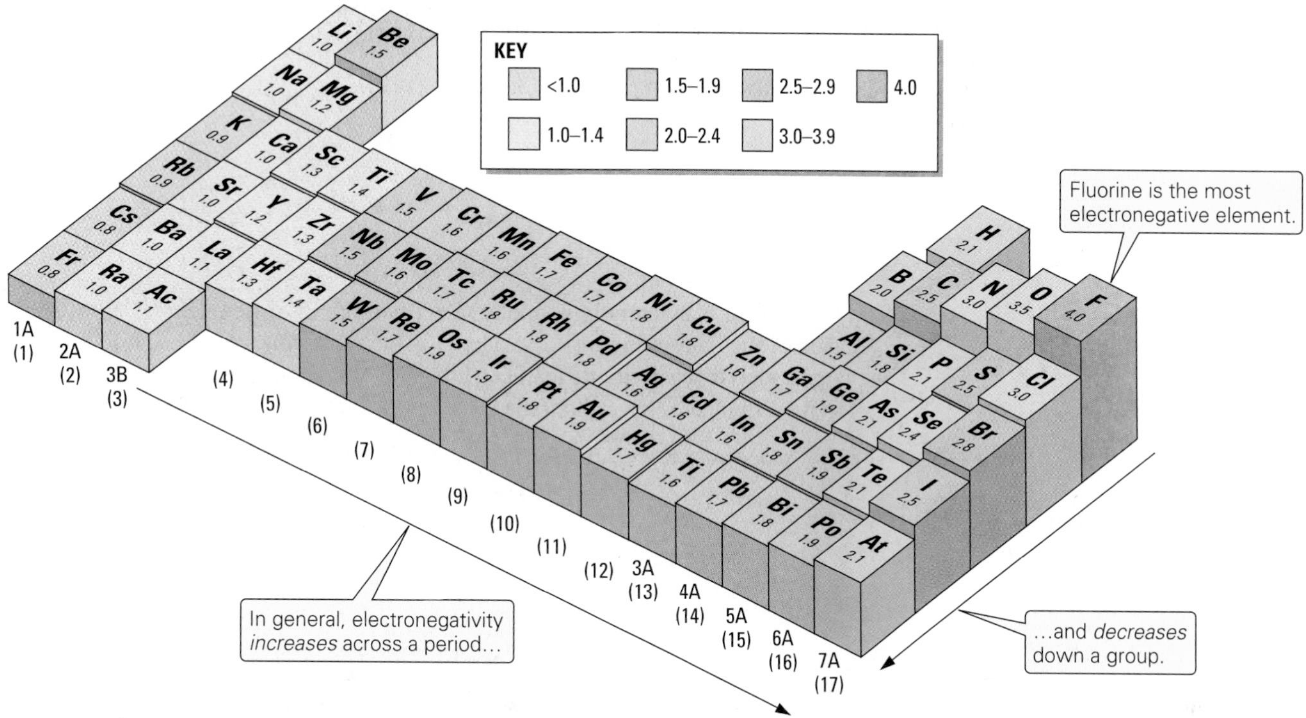


**Figure 12.9** Periodic trends in electronegativity values. Electronegativity values generally increase from left to right across any period of the table, and they decrease from the top to the bottom of any group.

electronegativity differences are zero and the bonds are nonpolar. In HF the electronegativity difference is 1.9 (4.0 for fluorine minus 2.1 for hydrogen). A bond between carbon and chlorine, for example, with an electronegativity difference of $3.0 - 2.5 = 0.5$, is more polar than an H—H bond, but less polar than an H—F bond.

The more electronegative element toward which the bonding electrons are displaced acts as the negative pole in a polar covalent bond. The less electronegative element is the positive pole. This is sometimes indicated by using an arrow rather than a simple dash, with the arrow pointing to the negative pole. In a bond between hydrogen and fluorine this is H $+\!\!\!\rightarrow$ F. Another representation is $\delta^-$ written in the region of the negative pole and $\delta^+$ in the area with the positive pole. The character $\delta$ is a lowercase Greek delta. In this use it represents a partial negative or partial positive charge—less than the full charge found on ions. Thus for hydrogen fluoride ${}^{\delta^+}\text{H—F}^{\delta^-}$.

**P/P/REVIEW:** Bond polarities are used to predict the polarities of molecules (Chapter 13). Molecular polarity is largely responsible for the physical properties of many compounds (Section 15.3).

## Active Example 12.2

Using data from Figure 12.9, arrange the following bonds in order of increasing polarity and place an arrowhead pointing toward the element that will act as the negative pole: H—O, H—S, H—P, H—C.

Locate the elements in the table, and calculate and write down the differences in electronegativity.

H—O

H—S

H—P

H—C

H—O: $3.5 - 2.1 = 1.4$
H—S: $2.5 - 2.1 = 0.4$
H—P: $2.1 - 2.1 = 0.0$
H—C: $2.5 - 2.1 = 0.4$

Now arrange the bonds in order from the least polar to the most polar.

H—P, H—S, H—C, H—O *or* H—P, H—C, H—S, H—O

The H—P bond is essentially nonpolar (electronegativity difference = 0.0 to the precision of the data in Figure 12.9). H—O is the most polar bond because it has the largest electronegativity difference (1.4). The other bonds, with equal electronegativity differences, have about the same polarity.

Now place an arrowhead on each bond that points toward the negative pole.

H—P H+→S H+→C H+→O

Since sulfur, carbon, and oxygen are all more electronegative than hydrogen, the electron density in these bonds is shifted away from hydrogen toward the other element. There is essentially no electronegativity difference in the H—P bond, so neither atom will act as a negative pole.

## 12.5 Multiple Bonds

**Goal** **10** Distinguish among single, double, and triple bonds, and identify these bonds in a Lewis diagram.

In special situations, evidence indicates that quadruple bonds, quintuple bonds, and sextuple bonds exist in compounds.

So far our consideration of covalent bonds has been limited to the sharing of one pair of electrons by two bonded atoms. Such a bond is called a **single bond.** In many molecules, we find two atoms bonded by two pairs of electrons; this is a **double bond.** When two atoms are bonded by three pairs of electrons, the bond is called a **triple bond.** All four electrons in a double bond and all six electrons in a triple bond are counted as valence electrons for each of the bonded atoms.

The most abundant substance containing a triple bond is nitrogen, $N_2$. Its Lewis diagram may be thought of as the combination of two nitrogen atoms, each with three unpaired electrons:

$$:\dot{\underset{\cdot}{N}}\cdot + \cdot\dot{\underset{\cdot}{N}}: \longrightarrow :N:::N: \quad or \quad :N{\equiv}N:$$

Counting the bonding electrons for both atoms, each nitrogen atom is satisfied with a full octet of electrons.

Experimental evidence supports the idea of **multiple bonds,** a general term that includes double and triple bonds. A triple bond is stronger and the distance between bonded atoms is shorter than the same measurements for a double bond between the same atoms, and a double bond is stronger and shorter than a single bond. Bond strength is measured as the energy required to break a bond. The greater the energy needed to break a bond, the stronger the bond. The triple bond in $N_2$ is among the strongest bonds known. This is one of the reasons elemental nitrogen is so stable and unreactive in the earth's atmosphere.

Nitrogen makes up 78% of the earth's atmosphere by volume.

# 12.6 Atoms That Are Bonded to Two or More Other Atoms

Using hydrogen and fluorine as examples, we have seen that two atoms that have a single unpaired valence electron are able to form a covalent bond by sharing those electrons. What if an atom has two unpaired valence electrons? Can it form two bonds with two different atoms? Yes. In fact, that is how a water molecule forms. A hydrogen atom forms a bond with one of the two unpaired valence electrons in an oxygen atom, and a second hydrogen atom does the same with the second unpaired oxygen electron:

$$\mathrm{H\cdot + \cdot\ddot{\underset{\cdot\cdot}{O}}\cdot + \cdot H \longrightarrow H:\ddot{\underset{\cdot\cdot}{O}}:H} \quad \textit{or} \quad \mathrm{H{-}\ddot{\underset{\cdot\cdot}{O}}{-}H}$$

This is the same as a hydrogen atom forming a bond with another hydrogen atom or with a fluorine atom.

A nitrogen atom has five valence electrons, three of which are unpaired. It therefore forms bonds with three hydrogen atoms to produce a molecule of ammonia, $NH_3$. Carbon has four valence electrons, only two of which are unpaired. This would lead us to expect that a carbon atom can form bonds with only two hydrogen atoms. In fact, all four electrons form bonds. The compound produced is methane, $CH_4$, the principal component of the natural gas burned as fuel in many homes. The Lewis diagrams of ammonia and methane are

All expectations should be checked experimentally. That carbon forms only two bonds is an example of a logical prediction that, when tested in the laboratory, is not confirmed. All organic life on the earth has as its basis the ability of the carbon atom to form four bonds.

```
   ..             H
                  |
 H—N—H         H—C—H
   |              |
   H              H
 ammonia       methane
```

Many polyatomic molecules contain multiple bonds. Ethylene, $C_2H_4$, the structural unit of the plastic polyethylene, has a double bond between two carbon atoms. There is a triple bond between carbon atoms in acetylene, $C_2H_2$, the fuel used in a welder's torch. The carbon atom in carbon dioxide, $CO_2$, is double bonded to two oxygen atoms. The Lewis diagrams for these compounds are

```
H       H
 \     /
  C = C        H—C≡C—H        :Ö=C=Ö:
 /     \
H       H
 ethylene       acetylene      carbon dioxide
```

Notice that if both lone pairs and bonding electron pairs are counted, the octet rule is satisfied for all atoms except hydrogen in all Lewis diagrams in this section. Hydrogen, as usual, duplicates the two-electron count of the noble gas helium.

## √ Target Check 12.4

*Which among the following have a double bond? Which have a triple bond? Which have a multiple bond?*

```
   H                                      :O:            H :O:
   |                                  ..  ||  ..         |  ||  ..
H—C—C≡N:          H  H             H—N—C—N—H        H—C—C—N—H
   |               |  |               |     |            |     |
   H            H—C=C—C≡N:            H     H            H     H
  (I)              (II)                (III)               (IV)
```

## √ Target Check 12.5

a) Is it possible, under the octet rule, for a single atom to be bonded by double bonds to each of three other atoms? Explain your answer.

b) What is the maximum number of atoms that can be bonded to the same atom and have that central atom conform to the octet rule? Explain.

## 12.7 Exceptions to the Octet Rule

### Odd-Electron Molecules

Not all substances conform to the octet rule. Two common oxides of nitrogen, NO and $NO_2$, have an odd number of electrons. It is therefore impossible to write Lewis diagrams for these compounds in which each atom is surrounded by eight valence electrons. Their Lewis diagrams are drawn with one single electron in a space where an electron pair would usually appear:

$$\text{N}{=}\text{O} \qquad \text{O}{=}\text{N}{-}\text{O}$$

### Molecules with More Than Four Electron Pairs Around the Central Atom

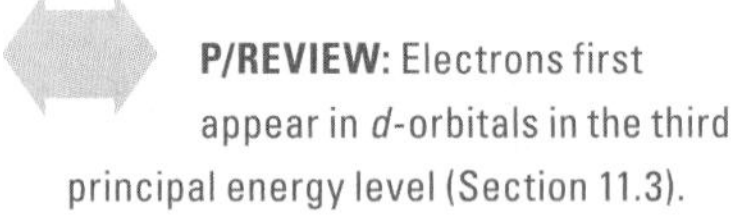

**P/REVIEW:** Electrons first appear in *d*-orbitals in the third principal energy level (Section 11.3).

All of the bonds we have encountered thus far have been formed by the valence electrons, the highest-energy *s*- and *p*-orbital electrons in the atom. Atoms of elements in the third period and higher can have more than four electron pairs surrounding them. This is accomplished by involving *d*-orbitals in the formation of covalent bonds. Phosphorus pentafluoride, $PF_5$, places five electron-pair bonds around the phosphorus atom; six pairs surround sulfur in $SF_6$:

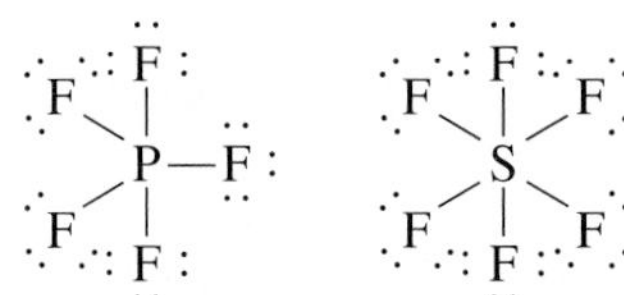

### Molecules with Fewer Than Four Electron Pairs Around the Central Atom

Two other substances for which satisfactory octet-rule diagrams can be drawn, but which are contradicted experimentally, are the fluorides of beryllium and boron. We might even expect $BeF_2$ and $BF_3$ to be ionic compounds, but laboratory evidence strongly supports covalent structures having the Lewis diagrams

$$\text{F}{-}\text{Be}{-}\text{F} \qquad \text{BF}_3$$

In these compounds the beryllium and boron atoms are surrounded by two and three pairs of electrons, respectively, rather than four.

### Oxygen

Certain molecules whose Lewis diagrams obey the octet rule do not have the properties that would be predicted. On paper, oxygen, $O_2$, appears to have an ideal double bond:

$$\text{O}{=}\text{O}$$

**P/REVIEW:** Oxygen and the fluorides of beryllium and boron are additional examples showing that all predictions must be confirmed experimentally. A specific step in the progress of chemistry begins and ends in the laboratory, but each ending is a new beginning, leading to new predictions and new experiments to confirm them. In this way our knowledge of chemistry increases.

But liquid oxygen is paramagnetic, meaning that it is attracted by a magnetic field. (See Fig. 12.10, which shows liquid oxygen being held in place between the poles of a magnet.) This is characteristic of molecules that have unpaired electrons, a fact that might suggest a Lewis diagram with each oxygen surrounded by seven electrons:

$$\text{O}{-}\text{O}$$

However, this conflicts with other evidence that the oxygen atoms are connected by something other than a single bond. In essence, it is impossible to write a single Lewis diagram that satisfactorily explains all the properties of molecular oxygen.

**Figure 12.10** Physical properties of oxygen. When pure gaseous oxygen is cooled to temperatures between −183°C and −218°C, it becomes a liquid (*left*). Liquid oxygen has a pale blue color (*center*). If the liquid is poured into the field of a strong magnet, some of it is trapped and held between the poles of the magnet (*right*). This paramagnetism is due to the unpaired electrons in the oxygen molecule.

## 12.8 Metallic Bonds

Goal **11** Describe how metallic bonding differs from ionic and covalent bonding.

Goal **12** Sketch a particulate-level illustration of the electron-sea model of metallic bonding.

Another familiar group of substances has properties quite different from those of either ionic compounds or molecular substances. These are the metals, the elements to the left of the stair-step line in the periodic table (Section 5.6).

Most metal atoms have one, two, or three valence electrons. These electrons are loosely held, which is why they are easily removed to form cations. When surrounded entirely by other metal atoms, there is no place for the valence electrons to go. They cannot transfer to half-filled *p*-electron orbitals in nonmetal atoms to form anions, and there is no way they can establish covalent bonds by sharing electron pairs with other atoms. A particulate-level model of metals consists of metal ions—the positively charged ions that result from the loss of the outermost *s* and *p* valence electrons—submersed in a freely moving "sea of electrons," an "ocean" of negative charge, as shown in Figure 12.11.

Tin (Z = 50) and lead (Z = 82) are notable exceptions to the common pattern of metal atoms having one, two, or three valence electrons. They have four valence electrons.

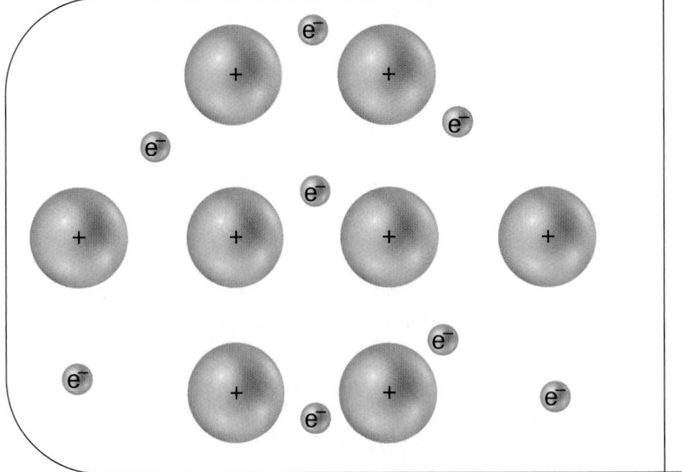

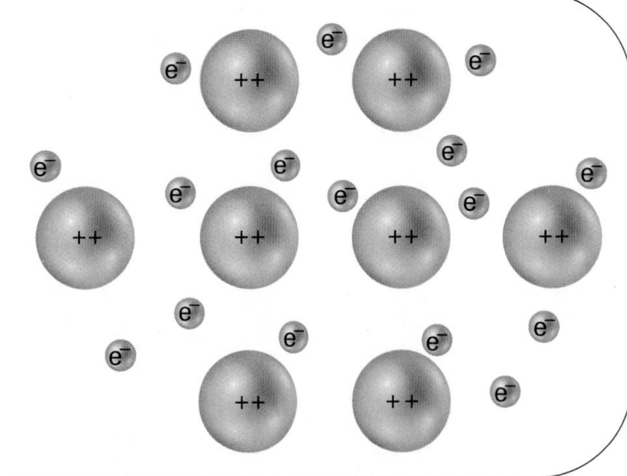


**Figure 12.11** The electron-sea model of a metallic crystal. The monatomic ions formed by the metal remain fixed in a definite crystal pattern, but the highest-energy valence electrons are relatively free to move, which explains the high electrical conductivity of metals. The drawing on the left might represent metallic sodium, or another Group 1A/1 element, with its highest-energy $ns^1$ electrons, and that on the right, magnesium, or another Group 2A/2 element, with its highest-energy $ns^2$ electrons.

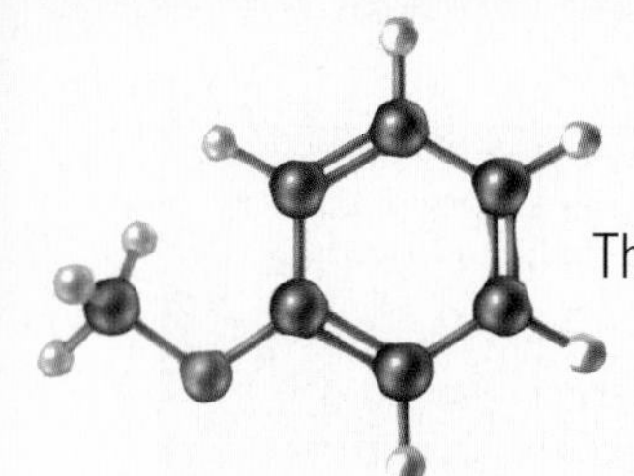

# Everyday Chemistry

## The Influence of Bonding on Macroscopic Properties

Look at the photographs of a diamond (Fig. 12.12[a]) and one type of pencil "lead" that contains graphite (Fig. 12.12[b]). (The original lead pencil was replaced by the graphite pencil. Modern pencils use a mixture of graphite and clay. Different hardnesses of pencils result from varying the mixture of graphite to clay.) Would you believe that both substances are pure carbon? Yes, it's true, diamonds and graphite are pure carbon. The particulate-level composition of each is nothing but carbon atoms. Why do they appear so very different? The answer lies in how the carbon atoms bond to one another.

Diamond has a particulate-level arrangement in which each carbon atom is covalently bonded to four other carbon atoms (Fig. 12.13). Look at a single carbon atom "ball" in the figure and count the number of bonding "sticks." Count the number of bonds to another atom. Note how each carbon atom in diamond has four bonds. This arrangement, where each atom is bonded to four atoms, is responsible for the physical properties of diamond. It is one of the hardest natural materials known.

Now look at the arrangement of carbon atoms in graphite as shown in Figure 12.14.

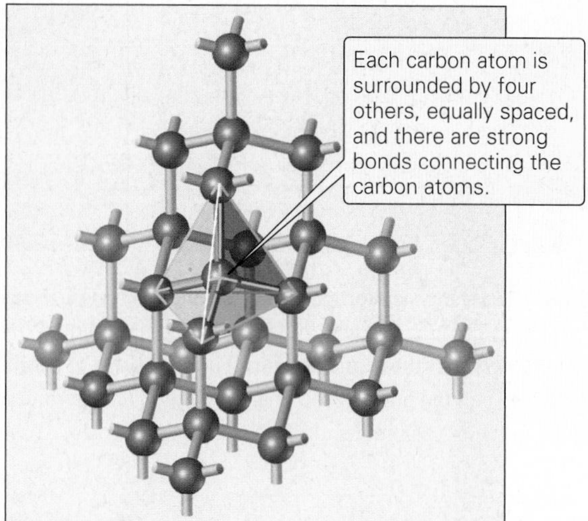

Each carbon atom is surrounded by four others, equally spaced, and there are strong bonds connecting the carbon atoms.


**Figure 12.13** A particulate-level model of diamond. Each carbon atom is bonded to four other carbon atoms. This bonding arrangement leads to the characteristic physical properties of diamond at the macroscopic level.

© Cengage Learning/Charles D. Winters

(a) A diamond. The physical properties of a diamond include a density of 3.5 g/cm$^3$, lack of color, and extreme hardness.

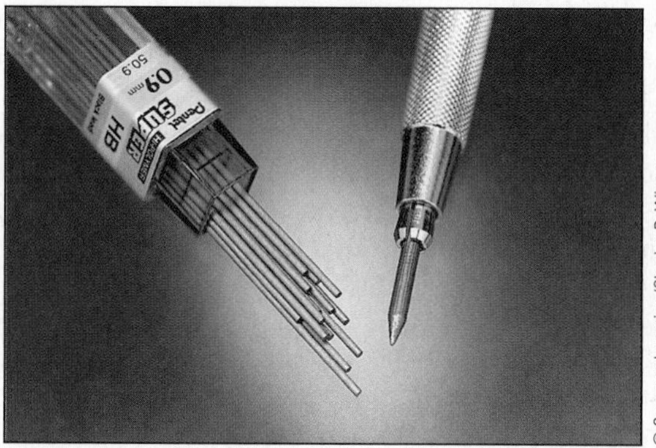


© Cengage Learning/Charles D. Winters

(b) Graphite. The physical properties of graphite include a density of 2.3 g/cm$^3$, black in color, and relative softness.

**Figure 12.12** These two forms of carbon are physically very different.

## Thinking About Your Thinking

### Mental Models

**Figure 12.15 is the starting point from which you can form a mental model of the subatomic structure of metals. Notice how the highest-energy (outermost) valence electron(s) in a metal atom is (are) separated from the remainder of the atom. These electrons have much more freedom of movement than those that remain localized near a specific atom in the crystal. The positively charged metal ions are held in fixed positions in the crystal because of their attraction to the negatively charged valence electrons that move among the ions.**

**The mobility of the valence electrons is what gives metals many of their characteristic macroscopic properties. Mentally imagine how the valence electrons can carry charge and be responsible for electric current. Even though a metal is electrically neutral, charged particles—electrons—can carry current through a sample.**

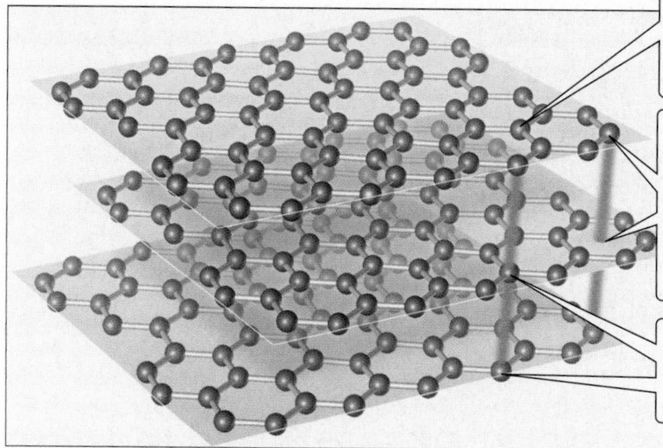

Each carbon atom is surrounded by three others in a flat sheet, and there are strong bonds connecting the carbon atoms within the same sheet.
The flat sheets stack to form the three-dimensional structure. Some carbon atoms in one sheet are aligned with the centers of rings in the next sheet.
Some carbon atoms in one sheet are aligned with carbon atoms in the next sheet.


**Figure 12.14** A particulate-level model of graphite. Each carbon atom is bonded to three other carbon atoms in a flat sheet. Very weak chemical forces loosely hold the sheets together until an external force peels them apart. This bonding arrangement leads to the characteristic physical properties of graphite at the macroscopic level.

How many bonds does each carbon atom have? You can see that each atom in graphite is bonded to three other atoms. Notice how all of the bonded atoms are in a single plane. Only very weak attractive forces exist between each "sheet" of bonded atoms in graphite. This bonding arrangement is what allows sheets of graphite to stick to paper when you write with a graphite pencil. The bonded atoms stay bonded together, but the force of pushing the pencil on paper is sufficient to slide these sheets apart.

A third form of pure carbon was discovered in the 1980s, and it was called buckminsterfullerene, or, as the molecules are more commonly known, buckyballs. Sixty carbon atoms are arranged in a cagelike structure (Fig. 12.15[a]), with a carbon atom at the equivalent of each corner of a soccer ball (Fig. 12.15[b]). Each atom is bonded to three other atoms to form the $C_{60}$ molecule shown as a model in Figure 12.15(a). This form of carbon is found in the soot made by burning carbon-containing substances in a low-oxygen environment.

The macroscopic properties of a substance depend on much more than just the number and type of atoms that make up the molecule. Bonding has a tremendous influence, too. If you ever have any doubt, compare the price of the two common forms of pure carbon, diamond and pencil "lead," and you'll quickly see the economic value of chemical bonds!

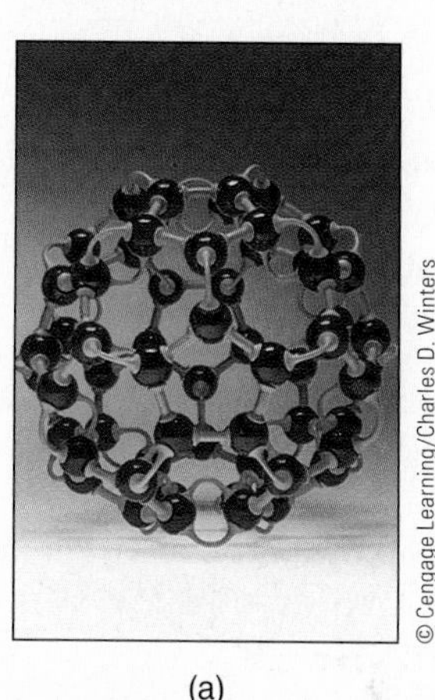


(a)

(b)

**Figure 12.15** (a) A particulate-level model of buckminster fullerene. Each carbon atom is bonded to three other carbon atoms in a cagelike sphere. The physical properties of bucky balls are very different from those of either diamond or graphite. (b) A soccer ball is another model for the particulate-level arrangement of carbon atoms in buckminsterfullerene. A carbon atom lies at the intersection of each set of three seams on the ball. The result is alternating five-membered (black) and six-membered (white) rings.

A **metallic bond** occurs because of the attractive forces between the positively charged metal ions and the negatively charged valence electrons that move among them. This is like the attractive forces caused by an electron pair spending most of its time between atoms in a covalent bond. In covalent bonds we say the bonding electrons are **localized** between two specific atoms. In contrast, electrons in a metallic bond are **delocalized** because the electrons do not stay near any single atom or pair of atoms.

The nature of the metallic bond explains many properties of metals. An electrical current is made up of moving electrons. Because electrons are free to move within the metallic crystal, metals are good conductors of electricity. Metals bend instead of break because their atoms can rearrange themselves in the electron sea.

Metals can combine with other metals in many different ways. These combinations are not compounds; they lack a constant composition. They are mixtures rather than pure substances because their properties are not constant but depend on the relative composition of the mix. Steel is made up of iron and other elements. Brass, a

**P/REVIEW:** The energy required to remove a valence electron from an atom is called its ionization energy. Figure 11.17 in Section 11.6 shows that the ionization energies of metals are lower than the ionization energies of nonmetals.

**Table 12.2** Composition of Some Common Alloys

| Alloy | Composition |
|---|---|
| 18 K gold | 75% gold, 12.5% silver, 12.5% copper |
| Brass | 60–95% copper, 5–40% zinc |
| Bronze | 90% copper, 10% tin |
| Carbon steel | 99% iron, 0.2–1.5% carbon |
| Pewter | 91% tin, 7.5% antimony, 1.5% copper |
| Stainless steel | About 70% iron, 18–20% chromium, 8–12% nickel, may have small quantities of other elements |
| Sterling silver | 92.5% silver, 7.5% copper |

**P/REVIEW:** Recall from Section 2.4 that constant physical and chemical properties are what distinguish pure substances from those that are impure—mixtures. Pure substances are either elements or compounds.

mixture of copper and zinc, and bronze, a combination of copper and tin, are common alloys (Table 12.2). A solid mixture of two or more elements that has macroscopic metallic properties is an **alloy.**

# Chapter 12 in Review

Most of the key terms and concepts and many others appear in the Glossary. Use your Glossary regularly.

**12.1 Monatomic Ions with Noble-Gas Electron Configurations**

Goal 1 Define and distinguish between cations and anions.

Goal 2 Identify the monatomic ions that are isoelectronic with a given noble-gas atom and write the electron configuration of those ions.

**Key Terms and Concepts: Anion, cation, chemical bond, hydronium ion**

**12.2 Ionic Bonds**

Goal 3 Use Lewis symbols to illustrate how an ionic bond can form between monatomic cations from Groups 1A, 2A, and 3A (1, 2, 13) and anions from Groups 5A, 6A, and 7A (15–17) of the periodic table.

**Key Terms and Concepts: Crystal, ionic bond, ionic compound**

**12.3 Covalent Bonds**

Goal 4 Describe, use, or explain each of the following with respect to forming a covalent bond: electron cloud, charge cloud, or charge density; valence electrons; half-filled electron orbital; filled electron orbital; electron sharing; orbital overlap; octet rule or rule of eight.

Goal 5 Use Lewis symbols to show how covalent bonds are formed between two nonmetal atoms.

Goal 6 Distinguish between bonding electron pairs and lone pairs.

**Key Terms and Concepts: Charge density, covalent bond, electron cloud, Lewis diagram (formula, structure), lone pair, molecular compound, molecule, octet rule (rule of eight), overlap**

**12.4 Polar and Nonpolar Covalent Bonds**

Goal 7 Distinguish between polar and nonpolar covalent bonds.

Goal 8 Predict which end of a polar bond between identified atoms is positive and which end is negative. (You may refer to a periodic table.)

Goal 9 Rank bonds in order of increasing or decreasing polarity based on periodic trends in electronegativity values or actual values, if given. (You may refer to a periodic table.)

**Key Terms and Concepts: Electronegativity, nonpolar and polar covalent bonds**

**12.5 Multiple Bonds**

Goal 10 Distinguish among single, double, and triple bonds, and identify these bonds in a Lewis diagram.

**Key Terms and Concepts: Single, double, triple, and multiple bonds**

**12.6 Atoms That Are Bonded to Two or More Other Atoms**

**12.7 Exceptions to the Octet Rule**

**12.8 Metallic Bonds**

Goal 11 Describe how metallic bonding differs from ionic and covalent bonding.

Goal 12 Sketch a particulate-level illustration of the electron-sea model of metallic bonding.

**Key Terms and Concepts: Alloy, electron-sea model, localized and delocalized electrons, metallic bond**

# Study Hints and Pitfalls to Avoid

How can you tell if the bond between atoms of two elements is most likely to be ionic, covalent, or metallic? Metal atoms have one, two, or three valence electrons. Monatomic cations form when metal atoms lose these electrons, usually reaching an outer electron configuration that is isoelectronic with a noble gas. Nonmetal atoms in Groups 5A/15 through 7A/17 have five, six, or seven valence electrons. These atoms form monatomic anions by gaining enough electrons to reach an electron configuration that matches that of a noble gas. These oppositely charged ions form ionic bonds between themselves, yielding the crystalline structure that is typical of an ionic compound. Conclusion: *Ionic bonds generally form between a metal and a nonmetal.*

Nonmetal atoms have too many valence electrons to lose them all and become cations, so an anion and a cation cannot both come from two nonmetal atoms. By sharing electrons, however, they form covalent bonds in which each atom reaches a noble-gas structure. The compounds formed in this way are molecular. Conclusion: *The bond between two nonmetals generally is covalent.*

To form a metallic bond, valence electrons must be shared by all "metal ions" in a crystal. Atoms that have one, two, or three valence electrons, when surrounded by other similar atoms, are apt to donate these electrons to a "sea of electrons" in which the metal ions are located. With a few exceptions, only *metals* have one, two, or three valence electrons. Conclusion: *The bonding among metal atoms is metallic.*

Note that hydrogen is a nonmetal that has only one valence electron rather than four, five, six, or seven. Nevertheless, a hydrogen atom reaches the noble-gas structure of helium with one additional electron. Hydrogen therefore forms either ionic or covalent bonds in the same way as other nonmetals.

You are not expected to memorize the electronegativities of the elements in Figure 12.9 (unless your instructor states otherwise). However, you should know that electronegativities are higher at the top of any column (group) and at the right of any row (period) in the periodic table. From this, you can often predict which of two bonds is more polar.

# Concept-Linking Exercises

*Write a brief description of the relationships among each of the following groups of terms or phrases. Answers to the Concept-Linking Exercises are given after answers to the Target Checks in Appendix III.*

1. Chemical bond, atoms, molecules, ionic compounds, valence electron
2. Ionic bond, electron transfer, crystal, cation, anion
3. Covalent bond, electron cloud, overlap, molecule
4. Polar bond, nonpolar bond, electronegativity, charge density, distribution of bonding electron charge
5. Metallic bond, electron-sea model, valence electrons
6. Ionic bond, covalent bond

# Small-Group Discussion Questions

*Small-Group Discussion Questions are for group work, either in class or under the guidance of a leader during a discussion section.*

1. Why do Group 1A/1 elements tend to form ions with a 1+ charge? Provide similar explanations for the common ions formed from third-period elements from Groups 2A/2, 3A/3, 5A/15, 6A/16, and 7A/17. Why do Group 8A/18 elements not exist as ions?
2. Consider the particulate-level depiction of the arrangement of ions in sodium chloride illustrated in Figure 12.2. Why are the sodium ions smaller than the chloride ions? Why are there equal numbers of sodium ions and chloride ions? Draw a sketch of the arrangement of ions in magnesium sulfide. How is it similar and how is it different from the sodium chloride illustration?
3. Be as specific as possible and state which orbitals overlap in forming each of the following covalent bonds: H—F, H—H, F—F, H—Cl, Cl—Cl. Explain your reasoning in each case.
4. What is the definition of *electronegativity*? What is the periodic trend in electronegativity values? In Section 11.6, you studied the periodic trends in a number of properties of elements. How are these other trends related to periodic trends in electronegativity values? Can any other periodic trend explain the periodic trend in electronegativity or vice versa?
5. Nitrogen molecules contain a triple bond. What orbitals from each nitrogen atom overlap to form the three bonds? Explain.
6. The Lewis diagrams for ammonia, methane, ethylene, acetylene, and carbon dioxide are given in Section 12.6. For each atom in each molecule, show how the octet rule is satisfied. For the hydrogen atoms, explain how the hydrogen atom achieves the electron configuration of a helium atom.
7. Describe the evidence that indicates that oxygen molecules do not have a double bond. Show how the Lewis diagram for oxygen with a double bond between the two atoms satisfies the octet rule. Explain how oxygen illustrates the limitations of Lewis diagrams and the need for experimental investigation of the structure of molecules.
8. One well-known macroscopic property of metals is their ability to conduct electricity. Explain how the particulate-level electron-sea model of metals provides an explanation for this macroscopic property.
9. Draw, sketch, and/or illustrate each of the following bond types at the particulate level: ionic, nonpolar covalent, polar covalent, metallic. Write a brief statement that describes the critical similarities and differences in the bond types.

# Questions, Exercises, and Problems

*Blue-numbered questions are answered in Appendix III. ■ denotes problems assignable in OWL. In the first section of Questions, Exercises, and Problems, similar exercises and problems are paired in consecutive odd-even number combinations.*

**Section 12.1: Monatomic Ions with Noble-Gas Electron Configurations**

1. Write the electron configuration for the ions of the third-period elements that form monatomic ions that are

isoelectronic with a noble-gas atom. Also identify the noble-gas atoms having those configurations.

2. ■ A monatomic ion with a charge of 1− has an electronic configuration of $1s^2 2s^2 2p^6 3s^2 3p^6$. (a) Is this ion a cation or an anion? (b) With what noble gas is it isoelectronic? (c) What is the symbol of the ion?

3. Identify by symbol two positively charged monatomic ions that are isoelectronic with argon.

4. ■ Considering only ions with charges of 1+, 2+, 1−, and 2− or neutral atoms, give the symbols for four species that are isoelectronic with the chloride ion, $Cl^-$.

5. Write the symbols of two ions that are isoelectronic with the barium ion.

6. ■ Considering only ions with charges of 1+, 2+, 1−, and 2− or neutral atoms, give the symbols for four species that are isoelectronic with Xe.

### Section 12.2: Ionic Bonds

7. Aluminum oxide is an ionic compound. Sketch the transfer of electrons from aluminum atoms to oxygen atoms that accounts for the chemical formula of the compound $Al_2O_3$.

8. Using Lewis symbols, show how atoms of sulfur and sodium form ionic bonds, leading to the correct formula of sodium sulfide, $Na_2S$.

9. When potassium and chlorine react and form an ionic compound, why is there only one chlorine atom for each potassium atom instead of two?

10. ■ Fill in the blanks with the smallest integers possible. When gallium (Z = 31) reacts with sulfur to form an ionic compound, each metal atom loses _____ electron(s) and each nonmetal atom gains _____ electron(s). There must be _____ gallium atom(s) for every _____ sulfur atom(s) in the reaction.

### Section 12.3: Covalent Bonds

11. What is the meaning of *orbital overlap* in the formation of a covalent bond?

12. What is an electron cloud?

13. Sketch the formation of two covalent bonds by an atom of sulfur in making a molecule of hydrogen sulfide, $H_2S$.

14. ■ How many covalent bond(s) would the element germanium (Z = 32) be expected to form in order to obey the octet rule? Use the octet rule to predict the formula of the compound that would form between germanium and hydrogen, if the molecule contains only one germanium atom and only single bonds are formed.

15. Circle the lone electron pairs in the Lewis diagram for hydrogen chloride: $H—\ddot{\underset{..}{Cl}}:$

16. ■ How many covalent bond(s) would the element tellurium (Z = 52) be expected to form in order to obey the octet rule? Use the octet rule to predict the formula of the compound that would form between tellurium and hydrogen, if the molecule contains only one tellurium atom and only single bonds are formed.

17. Show how atoms achieve the stability of noble-gas atoms in forming covalent bonds.

18. Does the energy of a system tend to increase, decrease, or remain unchanged as two atoms form a covalent bond?

### Section 12.4: Polar and Nonpolar Covalent Bonds

19. Compare the electron cloud formed by the bonding electron pair in a polar bond with that in a nonpolar bond.

20. What is meant by saying that a bond is polar or nonpolar? What bonds are completely nonpolar?

21. Refer to Figure 12.9 and list the following bonds in order of decreasing polarity: S—O, N—Cl, C—C.

22. ■ Consider the following bonds: Ge—Se, Br—Se, Br—Ge. Indicate the direction of polarity of each. Which bond is expected to be the most polar?

23. For each bond in Question 21, identify the positive pole, if any.

24. ■ Consider the following bonds: Te—Se, O—Te, O—Se. Indicate the direction of polarity of each. Which bond is expected to be the most polar?

25. Identify the trends in electronegativities in the periodic table.

26. ■ Arrange the following elements in order of increasing electronegativity: silicon, chlorine, sulfur, phosphorus.

### Section 12.5: Multiple Bonds

27. Atoms with double bonds and triple bonds can conform to the octet rule. Could an atom with a quadruple (four) bond obey that rule? Why or why not?

28. What is a multiple bond? Distinguish among single, double, and triple bonds.

### Section 12.6: Atoms That Are Bonded to Two or More Other Atoms

29. An atom, X, is bonded to another atom by a double bond. What is the largest number of *additional atoms*—don't count the one to which it is already bonded—to which X may be bonded and still conform to the octet rule? What is the minimum number? Justify your answers.

30. What is the maximum number of atoms to which a central atom in a molecule can bond and still conform to the octet rule? What is the minimum number?

31. A molecule contains a triple bond. Theoretically, what are the maximum and minimum numbers of atoms that can be in the molecule and still conform to the octet rule? Sketch Lewis diagrams that justify your answer.

32. Draw Lewis diagrams of *central* atoms of molecules showing all possible combinations of single, double, and triple bonds that can form around an atom that conforms to the octet rule. It is not necessary to show the atoms to which the central atom is bonded—just the bonds.

### Section 12.7: Exceptions to the Octet Rule

33. Because nitrogen has five valence electrons, it is sometimes difficult to fit a nitrogen atom into a Lewis diagram that obeys the octet rule. Why is this so? Without actually drawing them, can you tell which of the following species do not have a Lewis diagram that satisfies the octet rule? $N_2O$, $NO_2$, $NF_3$, NO, $N_2O_3$, $N_2O_4$, NOCl, $NO_2Cl$

34. Why is it impossible to draw a Lewis diagram that obeys the octet rule if the species has an odd number of electrons?

## Section 12.8: Metallic Bonds

35. What are the meanings of the terms *localized* and *delocalized,* when used to describe electrons in a compound?

36. Is a metallic bond more similar to an ionic bond or a covalent bond? Explain.

37. How would a particulate-level illustration differ if you were to draw a calcium crystal instead of a potassium crystal?

38. Sketch a particulate-level illustration, similar to Figure 12.15, that shows the electron-sea model for a potassium crystal.

39. Are alloys pure substances or mixtures? Are they compounds? Explain your answers.

40. What is an alloy? Give an example.

## GENERAL QUESTIONS

41. Distinguish precisely, and in scientific terms, the differences among items in each of the following pairs or groups.
    a) Ionic compound, molecular compound
    b) Ionic bond, covalent bond
    c) Lone pair, bonding pair (of electrons)
    d) Nonpolar bond, polar bond
    e) Single, double, triple, multiple bonds

42. Classify each of the following statements as true or false:
    a) A single bond between carbon and nitrogen is polar.
    b) A bond between phosphorus and sulfur will be less polar than a bond between phosphorus and chlorine.
    c) The electronegativity of calcium is less than the electronegativity of aluminum.
    d) Strontium (Z = 38) ions, $Sr^{2+}$, are isoelectronic with bromide ions, $Br^-$.
    e) The monatomic ion formed by selenium (Z = 34) is expected to be isoelectronic with a noble-gas atom.
    f) Most elements in Group 4A/14 do not normally form monatomic ions.
    g) Multiple bonds can form only between atoms of the same element.
    h) If an atom is triple-bonded to another atom, it may still form a bond with one additional atom.
    i) An atom that conforms to the octet rule can bond to no more than three other atoms if one bond is a double bond.
    j) Electrons are localized between atoms in a metal.
    k) Alloys are pure substances.

43. Explain why ionic bonds are called electron-*transfer* bonds and covalent bonds are known as electron-*sharing* bonds.

44. Explain why the total energy of a system changes in the formation of (a) an ionic bond and (b) a covalent bond.

45. Compare the bond between potassium and chlorine in potassium chloride with the bond between two chlorine atoms in chlorine gas. Which bond is ionic, and which is covalent? Describe how each bond forms.

46. ■ Considering bonds between the following pairs of elements, which are most apt to be ionic and which are most apt to be covalent: sodium and sulfur; fluorine and chlorine; oxygen and sulfur? Explain your choice in each case.

47. What is the electron configuration of the hydrogen ion, $H^+$? Explain your answer.

48. ■ Identify the pairs among the following that are *not* isoelectronic: (a) Ne and $Na^+$; (b) $S^{2-}$ and $Cl^-$; (c) $Mg^{2+}$ and Ar; (d) $K^+$ and $S^{2-}$; (e) $Ba^{2+}$ and $Te^{2-}$ (Te, Z = 52).

49. Which orbitals of each atom overlap in forming a bond between bromine and oxygen?

50. Is there any such thing as a completely nonpolar bond? If yes, give an example.

51. If you did not have an electronegativity table, could you predict the relative electronegativities of elements whose positions are (B)—(A) in the periodic table? What about elements whose positions are (X)—(Y)? In both cases, explain why or why not.

## MORE CHALLENGING PROBLEMS

52. ■ A monatomic ion with a 2− charge has the electron configuration $1s^2 2s^2 2p^6 3s^2 3p^6 4s^2 3d^{10} 4p^6$. (a) What neutral noble-gas atom has the same electron configuration? (b) What is the monatomic ion with a 2− charge that has this configuration? (c) Write the symbol of an ion with a 1+ charge that is isoelectronic with the species in (a) and (b).

53. ■ If the monatomic ions in Question 52 (b) and (c) combine to form a compound, what is the formula of that compound?

54. "The bond between a metal atom and a nonmetal atom is most apt to be ionic, whereas the bond between two nonmetal atoms is most apt to be covalent." Explain why this statement is true.

55. The metallic bond is neither ionic nor covalent. Explain, according to the octet rule, why this is so.

56. How do the energy and stability of bonded atoms and noble-gas electron configurations appear to be related in forming covalent and ionic bonds?

57. ■ Which bond, F—Si or O—P, is more polar? You may look at a full periodic table in answering this question, but do not look at any source of electronegativity values.

58. Arrange the following bonds in order of increasing polarity: Na—O; Al—O; S—O; K—O; Ca—O. If the polarity of any two bonds cannot be positively placed relative to each other based on periodic trends, explain why.

59. There are two iodides of arsenic (Z = 33), $AsI_3$ and $AsI_5$. A Lewis diagram that conforms to the octet rule can be drawn for one of these but not for the other. Draw the diagram that is possible. From that diagram, see if you can figure out how the second molecule might be formed by covalent bonds, even though it violates the octet rule. Draw the Lewis diagram for the second molecule.

60. How is it possible for central atoms in molecules to be surrounded by five or six bonding electron pairs when there are only four valence electrons from the *s* and *p* orbitals of any atom?

61. Suggest why $BF_3$ behaves as a molecular compound, whereas $AlF_3$ appears to be ionic.

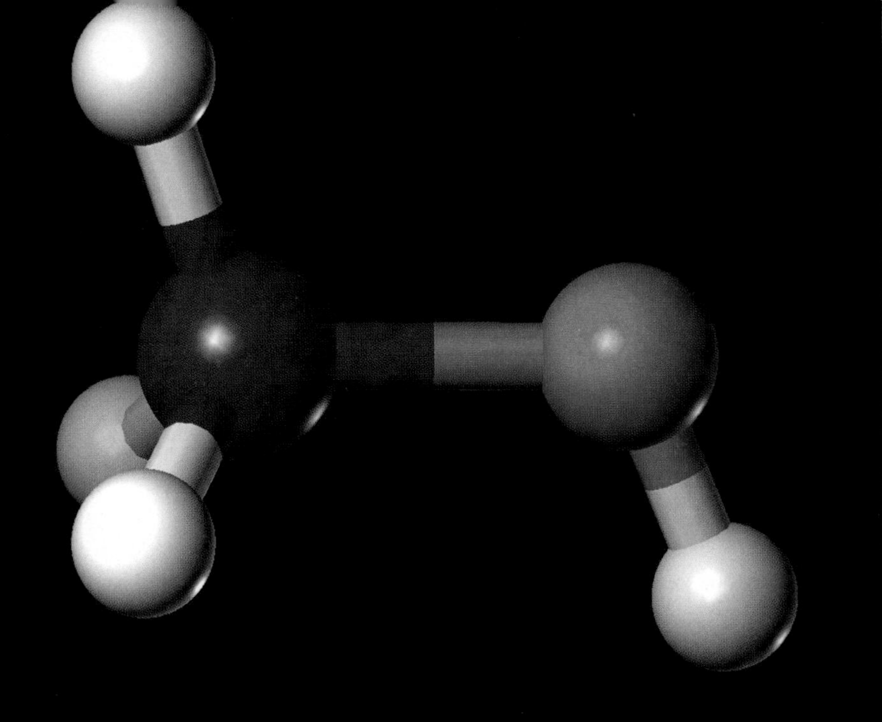

© Cengage Learning/George G. Stanley

This model represents the structure and shape of a molecule of methanol, $CH_3OH$. Hydrogen atoms are shown in white, the black sphere represents a carbon atom, and the red sphere, oxygen. The sticks connecting the balls represent single bonds. In this chapter you will learn to predict and sketch the three-dimensional shapes of molecules, starting with nothing more than their formulas.

13

# Structure and Shape

CONTENTS



You learned in Chapter 12 that atoms in molecular compounds and polyatomic ions are held together by covalent bonds. Lewis diagrams show, in two dimensions, how the atoms are connected. However, Lewis diagrams do not show how the atoms are *arranged* in three dimensions—the actual shape of the molecule. In this chapter you will learn how the distribution of electron pairs leads to the distribution of atoms, which, in turn, leads to the structure and shape of molecules. Determination of the shape of a molecule begins with the Lewis diagram, and in case it has been a while since you studied Lewis diagrams, we will review them briefly. Important terms are printed in italics.

*Lewis symbols* for elements, also called *electron-dot symbols,* were introduced in Section 11.5. The number of dots equal to the number of *valence electrons* is placed around the chemical symbol of the element. In Section 12.3 these symbols were used to show how two electrons are shared in forming a chemical bond between two atoms. The electrons can be shown as two dots, but usually a bonding pair appears as a dash between the symbols of the bonded atoms.

Online homework for this chapter may be assigned in OWL

Two atoms bonded to each other by one electron pair are connected by a *single bond.* Sometimes two atoms are bonded by more than one electron pair. This is shown by two or three dashes, which represent *double* or *triple bonds*, respectively. Bonded atoms may have valence electrons that are not used for bonding; these are represented by dots. In reaching a stable *octet of electrons,* these nonbonding electrons occur in pairs called lone pairs.

Once you are confident in your understanding of these Chapter 12 concepts, you are ready to learn how to draw Lewis diagrams.

## 13.1 Drawing Lewis Diagrams

**Goal** 1 Draw the Lewis diagram for any molecule or polyatomic ion made up of main group elements.

You may use the procedure that follows to sketch the Lewis diagram for any species that obeys the octet rule. Each step is illustrated by drawing the Lewis diagram for carbon tetrachloride, $CCl_4$.

***Learn It Now!*** Matching a main group element with its group number in the periodic table is a quick way to count the valence electrons in atoms of that element.

**Step 1** *Calculate the total number of valence electrons in the molecule or ion.* Note that the number of valence electrons for a main group element is the same as its column number in the periodic table (or the final digit of the column number if you are using IUPAC group numbers). If the species is an ion, the number of valence electrons must be adjusted to account for the charge on the ion. For each positive charge, subtract one electron; for each negative charge, add one electron.

In $CCl_4$ there are four valence electrons from carbon and seven from each chlorine:

$$4\ (C) + 4 \times 7\ (Cl) = 32$$

**PREVIEW:** Electronegativity values increase from left to right across any row of the periodic table and from the bottom to the top of any column. See Section 12.4. The atoms in a chemical formula are usually written from left to right in order of increasing electronegativity. Thus, the least electronegative element is usually the first one written in a formula. There are exceptions, however, so be cautious in assuming that the first element listed is the least electronegative.

**Step 2** *Determine the central atom(s) of the molecule or ion.* The central atom of the molecule or ion is usually the least electronegative atom of the species. The most common exception is the hydrogen atom. Hydrogen is never the central atom in a molecule.

Carbon is less electronegative than chlorine, so the central atom in $CCl_4$ is C. Carbon is always a central atom in a species that contains carbon.

**Step 3** *Draw a tentative diagram for the molecule or ion,* joining atoms by single bonds and placing electron dots around each symbol except hydrogen, so the total number of electrons for each atom is eight. In some cases, only one arrangement of atoms is possible. In others, two or more diagrams are possible. Ultimately, chemical or physical evidence must be used to decide which diagram is correct. A few general rules will help you make diagrams that are most likely to be correct:

a) A hydrogen atom always forms one bond and has no lone pairs of electrons. Hydrogen is always a **terminal atom in a Lewis diagram—an atom that is bonded to only one other atom.**

b) A carbon atom is always surrounded by four electron pairs: four single bonds, a double bond and two single bonds, two double bonds, etc. Carbon is always a **central atom in a Lewis diagram—an atom that is bonded to two or more other atoms.**

c) When several carbon atoms appear in the same molecule, they are often bonded to each other.

d) Make your diagram as balanced as possible. In particular, a compound or ion having two or more oxygen atoms and one atom of another nonmetal usually has the oxygen atoms arranged around the central nonmetal atom. If hydrogen is also present, it is usually bonded to an oxygen atom, which is then bonded to the nonmetal with a single bond: X—O—H, where X is the nonmetal.

The Active Examples that follow show you how to place hydrogen, oxygen, and other elements in Lewis diagrams.

$CCl_4$ is described by Steps 3b and 3d. The four chlorine atoms are placed around the carbon atom. Here is the tentative diagram:

```
        ..
       :Cl:
   ..   |   ..
  :Cl—  C  —Cl:
   ..   |   ..
       :Cl:
        ..
```

**Step 4** *Compare the number of valence electrons you have available from Step 1 to the number that you used in your tentative Lewis diagram. If they are not equal, replace two lone pairs with one bonding pair. Repeat if necessary.* When your tentative Lewis diagram has more electrons than are needed, erase two lone pairs of electrons, one from the central atom and one from a terminal atom, and replace them with one bonding electron pair between those two atoms.

You will see how to do this in Active Example 13.3. We call this a "cash-in-two-pairs-for-one" approach.

There are four bonds and 12 unshared pairs in the tentative diagram:

$$4 \times 2 \text{ (bonds)} + 12 \times 2 \text{ (lone pairs)} = 32$$

These 32 electrons are equal to the total number of electrons counted in Step 1. No further adjustment is necessary.

**Step 5** *Check to be sure that each atom other than hydrogen has four electron pairs and that hydrogen has only one electron pair.* Remember that a bonding pair is counted for both bonded atoms. This final step will ensure that you have drawn a correct Lewis diagram.

Checking each atom in the molecule:

```
   |       ..       |        ..      ..
  —C—     :Cl—     :Cl:    —Cl:     :Cl:
   |       ..       ..       ..       |
```

All atoms have four electron pairs. The Lewis diagram is now complete.

The steps in drawing a Lewis diagram are as follows:

**procedure**

**Drawing a Lewis Diagram**

**Step 1:** Count the total number of valence electrons. Adjust for charge on ions.

**Step 2:** Place the least electronegative atom(s) in the center of the molecule.

**Step 3:** Draw a tentative diagram. Join atoms by single bonds. Add unshared pairs to complete the octet around all atoms except hydrogen.

**Step 4:** Calculate the number of valence electrons in your tentative diagram and compare it with the actual number of valence electrons. If the tentative diagram has too many electrons, remove a lone pair from the central atom and from a terminal atom, and replace them with an additional bonding pair between those atoms. If the tentative diagram still has too many electrons, repeat the process.

**Step 5:** Check the Lewis diagram. Hydrogen atoms must have only one bond, and all other atoms should have a total of four electron pairs.

## Active Example 13.1

Draw Lewis diagrams for ammonia, $NH_3$, and phosphorus tribromide, $PBr_3$.

Start with the ammonia molecule. Count the valence electrons.

$5 \text{ (N)} + 3 \times 1 \text{ (H)} = 8$

Now determine the central atom and draw the tentative diagram.

*continued*

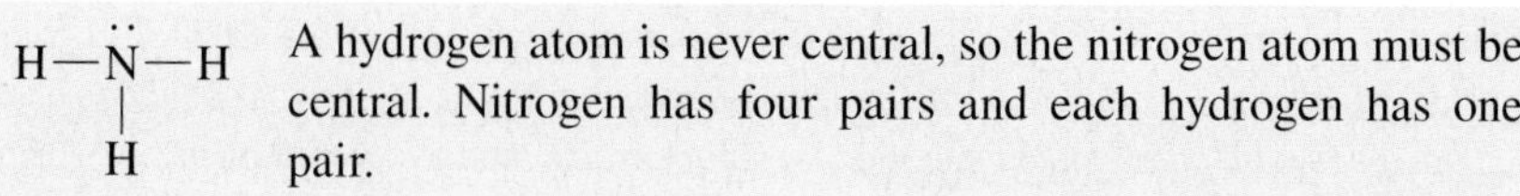

H—N—H (with one H below N) A hydrogen atom is never central, so the nitrogen atom must be central. Nitrogen has four pairs and each hydrogen has one pair.

Compare the number of electrons in your tentative diagram with the number available.

Eight electrons in the tentative diagram match the number available. The diagram is complete.

Now draw the Lewis diagram for $PBr_3$.

:Br—P—Br: (with :Br: below P)

5 (P) + 3 × 7 (Br) = 26 valence electrons

Phosphorus is less electronegative than bromine, so P is the central atom. The number of electrons in the tentative diagram matches the number available.

Notice the similarity between this diagram and the one for $NH_3$. The central elements are both in Group 5A/15 and have five valence electrons. The other element in each case is single-bonded to the central atom.

## Active Example 13.2

Write Lewis diagrams for the ClF molecule and the $ClO^-$ ion.

*Step 1* is to count the valence electrons for each species.

ClF: 7 (Cl) + 7 (F) = 14
$ClO^-$: 7 (Cl) + 6 (O) + 1 (charge) = 14

Chlorine and fluorine, both in Group 7A/17, have seven valence electrons. Oxygen in Group 6A/16 has six. In $ClO^-$ there is one additional electron to account for the 1− charge on the ion.

*Step 2* is to determine the central atom. With two atoms, the only possible diagram has them bonded to each other.

*Step 3* is to draw the tentative diagrams. Join the atoms by single bonds and complete the octet for each atom.

:Cl—F:    :Cl—O:

*Step 4* is to compare the actual number of valence electrons available to the number in the tentative diagrams. Each bond accounts for two of the total number of electrons. How many valence electrons are in each tentative diagram?

$1 \times 2 \text{ (bond)} + 6 \times 2 \text{ (lone pairs)} = 14$

The number of available valence electrons equals the number of electrons in the tentative diagrams. No modifications are necessary.

The final step, *Step 5*, is to check that each atom has an octet. The diagrams check. In both cases there is just the right number of electrons to complete the octet for both atoms. The diagrams are therefore complete. In the case of $ClO^-$ the diagram is enclosed in brackets and a charge is shown because it is an ion:

$$[:\ddot{\underset{..}{Cl}}-\ddot{\underset{..}{O}}:]^-$$

Did you notice how each step was the same for the two species in Active Example 13.2? This is because any two species that have (1) the same number of atoms and (2) the same number of valence electrons also have similar Lewis diagrams, whether they are molecules or polyatomic ions.

In all examples so far, covalent bonds have been formed when each atom contributes one electron to the bonding pair. This is not always the case. Many bonds are formed where one atom contributes both electrons and the other atom offers only an empty orbital.* To illustrate, an ammonium ion is produced when a hydrogen ion is bonded to the unshared electron pair of the nitrogen atom in an ammonia molecule:

$$H^+ + :\underset{\underset{H}{|}}{\overset{\overset{H}{|}}{N}}-H \longrightarrow \left[H-\underset{\underset{H}{|}}{\overset{\overset{H}{|}}{N}}-H\right]^+$$

ammonia    ammonium ion

The four bonds in an ammonium ion are identical. This shows that a bond formed from an electron pair and an empty orbital is the same as a bond formed by one electron from each atom.

## Active Example 13.3

Draw the Lewis diagram for the hydrogen carbonate ion, $HCO_3^-$.

In *Step 1*, you count the total number of valence electrons. Don't forget to account for the charge.

*continued*

*This is called a *coordinate covalent bond.*

1 (H) + 4 (C) + 3 × 6 (O) + 1 (charge) = 24

Now determine the central atom (*Step 2*).

Carbon is the central atom.

Hydrogen is always a terminal atom. Carbon is less electronegative than oxygen, so the central atom is C. Carbon will always be a central atom in a species that contains carbon.

*Step 3* is to draw the tentative diagram for $HCO_3^-$. Remember that oxygen is usually arranged around the central nonmetal atom and that hydrogen is usually bonded to oxygen.

```
     ..  ..  ..
H — O — C — O :
     ..  |   ..
        :O:
         ..
```

The hydrogen atom can be attached to any of the oxygen atoms. Hydrogen has one electron pair and all other atoms have four.

Now count the number of electrons in your tentative diagram.

There are 26 electrons.

The tentative diagram has four bonds and nine unshared pairs:
4 × 2 (bonds) + 9 × 2 (lone pairs) = 26

In *Step 1*, you determined that 24 valence electrons are available, so you have two more electrons in the tentative diagram than you need. Complete *Step 4* by removing one lone pair from the central atom (carbon) and one lone pair from either terminal oxygen atom. Then place another bonding pair between those atoms to form a double bond.

```
     ..      ..
H — O — C = O :
     ..  |
        :O:
         ..
```

Now check to be sure that all atoms except hydrogen have four electron pairs.

All atoms check.

Checking each atom in the ion:

```
        ..              .              |
H—     —O—     —C=    =O:            :O:
        ..      |       .             ..
```

The hydrogen has one electron pair, and all other atoms have four.

Lewis diagrams of ions are enclosed in square brackets with the charge indicated as a superscript on the upper right. Complete the diagram.

```
⎡     ..       .⎤ −
⎢ H—O—C=O:  ⎥
⎢     ..   |    .⎥
⎣         :O:   ⎦
           ..
```

The double bond in the above Active Example could have been placed between the carbon atom and either of the terminal oxygen atoms. When changing only the positions of electrons produces two or more equivalent Lewis diagrams for a molecule or ion, the diagrams represent **resonance structures.** On paper, the bonds between the central carbon and the outlying oxygens look different. In the molecule itself, they are identical. Moreover, the bond strengths and lengths are between those found in true single and double bonds connecting the same two atoms. The actual molecule is an average of the resonance structures, and it is called a **resonance hybrid.**

It is customary to place a two-headed arrow between resonance structures:

```
   ..      .             ..     ..
H—O—C=O:  ⟷  H—O—C—O:
   ..  |   .             ..  ‖  ..
      :O:                    .O.
       ..                    . .
```

However, further discussion of resonance is beyond the scope of an introductory text. Therefore, when we encounter a resonance structure, we will simply show one of the alternative diagrams.

## Active Example 13.4

Draw the Lewis diagram for the sulfite ion, $SO_3^{2-}$.

Do *Steps 1, 2,* and *3:* Get the total valence electron count, determine the central atom of the molecule, and draw a tentative diagram.

*continued*

6 (S) + 3 × 6 (O) + 2 (charge) = 26

$$:\ddot{O}-\ddot{S}-\ddot{O}: \text{ with } :\ddot{O}: \text{ bonded below S}$$

Sulfur is less electronegative than oxygen, so it is the central atom. The ion has a 2− charge, so two electrons must be added to the valence electrons of the atoms themselves. This diagram has the three oxygen atoms distributed around sulfur as the central (least electronegative) atom, conforming to *Steps 2* and *3d* in the earlier detailed procedure.

Now calculate the number of valence electrons in the tentative diagram, compare these with the number available, and modify the diagram if necessary.

3 × 2 (bonds) + 10 × 2 (unshared pairs) = 26

The number of electrons available is equal to the number in the tentative diagram. No further modification is necessary.

Finally, check the diagram to be sure that every atom has an octet and add the "finishing touch" to complete the diagram in the space near the beginning of the Active Example.

$$\left[:\ddot{O}-\ddot{S}-\ddot{O}: \text{ with } :\ddot{O}: \text{ bonded below S}\right]^{2-}$$

Every atom is surrounded by four electron pairs. The "finishing touch" is to surround the diagram with brackets and indicate the charge.

You might wish to compare the $SO_3^{2-}$ diagram with the diagram for $PBr_3$ in Active Example 13.1. Both species have four atoms and 26 electrons, so their diagrams are the same.

## Active Example 13.5

Draw the Lewis diagram for $SO_2$.

Complete the procedure through the point at which you compare the number of valence electrons available with the number in the tentative diagram.

$$:\ddot{O}-\ddot{S}-\ddot{O}:$$

Total electrons: 6 (S) + 2 × 6 (O) = 18
S is less electronegative than O, so it is the central atom. The tentative diagram has 20 valence electrons.

This time there are too many electrons in the tentative diagram. Modify the tentative diagram by removing two lone pairs and replacing them with one bonding pair.

$:\ddot{\underset{\cdot\cdot}{O}}-\underset{\cdot\cdot}{S}=\dot{\underset{\cdot}{O}}\cdot$ or $\cdot\dot{\underset{\cdot}{O}}=\underset{\cdot\cdot}{S}-\ddot{\underset{\cdot\cdot}{O}}:$

These Lewis diagrams are resonance structures. Either diagram is acceptable.

The rules we are following are readily applied to simple organic molecules, which always contain carbon and hydrogen atoms, and may contain atoms of other elements, notably oxygen. If oxygen is present in an organic compound, it usually forms two bonds. If you remember that carbon forms four bonds and hydrogen forms one, and that two or more carbon atoms often bond to each other (Steps 3a–c at the beginning of this section), your tentative diagrams are likely to be correct.

**P/REVIEW:** Organic chemistry is the chemistry of carbon compounds. Chapter 21 is an introduction to this branch of chemistry.

## Active Example 13.6

Write the Lewis diagram for propane, $C_3H_8$.

```
    H   H   H
    |   |   |
H — C — C — C — H
    |   |   |
    H   H   H
```

All 20 of the valence electrons—12 from the three carbon atoms plus 8 from the eight hydrogen atoms—are needed for the 10 single bonds in the diagram. There are no lone pairs.

## Active Example 13.7

Draw the Lewis diagram for acetylene, $C_2H_2$.

Take it through the point at which you compare the number of valence electrons available with the number in the tentative diagram.

Total electrons: $2 \times 4\ (C) + 2 \times 1\ (H) = 10$

The hydrogen atoms must be terminal, so the carbon atoms are central:

$H-\ddot{\underset{\cdot\cdot}{C}}-\ddot{\underset{\cdot\cdot}{C}}-H$ The tentative diagram has 14 valence electrons.

There are 4 too many electrons in the tentative diagram this time; however, the remedy is the same. Erase two lone pairs from the carbons and replace them with a bonding pair to reduce the electron count by 2. Then do it again to reduce the electron count by a total of 4. The result is a triple bond.

*continued*

H—C≡C—H

Twice "cashing in two pairs for one" yields a triple bond.

### Active Example 13.8

Draw a Lewis diagram for $C_2H_6O$.

Try to take this one all the way. Bond the carbon atoms to each other.

```
    H   H
    |   |   ..
H—C—C—O—H
    |   |   ..
    H   H
```

The tentative diagram has the correct number of valence electrons, 20. You may have switched the —O—H group with any of the hydrogen atoms in this diagram, and if so, your Lewis diagram is also correct. By our insisting that the carbons be bonded to each other, the oxygen had to go between a carbon and a hydrogen.

The compound in Active Example 13.8 is ethanol. If we had not insisted that the carbon atoms be bonded to each other, the oxygen atom might have been placed between them:

```
    H       H
    |   ..  |
H—C—O—C—H
    |   ..  |
    H       H
```

This is another well-known compound, dimethyl ether. Note that both compounds have the same molecular formula, $C_2H_6O$, but different structures. Compounds that have the same molecular formulas but different structures are called **isomers** of each other. Isomers are distinctly different substances; each isomer has its own unique set of properties. For example, ethanol boils at 78°C and is a liquid at room temperature, while dimethyl ether boils at −24°C and is a gas a room temperature.

## 13.2 Electron-Pair Repulsion: Electron-Pair Geometry

**Goal** 2 Describe the electron-pair geometry when a central atom is surrounded by two, three, or four electron pairs.

The shape of a molecule plays a major role in determining the macroscopic properties of a substance. We examine this role in other chapters in this book. To understand and predict the shape-property relationship, you first need to know what is responsible for molecular shape. This is the focus of this section and the next. Discussion in these sections is limited to molecules having only single bonds. We then expand our consideration to molecules with multiple bonds in Section 13.4.

No single theory or model yet developed succeeds in explaining all the molecular shapes observed in the laboratory. A theory that explains one group of molecules cannot explain another group. Each model has its advantages and limitations. Chemists therefore use them all within the areas to which they apply, fully recognizing that there is still much to learn about how atoms are assembled in molecules.

In this text we will explore one of the models used to explain **molecular geometry,** the more precise term used to describe the shape of a molecule. It is called the **valence shell electron-pair repulsion or VSEPR theory.** VSEPR theory applies primarily to substances in which a second-period atom is bonded to two, three, or four other atoms. You may wonder why we focus so much attention on so few elements. The answer is that the second period includes carbon, nitrogen, and oxygen. Carbon alone is present in about 95% of all known compounds, and a large percentage of those include oxygen or nitrogen or both. These elements warrant this kind of attention.

**P/REVIEW:** The forces electrically charged particles exert on each other are described in Section 2.7. A particle with a positive charge is attracted to a particle with a negative charge. Two particles with the same charge, both positive or both negative, repel each other.

The basic idea of VSEPR is that the electron pairs we draw in Lewis diagrams repel each other in real molecules. Therefore they distribute themselves in positions around the central atom that are as far away from each other as possible. These are the locations of lowest potential energy; they satisfy the "minimization of energy" tendency that, we have noted, is one driving force in nature. This arrangement of electron pairs is called **electron-pair geometry.** The electron pairs may be shared in a covalent bond, or they may be lone pairs; it makes no difference.

**P/REVIEW:** Minimization of energy is mentioned as a driving force for change in many places in this book, notably in Section 2.8.

Earlier, we drew Lewis diagrams in which carbon, nitrogen, or oxygen was the central atom. In all cases the central atom was surrounded by four pairs of electrons. In Section 12.7 we showed how beryllium and boron—also Period 2 elements—do not conform to the octet rule. The beryllium atom in $BeF_2$ is flanked by only two electron pairs; in $BF_3$ the boron atom has three electron pairs around it. Our question then is this: How do two, three, or four electron pairs distribute themselves around a central atom so they are as far apart as possible? This question is answered by identifying the **electron-pair angle,** the angle formed by any two electron pairs and the central atom.

When electron pairs are as far apart as possible, all electron-pair angles around the central atoms are equal. The electron-pair geometries that result from two, three, or four electrons pairs are shown in Figure 13.1. The bond angles are derived by geometry.

Can we find these angles in our familiar, large-scale world? Indeed we can. If balloons *of similar size and shape* are tied together, they naturally arrange themselves in the same way (Fig. 13.2). These arrangements are their minimum-energy positions.

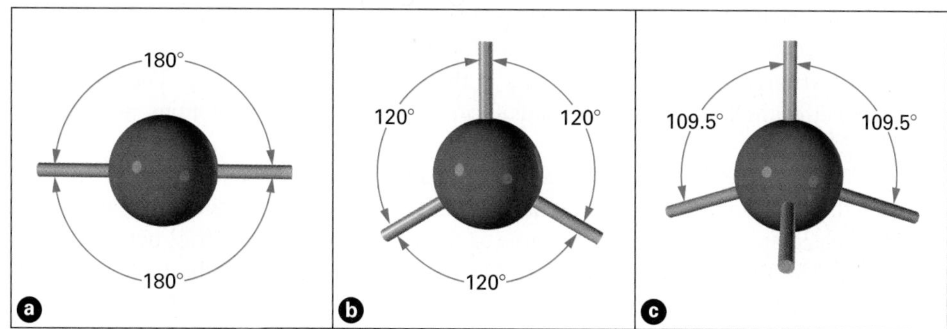


**Figure 13.1** Electron-pair geometry. Ball-and-stick models show the arrangement of two, three, and four electron pairs (sticks) around a central atom (ball). (a) According to the electron-pair repulsion principle, two sticks are as far from each other as possible when they are diametrically opposite each other. The geometry is linear, and the angle formed is 180°. (b) Three sticks are as far from each other as possible when equally spaced on a circumference of the ball. The sticks and the center of the ball are in the same plane and the angles are 120°. The geometry is trigonal (triangular) planar. (c) Four electron pairs are as far from each other as possible when arranged to form a tetrahedron. Each angle is 109.5°, which is called the tetrahedral angle.

**Figure 13.2** Electron-pair geometries for two to four electron pairs. If one ties together several balloons of similar size and shape, they will naturally assume the geometries shown.

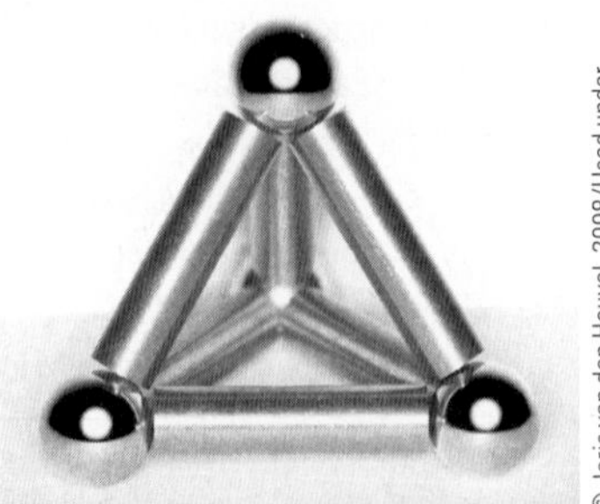

Figure 13.3 Tetrahedral models. The metal figure is a tetrahedron. Its four faces are identical equilateral triangles. The model of methane, $CH_4$, has a tetrahedral structure. The carbon atom is in the middle of the tetrahedron, and a hydrogen atom is found at each of the four corners.

**Table 13.1** Electron-Pair Geometries

| Electron Pairs | Geometry | Electron-Pair Angles |
|---|---|---|
| 2 | Linear | 180° |
| 3 | Trigonal (triangular) planar | 120° |
| 4 | Tetrahedral | 109.5° |

The balloons are like identically sized and shaped electron orbitals. Atom-size (particulate) properties are reproduced naturally on a larger (macroscopic) scale.

The geometries are summarized in Tables 13.1 and 13.2.

Table 13.1 and the caption to Figure 13.1 introduce the words ***tetrahedron*** and ***tetrahedral.*** A tetrahedron is the simplest regular solid. A regular solid is a solid figure with identical faces. A cube is a regular solid that has six identical squares as its faces. A tetrahedron has four identical equilateral triangles for its faces, as shown in Figure 13.3. This geometric figure appears in all molecules in which carbon forms single bonds with four other atoms. Notice that a tetrahedron is a three-dimensional figure. It can only be drawn in perspective on the two-dimensional plane of a book page.

## 13.3 Molecular Geometry

Goal **3** Given or having derived the Lewis diagram of a molecule or polyatomic ion in which a second-period central atom is surrounded by two, three, or four pairs of electrons, predict and sketch the molecular geometry around that atom.

Goal **4** Draw a wedge-and-dash diagram of any molecule for which a Lewis diagram can be drawn.

Molecular geometry describes the shape of a molecule and the arrangement of atoms around a central atom. You might think of it as an "atom geometry," in the same sense that the arrangement of electron pairs is the electron-pair geometry. Thus the **bond angle** is the angle between two bonds formed by the same central atom, as shown in Figure 13.4.

When all the electron pairs around a central atom are bonding pairs—that is, when there are no lone pairs—the bond angles are the same as the electron-pair angles. The molecular geometries are the same as the electron-pair geometries described previously. The same terms are also used to describe the shapes of the molecules. If a molecule contains one or two lone pairs, the bond angles are close to the electron-pair angles predicted by the VSEPR theory, but we need different terms to describe the molecular geometries of these molecules.

We now describe the molecular geometries for six combinations of electron pairs and atoms that are connected to the central atom by single bonds. These descriptions are illustrated and summarized in Table 13.2. Line references are to line numbers in that table.

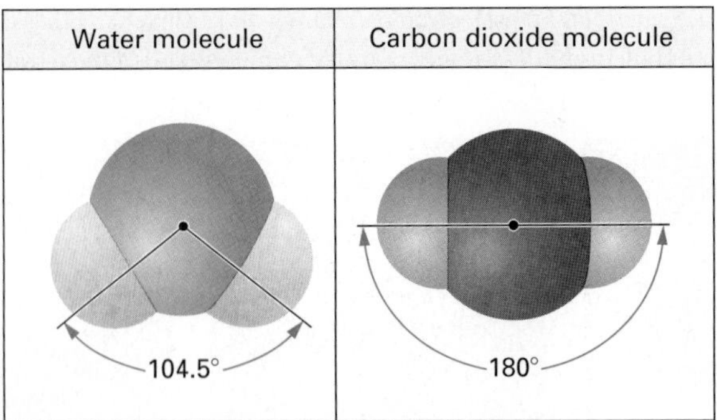


Figure 13.4 Bond angle. If an atom forms bonds with two other atoms, the angle between the bonds is the bond angle. In a water molecule *(left)*, the bonds form at an angle of 104.5°. In a carbon dioxide molecule *(right)*, the bonds lie in a straight line and the bond angle is 180°.

**Table 13.2 Electron-Pair and Molecular Geometries**

| Line | Electron Pairs | Bonded Atoms | Electron-Pair Geometry | Ball-and-Stick Model | Electron-Pair and Bond Angle | Molecular Geometry | Lewis Diagram | Ball-and-Stick Model | Space Filling Model | Example |
|---|---|---|---|---|---|---|---|---|---|---|
| 1 | 2 | 2 | Linear | | 180° | Linear | A—B—A | | | $BeF_2$ |
| 2 | 3 | 3 | Trigonal (triangular) planar | | 120° | Trigonal (triangular) planar | A, A, B, A (B bonded to three A) | | | $BF_3$ |
| 3 | 3 | 2 | Trigonal (triangular) planar | | 120° | Angular or bent | A=B(:)—A | | | $SO_2$ |
| 4 | 4 | 4 | Tetrahedral | | 109.5° | Tetrahedral | A—B—A with A above and A below B | | | $CH_4$ |
| 5 | 4 | 3 | Tetrahedral | | 109.5° | Trigonal (triangular) pyramid or pyramidal | A—B̈—A with A below B | | | $NH_3$ |
| 6 | 4 | 2 | Tetrahedral | | 109.5° | Bent or angular | A—B̈(:)—A or A—B̈: with A below B | | | $H_2O$ |

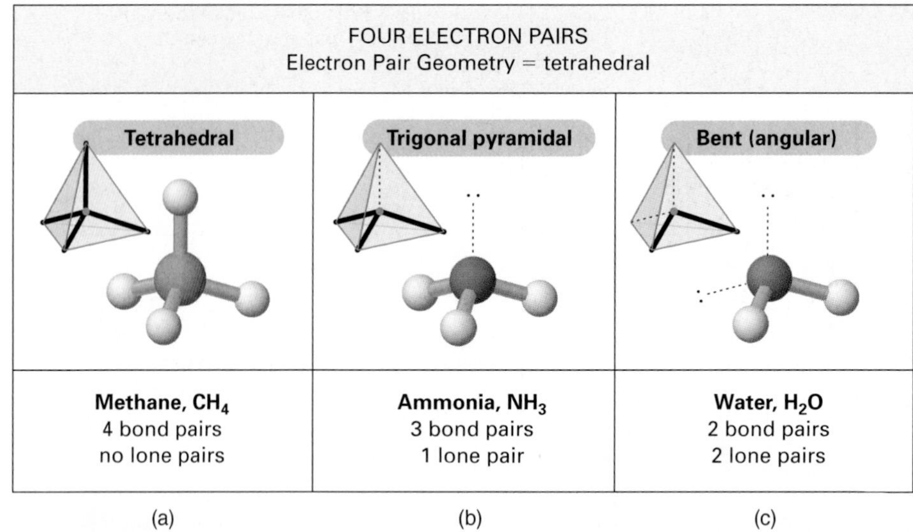


**Figure 13.5** Molecular geometries based on four electron pairs around the central atom. (a) Methane, $CH_4$, is a typical five-atom molecule. The four hydrogen atoms are at the corners of a tetrahedron, and the carbon atom is at its center. The molecule is three-dimensional. If the top hydrogen atom and the carbon atom are in the plane of the page, the front hydrogen atom in the base is closer to you than the page, and the other two hydrogen atoms are behind the page. (b) Ammonia, $NH_3$, is a four-atom molecule having the shape of a pyramid with a triangular base. It is like the $CH_4$ molecule without the top hydrogen atom. The nitrogen atom is in the plane of the page, the front hydrogen atom is in front of the page, and the other two hydrogen atoms are behind the page. Like the carbon atom in methane, the nitrogen atom in ammonia is surrounded by four electron pairs. (c) Water, $H_2O$, is a three-atom molecule with a bent (angular) shape. It is like the $CH_4$ molecule without the top and back-left hydrogens, or like the ammonia molecule without the back-left hydrogen and with the carbon or nitrogen atoms replaced by an oxygen atom. With only three atoms, the bond angle of the water molecule lies in two dimensions. Like the carbon atom in methane and the nitrogen atom in ammonia, the oxygen atom in water is surrounded by four electron pairs. These pairs are *not* in the same plane as the three atoms; one pair is above that plane and the other is beneath it.

**Two Electron Pairs, Two Bonded Atoms** Two electron pairs, both bonding, yield the same electron-pair and molecular geometries: **linear** (Line 1). A linear geometry has a 180° bond angle.

**Three Electron Pairs, Three Bonded Atoms** Three electron pairs, all bonding, yield the same electron-pair and molecular geometries: **trigonal (triangular) planar** (Line 2). Each bond angle is 120°.

Practicing chemists do not consistently use a single term to describe the shape of a molecule with three electron pairs and two bonded atoms. Follow the advice of your instructor as to whether you should use *angular* or *bent* to describe this shape.

**Three Electron Pairs, Two Bonded Atoms** The three electron pairs retain their trigonal planar geometry, but only two are bonded to atoms. The resulting molecular geometry is called **angular** (or **bent**) (Line 3), and the bond angle is 120°.*

**Four Electron Pairs, Four Bonded Atoms** Four electron pairs, all bonding, yield the same electron-pair and molecular geometries: tetrahedral (Line 4). The tetrahedral methane molecule, $CH_4$, looks like a tall pyramid with a triangular base (Fig. 13.5[a]). Each bond angle is 109.5°—the tetrahedral angle.

**Four Electron Pairs, Three Bonded Atoms** The four electron pairs retain their tetrahedral geometry, which is modified because only three of the electron pairs form bonds to other atoms (Line 5). The resulting shape is like a "squashed-down" pyramid, called **trigonal (triangular) pyramidal** (Fig. 13.5[b]).

*In this book, we will disregard minor changes in bond angles due to lone pairs.

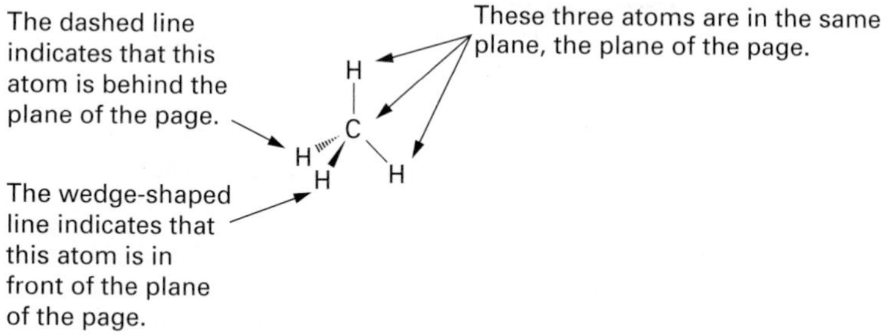

The dashed line indicates that this atom is behind the plane of the page.

These three atoms are in the same plane, the plane of the page.

The wedge-shaped line indicates that this atom is in front of the plane of the page.

**Figure 13.6** Conventions used for drawing wedge-and-dash diagrams.

**Four Electron Pairs, Two Bonded Atoms** Again, the tetrahedral electron-pair geometry is predicted (Line 6). The molecular geometry is bent (or angular) (Fig. 13.5[c]).

With the help of the six preceding paragraphs and their summaries in Table 13.2, you are now ready to predict and name some electron-pair and molecular geometries around a central atom and sketch three-dimensional representations of molecules. Chemists often use **wedge-and-dash diagrams** to indicate the three-dimensional structure of molecules.

Chemists also interchange *bent* and *angular* for four electron pairs/two bonded atoms. As always, follow your instructor's preference in describing this shape.

We will follow these conventions, which are illustrated in Figure 13.6:

summary

**Drawing Wedge-and-Dash Diagrams**

1. When two atoms are in the same plane as the page, they are connected with a solid line of uniform width.
2. When an atom is behind the plane of the page, it is connected to the central atom by a line that is dashed. The width of the dashed line increases as it moves away from the central atom.
3. When an atom is in front of the plane of the page, it is connected to the central atom by a line that is wedge-shaped. The width of the wedge-shaped line increases as it moves away from the central atom.

To predict molecular geometries, we suggest the following procedure:

procedure

**Predicting Molecular Geometries**

**Step 1:** Draw the Lewis diagram.

**Step 2:** Count the electron pairs around the central atom, both bonding and unshared.

**Step 3:** Determine electron-pair and molecular geometries. This is best done by reason rather than by memorization. Ask yourself, and picture the answer in your mind: "Where will the electron pairs go to be as far apart as possible?" There are three answers:

a) Two electron pairs: Electron-pair and molecular geometries are both linear. Bond angle is 180°.

b) Three electron pairs: Electron-pair geometry is trigonal planar. Bond angles are 120°.
   i) All electron pairs bonding: Molecular geometry is trigonal planar.
   ii) Two electron pairs bonding, one lone pair: Molecular geometry is angular (bent).

c) Four electron pairs: Electron-pair geometry is tetrahedral. Bond angles are tetrahedral (109.5°).
   i) All electron pairs bonding: Molecular geometry is tetrahedral.
   ii) Three electron pairs bonding, one lone pair: Molecular geometry is trigonal pyramidal.
   iii) Two electron pairs bonding, two lone pairs: Molecular geometry is bent (angular).

**Step 4:** Sketch the wedge-and-dash diagram. This should match the "mental picture" you formed in Step 3.

## Thinking About Your Thinking

### Mental Models

**Learning to form three-dimensional mental models of molecules and polyatomic ions is a skill you should develop while studying this chapter. The Lewis diagram is a convenient two-dimensional paper-and-pencil representation of the distribution of electrons in a molecule, but it has limited information. A goal to work toward is to be able to look at a Lewis diagram and then see a model of that species in your mind. Our emphasis on three-dimensional wedge-and-dash diagrams in the Active Examples that follow and in the end-of-chapter questions is meant to help you make the connection between Lewis diagrams and these mental models. If you have a molecular model kit, your models and the three-dimensional sketches you draw should match.**

## Active Example 13.9

Predict the electron-pair and molecular geometries of carbon tetrachloride, $CCl_4$.

The Lewis diagram is shown at the right. From this you should establish the number of electron pairs around the central atom and the number of atoms bonded to the central atom. Both geometries follow.

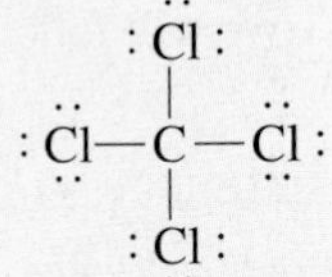

With four electron pairs around carbon, all bonded to other atoms, both geometries are tetrahedral.

Finally, draw a three-dimensional wedge-and-dash diagram of the molecule.

## Active Example 13.10

Describe the shape of a molecule of boron trihydride, $BH_3$, and sketch the molecule.

First draw the Lewis diagram. Remember that boron has only three valence electrons to contribute to covalent bonds. From the structure, write your description.

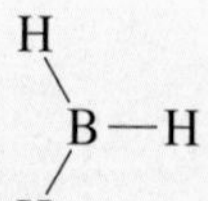

Three electron pairs yield both an electron-pair geometry and a molecular geometry that are trigonal planar, with 120° bond angles.

Complete the Active Example with a wedge-and-dash diagram of the molecule. Remember that all atoms are in the same plane.

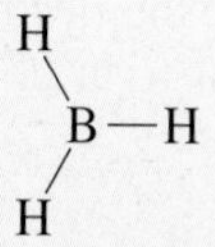

H
B—H
H

When all three atoms and all three bonds lie in the same plane, there is no need for wedges or dashes. It is simplest to put the entire molecule in the plane of the page.

## Active Example 13.11

Predict the electron-pair geometry and shape of a molecule of dichlorine oxide, $Cl_2O$. Draw a wedge-and-dash representation of the molecule.

Start with the electron-pair and molecular geometries. Draw the Lewis diagram and then name each.

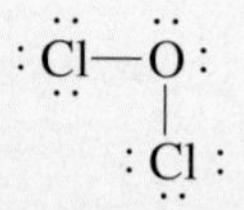

:Cl—O:
|
:Cl:

The electron-pair geometry is tetrahedral, and the molecular geometry is bent.

Oxygen has four electron pairs around it, yielding an electron-pair geometry that is tetrahedral. Only two of the electron pairs are bonded to other atoms, so the molecule is bent. The structure is similar to that of water. Even if you drew the correct Lewis diagram as

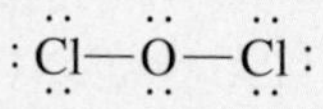

:Cl—O—Cl:

you need to recognize that the four electron pairs around the central O lead to a tetrahedral electron-pair geometry and a bent molecular geometry. It is a good idea to draw the Lewis diagram of the molecule showing the bent geometry.

*continued*

To draw the wedge-and-dash diagram, carefully consider the bent shape. It is preferable to have as many co-planar atoms as possible in your final sketch. Unshared pairs are not shown in a wedge-and-dash diagram.

Cl—O
    \
     Cl

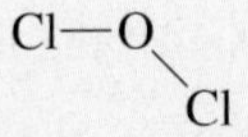

A bent molecular geometry includes the tetrahedral angle, 109.5°. All three atoms can be drawn in the same plane.

### Active Example 13.12

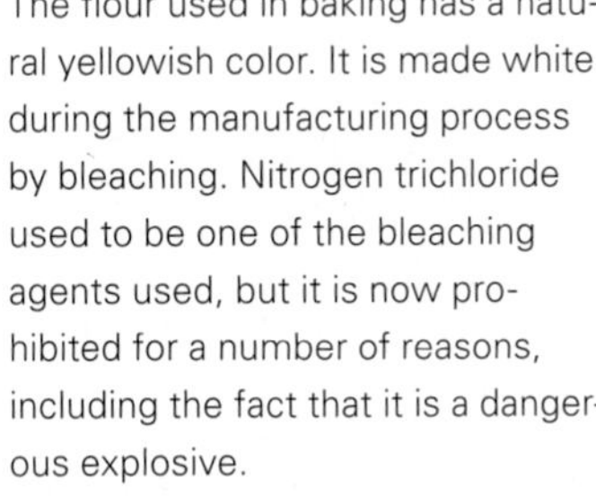

The flour used in baking has a natural yellowish color. It is made white during the manufacturing process by bleaching. Nitrogen trichloride used to be one of the bleaching agents used, but it is now prohibited for a number of reasons, including the fact that it is a dangerous explosive.

Draw the wedge-and dash diagram for nitrogen trichloride, $NCl_3$. Name the electron-pair and molecular geometries.

Take it all the way. Draw the Lewis diagram and, from that, state the electron-pair and molecular geometries. Then draw the wedge-and-dash diagram of the molecule.

:C̤̈l—N̈—C̤̈l:
      |
    :C̤̈l:

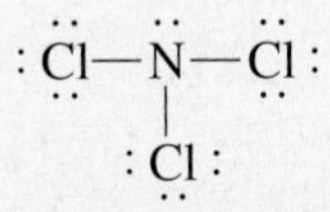

With four electron pairs surrounding the central atom, the electron-pair geometry is tetrahedral. The three bonded atoms yield a trigonal pyramid molecular geometry.

      N
Cl◀ ⋮ \
   Cl   Cl

The wedge-and-dash diagram reflects the tetrahedral electron-pair geometry with one lone pair of electrons.

## 13.4 The Geometry of Multiple Bonds

**Goal** 5 For a molecule with more than one central atom and/or multiple bonds, draw the Lewis diagram and predict and sketch the molecular geometry around each central atom, and draw a wedge-and-dash diagram of the molecule.

Experimental evidence shows that the two or three electron pairs in a multiple bond behave as a single electron pair in establishing molecular geometry. This appears if

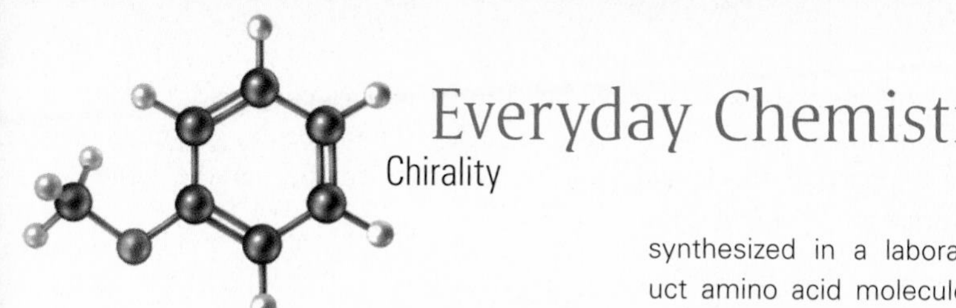

## Everyday Chemistry

### Chirality

The word *chiral* (pronounced ki-ral) refers to an object that cannot be superimposed on its mirror image. Your hands are an example of chiral objects. Hold your left hand next to your right hand. Your right hand is a mirror image of your left and vice versa. If you try to stack your hands on one another, you see that they cannot be superimposed, or matched up to one another (Fig. 13.7). In fact, the word *chiral* is derived from the Greek word *cheir,* which means *hand.* A chiral object is similar to a hand. Feet and ears are other body parts that are also chiral. Achiral (not chiral) objects include anything with a mirror image that is superimposable. Your pencil, a blank piece of paper, your pants, and perhaps your chair are examples of achiral objects near you now.

Nature has many other chiral objects. Seashells are chiral, and furthermore, they are almost always right-handed. Figure 13.8 shows a right-handed shell. Snail shells are also chiral. Vine plants will exhibit chirality as they grow up a pole. A number of humanmade objects are chiral. Many screws, shoes, propellers, and spiral notebook binders are chiral.

The chirality found in the macroscopic world also occurs naturally at the particulate level. Many molecules have chirality. Fascinatingly, as with the case of seashells, nature also has a preferred handedness for molecules. The molecular building blocks of the proteins in your body are known as amino acids. When synthesized in a laboratory, the product amino acid molecules are half left-handed and half right-handed. However, the amino acids in your body and in all living systems are exclusively left-handed.

Chiral drugs are an important issue in pharmacy and medicine. Ibuprofen, the active ingredient in the pain relievers Advil, Motrin, and Nuprin, is an equal mixture of left-handed and right-handed molecules. Yet only the left-handed molecule is medicinally active. The body converts the right-handed molecules into left-handed molecules. In other prescription drugs, one handedness will be active and the other will be inert or even harmful. A significant research effort in chemistry is presently directed toward synthesizing and purifying single-handed drugs and other compounds.

You will study chirality in more detail if you take a course in organic chemistry. Knowledge of a chemical formula alone is insufficient to completely identify many molecules. If you take chirality into consideration, even the Lewis diagram by itself does not tell the whole story! The structure and shape must be specified to the type of handedness for chiral molecules.

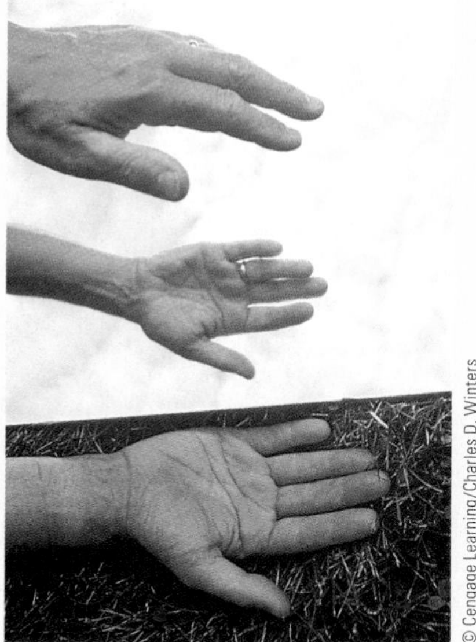


**Figure 13.7** Hands as an example of chiral objects. When you look at your left hand in a mirror, you see an image of your right hand. If you superimpose your palm-down left hand over your palm-down right hand, however, you see that they are not the same. Thus hands are nonsuperimposable mirror images of one another, and therefore they are chiral objects.

**Figure 13.8** A seashell as an example of a chiral object. The spiral pattern in this seashell has a handedness. This is a right-handed shell.

we compare beryllium difluoride, carbon dioxide, and hydrogen cyanide, whose Lewis diagrams are

$$:\ddot{\underset{..}{F}}-\text{Be}-\ddot{\underset{..}{F}}: \qquad \ddot{\underset{..}{O}}=\text{C}=\ddot{\underset{..}{O}} \qquad \text{H}-\text{C}\equiv\text{N}:$$

All three molecules are linear; their bond angles are 180°. The two electron pairs in $BeF_2$ are as far from each other as possible. According to the VSEPR principle, this is responsible for the 180° bond angle in that compound. Carbon is flanked by two double bonds in $CO_2$ and one single bond and one triple bond in HCN. Evidently, the second and third electron pairs in double and triple bonds don't affect the molecular geometry.

Further evidence supporting this conclusion comes from comparing the bond angles in boron trifluoride and formaldehyde:

```
          ..                 ..
          F:                 O:
         /                  //
 :F̤—B                H—C
         \                  \
          F:                 H
          ..
```

The shapes are both trigonal planar with 120° bond angles. This is the angle predicted for three electron pairs under the VSEPR theory.

We can conclude that the *number of regions of electron density* that surround a central atom determine the electron-pair geometry around that atom. A region of electron density can be a single, double, or triple bond, or a lone pair. No matter the number of pairs of bonding electrons between two atoms, each region of electron density is distributed as far away from other regions of electron density as possible, as predicted by VSEPR theory.

## Active Example 13.13

Determine the molecular geometry of ethylene, $C_2H_4$. You will need to describe the geometry around each carbon atom. Sketch a wedge-and-dash diagram of the molecule.

Start with the Lewis diagram and then determine and state the electron-pair and molecular geometries around each carbon atom.

```
H       H
 \     /
  C=C
 /     \
H       H
```

The electron-pair and molecular geometries are trigonal planar around each carbon atom. For each carbon atom, there are three regions of electron density, making the electron-pair geometry trigonal planar. Three bonded atoms makes the molecular geometry trigonal planar.

Now draw the wedge-and-dash diagram. Recall that all atoms with a trigonal *planar* electron-pair geometry lie in the same plane. Assume that all six atoms are in the same plane, which they are.

```
H       H
 \     /
  C=C
 /     \
H       H
```

With all atoms in the same plane, no wedges or dashes are needed. The Lewis diagram and the wedge-and-dash diagram are the same.

## Active Example 13.14

Describe the molecular geometry around each carbon atom in ethane, $C_2H_6$. Draw a wedge-and-dash diagram that illustrates the geometry.

Determine the number of valence electrons and draw the Lewis diagram.

Ethane is a component of natural gas. In industry, it is often converted to ethylene, and then to polyethylene, a plastic that has many uses, such as milk jugs, plastic bags, and food packaging.

$2 \times 4\ (C) + 6 \times 1\ (H) = 14$ valence electrons

```
    H   H
    |   |
H — C — C — H
    |   |
    H   H
```

What is the electron-pair and molecular geometry around each carbon atom?

Tetrahedral around each carbon atom.

Each carbon atom is surrounded by four regions of electron density, four bonded.

Finish the Active Example by constructing the wedge-and-dash diagram. You have two central atoms, each with a tetrahedral molecular geometry. The diagram is easiest to draw and interpret if you put both carbon atoms in the plane of the page.

```
  H        H
   \       ▸H
    C — C
 H◂       \
   H        H
```

We placed both carbon atoms and a hydrogen atom bonded to each carbon atom in the plane of the page to maximize the number of atoms in the plane of the page. That makes it necessary to draw one hydrogen atom out of the page and one hydrogen atom behind the page on each carbon atom. Variations on this diagram are acceptable as long as they correctly reflect the tetrahedral geometry around each carbon atom, as seen in this illustration:

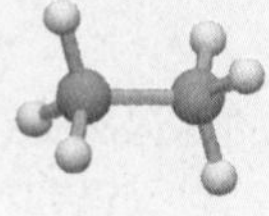

## 13.5 Polarity of Molecules

**Goal** 6 Given or having determined the Lewis diagram of a molecule, predict whether the molecule is polar or nonpolar.

**P/REVIEW:** A polar covalent bond is defined in Section 12.4 as a bond in which bonding electrons are shared unequally. The charge density of the bonding electrons is shifted toward the more electronegative atom.

We previously discussed the polarity of covalent bonds. Now that you have some idea about how atoms are arranged in molecules, you are ready to learn about the polarity of molecules themselves. **A polar molecule is one in which there is an asymmetrical* distribution of charge,** resulting in + and − poles. A simple example is the HF molecule. The fact that the bonding electrons are closer to the fluorine atom gives the fluorine end of the molecule a partial negative charge, while the hydrogen end acts as a positive pole. (See the illustration in the P/Review in the margin.) In general, any diatomic molecule in which the two atoms differ from each other will be at least slightly polar. Other examples are HCl and BrCl. In both of these molecules the more electronegative chlorine atom acts as a negative pole. In an electric field, polar molecules tend to line up with the more electronegative atoms pointing toward the plate with the positive charge and the less electronegative (more electropositive) atoms pointing toward the plate with the negative charge (Fig. 13.9).

When a molecule has more than two atoms, we must know something about the bond angles in order to decide whether the molecule is polar or nonpolar. Consider, for example, the two triatomic molecules $CO_2$ and $H_2O$. Despite the presence of two polar bonds, the linear $CO_2$ *molecule* is **nonpolar,** as illustrated in Active Figure 13.10(a). We say that the cancellation of the polar bonds results in no overall (net) regions of positive and negative charge (a dipole), or no net dipole moment (*moment* refers to the combined effect of a quantity and a force). Since the oxygen atoms are symmetric around the carbon atom, the two polar C=O bonds cancel each other.

In contrast, the bent water molecule is polar; the two polar bonds do not cancel each other because the molecule is not symmetrical around a horizontal axis (Active Fig. 13.10[b]). The bonding electrons spend more time near the more electronegative oxygen atom, which is the negative pole. The positive pole is midway between the two hydrogen atoms.

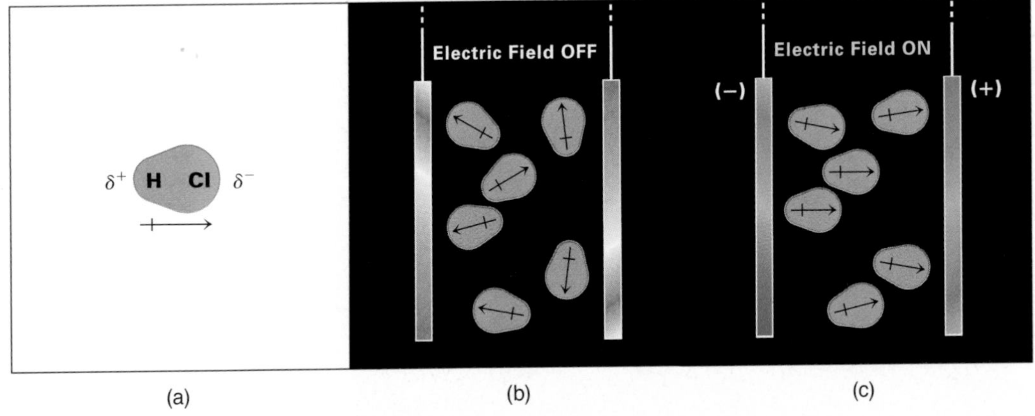


**Figure 13.9** Orientation of polar molecules in an electric field. (a) HCl is an example of a polar molecule. The partially positive end is shaded in red, and the partially negative side is blue. The arrow is drawn with a plus sign at the positive end and an arrowhead at the negative end. The arrow points in the direction of greatest charge density for the bonding electrons. (b) Two plates are connected through a switch to a source of an electric field. With the switch open, the orientation of the molecules is random (*left*). When the switch is closed (*right*), the molecules line up with the positive end (*red*) toward the negative plate and the negative pole (*blue*) toward the positive plate.

**Symmetry* refers to balance. As we use the word, a molecule has an *asymmetrical distribution of charge* if charge distribution is unbalanced. There is a point where positive charge appears to be concentrated and a different point where negative charge appears to be concentrated.

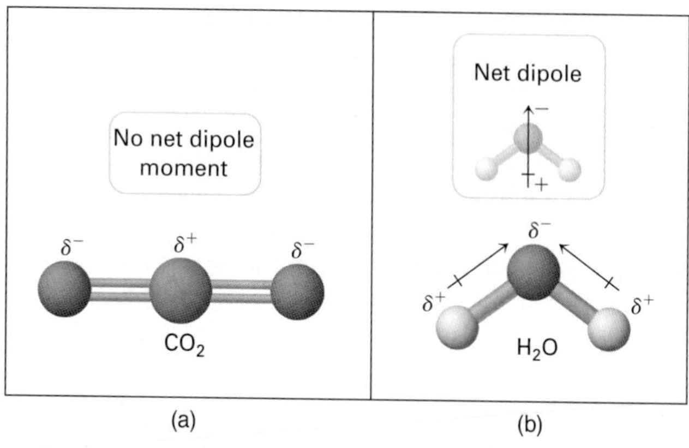


**Active Figure 13.10** Polarity of triatomic molecules. Molecular polarity depends on both bond polarity and molecular geometry. **Watch this active figure at *http://www.cengage.com/cracolice.***

Another molecule that is nonpolar despite the presence of polar bonds is $CCl_4$. The four C—Cl bonds are themselves polar, with the bonding electrons displaced toward the chlorine atoms. But because the four chlorines are symmetrically distributed about the central carbon atom (Fig. 13.11), the polar bonds cancel each other. If one of the chlorine atoms in $CCl_4$ is replaced by hydrogen, the symmetry of the molecule is destroyed. As an example, the chloroform molecule, $CHCl_3$, is polar. Figure 13.11 shows other similar tetrahedral molecules and an analysis of their polarities.

From these observations we can state an easy way to decide whether a simple molecule is polar or nonpolar. If the central atom has no lone pairs and all atoms bonded to it are identical, the molecule is nonpolar. If these conditions are not met, the molecule is polar.

## Thinking About Your Thinking

### Mental Models

**Forming mental models of molecules will help you understand the polarity concept. The challenge in this section is to combine the concept of polar and nonpolar bonds with that of polar and nonpolar molecules. Begin by imagining an asymmetric distribution of bonding electrons as illustrated in Section 12.4 and the P/Review at the beginning of this section. The bonding electrons are displaced toward the atom of the more electronegative element.**

**Now expand this mental model to a molecule with more than one nonpolar bond. Imagine a central atom bonded to four terminal atoms that are the same, as in Figure 13.11. The electrons in the bonds are shifted toward the atom of the more electronegative element, but the unequal distribution of charge is evenly spread out**

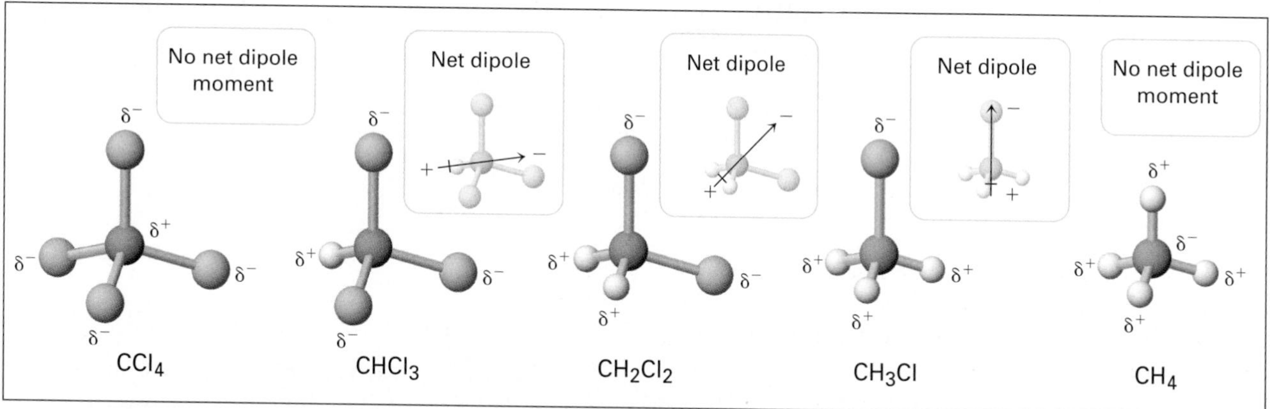


**Figure 13.11** Polar and nonpolar molecules. The carbon–chlorine bond is polar, with chlorine as the end toward which bonding electrons are displaced. The carbon–hydrogen bond is also polar, with carbon at the more negative end. $CCl_4$ and $CH_4$ are nonpolar because the dipoles from the polar bonds cancel due to the molecular symmetry. $CHCl_3$, $CH_2Cl_2$, and $CH_3Cl$ are polar because the dipoles from the polar bonds do not cancel.

**in opposing directions within the molecule. All of the pushes or pulls cancel one another out, and the molecule itself is nonpolar. Now imagine that you destroy the symmetry of the model in your mind by changing one of the terminal atoms to something different, as illustrated in Figure 13.11. Can you mentally visualize the lack of symmetry in the polar bonds? Think about how the molecule has an area that is electron-rich and an opposite area that is relatively electron-poor. This is a polar molecule.**

**Practice looking at the polar bonds within molecules and their net effect on overall molecular polarity whenever you draw Lewis diagrams or three-dimensional ball-and-stick diagrams.**

## Active Example 13.15

Is the $BF_3$ molecule polar? Is the $NH_3$ molecule polar?

The geometries of both of these molecules are described in Table 13.2. Consider $BF_3$ first. Sketch the Lewis diagram. Place arrows on each of the bonding electron pairs pointing to the more electronegative element. Is the molecule polar?

$BF_3$ is nonpolar.

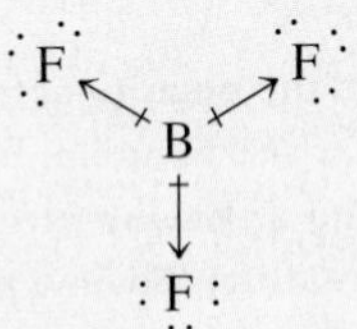


Even though fluorine is more electronegative than boron, the three fluorine atoms are arranged symmetrically around the boron atom. The polar bonds cancel.

Since the boron atom has no lone pairs and all atoms bonded to it are identical, the molecule must be nonpolar.

Now draw the Lewis diagram for $NH_3$, with the arrows pointing to the more electronegative element. Also sketch the wedge-and-dash diagram of the trigonal pyramidal structure of the molecule with the same arrows included. Is $NH_3$ polar or nonpolar?

$NH_3$ is polar.

δ−
H→N←H
H
δ+

N
H H H

The bonding electrons in ammonia are displaced toward the more electronegative nitrogen atom. The bonds do not cancel in the asymmetrical pyramidal shape, so the molecule is polar. The wedge-and-dash diagram, which illustrates the molecular shape, better suggests the charge displacement toward the nitrogen atom.

## 13.6 The Structure of Some Organic Compounds (Optional)

Goal 7 Distinguish between organic compounds and inorganic compounds.

Goal 8 Distinguish between hydrocarbons and other organic compounds.

Goal 9 On the basis of structure and the geometry of the identifying group, distinguish among alcohols, ethers, and carboxylic acids.

### The Bonding Capabilities of the Carbon Atom

**Organic chemistry** is the chemistry of carbon compounds. The vast majority of compounds that have been synthesized, characterized, and catalogued by chemists are carbon compounds. Carbonates, cyanides, oxides of carbon, and a few other carbon-containing compounds are exceptions that are often classified as inorganic.

The property of carbon that qualifies it to define a whole branch of chemistry is the bonding capability of the carbon atom. Carbon atoms have four valence electrons that enable them to form four covalent bonds with other atoms. The carbon atom is also the right size for these bonds to be of just the right strength—strong enough to form stable molecules but weak enough to undergo reactions. Most important is the fact that these carbon atoms may be bonded to other carbon atoms. As a result, many organic compounds have extremely long chains of carbon atoms. No other element has the bonding characteristics of carbon.

The structure of molecules is the primary concern of organic research chemists. They build on the previously discovered physical and chemical characteristics of a variety of structures. From this starting point, they "create," on paper or on a computer, Lewis diagrams and molecular models of molecules that should have certain desirable properties. Then, beginning with the structures of chemicals that are known and available, they figure out chemical reactions that change the existing structures to those that are wanted. This kind of research has led to many things we use daily, including synthetic fabrics, plastics, and medical products.

**P/REVIEW:** The structure of different kinds of organic compounds is explored in greater detail in Chapter 21.

### Hydrocarbons

**Hydrocarbons are compounds made up of only carbon and hydrogen.** One type of hydrocarbon is an **alkane,** in which each carbon atom forms four single bonds. Alkanes differ from each other by the number of carbon atoms that bond together to form a chain. Methane, $CH_4$, is the simplest alkane, with only one carbon atom. Ethane, $C_2H_6$, has two carbon atoms bonded to one another; propane, $C_3H_8$, has three carbon atoms in a chain; and butane, $C_4H_{10}$, has four carbon atoms bonded together. Compare the Lewis diagrams with the three-dimensional representations of the three compounds in Active Figure 13.12.

The formulas for ethane and propane can be written in two different ways. Ethane is $C_2H_6$ or $CH_3CH_3$. Propane is $C_3H_8$ or $CH_3CH_2CH_3$. $C_2H_6$ and $C_3H_8$ are molecular formulas. $CH_3CH_3$ and $CH_3CH_2CH_3$ are called **condensed structural formulas.**

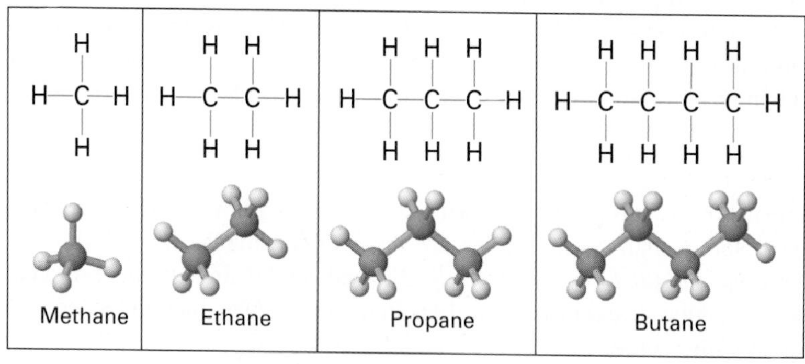


**Active Figure 13.12** Alkanes. An alkane is a hydrocarbon in which each carbon atom forms four bonds. **Watch this active figure at *http://www.cengage.com/cracolice.***

Figure 13.13 The isomers of butane and pentane.

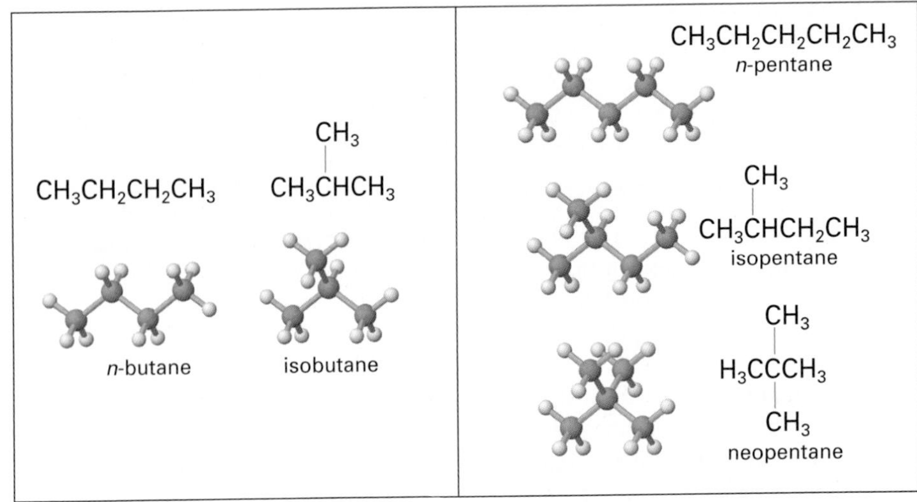


(a) Butane has two isomers, *n*-butane and isobutane.

(b) Pentane has three isomers, *n*-pentane, isopentane, and neopentane.

They suggest the structure of the molecule. Ethane, $CH_3CH_3$, is made up of two $—CH_3$ groups bonded together. Divide the Lewis diagram between the carbon atoms and that's exactly what you have. The condensed structural formula for propane, $CH_3CH_2CH_3$, suggests correctly that the molecule is made up of two $—CH_3$ groups with a $—CH_2—$ group between them. Any number of $—CH_2—$ groups can be placed between two $—CH_3$ groups.

Butane is the simplest alkane that exists as two distinct isomers, that is, as molecules with the same molecular formula but different arrangements of atoms (Fig. 13.13[a]). The two butanes are distinctly different chemical compounds, each having its own set of physical and chemical properties. Notice that the name *butane* is modified to distinguish between the isomers. Pentane, $C_5H_{12}$, exists as three isomers, as shown in Figure 13.13(b). Again, each molecule has its own distinct set of properties. Even though isomers are made from the same set of atoms, their differing arrangement into molecules makes them different substances.

As the number of carbon atoms in an alkane increases, the number of isomers increases dramatically. There are 5 isomers possible for six-carbon alkanes, 9 isomers possible for seven-carbon alkanes, and 75 isomers possible for ten-carbon alkanes. It is possible to draw more than 300,000 isomeric structures for $C_{20}H_{42}$ and more than 100 million for $C_{30}H_{62}$. As you may guess, not all of them have been prepared and identified! This does give us some idea, though, why there are many more known organic compounds than inorganic compounds.

## Thinking About Your Thinking

### Mental Models

**As the number of atoms in a molecule increases, you need more practice to become proficient at forming mental models after drawing a two-dimensional Lewis diagram. You will spend a good deal of time developing this skill in a future course. At this point in your study of chemistry, we will help you visualize organic compounds by showing particulate-level illustrations of most molecules along with their Lewis diagrams.**

Compare the Lewis diagrams with the three-dimensional representations of the molecules. Notice that the molecules are three-dimensional—*they do not look like their two-dimensional Lewis diagrams.* All bond angles are tetrahedral. Thus what appears to be a nice straight line of carbon atoms in a Lewis diagram is actually a zigzag chain of carbon atoms.

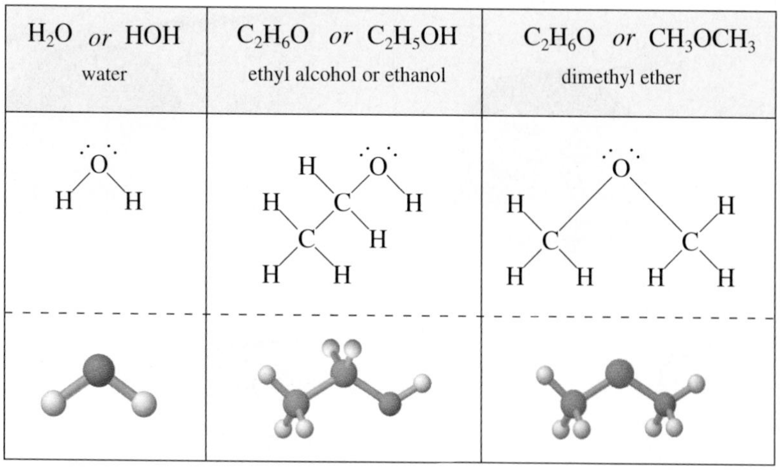

| $H_2O$ *or* HOH | $C_2H_6O$ *or* $C_2H_5OH$ | $C_2H_6O$ *or* $CH_3OCH_3$ |
|---|---|---|
| water | ethyl alcohol or ethanol | dimethyl ether |

Figure 13.14 Water, ethanol, and dimethyl ether. All molecules have a bent structure around an oxygen atom with two unshared electron pairs and single bonds to two other atoms.

## Alcohols and Ethers

In Active Example 13.8 you drew the Lewis diagram for ethanol, and you were shown the diagram for its isomer, dimethyl ether. These diagrams, plus the Lewis diagram for water, along with their three-dimensional drawings, are shown in Figure 13.14. All three molecules include an oxygen atom, with two lone pairs of electrons, that is single-bonded to two other atoms. This, as you learned in Section 13.3, yields a bent structure around the oxygen atom. In Section 13.5 you saw that a molecule with this structure is polar. This polarity is present in all three molecules, but to a diminishing extent from left to right.

The —OH part of an alcohol is called a **hydroxyl group.** Essentially, **an alcohol is an alkane in which a hydrogen atom is replaced by a hydroxyl group** (Fig. 13.15). An alcohol can also be thought of as a derivative of water in which one of the $H_2O$ hydrogens is replaced with a hydrocarbon group. The —OH group may be anyplace in the molecule, and there may be more than one. Another example of an alcohol is methanol, also called methyl alcohol or wood alcohol, which is used as a solvent, as a starting substance from which more complex molecules are produced, and as a gasoline antifreeze (Fig. 13.16[a]). Similarly, 2-propanol, often called isopropyl alcohol or rubbing alcohol, is also used as a solvent and a reactant in the manufacture of other substances, in addition to its most well-known use as a disinfectant and to cool and soothe skin (Fig. 13.16[b]). Ethylene glycol is an example of a diol, an alcohol with two hydroxyl groups. It is the major component in most automobile antifreeze solutions (Fig. 13.17).

The chemical properties of an alcohol are the chemical properties of the hydroxyl group. Similar properties are present in all alcohols, though to different degrees. The physical properties of the alcohols are also associated with the hydroxyl group, particularly among the smaller alcohol molecules. In large molecules, the properties of the hydrocarbon section usually have the greatest influence on the physical properties of the overall molecule. For example, alcohols with shorter hydrocarbon portions,

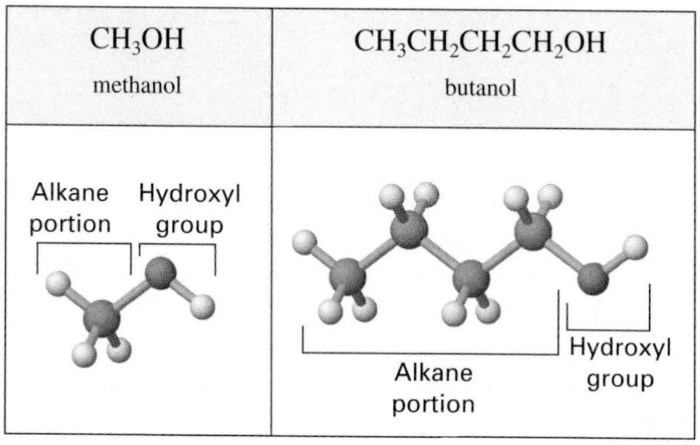

| $CH_3OH$ | $CH_3CH_2CH_2CH_2OH$ |
|---|---|
| methanol | butanol |
| Alkane portion; Hydroxyl group | Alkane portion; Hydroxyl group |

Figure 13.15 Alcohols. An alcohol is an alkane in which a hydroxyl group replaces a hydrogen atom.

Figure 13.16 Commercial products that include methanol and isopropyl alcohol.

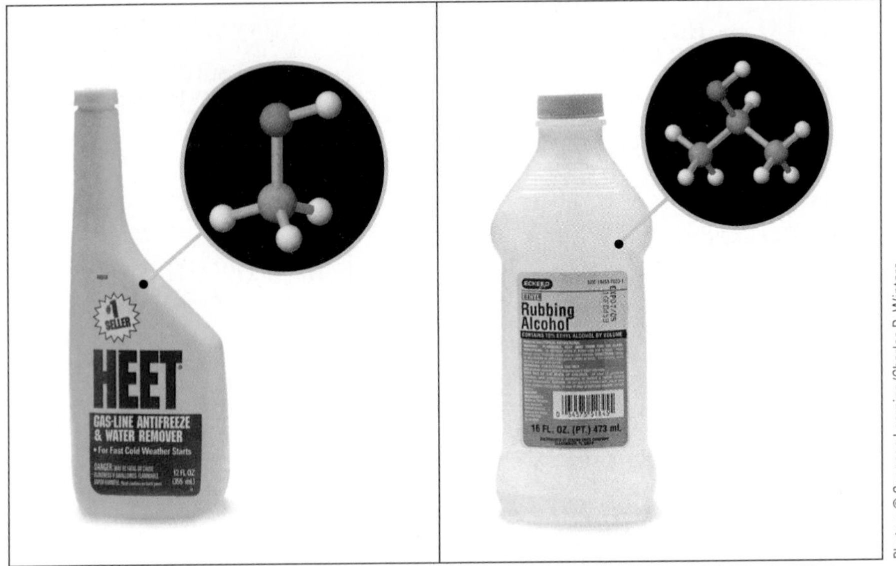


(a) Methanol, $CH_3OH$, is 99% of this gasoline antifreeze solution.

(b) Isopropyl alcohol, $CH_3CH(OH)CH_3$, is commonly sold as a 70% solution called rubbing alcohol.

Figure 13.17 Ethylene glycol, $HOCH_2CH_2OH$, is the major component in most automobile antifreeze solutions. The molecule has two hydroxyl groups, making it a dialcohol, or a diol.

such as $CH_3OH$ and $CH_3CH_2OH$, dissolve in water in all proportions, whereas alcohols with longer hydrocarbon portions, such as $CH_3CH_2CH_2CH_2CH_2CH_2CH_2CH_2OH$, are immiscible in water.

**An ether is a compound that has two hydrocarbon groups bonded to an oxygen atom.** Ethers can be thought of as water molecules in which both hydrogen atoms are replaced by hydrocarbon groups. Either or both of the $—CH_3$ groups in dimethyl ether (see Fig. 13.14) may be replaced by longer carbon chains. Two additional examples are shown in Figure 13.18.

## Carboxylic Acids

The formulas of formic acid, H—COOH, acetic acid, $CH_3$—COOH, and propanoic acid, $CH_3CH_2$—COOH, suggest the structure of the first three members of a series of **carboxylic acids,** which contain the **carboxyl group,** —COOH. Their Lewis diagrams and ball-and-stick models are shown in Figure 13.19. As with hydrocarbons, alcohols, and ethers, the carbon chain may extend indefinitely.

Acetic acid is the best known of the carboxylic acids. Given the structure of acetic acid, you can see the difference between the ionizable hydrogen, which is bonded to an

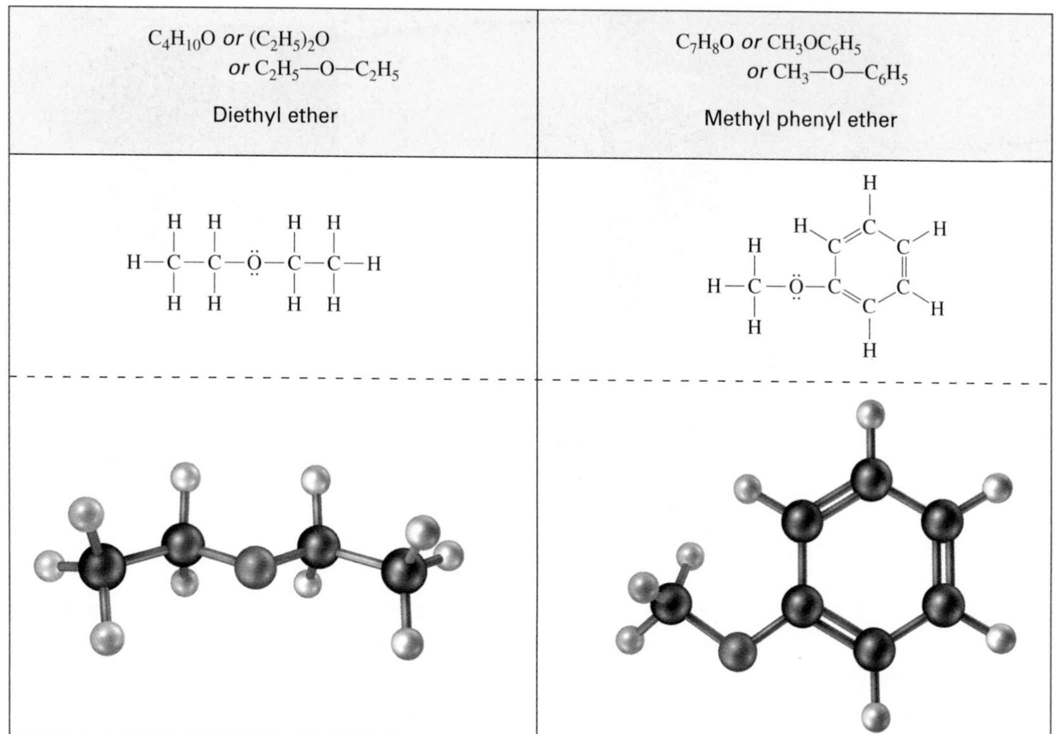


**Figure 13.18** Ethers. An ether is a compound that has two hydrocarbon groups bonded to an oxygen atom. In everyday language, the term *ether* refers to diethyl ether, which was once commonly used as an anesthetic. Scientists must use more precise language, because ether actually refers to an entire class of molecules.

| $HCOOH$ or $HCHO_2$ | $CH_3COOH$ or $HC_2H_3O_2$ | $C_2H_5COOH$ or $HC_3H_5O_2$ |
|---|---|---|
| Formic acid | Acetic acid | Propanoic acid |

**Figure 13.19** Carboxylic acids. A carboxylic acid is characterized by the presence of the carboxyl group, —COOH.

oxygen atom, and the other hydrogens, which are bonded to carbon atoms. The organic chemist's way of writing the ionization equation, along with Lewis diagrams, is

$$CH_3COOH \longrightarrow H^+ + CH_3COO^-$$

acetic acid ⟶ hydrogen ion + acetate ion

**P/REVIEW:** Acetic acid is given in this book as one of the "other acids" in Section 6.7. In that section, its ionization equation is given as $HC_2H_3O_2 \rightarrow H^+ + C_2H_3O_2^-$.

Many insects, including the red ant *Formica rufa* from which it was named, use formic acid as a chemical weapon, a communication system, and for protection against parasites.

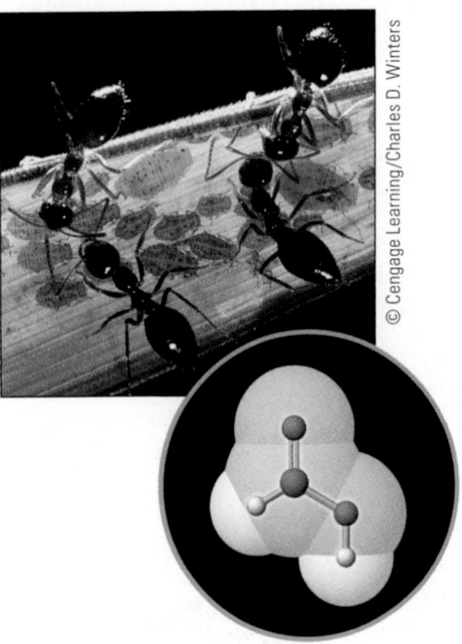

© Cengage Learning/Charles D. Winters

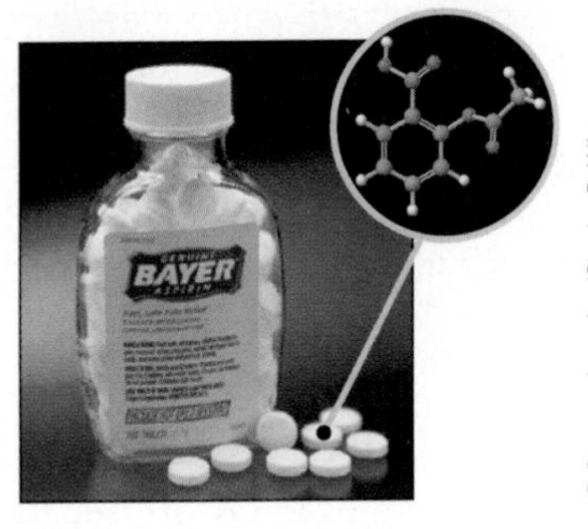



© Cengage Learning/Charles D. Winters

Aspirin is acetylsalicylic acid, a carboxylic acid.

Carboxylic acids are weak acids, which means that they ionize only slightly in water. To the extent they ionize, however, it is always the carboxylic hydrogen that reacts when they react with bases in neutralization reactions.

P/REVIEW: Writing the equation for the reaction between an acid and a hydroxide base is discussed in Section 8.9. It is the most common form of a neutralization reaction.

The geometry of the carboxyl group is shown in the preceding Lewis diagrams. The molecule is bent around the oxygen atom that is single-bonded to both the carbon atom and the ionizable hydrogen atom, forming a tetrahedral angle. In determining the geometry around the carbon atom, the double bond counts as if it were one of three single bonds. The geometry is trigonal planar with 120° bond angles. Bond angles around carbon atoms in the alkane part of the molecule are tetrahedral.

## Chapter 13 in Review

Most of the key terms and concepts and many others appear in the Glossary. Use your Glossary regularly.

### 13.1 Drawing Lewis Diagrams

Goal 1 Draw the Lewis diagram for any molecule or polyatomic ion made up of main group elements.

**Key Terms and Concepts: Central atom, isomer, resonance hybrid, resonance structure, terminal atom**

### 13.2 Electron-Pair Repulsion: Electron-Pair Geometry

Goal 2 Describe the electron-pair geometry when a central atom is surrounded by two, three, or four electron pairs.

**Key Terms and Concepts: Electron-pair angle, electron-pair geometry, molecular geometry, tetrahedron (tetrahedral), valence shell electron-pair repulsion (VSEPR) theory**

### 13.3 Molecular Geometry

Goal 3 Given or having derived the Lewis diagram of a molecule or polyatomic ion in which a second-period central atom is surrounded by two, three, or four pairs of electrons, predict and sketch the molecular geometry around that atom.

Goal 4 Draw a wedge-and-dash diagram of any molecule for which a Lewis diagram can be drawn.

**Key Terms and Concepts: Angular, bent, bond angle, linear, trigonal (triangular) planar, trigonal (triangular) pyramidal, wedge-and-dash diagrams**

### 13.4 The Geometry of Multiple Bonds

Goal 5 For a molecule with more than one central atom and/or multiple bonds, draw the Lewis diagram and predict and sketch the molecular geometry around each central atom, and draw a wedge-and-dash diagram of the molecule.

### 13.5 Polarity of Molecules

Goal 6 Given or having determined the Lewis diagram of a molecule, predict whether the molecule is polar or nonpolar.

**Key Terms and Concepts: Polar, nonpolar molecule**

### 13.6 The Structure of Some Organic Compounds (Optional)

Goal 7 Distinguish between organic compounds and inorganic compounds.

Goal 8 Distinguish between hydrocarbons and other organic compounds.

Goal 9 On the basis of structure and the geometry of the identifying group, distinguish among alcohols, ethers, and carboxylic acids.

**Key Terms and Concepts: Alcohol, alkane, carboxyl group, carboxylic acid, condensed structural formula, ether, hydrocarbon, hydroxyl group, organic chemistry**

# Study Hints and Pitfalls to Avoid

There are many molecules for which you can draw two or more Lewis diagrams that satisfy the octet rule. You can also draw incorrect diagrams that satisfy the octet rule. Your diagram is most likely to be correct if you remember that (1) hydrogen always forms one bond and carbon almost always forms four, (2) two or more oxygen atoms are distributed around a central atom, (3) an oxygen atom is between a hydrogen atom and another nonmetal atom, and (4) your diagram should be as symmetric as possible.

The most common errors in Lewis diagrams are bonding oxygen atoms to each other and surrounding a central atom with three or five electron pairs. There are some compounds in which oxygen atoms are bonded to each other, but not many. One of these appears in the question section that follows. The errors involving three or five electron pairs most often occur when double bonds are present. Always check your final diagram to be sure all atoms conform to the octet rule.

Two exceptions to the octet rule where central atoms are surrounded by fewer than four pairs of electrons, are compounds of beryllium and boron, with two and three electron pairs, respectively, found in stable compounds.

Wedge-and-dash diagrams are based on three basic structures: (1) a linear geometry with a 180° angle and all atoms in the same plane, (2) a trigonal planar geometry with 120° angles and all atoms in the same plane, and (3) a tetrahedral geometry with 109.5° angles between the atoms. The tetrahedral diagram is most easily drawn by starting with three co-planar atoms—the central atom and two of the terminal atoms at a 109.5° angle—and then drawing one atom coming forward out of the plane of the page with a wedge, and finally drawing one atom going behind the plane of the page with a dash. More complex molecules are drawn as combinations of the three basic structures. The most common error in wedge-and-dash diagrams is failing to account for lone pairs. Lone pairs are not shown in a wedge-and-dash diagram, yet they are there in the actual molecule, and they affect the molecular geometry.

Geometry places limits on the shapes of some molecules. If there are only two atoms, the geometry is linear; two points determine a line. If there are three atoms, they are either in a straight line (linear) or they are not (angular or bent). Four atoms take you into the *possibility* of a three-dimensional molecule. That is why you must distinguish between trigonal planar and trigonal pyramidal. The adjective *trigonal* is necessary because some elements in the third and later periods form square planar and square pyramidal structures.

To distinguish between polar and nonpolar molecules, test the molecule for two conditions that are required for nonpolarity. First, all atoms bonded to the central atom must be the same element. Second, there can be no lone pairs. If the molecule passes both tests, it is nonpolar; if it fails either test, it is polar.

# Small-Group Discussion Questions

*Small-Group Discussion Questions are for group work, either in class or under the guidance of a leader during a discussion section.*

1. Molecules with multiple carbon atoms sometimes exist in a configuration where the carbon atoms are in a closed loop, bonded to one another. Cyclopropane, $C_3H_6$, cyclobutane, $C_4H_8$, cyclopentane, $C_5H_{10}$, and cyclohexane, $C_6H_{12}$, are examples of this type of molecule, called cycloalkanes. Draw the Lewis diagram of each.
2. Five "*-ic* acids" form the basis for the nomenclature system presented in Chapter 6. Draw the Lewis diagram of each acid. Why are these molecules called oxyacids? What bonding arrangement is responsible for the acidic hydrogen? How are non-acidic hydrogens different?
3. How many distinctly different Lewis diagrams can you draw for each of the following formulas: $C_4H_{10}$, $C_5H_{12}$, $C_6H_{14}$?
4. How many resonance structures can be drawn for each of the following: ozone, $O_3$, carbonate ion, benzene, $C_6H_6$, and cyanate ion, $CNO^-$?
5. We suggested that you might think of molecular geometry as an "atom geometry." Why, then, do unshared electron pairs on a central atom influence molecular geometry, given that they are not bonded to an atom?
6. In the end-of-chapter Questions, Exercises, and Problems, you are asked to sketch the wedge-and-dash diagram for Questions 19–22. Sketch the wedge-and-dash diagrams for the remaining questions in that section, Questions 23–30.
7. Rank the following in order of increasing polarity: methane, $CH_4$, carbon disulfide, boron trifluoride, nitrogen trichloride. Explain your reasoning.
8. Draw the Lewis diagram of a four-carbon molecule that is an example of each of the following: hydrocarbon, alcohol, ether, carboxylic acid. Then draw a second example of a four-carbon molecule from each class that is different from your first example.
9. What is the role of the quantum model of the atom in the determination of molecular geometry? Explain in as much detail as possible, discussing the links and relationships among quantum mechanics, bonding, structure, and shape.

# Lewis Diagram Recognition Exercises

*Classify each of the following Lewis diagrams as either acceptable or unacceptable. For those that are unacceptable, explain why and draw an acceptable diagram.*

1. $BCl_3$

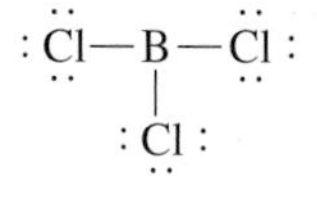

```
  ..       ..
: Cl — B — Cl :
  ..   |   ..
      :Cl:
       ..
```

2. $NO_3^-$

```
  ..            .. -
[ : N — O = O :  ]
    ..  |   ..
      : O :
        ..
```

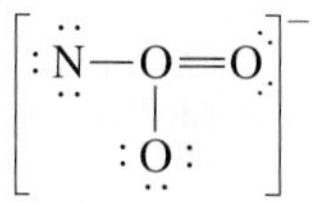

3. $HClO_3$

```
        H
        |
   :Ö—Cl—Ö:
        |
       :O:
        ¨
```

4. $C_3H_6$

```
     H        H
     |   ..   |
  H—C—C—C—H
     |        |
     H        H
```

5. HCN

```
  H—C═N̈:
```

6. $BeF_2$

```
  :F̈—Be—F̈:
```

7. $C_2H_6O$

```
      H
      |   ..
   H—C—O:
      |   |
      H   |
          |
   H—C—H
      |
      H
```

8. $CH_3Br$

```
      H
      |   ..
   H—C═Br
      |   ¨
      H
```

9. CO $C{\equiv}O$

10. $H_2O_2$ $\ddot{O}{=}H{=}\ddot{O}{-}H$

# Questions, Exercises, and Problems

*Blue-numbered questions are answered in Appendix III. ■ denotes problems assignable in OWL. In the first section of Questions, Exercises, and Problems, similar exercises and problems are paired in consecutive odd-even number combinations.*

### Section 13.1: Drawing Lewis Diagrams

*Write Lewis diagrams for each of the following sets of molecules.*

1. HI, $H_2O$, $NCl_3$
2. HF, $OF_2$, $NF_3$
3. $CO_2$, $SF_2$, $BrO_3^-$
4. ■ $NH_4^+$, $PH_3$, $NHF_2$
5. $BrO^-$, $H_2PO_4^-$, $ClO_4^-$
6. ■ $C_3H_8$, $C_3H_6$ (do not consider ring structures)
7. $CHFCl_2$, $CHF_3$, $CClI_3$
8. ■ $C_2H_6O$ (bond O to atoms of two different elements), $C_2H_6O$ (bond O to atoms of the same element)

*There are two or more acceptable diagrams for most species in Questions 9 through 18.*

9. $C_2H_2Cl_4$, $C_2H_2Cl_2F_2$, $C_3H_4Br_3I$
10. $C_2H_4Br_2$, $C_2H_4BrF$, $C_3H_5FBr_2$
11. $C_4H_8$, $C_2H_6O$, $C_3H_8O_2$
12. $C_3H_8$, $C_3H_6$, $C_3H_6O$
13. $C_6H_{14}$, $C_4H_6O$, $C_2H_2F_2$
14. $C_5H_{12}$, $C_4H_8O$, $C_4H_6$
15. Butanoic acid, $C_3H_7COOH$
16. Acetic acid, $CH_3COOH$
17. $NO_2^+$, $N_2O$, $NO^+$
18. Hydroxide ion, $OH^-$; water, $H_2O$; methanol, $CH_3OH$

### Section 13.2: Electron-Pair Repulsion: Electron-Pair Geometry

### Section 13.3: Molecular Geometry

### Section 13.4: The Geometry of Multiple Bonds

*Questions 19 through 30: For each molecule or ion, or for the atom specified in a molecule or ion, write the Lewis diagram, then describe (a) the electron-pair geometry and (b) the molecular geometry predicted by the valence shell electron-pair repulsion theory. Also sketch the wedge-and-dash diagram of each molecule or ion in Questions 19–22.*

19. $BCl_3$, $PH_3$, $H_2S$
20. ■ $BF_4^-$, $CCl_4$
21. $BrO^-$, $ClO_3^-$, $PO_4^{3-}$
22. ■ $PCl_3$, $SeOF_2$
23. Each carbon atom in $C_3H_7OH$
24. ■ Each carbon atom in $CH_3CH_2COOH$
25. Nitrogen atom in $C_2H_5NH_2$
26. ■ Each carbon atom and the oxygen atom in $CH_3CH_2OCH_2CH_3$
27. Each carbon atom in $C_2H_2$
28. ■ Each carbon atom and the nitrogen atom in $CH_3CONH_2$
29. Carbon atom in HCN
30. Carbon atom in $OCCl_2$
31. ■ The Lewis diagram of a certain compound has the element E as its central atom. The bonding and lone-pair electrons around E are shown. What is the molecular geometry around E?

```
    |
  —E—
    ¨
```

32. Draw a central atom with whatever combination of bonding and lone-pair electrons is necessary to yield an angular or bent structure around that atom.

*Questions 33 through 38: For each space-filling or ball-and-stick model shown, identify the electron-pair and molecular geometry. There are no "hidden" atoms in any of the models.*

33.

34. ■

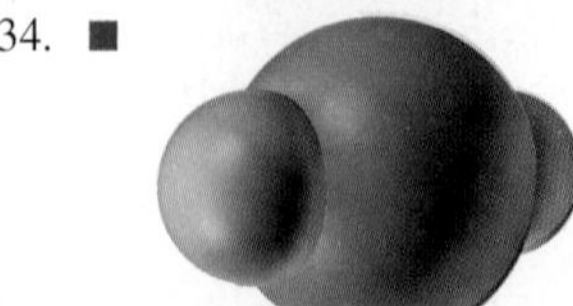

35.

36. ■

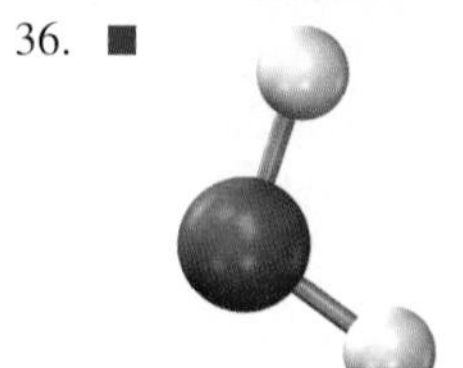

37.

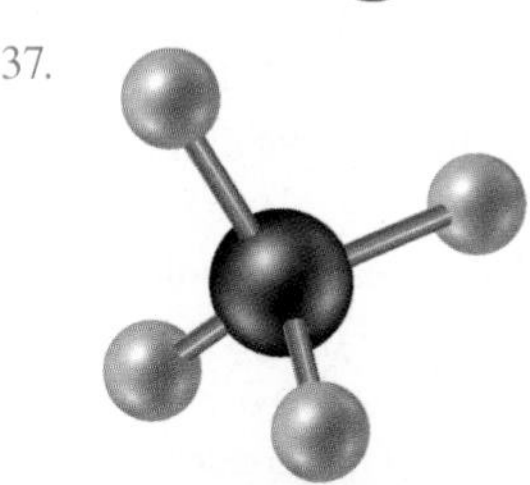

38.

*Questions 39 through 42: Estimate the value of the bond angles identified by letters (A, B, C, D) in each Lewis diagram. Explain your reasoning in each case.*

39. Aspirin, acetylsalicylic acid:

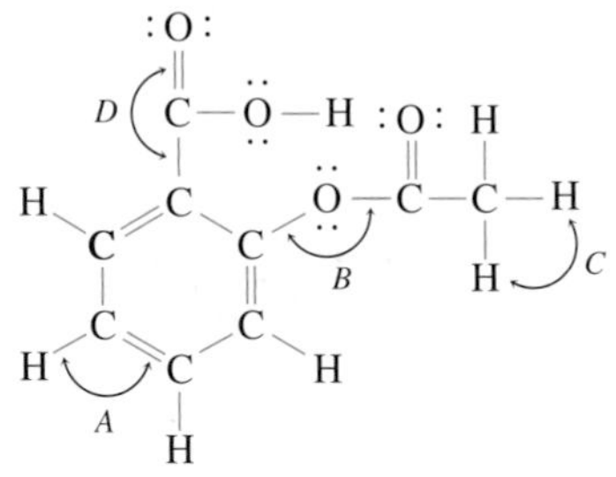

40. ■ Lactic acid:

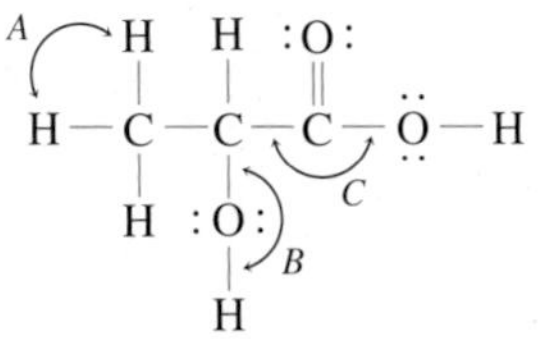

41. Cinnamaldehyde:

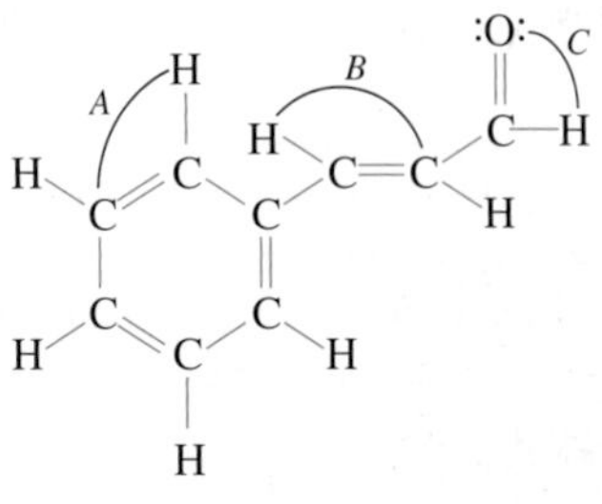

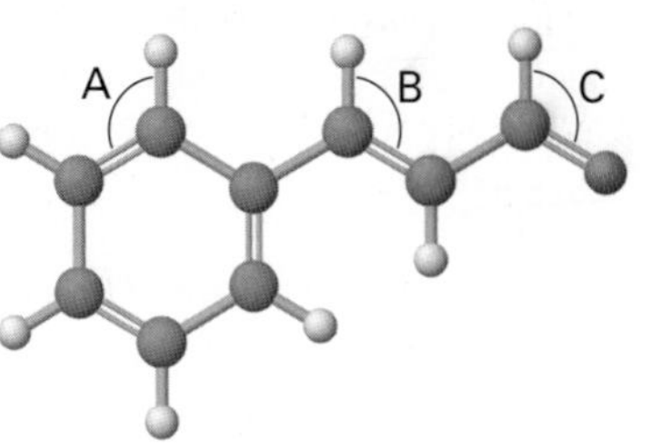


42. Amphetamine:

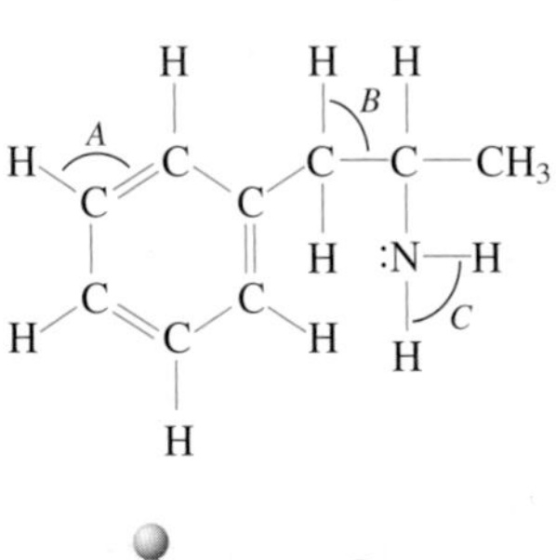

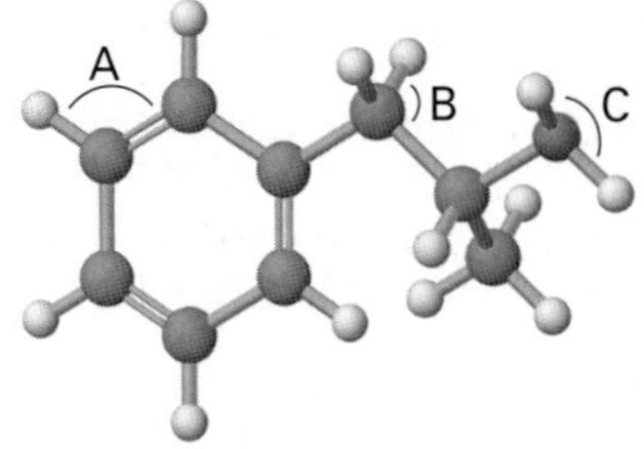


## Section 13.5: Polarity of Molecules

*Consider the following general Lewis diagrams for Questions 43 and 44.*

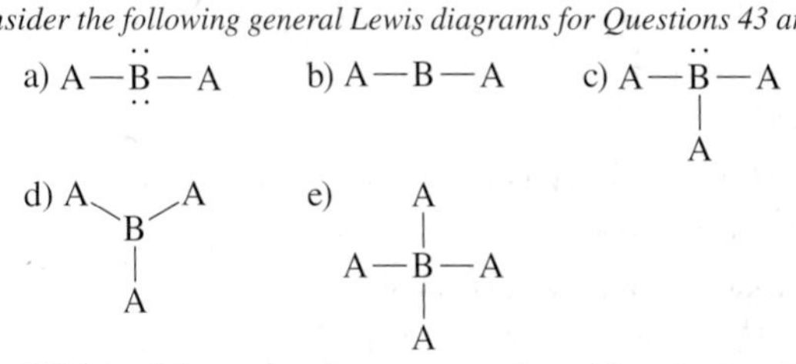

a) A—B—A (B with two lone pairs) b) A—B—A c) A—B—A with A below B (B with one lone pair)

d) A—B—A with A below B (trigonal) e) A—B—A with A above and below B

43. Which of the molecules are nonpolar with polar bonds?

44. Which of the molecules are polar?

45. Is the carbon tetrachloride molecule, $CCl_4$, which contains four polar bonds (electronegativity difference 0.5) polar or nonpolar? Explain.

46. ■ For each formula given, draw the Lewis diagram, determine the molecular geometry, and determine whether the molecule is polar or nonpolar.

a) $BF_3$ b) $NH_3$ c) $GeH_4$

47. Describe the shapes and compare the polarities of HF and HBr molecules. In each case, identify the end of the molecule that is more positive.

48. ■ For each formula given, draw the Lewis diagram, determine the molecular geometry, and determine whether the molecule is polar or nonpolar.

a) $BeF_2$ b) $H_2O$ c) $BF_3$

49. Sketch the Lewis diagram of the water molecule, paying particular attention to the bond angle and using arrows to indicate the polarity of each bond. Then sketch the methanol molecule, $HOCH_3$, again using arrows to show bond polarity. Predict the approximate shape of both molecules around the oxygen atom. Also predict relative polarities of the two molecules and explain your prediction.

50. ■ For each formula given, draw the Lewis diagram, determine the molecular geometry, and determine whether the molecule is polar or nonpolar.

a) $SiCl_4$ b) $SeH_2$ c) $CH_2O$

*Select the answers for Questions 51 through 54 from the group of Lewis diagrams that follows. Each question may have more than one answer.*

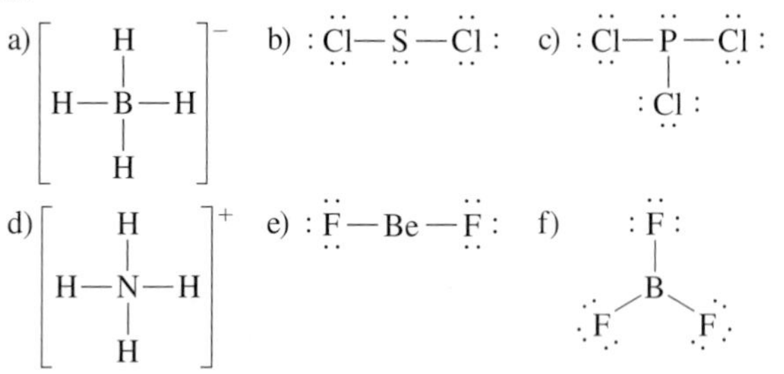

```
a) ⎡     H     ⎤⁻   b) :C̈l—S̈—C̈l:   c) :C̈l—P̈—C̈l:
   ⎢     |     ⎥                          |
   ⎢  H—B—H    ⎥                        :C̈l:
   ⎢     |     ⎥
   ⎣     H     ⎦

d) ⎡     H     ⎤⁺   e) :F̈—Be—F̈:   f)     :F̈:
   ⎢     |     ⎥                            |
   ⎢  H—N—H    ⎥                            B
   ⎢     |     ⎥                          /   \
   ⎣     H     ⎦                       :F̈:   :F̈:
```

51. Which species have tetrahedral shapes?

52. Which species are linear?

53. Which neutral molecules are polar?

54. ■ Identify all species that have trigonal planar geometries and all whose shapes are trigonal pyramidal.

**Section 13.6: The Structure of Some Organic Compounds (Optional)**

55. What distinguishes an organic compound from an inorganic compound?

56. What features of the carbon atom account for its ability to form so many chemical compounds?

57. Identify the hydrocarbons among the following: $CH_3OH$, $CH_3(CH_2)_6CH_3$, $C_6H_6$, $CH_2(NH_2)_2$, $C_8H_{18}$.

58. *Hydrocarbon* and *carbohydrate* are similar terms. Both are made by combining words or parts of words. Can you guess what those words or parts of words are in each case? Can a carbohydrate be an example of a hydrocarbon, or vice versa?

59. Distinguish between the *structure* of an organic molecule, as given by its Lewis diagram, and the three-dimensional *shape* of the molecule.

60. How are molecular structure and molecular shape shown in Lewis diagrams?

61. Why can a zigzag chain of carbon atoms be drawn as a straight line in a Lewis diagram?

62. What is a condensed structural formula? Why is it useful in drawing Lewis diagrams of organic molecules?

63. How are alcohols and ethers similar to water in structure and shape? What distinguishes between alcohols and ethers?

64. Name and describe the group that is present in all alcohols. What does the presence of this group contribute to an alcohol molecule and its properties?

65. How do carboxylic acids differ from other acids, such as hydrochloric, sulfuric, and carbonic acid?

66. What are carboxylic acids? Identify the group that is present in carboxylic acids. Write a Lewis diagram for the group and describe the geometry around all central atoms within the group.

**GENERAL QUESTIONS**

67. Distinguish precisely, and in scientific terms, the differences among items in each of the following pairs or groups.

a) Central atom, terminal atom

b) Molecular geometry, electron-pair geometry

c) Angular geometry, trigonal planar geometry

d) Trigonal planar geometry, trigonal pyramidal geometry

e) Trigonal pyramidal geometry, tetrahedral geometry

f) Polar molecule, nonpolar molecule

68. Classify each of the following statements as true or false:

a) Molecular geometry around an atom may or may not be the same as electron-pair geometry around the atom.

b) Electron-pair geometry is the direct effect of molecular geometry.

c) If the geometry of a molecule is linear, the molecule must have at least one double bond.

d) A molecule with a double bond cannot have a trigonal pyramidal geometry around the double-bonded atom.

e) A $CO_2$ molecule is linear, but an $SO_2$ molecule is bent.

f) A molecule is polar if it contains polar bonds.

g) A molecule with a central atom that has one lone pair of electrons is always polar.

h) A molecule with a central atom that has two lone pairs and two bonded pairs of electrons is always polar.

i) Carbon atoms normally form four bonds.

j) Hydrogen atoms never form double bonds.

69. One kind of $C_6H_{12}$ molecule has its carbon atoms in a ring. Draw the Lewis diagram.

70. Draw Lewis diagrams for these five acids of bromine: HBr, HBrO, $HBrO_2$, $HBrO_3$, $HBrO_4$.

71. Draw a wedge-and-dash representation of the molecule shown below, then criticize the statement, "Carbon atoms in the molecule lie in a straight line."

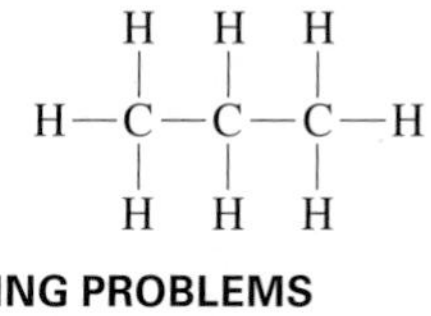

```
    H   H   H
    |   |   |
H — C — C — C — H
    |   |   |
    H   H   H
```

**MORE CHALLENGING PROBLEMS**

72. Benzene, $C_6H_6$, is a planar compound in which the carbon atoms are arranged in a hexagon (six-sided regular polygon). A Lewis diagram of this molecule has a geometry that accounts for the observed *shape* of the molecule, including bond angles. Draw this diagram. (Experimental evidence shows that the diagram does not describe the *structure* of benzene accurately.)

73. Describe the shapes of $C_2H_6$ and $C_2H_4$. In doing so, explain why one molecule is planar and the other molecule cannot be planar.

74. $NaCHO_2$ is the way an inorganic chemist is most likely to write the formula of sodium formate. Write the formula of the formate ion as the inorganic chemist is likely to write it. Now draw a Lewis diagram of the formate ion. Compare the diagram with the Lewis diagram of formic acid, $HCHO_2$. Is your formate ion diagram what you would expect from the ionization of formic acid?

75. Draw two different Lewis diagrams of $C_4H_6$.

76. $C_4H_{10}O$ is the formula of diethyl ether. The same group of atoms is attached on either side of the oxygen atom. Draw the Lewis diagram.

77. Compare Lewis diagrams for $CCl_4$, $SO_4^{2-}$, $ClO_4^-$, and $PO_4^{3-}$. Identify two things that are alike about these diagrams and the way they are drawn. From these generalizations, can you predict the Lewis diagram of $SeO_4^{2-}$ and $CI_4$?

78. What are the shapes of the following: (a) 1,2-dibromoethene, $C_2H_2Br_2$ (the bromine atoms are on different carbon atoms); (b) acetylene, $C_2H_2$; (c) *n*-butane, $CH_3CH_2CH_2CH_3$ (the carbon atoms are in a chain)?

# 14

## CONTENTS



In March 1999, the Breitling Orbiter 3 became the first balloon to make a nonstop flight around the world. The very first manned balloon flight occurred in 1783. Why was there a 216-year time lag between the first flight and the first successful circumnavigation of the globe? The answer lies in understanding the engineering and technology necessary to design a balloon capable of flying for 20 continuous days while covering a distance of 28,431 miles. In this chapter, you will learn about the relationship between the temperature and the density of gases, which, in part, governs the flight of hot air balloons.

# The Ideal Gas Law and Its Applications

In Chapter 4 you learned about several proportionalities among the volume, pressure, and temperature of a fixed amount of gas. In this chapter the amount becomes a fourth variable. The result is a single mathematical relationship, the Ideal Gas Law, that summarizes all the measurable properties of gases.

Online homework for this chapter may be assigned in OWL

## 14.1 Gases Revisited

These properties of gases were introduced in Section 4.1:

Gases may be compressed.

Gases expand to fill their containers uniformly.

All gases have low densities compared with those of liquids and solids.

Gases may be mixed in the same volume.

A gas exerts constant, uniform pressure in all directions on the walls of its container.

These properties are accounted for in the kinetic molecular theory (Section 4.2), which includes the ideal gas model. This model describes the particulate behavior of gases as follows:

Gases consist of molecular particles moving at any given instant in straight lines.

Molecules collide with each other and with the container walls without loss of energy.

Gas molecules are very widely spaced.

The actual volume of molecules is negligible compared to the space they occupy.

Gas molecules behave as independent particles; attractive forces between them are negligible.

Gas measurements (Section 4.3) and the units in which they are usually expressed include

pressure (P), expressed in torr or atmospheres (atm)

volume (V), expressed in liters (L)

temperature (T), expressed in degrees Celsius (°C) or kelvins (K)

amount (n), expressed in moles (Section 7.3)

The relationship that shows how a change in pressure, volume, or temperature of a gas sample (n is constant) affects the other two quantities is given in the equation for the Combined Gas Law (Section 4.6):

$$\frac{P_1V_1}{T_1} = k_c = \frac{P_2V_2}{T_2} \quad or \quad \frac{P_1V_1}{T_1} = \frac{P_2V_2}{T_2} \qquad \textbf{(4.19)}$$

Subscripts 1 and 2 refer to initial and final values of pressure, temperature, and volume, respectively. If Equation 4.19 is solved for $V_2$, we are able to calculate the volume of a gas at any temperature and pressure if we know its volume at any other temperature and pressure:

$$V_2 = V_1 \times \frac{P_1}{P_2} \times \frac{T_2}{T_1} \qquad \textbf{(4.20)}$$

Gas volumes in many problems are given at 0°C (273 K) and 1 bar. These conditions are known as standard temperature and pressure, or STP.

Photo by Kean Collection/Getty Images

Joseph Gay-Lussac (1778–1850). Gay-Lussac's investigations of the gaseous state of matter led to the Law of Combining Volumes.

## 14.2 Avogadro's Law

**Goal** 1 If pressure and temperature are constant, state how volume and amount of gas are related and explain phenomena or make predictions based on that relationship.

Early in the 19th century the French scientist Joseph Gay-Lussac noticed that when gases react with each other, the reacting volumes are always in the ratio of small whole numbers *if the volumes are measured at the same temperature and pressure.* This observation is known as the **Law of Combining Volumes.** It extends to gaseous products, too. An example appears in Figure 14.1.

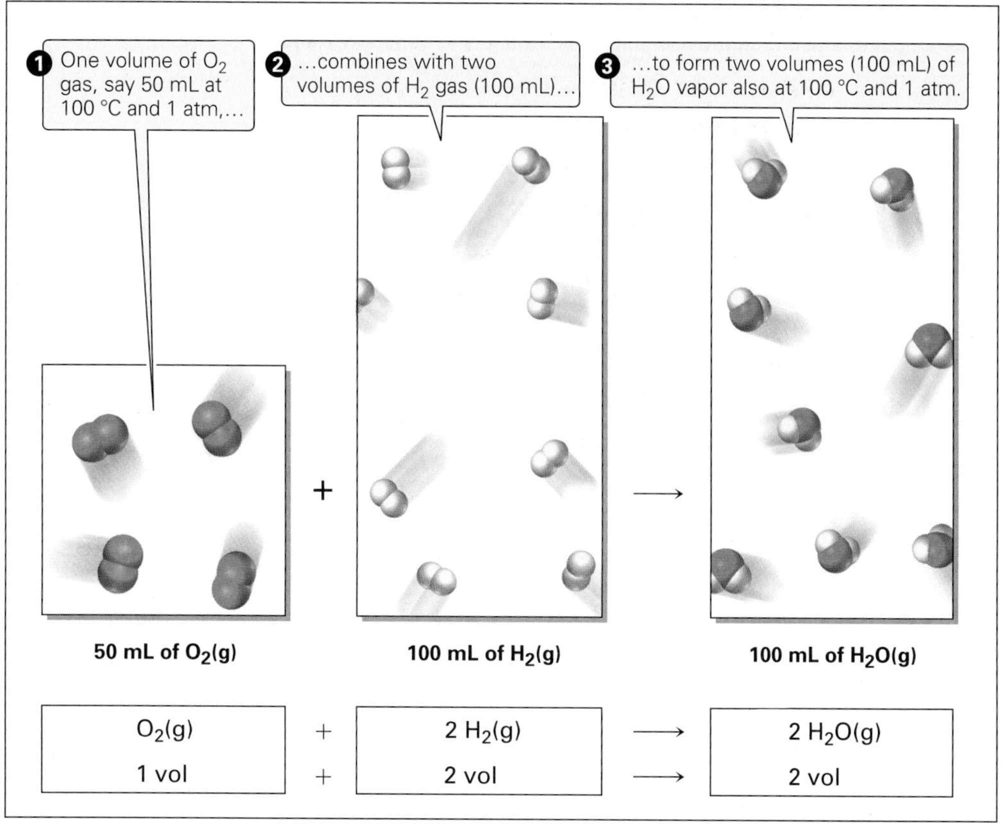


**Figure 14.1** Law of Combining Volumes. Experiments show that when gases at the same temperature and pressure react, the reacting volumes are in a ratio of small whole numbers. The small whole-number ratio extends to products, too. In the reaction, 1 volume of oxygen, $O_2$, reacts with 2 volumes of hydrogen, $H_2$, to form 2 volumes of gaseous water (steam), $H_2O$.

Shortly after Gay-Lussac's observations became known, the Italian scientist Amedeo Avogadro reasoned that they could be explained if **equal volumes of all gases *at the same temperature and pressure* contain the same number of molecules** (Fig. 14.2). If the reacting molecules react in a 1:1 ratio, and the reacting volumes also have a 1:1 ratio, then the equal volumes of the different gases must have the same number of molecules. If both ratios are 1:2, then the larger volume must have twice as many molecules as the smaller volume of the other gas. It follows that **$V \propto n$ at constant temperature and pressure**. These statements are known as **Avogadro's Law** (Fig. 14.3).

Count Amedeo Avogadro (1776–1856), depicted on a 1956 Italian postage stamp. The quote says, "Equal volumes of gas in the same conditions of temperature and of pressure contain the same number of molecules."

Avogadro's Law and Figure 14.2 give us a hint about why the formulas of hydrogen and chlorine are $H_2$ and $Cl_2$, not simply H and Cl. According to Avogadro's Law, one molecule of hydrogen divides to form two molecules of hydrogen chloride. It follows that the hydrogen molecule contains an even number of hydrogen atoms; if the starting number of hydrogen atoms divides into two equal parts, the starting number must be even. The same applies to chlorine. Other experiments prove that, for both elements, the even number is 2, not 4, 6, or some higher even number. Similar results show that nitrogen, oxygen, fluorine, bromine, and iodine form diatomic (two-atom) molecules with the formulas $N_2$, $O_2$, $F_2$, $Br_2$, and $I_2$.

## Thinking About Your Thinking

### Proportional Reasoning

**Avogadro's Law is an example of a direct proportionality. This means that the volume of a gas at constant temperature and pressure is equal to a constant times the number of gas particles in the container. Also, volume is zero when the number of particles is zero. If the number of particles is doubled, the volume is doubled; when the number of particles is tripled, the volume is tripled, etc. In dimensional-analysis**

**terms, a *PER* expression can be written that allows conversion between container volume and number of particles for a gas at constant T and P.**

## √ Target Check 14.1

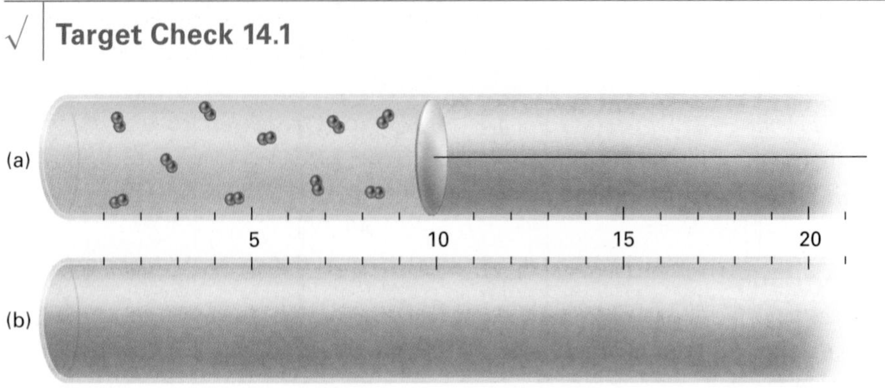


Figure 14.2 Avogadro's Law. Experimentally, it is found that, for gases at the same temperature and pressure, 1 liter of hydrogen reacts with 1 liter of chlorine to form 2 liters of hydrogen chloride. The logic behind Avogadro's reasoning appears in the subsequent particulate-level illustrations. In the first illustration, the same number of molecules in the equal-reacting volumes use all the molecules of both gases. In the second illustration, the number of molecules in the equal-reacting volumes is not the same. This means that one reactant will be completely consumed in the chemical reaction and some of the other reactant will be left over. That does *not* happen, however; it is contrary to experimental evidence. Therefore, equal volumes–equal molecules must be correct.

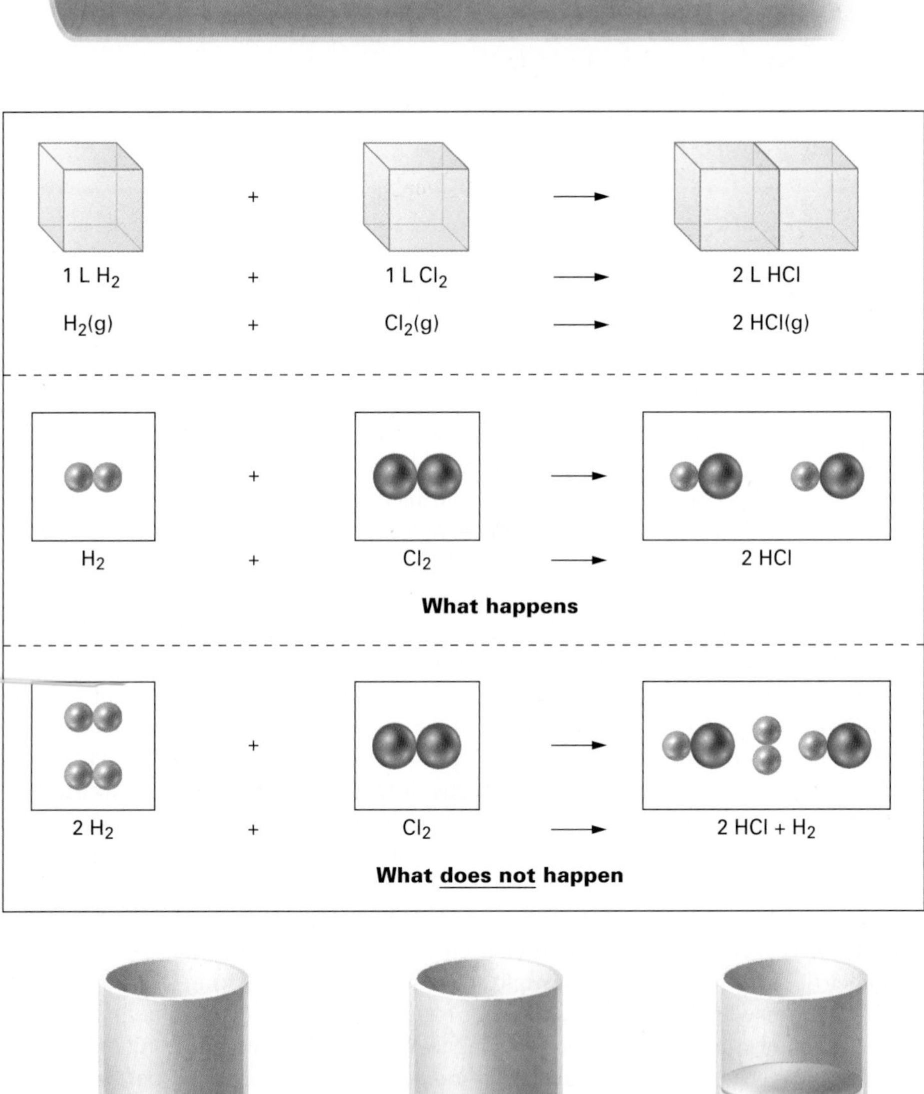


Figure 14.3 Avogadro's Law. When temperature and pressure conditions are constant, the volume of a container is directly proportional to the number of particles in the container.

*A horizontal cylinder (a) is closed at one end by a piston that moves freely left or right depending on the pressure exerted by the enclosed gas. The gas consists of 10 two-atom molecules. A reaction occurs in which 5 of the molecules separate into one-atom particles. In cylinder (b), sketch the position to which the piston would move as a result of the reaction. Pressure and temperature remain constant throughout the process. (Hint: How many total particles would be present after the reaction? Include them in your sketch.)*

## 14.3 The Ideal Gas Law

**Goal** 2 Explain how the ideal gas equation can be constructed by combining Charles's, Boyle's, and Avogadro's Laws, and explain how the ideal gas equation can be used to derive each of the three two-variable laws.

Having accumulated several different proportionalities between volume and the three other measurable properties of a gas, we now look for a single equation that ties them all together. We have seen from Charles's Law that $V \propto T$, from Boyle's Law that $V \propto \frac{1}{P}$, and from Avogadro's Law that $V \propto n$. If volume is proportional to three different quantities, it is logical to assume that it is proportional to their product, $n \times T \times \frac{1}{P}$ or

$$V \propto \frac{nT}{P} \qquad (14.1)$$

This assumption is indeed verified experimentally.

Inserting a proportionality constant, R, yields an equation:

$$V = R\frac{nT}{P} \qquad (14.2)$$

Rearranging gives the **ideal gas equation** in its most common form:

$$PV = nRT \qquad (14.3)$$

The relationship symbolized in this equation is called the **Ideal Gas Law.** If you are a science or engineering major, you will probably encounter Equation 14.3 in at least half a dozen other courses. Knowledge of the relationships among the four variables is exceptionally useful, and you should memorize the ideal gas equation.

Now that you have seen the logic in developing the Ideal Gas Law, you will find that is it convenient to derive Boyle's, Charles's, or Avogadro's Law, as necessary, by starting from the ideal gas equation. For example, if you have a fixed quantity of gas at constant temperature, n and T in the ideal gas equation are constants, making the entire nRT product a constant, which we can symbolize with k:

$$PV = nRT = k \qquad (14.4)$$

Rearranging to isolate V,

$$V = k \times \frac{1}{P} \quad \textit{or} \quad V \propto \frac{1}{P} \qquad (14.5)$$

Equation 14.5 is Boyle's Law. Figure 14.4 shows how the other laws are derived from the Ideal Gas Law through a similar approach.

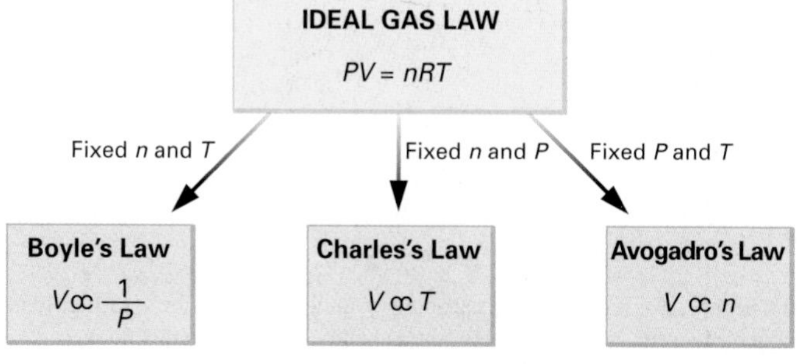


**Figure 14.4** Derivation of Boyle's, Charles's, and Avogadro's Laws from the ideal gas law.

The proportionality constant in the ideal gas equation, R, is the **universal gas constant.** Its value is the same for any gas or mixture of gases that behaves like an ideal gas.* It is significant to note that both Equation 14.3 and the value of R can be derived by applying the laws of physics to the ideal gas model. The fact that the same conclusion can be reached by theoretical calculations and by experiments makes us quite confident that the ideal gas model is correct.

The value of R may be found experimentally. . . .

## Active Example 14.1

A 0.592-mole sample of helium was found to occupy 14.7 liters when measured at 12°C and 0.942 atm. From these experimental data, calculate the value of R.

The relationship PV = nRT is the key to this problem. *PLAN* the problem. Use algebra to solve the equation for the wanted quantity, and then substitute the given information and find the answer. Watch your units.

*GIVEN:* P = 0.942 atm; V = 14.7 L; n = 0.592 mol; T = 12°C (285 K)

*WANTED:* R

*EQUATION:*
$$R = \frac{PV}{nT} = 0.942 \text{ atm} \times 14.7 \text{ L} \times \underbrace{\frac{1}{0.592 \text{ mol}} \times \frac{1}{285 \text{ K}}}_{\text{Remember that division is the same as "invert and multiply"}} = 0.0821 \text{ L} \cdot \text{atm/mol} \cdot \text{K} \quad \textbf{(14.6)}$$

The answer to Active Example 14.1 is an important constant. Your instructor may require you to memorize it.

## Active Example 14.2

At 23°C, a 656-mL sample of argon exerted a pressure of 839 torr. It was determined that the sample contained 0.0298 mole of the gas. Use these data to find the value of R in L · torr/mol · K.

Use the ideal gas equation to solve the problem. Be careful with the volume of the gas.

*Most real gases behave like an ideal gas, particularly under low-pressure and high-temperature conditions. We will not consider the differences between real and ideal gases in this book.

*GIVEN:* P = 839 torr; V = 656 mL (0.656 L); n = 0.0298 mol; T = 23°C (296 K)

*WANTED:* R in L · torr/mol · K

*EQUATION:* $$R = \frac{PV}{nT} = 839 \text{ torr} \times 0.656 \text{ L} \times \frac{1}{0.0298 \text{ mol}} \times \frac{1}{296 \text{ K}} = 62.4 \text{ L} \cdot \text{torr/mol} \cdot \text{K} \qquad \textbf{(14.7)}$$

The value given in Equation 14.7 is more convenient to use when pressure is expressed in torr.

A useful variation of the ideal gas equation replaces n, the number of moles, by the mass of a sample, m, divided by molar mass, MM. Thus:

$$n = \frac{\text{mass}}{\text{molar mass}} = \frac{m}{MM} = g \times \frac{\text{mol}}{g} = \text{mol}$$

$$PV = nRT = \frac{m}{MM}RT = \frac{mRT}{MM} \qquad \textbf{(14.8)}$$

Some instructors do not approve of Equations 14.7 and 14.8. If pressure is given in torr, they prefer that it be converted to atmospheres and then used with Equation 14.6. Others prefer that the conversion between grams and moles be done independently of the ideal gas equation rather than substituting m/MM. As usual, our recommendation is that you follow the directions of your instructor.

## 14.4 The Ideal Gas Equation: Determination of a Single Variable

**Goal** 3 Given values for all except one of the variables in the ideal gas equation, calculate the value of the remaining variable.

If you know values for all variables except one in the ideal gas equation, you can use algebra to find the value of the unknown variable. In fact, that's how you just calculated R. As with all problems to be solved algebraically, first solve the equation for the wanted quantity. Then substitute the quantities that are given, *including units,* and calculate the answer.

### Active Example 14.3

What volume will be occupied by 0.393 mole of nitrogen at 0.971 atm and 24°C?

First notice that you are given amount (n), pressure (P), and temperature (T). You are asked to find volume (V). Thus the mathematical connection between the *GIVEN* and *WANTED* is the ideal gas equation, solved for the wanted quantity. The given measurements are clearly identified, but another given, R, calls for a decision. Which value of R should you use? The choice is based on the units of pressure.

Use 0.0821 L · atm/mol · K, the value of R that includes atmospheres.

Complete the *PLAN* of the problem.

*continued*

Remember to always include units in solving problems by algebra. You will be performing many algebraic manipulations when solving problems involving the ideal gas equation. Your calculation will have many units. One term alone, R, includes four units. There is opportunity for routine errors in algebra. However, if you include units in your setups, you will catch those errors before you even pick up your calculator.

*GIVEN:* 0.393 mol nitrogen; 0.971 atm; 24°C (297 K); 0.0821 L · atm/mol · K

*WANTED:* L nitrogen *EQUATION:* $V = \frac{nRT}{P}$

Notice that Celsius degrees were changed immediately to kelvins so that the change would not be forgotten.

Now the known values may be substituted and the answer calculated.

$$V = \frac{nRT}{P} = 0.393\ \cancel{mol} \times \frac{0.0821\ L \cdot \cancel{atm}}{\cancel{mol} \cdot \cancel{K}} \times 297\ \cancel{K} \times \frac{1}{0.971\ \cancel{atm}} = 9.87\ L$$

### √ Target Check 14.2

*Calculate the pressure in atmospheres exerted by 2.18 moles of hydrogen in a 5.32-L tank at 15°C.*

One useful application of Equation 14.8, PV = (m/MM)RT, is in determining the molar mass of an unknown substance. If it is a liquid or a solid at room conditions, it may be heated beyond its boiling point to change it to a gas so that the gas laws will apply. The next Active Example illustrates the procedure.

## Active Example 14.4

A researcher vaporizes 1.67 grams of an unknown liquid at a temperature of 125°C. Its volume is 0.421 liter at 749 torr. Calculate the molar mass.

This time pressure is given in torr, so you use 62.4 L · torr/mol · K for R. Molar mass is the *WANTED* quantity, and it is the only unknown in the PV = mRT/MM form of the ideal gas equation (Equation 14.8). Complete the Active Example.

*GIVEN:* 1.67 g; 125°C (398 K); 0.421 L; 749 torr *WANTED:* Molar mass (g/mol)

*EQUATION:* $MM = \frac{mRT}{PV} = 1.67\ g \times \frac{62.4\ \cancel{L} \cdot \cancel{torr}}{mol \cdot \cancel{K}} \times 398\ \cancel{K} \times \frac{1}{749\ \cancel{torr}} \times \frac{1}{0.421\ \cancel{L}} = 132\ g/mol$

Air is a mixture, so you can't look up its molar mass. Air is also a gas; therefore, it obeys the gas laws. These laws are independent of the identity of the particles in the gas sample, whether they are all alike or are molecules of different gases. You can use the gas laws to find the "effective molar mass" of air—the mass of one mole of molecules of all the different gases in air—at a given temperature and pressure. Before doing so, take an educated guess. Air is about 98% nitrogen and oxygen. Ignore the other 2% and consider air to be 80% $N_2$ (28 g/mol) and 20% $O_2$ (32 g/mol). What do you predict for the effective molar mass of air?

### Active Example 14.5

At a time at which the temperature is 22°C and the pressure is 0.988 atm, the air in a 0.50-liter container is found to have a mass of 0.59 gram. Find the effective molar mass of air—the mass per mole of mixed molecules in the container.

This is just like Active Example 14.4. Solve it all the way.

*Given:* 22°C (295 K); 0.988 atm; 0.50 L; 0.59 g *Wanted:* molar mass (g/mol)

*Equation:* $MM = \frac{mRT}{PV} = 0.59 \text{ g} \times \frac{0.0821 \text{ L} \cdot \text{atm}}{\text{mol} \cdot \text{K}} \times 295 \text{ K} \times \frac{1}{0.988 \text{ atm}} \times \frac{1}{0.50 \text{ L}} = 29 \text{ g/mol}$

With oxygen at 32 g/mol and nitrogen at 28 g/mol, we would expect a mixture to have a value between them. Nitrogen is present in the larger amount, so we would expect the in-between value to be closer to 28 than 32. The result, 29 g/mol, is reasonable.

## 14.5 Gas Density and Molar Volume

### Gas Density

**Goal** 4 Calculate the density of a known gas at any specified temperature and pressure.

**Goal** 5 Given the density of a pure gas at specified temperature and pressure, or information from which it may be found, calculate the molar mass of that gas.

The customary unit for the density of a liquid or solid is grams per cubic centimeter, $g/cm^3$. The number of grams per cubic centimeter of a gas is so small, however, that gas densities are usually given in grams per liter, g/L. The quantities represented by these units, mass (m) and volume (V), both appear in Equation 14.8. The equation can therefore be solved for the density of a gas, m/V, in terms of pressure, temperature, molar mass, and the ideal gas constant:

$$\frac{m}{V} = D = \frac{(MM)P}{RT} \qquad \textbf{(14.9)}$$

The density of the carbon dioxide gas being expelled from this fire extinguisher is greater than the density of air. The carbon dioxide therefore flows downward, depriving the fire of the oxygen it needs to continue burning. Carbon dioxide is invisible; the white "smoke" is water vapor condensing from the air.

## Thinking About Your Thinking

### Proportional Reasoning

**Equation 14.9 relates the density of a gas to four quantities:**

$$D = \frac{(MM)P}{RT}$$

**If density is measured at constant temperature and pressure, three of those quantities are constant, T, P, and R:**

$$D = MM \times \frac{P}{RT} = MM \times \text{constant}$$

**Therefore, at constant T and P, density is directly proportional to the molar mass of the gas. The greater the density is, the greater the molar mass of the gas will be. In the laboratory, measurement of gas density (actually, measurement of mass and volume) is a method of determining molar mass.**

## Active Example 14.6

Use Equation 14.9 to calculate the density of nitrogen at 0.632 atm and 44°C.

*PLAN* the problem and calculate the answer.

*GIVEN:* 0.632 atm; 44°C (317 K)    *WANTED:* Density (assume g/L)

*EQUATION:* $D = \frac{(MM)P}{RT} = \frac{28.02 \text{ g}}{\cancel{\text{mol}}} \times 0.632 \cancel{\text{atm}} \times \frac{\cancel{\text{mol}} \cdot \cancel{\text{K}}}{0.0821 \text{ L} \cdot \cancel{\text{atm}}} \times \frac{1}{17 \cancel{\text{K}}} = 0.680 \text{ g/L}$

## Active Example 14.7

At 125°C and 749 torr, the density of an unknown gas is 3.97 g/L. Find the molar mass of that gas.

At first glance, it might seem that there is not enough information to solve this problem. Equation 14.9, m/V = (MM)P/RT, has six variables. Including the value of R, you have only four values: the ideal gas constant, temperature, pressure, and density. There are two unknowns. But are there? What is density? Mass per unit volume. Unit means 1. Density provides you with both the mass *and* the volume of the gas, 3.97 grams per *1* liter. So there is only one unknown. Solve the problem.

*GIVEN:* 125°C (398 K); 749 torr; 3.97 g/L    *WANTED:* molar mass (g/mol)

*EQUATION:* $MM = \frac{mRT}{PV} = \frac{3.97 \text{ g}}{\cancel{\text{L}}} \times \frac{62.4 \cancel{\text{L}} \cdot \cancel{\text{torr}}}{\text{mol} \cdot \cancel{\text{K}}} \times 398 \cancel{\text{K}} \times \frac{1}{749 \cancel{\text{torr}}} = 132 \text{ g/mol}$

© Cengage Learning/Charles D. Winters

(a)

Greg Gawlowski/Dembinski Photo Associates

(b)

Gas density and floating balloons. (a) A helium-filled balloon floats but air-filled balloons lie on the surface of the table. The density of helium is lower than that of air at the same temperature and pressure. (b) Hot air balloons float because the higher temperature of the air enclosed in the balloon makes it less dense than the surrounding air.

Active Example 14.7 may have looked familiar. It's the same problem as Active Example 14.4. The only difference is that in Active Example 14.7 *density* is given, and in Active Example 14.4 the numbers that make up the density are given: 1.67 g/0.421 L = 3.97 g/L. Notice Goal 4: "Calculate the density . . . *or information from which it may be found. . . .*"

© Cengage Learning/Charles D. Winters

**Figure 14.5** Molar volume at standard temperature and pressure (STP). The volume of the box in front of the student is 22.7 liters, the volume occupied by one mole of gas at 1 bar and 0°C.

## Molar Volume

Goal 6 Calculate the molar volume of any gas at any given temperature and pressure.

Goal 7 Given the molar volume of a gas at any specified temperature or pressure, or information from which the molar volume may be determined, and either the number of moles in or the volume of a sample of that gas, calculate the other quantity.

The ideal gas equation, PV = nRT, can be solved for the ratio of volume to moles, V/n:

$$MV \equiv \frac{V}{n} = \frac{RT}{P} \qquad \textbf{(14.10)}$$

Equation 14.10 is the defining equation for molar volume. Defining equations and their corresponding PER expressions are summarized in Section 3.8.

This V/n ratio is the **molar volume (MV) of a gas, the volume occupied by one mole of gas molecules.** The molecules may all be the same—a pure substance—or they may be a mixture of two or more gases. Equation 14.10 shows that molar volume depends on temperature and pressure; P and T are both variables in the equation. All gases have the same molar volume at any given temperature and pressure. Its value can be calculated for any temperature and pressure by substituting their values into the equation. For example, at the common reference conditions of standard temperature (0°C, or 273 K) and standard pressure (1 bar),

$$MV \equiv \frac{V}{n} = \frac{RT}{P} = \frac{0.0821\ \text{L}\cdot\cancel{\text{atm}}}{\text{mol}\cdot\cancel{\text{K}}} \times 273\ \cancel{\text{K}} \times \frac{1}{1\ \cancel{\text{bar}}} \times \frac{1.013\ \cancel{\text{bar}}}{\cancel{\text{atm}}} = 22.7\ \text{L/mol}$$

This value, **22.7 L/mol,** is used in many problems in which gas volumes are expressed at standard temperature and pressure, STP. *This is one to memorize.* A 22.7-L box is shown in Figure 14.5.

***Learn It Now!*** The STP molar volume of an ideal gas is 22.7 L/mol. This conversion factor provides you with a link between the macroscopic volume of a gas at 0°C and 1 bar and the particulate-level number of particles, grouped in moles.

There is not widespread agreement on standard temperature and pressure. IUPAC defines STP as 0°C and 1 bar, which leads to a molar volume of 22.7 L/mol for a gas at standard conditions. However, before 1982, IUPAC used 0°C and 1 atmosphere, which led to 22.4 L/mol as molar volume at standard conditions. As always, follow your instructor's guidelines as to the best choice for standard conditions and the resulting molar volume of a gas.

## Active Example 14.8

Calculate the molar volume of $SO_2$ at 8°C and 0.568 atm.

This is a plug-in to Equation 14.10, MV = V/n = RT/P. Complete the problem.

*GIVEN:* 8°C (281 K); 0.568 atm *WANTED:* MV (L/mol)

*EQUATION:* $MV = \frac{V}{n} = \frac{RT}{P} = \frac{0.0821\ L \cdot \cancel{atm}}{mol \cdot \cancel{K}} \times 281\ \cancel{K} \times \frac{1}{0.568\ \cancel{atm}} = 40.6\ L/mol$

Notice that the identity of the gas, $SO_2$, has no bearing on the problem. At a given temperature and pressure, all gases have the same molar volume.

Here again you come upon the macroscopic and particulate character of matter. Mass and volume are macroscopic; they can be seen and measured. Particles are too small to see or count directly, so their number is expressed in terms of moles.

Note the similarity between molar volume, L/mol, and molar mass, g/mol. You have used molar mass to convert between mass of a substance and number of moles. In the same way, you can use molar volume to convert between liters of a gas and number of moles. There is one *very* important difference, however: The molar mass of a substance is always the same, but the molar volume of a gas is *not*. Molar volume is pressure- and temperature-dependent. Be sure you use the molar volume at the given temperature and pressure when using molar volume for this conversion.

## Active Example 14.9

What volume is occupied by 4.21 moles of ethane, $C_2H_6$, (a) at standard temperature and pressure and (b) at the given temperature and pressure of Active Example 14.8, 8°C and 0.568 atm?

You can solve this problem with dimensional analysis, using the molar volume of a gas at STP, 22.7 L/mol. *PLAN* Part (a) and answer the question.

*GIVEN:* 4.21 mol *WANTED:* volume (assume L)

*PER:* $\frac{22.7\ L}{mol}$

*PATH:* mol $\longrightarrow$ L

$4.21\ \cancel{mol} \times \frac{22.7\ L}{\cancel{mol}} = 95.6\ L$

Now solve Part (b) using the molar volume you calculated in Active Example 14.8, which was 40.6 L/mol.

*GIVEN:* 4.21 mol  *WANTED:* volume (assume L)

*PER:* $\frac{40.6\text{ L}}{\text{mol}}$

*PATH:* mol $\longrightarrow$ L

$$4.21\ \cancel{\text{mol}} \times \frac{40.6\text{ L}}{\cancel{\text{mol}}} = 171\text{ L}$$

## 14.6 Gas Stoichiometry at Standard Temperature and Pressure

**Goal** 8 Given a chemical equation, or a reaction for which the equation can be written, and the mass or number of moles of one species in the reaction, or the STP volume of a gaseous species, find the mass or number of moles of another species, or the STP volume of another gaseous species.

In Chapter 10, the quantities in stoichiometry problems are expressed in mass units, usually grams. The quantities can also be measured in volume units when the reactants or products are gases. But remember, the volume of a gas depends on temperature and pressure. At standard temperature and pressure, the molar volume of all gases is 22.7 L/mol.

Being able to convert between gas volume at STP and number of moles means we can expand our stoichiometry pattern to include gas volume as a macroscopic quantity, as illustrated in Figure 14.6.

In this section, the "given temperature and pressure" are STP.

The calculation setup for mass stoichiometry (Equation 10.3) is modified here for gases at STP:

$$\cancel{\text{g Given}} \times \frac{\cancel{\text{mol Given}}}{\cancel{\text{g Given}}} \times \frac{\cancel{\text{mol Wanted}}}{\cancel{\text{mol Given}}} \times \frac{22.7\text{ L Wanted}}{\cancel{\text{mol Wanted}}} \qquad \textbf{(14.11)}$$

$$\cancel{\text{L Given}} \times \frac{\cancel{\text{mol Given}}}{22.7\ \cancel{\text{L Given}}} \times \frac{\cancel{\text{mol Wanted}}}{\cancel{\text{mol Given}}} \times \frac{\text{g Wanted}}{\cancel{\text{mol Wanted}}} \qquad \textbf{(14.12)}$$

To solve a stoichiometry problem that involves a gas whose volume is measured at STP, simply use 22.7 L/mol where you would have used molar mass if the amount had been given in grams.

### Active Example 14.10

What volume of hydrogen, measured at STP, can be released by 42.7 grams of zinc as it reacts with hydrochloric acid? The equation is $Zn(s) + 2\ HCl(aq) \rightarrow H_2(g) + ZnCl_2(aq)$.

The first two steps are just like mass-to-mass stoichiometry: Change the *GIVEN* mass to moles, and then change moles of *GIVEN* to moles of *WANTED*. Set up the problem that far, but do not solve it.

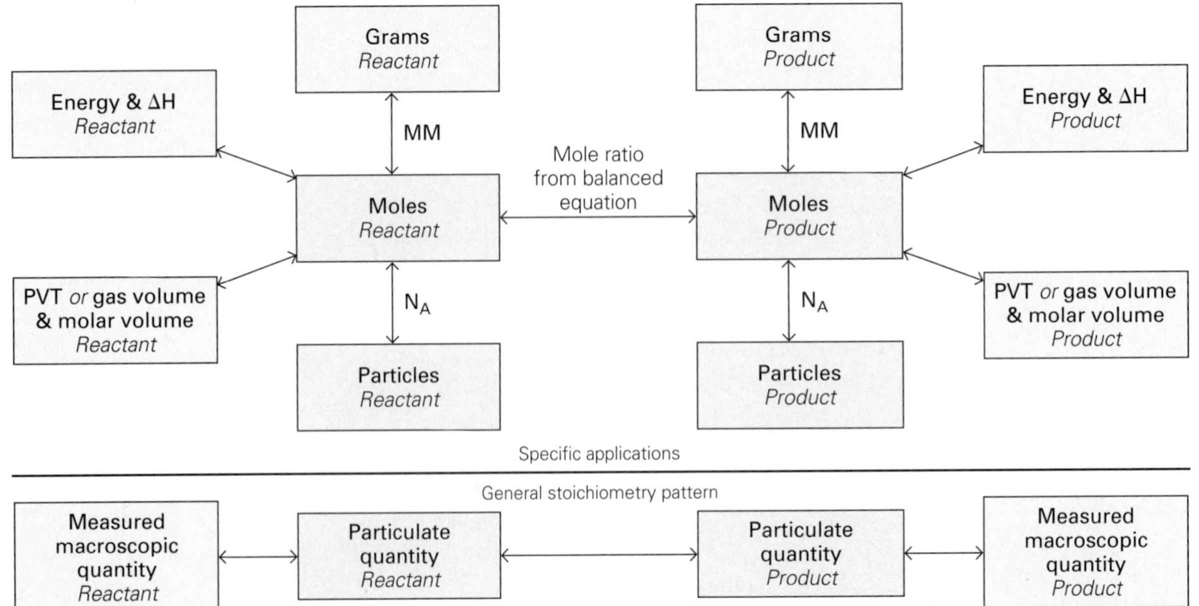


**Figure 14.6** Conversion among grams, energy and ΔH, or gas pressure, volume, and temperature, moles, and number of particles for two species in a chemical change. In Chapter 10, we introduced the general stoichiometry pattern with mass, measured in grams. We now add gas volume at a known temperature and pressure as another measured macroscopic quantity. In general, a measured macroscopic quantity is changed to the number of particles in the sample, measured in moles, and then moles of one species in the chemical change are converted to moles of another species using the mole ratio from the balanced equation. The resulting particulate quantity can be converted to a measured macroscopic quantity.

GIVEN: 42.7 g Zn  WANTED: volume $H_2$ (assume L)

PER: $\frac{1 \text{ mol Zn}}{65.38 \text{ g Zn}}$  $\frac{1 \text{ mol } H_2}{1 \text{ mol Zn}}$  $\frac{22.7 \text{ L}}{\text{mol } H_2}$

PATH: g Zn ⟶ mol Zn ⟶ mol $H_2$ ⟶ L $H_2$

$$42.7 \text{ g Zn} \times \frac{1 \text{ mol Zn}}{65.38 \text{ g Zn}} \times \frac{1 \text{ mol } H_2}{1 \text{ mol Zn}} \times \frac{\qquad}{\qquad} =$$

To this point in the problem you have moles of hydrogen. The molar volume of a gas at STP is 22.7 liters per mole. Use this conversion factor to complete the problem in the space provided above.

$$42.7 \text{ g Zn} \times \frac{1 \text{ mol Zn}}{65.38 \text{ g Zn}} \times \frac{1 \text{ mol } H_2}{1 \text{ mol Zn}} \times \frac{22.7 \text{ L } H_2}{\text{mol } H_2} = 14.8 \text{ L } H_2$$

**P/REVIEW:** In Section 10.2 you learned how to calculate the mass of one species in a chemical reaction with a known equation from a given mass of another species in three steps that we called a stoichiometry path: (1) Convert the mass of the given species to moles; (2) from the chemical equation, convert the moles of given species to moles of wanted species; (3) convert the moles of wanted species to grams. In this section you have learned how to solve the same problem if the given and/or wanted species is the volume of a gas at STP. In Section 14.7 or 14.8 you will learn how to solve the problem if the given and/or wanted species is the volume of the gas at nonstandard conditions.

**GAS STOICHIOMETRY OPTIONS** The next two sections offer alternative ways to solve gas stoichiometry problems at given temperatures and pressures.

**Option ❶**
In the **molar volume method (Section 14.7)**, the ideal gas equation is solved for the molar volume at the given temperature and pressure. The calculated molar volume is then used to solve the problem by the three steps in the stoichiometry path, just as 22.7 L/mol is used to solve a problem at STP. All problems are solved in the same way. The section describing the molar volume method is identified by a tan bar in the inside margin, as next to this paragraph.

**Option ❷**
In the **ideal gas equation method (Section 14.8)**, there are two procedures: (1) If the given quantity is a gas, the ideal gas equation is solved for n to change the given volume to moles. The problem is completed by the second and third steps in the stoichiometry path. (2) If the wanted quantity is a gas, the moles of wanted quantity are calculated by the first and second steps in the stoichiometry path. The ideal gas equation is then solved for V to convert the moles of gas to liters. The section describing the ideal gas equation method is identified by a green bar in the inside margin, as next to this paragraph.

***IMPORTANT:* Your instructor will probably assign one section and tell you to skip the other.** If you must choose between the sections, we suggest that you compare the Active Examples and end-of-book answers for both methods—the same problems are used—and choose the one that looks best for you. Whether your instructor chooses or you do, we recommend that you learn one method and disregard the other.

## 14.7 Gas Stoichiometry: Molar Volume Method (Option 1)

**Goal** 9 Given a chemical equation, or a reaction for which the equation can be written, and the mass or number of moles of one species in the reaction, or the volume of any gaseous species at a given temperature and pressure, find the mass or number of moles of any other species, or the volume of any other gaseous species at a given temperature and pressure.

We now consider gas stoichiometry problems at temperatures and pressures other than STP. At STP you know that molar volume is 22.7 L/mol, and you have used it to solve those problems. In this section you must first calculate the molar volume at the given temperature and pressure. Once you know the molar volume, you can solve all gas stoichiometry problems in exactly the same way.

### Active Example 14.11

What volume of hydrogen, measured at 739 torr and 21°C, can be released by 42.7 grams of zinc as it reacts with hydrochloric acid? The equation is $Zn(s) + 2\ HCl(aq) \rightarrow H_2(g) + ZnCl_2(aq)$.

*continued*

This is the same as Active Example 14.10, except that the temperature and pressure are not the standard values. So the first step is to find the molar volume at the conditions of the problem, 739 torr and 21°C. Complete the first step.

*Given:* 739 torr, 21°C (294 K) *Wanted:* Molar volume, L/mol

*Equation:* $MV \equiv \frac{V}{n} = \frac{RT}{P} = \frac{62.4\ L \cdot \cancel{torr}}{mol \cdot \cancel{K}} \times 294\ \cancel{K} \times \frac{1}{739\ \cancel{torr}} = 24.8\ L/mol$

Complete the problem exactly as you did in Active Example 14.10, but use your newly calculated molar volume.

*Given:* 42.7 g Zn *Wanted:* volume $H_2$ (assume L)

*Per:* $\frac{1\ mol\ Zn}{65.38\ g\ Zn}$ $\frac{1\ mol\ H_2}{1\ mol\ Zn}$ $\frac{24.8\ L\ H_2}{mol\ H_2}$

*Path:* $g\ Zn \longrightarrow mol\ Zn \longrightarrow mol\ H_2 \longrightarrow L\ H_2$

$$42.7\ \cancel{g\ Zn} \times \frac{1\ \cancel{mol\ Zn}}{65.38\ \cancel{g\ Zn}} \times \frac{1\ \cancel{mol\ H_2}}{1\ \cancel{mol\ Zn}} \times \frac{24.8\ L\ H_2}{\cancel{mol\ H_2}} = 16.2\ L\ H_2$$

Notice that the molar volume you used as a multiplier was calculated from RT/P. You could have multiplied the setup by RT/P without calculating the molar volume:

$$42.7\ \cancel{g\ Zn} \times \frac{1\ \cancel{mol\ Zn}}{65.38\ \cancel{g\ Zn}} \times \frac{1\ \cancel{mol}\ H_2}{1\ \cancel{mol\ Zn}} \times \frac{62.4\ L \cdot \cancel{torr}}{\cancel{mol} \cdot \cancel{K}} \times 294\ \cancel{K} \times \frac{1}{739\ \cancel{torr}} = 16.2\ L\ H_2$$

So if you want to multiply by molar volume, multiply by RT/P. If you want to divide by molar volume, multiply by the inverse, P/RT. Use this shortcut if you wish, and don't be concerned about getting the fraction right side up. Your units will tell you quickly if you have the wrong fraction.

**procedure**

**Solving a Gas Stoichiometry Problem**

**Step 1:** Use the ideal gas equation to find the molar volume at the given temperature and pressure: V/n = RT/P.

**Step 2:** Use the molar volume to calculate the wanted quantity by all three steps of the stoichiometry path.

## Active Example 14.12

How many liters of $CO_2$, measured at 744 torr and 131°C, will be produced by the complete burning of 16.2 grams of gaseous butane, $C_4H_{10}$? The equation is $2\ C_4H_{10}(g) + 13\ O_2(g) \rightarrow 8\ CO_2(g) + 10\ H_2O(\ell)$.

Complete the problem.

First, you calculate the molar volume at the given temperature and pressure:

*GIVEN:* 744 torr, 131°C (404 K) *WANTED:* Molar volume, L/mol

*EQUATION:* $MV \equiv \frac{V}{n} = \frac{RT}{P} = \frac{62.4\ L \cdot \cancel{torr}}{mol \cdot \cancel{K}} \times 404\ \cancel{K} \times \frac{1}{744\ \cancel{torr}} = 33.9\ L/mol$

Now you use the molar volume as a conversion factor in the final step:

*GIVEN:* 16.2 g $C_4H_{10}$ *WANTED:* L $CO_2$

*PER:* $\frac{1\ mol\ C_4H_{10}}{58.12\ g\ C_4H_{10}}$ $\frac{8\ mol\ CO_2}{2\ mol\ C_4H_{10}}$ $\frac{33.9\ L\ CO_2}{mol\ CO_2}$

*PATH:* $g\ C_4H_{10} \longrightarrow mol\ C_4H_{10} \longrightarrow mol\ CO_2 \longrightarrow L\ CO_2$

$$16.2\ \cancel{g\ C_4H_{10}} \times \frac{1\ \cancel{mol\ C_4H_{10}}}{58.12\ \cancel{g\ C_4H_{10}}} \times \frac{8\ \cancel{mol\ CO_2}}{2\ \cancel{mol\ C_4H_{10}}} \times \frac{33.9\ L\ CO_2}{\cancel{mol\ CO_2}} = 37.8\ L\ CO_2$$

Alternatively, using RT/P for molar volume, the setup is

$$16.2\ \cancel{g\ C_4H_{10}} \times \frac{1\ \cancel{mol\ C_4H_{10}}}{58.12\ \cancel{g\ C_4H_{10}}} \times \frac{8\ \cancel{mol}\ CO_2}{2\ \cancel{mol\ C_4H_{10}}} \times \frac{62.4\ L \cdot \cancel{torr}}{\cancel{mol} \cdot \cancel{K}} \times 404\ \cancel{K} \times \frac{1}{744\ \cancel{torr}} = 37.8\ L\ CO_2$$

## 14.8 Gas Stoichiometry: Ideal Gas Equation Method (Option 2)

**Goal** **9** Given a chemical equation, or a reaction for which the equation can be written, and the mass or number of moles of one species in the reaction, or the volume of any gaseous species at a given temperature and pressure, find the mass or number of moles of any other species, or the volume of any other gaseous species at a given temperature and pressure.

The stoichiometry path may be summarized as given quantity → mol given → mol wanted → wanted quantity. In a gas stoichiometry problem, the first or third step in the path is a conversion between moles and liters of gas at a given temperature and pressure. If you are given volume, you must convert to moles; if you find moles of wanted substance, you must convert to volume. These conversions are made with the ideal gas equation, PV = nRT. You have already made conversions like these. For example, in Active Example 14.3, you calculated the volume occupied by 0.393 mol $N_2$ at 24°C and 0.971 atm. You used the ideal gas equation solved for V.

The ideal gas equation method for solving gas stoichiometry problems combines use of the ideal gas equation (Sections 14.3 and 14.4) and the stoichiometry path (Section 10.2). Let's see how in the following Active Example.

## Active Example 14.13

What volume of hydrogen, measured at 739 torr and 21°C, can be released by 42.7 grams of zinc as it reacts with hydrochloric acid? The equation is $Zn(s) + 2\,HCl(aq) \rightarrow H_2(g) + ZnCl_2(aq)$.

This is the same as Active Example 14.10, except that the temperature and pressure are not the standard values. The first two steps are as before: Change the GIVEN mass to moles and then change moles of GIVEN to moles of WANTED. When temperature and pressure are given, however, you need to calculate the number of moles of hydrogen. Set up the problem and complete the calculation.

GIVEN: 42.7 g Zn WANTED: mol $H_2$

PER: $\frac{1 \text{ mol Zn}}{65.38 \text{ g Zn}}$ $\frac{1 \text{ mol H}_2}{1 \text{ mol Zn}}$

PATH: g Zn ⟶ mol Zn ⟶ mol $H_2$

$$42.7 \cancel{\text{g Zn}} \times \frac{1 \cancel{\text{mol Zn}}}{65.38 \cancel{\text{g Zn}}} \times \frac{1 \text{ mol H}_2}{1 \cancel{\text{mol Zn}}} = 0.653 \text{ mol H}_2$$

Now the problem has become, "What is the volume of 0.653 mol $H_2$, measured at 739 torr and 21°C?" This is just like Active Example 14.3. Solve the ideal gas equation for V, plug in the numbers, and calculate the answer. Be sure to use the appropriate value for R.

GIVEN: 0.653 mol $H_2$, 739 torr, 21°C (294 K) WANTED: L $H_2$

EQUATION: $V = \frac{nRT}{P} = 0.653 \cancel{\text{mol}} \times \frac{62.4 \text{ L} \cdot \cancel{\text{torr}}}{\cancel{\text{mol}} \cdot \cancel{\text{K}}} \times 294 \cancel{\text{K}} \times \frac{1}{739 \cancel{\text{torr}}} = 16.2 \text{ L H}_2$

In a gas stoichiometry problem, either the given quantity or the wanted quantity is a gas at specified temperature and pressure. The problem is usually solved in two steps, the order of which depends on whether the gas volume is the wanted quantity or the given quantity.

procedure

**Solving a Gas Stoichiometry Problem**

**Volume Given**

**Step 1:** Use the ideal gas equation to change given volume to moles: n = PV/RT.

**Step 2:** Use result in Step 1 to calculate wanted quantity using Steps 2 and 3 of the stoichiometry path.

**Volume Wanted**

**Step 1:** Calculate moles of wanted substance using Steps 1 and 2 of the stoichiometry path.

**Step 2:** Use ideal gas equation to change moles calculated above to volume: V = nRT/P.

## Active Example 14.14

How many liters of $CO_2$, measured at 744 torr and 131°C, will be produced by the complete burning of 16.2 grams of gaseous butane, $C_4H_{10}$? The equation is $2\ C_4H_{10}(g) + 13\ O_2(g) \rightarrow 8\ CO_2(g) + 10\ H_2O(\ell)$.

Follow the procedure outlined previously and calculate the answer.

*GIVEN:* 16.2 g $C_4H_{10}$, 744 torr, 131°C (404 K) *WANTED:* L $CO_2$

*PER:* $\frac{1\ \text{mol}\ C_4H_{10}}{58.12\ \text{g}\ C_4H_{10}}$ $\frac{8\ \text{mol}\ CO_2}{2\ \text{mol}\ C_4H_{10}}$

*PATH:* g $C_4H_{10}$ ⟶ mol $C_4H_{10}$ ⟶ mol $CO_2$

$$16.2\ \text{g}\ C_4H_{10} \times \frac{1\ \text{mol}\ C_4H_{10}}{58.12\ \text{g}\ C_4H_{10}} \times \frac{8\ \text{mol}\ CO_2}{2\ \text{mol}\ C_4H_{10}} = 1.11\ \text{mol}\ CO_2$$

*EQUATION:* $V = \frac{nRT}{P} = 1.11\ \text{mol} \times \frac{62.4\ \text{L} \cdot \text{torr}}{\text{mol} \cdot \text{K}} \times 404\ \text{K} \times \frac{1}{744\ \text{torr}} = 37.6\ \text{L}\ CO_2$

The two steps in the procedure can be combined so you can solve the problem in a single setup. The conversion factor between liters and moles is the molar volume at the temperature and pressure of the problem (see Section 14.5, particularly Active Example 14.9). From the ideal gas equation, molar volume is V/n, which, according to Equation 14.10, is RT/P. Thus if you need to change moles to liters, multiply by V/n in the form RT/P. To change liters to moles, divide by RT/P, or multiply by its inverse, P/RT. This can be done in the same setup as the two steps in the stoichiometry path. The single setup for Active Example 14.14 becomes

$$16.2\ \text{g}\ C_4H_{10} \times \frac{1\ \text{mol}\ C_4H_{10}}{58.12\ \text{g}\ C_4H_{10}} \times \frac{8\ \text{mol}\ CO_2}{2\ \text{mol}\ C_4H_{10}} \times \frac{RT}{P}$$

$$16.2\ \text{g}\ C_4H_{10} \times \frac{1\ \text{mol}\ C_4H_{10}}{58.12\ \text{g}\ C_4H_{10}} \times \frac{8\ \text{mol}\ CO_2}{2\ \text{mol}\ C_4H_{10}} \times \frac{62.4\ \text{L} \cdot \text{torr}}{\text{mol} \cdot \text{K}} \times 404\ \text{K} \times \frac{1}{744\ \text{torr}} = 37.6\ \text{L}\ CO_2$$

We don't encourage you to use this RT/P factor, but we don't discourage you, either. If you are comfortable with it, use it. If you do use it, be sure to include units in your setup.

## 14.9 Volume–Volume Gas Stoichiometry

Goal **10** Given a chemical equation, or a reaction for which the equation can be written, and the volume of any gaseous species at a given temperature and pressure, find the volume of any other gaseous species at a given temperature and pressure.

Avogadro's Law (Section 14.2) states that gas volume is directly proportional to the number of moles at constant temperature and pressure, $V \propto n$. This means that the ratio of volumes of gases in a reaction is the same as the ratio of moles, *provided that the gas volumes are measured at the same temperature and pressure.* The ratio of moles comes from the coefficients in the chemical equation. It follows that the coefficients give us a ratio of gas volumes, too. This is illustrated by the "volume equation" and its corresponding "molar equation" in Figure 14.1. The volumes of the individual gases are in the same ratio as the numbers of moles in the equation.

### Thinking About Your Thinking

Proportional Reasoning

**When gas volumes are measured at the same temperature and pressure, the ratio of volumes of gases in a reaction is the same as the ratio of moles as given by the coefficients in a balanced chemical equation. The gas volumes are directly proportional to one another. Any *PER* expression that relates mole ratios can also be written as a volume ratio for a gas-phase reaction when volumes are measured at the same T and P.**

This volume ratio is useful for stoichiometry problems when both the given and wanted quantities are gases measured at the same temperature and pressure. For example, consider the reaction $3\ H_2(g) + N_2(g) \rightarrow 2\ NH_3(g)$. Let's calculate the volume of ammonia that will be produced by the reaction of 5 liters of $N_2$, with both gases measured at STP. The equation coefficients, interpreted for gas volumes, tell us that 2 liters of ammonia are formed per 1 liter of nitrogen used. The one-step unit path is $L\ N_2 \rightarrow L\ NH_3$:

$$5\ \cancel{L\ N_2} \times \frac{2\ L\ NH_3}{1\ \cancel{L\ N_2}} = 10\ L\ NH_3$$

The full stoichiometry setup for the problem confirms this result:

$$5\ \cancel{L\ N_2} \times \frac{1\ \cancel{mol\ N_2}}{\cancel{22.7}\ \cancel{L\ N_2}} \times \frac{2\ \cancel{mol\ NH_3}}{1\ \cancel{mol\ N_2}} \times \frac{\cancel{22.7}\ L\ NH_3}{1\ \cancel{mol\ NH_3}} = 10\ L\ NH_3$$

Notice that the 22.7s cancel. Molar volumes will always cancel if the given and wanted gases are measured at the same temperature and pressure.

Be sure to recognize the restriction on the volume-to-volume conversion with coefficients from the equation: *Both gas volumes must be measured at the same temperature and pressure.* They don't have to be at STP, but they must be the same to make their molar volumes identical.

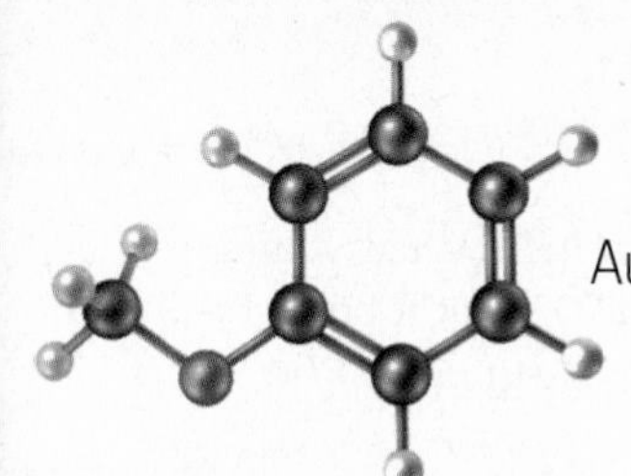

# Everyday Chemistry

## Automobile Air Bags

Since 1998 all cars sold in the United States have been required to have driver and passenger air bags. Research by the U.S. Department of Transportation shows that air bags reduce the risk of death in a frontal collision by about 30%. An air bag plus wearing lap and shoulder belts reduces the likelihood of moderate injury in front-end crashes by about 60%.

An air-bag system has three components. The bag itself is made of nylon fabric. Nylon was the first completely synthetic fiber chemists manufactured. The second component in the system is a sensor that sends a signal to inflate the bag when a front-end collision occurs with sufficient force. This is to prevent the bag from inflating because of a minor fender bender. The third component is the inflation system, which is an application of the principles of gas stoichiometry presented in this chapter. A solid mixture of sodium azide, $NaN_3$, potassium nitrate, $KNO_3$, and silicon dioxide, $SiO_2$, is in the core of the bag, surrounding an electronic igniter. When the sensor detects a collision, the igniter detonates the solids, which form gaseous nitrogen, inflating the bag. The primary source of nitrogen is the decomposition reaction of sodium azide: $2\ NaN_3 \rightarrow 2\ Na + 3\ N_2$.

Sodium is very reactive and cannot be allowed to remain in the bag after detonation. Potassium nitrate is added to the solid mixture to react with the sodium: $10\ Na + 2\ KNO_3 \rightarrow K_2O + 5\ Na_2O + N_2$. Notice that this reaction produces even more nitrogen gas.

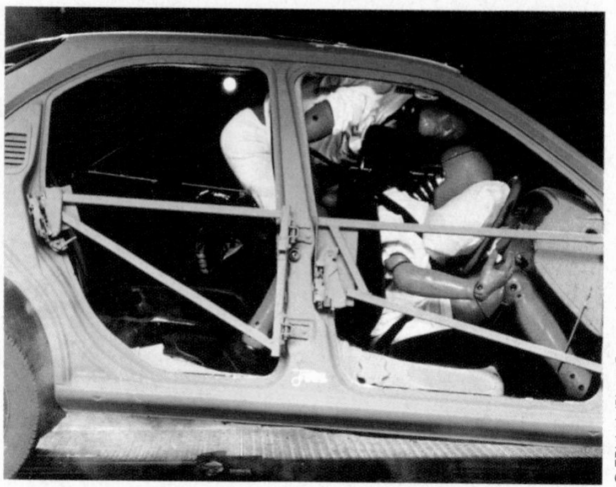

The crash test dummy in the driver's seat is being restrained by a seat belt and an inflated air bag. The dummy in the back seat, not wearing a seat belt, is being propelled through the cabin toward the front windshield.

© TRL Ltd./Photo Researchers, Inc.

The final reaction involves the potassium oxide and sodium oxide produced in the second reaction. The silicon dioxide in the original solid mixture is a type of sand, and when this sand reacts with Group 1A/1 oxides, a type of glass is produced. This glass is nonflammable and unreactive, so it is a safe final product. After the air bag is deployed, it deflates through small holes that allow the nitrogen gas to leak out, leaving behind a deflated bag containing inert substances.

Research is continuing to find the most effective method to employ side air bags to protect passengers from nonfrontal collisions. The primary challenge is the short time lag between the collision and the need for the inflated air bag. In a front-end collision, the relatively long distance between the front bumper and the passenger allows sufficient time for the bag to fully inflate before the passengers strike the dashboard or windshield. With side collisions, the distance from the car body to the passenger is much smaller.

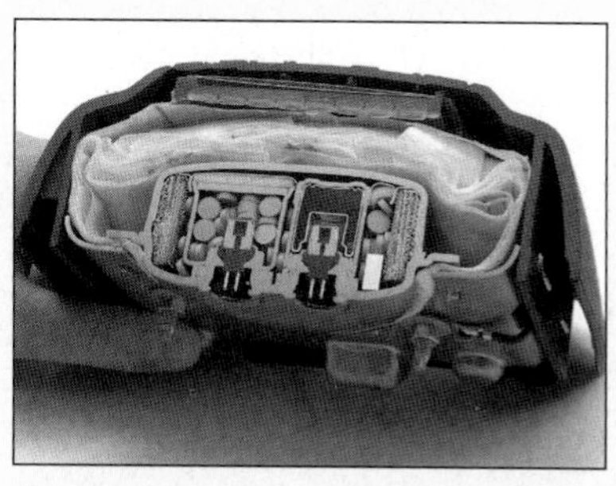

Autoliv ASP

❶ When a car decelerates in a collision, an electromechanical contact is made in the sensor unit. The gas generant (green solid) ignites, releasing nitrogen gas, and the folded nylon bag deploys by tearing through the thermoplastic cover.

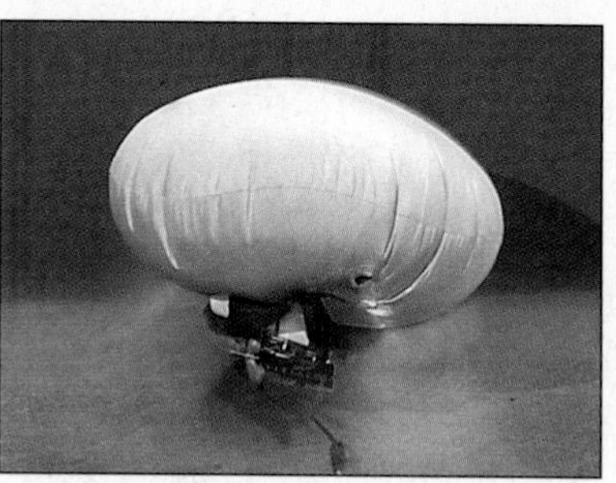

© Cengage Learning/Charles D. Winters

❷ Driver side air bags inflate with 35–70 L of $N_2$ gas, whereas passenger air bags hold about 60–160 L.

© Cengage Learning/Charles D. Winters

❸ The bag deflates within 0.2 s, the gas escaping through holes in the bottom of the bag.

## Active Example 14.15

A student burns 1.30 liters of gaseous ethylene, $C_2H_4$, completely. What volume of oxygen is required if both gas volumes are measured at STP? What if both volumes are measured at 22°C and 748 torr? The equation is $C_2H_4(g) + 3\ O_2(g) \rightarrow 2\ CO_2(g) + 2\ H_2O(g)$.

Solve the problem at STP.

*GIVEN:* 1.30 L $C_2H_4$ *WANTED:* volume $O_2$ (assume L)

*PER:* $\dfrac{3\ L\ O_2}{1\ L\ C_2H_4}$

*PATH:* L $C_2H_4 \longrightarrow$ L $O_2$

$$1.30\ L\ \cancel{C_2H_4} \times \frac{3\ L\ O_2}{1\ L\ \cancel{C_2H_4}} = 3.90\ L\ O_2$$

That completes the problem for STP. Now solve it when both gas volumes are measured at 22°C and 748 torr.

The answer is 3.90 L $O_2$, calculated by the same setup.

The volume ratio may be used whenever both gases are measured at the *same temperature and pressure.* It does not have to be STP.

More often than not, the given and wanted gas volumes are at different temperatures and pressures. The convenient L (given) → L (wanted) cannot be used—yet. First, you use the combined gas equation solved for $V_2$ (see Section 14.1, Equation 4.20) to change the volume of the given gas from its initial temperature and pressure to what that volume would be at the temperature and pressure of the wanted gas. Then both gases are at the same temperature and pressure, and you can use the L (given) → L (wanted) shortcut.

## Active Example 14.16

It is found that 1.75 L $O_2$, measured at 24°C and 755 torr, is used in burning sulfur. At one point in the exhaust hood the sulfur dioxide produced is at 165°C and 785 torr. Find the volume of $SO_2$ at those conditions. The equation is $S(s) + O_2(g) \rightarrow SO_2(g)$.

Begin by filling in a table of initial and final values.

| | Volume | Temperature | Pressure |
|---|---|---|---|
| Initial Value (1) | | | |
| Final Value (2) | | | |

| | Volume | Temperature | Pressure |
|---|---|---|---|
| Initial Value (1) | 1.75 L | 24°C; 297 K | 755 torr |
| Final Value (2) | $V_2$ | 165°C; 438 K | 785 torr |

Oxygen is the gas whose volume you know. You change the known oxygen volume to what it would be at the temperature and pressure at which the wanted gas volume is to be expressed.

Use the combined gas equation to find $V_2$.

$$V_2 = V_1 \times \frac{P_1}{P_2} \times \frac{T_2}{T_1} = 1.75\ \text{L} \times \frac{755\ \cancel{\text{torr}}}{785\ \cancel{\text{torr}}} \times \frac{438\ \cancel{\text{K}}}{297\ \cancel{\text{K}}} = 2.48\ \text{L}\ O_2$$

The problem now becomes, "What volume of $SO_2$ is produced by the reaction of 2.48 L $O_2$, with both gases at the same temperature and pressure?" This makes the volume-conversion factor the same as the mole ratio from the equation. PLAN and complete the problem.

GIVEN: 2.48 L $O_2$ WANTED: volume $SO_2$ (assume L)

PER: $\frac{1\ \text{L}\ SO_2}{1\ \text{L}\ O_2}$

PATH: L $O_2$ $\longrightarrow$ L $SO_2$

$$2.48\ \cancel{\text{L}\ O_2} \times \frac{1\ \text{L}\ SO_2}{1\ \cancel{\text{L}\ O_2}} = 2.48\ \text{L}\ SO_2$$

Notice that you did not have to calculate the numerical value of the volume of oxygen at the conditions at which the $SO_2$ volume is to be measured. The calculation setup is sufficient because it is equal to the calculated volume. The conversion from given substance to wanted can be tacked on as a final step:

$$1.75\ \cancel{\text{L}\ O_2} \times \frac{755\ \cancel{\text{torr}}}{785\ \cancel{\text{torr}}} \times \frac{438\ \cancel{\text{K}}}{297\ \cancel{\text{K}}} \times \frac{1\ \text{L}\ SO_2}{1\ \cancel{\text{L}\ O_2}} = 2.48\ \text{L}\ SO_2$$

# Chapter 14 in Review

Most of the key terms and concepts and many others appear in the Glossary. Use your Glossary regularly.

### 14.1 Gases Revisited

**Key Terms and Concepts: Combined Gas Law, standard temperature and pressure (STP)**

### 14.2 Avogadro's Law

Goal 1 If pressure and temperature are constant, state how volume and amount of gas are related and explain phenomena or make predictions based on that relationship.

**Key Terms and Concepts: Avogadro's Law, Law of Combining Volumes**

### 14.3 The Ideal Gas Law

Goal 2 Explain how the ideal gas equation can be constructed by combining Charles's, Boyle's, and Avogadro's Laws, and explain how the ideal gas equation can be used to derive each of the three two-variable laws.

**Key Terms and Concepts: Ideal gas equation, PV = nRT, Ideal Gas Law, universal gas constant, R**

### 14.4 The Ideal Gas Equation: Determination of a Single Variable

Goal 3 Given values for all except one of the variables in the ideal gas equation, calculate the value of the remaining variable.

### 14.5 Gas Density and Molar Volume

Goal 4 Calculate the density of a known gas at any specified temperature and pressure.

Goal 5 Given the density of a pure gas at specified temperature and pressure, or information from which it may be found, calculate the molar mass of that gas.

Goal 6 Calculate the molar volume of any gas at any given temperature and pressure.

Goal 7 Given the molar volume of a gas at any specified temperature or pressure, or information from which the molar volume may be determined, and either the number of moles in or the volume of a sample of that gas, calculate the other quantity.

**Key Terms and Concepts: 22.7 L/mol, gas density, molar volume (MV)**

### 14.6 Gas Stoichiometry at Standard Temperature and Pressure

Goal 8 Given a chemical equation, or a reaction for which the equation can be written, and the mass or number of moles of one species in the reaction, or the STP volume of a gaseous species, find the mass or number of moles of another species, or the STP volume of another gaseous species.

### 14.7 Gas Stoichiometry: Molar Volume Method (Option 1)

### 14.8 Gas Stoichiometry: Ideal Gas Equation Method (Option 2)

Goal 9 Given a chemical equation, or a reaction for which the equation can be written, and the mass or number of moles of one species in the reaction, or the volume of any gaseous species at a given temperature and pressure, find the mass or number of moles of any other species, or the volume of any other gaseous species at a given temperature and pressure.

**Key Terms and Concepts: Gas stoichiometry**

### 14.9 Volume–Volume Gas Stoichiometry

Goal 10 Given a chemical equation, or a reaction for which the equation can be written, and the volume of any gaseous species at a given temperature and pressure, find the volume of any other gaseous species at a given temperature and pressure.

**Key Terms and Concepts: Volume–volume gas stoichiometry**

# Study Hints and Pitfalls to Avoid

Your ability to solve the gas problems in this chapter depends largely on your algebra skills. Most students find it easiest to determine what is wanted and then solve the ideal gas equation for that variable. If the wanted quantity is density or molar volume, solve the equation for the combination of variables that represents the desired property. Then substitute the known variables, *including units,* and calculate the answer. Units are important: If they don't come out right, you know there is an error in the algebra.

Gas stoichiometry problems are usually solved in two steps. However, using RT/P for molar volume makes it possible to solve the problem in a single setup. If you're comfortable with this procedure, use it. If two separate steps are easier for you, solve the problem that way. Either way, be careful when you must divide by RT/P. Division appears in the setup as multiplication by P/RT. Again, the consistent use of units will help you to avoid a mistake.

If you have the impression that we regard the use of units in calculation setups as very important—well, you're right!

# Concept-Linking Exercises

*Write a brief description of the relationships among each of the following groups of terms or phrases. Answers to the Concept-Linking Exercises are given after answers to the Target Checks in Appendix III.*

1. Law of Combining Volumes, Avogadro's Law, chemical equation
2. Ideal Gas Law, ideal gas equation, universal gas constant
3. Moles, atmospheres, torr, liters, kelvins
4. 22.7, 62.4, 0.0821

## Small-Group Discussion Questions

*Small-Group Discussion Questions are for group work, either in class or under the guidance of a leader during a discussion section.*

1. Explain how kinetic molecular theory (Section 4.2) explains each of the five macroscopic characteristics of gases listed in Section 4.1 and summarized at the beginning of Section 14.1.
2. Define the Law of Combining Volumes and Avogadro's Law. Explain how the Law of Combining Volumes is explained by Avogadro's Law. In light of the Law of Conservation of Mass, how can 50 mL of oxygen combine with 100 mL of hydrogen to produce 100 mL of steam?
3. Derive the Ideal Gas Law from Boyle's Law, Charles's Law, and Avogadro's Law. Then show how the Ideal Gas Law yields each of the three two-variable gas laws when the other two variables are held constant.
4. What is the Fahrenheit temperature of 1.0-$ft^3$ sample of gas that has a pressure of 15 psi if the container holds 0.1 mole of the gas?
5. The appearance of the atmosphere of Jupiter has long fascinated astronomers and many hobbyists because of its Giant Red Spot and colorful cloud layer. The atmospheric pressure builds up from its outer boundary toward the surface. At 1 bar pressure, the temperature is −108°C, and the density of the atmospheric gas mixture is 0.16 $kg/m^3$. What is the average molar mass of the mixture? The atmosphere is mostly composed of hydrogen and helium. Use the average molar mass to find the percentage of each of the two major components of the Jovian atmosphere.
6. Compare and contrast molar volume and molar mass. How are they similar? How are they different?
7. Two premises of the ideal gas model are usually not precisely correct for real gases. One is the independence of the particles. In a real gas, the particles are attracted to one another, sticking together for brief periods of time. How do you suppose that affects gas pressure? Another characteristic of real gases is that they have particle volume. How does the container volume available to a real gas compare with the volume available to an ideal gas? What assumption about container volume do you make when you apply the ideal gas model?
8. Acetylene, the fuel for welder's torches, can be made by the reaction of calcium carbide and water at high temperature: $CaC_2(s) + 2\ H_2O(\ell) \rightarrow C_2H_2(g) + Ca(OH)_2(s)$. Determine the percent yield of acetylene for a laboratory-scale reaction of 5.30 grams of calcium carbide with excess water, producing 1595 mL of acetylene measured at the temperature and pressure of the lab, 25°C and 733 mm Hg.
9. Chlorine and sodium chlorite react to form chlorine dioxide and sodium chloride: $Cl_2(g) + 2\ NaClO_2(s) \rightarrow 2\ ClO_2(g) + 2\ NaCl(s)$. If 0.73 liter of chlorine at 22°C and $1.0 \times 10^3$ torr reacts with 7.8 grams of sodium chlorite, how many grams of chlorine dioxide will be produced? What volume or mass of reactant will remain unreacted?

## Questions, Exercises, and Problems

*Blue-numbered questions are answered in Appendix III. ■ denotes problems assignable in OWL. In the first section of Questions, Exercises, and Problems, similar exercises and problems are paired in consecutive odd-even number combinations.*

**Section 14.2: Avogadro's Law**

1. Compare the volumes of $1 \times 10^{23}$ hydrogen molecules, $1 \times 10^{23}$ oxygen molecules, and $2 \times 10^{23}$ nitrogen molecules, all at the same temperature and pressure.
2. ■ Which of the following gas samples would have the largest volume, if all samples are at the same temperature and pressure? (a) 263 grams of Xe; (b) $5 \times 10^{23}$ molecules of $H_2$; (c) 4.00 moles of $CO_2$; (d) they would all have the same volume.

**Section 14.4: The Ideal Gas Equation: Determination of a Single Variable**

3. Find the pressure in torr produced by 0.0888 mol of carbon dioxide in a 5.00-liter vessel at 36°C.
4. ■ A 0.917-mol sample of hydrogen gas at a temperature of 25.0°C is found to occupy a volume of 21.7 liters. What is the pressure of this gas sample in torr?
5. The pressure exerted by 6.04 mol of nitrogen monoxide at a temperature of 18°C is 17.2 atm. What is the volume of the gas in liters?
6. ■ A 0.512-mol sample of argon gas is collected at a pressure of 872 torr and a temperature of 18°C . What is the volume (L) of the sample?
7. A 784-mL hydrogen lecture bottle is left with the valve slightly open. Assuming no air has mixed with the hydrogen, how many moles of hydrogen are left in the bottle after the pressure has become equal to an atmospheric pressure of 752 torr at a temperature of 22°C?
8. ■ A sample of xenon gas collected at a pressure of 1.18 atm and a temperature of 18°C is found to occupy a volume of 26.7 liters. How many moles of Xe gas are in the sample?
9. At what temperature (°C) will 0.810 mol of chlorine in a 15.7-L vessel exert a pressure of 756 torr?
10. ■ A 0.142-mol sample of argon gas has a volume of 834 milliliters at a pressure of 4.83 atm. What is the temperature of the Ar gas sample on the Celsius scale?
11. How many moles of carbon monoxide must be placed into a 40.0-L tank to develop a pressure of 965 torr at 18°C?
12. ■ A sample of neon gas collected at a pressure of 0.946 atm and a temperature of 276 K is found to occupy a volume of 712 milliliters. How many moles of Ne gas are in the sample?
13. Find the volume of 0.621 mol of helium at −32°C and 0.771 atm.

14. ■ A sample of neon gas collected at a pressure of 531 mm Hg and a temperature of 291 K has a mass of 10.2 grams. What is the volume (L) of the sample?

### Section 14.5: Gas Density and Molar Volume

15. The STP density of an unknown gas is found to be 2.32 g/L. What is the molar mass of the gas?

16. ■ What is the density of a sample of oxygen gas at a pressure of 1.40 atm and a temperature of 49°C?

17. The "effective" molar mass of air is 29 g/mol. Use this value to calculate the density of air (a) at STP and (b) at 20°C and 751 torr.

18. ■ A 2.94-gram sample of an unknown gas is found to occupy a volume of 2.06 L at a pressure of 1.16 atm and a temperature of 46°C. What is the molar mass of the unknown gas?

19. If the density of an unknown gas at 41°C and 2.61 atm is 1.61 g/L, what is its molar mass?

20. ■ A sample of an unknown gas is found to have a density of 2.00 g/L at a pressure of 0.939 atm and a temperature of 40.0°C. What is the molar mass of the unknown gas?

21. The gas in an 8.07-liter cylinder at 13°C has a mass of 33.5 grams and exerts a pressure of 3.25 atm. Find the molar mass of the gas.

22. ■ A 0.201-mol sample of an unknown gas contained in a 5.00 L flask is found to have a density of 1.72 g/L. What is the molar mass of the unknown gas?

23. $NO_2$ and $N_2O_4$ both have the same empirical (simplest) formula. At a temperature and pressure at which both substances are gases, can you tell without calculating which gas is more dense? Explain.

24. ■ A 0.284-mol sample of Xe gas is contained in a 7.00-L flask at room temperature and pressure. What is the density of the gas, in grams/liter, under these conditions?

25. Compare the molar volumes of helium and neon at 30°C and 1.10 torr. What are their values in L/mol?

26. ■ What is the molar volume of carbon dioxide gas at a pressure of 0.974 atm and a temperature of 43°C?

27. Find the molar volume of acetylene, $C_2H_2$, at 21°C and 0.908 atm.

28. ■ The molar volume for oxygen gas at a pressure of 0.684 atm and a temperature of 31°C is 36.5 L/mol. What is the volume occupied by 1.31 moles of $O_2$ gas at the same temperature and pressure?

### Section 14.6: Gas Stoichiometry at Standard Temperature and Pressure

29. One small-scale laboratory method for preparing oxygen is to heat potassium chlorate in the presence of a catalyst: $2\ KClO_3(s) \rightarrow 2\ KCl(s) + 3\ O_2(g)$. Find the STP volume of oxygen that can be produced by 5.74 g $KClO_3$.

30. ■ What volume of carbon dioxide is produced at 0°C and 1 atm when 64.2 g of calcium carbonate reacts completely according to the reaction $CaCO_3(s) \rightarrow CaO(s) + CO_2(g)$?

31. The reaction used to produce chlorine in the laboratory is $2\ KMnO_4(aq) + 16\ HCl(aq) \rightarrow 2\ MnCl_2(aq) + 5\ Cl_2(aq) + 8\ H_2O(aq) + 2\ KCl(aq)$. Calculate the number of grams of potassium permanganate, $KMnO_4$, that are needed to produce 9.81 L of chlorine, measured at STP.

32. ■ How many grams of carbon (graphite) are required to react completely with 42.5 L of oxygen gas at 0°C and 1 atm according to the reaction $C(s) + O_2(g) \rightarrow CO_2(g)$?

### Section 14.7: Gas Stoichiometry: Molar Volume Method (Option 1)

### Section 14.8: Gas Stoichiometry: Ideal Gas Equation Method (Option 2)

*Questions 33–40 may be solved by the molar volume method (Section 14.7) or by the ideal gas equation method (Section 14.8). In the answer section, the setups are given first for the molar volume method and then for the ideal gas equation method. Check your work according to the section you studied.*

33. One source of sulfur dioxide used in making sulfuric acid comes from sulfide ores by the reaction $4\ FeS_2(s) + 11\ O_2(g) \rightarrow 2\ Fe_2O_3(s) + 8\ SO_2(g)$. How many liters of $SO_2$, measured at 983 torr and 214°C, are produced by the reaction of 598 g $FeS_2$?

34. ■ What volume of chlorine gas at 42.0°C and 1.33 atm is required to react completely with 41.5 g of phosphorus according to the reaction $P_4(s) + 6\ Cl_2(g) \rightarrow 4\ PCl_3(\ell)$?

35. How many grams of water must decompose by electrolysis to produce 23.9 L $H_2$, measured at 28°C and 728 torr?

36. ■ What volume of hydrogen gas at 29°C and 0.809 atm is produced when 31.8 g of iron reacts completely according to the reaction $Fe(s) + 2\ HCl(aq) \rightarrow FeCl_2(aq) + H_2(g)$?

37. The reaction chamber in a modified Haber process for making ammonia by the direct combination of its elements is operated at 575°C and 248 atm. How many liters of nitrogen, measured at these conditions, will react to produce $9.16 \times 10^3$ grams of ammonia?

38. ■ How many grams of iron are required to react completely with 5.49 L of oxygen gas at 41°C and 1.41 atm according to the reaction $2\ Fe(s) + O_2(g) \rightarrow 2\ FeO(s)$?

39. When properly detonated, ammonium nitrate explodes violently, releasing hot gases: $NH_4NO_3(s) \rightarrow N_2O(g) + 2\ H_2O(g)$. If the total volume of gas produced, both dinitrogen oxide and steam, is 82.3 L at 447°C and 896 torr, how many grams of $NH_4NO_3$ exploded?

40. ■ How many grams of sodium are needed to produce 25.3 L of hydrogen gas at 43°C and 0.886 atm according to the reaction $2\ Na(s) + 2\ H_2O(\ell) \rightarrow 2\ NaOH(aq) + H_2(g)$?

### Section 14.9: Volume–Volume Gas Stoichiometry

41. Sulfur burns to $SO_2$ with a beautiful deep blue–purple flame, but with a foul, suffocating odor: $2\ S + O_2 \rightarrow 2\ SO_2$. (a) How many liters of $O_2$ are needed to form 35.2 L $SO_2$, both gases being measured at 741 torr and 26°C? (b) What if only the $SO_2$ is at those conditions, but the $O_2$ is at 17°C and 847 torr?

42. ■ Consider the following gas phase reaction: $4\ NH_3(g) + 5\ O_2(g) \rightarrow 4\ NO(g) + 6\ H_2O(g)$. If 6.98 liters of $O_2(g)$ gas at 129°C and 0.836 atm are reacted, what volume of $H_2O(g)$ will be produced if it is collected at 74°C and 0.630 atm?

43. Gaseous chlorine dioxide, $ClO_2$, is used to bleach flour and in water treatment. It is produced by the reaction of chlorine with sodium chlorite: $Cl_2 + 2\ NaClO_2 \rightarrow 2\ ClO_2 + 2\ NaCl$. How many liters of $ClO_2$, measured at 0.961 atm and 31°C, will be produced by 283 L $Cl_2$ at 2.92 atm and 21°C?

44. ■ Consider the following gas phase reaction: $N_2(g) + 2\ O_2(g) \rightarrow 2\ NO_2(g)$. If 8.44 liters of $N_2(g)$ gas at 127°C and 0.638 atm is used, what volume of $NO_2(g)$ gas will be formed if it is collected at 187°C and 0.982 atm?

45. In the natural oxidation of hydrogen sulfide released by decaying organic matter, the following reaction occurs: $2\ H_2S + 3\ O_2 \rightarrow 2\ SO_2 + 2\ H_2O$. How many milliliters of hydrogen sulfide, measured at 19°C and 549 torr, will be used in a reaction that also uses 704 mL $O_2$ at 159 torr and 26°C?

46. ■ Consider the following gas phase reaction: $H_2(g) + Cl_2(g) \rightarrow 2\ HCl(g)$. If 8.36 liters of $Cl_2(g)$ gas at 113°C and 0.693 atm is used, what volume of HCl(g) gas will be formed if it is collected at 64°C and 0.355 atm?

## GENERAL QUESTIONS

47. Distinguish precisely, and in scientific terms, the differences among items in each of the following pairs or groups.
    a) Avogadro's Law, Law of Combining Volumes
    b) Gas density, molar volume, molar mass
    c) Ideal gas equation, combined gas equation

48. Determine whether each of the following statements is true or false:
    a) The molar volume of a gas at 1.06 atm and 212°C is less than 22.7 L/mol.
    b) The mass of 5.00 L of $NH_3$ is the same as the mass of 5.00 L of CO if both volumes are measured at the same temperature and pressure.
    c) At a given temperature and pressure, the densities of two gases are proportional to their molar masses.
    d) To change liters of a gas to moles, multiply by RT/P.

49. ■ Calculate the volume of 6.74 g $C_2H_4(g)$ at 41°C and 733 torr.

50. Find the mass of 57.9 liters of krypton at 775 torr and 6°C.

51. What is the density of $H_2S$ at 0.972 atm and 14°C?

52. ■ At 17°C and 0.835 atm, 16.2 liters of ammonia has a mass of 9.68 g. What is the molar volume of ammonia at those conditions? (This is easier than it may seem!)

53. ■ A 7.60-g sample of pure liquid is vaporized at 183°C and 179 torr. At these conditions it occupies 3.87 L. What is the molar mass of the substance?

## MORE CHALLENGING PROBLEMS

54. The density of nitrogen at 0.913 atm and 18°C is 1.07 g/L. Explain how this shows that the formula of nitrogen is $N_2$ rather than just N.

*The answer to Question 56 follows from the answer to Question 55. Answer Question 55 first.*

55. At a given temperature and pressure, what mathematical relationship exists between the density and molar mass of a gas? Explain your answer.

56. ■ Labels have become detached from cylinders of two gases, one of which is known to be propane, $C_3H_8$, and the other butane, $C_4H_{10}$. The densities of the two gases are compared at the same temperature and pressure. The density of gas A is 1.37 g/L, and the density of gas B is higher. Which gas is A and which is B? What is the density of B?

57. ■ An organic chemist has produced a solid she believes to be pure; she expects a molar mass of 346 g/mol. Using 4.08 grams of the solid, she melts and then boils it in a 3.36-L vacuum chamber at 117 torr and 243°C. She is disappointed; the molar mass is close to what she expected, but not close enough. (a) What molar mass did she find? (b) Her finding suggested to her what her mistake might have been. Does it suggest anything to you? [*Hint:* Part (b) is beyond the scope of Chapter 14, but you might have an inspiration if you have studied Chapter 13.]

© Hans Pfletschinger/Peter Arnold, Inc.

Liquid drops adopt a spherical shape because of a macroscopic measurable property known as surface tension. This property can be understood at the particulate level in terms of the strengths of the attractive forces among the particles that make up the liquid. Strong attractive forces lead to high surface tension.

# 15

# Gases, Liquids, and Solids

## CONTENTS



In Chapter 12 you studied chemical bonds, the attractive forces between atoms or ions that hold them together to form compounds. In this chapter we focus on covalently bonded molecules and shift our attention to the attractive forces *between* the molecules. The strength of these attractive forces is largely responsible for the physical properties of compounds, including melting and boiling points. Thus attractive forces between molecules determine whether a substance exists as a solid, liquid, or gas at a given temperature and pressure.

You first considered the particulate character of matter in the kinetic molecular theory in Section 2.2. The particles of a solid were described as being held in fixed position relative to each other. A degree of freedom is reached in the liquid state, in which particles move about among themselves, but still remain together at the bottom of the container that holds them. As a gas, the particles gain complete independence from each other and fly about randomly to fill their containers.

Online homework for this chapter may be assigned in OWL

The gaseous state is examined more fully in Chapter 4. In Section 4.1, five properties of gases are identified:

1. Gases may be compressed.
2. Gases fill their containers uniformly.
3. All gases have low densities.
4. Gases may be mixed.
5. A confined gas exerts constant pressure on the walls of its container uniformly in all directions.

In this chapter we discuss the relationships among gases, liquids, and solids and the energy changes that accompany a change of state. To understand some of these relationships you must first understand a bit more about mixtures of gases, Item 4 in the preceding list.

## 15.1 Dalton's Law of Partial Pressures

Goal **1** Given the partial pressure of each component in a mixture of gases, find the total pressure.

Goal **2** Given the total pressure of a gaseous mixture and the partial pressures of all components except one, or information from which those partial pressures can be obtained, find the partial pressure of the remaining component.

According to the ideal gas model described in Chapter 4, a gas is made up of tiny molecules that are widely separated from each other so they occupy the whole volume of the container that holds them. It is the vast open space between molecules that makes it possible for gases to mix. If Gas A is added to Gas B in a rigid container (constant volume), the A molecules distribute themselves throughout the open space between the B molecules. The particle volume of the A and B molecules is negligible compared to the macroscopic volume occupied by the gas as a whole. One thing changes, though: pressure. It goes up.

Figure 15.1 illustrates an experiment. First note that the volume of all three containers is the same. The gas pressure in the nitrogen vessel is 186 mm Hg, and the pressure in the oxygen vessel is 93 mm Hg, as shown. When the two samples are mixed into a single vessel, the pressure exerted by the combined gases is the sum of the pressures of the individual gases, 186 mm Hg + 93 mm Hg = 279 mm Hg. In effect, each gas continues to exert the same pressure it exerted before the gases were mixed. After mixing, the individual gases are exerting their pressures in the same container. The total pressure is the sum of the two individual pressures.

John Dalton, of atomic theory fame, summed up these observations in what is now known as **Dalton's Law of Partial Pressures: The total pressure exerted by a mixture of gases is the sum of the partial pressures of the gases in the mixture. The partial pressure of one gas in a mixture is the pressure that gas would exert if it alone occupied the same volume at the same temperature.** Mathematically, this is

$$P = p_1 + p_2 + p_3 + \cdots \qquad \textbf{(15.1)}$$

where P is the total pressure and $p_1$, $p_2$, $p_3$, . . . are the partial pressures of gases 1, 2, 3, . . . . Notice that we use an uppercase P for total pressure and a lowercase p for partial pressure.

**P/REVIEW:** John Dalton's atomic theory proposes that all matter is made up of tiny particles called atoms. His theory is described in Section 5.1.

### Active Example 15.1

In a gas mixture the partial pressure of methane is 154 torr; of ethane, 178 torr; and of propane, 449 torr. Find the total pressure exerted by the mixture.

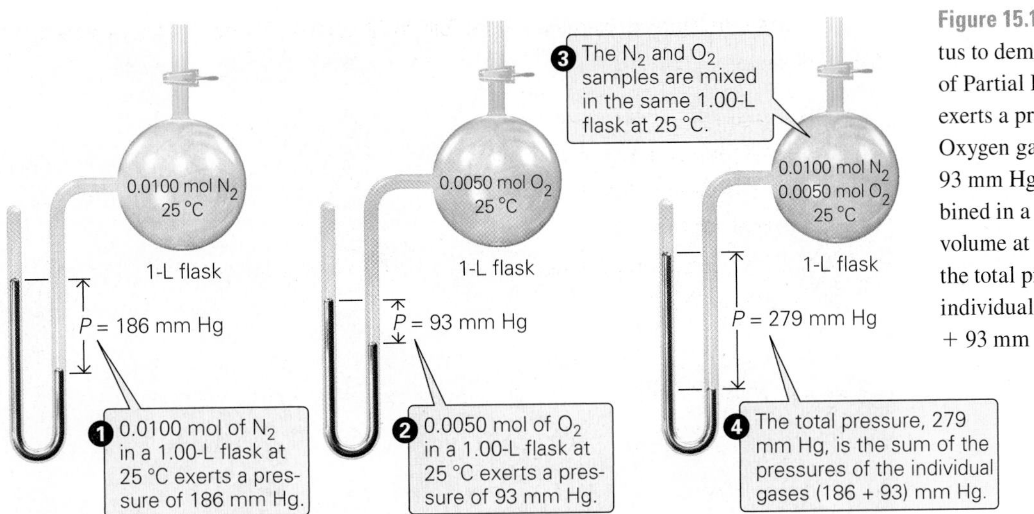
0.0100 mol $N_2$
25 °C
1-L flask
P = 186 mm Hg
1 0.0100 mol of $N_2$ in a 1.00-L flask at 25 °C exerts a pressure of 186 mm Hg.
0.0050 mol $O_2$
25 °C
1-L flask
P = 93 mm Hg
2 0.0050 mol of $O_2$ in a 1.00-L flask at 25 °C exerts a pressure of 93 mm Hg.
3 The $N_2$ and $O_2$ samples are mixed in the same 1.00-L flask at 25 °C.
0.0100 mol $N_2$
0.0050 mol $O_2$
25 °C
1-L flask
P = 279 mm Hg
4 The total pressure, 279 mm Hg, is the sum of the pressures of the individual gases (186 + 93) mm Hg.

**Figure 15.1** Experimental apparatus to demonstrate Dalton's Law of Partial Pressures. Nitrogen gas exerts a pressure of 186 mm Hg. Oxygen gas exerts a pressure of 93 mm Hg. If the gases are combined in a container of the same volume at the same temperature, the total pressure is the sum of the individual pressures, 186 mm Hg + 93 mm Hg = 279 mm Hg.

This is an application of Equation 15.1, $P = p_1 + p_2 + p_3 + \cdots$.

$P = p_1 + p_2 + p_3 = 154 \text{ torr} + 178 \text{ torr} + 449 \text{ torr} = 781 \text{ torr}$

Gases generated in the laboratory may be collected by bubbling them through water, as shown in Figure 15.2. As the oxygen bubbles rise through the water, they become saturated with water vapor. The gas collected is therefore actually a mixture of oxygen and water vapor. The pressure exerted by the mixture is the sum of the partial pressure of the oxygen and the partial pressure of the water vapor. As you will see in Section 15.4, water vapor pressure depends only on temperature. It may be found in reference books.

## Active Example 15.2

Oxygen is generated for a laboratory experiment by bubbling the gas through water, as illustrated in Figure 15.2. The total pressure of the oxygen saturated with water vapor is 755 torr. The temperature of the gas mixture is 22°C, and water vapor pressure at that temperature is 19.8 torr. What is the partial pressure of the oxygen?

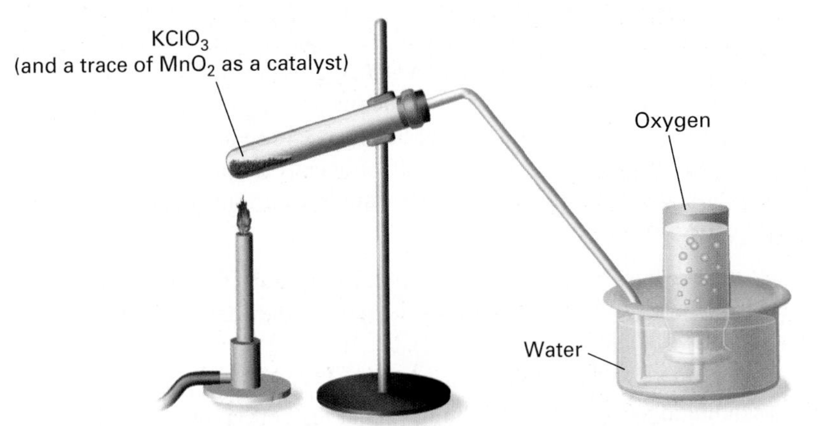
$KClO_3$
(and a trace of $MnO_2$ as a catalyst)
Oxygen
Water

**Figure 15.2** Laboratory preparation of oxygen. The heat applied to the test tube decomposes the $KClO_3$ into $O_2$ and KCl. The oxygen is directed into the bottom of an inverted bottle that initially is filled with water. As the oxygen accumulates in the bottle, it displaces the water until the bottle is filled with oxygen that initially is saturated with water vapor.

SCUBA (self contained underwater breathing apparatus) diving. The solubility of gases in the blood is proportional to their partial pressures. If air is compressed, the partial pressures of nitrogen and oxygen are increased, leading to problems such as nitrogen narcosis and oxygen toxicity. Helium is added to the breathing mixture for deep-sea diving.

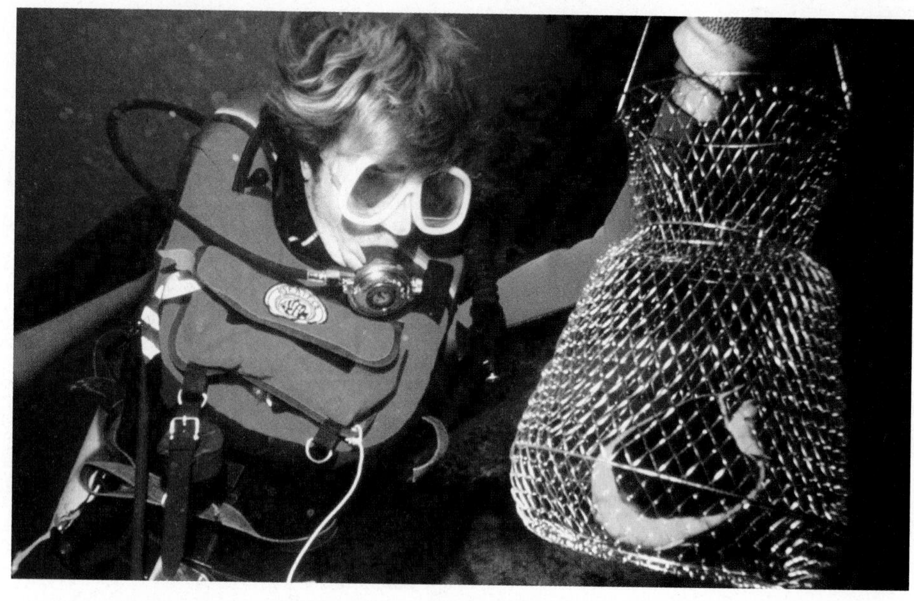

NOAA

As in Active Example 15.1, apply Equation 15.1, $P = p_1 + p_2 + p_3 + \cdots$.

$P = p_{H_2O} + p_{O_2}$;
rearranging, $p_{O_2} = P - p_{H_2O} = 755\ \text{torr} - 19.8\ \text{torr} = 735\ \text{torr}$

## 15.2 Properties of Liquids

**Goal 3** Explain the differences between the physical behavior of liquids and gases in terms of the relative distances among particles and the effect of those distances on intermolecular forces.

**Goal 4** For two liquids, given comparative values of physical properties that depend on intermolecular attractions, predict the relative strengths of those attractions; or, given a comparison of the strengths of the intermolecular attractions, predict the relative values of physical properties that the attractions cause.

The properties of liquids are easy to observe and describe—more so than the properties of gases. To understand liquid properties, however, it is helpful to compare the structure of a liquid with the structure of a gas. In Chapter 4 you learned that gas particles are so far apart that attractive and repulsive forces between the particles are negligible. These forces are electrostatic in character. They are inversely related to the distance between the particles; the smaller the distance is, the stronger the forces are. In a liquid, particles are very close to each other. Consequently, the intermolecular attractions in a liquid are strong enough to affect its physical properties.

We can now compare the properties of liquids with four properties of gases that were listed in Section 4.1:

1. *Gases may be compressed; liquids cannot.* Liquid particles are "touchingly close" to each other. There is no space between them, so they cannot be pushed closer, as in the compression of a gas.
2. *Gases expand to fill their containers; liquids do not.* The strong attractions between liquid particles hold them together at the bottom of a container.

3. *Gases have low densities; liquids have relatively high densities.* Density is mass per unit volume—mass divided by volume. If the particles of a liquid are close together compared with the particles of a gas, a given number of liquid particles will occupy a much smaller volume than the same number of particles occupies as a gas. A small denominator in the density ratio for a liquid means a higher value for the ratio.
4. *Gases may be mixed in a fixed volume; liquids cannot.* When one gas is added to another, the particles of the second gas occupy some of the space between the particles of the first gas. There is no space between particles of a liquid, so combining liquids must increase volume.

A liquid has several measurable properties whose values depend on intermolecular attractions, the tendency of the particles to stick together. In fact, if you think in terms of the "stick-togetherness" as equated with the strength of the intermolecular forces, you can usually predict relative values of these properties for two liquids. The greater the stick-togetherness is, the stronger the intermolecular forces are. We now identify five of these properties.

**Vapor Pressure** The open space above any liquid contains some particles in the gaseous, or vapor, state. This is due to evaporation (Fig. 15.3). The partial pressure exerted by these gaseous particles is called **vapor pressure.** If the gas space above the liquid is closed, the vapor pressure increases to a definite value called the **equilibrium vapor pressure.** In Section 15.4 you will study the mechanics by which that pressure is reached. Vapor pressure is inversely related to intermolecular attractions. If stick-togetherness is high between liquid particles, very few liquid particles "escape" into the gaseous state, so the vapor pressure is low.

**Molar Heat of Vaporization** It takes energy to overcome intermolecular attractions, separate liquid particles from each other, and keep them apart (Fig. 15.4). The energy required to change one mole of a liquid to its vapor, while at constant temperature and pressure, is called **molar heat of vaporization.** The energy released in the opposite process, as vapor condenses to the liquid phase, is called **heat of condensation.** The greater the stick-togetherness is, the greater the amount of energy needed or released.

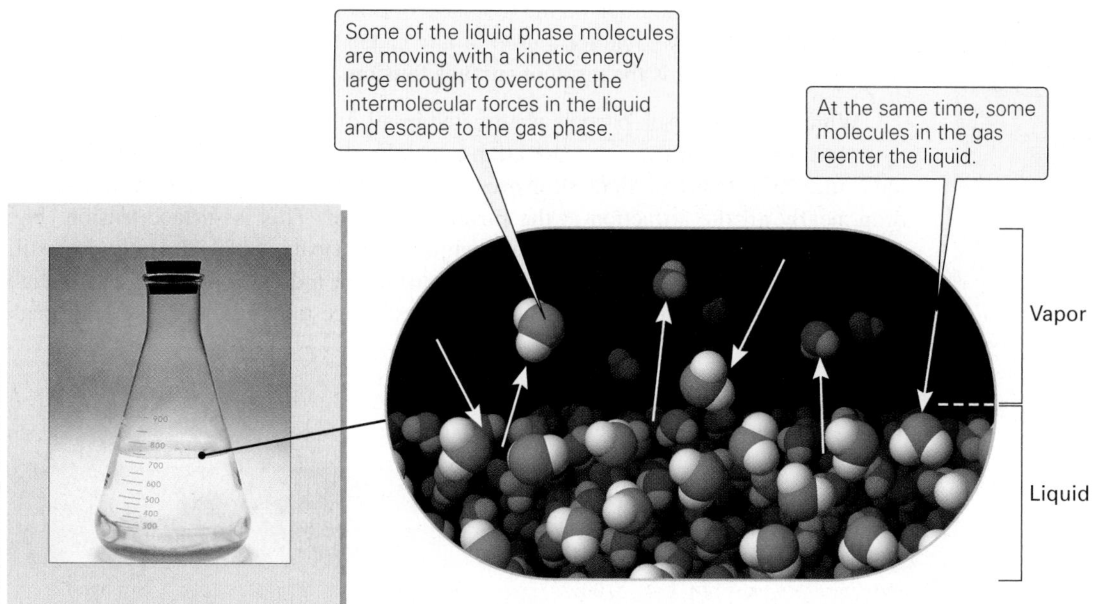

Some of the liquid phase molecules are moving with a kinetic energy large enough to overcome the intermolecular forces in the liquid and escape to the gas phase.
At the same time, some molecules in the gas reenter the liquid.
Vapor
Liquid
© Cengage Learning/Charles D. Winters


**Figure 15.3** Evaporation. Some molecules at the surface of the liquid phase have sufficient energy to escape and enter the vapor phase. Some molecules in the vapor phase will reenter the liquid phase when they make contact with the surface. The partial pressure exerted by the molecules in the vapor phase is called vapor pressure.

Figure 15.4 Heats of vaporization and condensation. Energy must be added to a system to overcome the attractive forces that are exerted among liquid molecules. When an equal quantity of vapor condenses to a liquid, an equal amount of energy is released.

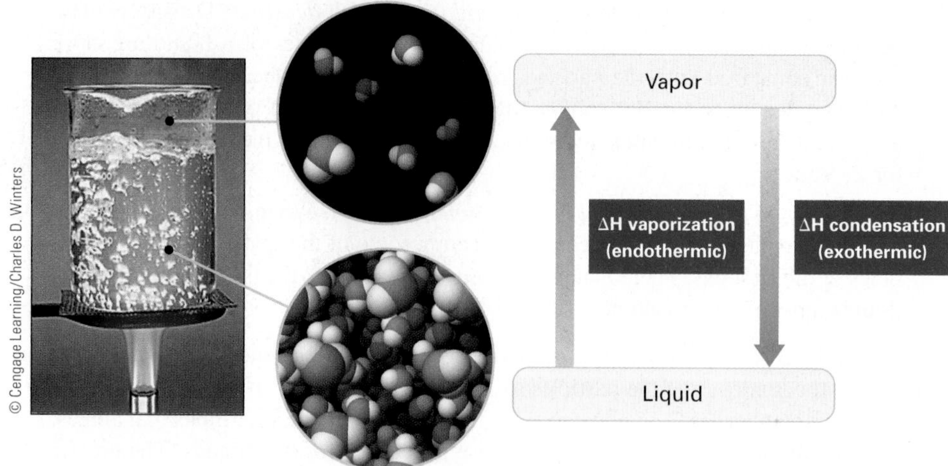


**Boiling Point** Liquids can be changed to gases by boiling. The boiling process is discussed in more detail in Section 15.5. At the **boiling point,** the average kinetic energy of the liquid particles is high enough to overcome the forces of attraction that hold the particles in the liquid state. When stick-togetherness is high, it takes more motion (a higher temperature) to separate the particles within the liquid, where boiling occurs.

The trends in vapor pressure, molar heat of vaporization, and boiling point are shown for several substances in Table 15.1.

**Viscosity** Particles in a liquid are free to move about relative to each other; they "flow." Some liquids flow more easily than others. Water, for example, can be poured much more freely than syrup, and syrup more readily than honey (Fig. 15.5). The ability of a liquid to flow is measured by its **viscosity**. Viscosity is an internal resistance to flow, and it is partially based on intermolecular attractions. When comparing particles of about the same size, more stick-togetherness means higher viscosity.

**Surface Tension** When a liquid is broken into "small pieces" it forms spherical drops (see the photograph on the first page of this chapter). A sphere has the smallest surface area possible for a drop of any given volume. This tendency toward a minimum surface is the result of **surface tension.**

Within a liquid, each particle is attracted in all directions by the particles around it. At the surface, however, the attraction is nearly all downward, pulling the surface molecules into a sort of tight skin over a standing liquid. Similarly, in a spherical drop, nearly all the attraction at the surface is inward. This is surface tension (Fig. 15.6). Its effect in water may be seen when a needle floats if placed gently on a still surface, or when small bugs run across the surface of a quiet pond (Fig. 15.7). High stick-togetherness at the surface means more resistance to anything that would break through or stretch that surface.

Figure 15.5 Viscosity. Honey pours relatively slowly, and thus it has a relatively high viscosity. This macroscopic characteristic indicates that there are relatively strong intermolecular attractions among the molecules that make up the mixture commonly known as honey.

**Table 15.1** Physical Properties of Liquids

| Substance | Vapor Pressure at 20°C (torr) | Heat of Vaporization (kJ/mol) | Normal Boiling Point (°C) | Intermolecular Forces |
|---|---|---|---|---|
| Mercury | 0.0012 | 59 | 357 | Strongest |
| Water | 17.5 | 41 | 100 | ↑ |
| Benzene | 75 | 31 | 80 | |
| Ether | 442 | 26 | 35 | |
| Ethane | 27,000 | 15 | −89 | Weakest |

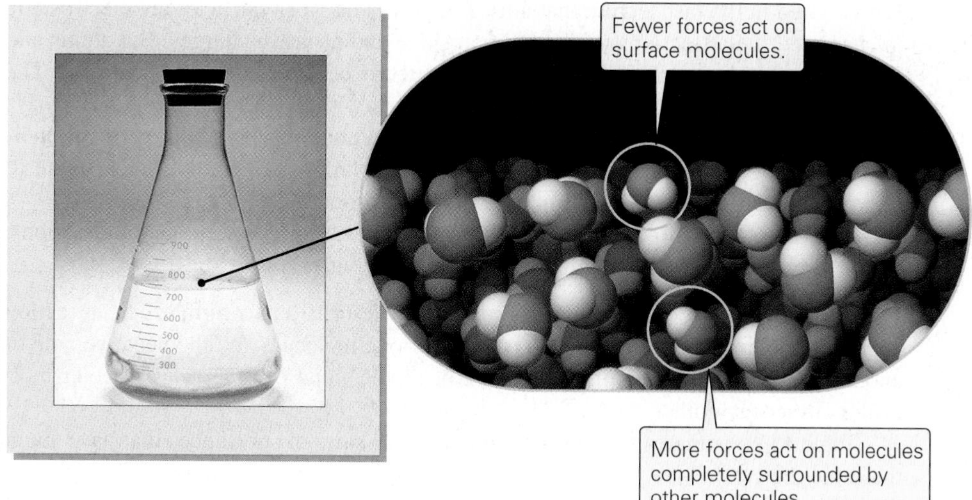
Fewer forces act on surface molecules.
More forces act on molecules completely surrounded by other molecules.
900
800
700
600
500
400
300

**Figure 15.6** Surface tension. Unbalanced attractive forces at the surface of a liquid pull surface molecules downward and sideways, but not up. Molecules within the water are attracted in all directions. Surface tension is a result of the difference in these forces.

summary

**Properties of Liquids and Intermolecular Attractions**

**Vapor pressure:** Liquids with relatively strong intermolecular attractions evaporate less readily, yielding lower vapor concentrations and therefore lower vapor pressures, than liquids with weak intermolecular forces.

**Molar heat of vaporization:** The molar heat of vaporization of a liquid with strong intermolecular attractions is higher than the molar heat of vaporization of a liquid with weak intermolecular attractions.

**Boiling point:** Liquids with strong intermolecular attractions require higher temperatures for boiling than liquids with weak intermolecular attractions.

**Viscosity:** Liquids with strong intermolecular attractions are generally more viscous than liquids with weak intermolecular attractions.

**Surface tension:** Liquids with strong intermolecular attractions have higher surface tension than liquids with weak intermolecular attractions.

**Figure 15.7** Insect walking on water. The surface tension of water creates a difficult-to-penetrate skin that will support small bugs or thin pieces of dense metals, such as a needle or razor blade. A bug literally runs on the water; it does not float in it.

### √ Target Check 15.1

a) What main difference between gases and liquids at the particulate level accounts for the large differences in their macroscopic properties?
b) Intermolecular attractions are stronger in A than in B, with all other potentially influencing factors being about equal. Which do you expect will have the higher surface tension, molar heat of vaporization, vapor pressure, boiling point, and viscosity?
c) X has a higher molar heat of vaporization than Y. Which do you expect will have a higher vapor pressure? Why?

## 15.3 Types of Intermolecular Forces

**Goal** 5 Identify and describe or explain induced dipole forces, dipole forces, and hydrogen bonds.

**Goal** 6 Given the structure of a molecule, or information from which it may be determined, identify the significant intermolecular forces present.

**Goal** 7 Given the molecular structures of two substances, or information from which they may be obtained, compare or predict relative values of physical properties that are related to them.

P/REVIEW: If you have drawn a Lewis diagram of a molecule (Section 13.1), you can predict that it is polar if either of these conditions exists: a central atom has a lone pair of electrons, or a central atom is bonded to atoms of different elements (see Section 13.5).

It was stated in the last section that attractive forces between particles are electrostatic in character; the attractions are between positive and negative charges. But atoms and molecules are electrically neutral. How can there be electrostatic attractions? The answer is that the *distribution* of electrical charge within the molecule is not always uniform. Some molecules are polar and some are nonpolar. In addition, some molecules are large and some are small. Molecular polarity and size both contribute to intermolecular attraction and therefore to physical properties.

Three kinds of **intermolecular forces** can be traced to electrostatic attractions: dipole forces, induced dipole forces, and hydrogen bonds.

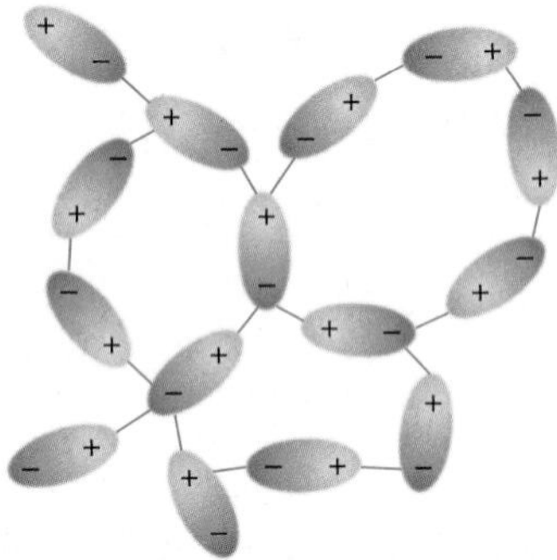

Figure 15.8 Dipole forces. Molecules tend to arrange themselves by bringing oppositely charged regions close to each other and forcing similarly charged regions far from each other.

1. *Dipole forces.* A polar molecule is sometimes described as a **dipole**. The attraction between dipoles is between the positive pole of one molecule and the negative pole of another. Figure 15.8 shows the alignment of dipoles, one of several ways polar molecules attract each other.

Table 15.2 compares the boiling points of four pairs of substances that have about the same molecular size, indicated approximately by their molar masses. In each pair the boiling point of the substance with polar molecules is higher than the boiling point of the nonpolar substance. This is because polar molecules have stronger intermolecular attractions than nonpolar molecules.

2. *Induced dipole forces.* Attractions between nonpolar molecules are called **induced dipole forces.** These are also called **dispersion forces, London forces,** or **London dispersion forces.** They are believed to be the result of shifting electron clouds within the molecules. If the electron movement in a molecule results in a temporary concentration of electrons at one side of the molecule, the molecule becomes a "temporary dipole." This is shown in the left molecule in the top pair in Figure 15.9. The electrons repel the electrons in the molecule next to it, pushing them to the far side of that molecule. The second molecule is thus "induced" to form a second temporary dipole (bottom pair in Fig. 15.9). As long as these dipoles exist—a very small fraction of a second in each case—there is a weak attraction between them.

The strength of induced dipole forces depends on the ease with which electron distributions can be induced to be distorted or "polarized." Large molecules, with many electrons and with electrons far removed from atomic nuclei, are more easily polarized than small molecules. Larger molecules are also generally heavier. Consequently,

Figure 15.9 Induced dipole forces. Electron clouds in molecules are constantly shifting. The temporary dipole in the left molecule in the top row "induces" the molecule next to it to become another temporary dipole (bottom row). The "instantaneous dipoles" are attracted to each other briefly. A small fraction of a second later, the clouds shift again and continue to interact with each other or with nearby molecules.

**Table 15.2** Boiling Points of Polar Versus Nonpolar Substances

| Formulas | Polar or Nonpolar | Molecular Mass | Boiling Point (°C) | Formulas | Polar or Nonpolar | Molecular Mass | Boiling Point (°C) |
|---|---|---|---|---|---|---|---|
| $N_2$ | Nonpolar | 28 | −196 | $GeH_4$ | Nonpolar | 77 | −88 |
| CO | Polar | 28 | −192 | $AsH_3$ | Polar | 78 | −63 |
| $SiH_4$ | Nonpolar | 32 | −112 | $Br_2$ | Nonpolar | 160 | 59 |
| $PH_3$ | Polar | 34 | −88 | ICl | Polar | 162 | 97 |

**Figure 15.10** Induced dipole forces and molecular size. Larger molecules, with a greater number of electrons, polarize more easily than smaller molecules. Bromine (*left*) exists as a liquid at room conditions, and iodine (*right*) is a solid. Both molecules are nonpolar and have induced dipole forces as the dominant intermolecular force. However, bromine has 70 electrons, whereas iodine has 106 electrons, so iodine polarizes more easily. Thus, the intermolecular attractive forces among iodine molecules are greater than among bromine molecules. This particulate-level difference is responsible for the macroscopic observations that iodine is a solid and bromine is a liquid.

intermolecular forces tend to increase with increasing molar mass among otherwise similar substances (Fig. 15.10). Notice in Table 15.2 the increase in boiling points for both polar and nonpolar molecules as molar mass increases.

3. *Hydrogen bonds.* Some polar molecules have intermolecular attractions that are much stronger than ordinary dipole forces. These molecules always have a hydrogen atom bonded to an atom that is small and highly electronegative and that has at least one unshared pair of electrons. Nitrogen, oxygen, and fluorine are generally the only elements whose atoms satisfy these requirements (see Fig. 15.11).

The covalent bond formed between the hydrogen atom and the atom of nitrogen, oxygen, or fluorine is strongly polar. The electron pair is shifted away from the hydrogen atom toward the more electronegative atom. This leaves the hydrogen nucleus—nothing more than a proton—as a small, highly concentrated region of positive charge at the edge of a molecule. The negative pole of another molecule, which is the region near an unshared electron pair on a nitrogen, oxygen, or fluorine atom, can get quite close to the hydrogen atom of the first molecule. This results in an extra-strong attraction between the molecules. This kind of intermolecular attraction is a **hydrogen bond.**

Notice that a hydrogen bond is an *intermolecular force,* an attraction between *different* molecules. It is not a covalent bond between atoms in the *same* molecule. The

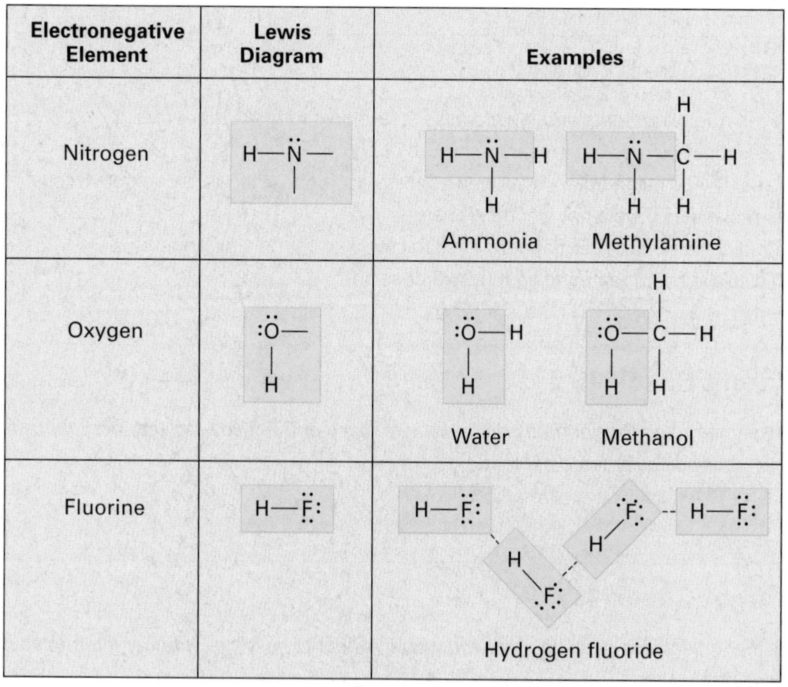

**P/REVIEW:** Electronegativity estimates the strength with which an atom attracts the pair of electrons that forms a bond between it and another atom. Covalent bonds are polar when there is a high electronegativity difference between the bonded atoms. See Section 12.4.

| Electronegative Element | Lewis Diagram | Examples | |
|---|---|---|---|
| Nitrogen | H—N̈— (with bond down) | H—N̈—H with H below; Ammonia | H—N̈—C—H with H above C and H below N and C; Methylamine |
| Oxygen | :Ö— with H below | :Ö—H with H below; Water | :Ö—C—H with H above C and H below O and C; Methanol |
| Fluorine | H—F: | H—F: ··· H—F: ··· H—F: ··· H—F: (dashed hydrogen bonds); Hydrogen fluoride | |

**Figure 15.11** Recognizing hydrogen bonding. Hydrogen bonds occur when a hydrogen atom is covalently bonded to a small atom that is highly electronegative and has one or more unshared electron pairs. Fluorine, oxygen, and nitrogen atoms fit this description. The hydrogen bond is between the atom of one of these elements in one molecule and the hydrogen atom of a nearby molecule. Hydrogen bonds between molecules are illustrated with dashed lines in the hydrogen fluoride example.

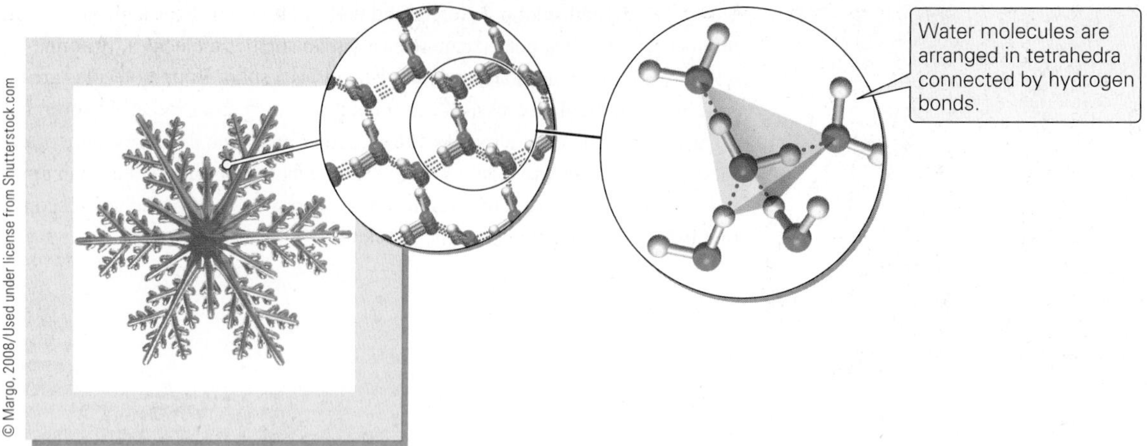


**Figure 15.12** Hydrogen bonding in water. Intermolecular hydrogen bonds are present between the electronegative oxygen region of one molecule and the electropositive hydrogen region of a second molecule.

dotted lines in Figure 15.12 represent hydrogen bonds between water molecules. While a hydrogen bond is much stronger than an ordinary dipole–dipole force, it is roughly one-tenth as strong as a covalent bond between atoms of the same two elements.

Of the three kinds of intermolecular attractions, hydrogen bonds are the strongest. When present between small molecules, hydrogen bonds are primarily responsible for the physical properties of a liquid. Dipole forces are next strongest, and induced dipole forces are the weakest of the three. Induced dipole forces are present between all molecules. In small molecules, induced dipole forces are important only when the other forces are absent. But between large molecules—molecules that contain many atoms or even few atoms that have many electrons—induced dipole forces are quite strong and often play the main role in determining physical properties.

Active Figure 15.13 summarizes the kinds of intermolecular forces and their effects on boiling points of similar compounds in three chemical families. We recommend that you study it carefully.

### √ Target Check 15.2

*Identify the true statements and rewrite false statements to make them true.*

a) Induced dipole forces are present only with nonpolar molecules.
b) All other things being equal, hydrogen bonds are stronger than dipole–dipole forces.
c) Polar molecules have a net electrical charge.
d) Intermolecular forces are magnetic in character.
e) $H_2O$ displays hydrogen bonding, but $H_2S$ does not.

### √ Target Check 15.3

*Determine the molecular geometry and polarity of each of the following and, from that, identify the major intermolecular force present.*

a) $CH_4$ b) $CO_2$ c) $OF_2$ d) HOCl

### √ Target Check 15.4

*Identify the molecule in each pair that you would expect to have the stronger intermolecular forces and state why.*

a) $CCl_4$ or $CBr_4$ b) $NH_3$ or $PH_3$

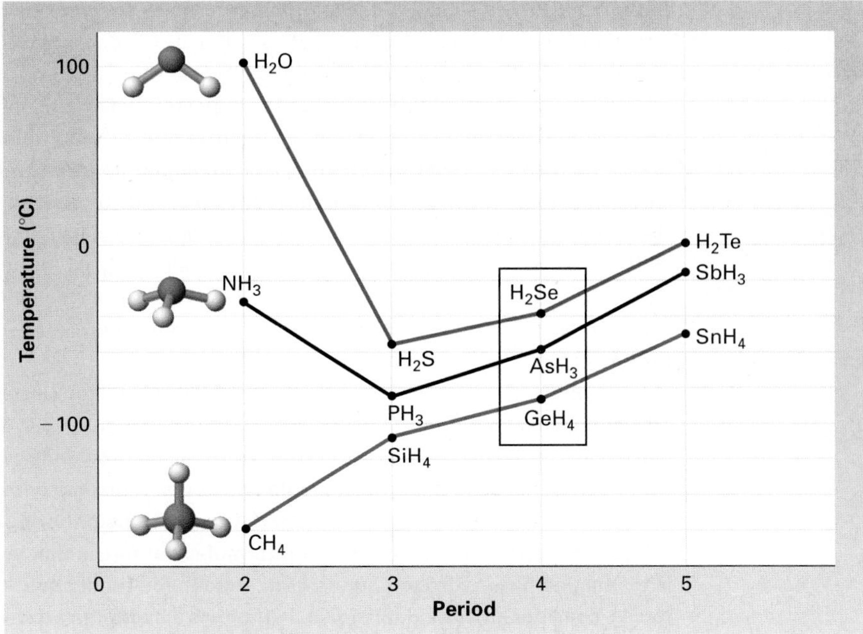


**Active Figure 15.13** Intermolecular attractions illustrated by boiling points of hydrogen-containing compounds. Liquids with strong intermolecular attractions usually boil at higher temperatures than liquids with weak intermolecular attractions. These attractions are caused by induced dipole forces, dipole forces, and hydrogen bonding. Holding two of these variables essentially constant and changing the third, we can see how each variable affects the attractions by comparing boiling points. **Watch this active figure at *http://www.cengage.com/cracolice.***

**Induced dipole forces** The Group 4A/14 hydrogen-containing compounds (blue line) all have tetrahedral molecular geometries. They are nonpolar, and they have no hydrogen bonding. The only intermolecular forces are induced dipole forces. The molecules differ only in molecular size (mass), ranging from $CH_4$, the smallest, to $SnH_4$, the largest. The boiling points of the four compounds increase as their molecular sizes increase. Except for $H_2O$ and $NH_3$, the same trend appears for Group 5A/15 (black line) and Group 6A/16 (red line) hydrogen-containing compounds. This suggests that, *other things being equal, intermolecular attractions increase as molecular size increases.*

**Dipole forces** The molecules in the rectangle for the Period 4 hydrogen-containing compounds—$H_2Se$, $AsH_3$, and $GeH_4$—are about the same size (nearly equal molar mass), and none of them has hydrogen bonding. They differ most in polarity. $GeH_4$ has tetrahedral molecules; it is nonpolar. The trigonal pyramidal molecules of $AsH_3$ are polar, but less so than bent $H_2Se$ molecules. The least polar compound, $GeH_4$, has the lowest boiling point, and the most polar compound, $H_2Se$, has the highest boiling point. The same trend appears with the Period 3 and Period 5 hydrogen-containing compounds. This indicates that, *other things being equal, intermolecular attractions increase as molecular polarity increases.*

**Hydrogen bonding** The high boiling points of $H_2O$ and $NH_3$ violate the trends in which small molecules boil at lower temperatures than large molecules that are otherwise similar. $H_2O$ and $NH_3$ are the only two substances shown that have hydrogen bonding. This indicates that, for small molecules in particular, *hydrogen bonding causes exceptionally strong intermolecular attractions.*

## Thinking About Your Thinking

### Mental Models

**Particulate-level mental models of what *causes* each of the intermolecular forces are needed for a complete understanding of this concept. First, be sure that you grasp the distinction between *intra*molecular forces, which are chemical bonds between atoms within a single molecule, and *inter*molecular forces, the attractive forces that operate between whole molecules and other particles. Your "mental movie" needs to be based on an electron cloud model of a molecule or atom, such as the quantum model shown in Figure 11.10. No matter how many atoms are used to make up the particle, consider only the outer electron cloud of the particle for this mental model.**

**Consider Figure 15.9. The top right electron cloud shows the most common state for a nonpolar molecule—one where the outer electrons are evenly distributed**

**throughout the molecule. Induced dipole forces are a result of a short-term uneven shifting of this electron cloud so that one area of the particle has a slight excess of electrons and the opposite side is slightly deficient in electrons, relative to the normal balanced distribution. This makes the molecule just a little polar, with very weak negative and positive regions. The charged region of this molecule then induces (hence the name of this force) its neighbor to adjust its electron distribution. If the electron density of a molecule is shifted to the right (top left particle in Fig. 15.9), the negative region repels the electrons in the left side of its neighbor, inducing it to shift its electron density to the right (bottom pair in Fig. 15.9). Now a slightly negative region is near a slightly positive region, and there is a weak attraction. If you can imagine this process, you have a good start at your mental model of induced dipole forces.**

**Now imagine a particle where the uneven distribution of electrons is permanent. This is the case in polar molecules, as illustrated in Figure 15.8. The darker-shaded end of each molecule is the portion with a greater electron density and the accompanying partial negative charge. The region of a polar molecule with a partial negative charge is attracted to the positively charged end of a neighboring polar molecule.**

**The third type of mental model for intermolecular forces that you need to develop is to imagine how hydrogen bonds form. Before you think about the intermolecular forces, consider an individual molecule that has a hydrogen atom covalently bonded to oxygen, such as water. The bonding electrons in the O—H bond are displaced toward the oxygen atom, which is relatively small and highly electronegative. This leaves the hydrogen atom nucleus—a positively charged proton—at the end of the molecule with very little surrounding electron density. That positively charged region will be strongly attracted to the oxygen-atom region of a neighboring molecule, given that the oxygen atom not only has two unshared pairs of electrons but also has its bonding electrons shifted toward it and away from the hydrogen atom. An intermolecular attractive force results.**

**The ability to imagine the particulate-level reasons for intermolecular attractive forces is a powerful tool in your "thinking as a chemist" knowledge stockpile. The more you understand this type of particulate-level behavior, the better you will understand the macroscopic properties that result from it.**

## 15.4 Liquid–Vapor Equilibrium

Goal 8 Describe or explain the equilibrium between a liquid and its own vapor and the process by which it is reached.

In Section 4.3 we discussed how temperature is a measure of the average kinetic energy of the particles in a sample. The range of kinetic energies at a certain temperature is shown in Figure 15.14. Kinetic energy is plotted horizontally, and the fraction, or percentage, of the sample having a given kinetic energy is plotted vertically. The area beneath the curve represents all of the particles in the sample, or 100% of the sample.

To evaporate, or vaporize, a molecule must be at the surface of a liquid. It also must have enough kinetic energy to overcome the attractions of other molecules that would hold it in the liquid state. If E in Figure 15.14 represents this minimum amount of kinetic energy—we will call it the *escape energy*—only those surface molecules having that energy or more can get away. The fraction, or percentage, of all surface molecules having that much energy is given by the area beneath the curve to the right of E.

The rate at which a particular liquid evaporates depends on two things, temperature and surface area. If we think in terms of a unit area and hold temperature constant, the vaporization rate is also constant. These conditions are assumed in the experiment described in Figure 15.15. Study that figure and its caption carefully before proceeding to the next paragraph, which is based on the illustration. The caption contains the information you need to satisfy Goal 8.

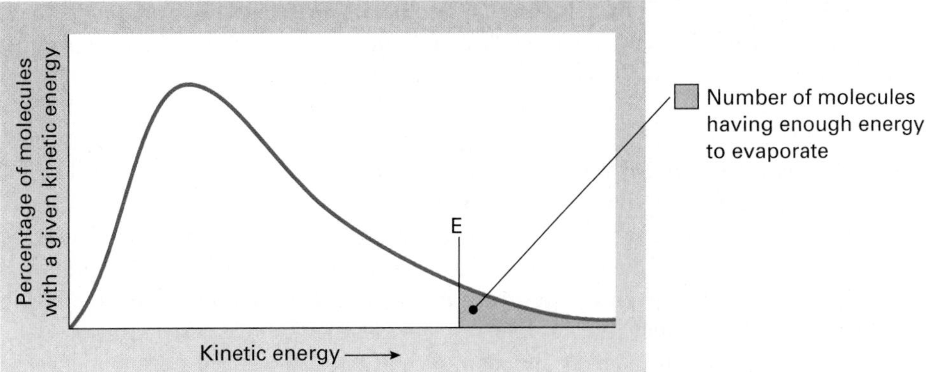


**Figure 15.14** Kinetic energy distribution curve for a liquid at a given temperature. E is the escape energy, the minimum kinetic energy a molecule must have to break from the surface and evaporate into the gas phase. The total area between the curve and the horizontal axis represents the total number of molecules in the sample. The tan area beneath the curve to the right of E represents the number of molecules having kinetic energy equal to or greater than E. At the temperature for which the graph is drawn, only a small fraction of all molecules has enough energy to evaporate.

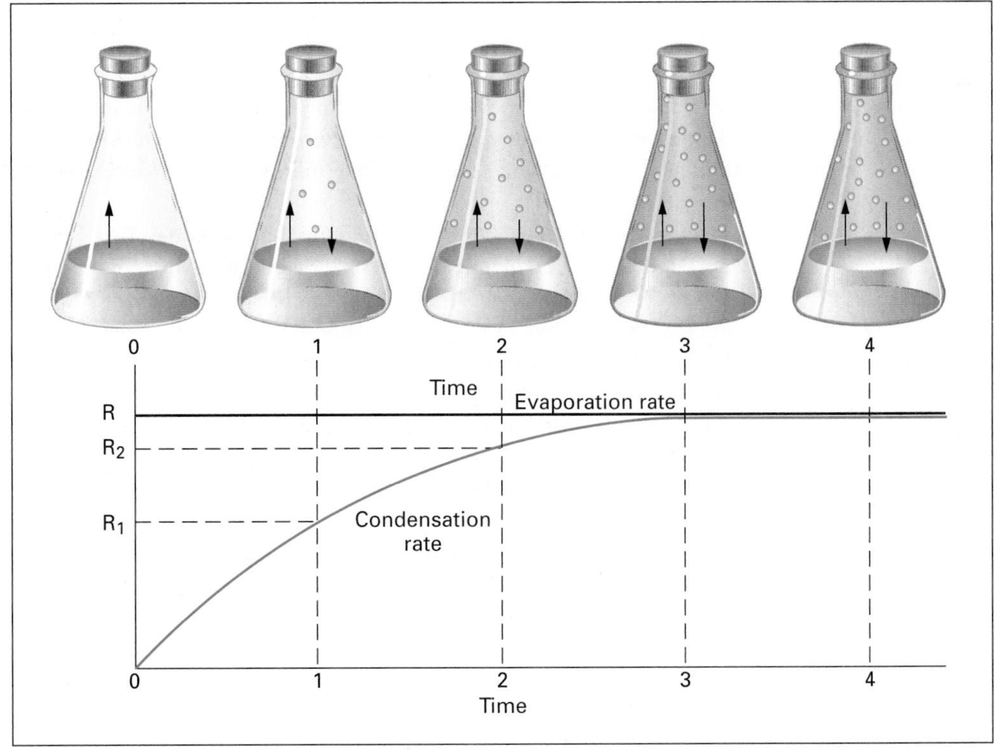


**Figure 15.15** A liquid with weak intermolecular forces, such as benzene, is placed in an Erlenmeyer flask, which is then closed with a rubber stopper. At constant temperature and surface area, the liquid begins to evaporate at a constant rate. This rate is represented by the fixed length of the arrows pointing upward from the liquid in each view of the flask. It is also shown as the horizontal line in the graph of evaporation rate versus time.

At first (Time 0), the movement of molecules is entirely in one direction, from the liquid to the vapor. As the concentration of molecules in the vapor builds up, an occasional molecule hits the surface and reenters the liquid. The change of state from gas to a liquid is called **condensation**. The rate of condensation per unit area at constant temperature depends on the concentration of molecules in the vapor state. At Time 1 there will be a small number of molecules in the vapor state, so the condensation rate ($R_1$) will be more than zero, but much less than the evaporation rate (R). This is shown by the condensation arrow being shorter than the evaporation arrow in the Time 1 flask.

As long as the rate of evaporation is greater than the rate of condensation, the vapor concentration will rise. Therefore, the rate of return from vapor to liquid rises with time (Time 2). Eventually, the rates of vaporization and condensation become equal (Time 3). The number of molecules moving from vapor to liquid in unit time just balances the number moving in the opposite direction. There is no further change in the vapor concentration, so the opposing rates remain equal (Time 4).

**P/REVIEW:** The reversibility of chemical changes was first mentioned in Section 8.7. The dynamic equilibrium described here is for a physical change. Another physical equilibrium is described in Section 16.3. Chemical equilibria are those for which rates of reversible reactions in the forward and reverse directions are equal. Chemical equilibrium is examined in some depth in Section 18.5.

When a system such as that described in Figure 15.15 reaches the point at which the **rates of change in opposite directions are equal,** the system is said to be in **equilibrium.** It is important to note that once equilibrium is reached (Times 3 and 4 in Fig. 15.15), the vapor concentration remains constant. On the *macroscopic* level, nothing is happening. On the *particulate* level, however, molecules are continually switching between the liquid and vapor states. Because of this constant activity, this kind of equilibrium is called a **dynamic equilibrium.**

When the vapor concentration becomes constant at equilibrium, vapor pressure also becomes constant. **The partial pressure exerted by a vapor in equilibrium with its liquid phase at a given temperature is the equilibrium vapor pressure of the substance at that temperature.**

Changes that occur in either direction, such as the change from a liquid to a vapor and the opposite change from a vapor to a liquid, are called **reversible changes;** if the change is chemical, it is a **reversible reaction.** Chemists write equations describing reversible changes with a double arrow, one pointing in each direction. For example, the reversible change between liquid benzene, $C_6H_6(\ell)$, and benzene vapor, $C_6H_6(g)$, is represented by the equation

$$C_6H_6(\ell) \rightleftharpoons C_6H_6(g) \qquad \textbf{(15.2)}$$

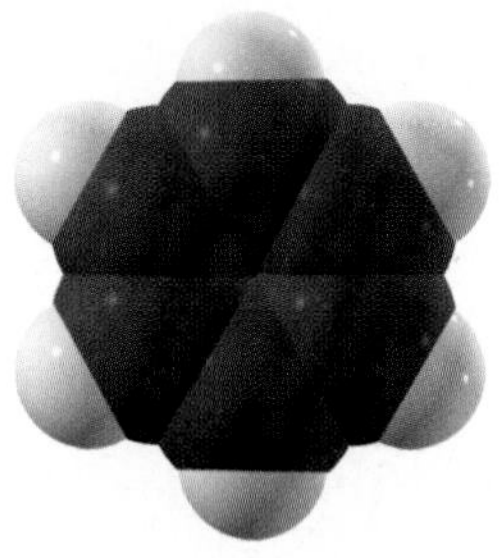

Benzene, $C_6H_6$, is an important substance in chemistry. Its three-dimensional structure is shown above. It is the starting point for producing many organic molecules, and it is a widely used solvent. It occurs naturally in coal tar. See Section 21.5.

## Thinking About Your Thinking

### Mental Models

**Developing a mental model of equilibrium processes is one of the most important thinking skills you will acquire in your chemistry coursework. Many chemical reactions that occur in living organisms and in the environment are reversible equilibrium processes.**

**Figure 15.15 is the starting point for your mental model. Turn that static series of "snapshots" into a dynamic "videoclip" in your mind. Imagine the particles evaporating at a constant speed throughout the process. Regardless of the point in the process, the number of particles evaporating per unit of time remains the same. Now add the condensing particles to your mental video. The condensation speed depends on the number of particles in the space above the liquid. Initially, it is at zero and it increases, rapidly at first, then more slowly as equilibrium is approached, until the speed at which the particles return to the liquid is equal to the speed at which they leave the liquid. Equilibrium is established when the rate of evaporation equals the rate of condensation.**

**When you encounter equilibrium questions on examinations, use your mental model as you develop a written answer. Imagine the scenario posed in the question and give a written description of what you "see" happening in your mind.**

## The Effect of Temperature

Goal | **9** Describe the relationship between vapor pressure and temperature for a liquid–vapor system in equilibrium; explain this relationship in terms of the kinetic molecular theory.

Many laboratory experiments show that vapor pressure increases as temperature rises. Active Figure 15.16 shows how the vapor pressures of several substances change with temperature. Notice how a relatively small increase in temperature causes a large increase in vapor pressure. The vapor pressure of water, for example, is 18 torr at 20°C (293 K) and 55 torr at 40°C (313 K). The vapor pressure more than triples (increases by 200%) while absolute temperature increases by only about 7%.

**P/REVIEW:** The absolute temperature scale is based on the Celsius degree and setting the lowest temperature possible at zero kelvin. $T_K = T_{°C} + 273$. See Section 4.3.

Figure 15.17 explains the effect of temperature on vapor pressure. The curve labeled $T_1$ gives the kinetic energy distribution at one temperature, and $T_2$ is the curve

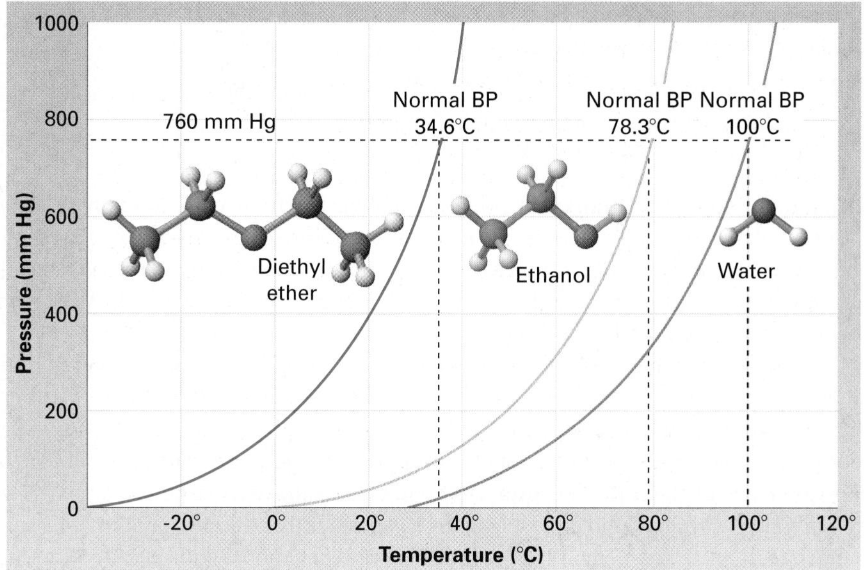


**Active Figure 15.16** Vapor pressures of three liquids at different temperatures. **Watch this active figure at *http://www.cengage.com/cracolice.***

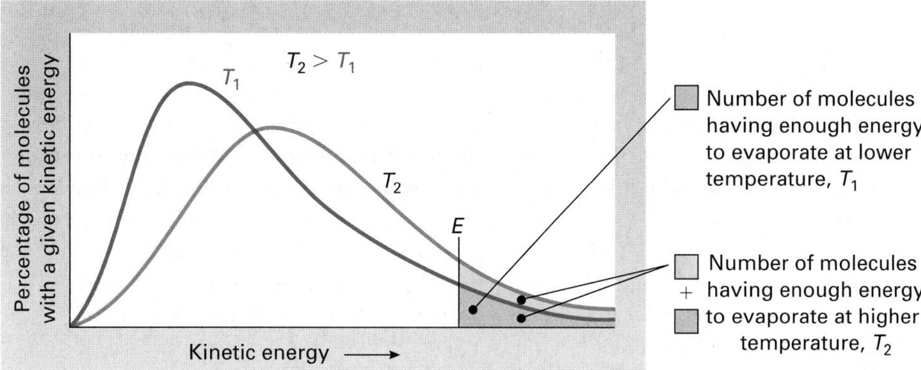


**Figure 15.17** Kinetic energy distribution curves for the same liquid at two temperatures. Curve $T_1$ is for the lower temperature. As temperature rises from $T_1$ to $T_2$, the average kinetic energy increases. The curve shifts to the right and is spread out over a wider range. The area beneath the curve to the right of the escape energy, E, represents the fraction of the total number of molecules that have enough kinetic energy to evaporate. At $T_1$ only the fraction representing the area under the marked part of the red curve has the escape energy, but at higher temperature, $T_2$, that fraction is represented by the marked area under the blue curve.

for a higher temperature. As in Figure 15.14, E is the escape energy, the minimum energy a molecule must have to evaporate into the gas phase. The area beneath the curve to the right of E is the fraction of the sample that has enough kinetic energy to evaporate. As illustrated, the area for $T_2$ is nearly twice the area for $T_1$. We conclude that it is an *increase in the number of molecules with enough energy to evaporate* that is responsible for higher vapor pressure at higher temperatures.

The term *humidity,* as it is commonly used in everyday language to describe weather, refers to what scientists call *relative humidity,* which is the ratio of the actual water vapor pressure to the equilibrium water vapor pressure at that temperature, $p_{H_2O}/p_{H_2O_{eq}}$. At high humidity, sweating is less effective at cooling our bodies because the evaporation rate and condensation rate of water in the air are nearly equal.

## √ Target Check 15.5

*Identify the true statements, and rewrite the false statements to make them true.*

a) A liquid–vapor equilibrium is reached when the amount of liquid is equal to the amount of vapor.
b) Rate of evaporation depends on temperature.
c) Equilibrium vapor pressure is higher at higher temperatures.

## 15.5 The Boiling Process

**Goal** **10** Describe the process of boiling and the relationships among boiling point, vapor pressure, and surrounding pressure.

When a liquid is heated in an open container, bubbles form, usually at the base of the container where heat is being applied. The first bubbles are often air, driven out of solution by an increase in temperature. Eventually, when a certain temperature is reached, vapor bubbles form throughout the liquid, rise to the surface, and break. When this happens, we say the liquid is **boiling.**

In order for a stable bubble to form in a boiling liquid, the vapor pressure within the bubble must be high enough to push back the surrounding liquid and the atmosphere above the liquid (see Fig. 15.18). The minimum temperature at which this can occur is called the **boiling point: The boiling point is the temperature at which the vapor pressure of the liquid is equal to the pressure above its surface.** Actually, the vapor pressure within a bubble must be a tiny bit greater than the surrounding pressure, which suggests that bubbles probably form in local "hot spots" within the boiling liquid. The boiling temperature at one atmosphere—the temperature at which the vapor pressure is equal to one atmosphere—is called the **normal boiling point.** Figure 15.16 shows that the normal boiling point of water is 100°C; of ethyl alcohol, 78.3°C, and of ethyl ether, 34.6°C.

According to the definition, the boiling point of a liquid depends on the pressure above it. If that pressure is reduced, the temperature at which the vapor pressure equals the lower surrounding pressure comes down also, and the liquid will boil at that lower temperature. This is why liquids boil at reduced temperatures at higher altitudes. In mile-high Denver, where atmospheric pressure is typically about 630 torr, water boils at 95°C. It is possible to boil water at room temperature by lowering the pressure in the space above it. When pressure is reduced to 20 torr, water boils at 22°C, which is "room temperature." A method for purifying a compound that might decompose or react with oxygen at its normal boiling point is to boil it at reduced temperature in a partial vacuum and then condense the vapor.

It is also possible to *raise* the boiling point of a liquid by *increasing* the pressure above it. The pressure cooker used in the kitchen takes advantage of this effect. By allowing the pressure to build up within the cooker, it is possible to reach temperatures as high as 110°C without boiling off the water. At this temperature food cooks in about half the time required at 100°C.

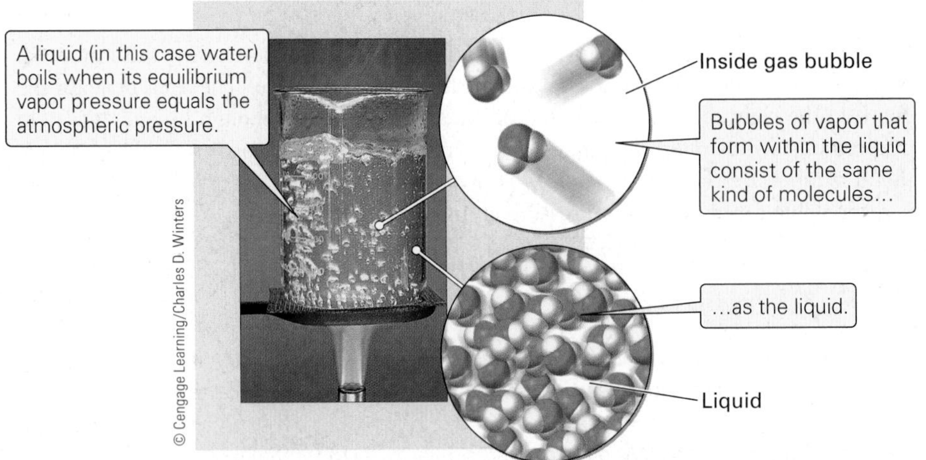
A liquid (in this case water) boils when its equilibrium vapor pressure equals the atmospheric pressure.
Inside gas bubble
Bubbles of vapor that form within the liquid consist of the same kind of molecules...
...as the liquid.
Liquid
© Cengage Learning/Charles D. Winters

**Figure 15.18** The boiling process. A liquid boils when its vapor pressure equals the pressure above it. The temperature at which this occurs is the boiling point.

## √ Target Check 15.6

*Are the following true or false?*

a) Water can be made to boil at 15°C.

b) Bubbles can form anyplace in a boiling liquid.

## 15.6 Water—An "Unusual" Compound

**Goal** 11 Describe the typical relative density relationship between the solid and liquid phase of a substance, and explain why water is an exception to this trend.

Through much of this book you have seen trends and regularities among physical and chemical properties. Many of these have been related to the periodic table. Predictions have been based on these trends. A prediction is not reliable, though, until it is confirmed in the laboratory. Sometimes a substance does not behave as it is expected to, and we have to look further, but most substances fit into regular patterns.

Water does not fit.

Water is so common, so much a part of our daily lives, that it is hard to think of it as being unusual. But in terms of trends, unusual is exactly what water is. One example appears in Active Figure 15.13. Beginning at tellurium (Z = 52) in Group 6A/16, the boiling points of the hydrogen-containing compounds drop as the molecules become smaller, as expected: −4°C for $H_2Te$, −42°C for $H_2Se$, and −62°C for $H_2S$. If the trend continued, the boiling point of $H_2O$ should be about −72°C. Instead, it is +100°C. And that is only one example of water's unusual behavior.

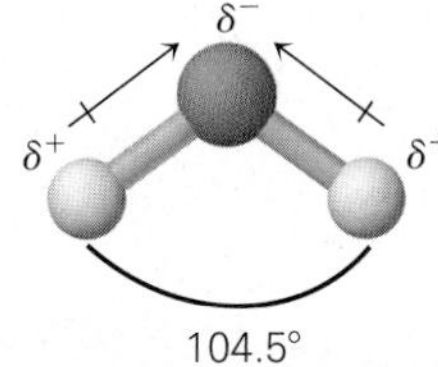


**Figure 15.19** The water molecule. The geometry of the water molecule and the polarity of its bonds make water molecules highly polar. In addition, water displays strong hydrogen bonding. These three factors account for exceptionally strong intermolecular attractions that influence many properties of water.

A close examination of the water molecule (Fig. 15.19) gives us some clues to explain this unique behavior. Aside from fluorine, oxygen is the most electronegative non-noble gas element there is. Therefore, the electrons forming each bond between hydrogen and oxygen are drawn strongly toward the oxygen atom, resulting in two very polar bonds—more polar than the bonds in other hydrogen-containing compounds in the group. Furthermore, the 104.5° bond angle makes a strong dipole. Finally, add hydrogen bonding, which is probably the most important contributor to strong intermolecular attractions in water.

Among molecules of comparable size, water has several other unusual properties. Exceptionally high surface tension and heat of vaporization are among them. Its vapor pressure is particularly low, even compared to larger molecules whose vapor pressures you would expect to be low. Check the compounds in Active Figure 15.16, for example. We don't usually think of water as being viscous, but it is viscous when compared with substances with similar structures. Water dissolves a wider variety of gaseous, liquid, and solid substances than most solvents. This is explained in Chapter 16. Finally, the mere fact that water is a liquid at room conditions is unusual. It is one of a very small number of compounds without carbon that exist as liquids at normal temperatures and pressures.

Water's most visible unusual property is that its solid form, ice, floats on its liquid form. Almost all substances expand—become less dense—when heated, and they contract—become more dense—when cooled. Water becomes more dense as it is cooled—until it reaches 4°C (under normal atmospheric conditions). Below 4°C, it becomes less dense. When water freezes, there is about a 9% increase in volume as the molecules arrange themselves into an "open" crystal structure. Compare this with the closer packing the molecules have in the liquid state (see Fig. 15.20). This expansion exerts enough force to break water pipes if the liquid is permitted to freeze in them.

Water is about the only substance we normally encounter in the solid, liquid, and gas phases. An example of water in the gas phase is the water vapor in air, which is the humidity mentioned earlier.

Water is a most unusual compound. Much of life on earth could not exist but for its molecular structure and the unique macroscopic properties that result.

**Figure 15.20** Fortunately for these penguins, water is one of the very few substances whose solid phase is *less* dense than the liquid. Water molecules in solid form (ice) are held in a crystal pattern that has voids between the molecules. When ice melts, the crystal collapses, the molecules are closer together, and the liquid is more dense than the solid. This is why ice floats in water, a solid–liquid property shared by few other substances.

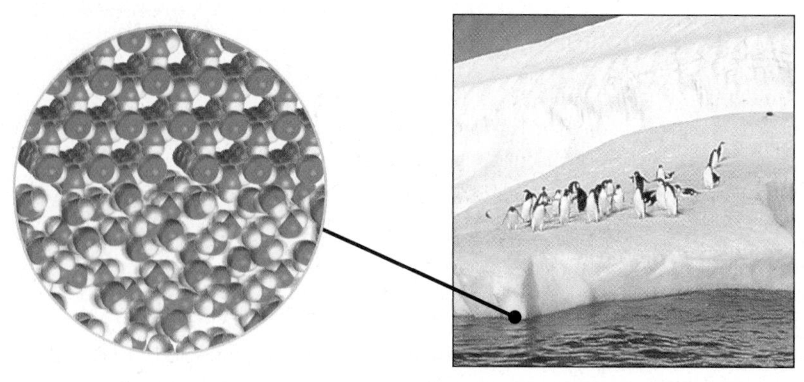

BIOS—V. Bretagnolle/Peter Arnold, Inc.

IBM Research/Peter Arnold, Inc.

A magnet levitates above a super conductor. Materials science researchers are studying methods for preparing new substances with superconducting properties.

## 15.7 The Solid State

**Goal** 12 Distinguish among amorphous, polycrystalline, and crystalline solids.

A rapidly growing area of chemistry is the study of the **solid state** of matter. Many new discoveries are being made, and the traditional subdivisions of chemistry are being reorganized to accommodate the growth of interest in solids. Why is there so much new interest in solids? Materials such as high-temperature superconductors that conduct electricity with no resistance and glass–ceramics that can be safely transferred directly from the stove to the refrigerator are examples of useful products that are driving research in the field known as **materials science.**

Solids can be classified based on their *macroscopic* properties. The differences on the macroscopic scale, however, result from differences at the *particulate* level. In particular, we can classify solids based on the forces holding the particles together. We will start by dividing solids into three general classifications based on the way the particles are arranged in the solid.

A solid whose particles are arranged in a geometric pattern that repeats itself over and over in three dimensions is a **crystalline solid** (Fig. 15.21[*right*] and Fig. 15.22[a]). Each particle occupies a fixed position in the crystal. It can vibrate about that site but cannot move past its neighbors. The high degree of order often leads to large crystals that have a precise geometric shape. In ordinary table salt we can distinguish small cubic crystals of sodium chloride. Large, beautifully formed crystals of minerals such as quartz ($SiO_2$) and fluorite ($CaF_2$) are found in nature.

In an **amorphous solid** such as glass, rubber, or plastic, there is no long-range ordering of the particles in the solid (Fig. 15.21[*left*] and Fig. 15.22[b]). Even though the arrangement around a particular point may appear to be in a regular pattern, the pattern does not repeat itself throughout the solid. From a structural standpoint, we may regard an amorphous solid as an intermediate between the solid and liquid states. In many amorphous solids the particles have some freedom to move with respect to one another. The elasticity of rubber and the tendency of glass to flow when subjected to stress over a long period of time suggest that the particles in these materials are not rigidly fixed in position.

**Polycrystalline solids** are intermediate in particulate organization, falling between disordered amorphous solids and orderly crystalline solids. The particles are arranged into small, orderly crystals, as in a crystalline solid, but then these crystals are randomly arranged to form the solid, as in an amorphous solid (Fig. 15.23). Examples of polycrystalline solids include steels, brasses, and aluminum.

Crystalline solids have characteristic physical properties that identify them. Sodium chloride, for example, melts sharply at 801°C. This is in striking contrast to

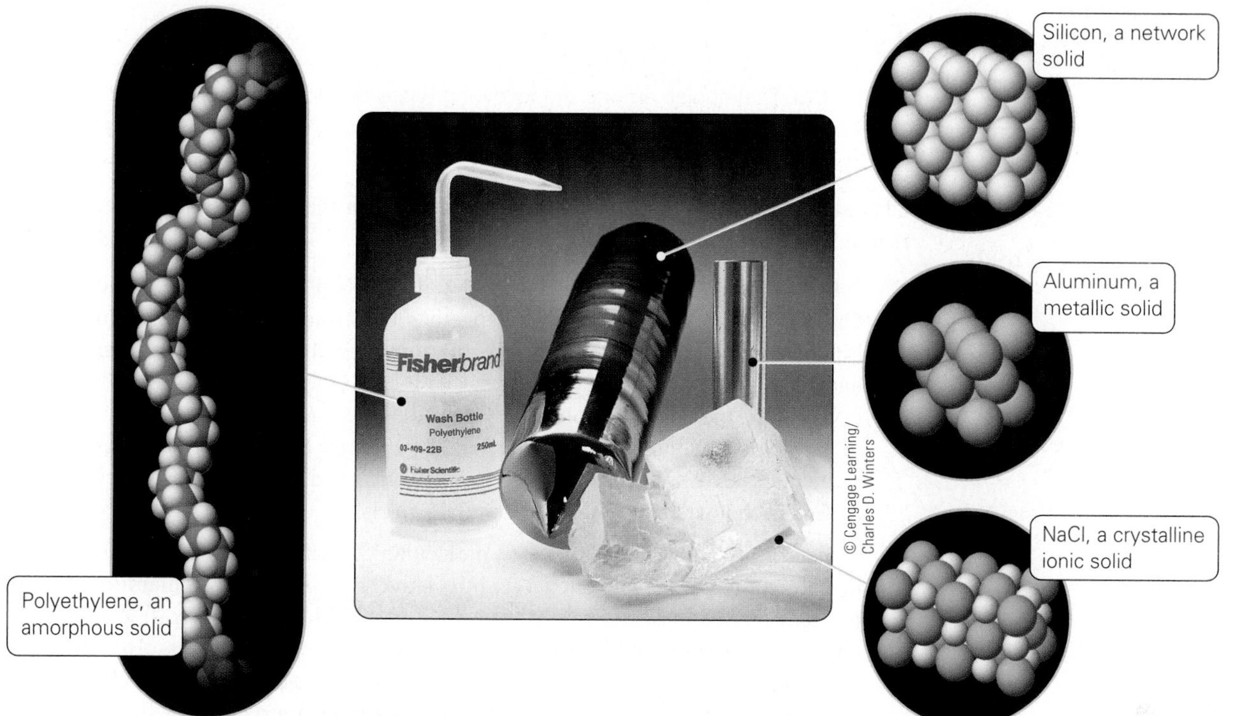


Figure 15.21 Amorphous (*left*) and crystalline (*right*) solids. An amorphous solid, such as the polyethylene bottle shown, has little long-range ordering at the particulate level. In contrast, each of the crystalline solids shown has an orderly, regularly repeating arrangement of the particles.

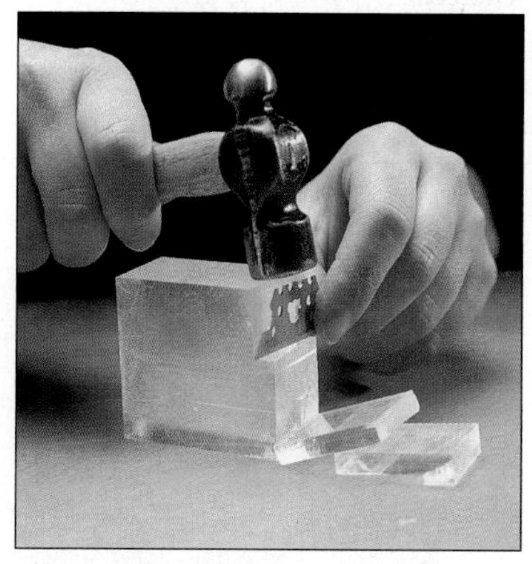

(a) Crystalline solids tend to cleave so that the smaller pieces have smooth and flat faces.

(b) Amorphous solids do not break along lines with predictable patterns.

Figure 15.22 Macroscopic properties of crystalline and amorphous solids. The presence or absence of a regular structure at the particulate level influences the macroscopic shape of the solid.

Figure 15.23 A polycrystalline solid. This microscopic-level photograph shows small, orderly crystals, randomly arranged in solid sphalerite. Sphalerite is the foremost ore of zinc, and it is mostly composed of zinc ions and sulfide ions, with a relatively small quantity of iron(II) ions.

glass, an amorphous solid, which first softens and then slowly liquifies over a wide range of temperatures.

### √ Target Check 15.7

*Identify the main structural differences among crystalline solids, polycrystalline solids, and amorphous solids.*

# 15.8 Types of Crystalline Solids

**Goal** 13 Distinguish among the following types of crystalline solids: ionic, molecular, covalent network, and metallic.

Crystalline solids can be divided into four classes on the basis of the types of forces that hold particles together in the crystal lattice.

**Ionic Crystals** Examples of ionic crystals are NaF, $CaCO_3$, AgCl, and $NH_4Br$. Oppositely charged ions are held together by strong electrostatic forces—ionic bonds. Ionic crystals are typically high-melting and frequently water-soluble, and in the solid state, they have very low electrical conductivities. When ionic crystals are melted or dissolved in water, the ions can move around, so the liquid or the solution conducts electricity readily. Figures 15.24 and 15.25 illustrate the particulate-level structure of two ionic crystals, and Figure 15.26 shows lead(II) sulfide at the macroscopic and particulate levels.

**Molecular Crystals** Examples of molecular crystals include $I_2$ and ICl. Small, discrete molecules are held together by relatively weak intermolecular forces of the types discussed in Section 15.3. Molecular crystals are typically soft, low-melting, and generally (but not always) insoluble in water. They usually dissolve in nonpolar or slightly polar organic solvents such as carbon tetrachloride or chloroform. Substances with a molecular crystal particulate-level structure, with rare exceptions, are nonconductors when pure, even in the liquid state. Sulfur has numerous forms at the molecular level, but the most common form of the solid is $S_8$ molecules, with eight atoms bonded in a ring (Fig. 15.27[a]). When the solid is heated to a liquid under certain conditions, the molecules rearrange into long chains and the substance has the macroscopic characteristics of a plastic (Fig. 15.27[b]).

**Covalent Network Solids** Examples of covalent network solids are diamond, C, and quartz, $SiO_2$. Atoms are covalently bonded to each other to form one large network of indefinite size. There are no small discrete molecules in network solids. In diamond, each carbon atom is covalently bonded to four other carbon atoms to

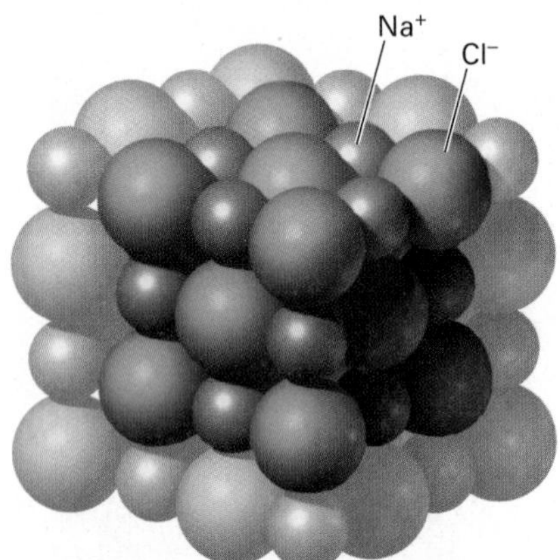


Figure 15.24 A particulate-level model of an ionic crystal. The gray spheres represent sodium ions and the green spheres are chloride ions. The macroscopic physical properties of ionic crystals result from the strong electrostatic attractions among the ions.

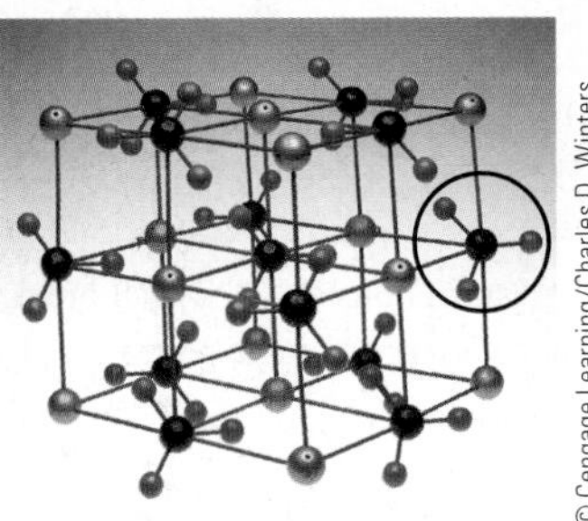



Figure 15.25 A particulate-level model of a calcium carbonate crystal. This is an example of an ionic crystal that has a polyatomic ion. The gray spheres represent calcium ions, $Ca^{2+}$. The black spheres with three red spheres attached are carbonate ions, $CO_3^{2-}$. The circle encloses one carbonate ion.

Figure 15.26 An ionic crystal. The chemical composition of the mineral galena is lead(II) sulfide.

**Figure 15.27** Two forms of sulfur. (a) One form of solid sulfur is an example of a molecular crystal in which distinct molecular units can be identified. (b) One liquid form consists of chains of sulfur atoms.

give a structure that repeats throughout the entire crystal. The structure of silicon dioxide resembles that of diamond in that the atoms are held together by a continuous series of covalent bonds. Each silicon atom is bonded to four oxygen atoms, and each oxygen is bonded to two silicon atoms. Figure 15.28 illustrates quartz crystals.

Covalent network solids, like ionic crystals, have high melting points. A very high temperature is needed to break the covalent bonds in crystals of quartz (about 1700°C) or diamond (about 3500°C). Covalent network solids are almost always insoluble in water or any common solvent. They are generally poor conductors of electricity in either the solid or liquid state.

**Metallic Crystals** Aluminum is an example of a metallic crystal (Fig. 15.29). A simple model of bonding in a metal consists of a crystal of positive ions through which valence electrons move freely. This so-called electron-sea model of a metallic crystal is illustrated in Figure 12.15. Positively charged ions form the backbone of the crystal; the electrons surrounding these ions are not tied down to any particular ion and therefore are not restricted to a particular location. It is because of these freely moving electrons that metals are excellent conductors of electricity.

**P/REVIEW:** Metallic bonding results from the attractive forces between the positively charged metal ions in the crystal and the negatively charged electrons in the "electron sea." See Section 12.8.

Some metallic crystals can appear to be amorphous at the macroscopic level, but when observed at the microscopic level, they are found to be polycrystalline. We

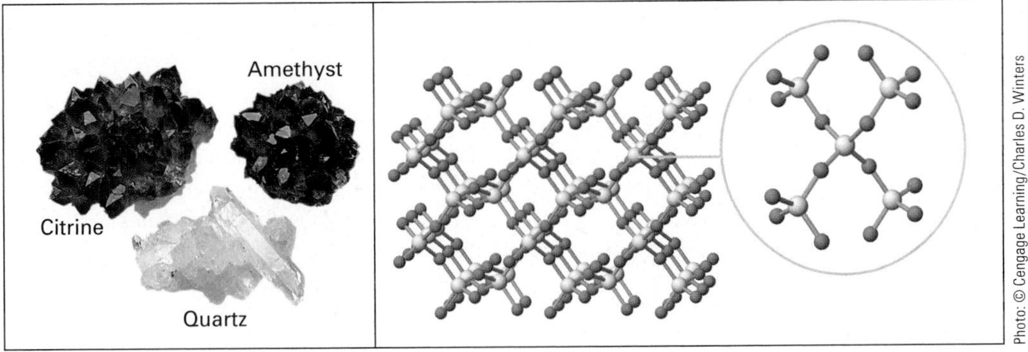


**Figure 15.28** Forms of quartz crystals. Pure quartz, $SiO_2$, is colorless. Amethyst and citrine are mostly composed of silicon and oxygen, but small amounts of iron impurities make the crystals colored. Quartz is an example of a covalent network solid. Atoms are covalently bonded in a regularly repeating pattern, but there are no individual molecules of fixed size.

© Cengage Learning/Charles D. Winters

**Figure 15.29** Aluminum metal. This is an example of a metallic crystal.

**Table 15.3** General Properties of Crystals

| Type | Examples | Properties |
|---|---|---|
| Ionic | $KNO_3$, NaCl, MgO | High-melting; generally water-soluble; brittle; conduct only when melted or dissolved in water |
| Molecular | $C_{10}H_8$, $I_2$ | Low-melting; usually more soluble in organic solvents than in water; nonconductors in pure state |
| Network | $SiO_2$, C | Very high-melting; insoluble in all common solvents; brittle; non- or semiconductors |
| Metallic | Cu, Fe | Wide range of melting points; insoluble in all common solvents; malleable, ductile; good electrical conductors |

can see the tiny individual crystals with a microscope. These individual crystals are arranged randomly.

The general properties of the four kinds of crystalline solids are summarized in Table 15.3.

## √ Target Check 15.8

*Identify the true statements, and rewrite the false statements to make them true.*

a) A high-melting solid that conducts electricity is probably a metal.
b) Covalent network solids are usually good conductors of electricity.
c) A solid that melts at 152°C is probably an ionic crystal.
d) A soluble molecular crystal is a nonconductor of electricity but a good conductor when dissolved.

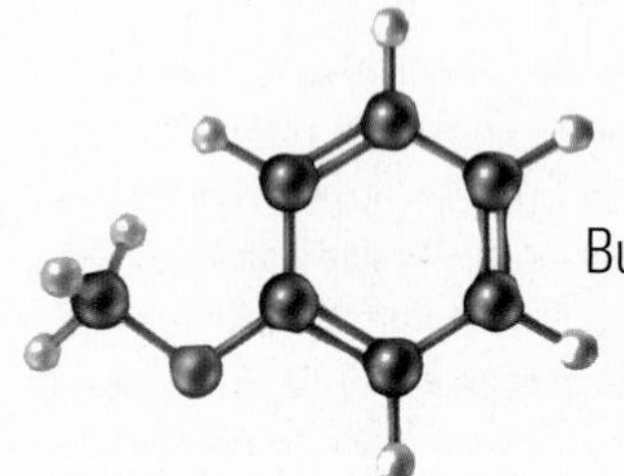

## Everyday Chemistry

### Buckyballs

A long tradition in chemistry holds that the discoverer or inventor of new substances gives the new class of material its nonsystematic name. There are classes of organic compounds called "propellanes" because their Lewis structures look like propellers, "basketanes," like baskets, and "barrelanes," like barrels.

In 1985, when chemists discovered (in soot, of all places) a stable $C_{60}$ cluster whose Lewis structure resembled a geodesic dome, they gave the substance the name buckminsterfullerene after Buckminster Fuller, the inventor of the geodesic dome (Fig. 15.30). Everybody now calls these clusters buckyballs because each cluster looks like a hollow soccer ball (Fig. 15.31).

Chemists were amazed because $C_{60}$ is a new allotrope, or form, of carbon. The graphite and diamond forms of carbon were discovered in prehistoric times, and a small amount of $C_{60}$ was apparently being made every time a sooty fire burned on the planet. Nobody noticed buckyballs because the $C_{60}$ was being oxidized (burned) by the same flame that created it.

Both graphite and diamond are high-melting network solids. A buckyball is a distinct cluster of carbon atoms, a molecule of carbon. The physical and chemical properties of buckyballs were completely undiscovered, and research laboratories around the world jumped into the fray.

Buckyballs have amazing physical properties. They are soft, like graphite. One research group is working to compress the hollow buckyball to about two-thirds of its original volume. Calculations predict that these squeezed buckyballs will be harder than diamond, the hardest known substance. An interesting experiment combining these ideas of softness and great hardness threw buckyballs against a steel surface at 17,000 miles per hour. Showing unprecedented resil-

*Continued on facing page*

© Xuanlu Wang, 2008/Used under license from Shutterstock.com

**Figure 15.30** Geodesic dome at Science World in Vancouver, British Columbia, Canada. The philosopher, architect, and engineer R. Buckminster Fuller invented the geodesic dome. The architectural advantage of its linked hexagon and pentagon structure is that it can cover a large area without internal supports. The similar molecular structure of buckminsterfullerene inspired its discovers to name the molecule in honor of Fuller.

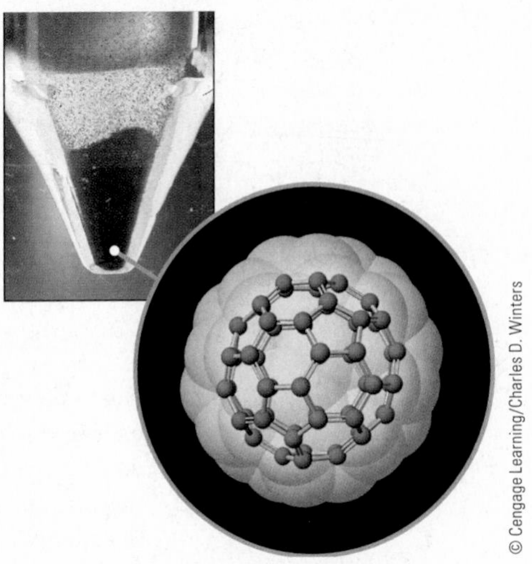

© Cengage Learning/Charles D. Winters

**Figure 15.31** Buckminsterfullerene, $C_{60}$. The photo shows buckminsterfullerene, a black powder, in the tip of a pointed glass tube. The molecular structure of this form of the element carbon is similar to that of a geodesic dome and a soccer ball.

ience, the buckyballs simply bounced back.

The interior of a buckyball is large enough to hold an atom of any element in the periodic table. Researchers wasted no time in putting different metals atoms in the center of buckyballs. Thus resulted a new family of superconductors. Other teams are working on using buckyballs as the source of tiny ball bearings, lightweight batteries, and even superconducting wires just one cluster thick. A group at DuPont (home of Teflon) made buckyballs with carbon–fluorine bonds (like Teflon) and is experimenting with "Teflonballs."

Buckyballs are simply one member of the family of fullerenes, which ranges from about $C_{30}$ up to at least $C_{600}$. Researchers have recently discovered "buckytubes," hollow needle-like tubes that nest within one another. Can we extend buckytubes to form buckyfibers? Are these the first elementary building blocks of a new carbon-based technology? We're just beginning to find out.

Buckminsterfullerene is a superb modern example of the value of fundamental research to science in general and to applied areas in particular. Buckminsterfullerene was discovered by chemists doing experiments trying to determine the role of carbon in space and in the distant stars. The result was dynamic new paths in chemistry, physics, and materials science here on earth.

The American chemists Robert F. Curl, Jr., and Richard E. Smalley and the British chemist Sir Harold W. Kroto were awarded the 1996 Nobel Prize in Chemistry for their discovery of the family of fullerenes.

F. Carter Smith/Corbis Sygma

Robert F. Curl Jr.

© Paul Seheult/Eye Ubiquitous/CORBIS

Sir Harold W. Kroto

© AP Photo/Bob Child

Richard E. Smalley

P/REVIEW: Energy units are discussed more fully in Section 10.7. You may wish to consult that section if you want more information.

P/REVIEW: The symbol $\Delta H$ is used for enthalpy of reaction, or heat of reaction. Thermochemical equations were introduced in Section 10.8. The Greek $\Delta$ represents change. The sign of a $\Delta$ quantity is described more formally in Section 15.10.

P/REVIEW: In Section 15.2, we described molar heat of vaporization, the energy required to change one mole of a liquid to a gas. It is expressed in kJ/mol. In this section we use the more convenient unit kJ/g to express heat of vaporization. Use the molar mass of a substance (g/mol) to convert between kJ/g and kJ/mol, if necessary.

In kJ/mol, $\Delta H_{vap}$ is the enthalpy change when one mole of substance is changed from a liquid to a vapor. This can be expressed in a thermochemical equation. For water, $H_2O(\ell) \rightarrow H_2O(g)$, $\Delta H$ = 40.7 kJ.

## 15.9 Energy and Change of State

**Goal** 14 Given two of the following, calculate the third: (a) mass of a pure substance changing between the liquid and vapor (gaseous) states, (b) heat of vaporization, (c) energy change.

**Goal** 15 Given two of the following, calculate the third: (a) mass of a pure substance changing between the solid and liquid states, (b) heat of fusion, (c) energy change.

Heat energy is added to or removed from a substance when it changes state. It takes energy to melt a solid, an endothermic change; energy is lost by a liquid when it freezes, an exothermic change. In the change between a liquid and a gas, vaporization is endothermic and condensation is exothermic. In this section you will learn how to calculate the energy change that accompanies a change of state.

The SI unit of energy is the **joule (J).** The joule is a small unit of energy. It is quite suitable for use with temperature changes (see the next section) and melting, but it is too small for the larger amounts of energy involved in boiling and chemical reactions. The kilojoule (kJ) is commonly used for these changes.

It has been found experimentally that the energy required to vaporize a substance, q, is proportional to the amount of substance. Amount may be expressed in moles or in grams. The proportionality is changed into an equation by means of a proportionality constant, $\mathbf{\Delta H_{vap}}$, known as the **heat of vaporization:**

$$q \propto m \xrightarrow{\text{the proportionality changes to an equation}} q = \Delta H_{vap} \times m \qquad \textbf{(15.3)}$$

Solving for $\Delta H_{vap}$ gives the defining equation for heat of vaporization:

$$\Delta H \equiv \frac{q}{m} \qquad \textbf{(15.4)}$$

The units of heat of vaporization follow from the equation, energy units per unit quantity. If quantity is expressed in moles, the units are kJ/mol; if expressed in grams, the units are kJ/g. In this chapter we will limit ourselves to weighable quantities and express all heats of vaporization in kJ/g. The heats of vaporization of several substances are given in Table 15.4.

**Table 15.4** Heats of Fusion and Heats of Vaporization

| Substance | Melting Point (°C) | Boiling Point (°C) | Heat of Fusion (J/g) | Heat of Vaporization (kJ/g) |
|---|---|---|---|---|
| Ag | 962 | 2162 | 105 | 2.4 |
| Al | 660 | 2519 | 397 | 10.9 |
| Au | 1064 | 2856 | 63 | 1.7 |
| Bi | 271 | 1564 | 52 | 0.8 |
| Cd | 321 | 767 | 56 | 0.9 |
| Cu | 1084 | 2927 | 206 | 4.7 |
| Fe | 1538 | 2861 | 247 | 6.2 |
| $H_2O$ | 0 | 100 | 333 | 2.26 |
| Hg | 39 | 357 | 11 | 0.3 |
| Na | 98 | 883 | 113 | 4.2 |
| NaCl | 801 | 1413 | 519 | — |
| Ni | 1455 | 2913 | 293 | 6.4 |
| Pb | 327 | 1749 | 23 | 0.9 |
| Zn | 420 | 907 | 112 | 1.8 |

When a vapor condenses to a liquid at the boiling point, the reverse energy change occurs. The energy change is then referred to as the **heat of condensation.** Values are the same as heats of vaporization, except that they are negative. This indicates that heat energy is removed *from* the substance (an exothermic change) instead of being added to it. In solving a problem, use the negative of heat of vaporization if a gas is condensing.

## Active Example 15.3

Calculate the heat of vaporization of a substance if 241 kJ is required to vaporize a 78.2-g sample.

This is a straightforward algebra problem in which the equation is solved for the wanted quantity, known values are substituted, and the answer is calculated. In this case the defining equation, $\Delta H_{vap} \equiv q/m$, is already solved for the wanted quantity. Set up and solve the problem completely.

GIVEN: 241 kJ; 78.2 g WANTED: $\Delta H_{vap}$ in kJ/g

EQUATION: $\Delta H_{vap} \equiv \frac{q}{m} = \frac{241\ \text{kJ}}{78.2\ \text{g}} = 3.08\ \text{kJ/g}$

**P/REVIEW:** In Section 3.8, and specifically in Active Example 3.23, we solve a density problem by an equation and also by dimensional analysis when the density is known. The same choice is available with change-of-state problems when $\Delta H$ is known.

The most common calculation is finding the energy added or removed in changing the state of a given mass of material. In previous examples of defining equations and PER relationships, the property defined has been used as a conversion factor for a dimensional-analysis setup. This time, however, we use the equation form, $q = m \times \Delta H_{vap}$ (Equation 15.3). This approach makes a neater package when change-of-state problems are combined with change-of-temperature problems, as they will be shortly.

## Active Example 15.4

How much energy is needed to vaporize 188 grams of a liquid if its heat of vaporization is 1.13 kJ/g?

All you have to do is plug the numbers and units into the equation. Set up and solve.

GIVEN: 188 g; $\Delta H_{vap} = 1.13$ kJ/g WANTED: Energy (assume kJ)

EQUATION: $q = m \times \Delta H_{vap} = 188\ \text{g} \times \frac{1.13\ \text{kJ}}{\text{g}} = 212\ \text{kJ}$

To melt a crystalline solid, energy must be applied to overcome the forces that hold the crystal structure together. Conceptually similar to heat of vaporization, **the heat of fusion, $\Delta H_{fus}$, of a substance is the energy required to melt one gram of that substance.** (*Fusion* is changing a solid to a liquid by the addition of heat energy; that is, the melting process.) Heats of fusion are generally much smaller than heats of vaporization, so the usual units are joules per gram, J/g. Some typical heats of fusion are given in Table 15.4.

Just as condensation is the opposite of vaporization, freezing is the opposite of melting. The amount of energy released in freezing a sample is identical to the amount of energy required to melt that sample. Accordingly, **heat of solidification** is numerically equal to heat of fusion, but the sign is negative.

The heat flow equation for the change between solid and liquid is like the equation for the liquid-to-gas change:

$$q = m \times \Delta H_{fus} \qquad \textbf{(15.5)}$$

Calculation methods are just like those in vaporization problems.

## Active Example 15.5

Calculate the heat flow when 135 grams of sodium freezes. Express the answer in both joules and kilojoules.

The heat of fusion of sodium may be found in Table 15.4. That and $q = m \times \Delta H_{fus}$ are all you need.

*Given:* 135 g Na; $\Delta H_{fus} = 113$ J/g (from Table 15.4)

*Wanted:* Heat flow (J and kJ)

*Equation:* $q = m \times (-\Delta H_{fus}) = 135 \text{ g Na} \times \frac{-113 \text{ J}}{\text{g Na}} \times \frac{1 \text{ kJ}}{1000 \text{ J}} = -15.3 \text{ kJ}$

$-1.53 \times 10^4$ J

The negative sign is applied to $\Delta H_{fus}$ because the metal is freezing, not melting.

**P/REVIEW:** The J → kJ change in Active Example 15.5 can be made by moving the decimal point three places to the left or by reducing the exponent by 3. Changing joules to kilojoules by moving the decimal point is an example of the larger/smaller rule (Section 3.2) applied to metric conversions (Section 3.4). A quantity may be expressed in a large number of small units (J) or a small number of large units (kJ).

# 15.10 Energy and Change of Temperature: Specific Heat

**Goal** **16** Given three of the following quantities, calculate the fourth: (a) energy change, (b) mass of a pure substance, (c) specific heat of the substance, (d) temperature change, or initial and final temperatures.

Let's look more closely at the algebraic sign of a change, a delta quantity. A Δ quantity is always calculated by subtracting the initial value from the final value. In this section we will work with a change in temperature, ΔT. By definition

$$\Delta T \equiv T_{final} - T_{initial} = T_f - T_i$$

If the temperature rises from 20°C to 25°C, ΔT is positive:

$$\Delta T \equiv T_f - T_i = 25°C - 20°C = 5°C$$

However, if the temperature falls from 25°C to 20°C, ΔT is negative:

$$\Delta T \equiv T_f - T_i = 20°C - 25°C = -5°C$$

You have seen that heat energy must be added if a substance is to melt or boil. Energy is released in the opposite processes of freezing or condensing. Such changes of state are constant-temperature processes. The steam and the boiling water in a kettle are both at 100°C, and they remain there as long as the water boils. The water and the crushed ice in a glass are both at 0°C, and they remain there until all of the ice melts. However, what happens when energy is added or lost but there is no change of state? The temperature changes.

Experiments indicate that the heat flow, q, in heating or cooling a substance is proportional to both the mass of the sample, m, and its temperature change, ΔT. Combining these proportionalities yields

$$q \propto m \text{ and } q \propto \Delta T \xrightarrow{\text{combine the proportionalities}} q \propto m \times \Delta T \qquad (15.6)$$

Introducing a proportionality constant, c,

$$q \propto m \times \Delta T \xrightarrow{\text{the proportionality changes to an equation}} q = m \times c \times \Delta T \qquad (15.7)$$

The units of c may be found by solving the equation for that quantity:

$$c = \frac{q}{m \times \Delta T} \qquad (15.8)$$

We therefore expect energy units divided by the product of mass units and temperature units. In this text, we will express values of c in J/g · °C. You may also encounter J/g · K in other sources.

**P/REVIEW:** When a variable is proportional to two or more other variables, it is proportional to the product of those variables. The proportionality is changed into an equation with a proportionality constant. See Sections 3.8 and 4.6.

## Thinking About Your Thinking

### Proportional Reasoning

**Two pairs of proportionalities are expressed in Equation 15.6: Heat flow is proportional to the amount of substance and to the temperature change of that substance. Consider each of these separately. Think about the difference between having a drop of boiling water coming into contact with your skin versus an entire cup of boiling water. The temperature of the two samples of water is the same, but the mass is not. The amount of heat that transfers from the water to your body is greater when more water is spilled; heat flow is proportional to mass.**

**Now hold constant the quantity variable and consider the effect of ΔT. A drop of 100°C water on your skin causes a more severe burn than a drop of 50°C water because the temperature change as your skin and the water exchange heat energy is greater for the 100°C drop. Heat flow is proportional to temperature change.**

The proportionality constant, c, is a property of a pure substance called its specific heat. **Specific heat is the heat flow required to change the temperature of one gram of a substance by one degree Celsius.** It measures the relative ease with which a substance may be heated or cooled. A substance with a low specific heat, such as aluminum, gains little energy in warming through a given temperature change compared with a substance with high specific heat, such as water. Similarly, when a slice of pizza wrapped in aluminum foil is removed from a hot oven, the foil cools much more rapidly than the pizza sauce (which is largely water). This is a result of the aluminum gaining less energy than the water when both were subjected to the same change in temperature. Specific heats of selected substances are given in Table 15.5.

A substance with high specific heat is best for retaining energy. In one type of solar heating system, rocks gain energy from the sun during the day and release that energy at night. The specific heat of the rocks used is relatively high. The rocks are able to gain a large amount of energy per unit of mass, which is later released when the sun no longer provides energy. At 4.18 J/g · °C, water has one of the highest specific heats of all substances. This is one reason that air temperatures near large bodies of water are usually higher in the winter and lower in the summer than nearby inland temperatures.

In specific-heat problems, you always are given or have available three of the four factors in Equation 15.7, q = m × c × ΔT, and you solve algebraically for the fourth.

**Table 15.5** Selected Specific Heats

| SUBSTANCE | c (J/g · °C) |
|---|---|
| **Elements** | |
| Aluminum(s) | 0.90 |
| Cadmium(s) | 0.23 |
| Cadmium(ℓ) | 0.27 |
| Carbon | |
| Diamond(s) | 0.51 |
| Graphite(s) | 0.71 |
| Cobalt(s) | 0.42 |
| Copper(s) | 0.38 |
| Gold(s) | 0.13 |
| Gold(ℓ) | 0.15 |
| Iron(s) | 0.45 |
| Iron(ℓ) | 0.45 |
| Lead(s) | 0.13 |
| Magnesium(s) | 1.02 |
| Silicon(s) | 0.71 |
| Silver(s) | 0.24 |
| Silver(ℓ) | 0.32 |
| Sulfur(s) | 0.74 |
| Zinc(s) | 0.39 |
| Zinc(ℓ) | 0.51 |
| **Compounds** | |
| Acetone(ℓ) | 2.17 |
| Benzene(ℓ) | 1.74 |
| Carbon tetrachloride(ℓ) | 0.85 |
| Ethanol(ℓ) | 2.44 |
| Methanol(ℓ) | 2.53 |
| Water | |
| Solid (ice) | 2.06 |
| Liquid | 4.18 |
| Gas (steam) | 2.00 |
| **Common Substances** | |
| Concrete(s) | 0.88 |
| Glass(s) | 0.84 |
| Granite(s) | 0.79 |
| Wood(s) | 1.76 |

© Don Mason/Corbis

Effect of high specific heat of water. The ocean stores energy during the day and releases it at night, moderating the air temperature of coastal cities, such as Seattle, which is shown in this photograph.

## Active Example 15.6

How much energy is required to raise the temperature of 475 grams of water for a pot of tea from 14°C to 95°C? Answer in both joules and kilojoules.

The equation $q = m \times c \times \Delta T$ is already solved for the wanted quantity, so the problem may be solved by direct substitution. For $\Delta T$, substitute $T_f - T_i$. As always, include units in your calculation setup.

*GIVEN:* 475 g; $T_i = 14°C$; $T_f = 95°C$; $c = 4.18\ J/g \cdot °C$

*WANTED:* q in J and kJ

*EQUATION:* $q = m \times c \times \Delta T = 475\ \cancel{g} \times \frac{4.18\ \cancel{J}}{\cancel{g} \cdot \cancel{°C}} \times (95 - 14)\cancel{°C} \times \frac{1\ kJ}{1000\ \cancel{J}} = 1.6 \times 10^2\ kJ$

$1.6 \times 10^5\ J$

The joule → kilojoule change can also be calculated by noting that the units get bigger, so the number becomes smaller, and thus the exponent is reduced by 3.

## 15.11 Change in Temperature Plus Change of State

Goal **17** Sketch, interpret, or identify regions in a graph of temperature versus energy for a pure substance over a temperature range from below the melting point to above the boiling point.

Goal **18** Given (a) the mass of a pure substance, (b) $\Delta H_{vap}$ and/or $\Delta H_{fus}$ of the substance, and (c) the average specific heat of the substance in the solid, liquid, and/or vapor state, calculate the total heat flow in going from one state and temperature to another state and temperature.

If you were to take some ice (water in the solid state) from a freezer, place it in a flask, and then apply heat steadily, five things would happen:

1. The ice would warm to its melting point.
2. The ice would melt at the melting point.
3. The water would warm to its boiling point.
4. The water would boil at the boiling point.
5. The steam would become hotter.

The heat flow for each of these steps can be calculated by the methods set forth in Sections 15.9 and 15.10. The specific heats of ice, water, and steam would be used for Steps 1, 3, and 5. The heat of fusion would be used for Step 2, and the heat of vaporization would be used for Step 4. The total heat flow would be the sum of the five separate heat flows.

Figure 15.32 illustrates this process by words, a graph, and the appropriate equation for each step. The shape of the temperature-versus-heat graph is typical for any pure substance. Refer to this illustration as you work through the next Active Example. It will help you to see clearly what is being done in each of the five steps. The general procedure for this kind of problem is

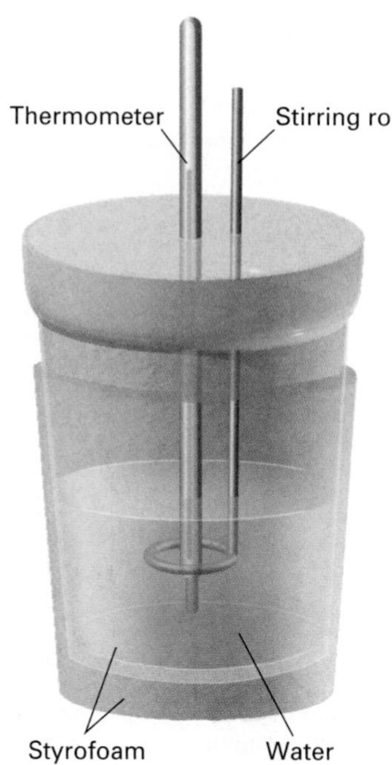


Data for heat-flow problems are obtained from calorimeters, highly insulated containers in which chemical or physical changes occur. This coffee-cup calorimeter is commonly used in introductory college chemistry laboratories.

**How to Calculate Total Heat Flow for a Change in Temperature Plus a Change of State**

**Step 1:** Sketch a graph having the shape shown in Figure 15.32. Mark the starting and ending points for the particular problem. Then mark the beginning and ending points of any change of state between the starting and ending points for the problem.

**Step 2:** Calculate the heat flow, q, for each sloped and horizontal portion of the graph between the starting and ending points.

**Step 3:** Add the heat flows calculated in Step 2. *Caution: Be sure the units are the same, either kilojoules or joules, for all numbers being added.*

## Active Example 15.7

Calculate the heat flow when 45.0 grams of steam, initially at 111°C, is changed to water at 25°C.

Begin by sketching the temperature–heat curve. Your sketch will be like Figure 15.32 but labeled specifically for the present Active Example. Use the letter A to label the initial temperature, 111°C, and the letters B and C to label the beginning and ending of the state changes between the initial temperature and the final temperature. Use the letter D to label the final temperature, 25°C. List each of the labeled temperatures of the y axis of your graph.

Point A is the starting point at 111°C. The horizontal line is drawn at 100°C, the boiling (and condensing) temperature of water. Points B and C are at the ends of the condensing line. Point D is liquid water at 25°C, the final temperature.

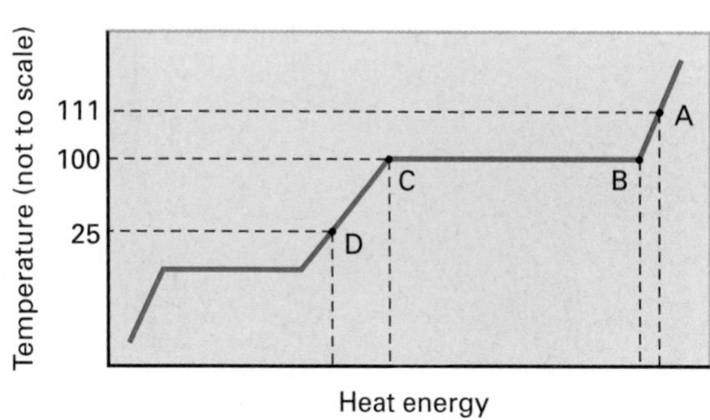


*continued*

TEMPERATURE-HEAT-ENERGY GRAPH FOR A PURE SUBSTANCE

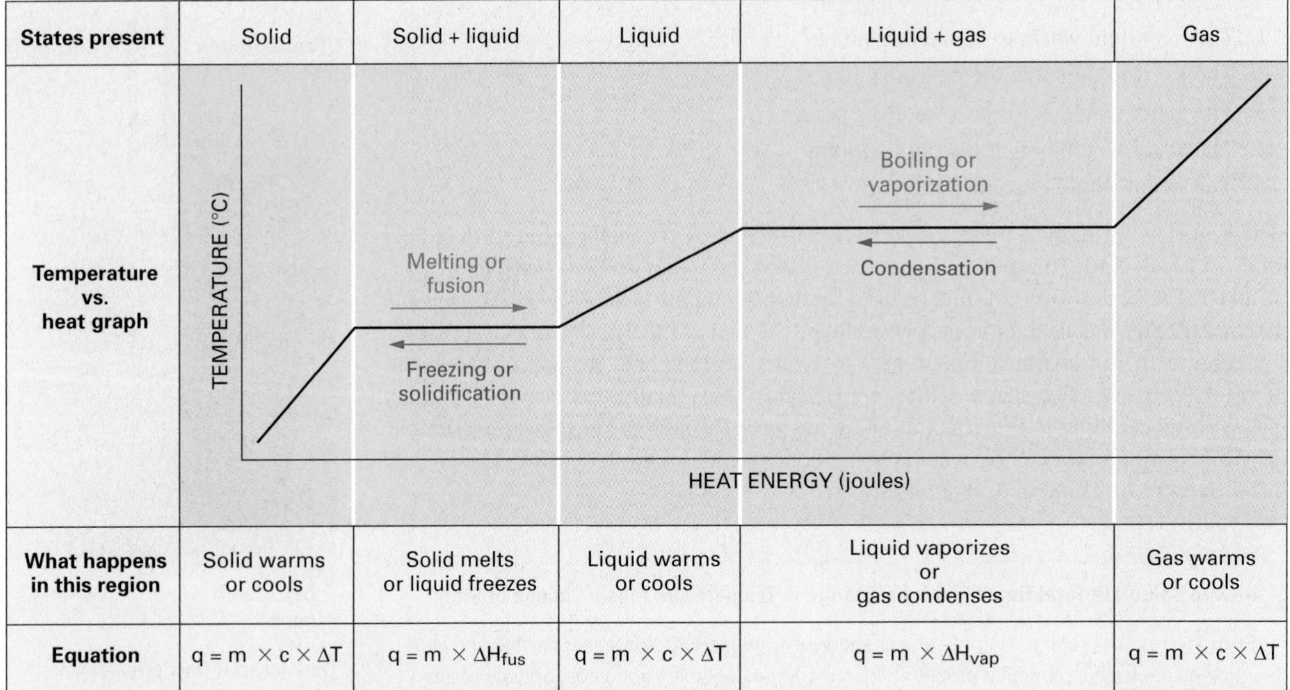

States present
Solid
Solid + liquid
Liquid
Liquid + gas
Gas
Temperature vs. heat graph
TEMPERATURE (°C)
HEAT ENERGY (joules)
Melting or fusion
Freezing or solidification
Boiling or vaporization
Condensation
What happens in this region
Solid warms or cools
Solid melts or liquid freezes
Liquid warms or cools
Liquid vaporizes or gas condenses
Gas warms or cools
Equation
q = m × c × ΔT
q = m × ΔHfus
q = m × c × ΔT
q = m × ΔHvap
q = m × c × ΔT


**Figure 15.32** Temperature–heat–energy graph for state and temperature changes. *Solid column:* When a solid below the freezing point is heated, temperature increases; when cooled, the temperature decreases. There is no change of state; the substance remains a solid. *Solid + liquid column:* At the melting point the solid melts as heat is added, or it freezes as heat is removed. Temperature remains constant during the change between solid and liquid. *Liquid column:* As heat is added to a liquid, its temperature increases; as heat is removed, the temperature goes down. There is no change of state; the substance remains a liquid. *Liquid + gas column:* At the boiling point the liquid boils as heat is added, or it condenses as heat is removed. Temperature remains constant during the change between liquid and gas. *Gas column:* As heat is added to a gas, its temperature increases; as heat is removed, the temperature goes down. There is no change of state; the substance remains a gas.

The heat flow must be calculated for cooling steam from point A to point B, 111°C to 100°C ($q_{A\text{ to }B}$). Then it must be calculated for condensing the steam at 100°C ($q_{B\text{ to }C}$). Finally, q must be calculated for cooling water from 100°C to 25°C ($q_{C\text{ to }D}$).

Calculate $q_{A\text{ to }B}$, cooling the 45.0 g of steam from 111°C to 100°C, using $q = m \times c \times \Delta T$. The value of c for steam is in Table 15.5. Express the final answer in kilojoules.

$$q_{A\text{ to }B} = m \times c \times \Delta T = 45.0\ \cancel{g} \times \frac{2.00\ \cancel{J}}{\cancel{g} \cdot \cancel{°C}} \times (100 - 111)\cancel{°C} \times \frac{1\ \text{kJ}}{1000\ \cancel{J}} = -0.99\ \text{kJ}$$

The minus sign indicates that energy is being lost by the system; the change is exothermic. Your q value will always have the correct sign as long as you carefully apply the $T_{final} - T_{initial}$ rule for a Δ quantity. The two significant figures in the answer come from $\Delta T = -11°C$.

Now find $q_{B \text{ to } C}$, changing 45.0 g of steam at 100°C to liquid water at 100°C. Remember, energy is lost in this process. You'll have to flip back to Table 15.4 to find the value of the heat of vaporization of water.

$$q_{B \text{ to } C} = m \times -\Delta H_{vap} = 45.0 \, \cancel{g} \times \frac{-2.26 \text{ kJ}}{\cancel{g}} = -102 \text{ kJ}$$

The sign of the heat of vaporization is changed to negative to get the heat of condensation.

Now cool the liquid water to 25°C, points C to D on your graph, using $q = m \times c \times \Delta T$ again, but with the value of the specific heat of liquid water.

$$q = m \times c \times \Delta T = 45.0 \, \cancel{g} \times \frac{4.18 \, \cancel{J}}{\cancel{g} \cdot \cancel{°C}} \times (25 - 100)\cancel{°C} \times \frac{1 \text{ kJ}}{1000 \, \cancel{J}} = -14 \text{ kJ}$$

You know the energy for the three steps. Complete the problem by finding the sum of the individual steps.

$$\Sigma q^* = q_{A \text{ to } B} + q_{B \text{ to } C} + q_{C \text{ to } D} = -0.99 \text{ kJ} + (-102 \text{ kJ}) + (-14 \text{ kJ}) = -117 \text{ kJ}$$

## Active Example 15.8

Calculate the total heat flow when 19.6 grams of ice, initially at −12°C, is heated to steam at 115°C.

Start with a graph of temperature versus heat energy. Label all temperatures that will be included in calculations. Use A for the initial temperature of the ice, −12°C, B for the solid at 0°C, C for the liquid at 0°C, and so on. Indicate the state of the water (e.g., solid, solid + liquid, liquid, etc.) on your graph.

*continued*

*The Greek letter sigma, Σ, is used to indicate "the sum of" two or more quantities.

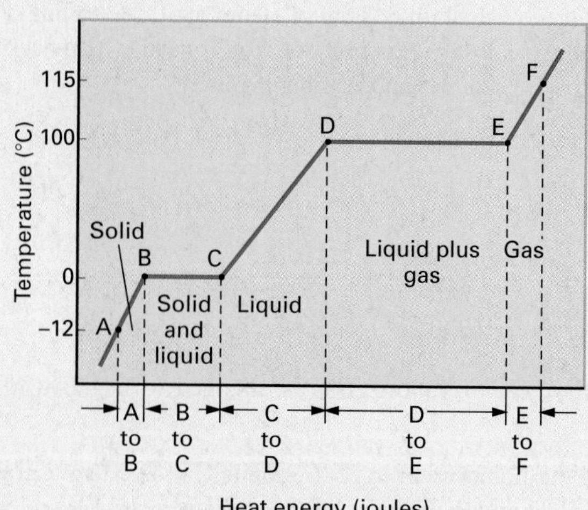


Determine $q_{A\text{ to }B}$, warming the ice from −12°C to its melting point, using $q = m \times c \times \Delta T$ and the value of the specific heat of ice. Use kilojoules as the final energy unit for all steps in the problem.

$$q_{A\text{ to }B} = m \times c \times \Delta T = 19.6\ \cancel{g} \times \frac{2.06\ \cancel{J}}{\cancel{g} \cdot \cancel{^\circ C}} \times [0 - (-12)]\cancel{^\circ C} \times \frac{1\ kJ}{1000\ \cancel{J}} = 0.48\ kJ$$

Your graph indicates that the next calculation is $q_{B\text{ to }C}$, melting ice at 0°C, using $q = m \times \Delta H_{fus}$. Look up the heat of fusion of $H_2O$ in Table 15.4 and complete this step. Remember that we are expressing all heat flows in kilojoules.

$$q_{B\text{ to }C} = m \times \Delta H_{fus} = 19.6\ \cancel{g} \times \frac{333\ \cancel{J}}{\cancel{g}} \times \frac{1\ kJ}{1000\ \cancel{J}} = 6.53\ kJ$$

Continuing to follow the graph, the next region is $q_{C\text{ to }D}$, warming the liquid water from 0°C to 100°C, with $q = m \times c \times \Delta T$. Do the calculation.

$$q_{C\text{ to }D} = m \times c \times \Delta T = 19.6\ \cancel{g} \times \frac{4.18\ \cancel{J}}{\cancel{g} \cdot \cancel{^\circ C}} \times (100 - 0)\cancel{^\circ C} \times \frac{1\ kJ}{1000\ \cancel{J}} = 8.19\ kJ$$

The next region on the graph is $q_{D\text{ to }E}$, vaporizing the liquid at 100°C, which requires the relationship $q = m \times \Delta H_{vap}$. Table 15.4 has values of heats of vaporization. Complete this step.

$$q_{D \text{ to } E} = m \times \Delta H_{vap} = 19.6 \text{ g} \times \frac{2.26 \text{ kJ}}{\text{g}} = 44.3 \text{ kJ}$$

The only region remaining on the graph is $q_{E \text{ to } F}$, heating the steam from 100°C to 115°C, once again requiring $q = m \times c \times \Delta T$. Determine this heat flow.

$$q_{E \text{ to } F} = m \times c \times \Delta T = 19.6 \text{ g} \times \frac{2.00 \text{ J}}{\text{g} \cdot °\text{C}} \times (115 - 100)°\text{C} \times \frac{1 \text{ kJ}}{1000 \text{ J}} = 0.59 \text{ kJ}$$

In *Step 3* of our procedure, the individual heat flows are added. All values are known in kilojoules, so use that unit and find the total heat flow, as requested in the problem statement.

$$\begin{aligned}\Sigma q &= q_{A \text{ to } B} + q_{B \text{ to } C} + q_{C \text{ to } D} + q_{D \text{ to } E} + q_{E \text{ to } F} \\ &= 0.48 \text{ kJ} + 6.53 \text{ kJ} + 8.19 \text{ kJ} + 44.3 \text{ kJ} + 0.59 \text{ kJ} = 60.1 \text{ kJ}\end{aligned}$$

# Chapter 15 in Review

Most of the key terms and concepts and many others appear in the Glossary. Use your Glossary regularly.

### 15.1 Dalton's Law of Partial Pressures

Goal 1 Given the partial pressure of each component in a mixture of gases, find the total pressure.

Goal 2 Given the total pressure of a gaseous mixture and the partial pressures of all components except one, or information from which those partial pressures can be obtained, find the partial pressure of the remaining component.

**Key Terms and Concepts: Dalton's Law of Partial Pressures**

### 15.2 Properties of Liquids

Goal 3 Explain the differences between the physical behavior of liquids and gases in terms of the relative distances among particles and the effect of those distances on intermolecular forces.

Goal 4 For two liquids, given comparative values of physical properties that depend on intermolecular attractions, predict the relative strengths of those attractions; or, given a comparison of the strengths of the intermolecular attractions, predict the relative values of physical properties that the attractions cause.

**Key Terms and Concepts: Boiling point, equilibrium vapor pressure, heat of condensation, molar heat of vaporization, surface tension, vapor pressure, viscosity**

### 15.3 Types of Intermolecular Forces

Goal 5 Identify and describe or explain induced dipole forces, dipole forces, and hydrogen bonds.

Goal 6 Given the structure of a molecule, or information from which it may be determined, identify the significant intermolecular forces present.

Goal 7 Given the molecular structures of two substances, or information from which they may be obtained, compare or predict relative values of physical properties that are related to them.

**Key Terms and Concepts: Dipole and dipole forces, dispersion forces, hydrogen bond, induced dipole forces, London (dispersion) forces**

### 15.4 Liquid–Vapor Equilibrium

Goal 8 Describe or explain the equilibrium between a liquid and its own vapor and the process by which it is reached.

Goal 9 Describe the relationship between vapor pressure and temperature for a liquid–vapor system in equilibrium; explain this relationship in terms of the kinetic molecular theory.

**Key Terms and Concepts: Condensation, dynamic equilibrium, equilibrium, equilibrium vapor pressure, reversible change, reversible reaction**

### 15.5 The Boiling Process

Goal 10 Describe the process of boiling and the relationships among boiling point, vapor pressure, and surrounding pressure.

**Key Terms and Concepts: Boiling, boiling point, normal boiling point**

**15.6 Water—An "Unusual" Compound**

Goal 11 Describe the typical relative density relationship between the solid and liquid phase of a substance, and explain why water is an exception to this trend.

**15.7 The Solid State**

Goal 12 Distinguish among amorphous, polycrystalline, and crystalline solids.

**Key Terms and Concepts: Amorphous solid, crystalline solid, materials science, polycrystalline solid, solid state**

**15.8 Types of Crystalline Solids**

Goal 13 Distinguish among the following types of crystalline solids: ionic, molecular, covalent network, and metallic.

**Key Terms and Concepts: Covalent network solid, ionic crystal, metallic crystal, molecular crystal**

**15.9 Energy and Change of State**

Goal 14 Given two of the following, calculate the third: (a) mass of a pure substance changing between the liquid and vapor (gaseous) states, (b) heat of vaporization, (c) energy change.

Goal 15 Given two of the following, calculate the third: (a) mass of a pure substance changing between the solid and liquid states, (b) heat of fusion, (c) energy change.

**Key Terms and Concepts: Heat of condensation, Heat of fusion ($\Delta H_{fus}$), heat of solidification, heat of vaporization ($\Delta H_{vap}$), joule (J)**

**15.10 Energy and Change of Temperature: Specific Heat**

Goal 16 Given three of the following quantities, calculate the fourth: (a) energy change, (b) mass of a pure substance, (c) specific heat of the substance, (d) temperature change, or initial and final temperatures.

**Key Terms and Concepts: Specific heat**

**15.11 Change in Temperature Plus Change of State**

Goal 17 Sketch, interpret, or identify regions in a graph of temperature versus energy for a pure substance over a temperature range from below the melting point to above the boiling point.

Goal 18 Given (a) the mass of a pure substance, (b) $\Delta H_{vap}$ and/or $\Delta H_{fus}$ of the substance, and (c) the average specific heat of the substance in the solid, liquid, and/or vapor state, calculate the total heat flow in going from one state and temperature to another state and temperature.

## Study Hints and Pitfalls to Avoid

In this chapter, we discuss the three common states of matter and the energy changes that occur when matter changes states. As you review this chapter, try to get the big picture. This is what will enable you to *explain* and *predict*, which are the key words in the early performance goals in this chapter.

The concept of a dynamic equilibrium is introduced in the liquid–vapor system. Take time to understand the idea of this equilibrium and you will find it much easier to understand the saturated solution equilibrium in Chapter 16. The ideas are almost the same. The concept comes up again in Chapter 18 on chemical equilibrium.

Recognize that combination specific-heat and change-of-state problems are a *group* of problems in which each problem is related to one short stretch on the horizontal axis of the temperature–heat curve (Fig. 15.32). It helps a lot to sketch the graph, as you were guided to do in the last Active Example of the chapter. Individually, the problems are relatively easy. Whatever is associated with a sloped line is to be solved by the specific-heat equation, $q = m \times c \times \Delta T$. The calculation associated with a horizontal (change-of-state) line is $q = m \times \Delta H_{vap}$ or $m \times \Delta H_{fus}$. When you get all the q's calculated, add them for the final answer. When you reach this point, remember that joules cannot be added to kilojoules! Be sure all heat flows are expressed in the same units before you add them, and be sure to apply the addition rule for significant figures.

## Concept-Linking Exercises

*Write a brief description of the relationships among each of the following groups of terms or phrases. Answers to the Concept-Linking Exercises are given after the answers to the Target Checks in Appendix III.*

1. Dipole forces, induced dipole forces, dispersion forces, London forces, London dispersion forces, hydrogen bonds, covalent bond, ionic bond
2. Stick-togetherness, vapor pressure, molar heat of vaporization, boiling point, viscosity, surface tension
3. Boiling point, normal boiling point, temperature, vapor pressure
4. Dynamic equilibrium, liquid–vapor equilibrium, evaporation rate, condensation rate, temperature, vapor pressure
5. Ionic crystal, molecular crystal, covalent network solid, metallic crystal
6. Amorphous solid, crystalline solid, polycrystalline solid
7. Heat of vaporization, heat of condensation, heat of fusion, heat of solidification, specific heat

## Small-Group Discussion Questions

*Small-Group Discussion Questions are for group work, either in class or under the guidance of a leader during a discussion section.*

1. Explain how the ideal gas model relates to Dalton's Law of Partial Pressures.
2. Give a particulate-level explanation for vapor pressure, heat of vaporization, boiling point, viscosity, and surface tension.

3. Draw Lewis diagrams and sketch models of electron clouds to illustrate each of the three types of intermolecular forces. Explain why each force occurs.

4. What are the differences among ionic bonds, metallic bonds, covalent bonds, and hydrogen bonds? Explain in as much detail as possible.

5. Sketch a plot of condensation rate versus time for a volatile liquid in a sealed container from the time the container is sealed until a dynamic liquid–vapor equilibrium is achieved. On the same axes, also plot evaporation rate versus time. Describe the factors that affect the condensation and evaporation rates.

6. Sketch a kinetic energy distribution curve for a liquid. Use the curve to explain why only some of the molecules in a liquid sample have enough energy to evaporate. Then draw a second curve on the same axes for the same liquid at a higher temperature. Use the curves to explain why vapor pressure is higher at higher temperatures.

7. Sketch a pressure-versus-temperature curve to illustrate the nature of the change in vapor pressure with temperature for a typical liquid. If this liquid has induced dipole forces as the primary intermolecular forces, how would a curve for a liquid with dipole forces compare? Explain and draw this curve on your graph. Also explain the relative vapor pressure and draw a curve for a liquid with hydrogen bonding as the primary intermolecular force.

8. Define the term *boiling point*. Explain how it is possible to boil water at temperatures below and above 100°C. How is boiling point related to a vapor-pressure-versus-temperature curve for a liquid?

9. Distinguish among crystalline solids, amorphous solids, and polycrystalline solids at the macroscopic and particulate levels. Also distinguish among ionic crystals, molecular crystals, covalent network solids, and metallic crystals at both the macroscopic and particulate levels.

10. A student conducts a coffee cup calorimetry experiment to determine the specific heat of an unknown metal. First, she measures 30.0 mL of room temperature water at 23.3°C into the calorimeter. She then adds 30.0 mL of water heated to 52.3°C to the calorimeter, and through graphical extrapolation, the instantaneous final temperature of the mixture is determined to be 36.2°C. She then measures 100.0 mL of water at 23.3°C to the dry calorimeter. She had a 72.45 g sample of the metal in a boiling water bath, which was at 98.7°C. She then transferred the metal to the water in the calorimeter, and the final temperature of the water and the metal was 25.9°C. Calculate the specific heat of the metal.

# Questions, Exercises, and Problems

*Blue-numbered questions are answered in Appendix III. ■ denotes problems assignable in OWL. In the first section of Questions, Exercises, and Problems, similar exercises and problems are paired in consecutive odd-even number combinations.*

### Section 15.1: Dalton's Law of Partial Pressures

1. A mixture of helium and argon occupies $1 \times 10^2$ L. The partial pressure of helium is 0.6 atm, and the partial pressure of argon is 0.4 atm. What are the partial volumes of the two gases? Explain your answer.

2. ■ A mixture of argon and carbon dioxide gases is maintained in a 8.05 L flask at a temperature of 43°C. If the partial pressure of argon is 0.326 atm and the partial pressure of carbon dioxide is 0.234 atm, what is the total pressure in the flask?

3. Atmospheric pressure is the total pressure of the gaseous mixture called air. Atmospheric pressure is 749 torr on a day that the partial pressures of nitrogen, oxygen, and argon are 584 torr, 144 torr, and 19 torr, respectively. What is the partial pressure of all the other gases in the air on that day?

4. ■ The stopcock connecting a 1.00 L bulb containing oxygen gas at a pressure of 540 torr and a 1.00 L bulb containing helium gas at a pressure of 776 torr is opened, and the gases are allowed to mix. Assuming that the temperature remains constant, what is the final pressure in the system?

5. A sample of "wet" hydrogen gas was collected over water (see Fig. 15.2). The sample was adjusted so that it was at room temperature and pressure, 19°C and 733 mm Hg. Water vapor pressure at 19°C is 16.5 mm Hg. What is the partial pressure of the "dry" hydrogen gas?

6. ■ Sodium metal reacts with water to produce hydrogen gas according to the following equation: $2\ Na(s) + 2\ H_2O(\ell) \rightarrow 2\ NaOH(aq) + H_2(g)$. In an experiment, the $H_2$ gas is collected over water in a vessel where the total pressure is 745 torr and the temperature is 20°C, at which temperature the vapor pressure of water is 17.5 torr. Under these conditions, what is the partial pressure of $H_2$? If the wet $H_2$ gas formed occupies a volume of 8.04 L, what number of moles of $H_2$ is formed?

### Section 15.2: Properties of Liquids

7. Why will two gases mix with each other more quickly than two liquids?

8. Why is the liquid state of a substance more dense than the gaseous state?

9. Why are intermolecular attractions stronger in the liquid state than in the gaseous state?

10. Explain why water is less compressible than air.

11. How do intermolecular attractive forces influence the boiling point of a pure substance?

12. How are intermolecular attractive forces and equilibrium vapor pressure related? Suggest a reason for this relationship.

13. Why does molar heat of vaporization depend on the strength of intermolecular forces?

14. What relationship exists between viscosity and intermolecular forces?

15. A tall glass cylinder is filled to a depth of 1 meter with water. Another tall glass cylinder is filled to the same

depth with syrup. Identical ball bearings are dropped into each tube at the same instant. In which tube will the ball bearings reach the bottom first? Explain your prediction in terms of viscosity and intermolecular attractive forces.

16. Which liquid is more viscous, water or motor oil? In which liquid do you suppose the intermolecular attractions are stronger? Explain.

17. If water spills on a laboratory desktop, it usually spreads over the surface, wetting any papers or books in its path. If mercury spills, it neither spreads nor makes paper wet, but forms little drops that are easy to combine into pools by pushing them together. Suggest an explanation for these facts in terms of the apparent surface tension and intermolecular attractive forces in mercury and water.

18. A drop of honey and a drop of water of identical volumes are placed on a plate. The water drop forms a large shallow pool, but the honey drop forms a circular blob with a much smaller diameter. Compare the surface tensions of the two liquids. Which liquid has stronger intermolecular attractive forces? Explain.

19. The level at which a duck floats on water is determined more by the thin oil film that covers its feathers than by a body density that is lower than the density of water. The water does not "mix" with the oil, and therefore does not penetrate the feathers. If, however, a few drops of wetting agent are placed in the water near the duck, the poor duck will sink to its neck. State the effect of a wetting agent on surface tension and intermolecular attractions of water.

20. The cleansing ability of soap depends in large part on its ability to change the surface tension in water. How do you suppose soap affects the surface tension of water? Explain.

*Questions 21 through 24: The table below gives the normal boiling and melting points for three nitrogen oxides.*

| | NO | $N_2O$ | $NO_2$ |
|---|---|---|---|
| Boiling point | −152°C | −88.5°C | +21.2°C |
| Melting point | −164°C | −90.8°C | −11.2°C |

21. Which of the three oxides would you expect to have the highest molar heat of vaporization? Explain how you reached your conclusion.

22. Samples of the three substances are side by side at a temperature at which all are solids. The temperature is raised gradually. List the compounds in the order at which they will begin to melt.

23. Which of the three oxides would you expect to have a measurable vapor pressure at −90°C? Explain your answer.

24. Which of the three oxides would you expect to have the lowest viscosity at −90°C? Justify your conclusion.

**Section 15.3: Types of Intermolecular Forces**

25. Other things being equal, which produces stronger intermolecular attractions, induced dipole forces or dipole forces? What "other things being *un*equal" would reverse this order of attractions?

26. Suggest a molecular structure in which hydrogen bonding may be present, but its contribution to intermolecular attractions is less than the contribution of induced dipole forces. Justify your suggestion.

27. What are the principal intermolecular forces in each of the following compounds: $NH(CH_3)_2$, $CH_2F_2$, $C_3H_8$?

28. ■ Identify the principal intermolecular forces in each of the following compounds: NOCl, $NH_2Cl$, $SiCl_4$.

29. Compare dipole forces and hydrogen bonds. How are they different, and how are they similar?

30. Given an ionic compound and a polar molecular compound of about the same molar mass, which is likely to have the higher melting point? Why? In explaining your answer, identify the interparticle forces present and the roles they play.

*Questions 31 through 34: On the basis of molecular size, molecular polarity, and hydrogen bonding, predict for each pair of compounds the one that has the higher boiling point. State the reason for your choice. Assume molecular size is related to molar mass.*

31. $CH_4$ and $NH_3$

32. $CH_4$ and $CCl_4$

33. Ar and Ne

34. $H_2S$ and $PH_3$

35. What feature of the hydrogen atom, when bonded to an appropriate second element, is largely responsible for the strength of hydrogen bonding between molecules?

36. Identify elements to which hydrogen atoms must be bonded if hydrogen bonding is to be a significant intermolecular attractive force. How are these elements different from other elements to which hydrogen might be bonded? Explain why the difference is important.

37. Of the three types of intermolecular forces, which one(s) (a) increase with molecular size; (b) account for the high melting point, boiling point, and other abnormal properties of water?

38. Of the three types of intermolecular forces, which one(s) operate(s) (a) in all molecular substances; (b) between all polar molecules?

39. Identify the intermolecular forces present in each of the following:

```
a)   H  H                 b)   H
     |  |                      |
  H—C—C—O                  H—C—O     H
     |  |   \                  |   \  /
     H  H    H                 H    C
                                   / \
                                  H   H

c)   H  H  H              d)      H
     |  |  |                      |
  H—C—C—C—H               H—N—C—H
     |  |  |                  |  |
     O  O  O                  H  H
    /  /  /
   H  H  H

e)   H  H
     |  |
  H—C—C—H
     |  |
     H  H
```

40. Identify the intermolecular forces present in each of the following:

a) H—C≡N b) O=C=O

c) $PH_3$ (P bonded to three H) d) $CHCl_3$ (C bonded to H and three Cl) e) $BF_3$ (B bonded to three F)

41. Predict which compound, $CO_2$ or $CS_2$, has the higher melting and boiling points. Explain your prediction.

42. ■ Select the best answer to complete the following statement. The boiling point of ICl (97°C) is higher than the boiling point of $Br_2$ (59°C) because:

a) the molecular mass of ICl is 162.4 u, while that of $Br_2$ is 159.8 u

b) there is hydrogen bonding in ICl, but not in $Br_2$

c) ICl is an ionic compound, while $Br_2$ is a molecular compound

d) ICl is polar, while $Br_2$ is nonpolar

e) induced dipole forces are much stronger for ICl than for $Br_2$

43. Predict which compound, $CH_4$ or $CH_3F$, has the higher vapor pressure as a liquid at a given temperature. Explain your prediction.

44. Predict which compound, $SO_2$ or $CO_2$, has the higher vapor pressure as a liquid at a given temperature. Explain your prediction.

**Section 15.4: Liquid–Vapor Equilibrium**

45. What is the meaning of *equilibrium*?

46. Why do we describe a liquid–vapor equilibrium as a *dynamic* equilibrium?

47. Explain why the rate of evaporation from a liquid depends on temperature.

48. ■ Use the following vapor pressure data to answer the questions:

| | Liquid | Vapor Pressure (torr) | Temperature (°C) |
|---|---|---|---|
| A | $CH_3COOCH_3$ | 400 | 40.0 |
| B | $C_7H_{16}$ | 400 | 78.0 |

In which liquid are the intermolecular attractive forces the strongest? Explain. Will the vapor pressure of $CH_3COOCH_3$ at 78°C be higher or lower than 400 torr? Explain.

**Section 15.5: The Boiling Process**

49. ■ The vapor pressure of a certain compound at 20°C is 906 torr. Is the substance a gas or a liquid at an external pressure of 760 torr? Explain.

50. Define *boiling point*. Draw a vapor-pressure-versus-temperature curve and locate the normal boiling point on it.

51. Liquid feed water is delivered to modern boilers at a temperature well above the normal boiling point of water. Explain how this is possible.

52. Normally, a gas may be condensed by cooling it. Suggest a second method and explain why it will work.

53. Explain why low-boiling liquids usually have low molar heats of vaporization.

54. ■ Given the following vapor pressure data:

| | Liquid | Vapor Pressure (torr) | Temperature (°C) |
|---|---|---|---|
| A | $CS_2$ | 400 | 28.0 |
| B | $C_8H_{18}$ | 400 | 104.0 |

In which liquid are the intermolecular attractive forces the strongest? Explain. Which liquid would be expected to have the highest normal boiling point? Explain.

55. At 20°C the vapor pressure of substance M is 520 torr; of substance N, 634 torr. Which substance will have the lower boiling point? The lower molar heat of vaporization?

56. ■ The molar heat of vaporization of substance X is 34 kJ/mol; of substance Y, 27 kJ/mol. Which substance would be expected to have the higher normal boiling point? The higher vapor pressure at 25°C?

**Section 15.7: The Solid State**

57. Is ice a crystalline solid or an amorphous solid? On what properties do you base your conclusion?

58. Compare amorphous, crystalline, and polycrystalline solids in terms of structure. How do crystalline and amorphous solids differ in physical properties? Explain the difference.

**Section 15.8: Types of Crystalline Solids**

*Questions 59 and 60: The physical properties of four solids are tabulated below. In each case, state whether the solid is most likely to be ionic, molecular, metallic, or a covalent network solid.*

| | Solid | Melting Point | Water Solubility | Conductivity (Pure) | Type of Solid |
|---|---|---|---|---|---|
| 59. | A | 2000°C | Insoluble | Nonconductor | ________ |
| | B | 1050°C | Soluble | Nonconductor | ________ |
| 60. | C | 150°C | Insoluble | Nonconductor | ________ |
| | D | 1450°C | Insoluble | Excellent | ________ |

**Section 15.9: Energy and Change of State**

*See Table 15.4 for heats of fusion and vaporization.*

61. A student is to find the heat of vaporization of isopropyl alcohol (rubbing alcohol). She vaporizes 61.2 g of the liquid at its boiling point and measures the energy required as 44.8 kJ. What heat of vaporization does she report?

62. ■ It is observed that 58.2 kJ of energy are released when a 26.4 g sample of an unknown liquid condenses. What is the heat of vaporization of the unknown liquid in kJ/g?

63. Calculate the energy released as 227 grams of sodium vapor condenses.

64. ■ The following information is given for benzene, $C_6H_6$, at 1 atm:

boiling point = 80°C $\Delta H_{vap} = 0.393$ kJ/g

melting point = 6°C $\Delta H_{fus} = 127$ J/g

How many kJ of energy are needed to vaporize a 28.6-g sample of liquid benzene at its normal boiling point of 80°C?

65. 79.4 kJ was released by the condensation of a sample of ethyl alcohol. If $\Delta H_{vap} = 0.880$ kJ/g, what was the mass of the sample?

66. ■ The following information is given for magnesium at 1 atm:

boiling point = 1090°C  $\Delta H_{vap} = 5.42$ kJ/g
melting point = 649°C  $\Delta H_{fus} = 368$ J/g

Heat is added to a sample of solid magnesium at its normal melting point of 649°C. How many grams of magnesium will melt if 10.5 kJ of energy is added?

67. Acetone, $C_3H_6O$, is a highly volatile solvent sometimes used as a cleansing agent prior to vaccination. It evaporates quickly from the skin, making the skin feel cold. How much energy is absorbed by 23.8 g of acetone as it evaporates if its molar heat of vaporization is 32.0 kJ/mol?

68. ■ The following information is given for cadmium at 1 atm:

boiling point = 765°C  $\Delta H_{vap} = 0.890$ kJ/g
melting point = 321°C  $\Delta H_{fus} = 54.4$ J/g

What is the energy change, q in kJ, for the process of freezing a 22.4-g sample of liquid cadmium at its normal melting point of 321°C?

69. Calculate the energy lost when 3.30 kg of lead freezes.

70. How much energy is required to melt 35.4 g of gold?

71. ■ 36.9 g of an unknown metal releases 2.51 kJ of energy in freezing. What is the heat of fusion of that metal?

72. An energy input of 7.08 kJ is required to melt 46.9 g of naphthalene, which can be used as mothballs. What is the heat of fusion of naphthalene?

73. A piece of zinc releases 4.45 kJ while freezing. What is the mass of the sample?

74. ■ Calculate the number of grams of silver that can be changed from a solid to a liquid by 11.3 kJ.

### Section 15.10: Energy and Change of Temperature: Specific Heat

*See Table 15.5 for specific heat values.*

75. Samples of two different metals, A and B, have the same mass. Both samples absorb the same amount of energy. The temperature of A increases by 11°C, and the sample of B increases by 13°C. Which metal has the higher specific heat? Explain your reasoning.

76. ■ In the laboratory a student finds that it takes 27.6 joules to increase the temperature of 10.7 grams of gaseous xenon from 22.7 to 39.9 degrees Celsius. What is the measured specific heat of xenon?

77. Find the quantity of energy released (in joules) as 467 grams of zinc cools from 68°C to 31°C.

78. ■ How much energy is required to raise the temperature of 10.4 grams of solid magnesium from 22.4°C to 39.5°C?

79. How much energy (kJ) is released when 2.30 kilograms of gold is cooled from 88°C to 22°C?

80. ■ What is the energy change when the temperature of 10.1 grams of solid silver is decreased from 38.1°C to 23.6°C?

81. The mass of a handful of copper coins is 144 grams. The coins are at a temperature of 33°C. If they lose 1.47 kJ when they are tossed in a fountain and drop to the fountain's water temperature, what is that temperature?

82. ■ A sample of solid sulfur is heated with an electrical coil. If 105 joules of energy is added to a 12.4-gram sample initially at 24.4°C, what is the final temperature of the sulfur?

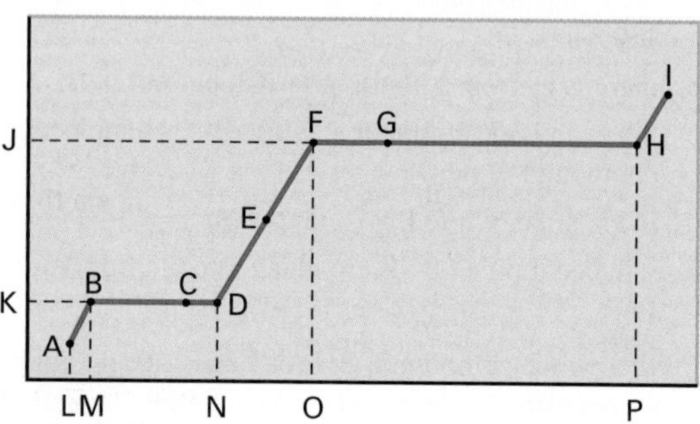


**Figure 15.33**

### Section 15.11: Change in Temperature Plus Change of State

*Questions 83 through 92: Figure 15.33 is a graph of temperature versus energy for a sample of a pure substance. Assume that letters J through P on the horizontal and vertical axes represent numbers and that expressions such as R − S or X + Y + Z represent arithmetic operations to be performed with those numbers.*

83. Identify by letter the boiling and freezing points in Figure 15.33.

84. What values are plotted, both vertically and horizontally?

85. Identify all points on the curve in Figure 15.33 where the substance is entirely gas.

86. Identify in Figure 15.33 all points on the curve where the substance is entirely liquid.

87. Identify in Figure 15.33 all points on the curve where the substance is partly solid and partly liquid.

88. Identify all points on the curve in Figure 15.33 where the substance is partly liquid and partly gaseous.

89. Describe the physical changes that occur as energy N − P is removed from the sample.

90. Describe what happens physically as the energy represented by N − M is added to the sample.

91. Using letters from the graph, show how you would calculate the energy required to boil the liquid at its boiling point.

92. Using letters from the graph, write the expression for the energy required to raise the temperature of the liquid from the freezing point to the boiling point.

93. A 127-gram piece of ice is removed from a refrigerator at −11°C. It is placed in a bowl where it melts and eventually warms to room temperature, 21°C. Calculate the amount of energy the sample has gained from the atmosphere.

94. ■ The following information is given for *n*-pentane at 1 atm:

boiling point = 36.20°C $\Delta H_{vap}$ (36.20°C) = 357.6 J/g
melting point = −129.7°C $\Delta H_{fus}$ (−129.7°C) = 116.7 J/g
specific heat gas = 1.650 J/g °C
specific heat liquid = 2.280 J/g °C

A 26.10-g sample of liquid *n*-pentane is initially at −51.40°C. If the sample is heated at constant pressure (P = 1 atm), how many kJ of energy are needed to raise the temperature of the sample to 65.30°C?

95. ■ A home melting pot is used for a metal casting hobby. At the end of a work period, the pot contains 689 grams of zinc at 552°C. How much energy will be released as the molten metal cools, solidifies, and cools further to room temperature, 21°C? Find the necessary data from the tables in the textbook.

96. ■ The following information is given for chromium at 1 atm:

boiling point = 2672°C $\Delta H_{vap}$ (2672°C) = 5874 J/g
melting point = 1857°C $\Delta H_{fus}$ (1857°C) = 281.5 J/g
specific heat gas = 0.4600 J/g °C
specific heat liquid = 0.9370 J/g °C

A 36.90-g sample of solid chromium is initially at 1837°C. If the sample is heated at constant pressure (P = 1 atm), how many kJ of heat are needed to raise the temperature of the sample to 2068°C?

97. A certain "white metal" alloy of lead, antimony, and bismuth melts at 264°C, and its heat of fusion is 29 J/g. Its average specific heat is 0.21 J/g °C as a liquid and 0.27 J/g °C as a solid. How much energy is required to heat 941 kg of the alloy in a melting pot from a starting temperature of 26°C to its operating temperature, 339°C?

98. ■ The following information is given for bismuth at 1 atm:

boiling point = 1627°C $\Delta H_{vap}$ (1627°C) = 822.9 J/g
melting point = 271.0°C $\Delta H_{fus}$ (271.0°C) = 52.60 J/g
specific heat gas = 0.1260 J/g °C
specific heat liquid = 0.1510 J/g °C

A 22.80-g sample of liquid bismuth at 553.0°C is poured into a mold and allowed to cool to 28.0°C. How many kJ of energy are released in this process?

## GENERAL QUESTIONS

99. Distinguish precisely, and in scientific terms, the differences among items in each of the following groups.
   a) Intermolecular forces, chemical bonds
   b) Vapor pressure, equilibrium vapor pressure
   c) Molar heat of vaporization, heat of vaporization
   d) Dipole forces, induced dipole forces, dispersion forces, London forces, hydrogen bonds
   e) Evaporation, vaporization, boiling, condensation
   f) Fusion, solidification
   g) Boiling point, normal boiling point
   h) Amorphous solid, crystalline solid, polycrystalline solid
   i) Ionic, molecular, covalent network, metallic crystals
   j) Heat of vaporization, heat of condensation
   k) Heat of fusion, heat of solidification
   l) Specific heat, heat of vaporization, heat of fusion

100. Classify each of the following statements as true or false.
   a) Intermolecular attractions are stronger in liquids than in gases.
   b) Substances with weak intermolecular attractions generally have low vapor pressures.
   c) Liquids with high molar heats of vaporization usually are more viscous than liquids with low molar heats of vaporization.
   d) A substance with a relatively high surface tension usually has a very low boiling point.
   e) All other things being equal, hydrogen bonds are weaker than induced dipole or dipole forces.
   f) Induced dipole forces become very strong between large molecules.
   g) Other things being equal, nonpolar molecules have stronger intermolecular attractions than polar molecules.
   h) The essential feature of a dynamic equilibrium is that the rates of opposing changes are equal.
   i) Equilibrium vapor pressure depends on the concentration of a vapor above its own liquid.
   j) The heat of vaporization is equal to the heat of fusion, but with opposite sign.
   k) The boiling point of a liquid is a fixed property of the liquid.
   l) If you break (shatter) an amorphous solid, it will break in straight lines, but if you break a crystal, it will break in curved lines.
   m) Ionic crystals are seldom soluble in water.
   n) Molecular crystals are nearly always soluble in water.
   o) The numerical value of heat of vaporization is always larger than the numerical value of heat of condensation.
   p) The units of heat of fusion are kJ/g · °C.
   q) The temperature of water drops while it is freezing.
   r) Specific heat is concerned with a change in temperature.

101. Identify the intermolecular attractions in $CH_3OH$ and $CH_3F$. Which of the two substances do you expect will have the higher boiling point and which will have the higher equilibrium vapor pressure? Justify your choices.

102. Under what circumstances might you find that a substance having only induced dipole forces is more viscous that a substance that exhibits hydrogen bonding?

## MORE CHALLENGING PROBLEMS

103. A liquid in a beaker is placed in an airtight cylinder. The liquid evaporates until equilibrium is established between the liquid and vapor states. The piston in the cylinder is suddenly adjusted to reduce the volume of the cylinder. If there is no change in temperature, what changes will occur, if any, in the rates of evaporation and condensation?

104. Three closed containers have identical volumes. A beaker containing a large quantity of ether, a highly volatile liquid, is placed in Container A. It evaporates until equilibrium is reached with a substantial amount of ether remaining. A beaker with a small amount of ether is placed in Container B. The ether all evaporates. A beaker with an intermediate amount of ether is placed in Container C. It evaporates until it reaches equilibrium with only a small amount of ether remaining. Compare the final ether vapor pressures in the three containers. Explain your answer.

105. ■ The equilibrium vapor pressure of water at 24°C is 22.4 torr. A sealed flask contains air at 24°C and 757 torr and a glass vial filled with liquid water. The vial is broken, allowing some of the water to evaporate. What is the maximum pressure this system can reach?

*Questions 106 and 107 are based on the apparatus shown in Active Figure 15.34. Study the caption that explains how vapor pressure is measured, explore the interactive version of the figure, and then answer the questions.*

106. Assume you are about to use the apparatus to determine the equilibrium vapor pressure of a volatile liquid. Would you expect the vapor pressure shown by the manometer to (a) increase uniformly until equilibrium is reached; (b) increase rapidly at first, and then slowly as equilibrium is reached; or (c) increase slowly at first and rapidly as equilibrium is approached? Justify your answer.

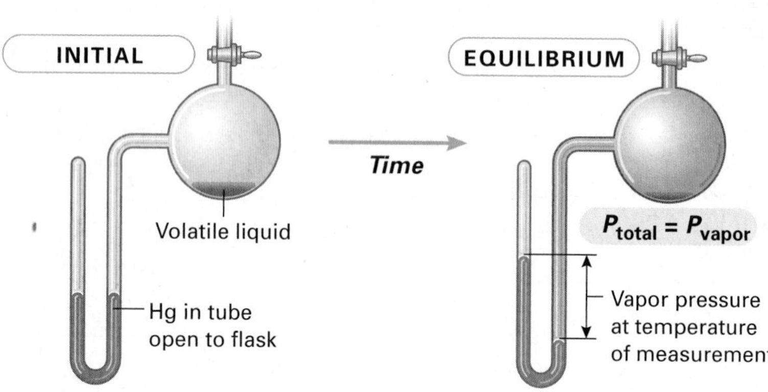


**Active Figure 15.34** Measurement of vapor pressure. Initially, the flask contains the liquid whose vapor pressure is to be measured. The flask, tubes, and manometer above the mercury in the right leg are all at atmospheric pressure. The mercury in the left leg of the manometer is also at atmospheric pressure, so the mercury levels are the same in the two legs. To measure vapor pressure, evaporation occurs until equilibrium is reached. Vapor causes an increase in pressure that is measured directly by the difference in the mercury levels of the two legs of the manometer. **Watch this active figure at *http://www.cengage.com/cracolice.***

107. Suppose that all of the liquid initially introduced to the flask evaporated. Why and how could this occur? Explain in terms of evaporation and condensation rates. How would the vapor pressure shown by the manometer compare with the equilibrium vapor pressure at the existing temperature? Is there further action you can take to complete the vapor pressure measurement, or is it necessary to start over? Justify your answer.

108. An industrial process requires boiling a liquid whose boiling point is so high that maintenance costs on associated pumping equipment are prohibitive. Suggest a way this problem might be solved.

109. ■ A calorimeter contains 72.0 grams of water at 19.2°C. A 141-gram piece of tin is heated to 89.0°C and dropped into the water. The entire system eventually reaches 25.5°C. Assuming all of the energy gained by the water comes from the cooling of the tin—no energy loss to the calorimeter or the surroundings—calculate the specific heat of the tin.

110. ■ A 54.1-g aluminum ice tray in a home refrigerator holds 408 g of water. Calculate the energy that must be removed from the tray and its contents to reduce the temperature from 17°C to 0°C, freeze the water, and drop the temperature of the tray and ice to −9°C. Assume the specific heat of aluminum remains constant over the temperature range involved.

111. The labels have come off the bottles of two white crystalline solids. You know one is sugar and the other is potassium sulfate. Suggest a safe test by which you could determine which is which.

112. The melting point of an amorphous solid is not always a definite value as it should be for a pure substance. Suggest a reason for this.

113. Why does dew form overnight?

114. ■ It is a hot summer day and Chris wants a glass of lemonade. There is none in the refrigerator, so a new batch is prepared from freshly squeezed lemons. When finished, there are 175 grams of lemonade at 23°C. That is not a very refreshing temperature, so it must be cooled with ice. But Chris doesn't like ice in lemonade! Therefore, just enough ice is used to cool the lemonade to 5°C. Of course, the ice will melt and reach the same temperature. If the ice starts at −8°C, and if the specific heat of lemonade is the same as that of water, how many grams of ice does Chris use? Assume there is no heat transfer to or from the surroundings. Answer in two significant figures.

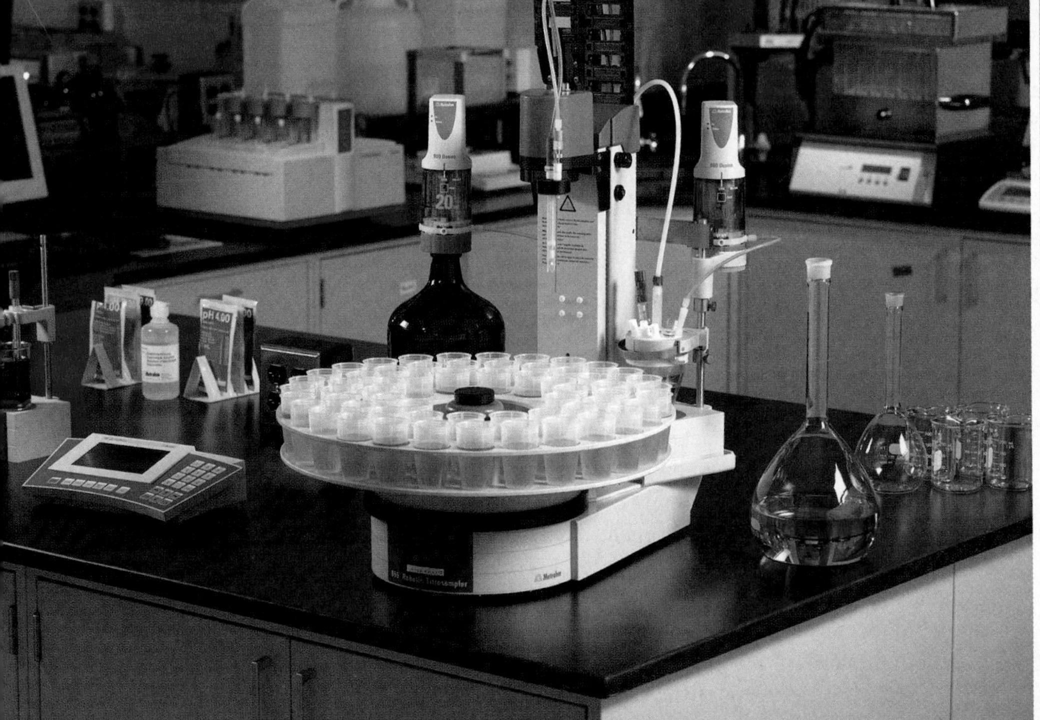

Courtesy of Metrohm USA, Inc.

# 16

Many chemical reactions commonly occur in solution, including those performed in research and instructional laboratories, industrial manufacturing plants, and even your home. This photograph shows an automated titration apparatus, which is often used in analytical chemistry laboratories. Titration is the very careful addition of one solution to another by a device that measures the volume of solution required to react with a measured amount of another dissolved substance. Titration is a tool used by many scientists including those who work in environmental analysis, the food and pharmaceutical industries, the health sciences, and agriculture.

# Solutions

In this chapter we examine solutions. Solutions are everywhere, both inside and outside the chemistry laboratory. All natural waters are solutions. What we call "fresh water" has a very low concentration of dissolved substances; the concentration is much higher with ocean water, or "salt water." "Hard water" has calcium and magnesium salts dissolved in it. Rainwater is nearly pure, but even rainwater is a solution of atmospheric gases in very low concentrations. The small amount of oxygen dissolved in rivers, lakes, and oceans is mighty important to fish, which cannot survive without it.

OWL

Online homework for this chapter may be assigned in OWL

## CONTENTS

**P/REVIEW:** Pure substances have definite, unchanging physical and chemical properties. The properties of a mixture, however, depend on how much of each component is in the mixture. As the composition changes, so do the properties. This is shown in Figure 2.9, Section 2.4.

Technically, the oceans and the atmosphere are not solutions because they are not homogeneous. However, laboratory-size samples of both satisfy the definition.

## 16.1 The Characteristics of a Solution

**Goal** 1 Define the term *solution,* and, given a description of a substance, determine if it is a solution.

**A solution is a homogeneous mixture.** This implies uniform distribution of solution components so that a sample taken from any part of the solution will have the same composition as any other sample of the same solution. Two solutions made up of the same substances may, however, have different compositions. A solution of ammonia in water, for example, may contain 1% ammonia by mass, or 2%, 5%, 20.3%, . . . up to the 29% solution called "concentrated ammonia." This leads to variable physical properties, which are determined by the mixture's composition.

A solution may exist in any of the three states—gas, liquid, or solid. Air is a gaseous solution made up of nitrogen, oxygen, argon, and other gases in small amounts. Like oxygen in water, dissolved carbon dioxide in carbonated beverages is a familiar liquid solution of a gas. Alcohol in water is an example of the solution of two liquids, and the oceans are liquid solutions of dissolved solids. Solid-state solutions are common in the form of metal alloys.

Particle size distinguishes solutions from other mixtures. Dispersed particles in solutions, which may be atoms, ions, or molecules, are very small—generally less than $5 \times 10^{-7}$ cm in diameter. Particles of this size do not settle on standing, and they are too small to see.

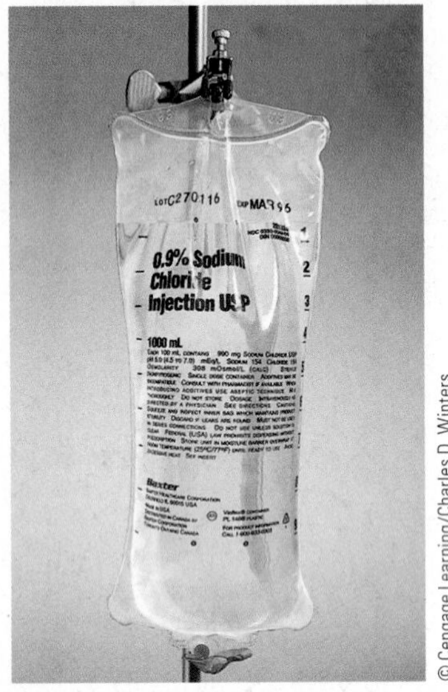



Dehydrated patients receive a sodium chloride (saline) solution that has a concentration of dissolved substances similar to the concentration in cells of the body.

### √ Target Check 16.1

*Identify true statements, and rewrite the false statements to make them true.*

a) At a given instant, a solution has a definite percentage composition.
b) The physical properties of a solution of A in B are variable.
c) A solution is always made up of two pure substances.
d) Different parts of a solution can be detected visually.

## 16.2 Solution Terminology

**Goal** 2 Distinguish among terms in the following groups: *solute* and *solvent; concentrated* and *dilute; solubility, saturated, unsaturated,* and *supersaturated; miscible* and *immiscible.*

In discussing solutions, we use a language of closely related and sometimes overlapping terms. We will now identify and define these terms.

**Solute and Solvent** When solids or gases are dissolved in liquids, the solid or gas is said to be the **solute** and the liquid the **solvent** (Fig. 16.1). More generally, the solute is taken to be the substance present in a relatively small amount. The medium in which the solute is dissolved is the solvent. The distinction is not precise, however. Water is capable of dissolving more than its own mass of some solids, but water continues to be called the solvent. In alcohol–water solutions, either liquid may be the more abundant, and in a given context, either might be called the solute or solvent.

**Concentrated and Dilute** A **concentrated** solution has a *relatively* large quantity of a specific solute per unit amount of solution, and a **dilute** solution has a *relatively* small quantity of the same solute per unit amount of solution (Fig. 16.2). The terms compare concentrations of two solutions of the *same solute and solvent.* They carry no other quantitative meaning.

**Solubility, Saturated, and Unsaturated** **Solubility** is a measure of how much solute will dissolve in a given amount of solvent at a given temperature. It is sometimes

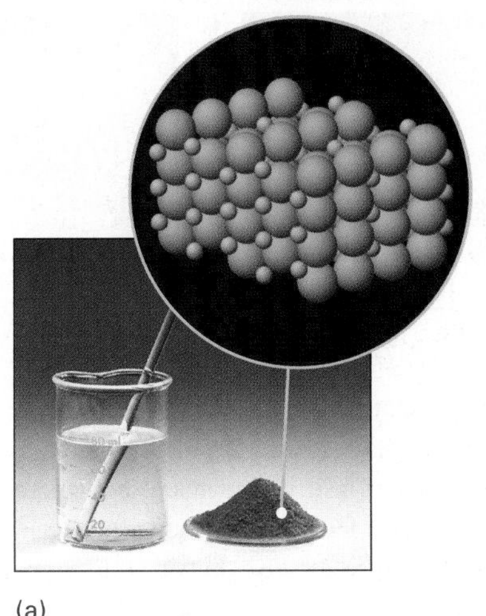

**Figure 16.1** Solute and solvent. (a) When a solid is dissolved in water, the solid is the solute and water is the solvent. The solute in this photograph is solid copper(II) chloride. (b) The solution formed is a homogeneous mixture. Hydrated copper(II) ions and hydrated chloride ions are mixed among water molecules. At the macroscopic level, the solution has the same appearance throughout.

expressed by giving the number of grams of solute that will dissolve in 100 grams of solvent. A solution whose concentration is at the solubility limit for a given temperature is a **saturated** solution. If the concentration of a solute is less than the solubility limit, the solution is **unsaturated.**

**Supersaturated Solutions** Under carefully controlled conditions, a solution can be produced in which the concentration of the solute is greater than the normal solubility limit. Such a solution is said to be **supersaturated.** A supersaturated solution of sodium acetate, for example, may be prepared by dissolving 80 grams of the salt in 100 grams of water at about 50°C. If the solution is cooled to room temperature without stirring, shaking, or other disturbance, all 80 grams of the solute will remain in solution, even though the solubility at 20°C is only 47 grams per 100 grams of water. The solution is unstable, however. A slight physical disturbance or the addition of a single crystal of solid sodium acetate can start crystallization. This proceeds quickly until the solution concentration reaches the saturated solution solubility limit (Fig. 16.3).

**Miscible and Immiscible** *Miscible* and *immiscible* are terms customarily limited to solutions of liquids in liquids. If two liquids dissolve in each other in all proportions, they are said to be **miscible** in each other (Active Fig. 16.4[a]). Alcohol and water, for example, are miscible liquids. Liquids that are insoluble in each other, such as oil and water, are **immiscible** (Active Fig. 16.4[b]). Some liquid pairs will mix appreciably with each other, but in limited proportions; they are said to be *partially miscible.*

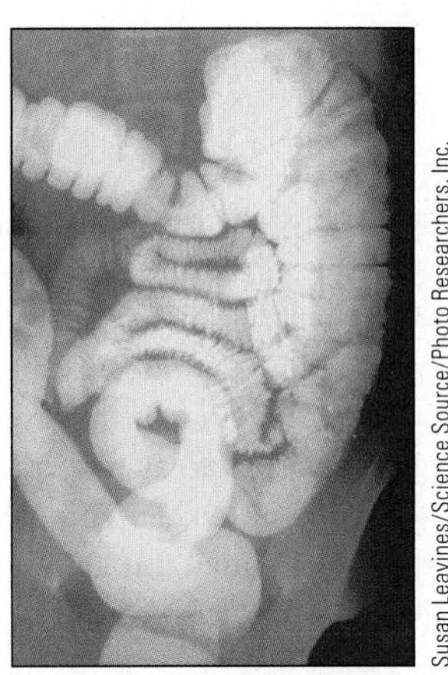


Although soluble barium-ion-containing compounds are toxic to humans, barium sulfate is used for x-rays of the digestive tract because its solubility is very low and it improves the visibility of the tract on the image.

**Figure 16.2** Concentrated and dilute. A copper wire reacts with an aqueous silver nitrate solution, forming silver metal and a copper(II) nitrate solution. The blue color of the solution is due to the hydrated copper(II) ions. (a) When the reaction has proceeded for a relatively short period of time, the color is light, indicating a dilute solution. (b) After a longer time span, the blue becomes much more intense, indicating a relatively large concentration of copper(II) ions. The solution is concentrated.

(a)

(b)

(c)

(d)

(e)



**Figure 16.3** Supersaturated solution. (a) A supersaturated solution of sodium acetate was prepared by dissolving solid sodium acetate in warm water. The solution was then allowed to cool without physical disturbance. Note that, at the macroscopic level, you cannot tell that the solution is supersaturated. (b) A single grain of sodium acetate is added to the solution to initiate crystallization. (c and d) Crystallization continues as long as the solution exceeds its solubility limit at the present temperature of the system. (e) When the solution reaches its solubility limit, a large quantity of solid sodium acetate is in the flask. At the macroscopic level, it appears as if the crystal is static. However, as you will learn in the next section, the solid continues to dissolve, but at a rate equal to the crystallization rate.

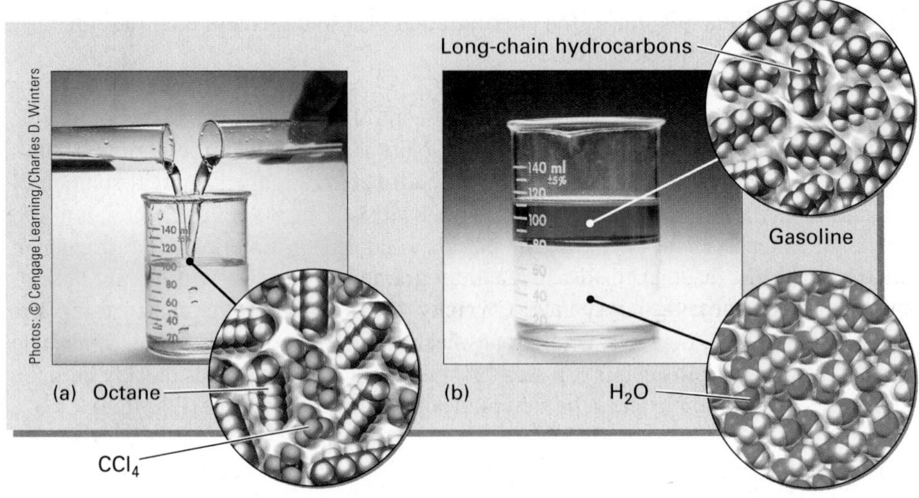


**Active Figure 16.4** Miscible and immiscible. (a) Octane and carbon tetrachloride dissolve in each other, and thus they are miscible. (b) Gasoline (with some red dye added for visibility) and water are insoluble in each other. They are immiscible. **Watch this active figure at *http://www.cengage.com/cracolice.***

> **√ Target Check 16.2**
>
> *0.100 g of A is dissolved in $1.00 \times 10^3$ mL of water, and 10.0 g of B is dissolved in $5.00 \times 10^2$ mL of water. Identify or explain why you cannot identify:*
>
> a) The solute and solvent in each solution
> b) The solution that is more likely to be saturated
> c) The "dilute" solution

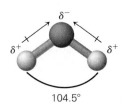

$\delta^-$ $\delta^+$ $\delta^+$ 104.5°

**Figure 16.5** The water molecule. The 104.5° H—O—H bond angle and the asymmetric distribution of the electrons bonding the hydrogen atoms (gray) to the oxygen atom (red) make water molecules highly polar. There is a partial negative charge near the oxygen atom, and there is a partial positive charge between the hydrogen atoms. The lowercase Greek delta, δ, is used to indicate a partial charge.

## 16.3 The Formation of a Solution

**Goal 3** Describe the formation of a saturated solution from the time excess solid solute is first placed into a liquid solvent.

**Goal 4** Identify and explain the factors that determine the time required to dissolve a given amount of solute or to reach equilibrium.

Most solutions of interest to a chemist, biologist, or geologist are water solutions. The water molecule has unique particulate-level properties that make it a good macroscopic-level solvent. Studies of the water molecule show that it is **polar** (Fig. 16.5). A polar molecule is one with an asymmetrical (unbalanced) distribution of charge, resulting

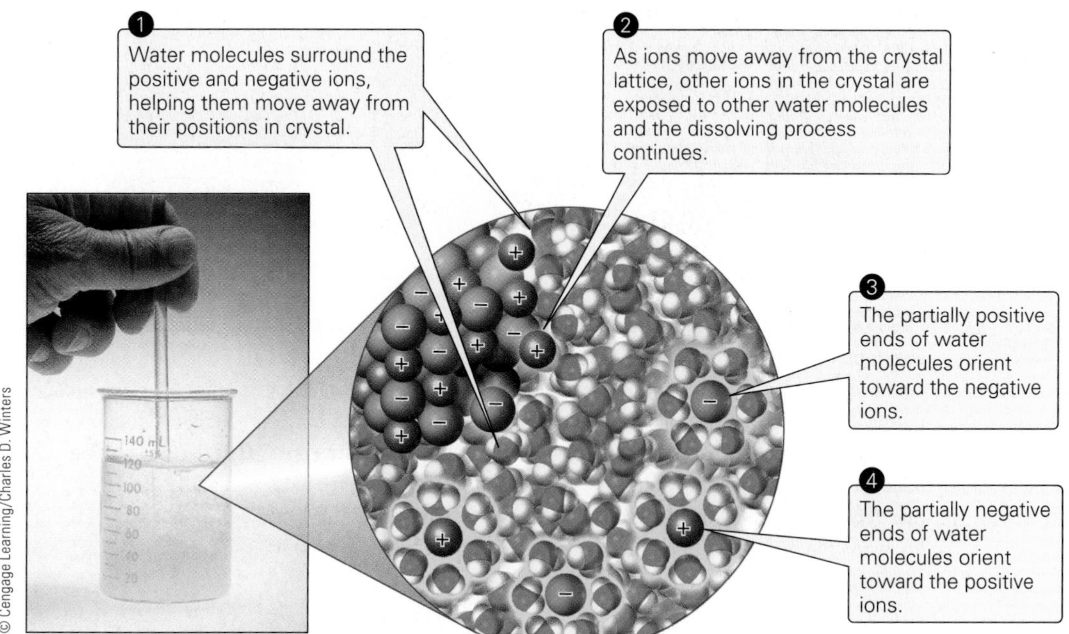

1
Water molecules surround the positive and negative ions, helping them move away from their positions in crystal.
2
As ions move away from the crystal lattice, other ions in the crystal are exposed to other water molecules and the dissolving process continues.
3
The partially positive ends of water molecules orient toward the negative ions.
4
The partially negative ends of water molecules orient toward the positive ions.
140 mL
120
100
80
60
40
20
© Cengage Learning/Charles D. Winters


**Figure 16.6** Dissolving an ionic solute in water. The negative ions of the solute are pulled from the crystal by relatively positive hydrogen regions in polar water molecules. Similarly, positive ions are attracted by the relatively negative oxygen region in the molecule. In solution, the ions are surrounded by those regions of the water molecules that have the opposite charge. These are hydrated ions.

in positive and negative poles. The negatively charged electrons of the water molecule spend more time near the oxygen atom, which causes that area of the molecule to have a partial negative charge. The area midway between the two hydrogen atoms exhibits a partial positive charge.

When a soluble ionic crystal is placed in water, the negatively charged ions at the crystal's surface are attracted by the positive region of the polar water molecules (Fig. 16.6). A "tug of war" for the negative ions begins. Water molecules tend to pull them from the crystal, while neighboring positive ions tend to hold them in the crystal. In a similar way, positive ions at the crystal's surface are attracted to the negative portion of the water molecules and are torn from the crystal. Once released, the ions are surrounded by the polar water molecules. Such ions are said to be **hydrated.**

**P/REVIEW:** Determination of molecular polarity (Section 13.5) follows from an analysis of bond polarity (Section 12.4) and molecular geometry (Section 13.3).

The dissolving process is reversible. As the dissolved solute particles move randomly through the solution, they come into contact with the undissolved solute or with each other and crystallize, that is, return to the solid state. For NaCl this process may be represented by the "reversible reaction" equation

$$NaCl(s) \rightleftharpoons Na^+(aq) + Cl^-(aq)$$

**P/REVIEW:** Reversible changes are those that can proceed in either direction, as suggested by the double arrow in the equation. In one direction, species on the left of the equation are reactants and species on the right are products; in the reverse direction, products are on the left and reactants are on the right. The change may be either chemical (Section 8.7) or physical (Section 15.4). Dissolving is a physical change. Though the formulas of the reactants and products look different, there has been no change at the particulate level. NaCl(s) is an assembly of $Na^+$ and $Cl^-$ ions in a crystal, as shown in Figure 12.2, Section 12.2.

The rate per unit of surface area at which a solute dissolves depends primarily on temperature. The rate of crystallization per unit area depends primarily on the concentration of the solute at the crystal surface. When the rates of dissolving and crystallization become the same, the solution is saturated at its solubility limit at the existing temperature. A **dynamic equilibrium** is reached. The process is described in the caption of Figure 16.7.

The time required to dissolve a given amount of solute—or to reach equilibrium, if excess solute is present—depends on several factors:

1. The dissolving process depends on surface area. A finely divided solid offers more surface area per unit of mass than a coarsely divided solid. Therefore, a finely divided solid dissolves more rapidly.
2. In a still solution, concentration builds up at the solute surface, causing a higher crystallization rate than would be present if the solute were uniformly distributed.

**P/REVIEW:** Equilibrium for crystallization/dissolving rates in a saturated solution is quite like the equilibrium for evaporation/condensation rates in a liquid–vapor equilibrium, which is discussed in Section 15.4. If Section 15.4 was not assigned to you earlier, you might find it helpful to study Figure 15.15 now.

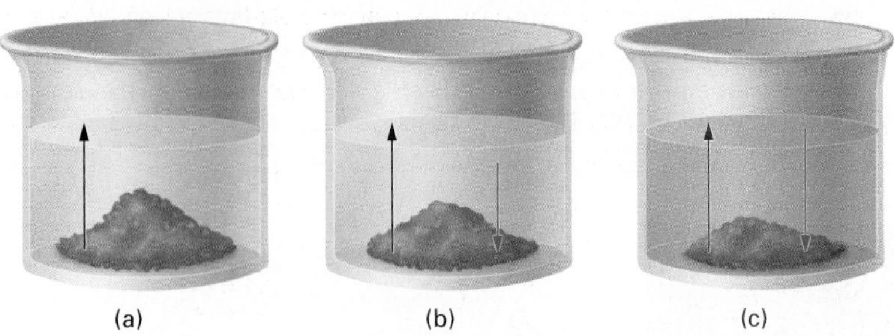


**Figure 16.7** Development of equilibrium in forming a saturated solution. If temperature is held constant, the rate of dissolving per unit of solute surface area is constant. This is shown by the equal-length black arrows in the three views. The crystallization rate per unit of surface area increases as the solution concentration at the surface increases. In (a), when dissolving has just begun, solution concentration is zero, and the crystallization rate is zero. In (b) the solution concentration has risen to yield a crystallization rate above zero, but less than the dissolving rate. This is shown by the different-length arrows. Concentration continues to increase until (c), when the crystallization rate has become equal to the dissolving rate. Equilibrium has been reached, and the solution is saturated.

Stirring or agitating the solution prevents this buildup and maximizes the net dissolving rate. The net dissolving rate is the rate of dissolving minus the rate of crystallization. The "macro" dissolving rate—the effective rate we can see—is the combined result of two invisible particulate happenings.

3. At higher temperatures, particle movement is more rapid, thereby speeding up all physical processes.

## Thinking About Your Thinking

### Mental Models

**You need to form two mental models as you study Section 16.3. The first is of the particulate-level process that occurs when a solid ionic solute is dissolved in water, as shown in Figure 16.6. First, imagine the solid ionic solid. Think about the positive ion–negative ion–positive ion–negative ion–etc. pattern in a simple ionic compound. Now imagine that this solid is in liquid water. A positive ion at the edge of the solid crystal will attract the negative portion of a polar water molecule, the end where the two lone pairs on the oxygen atom are exposed. If that ion breaks free from the crystal, the negative ends of a number of water molecules will quickly surround it. A similar process occurs for negative ions in the solid crystal and the positive ends of water molecules.**

**The second mental model that you need to develop is a "video in your mind" of the equilibrium process illustrated in Figure 16.7. Most importantly, you need to be able to picture each of the two processes, dissolving and crystallization, and the factors that affect each, in your mind. The dissolving rate is essentially constant under the conditions mentioned in the caption of Figure 16.7. The crystallization rate increases with increasing solution concentration. Picture both of these processes happening simultaneously. Equilibrium occurs when the opposing rates are equal.**

**When you come to a homework or exam question that involves dissolving an ionic solute in water or the saturated solution equilibrium process, call your mental model into your working memory, and then use your model to help answer the question. In this way, you are learning to think as a chemist.**

## √ Target Check 16.3

*Assume that temperature remains constant while a solute dissolves until the solution becomes saturated. For a unit area,*

a) Is the rate of dissolving when the solution is one-third saturated more than, equal to, or less than the rate of dissolving when the solution is two-thirds saturated?

b) Is the rate of crystallization when the solution is one-third saturated more than, equal to, or less than the rate of crystallization when the solution is two-thirds saturated?

c) Is the *net* rate of dissolving when the solution is one-third saturated more than, equal to, or less than the net rate of dissolving when the solution is two-thirds saturated?

## 16.4 Factors That Determine Solubility

**Goal** 5 Given the structural formulas of two molecular substances, or other information from which the strength of their intermolecular forces may be estimated, predict if they will dissolve appreciably in each other, and state the criteria on which your prediction is based.

**Goal** 6 Predict how and explain why the solubility of a gas in a liquid is affected by a change in the partial pressure of that gas over the liquid.

The extent to which a particular solute dissolves in a given solvent depends on three factors:

1. The strength of intermolecular forces within the solute, within the solvent, and between the solute and solvent
2. The partial pressure of a solute gas over a liquid solvent
3. The temperature

**P/REVIEW:** Intermolecular forces fall into three categories: induced dipole forces, dipole forces, and hydrogen bonds. All other things being equal, these are listed in order of increasing strength. Molecular structure is primarily responsible for the types of intermolecular forces in any given substance. See Section 15.3 for further information on intermolecular forces.

**Intermolecular Forces** Solubility, a macroscopic property, depends on intermolecular forces at the particulate level. Generally speaking, *if forces between molecules of A are about the same as the forces between molecules of B, A and B will probably dissolve in each other.* This is commonly paraphrased as *like dissolves like* (Fig. 16.8[a]). On the other hand, if the intermolecular forces between molecules of A are quite different from the forces between molecules of B, it is unlikely that they will dissolve in each other (Fig. 16.8[b]).

(a) Ethylene glycol, HO—$CH_2CH_2$—OH, is the primary ingredient in most automotive antifreeze solutions. At the particulate level, it has dipoles and exhibits hydrogen bonding, and thus it is soluble in water, which has similar intermolecular forces.

(b) Motor oil is a mixture that primarily consists of nonpolar hydrocarbon molecules. Induced dipole forces are the primary intermolecular forces, so motor oil is not soluble in water.

**Figure 16.8** Intermolecular forces and solubility.

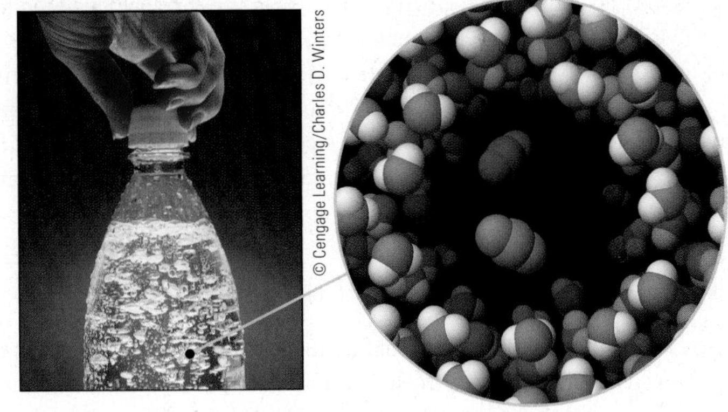


**Figure 16.9** Carbonated beverages. Bottling under a high carbon dioxide partial pressure increases the solubility of the gas in the liquid solution. When you open the bottle, the partial pressure of carbon dioxide above the solution drops to become equal to atmospheric $p_{CO_2}$, dramatically decreasing the solubility of the gas, which escapes from the solution in the form of gas-phase bubbles that consist of carbon dioxide molecules (and a relatively small quantity of water molecules).

Consider, for example, hexane, $C_6H_{14}$, and decane, $C_{10}H_{22}$. Each substance has only induced dipole forces. The forces are roughly the same for the two substances, which are soluble in each other. Neither, however, is soluble in water or methanol, $CH_3OH$, two liquids that exhibit strong hydrogen bonding. But water and methanol are soluble in each other, again supporting the correlation between solubility and similar intermolecular forces.

**Partial Pressure of Solute Gas Over Liquid Solution** Changes in partial pressure of a solute gas over a liquid solution have a pronounced effect on the solubility of that gas. This is sometimes startlingly apparent on opening a bottle of a carbonated beverage (Fig. 16.9). Such beverages are bottled under a carbon dioxide partial pressure that is slightly greater than one atmosphere, which increases the solubility of the gas. This is what is meant by "carbonated." As the pressure is released on opening, solubility decreases, which causes bubbles of carbon dioxide to escape from the solution.

**P/REVIEW:** The partial pressure of one gas in a mixture of gases is the pressure that one gas would exert if it alone occupied the same volume at the same temperature. See Section 15.1.

In most dilute solutions, the solubility of a gaseous solute in a liquid is directly proportional to the partial pressure of the gas over the surface of the liquid (Fig. 16.10). An equilibrium is reached that is similar to the liquid–vapor equilibrium described in Section 15.4 and the solid-in-liquid equilibrium discussed in Section 16.3. Neither the partial pressure nor the total pressure caused by *other* gases affects the solubility of the solute gas. This is what would be expected for an ideal gas, where all molecules are widely separated and completely independent.

Pressure has little or no effect on the solubility of solids or liquids in a liquid solvent. The gas pressure above a liquid surface has no effect on the equilibrium process that occurs as a solid solute dissolves in a liquid solvent.

**Temperature** Temperature exerts a major influence on most chemical equilibria, including solution equilibria. Consequently, solubility depends on temperature. Figure

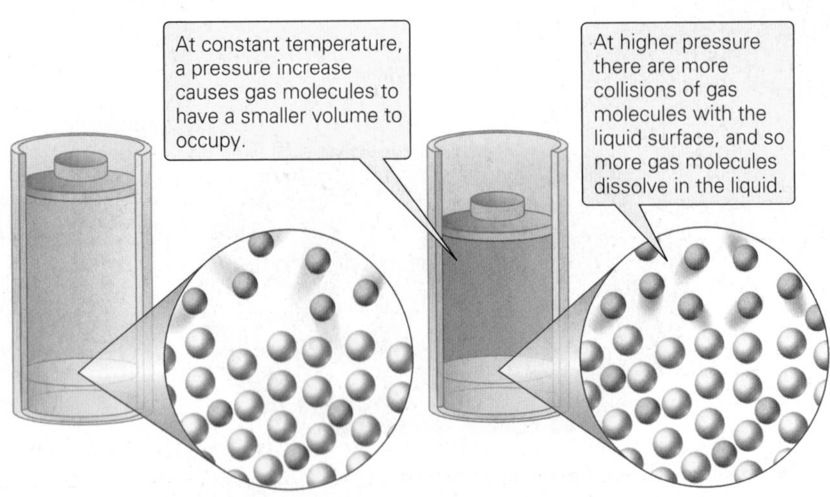

At constant temperature, a pressure increase causes gas molecules to have a smaller volume to occupy.
At higher pressure there are more collisions of gas molecules with the liquid surface, and so more gas molecules dissolve in the liquid.


**Figure 16.10** Gas solubility and partial pressure. The solubility of a gas in a liquid is directly proportional to its partial pressure: solubility $\propto$ partial pressure.

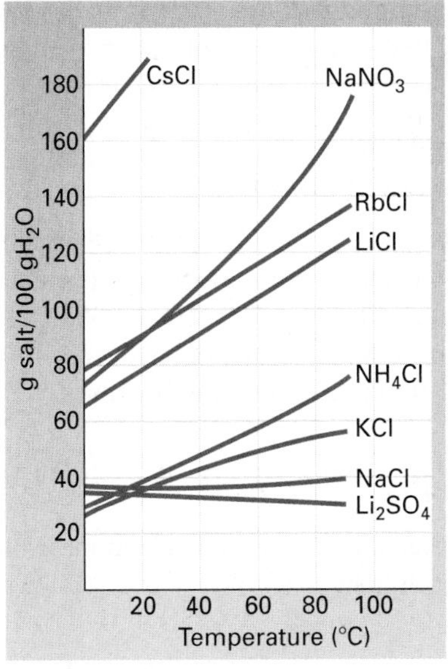


(a) Temperature–solubility curves for various salts in water.

(b) Ammonium chloride is dissolved in warm water.

(c) The solution is cooled by placing the test tube in an ice water bath, and solid ammonium chloride precipitates because it is less soluble at a lower temperature.



**Figure 16.11** Temperature dependence of the solubility of ionic compounds in water.

16.11 indicates that the solubility of most solids increases with rising temperature, but there are notable exceptions. The solubilities of gases in liquids, on the other hand, are generally lower at higher temperatures. The explanation of the relationship between temperature and solubility involves energy changes in the solution process, as well as other factors.

**summary**

**Factors That Determine Solubility**

1. Substances with similar intermolecular forces will usually dissolve in one another.
2. The greater the partial pressure of a gas over a liquid solution, the more soluble the gas.
3. The solubility of most solids in liquids increases with increasing temperature, but there are exceptions. The solubility of most gases in liquids decreases with increasing temperature.

### √ Target Check 16.4

*If you are given the structural formulas of two substances, what would you look for to predict if one will dissolve in the other?*

## 16.5 Solution Concentration: Percentage by Mass

**Goal 7** Given mass of solute and of solvent or solution, calculate percentage concentration.

**Goal 8** Given mass of solution and percentage concentration, calculate mass of solute and solvent.

The concentration of a solution tells us how much solute is present per given amount of solution or given amount of solvent. As a PER expression, concentration has the form of a fraction. Amount of solute appears in the numerator and may be in grams, moles, or equivalents (eq), a unit we will describe in Section 16.8. Quantity of solvent or solution is in the denominator and may be in mass or volume units. In general, concentration is

$$\frac{\text{quantity of solute}}{\text{quantity of solution}} \quad \textit{or} \quad \frac{\text{quantity of solute}}{\text{quantity of solvent}}$$

We begin with percentage by mass. Percentage concentration is based on the ratio g solute/g solution. We will refer to this as a **mass ratio.** The mass ratio is a decimal fraction that represents the grams of solute in 1 gram of solution. Therefore, 100 times the mass ratio is the grams of solute in 100 grams of solution—grams of solute per 100 grams of solution—which is the definition of **percentage concentration by mass.**

**P/REVIEW:** Equation 16.1 corresponds with Equation 7.2 in Section 7.6, where percentage calculations are discussed. When given mass quantities, you can find the percentage of any component with Equation 16.1. When the percentage of a component is known, it is a conversion factor between the mass of that component and the mass of the solution.

$$\%\text{ by mass} = \frac{\text{g solute}}{\text{g solution}} \times 100 = \frac{\text{g solute}}{\text{g solute} + \text{g solvent}} \times 100 \qquad \textbf{(16.1)}$$

Be careful about the denominator. If a problem gives the mass of solute and mass of solvent, be sure to add them to get the mass of the solution.

## Active Example 16.1

When 125 grams of a solution was evaporated to dryness, 42.3 grams of solute was recovered. What was the percentage of the solute?

This involves a direct substitution into Equation 16.1.

The concentrations of very dilute solutions are conveniently given in parts per million, ppm. A 0.0025% solution, which is 0.0025 parts per hundred, is equivalent to 25 ppm.

GIVEN: 125 g solution; 42.3 g solute WANTED: % by mass

EQUATION: $\%\text{ by mass} = \frac{\text{g solute}}{\text{g solution}} \times 100 = \frac{42.3\text{ g}}{125\text{ g}} \times 100 = 33.8\%$

## Active Example 16.2

You are to prepare $2.50 \times 10^2$ grams of 7.00% $Na_2CO_3$ solution. How many grams of sodium carbonate and how many milliliters of water do you use? (The density of water is 1.00 g/mL.)

This time you can use percentage (grams solute/100 grams solution) as a dimensional-analysis conversion factor to find the grams of $Na_2CO_3$. Calculate that quantity first.

GIVEN: $2.50 \times 10^2$ g 7.00% $Na_2CO_3$ solution  WANTED: g $Na_2CO_3$

PER: $\frac{7.00 \text{ g } Na_2CO_3}{100 \text{ g solution}}$

PATH: g solution $\longrightarrow$ g $Na_2CO_3$

$$2.50 \times 10^2 \cancel{\text{g solution}} \times \frac{7.00 \text{ g } Na_2CO_3}{100 \cancel{\text{g solution}}} = 17.5 \text{ g } Na_2CO_3$$

The mass of the solution is $2.50 \times 10^2$ g. The mass of the solute in the solution is 17.5 g. The rest is water. What is the mass of the water? What is the volume of that mass of water?

Concentrations of medicinal solutions are sometimes given in terms of weight/volume percent, the mass of solute per 100 mL of solution. The density of dilute solutions is very close to 1 g/mL, so the mass of 100 mL of solution is very close to 100 g. Thus, for solutions of 5% or less, mass percent and weight/volume percent are essentially equal.

g $H_2O$ = g solution − g solute = $2.50 \times 10^2$ g − 17.5 g $Na_2CO_3$ = 233 g $H_2O$

At 1.00 g/mL, 233 g $H_2O$ = 233 mL $H_2O$

## 16.6 Solution Concentration: Molarity

**Goal** **9** Given two of the following, calculate the third: moles of solute (or data from which it may be found), volume of solution, molarity.

In working with liquids, volume is easier to measure than mass. Therefore, a solution concentration based on volume is usually more convenient to use than one based on mass. **Molarity, M, is the moles of solute per liter of solution.** The defining equation is

$$M \equiv \frac{\text{moles solute}}{\text{liter solution}} = \frac{\text{mol}}{\text{L}} \qquad \textbf{(16.2)}$$

If a solution contains 0.755 mole of sulfuric acid per liter, we identify it as 0.755 M $H_2SO_4$. In words, it is "point seven-five-five molar sulfuric acid." In a calculation setup we would write "0.755 mol $H_2SO_4$/L."

Notice that molarity is a PER relationship. The amount of solute in a sample of solution is proportional to the size of the sample, and the molarity is the proportionality constant. If both sides of Equation 16.2 are multiplied by volume in L, the result is

$$\text{Volume} \times \text{Molarity} = V \times M = L \times \frac{\text{mol}}{\text{L}} = \text{mol} \qquad \textbf{(16.3)}$$

We will make good use of this equation in the solution dilution and solution stoichiometry sections later in this chapter.

All the calculation methods you have used before with PER relationships can be used with molarity. Notice that the units in the denominator are liters, but volume is often given in milliliters. If the volume is given in milliliters, you must convert it to liters.

Also notice that molarity, as expressed in Equations 16.2 and 16.3, includes one unit that is particulate (number of particles grouped in moles) and one unit that is macroscopic (volume, which can be measured). To be practical—to actually work with molarity in the laboratory—you must convert moles to a macroscopic unit, typically grams. The conversion factor, as you probably know from many uses by now, is molar mass in grams per mole.

# Everyday Chemistry

## The World's Oceans: The Most Abundant Solution

Oceans cover 71 percent of the earth's surface and contain all but 2.5% of the earth's water (Fig. 16.12). As land animals, we humans often fail to recognize the enormity of the oceans and the central role these aqueous solutions play in the earth's environment.

The composition of seawater varies depending on the location from which a sample is taken, but on average, it is a 3.5% solution of a variety of dissolved solids. To be precise, an entire ocean does not meet the criterion of being a homogeneous mixture and therefore is not a solution. Nonetheless, many environmental factors lead to a relatively thorough mixing effect in the world's oceans, and very large samples are, in fact, homogeneous. Scientists who have sampled seawater from across the world have verified this experimentally, finding that the samples are nearly identical when analyzed for the major solution components.

**Figure 16.12** Distribution of the earth's water. Almost all of it is found in the oceans. The total percentage of groundwater and fresh water is less than 1%.

Let's consider the concentration of seawater more carefully. By definition, percentage by mass is

$$\%\text{ by mass} = \frac{\text{g solute}}{\text{100 g solution}}$$

Seawater is therefore 3.5 grams of dissolved solids per 100 grams of solution. At typical ocean temperatures, the density of seawater is 1.0 g/mL. We can therefore determine the mass of dissolved solids per liter:

$$\frac{\text{3.5 g solids}}{\text{100 g seawater}} \times \frac{\text{1 g seawater}}{\text{1.0 mL seawater}} \times \frac{\text{1000 mL seawater}}{\text{L seawater}} = \frac{\text{35 g solids}}{\text{L seawater}}$$

Imagine a one-liter soft drink bottle filled with seawater. There is 35 grams (1.2 ounces) of dissolved solids in that bottle.

Chloride ions and sodium ions are by far the most abundant species dissolved in the oceans, making up about 85% of the dissolved solids. If you've ever gotten a mouthful of ocean water, you can recognize the salty taste characteristic of these two components of table salt. If we add magnesium ion, sulfate ion, calcium ion, and potassium ion to our list of dissolved solids in seawater, we account for 99% of the dissolved solids. Trace amounts of many other species are found in the oceans. Chemists have identified over 70 elements in seawater samples.

Sea salt is obtained by evaporating ocean water. One cubic foot of seawater yields more than two pounds of salt. In comparison, the same-size sample from a typical freshwater lake yields less than an ounce of dissolved solids.

The high concentration of solute particles in the oceans can be traced to

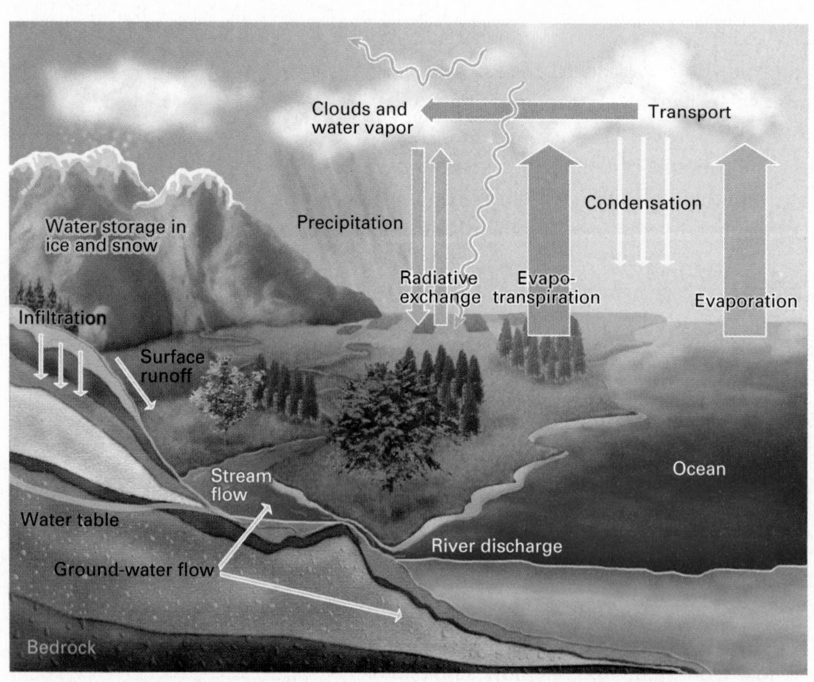


**Figure 16.13** The global water cycle. The relatively high concentration of dissolved solids in the oceans results from this process.

*Continued on facing page.*

the global water cycle. Water from the atmosphere falls to the earth as rain or snow. This water dissolves some of the solids in the earth's surface, carries the solids into the ocean, and then leaves them there as the water evaporates back into the atmosphere (Fig. 16.13). Ocean water evaporates and is carried over landmasses by the winds. When the water is released from the atmosphere in the form of rain or snow, some finds its way into rivers, which mix the water with dissolved solids and eventually run to the seas. As the cycle continues, dissolved ions are continually deposited in the oceans.

All natural waters are solutions, and the size of the world's oceans makes them the most abundant aqueous solution on the earth. Seawater is among the most concentrated of the planet's natural waters, and its concentration continues to gradually increase over time. Chemists and other scientists study the oceans and their interaction with all aspects of the environment. Many concepts from chemistry are central to this research.

Stone/Getty Images

The ocean is a liquid solution of many dissolved solids.

## Active Example 16.3

Calculate the molarity of a solution made by dissolving 15 g of NaOH in water and diluting to $1.00 \times 10^2$ mL.

Begin by *PLANNING* the problem.

Recall that *PLAN* means that you should write the *GIVEN*, *WANTED*, and *PER/PATH*. If there is an *EQUATION* needed to solve the problem, write that also.

*GIVEN:* 15 g NaOH; $1.00 \times 10^2$ mL *WANTED:* M

*PER:* $\frac{1 \text{ mol NaOH}}{40.00 \text{ g NaOH}}$

*PATH:* g NaOH ⟶ mol NaOH

*EQUATION:* $M \equiv \frac{\text{mol}}{\text{L}}$

To find molarity from its definition (Equation 16.2), you need to know the volume of solution, which is given, and the moles of NaOH. Grams are given. The grams-to-moles conversion must be made before you can use the defining equation. How many moles of NaOH are in the solution?

$$15 \text{ g } \cancel{\text{NaOH}} \times \frac{1 \text{ mol NaOH}}{40.00 \text{ g } \cancel{\text{NaOH}}} = 0.38 \text{ mol NaOH}$$

*continued*

You now have both moles and volume, the numerator and the denominator in the defining equation for molarity. Plug them into the equation *and do whatever else you must do.* Then calculate the answer.

EQUATION: $M \equiv \frac{\text{mol}}{\text{L}} = \frac{0.38 \text{ mol NaOH}}{1.00 \times 10^2 \text{ mL}} \times \frac{1000 \text{ mL}}{\text{L}} = 3.8 \text{ mol NaOH/L} = 3.8 \text{ M NaOH}$

The denominator was given in the statement of the problem, but in the wrong units. Milliliters must be converted to liters to be used in the defining equation for molarity.

## Active Example 16.4

How many grams of silver nitrate must be dissolved to prepare $5.00 \times 10^2$ mL of 0.150 M $AgNO_3$?

PLAN the problem. Molarity is used as a PER expression to convert between moles and volume in milliliters.

GIVEN: $5.00 \times 10^2$ mL 0.150 M $AgNO_3$ WANTED: g $AgNO_3$

PER: $\frac{1 \text{ L}}{1000 \text{ mL}}$ $\frac{0.150 \text{ mol AgNO}_3}{\text{L}}$ $\frac{169.9 \text{ g AgNO}_3}{\text{mol AgNO}_3}$

PATH: mL $\longrightarrow$ L $\longrightarrow$ mol $AgNO_3$ $\longrightarrow$ g $AgNO_3$

In this problem, molarity is the conversion factor in the second step of the PATH, changing the given volume to moles of silver nitrate. Moles are then changed to mass, using molar mass as the third conversion factor.

Complete the Active Example.

$$5.00 \times 10^2 \text{ mL} \times \frac{1 \text{ L}}{1000 \text{ mL}} \times \frac{0.150 \text{ mol AgNO}_3}{\text{L}} \times \frac{169.9 \text{ g AgNO}_3}{\text{mol AgNO}_3} = 12.7 \text{ g AgNO}_3$$

A clearer understanding of molarity can be gained by mentally "preparing" a solution. As an example, consider the preparation of 250.0 mL of 0.0100 M potassium permanganate solution. First, you determine the mass of solute needed:

$$250.0 \text{ mL} \times \frac{1 \text{ L}}{1000 \text{ mL}} \times \frac{0.0100 \text{ mol KMnO}_4}{\text{L}} \times \frac{158.04 \text{ g KMnO}_4}{\text{mol KMnO}_4} = 0.395 \text{ g KMnO}_4$$

Now weigh out the 0.395 g $KMnO_4$ into a 250-mL volumetric flask (Fig. 16.14[1]). Add *less than* 250 mL of water to the flask. After dissolving the solute (Fig. 16.14[2]), add water to the 250-mL mark on the neck of the flask (Fig. 16.14[3]). Finally, stopper and shake the flask to ensure thorough mixing (Fig. 16.14[4]). Notice that molarity is based on the volume of the *solution,* not the volume of the *solvent.* This is why the solute is dissolved in less than 250 mL of water and then diluted to that volume.

Combine ~240 mL of distilled $H_2O$ with 0.395 g (0.0025 mol) $KMnO_4$ in a 250.0 mL volumetric flask.

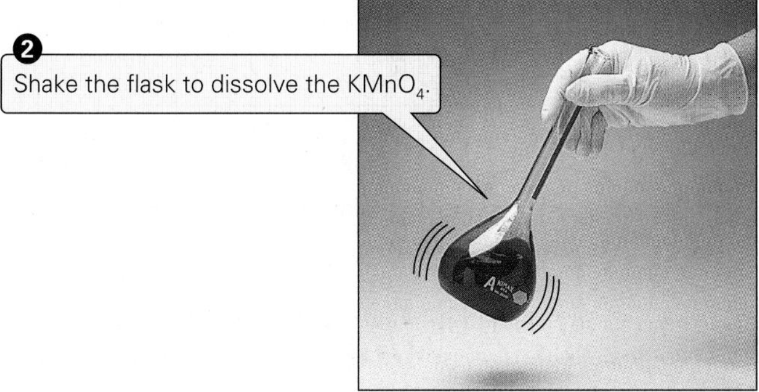


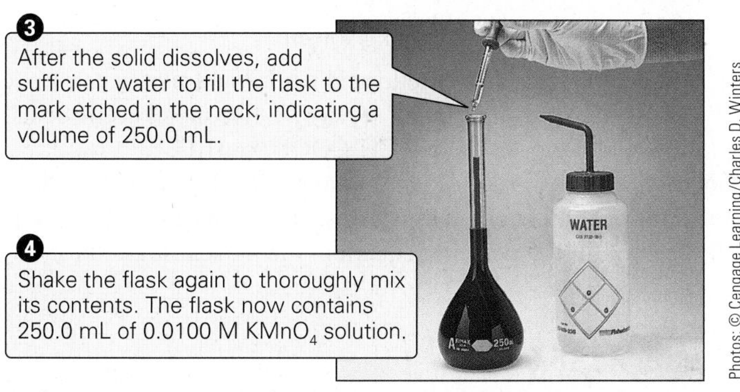

After the solid dissolves, add sufficient water to fill the flask to the mark etched in the neck, indicating a volume of 250.0 mL.
Shake the flask again to thoroughly mix its contents. The flask now contains 250.0 mL of 0.0100 M $KMnO_4$ solution.
Photos: © Cengage Learning/Charles D. Winters


**Figure 16.14** Preparing a solution of specified molarity.

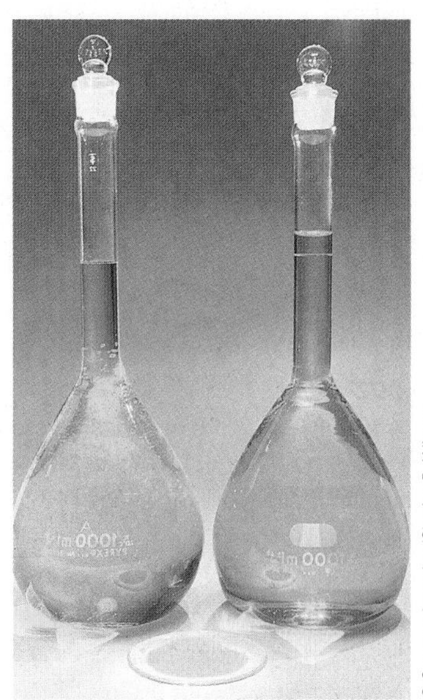


**Figure 16.15** Solution versus solvent volume. Each solution contains 0.10 mole of potassium chromate. The solution on the left was made by adding the solute and then adding enough water to make a *solution* volume of 1000 mL. Note that the volume of this solution is exactly 1000 mL because it precisely reaches the etch mark on the volumetric flask. This is the appropriate procedure to make a solution of specified molarity. The solution on the right was made by adding 1000 mL of water—a *solvent* volume of 1000 mL—to the solute. Note that the *solution* volume in the flask on the right exceeds 1000 mL.

Figure 16.15 contrasts a solution made based on solution volume with one made based on solvent volume. Remember, the denominator in the molarity fraction mol/L is *liters of solution.*

Using molarity as a conversion factor in changing volume to moles or moles to volume is one of its most important applications. Volume may be expressed in either liters or milliliters.

## Active Example 16.5

Find the volume of a 1.40-M solution that contains 0.287 mole of ammonia. Answer in both liters and milliliters.

PLAN and solve the problem.

*continued*

P/REVIEW: The alternative ways to solve Active Example 16.5, dimensional analysis or algebra, were offered with Active Example 3.23 (Section 3.8). In this chapter, we recommend the dimensional analysis approach because it fits as one step in solving the solution stoichiometry problems you will encounter in Section 16.11.

*GIVEN:* 0.287 mol $NH_3$; 1.40 M $NH_3$ *WANTED:* L and mL

*PER:* $\frac{1\text{ L}}{1.40\text{ mol NH}_3}$ $\frac{1000\text{ mL}}{\text{L}}$

*PATH:* mol $NH_3$ ⟶ L ⟶ mL

$$0.287\ \cancel{\text{mol NH}_3} \times \frac{1\ \cancel{\text{L}}}{1.40\ \cancel{\text{mol NH}_3}} \times \frac{1000\text{ mL}}{\cancel{\text{L}}} = 205\text{ mL}$$

$$= 0.205\text{ L}$$

You may have noticed that Active Example 16.5 could have been solved algebraically with Equation 16.2 or 16.3. Solving either equation for volume, L, yields

$$\text{L} = \frac{\text{mol}}{\text{M}} = \frac{0.287\text{ mol}}{1.40\text{ mol/L}} = 0.287\text{ mol} \times \frac{1\text{ L}}{1.40\text{ mol}} = 0.205\text{ L}$$

Molality is not a convenient concentration unit to work with—not nearly as convenient as molarity. Measuring volume of solution, the liters in moles per liter, is easy, whereas neither moles of solute nor kilograms of solvent, the units in molality, can be measured directly in a solution that is already prepared. So why not use molarity all the time and forget about molality? Molarity is temperature-dependent. As the volume of a solution changes on heating or cooling, its molarity changes—not a lot, but enough so that proper ties that can be related to molality cannot be satisfactorily related to molarity.

## 16.7 Solution Concentration: Molality (Optional)

**Goal** **10** Given two of the following, calculate the third: moles of solute (or data from which it may be found), mass of solvent, molality.

Many physical properties are related to solution concentration expressed as **molality, m, the number of moles of solute dissolved in one kilogram of solvent.** The defining equation is

$$\text{m} \equiv \frac{\text{mol solute}}{\text{kg solvent}} \qquad \textbf{(16.4)}$$

If a solution contains 0.755 mole of acetic acid per kilogram of water, we identify it as 0.755 m $CH_3COOH$. In words it is "point seven-five-five molal acetic acid." In a calculation setup, we would write "0.755 mol $CH_3COOH$/kg $H_2O$."

Like molarity, molality is defined as a ratio that leads to a *PER* relationship. Notice that the units in the denominator are kilograms, but the mass of solvent is usually given in grams. To be used in the defining equation, the solvent mass must be changed to kilograms.

### Active Example 16.6

Calculate the molality of a solution prepared by dissolving 15.0 grams of sugar, $C_{12}H_{22}O_{11}$, in $3.50 \times 10^2$ milliliters of water. (The density of water is 1.00 g/mL.)

The definition of molality is moles of solute per kilogram of solvent. Grams of $C_{12}H_{22}O_{11}$ will have to be converted to moles, and milliliters of $H_2O$ must be changed to kilograms. Go for both conversions.

GIVEN: 15.0 g $C_{12}H_{22}O_{11}$ WANTED: mol $C_{12}H_{22}O_{11}$

PER: $\dfrac{1 \text{ mol } C_{12}H_{22}O_{11}}{342.30 \text{ g } C_{12}H_{22}O_{11}}$

PATH: g $C_{12}H_{22}O_{11}$ ⟶ mol $C_{12}H_{22}O_{11}$

$$15.0 \text{ g } \cancel{C_{12}H_{22}O_{11}} \times \frac{1 \text{ mol } C_{12}H_{22}O_{11}}{342.30 \cancel{\text{ g } C_{12}H_{22}O_{11}}} = 0.0438 \text{ mol } C_{12}H_{22}O_{11}$$

GIVEN: $3.50 \times 10^2$ mL $H_2O$ WANTED: kg $H_2O$

PER: $\dfrac{1.00 \text{ g } H_2O}{\text{mL } H_2O}$ $\dfrac{1 \text{ kg } H_2O}{1000 \text{ g } H_2O}$

PATH: mL $H_2O$ ⟶ g $H_2O$ ⟶ kg $H_2O$

$$3.50 \times 10^2 \cancel{\text{ mL } H_2O} \times \frac{1.00 \cancel{\text{ g } H_2O}}{\cancel{\text{mL } H_2O}} \times \frac{1 \text{ kg } H_2O}{1000 \cancel{\text{ g } H_2O}} = 0.350 \text{ kg } H_2O$$

Complete the Active Example by substituting the calculated quantities into Equation 16.4, $m \equiv \dfrac{\text{mol solute}}{\text{kg solvent}}$.

GIVEN: 0.0438 mol $C_{12}H_{22}O_{11}$; 0.350 kg $H_2O$ WANTED: m (mol/kg)

EQUATION: $m \equiv \dfrac{\text{mol solute}}{\text{kg solvent}} = \dfrac{0.0438 \text{ mol } C_{12}H_{22}O_{11}}{0.350 \text{ kg } H_2O} = 0.125 \text{ m } C_{12}H_{22}O_{11}$

## Active Example 16.7

How many grams of KCl must be dissolved in $2.50 \times 10^2$ g $H_2O$ to make a 0.400 m solution?

This time you can use molality as a conversion factor in solving the problem by dimensional analysis. A unit adjustment is needed first, however. Complete the problem.

GIVEN: $2.50 \times 10^2$ g $H_2O$; 0.400 m solution WANTED: g KCl

PER: $\dfrac{1 \text{ kg } H_2O}{1000 \text{ g } H_2O}$ $\dfrac{0.400 \text{ mol KCl}}{\text{kg } H_2O}$ $\dfrac{74.55 \text{ g KCl}}{\text{mol KCl}}$

PATH: g $H_2O$ ⟶ kg $H_2O$ ⟶ mol KCl ⟶ g KCl

$$2.50 \times 10^2 \cancel{\text{ g } H_2O} \times \frac{1 \cancel{\text{ kg } H_2O}}{1000 \cancel{\text{ g } H_2O}} \times \frac{0.400 \cancel{\text{ mol KCl}}}{\cancel{\text{kg } H_2O}} \times \frac{74.55 \text{ g KCl}}{\cancel{\text{mol KCl}}} = 7.46 \text{ g KCl}$$

Many chemists never use normality to express concentration. Since advocates of SI units have been pushing to make SI the only system of units used worldwide, they argue that the mole is a base unit under the SI system, whereas the equivalent is not even recognized. (Historically, the equivalent has an undisputed place in the evolution of chemical understanding.) In addition, they argue, there is no problem that can be solved by normality that cannot be solved by molarity, so why have two systems? In the field—in the industrial laboratories where so many practicing chemists work—normality is so convenient that there is unyielding resistance to change. Moreover, many academic chemists support that position. This is one of the major areas of disagreement among chemists.

## 16.8 Solution Concentration: Normality (Optional)

Goal **11** Given an equation for a neutralization reaction, state the number of equivalents of acid or base per mole and calculate the equivalent mass of the acid or base.

Goal **12** Given two of the following, calculate the third: equivalents of acid or base (or data from which they may be found), volume of solution, normality.

A particularly convenient concentration in routine analytical work is **normality, N, the number of equivalents, eq, per liter of solution.** (The equivalent will be defined shortly.) The defining equation is

$$N \equiv \frac{\text{equivalents solute}}{\text{liter solution}} = \frac{\text{eq}}{\text{L}} \qquad \textbf{(16.5)}$$

If a solution contains 0.755 equivalent of phosphoric acid per liter, we identify it as 0.755 N $H_3PO_4$. In words it is "point seven-five-five normal phosphoric acid." In a calculation setup we would write "0.755 eq $H_3PO_4$/L."

The defining equation for normality is very similar to the equation for molarity (Equation 16.2). Many of the calculations are similar, too. But to understand the difference, we must see what an equivalent is.

**One equivalent of an acid is the quantity that yields one mole of hydrogen ions in a chemical reaction. One equivalent of a base is the quantity that reacts with one mole of hydrogen ions.** Because hydrogen and hydroxide ions combine on a one-to-one ratio, one mole of hydroxide ions is one equivalent of base.

According to these statements, both one mole of HCl and one mole of NaOH are one equivalent. They yield, respectively, one mole of $H^+$ ions and one mole of $OH^-$ ions. $H_2SO_4$, on the other hand, may have two equivalents per mole because it can release two moles of $H^+$ ions per mole of acid. Similarly, one mole of $Al(OH)_3$ may represent three equivalents because three moles of $OH^-$ may react.

The number of equivalents in a mole of an acid depends on a specific reaction, not just the number of moles of H's in a mole of the compound. What counts is the number of H's that react. By controlling reaction conditions, phosphoric acid can have one, two, or, theoretically, three equivalents per mole:

$$NaOH(aq) + H_3PO_4(aq) \rightarrow NaH_2PO_4(aq) + H_2O(\ell) \qquad \text{1 eq acid/mol} \qquad \textbf{(16.6)}$$

$$2\ NaOH(aq) + H_3PO_4(aq) \rightarrow Na_2HPO_4(aq) + 2\ H_2O(\ell) \qquad \text{2 eq acid/mol} \qquad \textbf{(16.7)}$$

$$3\ NaOH(aq) + H_3PO_4(aq) \rightarrow Na_3PO_4(aq) + 3\ H_2O(\ell) \qquad \text{3 eq acid/mol} \qquad \textbf{(16.8)}$$

There is one equivalent per mole of base in each of the preceding reactions. NaOH can have only one equivalent per mole because there is only one mole of $OH^-$ in one mole of NaOH. It is noteworthy that *the number of equivalents of acid and base in each reaction are the same.* In Equation 16.6, 1 mol $H_3PO_4$ gives up only 1 eq $H^+$, and it reacts with 1 mol NaOH, which is 1 eq NaOH. In Equation 16.7, 1 mol $H_3PO_4$ yields 2 eq $H^+$, and it reacts with 2 mol NaOH, which is 2 eq NaOH. In Equation 16.8 there are 3 eq $H^+$ and 3 mol NaOH, which is 3 eq NaOH.

### Active Example 16.8

State the number of equivalents of acid and base per mole in each of the following reactions: $2\ HBr + Ba(OH)_2 \rightarrow BaBr_2 + 2\ H_2O$; $H_3C_6H_5O_7 + 2\ KOH \rightarrow K_2HC_6H_5O_7 + 2\ H_2O$.

Remember, you are interested only in the number of moles of $H^+$ or $OH^-$ that *react,* not the number present, in one mole of acid or base. The formula of citric acid, $H_3C_6H_5O_7$, is written as an inorganic chemist is most apt to write it—with three ionizable hydrogens first. Insert the numbers of equivalents per mole in the spaces below.

| | eq acid/mol | eq base/mol |
|---|---|---|
| $2\ HBr + Ba(OH)_2 \rightarrow BaBr_2 + 2\ H_2O$ | | |
| $H_3C_6H_5O_7 + 2\ KOH \rightarrow K_2HC_6H_5O_7 + 2\ H_2O$ | | |

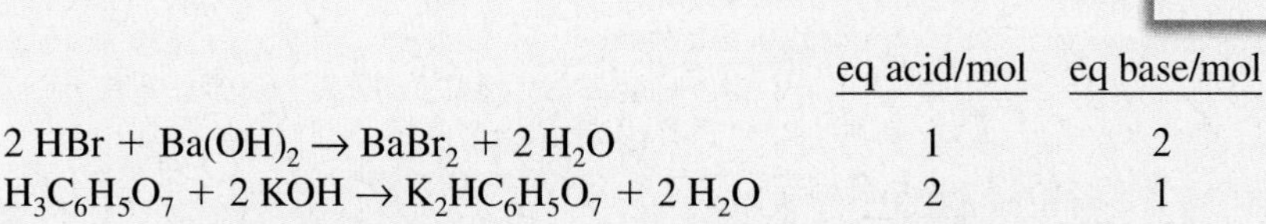

| | eq acid/mol | eq base/mol |
|---|---|---|
| $2\ HBr + Ba(OH)_2 \rightarrow BaBr_2 + 2\ H_2O$ | 1 | 2 |
| $H_3C_6H_5O_7 + 2\ KOH \rightarrow K_2HC_6H_5O_7 + 2\ H_2O$ | 2 | 1 |

In the first equation each mole of $Ba(OH)_2$ yields two $OH^-$ ions, so there are two eq/mol. Each mole of HBr produces one $H^+$, so there is one eq/mol. In the second equation, there could be one, two, or three equivalents of $H_3C_6H_5O_7$/mol, depending on how many ionizable hydrogens are released in the reaction. In this case, that number is two: $H_3C_6H_5O_7 \rightarrow 2\ H^+ + HC_6H_5O_7^{2-}$. The 2 $H^+$ ions released combine with the 2 $OH^-$ ions from 2 moles of KOH to form 2 $H_2O$ molecules. In KOH, there is only one $OH^-$ in a formula unit, so there can be only one eq/mol.

Notice that, as with the three phosphoric acid neutralizations, *the number of equivalents of acid and base in each equation in Active Example 16.8 is the same.* In the first reaction, 2 mol HBr is 2 eq, and 1 mol $Ba(OH)_2$ is 2 eq. The "same number of equivalents of all reactants" idea extends to the product species, too. Once you find the number of equivalents of one species in a reaction, you have the number of equivalents of *all* species.

The fact that the number of equivalents of *all* species in a chemical reaction is the same is *why* normality is a convenient concentration unit for analytical work. You will see how this becomes an advantage in Section 16.13.

It is sometimes convenient to find the **equivalent mass, g/eq,** of a substance, **the number of grams per equivalent.** Equivalent mass is similar to molar mass, g/mol. You can readily calculate equivalent mass by dividing molar mass by equivalents per mole:

$$\frac{\text{g/mol}}{\text{eq/mol}} = \frac{\text{g}}{\text{mol}} \times \frac{\text{mol}}{\text{eq}} = \text{g/eq} \qquad \textbf{(16.9)}$$

In ordinary acid–base reactions there are one, two, or three equivalents per mole. It follows that the equivalent mass of an acid or a base is the same as, one half of, or one third of the molar mass. The molar mass of phosphoric acid is 97.99 g/mol. For the three reactions of phosphoric acid (Equations 16.6, 16.7, and 16.8), the equivalent masses are

$$\text{Equation 16.6: } \frac{97.99\text{ g } H_3PO_4\text{/mol}}{1\text{ eq } H_3PO_4\text{/mol}} = \frac{97.99\text{ g } H_3PO_4}{1\text{ eq } H_3PO_4} = 97.99\text{ g } H_3PO_4\text{/eq } H_3PO_4$$

$$\text{Equation 16.7: } \frac{97.99\text{ g } H_3PO_4\text{/mol}}{2\text{ eq } H_3PO_4\text{/mol}} = \frac{97.99\text{ g } H_3PO_4}{2\text{ eq } H_3PO_4} = 49.00\text{ g } H_3PO_4\text{/eq } H_3PO_4$$

$$\text{Equation 16.8: } \frac{97.99\text{ g } H_3PO_4\text{/mol}}{3\text{ eq } H_3PO_4\text{/mol}} = \frac{97.99\text{ g } H_3PO_4}{3\text{ eq } H_3PO_4} = 32.66\text{ g } H_3PO_4\text{/eq } H_3PO_4$$

## Active Example 16.9

Calculate the equivalent masses of KOH, $Ba(OH)_2$, and $H_3C_6H_5O_7$ for the reactions in Active Example 16.8. Their molar masses are 56.11 g/mol KOH, 171.3 g/mol $Ba(OH)_2$, and 192.12 g/mol $H_3C_6H_5O_7$.

*continued*

$$\text{KOH:}\ \frac{56.11\ \text{g KOH}}{1\ \text{eq KOH}} = 56.11\ \text{g KOH/eq KOH}$$

$$\text{Ba(OH)}_2\text{:}\ \frac{171.3\ \text{g Ba(OH)}_2}{2\ \text{eq Ba(OH)}_2} = 85.65\ \text{g Ba(OH)}_2\text{/eq Ba(OH)}_2$$

$$\text{H}_3\text{C}_6\text{H}_5\text{O}_7\text{:}\ \frac{192.12\ \text{g H}_3\text{C}_6\text{H}_5\text{O}_7}{2\ \text{eq H}_3\text{C}_6\text{H}_5\text{O}_7} = 96.060\ \text{g H}_3\text{C}_6\text{H}_5\text{O}_7\text{/eq H}_3\text{C}_6\text{H}_5\text{O}_7$$

Just as molar mass makes it possible to convert in either direction between grams and moles, equivalent mass sets the path between grams and equivalents. In practice, it is often more convenient to use the fractional form for equivalent mass—the molar mass over the number of equivalents per mole. We use both setups in the next Active Example, but only the fractional setup thereafter. If your instructor emphasizes equivalent mass as a quantity, you should, of course, follow those instructions.

## Active Example 16.10

Calculate the number of equivalents in 68.5 g $Ba(OH)_2$.

The numbers you need are in Active Example 16.9. Complete the Active Example.

Using equivalent mass, $68.5\ \cancel{\text{g Ba(OH)}_2} \times \frac{1\ \text{eq Ba(OH)}_2}{85.65\ \cancel{\text{g Ba(OH)}_2}} = 0.800\ \text{eq Ba(OH)}_2$

Using the fractional setup, $68.5\ \cancel{\text{g Ba(OH)}_2} \times \frac{2\ \text{eq Ba(OH)}_2}{171.3\ \cancel{\text{g Ba(OH)}_2}} = 0.800\ \text{eq Ba(OH)}_2$

You are now ready to use the equivalent concept in normality problems.

## Active Example 16.11

Calculate the normality of a solution that contains 2.50 g NaOH in $5.00 \times 10^2$ mL of solution.

This is just like Active Example 16.3 except that moles in mol/L have been replaced by equivalents in eq/L. In that Active Example you first found the moles of NaOH. This time *PLAN* the problem and take the first step by finding the equivalents of NaOH.

*GIVEN:* 2.50 g NaOH; $5.00 \times 10^2$ mL solution *WANTED:* N (eq/L)

*PER:* $\frac{1 \text{ eq NaOH}}{40.00 \text{ g NaOH}}$

*PATH:* g NaOH ⟶ eq NaOH

$$2.50 \text{ g NaOH} \times \frac{1 \text{ eq NaOH}}{40.00 \text{ g NaOH}} = 0.0625 \text{ eq NaOH}$$

One mole of NaOH yields one mole of $OH^-$ ions, so there is one equivalent of NaOH per mole.

You now have both equivalents and volume, the numerator and the denominator in the defining equation for normality. Plug them into the equation and calculate the answer. Remember to "do whatever else you must do."

*EQUATION:* $N \equiv \frac{\text{eq}}{\text{L}} = \frac{0.0625 \text{ eq}}{5.00 \times 10^2 \text{ mL}} \times \frac{1000 \text{ mL}}{\text{L}} = 0.125 \text{ N NaOH}$

The denominator was given in the statement of the problem, but in the "wrong" units. Milliliters must be converted to liters to be used in the defining equation for normality.

Just as molarity provides a way to convert in either direction between moles of solute and volume of solution, normality offers a unit path between equivalents of solute and volume of solution.

## Active Example 16.12

How many equivalents are in 18.6 mL 0.856 N $H_2SO_4$?

*PLAN* and solve the problem.

*continued*

GIVEN: 18.6 mL 0.856 N $H_2SO_4$ WANTED: eq $H_2SO_4$

PER: $\frac{1 \text{ L}}{1000 \text{ mL}}$ $\frac{0.856 \text{ eq } H_2SO_4}{\text{L}}$

PATH: mL ⟶ L ⟶ eq $H_2SO_4$

$$18.6 \text{ mL} \times \frac{1 \text{ L}}{1000 \text{ mL}} \times \frac{0.856 \text{ eq } H_2SO_4}{\text{L}} = 0.0159 \text{ eq } H_2SO_4$$

Active Example 16.12 shows that normality is a conversion factor you can use to convert between volume and equivalents:

$$V \times N = L \times \frac{eq}{L} = eq \qquad \textbf{(16.10)}$$

We will use this fact in Section 16.13.

Knowing how to prepare a solution of specified normality is another valuable skill.

## Active Example 16.13

How many grams of phosphoric acid are needed to prepare $1.00 \times 10^2$ mL 0.350 N $H_3PO_4$ for use in the reaction $H_3PO_4 + NaOH \rightarrow NaH_2PO_4 + H_2O$?

This is like Active Example 16.4, except that moles have been replaced by equivalents. Complete the problem.

GIVEN: $1.00 \times 10^2$ mL 0.350 N $H_3PO_4$ WANTED: g $H_3PO_4$

PER: $\frac{1 \text{ L}}{1000 \text{ mL}}$ $\frac{0.350 \text{ eq } H_3PO_4}{\text{L}}$ $\frac{97.99 \text{ g } H_3PO_4}{\text{eq } H_3PO_4}$

PATH: mL ⟶ L ⟶ eq $H_3PO_4$ ⟶ g $H_3PO_4$

$$1.00 \times 10^2 \text{ mL} \times \frac{1 \text{ L}}{1000 \text{ mL}} \times \frac{0.350 \text{ eq } H_3PO_4}{\text{L}} \times \frac{97.99 \text{ g } H_3PO_4}{\text{eq } H_3PO_4} = 3.43 \text{ g } H_3PO_4$$

Only one of the three available hydrogens in $H_3PO_4$ reacts, so there is only 1 eq/mol.

## 16.9 Solution Concentration: A Summary

The best guarantee of success in working with solution concentration is a clear understanding of the units in which it is expressed. You may take these units from concentration ratios:

$$\frac{\text{quantity of solute}}{\text{quantity of solution}} \quad or \quad \frac{\text{quantity of solute}}{\text{quantity of solvent}}$$

Table 16.1 makes these ratios specific for percentage concentration, molarity, molality, and normality.

**Table 16.1** Summary of Solution Concentrations

| Name (Symbol) | Mathematical Form |
|---|---|
| Percentage (%) | $\frac{\text{g solute}}{\text{g solute} + \text{g solvent}} \times 100$ |
| Molarity (M) | $\frac{\text{mol solute}}{\text{L solution}}$ |
| Molality (m) | $\frac{\text{mol solute}}{\text{kg solvent}}$ |
| Normality (N) | $\frac{\text{eq solute}}{\text{L solution}}$ |

## Thinking About Your Thinking

### Proportional Reasoning

**All of the solution concentration units introduced in this chapter are direct proportionalities. Percentage concentration by mass is a direct proportionality between mass of solute and mass of solution; molarity, between moles of solute and liters of solution; molality, between moles of solute and kilograms of solvent; and normality, between equivalents of solute and liters of solution. These proportional relationships allow you to think of solution concentration units as conversion factors between the two units in the fraction. Do you know mass of solution and need mass of solute? Use percentage concentration. Do you know volume of solution and need moles of solute? Use molarity. Thinking about solution concentration units in this way allows you to become more skilled at solving quantitative problems.**

# 16.10 Dilution of Concentrated Solutions

Goal **13** Given any three of the following, calculate the fourth: (a) volume of concentrated solution, (b) molarity of concentrated solution, (c) volume of dilute solution, (d) molarity of dilute solution.

Some common acids and bases are available in concentrated solutions that are diluted to a lower concentration for use. To dilute a solution, you simply add more solvent. The number of moles of solute remains the same, but it is distributed over a larger volume, as illustrated in Figure 16.16. The number of moles is V × M (volume × molarity)

(a) Before

(b) After

**Figure 16.16** Dilution of a concentrated solution. Concentrated solutions are diluted by adding more *solvent* particles. The volume changes because of this addition. The number of *solute* particles, whether counted individually or grouped into moles, remains the same before (a) and after (b) the dilution. Thus, the number of moles of solute particles is a constant:

$$V_c \times M_c = n = V_d \times M_d$$

(Equation 16.3). Using subscript *c* for the *c*oncentrated solution and subscript *d* for the *d*ilute solution, we obtain

$$V_c \times M_c = V_d \times M_d \qquad \textbf{(16.11)}$$

Equation 16.11 has two important applications. They are illustrated in the next two Active Examples.

## Thinking About Your Thinking

### Mental Models

**Figure 16.16 is drawn to help you form a particulate-level mental model of what happens when a solution is diluted by adding more solvent. Note, in particular, that the number of solute particles remains the same before and after dilution.**

## Active Example 16.14

How many milliliters of commercial hydrochloric acid, which is 11.6 molar, should you use to prepare 5.50 liters of 0.500 M HCl?

Start by assigning each of the variables below:

$V_c =$      $M_c =$

$V_d =$      $M_d =$

$V_c$ = wanted      $M_c$ = 11.6 M

$V_d$ = 5.50 L      $M_d$ = 0.500 M

Solve Equation 16.11 for the wanted quantity, substitute, and calculate the answer.

EQUATION: $V_c = \frac{V_d \times M_d}{M_c} = \frac{5.50\ \text{L} \times 0.500\ \text{M}}{11.6\ \text{M}} \times \frac{1000\ \text{mL}}{\text{L}} = 237\ \text{mL}$

$= 0.237\ \text{L}$

Solution of the equation yields the answer 0.237 L, as shown. But the question asks for milliliters. The final conversion changes the answer to the wanted units.

Figure 16.17 shows the procedure for preparation of a dilute solution.

## Active Example 16.15

A student adds 50.0 mL $H_2O$ to 25.0 mL 0.881 M NaOH. What is the concentration of the diluted solution?

There is a little trick to this question, but if you read it carefully, you will not be trapped. Complete the Active Example.

**Figure 16.17** Procedure for diluting a concentrated solution.

*GIVEN:* $V_c = 25.0$ mL; $M_c = 0.881$ M; $V_d = 75.0$ mL *WANTED:* $M_d$

*EQUATION:* $$M_d = \frac{V_c \times M_c}{V_d} = \frac{25.0 \text{ mL} \times 0.881 \text{ M}}{75.0 \text{ mL}} = 0.294 \text{ M}$$

The tricky part of this Active Example is the volume of the diluted solution. The problem states that 50.0 mL is added to 25.0 mL, giving the total volume of 75.0 mL.

## 16.11 Solution Stoichiometry

**Goal** 14 Given the quantity of any species participating in a chemical reaction for which the equation can be written, find the quantity of any other species, either quantity being measured in (a) grams, (b) volume of solution at specified molarity, or (c) (if gases have been studied) volume of gas at given temperature and pressure.

For any reaction whose equation is known, the three steps for solving a stoichiometry problem are:

1. Convert the quantity of given species to number of particles, grouped in moles.
2. Convert the moles of given species to moles of wanted species.
3. Convert the moles of wanted species to the quantity units required.

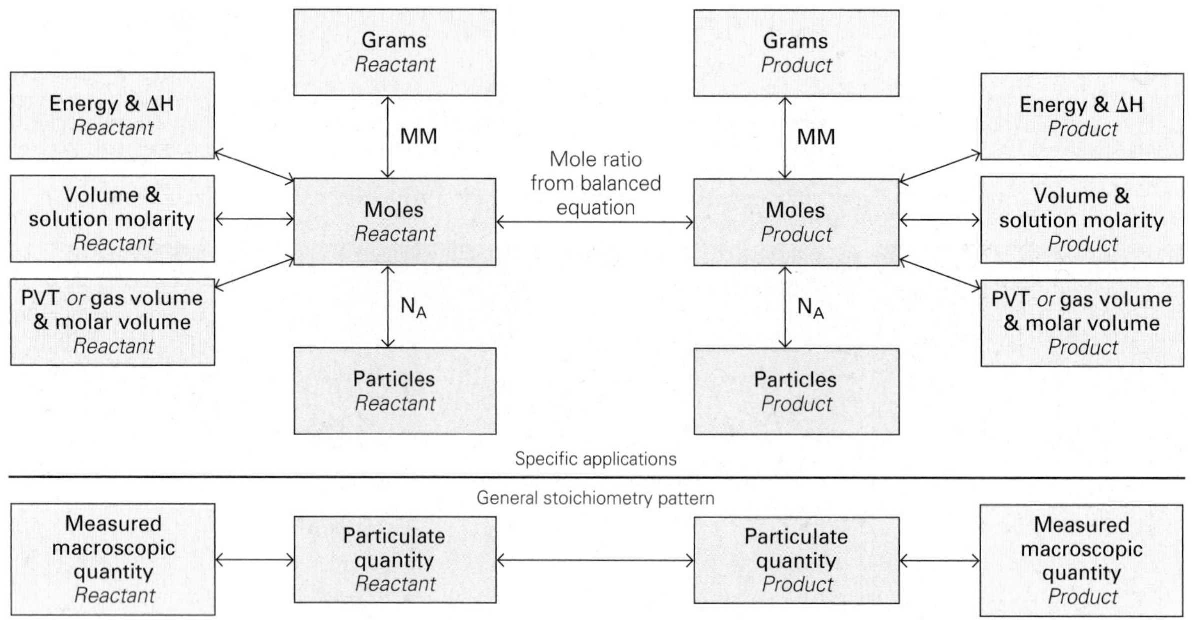


**Figure 16.18** Conversion among grams; energy and ΔH; volume and molarity; gas pressure, volume, and temperature; moles; and number of particles for two species in a chemical change. This figure shows all stoichiometric relationships introduced in this textbook. In general, a measured macroscopic quantity is changed to the number of particles in the sample, measured in moles, and then moles of one species in the chemical change are converted to moles of another species using the mole ratio from the balanced equation. The resulting particulate quantity can be converted to a measured macroscopic quantity.

In Chapter 10, the quantities of given and wanted species in Steps 1 and 3 were measured in grams. Later in Chapter 10, we looked at how quantity of energy was related to mass of reactant or product. In Chapter 14, either or both quantities were measured in (a) grams or (b) volume of gas at a known temperature and pressure. In this section, either or both quantities may be measured in (a) grams, (b) volume of a gas at a specified temperature and pressure, or (c) volume of solution of known concentration.

It is critical for you to see these connections as a general relationship in which the quantities of given or wanted substances may be expressed in any combination of mass, energy, volume of solution of known concentration, and volume of gas at known temperature and pressure. This general stoichiometric pattern is illustrated in Figure 16.18.

## Active Example 16.16

How many grams of lead(II) iodide will precipitate (Fig. 16.19) when excess potassium iodide solution is added to 50.0 mL of 0.811 M lead(II) nitrate?

PLAN the problem, including the reaction equation.

**Figure 16.19** Solid lead(II) iodide precipitates.

$Pb(NO_3)_2(aq) + 2\ KI(aq) \rightarrow PbI_2(s) + 2\ KNO_3(aq)$

*GIVEN:* 50.0 mL 0.811 M $Pb(NO_3)_2$ *WANTED:* g $PbI_2$

$$\text{PER: } \quad \frac{1\text{ L Pb(NO}_3)_2}{1000\text{ mL Pb(NO}_3)_2} \qquad \frac{0.811\text{ mol Pb(NO}_3)_2}{\text{L Pb(NO}_3)_2}$$

$$\text{PATH: mL Pb(NO}_3)_2 \longrightarrow \text{L Pb(NO}_3)_2 \longrightarrow$$

$$\frac{1\text{ mol PbI}_2}{1\text{ mol Pb(NO}_3)_2} \qquad \frac{461.0\text{ g PbI}_2}{\text{mol PbI}_2}$$

$$\text{mol Pb(NO}_3)_2 \longrightarrow \text{mol PbI}_2 \longrightarrow \text{g PbI}_2$$

The conversion from mL $Pb(NO_3)_2$ to mol $Pb(NO_3)_2$ was done in two steps in our *PER/PATH,* but you can do it in one step by thinking of molarity, mol/L, as its equivalent, mol/1000 mL.

Begin the setup with the first two conversions in the unit path to moles of $Pb(NO_3)_2$. Do not calculate the answer.

$$50.0\ \cancel{\text{mL Pb(NO}_3)_2} \times \frac{1\ \cancel{\text{L Pb(NO}_3)_2}}{1000\ \cancel{\text{mL Pb(NO}_3)_2}} \times \frac{0.811\text{ mol Pb(NO}_3)_2}{\cancel{\text{L Pb(NO}_3)_2}} \times$$

Once the volume of the solution is in liters, the defining equation for molarity (Equation 16.2, M ≡ mol/L) can be used as a conversion factor to find the number of moles.

The last two steps of the unit path complete the problem. Use the space above to finish the setup and write the final answer.

$$50.0\ \cancel{\text{mL Pb(NO}_3)_2} \times \frac{1\ \cancel{\text{L Pb(NO}_3)_2}}{1000\ \cancel{\text{mL Pb(NO}_3)_2}} \times \frac{0.811\ \cancel{\text{mol Pb(NO}_3)_2}}{\cancel{\text{L Pb(NO}_3)_2}} \times \frac{1\ \cancel{\text{mol PbI}_2}}{1\ \cancel{\text{mol Pb(NO}_3)_2}} \times \frac{461.0\text{ g PbI}_2}{\cancel{\text{mol PbI}_2}} = 18.7\text{ g PbI}_2$$

## Active Example 16.17

Calculate the number of milliliters of 0.842 M sodium hydroxide required to precipitate as copper(II) hydroxide all of the copper ions in 30.0 mL of 0.635 M copper(II) sulfate.

The first two steps this time are the same as in the last Active Example. Set up that far. We'll use the one-step shortcut volume-to-moles conversion mentioned in Active Example 16.16, thinking of molarity as mol/1000 mL.

*continued*

$2\ NaOH(aq) + CuSO_4(aq) \rightarrow Cu(OH)_2(s) + Na_2SO_4(aq)$

*GIVEN:* 30.0 mL 0.635 M $CuSO_4$ *WANTED:* mL 0.842 M NaOH

*PER:* $\frac{0.635 \text{ mol } CuSO_4}{1000 \text{ mL } CuSO_4}$ $\frac{2 \text{ mol NaOH}}{1 \text{ mol } CuSO_4}$

*PATH:* mL $CuSO_4$ ⟶ mol $CuSO_4$ ⟶ mol NaOH

$$30.0 \cancel{\text{mL } CuSO_4} \times \frac{0.635 \cancel{\text{mol } CuSO_4}}{1000 \cancel{\text{mL } CuSO_4}} \times \frac{2 \text{ mol NaOH}}{1 \cancel{\text{mol } CuSO_4}} \times$$

At this point you have the number of moles of NaOH. Its molarity may be used to change it to volume in milliliters, as in Active Example 16.5. Use the space above to complete the last step in the setup and write the final answer.

$$30.0 \cancel{\text{mL } CuSO_4} \times \frac{0.635 \cancel{\text{mol } CuSO_4}}{1000 \cancel{\text{mL } CuSO_4}} \times \frac{2 \cancel{\text{mol NaOH}}}{1 \cancel{\text{mol } CuSO_4}} \times \frac{1000 \text{ mL NaOH}}{0.842 \cancel{\text{mol NaOH}}} = 45.2 \text{ mL NaOH}$$

In Section 10.2, where you first learned how to solve chemical reaction problems, you read these words: "Sometimes it is more convenient to measure mass in larger or smaller units. Most analytical and academic chemists, for example, work in milligrams and millimoles of substance." That applies to volume, too. We have already been considering solution volume in milliliters. If we scale everything down to 1/1000 size, everything we have seen about liters and moles applies equally to milliliters and millimoles. Molarity then becomes millimoles/milliliter:

$$M \equiv \frac{\text{mol}}{\text{L}} = \frac{\cancel{\text{mol}}}{\cancel{\text{L}}} \times \frac{1000 \text{ mmol}}{\cancel{\text{mol}}} \times \frac{\cancel{\text{L}}}{1000 \text{ mL}} = \frac{\text{mmol}}{\text{mL}} \quad \textbf{(16.12)}$$

To illustrate the use of milliunits, we will repeat Active Example 16.17.

## Active Example 16.18

Calculate the number of milliliters of 0.842 M sodium hydroxide required to precipitate as copper(II) hydroxide all of the copper ions in 30.0 mL 0.635 M copper(II) sulfate.

No matter whether we set up the problem in base units or milliunits, the reaction equation, *GIVEN*, and *WANTED* remain the same:

$2\ NaOH(aq) + CuSO_4(aq) \rightarrow Cu(OH)_2(s) + Na_2SO_4(aq)$

*GIVEN:* 30.0 mL 0.635 M $CuSO_4$ *WANTED:* mL 0.842 M NaOH

The *PER/PATH* will be different, however. Instead of reading molarity as moles per liter, you will use mmol/mL. Similarly, instead of interpreting the coefficients from the balanced equation as a ratio of moles, think of it as a ratio of millimoles. Write the *PER/PATH* completely in milliunits.

*PER:* $\frac{0.635 \text{ mmol } CuSO_4}{\text{mL } CuSO_4}$ $\frac{2 \text{ mmol NaOH}}{1 \text{ mmol } CuSO_4}$ $\frac{1 \text{ mL NaOH}}{0.842 \text{ mmol NaOH}}$

*PATH:* mL $CuSO_4$ ⟶ mmol $CuSO_4$ ⟶ mmol NaOH ⟶ mL NaOH

Complete the problem with the setup and answer.

$$30.0 \text{ mL } CuSO_4 \times \frac{0.635 \text{ mmol } CuSO_4}{\text{mL } CuSO_4} \times \frac{2 \text{ mmol NaOH}}{1 \text{ mmol } CuSO_4} \times \frac{1 \text{ mL NaOH}}{0.842 \text{ mmol NaOH}} = 45.2 \text{ mL NaOH}$$

## 16.12 Titration Using Molarity

Goal **15** Given the volume of a solution that reacts with a known mass of a primary standard and the equation for the reaction, calculate the molarity of the solution.

Goal **16** Given the volumes of two solutions that react with each other in a titration, the molarity of one solution, and the equation for the reaction or information from which it can be written, calculate the molarity of the second solution.

One common laboratory operation in analytical chemistry is called **titration** (see Active Fig. 16.20). Titration is the very careful addition of one solution to another by means of a device that can measure delivered volume precisely, such as a **buret.** The buret accurately measures the volume of solution required to react with a carefully measured amount of another dissolved substance. When that precise volume has been reached, an **indicator** changes color, and the flow from the buret is stopped. Phenolphthalein is a typical indicator for acid–base titrations. It is colorless in an acid solution and pink in a basic solution.

Titration can be used to **standardize** a solution, which means finding its concentration for use in later titrations. Sodium hydroxide cannot be weighed accurately because it absorbs moisture from the air and increases in weight during the weighing process. Therefore, it is not possible to prepare a sodium hydroxide solution whose molarity is known precisely. Instead, the solution is standardized against a weighed quantity of something that can be weighed accurately. Such a substance is called a **primary standard.** Oxalic acid dihydrate, $H_2C_2O_4 \cdot 2\ H_2O$, is an example. When used to standardize sodium hydroxide, the equation is

$$H_2C_2O_4(aq) + 2\ NaOH(aq) \rightarrow Na_2C_2O_4(aq) + 2\ H_2O(\ell)$$

Notice that the water in the hydrate is not a part of the equation. When one mole of $H_2C_2O_4 \cdot 2\ H_2O$ dissolves, the hydrate water becomes a part of the solution and one mole of $H_2C_2O_4$ is available for reaction. The hydrate water must be taken into account in weighing the $H_2C_2O_4 \cdot 2\ H_2O$, however.

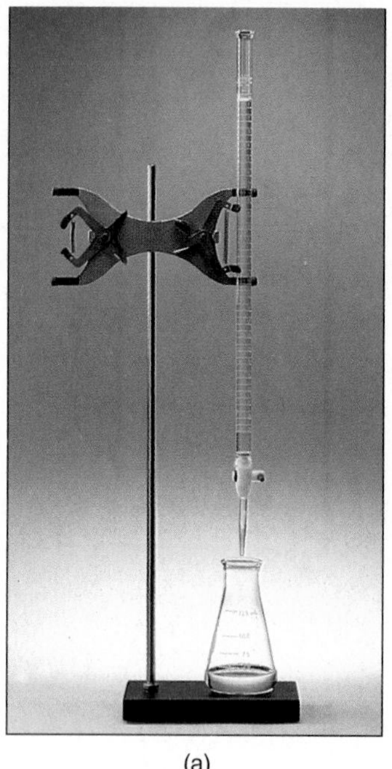

(a)

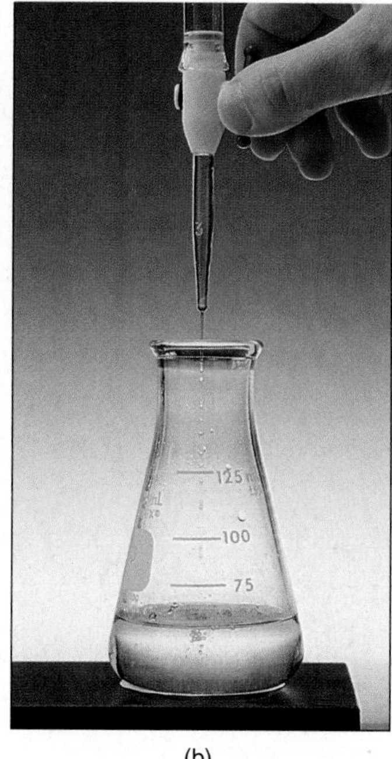

(b)

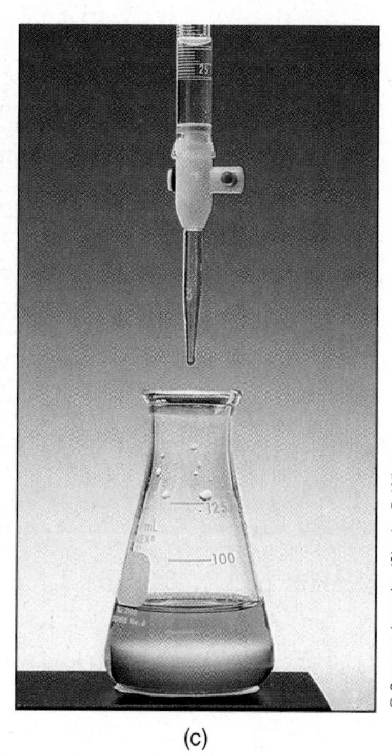

(c)



**Active Figure 16.20** Titrating from a buret into a flask. (a) The flask contains an acidic solution of unknown concentration. A small quantity of dye is added to the acid. The dye is colorless in acidic solution. The buret is filled with a basic solution of known concentration. (b) By careful control of the valve, the chemist may deliver liquid from the buret to the flask in a steady stream, drop by drop, or in a single drop. (c) When the volume of base delivered by the buret reacts completely with the acid in the flask, the indicator dye changes color, signaling the chemist to stop the flow of base solution and read the delivered volume. **Watch this active figure at *http://www.cengage.com/cracolice.***

## Active Example 16.19

A chemist dissolves 1.18 g $H_2C_2O_4 \cdot 2\ H_2O$ (126.07 g/mol) in water and titrates the solution with a solution of NaOH of unknown concentration. She determines that 28.3 mL NaOH(aq) is required to neutralize the acid. Calculate the molarity of the NaOH solution.

The general plan for any titration problem is outlined by Figure 16.18. You have the mass of the given species, which must be converted to moles. The moles of the given species are then converted to moles of the wanted species. This part of the Active Example is the same as the problems in Chapter 10. Plan the problem and solve for moles of NaOH.

$H_2C_2O_4(aq) + 2\ NaOH(aq) \rightarrow Na_2C_2O_4(aq) + 2\ H_2O(\ell)$

*GIVEN:* 1.18 g $H_2C_2O_4 \cdot 2\ H_2O$; 28.3 mL NaOH *WANTED:* mol NaOH

*PER:* $\dfrac{1\text{ mol } H_2C_2O_4 \cdot 2\ H_2O}{126.07\text{ g } H_2C_2O_4 \cdot 2\ H_2O}$ $\dfrac{2\text{ mol NaOH}}{1\text{ mol } H_2C_2O_4}$

*PATH:* g $H_2C_2O_4 \cdot 2\ H_2O \longrightarrow$ mol $H_2C_2O_4 \longrightarrow$ mol NaOH

$$1.18\text{ g }\cancel{H_2C_2O_4 \cdot H_2O} \times \frac{1\text{ mol }\cancel{H_2C_2O_4 \cdot 2\ H_2O}}{126.07\text{ g }\cancel{H_2C_2O_4 \cdot 2\ H_2O}} \times \frac{2\text{ mol NaOH}}{1\text{ }\cancel{\text{mol } H_2C_2O_4}} = 0.0187\text{ mol NaOH}$$

You now have the numerator for the fraction in the equation that defines molarity, and you almost have the denominator. Complete the problem. (If you're puzzled, remember the exact definition of molarity.)

*EQUATION:* $M \equiv \dfrac{\text{mol}}{\text{L}} = \dfrac{0.0187\text{ mol NaOH}}{28.3\ \cancel{\text{mL}}} \times \dfrac{1000\ \cancel{\text{mL}}}{1\text{ L}} = 0.661\text{ M NaOH}$

Molarity is moles per liter. Volume, given in milliliters, must be changed to liters.

As you gain experience in solving titration problems, you may find it more convenient not to calculate the intermediate answer, 0.0187 mol NaOH. Instead, you can use the setup, which represents moles of NaOH, and divide by (multiply by the inverse of) volume, which must be changed to liters:

$$1.18\text{ g }\cancel{H_2C_2O_4 \cdot 2\ H_2O} \times \frac{1\text{ mol }\cancel{HC_2O_4 \cdot 2\ H_2O}}{126.07\text{ g }\cancel{H_2C_2O_4 \cdot 2\ H_2O}} \times \frac{2\text{ mol NaOH}}{1\text{ }\cancel{\text{mol } H_2C_2O_4}} \times \frac{1}{28.3\ \cancel{\text{mL}}} \times \frac{1000\ \cancel{\text{mL}}}{\text{L}} = 0.661\text{ M NaOH}$$

Once a solution is standardized, we may use it to find the concentration of another solution. This procedure is widely used in industrial laboratories.

## Active Example 16.20

A 25.0-mL sample of an electroplating solution is analyzed for its sulfuric acid concentration. It takes 46.8 mL of the 0.661 M NaOH from Active Example 16.19 to neutralize the sample. Find the molarity of the acid.

This time the first stoichiometry step begins with a volume of solution of known molarity rather than the mass of a solid. Notice also that both *GIVEN* and *WANTED* quantities include solution volumes in milliliters. Does that suggest anything to you? Remember Active Example 16.18, which we solved using volume in mL and molarity as mmol/mL. It's easier that way. Complete the Active Example.

*continued*

$$2\ NaOH(aq) + H_2SO_4(aq) \rightarrow Na_2SO_4(aq) + 2\ H_2O(\ell)$$

GIVEN: 46.8 mL 0.661 M NaOH; 25.0 mL $H_2SO_4$ WANTED: M $H_2SO_4$

PER: $\frac{0.661 \text{ mmol NaOH}}{\text{mL NaOH}}$ $\frac{1 \text{ mmol } H_2SO_4}{2 \text{ mmol NaOH}}$

PATH: mL NaOH ⟶ mmol NaOH ⟶ mmol $H_2SO_4$

$$46.8 \text{ mL NaOH} \times \frac{0.661 \text{ mmol NaOH}}{\text{mL NaOH}} \times \frac{1 \text{ mmol } H_2SO_4}{2 \text{ mmol NaOH}} = 15.5 \text{ mmol } H_2SO_4$$

$$\text{EQUATION: } M \equiv \frac{\text{mol}}{\text{L}} = \frac{\text{mmol}}{\text{mL}} = \frac{15.5 \text{ mmol } H_2SO_4}{25.0 \text{ mL}} = 0.620 \text{ M } H_2SO_4$$

There are three significant figures in both the numerator and the denominator of the equation, so there are three significant figures in the result. Even though your calculator does not display the zero in "0.620 M $H_2SO_4$," you must include it in the answer.

## 16.13 Titration Using Normality (Optional)

Goal **17** Given the volume of a solution that reacts with a known mass of a primary standard and the equation for the reaction, calculate the normality of the solution.

Goal **18** Given the volumes of two solutions that react with each other in a titration and the normality of one solution, calculate the normality of the second solution.

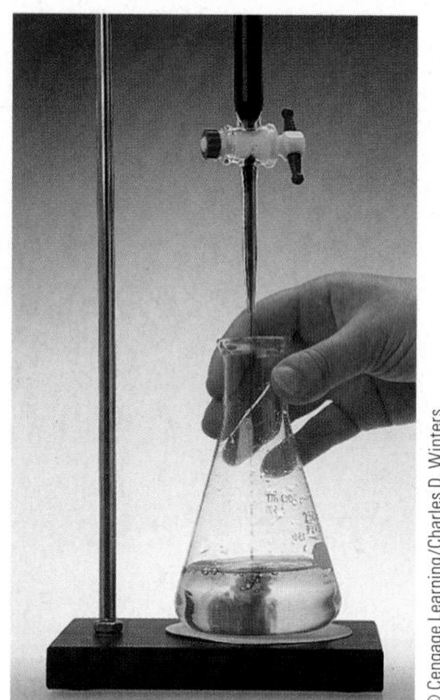

© Cengage Learning/Charles D. Winters

A purple solution containing permanganate ion, $MnO_4^-$, is titrated into an acidified solution that contains iron(II) ion. One product of the reaction is iron(III) ion, which has a pale yellow color in aqueous solution.

We noted earlier that normality is a convenient concentration unit in analytical work. This is because of the fact pointed out in Section 16.8: *The number of equivalents of all species in a reaction is the same.* Consequently, for an acid–base reaction,

$$\text{equivalents of acid} = \text{equivalents of base} \qquad \textbf{(16.13)}$$

There are two ways to calculate the number of equivalents (eq) in a sample of a substance. If you know the mass of the substance and its equivalent mass, use equivalent mass as a conversion factor to get equivalents, as in Active Example 16.10. If the sample is a solution and you know its volume and normality, multiply one by the other. $V \times N = eq$, according to Equation 16.10.

We illustrate normality calculations by repeating Active Examples 16.19 and 16.20, which were solved with molarity. In the first Active Example a primary standard is used to standardize a solution. In the second Active Example the standardized solution is used to find the concentration of another solution.

### Active Example 16.21

A chemist dissolves 1.18 g $H_2C_2O_4 \cdot 2\ H_2O$ (126.07 g/mol) in water and titrates the solution with 28.3 mL of a solution of NaOH of unknown concentration. Calculate the normality of the NaOH for the reaction $H_2C_2O_4(aq) + 2\ NaOH(aq) \rightarrow Na_2C_2O_4(aq) + 2\ H_2O(\ell)$.

Begin by planning the problem to the point where you find equivalents of $H_2C_2O_4$. Ordinarily, you include the molar mass of a substance as a PER expression in the plan. When working in normality, however, you use equivalent mass, g/eq, which can be expressed as a ratio of molar mass to equivalents per mole (Equation 16.9). We suggest you write it that way in your PLAN.

*GIVEN:* 1.18 g $H_2C_2O_4 \cdot 2\ H_2O$; 28.3 mL NaOH
*WANTED:* eq $H_2C_2O_4$

*PER:* $\frac{2 \text{ eq } H_2C_2O_4}{126.07 \text{ g } H_2C_2O_4 \cdot 2\ H_2O}$

*PATH:* $\text{g } H_2C_2O_4 \cdot 2\ H_2O \longrightarrow \text{eq } H_2C_2O_4$

The equation shows that both ionizable hydrogens from $H_2C_2O_4$ are used, so there are two equivalents per mole of acid.

To use the defining equation for normality, N ≡ eq/L, you must know, or be able to find, both the numerator and denominator quantities. The denominator results from a mL → L conversion from one of the givens. As for the numerator, you have already written the setup to find equivalents of $H_2C_2O_4$. And what is that equal to? It equals the number of equivalents of NaOH. The reaction has the same number of equivalents of all species. If you can find one, you've found them all. Using only what you need from the foregoing information, write the setup and calculate the number of equivalents of NaOH.

$$1.18 \cancel{\text{g } H_2C_2O_4 \cdot 2\ H_2O} \times \frac{2 \text{ eq NaOH or } H_2C_2O_4}{126.07 \cancel{\text{g } H_2C_2O_4 \cdot 2\ H_2O}} = 0.0187 \text{ eq NaOH}$$

Notice the "or" in the numerator. The number of equivalents is the same for all species in a reaction.

Complete the problem.

*EQUATION:* $\text{N} \equiv \frac{\text{eq}}{\text{L}} = \frac{0.0187 \text{ eq NaOH}}{28.3 \cancel{\text{mL}}} \times \frac{1000 \cancel{\text{mL}}}{\text{L}} = 0.661 \text{ N NaOH}$

Once you know the normality of one solution, you can use it to find the normality of another solution. Equation 16.10 (Section 16.8) indicates that the number of equivalents of a species in a reaction is the product of solution volume times normality. The number of equivalents of all species in a reaction is the same, so

$$V_1N_1 = \text{eq} = V_2N_2 \quad or \quad V_1N_1 = V_2N_2 \qquad \textbf{(16.14)}$$

Equations 16.14 and 16.15 are the two key equations that make normality so useful in a laboratory that runs the same titrations again and again.

where subscripts 1 and 2 identify the reacting solutions. Solving for the second normality, we obtain

$$N_2 = \frac{V_1N_1}{V_2} \qquad \textbf{(16.15)}$$

That's all it takes to calculate normality in this follow-up to Active Example 16.20.

## Active Example 16.22

A 25.0-mL sample of an electroplating solution is analyzed for its sulfuric acid concentration. It takes 46.8 mL of the 0.661 N NaOH from Active Example 16.21 to neutralize the sample. Find the normality of the acid.

*GIVEN:* $V_1 = 46.8\ mL; N_1 = 0.661\ N; V_2 = 25.0\ mL$ *WANTED:* $N_2$

*EQUATION:* $$N_2 = \frac{V_1N_1}{V_2} = \frac{46.8\ \cancel{mL} \times 0.661\ N}{25.0\ \cancel{mL}} = 1.24\ N\ H_2SO_4$$

The normality in Active Example 16.22 is twice the molarity in Active Example 16.20, 0.620 M. This is always the case when there are 2 eq/mol. For an X molar solution

$$\frac{X\ mol}{L} \times \frac{2\ eq}{mol} = 2X\ eq/L$$

When there is one equivalent per mole, the molarity and normality of the solution are the same:

$$\frac{X\ mol}{L} \times \frac{1\ eq}{mol} = X\ eq/L$$

## 16.14 Colligative Properties of Solutions (Optional)

Goal **19** Given (a) the molality of a solution, or data from which it may be found, (b) the normal freezing or boiling point of the solvent, and (c) the freezing- or boiling-point constant, find the freezing or boiling point of the solution.

Goal **20** Given the freezing-point depression or boiling-point elevation and the molality of a solution, or data from which they may be found, calculate the molal freezing-point constant or molal boiling-point constant.

Goal **21** Given (a) the mass of solute and solvent in a solution, (b) the freezing-point depression or boiling-point elevation, or data from which they may be found, and (c) the molal freezing/boiling-point constant of the solvent, find the approximate molar mass of the solute.

A pure solvent has distinct physical properties, as does any pure substance. Introducing a solute into the solvent affects these properties. The properties of the solution depend on the relative amounts of solvent and solute. It has been found experimentally that, in *dilute* solutions of certain solutes, the *change* in some of these properties is proportional to the molal concentration of the solute particles. These are called **colligative properties: solution properties that are determined only by the *number* of solute particles dissolved in a fixed quantity of solvent and not on the identity of the solute particles.**

Freezing and boiling points of solutions are colligative properties. Perhaps the best-known example of such a solution is the antifreeze used in automobile cooling systems. The solute that is dissolved in the radiator water reduces the freezing tempera-

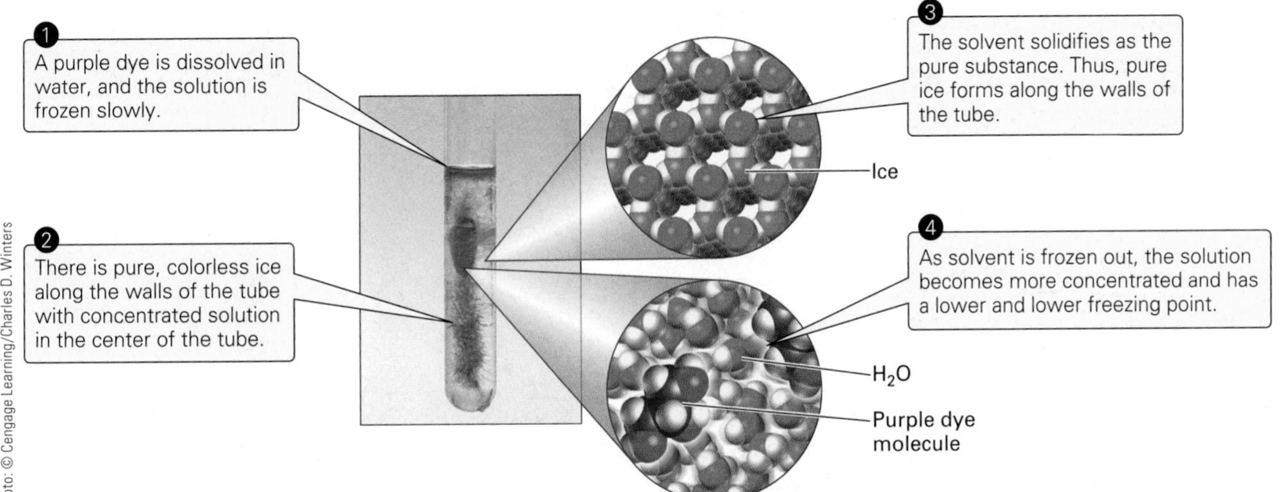

1
A purple dye is dissolved in water, and the solution is frozen slowly.
2
There is pure, colorless ice along the walls of the tube with concentrated solution in the center of the tube.
3
The solvent solidifies as the pure substance. Thus, pure ice forms along the walls of the tube.
Ice
4
As solvent is frozen out, the solution becomes more concentrated and has a lower and lower freezing point.
$H_2O$
Purple dye molecule
Photo: © Cengage Learning/Charles D. Winters


**Figure 16.21** Freezing-point depression. The change in freezing point of a solution is proportional to its molal concentration.

ture well below the normal freezing point of pure water. It also raises the boiling point above the normal boiling point.

The change in a freezing point is the **freezing-point depression, $\Delta T_f$** (Fig. 16.21), and the change in a boiling point is the **boiling-point elevation, $\Delta T_b$.** The two proportionalities and their corresponding equations are

$$\Delta T_f \propto m \xrightarrow{\text{the proportionality changes to an equation}} \Delta T_f = K_f \times m \qquad \textbf{(16.16)}$$

$$\Delta T_b \propto m \xrightarrow{\text{the proportionality changes to an equation}} \Delta T_b = K_b \times m \qquad \textbf{(16.17)}$$

The proportionality constants, $K_f$ and $K_b$, are, respectively, the **molal freezing-point depression constant** and the **molal boiling-point elevation constant.** The freezing- and boiling-point constants are properties of the *solvent,* no matter what the solute may be. The freezing-point constant for water is 1.86°C/m, and the boiling-point constant is 0.52°C/m.

Chemists often take some liberties in units and algebraic signs when solving freezing- and boiling-point problems. Technically, °C · kg solvent/mole solute are the units for $K_f$ or $K_b$. These units are usable, but they are awkward. The substitute, °C/m, is acceptable if we keep calculations based on Equations 16.16 and 16.17 separate from other calculations. We will follow that practice.

Again, technically, if the freezing point of the solvent is taken as the "initial" temperature and the always lower freezing point of the solution as the "final" temperature, $\Delta T_f$ and $K_f$ must be negative quantities. In some texts they are regarded as such. Most chemists, however, use the words depression and elevation to identify clearly the direction the temperature is changing, and then treat both constants and the magnitude of temperature change as positive numbers. We follow this practice, too.

## Active Example 16.23

Determine the freezing point of a solution of 12.0 g urea, $CO(NH_2)_2$, in $2.50 \times 10^2$ grams of water.

To use Equation 16.16, $\Delta T_f = K_f \times m$, you need to express the solution concentration in molality. The defining equation for molality, Equation 16.4, is $m \equiv$ mol solute/kg solvent (Section 16.7). Start by finding the moles of solute and the molality of the solution. Watch the units carefully.

*continued*

*GIVEN:* 12.0 g $CO(NH_2)_2$; $2.50 \times 10^2$ g $H_2O$ *WANTED:* mol $CO(NH_2)_2$; m $CO(NH_2)_2$

*PER:* $\frac{1 \text{ mol } CO(NH_2)_2}{60.06 \text{ g } CO(NH_2)_2}$

*PATH:* g $CO(NH_2)_2 \longrightarrow$ mol $CO(NH_2)_2$

$$12.0 \text{ g } \cancel{CO(NH_2)_2} \times \frac{1 \text{ mol } CO(NH_2)_2}{60.06 \text{ g } \cancel{CO(NH_2)_2}} = 0.200 \text{ mol } CO(NH_2)_2$$

$$m \equiv \frac{\text{mol solute}}{\text{kg solvent}} = \frac{0.200 \text{ mol } CO(NH_2)_2}{2.50 \times 10^2 \text{ g } \cancel{H_2O}} \times \frac{1000 \text{ g } \cancel{H_2O}}{\text{kg } H_2O} = 0.800 \text{ m } CO(NH_2)_2$$

Just as you changed milliliters of solution to liters to satisfy the defining equation for molarity, you must change grams of solvent into kilograms to satisfy the defining equation for molality.

Find the freezing-point depression by substitution into Equation 16.16, $\Delta T_f = K_f \times m$. $K_f$ for water is 1.86°C/m.

© Cengage Learning/Charles D. Winters

One reason for putting salt on icy streets in winter is that some dissolves in whatever liquid is present. This lowers the freezing temperature and melts at least some of the ice or turns it to slush.

*EQUATION:* $\Delta T_f = K_f m = \frac{1.86°C}{\cancel{m}} \times 0.800\ \cancel{m} = 1.49°C$

The freezing-point depression is 1.49°C. The normal freezing point of water is 0°C. Complete the Active Example by finding the freezing point of the solution.

$0°C - 1.49°C = -1.49°C$

Freezing-point depression and boiling-point elevation can be used to find the approximate molar mass of an unknown solute. The solution is prepared with measured masses of the solute and a solvent whose freezing- or boiling-point constant is known. The freezing-point depression or boiling-point elevation is found by experiment. The calculation procedure is as follows.

**procedure**

**How to Calculate the Molar Mass of a Solute from Freezing-Point Depression or Boiling-Point Elevation Data**

**Step 1:** Calculate molality from $m = \Delta T_f/K_f$ or $m = \Delta T_b/K_b$. Express as mol solute/kg solvent.

**Step 2:** Using molality as a conversion factor between moles of solute and kilograms of solvent, find the number of moles of solute.

**Step 3:** Use the defining equation for molar mass, MM ≡ g/mol, to calculate the molar mass of the solute.

## Active Example 16.24

The molal boiling-point elevation constant of benzene is 2.5°C/m. A solution of 15.2 g of unknown solute in 91.1 g benzene boils at a temperature 2.1°C higher than the boiling point of pure benzene. Estimate the molar mass of the solute.

Calculate the molality of the solution (*Step 1*).

GIVEN: $\Delta T_b = 2.1°C$; $K_b = 2.5°C/m$ WANTED: m (mol solute/kg benzene)

EQUATION: $$m = \frac{\Delta T_b}{K_b} = 2.1\cancel{°C} \times \frac{m}{2.5\cancel{°C}} = 0.84\ m = \frac{0.84 \text{ mol solute}}{\text{kg benzene}}$$

Now use this molality to convert from the given 91.9 grams of benzene to moles of solute (*Step 2*).

GIVEN: 91.1 g benzene WANTED: mol solute

PER: $\frac{1 \text{ kg benzene}}{1000 \text{ g benzene}}$ $\frac{0.84 \text{ mol solute}}{\text{kg benzene}}$

PATH: g benzene ⟶ kg benzene ⟶ mol solute

$$91.1\ \cancel{\text{kg benzene}} \times \frac{1\ \cancel{\text{kg benzene}}}{1000\ \cancel{\text{g benzene}}} \times \frac{0.84 \text{ mol solute}}{\cancel{\text{kg benzene}}} = 0.077 \text{ mol solute}$$

In *Step 3*, you use the defining equation to find molar mass.

GIVEN: 0.077 mol solute; 15.2 g solute WANTED: MM

EQUATION: $$MM \equiv \frac{g}{mol} = \frac{15.2 \text{ g}}{0.077 \text{ mol}} = 2.0 \times 10^2 \text{ g/mol}$$

# Chapter 16 in Review

Most of the key terms and concepts and many others appear in the Glossary. Use your Glossary regularly.

**16.1 The Characteristics of a Solution**

Goal 1 Define the term *solution,* and, given a description of a substance, determine if it is a solution.

**Key Terms and Concepts: Solution**

**16.2 Solution Terminology**

Goal 2 Distinguish among terms in the following groups: *solute* and *solvent; concentrated* and *dilute; solubility, saturated, unsaturated,* and *supersaturated; miscible* and *immiscible.*

**Key Terms and Concepts: Concentrated, dilute, immiscible, miscible, saturated, solubility, solute, solvent, supersaturated, unsaturated**

**16.3 The Formation of a Solution**

Goal 3 Describe the formation of a saturated solution from the time excess solid solute is first placed into a liquid solvent.

Goal 4 Identify and explain the factors that determine the time required to dissolve a given amount of solute or to reach equilibrium.

**Key Terms and Concepts: Dynamic equilibrium, hydrated, polar**

**16.4 Factors That Determine Solubility**

Goal 5 Given the structural formulas of two molecular substances, or other information from which the strength of their intermolecular forces may be estimated, predict if they will dissolve appreciably in each other, and state the criteria on which your prediction is based.

Goal 6 Predict how and explain why the solubility of a gas in a liquid is affected by a change in the partial pressure of that gas over the liquid.

**16.5 Solution Concentration: Percentage by Mass**

Goal 7 Given mass of solute and of solvent or solution, calculate percentage concentration.

Goal 8 Given mass of solution and percentage concentration, calculate mass of solute and solvent.

**Key Terms and Concepts: Mass ratio, percentage concentration by mass**

**16.6 Solution Concentration: Molarity**

Goal 9 Given two of the following, calculate the third: moles of solute (or data from which it may be found), volume of solution, molarity.

**Key Terms and Concepts: Molarity (M)**

**16.7 Solution Concentration: Molality (Optional)**

Goal 10 Given two of the following, calculate the third: moles of solute (or data from which it may be found), mass of solvent, molality.

**Key Terms and Concepts: Molality (m)**

**16.8 Solution Concentration: Normality (Optional)**

Goal 11 Given an equation for a neutralization reaction, state the number of equivalents of acid or base per mole and calculate the equivalent mass of the acid or base.

Goal 12 Given two of the following, calculate the third: equivalents of acid or base (or data from which they may be found), volume of solution, normality.

**Key Terms and Concepts: Equivalent, equivalent mass, normality (N)**

**16.9 Solution Concentration: A Summary**

**16.10 Dilution of Concentrated Solutions**

Goal 13 Given any three of the following, calculate the fourth: (a) volume of concentrated solution, (b) molarity of concentrated solution, (c) volume of dilute solution, (d) molarity of dilute solution.

**16.11 Solution Stoichiometry**

Goal 14 Given the quantity of any species participating in a chemical reaction from which the equation can be written, find the quantity of any other species, either quantity being measured in (a) grams, (b) volume of solution at specified molarity, or (c) (if gases have been studied) volume of gas at given temperature and pressure.

**16.12 Titration Using Molarity**

Goal 15 Given the volume of a solution that reacts with a known mass of a primary standard and the equation for the reaction, calculate the molarity of the solution.

Goal 16 Given the volumes of two solutions that react with each other in a titration, the molarity of one solution, and the equation for the reaction or information from which it can be written, calculate the molarity of the second solution.

**Key Terms and Concepts: Buret, indicator, primary standard, standardize, titration**

**16.13 Titration Using Normality (Optional)**

Goal 17 Given the volume of a solution that reacts with a known mass of a primary standard and the equation for the reaction, calculate the normality of the solution.

Goal 18 Given the volumes of two solutions that react with each other in a titration and the normality of one solution, calculate the normality of the second solution.

**16.14 Colligative Properties of Solutions (Optional)**

Goal 19 Given (a) the molality of a solution, or data from which it may be found, (b) the normal freezing or boiling point of the solvent, and (c) the freezing- or boiling-point constant, find the freezing or boiling point of the solution.

Goal 20 Given the freezing-point depression or boiling-point elevation and the molality of a solution, or data from which they may be found, calculate the molal freezing-point constant or molal boiling-point constant.

Goal 21 Given (a) the mass of solute and solvent in a solution, (b) the freezing-point depression or boiling-point elevation, or data from which they may be found, and (c) the molal freezing/boiling-point constant of the solvent, find the approximate molar mass of the solute.

**Key Terms and Concepts: Boiling-point elevation, colligative properties, freezing-point depression, molal boiling-point elevation constant, molal freezing-point depression constant**

## Study Hints and Pitfalls to Avoid

To solve solution problems easily, you must have a clear understanding of concentrations and the units in which they are expressed. Table 16.1 summarizes all the concentrations used in this chapter. Study carefully the concentrations that have been assigned to you. Then practice with the end-of-chapter questions until you have complete mastery of each performance goal.

To understand solution stoichiometry, you must first understand both fundamental stoichiometry concepts and solution concentrations. If you have difficulty solving solution stoichiometry problems, ask yourself if you thoroughly understand: (a) writing chemical formulas from names, (b) calculating molar masses from chemical formulas, (c) using molar mass to change from moles to mass and from mass to moles, (d) writing and balancing chemical equations, (e) using chemical equations to determine mole ratios, and (f) using molarity to convert from moles to volume and from volume to moles.

Once in a while a student is tempted to change between moles and liters by using 22.7 L/mol, or even worse, 22.7 mol/L. The number 22.7 is so convenient, and the units look like just what is needed. But they are not; 22.7 applies only to gases, not to solutions, and then only to gases at STP.

## Concept-Linking Exercises

*Write a brief description of the relationships among each of the following groups of terms or phrases. Answers to the Concept-Linking Exercises are given after the answers to the Target Checks in Appendix III.*

1. Solution, percentage composition, mixture, pure substance
2. Solubility, saturated, unsaturated, supersaturated
3. Solubility, intermolecular forces, partial pressure of a solute gas, temperature
4. Titration, indicator, standardization of a solution, primary standard, molarity

*The following exercises are from optional sections.*

5. Normality, equivalent, equivalent mass
6. Colligative properties, molality, freezing-point depression, molal freezing-point depression constant, boiling-point elevation, molal boiling-point elevation constant

## Small-Group Discussion Questions

*Small-Group Discussion Questions are for group work, either in class or under the guidance of a leader during a discussion section.*

1. Both the ocean and the atmosphere can be classified as solutions in some situations and not as solutions in other situations. Explain.
2. Give specific examples of solutions in each of the following states: (a) gas in gas, (b) gas in liquid, (c) liquid in liquid, (d) solid in liquid, (e) solid in solid. For each example, identify the solute and solvent.
3. What are the similarities and differences between miscibility and solubility?
4. Do you expect table salt to dissolve in water and/or carbon tetrachloride (both are liquids at room temperature)? For each liquid, describe the dissolving process or explain why dissolving does not occur. Sketch the dissolving process at the particulate level.
5. Develop a list of ten compounds, other than metals or metal alloys, that have intermolecular forces sufficiently different from that of water to make you confident that they will not dissolve in water. You may use your textbook to find compounds. Then develop a list of ten compounds that you predict will dissolve in water. Give a brief justification for each of your choices.
6. If you were to be charged with manufacturing a brand of sugar that will be marketed as fast dissolving, what characteristics would you give it? In general, what characteristics speed the dissolving process? How do those characteristics affect the solubility of sugar?
7. In Section 16.5, we state that for solutions of 5% or less, mass percent and weight/volume percent are essentially equal. A student conducts an experiment to test this statement. He finds the following densities at 20°C: water, 0.9982 g/mL; 10.000 g sodium chloride in 100.00 g solution, 1.0726 g/mL; 10.000 g sugar (sucrose) in 100.00 g solution, 1.0400 g/mL. Do these data confirm or refute the statement?
8. Write a set of instructions sufficiently detailed so that a person who has no knowledge of chemistry can prepare 250.0 mL of a 0.10 M solution of sodium nitrate. Assume that you have standard chemistry laboratory equipment and glassware available.
9. (Optional Section 16.7) What are the similarities and differences between molarity and molality? What are the advantages and disadvantages of each unit?
10. (Optional Section 16.8) Why is equivalent concentration a convenient unit for analytical work? Explain how phosphoric acid can have between 1 and 3 equivalents per mole.
11. Explain how each solution concentration unit assigned in your course fits the general definition of a concentration ratio: quantity of solute per quantity of solution or quantity of solute per quantity of solvent. Also explain how each unit is a direct proportionality.
12. Write a set of instructions sufficiently detailed so that a person who has no knowledge of chemistry can prepare 100.0 mL of a 0.10 M solution of hydrochloric acid from a 12 M solution. Assume that you have standard chemistry laboratory equipment and glassware available.
13. Figure 16.18 summarizes stoichiometry, as it is presented in this introductory course. Four different measured macroscopic quantities can be changed to the same set of four

quantities for any pair of substances involved in a chemical change. This yields 16 different combinations per pair of substances. Divide the 16 combinations among your group members, and find or write a problem for each of the combinations (for example, mass to gas volume and solution volume to quantity of energy).

14. Consider the titration of 0.100 M sodium hydroxide into 10.00 mL of 0.100 M sulfuric acid. Sketch a buret and a flask. Illustrate the titration process by illustrating the buret and flask (a) before the titration starts, (b) half way to neutralizing the acid, and (c) at the completion of the titration. Next, calculate the molar concentration of the acid at points (b) and (c). Finally, construct particulate-level sketches of a tiny portion of the contents of the flask at each of the three points.

15. (Optional Section 16.13) For the same titration described in question 14, calculate the normality of the acid at points (b) and (c), assuming that the acid is completely neutralized at the end of the titration.

16. (Optional Section 16.14) Define the term *colligative property*. Give two examples of solution properties that are colligative and two that are not colligative. Use dimensional analysis to determine the units of the molal boiling-point elevation constant without using molality.

## Questions, Exercises, and Problems

*Blue-numbered questions are answered in Appendix III. ■ denotes problems assignable in OWL. In the first section of Questions, Exercises, and Problems, similar exercises and problems are paired in consecutive odd-even number combinations.*

*Many questions in this chapter are written with the assumption that you have studied Chapter 6 and can write the required formulas from their chemical names. If this is not the case, we have placed a list of all chemical formulas needed to answer Chapter 16 questions at the end of the Questions, Exercises, and Problems.*

*If you have studied Chapter 6 and you get stuck while answering a question because you cannot write a formula, we urge you to review the appropriate section of Chapter 6 before continuing. Avoid the temptation to "just peek" at the list of formulas. Developing skill in chemical nomenclature is an important part of your learning process in this course, and in many cases, you truly learn the material from Chapter 6 as you apply it here in Chapter 16 and throughout your study of chemistry.*

### Section 16.1: The Characteristics of a Solution

1. Mixtures of gases are always true solutions. True or False? Explain why.

2. Every pure substance has a definite and fixed set of physical and chemical properties. A solution is prepared by dissolving one pure substance in another. Is it reasonable to expect that the solution will also have a definite and fixed set of properties that are different from the properties of either component? Explain your answer.

3. Can you see particles in a solution? If yes, give an example, and if no, say why.

4. What kinds of solute particles are present in a solution of an ionic compound? Of a molecular compound?

### Section 16.2: Solution Terminology

5. Distinguish between the solute and solvent in each of the following solutions: (a) saltwater [NaCl(aq)]; (b) sterling silver (92.5% Ag, 7.5% Cu); (c) air (about 80% $N_2$, 20% $O_2$). On what do you base your distinctions?

6. Explain why the distinction between solute and solvent is not clear for some solutions.

7. Would it be proper to say that a saturated solution is a concentrated solution? Or that a concentrated solution is a saturated solution? Point out the distinctions between these sometimes confused terms.

8. Solution A contains 10 g of solute dissolved in 100 g of solvent, while solution B has only 5 g of a different solute per 100 g of solvent. Under what circumstances can solution A be classified as dilute and solution B as concentrated?

9. What happens if you add a very small amount of solid salt (NaCl) to each beaker described below? Include a statement comparing the *amount* of solid eventually found in the beaker with the amount you added: (a) a beaker containing *saturated* NaCl solution, (b) a beaker with *unsaturated* NaCl solution, (c) a beaker containing *supersaturated* NaCl solution.

10. Suggest simple laboratory tests by which you could determine whether a solution is unsaturated, saturated, or supersaturated. Explain why your suggestions would distinguish among the different classifications.

11. In stating solubility, an important variable must be specified. What is that variable, and how does solubility of a solid solute usually depend on it?

12. Suggest units in which solubility might be expressed other than grams per 100 g of solvent.

13. When acetic acid, a clear, colorless liquid, and water are mixed, a clear, uniform, colorless liquid results. Is acetic acid soluble in water? Is acetic acid miscible in water? Explain your answers.

14. ■ (a) The solubility of lead(II) chloride in water is 4.50 grams per liter. If a lead(II) chloride solution had a concentration of 6.35 grams per liter, would it be saturated, supersaturated, or unsaturated? (b) The solubility of calcium sulfate in water is 0.667 grams per liter. If a calcium sulfate solution had a concentration of $3.47 \times 10^{-3}$ grams per liter, would it be relatively concentrated or dilute? (c) The liquid acetone and water dissolve in each other in all proportions. Are liquid acetone and water said to be miscible, immiscible, or partially miscible?

### Section 16.3: The Formation of a Solution

15. How is it that both cations and anions, positively charged ions and negatively charged ions, can be hydrated by the same substance, water?

16. What does it mean to say that a solute particle is hydrated?

17. Explain why the dissolving process is reversible.

18. Describe the forces that promote the dissolving of a solid solute in a liquid solvent.

19. Describe the changes that occur between the time excess solute is placed into water and the time the solution becomes saturated.

20. Compare and contrast a dynamic equilibrium on both the particulate and the macroscopic levels. Pay particular attention to explaining why the equilibrium is dynamic.

21. At what time during the development of a saturated solution is the rate at which ions move from the aqueous phase to the solid phase [solute(aq) $\rightarrow$ solute(s)] greater than the rate from the solid phase to the aqueous phase [solute(s) $\rightarrow$ solute(aq)]?

22. Compare the rate at which ions pass from the solid phase to the aqueous phase with the rate at which they pass from the aqueous phase to the solid phase when the solution is unsaturated.

23. Why can't you prepare a supersaturated solution by adding more solute and stirring until it dissolved?

24. Why do people stir coffee after putting sugar into it? Would putting the coffee and sugar into a closed container and shaking be as effective as stirring?

25. Identify three ways in which you can reduce the amount of time required to dissolve a given amount of solute in a fixed quantity of solvent.

26. Explain how each action in Question 25 speeds the dissolving process.

### Section 16.4: Factors That Determine Solubility

*Questions 27 through 30: Structural diagrams for several substances are given in Table 16.2.*

27. Which of the following solutes do you expect to be more soluble in water than in cyclohexane: (a) formic acid, (b) benzene, (c) methylamine, (d) tetrafluoromethane? Explain your choice(s).

28. Which of the following solutes do you expect to be more soluble in water than in cyclohexane: (a) dimethyl ether, (b) hexane, (c) tetrachloroethene, (d) hydrogen fluoride? Explain your choice(s).

29. Which compound, glycerine or hexane, do you expect would be more miscible in water? Why?

30. Suppose you have a spot on some clothing and water will not take it out. If you have ethanol and cyclohexane available, which would you choose as the more promising solvent to try? Why?

31. On opening a bottle of carbonated beverage, many bubbles are released and you hear the sound of escaping gas. This suggests that the beverage is bottled under high pressure. Yet, for safety reasons, the pressure cannot be much more than one atmosphere. What gas do you suppose is in the small space between the beverage and the cap of a bottle of carbonated beverage before the cap is removed?

32. ■ Determine whether each of the following statements is true or false:

a) The rate of crystallization when a solution is 2/3 saturated is higher than the rate of crystallization when a solution is 1/3 saturated.

**Table 16.2** Lewis Diagrams for Questions 27 to 30

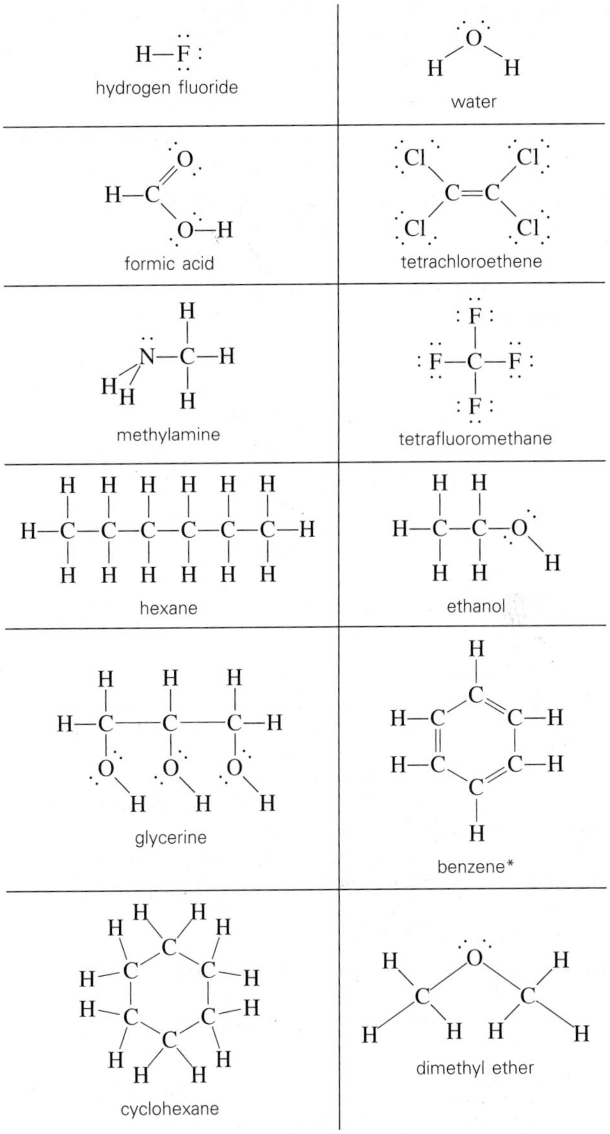

| | |
|---|---|
| H—F: <br> hydrogen fluoride | water |
| formic acid | tetrachloroethene |
| methylamine | tetrafluoromethane |
| hexane | ethanol |
| glycerine | benzene* |
| cyclohexane | dimethyl ether |

*This diagram is not consistent with all of the properties of benzene, but it is adequate to predict benzene's ability to dissolve other substances or to be dissolved by other solvents.

b) The solubility of a solid in a liquid always increases with temperature.

c) The solubility of a gas in a liquid increases as the partial pressure of that gas over the surface of the liquid increases.

### Section 16.5: Solution Concentration: Percentage by Mass

33. Calculate the percentage concentration of a solution prepared by dissolving 2.32 g of calcium chloride in 81.0 g of water.

34. ■ If 20.2 grams of an aqueous solution of calcium nitrate, $Ca(NO_3)_2$, contains 2.46 grams of calcium nitrate, what is the percentage by mass of calcium nitrate in the solution?

35. How many grams of ammonium nitrate must be weighed out to make 415 g of a 58.0% solution? In how many milliliters of water should it be dissolved?

36. ■ How many grams of $CuSO_4$ are there in 219 grams of an aqueous solution that is 20.2% by weight $CuSO_4$?

**Section 16.6: Solution Concentration: Molarity**

37. Potassium iodide is the additive in "iodized" table salt. Calculate the molarity of a solution prepared by dissolving 2.41 g of potassium iodide in water and diluting to 50.0 mL.

38. ■ A student weighs out a 4.80 g sample of $AlBr_3$, transfers it to a 100-mL volumetric flask, adds enough water to dissolve it, and then adds water to the 100-mL mark. What is the molarity of aluminum bromide in the resulting solution?

39. ■ A student dissolves 18.0 g of anhydrous nickel chloride in water and dilutes it to 90.0 mL. He also dissolves 30.0 g of nickel chloride hexahydrate in water and dilutes it to 90.0 mL. Identify the solution with the higher molar concentration and calculate its molarity.

40. The chemical name for the "hypo" used in photographic developing is sodium thiosulfate. It is sold as a pentahydrate, $Na_2S_2O_3 \cdot 5\ H_2O$. What is the molarity of a solution prepared by dissolving $1.2 \times 10^2$ g of this compound in water and diluting the solution to $1.25 \times 10^3$ mL?

41. Large quantities of silver nitrate are used in making photographic chemicals. Find the mass that must be used in preparing $2.50 \times 10^2$ mL of 0.058 M silver nitrate.

42. ■ How many grams of zinc iodide, $ZnI_2$, must be dissolved to prepare $2.50 \times 10^2$ mL of a 0.150 M aqueous solution of the salt?

43. Potassium hydroxide is used in making liquid soap. How many grams would you use to prepare 2.50 L of 1.40 M potassium hydroxide?

44. ■ You need to make an aqueous solution of 0.123 M manganese(II) acetate (acetate ion, $C_2H_3O_2^-$) for an experiment in lab, using a 250-mL volumetric flask. How much solid manganese(II) acetate should you add?

45. What volume of concentrated sulfuric acid, which is 18 molar, is required to obtain 5.19 mole of the acid?

46. ■ How many milliliters of a 0.196 M aqueous solution of manganese(II) nitrate, $Mn(NO_3)_2$, must be taken to obtain 6.17 grams of the salt?

47. If 0.132 M sodium chloride is to be the source of 8.33 g of dissolved solute, what volume of solution is needed?

48. Calculate the volume of concentrated ammonia solution, which is 15 molar, that contains 75.0 g of ammonia.

49. Calculate the moles of silver nitrate in 55.7 mL of 0.204 M silver nitrate.

50. How many moles of solute are in 65.0 mL of 2.20 M sodium hydroxide?

51. Despite its intense purple color, potassium permanganate is used in bleaching operations. How many moles are in 25.0 mL of 0.0841 M $KMnO_4$?

52. A student uses 29.3 mL of 0.482 M sulfuric acid to titrate a base of unknown concentration. How many moles of sulfuric acid react?

53. The density of 3.30 M $KNO_3$ is 1.15 g/mL. What is its percentage concentration?

54. ■ An aqueous solution of 6.02 M hydrochloric acid, HCl, has a density of 1.10 g/mL. What is the percent concentration by mass of HCl in the solution?

**Section 16.7: Solution Concentration: Molality (Optional)**

55. Calculate the molal concentration of a solution of 44.9 g of naphthalene, $C_{10}H_8$, in 175 g of benzene, $C_6H_6$.

56. ■ What is the molality of a solution prepared by dissolving 18.8 g $NH_4I$ in $4.50 \times 10^2$ mL of water? Assume that the density of water is 1.00 g/mL.

57. Diethylamine, $(CH_3CH_2)_2NH$, is highly soluble in ethanol, $C_2H_5OH$. Calculate the number of grams of diethylamine that would be dissolved in $4.00 \times 10^2$ g of ethanol to produce 4.70 m $(CH_3CH_2)_2NH$.

58. ■ How many grams of aluminum nitrate, $Al(NO_3)_3$, must be dissolved in $5.00 \times 10^2$ grams of water to prepare a 0.121 m solution of the salt?

59. How many milliliters of water are needed to dissolve 97.7 mg sodium chloride in the preparation of a $2.80 \times 10^{-3}$ m solution?

60. ■ How many grams of water are required to dissolve 12.7 grams of iron(III) chloride, $FeCl_3$, in order to produce a 0.172 m solution?

**Section 16.8: Solution Concentration: Normality (Optional)**

61. What is equivalent mass? Why can you state positively the equivalent mass of LiOH, but not of $H_2SO_4$?

62. Explain why the number of equivalents in a mole of acid or base is not always the same.

63. State the number of equivalents in one mole of $HNO_2$; in one mole of $H_2SeO_4$ in the reaction $H_2SeO_4 \rightarrow H^+ + HSeO_4^-$. (Se is selenium, Z = 34.)

64. ■ Give the number of equivalents of acid and base per mole in each of the following reactions:

a) $HNO_3 + KOH \rightarrow KNO_3 + H_2O$

b) $H_2S + 2\ NaOH \rightarrow Na_2S + 2\ H_2O$

65. State the maximum number of equivalents per mole of $Cu(OH)_2$; per mole of $Fe(OH)_3$.

66. What is the maximum number of equivalents in one mole of $Zn(OH)_2$ and RbOH? (Rb is rubidium, Z = 37.)

67. Calculate the equivalent masses of $HNO_2$ and $H_2SeO_4$ in Question 63.

68. ■ Consider the reaction $HBr + KOH \rightarrow KBr + H_2O$. (a) What is the equivalent mass of KOH? (b) Calculate the number of equivalents in 29.5 grams of KOH.

69. What are the equivalent masses of $Cu(OH)_2$ and $Fe(OH)_3$ in Question 65?

70. ■ Consider the reaction $H_3PO_4 + 3\ KOH \rightarrow K_3PO_4 + 3\ H_2O$. (a) What is the equivalent mass of $H_3PO_4$? (b) Calculate the number of equivalents in 38.4 grams of $H_3PO_4$.

71. What is the normality of the solution made when 2.25 g potassium hydroxide is dissolved in water and diluted to $2.50 \times 10^2$ mL?

72. ■ Calculate the normality of a solution that contains 3.46 grams of $H_3PO_4$ in 909 mL of solution for the reaction $H_3PO_4 + 3\ KOH \rightarrow K_3PO_4 + 3\ H_2O$.

73. $NaHSO_4$ is used as an acid in the reaction $HSO_4^- \rightarrow H^+ + SO_4^{2-}$. What mass of $NaHSO_4$ must be dissolved in $7.50 \times 10^2$ mL of solution to produce 0.200 N $NaHSO_4$?

74. If 9.79 g of $NaHCO_3$ is dissolved in $5.00 \times 10^2$ mL of solution, what is the normality in the reaction $NaHCO_3 + HCl \rightarrow NaCl + H_2O + CO_2$?

75. A student dissolves 6.69 g $H_2C_2O_4$ in water, dilutes to $2.00 \times 10^2$ mL, and uses it in a reaction in which it ionizes as follows: $H_2C_2O_4 \rightarrow H^+ + HC_2O_4^-$. What is the normality of the solution?

76. ■ It is desired to prepare 600 mL of 0.150 normal $Ba(OH)_2$ for use in the reaction $2\ HCl + Ba(OH)_2 \rightarrow BaCl_2 + 2\ H_2O$. How many grams of $Ba(OH)_2$ are needed?

77. What is the molarity of (a) 0.965 N sodium hydroxide, (b) 0.237 N $H_3PO_4$ in $H_3PO_4 + 2\ NaOH \rightarrow Na_2HPO_4 + 2\ H_2O$?

78. ■ Consider the reaction $H_2CO_3 + 2\ NaOH \rightarrow Na_2CO_3 + 2\ H_2O$. (a) What is the normality of a $6.110 \times 10^{-2}$ M $H_2CO_3$ solution? (b) What volume of this solution would contain 0.345 eq?

79. How many equivalents of solute are in 73.1 mL of 0.834 N NaOH?

80. How many equivalents are in 2.25 L of 0.871 N $H_2SO_4$?

81. What volume of 0.492 N $KMnO_4$ contains 0.788 eq?

82. Calculate the volume of 0.371 N HCl that contains 0.0385 eq.

#### Section 16.10: Dilution of Concentrated Solutions

83. What is the molarity of the acetic acid solution if 45.0 mL of 17 M $HC_2H_3O_2$ is diluted to 1.5 L?

84. ■ In the laboratory, a student dilutes 12.9 mL of a 10.9 M perchloric acid solution to a total volume of $2.50 \times 10^2$ mL. What is the concentration of the diluted solution?

85. How many milliliters of concentrated nitric acid, 16 M $HNO_3$, will you use to prepare $7.50 \times 10^2$ mL of 0.69 M $HNO_3$?

86. ■ How many milliliters of 9.76 M hydrochloric acid solution should be used to prepare 2.00 L of 0.400 M HCl?

87. Calculate the volume of 18 M $H_2SO_4$ required to prepare 3.0 L of 2.9 N $H_2SO_4$ for the reactions in which the sulfuric acid is completely ionized.

88. ■ In the laboratory, a student adds 54.6 mL of water to 14.4 mL of a 0.791 M hydroiodic acid solution. What is the concentration of the diluted solution?

89. Calculate the normality of a solution prepared by diluting 15.0 mL of 15 M $H_3PO_4$ to $2.50 \times 10^2$ mL. The solution will be used in the reaction $H_3PO_4 + 2\ NaOH \rightarrow Na_2HPO_4 + 2\ H_2O$.

90. If 25.0 mL of 15 M $HNO_3$ is diluted to $4.00 \times 10^2$ mL, what is the normality of the diluted solution?

#### Section 16.11: Solution Stoichiometry

91. Calculate the grams of magnesium hydroxide that will precipitate from 25.0 mL of 0.398 M magnesium chloride by the addition of excess sodium hydroxide solution.

92. ■ How many grams of $Cu(OH)_2$ will precipitate when excess KOH solution is added to 56.0 mL of 0.522 M $CuBr_2$ solution?

93. Calculate the mass of calcium phosphate that will precipitate when excess sodium phosphate solution is added to 100.0 mL of 0.130 M calcium nitrate.

94. ■ How many milliliters of 0.464 M $HNO_3$ are needed to neutralize 7.84 g of $MgCO_3$?

95. How many milliliters of 1.50 M NaOH must react with aluminum to yield 2.00 L of hydrogen, measured at 22°C and 789 torr, by the reaction $2\ Al + 6\ NaOH \rightarrow 2\ Na_3AlO_3 + 3\ H_2$? Assume complete conversion of reactants to products.

96. Calculate the volume of chlorine, measured at STP, that can be recovered from 50.0 mL of 1.20 M HCl by the reaction $MnO_2 + 4\ HCl \rightarrow MnCl_2 + 2\ H_2O + Cl_2$, assuming complete conversion of reactants to products.

97. What volume of 0.842 M NaOH would react with 8.74 g of sulfamic acid, $NH_2SO_3H$, a solid acid with one replaceable hydrogen?

98. ■ Calculate the number of milliliters of 0.563 M $Ba(OH)_2$ required to precipitate as $Ca(OH)_2$ all of the $Ca^{2+}$ ions in 111 mL of 0.658 M $CaI_2$ solution.

#### Section 16.12: Titration Using Molarity

99. The equation for a reaction by which a solution of sodium carbonate may be standardized is $2\ HC_7H_5O_2 + Na_2CO_3 \rightarrow 2\ NaC_7H_5O_2 + H_2O + CO_2$. A student determines that 5.038 g of $HC_7H_5O_2$ uses 51.89 mL of the sodium carbonate solution in the titration. Find the molarity of the sodium carbonate.

100. ■ Potassium hydrogen phthalate is a solid, monoprotic acid frequently used in the laboratory as a primary standard. It has the unwieldy formula of $KHC_8H_4O_4$. This is often written in shorthand notation as KHP. If 25.0 mL of a potassium hydroxide solution are needed to neutralize 2.26 grams of KHP, what is the molarity of the potassium hydroxide solution?

101. A student is to titrate solid maleic acid, $H_2C_4H_2O_4$ (two replaceable hydrogens) with a KOH solution of unknown concentration. She dissolves 1.45 g of maleic acid in water and titrates 50.0 mL of the base to neutralize the acid. What is the molarity of the KOH solution?

102. ■ Oxalic acid dihydrate is a solid, diprotic acid that can be used in the laboratory as a primary standard. Its formula is $H_2C_2O_4 \cdot 2\ H_2O$. A student dissolves 0.750 grams of $H_2C_2O_4 \cdot 2\ H_2O$ in water and titrates the resulting solution with a solution of sodium hydroxide of unknown concentration. If 28.0 mL of the sodium hydroxide solution are required to neutralize the acid, what is the molarity of the sodium hydroxide solution?

103. A student finds that 37.80 mL of a 0.4052 M $NaHCO_3$ solution is required to titrate a 20.00-mL sample of sulfuric acid solution. What is the molarity of the acid? The reaction equation is $H_2SO_4 + 2\ NaHCO_3 \rightarrow Na_2SO_4 + 2\ H_2O + 2\ CO_2$.

104. ■ The molarity of an aqueous solution of potassium hydroxide is determined by titration against a 0.138 M

hydrobromic acid solution. If 27.4 mL of the base are required to neutralize 21.6 mL of hydrobromic acid, what is the molarity of the potassium hydroxide solution?

**Section 16.13: Titration Using Normality (Optional)**

105. What is the normality of the sodium carbonate solution in Question 99?

106. What is the normality of the base in Question 100?

107. What is the normality of the base in Question 101?

108. Calculate the normality of the base in Question 102. Set up the problem completely from the data, not from the answer to Question 102.

109. Calculate the normality of a solution of sodium carbonate if a 25.0-mL sample requires 39.8 mL of 0.405 N sulfuric acid in a titration.

110. What is the normality of an acid if 12.8 mL is required to titrate 15.0 mL of 0.882 N sodium hydroxide?

111. A chemist finds that 42.2 mL of 0.402 N sodium hydroxide is required to titrate 50.0 mL of a solution of tartaric acid ($H_2C_4H_4O_6$) of unknown concentration. Find the normality of the acid.

112. ■ The normality of an aqueous solution of perchloric acid is determined by titration with a 0.248 N potassium hydroxide solution. If 34.3 mL of potassium hydroxide are required to neutralize 29.3 mL of the acid, what is the normality of the perchloric acid solution?

113. When sodium hydroxide is titrated into phosphoric acid, 16.3 mL of 0.208 N sodium hydroxide is required for 20.0 mL of the phosphoric acid solution. Calculate the normality of the acid.

114. ■ Oxalic acid dihydrate is a solid, diprotic acid that can be used in the laboratory as a primary standard. Its formula is $H_2C_2O_4 \cdot 2\,H_2O$. A student dissolves 0.523 grams of $H_2C_2O_4 \cdot 2\,H_2O$ in water and titrates the resulting solution with a solution of potassium hydroxide of unknown concentration. If 32.8 mL of the potassium hydroxide solution are required to neutralize the acid, what is the normality of the potassium hydroxide solution?

115. A chemist dissolves 1.21 g of an organic compound that functions as a base in reaction with sulfuric acid in water and titrates it with 0.170 N sulfuric acid. What is the equivalent mass of the base if 30.7 mL of acid is required in the titration?

116. If 15.6 mL of 0.562 N sodium hydroxide is required to titrate a solution prepared by dissolving 0.631 g of an unknown acid, what is the equivalent mass of the acid?

**Section 16.14: Colligative Properties of Solutions (Optional)**

117. Is the partial pressure exerted by one component of a gaseous mixture at a given temperature and volume a colligative property? Justify your answer, pointing out in the process what classifies a property as "colligative."

118. The specific gravity of a solution of KCl is greater than 1.00. The specific gravity of a solution of $NH_3$ is less than 1.00. Is specific gravity a colligative property? Why, or why not?

119. A student dissolves 27.2 g of aniline, $C_6H_5NH_2$, in $1.20 \times 10^2$ g of water. At what temperatures will the solution freeze and boil?

120. ■ The boiling point of benzene, $C_6H_6$, is 80.10°C at 1 atmosphere. $K_b$ for benzene is 2.53°C/m. A nonvolatile molecular substance that dissolves in benzene is testosterone. If 10.14 grams of testosterone, $C_{19}H_{28}O_2$ (288.4 g/mol), is dissolved in 231.0 grams of benzene, what are the molality and the boiling point of the solution?

121. Calculate the freezing point of a solution of 2.12 g of naphthalene, $C_{10}H_8$, in 32.0 g of benzene, $C_6H_6$. Pure benzene freezes at 5.50°C, and its $K_f$ = 5.10°C/m.

122. ■ The freezing point of water is 0.00°C at 1 atmosphere. $K_f$ for water is 1.86°C/m. A molecular substance that dissolves in water is antifreeze (ethylene glycol). If 11.35 grams of antifreeze, $CH_2OHCH_2OH$ (62.10 g/mol), is dissolved in 272.3 grams of water, what are the molality and the boiling point of the solution?

123. What is the molality of a solution of an unknown solute in acetic acid if it freezes at 14.1°C? The normal freezing point of acetic acid is 16.6°C, and $K_f$ = 3.90°C/m.

124. ■ When 14.56 grams of TNT, $C_7H_5N_3O_6$ (227.1 g/mol), is dissolved in 264.3 grams of an organic solvent, the boiling point of the resulting solution is 0.200°C higher than that of the pure solvent. What are the molality of the solution and the value of $K_b$ for the solvent?

125. A solution of 16.1 g of an unknown solute in $6.00 \times 10^2$ g of water boils at 100.28°C. Find the molar mass of the solute.

126. ■ The boiling point of benzene, $C_6H_6$, is 80.10°C at 1 atmosphere. $K_b$ for benzene is 2.53°C/m. In a laboratory experiment, students synthesized a new compound and found that when 11.5 grams of the compound were dissolved in 246 grams of benzene, the solution began to boil at 80.43°C. The compound was also found to be a nonvolatile molecular compound. What is the molecular mass that they determined for this compound?

127. When 12.4 g of an unknown solute is dissolved in 90.0 g of phenol, the freezing point depression is 9.6°C. Calculate the molar mass of the solute if $K_f$ = 3.56°C/m for phenol.

128. ■ The freezing point of water is 0.00°C at 1 atmosphere. $K_f$ for water is 1.86°C/m. In a laboratory experiment, students synthesized a new compound and found that when 11.2 grams of the compound were dissolved in $2.80 \times 10^2$ grams of water, the solution began to freeze at −1.12°C. The compound was also found to be a nonvolatile molecular compound. What is the molecular mass that they determined for this compound?

129. The normal freezing point of an unknown solvent is 28.7°C. A solution of 11.4 g of ethanol, $C_2H_5OH$, in $2.00 \times 10^2$ g of the solvent freezes at 22.5°C. What is the molal freezing point constant of the solvent?

130. ■ When 19.77 grams of glucose, $C_6H_{12}O_6$ (180.2 g/mol), is dissolved in 225.6 grams of an organic solvent, the freezing point of the resulting solution is 1.06°C lower than that of the pure solvent. What is the molality of the solution? What is the value of $K_f$ for the solvent?

## General Questions

131. Distinguish precisely, and in scientific terms, the differences among items in each of the following pairs or groups.
    a) Solute, solvent
    b) Concentrated, dilute
    c) Saturated, unsaturated, supersaturated
    d) Soluble, miscible
    e) (Optional) molality, molarity, normality
    f) (Optional) molar mass, equivalent mass
    g) (Optional) freezing-point depression, boiling-point elevation
    h) (Optional) molal freezing-point constant, molal boiling-point constant

132. Determine whether each of the following statements is true or false:
    a) The concentration is the same throughout a beaker of solution.
    b) A saturated solution of solute A is always more concentrated than an unsaturated solution of solute B.
    c) A solution can never have a concentration greater than its solubility at a given temperature.
    d) A finely divided solute dissolves faster because more surface area is exposed to the solvent.
    e) Stirring a solution increases the rate of crystallization.
    f) Crystallization ceases when equilibrium is reached.
    g) All solubilities increase at higher temperatures.
    h) Increasing air pressure over water increases the solubility of nitrogen in the water.
    i) An ionic solute is more likely to dissolve in a nonpolar solvent than in a polar solvent.
    j) (Optional) The molarity of a solution changes slightly with temperature, but the molality does not.
    k) (Optional) If an acid and a base react on a two-to-one mole ratio, there are twice as many equivalents of acid as there are of base in the reaction.
    l) The concentration of a primary standard is found by titration.
    m) Colligative properties of a solution are independent of the kinds of solute particles, but they are dependent on particle concentration.

133. When you heat water on a stove, small bubbles appear long before the water begins to boil. What are they? Explain why they appear.

134. Antifreeze is put into the water in an automobile to prevent it from freezing in winter. What does the antifreeze do to the boiling point of the water, if anything?

135. Does percentage concentration of a solution depend on temperature?

## More Challenging Problems

136. Suggest a way to separate two miscible liquids. Do you know of any widespread industrial process in which this is done?

137. In Chapter 2 we explain that physical properties must be employed to separate components of a mixture. Suggest a way to separate two immiscible liquids.

138. The text gives no explanation of forces that must be overcome during the dissolving process. Can you imagine what the forces might be?

139. Silver acetate has a solubility of 2.52 g/100 g water at 80°C and 1.02 g/100 g water at 20°C. How do you prepare a supersaturated solution of silver acetate? Why does crystallization not occur as soon as the ion concentration is greater than the concentration of a saturated solution?

140. Consider a sample of pure sugar that has been finely powdered. What advantage does it have over granular (crystalline) sugar in making bakery goods? Would there be any advantage or disadvantage in using it for sweetening coffee? Explain.

141. ■ The density of 18.0% HCl is 1.09 g/mL. Calculate its molarity.

142. ■ Methyl ethyl ketone, $C_4H_8O$, is a solvent popularly known as MEK that is used to cement plastics. How many grams of MEK must be dissolved in $1.00 \times 10^2$ mL of benzene, density 0.879 g/mL, to yield a 0.254 molal solution?

143. Calculate the mass of $H_2C_2O_4 \cdot 2\,H_2O$ required to make $2.50 \times 10^2$ mL of 0.500 N $H_2C_2O_4$ that will be used in the reaction $H_2C_2O_4 + 2\,OH^- \rightarrow C_2O_4^{2-} + 2\,H_2O$.

144. Sodium carbonate decahydrate is used as a base in the reaction $CO_3^{2-} + 2\,H^+ \rightarrow CO_2 + H_2O$. Calculate the mass of the hydrate needed to prepare $1.00 \times 10^2$ mL of 0.500 N sodium carbonate.

145. ■ The iron(III) ion content of a solution may be found by precipitating it as iron(III) hydroxide and then decomposing the hydroxide to iron(III) oxide by heat. How many grams of iron(III) oxide can be collected from 35.0 mL of 0.516 M iron(III) nitrate?

146. A student adds 25.0 mL of 0.350 M sodium hydroxide to 45.0 mL of 0.125 M copper(II) sulfate. How many grams of copper(II) hydroxide will precipitate?

147. ■ A laboratory technician combines 25.0 mL of 0.269 M nickel chloride with 30.0 mL 0.260 M potassium hydroxide. How many grams of nickel hydroxide can precipitate?

148. ■ An analytical procedure for finding the chloride ion concentration in a solution involves the precipitation of silver chloride: $Ag^+ + Cl^- \rightarrow AgCl$. What is the molarity of the chloride ion if 16.80 mL of 0.629 M silver nitrate (the source of silver ion) is needed to precipitate all of the chloride in a 25.00-mL sample of the unknown?

149. Calculate the hydroxide ion concentration in a 20.00-mL sample of an unknown if 14.75 mL 0.248 M sulfuric acid is used in a neutralization reaction.

150. ■ A 694-mg sample of impure sodium carbonate was titrated with 41.24 mL of 0.244 M hydrochloric acid. Calculate the percentage of sodium carbonate in the sample.

151. A student received a 599-mg sample of a mixture of sodium hydrogen phosphate and sodium dihydrogen phosphate. She is to find the percentage of each compound in the sample. After dissolving the mixture, she titrated it with 19.58 mL 0.201 M sodium hydroxide. If the only reaction is $NaH_2PO_4 + NaOH \rightarrow Na_2HPO_4 + H_2O$, find the required percentages.

152. ■ A chemist combines 60.0 mL of 0.322 M potassium iodide with 20.0 mL of 0.530 M lead(II) nitrate. (a) How many grams of lead(II) iodide will precipitate? (b) What is the final molarity of the potassium ion? (c) What is the final molarity of the lead(II) or iodide ion, whichever one is in excess?

153. A solution is defined as a homogeneous mixture. Is a small sample of air a solution? Is the atmosphere a solution?

154. If you know either the percentage concentration of a solution or its molarity, what additional information must you have before you can convert to the other concentration?

# Formulas

37. KI

39. $NiCl_2 \cdot 6\ H_2O$

41. $AgNO_3$

43. KOH

44. $Mn(C_2H_3O_2)_2$

45. $H_2SO_4$

47. NaCl

48. $NH_3$

49. $AgNO_3$

50. NaOH

52. $H_2SO_4$

59. NaCl

71. KOH

91. $Mg(OH)_2$, $MgCl_2$, NaOH

93. $Ca_3(PO_4)_2$, $Na_3PO_4$, $Ca(NO_3)_2$

100. KOH

102. NaOH

104. KOH, HBr

109. $Na_2CO_3$, $H_2SO_4$

110. NaOH

111. NaOH

112. $HClO_4$, KOH

113. NaOH, $H_3PO_4$

114. KOH

115. $H_2SO_4$

116. NaOH

144. $Na_2CO_3 \cdot 10\ H_2O$

145. $Fe(OH)_3$, $Fe_2O_3$, $Fe(NO_3)_3$

146. NaOH, $CuSO_4$, $Cu(OH)_2$

147. $NiCl_2$, KOH, $Ni(OH)_2$

148. $AgNO_3$

149. $OH^-$, $H_2SO_4$

150. $Na_2CO_3$, HCl

152. KI, $Pb(NO_3)_2$, $PbI_2$

# 17

## CONTENTS



Did you know that your local grocery store is a major industrial outlet for acids and bases? It's a fact. They don't come in the kinds of bottles sold to college chemistry laboratories, but they are acids and bases nonetheless. If you doubt it, read some of the labels. The substances in the picture on the left are all acidic; those on the right are basic.

# Acid–Base (Proton–Transfer) Reactions

Originally, the word ***acid*** described something that had a sour, biting taste. The tastes of vinegar (acetic acid) and lemon juice (citric acid) are typical examples. Substances with such tastes have other common properties, too. They impart certain colors to some organic substances, such as litmus, which is red in acid solutions; they react with carbonate ions and release carbon dioxide; they react with and neutralize a base; and they release hydrogen when they react with particular metals.

Online homework for this chapter may be assigned in OWL

Traditionally, a **base** is something that tastes bitter. Bases impart a different color to some organic substances (blue to litmus); they neutralize acids; and they form precipitates when added to solutions of most metal ions. Also, bases feel slippery, or "soapy."

To understand these two distinct groups, we must ask, what features in their chemical structure and composition are responsible for their characteristic properties? This is the goal of Chapter 17.

## 17.1 The Arrhenius Theory of Acids and Bases (Optional)

Goal **1** Distinguish between an acid and a base according to the Arrhenius theory of acids and bases.

The E. F. Smith Memorial Collection in the History of Chemistry, Dept. of Special Collections, Van Pelt-Dietrich Library, University of Pennsylvania

Svante August Arrhenius (1859–1927). Arrhenius was awarded the 1903 Nobel Prize in Chemistry for proposing that ionic compounds can divide into oppositely charged particles in solution.

In 1884, Svante Arrhenius observed that all substances called acids contain hydrogen ions, $H^+$. Bases, on the other hand, always contained hydroxide ions, $OH^-$. An acid was thus identified as a **substance whose water solution contains more hydrogen ions than hydroxide ions,** and a base is a **substance whose water solution contains more hydroxide ions than hydrogen ions.** According to the **Arrhenius theory** of acids and bases, the properties of an acid are the properties of the hydrogen ion, and the properties of a base are the properties of the hydroxide ion.

Among the properties of acids and bases is their ability to neutralize each other. The net ionic equation for the reaction between a strong acid, HCl, and a strong base, NaOH, is a molecule-forming ion combination that yields water:

$$H^+(aq) + OH^-(aq) \rightarrow H_2O(\ell)$$

If the acid is weak, the hydroxide ion essentially pulls a hydrogen ion right out of the un-ionized molecule:

$$HC_2H_3O_2(aq) + OH^-(aq) \rightarrow H_2O(\ell) + C_2H_3O_2^-(aq)$$

The carbon dioxide from the reaction of an acid with a carbonate compound is the end result of the combination of hydrogen and carbonate ions. The expected product is carbonic acid, $H_2CO_3$. However, in Section 9.9 you learned that this is one of the "unstable products" of an ion combination. It decomposes into carbon dioxide and water:

$$2\,H^+(aq) + CO_3^{2-}(aq) \rightarrow CO_2(g) + H_2O(\ell)$$

Hydrogen ions of acids and hydrogen gas occupy a unique position in the activity series (Section 9.5). All other members of this series are metals and metal ions that engage in single-replacement-type redox reactions. The reaction of the hydrogen ion appears in the net ionic equation for the reaction of calcium with hydrobromic acid:

$$Ca(s) + 2\,H^+(aq) \rightarrow H_2(g) + Ca^{2+}(aq)$$

Table 9.3 shows that only the hydroxides of barium and the alkali metals are soluble, which is why these solutions are the only strong Arrhenius bases. It also explains why precipitates form when a strong base is combined with the salt of most metals. The net ionic equation for the precipitation of magnesium hydroxide is typical:

$$Mg^{2+}(aq) + 2\,OH^-(aq) \rightarrow Mg(OH)_2(s)$$

### √ Target Check 17.1

*According to the Arrhenius theory of acids and bases, how do you recognize an acid and a base?*

## 17.2 The Brønsted–Lowry Theory of Acids and Bases

**Goal 2** Given the equation for a Brønsted–Lowry acid–base reaction, explain how or why it can be so classified.

**Goal 3** Given the formula of a Brønsted–Lowry acid and the formula of a Brønsted–Lowry base, write the net ionic equation for the reaction between them.

When we use Lewis diagrams to examine the simple hydrogen–hydroxide ion concept of acids and bases, we find two things. First, all such reactions can be interpreted as a transfer of a proton (hydrogen ion) from the acid to a hydroxide ion. Second, substances other than hydroxide ions can remove protons from acids. These observations were announced independently by Johannes N. Brønsted and Thomas M. Lowry in 1923. Their proposal is known by both their names. According to the **Brønsted–Lowry theory of acids and bases, an acid–base reaction is a proton-transfer reaction in which the proton is transferred from the acid to the base. An acid is a compound from which a proton can be removed, and a base is a compound that can remove a proton from an acid.** According to this theory, anything that can take away a proton is a base. The hydroxide ion is only the most common example.

The Lewis diagram showing the formation of the hydronium ion when a hydrogen-bearing molecule reacts with water (Section 6.5) illustrates a proton-transfer reaction. Using nitric acid as an example, we have

Acid: $H^+$ source — Base: $H^+$ remover

$$HNO_3(aq) + H_2O(\ell) \longrightarrow H_3O^+(aq) + NO_3^-(aq)$$

The proton from the nitric acid molecule is transferred to one of the unshared electron pairs of the water molecule. The $HNO_3$ molecule is the acid, the proton source. The remover of the proton is the water molecule; water is a base in this reaction.

Another example of an acid–base reaction is the transfer of a proton from water to ammonia:

Base: $H^+$ remover — Acid: $H^+$ source

$$NH_3(g) + H_2O(\ell) \rightleftharpoons NH_4^+(aq) + OH^-(aq)$$

Johannes N. Brønsted (1879–1947). Brønsted was primarily interested in studying thermochemistry, in addition to his more famous work with acids and bases.

Special Collections, Van Pelt Library, University of Pennsylvania

**P/REVIEW:** As the term *hydrogen ion* is used here, it is a proton. The most common hydrogen atom has one proton and one electron. If you remove the electron, the proton is left. Its formula is $H^+$, which is a hydrogen ion. When in a water solution, this ion is thought to be hydrated, that is, chemically bonded to one or more water molecules. It is represented by the formula $H_3O^+$ and called a hydronium ion. See Section 6.5.

Thomas M. Lowry (1874–1936). Lowry was the first professor of physical chemistry at Cambridge University.

Special Collections, Van Pelt Library, University of Pennsylvania

This time water is an acid. It has a proton that can be taken by ammonia, the base. A substance that can behave as an acid in one case and a base in another, as water does in the two preceding reactions, is said to be **amphoteric.**

The double arrow in the equation for the ammonia–water reaction indicates that the reaction is reversible. It suggests correctly that the reaction reaches a chemical equilibrium. When a reversible reaction is read from left to right, the **forward reaction,** or the reaction in the **forward direction,** is described; from right to left, the change is the **reverse reaction,** or in the **reverse direction.** In 1 M $NH_3(aq)$, less than 1% of the ammonia is changed to ammonium ions. $NH_3$ is the major species in the solution, and $NH_4^+$ and $OH^-$ are the minor species. The reverse reaction is therefore said to be favored.

**P/REVIEW:** A reversible reaction is one in which the products of the reaction can react with themselves to reproduce the original reactants. A double arrow in the equation for a reaction identifies it as reversible. Sometimes a longer arrow is used to point in the favored direction—to the major species in the reaction, those present in higher concentration—while the shorter arrow points to the minor species, those with lower concentrations.

In writing net ionic equations, only the major species are shown (see Section 9.4). Reversible reactions are introduced in Section 8.7, and they are used in discussing liquid–vapor equilibria in Section 15.4. Other important applications of reversible reactions appear in Chapters 18 and 19.

$HNO_3$ is an acid, a proton source, in our first example reaction. $NH_3$ is a base, a proton remover, in the second reaction. Do you suppose that if we were to put $HNO_3$ and $NH_3$ together, the $HNO_3$ would release its proton to $NH_3$? The answer is yes. If $NH_3$ can take a proton from water, and if water can take a proton from $HNO_3$, then $NH_3$ can take a proton from $HNO_3$. Conventional and Lewis diagram equations describe the reaction:

Acid: $H^+$ source    Base: $H^+$ remover

$$HNO_3(aq) + NH_3(g) \rightleftharpoons NH_4^+(aq) + NO_3^-(aq)$$

Note that this is an acid–base reaction that does not directly involve either water or the hydroxide ion.

Reviewing the three acid–base reactions presented, we find that all fit into the general equation for a Brønsted–Lowry proton-transfer reaction:

$$\underset{\text{base proton remover}}{B} + \underset{\text{acid proton source}}{HA} \rightarrow HB^+ + A^-$$

In this equation, note that the charges are not “absolute” charges. They indicate, rather, that the acid species, in losing a proton, leaves a species having a charge one less than the acid, and that the base, in gaining a proton, increases by one in charge.

## Thinking About Your Thinking

### Mental Models

**Forming mental models of proton-transfer reactions is a skill to develop as you study chemistry. This type of reaction literally occurs thousands of times per second in your body. Many common biological molecules are classified as acids and bases.**

**Lewis diagrams are good models to use when considering proton-transfer reactions. Recall that a proton is symbolized as $H^+$, which is a hydrogen ion. The most common hydrogen atom isotope consists of a single proton (no neutron) and a single electron. If the electron is lost, what remains is nothing but a proton. This proton is what is transferred in acid–base reactions.**

**Take another look at the Lewis diagrams in the reactions illustrated in this section, this time looking for the common features of the bases, and in particular, the atom in the basic species that receives the proton. Do you see that the proton remover always has an unshared electron pair? Those negatively charged electrons are also part of a highly electronegative atom. Thus, the area in the species that acts as a base, or proton remover, has a greater negative charge than the remainder of the species. It attracts a positive charge.**

**Now revisit the Lewis diagrams of the acids in this section. Each has a hydrogen atom bonded to a highly electronegative atom. These bonding electrons are less attracted to the hydrogen atom than to the other atom in the bond. That "allows" the positively charged proton to "look for" a better source of opposite charge to be attracted to than the one it now has. When one is found, a proton-transfer reaction occurs.**

**If you keep in mind that an acid is a proton source, a base is a proton remover, and an acid–base reaction is a proton-transfer reaction, you will have an excellent mental model of proton-transfer reactions.**

### √ Target Check 17.2

*What is the difference between a Brønsted–Lowry acid and an Arrhenius acid? Between a Brønsted–Lowry base and an Arrhenius base? Are all Brønsted–Lowry bases also Arrhenius bases? Are all Arrhenius bases also Brønsted–Lowry bases? Explain.*

### √ Target Check 17.3

*Compare and contrast proton-transfer reactions with Brønsted–Lowry acid–base reactions.*

## 17.3 The Lewis Theory of Acids and Bases (Optional)

**Goal** 4 Distinguish between a Lewis acid and a Lewis base. Given the structural formula of a molecule or ion, state if it can be a Lewis acid, a Lewis base, or both, and explain why.

**Goal** 5 Given the structural equation for a Lewis acid–base reaction, explain how or why it can be so classified.

Look at the Lewis diagrams of the bases in Section 17.2. In each case the structural feature of the base that permits it to remove the proton is an unshared pair of electrons. A hydrogen ion has no electron to contribute to the bond, but it is able to accept an electron pair to form the bond. According to the **Lewis theory of acids and bases, a Lewis base is an electron-pair donor, and a Lewis acid is an electron-pair acceptor.**

A Lewis acid is not limited to a hydrogen ion. A common example is boron trifluoride, which behaves as a Lewis acid by accepting an unshared pair of electrons from ammonia:

$$\mathrm{BF_3} + \mathrm{:NH_3} \longrightarrow \mathrm{F_3B{-}NH_3}$$

$$\mathrm{A} + \mathrm{:B} \longrightarrow \mathrm{A{-}B}$$

Lewis Acid (electron–pair acceptor) — Lewis Base (electron–pair donor)

Other examples of Lewis acid–base reactions appear in the formation of some complex ions and in organic reactions. You will study these in more advanced courses.

### √ Target Check 17.4

*By the Brønsted–Lowry theory, water can be either an acid or a base. Can water be a Lewis acid? A Lewis base? Explain.*

**summary**

**Identifying Features of Acids and Bases**

The identifying features of acids and bases according to the three acid–base theories are summarized below:

| Theory | Acid | Base |
|---|---|---|
| Arrhenius | Hydrogen ion | Hydroxide ion |
| Brønsted–Lowry | Proton source | Proton remover |
| Lewis | Electron-pair acceptor | Electron-pair donor |

Our principal interest in acid–base chemistry is in aqueous solutions, where the Brønsted–Lowry theory prevails. The balance of this chapter is limited to the proton-transfer concept of acids and bases.

## 17.4 Conjugate Acid–Base Pairs

Goal 6 Define and identify conjugate acid–base pairs.

Goal 7 Given the formula of an acid or a base, write the formula of its conjugate base or acid.

We noted that reaction between ammonia and water reaches equilibrium. The fact is that most acid–base reactions reach equilibrium. Accordingly, the general Brønsted–Lowry acid–base proton-transfer reaction can be written with a double arrow to show that it is reversible. Look carefully at the reverse reaction, for which the arrow and the labels for the reactants are printed in red:

$$\underset{\text{base, proton receiver}}{\mathrm{B}} + \underset{\text{acid, proton source}}{\mathrm{HA}} \rightleftharpoons \underset{\text{acid, proton source}}{\mathrm{HB^+}} + \underset{\text{base, proton receiver}}{\mathrm{A^-}}$$

Is not $HB^+$ donating a proton to $A^-$ in the reverse reaction? In other words, $HB^+$ is an acid in the reverse reaction, and $A^-$ is a base. From this we see that the products of any proton-transfer acid–base reaction are *another* acid and base for the reverse reaction.

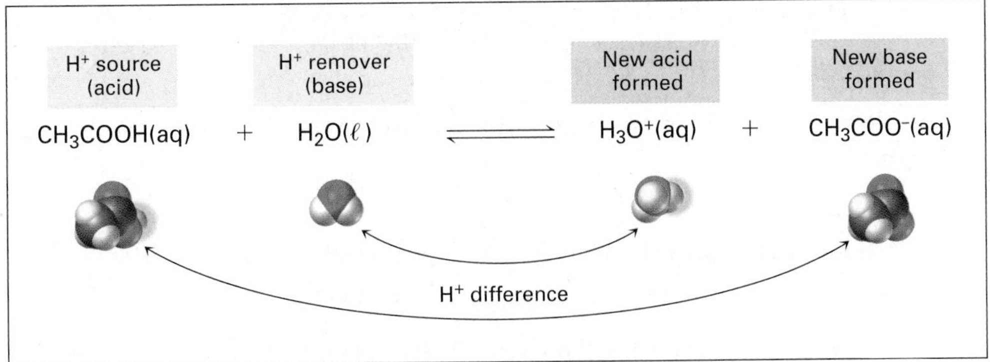


**Active Figure 17.1** Conjugate acid–base pairs. For a reversible proton-transfer reaction, characterized by the exchange of a single proton, $H^+$, there are two conjugate acid–base pairs. **Watch this active figure at *http://www.cengage.com/cracolice.***

Combinations such as base B and acid $HB^+$ as well as acid HA and base $A^-$ that result from an acid losing a proton or a base gaining one are called **conjugate acid–base pairs** (see Active Figure 17.1). In the preceding forward reaction, the acid, HA, releases a proton to the base, B, leaving the acid anion, $A^-$, and $HB^+$. HA and $A^-$ are a conjugate acid–base pair. In the reverse reaction, acid $HB^+$ is the proton source for base $A^-$, restoring the original reactants, B and HA. This time $HB^+$ and B are a conjugate acid–base pair. Thus, we see that, for every reversible proton-transfer reaction, two conjugate acid–base pairs can be identified, one for the forward reaction and one for the reverse.

Let's apply these ideas to reaction of ammonia with water:

$$NH_3(aq) + H_2O(\ell) \rightleftharpoons NH_4^+(aq) + OH^-(aq)$$

BASE acid ACID base

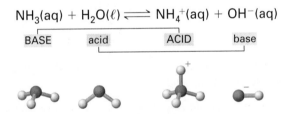

In the forward direction, $H_2O$ is an acid and $OH^-$ is its conjugate base. In the reverse direction, $NH_4^+$ is the acid and $NH_3$ is its conjugate base.

You can write the formula of the conjugate base of any acid simply by removing a proton. If $HCO_3^-$ acts as an acid, its conjugate base is $CO_3^{2-}$. You can also write the formula of the conjugate acid of any base by adding a proton. If $HCO_3^-$ is a base, its conjugate acid is $H_2CO_3$. Notice that $HCO_3^-$ is amphoteric. Any amphoteric substance has both a conjugate base and a conjugate acid.

## Active Example 17.1

Write the formula of the conjugate base of $H_3PO_4$, and write the formula of the conjugate acid of $C_7H_5O_2^-$.

In (b), don't let $C_7H_5O_2^-$ confuse you just because it is unfamiliar. Just do what must be done to find the formula of a conjugate acid.

*continued*

$H_3PO_4 \rightarrow H^+ + H_2PO_4^-$, the conjugate base of $H_3PO_4$

$C_7H_5O_2^- + H^+ \rightarrow HC_7H_5O_2$, the conjugate acid of $C_7H_5O_2^-$

Remove a proton to get a conjugate base, and add one to get a conjugate acid.

## Active Example 17.2

Nitrous acid engages in a proton-transfer reaction with formate ion, $CHO_2^-$:

$$HNO_2(aq) + CHO_2^-(aq) \rightleftharpoons NO_2^-(aq) + HCHO_2(aq)$$

Answer the questions about this reaction in the steps that follow.

For the forward reaction, identify the acid and the base.

$HNO_2$ is the acid; it releases a proton to $CHO_2^-$, the base, or proton remover.

Identify the acid and base for the reverse reaction.

$HCHO_2$ is the acid; it releases a proton to $NO_2^-$, the base, or proton remover.

Identify the conjugate of $HNO_2$. Is it a conjugate acid or a conjugate base?

$NO_2^-$ is the conjugate *base* of $HNO_2$. It is the base that remains after the proton has been released by the acid.

Identify the other conjugate acid–base pair and classify each species as the acid or the base.

$CHO_2^-$ and $HCHO_2$ make up the other conjugate acid–base pair. $CHO_2^-$ is the base, and $HCHO_2$ is the conjugate acid—the species produced when the base takes a proton.

## 17.5 Relative Strengths of Acids and Bases

**Goal** 8 Given a table of the relative strengths of acids and bases, arrange a group of acids or a group of bases in order of increasing or decreasing strength.

Section 9.3 distinguished between the relatively few strong acids and the many weak acids. **Strong acids** are those that ionize almost completely, whereas **weak acids** ionize but slightly. Hydrochloric acid is a strong acid; 0.10 M HCl is almost 100% ionized. Acetic acid is a weak acid; only 1.3% of the molecules ionize in 0.10 M $HC_2H_3O_2$.

In a Brønsted–Lowry sense an acid behaves as an acid by releasing protons. The more readily protons are surrendered, the stronger the acid. A base behaves as a base by removing protons. The stronger the ability to remove protons is, the stronger the base will be.

Look at the ionization equations for hydrochloric and acetic acids, one written above the other, with the stronger acid first:

| | | |
|---|---|---|
| Strong acid | $HCl \rightleftharpoons H^+ + Cl^-$ | ? base |
| Weak acid | $HC_2H_3O_2 \rightleftharpoons H^+ + C_2H_3O_2^-$ | ? base |

The conjugate bases of the acids are on the right-hand sides of the equations. What is the relative strength of the two bases? Which is stronger, $Cl^-$ or $C_2H_3O_2^-$? If a chloride ion gained a proton, it would form HCl, a strong acid that would immediately lose that proton. The $Cl^-$ ion has a weak attraction for protons. It is a **weak base.** If the acetate ion gained a proton, it would form $HC_2H_3O_2$, a weak acid that holds its proton tightly. The $C_2H_3O_2^-$ ion has a strong attraction for protons. It is a **strong base.** We can therefore complete the comparison between these acids and their conjugate bases:

| | | |
|---|---|---|
| Strong acid | $HCl \rightleftharpoons H^+ + Cl^-$ | Weak base |
| Weak acid | $HC_2H_3O_2 \rightleftharpoons H^+ + C_2H_3O_2^-$ | Strong base |

Table 17.1 lists many acids and bases in this way. Acid strength decreases from top to bottom, and the base strength increases. By referring to this table, we can compare the relative strengths of different acids and bases.

**Table 17.1** Relative Strengths of Acids and Bases

| Acid Name | Acid Formula | | Base Formula |
|---|---|---|---|
| Hydroiodic | HI | $\rightleftharpoons$ | $H^+ + I^-$ |
| Hydrobromic | HBr | $\rightleftharpoons$ | $H^+ + Br^-$ |
| Perchloric | $HClO_4$ | $\rightleftharpoons$ | $H^+ + ClO_4^-$ |
| Hydrochloric | HCl | $\rightleftharpoons$ | $H^+ + Cl^-$ |
| Sulfuric | $H_2SO_4$ | $\rightleftharpoons$ | $H^+ + HSO_4^-$ |
| Chloric | $HClO_3$ | $\rightleftharpoons$ | $H^+ + ClO_3^-$ |
| Nitric | $HNO_3$ | $\rightleftharpoons$ | $H^+ + NO_3^-$ |
| Hydronium ion | $H_3O^+$ | $\rightleftharpoons$ | $H^+ + H_2O$ |
| Oxalic | $H_2C_2O_4$ | $\rightleftharpoons$ | $H^+ + HC_2O_4^-$ |
| Sulfurous | $H_2SO_3$ | $\rightleftharpoons$ | $H^+ + HSO_3^-$ |
| Hydrogen sulfate ion | $HSO_4^-$ | $\rightleftharpoons$ | $H^+ + SO_4^{2-}$ |
| Phosphoric | $H_3PO_4$ | $\rightleftharpoons$ | $H^+ + H_2PO_4^-$ |
| Hydrofluoric | HF | $\rightleftharpoons$ | $H^+ + F^-$ |
| Nitrous | $HNO_2$ | $\rightleftharpoons$ | $H^+ + NO_2^-$ |
| Formic (methanoic) | $HCHO_2$ | $\rightleftharpoons$ | $H^+ + CHO_2^-$ |
| Benzoic | $HC_7H_5O_2$ | $\rightleftharpoons$ | $H^+ + C_7H_5O_2^-$ |
| Hydrogen oxalate ion | $HC_2O_4^-$ | $\rightleftharpoons$ | $H^+ + C_2O_4^{2-}$ |
| Acetic (ethanoic) | $HC_2H_3O_2$ | $\rightleftharpoons$ | $H^+ + C_2H_3O_2^-$ |
| Propionic (propanoic) | $HC_3H_5O_2$ | $\rightleftharpoons$ | $H^+ + C_3H_5O_2^-$ |
| Carbonic | $H_2CO_3$ | $\rightleftharpoons$ | $H^+ + HCO_3^-$ |
| Hydrosulfuric | $H_2S$ | $\rightleftharpoons$ | $H^+ + HS^-$ |
| Dihydrogen phosphate ion | $H_2PO_4^-$ | $\rightleftharpoons$ | $H^+ + HPO_4^{2-}$ |
| Hydrogen sulfite ion | $HSO_3^-$ | $\rightleftharpoons$ | $H^+ + SO_3^{2-}$ |
| Hypochlorous | HClO | $\rightleftharpoons$ | $H^+ + ClO^-$ |
| Boric | $H_3BO_3$ | $\rightleftharpoons$ | $H^+ + H_2BO_3^-$ |
| Ammonium ion | $NH_4^+$ | $\rightleftharpoons$ | $H^+ + NH_3$ |
| Hydrocyanic | HCN | $\rightleftharpoons$ | $H^+ + CN^-$ |
| Hydrogen carbonate ion | $HCO_3^-$ | $\rightleftharpoons$ | $H^+ + CO_3^{2-}$ |
| Monohydrogen phosphate ion | $HPO_4^{2-}$ | $\rightleftharpoons$ | $H^+ + PO_4^{3-}$ |
| Hydrogen sulfide ion | $HS^-$ | $\rightleftharpoons$ | $H^+ + S^{2-}$ |
| Water | HOH | $\rightleftharpoons$ | $H^+ + OH^-$ |
| Hydroxide ion | $OH^-$ | $\rightleftharpoons$ | $H^+ + O^{2-}$ |

Acid strength increasing (upward arrow); Base strength increasing (downward arrow)

## Thinking About Your Thinking

### Classification

**The classification thinking skill, as we discuss it in this book, is the understanding that different classification systems can be used for different purposes. In Chapter 9 we introduced the strong acid/strong base versus weak acid/weak base concept. This black-and-white view put all acids and bases into one or the other of these groups. In this section, we expand the acid and base strengths to include "shades of gray." We now recognize that acid and base strengths can be compared with each other. Instead of thinking with the absolute terms strong or weak, we think in the relative terms strong*er* or weak*er*.**

### Active Example 17.3

Use Table 17.1 to list the following acids in order of decreasing strength (strongest first): $HC_2O_4^-$, $NH_4^+$, $H_3PO_4$.

Find the three acids among those shown in the table and list them from the strongest (first) to the weakest (last).

$H_3PO_4$, $HC_2O_4^-$, $NH_4^+$

### Active Example 17.4

Using Table 17.1, list the following bases in order of decreasing strength (strongest first): $HC_2O_4^-$, $SO_3^{2-}$, $F^-$.

$SO_3^{2-}$, $F^-$, $HC_2O_4^-$

The ion $HC_2O_4^-$ appears in both Active Examples 17.3 and 17.4, first as an acid and second as a base. It is the intermediate ion in the two-step ionization of oxalic acid, $H_2C_2O_4$. The $HC_2O_4^-$ ion is amphoteric.

## 17.6 Predicting Acid–Base Reactions

Goal 9 Given the formulas of a potential Brønsted–Lowry acid and a Brønsted–Lowry base, write the equation for the possible proton-transfer reaction between them.

Goal 10 Given a table of the relative strengths of acids and bases and information from which a proton-transfer reaction equation between two species in the table may be written, write the equation and predict the direction in which the reaction will be favored.

A chemist likes to know if an acid–base reaction will occur when certain reactants are brought together. Obviously, there must be a potential proton source and remover—

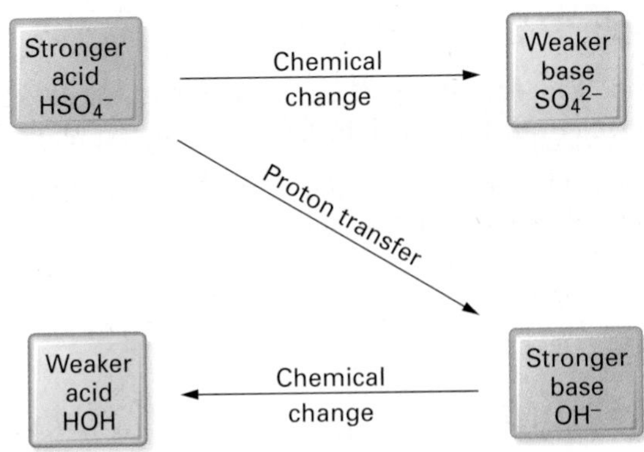


**Figure 17.2** Predicting acid–base reactions from positions in Table 17.1. The spontaneous chemical change always transfers a proton from the stronger acid to the stronger base, both shown in pink. The products of the reaction, the weaker acid and the weaker base, are shown in blue. The favored direction, forward or reverse, is the one that has the weaker acid and base as products.

there can be no proton-transfer reaction without both. From there the decision is based on the relative strengths of the conjugate acid–base pairs. The stronger acid and base are the most reactive. They do what they must do to behave as an acid and a base. The weaker acid and base are more stable—less reactive. It follows that *the stronger acid will always surrender a proton to the stronger base, yielding the weaker acid and base as favored species at equilibrium.* Figure 17.2 summarizes the proton transfer from the stronger acid to the stronger base, from the standpoint of positions in Table 17.1.

Hydrogen sulfate ion, $HSO_4^-$, is a relatively strong acid that holds its proton weakly. Hydroxide ion, $OH^-$, is a strong base that attracts a proton strongly. If an $HSO_4^-$ ion finds an $OH^-$ ion, the proton will transfer from $HSO_4^-$ to $OH^-$:

$$HSO_4^-(aq) + OH^-(aq) \rightleftharpoons SO_4^{2-}(aq) + HOH(\ell)$$

Now identify the conjugate acid–base pairs. In the forward direction the acid is $HSO_4^-$. Its conjugate base for the reverse reaction is $SO_4^{2-}$. Similarly, $OH^-$ is the base in the forward reaction, and HOH is the conjugate acid for the reverse reaction. Let's show the acid–base roles for the different directions with the letters A for acid and B for base:

$$\underset{A}{HSO_4^-(aq)} + \underset{B}{OH^-(aq)} \rightleftharpoons \underset{B}{SO_4^{2-}(aq)} + \underset{A}{HOH(\ell)}$$

Now compare the two acids in strength. $HSO_4^-$ is nearer the top of the list in Table 17.1, a much stronger acid than water. We therefore label $HSO_4^-$ with SA for stronger acid and water with WA for weaker acid. Similarly, compare the bases: $OH^-$ is a stronger base (SB) than $SO_4^{2-}$ (WB):

$$\underset{SA}{HSO_4^-(aq)} + \underset{SB}{OH^-(aq)} \rightleftharpoons \underset{WB}{SO_4^{2-}(aq)} + \underset{WA}{HOH(\ell)}$$

As you see, the weaker combination is on the right-hand side in this equation. This indicates that the reaction is favored in the forward direction. The proton transfers spontaneously from the strong proton source to the strong proton remover. The products that are in greater abundance are the weaker conjugate base and conjugate acid.

The following procedure is recommended for predicting acid–base reactions.

**procedure**

**How to Predict the Favored Direction of an Acid-Base Reaction**

**Step 1:** For a given pair of reactants, write the equation for the transfer of *one* proton from one species to the other. (Do not transfer two protons.)

**Step 2:** Label the acid and base on each side of the equation.

**Step 3:** Determine which side of the equation has *both* the weaker acid and the weaker base (they must both be on the same side). That side identifies the products in the favored direction.

## Active Example 17.5

Write the net ionic equation for the reaction between hydrofluoric acid and the sulfite ion, and predict which side will be favored at equilibrium.

The first step is to write the equation for the single-proton transfer reaction between HF and $SO_3^{2-}$. Complete this step.

$HF(aq) + SO_3^{2-}(aq) \rightleftharpoons F^-(aq) + HSO_3^-(aq)$

Next, you need to identify the acid and base on each side of the equation. Do so with letters A and B, as in the preceding discussion.

$$\underset{A}{HF(aq)} + \underset{B}{SO_3^{2-}(aq)} \rightleftharpoons \underset{B}{F^-(aq)} + \underset{A}{HSO_3^-(aq)}$$

In each case the acid is the species with a proton to donate. It is transferred from acid HF to base $SO_3^{2-}$ in the forward reaction and from acid $HSO_3^-$ to base $F^-$ in the reverse reaction.

Finally, determine which reaction, forward or reverse, is favored at equilibrium. It is the side with the weaker acid and base. Refer to Table 17.1.

The forward reaction is favored at equilibrium.

$HSO_3^-$ is a weaker acid than HF, and $F^-$ is a weaker base than $SO_3^{2-}$. These species are the products in the favored direction.

Up to this point most attention has been given to the direction in which an equilibrium is favored. This does not mean we can ignore the unfavored direction. Consider, for example, the reaction $NH_3(aq) + HOH(\ell) \rightleftharpoons NH_4^+(aq) + OH^-(aq)$. Although this reaction proceeds only slightly in the forward direction, many of the properties of household ammonia—its cleaning power, in particular—depend on the presence of $OH^-$ ions.

**Redox-Before-Acid–Base Option** If you have already studied Chapter 19, you know about the electron-transfer character of oxidation–reduction reactions. There are several similarities between those reactions and the proton-transfer reactions of Chapter 17. These are identified in the next section. If you have not yet studied Chapter 19, you may omit Section 17.7, which is repeated at the appropriate place in Chapter 19.

## 17.7 Acid–Base Reactions and Redox Reactions Compared

**Goal** **11** (If Chapter 19 has been studied) Compare and contrast acid–base reactions with redox reactions.

At this point it may be useful to pause briefly and point out how acid–base reactions resemble redox reactions.

1. An acid–base reaction is a transfer of protons; a redox reaction is a transfer of electrons.
2. In both cases the reactants are given special names to indicate their roles in the transfer process. An acid is a proton source; a base is a proton remover. A reducing agent is an electron source; an oxidizing agent is an electron remover.
3. Just as certain species can either provide or remove protons (for example, $HCO_3^-$ and $H_2O$) and thereby behave as an acid in one reaction and a base in another, certain species can either remove or provide electrons, acting as an oxidizing agent in one reaction and a reducing agent in another. An example is the $Fe^{2+}$ ion, which can oxidize Zn atoms to $Zn^{2+}$ in the reaction

   $$Fe^{2+}(aq) + Zn(s) \rightarrow Fe(s) + Zn^{2+}(aq)$$

   $Fe^{2+}$ can also reduce $Cl_2$ molecules to $Cl^-$ ions in another reaction:

   $$Cl_2(g) + 2\ Fe^{2+}(aq) \rightarrow 2\ Cl^-(aq) + 2\ Fe^{3+}(aq)$$

4. Just as acids and bases may be classified as "strong" or "weak" depending on how readily they remove or provide protons, the strengths of oxidizing and reducing agents may be compared according to their tendencies to attract or release electrons.
5. Just as most acid–base reactions in solution reach a state of equilibrium, most aqueous redox reactions also reach equilibrium. Just as the favored side of an acid–base equilibrium can be predicted from acid–base strength, the favored side of a redox equilibrium also can be predicted from oxidizing agent–reducing agent strength.

## 17.8 The Water Equilibrium

**Goal** **12** Given the hydrogen or hydroxide ion concentration of water or a water solution, calculate the other value.

In the remaining sections of this chapter, you will be multiplying and dividing exponentials, taking the square root of an exponential, and working with logarithms. We will furnish brief comments on these operations as we come to them. For more detailed instructions, see Appendix I, Parts A and C.

One of the most critical equilibria in all of chemistry is represented by the next-to-last line in Table 17.1, the ionization of water. Careful control of tiny traces of hydrogen and hydroxide ions marks the difference between success and failure in an untold number of industrial chemical processes. In biochemical systems, these concentrations are vital to survival.

Although pure water is generally regarded as a nonconductor, a sufficiently sensitive detector shows that even water contains a tiny concentration of ions. These ions come from the ionization of the water molecule:

$$2\ H_2O(\ell) \rightleftharpoons H_3O^+(aq) + OH^-(aq)$$

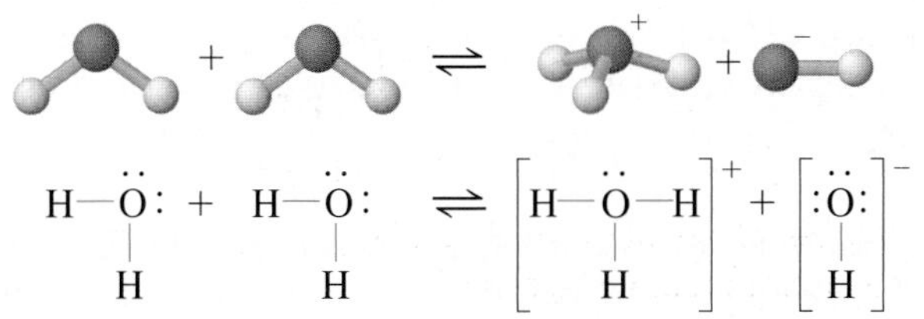

$$H{-}\ddot{\underset{|}{\overset{}{O}}}{:}\ + H{-}\ddot{O}{:} \rightleftharpoons [H{-}\ddot{O}{-}H]^+ + [{:}\ddot{O}{:}]^-$$

(each O also bonded to an H below: $H_2O + H_2O \rightleftharpoons H_3O^+ + OH^-$)

**PREVIEW:** The equilibrium constant is defined in detail in Section 18.7. The value of an equilibrium constant is fixed at a specified temperature.

Each chemical equilibrium has an **equilibrium constant** that is calculated from the concentrations of one or more species in the equation. The equilibrium constant for water, or **water constant, $K_w$,** at 25°C is

$$K_w = [H^+][OH^-] = 1.0 \times 10^{-14}$$

Enclosing a chemical symbol in square brackets is one way to represent the moles-per-liter (molar) concentration of that species. Thus, **$[H^+]$** and **$[OH^-]$** are the molar concentrations of the hydrogen and hydroxide ions, respectively.

$K_w = 1.0 \times 10^{-14}$ only at 25°C. In this book, we assume that the temperature is at 25°C for all calculations that involve the water constant.

At first we will consider $K_w = 10^{-14}$ and work only with concentrations that can be expressed with whole-number exponents. In the next section, after you have become familiar with the mathematical procedures, concentrations will be written in the usual exponential notation form, including coefficients.

The stoichiometry of the equilibrium constant expression for water indicates that the theoretical concentrations of hydrogen and hydroxide ions in pure water must be equal.* If $x = [H^+] = [OH^-]$, then substituting into the $K_w$ expression gives

To take the square root of an exponential, divide the exponent by 2.

$$x^2 = 10^{-14}$$

$$x = \sqrt{10^{-14}} = 10^{-7} \text{ moles/liter}$$

## Thinking About Your Thinking

Equilibrium

**The water equilibrium has the general mathematical form xy = constant. Compare $K_w = [H^+][OH^-]$ to this general form. A characteristic of equilibrium relationships of the type xy = constant is that when x increases, y must decrease, and vice versa. The variables are inversely proportional.**

**Water or water solutions in which $[H^+] = [OH^-] = 10^{-7}$ M are neutral solutions, neither acidic nor basic. A solution in which $[H^+] > [OH^-]$ is acidic; a solution in which $[OH^-] > [H^+]$ is basic.**

$K_w = [H^+][OH^-]$ indicates an inverse relationship between $[H^+]$ and $[OH^-]$; if one concentration goes up, the other goes down. In fact, if we know either the hydrogen or hydroxide ion concentration, we can find the other by solving $K_w = [H^+][OH^-]$ for the unknown.

## Active Example 17.6

Find the hydroxide ion concentration in a solution in which $[H^+] = 10^{-5}$ M. Is the solution acidic or basic?

To solve this problem, apply the relationship $K_w = [H^+][OH^-] = 1.0 \times 10^{-14}$.

*GIVEN:* $[H^+] = 10^{-5}$ M *WANTED:* $[OH^-]$

*EQUATION:* $[OH^-] = \frac{K_w}{[H^+]} = \frac{10^{-14}}{10^{-5}} = 10^{-14-(-5)} = 10^{-9}$ M

To divide exponentials, subtract the denominator exponent from the numerator exponent. With negative exponents, the more negative the exponent, the smaller the value.

*Natural water is not pure. Dissolved minerals from the ground and gases from the atmosphere may cause variations as large as two orders of magnitude from expected hydrogen and hydroxide ion concentrations.

Now, is the solution acidic or basic? State the answer and explain.

Since $[H^+] = 10^{-5} > 10^{-9} = [OH^-]$, the solution is acidic.

## 17.9 pH and pOH (Integer Values Only)

**Goal** **13** Given any one of the following, calculate the remaining three: hydrogen or hydroxide ion concentration expressed as 10 raised to an integral power or its decimal equivalent, pH, and pOH expressed as an integer.

Rather than express very small $[H^+]$ and $[OH^-]$ values in negative exponentials, chemists use base-10 **logarithms** in the form of "p" numbers. By this system, if Q is a number, then

$$pQ = -\log Q$$

Applied to $[H^+]$ and $[OH^-]$, this becomes

$$\mathbf{pH = -\log [H^+]} \quad \text{and} \quad \mathbf{pOH = -\log [OH^-]}$$

If the pH or pOH of a solution is known, the corresponding concentration is found by taking the antilogarithm of the negative of the "p" number:

$$[H^+] = \text{antilog}\ (-pH) = 10^{-pH};\ [OH^-] = \text{antilog}\ (-pOH) = 10^{-pOH}$$

From these logarithmic relationships it follows that, for a concentration expressed as an exponential, a "p" number is written simply by changing the sign of the exponent:

$$\text{If } [H^+] = 10^{-x}, \text{ then pH} = x \quad \text{and} \quad \text{if } [OH^-] = 10^{-y}, \text{ then pOH} = y$$

Conversely, given the "p" number, the concentration is 10 raised to the opposite of that number:

$$\text{If pH} = z, \text{ then } [H^+] = 10^{-z} \quad \text{and} \quad \text{if pOH} = w, \text{ then } [OH^-] = 10^{-w}$$

A logarithm is an exponent. If N is a number and if $N = 10^x$, then log $N = \log 10^x = x$. An **antilogarithm** (antilog) is a number whose logarithm is known. If y is the logarithm of number M, then antilog $y = 10^y = M$. In the first instance you are changing from a number to its logarithm; in the second you are changing from the logarithm to its number. Logarithms and how to use a calculator to solve logarithm problems are discussed in Parts A and C of Appendix I.

### Active Example 17.7

The hydrogen ion concentration of a solution is $10^{-9}$. What is its pH?

The pOH of a solution is 4. What is its hydroxide ion concentration?

$[H^+] = 10^{-pH} = 10^{-9}$ M. pH = 9, the negative of the exponent of 10.

$[OH^-] = 10^{-pOH} = 10^{-4}$ M. The exponent of 10 is the negative of the pOH.

The nature of the equilibrium constant and the logarithmic relationship between pH and pOH yield a simple equation that ties the two together:

$$pH + pOH = 14$$

If you know any one of the group consisting of pH, pOH, $[OH^-]$, or $[H^+]$, you can calculate the others. Figure 17.3 is a **"pH loop"** that summarizes these calculations.

**Figure 17.3** The "pH loop." Given the value for any corner of the pH loop, all other values may be calculated by progressing around the loop in either direction. Conversion equations are shown for each step.

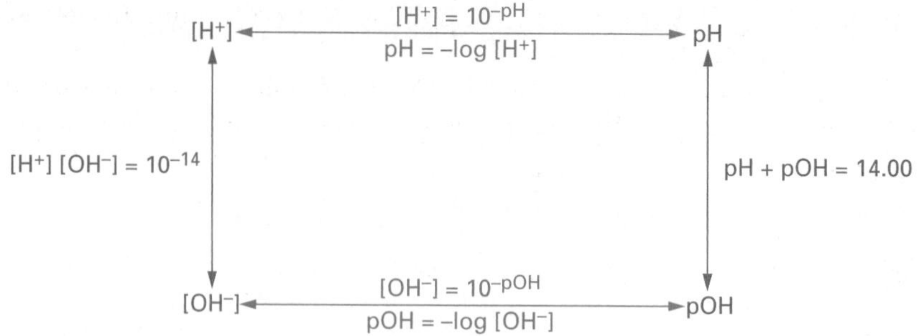


## Active Example 17.8

What are the pH, pOH, $[OH^-]$, and $[H^+]$ of 0.01 M sodium hydroxide?

To solve this problem, you need to work your way through the pH loop illustrated in Figure 17.3. Start by stating the hydroxide ion concentration in a 0.01 M solution of sodium hydroxide. Explain your reasoning.

If 0.01 mol of NaOH is dissolved in 1 L of solution, $NaOH(s) \rightarrow Na^+(aq) + OH^-(aq)$, it forms 0.01 mol of $Na^+(aq)$ and 0.01 mol of $OH^-(aq)$, so the concentration of the hydroxide ion is 0.01 molar, 0.01 mol/L: $[OH^-] = 0.01\ M = 10^{-2}\ M$.

Next, determine the pOH with the relationship $pOH = -\log [OH^-]$.

$pOH = -\log [OH^-] = -\log 10^{-2} = 2$

Now that you have pOH, you can find pH with the relationship $pH + pOH = 14$.

$pH = 14 - pOH = 14 - 2 = 12$

Finally, you can find the hydrogen ion concentration from $[H^+] = 10^{-pH}$.

$[H^+] = 10^{-pH} = 10^{-12}\ M$

Two things are worth noting about Active Example 17.8. First, if we extend the problem by one more step, we complete the full pH loop. We began with $[OH^-]$ and went counterclockwise through pOH, pH, and $[H^+]$. The $[H^+]$ of $10^{-12}$ can be converted to $[OH^-]$:

$$[OH^-] = \frac{K_w}{[H^+]} = \frac{10^{-14}}{10^{-12}} = 10^{-2}\ M$$

This is the same as the starting $[OH^-]$. Completing the loop may therefore be used to check the correctness of other steps in the process.

The second observation from Active Example 17.8 is that the loop may be circled in either direction. Starting with $[OH^-] = 10^{-2}$ and moving clockwise, we obtain

$$[H^+] = \frac{K_w}{[OH^-]} = \frac{10^{-14}}{10^{-2}} = 10^{-12}\ M$$

It follows that pH = 12 and pOH = 14 − 2 = 2, the same results reached by circling the loop in the opposite direction.

You should now be able to make a complete trip around the pH loop on your own.

## Active Example 17.9

The pH of a solution is 3. Calculate the pOH, $[H^+]$, and $[OH^-]$ in any order. Confirm your result by calculating the starting pH—by completing the loop.

You may go either way around the loop, but complete it whichever way you choose, making sure you return to the starting point.

| Counterclockwise | Clockwise |
|---|---|
| From pH = 3, $[H^+] = 10^{-3}$ M | pOH = 14 − 3 = 11 |
| $[OH^-] = \frac{10^{-14}}{10^{-3}} = 10^{-11}$ M | From pOH = 11, $[OH^-] = 10^{-11}$ M |
| From $[OH^-] = 10^{-11}$ M, pOH = 11 | $[H^+] = \frac{10^{-14}}{10^{-11}} = 10^{-3}$ M |
| pH = 14 − 11 = 3 | From $[H^+] = 10^{-3}$ M, pH = 3 |

Most of the solutions we work with in the laboratory and in biochemical systems have pH values between 1 and 14 (Fig. 17.4). This corresponds to $H^+$ concentrations between $10^{-1}$ and $10^{-14}$ M, as shown in Table 17.2.

Let's pause for a moment to develop a feeling for what pH means. pH is a measure of acidity. It is an inverse sort of measurement; the higher the acidity is, the lower the pH will be, and vice versa. Table 17.3 brings out this relationship.

On examining Table 17.3, we see that each pH unit represents a hydrogen-ion-concentration factor of 10. Thus, a solution of pH = 2 is 10 times as acidic as a solution with pH = 3, and 100 times as acidic as the solution of pH = 4. In general, the relative acidity in terms of $[H^+]$ is $10^x$, where x is the difference between the two pH measurements. From this we conclude that a 0.1 M solution of a strong acid, with pH = 1, is one million times as acidic as a neutral solution, with pH = 7. (One million is based on the pH difference, 7 − 1 = 6. As an exponential, $10^6$ = 1,000,000.)

If you understand the idea behind pH, you should be able to make some comparisons.

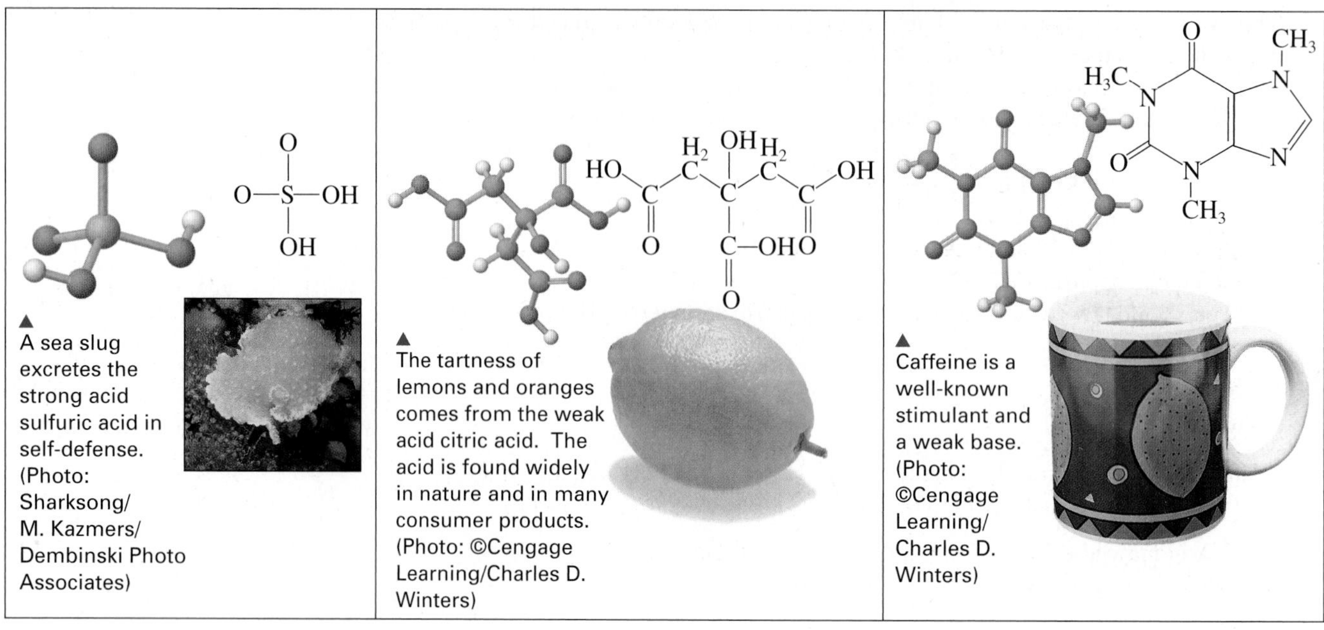

▲ A sea slug excretes the strong acid sulfuric acid in self-defense. (Photo: Sharksong/M. Kazmers/Dembinski Photo Associates)

▲ The tartness of lemons and oranges comes from the weak acid citric acid. The acid is found widely in nature and in many consumer products. (Photo: ©Cengage Learning/Charles D. Winters)

▲ Caffeine is a well-known stimulant and a weak base. (Photo: ©Cengage Learning/Charles D. Winters)

**Figure 17.4** Acids and bases in biochemical systems.

**Table 17.2** pH Values of Common Solutions

| Liquid | pH |
|---|---|
| Human gastric juices | 1.0–3.0 |
| Lemon juice | 2.2–2.4 |
| Vinegar | 2.4–3.4 |
| Carbonated drinks | 2.0–4.0 |
| Orange juice | 3.0–4.0 |
| Black coffee | 3.7–4.1 |
| Tomato juice | 4.0–4.4 |
| Cow's milk | 6.3–6.6 |
| Human blood | 7.3–7.5 |
| Seawater | 7.8–8.3 |
| Saturated $Mg(OH)_2$ | 10.5 |
| Household ammonia | 10.5–11.5 |
| 0.1 M $Na_2CO_3$ | 11.7 |
| 1 M NaOH | 14.0 |

**Table 17.3** pH and Hydrogen Ion Concentration

| $[H^+]$ | $[H^+]$ | pH | Acidity or Basicity* |
|---|---|---|---|
| 1.0 | $10^0$ | 0 | Strongly acid pH < 4 |
| 0.1 | $10^{-1}$ | 1 | |
| 0.01 | $10^{-2}$ | 2 | |
| 0.001 | $10^{-3}$ | 3 | |
| 0.0001 | $10^{-4}$ | 4 | |
| 0.0001 | $10^{-4}$ | 4 | Weakly acid $4 \le pH < 6$ |
| 0.00001 | $10^{-5}$ | 5 | |
| 0.000001 | $10^{-6}$ | 6 | |
| 0.000001 | $10^{-6}$ | 6 | Neutral (or near neutral) $6 \le pH < 8$ |
| 0.0000001 | $10^{-7}$ | 7 | |
| 0.00000001 | $10^{-8}$ | 8 | |
| 0.00000001 | $10^{-8}$ | 8 | Weakly basic $8 \le pH < 10$ |
| 0.000000001 | $10^{-9}$ | 9 | |
| 0.0000000001 | $10^{-10}$ | 10 | |
| 0.0000000001 | $10^{-10}$ | 10 | Strongly basic $10 \le pH$ |
| 0.00000000001 | $10^{-11}$ | 11 | |
| 0.000000000001 | $10^{-12}$ | 12 | |
| 0.0000000000001 | $10^{-13}$ | 13 | |
| 0.00000000000001 | $10^{-14}$ | 14 | |

*Ranges of acidity and basicity are arbitrary.

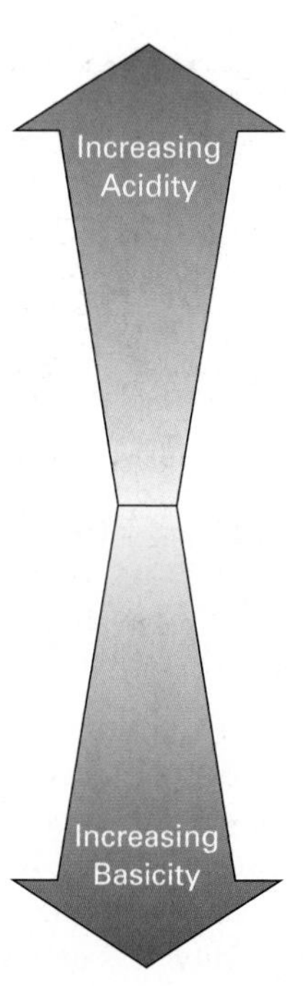


## Active Example 17.10

Arrange the following solutions in order of decreasing acidity (that is, highest $[H^+]$ first, lowest last): Solution A, pH = 8; Solution B, pOH = 4; Solution C, $[H^+] = 10^{-6}$ M; Solution D, $[OH^-] = 10^{-5}$ M.

To list these solutions in any order, all values should be converted to the same basis: pH, pOH, $[H^+]$, or $[OH^-]$. The question asks for a list in terms of acidity, $[H^+]$. Therefore, go around the loop and find $[H^+]$ for each solution:

A: $[H^+] = 10^{-pH} = 10^{-8}$ M

B: $pH = 14 - pOH = 10; [H^+] = 10^{-pH} = 10^{-10}$ M

C: $[H^+] = 10^{-6}$ M (given)

D: $[H^+] = \dfrac{K_w}{[OH^-]} = \dfrac{10^{-14}}{10^{-5}} = 10^{-9}$ M

In arranging these $[H^+]$ values in decreasing order, remember that the exponents are negative.

| Most acidic | $10^{-6}$ | > | $10^{-8}$ | > | $10^{-9}$ | > | $10^{-10}$ | Least acidic |
|---|---|---|---|---|---|---|---|---|
| | C | | A | | D | | B | |

Various methods are used to measure pH in the laboratory. Acid–base indicators (Fig. 17.5) were first mentioned in the introduction to this chapter. Each indicator is effective over a specific pH range. Paper strips impregnated with an indicator dye that changes color gradually over a limited pH range are used for rough measurements (Fig. 17.6[a]). More accurate measurements are made with pH meters (Fig. 17.6[b]).

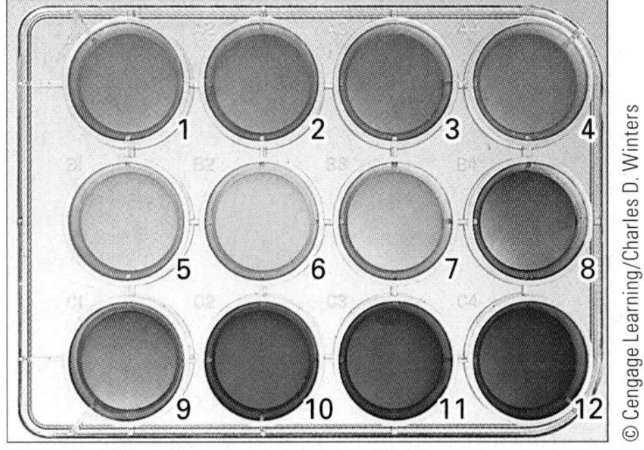




**Figure 17.5** A universal indicator is a solution that shows a wide range of colors as pH varies.

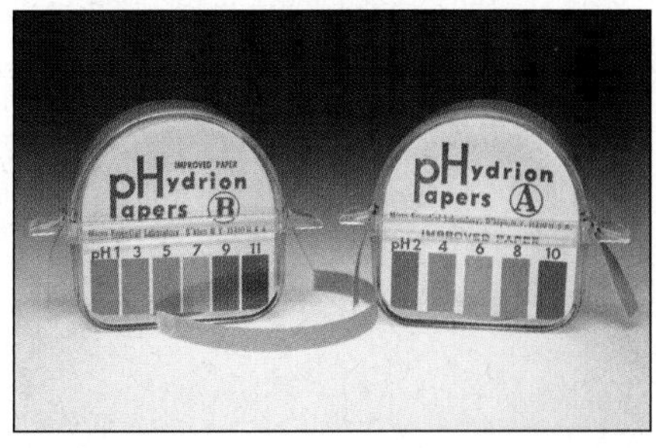


(a)

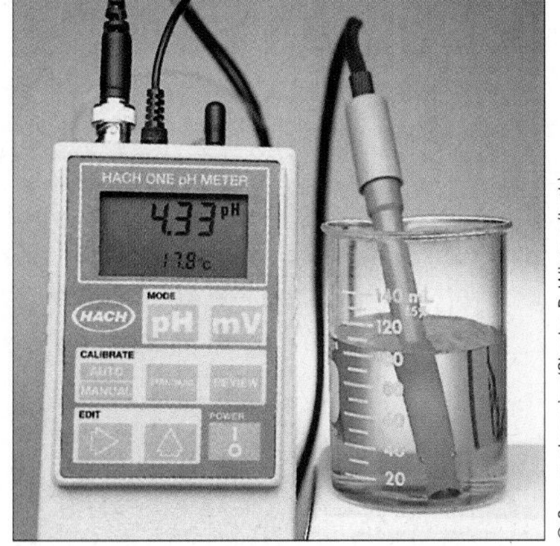


(b)



**Figure 17.6** Measurement of pH. (a) The color imparted to papers impregnated with certain dyes can be used for approximate measurements of pH. (b) A pH meter is a voltmeter calibrated to measure pH.

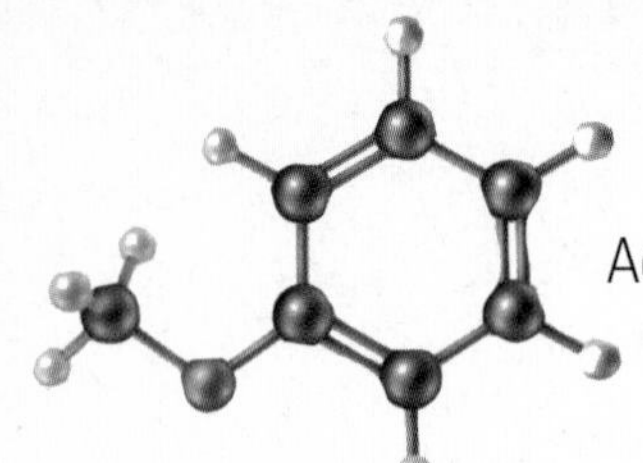

# Everyday Chemistry

## Acid–Base Reactions

The introductory photograph for this chapter shows a few items commonly found in U.S. households. They are organized into two groups. The items on the left, cola, lemon, and cherries, all have acidic solutions. The items on the right, bleach, lye, and borax, all dissolve in water to form basic solutions. This type of categorization also applies to most other household foods and cleaners. Foods tend to be acids, and cleaning solutions tend to be bases. Let's consider some acid–base reactions that occur with common acidic or basic substances.

The pH of the fluid in your stomach is about 1, which is much more acidic than the foods that you eat (see Table 17.2 for some food pH values). The acidity level of the contents of your stomach is primarily controlled by the natural formation of hydrochloric acid, which is believed to be in the stomach as an antibacterial agent and possibly as a predigestion agent for some foods. Under some conditions, such as when you ingest certain foods and beverages, the acidic solution in your stomach can pass into the esophagus, the tube that carries food into your stomach. When this happens, you experience a burning sensation behind your breastbone commonly known as heartburn.

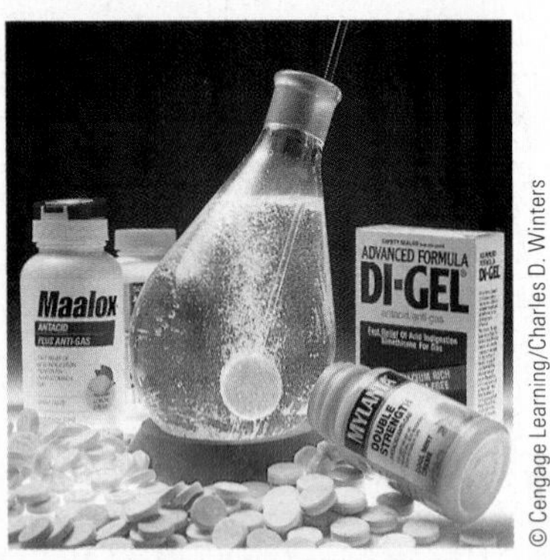

© Cengage Learning/Charles D. Winters

Common commercial antacids. Sales of antacids in the United States are estimated to exceed one billion dollars per year. About 25 million adults in the United States suffer from heartburn daily, and over 60 million are estimated to have at least one heartburn incident per month.

Antacids are used for heartburn relief. They are bases (a base is an anti-acid, or the chemical opposite of an acid), and they work via an acid–base reaction that neutralizes some of the acid in your esophagus and stomach. The active ingredient in Tums and other similar products is calcium carbonate:

$$CaCO_3(s) + 2\ H^+(aq) \rightarrow Ca^{2+}(aq) + H_2O(\ell) + CO_2(g)$$

Alka-Seltzer contains sodium bicarbonate:

$$NaHCO_3(s) + H^+(aq) \rightarrow Na^+(aq) + H_2O(\ell) + CO_2(g)$$

Rolaids used to feature dihydroxyaluminum sodium carbonate as its active ingredient:

$$NaAl(OH)_2CO_3(s) + 4\ H^+(aq) \rightarrow Na^+(aq) + Al^{3+}(aq) + 3\ H_2O(\ell) + CO_2(g)$$

but now they are usually made of calcium carbonate or a combination of calcium carbonate and magnesium hydroxide.

Note that no matter the antacid, an acid–base reaction occurs that yields the unstable "$H_2CO_3$," which decomposes to $H_2O(\ell)$ and $CO_2(g)$ (see Section 9.9). After ingesting these antacids, you normally experience a "burp" by which you expel the carbon dioxide gas.

Cleaning solutions are usually bases. This is because they are designed to decompose fatty molecules, which do not dissolve in water, into ions, which do dissolve in water. A typical fat consists of a long carbon–hydrogen chain with a —COOH group of atoms at the end.

© Cengage Learning/Charles D. Winters

Alka-Seltzer. The solid tablets contain both citric acid and sodium hydrogen carbonate, which cannot react until they are dissolved in solution. Once dissolved, an acid–base reaction takes place, forming bubbles of gaseous carbon dioxide.

$CH_3(CH_2)_{12}COOH$ is an example. In this molecule, 14 carbon atoms are bonded to one another in a long chain. The first 13 are bonded to hydrogen atoms and their neighboring carbon atoms. At the end of the chain, the last carbon atom is double-bonded to an oxygen atom and single-bonded to an —O—H group. This last hydrogen atom is acidic and takes part in acid–base reactions. Thus, the reaction between this fatty molecule, and the hydroxide ion in a cleaning solution is

$$CH_3(CH_2)_{12}COOH(s) + OH^-(aq) \rightarrow CH_3(CH_2)_{12}COO^-(aq) + H_2O(\ell)$$

Not only is the product conjugate base of the reactant fatty acid molecule now soluble in water, but it also actually aids in cleaning!

Proton-transfer reactions are among the most common of all reaction types. Although the transfer of a proton is a very simple particulate-level process, its macroscopic applications are countless. You will learn about many additional acid–base reactions throughout your studies in biology and chemistry.

## 17.10 Noninteger pH–[$H^+$] and pOH–[$OH^-$] Conversions (Optional)

Goal **14** Given any one of the following, calculate the remaining three: hydrogen ion concentration, hydroxide ion concentration, pH, and pOH.

Real-world solutions do not come neatly packaged in concentrations that can be expressed as whole-number powers of 10. [$H^+$] is apt to have a value such as $2.7 \times 10^{-4}$ M; the pH of a solution is likely to be 3.57. A chemist must be able to convert from each of these to the other.

A pH number is a logarithm. Table 17.4 shows the logarithms of 3.45 multiplied by five different powers of 10: 0, 1, 2, 8, and 12. One column shows the value of the logarithm to seven decimals. Another shows the logarithms rounded off to the correct number of significant figures. Notice three things:

1. The mantissa of the logarithm—the number to the right of the decimal in the Value column—is always the same, 0.5378191. This is the logarithm of 3.45, the coefficient for each entry in the Exponential Notation column in Table 17.4.
2. The characteristic of the logarithm—the number to the left of the decimal in the Value column—is the same as the exponent in the Exponential Notation column.
3. All numbers in the first two columns have three significant figures. This appears in both the decimal form of the number and in the coefficient when the number is written in exponential notation.

Item 2 shows that the digits to the left of the decimal in a logarithm—the characteristic—are related only to the exponent of the number when it is written in exponential notation. They have nothing to do with the coefficient of the number, which is where significant figures are expressed. Therefore, *in a logarithm, the digits to the left of the decimal are not counted as significant figures. Counting significant figures in a logarithm begins at the decimal point.*

The significant figures in a number written in exponential notation are the significant figures in the coefficient. These show up in the *mantissa* of the logarithm. To be correct in significant figures, the coefficient and the mantissa must have the same number of digits. The correctly rounded-off logarithms are in the right-hand column of Table 17.4. All numbers in that column are written in three significant figures. Only the digits after the decimal point are significant.

In working with pH, you will be finding logarithms of numbers smaller than 1. These logarithms are negative. The sign is changed to positive when the logarithm is written as a "p" value. Try one. Find log $3.45 \times 10^{-6}$. Enter the number into your calculator, and press the "log" key. The display should read −5.462180905. If $3.45 \times 10^{-6}$ represented [$H^+$], the pH would be the opposite of −5.462180905, or 5.462 rounded off to three significant figures and with the sign changed.*

**Table 17.4** Logarithms and Exponential Notation

| Number | | Logarithm | |
|---|---|---|---|
| **Decimal Form** | **Exponential Notation** | **Value** | **Rounded Off** |
| 3.45 | $3.45 \times 10^0$ | 0.5378191 | 0.538 |
| 34.5 | $3.45 \times 10^1$ | 1.5378191 | 1.538 |
| 345 | $3.45 \times 10^2$ | 2.5378191 | 2.538 |
| 345,000,000 | $3.45 \times 10^8$ | 8.5378191 | 8.538 |
| 3,450,000,000,000 | $3.45 \times 10^{12}$ | 12.5378191 | 12.538 |

*The mantissa appears to be different here from what it was in Table 17.4, but really it is not when you trace its origin: $\log (3.45 \times 10^{-6}) = \log 3.45 + \log 10^{-6} = 0.538 + (-6) = -5.462$.

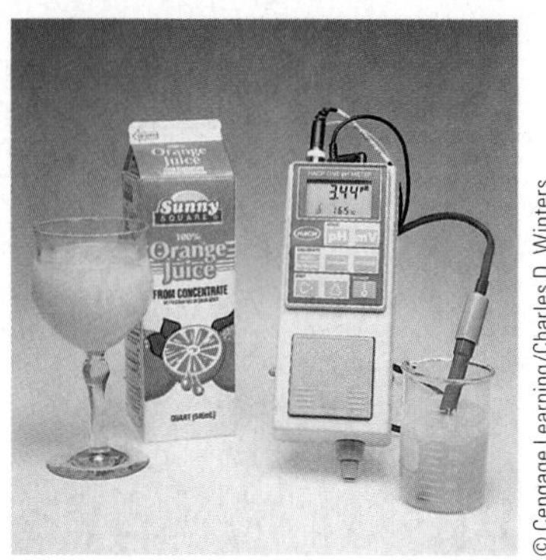




Measuring the pH of orange juice. Real-world solutions almost always have noninteger pH values.

In the Active Example that follows, the calculator sequence is given in detail. Although calculator keys may be marked differently, the procedure is essentially the same for most common calculators.

## Active Example 17.11

Calculate the pH of a solution if $[H^+] = 2.7 \times 10^{-4}$ M.

***Solution***

$$pH = -\log(2.7 \times 10^{-4}) = 3.57$$

| Press | Display |
|---|---|
| 2.7 | 2.7 |
| EE | 2.7 00 |
| 4 | 2.7 04 |
| +/− | 2.7−04 |
| log | −3.568636236 |
| +/− | 3.568636236 |

The answer should be rounded off to two significant figures, 3.57.

## Active Example 17.12

Find the pOH of a solution if its hydroxide ion concentration is $7.9 \times 10^{-5}$ M.

$$pOH = -\log(7.9 \times 10^{-5}) = 4.10 \text{ (two significant figures)}$$

The "p" value for any concentration is found by taking the log of the concentration and changing the sign. The two significant figures in the given concentration lead to a two-significant-figure "p" value, which is two digits after the decimal point.

## Active Example 17.13

The pOH of a solution is 6.24. Find $[OH^-]$.

$$[OH^-] = \text{antilog}(-6.24) = 10^{-6.24} = 5.8 \times 10^{-7} \text{ M}$$

| Press | Display |
|---|---|
| 6.24 | 6.24 |
| +/− | −6.24 |
| $10^x$ | 5.754399373−07 |

The pOH is given to two significant figures, so the hydroxide ion concentration is rounded to two significant figures to yield the final answer, $5.8 \times 10^{-7}$ M.

## Active Example 17.14

Find the hydrogen ion concentration of a solution if its pH is 11.62.

$[H^+] = \text{antilog}\,(-11.62) = 10^{-11.62} = 2.4 \times 10^{-12}\ M$

## Active Example 17.15

$[OH^-] = 5.2 \times 10^{-9}$ M for a certain solution. Calculate, in order, pOH, pH, and $[H^+]$, and then complete the pH loop by recalculating $[OH^-]$ from $[H^+]$.

$pOH = -\log\,(5.2 \times 10^{-9}) = 8.28$

$pH = 14.00 - 8.28 = 5.72$

$[H^+] = \text{antilog}\,(-5.72) = 10^{-5.72} = 1.9 \times 10^{-6}\ M$

$$[OH^-] = \frac{1.0 \times 10^{-14}}{1.9 \times 10^{-6}} = 5.3 \times 10^{-9}\ M$$

The variation between $5.2 \times 10^{-9}$ M and $5.3 \times 10^{-9}$ M comes from rounding off in expressing intermediate answers. If the calculator sequence is completed without rounding off, the loop returns to $5.2 \times 10^{-9}$ M for $[OH^-]$.

# Chapter 17 in Review

Most of the key terms and concepts and many others appear in the Glossary. Use your Glossary regularly.

### 17.1 The Arrhenius Theory of Acids and Bases (Optional)

Goal 1 Distinguish between an acid and a base according to the Arrhenius theory of acids and bases.

**Key Terms and Concepts: Acid, Arrhenius theory, base**

### 17.2 The Brønsted–Lowry Theory of Acids and Bases

Goal 2 Given the equation for a Brønsted–Lowry acid–base reaction, explain how or why it can be so classified.

Goal 3 Given the formula of a Brønsted–Lowry acid and the formula of a Brønsted–Lowry base, write the net ionic equation for the reaction between them.

**Key Terms and Concepts: Amphoteric, Brønsted–Lowry theory, favored direction (equilibrium), forward or reverse reaction or direction, proton source or remover, proton-transfer reaction**

### 17.3 The Lewis Theory of Acids and Bases (Optional)

Goal 4 Distinguish between a Lewis acid and a Lewis base. Given the structural formula of a molecule or ion, state if it can be a Lewis acid, a Lewis base, or both, and explain why.

Goal 5 Given the structural equation for a Lewis acid–base reaction, explain how or why it can be so classified.

**Key Terms and Concepts: Electron-pair donor or acceptor, Lewis theory**

### 17.4 Conjugate Acid–Base Pairs

Goal 6 Define and identify conjugate acid–base pairs.

Goal 7 Given the formula of an acid or a base, write the formula of its conjugate base or acid.

**Key Terms and Concepts: Conjugate acid–base pair, conjugate acid, conjugate base**

**17.5 Relative Strengths of Acids and Bases**

Goal 8 Given a table of the relative strengths of acids and bases, arrange a group of acids or a group of bases in order of increasing or decreasing strength.

**Key Terms and Concepts: Strong or weak acid or base**

**17.6 Predicting Acid–Base Reactions**

Goal 9 Given the formulas of a potential Brønsted–Lowry acid and a Brønsted–Lowry base, write the equation for the possible proton-transfer reaction between them.

Goal 10 Given a table of the relative strengths of acids and bases and information from which a proton-transfer reaction equation between two species in the table may be written, write the equation and predict the direction in which the reaction will be favored.

**17.7 Acid–Base Reactions and Redox Reactions Compared**

Goal 11 (If Chapter 19 has been studied) Compare and contrast acid–base reactions with redox reactions.

**17.8 The Water Equilibrium**

Goal 12 Given the hydrogen or hydroxide ion concentration of water or a water solution, calculate the other value.

**Key Terms and Concepts: Acidic, basic, or neutral solution, equilibrium constant, $[H^+]$, $[OH^-]$, water constant ($K_w$)**

**17.9 pH and pOH (Integer Values Only)**

Goal 13 Given any one of the following, calculate the remaining three: hydrogen or hydroxide ion concentration expressed as 10 raised to an integral power or its decimal equivalent, pH, and pOH expressed as an integer.

**Key Terms and Concepts: Logarithm, antilogarithm, pH loop, pH, pOH**

**17.10 Noninteger pH–$[H^+]$ and pOH–$[OH^-]$ Conversions (Optional)**

Goal 14 Given any one of the following, calculate the remaining three: hydrogen ion concentration, hydroxide ion concentration, pH, and pOH.

# Study Hints and Pitfalls to Avoid

You can always write the conjugate base of an acid by removing one H from the acid formula and reducing the acid charge by 1. Conversely, to write the formula of the conjugate acid of a base, add one H to the base formula and increase the charge by 1.

When writing the equation for a Brønsted–Lowry acid–base reaction, transfer only one proton to get the correct conjugate acid and base on the opposite side of the equation. Once the equation is written, each conjugate acid–base pair has the acid on one side and the base on the other side. The acid and base on the same side of the equation are *not* a conjugate pair.

A "p" number is the opposite (in sign) of the exponent of 10 when an ion concentration is given in exponential form. The exponents are always negative in common solutions, so "p" numbers are positive. That gives an inverse character to the concentration of an ion and its "p" number. For $H^+$, the larger the $[H^+]$, the more acidic the solution and the smaller the pH.

Be careful of negative exponents. The more negative the exponent, the smaller the value. Even though 4 is larger than 3, $-4$ is smaller than $-3$. Therefore, $10^{-4}$ is smaller than $10^{-3}$.

# Concept-Linking Exercises

*Write a brief description of the relationships among each of the following groups of terms or phrases. Answers to the Concept-Linking Exercises are given after the answers to the Target Checks in Appendix III. Terms from optional Section 17.3 are in italics. Exercise 11 covers terms related to logarithms, and it is also optional.*

1. Arrhenius acid, Brønsted–Lowry acid, *Lewis acid*
2. Arrhenius base, Brønsted–Lowry base, *Lewis base*
3. Brønsted–Lowry acid–base theory, proton-transfer reaction, proton source, proton remover, amphoteric
4. Reversible, forward direction, reverse direction, left, right, favored direction
5. *Lewis acid–base theory, electron-pair donor, electron-pair acceptor*
6. Conjugate acid, conjugate base, conjugate acid–base pair
7. Strong acid, weak acid
8. Water equilibrium, $10^{-14}$, $[H^+]$, water constant, $[OH^-]$, $K_w$
9. pH, pOH, $[H^+]$, $[OH^-]$
10. Logarithm, antilogarithm
11. *Characteristic, mantissa, coefficient, exponential, logarithm*

# Small-Group Discussion Questions

*Small-Group Discussion Questions are for group work, either in class or under the guidance of a leader during a discussion section.*

1. Although there are no "ammonium hydroxide" particles in a container of aqueous ammonia, bottles are often labeled as such. This traditional label likely originated from the Arrhenius acid–base model. Explain why Arrhenius theory requires hydroxide ions in solution, and discuss why it would lead to this name.
2. Hydrogen chloride is a stable, gaseous species at common conditions. When the gas is bubbled through water, a solution of hydrochloric acid is formed in which the major species are hydrated hydrogen and chloride ions. How can it be that hydrogen chloride gas is "content" to exist as a molecular compound until it comes into contact with water? Answer in terms of the Brønsted–Lowry model. Use the hydrogen chloride and water reaction as an illus-

tration and explain the Brønsted–Lowry definitions of *acid* and *base*.

3. Classify each of the following categories as being either Lewis acids or Lewis bases. Give specific examples of at least three species from each category. Illustrate Lewis acid–base reactions from at least one species from each category. Can any category contain both Lewis acids and Lewis bases? A molecule with a double bond, positive ions, a molecule with an unoccupied orbital in the same principal energy level as its valence electrons, anions, a molecule with a polar double bond, a molecule with an unshared electron pair.

4. Write the formula of both the conjugate acid *and* the conjugate base of each of the following: water, ammonia, hydroxide ion, hydrogen sulfate ion.

5. Classify each of the acids in Table 17.1 as strong or weak, according to the classification scheme introduced in Chapter 9. What acid demarcates the division between strong and weak? Rewrite the reactions in the vicinity of the division between strong and weak using a water molecule on the left side of the equation; for example, the reaction with HI would be rewritten $HI + H_2O \rightleftharpoons H_3O^+ + I^-$. Explain why the acid you identified as separating the strong acids from the weak acids acts as the separation point.

6. Sketch particulate-level illustrations of a strong acid solution and a weak acid solution.

7. Sketch particulate-level illustrations of a strong base solution and a weak base solution.

8. At 30°C, the value of $K_w$, the water constant, is $1.5 \times 10^{-14}$. Determine the pH, pOH, $[H^+]$, and $[OH^-]$ for water at 30°C.

9. Rank the following solutions from lowest to highest pH: pure water, natural rainwater, seawater, vinegar, carbonated beverages, sodium hydroxide solution. Explain your reasoning for each solution.

10. What is the pH of a solution with a hydrogen ion concentration of 1.0 M?

# Questions, Exercises, and Problems

*Blue-numbered questions are answered in Appendix III.* ■ *denotes problems assignable in OWL. In the first section of Questions, Exercises, and Problems, similar exercises and problems are paired in consecutive odd-even number combinations.*

### Section 17.1: The Arrhenius Theory of Acids and Bases (Optional)

### Section 17.2: The Brønsted–Lowry Theory of Acids and Bases

### Section 17.3: The Lewis Theory of Acids and Bases (Optional)

1. Identify at least two of the classical properties of acids and two of bases. For one acid property and one base property, show how it is related to the ion associated with an acid or base.

2. ■ In the following net ionic equations, identify each reactant as either a Brønsted–Lowry acid or a Brønsted–Lowry base. Also identify the proton source and the proton remover.

   a) $HCN(aq) + H_2O(\ell) \rightarrow CN^-(aq) + H_3O^+(aq)$

   b) $CH_3NH_2(aq) + H_2O(\ell) \rightarrow CH_3NH_3^+(aq) + OH^-(aq)$

3. Distinguish between an Arrhenius base and a Brønsted–Lowry base. Are the two concepts in agreement? Justify your answer.

4. ■ a) Write a net ionic equation to show that hypochlorous acid behaves as a Brønsted–Lowry acid in water.

   b) The substance triethanolamine is a weak nitrogen-containing base like ammonia. Write a net ionic equation to show that triethanolamine, $C_6H_{15}O_3N$, behaves as a Brønsted–Lowry base in water.

5. Explain or illustrate by an example what is meant by identifying a Lewis acid as an electron-pair acceptor and a Lewis base as an electron-pair donor.

6. ■ Classify each of the following substances into the appropriate category: (1) Lewis acid; (2) Lewis base; (3) substance that can act as either a Lewis acid or Lewis base; (4) substance that is neither a Lewis acid nor a Lewis base. (a) $CCl_4$, (b) $F^-$, (c) $Cu^{2+}$, (d) $Fe^{3+}$, (e) $O_2$

7. When diethyl ether reacts with boron trifluoride, a covalent bond forms between the molecules. Describe the reaction from the standpoint of the Lewis acid–base theory, based on the following "structural" equation:

$$BF_3 + :\ddot{O}(C_2H_5)_2 \longrightarrow F_3B{-}\ddot{O}(C_2H_5)_2$$

8. Aluminum chloride, $AlCl_3$, behaves more as a molecular compound than an ionic one. This is illustrated in its ability to form a fourth covalent bond with a chloride ion: $AlCl_3 + Cl^- \rightarrow AlCl_4^-$. From the Lewis diagram of the aluminum chloride molecule and the electron configuration of the chloride ion, show that this is an acid–base reaction in the Lewis sense, and identify the Lewis acid and the Lewis base.

### Section 17.4: Conjugate Acid–Base Pairs

9. Give the formula of the conjugate base of HF; of $H_2PO_4^-$. Give the formula of the conjugate acid of $NO_2^-$; of $H_2PO_4^-$.

10. ■ Write the formulas of each of the following: (a) the conjugate base of $CH_3COOH$, (b) the conjugate acid of $CN^-$, (c) the conjugate base of $H_2S$, (d) the conjugate acid of $HCO_3^-$.

11. For the reaction $HSO_4^-(aq) + C_2O_4^{2-}(aq) \rightleftharpoons SO_4^{2-}(aq) + HC_2O_4^-(aq)$, identify the acid and the base on each side of the equation—that is, the acid and base for the forward reaction and the acid and base for the reverse reaction.

12. ■ In the following net ionic equation, identify each species as either a Brønsted–Lowry acid or a Brønsted–Lowry base: $HF(aq) + CN^-(aq) \rightarrow F^-(aq) + HCN(aq)$. Identify the conjugate of each reactant and state whether it is a conjugate acid or a conjugate base.

13. Identify the conjugate acid–base pairs in Question 11.

14. ■ In the following net ionic equation, identify each species as either a Brønsted–Lowry acid or a Brønsted–Lowry base: $CH_3COO^-(aq) + HS^-(aq) \rightarrow CH_3COOH(aq) + S^{2-}(aq)$. Identify the conjugate of each reactant and state whether it is a conjugate acid or a conjugate base.

15. For the reaction $HNO_2(aq) + C_3H_5O_2^-(aq) \rightleftharpoons NO_2^-(aq) + HC_3H_5O_2(aq)$, identify both conjugate acid–base pairs.

16. Identify both conjugate acid–base pairs in the reaction $HClO(aq) + CO_3^{2-}(aq) \rightleftharpoons ClO^-(aq) + HCO_3^-(aq)$.

17. Identify the conjugate acid–base pairs in $NH_4^+(aq) + HPO_4^{2-}(aq) \rightleftharpoons NH_3(aq) + H_2PO_4^-(aq)$.

18. Identify the conjugate acid–base pairs in $H_2PO_4^-(aq) + HCO_3^-(aq) \rightleftharpoons HPO_4^{2-}(aq) + H_2CO_3(aq)$.

### Section 17.5: Relative Strengths of Acids and Bases

*Refer to Table 17.1 when answering questions in this section.*

19. What is the difference between a strong base and a weak base, according to the Brønsted–Lowry concept? Give two examples of strong bases and two examples of weak bases.

20. What is the difference between a strong acid and a weak acid, according to the Brønsted–Lowry concept? Give two examples of strong acids and two examples of weak acids.

21. List the following acids in order of their increasing strength (weakest acid first): $HC_2O_4^-$, $H_2SO_3$, $H_2O$, $HClO$.

22. ■ Select the specified acid from each of the following groups:
    a) Strongest of acetic acid, $CH_3COOH$; hydrogen carbonate ion, $HCO_3^-$; sulfurous acid, $H_2SO_3$
    b) Weakest of hydrofluoric acid, HF; phosphoric acid, $H_3PO_4$; sulfurous acid, $H_2SO_3$

23. List the following bases in order of their decreasing strength (strongest base first): $CN^-$, $H_2O$, $HSO_3^-$, $ClO^-$, $Cl^-$.

24. ■ Select the specified *conjugate base* from each of the following groups of acids:
    a) Weakest conjugate base of phosphoric acid, $H_3PO_4$; acetic acid, $CH_3COOH$; hydrogen carbonate ion, $HCO_3^-$
    b) Strongest conjugate base of: hydrosulfuric acid, $H_2S$; sulfurous acid, $H_2SO_3$; hydrocyanic acid, HCN

### Section 17.6: Predicting Acid–Base Reactions

*For each acid and base given in this section, complete a proton-transfer equation for the transfer of one proton. Using Table 17.1, predict the direction in which the resulting equilibrium will be favored.*

25. $HC_3H_5O_2(aq) + PO_4^{3-}(aq) \rightleftharpoons$

26. ■ $HCN(aq) + CH_3COO^-(aq) \rightleftharpoons$

27. $HSO_4^-(aq) + CO_3^{2-}(aq) \rightleftharpoons$

28. ■ $F^-(aq) + HCO_3^-(aq) \rightleftharpoons$

29. $H_2CO_3(aq) + NO_3^-(aq) \rightleftharpoons$

30. $H_3PO_4(aq) + CN^-(aq) \rightleftharpoons$

31. $NO_2^-(aq) + H_3O^+(aq) \rightleftharpoons$

32. $H_2BO_3^-(aq) + NH_4^+(aq) \rightleftharpoons$

33. $HSO_4^-(aq) + HC_2O_4^-(aq) \rightleftharpoons$

34. $HPO_4^{2-}(aq) + HC_2H_3O_2(aq) \rightleftharpoons$

### Section 17.7: Acid–Base Reactions and Redox Reactions Compared

35. Explain how a strong acid is similar to a strong reducing agent. Explain how a strong base compares with a strong oxidizing agent.

36. Show how redox and acid–base reactions parallel each other—how they are similar, but also what makes them different.

### Section 17.8: The Water Equilibrium

37. Of what significance is the very small value of $10^{-14}$ for $K_w$, the ionization equilibrium constant for water?

38. How is it that we can classify water as a nonconductor of electricity and yet talk about the ionization of water? If it ionizes, why does it not conduct?

39. $[H^+] = 10^{-5}$ M and $[OH^-] = 10^{-9}$ M in a certain solution. Is the solution acidic, basic, or neutral? How do you know?

40. ■ An aqueous solution has a hydrogen ion concentration of $10^{-4}$ M. (a) What is the hydroxide ion concentration in this solution? (b) Is this solution acidic, basic, or neutral?

41. What is $[OH^-]$ in 0.01 M HCl? (*Hint:* Begin by finding $[H^+]$ in 0.01 M HCl.)

42. ■ An aqueous solution has a hydroxide ion concentration of $10^{-11}$ M. (a) What is the hydrogen ion concentration in this solution? (b) Is this solution acidic, basic, or neutral?

### Section 17.9: pH and pOH (Integer Values Only)

43. In which classification, strongly acidic, weakly acidic, strongly basic, weakly basic, and neutral, or close to neutral, does each of the following solutions belong: (a) pH = 7, (b) pH = 9, (c) pOH = 3?

44. Identify the approximate ranges of the pH scale that could be classified as strongly acidic, weakly acidic, strongly basic, weakly basic, and neutral, or close to neutral.

45. If the pH of a solution is 8.6, is the solution acidic or basic? How do you reach your conclusion? List in order the pH values of a solution that is neutral, one that is basic, and one that is acidic.

46. Select any integer from 1 to 14 and explain what is meant by saying that this number is the pH of a certain solution.

*Questions 47 through 54: The pH, pOH, $[H^+]$, or $[OH^-]$ is given. Find each of the other values. Classify each solution as strongly acidic, weakly acidic, neutral (or close to neutral), weakly basic, or strongly basic.*

47. pH = 5

48. ■ pOH = 6

49. $[OH^-] = 10^{-1}$ M

50. ■ $[H^+] = 0.1$ M

51. pOH = 4

52. ■ $[OH^-] = 10^{-2}$ M

53. $[H^+] = 10^{-9}$ M

54. ■ pH = 4

### Section 17.10: Noninteger pH–[H⁺] and pOH–[OH⁻] Conversions (Optional)

*Questions 55 through 62: The pH, pOH, $[H^+]$, or $[OH^-]$ of a solution is given. Find each of the other values.*

55. $[OH^-] = 2.5 \times 10^{-10}$ M

56. ■ pH = 6.62

57. pH = 4.06

58. ■ $[OH^-] = 1.1 \times 10^{-11}$ M

59. $[H^+] = 2.8 \times 10^{-1}$ M

60. ■ pOH = 5.54

61. pOH = 7.40

62. ■ $[H^+] = 7.2 \times 10^{-2}$ M

### General Questions

63. Distinguish precisely, and in scientific terms, the differences among items in each of the following pairs.
   a) Acid and base—by Arrhenius theory
   b) Acid and base—by Brønsted–Lowry theory
   c) Acid and base—by Lewis theory
   d) Forward reaction, reverse reaction
   e) Acid and conjugate base, base and conjugate acid
   f) Strong acid, weak acid
   g) Strong base, weak base
   h) $[H^+]$, $[OH^-]$
   i) pH and pOH

64. Classify each of the following statements as true or false:
   a) All Brønsted–Lowry acids are Arrhenius acids.
   b) All Arrhenius bases are Brønsted–Lowry bases, but not all Brønsted–Lowry bases are Arrhenius bases.
   c) $HCO_3^-$ is capable of being amphoteric.
   d) $HS^-$ is the conjugate base of $S^{2-}$.
   e) If the species on the right side of an ionization equilibrium are present in greater abundance than those on the left, the equilibrium is favored in the forward direction.
   f) $NH_4^+$ cannot act as a Lewis base.
   g) Weak bases have a weak attraction for protons.
   h) The stronger acid and the stronger base are always on the same side of a proton-transfer reaction equation.
   i) A proton-transfer reaction is always favored in the direction that yields the stronger acid.
   j) A solution with pH = 9 is more acidic than one with pH = 4.
   k) A solution with pH = 3 is twice as acidic as one with pH = 6.
   l) A pOH of 4.65 expresses the hydroxide ion concentration of a solution in three significant figures.

65. Theoretically, can there be a Brønsted–Lowry acid–base reaction between $OH^-$ and $NH_3$? If not, why not? If yes, write the equation.

66. Explain what *amphoteric* means. Give an example of an amphoteric substance, other than water, that does not contain carbon.

67. ■ Very small concentrations of ions other than hydrogen and hydroxide are sometimes expressed with "p" numbers. Calculate pCl in a solution for which $[Cl^-] = 7.49 \times 10^{-8}$.

68. Can a substance that does not have hydrogen atoms be a Brønsted–Lowry acid? Explain.

69. Why do you suppose chemists prefer the pH scale to expressing hydrogen ion concentration directly? In other words, what is the advantage of saying pH = 4 rather than $[H^+] = 10^{-4}$?

70. ■ What is the bromide ion concentration in a solution if a report gives the pBr as 7.2?

### More Challenging Questions

71. Suggest a reason why the acid strength decreases with each step in the ionization of phosphoric acid: $H_3PO_4 \rightarrow H_2PO_4^- \rightarrow HPO_4^{2-}$.

72. Theoretically, can there be a Brønsted–Lowry acid–base reaction between $SO_4^{2-}$ and $F^-$? If not, why not? If yes, write the equation.

73. ■ Sodium carbonate is among the most widely used industrial bases. How can it be a base when it does not contain a hydroxide ion? Write the equation for a reaction that demonstrates its character as a base.

74. ■ Nonmetal oxides can react with water to form acids. For example, carbon dioxide reacts with water to form carbonic acid: $CO_2 + H_2O \rightarrow H_2CO_3$. What acid forms as the result of the reaction of sulfur trioxide and water? Write the equation for the reaction.

75. ■ Metal oxides can react with water to form bases. Write an equation to show how calcium oxide can react with water to form a base. Name the base.

76. Nitrogen dioxide is emitted in automobile exhaust. Explain how this can contribute to acid rain, which is rain with a low pH.

# 18

One condition always found in a liquid–vapor, solute–solution, or chemical equilibrium is that the change is reversible. The solution in this test tube changes color depending on its temperature. When heated, the solution turns blue. When cooled, the solution turns pink. The solution will change color over and over again when moved between the cold and warm water baths. The macroscopic-level color change occurs because the temperature change causes a particulate-level chemical change. In this chapter, you will learn about the effect of temperature—and other conditions—on a chemical equilibrium.

# Chemical Equilibrium

If you have not yet studied Chapters 15 and 16, or if it has been some time since you studied Chapters 15 and 16, you should review Sections 15.4 and 16.3 before beginning this chapter. Pay particular attention to Figures 15.15 and 16.7 and their captions.

In studying the liquid–vapor equilibrium in Chapter 15 and the solute–solution equilibrium in Chapter 16, you learned that equilibrium exists when forward and reverse rates are the same in a reversible *physical* change. Similarly, in this chapter you will see that chemical equilibrium exists when the forward and reverse rates are the same in a reversible *chemical* change.

Online homework for this chapter may be assigned in OWL

## CONTENTS

## 18.1 The Character of an Equilibrium

Goal 1 Identify a chemical equilibrium by the conditions it satisfies.

A careful review of the liquid–vapor equilibrium in Section 15.4 and the solution equilibrium in Section 16.3 reveals four conditions that are true for every equilibrium:

**1.** *The change is reversible and can be represented by an equation with a double arrow.* In a reversible change, the substances on the left side of the equation produce the substances on the right, and those on the right change back into the substances on the left. These are the forward and reverse reactions, respectively.

**2.** *The equilibrium system is **"closed"**—closed in the sense that no substance can enter or leave the immediate vicinity of the equilibrium.* All substances on either side of an equilibrium equation must remain to form the substances on the other side.

**3.** *The equilibrium is* ***dynamic.*** The reversible changes occur continuously, even though there is no appearance of change. By contrast, items in a static equilibrium are stationary, without motion, as an object hanging on a string.

**4.** *The things that are* equal *in an equilibrium are the forward rate of change (from left to right in the equation) and the reverse rate of change (from right to left).* Note in particular that the amounts of substances present in an equilibrium are not necessarily equal.

### √ Target Check 18.1

*Can a solution in a beaker that is open to the atmosphere contain a chemical equilibrium? Explain.*

## 18.2 The Collision Theory of Chemical Reactions

Goal 2 Distinguish between reaction-producing molecular collisions and molecular collisions that do not yield reactions.

If two molecules are to react chemically, it is reasonable to expect that they must come into contact with each other. What we see as a chemical reaction is the overall effect of a huge number of individual collisions between reacting particles. This view of chemical change is the **collision theory of chemical reactions.**

Figure 18.1 examines three kinds of molecular collisions for the imaginary reaction $A_2 + B_2 \rightarrow 2\ AB$. If the collision is to produce a reaction, the bond between A atoms in $A_2$ must be broken; similarly, the bonds in the $B_2$ molecules must be broken. It takes energy to break these bonds. This energy comes from the kinetic energy of the molecules just before they collide. In other words, there must be a violent, bond-breaking collision. This is most apt to occur if the molecules are moving at high speed. Figure 18.1(a) pictures a reaction-producing collision.

### Thinking About Your Thinking

Mental Models

**Figure 18.1 is a great starting point from which you can build a mental model of the collision theory of chemical reactions. Be sure to distinguish between successful and unsuccessful collisions. Figure 18.1(a) illustrates a successful collision. Try**

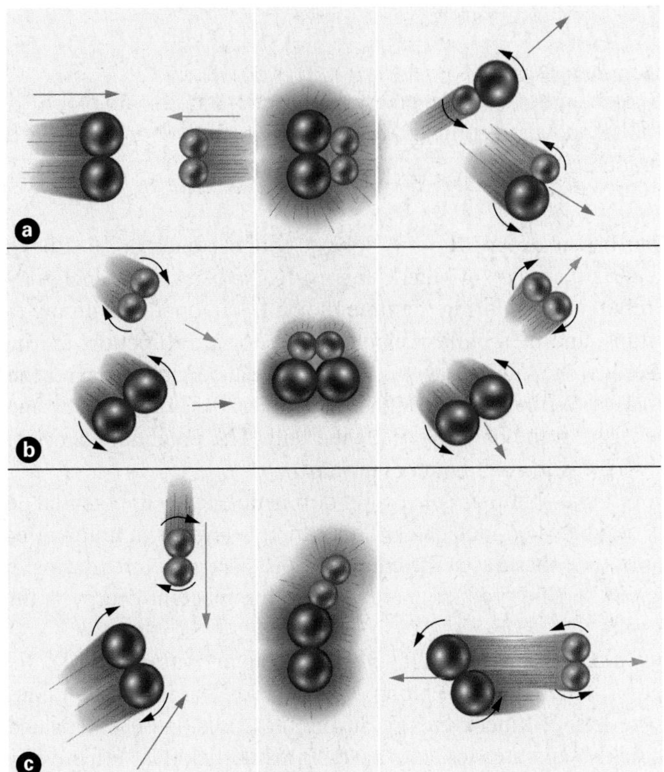


**Figure 18.1** Molecular collisions and chemical reactions. (a) Reaction-producing collision between molecules $A_2$ and $B_2$, which have sufficient energy and proper orientation. (b) Collision has proper orientation but not enough kinetic energy. There is no reaction. (c) Glancing collision has enough kinetic energy, but poor orientation. There is no reaction.

**to picture this in motion in your mind. There are two essential requirements for a reaction-producing collision. First, the particles must have enough kinetic energy to overcome the electrical repulsions that occur as the negatively charged electrons on the edge of one molecule repel the electrons of the other molecule. Second, the reactant particles must approach one another in an orientation that "lines up" the constituent atoms so that the product molecules will be formed.**

**In Figure 18.1(b), notice the middle frame in particular. Can you see how the orientation of the collision is such that the product molecules will form? If such a collision is not successful, it must be because the kinetic energy of the molecules is not great enough to overcome the electrostatic repulsions. Figure 18.1(c) shows another ineffective collision. It does not produce product molecules because the alignment of the colliding particles is such that the product cannot form. Be sure to imagine what these unsuccessful collisions would look like in three dimensions.**

Not all collisions result in a reaction; in fact, most do not. If the colliding molecules do not have enough kinetic energy to break the bonds, the original molecules simply repel each other with the same identity they had before the collision (Fig. 18.1[b]). Sometimes they may have enough kinetic energy, but only "sideswipe" each other and move off unchanged (Fig. 18.1[c]). Other sufficiently energetic collisions may have an orientation that pushes atoms in the original molecules closer together rather than pulling them apart.

To summarize: For an individual collision to result in a reaction, the particles must have (1) enough kinetic energy and (2) the proper orientation. The rate of a particular reaction depends on the frequency of effective collisions.

## √ Target Check 18.2

*Draw three kinds of molecular collisions that will not result in a chemical change.*

## 18.3 Energy Changes During a Molecular Collision

**Goal** 3 Sketch and/or interpret an energy-reaction coordinate graph. Identify the (a) activated complex region, (b) activation energy, and (c) $\Delta E$ for the reaction.

**P/REVIEW:** The change from kinetic energy to potential energy and then back to kinetic energy for a bouncing ball is an example of the Law of Conservation of Energy (Section 2.9).

When a rubber ball is dropped to the floor, it bounces back up. Just before it hits the floor, it has a certain amount of kinetic energy. It also has kinetic energy as it leaves the floor on the rebound. But during the time the ball is in contact with the floor, it slows down, even stops, and then builds velocity in an upward direction. During the period of reduced velocity, the kinetic energy, $\frac{1}{2} mv^2$, is reduced, and even reaches zero at the turnaround instant. While the collision is in progress, the initial kinetic energy changes to potential energy in the partially flattened ball. The potential energy changes back into kinetic energy as the ball bounces upward.

It is believed that a similar conversion of kinetic energy to potential energy occurs during a collision between molecules. This can be shown by a graph of energy versus a reaction coordinate that traces the energy of the system before, during, and after the collision (Fig. 18.2). The product energy minus the reactant energy is the $\Delta E$ for the reaction, $\Delta E = E_{products} - E_{reactants}$.

When two molecules are colliding, they form an **activated complex** that has a high potential energy at the top of the hump of the curve. The increase in potential energy comes from the loss of kinetic energy during the collision. The activated complex is unstable. It quickly separates into two parts. If the collision is "effective" in producing a reaction, the parts will be product molecules; if the collision is ineffective, they will be the original reactant molecules.

The hump in Figure 18.2 is a **potential energy barrier** that must be surpassed before a collision can be effective. Surpassing the barrier is like rolling a ball over a hill. If the ball has enough kinetic energy to get to the top, it will roll down the other side (Fig. 18.3[a]). This corresponds to an effective collision in which the colliding molecules have enough kinetic energy to get over the potential energy barrier. However, if the ball does not have enough kinetic energy to reach the top of the hill, it rolls back to where it came from (Fig. 18.3[b]). If the colliding particles do not have enough kinetic energy to meet the potential energy requirement, the collision is ineffective.

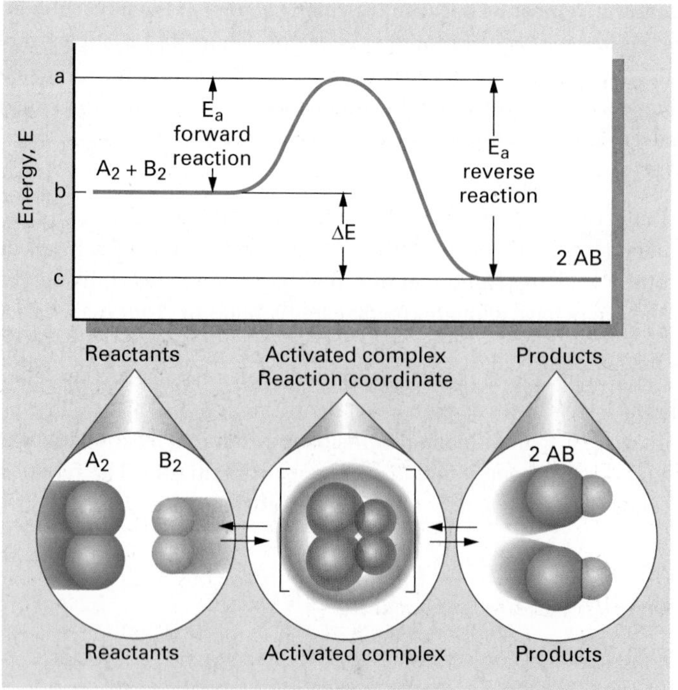


**Figure 18.2** Energy-reaction graph for the reaction $A_2 + B_2 \rightarrow 2\ AB$. $E_a$ represents the activation energy for the forward and reverse reactions, as indicated. (Energy values a, b, and c are reference points in an end-of-chapter question and a Thinking About Your Thinking feature.)

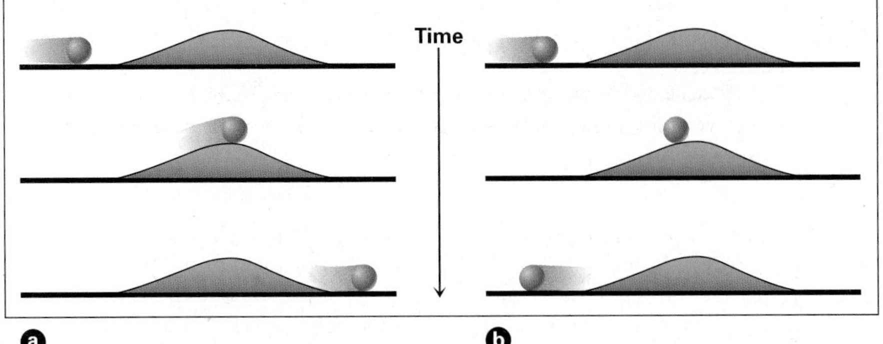


(a) In (a) the ball has more than enough kinetic energy to reach the top and roll down the other side.

(b) In (b) all the original kinetic energy is changed to potential energy before the ball reaches the top, so it rolls back down the same side.

**Figure 18.3** Potential energy barrier. As the ball rolls toward the hill (potential energy barrier), its speed (kinetic energy) determines whether or not it will pass over the hill.

## Thinking About Your Thinking

### Mental Models

**Figure 18.2 is designed to help you see the relationship between a graph, which is a mathematical construct, and what actually happens at the particulate level that results in the relationship expressed in the graph. This type of thinking frequently occurs in the sciences, and thus it is among the best ways of understanding scientific concepts. The key is to match your mental model of the actual process to the graph.**

**The horizontal portion of the curve at energy value b represents the total energy of the separate reactants $A_2$ and $B_2$. This is analogous to the ball rolling along the flat surface in Figure 18.3. The rising portion of the curve in Figure 18.2 represents increasing energy. As the two molecules approach one another, the potential energy of the system increases because of electrostatic repulsions. Similarly, in Figure 18.3, the potential energy of the ball increases as it moves up the hill. Imagine the molecules coming together and how that increases the potential energy as the negative edge of one molecule repels the negative edge of the other.**

**When the system energy reaches its maximum, energy a on the Figure 18.2 graph, the reacting particles have moved close enough together to form the activated complex, the highest-energy species that exists in this reaction profile. This is similar to the potential energy of the ball at the top of the hill in Figure 18.3(a). As the product molecules move apart, the energy of the system decreases, shown on the graph as the drop from energy a to energy c. This is like the ball rolling down the other side of the hill in Figure 18.3(a). The ball's potential energy is changed to kinetic energy. As you look at the downward slope on the graph, imagine the product molecules moving apart so that the electrostatic repulsions are no longer significant.**

The minimum kinetic energy needed to produce an effective collision is called **activation energy.** The difference between the energy at the peak of the Figure 18.2 curve and the reactant energies is the activation energy for the forward reaction. Similarly, the difference between the energy at the peak and the product energies is the activation energy for the reverse reaction.

**P/REVIEW:** Activation energy is similar to the escape energy in the evaporation of a liquid, as described in Section 15.4. Only molecules with more than a certain minimum kinetic energy are able to tear away from the bulk of the liquid and change to the vapor state.

### √ Target Check 18.3

*In what sense is activation energy a "barrier"?*

# 18.4 Conditions That Affect the Rate of a Chemical Reaction

## The Effect of Temperature on Reaction Rate

**Goal** | **4** State and explain the relationship between reaction rate and temperature.

**P/REVIEW:** The vapor pressure of a liquid rises as temperature increases. The boiling point of a liquid is the temperature at which its vapor pressure is equal to the pressure above the liquid. A higher pressure over the liquid requires a higher vapor pressure—and therefore a higher temperature—to boil the liquid. See Section 15.5.

Chemical reactions are faster at higher temperatures. This can be seen in the kitchen in several ways. Food is refrigerated to slow down the chemical changes that occur in spoiling. A pressure cooker reduces the time needed to cook some items in boiling water because water boils at a higher temperature under increased pressure. The opposite effect is seen in open cooking at high altitudes, where reduced atmospheric pressure allows water to boil at lower temperatures. Here cooking is slower, the result of reduced reaction rates.

Figure 18.4 explains the effect of temperature on reaction rates. The curve labeled $T_1$ gives the kinetic energy distribution among the particles in a sample at one temperature, and $T_2$ represents the distribution at a higher temperature. E is the activation energy, the minimum kinetic energy a particle must have to enter into a reaction-producing collision. It is the same at both temperatures. Only the fraction of the particles in the sample represented by the area beneath the curve to the right of E is able to react. Compare the areas to the right of E. As illustrated, the fraction that is able to react is about $2^1/2$ times as much for $T_2$ as for $T_1$. The rate of reaction is therefore higher at the higher temperature.

## The Effect of a Catalyst on Reaction Rate

**Goal** | **5** Using an energy-reaction coordinate graph, explain how a catalyst affects reaction rate.

**P/REVIEW:** In principle, the plot in Figure 18.4 is the same curve as in Figure 15.17 (Section 15.4), which we used to explain the effect of temperature on the vapor pressure of a liquid. The total area beneath the curve represents the entire sample, so it is the same at all temperatures. At higher temperatures the curve flattens and shifts to the right. The *average* kinetic energy that corresponds with temperature is found along the horizontal axis.

Driving from one city to another over rural roads and through towns takes a certain amount of time. If an interstate highway were built between the cities, you would have available an alternative route that would be much faster. This is what a **catalyst** does. It provides an alternative "route" for reactants to change to products. The activation energy with the catalyst is lower than the activation energy without the catalyst. The result is that a larger fraction of the molecules in a sample are able to enter into reaction-producing collisions, so the reaction rate increases. This is illustrated in Figure 18.5.

Catalysts exist in several different forms, and the precise function of many catalysts is not clearly understood. Some catalysts are mixed in the reacting chemicals,

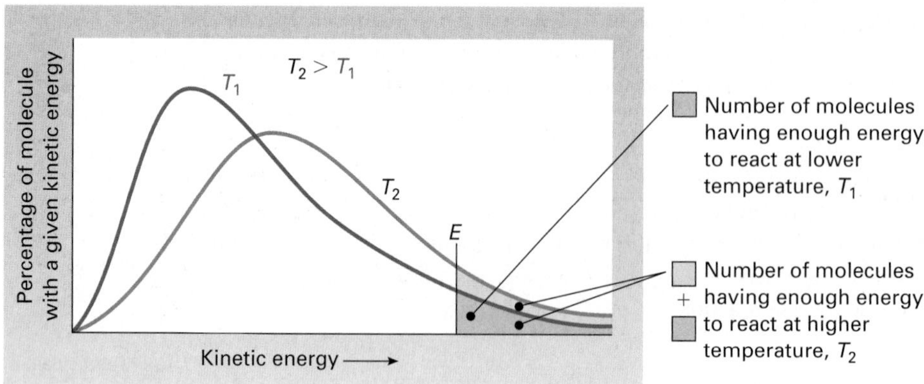


**Figure 18.4** Kinetic energy distribution curves at two temperatures. E is the activation energy, the minimum kinetic energy required for a reaction-producing collision. Only the fraction of molecules represented by the total area beneath the curve to the right of the activation energy has enough kinetic energy to react. At lower temperature $T_1$, that area is shaded in pink, and at higher temperature $T_2$, that area is the sum of the blue and pink shading. The larger fraction of molecules that is able to react, represented by the blue area, is responsible for the higher reaction rate at higher temperatures.

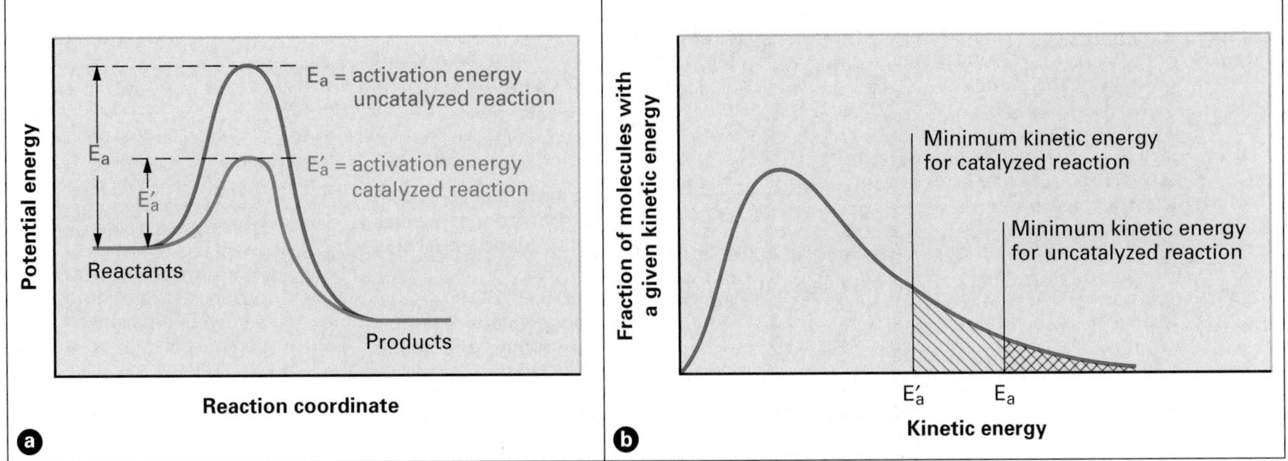


**Figure 18.5** (a) The effect of a catalyst on activation energy and reaction rate. A catalyst provides a way for a reaction to occur with a lower activation energy (blue curve) than the same reaction has without a catalyst (red curve). Molecules with a lower kinetic energy are therefore able to pass over the potential energy barrier. (b) The crosshatched area in each color represents the fraction of the total sample with enough energy to engage in reaction-producing collisions. The catalyzed area (blue) is much larger than the uncatalyzed area (red), so the catalyzed reaction rate is faster.

whereas others do no more than provide a surface on which the reaction may occur. In either case, the catalyst is not permanently affected by the reaction. Some catalysts undergo a chemical change, but during the course of the reaction they are regenerated in exactly the same amount that was present at the start.

A catalytic reaction that appears often in beginning chemistry laboratories is making oxygen by decomposing hydrogen peroxide, $H_2O_2$. Manganese(IV) oxide, which is commonly referred to by its non-systematic name, manganese dioxide, $MnO_2$, is mixed in as a catalyst (Fig. 18.6[a]). A well-known industrial process is the catalytic cracking of crude oil, in which large hydrocarbon molecules are broken down into simpler and more useful products in the presence of a catalyst (Fig. 18.7). Biological reactions are controlled by catalysts called *enzymes* (Fig. 18.6[c]).

Some substances interfere with a normal reaction path from reactants to products, forcing the reaction to a higher activation energy route that is slower. Such substances are called negative catalysts, or **inhibitors.** Inhibitors are used to control the rates of particular industrial reactions. Sometimes negative catalysts can have disastrous results, as when mercury poisoning prevents the normal biological functions of enzymes.

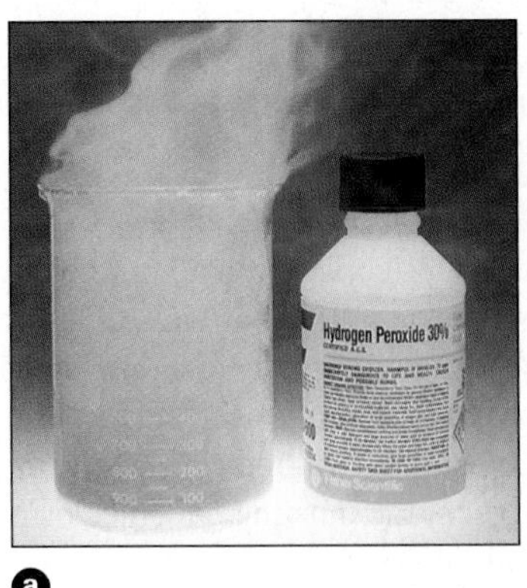

a

b

c



**Figure 18.6** Catalytic decomposition of hydrogen peroxide, $2\ H_2O_2 \xrightarrow{\text{catalyst}} 2\ H_2O + O_2$. (a) Solid $MnO_2$ catalyzes the decomposition of a 30% solution of $H_2O_2$. The chemical change is accompanied by the release of so much heat that some of the product $H_2O$ appears as steam. (b) The bombardier beetle shoots a hot liquid solution from a "gun" at the rear of its body. The beetle has a gland that produces hydrogen peroxide, which is mixed with the substance that catalyzes its decomposition when it becomes necessary for the insect to defend itself. (c) A biological catalyst—an enzyme—that speeds the decomposition of hydrogen peroxide solution occurs naturally in potatoes. You can see oxygen gas bubbling to the surface.

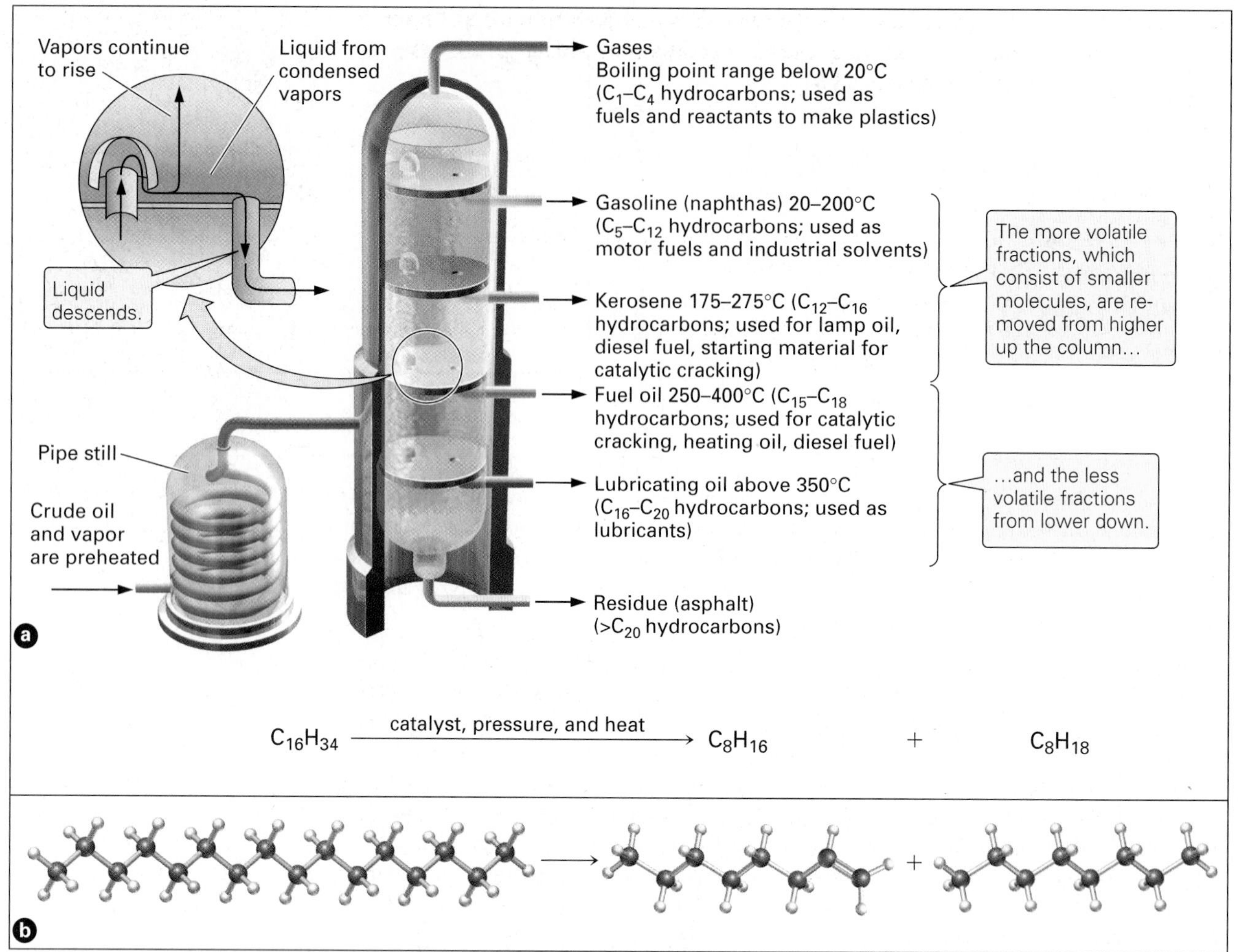


**Figure 18.7** Catalytic cracking. (a) Crude oil, a complex mixture, is distilled into simpler mixtures (fractions) based on their boiling points. In general, there is more demand for the gasoline fraction than for kerosene. (b) After distillation, the larger (12 to 16 carbons) molecules in the kerosene fraction can be decomposed into smaller (5 to 12 carbons) molecules that make up gasoline. This is accomplished by catalytic cracking, a process that uses a catalyst.

## The Effect of Concentration on Reaction Rate

**Goal** | **6** Identify and explain the relationship between reactant concentration and reaction rate.

If a reaction rate depends on the frequency of effective collisions, the influence of concentration is readily predictable. The more particles there are in a given volume, the more frequently collisions will occur and the more rapidly the reaction will take place.

The effect of concentration on reaction rate is easily seen in the rate at which objects burn in air compared with the rate of burning in an atmosphere of pure oxygen. If a burning splint is thrust into pure oxygen, the burning becomes brighter, more vigorous, and much faster. In fact, the typical laboratory test for oxygen is to ignite a splint, blow it out, and then, while there is still a faint glow, place it in oxygen. It immediately bursts back into full flame and burns vigorously (Fig. 18.8). Charcoal, phosphorus, and other substances may be used in place of the splint.

We have identified three factors that influence the rate of a chemical reaction. In relating these factors to equilibrium considerations, we will examine only concentration and temperature. These variables affect forward and reverse reaction rates differently. A catalyst, on the other hand, has the same effect on both forward and reverse rates. Therefore, a catalyst does not alter chemical equilibrium. A catalyst does cause a system to reach equilibrium more quickly.

**Figure 18.8** The glowing splint test for oxygen.

### √ Target Check 18.4

a) What happens to a reaction rate as temperature drops? Give two explanations for the change. State which one is more important and explain why.
b) How does a catalyst affect reaction rates?
c) Compare reaction rates when a given reactant concentration is high with the rate when the concentration is low. Explain the difference.

## 18.5 The Development of a Chemical Equilibrium

**Goal** 7 Trace and explain the changes in concentrations of reactants and products that lead to a chemical equilibrium.

The role of concentration in chemical equilibrium may be illustrated by tracing the development of an equilibrium. The forward reaction in $A_2 + B_2 \rightarrow 2\ AB$ is assumed to take place by the simple collision of $A_2$ and $B_2$ molecules, which separate as two AB molecules. The reverse reaction is exactly the reverse process: Two AB molecules collide and separate as one $A_2$ molecule and one $B_2$ molecule.

Figure 18.9 shows a graph of forward and reverse reaction rates versus time. Initially, at Time 0, pure $A_2$ and $B_2$ are introduced to the reaction chamber. At the initial concentrations of $A_2$ and $B_2$ the forward reaction begins at rate $F_0$. Initially there are no AB molecules present, so the reverse reaction cannot occur. At Time 0 the reverse reaction rate, $R_0$, is zero. These points are plotted on the graph.

As soon as the reaction begins, $A_2$ and $B_2$ are consumed, thereby reducing their concentrations in the reaction vessel. As these reactant concentrations decrease, the forward reaction rate declines. Consequently, at Time 1, the forward reaction rate drops to $F_1$. During the same interval, some AB molecules are produced by the forward reaction, and the concentration of AB becomes greater than zero. Therefore, the reverse reaction begins, with the reverse rate rising to $R_1$ at Time 1.

At Time 1 the forward rate is greater than the reverse rate. Therefore, $A_2$ and $B_2$ are consumed by the forward reaction more rapidly than they are produced by the reverse reaction. The net change in the concentrations of $A_2$ and $B_2$ is therefore downward, causing a further reduction in the forward rate at Time 2. Conversely, the reverse reaction uses AB more slowly than the forward reaction produces it. The net change in the concentration of AB is thus an increase. This, in turn, raises the reverse reaction rate at Time 2.

Similar changes occur over successive intervals until the forward and reverse rates eventually become equal. At this point a dynamic equilibrium is established. From this analysis we may state the following generalization:

> *For any reversible reaction in a closed system, whenever the opposing reactions are occurring at different rates, the faster reaction will gradually become slower, and the slower reaction will become faster. Finally, the reaction rates become equal, and equilibrium is established.*

**P/REVIEW:** The development of a liquid–vapor pressure equilibrium is described in Section 15.4. An equilibrium between excess solute and a saturated solution is examined in Section 16.3. Both equilibria involve physical changes in which one of the two opposing rates remains constant until the other catches up with it. Here we study how a chemical equilibrium develops. This time both forward and reverse rates change as equilibrium is reached.

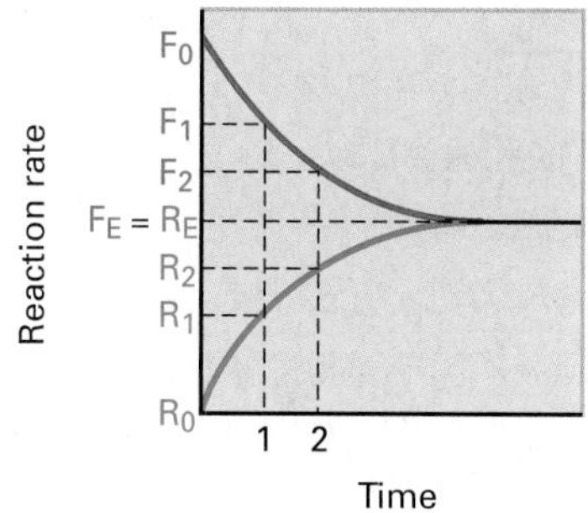


**Figure 18.9** Changes in reaction rates during the development of a chemical equilibrium. The forward rate is represented by the upper curve in red, and the reverse rate is shown by the lower curve in blue.

## 18.6 Le Chatelier's Principle

In the last section you saw how concentrations and reaction rates jockey with each other until the rates become equal and equilibrium is reached. In this section we start with a

Oesper Collection in the History of Chemistry, University of Cincinnati

Henri Louis Le Chatelier (1850–1936). In addition to being a talented scientist, Le Chatelier was an educational reformer and an excellent teacher.

system already at equilibrium and see what happens to it when the equilibrium is upset. To "upset" an equilibrium, you must somehow make the forward and reverse reaction rates unequal, at least temporarily. One way to do this is to change the concentration of at least one substance in the system. A gaseous equilibrium can often be upset simply by changing the volume of the container. If you change the temperature of a chemical equilibrium, you will also make the forward and reverse reaction rates unequal.

How an equilibrium responds to a disturbance can be predicted from the concentration and temperature effects already considered. The predictions may be summarized in **Le Chatelier's Principle,** which says that **if an equilibrium system is subjected to change, processes occur that tend to partially counteract the initial change, thereby bringing the system to a new position of equilibrium.**

We will now see how Le Chatelier's Principle explains three different equilibrium changes.

## The Concentration Effect

**Goal 8** Given the equation for a chemical equilibrium, predict the direction in which the equilibrium will shift because of a change in the concentration of one species.

The reaction of hydrogen and iodine to produce hydrogen iodide comes to equilibrium with hydrogen iodide as the favored species, that is, the species having the higher concentration. The equal forward and reverse reaction rates are shown by the equal-length arrows in the equation

$$H_2(g) + I_2(g) \rightleftharpoons 2\ HI(g)$$

The sizes of the formulas represent the relative concentrations of the different species in the reaction. [HI] is greater than $[H_2]$ and $[I_2]$, which are equal.

**P/REVIEW:** Enclosing the formula of a substance in brackets represents its concentration in moles per liter. Brackets are used for hydrogen and hydroxide ion concentrations, $[H^+]$ and $[OH^-]$, in Section 17.8.

If more HI is forced into the system, [HI] is increased. This raises the rate of the reverse reaction, in which HI is a reactant. This is indicated by the longer arrow from right to left:

$$H_2(g) + I_2(g) \leftharpoondown\!\!\!\rightharpoonup 2\ HI(g)$$

Because the rates are no longer equal, the equilibrium is destroyed.

Now the changes described in italics at the end of the last section begin. The unequal reaction rates cause the system to shift in the direction of the faster rate—to the left, or in the reverse direction. As a result, $H_2$ and $I_2$ are made by the reverse reaction faster than they are used by the forward reaction. Their concentrations increase, so the forward rate increases. Simultaneously, HI is used faster than it is produced, reducing both [HI] and the reverse reaction rate. Eventually, the rates become equal (note arrow lengths) at an intermediate value, and a new equilibrium is reached:

$$H_2(g) + I_2(g) \rightleftharpoons 2\ HI(g)$$

The sizes of the formulas indicate that all three concentrations are larger than they were originally, although [HI] has come down from the maximum it reached just after it was added.

The preceding example shows *why* an equilibrium shifts when it is disturbed in terms of concentrations and reaction rates. Le Chatelier's Principle makes it possible to predict the direction of a shift, forward or reverse, without such a detailed analysis. Just remember that the shift is always in the direction that tries to return the substance disturbed to its original condition. Active Figure 18.10 illustrates the Le Chatelier concentration effect.

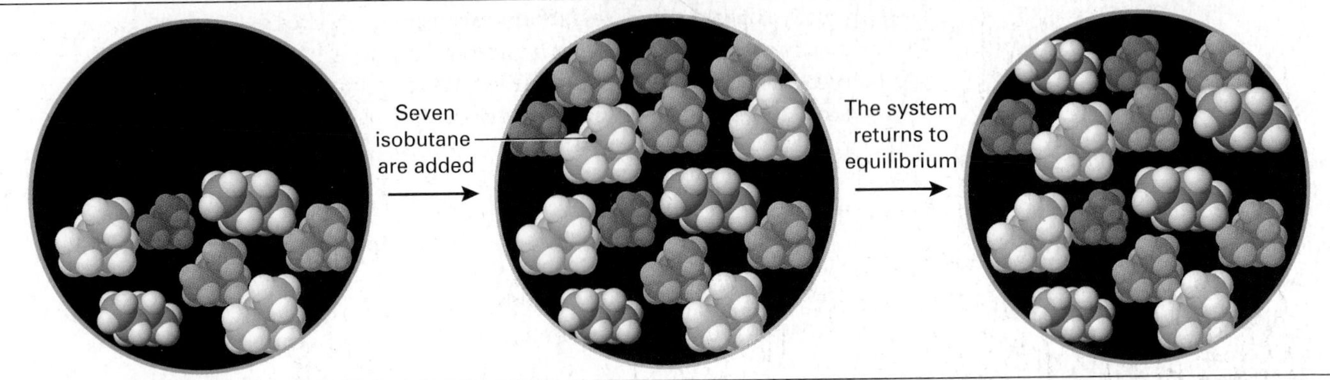
Seven
isobutane
are added
The system
returns to
equilibrium
1 An equilibrium mixture of 5 isobutane molecules and 2 butane molecules.
2 Seven isobutane molecules are added, so the system is no longer at equilibrium.
3 A net of 2 isobutane molecules has changed to butane molecules, to once again give an equilibrium mixture where the ratio of isobutane to butane is 5 to 2 (or 2.5/1).

**Active Figure 18.10** The concentration effect. The shift is in the direction that returns the disturbed concentration ratio to its equilibrium value. **Watch this active figure at *http://www.cengage.com/cracolice.***

## Active Example 18.1

The system $N_2(g) + 3\ H_2(g) \rightleftharpoons 2\ NH_3(g)$ is at equilibrium. Use Le Chatelier's Principle to predict the direction in which the equilibrium will shift if ammonia is withdrawn from the reaction chamber.

The equilibrium disturbance is clearly stated: Ammonia is withdrawn. In which direction, forward or reverse, must the reaction shift to *counteract* the removal of ammonia—that is, to *produce* more ammonia to replace some of what has been taken away? Choose one and explain your reasoning.

The shift will be in the *forward* direction.

Ammonia is the product of the forward reaction and therefore will be *partially* restored to its original concentration by a shift in the forward direction.

## Active Example 18.2

Predict the direction, forward or reverse, of a Le Chatelier shift in the equilibrium $CH_4(g) + 2\ H_2S(g) \rightleftharpoons 4\ H_2(g) + CS_2(g)$ caused by each of the following: (a) increase $[H_2S]$, (b) reduce $[CS_2]$, (c) increase $[H_2]$. Briefly justify each answer.

*continued*

a) Forward, counteracting partially the increase in $[H_2S]$ by consuming some of it.
b) Forward, restoring some of the $CS_2$ removed.
c) Reverse, consuming some of the added $H_2$.

## The Volume Effect

**Goal** **9** Given the equation for a chemical equilibrium involving one or more gases, predict the direction in which the equilibrium will shift because of a change in the volume of the system.

A change in the volume of an equilibrium system that includes one or more gases changes the concentration of those gases. Usually—there is one exception, as you will see shortly—there is a Le Chatelier shift that partially offsets the initial change. Both the change and the adjustment involve the pressure caused by the entire system.

According to the kinetic molecular theory and the model of an ideal gas (Section 4.2), the pressure exerted by a gas is the combined result of billions upon billions of collisions of molecules hitting the walls of the container that holds the gas. If the frequency of these collisions increases, pressure increases; if frequency is reduced, pressure is reduced. One way to increase the frequency is to reduce the volume of the container. The molecules have shorter distances to travel before hitting the walls, so the frequency goes up. Conversely, if volume is increased, frequency and, therefore, pressure are reduced. The relationship is an inverse proportionality, expressed by Boyle's Law, which you studied in Section 4.5: $P \propto 1/V$.

**P/REVIEW:** If you have studied Chapter 14, you will recognize that the text discussion leads to the ideal gas equation (Section 14.4), $PV = nRT$. Solving for P at constant temperature gives

$$P = RT \times \frac{n}{V} = k \times \frac{n}{V}$$

where k is a constant. If k is a proportionality constant, it follows that

$$P \propto \frac{n}{V}$$

which expresses the concentration of the gas.

There is another way to change the frequency of collisions between molecules and the walls of a container: Change the number of gas molecules in the container. More particles mean more collisions in a given period of time, and therefore higher pressure; fewer particles yield lower pressure. It can be demonstrated that this relationship is a direct proportionality: $P \propto n$, where n represents the number of gas molecules in terms of moles.

Whenever one quantity (P) is proportional to two other quantities (1/V and n), it is proportional to the product of those quantities:

$$P \propto n \times \frac{1}{V} \qquad P \propto \frac{n}{V}$$

The ratio n/V is an expression of the concentration of the gas in moles per liter.

If the volume of a gas is reduced, the denominator in the n/V concentration ratio becomes smaller and the fraction becomes larger; thus, pressure increases. Le Chatelier's Principle calls for a shift that will partially counteract the change—to reduce pressure. What change can reduce the pressure of the system at the new volume? The numerator in the n/V concentration ratio must be reduced; there must be fewer gaseous molecules in the system. In general:

**P/REVIEW:** The proportional relationship between equation coefficients and the number of gaseous molecules is used in stoichiometry problems when converting between volumes of different gases measured at the same temperature and pressures (Section 14.9).

*If a gaseous equilibrium is compressed, the increased pressure will be partially relieved by a shift in the direction of fewer gaseous molecules; if the system is expanded, the reduced pressure will be partially restored by a shift in the direction of more gaseous molecules (Fig. 18.11).*

The coefficients of gases in an equation are in the same proportion as the number of gaseous molecules. We use this fact in predicting Le Chatelier shifts caused by volume changes. The following Active Example shows how.

### Active Example 18.3

Predict the direction of the shift resulting from an expansion in the volume of the equilibrium $2\ SO_2(g) + O_2(g) \rightleftharpoons 2\ SO_3(g)$.

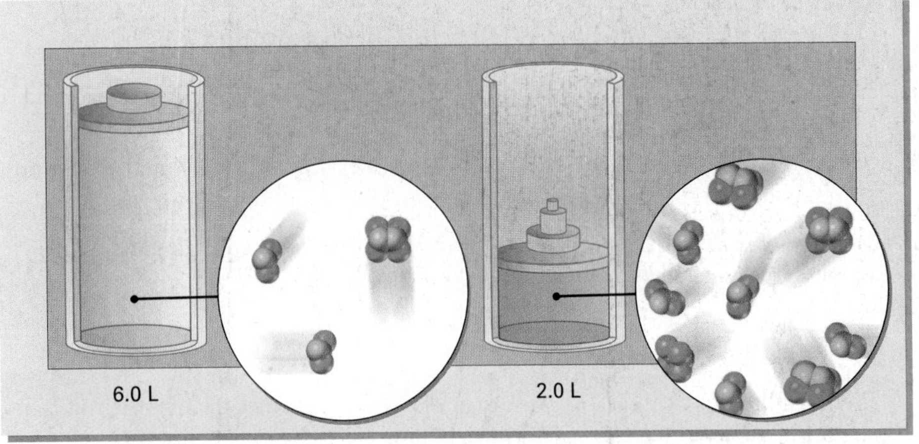


**Figure 18.11** The volume effect. The system $N_2O_4(g) \rightleftharpoons 2\ NO_2(g)$ is initially at equilibrium under conditions at which the ratio of $N_2O_4$ to $NO_2$ is 1:2 (*left*). The system is compressed to $^1/_3$ of the original volume (*right*). Le Chatelier's Principle predicts that the system will respond by shifting in the direction of fewer gaseous molecules, which is in the reverse direction. This shift is verified by counting the molecules; the ratio of $N_2O_4$ to $NO_2$ is now 4:5. As $NO_2$ reacts to form $N_2O_4$, the total number of particles in the system is reduced, partially offsetting the increased pressure.

***Solution***

First, will total pressure increase or decrease as a result of expansion? Boyle's Law indicates that pressure is inversely proportional to volume, so the total pressure will be less at the larger volume. The Le Chatelier shift must make up some of that lost pressure. How? By changing the number of gaseous molecules in the system. Will it take more molecules or fewer to raise pressure? Since pressure is proportional to concentration, $P \propto n/V$, the number of molecules, n, must increase to raise the pressure.

Now examine the reaction equation. Notice that a forward shift finds *three* reactant molecules, two $SO_2$ and one $O_2$, forming *two* $SO_3$ product molecules. The reverse shift has *two* reactant molecules yielding *three* product molecules. To increase the total number of molecules, then, the reaction must shift in the *reverse* direction: 2 molecules $\rightarrow$ 3 molecules.

## Active Example 18.4

The volume occupied by the equilibrium $SiF_4(g) + 2\ H_2O(g) \rightleftharpoons SiO_2(s) + 4\ HF(g)$ is reduced. Predict the direction of the shift in the position of equilibrium.

Will the shift be in the direction of more gaseous molecules or fewer? Explain.

Fewer.

If volume is reduced, pressure increases. Increased pressure is counteracted by fewer molecules.

Now predict the direction of the shift and justify your prediction by stating the numerical change in molecules from the equation.

*continued*

The shift is in the reverse direction. Molecules change from 4 on the right to 3 on the left.

Note the molecule change is 4 to 3, not 5 to 3. Only gaseous molecules are involved in pressure adjustments, so the $SiO_2(s)$ doesn't count.

## Active Example 18.5

Returning to the familiar $H_2(g) + I_2(g) \rightleftharpoons 2\ HI(g)$, predict the direction of the shift that will occur because of a volume increase.

Take it all the way, but be careful.

There will be no shift because each side of the equation has two gaseous molecules. When the number of gaseous molecules is the same, neither the number of molecules nor the pressure can be changed by a shift in equilibrium. Increasing or decreasing the volume has no effect on the equilibrium.

## The Temperature Effect

**Goal** **10** Given a thermochemical equation for a chemical equilibrium, or information from which it can be written, predict the direction in which the equilibrium will shift because of a change in temperature.

A change in temperature of an equilibrium will change both forward and reverse reaction rates, but the rate changes are not equal. The equilibrium is therefore destroyed temporarily. The events that follow are again predictable by Le Chatelier's Principle.

A thermochemical equation is one that includes a change in energy. It can be written in two ways: with the enthalpy-of-reaction term, $\Delta H$, to the right of the conventional equation or with the energy term included as if it were a reactant or product (Section 10.8). Including the energy term in the thermochemical equation, rather than showing $\Delta H$ separately, makes it easier to predict the Le Chatelier effect of a change in temperature. In the equation, we can think of energy as we would a substance being "added" or "removed."

## Active Example 18.6

If the temperature of the equilibrium $PCl_5(g) \rightleftharpoons PCl_3(g) + Cl_2(g) + 92.5\ kJ$ is increased, predict the direction of the Le Chatelier shift.

In order to raise the temperature of something, we must heat it. We therefore interpret an increase in the temperature as the "addition of heat," and a lowering of temperature as the "removal of heat." In applying Le Chatelier's Principle to a thermochemical equation, we may regard the heat in much the same manner as we regard any chemical species in the equation. Accordingly, if heat is *added* to the equilibrium system shown, in which direction must it shift to *use up,* or

*consume,* some of the heat that was added? (If chlorine were *added,* in which direction would the equilibrium shift to use up some of the chlorine?) Forward or reverse? Explain.

The equilibrium must shift in the *reverse* direction to use up some of the added heat. An endothermic reaction consumes heat. As the equation is written, the reverse reaction is endothermic.

## Active Example 18.7

The thermal decomposition of limestone reaches the following equilibrium: $CaCO_3(s) + 176\text{ kJ} \rightleftharpoons CaO(s) + CO_2(g)$. Predict the direction this equilibrium will shift if the temperature is reduced: forward or reverse. Justify your answer.

Reverse.

Reduction in temperature is interpreted as the removal of heat. The reaction will respond to replace some of the heat removed—as an exothermic reaction. Heat is produced as the reaction proceeds in the reverse direction.

Figure 18.12 gives a visible example of Le Chatelier's Principle as it relates to temperature.

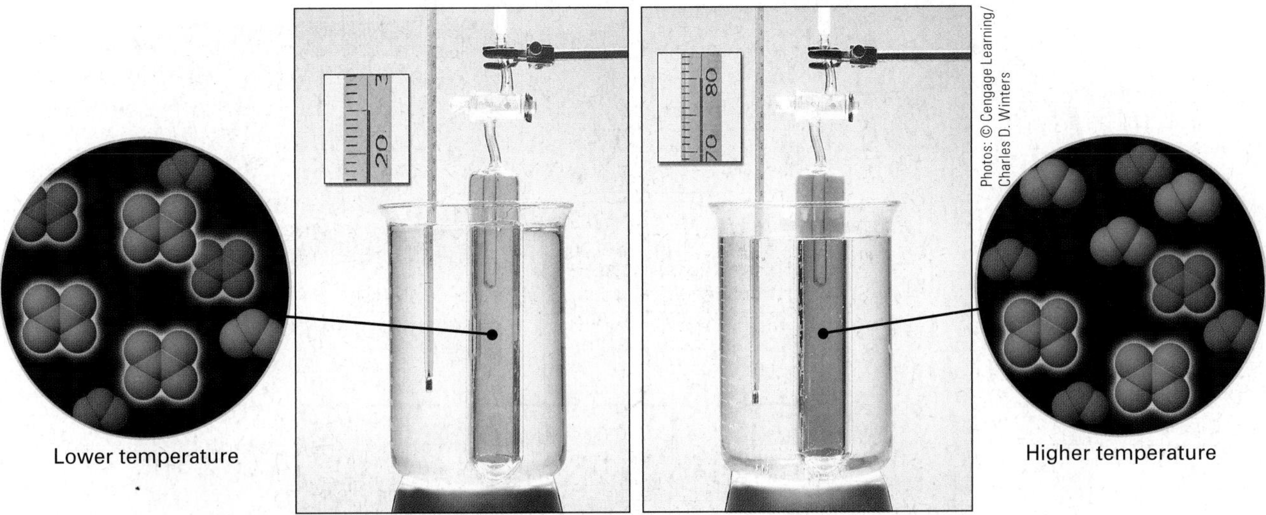


**Figure 18.12** The gas-phase equilibrium for the reaction $2\ NO_2(g) \rightleftharpoons N_2O_4(g) + Q$, where Q represents heat. The flasks contain the same total amount of gas. $NO_2$ is brown, while $N_2O_4$ is colorless. The left tube, at 25°C, contains very little brown gas, indicating that the equilibrium is shifted in the forward direction. The concentration of $N_2O_4$ molecules is relatively high. At higher temperature (right tube) the gas is much darker because brown $NO_2$ is formed as the reaction shifts in the reverse direction. The $N_2O_4$ concentration has decreased.

# Everyday Chemistry

## Fertilization of the World's Crops

Have you ever wondered how bread is made? Have you wondered how fruits and vegetables are grown? Have you ever wondered where taco shells come from? The ultimate source of all of these foods is farms. Each time you eat a plant product, unless it came from someone's backyard, there's a farmer responsible for planting and growing the crop that ultimately became your food.

The molecules that make up plants are composed mostly of carbon, hydrogen, and oxygen atoms. Plants obtain these atoms from chemical reactions in which the reactants are atmospheric carbon dioxide and water from the air or ground. A smaller, but still essential, component of the molecules is nitrogen. Chemistry plays a central role in providing a usable source of nitrogen for agriculture.

Since the atmosphere is about 78% nitrogen, it is logical to turn to the air for this element. Alas, the process is not that simple. Elemental nitrogen, $N_2$, has a very strong and stable triple bond between the two atoms. This makes it very unreactive. The challenge for chemists is to break the bond and put the nitrogen atoms into other substances that more easily release them to form molecules in plants.

Fritz Haber (1868–1934)

Farmers are well aware of natural sources of nitrogen fertilizers. Animal waste, composted household garbage and plant clippings, and blood meal (dried blood from slaughtered animals) have long been known as effective fertilizers. However, beginning in about 1900, the world's demand for fertilizer began to exceed the supply. Local natural sources, plus a large guano deposit in Chile, were no longer adequate to supply nitrogen-based fertilizer to meet the growing worldwide demand.

Fritz Haber, a German-born chemist, solved the problem. The Haber synthesis, as it is now called, is based on the equilibrium

$$N_2(g) + 3\,H_2(g) \rightleftharpoons 2\,NH_3(g) + 92\text{ kJ}$$

Nitrogen was obtained by distilling the air. Hydrogen was made by blowing steam over a bed of hot coke (mostly made of carbon), which produced hydrogen and carbon monoxide.

Haber's unique insight was to solve the problem of how to get the nitrogen-plus-hydrogen reaction to work. The key was finding the proper combination of temperature and pressure, as well as the right catalyst, to make the synthesis succeed. Haber had learned a great deal about energy relationships in chemical change, the combustion of hydrocarbons, oxidation–reduction reactions, and chemical equilibria before he began his quest to perform the ammonia synthe-

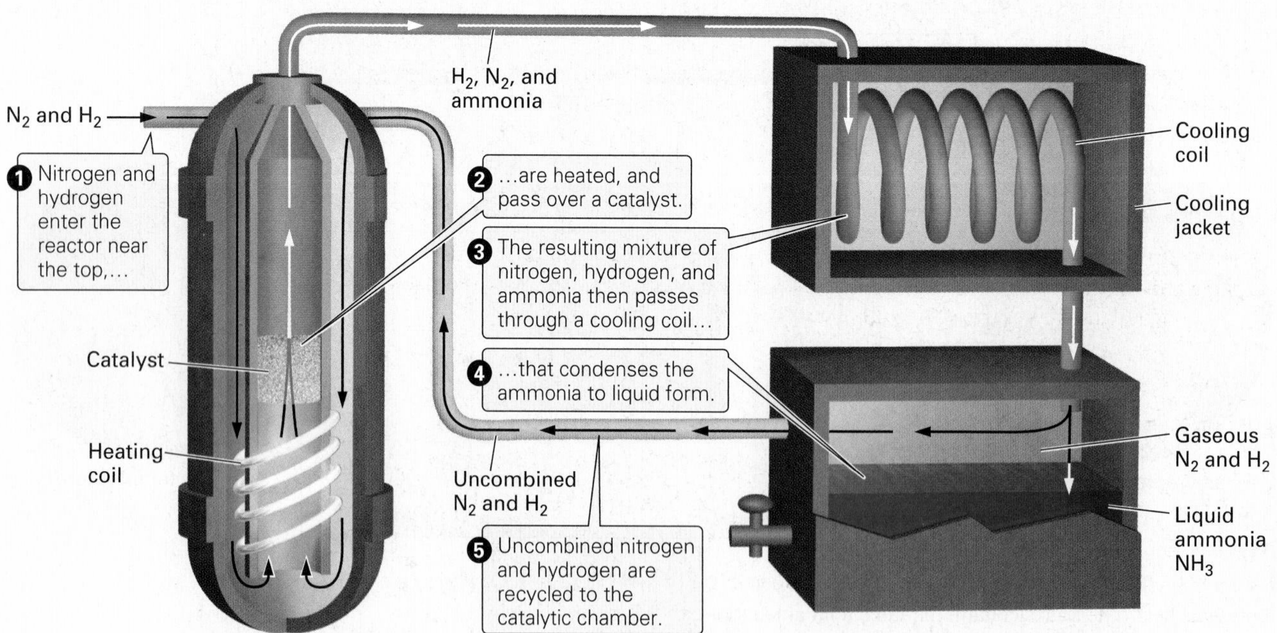


**Figure 18.13** Schematic diagram of the process for the industrial production of ammonia from nitrogen and hydrogen.

*Continued on facing page.*

sis. He then applied his copious knowledge and searched for the proper conditions through extensive experimentation, finally finding that an iron catalyst, a temperature of about 500°C, and a pressure of about 200 atm produced adequate yields under safe conditions (Fig. 18.13). Haber was awarded the 1918 Nobel Prize in Chemistry for his work.

An ammonia solution can be sprayed on soil directly as a relatively inexpensive fertilizer. The ammonia can also be further changed into ammonium sulfate, ammonium phosphate, or ammonium nitrate, which are other common fertilizers. Modern society would not be the same without Fritz Haber's contribution of a way to fertilize the world's crops.

## 18.7 The Equilibrium Constant

**Goal** **11** Given any chemical equilibrium equation, or information from which it can be written, write the equilibrium constant expression.

If 1.000 mole of $H_2(g)$ and 1.000 mole of $I_2(g)$ are introduced into a 1.000-liter reaction vessel, they will react to produce the following equilibrium:

$$H_2(g) + I_2(g) \rightleftharpoons 2\ HI(g)$$

At a temperature of 440°C, analysis at equilibrium will show hydrogen and iodine concentrations of 0.218 mol/L and a hydrogen iodide concentration of 1.564 mol/L.

Working in the opposite direction, if gaseous hydrogen iodide is introduced to a reaction chamber at 440°C at an initial concentration of 2.000 mol/L, it will decompose by the reverse reaction. The system will eventually come to equilibrium with $[H_2] = [I_2] = 0.218$ mol/L and $[HI] = 1.564$ mol/L, exactly the same equilibrium concentrations as in the first example. This illustrates the experimental fact that the position of an equilibrium when under the same conditions is the same, regardless of the direction from which it is approached.

As a third example, if the initial hydrogen iodide concentration is half as much, 1.000 mol/L, the equilibrium concentrations will be half what they are above: $[H_2] = [I_2] = 0.109$ mol/L and $[HI] = 0.782$ mol/L.

If the experiment is repeated at the same temperature, starting this time with hydrogen at 1.000 mol/L and iodine at 0.500 mol/L, the equilibrium concentrations will be $[H_2] = 0.532$ mol/L, $[I_2] = 0.032$ mol/L, and $[HI] = 0.936$ mol/L.

The data of these four equilibria are summarized in Table 18.1. Analysis of these data leads to the observation that, at equilibrium, the ratio $\frac{[HI]^2}{[H_2][I_2]}$ has a value of 51.5 for all four equilibria. For that matter, this ratio of equilibrium concentrations is always the same regardless of the initial concentrations of hydrogen, iodine, and hydrogen iodide, as long as the temperature is held at 440°C.

**Table 18.1** Equilibria of $H_2(g) + I_2(g) \rightleftharpoons 2\ HI(g)$ at 440°C

| Experiment Number | Initial | | | Equilibrium | | | |
|---|---|---|---|---|---|---|---|
| | $[H_2]$ | $[I_2]$ | $[HI]$ | $[H_2]$ | $[I_2]$ | $[HI]$ | $\frac{[HI]^2}{[H_2][I_2]}$ |
| 1 | 1.000 | 1.000 | 0 | 0.218 | 0.218 | 1.564 | 51.5 |
| 2 | 0 | 0 | 2.000 | 0.218 | 0.218 | 1.564 | 51.5 |
| 3 | 0 | 0 | 1.000 | 0.109 | 0.109 | 0.782 | 51.5 |
| 4 | 1.000 | 0.500 | 0 | 0.532 | 0.032 | 0.936 | 51.5 |

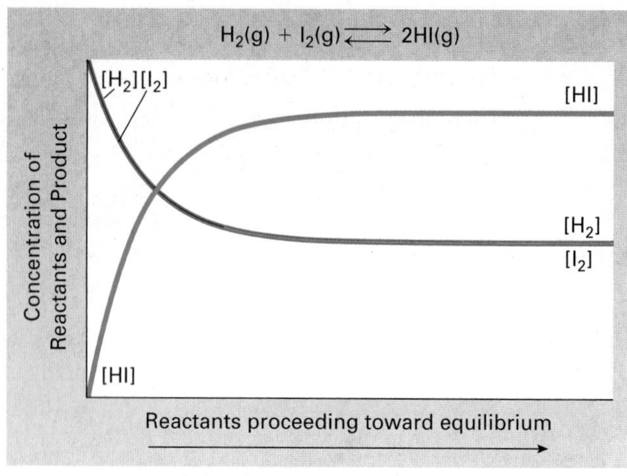


**Active Figure 18.14** The equilibrium constant. The ratio $[HI]^2/[H_2][I_2]$ is the same at equilibrium at any given temperature regardless of the initial concentrations of any species in the system. **Watch this active figure at *http://www.cengage.com/cracolice.***

Data from countless other equilibrium systems show a similar regularity that defines the **equilibrium constant, K: For any equilibrium at a given temperature, the ratio of the product of the concentrations of the species on the right side of the equilibrium equation, each raised to a power equal to its coefficient in the equation, to the corresponding product of the concentrations of the species on the left side of the equation, each raised to a power equal to its coefficient in the equation, is a constant.*** Thus, for $H_2(g) + I_2(g) \rightleftharpoons 2\ HI(g)$ at 440°C,

$$K = \frac{[HI]^2}{[H_2][I_2]} = 51.5$$

Active Figure 18.14 gives you an opportunity to explore the equilibrium constant concept as applied to the hydrogen–iodine–hydrogen iodide system.

The definition of K sets the procedure by which any equilibrium constant expression may be written. For the general equilibrium $a\ A + b\ B \rightleftharpoons c\ C + d\ D$ where A, B, C, and D are chemical formulas and a, b, c, and d are their coefficients in the equilibrium equation:

1. Write in the *numerator* the concentration of each species on the *right-hand* side of the equation. $K = \dfrac{[C][D]}{}$
2. For each species on the right-hand side of the equation, use its coefficient in the equation as an exponent. $K = \dfrac{[C]^c[D]^d}{}$
3. Write in the *denominator* the concentration of each species on the *left-hand* side of the equation. $K = \dfrac{[C]^c[D]^d}{[A][B]}$
4. For each species on the left-hand side of the equation, use its coefficient in the equation as an exponent. $K = \dfrac{[C]^c[D]^d}{[A]^a[B]^b}$

If the equation were written in reverse, $c\ C + d\ D \rightleftharpoons a\ A + b\ B$, the equilibrium constant would be the reciprocal of the preceding constant, $K = \dfrac{[A]^a[B]^b}{[C]^c[D]^d}$. Every equilibrium constant expression must be associated with a specific equilibrium equation.

*This development of the equilibrium constant expression can be duplicated with real laboratory data. Moreover, the same expression can be reached by a rigorous theoretical derivation. Theory and experiment support each other completely in this area.

## Active Example 18.8

Write equilibrium constant expressions for each of the following equilibria:

a) $2\ HI(g) \rightleftharpoons H_2(g) + I_2(g)$

b) $HI(g) \rightleftharpoons \frac{1}{2}\ H_2(g) + \frac{1}{2}\ I_2(g)$

c) $2\ Cl_2(g) + 2\ H_2O(g) \rightleftharpoons 4\ HCl(g) + O_2(g)$

Iodine gas has a blue-violet color.

a) $K = \dfrac{[H_2][I_2]}{[HI]^2}$

Notice that this is the reciprocal of the equilibrium constant developed in the text, when the equation was written $H_2(g) + I_2(g) \rightleftharpoons 2\ HI(g)$. Its numerical value at 440°C is 1/51.5, or 0.0194.

b) $K = \dfrac{[H_2]^{1/2}[I_2]^{1/2}}{[HI]}$

The value of this equilibrium constant is *not* the same as in (a), 0.0194. It is, in fact, the square root of 0.0194, or 0.139. This emphasizes why we must associate any equilibrium constant expression with a specific chemical equation.

c) $K = \dfrac{[HCl]^4[O_2]}{[Cl]^2[H_2O]^2}$

The procedure for writing the equilibrium constant expression is the same no matter how complex the equation may be.

So far all equilibrium constant expressions have been for equilibria in which all substances are gases. An equilibrium may also have solids, liquids, or dissolved substances as part of its equation. Solute concentrations are variable, and they appear in equilibrium constant expressions just like the concentrations of gases. If a liquid solvent or a solid is part of an equilibrium, however, its concentration is essentially constant. Its concentration is therefore omitted in the equilibrium constant expression. (We can, if you wish, say that its constant value is "included" in the value of K.) Remember:

*When writing an equilibrium constant expression, use only the concentrations of gases, (g), or dissolved substances, (aq). Do not include solids, (s), or liquids, (ℓ).*

In Section 17.8 this rule was applied to the ionization of water:

$$H_2O(\ell) \rightleftharpoons H^+(aq) + OH^-(aq) \qquad K_w = [H^+][OH^-]$$

$K_w$ has no denominator because the species on the left-hand side of the equilibrium equation is a liquid. This example also shows the common practice of using a subscript to identify a constant for a particular kind of equilibrium. We will describe other subscripts shortly.

## Active Example 18.9

Write the equilibrium constant expression for each of the following:

a) $CaCO_3(s) \rightleftharpoons CaO(s) + CO_2(g)$

b) $Li_2CO_3(s) \rightleftharpoons 2\ Li^+(aq) + CO_3^{2-}(aq)$

c) $4\ H_2O(g) + 3\ Fe(s) \rightleftharpoons 4\ H_2(g) + Fe_3O_4(s)$

d) $HF(aq) \rightleftharpoons H^+(aq) + F^-(aq)$

e) $NH_3(aq) + H_2O(\ell) \rightleftharpoons NH_4^+(aq) + OH^-(aq)$

a) $K = [CO_2]$ b) $K = [Li^+]^2\,[CO_3^{2-}]$

c) $K = \dfrac{[H_2]^4}{[H_2O]^4}$ d) $K = \dfrac{[H^+][F^-]}{[HF]}$ e) $K = \dfrac{[NH_4^+][OH^-]}{[NH_3]}$

## 18.8 The Significance of the Value of K

**Goal** **12** Given an equilibrium equation and the value of the equilibrium constant, identify the direction in which the equilibrium is favored.

By definition, an equilibrium constant is a ratio—a fraction. The numerical value of an equilibrium constant may be very large, very small, or anyplace in between. Even though there is no defined intermediate range, equilibria with constants between 0.01 and 100 ($10^{-2}$ to $10^2$) will have appreciable quantities of all species present at equilibrium.

To see what is meant by "very large" or "very small" K values, consider an equilibrium similar to the hydrogen iodide system studied in Section 18.7. If we substitute chlorine for iodine, the equilibrium equation is $H_2(g) + Cl_2(g) \rightleftharpoons 2\ HCl(g)$. At 25°C,

$$K = \frac{[HCl]^2}{[H_2][Cl_2]} = 2.4 \times 10^{33}$$

Chlorine gas has a greenish-yellow color.

This is a very large number—ten billion times larger than the number of particles in a mole! The only way an equilibrium constant ratio can become so huge is for the concentration of one or more reacting species to be very close to zero. If the denominator of a ratio is nearly zero, the value of the ratio will be very large. A near-zero denominator and large K mean the equilibrium is favored overwhelmingly in the forward direction.

By contrast, if the equilibrium constant is very small, it means the concentration of one or more of the species on the right-hand side of the equation is nearly zero. This puts a near-zero number in the numerator of K, and the equilibrium is strongly favored in the reverse direction.

### Thinking About Your Thinking

Equilibrium

**The concept discussed in Section 18.8 requires the equilibrium thinking skill. The value of the equilibrium constant, K, sets the *ratio* of product to reactant concentrations for an equilibrium system.**

summary

**The Significance of the Value of K**

If an equilibrium constant is very large (> 100), the forward reaction is favored; if the constant is very small (< 0.01), the reverse reaction is favored. If the constant is neither large nor small, appreciable quantities of all species are present at equilibrium.

**√ Target Check 18.5**

*For the following reactions, determine whether (1) the forward reaction is favored, (2) the reverse reaction is favored, or (3) appreciable quantities of all species are present at equilibrium:*

a) $HBr(aq) \rightleftharpoons H^+(aq) + Br^-(aq)$ $\quad K = 1 \times 10^9$

b) $Sn^{2+}(aq) + 4\ Cl^-(aq) \rightleftharpoons SnCl_4^{2-}(aq)$ $\quad K = 3.0 \times 10^1$

## 18.9 Equilibrium Calculations (Optional)

Equilibrium calculations cover a wide range of problem types. A thorough understanding of these calculations is essential to understanding many chemical phenomena in the laboratory, in industry, and in living organisms. We will sample only a few in this section.

You should write two things before attempting to solve any equilibrium problem. First is the equilibrium equation. Second is the equilibrium constant expression.

The numbers used in the equilibrium constant expression are concentrations in moles per liter. These concentrations are sometimes given, but sometimes you must figure them out from the available information. The figuring-out process uses the mole relationships expressed in the coefficients in an equation. For example, suppose you were to prepare two 500-mL (0.5-L) solutions. In the first, you dissolve 0.1 mole of NaCl; in the second you dissolve 0.1 mole of $CaCl_2$. The molarities of both solutions would be

$$M \equiv \frac{\text{mol}}{\text{L}} = \frac{0.1\ \text{mol}}{0.5\ \text{L}} = 0.2\ \text{mol/L}$$

But the concentrations used in equilibrium problems are usually ion concentrations. What are the $Na^+$, $Ca^{2+}$, and $Cl^-$ concentrations in the two solutions? To answer these questions, we must examine the dissolving equations:

$$NaCl(s) \rightarrow Na^+(aq) + Cl^-(aq) \quad \text{and} \quad CaCl_2(s) \rightarrow Ca^{2+}(aq) + 2\ Cl^-(aq)$$

The NaCl equation shows that one mole of NaCl yields one mole of $Na^+$ ion and one mole of $Cl^-$ ion. Therefore, the ion concentrations are the same as the solute concentration:

$$\frac{0.2\ \cancel{\text{mol NaCl}}}{\text{L}} \times \frac{1\ \text{mol Na}^+}{1\ \cancel{\text{mol NaCl}}} = 0.2\ \text{mol Na}^+/\text{L} \quad \text{and} \quad \frac{0.2\ \cancel{\text{mol NaCl}}}{\text{L}} \times \frac{1\ \text{mol Cl}^-}{1\ \cancel{\text{mol NaCl}}} = 0.2\ \text{mol Cl}^-/\text{L}$$

The equation for dissolving $CaCl_2$ shows that one mole of $CaCl_2$ yields one mole of $Ca^{2+}$ and *two* moles of $Cl^-$. Therefore the chloride ion concentration should be *twice* as large as the calcium ion concentration:

$$\frac{0.2\ \cancel{\text{mol CaCl}_2}}{\text{L}} \times \frac{1\ \text{mol Ca}^{2+}}{1\ \cancel{\text{mol CaCl}_2}} = 0.2\ \text{mol Ca}^{2+}/\text{L} \quad \text{and} \quad \frac{0.2\ \cancel{\text{mol CaCl}_2}}{\text{L}} \times \frac{2\ \text{mol Cl}^-}{1\ \cancel{\text{mol CaCl}_2}} = 0.4\ \text{mol Cl}^-/\text{L}$$

In the Active Examples that follow, we will use "twice as large" and similar reasoning without showing calculation setups.

Let's also look at the units of equilibrium constants. For reasons beyond the scope of this course, a rigorous treatment of equilibrium constant calculations results in an equilibrium constant with no units. In practice, determining the value of the equilibrium constant from solution concentrations in moles per liter or from gas partial pressures simplifies the calculations and generally results in an insignificant error for most situations. Therefore, we omit units on K values.

Now you are ready to solve some equilibrium constant problems.

## Solubility Equilibria

Goal **13** Given the solubility product constant or the solubility of a slightly soluble compound (or data from which the solubility can be found), calculate the other value.

Goal **14** Given the solubility product constant of a slightly soluble compound and the concentration of a solution having a common ion, calculate the solubility of the slightly soluble compound in the solution.

In Section 9.7 you used solubility Table 9.3 and the solubility guidelines in Active Figure 9.13 to predict whether or not a precipitate would form when ionic solutions are combined. The footnote to Table 9.3 said that, for purposes of writing net ionic equations, a compound that did not dissolve to a concentration of 0.1 mole per liter of water would be considered insoluble. However, no ionic compound is completely insoluble. It is appropriate, then, to refer to "low-solubility solids" rather than insoluble compounds.

The equilibrium equation for dissolving a low-solubility compound is very similar to the equation for the ionization of water. (See the discussion in Section 17.8.) For silver chloride, for example, the equilibrium and K equations are

$$AgCl(s) \rightleftharpoons Ag^+(aq) + Cl^-(aq) \qquad K_{sp} = [Ag^+][Cl^-]$$

The rules for writing an equilibrium constant expression from Section 18.7 indicate that solids and liquids are not included.

The equilibrium constant for a low-solubility compound is the **solubility product constant, $K_{sp}$.** A $K_{sp}$ expression has no denominator because the only species on the left side of the equation is a solid.

### Active Example 18.10

The chloride ion concentration of a saturated solution of silver chloride is $1.3 \times 10^{-5}$ M. Calculate $K_{sp}$ for silver chloride.

The stalagmites and stalactites found in many caves form as dissolved minerals precipitate from solution. These minerals are low-solubility compounds.

*Plan* this problem.

*Given:* $[Cl^-] = 1.3 \times 10^{-5}$ M  *Wanted:* $K_{sp}$

*Equations:* $AgCl(s) \rightleftharpoons Ag^+(aq) + Cl^-(aq)$; $K_{sp} = [Ag^+][Cl^-]$

As noted earlier, the equilibrium equation and the equilibrium constant expression are needed for all equilibrium problems.

One of the two concentrations you need to calculate $K_{sp}$ is given. But you need $[Ag^+]$, too. What is it? (*Hint:* Think about the NaCl discussion at the beginning of Section 18.9.)

$[Ag^+] = 1.3 \times 10^{-5}$ M

The equilibrium equation shows that equal numbers of moles of silver and chloride ions are released when silver chloride dissolves. Their concentrations are therefore equal.

You have the equation, and you have the numbers to use. Complete the problem.

$$K_{sp} = [Ag^+][Cl^-] = (1.3 \times 10^{-5})(1.3 \times 10^{-5}) = (1.3 \times 10^{-5})^2 = 1.7 \times 10^{-10}$$

## Active Example 18.11

The solubility of magnesium fluoride is 73 mg/L. What is its $K_{sp}$?

This time the solution concentration is given in mg/L. It must be converted to molarity, mol/L or mmol/mL. Take your choice. Our solution *Plan* will be in mol/L, but we'll show both calculation setups.

*continued*

*Given:* 73 mg $MgF_2$; 1.0 L *Wanted:* M (mol/L)

*Per:* $\frac{1\ g}{1000\ mg}$ $\frac{1\ mol\ MgF_2}{62.31\ g\ MgF_2}$

*Path:* mg $MgF_2$ ⟶ g $MgF_2$ ⟶ mol $MgF_2$

$$73\ \cancel{mg\ MgF_2} \times \frac{1\ g}{1000\ \cancel{mg}} \times \frac{1\ mol\ MgF_2}{62.31\ \cancel{g\ MgF_2}} = 1.2 \times 10^{-3}\ mol\ MgF_2$$

*Equation:* $M \equiv \frac{mol}{L} = \frac{1.2 \times 10^{-3}\ mol\ MgF_2}{1.0\ L} = 1.2 \times 10^{-3}\ M\ MgF_2$

In milliunits,

*Equation:* $M \equiv \frac{mmol}{mL} = \frac{73\ \cancel{mg\ MgF_2}}{1000\ mL} = \frac{1\ mmol\ MgF_2}{62.31\ \cancel{mg\ MgF_2}} = 1.2 \times 10^{-3}\ M\ MgF_2$

Now you know the molarity of the solution in moles of $MgF_2$ per liter. But you need the ion concentrations. Think back to the examination of the calcium chloride solution at the beginning of Section 18.9. How were those ion concentrations determined? What are both the magnesium and fluoride ion concentrations in $1.2 \times 10^{-3}$ M $MgF_2$? Before answering, write the equilibrium equation for $MgF_2$ and the $K_{sp}$ expression equation. They will help you see where you're going.

$MgF_2(s) \rightleftharpoons Mg^{2+}(aq) + 2\ F^-(aq)$ $K_{sp} = [Mg^{2+}][F^-]^2$
$[Mg^{2+}] = 1.2 \times 10^{-3}$ M
$[F^-] = 2 \times [Mg^{2+}] = 2 \times 1.2 \times 10^{-3}\ M = 2.4 \times 10^{-3}$ M

According to the equilibrium equation, one mole of $Mg^{2+}$ is produced for every mole of $MgF_2$ that dissolves. The molarity of $Mg^{2+}$ is therefore equal to the molarity of the solution. The equation also shows that twice as many fluoride ions are produced as magnesium ions. It follows that the molarity of $F^-$ is twice the molarity of $Mg^{2+}$.

You now have both ion concentrations. Substitute into the $K_{sp}$ equation and calculate the answer.

*Equation:* $K_{sp} = [Mg^{2+}][F^-]^2 = (1.2 \times 10^{-3})(2.4 \times 10^{-3})^2 = 6.9 \times 10^{-9}$

Solubility product constants have already been determined for most common salts. Their values may be found in handbooks. They are used in several kinds of problems, one of which is the reverse of the last two Active Examples.

## Active Example 18.12

Calculate the solubility of zinc carbonate in (a) mol/L and (b) g/100 mL. $K_{sp} = 1.4 \times 10^{-11}$ for $ZnCO_3$.

Begin with the equilibrium and $K_{sp}$ equations.

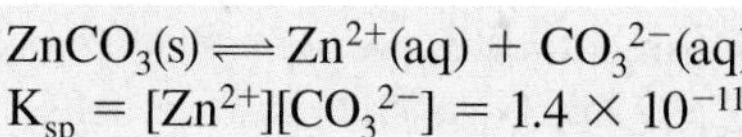

$ZnCO_3(s) \rightleftharpoons Zn^{2+}(aq) + CO_3^{2-}(aq)$
$K_{sp} = [Zn^{2+}][CO_3^{2-}] = 1.4 \times 10^{-11}$

If $ZnCO_3$ is the only source of both $Zn^{2+}$ and $CO_3^{2-}$ ions, what can you say about their concentrations?

$[Zn^{2+}] = [CO_3^{2-}]$

The equation shows that equal numbers of moles of $Zn^{2+}$ and $CO_3^{2-}$ are formed when $ZnCO_3$ dissolves. Therefore, their concentrations must be equal.

Algebra is applied to find the solubility of $ZnCO_3$. Let the letter s represent $[Zn^{2+}]$ at equilibrium: $s = [Zn^{2+}]$. Because $[Zn^{2+}] = [CO_3^{2-}]$, s is also equal to $[CO_3^{2-}]$. Now substitute s for the two ion concentrations in the $K_{sp}$ equation and calculate its value. (This will require using a procedure on your calculator that you have not performed in earlier problems in this book. The procedure is described in Appendix I.A.)

$K_{sp} = [Zn^{2+}][CO_3^{2-}] = s \times s = s^2 = 1.4 \times 10^{-11}$
$s = \sqrt{1.4 \times 10^{-11}} = 3.7 \times 10^{-6}\ M = [Zn^{2+}] = [CO_3^{2-}]$

On most calculators you find the square root of a number by entering the number and pressing the $\sqrt{\ }$ key.

You now know the solubility of $ZnCO_3$ in moles per liter, which answers part (a). To find it in grams per 100 mL, as requested in part (b), you must ask yourself, how many grams of $ZnCO_3$ are in 100 mL of $3.7 \times 10^{-6}$ M $ZnCO_3$? *PLAN*, set up, and solve the problem.

*GIVEN:* 100 mL  *WANTED:* g $ZnCO_3$

*PER:* $\frac{1\ L}{1000\ mL}$  $\frac{3.7 \times 10^{-6}\ mol\ ZnCO_3}{L}$  $\frac{125.39\ g\ ZnCO_3}{mol\ ZnCO_3}$

*PATH:* mL ⟶ L ⟶ mol $ZnCO_3$ ⟶ g $ZnCO_3$

$$100\ \cancel{mL} \times \frac{1\ \cancel{L}}{1000\ \cancel{mL}} \times \frac{3.7 \times 10^{-6}\ \cancel{mol\ ZnCO_3}}{L} \times \frac{125.39\ g\ ZnCO_3}{\cancel{mol\ ZnCO_3}} = 4.6 \times 10^{-5}\ g\ ZnCO_3$$

The solubility of zinc carbonate is $3.7 \times 10^{-6}$ mol/L or $4.6 \times 10^{-5}$ g/100 mL.

Suppose a soluble carbonate, such as $Na_2CO_3$, is added to the saturated solution of zinc carbonate in Active Example 18.12. What happens to the solubility of $ZnCO_3$? What does Le Chatelier's Principle predict? $[CO_3^{2-}]$ would no longer be equal to $[Zn^{2+}]$ because the $CO_3^{2-}$ ion would be coming from two sources. According to Le Chatelier's Principle, the equilibrium should shift in the direction that would use up some of the added $CO_3^{2-}$. That is the reverse direction. Less $ZnCO_3$ would dissolve; its solubility would be reduced. Let's see if that prediction is confirmed by calculation.

## Active Example 18.13

Calculate the solubility of $ZnCO_3$ in 0.010 M $Na_2CO_3$. Answer in moles per liter.

The product of the concentrations of the zinc and carbonate ions is constant, $1.4 \times 10^{-11}$, whether the ions come from the same source or different sources. Assuming that the $Na_2CO_3$ is completely dissolved, what is $[CO_3^{2-}]$ in 0.010 M $Na_2CO_3$? (*Hint:* Look back on the $Ca^{2+}$ ion discussion at the beginning of Section 18.9.)

$[CO_3^{2-}] = 0.010$ M in 0.010 M $Na_2CO_3$

In $Na_2CO_3(s) \rightleftharpoons 2\ Na^+(aq) + CO_3^{2-}(aq)$, the number of moles of carbonate ion in solution is the same as the number of moles of sodium carbonate dissolved.

Let's look again at the equilibrium and solubility product equations from Active Example 18.12:

$$ZnCO_3(s) \rightleftharpoons Zn^{2+}(aq) + CO_3^{2-}(aq) \qquad K_{sp} = [Zn^{2+}][CO_3^{2-}] = 1.4 \times 10^{-11}$$

$Zn^{2+}$ ion comes only from $ZnCO_3$. $[Zn^{2+}]$ is therefore the same as the solubility of zinc carbonate. You already have values for $K_{sp}$ and $[CO_3^{2-}]$. The zinc ion concentration is the only unknown in the solubility product equation. Go for it!

Notice that we did not include in $[CO_3^{2-}]$ any carbonate ion that came from the small amount of $ZnCO_3$ that dissolved. In water, that concentration was only 0.0000037 mol/L, and Le Chatelier's Principle predicted an even smaller amount in the $Na_2CO_3$ solution. By rules of significant figures, carbonate from $ZnCO_3$ is negligible when added to carbonate from $Na_2CO_3$: 0.010 + less than 0.0000037 = 0.010.

GIVEN: $K_{sp} = 1.4 \times 10^{-11}$; $[CO_3^{2-}] = 0.010$ M  WANTED: $[Zn^{2+}]$

$$[Zn^{2+}] = \frac{K_{sp}}{[CO_3^{2-}]} = \frac{1.4 \times 10^{-11}}{0.010} = 1.4 \times 10^{-9}\ M$$

Solving the solubility product equation for $[Zn^{2+}]$ and substituting known values of $K_{sp}$ and $[CO_3^{2-}]$ yields a zinc carbonate solubility that is, indeed, smaller than its solubility in water: $3.7 \times 10^{-6}$ M, from Active Example 18.12. Le Chatelier's Principle is confirmed.

Reducing the concentration of an ion or the solubility of a compound by adding an ion that is already present is an example of the **common ion effect.**

## Ionization Equilibria

NOTE: Calculations in this subsection include converting between noninteger pH values and $[H^+]$ as described in optional Section 17.10. If you have not yet studied Section 17.10, we recommend that you skip to the next subsection, "Gaseous Equilibria."

**Goal 15** Given the formula of a weak acid, HA, write the equilibrium equation for its ionization and the expression for its acid constant, $K_a$.

**Goal 16** Given any two of the following three values for a weak acid, HA, calculate the third: (a) the initial concentration of the acid; (b) the pH of the solution, the percentage dissociation of the acid, or $[H^+]$ or $[A^-]$ at equilibrium; (c) $K_a$ for the acid.

**Goal 17** For a weak acid, HA, given $K_a$, $[H^+]$, and $[A^-]$, or information from which they may be obtained, calculate the pH of the buffer produced.

**Goal 18** Given $K_a$ for a weak acid, HA, determine the ratio between [HA] and $[A^-]$ that will produce a buffer of specified pH.

In Section 9.3 you learned that weak acids ionize only slightly when dissolved in water. If HA is the formula of a weak acid, its ionization equation and equilibrium constant expression are

$$HA(aq) \rightleftharpoons H^+(aq) + A^-(aq) \qquad K_a = \frac{[H^+][A^-]}{[HA]}$$

The equilibrium constant is the **acid constant, $K_a$**. The undissociated molecule is the major species in the solution and the $H^+$ ion and the conjugate base of the acid, $A^-$, are the minor species.

The ionization of a weak acid is usually so small that it is negligible compared with the initial concentration of the acid. For example, if a 0.12 M acid is 3.0% ionized, the amount ionized is $0.030 \times 0.12 = 0.0036$ mol/L. When this is subtracted from the initial concentration and rounded off according to the rules of significant figures, the result is $0.12 - 0.0036 = 0.1164 = 0.12$. (This is similar to the negligible addition noted in the margin at the end of Active Example 18.13.)

In more advanced courses you will learn how to determine whether or not the ionization has a negligible effect on the initial concentration. In this book we assume that all ionization concentrations are negligible *when subtracted from the initial concentration*. The ion concentrations by themselves, however, are not negligible.

If the ionization of HA is the only source of $H^+$ and $A^-$ in the equilibrium $HA(aq) \rightleftharpoons H^+(aq) + A^-(aq)$, then $[H^+] = [A^-]$. (This is the same as $[Ag^+]$ being equal to $[Cl^-]$ in Active Example 18.10.) This makes it possible to calculate the percentage ionization and $K_a$ from the pH of a weak acid whose molarity has been determined by titration.

**P/REVIEW:** The major species in a solution are those that are present in higher concentrations, and the minor species have relatively low concentrations (Section 9.3). The conjugate base of an acid is the species that remains after the acid has lost its proton, $H^+$ (Section 17.4).

$[H^+] = 10^{-pH}$. Converting between noninteger pH values and $[H^+]$ is discussed in Section 17.10. Briefly, one calculator procedure for changing pH to $[H^+]$ is as follows:

1. Enter the pH.
2. Change its sign to minus.
3. Press $10^x$ or INV log.

Another procedure is to raise 10 to the −pH power. If neither procedure works with your calculator, see Section 17.10, Appendix I.A, or your calculator instruction book.

### Active Example 18.14

A 0.13 M solution of an unknown acid has a pH of 3.12. Calculate the percentage ionization and $K_a$.

The acid constant, $K_a = \frac{[HA^+][A^-]}{[HA]}$, gives you the equation for $K_a$. Complete the *PLAN* for the problem.

*GIVEN:* [HA] = 0.13 M; pH = 3.12 *WANTED:* $K_a$ and % ionization

Begin by finding the value of $[H^+]$ you need for the $K_a$ equation.

*continued*

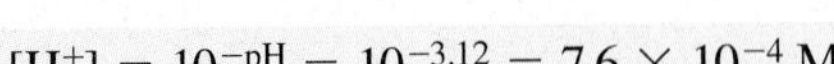

$[H^+] = 10^{-pH} = 10^{-3.12} = 7.6 \times 10^{-4}$ M

What else that you need has a value of $7.6 \times 10^{-4}$ M?

$[A^-] = 7.6 \times 10^{-4}$ M

Because the ionization of HA is the only source for both $H^+$ and $A^-$, the ionization equation tells us that their concentrations are equal.

You now have all the numbers for calculating $K_a$.

**P/REVIEW:** In Section 7.6, Equation 7.2 gives the general formula for calculating the percentage of any part, A, in the total of all parts:

$$\% \text{ of A} = \frac{\text{Parts of A}}{\text{total parts}} \times 100$$

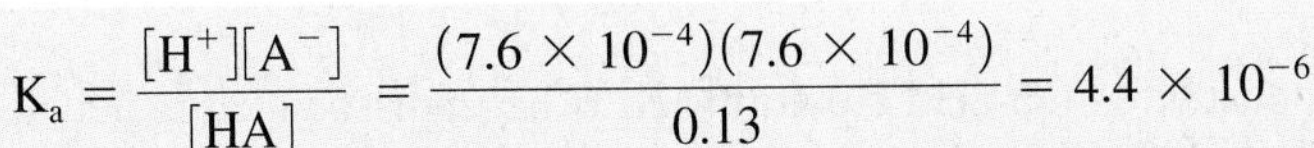

$$K_a = \frac{[H^+][A^-]}{[HA]} = \frac{(7.6 \times 10^{-4})(7.6 \times 10^{-4})}{0.13} = 4.4 \times 10^{-6}$$

Now you can calculate the percentage ionization. Of the 0.13 mole of HA initially in one liter of solution, only $7.6 \times 10^{-4}$ mol ionized. What percentage of 0.13 is $7.6 \times 10^{-4}$?

Measuring the pH of cola. Note that the solution is quite acidic. Nonetheless, the weak acid in the solution is only slightly ionized.

$$\frac{7.6 \times 10^{-4}}{0.13} \times 100 = 0.58\% \text{ ionized}$$

Just as the $K_{sp}$ values of low-solubility solids are listed in handbooks, so are the $K_a$ values of most weak acids. They can be used to determine $[H^+]$ and the pH of solutions of those acids. If the ionization of the acid is the only source of $H^+$ and $A^-$ ions, we can start with the acid constant equation, multiply both sides by [HA], and substitute $[H^+]$ for its equal $[A^-]$:

$$K_a = \frac{[H^+][A^-]}{[HA]} \xrightarrow{\text{Multiply both sides by [HA]}} K_a[HA] = [H^+][A^-] \xrightarrow{\text{Substitute } [H^+] \text{ for } [A^-]}$$

$$K_a[HA] = [H^+]^2 \xrightarrow{\text{Take the square root of each side}} [H^+] = \sqrt{K_a[HA]}$$

## Active Example 18.15

What is the pH of a 0.20 molar solution of the acid in Active Example 18.14, for which $K_a = 4.4 \times 10^{-6}$?

$[H^+]$ may be found by direct substitution into $[H^+] = \sqrt{K_a[HA]}$. Additionally, you will use pH = $-\log [H^+]$. *PLAN* and solve the problem.

*GIVEN:* $K_a = 4.4 \times 10^{-6}$; [HA] = 0.20 M *WANTED:* pH

*EQUATION:* $[H^+] = \sqrt{K_a[HA]} = \sqrt{(4.4 \times 10^{-6})(0.20)} = 9.4 \times 10^{-4}$ M

*EQUATION:* $pH = -\log [H^+] = -\log (9.4 \times 10^{-4}) = 3.03$

Just as we can force solubility equilibria in the reverse direction by adding a common ion, so can we force a weak acid equilibrium by adding a soluble salt of the acid. If we add $A^-$ to the equilibrium in $HA(aq) \rightleftharpoons H^+(aq) + A^-(aq)$, $[A^-]$ increases and the equilibrium is shifted to the left, reducing $[H^+]$. To find the pH of such a solution, we solve $K_a = \frac{[H^+][A^-]}{[HA]}$ for $[H^+]$:

$$[H^+] = K_a \times \frac{[HA]}{[A^-]}$$

Neither [HA] nor $[A^-]$ is changed significantly by the ionization of HA, which is even smaller than its ionization in water.

## Active Example 18.16

Find the pH of a 0.20 M solution of the acid in Active Examples 18.14 and 18.15 ($K_a = 4.4 \times 10^{-6}$) if the solution is also 0.15 M in $A^-$.

Direct substitution into $[H^+] = K_a \times \frac{[HA]}{[A^-]}$ gives $[H^+]$, and pH follows. Complete the problem.

$$HA(aq) \rightleftharpoons H^+(aq) + A^-(aq) \qquad K_a = \frac{[H^+][A^-]}{[HA]}$$

*GIVEN:* $K_a = 4.4 \times 10^{-6}$; [HA] = 0.20 M; $[A^-]$ = 0.15 M *WANTED:* pH

*EQUATION:* $[H^+] = K_a \times \frac{[HA]}{[A^-]} = 4.4 \times 10^{-6} \times \frac{0.20}{0.15} = 5.9 \times 10^{-6}$ M

*EQUATION:* $pH = -\log [H^+] = -\log (5.9 \times 10^{-6}) = 5.23$

The solution in Active Example 18.16 is a **buffer solution,** or, more simply, a **buffer.** A buffer is a solution that resists changes in pH because it contains relatively high concentrations of both a weak acid and a weak base. The acid is able to consume any

$OH^-$ that may be added, and the base can absorb $H^+$, both without significant change in either [HA] or [$A^-$]. For example, if 0.001 mol of HCl was dissolved in one liter of water, the pH would be 3. If the same amount of HCl was added to a liter of the buffer in Active Example 18.16, it would react with 0.001 mol of the $A^-$ present. The new concentration of $A^-$ would be $0.15 - 0.001 = 0.15$, unchanged according to the rules of significant figures. An additional 0.001 mol of HA would be formed in the reaction, but that added to 0.20 is still 0.20. In other words, the [HA]/[$A^-$] ratio would be unchanged, so the [$H^+$] and pH would also be unchanged.

This all suggests that a buffer can be tailor-made for any pH simply by adjusting the [HA]/[$A^-$] ratio to the proper value. Solving $K_a = \frac{[H^+][A^-]}{[HA]}$ for that ratio gives

$$K_a = \frac{[H^+][A^-]}{[HA]} \xrightarrow{\text{Divide both sides by } [H^+]} \frac{K_a}{[H^+]} = \frac{[A^-]}{[HA]} \xrightarrow{\text{Take the inverse of both sides}} \frac{[HA]}{[A^-]} = \frac{[H^+]}{K_a}$$

## Active Example 18.17

What [HA]/[$A^-$] ratio is necessary to produce a buffer with a pH of 5.00 if $K_a = 4.4 \times 10^{-6}$?

Convert the pH to [$H^+$], then plug that value and $K_a$ into $\frac{[HA]}{[A^-]} = \frac{[H^+]}{K_a}$ to get the answer. You don't even need a calculator for the pH → [$H^+$] conversion.

$$\frac{[HA]}{[A^-]} = \frac{[H^+]}{K_a} = \frac{10^{-5.00}}{4.4 \times 10^{-6}} = 2.3$$

If pH is 5.00, $[H^+] = 10^{-pH} = 10^{-5.00}$ M. This could be written simply as $10^{-5}$, but the two zeros are held to show that all numbers are to two significant figures.

## Gaseous Equilibria

**Goal** | **19** Given equilibrium concentrations of species in a gas-phase equilibrium, or information from which they can be found, and the equation for the equilibrium, calculate the equilibrium constant.

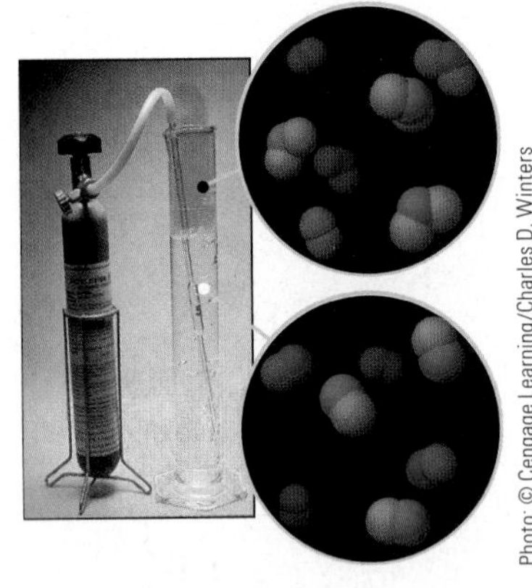

NO reacts with $O_2$ to form $NO_2$. Gaseous NO from the cylinder is bubbled through water so that you can see that it is colorless. When the NO bubbles meet the colorless $O_2$ in the air above the surface of the water, NO and $O_2$ react to form dark reddish-brown $NO_2$.



All the equilibria considered thus far in this section have been in aqueous solution. When an equilibrium involves only gases, the calculation principles and stoichiometric reasoning are the same as in solution equilibria. However, the changes in starting concentrations are not negligible. It often helps to trace these changes by assembling them into a table. The columns are headed by the species in the equilibrium just as they appear in the reaction equation. The three lines give the initial concentration of each substance, the change in the concentration as the system reaches equilibrium, and the equilibrium concentration.

## Active Example 18.18

A student places 0.052 mole of NO and 0.054 mole of $O_2$ in a 1.00-L vessel at a certain temperature. They react until equilibrium is reached according to the equation $2\,NO(g) + O_2(g) \rightleftharpoons 2\,NO_2(g)$. At equilibrium, $[NO_2] = 0.028$ M. Calculate K.

From the equilibrium equation, $K = \frac{[NO_2]^2}{[NO]^2\,[O_2]}$. We have set up a table.

Begin by inserting all given data. Assume that the initial concentration of $NO_2$ is zero, as implied but not directly specified by the problem statement. You'll be able to make four entries into the table.

| | $2\,NO(g)$ | + | $O_2(g)$ | $\rightleftharpoons$ | $2\,NO_2(g)$ |
|---|---|---|---|---|---|
| mol/L at start | | | | | |
| mol/L change, + or − | | | | | |
| mol/L at equilibrium | | | | | |

| | $2\,NO(g)$ | + | $O_2(g)$ | $\rightleftharpoons$ | $2\,NO_2(g)$ |
|---|---|---|---|---|---|
| mol/L at start | 0.052 | | 0.054 | | 0.000 |
| mol/L change, + or − | | | | | |
| mol/L at equilibrium | | | | | 0.028 |

To get any entry in the change (middle) line, find a species whose initial and final concentrations are known. In this case, $[NO_2]$ starts at 0.000 M and reaches 0.028 M. Its change is therefore +0.028 M. Insert that value into the table above.

| | $2\,NO(g)$ | + | $O_2(g)$ | $\rightleftharpoons$ | $2\,NO_2(g)$ |
|---|---|---|---|---|---|
| mol/L at start | 0.052 | | 0.054 | | 0.000 |
| mol/L change, + or − | | | | | +0.028 |
| mol/L at equilibrium | | | | | 0.028 |

You can use stoichiometry to find the changes in the other two species. The coefficients in the equation show that the reaction that produces 0.028 mol $NO_2$ uses 0.028 mol NO (both coefficients are the same, 2) and half as much $O_2$ (its coefficient is half of the others, 1), or 0.014 mol $O_2$. Add these values to the table above.

| | $2\,NO(g)$ | + | $O_2(g)$ | $\rightleftharpoons$ | $2\,NO_2(g)$ |
|---|---|---|---|---|---|
| mol/L at start | 0.052 | | 0.054 | | 0.000 |
| mol/L change, + or − | −0.028 | | −0.014 | | +0.028 |
| mol/L at equilibrium | | | | | 0.028 |

The final concentrations of the reactants are found by subtracting the amounts used from the starting concentrations—or adding them algebraically, as they appear in the table. Complete the table above.

| | $2\,NO(g)$ | + | $O_2(g)$ | $\rightleftharpoons$ | $2\,NO_2(g)$ |
|---|---|---|---|---|---|
| mol/L at start | 0.052 | | 0.054 | | 0.000 |
| mol/L change, + or − | −0.028 | | −0.014 | | +0.028 |
| mol/L at equilibrium | 0.024 | | 0.040 | | 0.028 |

The equilibrium constant, K, may now be calculated by substituting the equilibrium concentrations from the third line of the table into the equilibrium constant expression.

*continued*

$$K = \frac{[NO_2]^2}{[NO]^2\,[O_2]} = \frac{(0.028)^2}{(0.024)^2\,(0.040)} = 34$$

## Active Example 18.19

A researcher injects 2.00 mole of hydrogen iodide gas into a 1.00-L reaction vessel. It decomposes according to the equation $2\ HI(g) \rightleftharpoons H_2(g) + I_2(g)$. At equilibrium, 1.60 mole of HI remains. Calculate the value of the equilibrium constant at the temperature of this experiment.

Complete as much of the following table as possible, based on the given data.

| | $2\ HI(g)$ | $\rightleftharpoons$ | $H_2(g)$ | + | $I_2(g)$ |
|---|---|---|---|---|---|
| mol/L at start | | | | | |
| mol/L change, + or − | | | | | |
| mol/L at equilibrium | | | | | |

| | $2\ HI(g)$ | $\rightleftharpoons$ | $H_2(g)$ | + | $I_2(g)$ |
|---|---|---|---|---|---|
| mol/L at start | 2.00 | | 0 | | 0 |
| mol/L change, + or − | | | | | |
| mol/L at equilibrium | 1.60 | | | | |

The initial concentration minus the change must be equal to the equilibrium concentration. This fact will allow you to determine the mol/L change for HI(g). You can then use the reaction stoichiometry to complete the mol/L change line for hydrogen and iodine. Insert those three values into the table above.

| | $2\ HI(g)$ | $\rightleftharpoons$ | $H_2(g)$ | + | $I_2(g)$ |
|---|---|---|---|---|---|
| mol/L at start | 2.00 | | 0 | | 0 |
| mol/L change, + or − | −0.40 | | +0.20 | | +0.20 |
| mol/L at equilibrium | 1.60 | | | | |

The +0.20 mol/L change for $H_2(g)$ and $I_2(g)$ comes from the coefficients in the balanced chemical equation:

$$0.40\ \text{mol/L change for HI} \times \frac{1\ \text{mol}\ H_2(g)\ \text{or}\ I_2(g)}{2\ \text{mol}\ HI(g)} = 0.20\ \text{mol/L change for}\ H_2(g)\ \text{and}\ I_2(g)$$

Complete the problem by adding the mol/L at equilibrium for the product species. Once you've determined the equilibrium concentrations, you can write the K expression, substitute the equilibrium concentrations, and solve for the value of K.

| | $2 HI(g)$ | $\rightleftharpoons$ | $H_2(g)$ | + | $I_2(g)$ |
|---|---|---|---|---|---|
| mol/L at start | 2.00 | | 0 | | 0 |
| mol/L change, + or − | −0.40 | | +0.20 | | +0.20 |
| mol/L at equilibrium | 1.60 | | 0.20 | | 0.20 |

$$K = \frac{[H_2][I_2]}{[HI]^2} = \frac{(0.20)(0.20)}{(1.60)^2} = 0.016$$

# Chapter 18 in Review

Most of the key terms and concepts and many others appear in the Glossary. Use your Glossary regularly.

**18.1 The Character of an Equilibrium**

Goal 1 Identify a chemical equilibrium by the conditions it satisfies.

**Key Terms and Concepts: Closed system, dynamic equilibrium, static equilibrium**

**18.2 The Collision Theory of Chemical Reactions**

Goal 2 Distinguish between reaction-producing molecular collisions and molecular collisions that do not yield reactions.

**Key Terms and Concepts: Collision theory of chemical reactions**

**18.3 Energy Changes During a Molecular Collision**

Goal 3 Sketch and/or interpret an energy-reaction coordinate graph. Identify the (a) activated complex region, (b) activation energy, and (c) $\Delta E$ for the reaction.

**Key Terms and Concepts: Activated complex, activation energy, potential energy barrier**

**18.4 Conditions That Affect the Rate of a Chemical Reaction**

Goal 4 State and explain the relationship between reaction rate and temperature.
Goal 5 Using an energy-reaction coordinate graph, explain how a catalyst affects reaction rate.
Goal 6 Identify and explain the relationship between reactant concentration and reaction rate.

**Key Terms and Concepts: Catalyst, inhibitor, negative catalyst**

**18.5 The Development of a Chemical Equilibrium**

Goal 7 Trace and explain the changes in concentrations of reactants and products that lead to a chemical equilibrium.

**18.6 Le Chatelier's Principle**

Goal 8 Given the equation for a chemical equilibrium, predict the direction in which the equilibrium will shift because of a change in the concentration of one species.
Goal 9 Given the equation for a chemical equilibrium involving one or more gases, predict the direction in which the equilibrium will shift because of a change in the volume of the system.
Goal 10 Given a thermochemical equation for a chemical equilibrium, or information from which it can be written, predict the direction in which the equilibrium will shift because of a change in temperature.

**Key Terms and Concepts: Le Chatelier's Principle, shift of an equilibrium**

**18.7 The Equilibrium Constant**

Goal 11 Given any chemical equilibrium equation, or information from which it can be written, write the equilibrium constant expression.

**Key Terms and Concepts: Equilibrium constant (K)**

**18.8 The Significance of the Value of K**

Goal 12 Given an equilibrium equation and the value of the equilibrium constant, identify the direction in which the equilibrium is favored.

**18.9 Equilibrium Calculations (Optional)**

Goal 13 Given the solubility product constant or the solubility of a slightly soluble compound (or data from which the solubility can be found), calculate the other value.
Goal 14 Given the solubility product constant of a slightly soluble compound and the concentration of a solution having a common ion, calculate the solubility of the slightly soluble compound in the solution.
Goal 15 Given the formula of a weak acid, HA, write the equilibrium equation for its ionization and the expression for its acid constant, $K_a$.
Goal 16 Given any two of the following three values for a weak acid, HA, calculate the third: (a) the initial concentration of the acid; (b) the pH of the solution, the percentage dissociation of the acid, or $[H^+]$ or $[A^-]$ at equilibrium; (c) $K_a$ for the acid.
Goal 17 For a weak acid, HA, given $K_a$, $[H^+]$, and $[A^-]$, or information from which they may be obtained, calculate the pH of the buffer produced.
Goal 18 Given $K_a$ for a weak acid, HA, determine the ratio between [HA] and $[A^-]$ that will produce a buffer of specified pH.
Goal 19 Given equilibrium concentrations of species in a gas-phase equilibrium, or information from which they can be found, and the equation for the equilibrium, calculate the equilibrium constant.

**Key Terms and Concepts: Acid constant ($K_a$), buffer, buffer solution, common ion effect, solubility product constant ($K_{sp}$)**

## Study Hints and Pitfalls to Avoid

We are used to thinking about "equal" in terms of amounts of physical things. Therefore, it is understandable that students sometimes resist the idea of *equilibrium* being applied to a system in which amounts may differ by many orders of magnitude. In $A \rightleftharpoons B$ there may be millions of times more A than B in the system, but if the *rate* at which A becomes B is the same as the rate at which B becomes A, there is an equilibrium between them.

Whenever you are trying to analyze or understand an equilibrium, write the equilibrium equation. It gives you a visual image of what's happening. If you are solving a quantitative equilibrium problem, also write the equilibrium constant expression. With those equations in black and white before you, it's much easier to reason through a Le Chatelier shift or to know where to put the numbers you must use to calculate an answer.

What is temperature, how does it affect an equilibrium, and why does a small change in temperature have so drastic an effect on reaction rate? An understanding of Figure 18.4 will help you answer these questions. Realize that temperature is a measure of kinetic energy and that kinetic energy is measured along the horizontal axis of the graph, not the vertical axis. Therefore, the curve that stretches out farther to the right represents the higher energy, the higher temperature. This flattens the curve somewhat, making it look "lower"; but that "lower" is not lower temperature. The main reason higher temperatures speed reaction rates is the resultant increase in the portion of the sample with enough kinetic energy to react, represented by the shaded portion of the graph under the curve and far to the right.

Consider this trick question: If you heat an equilibrium that is endothermic as the equation is written, the forward reaction rate increases. What happens to the reverse rate? Does it increase, decrease, or remain the same? (The answer is in Appendix III, immediately after the Chapter 18 Answers to Blue-Numbered Questions, Exercises, and Problems.)

Sometimes students hesitate to find one ion concentration by doubling another and then squaring the higher concentration when calculating K. You did this with the fluoride ion concentration from $MgF_2$ in Active Example 18.11: $[F^-]$ is twice $[Mg^{2+}]$, and $[F^-]$ must be squared in calculating $K_{sp}$, which is $[Mg^{2+}][F^-]^2$. Finding $[F^-]$ is an independent step; so is squaring it when substituting into the $K_{sp}$ expression. It is no coincidence that $[F^-]$ is twice $[Mg^{2+}]$, but that is true if both ions come only from the solute. If they came from different sources, they would not have that 2:1 relationship—but you would still square $[F^-]$ in using $K_{sp}$.

In all equilibrium constant problems, find the values of the concentrations first and then substitute as the K expression requires. Keep the steps separate.

## Concept-Linking Exercises

*Write a brief description of the relationships between each of the following groups of terms or phrases. Answers to the Concept-Linking Exercises are given after answers to the Target Checks in Appendix III.*

1. Reversible reaction, equilibrium, forward rate, reverse rate
2. Collision theory of chemical reactions, energy-reaction graph, activated complex, activation energy
3. Temperature, catalyst, concentration, reaction rate
4. Le Chatelier's Principle, temperature, gas volume, concentration
5. Equilibrium constant, K, $K_{sp}$, $K_a$

## Small-Group Discussion Questions

*Small-Group Discussion Questions are for group work, either in class or under the guidance of a leader during a discussion section.*

1. You place two equally filled glasses of water on a table in your kitchen. You leave one open and cover the other with tightly sealed plastic wrap. The next day, the open glass contains less water than the covered glass. Do you have a liquid–vapor equilibrium in either system? Explain, citing the four conditions that are true for every equilibrium.

2. A quantitative model for the collision theory of chemical reactions can be constructed. The rate constant for a reaction is a proportionality constant that influences the speed of the reaction. It can be broken down into three factors. Each of these three factors, in turn, affect the speed of a reaction. In symbols, $k = p \times Z \times f$, where k is the rate constant. The factor p accounts for the orientation of the colliding molecules. The Z factor accounts for the number of collisions per unit time, when comparing equal concentrations of reactants. The f factor accounts for the fractions of collisions with sufficient energy for a reaction-producing collision. For each of the three factors, draw a particulate-level depiction of a favorable situation versus an unfavorable situation for a high-speed reaction process. Explain each pair of sketches in words.

3. Consider a simple equilibrium system $A \rightleftharpoons B$. The reaction proceeds in the forward direction at the rate of 10% of the number molecules per minute; that is, 10% of the total number of molecules of A change to B each minute. The reverse reaction also proceeds at the same rate, 10% of the molecules per minute. Thus, if we start with 100 molecules of A and no B, after 1 minute, 10 molecules of A change to B, but 10% of the 10 molecules of B change back to A, which is 1 molecule. Therefore, after 1 minute, the system will consist of 91 A molecules and 9 B molecules. Continue this analysis for this system, recording the number of A and B molecules after each minute until the system comes to equilibrium. What happens to the system at equilibrium? What if you start with 100 B molecules and no A molecules? How does this equilibrium system compare with the one where you started with 100 molecules of A?

4. Consider the chemical system illustrated in Active Figure 18.14. (a) If the initial concentrations of hydrogen and iodine are 1.0 M, estimate the value of the equilibrium constant for the reaction as written. (b) Is the forward rate of reaction or the reverse rate of reaction greater at each of the following points: (i) before the two concentration lines cross, (ii) at the time the two lines cross, and (iii) after the two lines cross?

5. Derive a mathematical statement that expresses the relationship among the values of the equilibrium constants for the following reactions: (a) $3\ H_2(g) + N_2(g) \rightleftharpoons 2\ NH_3(g)$, (b) $2\ NH_3(g) \rightleftharpoons 3\ H_2(g) + N_2(g)$; and (c) $3/2\ H_2(g) + 1/2\ N_2(g) \rightleftharpoons NH_3(g)$.

6. For systems not at equilibrium, it can be useful to calculate a concentration ratio known as the reaction quotient, Q. The reaction quotient has the same form as the equilibrium constant. For the general reaction $a\ A + b\ B \rightleftharpoons c\ C + d\ D$, $Q = \frac{[C]^c\,[D]^d}{[A]^a\,[B]^b}$. Q = K only at equilibrium. If Q is larger than K, in which direction will the reaction shift to reach equilibrium? What if Q is smaller than K? Explain both answers.

7. If you breathe into a bag, you will feel light headed and possibly pass out if you continue long enough. Explain this phenomenon, given the fact that carbonic acid in your blood is part of an equilibrium system that involves exhaled carbon dioxide: $H_2CO_3 \rightleftharpoons H_2O + CO_2$.

8. Consider the reaction $C(s) + H_2O(g) + \text{heat} \rightleftharpoons CO(g) + H_2(g)$. State how each of the following will affect the equilibrium concentrations of each reactant and product: (a) remove steam, (b) add hydrogen, (c) increase the temperature, (d) expand the volume of the container, (e) add carbon, (f) introduce a catalyst, (g) add helium to the container (helium does not react with any species in the equilibrium mixture).

9. Compare and contrast the macroscopic characteristics of an equilibrium system with the particulate-level characteristics of the system. Choose at least one physical equilibrium and one chemical equilibrium system to illustrate your comparison.

# Questions, Exercises, and Problems

*Blue-numbered questions whose numbers are answered in Appendix III. ■ denotes problems assignable in OWL. In the first section of Questions, Exercises, and Problems, similar exercises and problems are paired in consecutive odd-even number combinations.*

**Section 18.1: The Character of an Equilibrium**

1. What does it mean when an equilibrium is described as *dynamic*? Compare an equilibrium that is dynamic with one that is static.

2. What things are equal in an equilibrium? Give an example.

3. Undissolved table salt is in contact with a saturated salt solution in (a) a sealed container and (b) an open beaker. Which system, if either, can reach equilibrium? Explain your answer.

4. What is meant by saying that an equilibrium is confined to a "closed system"?

5. A garden in a park has a fountain that discharges water into a pond. The pond overflows into a stream that cascades to the bottom of a small pool. The water is then pumped up into the fountain. Is this system a dynamic equilibrium? Explain.

6. A river flows into a lake formed by a dam. Water flows through the dam's spillways as the river continues downstream. The water level of the lake is constant. Is this system a dynamic equilibrium? Explain.

**Section 18.2: The Collision Theory of Chemical Reactions**

7. Explain why a molecular collision can be sufficiently energetic to cause a reaction, yet no reaction occurs as a result of that collision.

8. According to the collision theory of chemical reactions, what two conditions must be satisfied if a molecular collision is to result in a reaction?

**Section 18.3: Energy Changes During a Molecular Collision**

*Assume heat to be the only form of reaction energy in the following questions. This makes $\Delta E$ equal to the $\Delta H$ discussed in Section 10.8.*

9. Sketch an energy-reaction coordinate graph for an endothermic reaction. Include a, b, and c points on the vertical axis and use them for the algebraic expressions of $\Delta E$ and the activation energy.

10. ■ In the reaction for which Figure 18.2 is the energy-reaction coordinate graph, is $\Delta E$ for the reaction positive or negative? Is the reaction exothermic or endothermic? Use the letters a, b, and c on the vertical axis of Figure 18.2 to state algebraically the $\Delta E$ and activation energy of the reaction.

11. Assuming the reaction described by Figure 18.2 to be reversible, compare the signs of the activation energies for the forward and reverse reactions. Which is positive, which is negative, or are they the same, and if so, are they positive or negative?

12. Assume that the reaction described by Figure 18.2 is reversible. Compare the magnitude of the activation energies for the forward and reverse reactions described by Figure 18.2. Which is greater, or are they equal?

13. What is an "activated complex"? Why is it that we cannot list the physical properties of the species represented as an activated complex?

14. Explain the significance of activation energy. For two reactions that are identical in all respects except activation energy, identify the reaction that would have the higher rate and tell why.

**Section 18.4: Conditions That Affect the Rate of a Chemical Reaction**

15. State the effect of a temperature increase and a temperature decrease on the rate of a chemical reaction. Explain each effect.

16. "At a given temperature, only a small fraction of the molecules in a sample has sufficient kinetic energy to engage in a chemical reaction." What is the meaning of that statement?

17. Suppose that two substances are brought together under conditions that cause them to react and reach equilibrium. Suppose that in another vessel the same substances and a catalyst are brought together, and again equilibrium is reached. How are the processes alike, and how are they different?

18. What is a catalyst? Explain how a catalyst affects reaction rates.

19. For the hypothetical reaction $A + B \rightarrow C$, what will happen to the reaction rate if the concentration of A is increased *and* the concentration of B is decreased? Explain.

20. For the hypothetical reaction $A + B \rightarrow C$, what will happen to the rate of reaction if the concentration of A is increased without changing the concentration of B? What will happen if the concentration of B is decreased without changing the concentration of A? Explain why in both cases.

**Section 18.5: The Development of a Chemical Equilibrium**

*If nitrogen and hydrogen are brought together at the proper temperature and pressure, they will react until they reach equilibrium:* $N_2(g) + 3\ H_2(g) \rightleftharpoons 2\ NH_3(g)$. *Answer the following questions with regard to the establishment of that equilibrium.*

21. When will the reverse reaction rate be at a maximum: at the start of the reaction, after equilibrium has been reached, or at some point in between?

22. When will the forward reaction rate be at a maximum: at the start of the reaction, after equilibrium has been reached, or at some point in between?

23. On a single set of coordinate axes, sketch graphs of the forward reaction rate versus time and the reverse reaction rate versus time from the moment the reactants are mixed to a point beyond the establishment of equilibrium.

24. What happens to the concentrations of each of the three species between the start of the reaction and the time equilibrium is reached?

**Section 18.6: Le Chatelier's Principle**

25. If the system $2\ SO_2(g) + O_2(g) \rightleftharpoons 2\ SO_3(g)$ is at equilibrium and the concentration of $O_2$ is reduced, predict the direction in which the equilibrium will shift. Justify or explain your prediction.

26. ■ Consider the following system at equilibrium at 500 K: $PCl_3(g) + Cl_2(g) \rightleftharpoons PCl_5(g)$. When some $Cl_2(g)$ is removed from the equilibrium system at constant temperature, in what direction must the reaction run? What will happen to the concentration of $PCl_3$?

27. If additional oxygen is pumped into the equilibrium system $4\ NH_3(g) + 5\ O_2(g) \rightleftharpoons 4\ NO(g) + 6\ H_2O(g)$, in which direction will the reaction shift? Justify your answer.

28. ■ Consider the following system at equilibrium at 723 K: $N_2(g) + 3\ H_2(g) \rightleftharpoons 2\ NH_3(g)$. When some $NH_3(g)$ is removed from the equilibrium system at constant temperature, in what direction must the reaction run? What will happen to the concentration of $H_2$?

29. Predict the direction of the shift for the equilibrium $Cu(NH_3)_4^{2+}(aq) \rightleftharpoons Cu^{2+}(aq) + 4\ NH_3(aq)$ if the concentration of ammonia were reduced. Explain your prediction.

30. ■ Consider the following system at equilibrium at 600 K: $COCl_2(g) \rightleftharpoons CO(g) + Cl_2(g)$. When some $COCl_2(g)$ is added to the equilibrium system at constant temperature, in what direction must the reaction run? What will happen to the concentration of CO?

31. A container holding the equilibrium $4\ H_2(g) + CS_2(g) \rightleftharpoons CH_4(g) + 2\ H_2S(g)$ is enlarged. Predict the direction of the Le Chatelier shift. Explain.

32. ■ Consider the following system at equilibrium at 298 K: $2\ NOBr(g) \rightleftharpoons 2\ NO(g) + Br_2(g)$. If the volume of the equilibrium system is suddenly decreased at constant temperature, in what direction must the reaction run? What will happen to the number of moles of $Br_2$?

33. In what direction will $CO(g) + H_2O(g) \rightleftharpoons CO_2(g) + H_2(g)$ shift as a result of a reduction in volume? Explain.

34. ■ Consider the following system at equilibrium at $1.15 \times 10^3$ K: $2\ SO_2(g) + O_2(g) \rightleftharpoons 2\ SO_3(g)$. If the volume of the equilibrium system is suddenly increased at constant temperature, in what direction must the reaction run? What will happen to the number of moles of $O_2$?

35. Which direction of the equilibrium $2\ NO_2(g) \rightleftharpoons N_2O_4(g) + 59.0$ kJ will be favored if the system is cooled? Explain.

36. ■ Consider the following system, for which $\Delta H = 87.9$ kJ, at equilibrium at 500 K: $PCl_5(g) + 87.9\ kJ \rightleftharpoons PCl_3(g) + Cl_2(g)$. If the temperature of the equilibrium system is suddenly increased, in what direction must the reaction run? What will happen to the concentration of $Cl_2$?

37. If your purpose were to increase the yield of $SO_3$ in the equilibrium $SO_2(g) + NO_2(g) \rightleftharpoons SO_3(g) + NO(g) + 41.8$ kJ, would you use the highest or lowest operating temperature possible? Explain.

38. ■ Consider the following system, for which $\Delta H = -10.4$ kJ, at equilibrium at 698 K: $H_2(g) + I_2(g) \rightleftharpoons 2\ HI(g)$. If the temperature of the equilibrium system is suddenly decreased, in what direction must the reaction run? What will happen to the concentration of $I_2$?

39. The solubility of calcium hydroxide is low; it reaches about $2.4 \times 10^{-2}$ M at saturation. In acid solutions, with many $H^+$ ions present, calcium hydroxide is quite soluble. Explain this fact in terms of Le Chatelier's Principle. (*Hint:* Recall what you know of reactions in which molecular products are formed.)

40. ■ Consider the following system, for which $\Delta H = 18.8$ kJ, at equilibrium at 350 K: $CH_4(g) + CCl_4(g) \rightleftharpoons 2\ CH_2Cl_2(g)$. The production of $CH_2Cl_2(g)$ is favored by which of the following: (a) increasing the temperature, (b) increasing the pressure (by changing the volume), (c) increasing the volume, (d) adding $CH_2Cl_2$, (e) removing $CCl_4$?

**Section 18.7: The Equilibrium Constant**

*For each equilibrium equation shown, write the equilibrium constant expression.*

41. $CO(g) + H_2O(g) \rightleftharpoons CO_2(g) + H_2(g)$
42. ■ $NH_4HS(s) \rightleftharpoons NH_3(g) + H_2S(g)$
43. $C(s) + H_2O(g) \rightleftharpoons CO(g) + H_2(g)$
44. ■ $2\ SO_2(g) + O_2(g) \rightleftharpoons 2\ SO_3(g)$
45. $Zn_3(PO_4)_2(s) \rightleftharpoons 3\ Zn^{2+}(aq) + 2\ PO_4^{3-}(aq)$
46. ■ $HF(aq) + H_2O(\ell) \rightleftharpoons H_3O^+(aq) + F^-(aq)$
47. $HNO_2(aq) + H_2O(\ell) \rightleftharpoons H_3O^+(aq) + NO_2^-(aq)$

48. ■ $(CH_3)_3N(aq) + H_2O(\ell) \rightleftharpoons (CH_3)_3NH^+(aq) + OH^-(aq)$

49. $Cu(NH_3)_4^{2+}(aq) \rightleftharpoons Cu^{2+}(aq) + 4\ NH_3(aq)$

50. ■ $Mg(OH)_2(s) \rightleftharpoons Mg^{2+}(aq) + 2\ OH^-(aq)$

51. The equilibrium between nitrogen monoxide, oxygen, and nitrogen dioxide may be expressed in the equation $2\ NO(g) + O_2(g) \rightleftharpoons 2\ NO_2(g)$. Write the equilibrium constant expression for this equation. Then express the same equilibrium in at least two other ways, and write the equilibrium constant expression for each. Are the constants numerically equal? Cite some evidence to support your answer.

52. "The equilibrium constant expression for a given reaction depends on how the equilibrium equation is written." Explain the meaning of that statement. You may, if you wish, use the equilibrium equation $N_2(g) + 3\ H_2(g) \rightleftharpoons 2\ NH_3(g)$ to illustrate your explanation.

### Section 18.8: The Significance of the Value of K

53. If sodium cyanide solution is added to silver nitrate solution, the following equilibrium will be reached: $Ag^+(aq) + 2\ CN^-(aq) \rightleftharpoons Ag(CN)_2^-(aq)$. For this equilibrium $K = 5.6 \times 10^{18}$. In which direction is the equilibrium favored? Justify your answer.

54. ■ For the following equilibrium system, $K = 4.86 \times 10^{-5}$ at 298 K: $HClO(aq) + F^-(aq) \rightleftharpoons ClO^-(aq) + HF(aq)$. Assuming that you start with equal concentrations of HClO and $F^-$, and that no $ClO^-$ or HF is initially present, which of the following best describes the equilibrium system: (a) the forward reaction is favored at equilibrium; (b) the reverse reaction is favored at equilibrium; (c) appreciable quantities of all species are present at equilibrium?

55. A certain equilibrium has a very small equilibrium constant. In which direction, forward or reverse, is the equilibrium favored? Explain.

56. ■ Acetic acid, $HC_2H_3O_2$, is a soluble weak acid. When placed in water it partially ionizes and reaches equilibrium. Write the equilibrium equation for the ionization. Will the equilibrium constant be large or small? Justify your answer.

*Questions 57 and 58: In Chapter 9 we discussed how to identify major and minor species and how to write net ionic equations. These skills are based on the solubility of ionic compounds, the strengths of acids, and the stability of certain ion combinations. Use these ideas to predict the favored direction of each equilibrium given. In each case state whether you expect the equilibrium concentration to be large or small.*

57. a) $H_2SO_3(aq) \rightleftharpoons H_2O(\ell) + SO_2(aq)$
    b) $H^+(aq) + C_2H_3O_2^-(aq) \rightleftharpoons HC_2H_3O_2(aq)$

58. ■ a) $HCl(aq) \rightleftharpoons H^+(aq) + Cl^-(aq)$
    b) $BaSO_4(s) \rightleftharpoons Ba^{2+}(aq) + SO_4^{2-}(aq)$

### Section 18.9: Equilibrium Calculations (Optional)

59. $Co(OH)_2$ dissolves in water to the extent of $3.7 \times 10^{-6}$ mol/L. Find its $K_{sp}$.

60. ■ A student measures the molar solubility of magnesium fluoride in a water solution to be $1.19 \times 10^{-3}$ M. Based on her data, what is the solubility product constant for this compound?

61. If 250 mL of water will dissolve only 8.7 mg of silver carbonate, what is the $K_{sp}$ of $Ag_2CO_3$?

62. ■ A student measures the solubility of lead(II) sulfate in a water solution to be $3.95 \times 10^{-3}$ grams /100 mL. Based on his data, what is the solubility product constant for this compound?

63. Find the moles per liter and grams per 100 mL solubility of silver iodate, $AgIO_3$, if its $K_{sp} = 2.0 \times 10^{-8}$.

64. $K_{sp} = 8.7 \times 10^{-9}$ for $CaCO_3$. Calculate its solubility in (a) moles per liter and (b) grams per 100 mL.

65. Find the solubility (mol/L) of $Mn(OH)_2$ if its $K_{sp} = 1.0 \times 10^{-13}$.

66. ■ $K_{sp}$ for manganese(II) hydroxide is $4.6 \times 10^{-14}$. What is the molar solubility of manganese(II) hydroxide in a water solution?

67. How many grams of calcium oxalate will dissolve in $2.5 \times 10^2$ mL of 0.22 M $Na_2C_2O_4$ if $K_{sp} = 2.4 \times 10^{-9}$ for $CaC_2O_4$?

68. ■ $K_{sp}$ for silver hydroxide is $2.0 \times 10^{-8}$. Calculate the molar solubility of silver hydroxide in 0.103 M NaOH.

69. The pH of 0.22 M $HC_4H_5O_3$ (acetoacetic acid) is 2.12. Find its $K_a$ and percent ionization.

70. ■ In the laboratory, a student measured the pH of a 0.48 M aqueous solution of nitrous acid to be 1.85. From these data, calculate the percentage ionization and the $K_a$ value for this acid.

71. Find the pH of 0.35 M $HC_2H_3O_2$ ($K_a = 1.8 \times 10^{-5}$).

72. ■ What is the pH of a 0.501 M aqueous solution of hypochlorous acid, for which $K_a = 3.5 \times 10^{-8}$?

73. A student dissolves 24.0 g of sodium acetate, $NaC_2H_3O_2$, in $5.00 \times 10^2$ mL of 0.12 M $HC_2H_3O_2$ ($K_a = 1.8 \times 10^{-5}$). Calculate the pH of the solution.

74. ■ What is the pH of a solution that contains 0.18 M sodium acetate and 0.21 M acetic acid? $K_a$ for acetic acid is $1.8 \times 10^{-5}$.

75. Find the ratio $[HC_2H_3O_2]/[C_2H_3O_2^-]$ that will yield a buffer in which pH = 4.25 ($K_a = 1.8 \times 10^{-5}$).

76. ■ What concentration ratio of hydrocyanic acid to cyanide ion, $[HCN]/[CN^-]$, will produce a buffer solution with a pH of 9.71? $K_a$ for hydrocyanic acid is $4.0 \times 10^{-10}$.

77. A student introduces 0.351 mol of CO and 1.340 mol of $Cl_2$ into a reaction chamber having a volume of 3.00 L. When equilibrium is reached according to the equation $CO(g) + Cl_2(g) \rightleftharpoons COCl_2(g)$, there are 1.050 mol of $Cl_2$ in the chamber. Calculate K.

78. ■ A student studied the following reaction in the laboratory at 530 K: $PCl_3(g) + Cl_2(g) \rightleftharpoons PCl_5(g)$. When he introduced $9.0 \times 10^{-2}$ moles of $PCl_3(g)$ and 0.11 moles of $Cl_2(g)$ into a 1.00 liter container, he found the equilibrium concentration of $PCl_5(g)$ to be $5.3 \times 10^{-2}$ M. Use these data to calculate the equilibrium constant, K, for this reaction.

### General Questions

79. Distinguish precisely, and in scientific terms, the differences among items in each of the following pairs.
    a) Reaction, reversible reaction
    b) Open system, closed system
    c) Dynamic equilibrium, static equilibrium
    d) Activated complex, activation energy
    e) Catalyzed reaction, uncatalyzed reaction

f) Catalyst, inhibitor

g) Buffered solution, unbuffered solution

80. Classify each of the following statements as true or false.

a) Some equilibria depend on a steady supply of a reactant in order to maintain the equilibrium.

b) Both forward and reverse reactions continue after equilibrium is reached.

c) Every time reactant molecules collide, there is a reaction.

d) Potential energy during a collision is greater than potential energy before or after the collision.

e) The properties of an activated complex are between those of the reactants and products.

f) Activation energy is positive for both the forward and reverse reactions.

g) Kinetic energy is changed to potential energy during a collision.

h) An increase in temperature speeds the forward reaction but slows the reverse reaction.

i) A catalyst changes the steps by which a reaction is completed.

j) An increase in concentration of a substance on the right-hand side of an equation speeds the reverse reaction rate.

k) An increase in the concentration of a substance in an equilibrium increases the reaction rate in which the substance is a product.

l) Reducing the volume of a gaseous equilibrium shifts the equilibrium in the direction of fewer gaseous molecules.

m) Raising temperature results in a shift in the forward direction of an endothermic equilibrium.

n) The value of an equilibrium constant depends on temperature.

o) A large K indicates that an equilibrium is favored in the reverse direction.

81. At Time 1 two molecules are about to collide. At Time 2 they are in the process of colliding, and their form is that of the activated complex. Compare the sum of their kinetic energies at Time 1 with the kinetic energy of the activated complex at Time 2. Explain your conclusions.

82. List three things you might do to increase the rate of the reverse reaction for which Figure 18.2 is the energy-reaction coordinate graph.

### More Challenging Quest ions

83. ■ The Haber process for making ammonia by direct combination of the elements is described by the equation $N_2(g) + 3\ H_2(g) \rightleftharpoons 2\ NH_3(g) + 92$ kJ. If a manufacturer wants to make the greatest amount of ammonia in the least time, removing product as the reaction proceeds, is the manufacturer more likely to conduct the reaction at (a) high pressure or low pressure, (b) high temperature or low temperature? Explain your choice in each case.

84. Under proper conditions the reaction in Question 83 will reach equilibrium. Is the manufacturer likely to conduct the reaction under those conditions, that is, at equilibrium? Explain.

85. The reaction in Question 83 has a yield of about 98% at 200°C and 1000 atm. Commercially, the reaction is performed at about 500°C and 350 atm, where the yield is only about 30%. Suggest why operation at the lower yield is economically more favorable.

86. The solubility of calcium hydroxide is low enough to be listed as "insoluble" in solubility tables, but it is much more soluble than most of the other ionic compounds that are similarly classified. Its $K_{sp}$ is $5.5 \times 10^{-6}$.

a) Write the equation for the equilibrium to which the $K_{sp}$ is related.

b) If you had such an equilibrium, name at least two substances or general classes of substances that might be added to (1) reduce the solubility of $Ca(OH)_2$ and (2) increase its solubility. Justify your choices.

c) Without adding calcium or hydroxide ion, name a substance or class of substances that would, if added, (1) increase $[OH^-]$ and (2) reduce $[OH^-]$. Justify your choices.

87. ■ The table below lists several "disturbances" that may or may not produce a Le Chatelier shift in the equilibrium $4\ NH_3(g) + 7\ O_2(g) \rightleftharpoons 6\ H_2O(g) + 4\ NO_2(g) +$ energy. If the disturbance is an immediate change in the concentration of any species in the equilibrium, place in the concentration column of that substance an *I* if the change is an increase or a *D* if it is a decrease. If a shift will result, place *F* in the shift column if the shift is in the forward direction or *R* if it is in the reverse direction. Then determine what will happen to the concentrations of the other species because of the shift, and insert *I* or *D* for increase or decrease. If there is no Le Chatelier shift, write *None* in the Shift column and leave the other columns blank.

| Disturbance | Shift | $[NH_3]$ | $[O_2]$ | $[H_2O]$ | $[NO_2]$ |
|---|---|---|---|---|---|
| Add $NO_2$ | | | | | |
| Reduce temperature | | | | | |
| Add $N_2$ | | | | | |
| Remove $NH_3$ | | | | | |
| Add a catalyst | | | | | |

88. Some systems at equilibrium are exothermic, and some are endothermic. Is this statement always true, sometimes true, or never true? Explain your answer.

89. The equilibrium constant, K, can have many values for any equilibrium. Why or how?

90. An all-gas-phase equilibrium can be reached between sulfur dioxide and oxygen on one side of the equation and sulfur trioxide on the other. Write two equilibrium constant expressions for this equilibrium, one of which has a value greater than 1 and one with a value less than 1. It is not necessary to say which is which.

91. "Hard" water has a high concentration of calcium and magnesium ions. Focusing on the calcium ion, a common home-water-softening process is based on a reversible chemical change that can be expressed by $Na_2Ze(s) + Ca^{2+}(aq) \rightleftharpoons CaZe(s) + 2\ Na^+(aq)$. $Na_2Ze$ represents a solid resin that is like an ionic compound between sodium ions and *zeolite ions,* a complex arrangement of silicate and aluminate groups; CaZe is the corresponding calcium compound. When, during the day, is this system most likely to reach equilibrium? Why doesn't it reach equilibrium and stay there? Periodically it is necessary to "recharge" the water softener by running salt water, NaCl(aq), through it. Why is this necessary? What concept discussed in this chapter does the recharging process illustrate?

# 19



All common batteries convert chemical energy into electrical energy by means of an oxidation–reduction reaction. A battery can be made by placing a strip of copper and a strip of zinc into a lemon! This lemon battery produces enough current to power a small clock. In this chapter, you will learn how a chemical reaction can generate an electric current.

# Oxidation–Reduction (Redox) Reactions

In Chapter 19 you will study oxidation–reduction reactions. You will see that redox reactions, as they are also known, are electron-transfer reactions. As you study this chapter, look for parallels to the proton-transfer reactions you studied in Chapter 17. You will find that this exercise will not only help you learn about redox reactions, but it will also help you better understand acid–base reactions.

Online homework for this chapter may be assigned in OWL

## CONTENTS

## 19.1 Electron-Transfer Reactions

Goal 1 Describe and explain oxidation and reduction in terms of electron transfer.

Goal 2 Given an oxidation half-reaction equation and a reduction half-reaction equation, combine them to form a net ionic equation for an oxidation–reduction reaction.

In the system in Figure 19.1, a strip of zinc is immersed in a solution containing zinc ions, and a piece of copper is placed in a solution that includes copper(II) ions. The solutions are connected by a **salt bridge** containing a solution of an ionic compound whose ions are not involved in the net chemical change. The two electrodes are connected by a wire. A lightbulb in the external circuit is lit because of a flow of electrons from the zinc electrode to the copper electrode, and its brightness is proportional to the force that moves the electrons through the circuit.

Where do the electrons entering the lightbulb come from, and where do they go when they leave the bulb? Four measurable observations answer those questions. After the cell has operated for a period of time (1) the mass of the zinc electrode decreases, (2) the $Zn^{2+}$ concentration increases, (3) the mass of the copper electrode increases, and (4) the $Cu^{2+}$ concentration decreases. The first two observations indicate that neutral zinc atoms lose two electrons to become zinc ions. Stated another way, zinc atoms are being divided into zinc ions and two electrons:

$$Zn(s) \rightarrow Zn^{2+}(aq) + 2\ e^- \qquad \textbf{(19.1)}$$

The electrons flow through the wire and the lightbulb to the copper electrode, where they join a copper ion to become a copper atom:

$$Cu^{2+}(aq) + 2\ e^- \rightarrow Cu(s) \qquad \textbf{(19.2)}$$

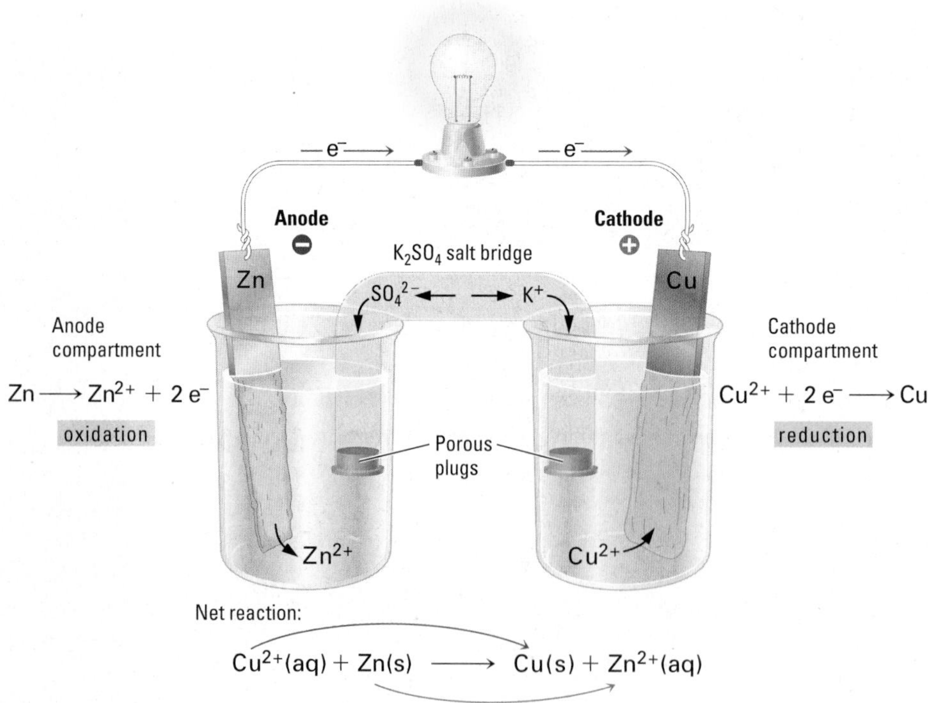


**Figure 19.1** An electrochemical cell. In this system, a chemical reaction generates an electric current. Zinc atoms change to zinc ions, releasing two electrons, which react with copper(II) ions to form copper atoms.

The chemical change that occurs at the zinc electrode is oxidation: **oxidation is defined as the loss of electrons.** The reaction is described as a **half-reaction** because it cannot occur by itself. There must be a second half-reaction. The electrons lost by the substance **oxidized** must have someplace to go. In this case they go to the copper ion, which is **reduced. Reduction is a gain of electrons.**

Equations 19.1 and 19.2 are **half-reaction equations.** If the half-reaction equations are added algebraically, the result is the net ionic equation for the oxidation–reduction (redox) reaction:

$$Zn(s) \rightarrow Zn^{2+}(aq) + \cancel{2\,e^-} \qquad \textbf{(19.1)}$$

$$Cu^{2+}(aq) + \cancel{2\,e^-} \rightarrow Cu(s) \qquad \textbf{(19.2)}$$

$$Zn(s) + Cu^{2+}(aq) \rightarrow Cu(s) + Zn^{2+}(aq) \qquad \textbf{(19.3)}$$

This chemical change is an **electron-transfer reaction.** Electrons have been transferred from zinc atoms to copper(II) ions. Notice that although no electrons appear in the final equation, the electron-transfer character of the reaction is quite clear in the half-reactions. Notice also that the number of electrons lost by one species is exactly equal to the number of electrons gained by the other.

If there is no need for the electrical energy that can be derived from this cell, the same reaction can be performed by simply dipping a strip of zinc into a solution containing copper(II) ions (Active Figure 19.2). A coating of copper atoms quickly forms on the surface of the zinc. If the copper atoms are washed off the zinc and the zinc is weighed, its mass will be less than it was at the beginning. The concentration of copper ions in the solution goes down, and zinc ions appear. The half-reaction and net ionic equations are exactly as they are for the system connected to a lightbulb.

This same reaction was used in Active Example 9.6: "A reaction occurs when a piece of zinc is dipped into copper(II) nitrate. Write the conventional, total ionic, and net ionic equations." The conventional equation is a "single-replacement" equation:

$$Zn(s) + Cu(NO_3)_2(aq) \rightarrow Cu(s) + Zn(NO_3)_2(aq) \qquad \textbf{(19.4)}$$

Equation 19.3 is the net ionic equation produced in Active Example 9.6.

All of the single-replacement redox reactions encountered in Chapters 8 and 9 can be analyzed in terms of half-reactions. For example,

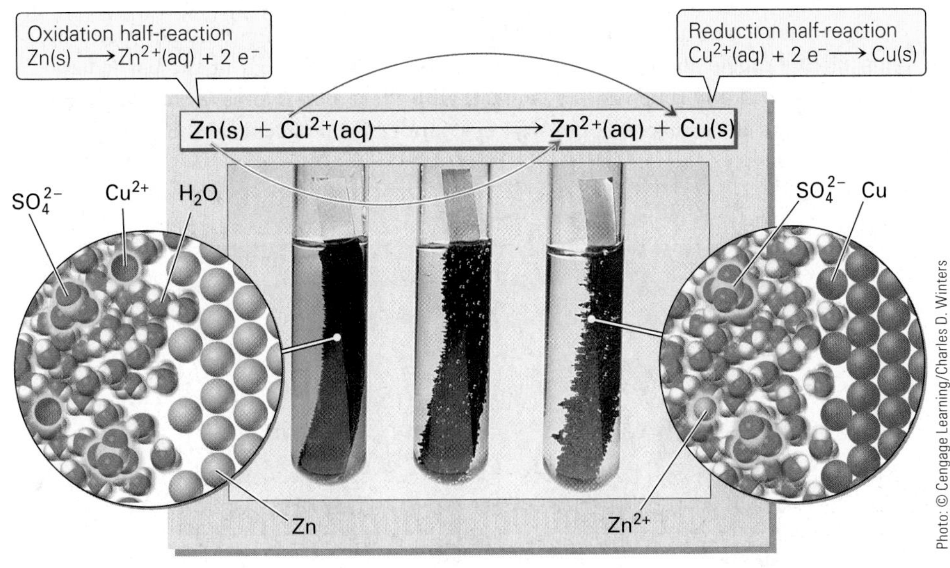


**Active Figure 19.2** Zinc reacts with a solution of copper(II) sulfate. A zinc strip is dipped into a blue copper(II) sulfate solution (*left*). Notice the copper coating that forms immediately. As the reaction progresses, copper metal accumulates on the surface of the zinc strip (*center*). The hydrated copper(II) ions impart a blue color to the solution that fades as they react to form copper atoms (*left to right*). Hydrated zinc and sulfate ions are colorless in solution. ***Watch this active figure at http://www.cengage.com/cracolice.***

**Figure 19.3** Iron metal is added to a hydrochloric acid solution. Hydrogen gas is bubbling from the solution.

**1.** The evolution of hydrogen gas on adding iron to hydrochloric acid (Section 9.5) (Fig. 19.3):

$$\begin{array}{ll} \text{Reduction:} & 2\,H^+(aq) + \cancel{2e^-} \rightarrow H_2(g) \\ \text{Oxidation:} & Fe(s) \rightarrow Fe^{2+}(aq) + \cancel{2e^-} \\ \hline \text{Redox:} & 2\,H^+(aq) + Fe(s) \rightarrow H_2(g) + Fe^{2+}(aq) \end{array} \quad \textbf{(19.5)}$$

**2.** The preparation of bromine by bubbling chlorine gas through a solution of sodium bromide (Fig. 19.4):

$$\begin{array}{ll} \text{Reduction:} & Cl_2(g) + \cancel{2e^-} \rightarrow 2\,Cl^-(aq) \\ \text{Oxidation:} & 2\,Br^-(aq) \rightarrow Br_2(\ell) + \cancel{2e^-} \\ \hline \text{Redox:} & Cl_2(g) + 2\,Br^-(aq) \rightarrow 2\,Cl^-(aq) + Br_2(\ell) \end{array} \quad \textbf{(19.6)}$$

**3.** The reduction of silver ion (Active Example 8.9 in Section 8.8 and Active Example 9.8 in Section 9.5) by placing copper into a silver nitrate solution (Fig. 19.5):

$$\begin{array}{ll} \text{Reduction:} & 2\,Ag^+(aq) + \cancel{2e^-} \rightarrow 2\,Ag(s) \\ \text{Oxidation:} & Cu(s) \rightarrow Cu^{2+}(aq) + \cancel{2e^-} \\ \hline \text{Redox:} & 2\,Ag^+(aq) + Cu(s) \rightarrow 2\,Ag(s) + Cu^{2+}(aq) \end{array} \quad \textbf{(19.7)}$$

The development of Equation 19.7 needs special comment. The usual reduction equation for silver ion is $Ag^+(aq) + e^- \rightarrow Ag(s)$. Because two moles of electrons are lost in the oxidation reaction, *two moles of electrons must be gained in the reduction reaction.* As has already been noted, the number of electrons lost by one species must equal the number gained by the other species. It is therefore necessary to multiply the usual $Ag^+$ reduction equation by 2 to bring about this equality in electrons gained and lost. They then cancel when the half-reaction equations are added.

## Thinking About Your Thinking

### Mental Models

**As with the proton-transfer reactions of Chapter 17, working on the formation of a mental model of electron-transfer reactions is a critical step toward thinking as a chemist. Both types of transfer reactions abound in biological systems. Students who study biochemistry are often amazed when they learn that nearly every chemical change that occurs in a living cell can be classified as either a proton-transfer or an electron-transfer reaction.**

**We can use Figure 19.1 and Active Figure 19.2 as a starting point to help you form your mental movie about how electron-transfer reactions occur at the particulate level. Look back and refamiliarize yourself with these two figures with a focus on what happens at the macroscopic level in Figure 19.1 and a focus on the particulate**

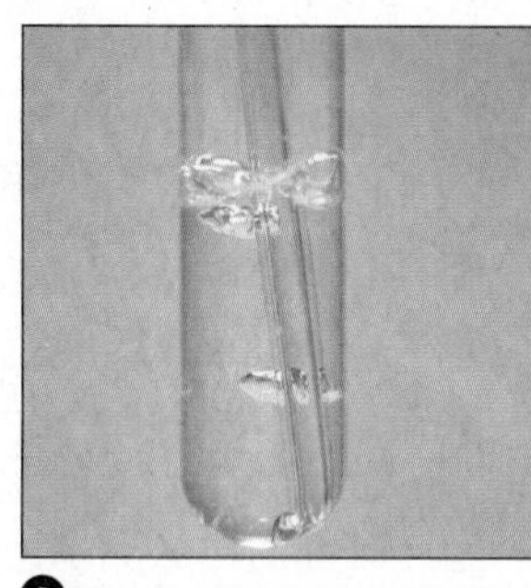
a

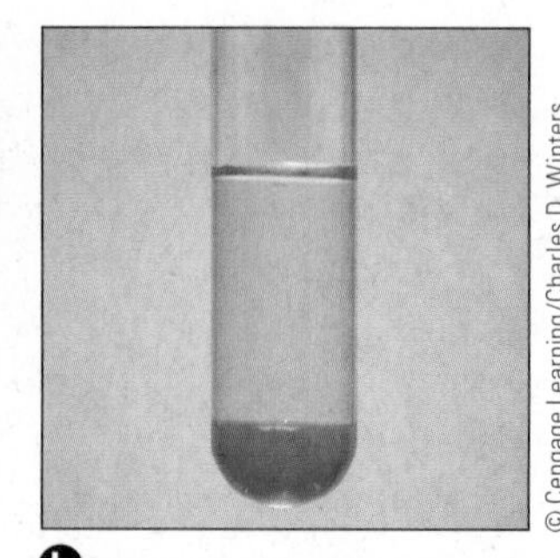
b



**Figure 19.4** Preparation of bromine. (a) Chlorine gas is bubbled through a solution of sodium bromide. (b) The liquid bromine product is extracted by adding carbon tetrachloride. The top water layer and the bottom carbon tetrachloride layer both contain dissolved bromine, but it is more concentrated in the bottom.

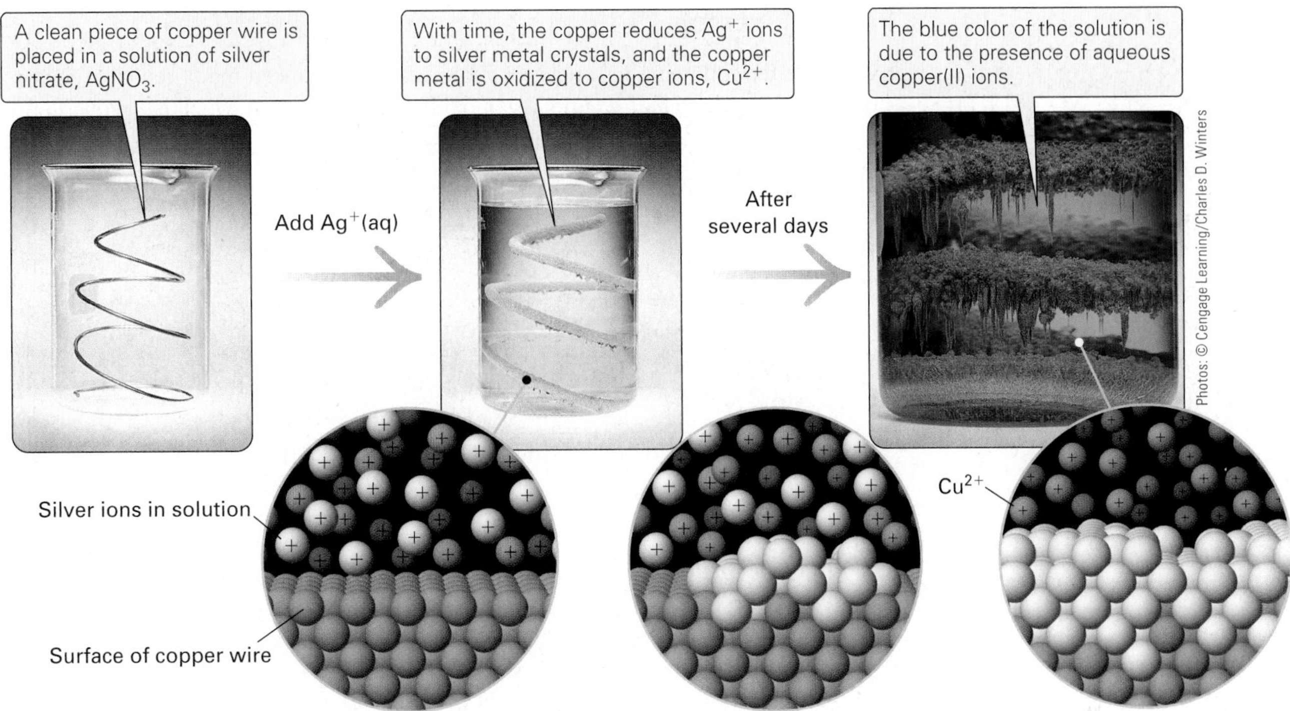
A clean piece of copper wire is placed in a solution of silver nitrate, AgNO3.
With time, the copper reduces Ag+ ions to silver metal crystals, and the copper metal is oxidized to copper ions, Cu2+.
The blue color of the solution is due to the presence of aqueous copper(II) ions.
Add Ag+(aq)
After several days
Silver ions in solution
Surface of copper wire
Cu2+
Photos: © Cengage Learning/Charles D. Winters

**Figure 19.5** The reduction of silver ion to solid silver and the oxidation of solid copper to copper(II) ion.

**level changes in Active Figure 19.2. Now recall the electron-sea model of metallic bonding (Section 12.8), by which a metal is pictured as a crystal structure of metal ions immersed in a sea of electrons. Imagine the metallic zinc electrode in this way: zinc ions in a sea of electrons.**

**If one of those $Zn^{2+}$ ions moves from the electrode into solution, its electrons are left behind. The charge balance in the metal is upset. There is more negative charge than positive. Negatively charged electrons repel other negatively charged electrons, creating a "flow" of electrons from the zinc electrode, through the connecting wires and voltmeter, to the metallic copper electrode. These give the copper electrode a small negative charge that attracts it to the copper(II) ions in the solution. Each copper(II) ion is joined by two electrons to become a copper atom that is deposited on the electrode.**

**Since a positively charged copper(II) ion is removed from solution, the cell compartment has a charge imbalance, with an excess negative charge. This can be corrected by the flow of negatively charged $SO_4^{2-}$ ions out of the compartment and into the salt bridge, or with a positively charged ion (or ions) from the salt bridge moving into the compartment. In either case, the electrical charge in the salt bridge becomes out of balance, with a negative charge.**

**The compartment with zinc ions and sulfate ions compensates for the negative charge in the salt bridge. Either negative ions from the salt bridge enter this compartment or positive ions from the compartment enter the salt bridge. In both cases, the charge balance in the zinc sulfate solution is upset, with too many negative charges. This induces the formation of a zinc ion from the zinc electrode, which takes us back to the beginning of the cycle.**

**Run the cycle through your mind a few times until it becomes very clear. In particular, be sure that your mental model includes a flow of electrons (actually, more of a "jiggle" from neighbor to neighbor) through the metal electrodes and the connecting wires and voltmeter. This charge is compensated for by a flow of ions (again, it's actually a "jiggling" of neighbors) through the aqueous solutions and into and out of the salt bridge.**

### Active Example 19.1

Combine the following half-reactions to produce a balanced redox reaction equation. Indicate which half-reaction is an oxidation reaction and which is a reduction.

$$Co^{2+}(aq) + 2\ e^- \rightarrow Co(s)$$
$$Sn(s) \rightarrow Sn^{2+}(aq) + 2\ e^-$$

Reduction: $Co^{2+}(aq) + \cancel{2e^-} \rightarrow Co(s)$

Oxidation: $Sn(s) \rightarrow Sn^{2+}(aq) + \cancel{2e^-}$

Redox: $Co^{2+}(aq) + Sn(s) \rightarrow Co(s) + Sn^{2+}(aq)$

### Active Example 19.2

Combine the following half-reactions to produce a balanced redox equation. Identify the oxidation half-reaction and the reduction half-reaction.

$$Fe^{2+}(aq) \rightarrow Fe^{3+}(aq) + e^-$$
$$Al^{3+}(aq) + 3\ e^- \rightarrow Al(s)$$

Oxidation: $3\ Fe^{2+}(aq) \rightarrow 3\ Fe^{3+}(aq) + \cancel{3e^-}$

Reduction: $Al^{3+}(aq) + \cancel{3e^-} \rightarrow Al(s)$

Redox: $Al^{3+}(aq) + 3\ Fe^{2+}(aq) \rightarrow Al(s) + 3\ Fe^{3+}(aq)$

In this Active Example it is necessary to multiply the oxidation half-reaction equation by 3 in order to balance the electrons gained and lost.

Another reaction involving iron and aluminum introduces an additional technique.

### Active Example 19.3

Arrange and modify the following half-reactions as necessary, so they add up to produce a balanced redox equation. Identify the oxidation half-reaction and the reduction half-reaction.

$$Fe^{2+}(aq) + 2\ e^- \rightarrow Fe(s) \qquad Al(s) \rightarrow Al^{3+}(aq) + 3\ e^-$$

This will extend you a bit when it comes to balancing electrons. Two electrons are transferred for each atom of iron and three per atom of aluminum. In what ratio must the atoms be used to equate the electrons gained and lost? Rewrite, multiply, and add the half-reaction equations accordingly.

| | |
|---|---|
| Reduction: | $3\ Fe^{2+}(aq) + \cancel{6e^-} \rightarrow 3\ Fe(s)$ |
| Oxidation: | $2\ Al(s) \rightarrow 2\ Al^{3+}(aq) + \cancel{6e^-}$ |
| Redox: | $3\ Fe^{2+}(aq) + 2\ Al(s) \rightarrow 2\ Al^{3+}(aq) + 3\ Fe(s)$ |

In this Active Example, electrons are transferred two at a time in the iron half-reaction and three at a time in the aluminum half-reaction. The simplest way to equate these is to take the iron half-reaction three times and the aluminum half-reaction twice. This gives six electrons for both half-reactions. The number of electrons lost in the oxidation half-reaction must always equal the number of electrons gained in the reduction half-reaction.

## 19.2 Voltaic and Electrolytic Cells

Goal 3 Distinguish among electrolytic cells, voltaic cells, and galvanic cells.

Goal 4 Describe and identify the parts of an electrolytic or voltaic (galvanic) cell and explain how it operates.

Figure 19.6 illustrates the same copper-zinc system first shown in Figure 19.1, but the lightbulb of Figure 19.1 has been replaced with a voltmeter, a device that measures the force pushing electrons through the wire. This system is an example of a **voltaic cell,** which is also sometimes called a **galvanic cell.** A voltaic cell is one in which a spontaneous chemical reaction occurs, producing electricity and supplying it to an external circuit.

All voltaic cells are characterized by an electron-transfer reaction that occurs on its own without an external energy source. The **anode** is the electrode at which oxidation occurs. Since an oxidation reaction yields electrons, the anode has a negative charge. Reduction occurs at the **cathode,** which has a positive charge. Electrons flow through the external circuit from anode to cathode. The two **half-cells** are connected

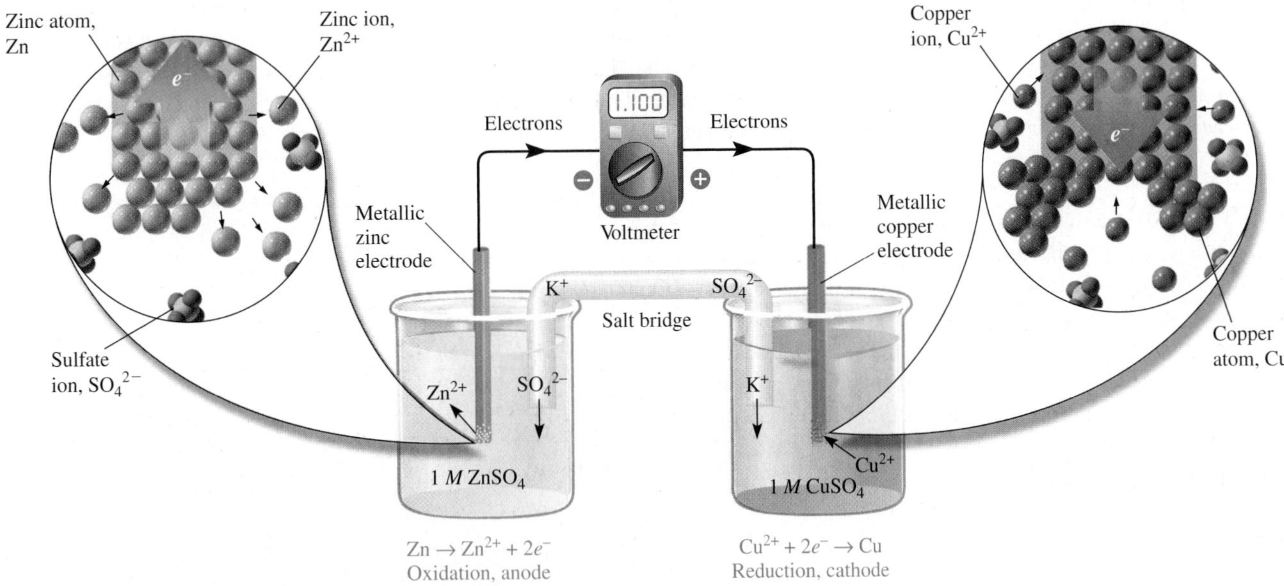


**Figure 19.6** A voltaic cell, also called a galvanic cell, consists of two different electrodes immersed in two different ionic solutions, the electrolytes. It produces a spontaneous flow of electricity through both the electrolyte and the external circuit. Chemical changes occur at the electrodes of both half-cells. The electrode at which oxidation occurs in the anode, and the electrode at which reduction occurs in the cathode.

Figure 19.7 Nickel-cadmium (ni-cad) batteries. The fundamental components of these batteries are nickel oxide, nickel hydroxide, and nickel metal, cadmium hydroxide and cadmium metal, and potassium hydroxide.

Figure 19.8 Lithium batteries. These batteries are typically made from lithium metal and manganese(IV) oxide.

by a salt bridge that maintains electrical neutrality in the solutions by allowing the passage of ions while preventing the mixing of the solutions.

The familiar alkaline battery used for flashlights, toys, and other electrical devices is a voltaic cell. When size is critical, as in calculators and watches, a mercury cell may be used. All of these cells "run down" and must be replaced when the chemical reactions in them reach equilibrium. The "ni-cad" (nickel–cadmium) voltaic cell runs down, too, but unlike the others, it can be recharged (Fig. 19.7). Lithium batteries, although relatively expensive, are now frequently used in portable consumer electronic devices because of their light weight (Fig. 19.8). Lithium has the lowest density of any metal. Another familiar rechargeable battery is the lead storage battery used in automobiles.

**P/REVIEW:** The operation of an electrolytic cell is described in Section 9.1. An electrolyte must contain charged ions for the cell to conduct a current when it is connected to an outside source of electricity.

An **electrolytic cell** is made up of a container holding an ionic liquid or an ionic solution called an **electrolyte** and two **electrodes** (Fig. 19.9). When the electrodes are connected to an outside source of electricity, they become charged, one positively and one negatively. Ions in the electrolyte move to the oppositely charged electrode, where chemical reactions occur. The movement of ions is an electric current. The whole process is called **electrolysis.**

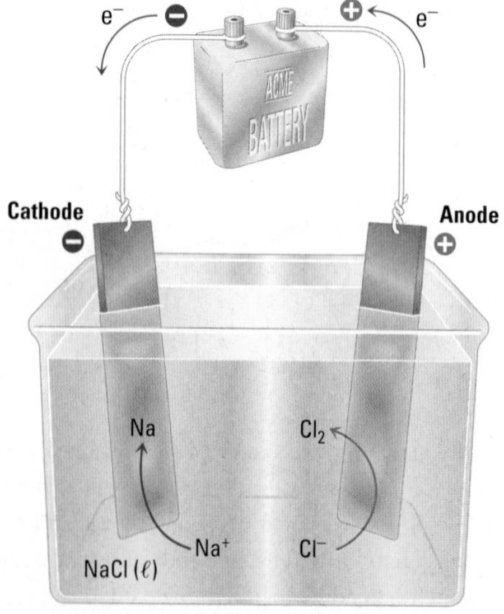


Figure 19.9 An electrolytic cell. This cell contains sodium chloride heated past its melting point, forming liquid or molten sodium chloride. An external energy source, such as a battery, causes a flow of electric current through the cell and an external circuit. The cells consist of two identical electrodes immersed in liquid. The sodium ions in the liquid are attracted to the cathode, where they react with electrons, forming sodium metal. The chloride ions are attracted to the anode, where they are oxidized to chlorine gas.

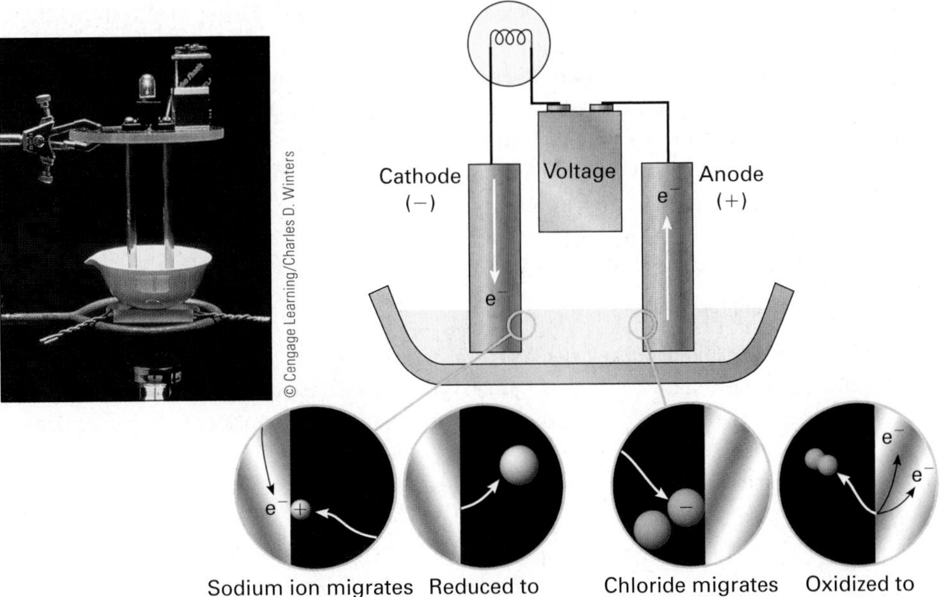


**Figure 19.10** Electrolysis of sodium chloride. Solid sodium chloride is heated past its melting point. A battery causes a flow of electric current, pumping electrons to the cathode, where sodium ions are reduced to sodium metal. Electrons are pumped from the anode, where chloride ions are oxidized to chlorine gas.

With the exception of barium, electrolytic cells are used to industrially produce all of the Group 1A/1 and 2A/2 metals. Sodium chloride is electrolyzed commercially in an apparatus called the Downs cell to produce sodium and chlorine. Figure 19.10 shows the small-scale electrolysis of sodium chloride. Chlorine also comes from the electrolysis of sodium chloride *solutions,* after which the used electrolyte is evaporated to recover sodium hydroxide. Many common objects are made of metals that are electroplated with copper, nickel, chromium, zinc, tin, silver, gold, and other elements (Fig. 19.11). The electrodeposits not only add beauty to the final product, but they also protect the underlying base metal from corrosion.

**Figure 19.11** Electroplating. An electrolytic cell is used to add a coating of a relatively expensive metal to a less expensive base metal. The base metal for the Oscar statues shown here is brittanium, an alloy (solid solution) of tin, copper, and antimony. They are then sequentially electroplated with copper, nickel, silver, and finally, 24-karat gold.

### √ Target Check 19.1

*Are the batteries used to power portable MP3 players best classified as electrolytic, voltaic, or galvanic cells? Explain.*

## 19.3 Oxidation Numbers and Redox Reactions

**Goal** 5 Given the formula of an element, molecule, or ion, assign an oxidation number to each element in the formula.

**Goal** 6 Describe and explain oxidation and reduction in terms of change in oxidation numbers.

The redox reactions that we have discussed up to this point have been relatively simple ones involving only two reactants. With Equations 19.3 and 19.5 through 19.7, we can see at a glance which species has gained and which has lost electrons. Some oxidation–reduction reactions are not so readily analyzed. Consider, for example, a reaction that is sometimes used in the general chemistry laboratory to prepare chlorine gas from hydrochloric acid:

$$MnO_2(s) + 4\,H^+(aq) + 2\,Cl^-(aq) \rightarrow Mn^{2+}(aq) + Cl_2(g) + 2\,H_2O(\ell) \quad \textbf{(19.8)}$$

or the reaction, in a lead storage battery that produces the electrical spark to start an automobile:

$$Pb(s) + PbO_2(s) + 4\,H^+(aq) + 2\,SO_4^{2-}(aq) \rightarrow 2\,PbSO_4(s) + 2\,H_2O(\ell) \quad \textbf{(19.9)}$$

© Cengage Learning/Charles D. Winters

Laboratory preparation of chlorine. In this reaction, NaCl, $K_2Cr_2O_7$, and $H_2SO_4$ react, forming $Cl_2$ as one product.

It is by no means obvious which species are gaining and losing electrons in these reactions.

"Electron bookkeeping" in redox reactions like Equations 19.8 and 19.9 is accomplished by using **oxidation numbers, a number assigned to each element in a species that is used to keep track of electrons,** which were introduced in Section 6.4. By following a set of rules, we may assign oxidation numbers to each element in a molecule or ion. The rules are as follows:

summary

**Oxidation Number Rules**

1. The oxidation number of any elemental substance is 0 (zero).
2. The oxidation number of a monatomic ion is the same as the charge on the ion.
3. The oxidation number of combined oxygen is −2, except in peroxides (−1), superoxides ($-\frac{1}{2}$), and $OF_2$ (+2).
4. The oxidation number of combined hydrogen is +1, except as a monatomic hydride ion, $H^-$.
5. In any molecular or ionic species, the sum of the oxidation numbers of all atoms in a molecule or polyatomic ion is equal to the charge on the species.

## Active Example 19.4

What are the oxidation numbers of the elements in $CO_2$?

Oxidation Rules 1, 2, and 4 do not apply. Rule 3 gives one of the two oxidation numbers required. What is it?

Oxygen, −2

Carbon can exist in several different oxidation states. You can decide which one by applying Rule 5. What is that oxidation number?

+4

Note that Rule 5 requires that the sum of the oxidation numbers of *atoms* in the formula unit be equal to the charge on the unit, which is 0 for $CO_2$. There are two oxygen atoms, each at −2. The total contribution of oxygen is 2(−2), or −4. The sum of −4 plus the oxidation number of carbon is equal to 0. Carbon must therefore be +4.

There is a mechanical way to reach the same conclusion that you might find helpful in more complicated examples. Applied to $CO_2$:

| | |
|---|---|
| Write the formula with space between the symbols of elements or ions. Place the oxidation number of each element or ion beneath its symbol. Use n for the unknown oxidation number. | C $O_2$<br>n −2 |
| Multiply each oxidation number by the number of atoms of that element in the formula unit. | C $O_2$<br>n 2(−2) |
| Add the oxidation numbers, set them equal to the charge on the species, and solve for the unknown oxidation number. In this case n = +4. | C $O_2$<br>n + 2(−2) = 0<br>n = +4 |

## Active Example 19.5

Find the oxidation number of (a) S in $SO_4^{2-}$ and (b) Cr in $HCr_2O_7^-$.

(a) S $O_4$
$n + 4(-2) = -2$
$n = +6$

(b) H $Cr_2$ $O_7$
$1 + 2(n) + 7(-2) = -1$
$n = +6$

Let's glance back at some of the equations in Section 19.2 to see if there isn't a regularity between oxidation and reduction and the change in oxidation number. Table 19.1 summarizes the changes in Equations 19.5 and 19.6 and in Active Example 19.2. Notice that for every oxidation half-reaction, the oxidation number increases. Conversely, for every reduction half-reaction, the oxidation number goes down.

We can now state a broader definition of oxidation and reduction. **Oxidation is an increase in oxidation number; reduction is a decrease in oxidation number.** These definitions are more useful in identifying the elements oxidized and reduced when the electron transfer is not apparent. All you must do is find the elements that change their oxidation numbers and determine the direction of each change. One element must increase, and the other must decrease. (This corresponds with one species losing electrons while another gains.)

There are some techniques that enable you to spot quickly an element that changes oxidation number or to dismiss quickly some elements that do not change. These are as follows:

1. An element that is in its elemental state must change. As an element on one side of the equation, its oxidation number is 0; as anything other than an element on the other side, its oxidation number is *not* 0 (Active Figure 19.12).

2. In other than elemental form, hydrogen is +1 and oxygen is −2. Unless they are elements on one side, they do not change. In more advanced courses you will have to be alert to the hydride, peroxide, and superoxide exceptions noted in the oxidation-number rules.

3. A Group 1A/1 or 2A/2 element has only one oxidation state other than 0. If it does not appear as an element, it does not change. This observation is helpful when you must find the element oxidized or reduced in a conventional equation.

Additional exceptions to the oxidation number rules may also be introduced in more advanced courses. Follow the advice of your instructor about which exceptions, if any, you are to learn in the course in which this textbook is being used.

We will now use these ideas and your ability to assign oxidation numbers to find the elements oxidized and reduced in Equation 19.8:

$$MnO_2(s) + 4\,H^+(aq) + 2\,Cl^-(aq) \rightarrow Mn^{2+}(aq) + Cl_2(g) + 2\,H_2O(\ell)$$

Chlorine is an element on the right, so it must be something else on the left. It is—the chloride ion, $Cl^-$. The oxidation number change is −1 to 0, an *increase,* so chlorine is *oxidized.*

**Table 19.1 Summary of Selected Oxidation–Reduction Reactions**

| Source | Oxidation Half-Reaction | Oxidation Number Change | Reduction Half-Reaction | Oxidation Number Change |
|---|---|---|---|---|
| Eq. 19.5 | $Fe \rightarrow Fe^{2+} + 2\,e^-$ | $0 \rightarrow +2$; increase | $2\,H^+ + 2\,e^- \rightarrow H_2$ | $+1 \rightarrow 0$; reduction |
| Eq. 19.6 | $2\,Br^- \rightarrow Br_2 + 2\,e^-$ | $-1 \rightarrow 0$; increase | $Cl_2 + 2\,e^- \rightarrow 2\,Cl^-$ | $0 \rightarrow -1$; reduction |
| Ex. 19.2 | $Fe^{2+} \rightarrow Fe^{3+} + e^-$ | $+2 \rightarrow +3$; increase | $Al^{3+} + 3\,e^- \rightarrow Al$ | $+3 \rightarrow 0$; reduction |

Photos: © Cengage Learning/ Charles D. Winters

**Active Figure 19.12** Oxidation numbers of reacting elements must change. (a) The oxidation numbers of aluminum and bromine are zero. (b) In a vigorous oxidation–reduction reaction, $2\ Al(s) + 3\ Br_2(\ell) \rightarrow Al_2Br_6(s)$, aluminum is oxidized and bromine is reduced. (c) The product of the reaction is solid $Al_2Br_6$, in which the oxidation number of aluminum is +3, and of bromine, −1. **Watch this active figure at *http://www.cengage.com/cracolice.***

Neither hydrogen nor oxygen appear as elements, so we conclude that they do not change oxidation state. That leaves manganese. Its oxidation state is +4 in $MnO_2$ on the left and +2 as $Mn^{2+}$ on the right. This is a *decrease,* from +4 to +2, so manganese is *reduced.*

## Active Example 19.6

Determine the element oxidized and the element reduced in a lead storage battery, Equation 19.9: $Pb(s) + PbO_2(s) + 4\ H^+(aq) + 2\ SO_4^{2-}(aq) \rightarrow 2\ PbSO_4(s) + 2\ H_2O(\ell)$.

To make this easier, we'll rewrite the equation in the answer space. Beneath the equation, write oxidation numbers for as many elements as necessary until you come up with the pair that changes. Then identify the oxidation and reduction changes. (Be careful. This one is a bit tricky.)

$$Pb(s) + PbO_2(s) + 4\ H^+(aq) + 2\ SO_4^{2-}(aq) \rightarrow 2\ PbSO_4(s) + 2\ H_2O(\ell)$$

Lead is both oxidized (0 in Pb to +2 in $PbSO_4$) and reduced (+4 in $PbO_2$ to +2 in $PbSO_4$).

The oxidation of lead can be spotted quickly because it is an element on the left. You might have thought sulfur to be the element reduced, but its oxidation state is +6 in the sulfate ion whether the ion is by itself on the left or part of a solid ionic compound on the right.

Even though the oxidation number concept is very useful for keeping track of what the electrons are doing in a redox reaction, we should emphasize that it has been invented to meet a need. It has no experimental basis. Unlike the charge of a monatomic ion, the oxidation number of an atom in a molecule or polyatomic ion cannot be measured in the laboratory. It is all very good to talk about "+4 manganese" in $MnO_2$ or "+6 sulfur" in the $SO_4^{2-}$ ion, but take care not to fall into the trap of thinking that the elements in these species actually carry positive charges equal to their oxidation numbers.

**summary**

**Definitions of Oxidation and Reduction**

| | Oxidation | Reduction |
|---|---|---|
| Change in electrons | Loss of electrons | Gain of electrons |
| Change in oxidation number | Increase | Decrease (reduction) |

## √ Target Check 19.2

*What is true about the oxidation numbers of each of the following: (a) elemental substance, (b) oxygen in a compound (some exceptions), (c) molecule, (d) monatomic ion, (e) hydrogen in a compound (one exception), (f) polyatomic ion?*

# 19.4 Oxidizing Agents and Reducing Agents

**Goal** 7 Given a redox equation, identify the oxidizing agent, the reducing agent, the element oxidized, and the element reduced.

The two essential reactants in a redox reaction are given special names to indicate the roles they play. The species that removes electrons, that is, the species that is itself reduced, is referred to as an **oxidizing agent.** The species from which the electrons are removed so reduction of another element can occur is called a **reducing agent.** Thus, the reducing agent is itself oxidized. For example, in Equation 19.5,

$$2\ H^+(aq) + Fe(s) \rightarrow H_2(g) + Fe^{2+}(aq)$$

$H^+$ has taken electrons from Fe—it has *oxidized* Fe to $Fe^{2+}$—and is therefore the oxidizing agent. Conversely, Fe is the source of electrons taken by $H^+$—it has allowed $H^+$ to be *reduced* to $H_2$—and is therefore the reducing agent. In Equation 19.8,

$$MnO_2(s) + 4\ H^+(aq) + 2\ Cl^-(aq) \rightarrow Mn^{2+}(aq) + Cl_2(g) + 2\ H_2O(\ell)$$

$Cl^-$ is the reducing agent, reducing manganese from +4 to +2. The oxidizing agent is $MnO_2$—the whole compound, not just the Mn; it oxidizes chlorine from −1 to 0.

The following Active Example summarizes the redox concepts.

Sometimes oxidizing agents are simply referred to as *oxidizers,* and reducing agents may also be called *reducers.*

## Active Example 19.7

Consider the redox equation

$$5\ NO_3^-(aq) + 3\ As(s) + 2\ H_2O(\ell) \rightarrow 5\ NO(g) + 3\ AsO_4^{3-}(aq) + 4\ H^+(aq)$$

a) Determine the oxidation number in each species:

N: _____ in $NO_3^-$ and _____ in NO.

As: _____ in As and _____ in $AsO_4^{3-}$. H: _____ in $H_2O$ and _____ in $H^+$.

O: _____ in $NO_3^-$, _____ in $H_2O$, _____ in NO, and _____ in $AsO_4^{3-}$.

b) Identify: (1) the element oxidized _____ (2) the element reduced _____

(3) the oxidizing agent _____ (4) the reducing agent _____

*continued*

a) N: +5 in $NO_3^{3-}$ and +2 in NO.
As: 0 in As and +5 in $AsO_4^{3-}$.
H: +1 in both $H_2O$ and $H^+$.
O: −2 in all species.

b) (1) As is oxidized, increasing in oxidation number from 0 to +5.
(2) N is reduced, decreasing in oxidation number from +5 to +2.
(3) $NO_3^-$ is the oxidizing agent, removing electrons from As.
(4) As is the reducing agent, furnishing electrons to $NO_3^-$.

## 19.5 Strengths of Oxidizing Agents and Reducing Agents

Goal 8 Distinguish between strong and weak oxidizing agents.

Goal 9 Given a table of the relative strengths of oxidizing and reducing agents, arrange a group of oxidizing agents or a group of reducing agents in order of increasing or decreasing strength.

An oxidizing agent earns its title by its ability to take electrons from another substance. A **strong oxidizing agent** has a strong attraction for electrons. Conversely, a **weak oxidizing agent** attracts electrons only slightly. The strength of a reducing agent is measured by its ability to give up electrons. A **strong reducing agent** releases electrons readily, whereas a **weak reducing agent** holds on to its electrons.

Table 19.2 lists oxidizing agents in order of decreasing strength on the left side of the equation and lists reducing agents in order of increasing strength on the right side. The strongest oxidizing agent shown is fluorine, $F_2$, located at the top of the left column. Chlorine, $Cl_2$, listed below fluorine, is used as a disinfectant in water supplies because of its ability to oxidize harmful organic matter. Notice that all equations in Table 19.2 are written as reduction half-reactions.

**Acid–Base-Before-Redox Option** If you have already studied Chapter 17, you might find it interesting to compare Table 19.2 with Table 17.1 in Section 17.5. Table 17.1 lists acids in order of decreasing strength on the left and bases in order of increasing strength on the right. In Table 17.1, the substances are listed according to their tendencies to release and take protons; in Table 19.2 the substances are listed according to their tendencies to take or give electrons. Just as Table 17.1 enables us to write acid–base reaction equations and predict the direction that will be favored at equilibrium, Table 19.2 enables us to do the same for redox reactions. These and other similarities between redox (electron-transfer) and acid–base (proton-transfer) reactions are summarized in Section 19.7.

### √ Target Check 19.3

*Why does a strong oxidizing agent become a weak reducing agent when it gains an electron?*

## 19.6 Predicting Redox Reactions

Goal 10 Given a table of the relative strengths of oxidizing and reducing agents, and information from which an electron-transfer reaction equation between two species in the table may be written, write the equation and predict the direction in which the reaction will be favored.

In Section 9.5 you wrote net ionic equations for possible single-replacement redox reactions. You used the activity series given in Table 9.2—reprinted in the margin on page 596—to predict whether or not the reaction would occur. In Table 9.2, the ele-

**Table 19.2 Relative Strengths of Oxidizing and Reducing Agents**

| STRENGTH | Oxidizing Agent | Reducing Agent | STRENGTH |
|---|---|---|---|
| Increasing ↑ | $F_2(g) + 2e^-$ | $\rightleftharpoons 2F^-(aq)$ | Decreasing ↑ |
| | $Au^+(aq) + e^-$ | $\rightleftharpoons Au(s)$ | |
| | $Au^{3+}(aq) + 3e^-$ | $\rightleftharpoons Au(s)$ | |
| | $Cl_2(g) + 2e^-$ | $\rightleftharpoons 2Cl^-(aq)$ | |
| | $O_2(g) + 4H^+(aq) + 4e^-$ | $\rightleftharpoons 2H_2O(\ell)$ | |
| | $Br_2(\ell) + 2e^-$ | $\rightleftharpoons 2Br^-(aq)$ | |
| | $NO_3^-(aq) + 4H^+ + 3e^-$ | $\rightleftharpoons NO(g) + 2H_2O(\ell)$ | |
| | $Ag^+(aq) + e^-$ | $\rightleftharpoons Ag(s)$ | |
| | $Fe^{3+}(aq) + e^-$ | $\rightleftharpoons Fe^{2+}(aq)$ | |
| | $I_2(s) + 2e^-$ | $\rightleftharpoons 2I^-(aq)$ | |
| | $Cu^+(aq) + e^-$ | $\rightleftharpoons Cu(s)$ | |
| | $Cu^{2+}(aq) + 2e^-$ | $\rightleftharpoons Cu(s)$ | |
| | $Cu^{2+}(aq) + e^-$ | $\rightleftharpoons Cu^+(aq)$ | |
| | $2H^+(aq) + 2e^-$ | $\rightleftharpoons H_2(g)$ | |
| | $Pb^{2+}(aq) + 2e^-$ | $\rightleftharpoons Pb(s)$ | |
| | $Sn^{2+}(aq) + 2e^-$ | $\rightleftharpoons Sn(s)$ | |
| | $Ni^{2+}(aq) + 2e^-$ | $\rightleftharpoons Ni(s)$ | |
| | $Co^{2+}(aq) + 2e^-$ | $\rightleftharpoons Co(s)$ | |
| | $Cd^{2+}(aq) + 2e^-$ | $\rightleftharpoons Cd(s)$ | |
| | $Fe^{2+}(aq) + 2e^-$ | $\rightleftharpoons Fe(s)$ | |
| | $Zn^{2+}(aq) + 2e^-$ | $\rightleftharpoons Zn(s)$ | |
| | $Al^{3+}(aq) + 3e^-$ | $\rightleftharpoons Al(s)$ | |
| | $Mg^{2+}(aq) + 2e^-$ | $\rightleftharpoons Mg(s)$ | |
| | $Na^+(aq) + e^-$ | $\rightleftharpoons Na(s)$ | |
| | $Ca^{2+}(aq) + 2e^-$ | $\rightleftharpoons Ca(s)$ | |
| | $Sr^{2+}(aq) + 2e^-$ | $\rightleftharpoons Sr(s)$ | |
| | $Ba^{2+}(aq) + 2e^-$ | $\rightleftharpoons Ba(s)$ | |
| | $Rb^+(aq) + e^-$ | $\rightleftharpoons Rb(s)$ | |
| | $K^+(aq) + e^-$ | $\rightleftharpoons K(s)$ | |
| Decreasing ↓ | $Li^+(aq) + e^-$ | $\rightleftharpoons Li(s)$ | Increasing ↓ |

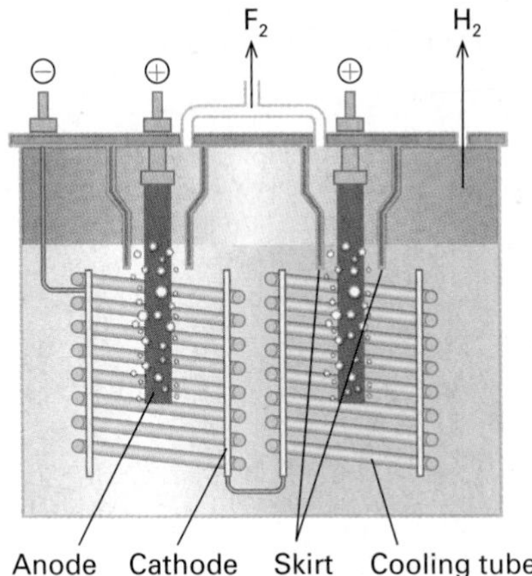


Industrial production of fluorine. Gaseous fluorine and hydrogen are the products of the electrolysis of a molten mixture of potassium fluoride and hydrogen fluoride. Great caution must be used in manufacturing, storing, and using fluorine due to its strength as an oxidizing agent.

One way to measure the relative strengths of oxidizing agents is to use them as electrodes in a galvanic cell and observe the voltage produced, as shown in Figure 19.6. In more advanced courses, Table 19.2 has an additional column in which voltmeter readings are recorded.

ments are listed in order of decreasing reactivity; the most active element is at the top. Any element in the activity series will react with and replace the dissolved ion of any element beneath it in the series.

Most metals in Table 9.2 are all included in the longer list of reducing agents in the right-hand column of Table 19.2. Moreover, their order, in terms of reactivity, is exactly the same in the two tables. (The order appears to be inverted because the reactivity *increases* as you go down the column in Table 19.2, and it is just the opposite of Table 9.2. The most active metal in Table 19.2 is at the bottom.) Now you can understand that the arrangement of the activity series is determined by the ease with which atoms of the element release electrons and function as reducing agents.

The reaction of solid iron with chlorine gas, forming iron(III) chloride. In this redox reaction, electrons are transferred from iron to chlorine. Chlorine is a stronger oxidizing agent than the iron(III) ion, and iron is a stronger reducing agent than chloride ion.

## Thinking About Your Thinking

### Classification

**Different classification systems can be used for different purposes. As you categorize your knowledge about chemistry, you need to reorganize your thinking when new information leads to a more efficient system. You will also find that there are strengths and weaknesses of various classification systems, and some may work better in one context while others work better in another context.**

**Table 9.2 Activity Series**

| Element | Activity |
|---|---|
| Li<br>K<br>Ba<br>Sr<br>Ca<br>Na | Will replace $H_2$ from liquid water, steam, or acid |
| Mg<br>Al<br>Mn<br>Zn<br>Cr | Will replace $H_2$ from steam or acid |
| Fe<br>Ni<br>Sn<br>Pb | Will replace $H_2$ from acid |
| $H_2$<br>Sb<br>Cu<br>Hg<br>Ag<br>Pd<br>Pt<br>Au | Will not replace $H_2$ from liquid water, steam, or acid |

**Section 19.6 provides new information that will cause you to reorganize how you think about the activity series introduced in Section 9.5. You can now see that the activity series is just a subset of elements derived from the Relative Strengths of Oxidizing and Reducing Agents, Table 19.2. The activity series is a list of reducing agents, listed from strongest to weakest. The information found in the activity series is all taken from the table of relative strengths.**

The next Active Example shows how to use Table 19.2 to write redox reaction equations.

## Active Example 19.8

Write the net ionic equation for the redox reaction between the cobalt(II) ion, $Co^{2+}$, and metallic silver, Ag.

In any redox reaction there must be an electron source (reducing agent) and an electron taker (oxidizing agent). Look at Table 19.2. Find $Co^{2+}$ among the oxidizing agents and Ag among the reducing agents. Take the half-reaction for the reduction of cobalt(II) ion directly from the table and write it down.

Reduction: $Co^{2+}(aq) + 2\ e^- \rightleftharpoons Co(s)$

To obtain the *oxidation* half-reaction for silver, you need to *reverse* the reduction half-reaction found in the table. Write it down.

Oxidation: $Ag(s) \rightleftharpoons Ag^+(aq) + e^-$

Multiply the oxidation equation by 2 to equalize electrons gained and lost and add it to the reduction equation to complete the net ionic equation.

| | |
|---|---|
| Reduction: | $Co^{2+}(aq) + \cancel{2e^-} \rightleftharpoons Co(s)$ |
| 2 × Oxidation: | $2\ Ag(s) \rightleftharpoons 2\ Ag^+(aq) + \cancel{2e^-}$ |
| Redox: | $Co^{2+}(aq) + 2\ Ag(s) \rightleftharpoons Co(s) + 2\ Ag^+(aq)$ |

The double arrows in Active Example 19.8 indicate that the reaction is reversible, as are most other redox reactions that occur in aqueous solution. When a reversible reaction equation is read from left to right, the **forward reaction,** or the reaction in the **forward direction,** is described; from right to left, the change is the **reverse reaction,** or in the **reverse direction.**

In an aqueous solution redox reaction, a strong oxidizing agent takes electrons from a strong reducing agent to produce weaker oxidizing and reducing agents. The reversible reactions proceed in both directions until equilibrium is reached. At that time, the concentrations of the weaker oxidizing and reducing agents are greater than the concentrations of the stronger oxidizer and reducer. The reaction is said to be **favored** in the direction pointing to the weaker oxidizing and reducing agents.

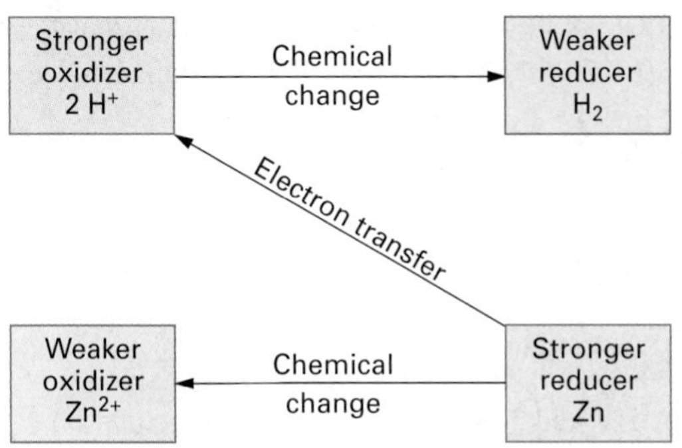


**Figure 19.13** Predicting redox reactions from positions in Table 19.2. The spontaneous chemical change always transfers one or more electrons from the stronger reducing agent to the stronger oxidizing agent, both shown in pink. The products of the reaction, the weaker oxidizing and reducing agents, are shown in blue. The favored direction, forward or reverse, has the weaker oxidizing and reducing agents as products.

Table 19.2 enables us to predict the favored direction of a redox reaction. This is illustrated in Figure 19.13. Applied to the reaction $2\ H^+(aq) + Fe(s) \rightleftharpoons H_2(g) + Fe^{2+}(aq)$ (Equation 19.5), the positions in the table establish $H^+$ and Fe as the stronger oxidizing agent and reducing agent, respectively. The reaction is favored in the forward direction, yielding the weaker oxidizing agent and reducing agent, $Fe^{2+}$ and $H_2$, respectively.

**P/REVIEW:** Figure 19.13 is strikingly similar to Figure 17.2 in Section 17.6, which describes the transfer of protons in an acid–base reaction.

## Active Example 19.9

In which direction, forward or reverse, will the redox reaction in Active Example 19.8 be favored? State the correct direction and explain. $Co^{2+}(aq) + 2\ Ag(s) \rightleftharpoons Co(s) + 2\ Ag^+(aq)$.

The reverse direction will be favored.

$Ag^+$ is a stronger oxidizing agent than $Co^{2+}$ and is therefore able to take electrons from cobalt atoms. Also, cobalt atoms are a stronger reducing agent than silver atoms, and therefore readily release electrons to $Ag^+$. The favored direction is toward the weaker reducing and oxidizing agents, Ag and $Co^{2+}$.

© Cengage Learning/Charles D. Winters

Metallic copper does not react with a strong acid to release hydrogen because copper and hydrogen ion are weaker reducing and oxidizing agents than hydrogen gas and copper ion, respectively.

## Active Example 19.10

Write the redox reaction between metallic copper and a strong acid, $H^+$, to form copper(II) ions and hydrogen gas and indicate the direction that is favored.

Reduction: $2\ H^+(aq) + 2e^- \rightleftharpoons H_2(g)$

Oxidation: $Cu(s) \rightleftharpoons Cu^{2+}(aq) + 2e^-$

Redox: $2\ H^+(aq) + Cu(s) \rightleftharpoons H_2(g) + Cu^{2+}(aq)$

The reverse reaction is favored.

Even though copper, like lead, does not react with a strong acid, it does react with nitric acid (see the Chapter 9 opening photo). This time copper and the nitrate ion are stronger reducing and oxidizing agents than nitrogen monoxide and copper ion, respectively. The dark gas is not nitrogen monoxide, but nitrogen dioxide, which forms when nitrogen monoxide comes into contact with oxygen in the air.

One of the properties of acids listed in Section 17.1 is their ability to release hydrogen gas on reaction with certain metals. Judging from Active Example 19.10, copper is not among those metals. The metals that do release hydrogen are the reducers below hydrogen in the right column of Table 19.2. But there is more to the reactions between metals and acids than meets the eye.

### Active Example 19.11

Write the equation between lead and nitric acid and predict which direction is favored.

Lead is in Table 19.2, but you will search in vain for $HNO_3$. The major species present in a nitric acid solution, $H^+$ and $NO_3^-$, are found in one equation, though (recall that nitric acid is a strong acid). We will comment on the imbalance between hydrogen ions and nitrate ions shortly. This reaction summarizes our equation-writing methods to this point. Take it all the way.

2 × Reduction: $2\ NO_3^-(aq) + 8\ H^+(aq) + \cancel{6\ e^-} \rightleftharpoons 2\ NO(g) + 4\ H_2O(\ell)$

3 × Oxidation: $3\ Pb(s) \rightleftharpoons 3\ Pb^{2+}(aq) + \cancel{6\ e^-}$

Redox: $2\ NO_3^-(aq) + 8\ H^+(aq) + 3\ Pb(s) \rightleftharpoons 2\ NO(g) + 4\ H_2O(\ell) + 3\ Pb^{2+}(aq)$

The forward reaction is favored.

Don't worry about those missing nitrate ions, the six unaccounted for from the eight moles of $HNO_3$ that furnished the 8 $H^+$. They are present as spectators.

**Acid–Base-Before-Redox Option** If you have already studied Chapter 17, you probably have noticed the several similarities between electron-transfer reactions (redox) and proton-transfer reactions (acid–base) that were mentioned earlier. These are summarized in Section 19.7, which is repeated at the appropriate place in Chapter 17.

## 19.7 Redox Reactions and Acid–Base Reactions Compared

Goal 11 Compare and contrast redox reactions with acid–base reactions.

At this point it may be useful to pause briefly and point out how acid–base reactions resemble redox reactions.

**1.** An acid–base reaction is a transfer of protons; a redox reaction is a transfer of electrons.

**2.** In both cases the reactants are given special names to indicate their roles in the transfer process. An acid is a proton source; a base is a proton remover. A reducing agent is an electron source; an oxidizing agent is an electron remover.

**3.** Just as certain species can either provide or remove protons (for example, $HCO_3^-$ and $H_2O$) and thereby behave as an acid in one reaction and as a base in another, certain species can either remove or provide electrons, acting as an oxidizing agent in one reaction and as a reducing agent in another. An example is the $Fe^{2+}$ ion, which can oxidize Zn atoms to $Zn^{2+}$ in the reaction

$$Fe^{2+}(aq) + Zn(s) \rightarrow Fe(s) + Zn^{2+}(aq)$$

$Fe^{2+}$ can also reduce $Cl_2$ molecules to $Cl^-$ ions in another reaction:

$$Cl_2(g) + 2\ Fe^{2+}(aq) \rightarrow 2\ Cl^-(aq) + 2\ Fe^{3+}(aq)$$

**4.** Just as we classify acids and bases as "strong" or "weak" depending on how readily they provide or remove protons, we can compare the strengths of oxidizing and reducing agents according to their tendencies to attract or release electrons.

**5.** Just as most acid–base reactions in solution reach a state of equilibrium, most aqueous redox reactions reach equilibrium. Just as we can predict the favored side of an acid–base equilibrium from acid–base strength, we can predict the favored side of a redox equilibrium from oxidizing agent–reducing agent strength.

## 19.8 Writing Redox Equations (Optional)

**Goal** **12** Given the before and after formulas of species containing elements that are oxidized and reduced in an acidic solution, write the oxidation and reduction half-reaction equations and the net ionic equation for the reaction.

Thus far we have considered only redox reactions for which the oxidation and reduction half-reactions are known. We are not always this fortunate. Sometimes we know only the species that contain the elements actually oxidized or reduced. We call the half-reaction equation that contains only these oxidized and reduced species a **skeleton equation.** Considering nitric acid, for example, suppose we know only that the product of the reduction of nitric acid is NO(g). How do we get from this information to the reduction half-reaction in Table 19.2?

The steps for writing a half-reaction equation in an acidic solution are listed below. Each step is illustrated for the $NO_3^-$-to-NO change in Active Example 19.11.

### procedure

**Writing Redox Equations for Half-Reactions in Acidic Solutions**

1. After identifying the element oxidized or reduced, write a skeleton half-reaction equation with the element in its original form (element, monatomic ion, or part of a polyatomic ion or compound) on the left and in its final form on the right:

$$NO_3^-(aq) \rightarrow NO(g)$$

2. Balance the element oxidized or reduced.

   Nitrogen is already balanced.

3. Balance elements other than hydrogen or oxygen, if any.

   There are none.

4. Balance oxygen by adding water molecules where necessary.

   There are three oxygens on the left and one on the right. Two water molecules are needed on the right:

$$NO_3^-(aq) \rightarrow NO(g) + 2\ H_2O(\ell)$$

5. Balance hydrogen by adding $H^+$ where necessary.

   There are four hydrogens on the right and none on the left. Four hydrogen ions are needed on the left:

$$4\ H^+(aq) + NO_3^-(aq) \rightarrow NO(g) + 2\ H_2O(\ell)$$

6. Balance charge by adding electrons to the more positive side.

   Total charge on the left is $+4 + (-1) = +3$; on the right it is zero. Three electrons are needed on the left:

$$3\ e^- + 4\ H^+(aq) + NO_3^-(aq) \rightarrow NO(g) + 2\ H_2O(\ell)$$

7. Recheck the equation to be sure it is balanced in both atoms and charge.

Notice that these instructions are for redox half-reactions in *acidic* solutions. The procedure is somewhat different with basic solutions, but we will omit that procedure in this introductory text.

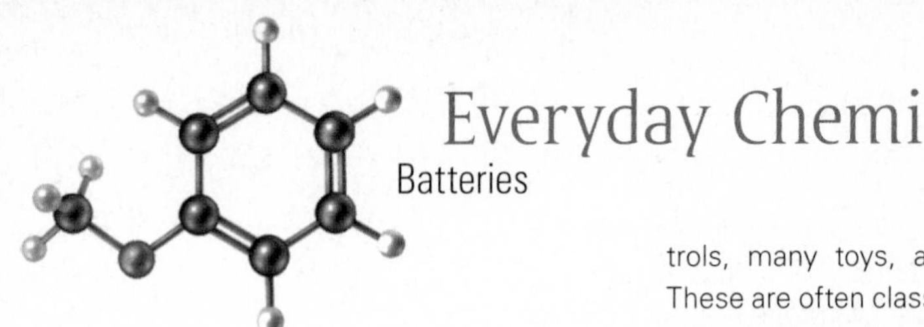

# Everyday Chemistry

## Batteries

Batteries are among the most common everyday applications of chemistry. Watches and clocks, MP3 players, toys, calculators, computers, smoke detectors, and automobiles are among the many common products that rely on battery power. Most citizens of industrialized nations can recall times when they became very aware of the role of batteries in their lives—those times when they encountered a dead battery!

Batteries are voltaic (or galvanic) cells. (Technically, the term *battery* only applies to a set of voltaic cells, but in common use, it describes both an individual cell and a set.) The same basic principles govern the operation of most common batteries. The reaction that occurs within a battery is an electron-transfer, or redox, reaction. As with all redox reactions, an oxidizing agent and a reducing agent are required, as well as electrodes and an electrolyte. The relative strengths of the oxidizing agent and reducing agent determine the characteristics of the battery.

Alkaline batteries (Fig. 19.14) are the power source in television remote controls, many toys, and digital cameras. These are often classified by their electrical potential and shape and size, such as the round AA or AAA 1.5-volt batteries or the rectangular 9-volt batteries. The anode of an alkaline battery is a gel that contains zinc in contact with a potassium hydroxide solution. The presence of the hydroxide ion is what gives these batteries their name. The cathode is a mixture of manganese(IV) oxide and graphite. The reactions are

Anode:

$$Zn(s) + 2\,OH^-(aq) \rightarrow ZnO(s) + H_2O(\ell) + 2\,e^-$$

Cathode:

$$2\,MnO_2(s) + H_2O(\ell) + 2\,e^- \rightarrow Mn_2O_3(s) + 2\,OH^-(aq)$$

Redox:

$$Zn(s) + 2\,MnO_2(s) \rightarrow ZnO(s) + Mn_2O_3(s)$$

Lead storage batteries (Fig. 19.15) are most frequently used to start cars. The most common configuration within such a battery is a series of six cells, each with a potential of 2 volts, for a total battery potential of 12 volts. The name *lead battery* is derived from both the anode, which is a collection of lead plates, and the cathode, lead(IV) oxide plates. The relatively high density of lead, 11.3 g/cm$^3$, makes lead storage batteries heavy. The electrolyte in these batteries is a sulfuric acid solution. The oxidation–reduction reactions are

Anode: $$Pb(s) + SO_4^{2-}(aq) \rightarrow PbSO_4(s) + 2\,e^-$$

Cathode:

$$PbO_2(s) + 4\,H^+(aq) + SO_4^{2-}(aq) + 2\,e^- \rightarrow PbSO_4(s) + 2\,H_2O(\ell)$$

Redox:

$$Pb(s) + PbO_2(s) + 4\,H^+(aq) + 2\,SO_4^{2-}(aq) \rightarrow 2\,PbSO_4(s) + 2\,H_2O(\ell)$$

One of the most important features of a lead storage battery is its ability to be recharged. A running engine powers an alternator that is designed to pump electrons through the battery, forcing the chemical reaction to run backward, regenerating the reactants:

$$2\,PbSO_4(s) + 2\,H_2O(\ell) \rightarrow Pb(s) + PbO_2(s) + 4\,H^+(aq) + 2\,SO_4^{2-}(aq)$$

From this point forward in your life, whenever you encounter a battery, we hope you will "see" the voltaic cell that lies beneath the metal or plastic casing. The electric power provided to your MP3 player, car starter, or any other battery-powered device comes from electron-transfer reactions.

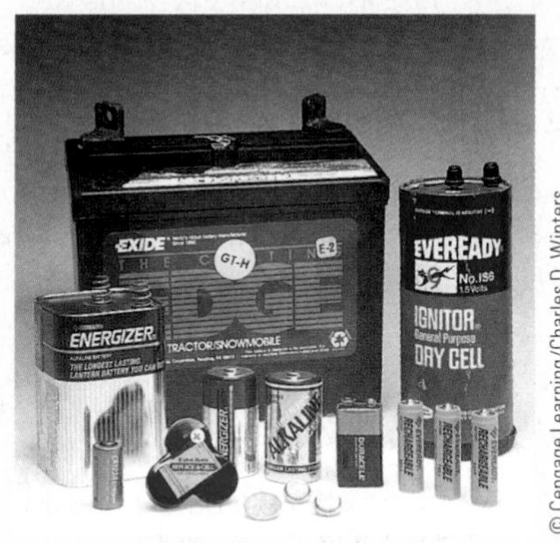


There are many sizes and shapes of batteries. The chemical reactions used to provide electrical energy vary, too. Nonetheless, all batteries are based on electron-transfer reactions.

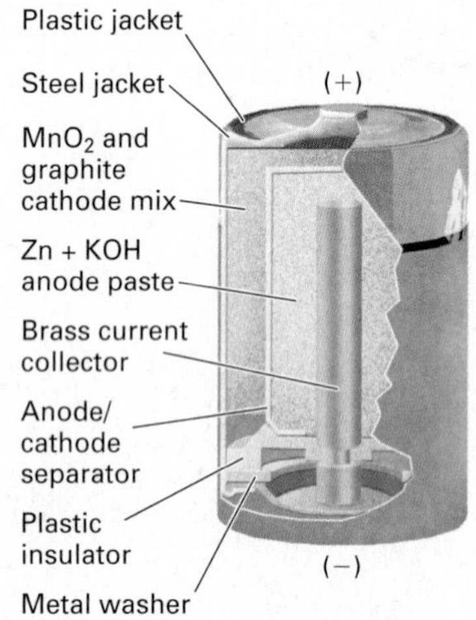


**Figure 19.14** A cutaway view of an alkaline battery. It is estimated that over 10,000,000,000 alkaline batteries are produced each year.

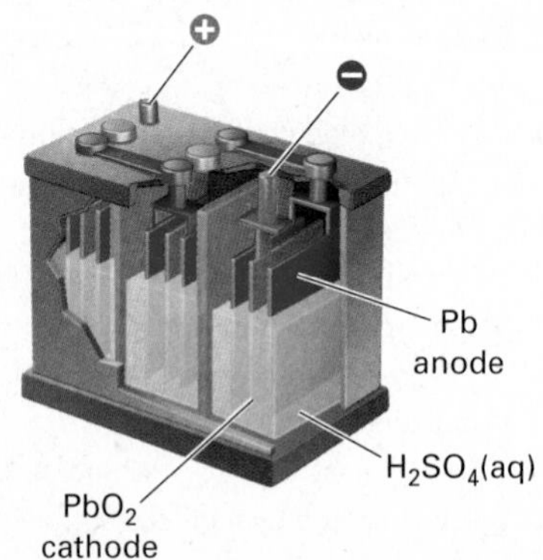


**Figure 19.15** A cutaway view of a lead storage battery. Nearly every automobile in the world uses a lead battery to power its starter.

When you have both half-reaction equations, proceed as in the earlier Active Examples.

## Active Example 19.12

Write the net ionic equation for the redox reaction between iodide and sulfate ions in an acidic solution. The products are iodine and sulfur. The skeleton equation is $I^-(aq) + SO_4^{2-}(aq) \rightarrow I_2(s) + S(s)$.

First, assign oxidation numbers to each element, and then identify the element reduced and the element oxidized.

Sulfur is reduced (ox. no. change +6 to 0) and iodine is oxidized (−1 to 0).

Balance atoms first, then charges, in the oxidation half-reaction, $I^-(aq) \rightarrow I_2(s)$.

$2\ I^-(aq) \rightarrow I_2(s) + 2\ e^-$ (This one happens to be in Table 19.2.)

Now for the reduction half-reaction, $SO_4^{2-}(aq) \rightarrow S(s)$, sulfur is already in balance. The only other element is oxygen. According to Step 4, you balance oxygen by adding the necessary water molecules. Complete that step.

$SO_4^{2-}(aq) \rightarrow S(s) + 4\ H_2O(\ell)$

Four oxygen atoms in a sulfate ion require four water molecules.

Next comes the hydrogen balancing, using $H^+$ ions.

$8\ H^+(aq) + SO_4^{2-}(aq) \rightarrow S(s) + 4\ H_2O(\ell)$

Finally, add to the positive side the electrons that will bring the charges into balance.

$6\ e^- + 8\ H^+(aq) + SO_4^{2-}(aq) \rightarrow S(s) + 4\ H_2O(\ell)$

On the left there are eight positive charges from hydrogen ion and two negative charges from sulfate ion, a net of 6+. On the right the net charge is zero. Charge is balanced by adding 6 electrons to the left (positive) side.

Now that you have the two half-reaction equations, finish writing the net ionic equation as you did before.

*continued*

Ottmar Bierwagen/Spectrum Stock

Large deposits of sulfur have been discovered in the United States, Mexico, and Poland. After the element is mined, it is shaped into large blocks for shipment.

$$\begin{array}{ll} \text{Reduction:} & \cancel{6e^-} + 8\,H^+(aq) + SO_4^{2-}(aq) \rightarrow S(s) + 4\,H_2O(\ell) \\ 3 \times \text{Oxidation:} & 6\,I^-(aq) \rightarrow 3\,I_2(s) + \cancel{6e^-} \\ \hline \text{Redox:} & 8\,H^+(aq) + SO_4^{2-}(aq) + 6\,I^-(aq) \rightarrow S(s) + 4\,H_2O(\ell) + 3\,I_2(s) \end{array}$$

Let's check to make sure the equation is, indeed, balanced:
—the atoms balance (six I, one S, four O, and eight H atoms on each side);
—the charge balances $[+8 + (-2) + 6(-1) = 0 + 0 + 0]$.

## Active Example 19.13

The permanganate ion, $MnO_4^-$, is a strong oxidizing agent that oxidizes chloride ion to chlorine in an acidic solution. Manganese ends up as a monatomic manganese(II) ion. Write the net ionic equation for the redox reaction.

This is a challenging Active Example, but watch how it falls into place when you follow the procedure that has been outlined. To be sure you have correctly interpreted the question, begin by writing an unbalanced skeleton equation. Put the identified reactants on the left side and the identified products on the right side.

$MnO_4^-(aq) + Cl^-(aq) \rightarrow Cl_2(g) + Mn^{2+}(aq)$

The oxidation half-reaction is easiest. Write it next.

$2\,Cl^-(aq) \rightarrow Cl_2(g) + 2\,e^-$

With a switch from iodine to chlorine, this is the same oxidation half-reaction as in the last Active Example.

Now write the formulas of the starting and ending species for the reduction half-reaction on opposite sides of the arrow.

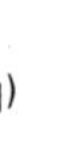

A purple solution containing permanganate ion is poured into a colorless acidic solution that contains chloride ion.

$MnO_4^-(aq) \rightarrow Mn^{2+}(aq)$

Can you take the skeleton reduction equation to a complete half-reaction equation? First do oxygen, then hydrogen, and then balance charge.

$8\ H^+(aq) + MnO_4^-(aq) + 5\ e^- \rightarrow Mn^{2+}(aq) + 4\ H_2O(\ell)$

Oxygen: $MnO_4^-(aq) \rightarrow Mn^{2+}(aq) + 4\ H_2O(\ell)$ (four waters for four oxygen atoms)

Hydrogen: $8\ H^+(aq) + MnO_4^-(aq) \rightarrow Mn^{2+}(aq) + 4\ H_2O(\ell)$ (eight $H^+$ for four waters)

Charge: $+8 + (-1) = +7$ on the left; $+2 + 0 = +2$ on the right. Charge is balanced by adding five negatives on the left, or five electrons in the final answer.

Now rewrite and combine the half-reaction equations for the net ionic equation.

2 × Reduction: $16\ H^+(aq) + 2\ MnO_4^-(aq) + \cancel{10\ e^-} \rightarrow 2\ Mn^{2+}(aq) + 8\ H_2O(\ell)$

5 × Oxidation: $10\ Cl^-(aq) \rightarrow 5\ Cl_2(g) + \cancel{10\ e^-}$

Redox: $16\ H^+(aq) + 2\ MnO_4^-(aq) + 10\ Cl^-(aq) \rightarrow 2\ Mn^{2+}(aq) + 8\ H_2O(\ell) + 5\ Cl_2(g)$

Checking:

—atoms balance (16 H, 2 Mn, 8 O, and 10 Cl);

—charges balance $[+16 + (-2) + (-10) = +4 = +4 + 0 + 0]$

Can you imagine a trial-and-error approach to an equation such as this?

# Chapter 19 in Review

Most of the key terms and concepts and many others appear in the Glossary. Use your Glossary regularly.

**19.1 Electron-Transfer Reactions**

Goal 1 Describe and explain oxidation and reduction in terms of electron transfer.

Goal 2 Given an oxidation half-reaction equation and a reduction half-reaction equation, combine them to form a net ionic equation for an oxidation–reduction reaction.

**Key Terms and Concepts: Electron-transfer reaction, half-reaction, half-reaction equation, oxidation, oxidized, reduced, reduction, salt bridge**

**19.2 Voltaic and Electrolytic Cells**

Goal 3 Distinguish among electrolytic cells, voltaic cells, and galvanic cells.

Goal 4 Describe and identify the parts of an electrolytic or voltaic (galvanic) cell and explain how it operates.

**Key Terms and Concepts: Anode, cathode, electrodes, electrolysis, electrolyte, electrolytic cell, half-cell, voltaic (galvanic) cell**

**19.3 Oxidation Numbers and Redox Reactions**

Goal 5 Given the formula of an element, molecule, or ion, assign an oxidation number to each element in the formula.

Goal 6 Describe and explain oxidation and reduction in terms of change in oxidation numbers.

**Key Terms and Concepts: Oxidation, oxidation number, reduction**

**19.4 Oxidizing Agents and Reducing Agents**

Goal 7 Given a redox equation, identify the oxidizing agent, the reducing agent, the element oxidized, and the element reduced.

**Key Terms and Concepts: Oxidizing agent, reducing agent**

**19.5 Strengths of Oxidizing Agents and Reducing Agents**

Goal 8 Distinguish between strong and weak oxidizing agents.

Goal 9 Given a table of the relative strengths of oxidizing and reducing agents, arrange a group of oxidizing agents or a group of reducing agents in order of increasing or decreasing strength.

**Key Terms and Concepts: Strong and weak oxidizing and reducing agents**

**19.6 Predicting Redox Reactions**

Goal 10 Given a table of the relative strengths of oxidizing and reducing agents, and information from which an electron-transfer reaction equation between two species in the table may be written, write the equation and predict the direction in which the reaction will be favored.

**Key Terms and Concepts: Favored direction, forward or reverse reaction or direction**

**19.7 Redox Reactions and Acid–Base Reactions Compared**

Goal 11 Compare and contrast redox reactions with acid–base reactions.

**19.8 Writing Redox Equations (Optional)**

Goal 12 Given the before and after formulas of species containing elements that are oxidized and reduced in an acidic solution, write the oxidation and reduction half-reaction equations and the net ionic equation for the reaction.

**Key Terms and Concepts: Skeleton equation**

# Study Hints and Pitfalls to Avoid

Oxidation and reduction are so closely related and similarly defined that they are easy to confuse. *Oxidation* is not a common word outside its chemical sense, but *reduction* is—and we can take advantage of it:

> If something is reduced, it gets smaller. If an element is reduced, its oxidation number becomes smaller. Oxidation is the opposite. Watch out for negative numbers that get smaller. "Getting smaller" means becoming more negative, as from −1 to −3.

"Becoming more negative" helps in the gain-or-loss-of-electron definition, too. "Becoming more negative" means getting more negative charge, or gaining negatively charged electrons. Oxidation is the opposite, that is, losing electrons.

Students sometimes summarize the relationship between species oxidized/reduced with oxidizing/reducing agents by saying, "Whatever is oxidized is the reducing agent, and whatever is reduced is the oxidizing agent." *Caution:* This is true for a *monatomic species only.* If the element being oxidized/reduced is a part of a polyatomic species, the entire compound or ion is the reducing/oxidizing agent.

There are several ways to balance complicated redox equations. We have shown you only one, and that is only for acidic solutions. If your instructor prefers another method, by all means use it. Whatever method you use, it takes practice to perfect it. You may question this while learning, but many students report that once they get the hang of it, balancing redox equations is fun!

# Concept-Linking Exercises

*Write a brief description of the relationships among each of the following groups of terms or phrases. Answers to the Concept-Linking Exercises are given after answers to the Target Checks in Appendix III.*

1. Electrolytic cell, electrolyte, voltaic cell, electrolysis, galvanic cell, electrode
2. Oxidation, reduction, oxidation half-reaction, reduction half-reaction
3. Oxidation, reduction, oxidation number
4. Oxidizing agent, reducing agent, oxidizer, reducer
5. Strong oxidizing agent, weak oxidizing agent, strong reducing agent, weak reducing agent

# Small-Group Discussion Questions

*Small-Group Discussion Questions are for group work, either in class or under the guidance of a leader during a discussion section.*

1. The oxidation–reduction process can be described as burning and "unburning." Determine which is which, and explain why these terms can be applied.
2. A voltaic cell is constructed with metal M in a solution containing $M^{2+}$ ions and Metal Me in a solution containing $Me^{2+}$ ions. The metals are attached via a conducting wire, and the solutions are connected by a salt bridge. As the cell runs, the color of the solution containing $M^{2+}$ gets lighter, and the color of the solution containing $Me^{2+}$ gets darker. (a) Sketch this cell. (b) Write an equation for each half-reaction. (c) On your sketch, indicate the direction in which electrons travel through the wire. (d) On your sketch, indicate the direction in which anions travel through the salt bridge and solutions. (e) Label the anode and cathode on your sketch, and indicate whether each is positive or negative. (f) What is oxidized and what is reduced? (g) What are the oxidizing and reducing agents? (h) Describe what happens to the mass of each metal as the cell runs.
3. Figure 19.9 is a macroscopic-level illustration of the electrolysis of liquid sodium chloride. Sketch a particulate-level illustration of the chemical changes that occur as this cell runs.
4. Vitamins such as vitamins C and E are often described as being good for you because they are antioxidants. Which of the terms that were introduced in this chapter (oxidized, reduced, oxidizing agent, reducing agent) describe the functions of antioxidants in the body? Explain.
5. Experiments are conducted with four metals, A, B, C, and D, and solutions of their ions. When solid D is added to solutions of the ions of the other metals, metallic A, B, and C are formed. When metal A is added to a solution containing ions of C, ions of A form along with metallic C. When hydrochloric acid is dropped on each metal, B and D react to yield hydrogen gas. A and C do not react with hydrochloric acid. List the metals A, B, C, and D in order of their strength as reducing agents, and explain your reasoning.
6. Steel objects can be coated with zinc in a process called galvanization. In terms of the relative strengths of oxidizing and reducing agents, why is this done?
7. A fuel cell is an electrochemical cell that has a continual supply of reactants. An example is a hydrogen–oxygen fuel cell, which has the half-reaction $H_2(g) \rightarrow 2\ H^+(aq) + 2\ e^-$ occur at the anode and the half-reaction $O_2(g) + 4\ H^+(aq) + 4\ e^- \rightarrow 2\ H_2O(\ell)$ occur at the cathode. (a) What are the fuels in a hydrogen-oxygen fuel cell? (b) Write the net ionic equation for the overall reaction that occurs in a fuel cell? (c) How does a fuel cell resemble a battery? How is it different? (d) What are the oxidation and reduction half-reactions in a hydrogen–oxygen fuel cell? (e) What advantage do the products of a hydrogen–oxygen fuel cell have over the products emitted by a conventional gasoline-burning engine? (f) What are the disadvantages of a hydrogen–oxygen fuel cell as compared with a gasoline engine?
8. Plants produce simple sugar by using energy from the sun in a process known as photosynthesis, which can be sum-

marized as $6\ CO_2 + 6\ H_2O + \text{energy} \rightarrow C_6H_{12}O_6 + 6\ O_2$. Animals metabolize simple sugar as a source of energy: $C_6H_{12}O_6 + 6\ O_2 \rightarrow 6\ CO_2 + 6\ H_2O + \text{energy}$. Explain the photosynthesis and metabolism of simple sugar in terms of oxidation and reduction. How would life on earth be altered if photosynthesis did not occur?

9. In Figure 19.9, which illustrates the electrolysis of liquid sodium chloride, we show a battery as a power source. It would have been more intuitive to show the apparatus plugged into a standard electrical outlet, but such a power source does not work for electrolysis because it supplies alternating current. Why can't alternating current be used for electrolytic cells? What do you suppose is used in industry as a power source for large-scale electroplating?

# Questions, Exercises, and Problems

*Blue-numbered questions are answered in Appendix III.* ■ *denotes problems assignable in OWL. In the first section of Questions, Exercises, and Problems, similar exercises and problems are paired in consecutive odd-even number combinations.*

### Section 19.1: Electron-Transfer Reactions

1. Using any example of a redox reaction, explain why such reactions are described as electron-transfer reactions.

2. ■ Identify the species oxidized and the species reduced in the following electron-transfer reaction: $Mg(s) + Co^{2+}(aq) \rightarrow Mg^{2+}(aq) + Co(s)$. As the reaction proceeds, electrons are transferred from _____ to _____.

3. Classify each of the following half-reaction equations as oxidation or reduction half-reactions:
   a) $Zn \rightarrow Zn^{2+} + 2\ e^-$
   b) $2\ H^+ + 2\ e^- \rightarrow H_2$
   c) $Fe^{2+} \rightarrow Fe^{3+} + e^-$
   d) $NO + 2\ H_2O \rightarrow NO_3^- + 4\ H^+ + 3\ e^-$

4. ■ Classify each of the following half-reaction equations as oxidation or reduction half-reactions:
   a) $2\ Cl^- \rightarrow Cl_2 + 2\ e^-$
   b) $Na \rightarrow Na^+ + e^-$
   c) $Sn^{2+} \rightarrow Sn^{4+} + 2\ e^-$
   d) $O_2 + 4\ H^+ + 4\ e^- \rightarrow 2\ H_2O$

*For the next four questions, classify the equation given as an oxidation or a reduction half-reaction equation.*

5. Dissolving ozone in water: $O_3 + H_2O + 2\ e^- \rightarrow O_2 + 2\ OH^-$

6. Tarnishing of silver: $2\ Ag + S^{2-} \rightarrow Ag_2S + 2\ e^-$

7. Dissolving gold (Z = 79): $Au + 4\ Cl^- \rightarrow AuCl_4^- + 3\ e^-$

8. One side of an automobile battery: $PbO_2 + SO_4^{2-} + 4\ H^+ + 2\ e^- \rightarrow PbSO_4 + 2\ H_2O$

9. ■ Combine the following half-reaction equations to produce a balanced redox equation:

$$Cr \rightarrow Cr^{3+} + 3\ e^-;\ Cl_2 + 2\ e^- \rightarrow 2\ Cl^-$$

10. ■ For the electron-transfer reaction $2\ Ag(s) + Br_2(\ell) \rightarrow 2\ Ag^+(aq) + 2\ Br^-(aq)$, write the oxidation half-reaction and the reduction half-reaction.

11. The half-reactions that take place at the electrodes of an alkaline cell, widely used in MP3 players, calculators, and other devices, are

$NiOOH + H_2O + e^- \rightarrow Ni(OH)_2 + OH^-$ and
$Cd + 2\ OH^- \rightarrow Cd(OH)_2 + 2\ e^-$

Which equation is for the oxidation half-reaction? Write the overall equation for the cell.

12. ■ Identify each of the following half-reactions as either an oxidation half-reaction or a reduction half-reaction: $Fe^{2+}(aq) \rightarrow Fe^{3+}(aq) + e^-$; $Cl_2(g) + 2\ e^- \rightarrow 2\ Cl^-(aq)$. Write a balanced equation for the overall redox reaction.

### Section 19.2: Voltaic and Electrolytic Cells

13. List as many things in your home as you can that are operated by voltaic cells.

14. How does electrolysis differ from the passage of electric current through a wire?

15. Can a galvanic cell operate an electrolytic cell? Explain.

16. Can an electrolytic cell operate a voltaic cell? Explain.

### Section 19.3: Oxidation Numbers and Redox Reactions

*For the next four questions, give the oxidation number of the element whose symbol is underlined.*

17. $\underline{Al}^{3+}$, $\underline{S}^{2-}$, $\underline{S}O_3^{2-}$, $Na_2\underline{S}O_4$

18. ■ $\underline{C}O_2$, $\underline{Cr}_2O_7^{2-}$, $\underline{Cl}^-$

19. $\underline{N}_2O_3$, $\underline{N}O_3^-$, $\underline{Cr}O_4^{2-}$, $NaH_2\underline{P}O_4$

20. ■ $H\underline{As}O_2$, $\underline{N}H_2OH$, $\underline{Ni}$

*In the next six questions, (1) identify the element experiencing oxidation or reduction, (2) state "oxidized" or "reduced," and (3) show the change in oxidation number. Example: $2\ Cl^- \rightarrow Cl_2 + 2\ e^-$. Chlorine oxidized from −1 to 0.*

21. a) $Br_2 + 2\ e^- \rightarrow 2\ Br^-$
    b) $Pb^{2+} + 2\ H_2O \rightarrow PbO_2 + 4\ H^+ + 2\ e^-$

22. ■ a) $Cu^{2+} + 2\ e^- \rightarrow Cu$
    b) $Co^{3+} + e^- \rightarrow Co^{2+}$

23. a) $8\ H^+ + IO_4^- + 8\ e^- \rightarrow I^- + 4\ H_2O$
    b) $4\ H^+ + O_2 + 4\ e^- \rightarrow 2\ H_2O$

24. a) $H_2O + SO_3^{2-} \rightarrow SO_4^{2-} + 2\ H^+ + 2\ e^-$
    b) $PH_3 \rightarrow P + 3\ H^+ + 3\ e^-$

25. a) $NO_2 + H_2O \rightarrow NO_3^- + 2\ H^+ + e^-$
    b) $2\ Cr^{3+} + 7\ H_2O \rightarrow Cr_2O_7^{2-} + 14\ H^+ + 6\ e^-$

26. a) $2\ HF \rightarrow F_2 + 2\ H^+ + 2\ e^-$
    b) $MnO_4^{2-} + 2\ H_2O + 2\ e^- \rightarrow MnO_2 + 4\ OH^-$

### Section 19.4: Oxidizing Agents and Reducing Agents

27. ■ Identify the oxidizing and reducing agents in $Cl_2$ + 2 $Br^-$ → 2 $Cl^-$ + $Br_2$.

28. ■ For the redox reaction $SO_4^{2-}$ + $Ni^{2+}$ → $SO_2$ + $NiO_2$, assign oxidation numbers and use them to identify the element oxidized, the element reduced, the oxidizing agent, and the reducing agent.

29. What is the oxidizing agent in the equation for the storage battery: Pb + $PbO_2$ + 4 $H^+$ + 2 $SO_4^{2-}$ → 2 $PbSO_4$ + 2 $H_2O$? What does it oxidize? Name the reducing agent and the species it reduces.

30. ■ For the redox reaction 3 $Mn^{2+}$ + 2 $HNO_3$ + 2 $H_2O$ → 3 $MnO_2$ + 2 NO + 6 $H^+$, assign oxidation numbers and use them to identify the element oxidized, the element reduced, the oxidizing agent, and the reducing agent.

### Section 19.5: Strengths of Oxidizing Agents and Reducing Agents

31. Which is the stronger reducer, Zn or $Fe^{2+}$? On what basis do you make your decision? What is the significance of one reducer being stronger than another?

32. ■ Consider the following half-reactions: $F_2(g) + 2\,e^- \rightleftharpoons 2\,F^-(aq)$; $Cu^{2+}(aq) + e^- \rightleftharpoons Cu^+(aq)$; $Fe^{2+}(aq) + 2\,e^- \rightleftharpoons Fe(s)$. Identify the strongest and weakest oxidizing agents and the strongest and weakest reducing agents.

33. Arrange the following oxidizers in order of increasing strength, that is, the weakest oxidizing agent first: $Na^+$, $Br_2$, $Fe^{2+}$, $Cu^{2+}$.

34. ■ Consider the following half-reactions: $Cd^{2+}(aq) + 2\,e^- \rightleftharpoons Cd(s)$; $Pb^{2+}(aq) + 2\,e^- \rightleftharpoons Pb(s)$; $Cl_2(g) + 2\,e^- \rightleftharpoons 2\,Cl^-(aq)$. Identify the strongest and weakest oxidizing agents and the strongest and weakest reducing agents.

### Section 19.6: Predicting Redox Reactions

*In this section, write the redox equation for the redox reactants given, using Table 19.2 as a source of the required half-reactions. Then predict the direction in which the reaction will be favored at equilibrium.*

35. $Br_2 + I^- \rightleftharpoons$

36. ■ $Ni^{2+} + Cd \rightleftharpoons$

37. $H^+ + Br^- \rightleftharpoons$

38. ■ $Cl_2 + Fe \rightleftharpoons$

39. $NO + H_2O + Fe^{2+} \rightleftharpoons$

40. ■ $H^+ + Mg \rightleftharpoons$

### Section 19.7: Redox Reactions and Acid–Base Reactions Compared

41. Explain how a strong acid is similar to a strong reducing agent. Explain how a strong base compares with a strong oxidizing agent.

42. Show how redox and acid–base reactions parallel each other—how they are similar, but also what makes them different.

### Section 19.8: Writing Redox Equations (Optional)

*In this section, each "equation" identifies an oxidizer and a reducer, as well as the oxidized and reduced products of the redox reaction. Write separate oxidation and reduction half-reaction equations, assuming that the reaction takes place in an acidic solution, and add them to produce a balanced redox equation.*

43. $S_2O_3^{2-} + Cl_2 \rightarrow SO_4^{2-} + Cl^-$

44. ■ $Zn + Sb_2O_5 \rightarrow Zn^{2+} + SbO^+$

45. $Sn + NO_3^- \rightarrow H_2SnO_3 + NO_2$

46. ■ $Ni^{2+} + Sn^{2+} \rightarrow NiO_2 + Sn$

47. $C_2O_4^{2-} + MnO_4^- \rightarrow CO_2 + Mn^{2+}$

48. ■ $Cl^- + NO_3^- \rightarrow ClO_3^- + NO$

49. $Cr_2O_7^{2-} + NH_4^+ \rightarrow Cr_2O_3 + N_2$

50. ■ $SO_4^{2-} + Hg \rightarrow SO_2 + Hg^{2+}$

51. $As_2O_3 + NO_3^- \rightarrow AsO_4^{3-} + NO$

52. ■ $Pb^{2+} + Zn^{2+} \rightarrow Zn + PbO_2$

### General Questions

53. Distinguish precisely, and in scientific terms, the differences among items in each of the following pairs.
    a) Oxidation, reduction (in terms of electrons)
    b) Half-reaction equation, net ionic equation
    c) Oxidation, reduction (in terms of oxidation numbers)
    d) Oxidizing agent (oxidizer), reducing agent (reducer)
    e) Electron-transfer reaction, proton-transfer reaction
    f) Strong oxidizing agent, weak oxidizing agent
    g) Strong reducing agent, weak reducing agent
    h) Atom balance, charge balance (in equations)

54. ■ Classify each of the following statements as true or false:
    a) Oxidation and reduction occur at the electrodes in a voltaic cell.
    b) The sum of the oxidation numbers in a molecular compound is zero, but in an ionic compound that sum may or may not be zero.
    c) The oxidation number of oxygen is the same in all of the following: $O^{2-}$, $HClO_3$, $S_2O_3^{2-}$, $NO_2$.
    d) The oxidation number of alkali metals is always −1.
    e) A substance that gains electrons is oxidized.
    f) A strong reducing agent has a strong attraction for electrons.
    g) The favored side of a redox equilibrium equation is the side with the weaker oxidizer and reducer.

55. One of the properties of acids listed in Chapter 17 is "the ability to react with certain metals and release hydrogen." Why is this property limited to certain metals? Identify two metals that do not release hydrogen from an acid and two that do.

56. The questions based on Section 19.8 identify reactants that engage in a redox reaction and their oxidized and reduced products. You were to write the redox equation. Nowhere among the reactants or products do you find a water molecule or a hydrogen ion. Yet, when writing the equation, you added these species. Why is this permissible?

57. There is a fundamental difference between the electrolytic cell in Figure 19.9 and the voltaic cell in Figure 19.6. What is that difference?

58. It is sometimes said that in a redox reaction the oxidizing agent is reduced and the reducing agent is oxidized. Is this statement (a) always correct, (b) never correct, or (c) sometimes correct? If you select (b) or (c), give an example in which the statement is incorrect.

#### More Challenging Questions

59. As an example of an electrolytic cell, the text states: "Sodium chloride is electrolyzed commercially in an apparatus called the Downs cell to produce sodium and chlorine." This is a high-temperature operation; the electrolyte is molten NaCl. Write the half-reaction equations for the changes taking place at each electrode. Is the electrode at which sodium is produced the anode or the cathode?

60. Examine Figure 19.16. Assume that the diagram represents one of the earliest known electroplating systems in which both electrodes are copper and the electrolyte is an acidic solution of copper(II) sulfate. The net operation of this system effectively transfers copper atoms from one electrode to the other.

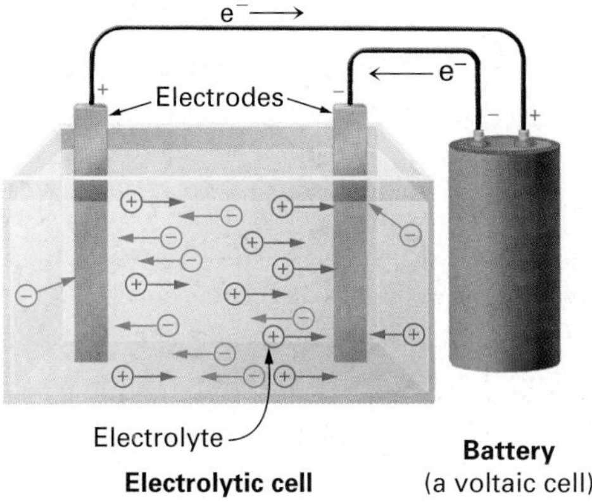


Figure 19.16

a) Write the oxidation and reduction half-reaction equations for the reactions that occur at the copper electrodes in Figure 19.16. Number your equations (1) and (2). (*Hint:* Read the introduction to this question very carefully. It contains what you need to write these equations.)
b) State which half-reaction equation, (1) or (2), represents the change at the electrode labeled "+" in Figure 19.16 and which is the change at the "−" electrode.
c) Which electrode, "+" or "−," is the anode and which is the cathode in Figure 19.16?
d) What ion or ions are carrying the charge through the electrolyte? From which electrode and to which electrode are they moving?
e) Ordinarily, this system operates at close to 100% efficiency, which means that 100 grams of copper dissolved at one electrode yields 100 grams of copper deposited at the other electrode. Sometimes, however, bubbles appear at one electrode, and the masses of copper dissolved and deposited are not quite equal. Considering everything in the solution, can you (1) identify the bubbles, (2) identify the electrode at which they appear, (3) write a half-reaction equation for what is occurring at that electrode, and (4) summarize in one sentence your answers to (1), (2), and (3)?

61. Marine equipment made of iron or copper alloys and through which seawater passes sometimes has inexpensive and replaceable zinc plugs sticking into the water stream. Can you suggest a reason for this? Explain your answer.

62. Have you ever put a piece of metal in your mouth, perhaps a paper clip or the end of a metal pencil, touched it to a metal filling in your teeth, and "tasted" the electricity produced? Whether you have done this or not, it does happen. Can you explain this phenomenon?

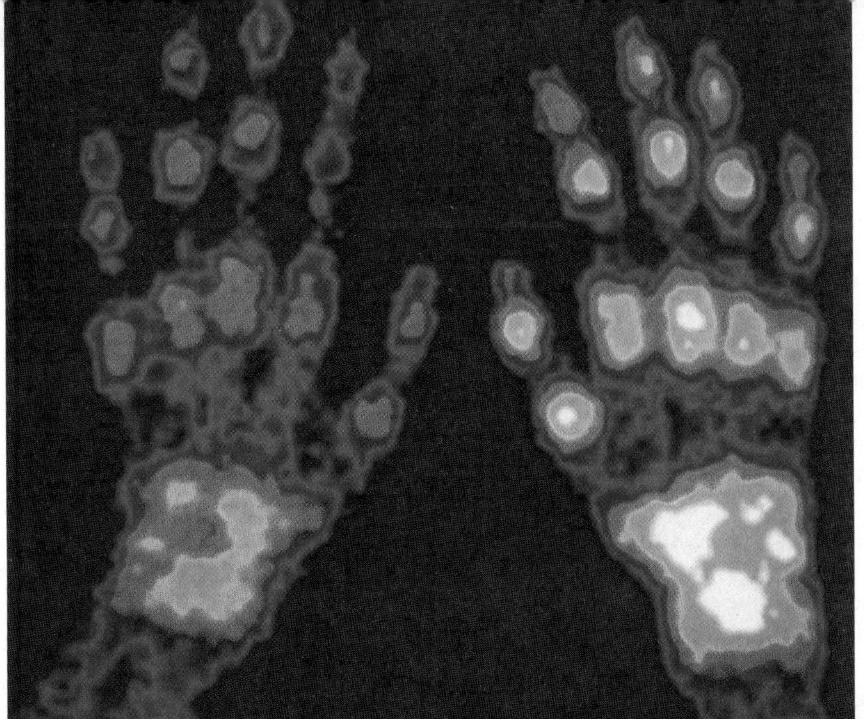

CNRI/Science Photo Library/Photo Researchers, Inc.

Nuclear chemistry has many medicinal applications. This is a gamma camera scan of the hands of a person with extensive arthritis. A radioactive substance was injected into the hands before the scan. This substance concentrates in inflamed tissue. The brighter areas indicate greater radiation emission. Notice how the right hand is more severely affected than the left.

# 20

# Nuclear Chemistry

CONTENTS



The study of the reactions of atomic nuclei by chemists is known as nuclear chemistry. In this chapter, you will learn about the nucleus and nuclear reactions. In practice, scientists in many fields study the nucleus. Much of the research in 21st-century physics has been directed toward understanding the forces that hold the nucleus together. Geologists use nuclear reactions to understand changes that occur on the earth. Astronomers study the nuclear reactions that occur in stars. The interaction of the radiation emitted from nuclear reactions with living systems is considered to be part of biology and medicine. You will see many examples in this chapter demonstrating that chemistry is truly the central science (see Fig. 1.6).

Online homework for this chapter may be assigned in OWL

© Bettmann/Corbis

Antoine Henri Becquerel (1852–1908) shared the 1903 Nobel Prize in Physics with Marie and Pierre Curie for his discovery of spontaneous radioactivity.

## 20.1 The Dawn of Nuclear Chemistry

**Goal** 1 Define radioactivity.

*Serendipity.* This pleasant-sounding word refers to finding valuable things you are not looking for—an accidental discovery, in other words. Serendipity has been a part of many scientific discoveries, but what happened to Henri Becquerel in 1896 stands above them all. What he stumbled across affects your life and the life of every living organism on this planet. Becquerel discovered nuclear chemistry.

Becquerel became interested in the penetrating power of X-rays soon after they were discovered—also accidentally—by Wilhelm Roentgen in 1895. Becquerel was also interested in phosphorescence, the phenomenon by which substances called phosphors glow after being exposed to light. He wondered if phosphorescent light could penetrate black paper as X-rays can. His plan was to put a uranium-containing phosphor on top of unexposed film wrapped in black paper and place them in sunlight. The paper would prevent the film from being exposed by the sunlight, so that if it were exposed at all, it would have to be from phosphorescent rays passing through the paper.

Alas, the day Becquerel chose for his experiment was cloudy. After waiting in vain for the sun to come out, he put his assembled material into a drawer to await the next sunny day. After several overcast days, Becquerel decided to develop the film to see if the initial cloudy-day experiment had caused even a trace of phosphorescent light to penetrate the paper. To his amazement, the film was highly exposed! The only explanation for this was that some sort of rays were leaving the uranium compound continuously, passing through the paper, and exposing the film. Sunlight and phosphorescence had nothing to do with the result.

**P/REVIEW:** Rutherford used alpha particles in his investigation of the atom (Section 5.3). Alpha particles are one of the three types of radioactive emissions discussed in the next section.

This is how Becquerel discovered **radioactivity, the spontaneous emission of particles and/or electromagnetic radiation resulting from the decay, or breaking up, of an atomic nucleus.** Of course, Becquerel did not know about the nucleus in 1896. In fact, Ernest Rutherford used one of those rays coming from a radioactive source in his 1911 experiment that led to the discovery of the nucleus (see the caption of Figure 5.3).

## 20.2 Radioactivity

**Goal** 2 Name, identify from a description, or describe three types of radioactive emissions.

Overall, there are at least 88 naturally occurring elements. (Elements with atomic numbers greater than 92 are not found or are found only in trace amounts in nature. They have been made in laboratories and, in some cases, on a commercial production level.) The nuclei of atoms of all the elements are referred to generally as **nuclides.*** All the isotopes of the elements in nature give us more than 300 different nuclides. Of these, 266 have been discovered that are stable; they should last forever.

**P/REVIEW:** If Sy is the symbol of an element, the nuclear symbol of an isotope of the element is

$^{\text{mass number}}_{\text{atomic number}}\text{Sy}$

The name of an isotope is the name of the element followed by the mass number, as $^{238}_{92}U$ is uranium-238. (See Section 5.4.)

The remaining 60-plus natural nuclides are radioactive. They are called **radionuclides.** Radionuclides are not stable. They are constantly changing, either into isotopes of the same element or into nuclides of different elements. When that happens, energy and a subatomic particle are given off. The new nuclide may also be radioactive, leading to further decay, or it may be a stable isotope of some element.

Three products from radioactive decay have been identified. If a beam consisting of all three products is aimed into an electric field, as in Figure 20.1, the products sepa-

*Some authors include the electrons outside the nucleus in the term *nuclide.*

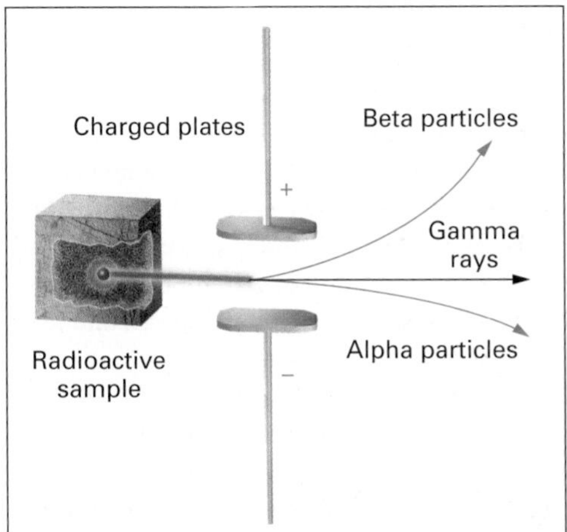


**Figure 20.1** Alpha ($\alpha$), beta ($\beta$), and gamma ($\gamma$) radioactive emissions. Alpha and beta particles are deflected by an electric field, but gamma rays are not. The alpha particles are deflected toward the negative plate, which means that they have a positive charge. The beta particles are deflected toward the positive plate to a greater degree than the alpha particles, indicating that they have less mass than alpha particles and a negative charge.

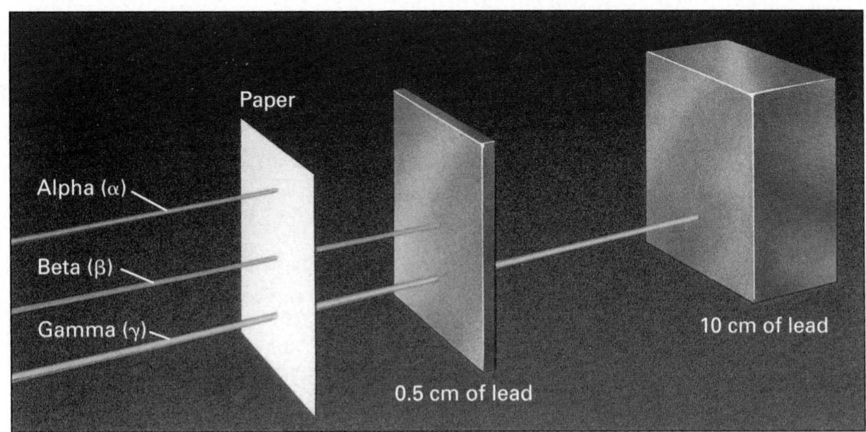


**Figure 20.2** Penetrating ability of radioactive emissions. Relatively large and highly charged alpha particles can be stopped by a sheet of paper. Beta particles penetrate paper, but they are stopped by moderately thick sheets of metals such as lead (0.5 cm = 0.2 in.). Gamma radiation has no mass or charge, and it penetrates paper and sheets of metals, but it can be stopped by thick blocks of metals (10 cm = 4 in.).

rate. One product, called an **alpha particle,** or **$\alpha$-particle,*** is attracted to the negatively charged plate, indicating that it has a positive charge. The $\alpha$-particle has little penetrating power; it can be stopped by the outer layer of skin or a few sheets of paper (Fig. 20.2). Alpha particles are now known to be nuclei of helium atoms, having the nuclear symbol ${}^{4}_{2}He$. Since alpha particles are just the nuclei of helium atoms, without the two electrons of a neutral helium atom, they have a 2+ charge. The emission of an alpha particle is an **alpha decay reaction,** or simply an **alpha decay.**

The second kind of radioactive emission also turns out to be a beam of particles, but these particles are negatively charged and therefore attracted to the positively charged plate (see Fig. 20.1). Called **beta particles,** or **$\beta$-particles,** they have been identified as electrons. The nuclear symbol for a beta particle is ${}^{0}_{-1}e$, indicating zero mass number and a 1− charge. $\beta$-particles have considerably more penetrating power than $\alpha$-particles, but they can be stopped by a sheet of lead or aluminum about 5 mm thick (see Fig. 20.2). The emission of a beta particle is a **beta decay reaction,** or **beta decay.**

The third kind of radiation is the **gamma ray,** or **$\gamma$-ray.** Gamma rays are not particles but very-high-energy electromagnetic rays, similar to X-rays. Because of their high energy, gamma rays have high penetrating power. They can be stopped only by thick layers of lead or heavy concrete walls, as shown in Figure 20.2. Because they do not have an electric charge, gamma rays are not deflected by an electric field (see Fig. 20.1).

*$\alpha$, $\beta$, and $\gamma$ are the Greek lowercase letters alpha, beta, and gamma.

**P/REVIEW:** The electromagnetic spectrum, which was described in Section 11.1 and Figure 11.1, includes gamma-rays, X-rays, ultraviolet and infrared rays, visible light, microwaves, and radio and television waves.

## Thinking About Your Thinking

### Mental Models

**Forming mental models of the three common products of radioactive decay will help you think about and understand radioactivity. An alpha particle is a helium nucleus, two protons and two neutrons (and no electrons). It therefore has a 2+ charge and is attracted to negatively charged objects. It is relatively bulky and highly charged for a product of radioactive decay, so it has the least penetrating power. A beta particle is an electron, with a 1− charge. It is attracted to positively charged objects, and its penetrating power is greater than that of an alpha particle because it has less of a charge and it is much smaller. Finally, gamma radiation is a packet of electromagnetic radiation, similar to visible light, but with higher energy. It has no charge, so it is not influenced by an electrically charged object. It has very high penetrating power because it is not a particle, but rather a form of energy. If you can now picture each type of radioactive emission in your mind, you have a good start at understanding radioactivity. Use your mental models as you answer questions and solve problems at the end of the chapter.**

Radioactive substances can be harmful, but some have become valuable tools in industry, research, and medicine. In the following sections we will discuss briefly some of these applications.

### √ Target Check 20.1

a) Write the nuclear symbol and electrical charge of an alpha particle and a beta particle.
b) List alpha, beta, and gamma rays in order of decreasing penetrating power.

## 20.3 The Detection and Measurement of Radioactivity

**Goal** 3 Identify the function of a Geiger counter and describe how it operates.

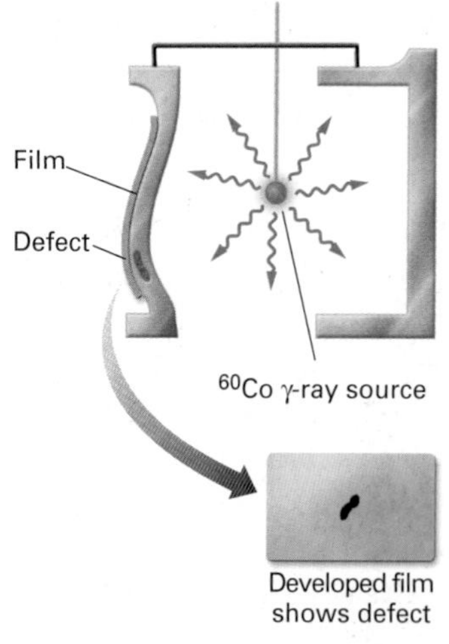


**Figure 20.3** A practical use of photographic film to detect radiation. A gamma-ray source is placed on one side of a cast metal part and film is placed on the other. If the part has a defect not visible to the eye, the gamma ray will reveal the defect by exposing the film more strongly at that location.

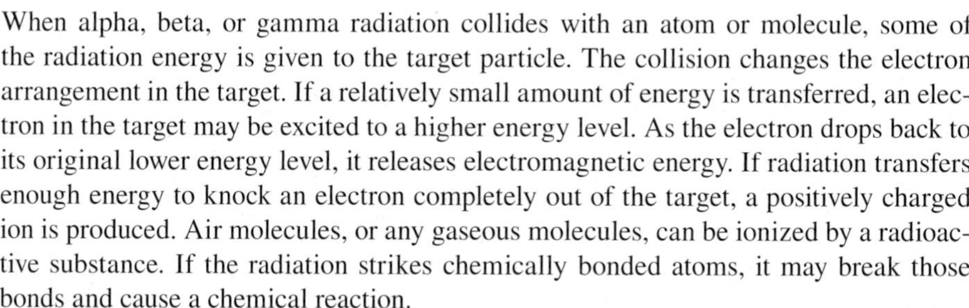

When alpha, beta, or gamma radiation collides with an atom or molecule, some of the radiation energy is given to the target particle. The collision changes the electron arrangement in the target. If a relatively small amount of energy is transferred, an electron in the target may be excited to a higher energy level. As the electron drops back to its original lower energy level, it releases electromagnetic energy. If radiation transfers enough energy to knock an electron completely out of the target, a positively charged ion is produced. Air molecules, or any gaseous molecules, can be ionized by a radioactive substance. If the radiation strikes chemically bonded atoms, it may break those bonds and cause a chemical reaction.

There are several ways to detect radioactivity. Perhaps the most obvious, but not necessarily the most convenient, is exposing photographic film, the very property that led to its discovery (Fig. 20.3). Another is the cloud chamber, an enclosed container that contains air and a supersaturated vapor, usually water. As ionizing radiation passes through the cloud chamber, some of the air is ionized. Water vapor condenses on the ions, leaving a cloudlike track that can be seen and photographed.

The **Geiger-Müller counter** (often shortened to **Geiger counter**) is the best-known instrument for measuring ionizing radiation. It consists of a tube filled with argon gas, as shown in Active Figure 20.4. The gas is ionized by radiation passing through a thin glass window, permitting an electrical discharge between two electrodes. The current may be measured quantitatively on a meter. Some Geiger counters emit a click when radiation is detected. Geiger counters are used to measure alpha and beta radiation.

Geiger counters do not measure gamma rays effectively because there are not enough gas particles to guarantee interaction with a neutral gamma ray. A **scintilla-**

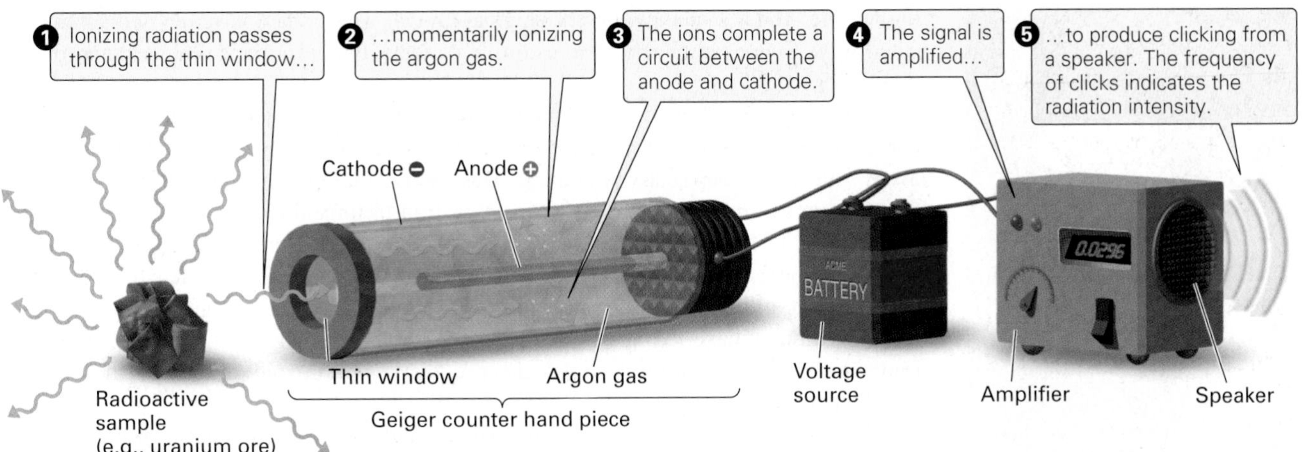
1 Ionizing radiation passes through the thin window...
2 ...momentarily ionizing the argon gas.
3 The ions complete a circuit between the anode and cathode.
4 The signal is amplified...
5 ...to produce clicking from a speaker. The frequency of clicks indicates the radiation intensity.
Cathode
Anode
ACME BATTERY
Thin window
Argon gas
Voltage source
Amplifier
Speaker
Radioactive sample (e.g., uranium ore)
Geiger counter hand piece

**Active Figure 20.4** How a Geiger counter works. **Watch this active figure at *http://www.cengage.com/cracolice.***

**tion counter** (Fig. 20.5) uses a transparent solid, which has a higher particle density than a gas. Higher density makes interaction with a gamma ray more likely. Particles in the solid absorb energy and then release some of it in flashes of light, which can be counted. This light emission, which may continue for some time after exposure to gamma rays stops, is **phosphorescence.** It is what Becquerel was looking for when he discovered radioactivity.

A common example of a scintillation counter is a watch with luminous hands. Watch hands used to be painted with a mixture of radium salts, which are radioactive, and zinc sulfide. Zinc sulfide is a phosphor that glows as the radium atoms decay. This kind of watch emitted both alpha and beta particles and will make a Geiger counter click. Newer luminous watches do not use radium salts, but instead use compounds containing tritium, ${}^{3}_{1}H$, which are sealed in microscopic glass vials glued to the watch hands and face. Tritium is a low-energy beta emitter, which does not trigger a Geiger-counter reading.

Two instruments used in medicine for detecting radioactivity are the gamma camera and the scanner (Fig. 20.6). The gamma camera is placed over a target area and takes a snapshot of it. The scanner moves while taking many pictures, each picture showing a "slice" of the area under study. These pictures may be combined to give a three-dimensional view of the interior of an organ. The photograph on the first page of this chapter shows a gamma camera scan, which is also known as a scintigram. These processes are simple, usually cause no discomfort to the patient, and are often used instead of exploratory surgery.

**Figure 20.5** Scintillation counter. This instrument is especially useful for detecting gamma radiation because it uses a high-density transparent solid, in contrast with the low-density gas in a Geiger counter. Radiation interacts with the solid, resulting in flashes of light, which are counted by a detector.

### √ Target Check 20.2

*How do Geiger counters and scintillation counters "measure" radioactivity? Precisely, what do they count? Do they count "radiations" or do they count something that is proportional to radiation? Suggest units for radioactivity as it would be measured by either of these counters.*

## 20.4 The Effects of Radiation on Living Systems

**Goal** 4 Explain how exposure to radiation may harm or help living systems.

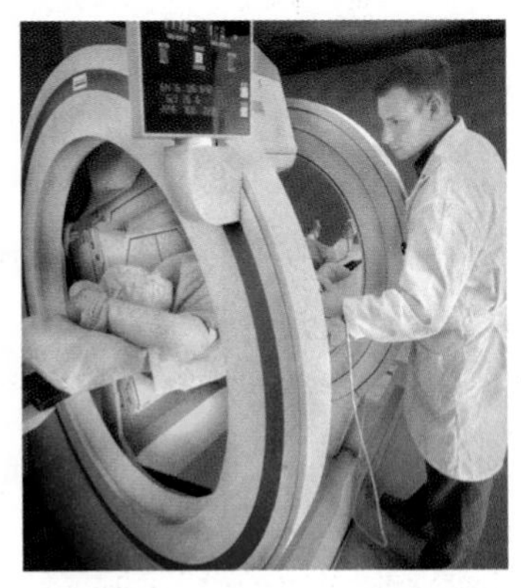

**Figure 20.6** Gamma camera scanner. Two gamma cameras scan the patient, one from above and one from below. In this model, the table moves past the cameras to scan the whole body. A typical bone scan takes about 20 minutes.

Shortly after radioactivity was discovered, people thought that radiation had certain curative powers. Radium compounds were made, and radium solutions were bottled and sold for drinking and bathing. That was before people knew the harmful effects of radiation exposure—and some early users of these "cures" paid a dear price for acquiring that

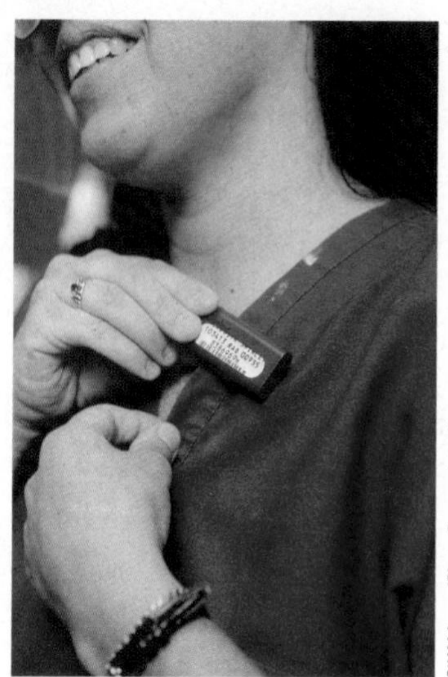


Film badge. Workers in industries where exposure to radiation is known to be above background levels are required to monitor their cumulative exposure dosage.

knowledge. Today's medical practitioners are much wiser. They have devised sophisticated ways to use radionuclides to examine patients, diagnose their illnesses, and treat their disorders.

The harmful effects of radiation on living systems come from its ability to break chemical bonds and thereby destroy healthy tissue. Obviously, the wise course is to avoid exposure to destructive radiation! Destructive radiation also has its good side: It destroys *unhealthy* tissue, too. Selective radiation of unhealthy tissue can eliminate it or reduce it. This is the key to radiation therapy (Fig. 20.7). The trick is to do it without damaging surrounding healthy tissue as well, or at least to be sure that the benefits of destroying unhealthy tissue outweigh the risks of destroying the good tissue. Modern practitioners have learned to do this fairly well, but there is still much room for improvement in targeting only the cancerous cells while avoiding the destruction of healthy cells.

The unit most commonly used to express radiation exposure is the **rem.*** More than 600 rems in one dose are fatal. The radiation therapy used in fighting cancers gives a *total* dose *over time* in the 4000- to 7000-rem range. This radiation indeed damages healthy tissue surrounding a tumor, but healthy cells are more capable of repairing themselves than are malignant cells.

Most of the radiation you encounter is measured not in rems, but in *milli*rems, mrem. For example, a chest X-ray is about 25 mrem, a complete diagnostic gastrointestinal X-ray series is about 2000 mrem, and a dental X-ray is about 0.5 mrem. The federal standard for occupational exposure is 5000 mrem/year.

Sources of radiation are everywhere. Radiation is emitted from rocks, soil, water, and even atoms in us. Although the atmosphere acts as a filter between us and the rest of the universe, we are constantly bombarded by radiation from sources outside the earth's atmosphere. Indeed, the average exposure to natural sources of radiation is estimated at 360 mrem/year. Figure 20.8 shows where this radiation comes from. Although it is not shown in the figure, exposure to the largest source of "natural" radiation (by far!) is voluntary for about 20% of Americans: tobacco smoke. At 1300 mrem per year, this is more than three times the dose for nonsmokers. Unfortunately for smokers, the alpha emitter $^{210}_{84}Po$ is found in the tobacco plant, and when inhaled, alpha particles can be very harmful.

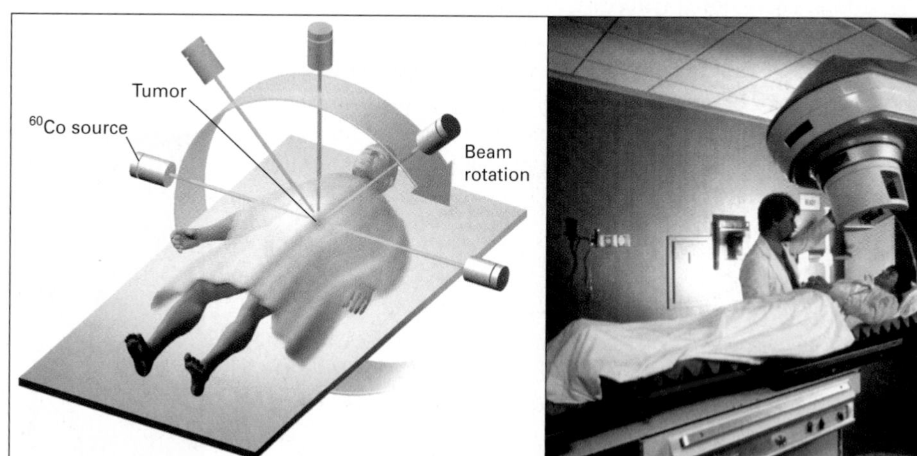

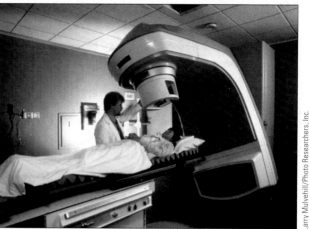

**Figure 20.7** Radiation therapy. A cobalt-60 radiation source emits gamma radiation, which is directed at a cancerous tumor. The beam is rotated to minimize the destruction of healthy tissue and concentrate the radiation at the tumor.

*The rem comes from **r**oentgen **e**quivalent in **m**an. The roentgen, R, is the amount of ionizing radiation that generates $2.09 \times 10^9$ ion pairs in one $cm^3$ of dry air. In SI units, 1 R = 0.00993 J/kg of air.

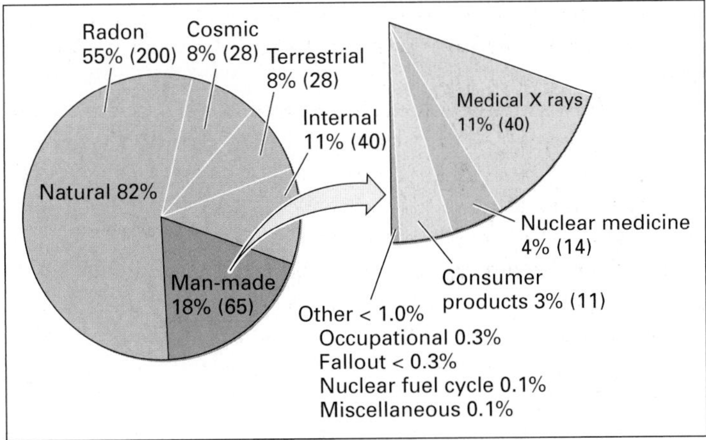


**Figure 20.8** Typical radiation experienced by people in the United States each year. The number of millirems per year for each source is in parentheses.

If you chose to live in a radiation-proof home to escape radiation originating from outer space, you would still have to contend with the radiation that is naturally in your body. If you weigh 150 pounds, you have about 225 grams of potassium ions in your body. Potassium ions participate in nerve conduction and the contraction of muscles, including the heart. Without potassium ions, you don't live. The natural abundance of ${}^{40}_{19}K$ is 0.0118% of all potassium ions. A 150-pound person therefore has about $4.1 \times 10^{20}$ radioactive ${}^{40}_{19}K$ atoms in his or her body. (You might like to confirm this statement by calculation. We'll show the calculation setup after the Target Check answers in Appendix III.)

### √ Target Check 20.3

*What is the easiest way you can protect yourself from the harmful effects of "natural" sources of radiation?*

## 20.5 Half-Life

**Goal** 5 Describe or illustrate what is meant by the half-life of a radioactive substance.

**Goal** 6 Given the starting quantity of a radioactive substance, Figure 20.9, and two of the following, calculate the third: half-life, elapsed time, quantity of isotope remaining.

The rate at which a radioactive substance decays is measured by its **half-life,** the time it takes for one half of the radioactive atoms in a sample to decay. Each radionuclide has its own unique half-life, commonly written $t_{1/2}$. The units of $t_{1/2}$ are time units per half-life, or time/half-life. Time may be expressed in seconds, minutes, hours, days, or years.

Figure 20.9 shows a graph of the fraction of an original sample that remains (vertical axes) after a number of half-lives (horizontal axis). The vertical axis at the left has a conventional scale, giving values in decimal fractions. The scale values on the axis on the right are the fractions that remain after each half-life period: 1/2 after the first half-life; 1/2 of 1/2, or 1/4 after the second; 1/2 of 1/4, or 1/8 after the third; and so forth. Thus, the fraction of a sample that is still present after n half-lives is $(1/2)^n$. If S is the starting quantity and R is the amount that remains after n half-lives, then

$$R = S \times (1/2)^n \qquad \textbf{(20.1)}$$

**Figure 20.9** Half-life decay curve for a radioactive substance.

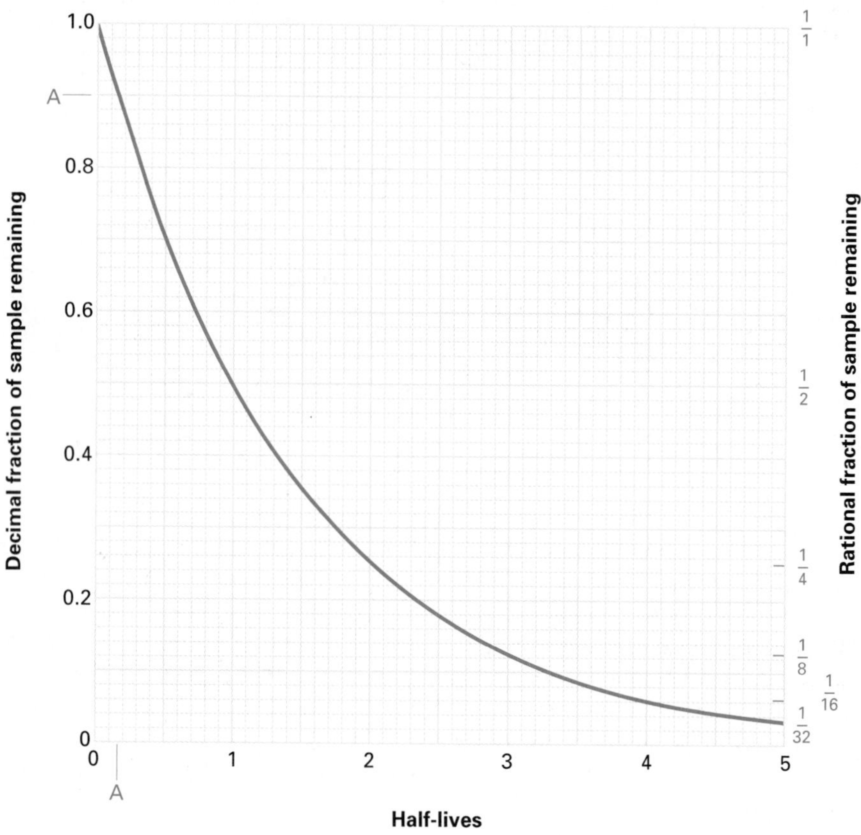


## Active Example 20.1

The half-life of ${}^{210}_{83}Bi$ is 5.0 days. If you begin with 16 grams of ${}^{210}_{83}Bi$, how many grams will you have left 25 days (5 half-lives) later?

This is a straightforward substitution into Equation 20.1. Recall that to raise a number to a power with a calculator, enter the number (0.5 for 1/2), press first $y^x$ and then =. Solve the problem.

*GIVEN:* 16 g ${}^{210}_{83}Bi$; 5 half-lives *WANTED:* g ${}^{210}_{83}Bi$ remaining

*EQUATION:* $R = S \times (0.5)^n = 16 \text{ g } {}^{210}_{83}Bi \times (0.5)^5 = 0.50 \text{ g } {}^{210}_{83}Bi$

## Active Example 20.2

The half-life of ${}^{45}_{19}K$ is 20 minutes. If you have a sample containing $2.1 \times 10^3$ micrograms ($\mu g$) of this isotope at noon, how many micrograms will remain at 3 o'clock in the afternoon?

This is a two-step problem. The number of micrograms of ${}^{45}_{19}K$ that remains is calculated by Equation 20.1. To use that equation, however, you need the number of half-lives, n. Because n is proportional to time, that number can be calculated by dimensional analysis. Plan the problem that far, set it up, and calculate the number of half-lives.

*Given:* 3 hours (noon to 3 p.m.); 20 min/half-life

*Wanted:* half-lives (n)

*Per:* $\frac{60 \text{ min}}{\text{hr}}$ $\qquad$ $\frac{1 \text{ half-life}}{20 \text{ min}}$

*Path:* hr ⟶ min ⟶ half-lives

$$3\ \cancel{\text{hr}} \times \frac{60\ \cancel{\text{min}}}{\cancel{\text{hr}}} \times \frac{1 \text{ half-life}}{20\ \cancel{\text{min}}} = 9 \text{ half-lives} = \text{n}$$

Now the problem is just like Active Example 20.1. Calculate the answer.

*Given:* $S = 2.1 \times 10^3\ \mu g\ ^{45}_{19}K$; $n = 9$ half-lives $\quad$ *Wanted:* $\mu g\ ^{45}_{19}K$ remaining (R)

*Equation:* $R = S \times (0.50)^n = 2.1 \times 10^3\ \mu g\ ^{45}_{19}K \times (0.5)^9 = 4.1\ \mu g\ ^{45}_{19}K$

To find the half-life of a radioactive isotope, you must determine starting and remaining quantities over a measured period of time. One way to interpret these data is to express the numbers as the fraction of the sample remaining, R/S. This is the vertical axis in Figure 20.9. Start with the fraction on the vertical axis, project horizontally to the curve, and then project vertically to the number of half-lives on the horizontal axis. Divide time by half-lives to get the half-life of the substance: time/half-lives = $t_{1/2}$.

You will use this procedure to solve the next Active Example.

## Active Example 20.3

The mass of radioactive $^{125}_{51}Sb$ in a sample is found to be 8.623 grams. The sample is set aside for 157 days, which is 0.430 year. At that time the sample contains 7.762 grams of $^{125}_{51}Sb$. Find the half-life of $^{125}_{51}Sb$ in years.

First, find the fraction of the isotope remaining after 0.430 year. The sample began with 8.623 grams, but only 7.762 grams remain. What fraction of 8.623 grams is 7.762 grams? Calculate that value.

*continued*

*GIVEN:* S = 8.632 g; R = 7.762 g
*WANTED:* Fraction remaining, R/S
R/S = 7.762 g/8.623 g = 0.9002

Now use Figure 20.9 to determine how many half-lives have passed when 0.9002 of the original amount—90% of the starting mass—remains.

0.16 half-life has passed.

"A" on the vertical axis of the graph corresponds to 0.90, the fraction of radionuclide remaining. Moving horizontally to the curve and projecting down to the horizontal axis—again "A"—gives 0.16 half-life.

You now have the number of years, 0.430 years, and the number of half-lives, 0.16 half-life. Calculate the years per half-life.

0.430 year/0.16 half-life = 2.7 years/half-life

You may wonder how you can solve Active Example 20.3 without Figure 20.9. Specifically, this means finding the number of half-lives, n, algebraically. The procedure is as follows:

Solve Equation 20.1 for $(0.5)^n$: $(0.5)^n = R/S$

Take the logarithm of both sides: $n \log 0.5 = \log (R/S)$

Solve for n: $$n = \frac{\log (R/S)}{\log 0.5}$$

Substitute values and calculate: $$n = \frac{\log(7.762/8.623)}{\log 0.5}$$

$$= \frac{\log 0.9002}{\log 0.5} = 0.1517$$

Then, 0.430 year/0.1517 half-life = 2.83 years/half-life. The small difference between the final values comes from the two-significant-figure reading of 0.16 half-life from Figure 20.9.

© Bettmann/Corbis

Willard Libby (1908–1980) was awarded the 1960 Nobel Prize in Chemistry for developing radio carbon dating.

This half-life rate of decay of radioactive substances is the basis of **radiocarbon dating,** by which scientists estimate the age of fossils. Carbon is found in all living organisms. Most carbon atoms are carbon-12; however, a small portion of the carbon in atmospheric carbon dioxide is carbon-14, a radioactive isotope with a half-life of 5.73 $\times 10^3$ years. When a plant or animal is alive, it takes in this isotope from its environment, while the same isotope in the organism is disappearing by nuclear disintegration. A "steady-state" situation exists while the organism lives, maintaining a constant ratio—the ratio found in the atmosphere—of carbon-14 to carbon-12. When the organism dies, the disintegration of carbon-14 continues, but its intake stops. This leads to a gradual reduction in the ratio of ${}^{14}_{6}C$ to ${}^{12}_{6}C$. By measuring the ratio and the amount of ${}^{14}_{6}C$ now present in a sample, we can calculate the ${}^{14}_{6}C$ present when the organism died. Figure 20.9 is then used to calculate the age of the sample.

## Active Example 20.4

Oetzi the Iceman (Figure 20.10) is a Neolithic hunter whose frozen remains were found in the Similaun Glacier on the Austrian/Italian border in the Tyrolean Alps.

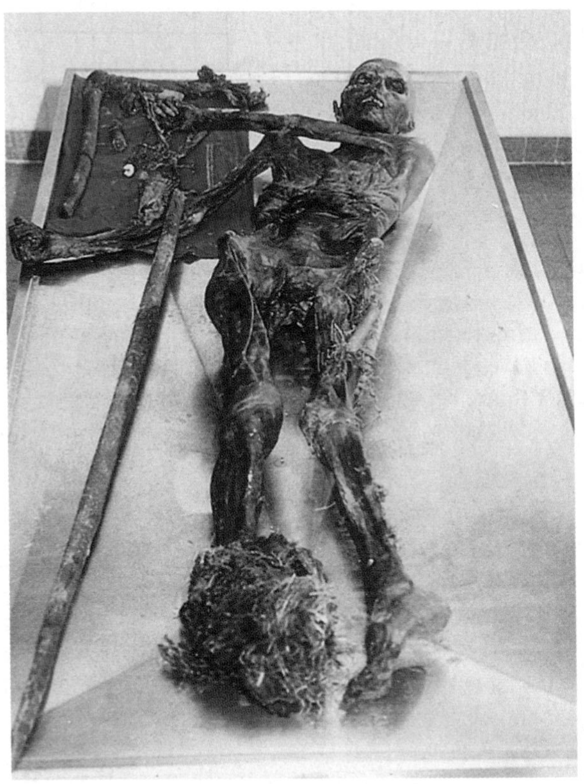


**Figure 20.10** Oetzi the Iceman. He was found in a glacier in the Tyrolean Alps near the Austrian/Italian border in 1991. He is believed to have lived in about 3300 B.C. He was so well preserved that scientists were able to examine the contents of his digestive system, finding that he was an omnivore.

Analysis of the Iceman shows that for every 15.3 units of $^{14}_{6}C$ present at the time of his death, 8.1 remain today. How old is the Iceman?

The quantities 15.3 units and 8.1 units identify starting and remaining amounts of radioactive isotope in the sample. Use these quantities and Figure 20.9 to determine the number of half-lives that have elapsed since the death of the Iceman.

GIVEN: S = 15.3 units; R = 8.1 units
WANTED: n (number of half-lives)
R/S = 8.1 units ÷ 15.3 units = 0.53

Matching number of half-lives from Figure 20.9: 0.93 half-life

The half-life of carbon-14 is $5.73 \times 10^3$ years. How many years are in 0.93 half-life?

*continued*

GIVEN: 0.93 half-life  WANTED: years

PER: $\frac{5.73 \times 10^3 \text{ years}}{\text{half-life}}$

PATH: half-life $\longrightarrow$ years

$$0.93 \text{ }\cancel{\text{half-life}} \times \frac{5.73 \times 10^3 \text{ years}}{\cancel{\text{half-life}}} = 5.3 \times 10^3 \text{ years}$$

Carbon dating has produced evidence of modern humans' presence on the earth as long ago as 200,000 years, although modern behavior and culture evolved much more recently. Similar dating techniques are also applied to mineral deposits. Analyses of geological deposits have identified rocks with an estimated age of 3.5 billion to 4.3 billion years, the latter figure being close to scientists' estimate of the age of the earth, 4.5 billion years. The oldest moon rocks also indicate an age of about 4.5 billion years.

## 20.6 Natural Radioactive Decay Series—Nuclear Equations

Goal 7 Describe a natural radioactive decay series.

Goal 8 Given the identity of a radioactive isotope and the particle it emits, write a nuclear equation for the emission.

**P/REVIEW:** The number of protons in an atom determines the element to which it belongs. All atoms of a specific element have the same number of protons, which is the same as the atomic number (see Section 5.4).

When a radionuclide emits an alpha or beta particle, there is a **transmutation** of an element, that is, a change from one element to another. This means that the remaining nuclide has a different atomic number—a different number of protons—from the original nuclide. A nuclear change—actually a nuclear reaction—has occurred. The original substance (nuclide) has been destroyed and a new substance (nuclide) has been formed.

The emission of a gamma ray does not change the elemental identity of the nucleus, even though energy is released. In that sense, a gamma emission, by itself, is not a nuclear change. Therefore, for the remainder of this chapter, we will consider only alpha and beta emissions in nuclear reactions.

Just as chemists write chemical equations to describe chemical changes, they write nuclear equations to describe nuclear changes. A nuclear equation shows the reactant nuclides or particles on the left and the product nuclides or particles on the right. The first step in the natural radioactive series observed by Becquerel is an alpha decay reaction. In it, a $^{238}_{92}U$ nucleus disintegrates, or decays, into a $^{4}_{2}He$ nucleus (alpha particle) and a $^{234}_{90}Th$ nucleus. The nuclear equation is

$$^{238}_{92}U \rightarrow {}^{4}_{2}He + {}^{234}_{90}Th \qquad \textbf{(20.2)}$$

Notice that this equation is balanced in both mass number and atomic number. The total number of neutrons and protons is 238, the mass number of the uranium isotope. The total mass number of the two products is 234 + 4, again 238. This accounts for all the protons and neutrons in the reactant, $^{238}_{92}U$. In terms of protons only, the 92 in a uranium nucleus are accounted for by 90 in the thorium nucleus plus 2 in the helium nucleus. *A nuclear equation is balanced if the sums of the mass numbers on the two sides of the equation are equal and if the sums of the atomic numbers are equal.*

The $^{234}_{90}Th$ nucleus resulting from the disintegration of uranium-238 is also radioactive. In a beta-decay reaction, it emits a beta particle, $^{0}_{-1}e$, and produces an isotope of protactinium, $^{234}_{91}Pa$:

$$^{234}_{90}Th \rightarrow {}^{234}_{91}Pa + {}^{0}_{-1}e \qquad \textbf{(20.3)}$$

In a beta-particle emission, the mass numbers of the reactant and product isotopes are the same, while the atomic number increases by 1. Although the actual process is more complex, it appears as if a neutron divides into a proton and an electron, and the electron is ejected.

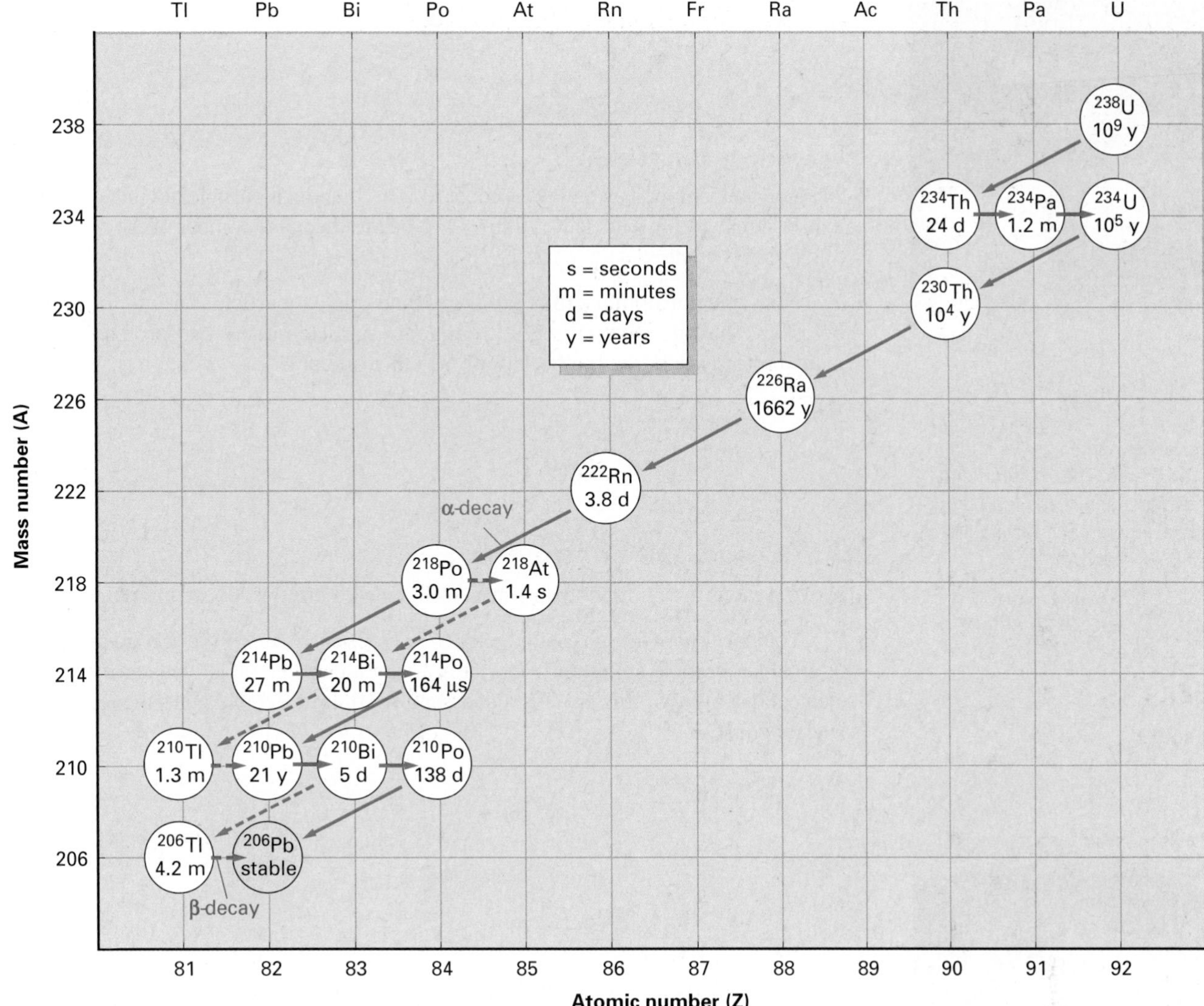


**Figure 20.11** Radioactive decay series. This series begins with $^{238}_{92}U$ and, after eight alpha emissions and six beta emissions, produces $^{206}_{82}Pb$ as a stable end product.

The two disintegrations described in Equations 20.2 and 20.3 are only the first two of 14 steps that begin with $^{238}_{92}U$. There are eight $\alpha$-particle emissions and six $\beta$-particle emissions, leading ultimately to a stable isotope of lead, $^{206}_{82}Pb$. This entire **natural radioactive decay series** is described in Figure 20.11. There are other natural disintegration series. One begins with $^{232}_{90}Th$ and ends with $^{208}_{82}Pb$, and another passes from $^{235}_{92}U$ to $^{207}_{82}Pb$.

## Active Example 20.5

Write the nuclear equation for the changes that occur in the uranium-238 disintegration series when $^{226}_{88}Ra$ ejects an $\alpha$-particle. Ra is the symbol for radium, one of the elements discovered by Pierre and Marie Curie in their study of radioactivity.

In a nuclear equation, one product will be the particle ejected. The mass number of the other product will be such that, when added to the mass number of the ejected particle, the total will be the mass number of the original isotope. What is the mass number of the second product of the emission of an alpha particle from a $^{226}_{88}Ra$ nucleus?

*continued*

© Cengage Learning/Charles D. Winters

Radon gas detection kit. Radon is believed to be the cause of 20,000 lung cancer deaths per year in the United States. The Environmental Protection Agency and the Surgeon General recommend testing all homes for radon.

222

The reactant isotope has a mass number of 226. It emits a particle having a mass number of 4. This leaves 226 − 4 = 222 as the mass number of the remaining particle.

Now find the atomic number of the second product particle. The atomic number of the starting isotope is 88. It emitted a particle having two protons. How many protons are left in the nucleus of the other product?

86

If two protons are emitted from a nucleus having 88 protons, 86 will remain.

You now know the mass number and the atomic number of the second product of an alpha-particle emission from $^{226}_{88}Ra$. Using a periodic table, you can find the elemental symbol of this product and assemble all three symbols into the required nuclear equation.

$$^{226}_{88}Ra \rightarrow {}^{4}_{2}He + {}^{222}_{86}Rn$$

AIP Emilio Segre Visual Archives, W. F. Muggers Collection

Marie Curie (1867–1934), who was awarded the 1903 Nobel Prize in Physics (with husband Pierre and Henri Becquerel) for the discovery of radioactivity and the 1911 Nobel Prize in Chemistry for the isolation of pure radium. She was the mother of Irene and Eve (Irene Curie-Joliot shared with her husband, Frederic, the 1935 Nobel Prize in Chemistry). Marie Curie was the first woman to teach at the Sorbonne (1906) and the first woman to be appointed full professor at the Sorbonne (1908).

The second product in Active Example 20.5, $^{222}_{86}Rn$, is a radioactive isotope of the gas radon. Radon-222 further decays by emitting an alpha particle. In recent years there has been concern about radon-222 that enters building basements from the soil outside. Exposure to radon gas may increase your chances of developing lung cancer, particularly if you smoke.

## Active Example 20.6

Write the nuclear equation for the emission of a $\beta$-particle from $^{210}_{83}Bi$.

The method is the same. Remember that the beta particle, $^{0}_{-1}e$, has zero mass number and an effective atomic number of −1. Both mass number and atomic number must be conserved in the equation.

$$^{210}_{83}Bi \rightarrow {}^{0}_{-1}e + {}^{210}_{84}Po$$

In the emission of a $\beta$-particle, the mass number of the radioactive isotope and the product isotope are the same. The product isotope has an atomic number greater by one than the radioactive isotope, an increase of one proton. Po is the symbol that corresponds to atomic number 84. The element is polonium, the other element discovered by the Curies in their investigation of radioactivity. The name of the element was selected to honor Madame Curie's native Poland.

## 20.7 Nuclear Reactions and Ordinary Chemical Reactions Compared

Goal **9** List or identify four ways in which nuclear reactions differ from ordinary chemical reactions.

Now that you have seen the nature of a nuclear change and the type of equation by which it is described, we will pause to compare nuclear reactions with the others you have studied. There are four areas of comparison:

1. In ordinary chemical reactions the chemical properties of an element depend only on the electrons outside the nucleus, and the properties are essentially the same for all isotopes of the element. The nuclear properties of the various isotopes of an element are quite different, however. In the radioactive decay series beginning with uranium-238, ${}^{234}_{90}Th$ emits a $\beta$-particle, whereas a bit farther down the line ${}^{230}_{90}Th$ ejects an $\alpha$-particle. Both ${}^{214}_{82}Pb$ and ${}^{210}_{82}Pb$ are $\beta$-particle emitters toward the end of the series, while the final product, ${}^{206}_{82}Pb$, has a stable nucleus, emitting neither alpha nor beta particles nor gamma rays.
2. Radioactivity is independent of the state of chemical combination of the radioactive isotope. The decomposition of ${}^{210}_{83}Bi$ occurs for atoms of that particular isotope whether they are in pure elemental bismuth, combined in bismuth chloride, $BiCl_3$, bismuth sulfate, $Bi_2(SO_4)_3$, or any other bismuth compound, or if they happen to be present in the low-melting alloy used in sprinkler systems for fire protection in large buildings.
3. Nuclear reactions result in the formation of different elements because of changes in the number of protons in the nucleus of an atom. In ordinary chemical reactions the atoms keep their identities while changing from one compound as a reactant to another as a product.
4. Both nuclear and ordinary chemical changes involve energy, but the amount of energy for a given amount of reactant in a nuclear change is enormous—greater by several orders of magnitude, or multiples of ten—compared with the energies of ordinary chemical reactions.

## 20.8 Nuclear Bombardment and Induced Radioactivity

Goal **10** Define or identify nuclear bombardment reactions.

Goal **11** Distinguish natural radioactivity from induced radioactivity produced by bombardment reactions.

Goal **12** Define or identify transuranium elements.

In natural radioactive decay we find an example of the alchemist's get-rich-quick dream of converting one element to another. But the natural process for uranium does not yield the gold coveted by the alchemist; rather, it produces the element lead, with which the dreamer wanted to begin his transmutation. The question remained after radioactivity was discovered: Can we initiate the transmutation of one ordinarily stable element into another?

In 1919 Ernest Rutherford answered "Yes" to that question. He found that he could bombard the nucleus of a nitrogen atom with a beam of alpha particles from a radioactive source, producing an atom of oxygen-17 and a hydrogen atom:

$${}^{14}_{7}N + {}^{4}_{2}He \rightarrow {}^{17}_{8}O + {}^{1}_{1}H$$

The oxygen isotope produced is stable; the experiment did not yield any radioactive isotopes. Similar experiments were conducted with other elements, using high-speed alpha particles as atomic "bullets." Scientists found that most of the elements up to

Hulton Archive/Getty Images

Marie, Irene, and Pierre Curie (*left to right*). Irene shared the 1935 Nobel Prize in Chemistry. Marie and Pierre shared the 1903 Nobel Prize in Physics, and Marie won the 1911 Nobel Prize in Chemistry.

potassium can be changed to other elements by **nuclear bombardment.** None of the isotopes produced were radioactive.

One experiment during this period was first thought to yield a nuclear particle that had emitted some sort of high-energy radiation, perhaps a gamma ray. In 1932 James Chadwick correctly interpreted the experiment, and in doing so he became the first person to identify the neutron. The reaction comes from bombarding a beryllium atom with a high-energy $\alpha$-particle:

$$^{9}_{4}\text{Be} + {}^{4}_{2}\text{He} \rightarrow {}^{12}_{6}\text{C} + {}^{1}_{0}\text{n}$$

$^{1}_{0}\text{n}$ is the nuclear symbol for the neutron, with zero charge and a mass number of 1.

Two years later, in 1934, Irene Curie, a daughter of Pierre and Marie Curie, and her husband, Frederic Joliot, used high-energy $\alpha$-particles to produce the first synthetic radionuclide. Their target was boron-10; the product was a radioactive nitrogen nucleus:

$$^{10}_{5}\text{B} + {}^{4}_{2}\text{He} \rightarrow {}^{13}_{7}\text{N} + {}^{1}_{0}\text{n}$$

Because this radionuclide is not found in nature, its decay is an example of **induced or artificial radioactivity.** When $^{13}_{7}\text{N}$ decays, it emits a particle having the mass of an electron and a charge equal to that of an electron, except that it is positive. This "positive electron" is called a **positron,** and it is represented by the symbol $^{0}_{1}\text{e}$. The decay equation is

$$^{13}_{7}\text{N} \rightarrow {}^{13}_{6}\text{C} + {}^{0}_{1}\text{e}$$

Today hundreds of radionuclides have been produced in laboratories all over the world. Many of these isotopes have been made in different kinds of **particle accelerators,** which use electric fields to increase the kinetic energy of the charged particles that bombard nuclei (Fig. 20.12). Particle accelerators are manufactured in two basic designs, linear and circular. Among the earliest and best-known accelerators is the cyclotron, so named because of its circular shape. It was invented by Ernest Lawrence at the University of California, Berkeley, who won the 1939 Nobel Prize in Physics for his efforts.

One of the more exciting areas of research with bombardment reactions has been the production of elements that do not exist in nature. Except in trace quantities, no natural elements having atomic numbers greater than 92 have ever been discovered. In 1940 it was found that uranium-238 is capable of capturing a neutron:

$$^{238}_{92}\text{U} + {}^{1}_{0}\text{n} \rightarrow {}^{239}_{92}\text{U} \qquad \textbf{(20.4)}$$

The newly formed isotope is unstable, progressing through two successive $\beta$-particle emissions, yielding isotopes of the elements having atomic numbers 93 and 94:

$$^{239}_{92}\text{U} \rightarrow {}^{0}_{-1}\text{e} + {}^{239}_{93}\text{Np (neptunium)} \qquad \textbf{(20.5)}$$

$$^{239}_{93}\text{Np} \rightarrow {}^{0}_{-1}\text{e} + {}^{239}_{94}\text{Pu (plutonium)} \qquad \textbf{(20.6)}$$

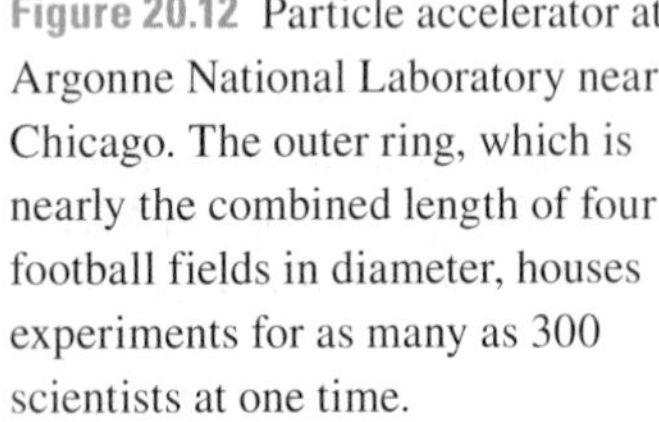

**Figure 20.12** Particle accelerator at Argonne National Laboratory near Chicago. The outer ring, which is nearly the combined length of four football fields in diameter, houses experiments for as many as 300 scientists at one time.

Courtesy of Argonne National Laboratory

Neptunium, plutonium, and all the other synthetic elements having atomic numbers greater than 92 are called the **transuranium elements.** All transuranium isotopes are radioactive, and some with very short half-lives have only been briefly isolated in extremely small amounts. Some of the bombardments yielding transuranium products use relatively high-mass isotopes as bullets. For example, einsteinium-247 is produced by bombarding uranium-238 with ordinary nitrogen nuclei:

$$^{238}_{92}U + ^{14}_{7}N \rightarrow ^{247}_{99}Es + 5\,^{1}_{0}n$$

### √ Target Check 20.4

a) What is produced in a nuclear bombardment reaction?

b) What property must a particle have if it is to be used in a particle accelerator?

## 20.9 Uses of Radionuclides

**Goal** 13 Identify and describe uses for synthetic radionuclides.

Today there are hundreds, possibly thousands, of uses for synthetic radionuclides. The best known of these are in medicine. People are not usually aware of others, although at times they may be close at hand. For example, do you have a smoke detector in your home? Battery-powered smoke detectors use a chip of americium-241, $^{241}_{95}Am$. The americium ionizes the air in the detector, which causes a small current to flow through the air. When smoke enters, it breaks the circuit and sets off the alarm, which is powered by a battery (Fig. 20.13). With a half-life of 458 years, the americium doesn't need changing every year as the battery does.

Artificial radionuclides are used in food preservation (Fig. 20.14). Worldwide, more than 40 classes of foods are irradiated with gamma rays from cobalt-60, $^{60}_{27}Co$,

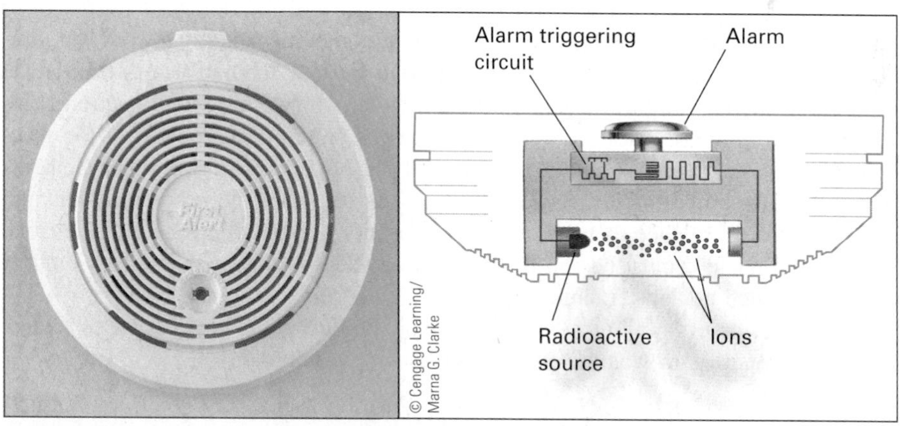


**Figure 20.13** Smoke detector. The radioactive source in a smoke detector is americium-241 oxide, embedded in a gold foil matrix. Americium is an alpha and gamma emitter. The radiation ionizes the air in the ionization chamber, producing an electric current. When smoke particles interact with the ions, the current is reduced, and the alarm is triggered.

**Figure 20.14** Using emissions from artificial radionuclides for food preservation. The roll on the right was irradiated to preserve it. Both rolls were then left for two weeks, and the roll on the left spoiled.

or cesium-137, $^{137}_{55}Cs$. This process retards the growth of bacteria, molds, and yeasts in foods, just as heat pasteurization extends the shelf life of milk. Higher doses of gamma radiation sterilize foodstuffs, killing insects as well as bacteria, molds, and yeasts.

Industrial applications of radionuclides include studies of piston wear and corrosion resistance. Petroleum companies use radionuclides to monitor the progress of some oils through pipelines. The thickness of thin sheets of metal, plastic, and paper is subject to continuous production control through the use of a Geiger counter to measure the amount of radiation that passes through the sheet; the thinner the sheet is, the more radiation that the counter will detect. Quality control laboratories can detect small traces of radioactive elements in a metal part.

Scientific research is another major application of radionuclides. Chemists use "tagged" atoms as *radioactive tracers* to study the mechanism, or series of individual steps, in complicated reactions. For example, by using water containing radioactive oxygen, scientists have determined that the oxygen in the glucose, $C_6H_{12}O_6$, formed in photosynthesis

$$6\ CO_2(g) + 6\ H_2O(\ell) \rightarrow C_6H_{12}O_6(s) + 6\ O_2(g)$$

comes entirely from the carbon dioxide and that all oxygen from water is released as oxygen gas. Archaeologists use neutron bombardment to produce radioactive isotopes in an artifact, which makes it possible to analyze the item without destroying it. Biologists employ radioactive tracers in the water absorbed by the roots of plants to study the rate at which the water is distributed throughout the plant system. These are but a few of the many ingenious applications that have been devised for this useful tool of science.

## 20.10 Nuclear Fission

**Goal** **14** Define or identify a nuclear fission reaction.

**Goal** **15** Define or identify a chain reaction.

In 1938, during the period when Nazi Germany was moving steadily toward war, dramatic and far-reaching events were taking place in German laboratories. A team made up of Otto Hahn, Fritz Strassman, and Lise Meitner was working with neutron bombardment of uranium. They were finding surprises among the products of the reaction. Namely, the products contained atoms of barium, krypton, and other elements far removed in both atomic mass and atomic number from the uranium atoms and neutrons used to produce them. The only explanation was, at that time, unbelievable; the uranium nucleus must be splitting into two nuclei of smaller mass. This kind of reaction is called **nuclear fission.**

© Corbis

Lise Meitner (1878–1968), one of a team of scientists who discovered nuclear fission. Meitner's work was recognized by naming element 109 meitnerium, Mt, in her honor.

In the fission of uranium-235 there are many products; it is not possible to write a single equation to show what happens. A representative equation is

$$^{235}_{92}U + ^{1}_{0}n \rightarrow ^{94}_{38}Sr + ^{139}_{54}Xe + 3\ ^{1}_{0}n \qquad \textbf{(20.7)}$$

Notice that it takes a neutron to initiate the reaction. Notice also that the reaction produces *three* neutrons. If one or two of these collide with other fissionable uranium nuclei, there is the possibility of another fission or two. And the neutrons from those reactions can trigger others, repeatedly, as long as the supply of nuclei lasts. This is what is meant by a **chain reaction** (Fig. 20.15), in which a nuclear product of the reaction becomes a nuclear reactant in the next step, thereby continuing the process.

The number of neutrons produced in the fission of $^{235}_{92}U$ varies with each reaction. Some reactions yield two neutrons per uranium atom; others, like that above, yield three, and still others produce four or more. The average is about 2.5. If the quantity of uranium, or any other fissionable isotope, is large enough that most of the neutrons produced are captured within the sample, rather than escaping to the surroundings, the chain reaction will continue. The minimum quantity required for this purpose is called the **critical mass.**

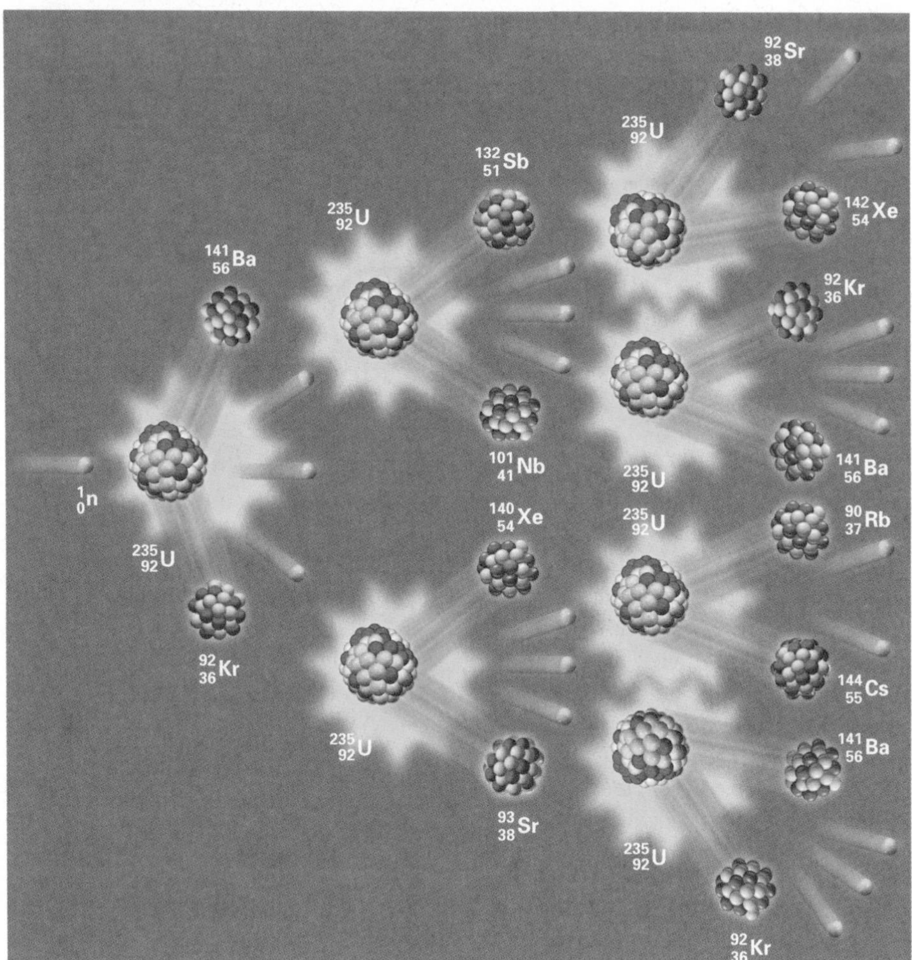


**Figure 20.15** Chain reaction started by the capture of a stray neutron. Each pair of isotopes shown is only one of the many different pairs that can be produced.

## Thinking About Your Thinking

### Mental Models

**Figure 20.15 is the starting point from which you can form a mental model of a nuclear chain reaction. On the left, you see the neutron, $^{1}_{0}n$, that is used to initiate the reaction. When it reacts with the uranium-235 nucleus, notice how two smaller nuclei (Ba and Kr) are produced and notice in particular the three neutrons that are also a product of the nuclear reaction. Now visually follow one neutron, the one moving down and to the right. It collides with another uranium-235 nucleus, producing two additional smaller nuclei and three additional neutrons. One neutron from an outside source gave rise to three neutrons, and these caused the formation of six neutrons (count the neutrons in the middle of the illustration), and these will in turn yield about 10 more neutrons, and so on. Now try to picture this process in your mind, as it would happen in three dimensions. As a test of your model, try to answer this question: Would a chain reaction be more likely to sustain itself in a sample in the shape of a sphere or in a sample of equal mass shaped like a sheet of paper? If you can mentally visualize the answer to this question, congratulations! You are well on your way to understanding nuclear fission.**

Uranium-235 is capable of sustaining a chain reaction, but it makes up only 0.7% of all naturally occurring uranium. Therefore, it is not a very satisfactory source of nuclear fuel (Fig. 20.16). An alternative is the plutonium isotope, $^{239}_{94}Pu$, produced from $^{238}_{92}U$, the most abundant uranium isotope (Equations 20.4 to 20.6). $^{239}_{94}Pu$ has a long half-life (24,110 years) and is fissionable. It has been used in the production of atomic bombs and is also used in some nuclear power plants to generate electrical energy. It is made in a **breeder reactor,** the name given to a device whose purpose is to produce fissionable fuel from nonfissionable isotopes.

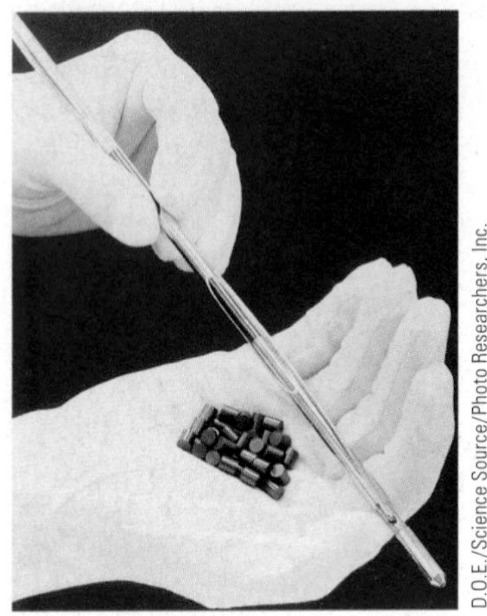

D.O.E./Science Source/Photo Researchers, Inc.

**Figure 20.16** Nuclear fuel. After mining, uranium is usually enriched to 3.5% $^{235}_{92}U$ in the form of uranium oxide, $UO_2$, powder. The powder is pressed into pellets, which are then inserted into tubes called fuel rods.

√ **Target Check 20.5**

*How do the products of a fission reaction compare with the reactants?*

## 20.11 Electrical Energy from Nuclear Fission

**Goal** **16** Describe how a nuclear power plant differs from a fossil-fueled power plant.

Except what is produced by hydroelectric plants located on major rivers, most electrical energy comes from generators driven by steam. Traditionally, the steam comes from boilers fuelled by oil, gas, or coal. The fast-dwindling supplies of these fossil fuels, and the uncertainties surrounding the availability and cost of petroleum from the countries where it is abundant, have once again brought attention to nuclear fission as an alternative energy source.

A diagram of a nuclear power plant is shown in Figure 20.17. The turbine, generator, and condenser are similar to those found in any fuel-burning power plant. The nuclear fission reaction has three main components: the fuel elements, control rods, and moderator. The fuel elements are simply long trays that hold fissionable material in the reactor. As the fission reaction proceeds, fast-moving neutrons are released. These neutrons are slowed down by a moderator, which is water in the reactor illustrated. When the slower neutrons collide with more fissionable material, the reaction continues. The reaction rate is governed by cadmium or boron control rods, which absorb excess neutrons. At times of peak power demand, the control rods are largely withdrawn from the reactor, permitting as many neutrons as necessary to find fissionable nuclei. When demand drops, the control rods are pushed in, absorbing neutrons and limiting the reaction.

The building and continued use of nuclear power plants faces some opposition in the United States. The threat of an accident that might release large amounts of radia-

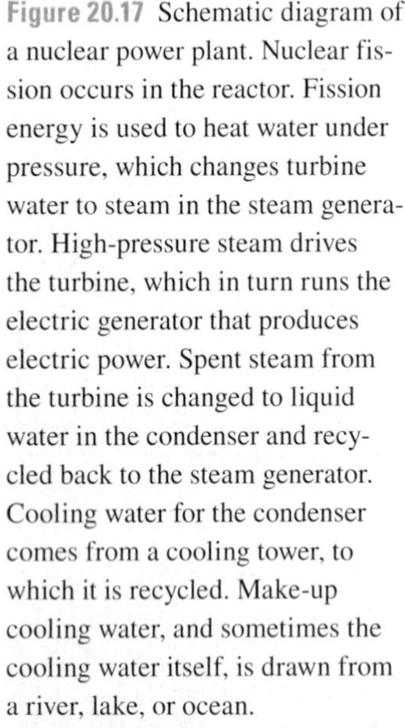

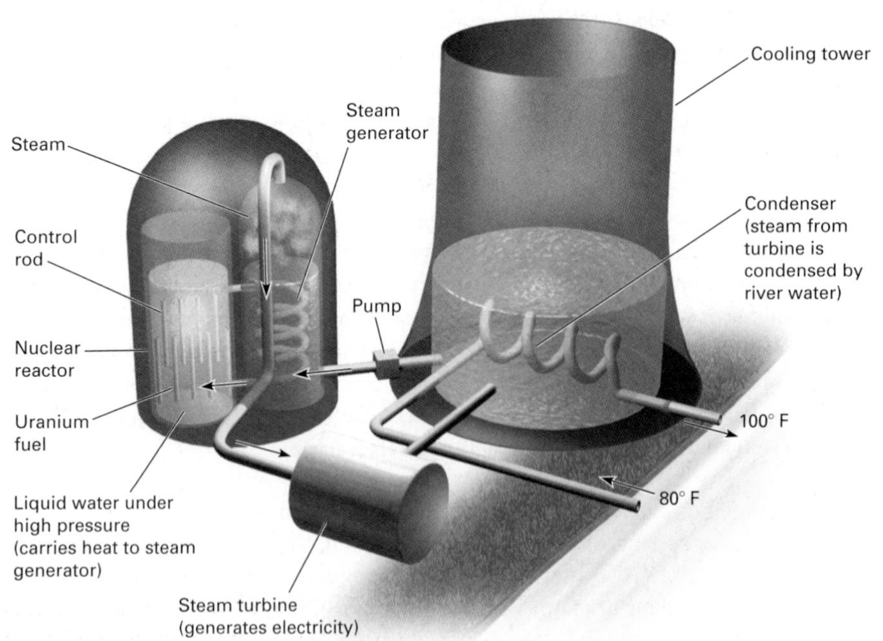


**Figure 20.17** Schematic diagram of a nuclear power plant. Nuclear fission occurs in the reactor. Fission energy is used to heat water under pressure, which changes turbine water to steam in the steam generator. High-pressure steam drives the turbine, which in turn runs the electric generator that produces electric power. Spent steam from the turbine is changed to liquid water in the condenser and recycled back to the steam generator. Cooling water for the condenser comes from a cooling tower, to which it is recycled. Make-up cooling water, and sometimes the cooling water itself, is drawn from a river, lake, or ocean.

Phil Degginger/Stone/Getty Images

Nuclear power plant. The cooling towers are the most visibly prominent feature in nuclear power plants.

Science Photo Library/Photo Researchers, Inc.

Vitrification is one technique for storing nuclear waste. The liquid waste is heated to a high temperature and mixed with substances that form a glassy solid when cooled. This prevents problems associated with storing liquids in tanks that may leak.

tion over a densely populated area is the major concern. This fear became an actuality in Chernobyl, Republic of Ukraine, in 1986 when two water-cooling systems failed. The chain of events that followed, including fire and a nonnuclear explosion, led to the release of radioactive gases that spread over much of Europe and into Asia. Cooling-system problems were also behind the Three Mile Island accident in Pennsylvania in 1979. In this incident, all radioactive substances were safely held within the reactor containment building, a safety feature in the design of U.S. power plants that is generally absent in the plants in the former USSR.

Whether getting energy from nuclear power plants is good or bad, ending the practice in the United States will not eliminate the danger. As Chernobyl demonstrated, the risk of accident is global and cannot be eliminated by a single nation. Recent statistics show that nuclear power plants produce about 20% of the electricity in the United States, 78% of France's electricity, and 34% of Japan's electricity (Fig. 20.18). Even if Americans chose to reduce their energy demands by 20%, other nations have chosen nuclear power as their way to become energy-independent of imported oil. Nuclear power, for better or worse, is here to stay.

Even if a nuclear accident never occurs, there is still the problem of how and where to dispose of the dangerous radioactive wastes from nuclear reactors. One method is to collect them in large containers that may be buried in caves deep below the surface of the earth. People who live and work near such disposal sites or along the routes by which the waste is transported are seldom enthusiastic about this solution.

Finally, there is fear that some irresponsible government may use nuclear fuel to manufacture nuclear weapons, spreading the threat of atomic warfare. More frightening is the possibility that some terrorist group might steal the materials needed to build a bomb. Although these threats cannot be removed from the earth today, perhaps they would be lessened if the large-scale production of nuclear fuel for electric power were eliminated.

On the other side of all these concerns is, of course, the question, "If we do not build and operate nuclear power plants, how else will we meet the energy needs of the coming decades?" Perhaps the next section offers one answer, if it can be reached at all.

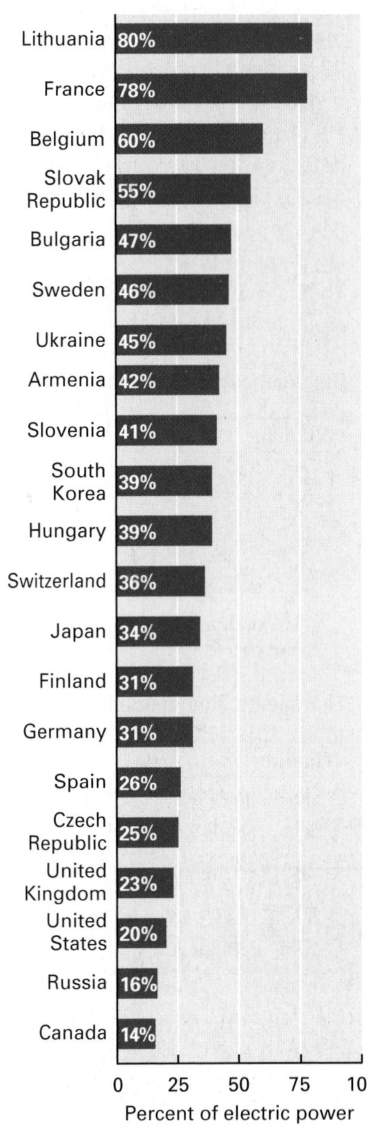


**Figure 20.18** Percentage of electricity from nuclear power by country. The United States relies on nonnuclear sources of electrical energy much more than many of the other nations of the world.

## 20.12 Nuclear Fusion

**Goal** **17** Define or identify a nuclear fusion reaction.

There is nothing new about nuclear energy. Humans did not invent it. In fact, without knowing it, humanity has been dependent on its benefits since before the beginning of

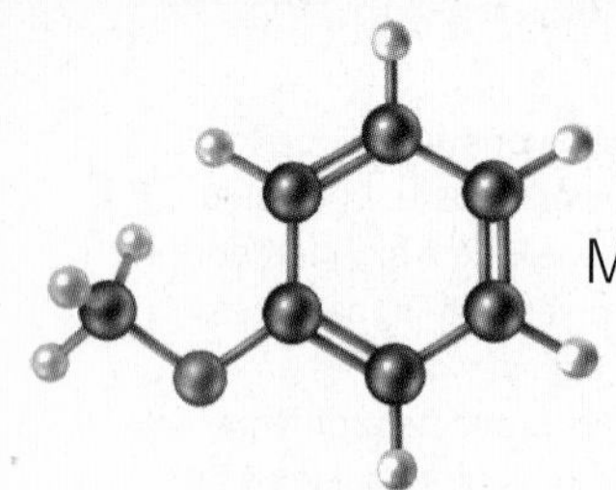

# Everyday Chemistry

## Medicine and Radionuclides

Hospitals and larger medical clinics typically have a Department of Nuclear Medicine. This department is responsible for the production, use, and disposal of radioactive materials used at the medical facility. Medical uses of radionuclides fall into two broad categories, diagnostic and therapeutic. A large hospital could use as many as 47 different radionuclides in as many as 194 diagnostic procedures and 29 therapeutic procedures.

Nearly all diagnostic radionuclides emit gamma rays, which are easy to detect. A gamma ray is like a nuclear needle; it makes a quick, exceedingly narrow passage through the body, limiting damage to a small number of cells. This is one criterion such radionuclides must meet. These radionuclides must also have a short half-life to limit the time of the patient's exposure to radiation. The mechanism by which the body eliminates the radionuclide must be known. Finally, the chemical behavior of the radionuclide must not interfere with normal body functions.

The goals of therapeutic uses of radionuclides differ from the goals of diagnostic uses. Therapeutic radionuclides are used to destroy abnormal, usually cancerous, cells as selectively as possible. Cell poisons such as radiation destroy abnormal cells more rapidly than normal cells because abnormal cells divide more quickly than normal cells. Therapeutic radionuclides are usually alpha or beta emitters. These decay particles cause heavy damage confined to a small area, owing to their low penetrating power. In the body, an alpha or beta emitter is a nuclear bull in a cellular china shop.

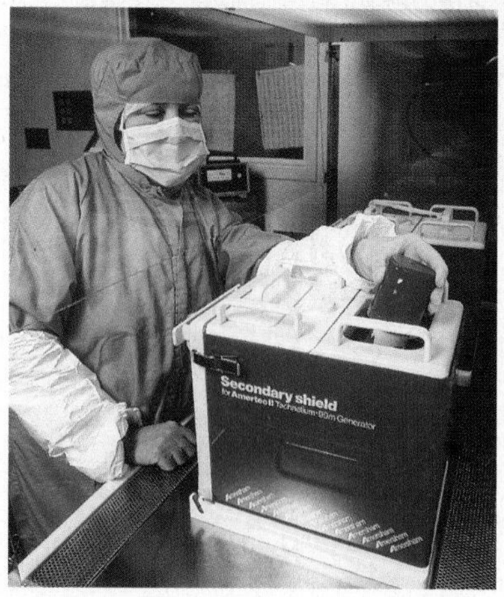


David Parker/Science Photo Library/Photo Researchers, Inc.

Radionuclide generator. The most widely used radionuclide generator converts molybdenum-99 to technetium-99m, where the m indicates that the isotope is *m*etastable, which means that the nucleus is in a temporary high-energy state. The excess energy is given off in the form of gamma radiation, which is used for diagnostic procedures.

**Diagnostic Radionuclides**

| Nuclide | | Half-Life | Emitted Particles | Uses |
|---|---|---|---|---|
| $^{51}_{24}Cr$ | chromium-51 | 28 days | gamma | spleen imaging |
| $^{59}_{26}Fe$ | iron-59 | 45 days | beta, gamma | bone marrow function |
| $^{99m}_{43}Tc$ | technetium-99 | 6 hours | gamma | bone, brain, liver, spleen imaging |
| $^{131}_{53}I$ | iodine-131 | 8 days | beta, gamma | thyroid functioning |
| $^{201}_{81}Tl$ | thallium-201 | 13 days | gamma | heart imaging |

**Therapeutic Radionuclides**

| Nuclide | | Half-Life | Emitted Particles | Uses |
|---|---|---|---|---|
| $^{32}_{15}P$ | phosphorus-32 | 14 days | beta | treatment of some leukemias, widespread carcinomas |
| $^{60}_{27}Co$ | cobalt-60 | 5.3 years | beta, gamma | external radiation source for cancer treatment |
| $^{90}_{39}Y$ | yttrium-90 | 64 hours | beta, gamma | implanted in tumors |
| $^{131}_{53}I$ | iodine-131 | 8 days | beta, gamma | treatment of thyroid cancer |
| $^{226}_{88}Ra$ | radium-226 | 1620 years | beta, gamma | implanted in tumors |

Thyroid gland imaging. (a) A healthy thyroid gland, located in the lower neck, is shaped like a bow tie. This image was produced by injecting the patient with a solution containing technetium-99m and scanning the neck with a gamma camera. (b) Hyperthyroidism, an overactive thyroid gland, is a common disease that has many causes. It can be treated with drugs, radioactive iodine treatment, or surgery.

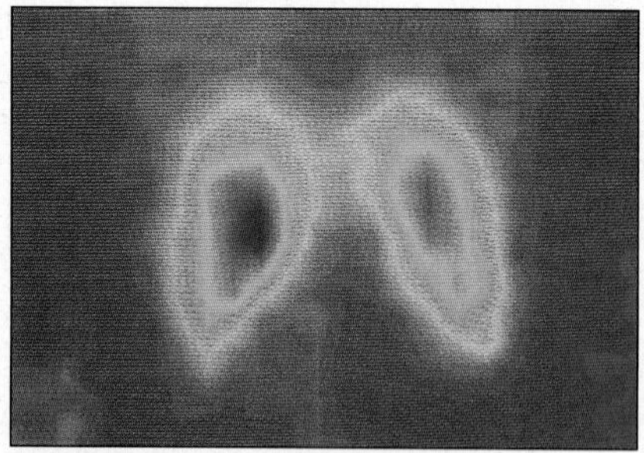

(a) Healthy human thyroid gland.

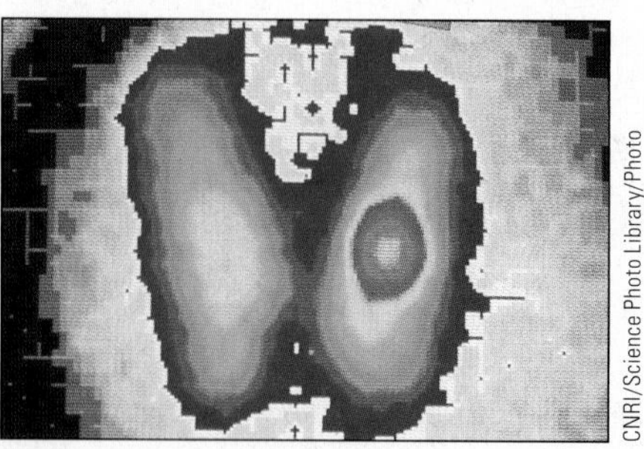

(b) Thyroid gland showing effect of hyperthyroidism.

CNRI/Science Photo Library/Photo Researchers, Inc.

recorded time. In its common form, though, we do not call it nuclear energy. We call it solar energy—the energy that comes from the sun.

The energy the earth derives from the sun comes from a type of nuclear reaction called **nuclear fusion, in which two small nuclei combine to form a larger nucleus.** The smaller nuclei are "fused" together, you might say. The typical fusion reaction believed to be responsible for the energy radiated by the sun is represented by the equation

$$^{2}_{1}H + ^{3}_{1}H \rightarrow ^{4}_{2}He + ^{1}_{0}n$$

Fusion processes are, in general, more energetic than fission reactions. The fusion of one gram of hydrogen in the above reaction yields about four times as much energy as the fission of an equal mass of uranium-235. So far, people have been able to produce only one kind of large-scale fusion reaction, the explosion of a hydrogen bomb.

Research efforts are being made to develop a nuclear fusion reactor as a source of useful energy. It has several advantages over fission. It yields more energy per given quantity of fuel. The isotopes required for fusion are far more abundant than those needed for fission. Best of all, fusion yields no radioactive waste, removing both the need for extensive disposal systems and the danger of an accidental release of radiation to the atmosphere.

The main obstacle to be overcome before energy can be obtained from fusion is the extremely high temperature needed to start and sustain the reaction. The trigger for a hydrogen bomb is the heat generated by an atomic bomb. Furthermore, no substance known can hold fusion reactants at the needed temperature. Experiments on magnetic containment have been conducted, in which the fuel is suspended in a magnetic field. Energy is then added by pulsing laser beams into the magnetic bottle. Temperatures as high as 40 million degrees Celsius have been reported, but they could not be sustained.

Once the technological obstacles to using energy from fusion are overcome, time remains a serious problem. Even with continued research, it will be well into this century before a significant proportion of our energy needs could possibly be met by fusion.

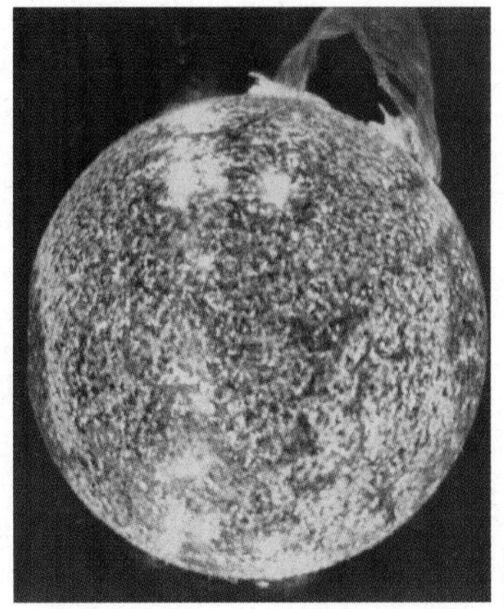
NASA

The sun. The fundamental energy source of the sun is a series of reactions that can be summarized as the conversion of hydrogen into helium.

### √ Target Check 20.6

*How do the products of a fusion reaction compare with the reactants?*

## Chapter 20 in Review

Most of the key terms and concepts and many others appear in the Glossary. Use your Glossary regularly.

**20.1 The Dawn of Nuclear Chemistry**

Goal 1 Define radioactivity.

**Key Terms and Concepts: Radioactivity**

**20.2 Radioactivity**

Goal 2 Name, identify from a description, or describe three types of radioactive emissions.

**Key Terms and Concepts: Alpha decay, alpha decay reaction, alpha particle or $\alpha$-particle, alpha ray or $\alpha$-ray, beta decay, beta decay reaction, beta particle or $\beta$-particle, beta ray or $\beta$-ray, gamma ray or $\gamma$-ray, nuclide, radionuclide**

**20.3 The Detection and Measurement of Radioactivity**

Goal 3 Identify the function of a Geiger counter and describe how it operates.

**Key Terms and Concepts: Cloud chamber, Geiger counter, Geiger-Müller counter, phosphorescence, scintillation counter**

**20.4 The Effects of Radiation on Living Systems**

Goal 4 Explain how exposure to radiation may harm or help living systems.

**Key Terms and Concepts: rem**

**20.5 Half-Life**

Goal 5 Describe or illustrate what is meant by the half-life of a radioactive substance.

Goal 6 Given the starting quantity of a radioactive substance, Figure 20.9, and two of the following, calculate the third: half-life, elapsed time, quantity of isotope remaining.

**Key Terms and Concepts: Half-life, radiocarbon dating**

**20.6 Natural Radioactive Decay Series—Nuclear Equations**

Goal 7 Describe a natural radioactive decay series.

Goal 8 Given the identity of a radioactive isotope and the particle it emits, write a nuclear equation for the emission.

**Key Terms and Concepts: Natural radioactive decay series, transmutation**

**20.7 Nuclear Reactions and Ordinary Chemical Reactions Compared**

Goal 9 List or identify four ways in which nuclear reactions differ from ordinary chemical reactions.

**20.8 Nuclear Bombardment and Induced Radioactivity**

Goal 10 Define or identify nuclear bombardment reactions.

Goal 11 Distinguish natural radioactivity from induced radioactivity produced by bombardment reactions.

Goal 12 Define or identify transuranium elements.

**Key Terms and Concepts: Induced (artificial) radioactivity, nuclear bombardment, particle accelerator, positron, transuranium element**

**20.9 Uses of Radionuclides**

Goal 13 Identify and describe uses for synthetic radionuclides.

**20.10 Nuclear Fission**

Goal 14 Define or identify a nuclear fission reaction.

Goal 15 Define or identify a chain reaction.

**Key Terms and Concepts: Breeder reactor, chain reaction, critical mass, nuclear fission**

**20.11 Electrical Energy from Nuclear Fission**

Goal 16 Describe how a nuclear power plant differs from a fossil-fueled power plant.

**20.12 Nuclear Fusion**

Goal 17 Define or identify a nuclear fusion reaction.

**Key Terms and Concepts: Nuclear fusion**

## Study Hints and Pitfalls to Avoid

Much of the information in this chapter is conceptual in nature. It is important to carefully learn the meaning of terms and phrases. Once you have mastered these, your study strategy should focus on learning the relationships among the concepts.

Many seemingly different problem types arise from Equation 20.1 and the graph in Figure 20.9. Practicing many end-of-chapter problems will help you learn the variations of the problems based on this equation and graph.

It is important to understand that writing a nuclear equation is different from writing a conventional chemical equation. The key to writing a nuclear equation is balancing both the sums of the mass numbers and the sums of the atomic numbers. In order to do this, you must memorize the mass number and atomic number of the alpha and beta particles.

The terms *fission* and *fusion* can be easy to reverse in one's mind. *Fission* was originally assigned to the process because it comes from the Latin word meaning "to split." *Fusion* describes "fusing together" of nuclei.

## Concept-Linking Exercises

*Write a brief description of the relationships among each of the following groups of terms or phrases. Answers to the Concept-Linking Exercises are given after the answers to the Target Checks in Appendix III.*

1. Alpha particles, beta particles, gamma rays, charge, penetrating power
2. Cloud chamber, Geiger counter, scintillation counter
3. Radiocarbon dating, half-life, carbon-14, carbon-12
4. Natural radioactive decay series, uranium, lead, alpha emission, beta emission
5. Nuclear bombardment, particle accelerator, transuranium elements
6. Nuclear fission, nuclear fusion, nuclear power plant, solar energy
7. Chain reaction, critical mass, breeder reactor

## Small-Group Discussion Questions

*Small-Group Discussion Questions are for group work, either in class or under the guidance of a leader during a discussion section.*

1. If a nuclide has a neutron-to-proton ratio that is high relative to the ratio typically found in stable nuclides, is it most likely to decay by alpha, beta, or gamma emission? How will a nuclide with a low neutron-to-proton ratio be likely to decay?
2. What are the limitations on carbon dating? For example, can it be used to date such artifacts as wooden caskets, ancient coins, animal bones, rocks, or scrolls? Can carbon dating be used for the bones of an animal that died ten years ago? Can it be used for objects that are millions of years old?
3. Draw particulate-level sketches of the following processes: (a) the radioactive decay of tritium (hydrogen-3) via beta emission, (b) the nuclear bombardment that led to the discovery of the neutron, where alpha particles bombarded beryllium-9, yielding carbon-12 and a neutron, and (c) the representative process for nuclear fusion, the reaction of hydrogen-2 and hydrogen-3 to yield helium-4 and a neutron.
4. The natural radioactive decay series that begins with uranium-235 and ends with lead-207 undergoes 11 reactions: alpha decay (hereafter $\alpha$), beta decay (hereafter $\beta$), $\alpha$, $\beta$, $\alpha$, $\alpha$, $\alpha$, $\alpha$, $\beta$, $\alpha$, and $\beta$. Write the equation for each step.

5. Two types of radioactive decay not mentioned in the chapter are positron emission and electron capture. A positron has a charge opposite of an electron but the same mass. It has the symbol ${}^{0}_{+1}e$. Electron capture is the capture of an electron by a nucleus. It is believed that the captured electron usually comes from the $n = 1$ or $n = 2$ levels. When electron capture occurs, the electron is a reactant. Write balanced nuclear equations for the following processes. (a) Polonium-207 undergoes positron decay. (b) Carbon-11 decays by positron emission. (c) Oxygen-15 undergoes positron emission. (d) Potassium-38 emits a positron as it decomposes. (e) Rubidium-81 captures an orbital electron. (f) Beryllium-7 captures an extranuclear electron. (g) Gallium-67 decays by electron capture. (h) Iron-55 captures an orbital electron.

6. The famous Einstein equation $E = \Delta m \times c^2$ can be used to calculate the quantity of energy needed to separate a nucleus into its constituent parts. Consider the decomposition of one mole of hydrogen-2 (commonly called deuterium) nuclei: ${}^{2}_{1}H$ (2.01410 g/mol) $\rightarrow$ ${}^{1}_{1}H$ (1.007825 g/mol) + ${}^{1}_{0}n$ (1.008665 g/mol). (a) Calculate the mass defect, $\Delta m$, in g/mol. (b) Calculate the quantity of energy needed to separate the proton and neutron of a deuterium nucleus. (*Hint:* $1\ J = 1\ kg \cdot m^2/s^2$). (c) What mass of hydrogen would need to be burned according to $2\ H_2(g) + O_2(g) \rightarrow 2\ H_2O(g) +$ 484 kJ in order to provide enough energy to decompose one mole of deuterium into subatomic particles? (d) What volume of hydrogen gas at room temperature is required to decompose one mole of deuterium into subatomic particles? (e) Is deuterium stable under ordinary conditions? Explain.

7. (a) Pure uranium metal is 0.7% uranium-235, with the rest being mostly uranium-238. Uranium-238 does not ordinarily undergo fission. Use this information to explain why natural uranium deposits do not undergo chain reactions. (b) The key to building a uranium bomb is to separate uranium-235 from the other naturally occurring isotopes. In the World War II era, that was accomplished by reacting uranium with fluorine to make gaseous uranium hexafluoride. The compound was then allowed to diffuse through a small opening into an empty chamber. Which uranium compound would move fastest in the gaseous state? Which side of a chamber, the one in which the gas was originally placed, or the one on the other side of the hole, would become enriched in uranium-235? Explain.

8. Fission of one mole of uranium-235 releases $2 \times 10^{10}$ kJ. Burning one ton of coal releases $2 \times 10^{7}$ kJ. What weight of coal needs to be burned to equal the energy output of the fission of one pound of uranium?

9. List advantages and disadvantages of both fossil-fueled and nuclear-powered electrical power plants. Which is the better choice?

10. In Chapter 2 we described the Law of Conservation of Mass. Here in Chapter 20 we illustrate nuclear changes where the identity of an element is changed and matter is converted into energy. Why do scientists believe that the Law of Conservation of Mass is true?

## Questions, Exercises, and Problems

*Blue-numbered questions are answered in Appendix III. ■ denotes problems assignable in OWL. In the first section of Questions, Exercises, and Problems, similar exercises and problems are paired in consecutive odd-even number combinations.*

#### Section 20.1: The Dawn of Nuclear Chemistry

#### Section 20.2: Radioactivity

1. What is a nuclide? How does a nuclide differ from an isotope?

2. ■ (a) What is the nuclear symbol for a beta particle? (b) What is the name for the Greek letter $\gamma$? (c) Of the radiations alpha, beta, and gamma, which is the most penetrating and which is the least penetrating?

3. *Decay* is a term used to describe what happens to a radioactive nucleus. What does *decay* mean in this sense?

4. ■ Which of the following characterizes an alpha ray (choose all that apply)? (a) It is composed of electrons; (b) it is composed of helium nuclei; (c) it is electromagnetic radiation; (d) it carries a positive charge; (e) it is a product of natural radioactive decay.

5. Compare the three forms of radioactive emissions in terms of mass number, electrical charge, and penetrating power.

6. ■ Which of the following characterizes a beta ray (choose all that apply)? (a) It is electromagnetic radiation; (b) it is composed of electrons; (c) it is attracted to the negatively charged plate in an electric field; (d) it is a product of natural radioactive decay; (e) it carries a negative charge.

#### Section 20.3: The Detection and Measurement of Radioactivity

7. What happens, or might happen, when an emission from a radioactive substance collides with an atom or molecule? Is this harmful? Explain.

8. Radiation is sometimes described as "ionizing radiation." What does this mean? Is all radiation ionizing? Justify your answer.

9. What is a Geiger-Müller counter (or, simply, Geiger counter)? How does it work?

10. Identify some properties of radioactive emissions that are used in detecting and measuring them.

11. How do Geiger and scintillation counters differ in how they tell an observer that an object is radioactive? Can either or both be used to measure radiation as well as detect it? If so, precisely what is measured?

12. What is a scintillation counter? How does it work?

13. How do gamma cameras and scanners record the presence of radiation?

14. Distinguish between a gamma camera and a scanner. What is their principal advantage in medical applications compared with nonradiological procedures for similar purposes?

### Section 20.4: The Effects of Radiation on Living Systems

15. Identify the greatest source of background radiation for U.S. citizens. What are the second and third greatest sources?

16. Is background radiation dangerous? Should we be concerned about it? If so, what can you do about it?

### Section 20.5: Half-Life

17. The radioactivity of a sample has dropped to 1/4 of its original intensity. How many half-lives have passed?

18. What is meant by the half-life of a radioactive substance?

19. What fraction of a radionuclide remains after the passage of seven half-lives?

20. ■ How many half-lives have passed when a radioactive substance has lost $\frac{15}{16}$ of its radioactivity?

21. Calculate the mass of radionuclide in a sample that will be left after 33 minutes if the sample originally has 12.9 grams of that radionuclide. The half-life of the radionuclide is 11.0 minutes.

22. ■ Chromium-51 is a radioisotope that is used to assess the lifetime of red blood cells. The half-life of chromium-51 is 27.7 days. If you begin with 86.9 milligrams of chromium-51, how many milligrams will you have left after 55.4 days have passed?

23. One of the more hazardous radioactive isotopes in the fallout of atomic bombs is strontium-90, for which the half-life is 28 years. If 654 g $^{90}_{38}Sr$ fall on a family farm on the day a child is born in 2006, how many grams will still be on the land when the farmer's granddaughter is born in 2062? How about when the granddaughter marries on the same farm in 2082?

24. ■ $^{214}_{83}Bi$ is a radioactive nuclide of the element bismuth with a half-life of 19.7 minutes. What percentage of a stored sample of this isotope would be lost due to radioactive decay in a 551-second period?

25. The half-life of $^{208}_{81}Tl$ is 3.1 minutes. A 84.6-gram sample is studied in the laboratory.
    a) How many grams of the isotope will remain after 12 minutes?
    b) In how many minutes will the mass of $^{208}_{81}Tl$ be 3.48 grams?

26. ■ If the mass of a sample of radioactive cesium-137 decays from 71.9 milligrams to 20.8 milligrams in 54.0 years, what is the half-life of cesium-137?

27. Uranium-235, the uranium isotope used in making the first atomic bomb, is the starting point of a natural radioactivity series. The next isotope in the series is thorium-231. At the beginning of a test period, a sample contained 9.53 grams of the thorium isotope. After 83.2 hours only 1.05 grams of the original isotope remained. What is the half-life of thorium-231?

28. ■ The half-life of the radioactive isotope thorium-234 is 24.1 days. How long will it take for the mass of a sample of thorium-234 to decay from 84.5 milligrams to 20.9 milligrams?

29. While excavating for the foundation of a new building, a contractor uncovered human skeletons in what turned out to be a burial ground from an ancient civilization. They were taken to a nearby university and submitted to radiocarbon dating analysis. It was found that the bones emit radiation at a rate of 55% of the rate of a living organism. How many years ago did the specimen die? (Use $5.73 \times 10^3$ years as the half-life of carbon-14.)

30. ■ The radioactive isotope carbon-14 is used for radiocarbon dating. The half-life of carbon-14 is $5.73 \times 10^3$ years. A wooden artifact in a museum has a $^{14}C$ to $^{12}C$ ratio that is 0.775 times that found in living organisms. Estimate the age of the artifact.

### Section 20.6: Natural Radioactive Decay Series—Nuclear Equations

31. What happens to the nucleus of an atom that experiences an alpha decay reaction? Compare the final nuclide with the original nuclide. Does the element undergo transmutation?

32. What happens to the nucleus of an atom that experiences a beta-decay reaction? Compare the final nuclide with the original nuclide. Does the element undergo transmutation?

33. Write nuclear equations for the beta emissions of $^{228}_{89}Ac$ and $^{212}_{83}Bi$.

34. ■ You may use the Table of Elements, as necessary, to answer these questions.
    a) When the nuclide bismuth-214 undergoes alpha decay, what are the name and symbol of the product nuclide?
    b) When the nuclide iron-59 undergoes beta decay, what are the name and symbol of the product nuclide?

35. Write nuclear equations for the alpha decay of $^{216}_{84}Po$ and $^{234}_{92}U$.

36. ■ You may use the Table of Elements, as necessary, to answer these questions. Write balanced nuclear equations for the following.
    a) The nuclide radon-222 undergoes alpha emission.
    b) The nuclide bismuth-214 undergoes beta emission.

### Section 20.7: Nuclear Reactions and Ordinary Chemical Reactions Compared

37. Why is it possible to speak of the "chemical properties of lead," but not the "nuclear chemical properties of lead"?

38. How do the chemical properties of carbon-12 compare with the chemical properties of carbon-14? If there is a difference, explain why.

39. The radioactivity of a sample of dirt containing uranium compounds records 5000 counts per minute when measured with a Geiger counter. The sample is treated physically to isolate the uranium compound, which is then decomposed chemically into pure uranium. If you disregard any loss of radioactivity because of decay during the purification process, will the pure uranium still radiate

at 5000 counts per minute, or will it be more or less than 5000? Explain your answer.

40. A fundamental idea of Dalton's atomic theory is that atoms of an element can be neither created nor destroyed. We now know that this is not always true. Specifically, it is not true for uranium and lead atoms as they appear in nature. Are the numbers of these atoms increasing or decreasing? Explain.

**Section 20.8: Nuclear Bombardment and Induced Radioactivity**

41. Distinguish between nuclear reactions that begin spontaneously and those that begin with nuclear bombardment. What is nuclear bombardment?

42. What distinguishes induced radioactivity from natural radioactivity?

43. Which of the following particles can be accelerated in particle accelerators and which cannot: electrons, protons, neutrons, positrons, alpha particles? Which property of the particle(s) governed your choice?

44. What does a particle accelerator do, and how does it do it?

45. Compare the atomic numbers of all elements that are naturally radioactive with the atomic numbers of elements that exhibit artificial radioactivity.

46. What are the transuranium elements? What property is associated with all transuranium elements? Do you know of any practical application of transuranium elements, or are they mostly laboratory curiosities, useful primarily in research?

47. Look at the periodic table. Are elements in the lanthanide series of elements transuranium elements? What about elements in the actinide series?

48. Is the element with atomic number 118 a transuranium element? Do you expect it to be radioactive? What sort of chemical properties do you expect it to have?

49. Complete each nuclear bombardment equation by supplying the nuclear symbol for the missing species.
   a) ${}^{44}_{20}\text{Ca} + {}^{1}_{1}\text{H} \rightarrow ? + {}^{1}_{0}\text{n}$
   b) ${}^{252}_{98}\text{Cf} + {}^{10}_{5}\text{B} \rightarrow 5\,{}^{1}_{0}\text{n} + ?$
   c) ${}^{106}_{46}\text{Pd} + {}^{4}_{2}\text{He} \rightarrow {}^{109}_{47}\text{Ag} + ?$

50. ■ Complete each nuclear bombardment equation by supplying the nuclear symbol for the missing species.
   a) $? + {}^{4}_{2}\text{He} \rightarrow {}^{12}_{6}\text{C} + {}^{1}_{0}\text{n}$
   b) ${}^{114}_{48}\text{Cd} + {}^{2}_{1}\text{H} \rightarrow ? + {}^{1}_{1}\text{H}$
   c) ${}^{235}_{92}\text{U} + {}^{1}_{0}\text{n} \rightarrow ? + {}^{92}_{38}\text{Sr} + 2\,{}^{1}_{0}\text{n}$

**Section 20.10: Nuclear Fission**

**Section 20.11: Electrical Energy from Nuclear Fission**

**Section 20.12: Nuclear Fusion**

51. How are fission reactions like fusion reactions, and how are they different?

52. Can radioactive decay be classified as nuclear fission? Why or why not?

53. What is a chain reaction? What essential feature must be present in a nuclear reaction before it can become a chain reaction?

54. Equation 20.7 is referred to as a "representative equation," indicating that it is only one of several possible equations for the fission of ${}^{235}_{92}\text{U}$ when it is bombarded by a single neutron. Another fission of the same reactants yields the nuclides ${}^{144}_{55}\text{Cs}$ and ${}^{90}_{37}\text{Rb}$. Write the equation for that reaction.

55. Can a fusion reaction be a chain reaction? Why or why not?

56. Starting a chain reaction is one thing; keeping it going is another. What is required if a chain reaction is to continue? By what term is this requirement identified?

57. What advantages do fusion reactions have over fission reactions as a source of nuclear power? If fusion reactions are more desirable than fission reactions, why don't we use them instead of fission reactions?

58. List some of the advantages and disadvantages of nuclear power plants compared with other sources of electrical energy. In your opinion, do the advantages outweigh the disadvantages?

**General Questions**

59. Distinguish precisely, and in scientific terms, the differences among items in each of the following pairs or groups.
   a) Alpha, beta, and gamma radiation
   b) X-rays, $\gamma$-rays
   c) $\alpha$-particle, $\beta$-particle
   d) Natural and induced radioactivity
   e) Chemical reaction, nuclear reaction
   f) Isotope, nuclide, radionuclide
   g) Element, transuranium element
   h) Nuclear fission, nuclear fusion
   i) Atomic bomb, hydrogen bomb

60. Classify each of the following statements as true or false.
   a) A radioactive atom decays in the same way whether or not the atom is chemically bonded in a compound.
   b) The chemical properties of a radioactive atom of an element are different from the chemical properties of a nonradioactive atom of the same element.
   c) $\alpha$-particles have more penetrating power than $\beta$-particles.
   d) $\alpha$- and $\beta$-emissions are particles, but a $\gamma$-ray is an "energy ray."
   e) Radioactivity is a nuclear change that has no effect on the electrons in nearby atoms.
   f) The number of protons in a nucleus changes when it emits a beta particle.
   g) The mass number of a nucleus changes in an alpha emission but not in a beta emission.
   h) Synthetic radionuclides have no application in everyday life or industry; they are used only for scientific research purposes.
   i) The first transmutations were achieved by the alchemists.

j) Radioisotopes can be made by bombarding a nonradioactive isotope with atomic nuclei or subatomic particles.

k) The atomic numbers of products of a fission reaction are smaller than the atomic number of the original nucleus.

l) Nuclear power plants are a safe source of electrical energy.

m) The main obstacle to developing nuclear fusion as a source of electrical energy is a shortage of nuclei to serve as "fuel."

61. A major form of fuel for nuclear reactors used to produce electrical energy is a fissionable isotope of plutonium. Plutonium is a transuranium element. Why is this element used instead of a fissionable isotope that occurs in nature?

62. Why is half-life used for measuring rate of decay rather than the time required for the complete decay of a radioactive isotope?

63. ■ A ton of high-grade coal has an energy output of about $2.5 \times 10^7$ kJ. The energy released in the fission of one mole of $^{235}_{92}U$ is about $2.0 \times 10^{10}$ kJ. How may tons of coal could be replaced by one pound of uranium-235, assuming the materials and the technology were available?

**More Challenging Questions**

64. Loss of mass is not a satisfactory way to express rate of decay, but we used it in Active Examples 20.1 through 20.3 because it is the easiest to visualize. However, we chose the words for these Active Examples very precisely; carefully read, they are correct. Now, why is loss of mass usually unsatisfactory as a measure of rate of decay? Suggest a better way, and explain its advantage.

65. Suppose you have a radionuclide, A, that goes through a two-step decay sequence, first to B and then to C, which is stable. Suppose also that the half-life from A to B is six days, and the half-life from B to C is one day. Predict by listing in declining order, greatest to smallest, the amounts of A, B, and C that will be present (a) at the end of 6 days and (b) at the end of 12 days. Explain your prediction.

66. ■ A fragment of cloth found just outside Jerusalem is believed to have been used by some person at about the beginning of the Christian era. Analysis shows that radiation from the fragment is 22.7 units, whereas radiation from a living specimen is 29.0 units when measured with the same instrument. Is it possible that the fragment might have come from the period believed? Justify your answer.

67. A sample of pure calcium chloride is prepared in a laboratory. A small but measurable amount of the calcium in the compound is made up of calcium-47 atoms, which are beta emitters with a half-life of 4.35 days. The compound is securely stored for a week in an inert atmosphere. When it is used at the end of that period, it is no longer pure. Why? With what element would you expect it to be contaminated?

68. ■ Two uranium sulfides have the formulas US and $US_2$. A laboratory worker prepares 50.0 grams of each compound. If the uranium in each compound has all the isotopes of uranium in their normal distribution in nature, which compound, if either, will exhibit the greater amount of radioactivity? If 0.5 mole of each compound is prepared, which compound, if either, will be more radioactive? Explain both answers.

69. Compare Equation 20.7 with the equation you wrote to answer Question 54. Which of the two reactions is more likely to contribute to a chain reaction? Explain.

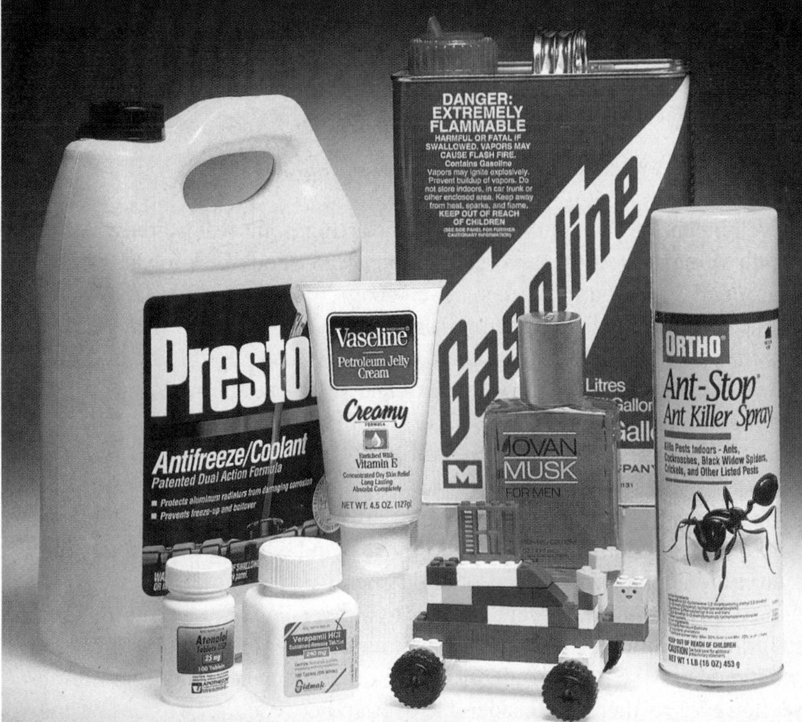




# 21

The starting material for everything you see in this photograph other than the metal cans is either natural gas, petroleum, or coal. Many common products that we use every day in modern society come from these substances. How can so many diverse substances be made from such a small collection of naturally occurring starting materials? How does the transformation take place? We will investigate the answers to these questions in this chapter.

# Organic Chemistry

Chapter 21 is a brief survey of organic chemistry. While some instructors might seek mastery of specific chemical concepts in some or all areas of this chapter, other instructors may believe that most topics are presented too briefly to set for them the kind of performance goals used in other chapters. Thus we offer two alternatives. Goals are introduced throughout the chapter in the usual manner. We also offer the following chapter-wide performance goals:

Goal A Distinguish between organic and inorganic chemistry.

Goal B Define the term *hydrocarbon.*

Goal C Distinguish between saturated and unsaturated hydrocarbons.

Goal D Write, recognize, or otherwise identify (a) the structural unit, or functional group, (b) the general formula, and (c) the molecular or structural formulas and/or names of specific examples of the following classes of organic compounds: alkanes, alkenes, alkynes, cycloalkanes, aromatic hydrocarbons, alcohols, ethers, aldehydes, ketones, carboxylic acids, esters, amines, and amides.

Goal E Define and give examples of isomerism.

Goal F Define and give examples of monomers and polymers.

Online homework for this chapter may be assigned in OWL

## CONTENTS

If your instructor does not express a preference between the two types of performance goals, we suggest you use the specific in-chapter performance goals to guide your study.

## 21.1 The Nature of Organic Chemistry

Goal 1 Distinguish between organic and inorganic compounds.

In the early development of chemistry, the logical starting point was a study of substances that occur in nature. As in the organization of any body of knowledge, substances were grouped by certain common macroscopic characteristics. One system assigned substances to groups labeled animal, vegetable, and mineral. Minerals and the compounds that may be derived from them originally made up the area of study commonly known as inorganic chemistry. Animal and vegetable substances were considered part of organic chemistry, which was originally defined as the chemistry of living organisms, including those compounds directly derived from living organisms by natural processes of decay.

The Oesper Collection in the History of Chemistry/University of Cincinnati

Friedrich Wöhler (1800–1882) can be considered as the founding father of the science of organic chemistry. He showed that a "life force" was not required for producing compounds found in living organisms; they could be synthesized in the laboratory.

In 1828 Friedrich Wöhler heated ammonium cyanide, which everyone agreed was not organic, and obtained urea, which is found in urine, and which everyone agreed *is* organic. The definitions of inorganic and organic chemistry had to be changed. The definitions changed from the macroscopic level—living versus nonliving—to the particulate level. Today, **organic chemistry is known as the chemistry of carbon compounds.** Inorganic chemistry is the study of all the other elements. These definitions are very loose, however, as there is considerable overlap between the two subdivisions of chemistry. In fact, carbon-containing compounds without hydrogen are still classified as inorganic. Carbonates, cyanides, and oxides of carbon are examples.

Organic chemistry is not "different" from inorganic chemistry. All of the chemical principles that apply to inorganic compounds, such as bonding, reaction rates, and equilibrium, apply equally to organic compounds. What truly makes organic chemistry unique is the ability of carbon atoms to bond to one another strongly to form long chain and ring systems. This results in the theoretical possibility of an almost unlimited number of organic compounds. This theoretical possibility is borne out in practice, too. In 1965 the Chemical Abstracts Service* began assigning a registry number to each new substance reported. The total number of classified chemical compounds has now passed 36 million, and the majority of these substances contain carbon. About 4000 new substances are registered each day.

### √ Target Check 21.1

a) Which of the following compounds are classified as organic, and which are classified as inorganic? $CH_4$, NaCl, $CH_3NH_2$, CO

b) The term *organic* has several meanings in everyday language, all of which are different from its meaning in chemistry. State an everyday definition of the term *organic,* and give an example of how it is used. Compare this with the chemical definition of the term.

## 21.2 The Molecular Structure of Organic Compounds

Goal 2 Given Lewis diagrams, ball-and-stick models, or space-filling models of two or more organic molecules with the same molecular formula, distinguish between isomers and different orientations of the same molecule.

P/REVIEW: Table 21.1 is derived largely from Table 13.2, Section 13.3. The earlier table includes illustrations that show how molecular structure is related to electron-pair geometry and the number of atoms bonded to the central atom.

Table 21.1 summarizes the covalent bonding properties of carbon, hydrogen, oxygen, nitrogen, and the halogens, the elements most frequently found in organic compounds. All the molecular geometries are indicated—the three-dimensional shapes and, where constant, the actual bond angles. Of particular significance is the number of covalent

---

**Chemical Abstracts* is a weekly summary of articles that have appeared in original research journals. It is published by the American Chemical Society.

**Table 21.1** Bonding in Organic Compounds

| Element | Number of Bonds* | Molecular Bond Geometry | | | |
|---|---|---|---|---|---|
| | | *Single Bond* | *Double Bond* | *Double Bond* | *Triple Bond* |
| Carbon | 4 | C (four single bonds) Tetrahedral: 109.5° angles | C= Trigonal Planar: 120° angles | =C= Linear: 180° angle | —C≡ Linear: 180° angle |
| Hydrogen | 1 | H— | | | |
| Halogens | 1 | :X— | | | |
| Oxygen | 2 | :O: Bent structure | O= | | |
| Nitrogen | 3 | :N Trigonal pyramidal structure | :N= Angular structure | | :N≡ |

*Number of bonds to which an atom of the element shown can contribute *one* electron.

bonds that atoms of the different elements usually form. This is determined by the electron configuration of the atom. The bonding relationships of these elements are basic to your understanding of the structure of organic compounds.

When carbon forms four single bonds, they are arranged tetrahedrally around the carbon atom; the molecular geometry is tetrahedral (Fig. 21.1). Recall from Chapter 13 that it is not possible to represent this three-dimensional shape accurately in a two-dimensional sketch. Thus the four bonds radiating from each carbon atom in

```
   x  z  y
   |  |  |
x—C—C—C—y
   |  |  |
   x  z  y
```

all form tetrahedral bond angles (109.5°). Furthermore, all three x positions are geometrically equal, the three y positions are equal, and the two z positions are equal. Also notice that if you flip the molecule end-to-end, the x positions are equivalent to the y positions.

Let's examine this relationship between two-dimensional sketches and three-dimensional molecules a bit more carefully. Compare the following two structural diagrams for $C_3H_6BrCl$ and the ball-and-stick diagrams beneath them:

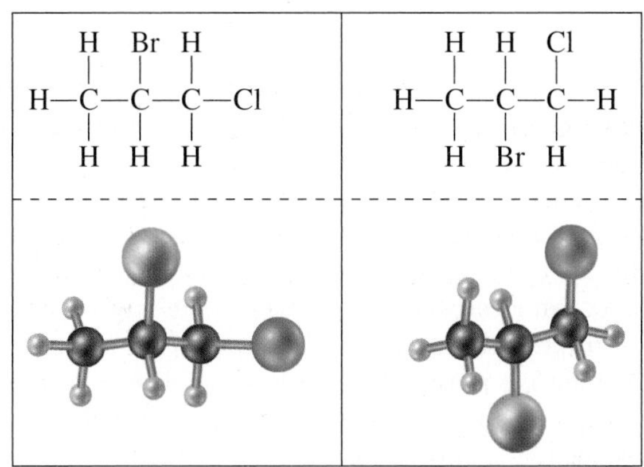


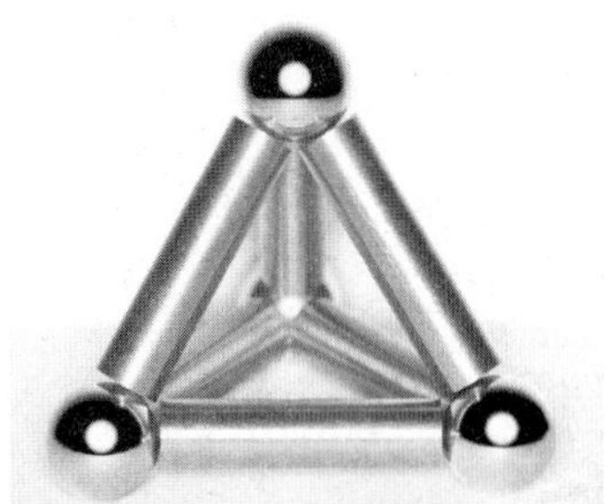

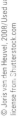

**Figure 21.1** Tetrahedral models. The metal figure is a tetrahedron. Its four faces are identical equilateral triangles. The model of methane, $CH_4$, has a tetrahedral structure. The carbon atom is in the middle of the tetrahedron and a hydrogen atom is found at each of the four vertices.

Are these the same molecule or isomers? One type of isomer is composed of compounds with the same molecular formula but different atom connectivities. The two compounds have the same molecular formulas, but do they have different atom connectivities? One way to answer this question is to consider whether a ball-and-stick model of the molecule would have to be taken apart and reassembled to form the other compound. Careful examination of the model diagrams shows that they have the same structure; the atoms are bonded in exactly the same way. The diagrams are different views of the same molecule.

The molecule

```
    H  Cl  H
    |   |   |
H—C—C—C—H
    |   |   |
    H  Br  H
```

is an isomer of the molecule above. Its molecular formula is still $C_3H_6BrCl$, but this time the chlorine and bromine atoms are both bonded to the central carbon—it differs in atom connectivity.

**√ Target Check 21.2**

*Thus far in this section, we have shown two isomers of the molecule $C_3H_6BrCl$. There are three more isomers. Draw their structural diagrams.*

## Thinking About Your Thinking

### Mental Models

**There are many structural diagrams and drawings of molecular models throughout this chapter. The mental translation between a structural diagram and the three-dimensional molecule it represents is essential to understanding organic chemistry. As you study this chapter, try to imagine a model of each molecule for which a structural diagram is drawn. If you have a model kit, build some of the molecules and rotate the model in your hands, looking at it from many angles. Practice rotating mental models of molecules in your mind. The more clearly you can see organic molecules in your mind, the better you will be at understanding this area of chemistry.**

# SECTIONS 21.3–21.9: HYDROCARBONS

Goal | **3** Distinguish between saturated and unsaturated hydrocarbons (or other compounds).

Goal | **4** Distinguish between alkanes, alkenes, and alkynes (Sections 21.3 and 21.4).

The simplest organic compounds are the **hydrocarbons, binary compounds of carbon and hydrogen.** Hydrocarbons are divided into two major classes, the aliphatic hydrocarbons and the aromatic hydrocarbons. We will examine three groups of aliphatic hydrocarbons: alkanes, alkenes, and alkynes. The features that distinguish aliphatic and aromatic hydrocarbons will be identified when we study the latter group. As you will see shortly, these classifications are all based on molecular structure.

Table 21.1 shows that a carbon atom is able to form four bonds to a maximum of four other atoms. A hydrocarbon in which each carbon atom is bonded to that maximum of four other atoms is **saturated** in the sense that there is no room for the carbon atom to form a bond to another atom. The table also shows that if a carbon atom is double- or triple-bonded, it is bonded to three or two other atoms. This is fewer than the maximum number of atoms with which the carbon atom is capable of bonding. Such a hydrocarbon is **unsaturated.**

## 21.3 Saturated Hydrocarbons: The Alkanes and Cycloalkanes

Goal 5 Given a formula of a hydrocarbon or information from which it can be written, determine whether the compound can be a normal alkane or a cycloalkane.

Goal 6 Given the name (or structural diagram) of a normal, branched, or halogen-substituted alkane, write the structural diagram (or name).

Goal 7 Given the name (or structural diagram) of a cycloalkane, write the structural diagram (or name).

In an **alkane, each carbon atom forms single bonds to four other atoms and there are no multiple bonds.** Most of the carbon atoms in an alkane are arranged in a continuous chain in which all bond angles are 109.5°, the tetrahedral angle. If *all* carbon atoms in the molecule are in a continuous chain, the compound is a **normal alkane.** In other alkanes some carbon atoms appear as "branches" off the main chain. They are isomers of the normal alkane having the same number of carbon atoms.

**P/REVIEW:** Isomers are distinctly different compounds, each with its own set of physical and chemical properties. See Section 13.1.

The first 10 alkanes are shown in Table 21.2. Careful examination of the formulas shows that they have the form $C_nH_{2n+2}$, where n is the number of carbon atoms. Notice also that the difference from one alkane to the next is one carbon and two hydrogen atoms, a $—CH_2—$ structural unit. A series of compounds in which each member differs from the members before it by a $—CH_2—$ unit is called a **homologous series.**

There are three ways to write the formula of an alkane. Using octane, the alkane with eight carbons, as an example, there is the molecular formula, $C_8H_{18}$. A molecular formula gives no information as to how the atoms are arranged. A Lewis diagram that omits unshared electron pairs, commonly referred to as a **structural formula** or **structural diagram,** shows that arrangement. A compromise between them is the **line formula,** which includes each $CH_2$ unit in the alkane chain. The line formula for octane is $CH_3CH_2CH_2CH_2CH_2CH_2CH_2CH_3$. A long line formula is usually shortened by grouping the $CH_2$ units: $CH_3(CH_2)_6CH_3$. This is called a **condensed formula.**

Table 21.2 also serves to introduce organic nomenclature. Notice that each alkane is named by combining a prefix and a suffix. The prefix indicates the number of carbon atoms in the compound, as shown in the third and fourth columns. The suffix identifying an alk*ane* is *-ane.* Thus, the name of methane comes from combining the prefix *meth-,* indicating one carbon, with the suffix *-ane,* indicating an alkane. The same prefixes are used to name other organic compounds and groups as well.

**Table 21.2** The Alkane Series

| Molecular Formula | Name | Number of Carbon Atoms | Prefix | Physical State at 25°C |
|---|---|---|---|---|
| $CH_4$ | Methane | 1 | Meth- | Gas |
| $C_2H_6$ | Ethane | 2 | Eth- | Gas |
| $C_3H_8$ | Propane | 3 | Prop- | Gas |
| $C_4H_{10}$ | Butane | 4 | But- | Gas |
| $C_5H_{12}$ | Pentane | 5 | Pent- | Liquid |
| $C_6H_{14}$ | Hexane | 6 | Hex- | Liquid |
| $C_7H_{16}$ | Heptane | 7 | Hept- | Liquid |
| $C_8H_{18}$ | Octane | 8 | Oct- | Liquid |
| $C_9H_{20}$ | Nonane | 9 | Non- | Liquid |
| $C_{10}H_{22}$ | Decane | 10 | Dec- | Liquid |

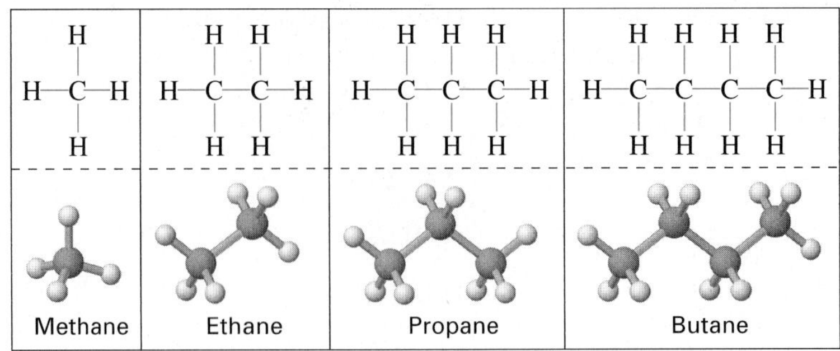


**Active Figure 21.2** Lewis diagrams and ball-and-stick models of the four simplest alkanes. There is a tetrahedral orientation of all four bonds around each carbon atom. **Watch this active figure at *http://www.cengage.com/cracolice.***

Although there is only 1 possible structure for methane, ethane, and propane, as shown in Active Figure 21.2, there are 2 possible isomers of butane, shown in Figure 21.3(a). Pentane has 3 isomers (Fig. 21.3[b]). There are 5 isomeric hexanes, 9 heptanes, and 75 possible decanes. It is possible to draw over 300,000 isomeric structures for $C_{20}H_{42}$ and more than 100 million for $C_{30}H_{62}$. It is not surprising that many of these compounds have not been prepared and identified! This does give you some idea, though, why there are so many organic compounds.

## Alkyl Groups

An **alkyl group is an alkane from which one hydrogen atom has been removed.** If, on paper, we remove a hydrogen atom from methane, $CH_4$, we get —$CH_3$, where the dash indicates a bond that the alkyl is able to form with another atom. This —$CH_3$ group, appearing in the structural formula of a compound (attached to a carbon atom other than a terminal carbon), is called a **methyl group.** The term is made up of the prefix *meth-* for one carbon (Table 21.2) and the suffix *-yl,* which is applied to alk*yl* groups. If we compare two compounds,

```
    H   H   H   H   H               H   H       H       H   H
    |   |   |   |   |               |   |       |       |   |
H — C — C — C — C — C — H   and H — C — C ————— C ————— C — C — H
    |   |   |   |   |               |   |       |       |   |
    H   H  [H]  H   H               H   H       |       H   H
                                                |
                                            H — C — H
                                                |
                                                H
```

we see that the shaded H in the first compound has been replaced by a —$CH_3$ group, or methyl group, in the second. If the replacement group has two carbon atoms

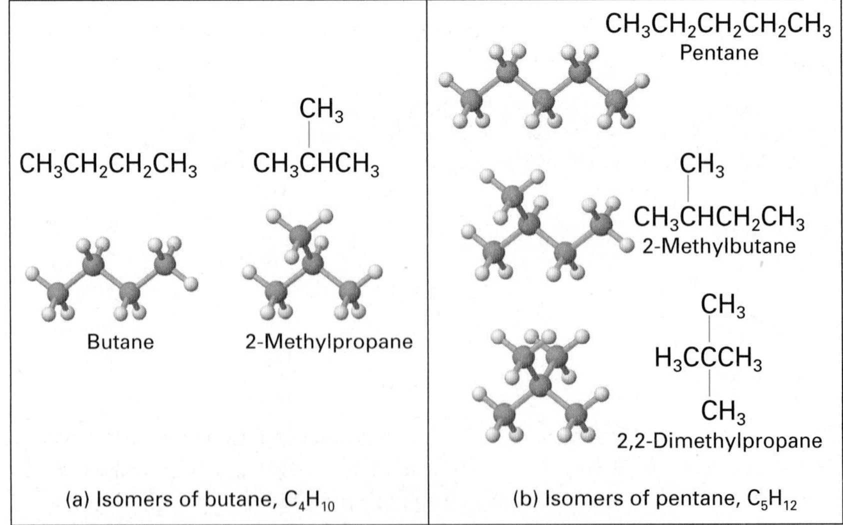


**Figure 21.3** Lewis diagrams and ball-and-stick models of isomers of butane and pentane. (a) Each isomer of butane is composed of four carbon atoms and ten hydrogen atoms. (b) Each isomer of pentane contains five carbon atoms and twelve hydrogen atoms.

```
   H  H     H     H  H
   |  |     |     |  |
H—C—C——C——C—C—H
   |  |     |     |  |
   H  H     |     H  H
            |
         H—C—H
            |
         H—C—H
            |
            H
```

it is an ethyl group, $—C_2H_5$, one hydrogen short of ethane, $C_2H_6$. All the alkyl groups are similarly named.

Frequently, we wish to show a bonding situation in which *any* alkyl group may appear. The letter R is used for this purpose. Thus R—OH could be $CH_3OH$, $C_2H_5OH$, $C_3H_7OH$, or any other alkyl group attached to an —OH group.

## Naming the Alkanes by the IUPAC System

We are now ready to describe the International Union of Pure and Applied Chemistry (IUPAC) system of naming isomers of the alkanes, as well as other compounds we will encounter shortly. The system follows a set of rules:

**procedure**

### Naming Alkanes

**Step 1: Identify as the parent alkane the longest continuous chain.** For example, in the compound having the skeleton structure

```
C—C—C—C—C
         |
         C
         |
         C
```

the longest chain is six carbons long, a hexane, not five as you might first expect. This is readily apparent if we number the carbon atoms in the original representation of the structure and in an equivalent layout:

```
                                    C
6   5   4   3               6   5   4   |3  2   1
C—C—C—C—C               C—C—C—C—C—C
            |
            C2
            |
            C1
```

**Step 2: Identify by number the carbon atom to which the alkyl group (or other species) is bonded to the chain.** In the example compound this is the *third* carbon, as shown. Notice that counting always begins at the end of the chain that places the branch on the *lowest*-number carbon atom possible.

**Step 3: Identify the branched group (or other species).** In this example, the branch is a methyl group, $—CH_3$.

```
             H
             |
C—C—C—C—C—H
         |   |
         C   H
         |
         C
```

These three items of information are combined to produce the name of the compound, *3-methylhexane.* The 3 comes from the third carbon (*Step 2*); methyl comes from the branch group (*Step 3*); and hexane is the parent alkane (*Step 1*).

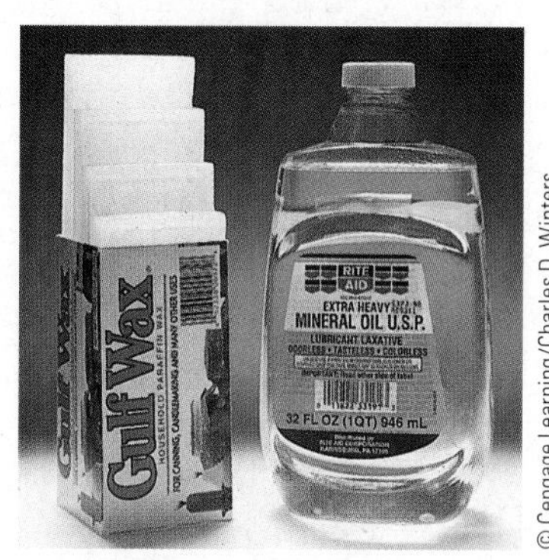



Paraffin wax and mineral oil. Paraffin wax is a mixture of straight-chain alkane molecules with 20 or more carbon atoms. Mineral oil is also a mixture mainly of alkane molecules with 17 to 50 carbon atoms, but it includes other hydrocarbons as well.

Sometimes the same branch appears more than once in a single compound. This situation is governed by the following rule:

**Step 4: If the same alkyl group, or other species, appears more than once, indicate the number of appearances by di-, tri-, tetra-, etc., and show the location of each branch by number.** For example

```
        C   C
        |   |
C—C—C—C—C
```

is 2,3-dimethylpentane. To write the structural formula for 2,2,5-trimethylhexane, we would establish a six-carbon skeleton and attach methyl groups as required, two to the second carbon and one to the fifth:

```
    C           C
    |           |
C—C—C—C—C—C
    |
    C
```

When attached to a hydrocarbon, halogen (Group 7A/17) atoms are named fluoro- (—F), chloro- (—Cl), bromo- (—Br), and iodo- (—I).

Twice above we have referred to "other species"—species other than an alkyl group that might be attached to a hydrocarbon. The most common species other than an alkyl is a halogen atom. If two chlorine atoms take the places of hydrogen atoms on the second and third carbons of *n*-pentane, for example, or take the places of the methyl groups in 2,3-dimethylpentane above, the compound would be 2,3-dichloropentane. This leads to the next nomenclature rule:

**Step 5:** If two or more different alkyl groups, or other species, are attached to the parent chain, they are named in alphabetical order. By this rule the compound

```
    Cl  Br
    |   |
C—C—C—C—C
```

is 3-bromo-2-chloropentane. The skeleton diagram for 2,2-dibromo-4-chloroheptane is

```
    Br      Cl
    |       |
C—C—C—C—C—C—C
    |
    Br
```

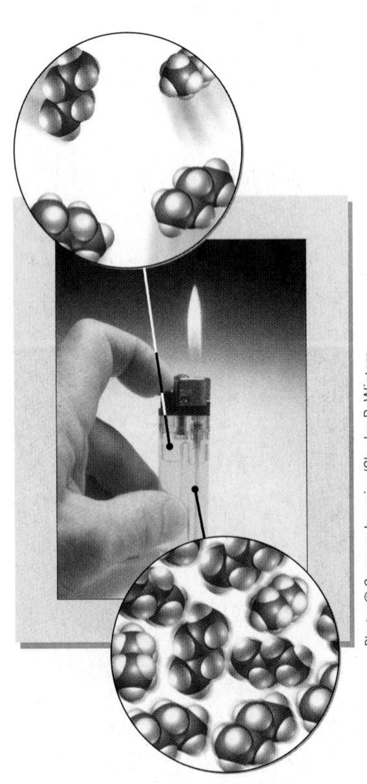

Photo: © Cengage Learning/Charles D. Winters

**Figure 21.4** A disposable butane lighter. Butane boils at −20.5°C at atmospheric pressure; it is pressurized in lighters so that some of the fuel exists in the liquid state.

## Physical Properties of the Alkanes

Alkane molecules are all nonpolar. In Chapter 15 we stated that intermolecular forces between nonpolar molecules increase with increasing molecular size. As a result, larger molecules have higher melting and boiling points. Among normal alkanes—those with continuous chains—compounds having fewer than five carbon atoms have the weakest intermolecular attractions. They have low boiling points and are gases at room temperature. All are used as fuels; methane is the main constituent of natural gas, and butane is often used as lighter fluid (Fig. 21.4).

Intermolecular forces are stronger between larger alkanes from $C_5H_{12}$ to $C_{17}H_{36}$. These higher-boiling compounds are liquids at room temperature (Fig. 21.5). Several of the lower-molar-mass liquid alkanes containing 5 to 12 carbon atoms are present in gasoline. Diesel fuel and lubricating oils are made up largely of higher-molar-mass liquid alkanes. Alkanes with molar masses greater than 300 are normally solids at room temperature.

## Cycloalkanes

If a hydrogen atom is removed from each end carbon of a normal alkane, the two end carbons can bond to each other to form a ring or cycle of carbon atoms. This ring compound, in which all carbon atoms are saturated, is a **cycloalkane.** Cycloalkanes can form with a minimum of three carbon atoms; however, the resulting 60° bond angles are severely strained from the normal tetrahedral angle, so cyclopropane is unstable. Cyclobutane, the four-carbon ring system, is also unstable (Fig. 21.6). The more common alkanes are those whose bond angles are close to or equal to the tetrahedral angle, such as cyclopentane, the five-membered ring system, and cyclohexane, the six-membered ring system.

**P/REVIEW:** In Sections 15.2 and 15.3 we discussed the relationships between intermolecular forces and physical properties. Substances with strong intermolecular forces tend to have higher boiling points, heats of vaporization, viscosity, and surface tension and lower vapor pressures. The forces, in increasing strength, were classified as induced dipole forces, dipole forces, and hydrogen bonds. Induced dipole forces increase with molecular size, and with large molecules they may exert more influence on physical properties than dipole forces and hydrogen bonds.

Three structural diagrams and a ball-and-stick diagram of cyclohexane are

H H H H
H—C C—H
H—C C—H
H H H H

$CH_2$
$H_2C$ $CH_2$
$H_2C$ $CH_2$
$CH_2$

The left diagram is the most complete. The middle diagram illustrates the common practice of using $—CH_2—$ when a carbon bonded to two hydrogen atoms is also bonded to two other atoms, as in a saturated chain. The diagram at the right is a skeleton diagram in which the vertex of each angle shows the relative location of a carbon atom. It is understood in such diagrams that each carbon atom forms as many additional bonds to hydrogen atoms as are necessary to bring its total number of bonds to four. All structural diagrams are equivalent to the ball-and-stick diagram.

To name a cycloalkane, apply the prefix *cyclo*-to the name of the open-chain alkane with the same number of carbon atoms. The diagrams above have six carbon atoms in the ring, hence the name cyclohexane. If one or more alkyl groups or other species, such as a halogen, takes the place of one of the invisible hydrogen atoms in the ring, alkane nomenclature rules for branched species apply. Thus

$CH_2$ $CH_3$
$H_2C$ CH
$H_2C$ $CH_2$
$CH_2$

is methylcyclohexane. Notice that the methyl carbon in the skeleton diagram is not shown. Like the vertex in the polygon, a bond with no symbol at its end is understood to be a carbon atom that forms additional bonds up to a total of four.

**Figure 21.5** Pentane, $C_5H_{12}$, boils at 36°C, so it is a liquid at room temperature and atmospheric pressure.

When two or more substitutions appear in a cycloalkane, their locations are determined by number. The ring carbon to which one substituent is bonded is number 1. Other numbers are assigned to give the lowest numbers possible to the other locations. Thus, the following compound is 1,3-dimethylcyclohexane, not 1,5-dimethylcyclohexane:

$CH_3$
CH
$H_2C$ $CH_2$
$H_2C$ CH
$CH_2$ $CH_3$

**P/REVIEW:** Figure 18.7 in Section 18.4 illustrates the distillation of crude oil into fractions based on boiling point ranges. The gasoline fraction includes $C_5—C_{12}$ hydrocarbons.

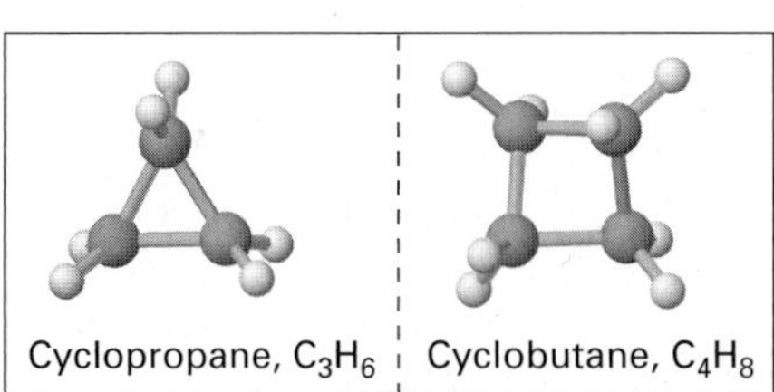


**Figure 21.6** Unstable cycloalkanes. The normal tetrahedral bond angle for a carbon–carbon bond is 109.5°. In cyclopropane, the C—C bond angles are 60°. Orbital overlap is poor, so the bonds are relatively weak and the molecule is very reactive. Cyclobutane has C—C bond angles of 88° because the carbons are not in the same plane (they would be 90° if the carbon atoms were co-planar). Again, the large deviation from the normal tetrahedral bond angle causes the molecule to be relatively unstable.

### √ Target Check 21.3

a) Identify the alkanes among $C_7H_{16}$, $C_5H_{10}$, $C_{11}H_{22}$, $C_9H_{20}$.
b) Write the formula of the alkyl group derived from pentane.
c) Write a structural diagram of 3,3-difluoro-4-iododecane.
d) Write a structural diagram of 1,1-diethylcyclohexane.

## 21.4 Unsaturated Hydrocarbons: The Alkenes and Alkynes

Goal 8 Given a formula of a hydrocarbon or information from which it can be written, determine whether the compound can be an alkene or an alkyne.

Goal 9 Given the name (or structural diagram) of an alkyne or an alkene, write the structural diagram (or name).

Goal 10 Identify and distinguish between *cis* and *trans* geometric isomers.

### Structure and Nomenclature

In a saturated hydrocarbon, each carbon atom is bonded to four other atoms. Hydrocarbons in which two or more carbon atoms are (1) connected by a double or triple bond and (2) bonded to fewer than four other atoms are unsaturated.

If one hydrogen atom, complete with its electron, is removed from each of two adjacent carbon atoms in an alkane (A), each carbon is left with a single unpaired electron (B). These electrons may then form a second bond between the two carbon atoms (C):

Words or chemical symbols are sometimes placed above or above and below the arrow of an equation to indicate a substance whose presence or removal is necessary for a reaction to proceed or to identify a reaction condition.

$$\underset{\text{A}}{\text{H}_3\text{C—CH}_3} \xrightarrow{-2\ \text{H}\cdot} \underset{\text{B}}{\text{H—}\dot{\text{C}}\text{H—}\dot{\text{C}}\text{H—H}} \longrightarrow \underset{\text{C}}{\text{H—CH=CH—H}}$$

Each carbon atom is now bonded to three other atoms. **An aliphatic hydrocarbon that contains at least one carbon-carbon double bond is called an alkene.** Figure 21.7(a) illustrates the simplest alkene, which has the common name ethylene (*eth-* = 2 carbon atoms + *-ylene* = alkene).

Removal of another hydrogen atom from each of the double-bonded carbon atoms in an alkene yields a triple bond:

$$\text{H—CH=CH—H} \xrightarrow{-2\ \text{H}\cdot} \text{H—}\dot{\text{C}}\text{=}\dot{\text{C}}\text{—H} \longrightarrow \text{H—C}\equiv\text{C—H}$$

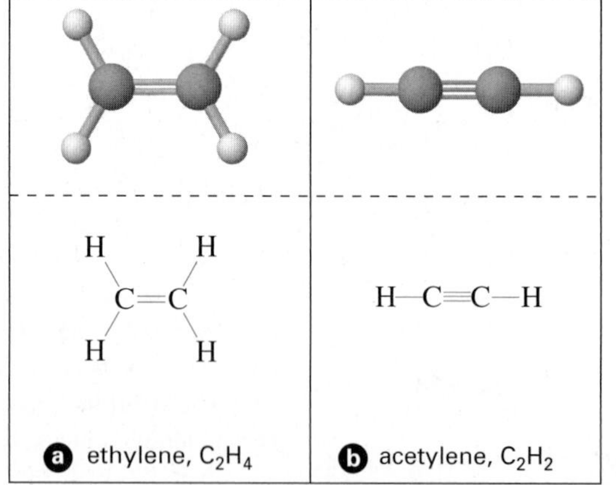


**Figure 21.7** Ball-and-stick models of the first members of the alkene (a) and alkyne (b) hydrocarbon series, ethylene, $C_2H_4$, and acetylene, $C_2H_2$.

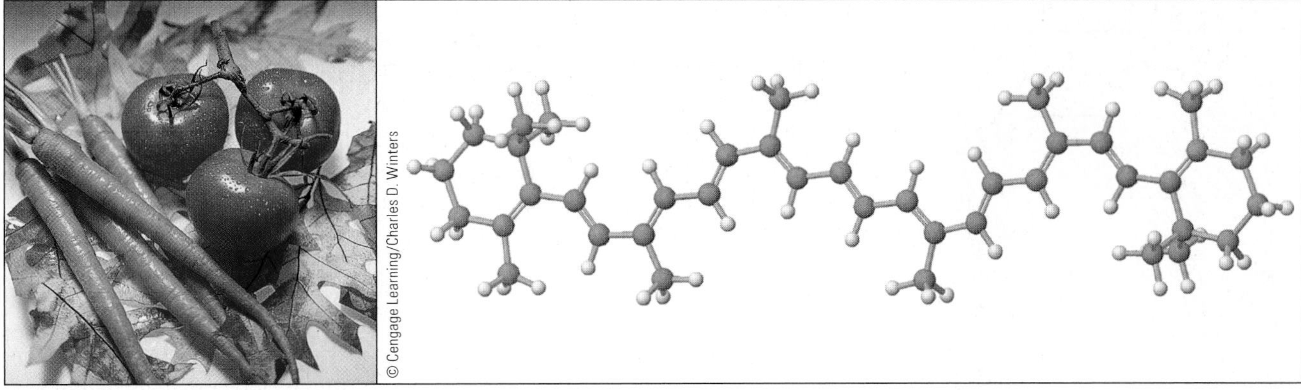


**Figure 21.8** Carotene. An example of a molecule with multiple double bonds is carotene, which has 11 carbon–carbon double bonds. The carotene molecule plays an indirect role in photosynthesis by transferring the energy it receives from sunlight to the chlorophyll molecule. Carotene exists in two forms, alpha-carotene and beta-carotene. Beta-carotene is found in yellow, orange, and green fruits, vegetables, and plants, such as these carrots, tomatoes, and leaves.

Each carbon atom is now bonded to two other atoms. **An aliphatic hydrocarbon in which two carbon atoms are triple-bonded to each other is called an alkyne.** Acetylene, the common name for the most common alkyne, $C_2H_2$, is illustrated in Figure 21.7(b).

Both the alkenes and the alkynes make up a new homologous series. Just as with the alkanes, each series may be extended by adding $-CH_2-$ units. Longer chains may have more than one multiple bond, but we will not consider such compounds in this text (Fig. 21.8). The general formula for an alkene is $C_nH_{2n}$, and for an alkyne, $C_nH_{2n-2}$.

Table 21.3 gives the names and formulas of some of the simpler unsaturated hydrocarbons. The IUPAC nomenclature system for the alkenes matches that of the alkanes. The suffix designating the alk*ene* hydrocarbon series is *-ene,* just as *-ane* identifies an alk*ane*. For example, pentene is $C_5H_{10}$, hexene is $C_6H_{12}$, and octene is $C_8H_{16}$. The

**Table 21.3** Unsaturated Hydrocarbons

| | | Formulas | | Names | |
|---|---|---|---|---|---|
| **Hydrocarbon Series** | **n** | ***Molecular*** | ***Structural*** | ***IUPAC*** | ***Common*** |
| Alkenes, $C_nH_{2n}$ | 2 | $C_2H_4$ | $H_2C=CH_2$ | Ethene | Ethylene |
| | 3 | $C_3H_6$ | $H_2C=CH-CH_3$ | Propene | Propylene |
| | 4 | $C_4H_8$ | $H_2C=CH-CH_2-CH_3$ | 1-Butene | Butylene |
| Alkynes, $C_nH_{2n-2}$ | 2 | $C_2H_2$ | $H-C\equiv C-H$ | Ethyne | Acetylene |
| | 3 | $C_3H_4$ | $H-C\equiv C-CH_3$ | Propyne | — |

**Figure 21.9** Oxyacetylene torch. Acetylene is the fuel in a welder's torch. It reacts with the oxygen in the air in a highly exothermic reaction, producing a 3000°C flame, which is hot enough to cut through and weld steel.

common names for the alkenes are produced similarly, except that the suffix is *-ylene*. These names for the smaller alkenes are often used instead of the IUPAC names: $C_2H_4$ is often called ethylene, $C_3H_6$ is propylene, and $C_4H_8$ is butylene.

Acetylene, $C_2H_2$, is the first member of the alkyne series (Fig. 21.9). Despite its *-ene* ending, acetylene is an alkyne, not an alkene. The IUPAC system is used for all alkynes except acetylene. By the IUPAC system, the ending *-yne* is used to indicate the presence of a triple bond. So, for the alkynes with two, three, and four carbon atoms, the IUPAC names are ethyne, propyne and butyne, respectively.

## Isomerism Among the Unsaturated Hydrocarbons

There are two possible isomers of butyne, shown below. They differ by the position of the multiple bond. The melting and boiling points are given below the structures to show that these are indeed different molecules.

$$-\overset{1}{C}\equiv\overset{2}{C}-\overset{3}{C}-\overset{4}{C}- \qquad -\overset{1}{C}-\overset{2}{C}\equiv\overset{3}{C}-\overset{4}{C}-$$

| | 1-butyne | 2-butyne |
|---|---|---|
| melting point | −122.5°C | −32.3°C |
| boiling point | +8.1°C | +27°C |

There are three distinct isomers of butene. Depending on the location of the double bond, we have the two structures:

$$-\overset{1}{C}=\overset{2}{C}-\overset{3}{C}-\overset{4}{C}- \qquad -\overset{1}{C}-\overset{2}{C}=\overset{3}{C}-\overset{4}{C}-$$

1-butene 2-butene

The third isomer is the result of *cis-trans* isomerism, as discussed in the next paragraph.

The part of a molecule that is on either side of a single bond may rotate freely around that bond as an axis. There is restricted rotation around a double bond (Fig. 21.10). This leads to two possible arrangements around a double bond. The two methyl groups can be on the *same* side of the double bond, as in *cis*-2-butene, or on opposite sides, as in *trans*-2-butene. *Cis*- and *trans*- are prefixes meaning, respectively, "on this side" and "across." Remember *trans* by associating it with a word such as transcontinental, meaning across a continent. Another name for *cis-trans* isomers is **geometric isomers.**

The three butene isomers are shown below, with their melting and boiling points, to show they are different substances.

$$-\overset{1}{C}=\overset{2}{C}-\overset{3}{C}-\overset{4}{C}-$$

| | 1-butene | *cis*-2-butene | *trans*-2-butene |
|---|---|---|---|
| melting point | −185.4°C | −138.9°C | −105.6°C |
| boiling point | −6.3°C | +3.7°C | +0.9°C |

Figure 21.11 shows another example of *cis-trans* isomers and the differences in their physical properties.

The compound 1-butene shows that the carbon chain is numbered through the multiple bond to give the multiple bond the lowest possible number. The compound

$$-\overset{5}{C}-\overset{4}{C}-\overset{3}{C}=\overset{2}{C}-\overset{1}{C}-$$

is 2-pentene, because the double bond is attached to the *second* carbon atom, counting from the right.

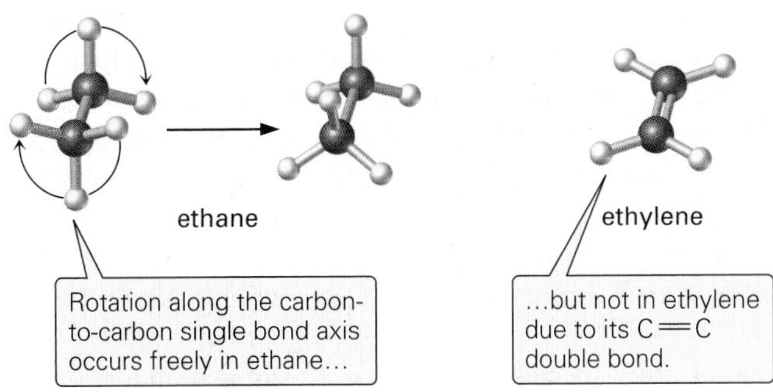


**Figure 21.10** Restriction of rotation about a double bond. In general, the groups of atoms on either side of a single bond have the ability to rotate freely about the bond, but the presence of a double bond essentially locks the atoms in place, making geometric isomerism possible.

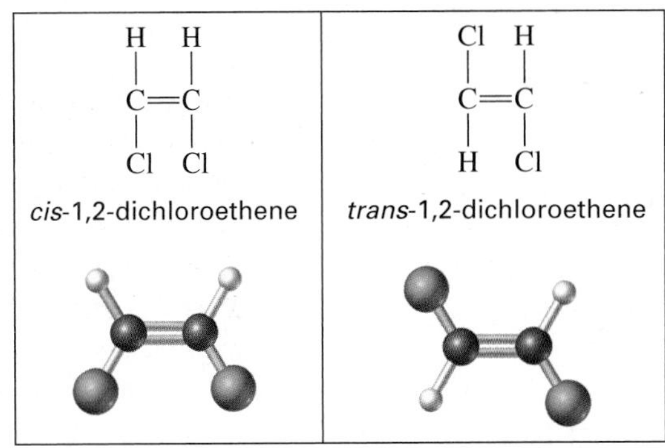


**Figure 21.11** Physical properties of *cis*- and *trans*-1,2-dichloroethene. The boiling point of the *cis* isomer is 60°C, and its melting point is −80°C. The boiling and melting points of the *trans* isomer are 48°C and −50°C, respectively. If rotation could occur around the double bond, the boiling and melting point of the isomers would be the same. Since the physical properties of the two isomers are different, there must be two different molecules. Rotation about the double bond must be restricted to account for the two isomers because atom connectivity is otherwise the same.

### √ Target Check 21.4

a) Identify the alkenes among the following: $C_4H_6$, $C_2H_6$, $C_7H_{12}$, $C_8H_{16}$.
b) Write a structural formula for *trans*-difluoroethene, $C_2H_2F_2$.
c) Identify the alkynes among the following: $C_4H_6$, $C_2H_6$, $C_7H_{12}$, $C_8H_{16}$.
d) How many isomeric straight-chain pentynes are possible? Write a structural formula for 2-pentyne.

## 21.5 Aromatic Hydrocarbons

**Goal 11** Distinguish between aliphatic and aromatic hydrocarbons.

**Goal 12** Given the name (or structural diagram) of an alkyl- or halogen-substituted benzene compound, write the structural diagram (or name).

Initially, aliphatic compounds were associated with oils and fats, which contain long carbon chains. By contrast, the term ***aromatic*** was associated with a series of compounds found in such pleasant-smelling substances as oil of cloves, vanilla, wintergreen, and cinnamon. Ultimately, chemists found that the key structure in aromatic hydrocarbons is the benzene ring. Now **any hydrocarbon that does not contain a benzene ring—a hydrocarbon that is not an aromatic hydrocarbon—is an aliphatic hydrocarbon.**

The simplest aromatic hydrocarbon is benzene, $C_6H_6$. Chemists struggled for decades to find a structural diagram for benzene that is consistent with its physical and chemical properties, but without success. Common forms and its space-filling model are

The structure at the left satisfies the octet rule and predicts a planar molecule with 120° bond angles, which benzene has, as shown in the model. However, the bonds in benzene are identical, rather than being alternating single and double bonds. The diagrams in the center describe the "second-bond" bonding electrons of the double bonds as being delocalized, belonging to the molecule as a whole rather than to specific bonds. It is this type of diagram that is most commonly used. As with cycloalkanes, each corner of the benzene molecule has a carbon atom, but it forms only one bond to an atom outside the carbon ring. Unless indicated otherwise, this atom is assumed to be hydrogen.

An alkyl group, halogen, or other species may replace a hydrogen on a benzene ring:

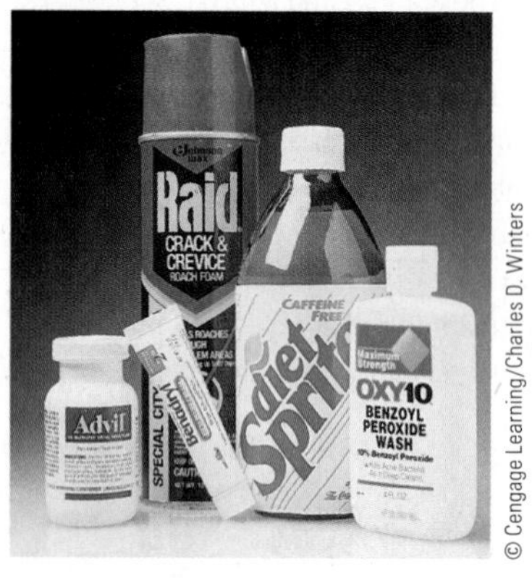



Consumer products containing compounds that contain a benzene ring. The ibuprofen in Advil, the propoxur in Raid, the diphenhydramine hydrochloride in Benadryl, the sodium benzoate in Sprite, and the benzoyl peroxide in Oxy-10 all have at least one benzene ring in their molecular structures.

$CH_3$ Br

methylbenzene
toluene

bromobenzene

If two bromines substitute for hydrogens on the same ring, we must consider three possible isomers:

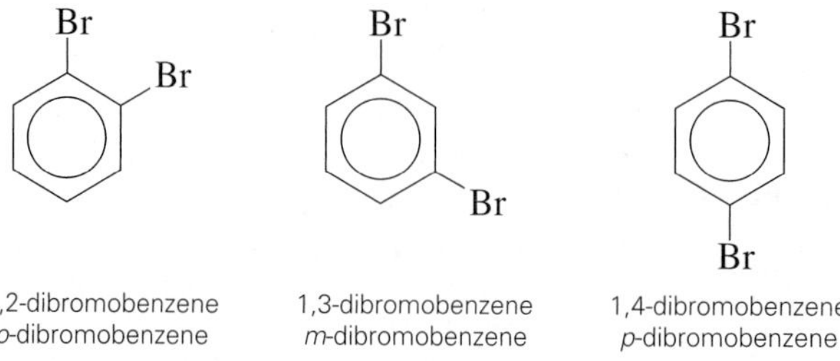


1,2-dibromobenzene
*o*-dibromobenzene

1,3-dibromobenzene
*m*-dibromobenzene

1,4-dibromobenzene
*p*-dibromobenzene

Two names are given for each isomer. The number system is the same as the system used to identify positions on a cyclohexane ring. It is more formal and serves any number of substituents. The other system uses the names *ortho*-dibromobenzene, *meta*-dibromobenzene, and *para*-dibromobenzene. *Ortho-*, *meta-*, and *para-* are prefixes commonly used when two hydrogens have been replaced from the benzene ring. Relative to position X, the other positions are shown here:

X
*o* *o*
*m* *m*
*p*

**P/REVIEW:** Substances with intermolecular attractions that are roughly equal are most apt to be soluble in each other (Section 16.4). Similar molecular polarity contributes to similar intermolecular attractions.

The physical properties of benzene and its derivatives are quite similar to those of other hydrocarbons. The compounds are nonpolar, insoluble in polar solvents such as water, but generally soluble in nonpolar solvents. In fact, derivatives of benzene are widely used as the solvent for many nonpolar organic compounds. Like other hydrocarbons of comparable molar mass, benzene is a liquid at room temperature.

### √ Target Check 21.5

a) Write the structural formula of 1,3,5-trifluorobenzene.

b) Many people buy mothballs that are made of *p*-dichlorobenzene. Write the structural formula for this substance.

## 21.6 Summary of the Hydrocarbons

The five categories of hydrocarbons we have considered are summarized in Table 21.4.

**Table 21.4** Summary of Hydrocarbons

| Type | Name | Formula | Saturation | Representative Structure |
|---|---|---|---|---|
| Aliphatic open-chain | Alkane | $C_nH_{2n+2}$ | Saturated | $-\overset{\vert}{\underset{\vert}{C}}-$ |
| | Alkene | $C_nH_{2n}$ | Unsaturated | $>C{=}C<$ |
| | Alkyne | $C_nH_{2n-2}$ | Unsaturated | $-C{\equiv}C-$ |
| Aliphatic cyclic | Cycloalkane | $C_nH_{2n}$ | Saturated | |
| Aromatic | — | — | Unsaturated | |

## 21.7 Sources and Preparation of Hydrocarbons

Almost all hydrocarbons are derived from fossil fuels: coal, natural gas, and petroleum. These substances are natural products that have resulted from the decay of plants and animals that lived millions of years ago. We are familiar with natural gas as the most common fuel for home heating and cooking with gas stoves. This fuel is primarily methane, $CH_4$, plus a smaller but significant portion of ethane, $C_2H_6$. Coal used to be a major source of benzene; benzene is actually a by-product of the preparation of coke from coal (Fig. 21.12). Other aromatic hydrocarbons are also recovered from the same process. Today, coal is primarily used as a fuel, and most benzene comes from petroleum.

**Figure 21.12** Coal. Although coal is an important starting material from which many products of the chemical industry are made, its primary use is as a fuel in electrical power plants.

**P/REVIEW:** A *catalytic* process is one that occurs in the presence of a catalyst. In Section 18.4 a catalyst is identified as a substance that speeds the rate of reaction without being permanently affected.

Petroleum is by far the largest source of the vast number of products broadly known as petrochemicals. Raw petroleum is a mixture of hydrocarbons containing up to 40 carbon atoms per molecule. These large molecules are not useful in their natural form, but they are broken into smaller molecules in petroleum refineries (Fig. 21.13; also see Fig. 18.7). *Catalytic cracking* essentially "cracks" the long carbon chains into shorter molecules of 5 to 10 carbon atoms. *Fractional distillation* separates hydrocarbons into "fractions" that boil at different temperatures. Alkanes of up to 4 or 5 carbon atoms per molecule may be obtained in pure form by this method. The boiling points of larger alkanes are too close for their complete separation, so chemical methods must be used to obtain pure products.

Courtesy of the American Petroleum Institute

**Figure 21.13** Petroleum refinery towers. Raw petroleum, or crude oil, is the starting point from which a huge number of products are made. The first step in the refining process is to distill the oil mixture into other mixtures based on boiling point ranges. These fractions are then further refined and processed.

Alkanes are prepared by several industrial and laboratory methods. One of the more important is the catalytic **hydrogenation** of an alkene. Hydrogenation is the reaction of a substance with hydrogen. The general reaction of the hydrogenation of an alkene is

$$C_nH_{2n} + H_2 \xrightarrow{\text{catalyst}} C_nH_{2n+2}$$

Unsaturated hydrocarbons are often prepared commercially from compounds derived originally from alkanes. Alkenes, for example, are produced from the *dehydration* of alcohols (Section 21.10) or the *dehydrohalogenation* of an alkyl halide. These two impressive terms describe very similar processes that are quite simple, at least in principle. Dehydration is the removal of a water molecule; dehydrohalogenation is the removal of a hydrogen atom and a halogen atom. For example, a water molecule may be separated from propyl alcohol, $C_3H_7OH$, to make propylene:

$$\begin{array}{c} \text{H}\ \ \text{H}\ \ \text{H} \\ |\ \ \ \ |\ \ \ \ | \\ \text{H—C—C—C—H} \\ |\ \ \ \ |\ \ \ \ | \\ \text{H}\ \ \text{H}\ \ \text{OH} \\ \text{propyl alcohol} \end{array} \xrightarrow{H_2SO_4} \begin{array}{c} \text{H}\ \ \text{H}\ \ \text{H} \\ |\ \ \ \ |\ \ \ \ | \\ \text{H—C—C=C—H} \\ | \\ \text{H} \\ \text{propylene} \end{array} + \begin{array}{c} \text{HOH} \\ \text{water} \end{array}$$

An alkyl halide is an alkane in which a halogen atom has been substituted for a hydrogen atom; viewed in another way, an alkyl halide is an alkyl group bonded to a halogen. The molecule is attacked with base in the presence of an alcohol to produce an alkene:

$$\begin{array}{c} \text{H}\ \ \text{H}\ \ \text{H} \\ |\ \ \ \ |\ \ \ \ | \\ \text{H—C—C—C—H} \\ |\ \ \ \ |\ \ \ \ | \\ \text{H}\ \ \text{H}\ \ \text{X} \\ \text{propyl halide} \end{array} + \begin{array}{c} \text{KOH} \\ \text{base} \end{array} \xrightarrow{\text{alcohol}} \begin{array}{c} \text{H}\ \ \text{H}\ \ \text{H} \\ |\ \ \ \ |\ \ \ \ | \\ \text{H—C—C=C—H} \\ | \\ \text{H} \\ \text{propene} \end{array} + \begin{array}{c} \text{KX} \\ \text{salt} \end{array} + \begin{array}{c} \text{HOH} \\ \text{water} \end{array}$$

Acetylene is the only alkyne produced commercially in large quantities. It is manufactured in a two-step process in which calcium oxide reacts with coke (carbon) at high temperatures to produce calcium carbide and carbon monoxide:

$$CaO(s) + 3\ C(s) \rightarrow CaC_2(s) + CO(g)$$

Calcium carbide then reacts with water to produce acetylene:

$$CaC_2(s) + 2\ H_2O(\ell) \rightarrow C_2H_2(g) + Ca(OH)_2(s)$$

## 21.8 Chemical Reactions of Hydrocarbons

**Goal** **13** Given the reactants in an addition or substitution reaction between (a) an alkane, alkene, alkyne, or benzene and (b) a hydrogen or halogen molecule, predict the products of the reaction.

The combustibility—ability to burn in air—of the hydrocarbons is probably the chemical reaction most important to modern society (Fig. 21.14). As components of liquid and gaseous fuels, hydrocarbons are among the most heavily processed and distributed chemical products in the world. When burned in an excess of air, the end products are water and carbon dioxide. For example,

$$CH_4(g) + 2\,O_2(g) \rightarrow 2\,H_2O(\ell) + CO_2(g)$$

**Figure 21.14** Combustion of natural gas. Hydrocarbons, such as natural gas, burn in air, releasing a relatively large amount of energy. This energy is then used to generate electricity, operate engines, heat factories, businesses, and homes, and, in general, to provide energy for most of the conveniences of modern society.

One major distinction separates the chemical properties of saturated hydrocarbons from those of the unsaturated hydrocarbons. By opening a multiple bond in an alkene or alkyne, the compound is capable of reacting by **addition,** simply by adding atoms of some element to the molecule. By contrast, an alkane molecule is literally saturated; there is no more room for an atom to join the molecule without first removing a hydrogen atom. A reaction in which a hydrogen atom in an alkane is replaced by an atom of another element is called a **substitution reaction.**

Both alkanes and alkenes undergo **halogenation reactions**—reaction with a halogen. These reactions serve to show the difference between addition and substitution reactions:

**Addition Reaction**

$$\begin{array}{ccccccccc} & & H & & H & & H & & \\ & & | & & | & & | & & \\ H & - & C & - & C & = & C & - & H \\ & & | & & & & & & \\ & & H & & & & & & \end{array} + Cl_2 \xrightarrow{CH_2Cl_2} \begin{array}{ccccccccc} & & H & & H & & H & & \\ & & | & & | & & | & & \\ H & - & C & - & C & - & C & - & H \\ & & | & & | & & | & & \\ & & H & & Cl & & Cl & & \end{array}$$

propene chlorine 1,2-dichloropropane

**Substitution Reaction**

$$\begin{array}{ccccccccc} & & H & & H & & H & & \\ & & | & & | & & | & & \\ H & - & C & - & C & - & C & - & H \\ & & | & & | & & | & & \\ & & H & & H & & H & & \end{array} + Cl_2 \xrightarrow{\text{heat or light}} \begin{array}{ccccccccc} & & H & & H & & H & & \\ & & | & & | & & | & & \\ H & - & C & - & C & - & C & - & H \\ & & | & & | & & | & & \\ & & H & & H & & Cl & & \end{array} + HCl$$

propane chlorine 1-chloropropane hydrogen chloride

The substituted chlorine atom may appear on either an end carbon atom or the middle carbon; the actual product is usually a mixture of 1-chloropropane and 2-chloropropane.

Normally, addition reactions are more readily accomplished than substitution reactions. This is hinted at in the reaction conditions specified above. The addition of a halogen to an alkene will occur easily at room temperature, whereas the substitution of a halogen for a hydrogen in an alkane requires either high temperature or ultraviolet light. This shows that unsaturated hydrocarbons are more reactive than saturated hydrocarbons.

Hydrogenation is also an addition reaction. We have already indicated that the hydrogenation of an alkene may be used to produce an alkane (Fig. 21.15). Hydrogenation of an alkyne is a stepwise process, which may often be controlled to give the intermediate alkene as a product:

$$R-C\equiv C-R' \xrightarrow[\text{catalyst}]{H_2} \begin{array}{ccccccc} & & H & & H & & \\ & & | & & | & & \\ R & - & C & = & C & - & R' \end{array} \xrightarrow[\text{catalyst}]{H_2} \begin{array}{ccccccc} & & H & & H & & \\ & & | & & | & & \\ R & - & C & - & C & - & R' \\ & & | & & | & & \\ & & H & & H & & \end{array}$$

alkyne alkene alkane

**Figure 21.15** Hydrogenation of an alkene produces an alkane. Vegetable oils have double bonds that are converted to saturated solid cooking fats by hydrogenation.

$R'$ in the diagram may be the same alkyl as R, or it may be a different alkyl.

Perhaps the most significant—and surprising—chemical property of benzene is that, despite its high degree of unsaturation, *it does not normally engage in addition reactions.* The classic 19th-century chemical tests for double bonds were reaction with bromine and with potassium permanganate to give addition products (Fig. 21.16). Benzene gives neither reaction. The most important reaction of benzene itself is the substitution reaction in

Figure 21.16 Addition reaction of bromine with alkenes. Chemists used to test for the presence of double bonds by reaction with bromine. Since bromine easily undergoes an addition reaction with alkenes, the disappearance of the red-brown color of bromine indicates that it has reacted. These photographs show that the bromine color disappears in the presence of bacon. We conclude that this is evidence of alkenes in bacon.

A few minutes $\longrightarrow$

which one hydrogen is displaced from the benzene ring. Several substances may be used for substitution, including the halogens:

$$\text{C}_6\text{H}_6 \text{ (benzene)} + Cl_2 \xrightarrow{FeCl_3} \text{C}_6\text{H}_5\text{Cl (chlorobenzene)} + HCl$$

Substitutions with nitric and sulfuric acids yield, respectively, nitrobenzene and benzenesulfonic acid. Second substitutions on the same ring are possible, although more difficult to bring about. Substitution reactions may also be performed on benzene derivatives, such as toluene, yielding isomers of nitrotoluene, for example. A triple nitro substitution produces 2,4,6-trinitrotoluene, better known simply as TNT.

## √ Target Check 21.6

*Complete the following equations.*

a)
$$\begin{array}{c} \text{H}\quad\text{H}\quad\text{H}\quad\text{H} \\ |\quad\;\;|\quad\;\;|\quad\;\;| \\ \text{H—C—C—C=C—H} \\ |\quad\;\;| \\ \text{H}\quad\text{H} \end{array} + Br_2 \xrightarrow{CH_2Cl_2}$$

b)
$$\begin{array}{c} \text{H}\quad\text{H}\quad\text{H}\quad\text{H}\quad\text{H}\quad\text{H} \\ |\quad\;\;|\quad\;\;|\quad\;\;|\quad\;\;|\quad\;\;| \\ \text{H—C—C—C—C—C—C—H} \\ |\quad\;\;|\quad\;\;|\quad\;\;|\quad\;\;|\quad\;\;| \\ \text{H}\quad\text{H}\quad\text{H}\quad\text{H}\quad\text{H}\quad\text{H} \end{array} + Cl_2 \xrightarrow{\text{light}}$$

c)
$$\text{C}_6\text{H}_6 + Cl_2 \xrightarrow{FeCl_3} \qquad + HCl$$

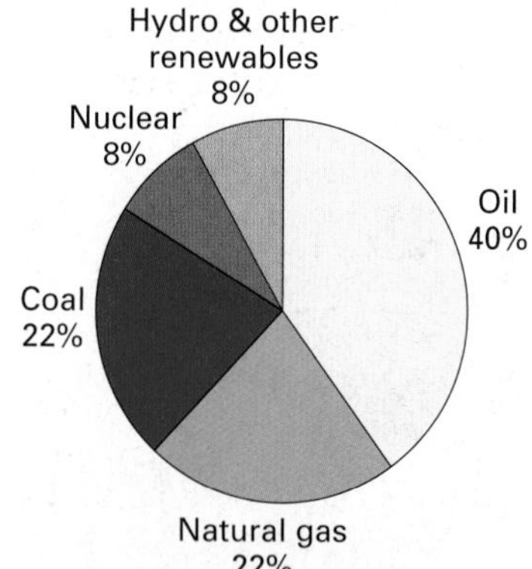


**2006 energy consumption = 99.9 quadrillion Btu ($99.9 \times 10^{15}$ Btu)**

Figure 21.17 United States energy sources in 2006. Oil, natural gas, and coal account for 84% of the total energy consumption.

## 21.9 Uses of Hydrocarbons

It is impossible to overstate the importance of hydrocarbons in modern society (Fig. 21.17). Alkanes move nations, both literally and figuratively—literally in transportation and figuratively in world politics. The oil crises of the 1970s and Operation Desert Storm in 1991 are events in which petroleum had a major impact on all industrialized nations. We rely heavily on petroleum products for the energy to heat homes, generate electricity, move people and goods, and manufacture almost everything we use, including many things derived directly from alkanes.

How close to you right now are hydrocarbons or products derived from hydrocarbons? Well, unless you're sitting naked as you study this book, you're probably wear-

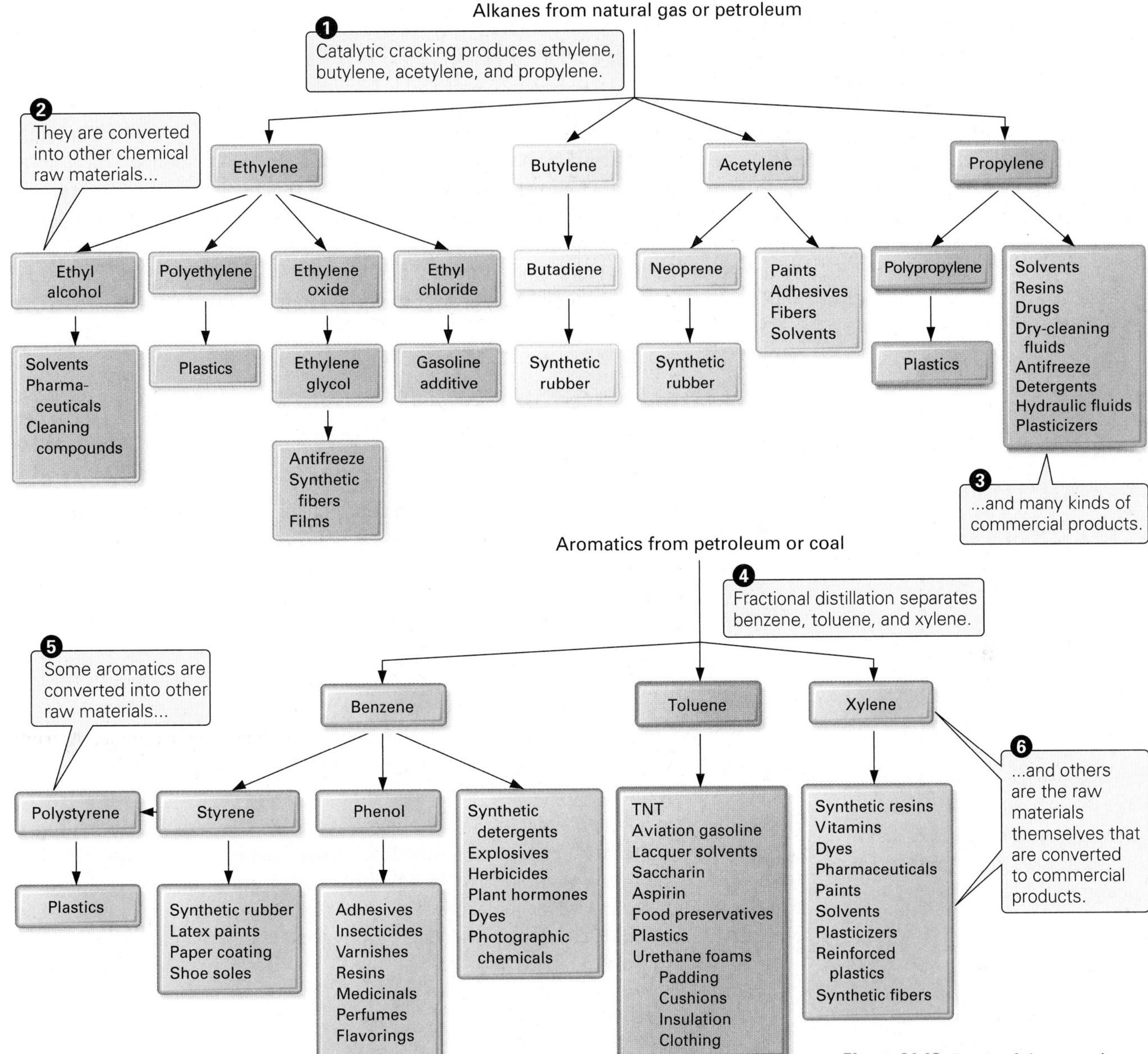


**Figure 21.18** Some of the organic chemicals obtained from fossil fuels and their uses as raw materials.

ing some. If everything you have on is not made of wool, cotton, silk, rubber, hemp, bone, or leather, you are almost certainly wearing something synthetic that started as a hydrocarbon. And even if you are wearing clothes derived entirely from natural products, the processes by which the natural fibers were converted to wearable clothing used hydrocarbons in some form.

Look around you. How many things do you see that are made of or derived from hydrocarbons? To help you, look at Figure 21.18. The chart starts with alkanes, takes you through other hydrocarbons, and ends at the useful products in the bottom box in each column. Can you not see, touch, or smell at least one of those end products at this very instant?

Where on the earth can you go to be totally separate from hydrocarbons? Almost nowhere. How about deep in dense woods? No, the wonderful smell of a pine forest is from naturally occurring hydrocarbons called (for good reasons) *pinenes*. On an all-wooden boat on an ocean, out of sight of land? Perhaps, if the boat isn't painted with an oil-based paint commonly used on marine vessels. In the middle of a desert, on top of Mount Everest, or at the North or South Pole? Perhaps, but wherever it is, you'd better be dressed in clothing that was manufactured in the 18th century!

# SECTIONS 21.10–21.14: ORGANIC COMPOUNDS CONTAINING OXYGEN AND NITROGEN

After carbon and hydrogen, the third element most commonly found in organic compounds is oxygen. Capable of forming two bonds (see Table 21.1), an oxygen atom serves as a connecting link between two other atoms or, double-bonded, usually to carbon, as a terminal atom in a **functional group.** A functional group is **an atom or a group of atoms that establishes the identity of a class of compounds and determines its chemical properties.** In the next five sections, we will look at several functional groups that contain oxygen atoms, and we will look briefly at two classes of nitrogen-bearing compounds that are very important in living organisms.

## 21.10 Alcohols and Ethers

Goal **14** Identify the structural formulas of the functional groups that distinguish alcohols and ethers.

Goal **15** Given the molecular structures of alcohols or ethers, or information from which they may be obtained, predict relative values of boiling points or solubility in water.

Goal **16** Given the name (or structural diagram) of an alcohol or ether, write the structural diagram (or name).

Goal **17** Given the reactants (or products) of a dehydration reaction between two alcohols, predict the products (or reactants) of the reaction.

Although the unshared electron pairs are not shown on the oxygen atoms in these structural diagrams, they play a key role in governing the physical and chemical properties of alcohols and ethers.

### Structures and Names of Alcohols and Ethers

Beginning with a water molecule in the diagrams below, if we remove a hydrogen atom, we have a **hydroxyl group.** This functional group identifies an **alcohol,** which is formed when the hydrogen atom is replaced with an alkyl group, R. Removal of both hydrogen atoms from a water molecule leaves the functional group of an **ether.** The ether molecule is completed when two alkyl groups bond to the oxygen.

```
  O          O           O          O           O
 / \        / \         / \        / \         / \
H   H          H       R   H                  R   R'
water    hydroxyl group  alcohol  ether group   ether
```

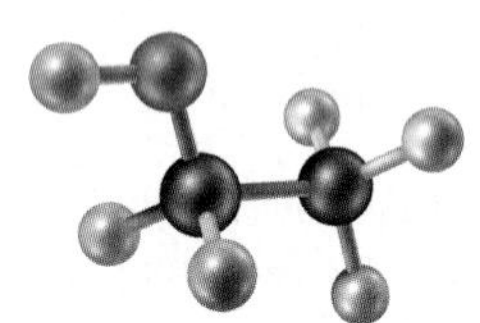

Figure 21.19 Models of ethanol (also called ethyl alcohol), $C_2H_5OH$.

Figures 21.19 and 21.20 are models of an alcohol and an ether, respectively. See if you can locate the functional group in each model. Alcohols are sometimes described by their common names, which originate in the name of the alkyl group to which the hydroxyl group is bonded. This system names the alkyl group, followed by "alcohol." Thus, $CH_3OH$ is *methyl alcohol* and $C_2H_5OH$ is *ethyl alcohol.* Under IUPAC nomenclature rules for alcohols, the *e* at the end of the corresponding alkane is replaced with the suffix *-ol* and the result is the name of the alcohol. Thus, methyl alcohol becomes *methanol,* and ethyl alcohol becomes *ethanol.*

Propyl alcohol has two isomers:

```
    H  H  H             H     H     H
    |  |  |             |     |     |
 H—C—C—C—H          H—C—–—C—–—C—H
    |  |  |             |     |     |
    H  H  OH            H     OH    H
  n-propyl alcohol     isopropyl alcohol
    1-propanol            2-propanol
```

These isomers are distinguished by stating the number of the carbon atom to which the hydroxyl group is bonded. Accordingly, *n*-propyl alcohol becomes *1-propanol* and isopropyl alcohol is formally designated *2-propanol.*

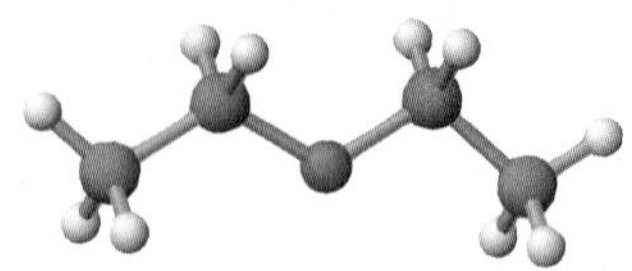

Figure 21.20 Model of diethyl ether, $C_2H_5OC_2H_5$.

All ethers are called "ether" and are identified specifically by naming first the two alkyl groups that are bonded to the functional group. If the groups are identical, the prefix *di-* may be used, as in diethyl ether.

## Sources and Preparation of Alcohols and Ethers

**Hydration of Alkenes** The major industrial source of several of our most important alcohols is the hydration of alkenes obtained from cracking petroleum. Beginning with ethylene, for example, the reaction may be summarized

$$\underset{\text{ethylene}}{H_2C{=}CH_2} + HOH \longrightarrow \underset{\text{ethanol}}{H{-}CH_2{-}CH_2{-}OH}$$

**Fermentation of Carbohydrates** Making ethanol by fermenting sugars in the presence of yeast is probably the oldest synthetic chemical process known (Fig. 21.21):

$$\underset{\text{glucose (sugar)}}{C_6H_{12}O_6} \xrightarrow{\text{yeast}} 2\,CO_2 + \underset{\text{ethanol}}{2\,C_2H_5OH}$$

A solution that is 95% ethyl alcohol (190 proof) may be obtained from the final mixture by fractional distillation. The mixture also yields two other products of commercial value today: 1-butanol and acetone (Section 21.11).

Under properly controlled conditions, ethers can be prepared by dehydrating alcohols. At 140°C, and with constant alcohol addition to replace the ether as it distills from the mixture, diethyl ether is formed from two molecules of ethanol:

$$\underset{\text{ethanol}}{CH_3{-}CH_2{-}O{-}H} + \underset{\text{ethanol}}{H{-}O{-}CH_2{-}CH_3} \longrightarrow \underset{\text{diethyl ether}}{CH_3{-}CH_2{-}O{-}CH_2{-}CH_3} + HOH$$

## Physical and Chemical Properties of Alcohols and Ethers

The structural similarity between water and alcohols suggests similar intermolecular forces and therefore similar physical properties. One- to three-carbon-atom alcohols

Courtesy of Professor James M. Bobbit, Twenty Mile Vineyard

**Figure 21.21** Wine production. Yeast metabolizes sugar to obtain energy, producing ethanol as a product of the chemical change. Exposure to oxygen must be avoided during fermentation to prevent oxidation of the ethanol to acetic acid (Section 21.12). Home fermenting and brewing carboys are fitted with an airlock to allow carbon dioxide to escape while preventing oxygen from entering.

P/REVIEW: Hydrogen bonding is the attraction between molecules in which a hydrogen atom is bonded to a nitrogen, oxygen, or fluorine atom that has at least one unshared pair of electrons (Section 15.3).

are liquids with boiling points ranging from 65°C to 97°C. These are comparable to the boiling point of water but well above the boiling points of alkanes of about the same molar mass. This is largely because of hydrogen bonding. Hydrogen bonding also accounts for the complete miscibility (solubility) between one- to three-carbon-atom alcohols and water. Solubility drops off sharply as the alkyl chain lengthens and the molecule assumes more of the character of the parent alkane. As usual, boiling points rise with increasing molecular size.

Ether molecules are less polar than alcohol molecules, and there is no opportunity for hydrogen bonding between them. Intermolecular attractions are therefore lower, as are their respective boiling points. Up to three carbons, ethers are gases at room conditions. Diethyl ether, with four carbons, is a volatile liquid that boils at 35°C. The solubility of ether molecules in water is about the same as the solubility of isomeric alcohols, probably due to the hydrogen bonding between the ether molecules and water molecules.

The chemical properties of alcohols are essentially the chemical properties of the functional group, —OH. In some reactions the C—OH bond is broken, separating the entire hydroxyl group. In other reactions the O—H bond in the hydroxyl group is broken.

Ethers are relatively unreactive compounds, being quite resistant to attack by active metals, strong bases, and oxidizing agents. They are, however, highly flammable and must be handled cautiously in the laboratory.

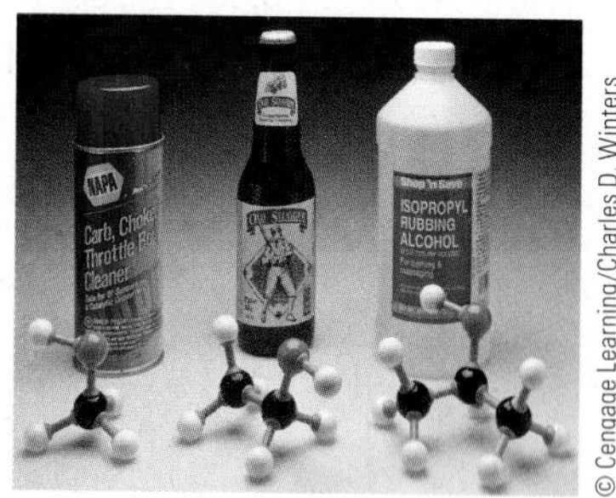




**Figure 21.22** Some common alcohols. Carburetor cleaner contains methanol, or methyl alcohol, $CH_3OH$. The alcohol in alcoholic beverages is ethanol, or ethyl alcohol, $C_2H_5OH$. Rubbing alcohol is 2-propanol, or isopropyl alcohol, $C_3H_7OH$.

## Common Alcohols and Ethers

Figure 21.22 shows some common alcohols. Methanol, $CH_3OH$, is an industrial chemical with production measured in the billions of pounds annually. Its major use is as a building block for other molecules. It is also used as an alternative automotive fuel because its combustion emits less pollution than gasoline and it can be made from sources other than petroleum. Methanol is found in consumer products such as windshield washer fluids, windshield deicers, and paint strippers. Taken internally, methanol can be a deadly poison, and even in doses as small as a few teaspoons, it can cause blindness. Fortunately, prompt medical treatment after methanol ingestion can prevent its tragic effects.

In addition to its uses in beverages, ethanol is used in organic solvents and in the preparation of various organic compounds such as chloroform and diethyl ether. Its production is also measured in the billions of pounds annually.

Other widely used alcohols include isopropyl alcohol, which is sold as rubbing alcohol, and *n*-butanol, used in lacquers in the automobile industry. Alcohols containing more than one hydroxyl group are also common. Permanent antifreeze in automobiles is ethylene glycol, which has two hydroxyl groups in the molecule (Fig. 21.23). Glycerin, or glycerol, a trihydroxyl alcohol, has many uses in the manufacture of drugs, cosmetics, explosives, and other chemicals (Fig. 21.24).

The word *ether* generally makes people think of the anesthetic that is so identified. This compound is diethyl ether, or simply ethyl ether; its line formula is $C_2H_5—O—C_2H_5$. Recently, its isomer, methyl propyl ether (Neothyl), $CH_3—O—C_3H_7$, has been gaining popularity as an anesthetic. It has fewer objectionable aftereffects than diethyl ether. Diethyl ether is used as a solvent for dissolving fats from foods and animal tissue in the laboratory. Because of its great combustibility, diethyl ether is also used as a cold-weather starting fluid for automobiles.

**Figure 21.23** Engine antifreeze and coolant. The substance marketed as "permanent antifreeze" (because it can be used in both the winter and summer) is a solution with ethylene glycol, $HO—CH_2CH_2—OH$, as its primary component. A molecule with two hydroxyl groups is called a diol.

### √ Target Check 21.7

a) Write the name and structural formula of the functional group that identifies an alcohol.
b) Write the structural formula of the functional group that identifies an ether.
c) Draw line formulas for all compounds with the formula $C_3H_8O$. Identify these as alcohols or ethers.

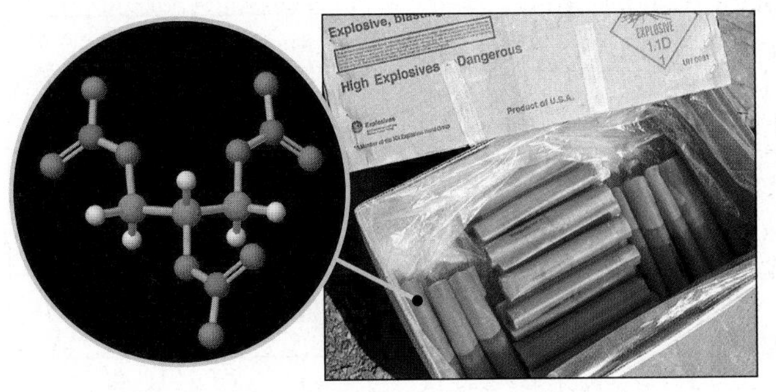




**Figure 21.24** Nitroglycerin. Glycerin, also called glycerol or 1,2,3-propanetriol, reacts with nitric acid to yield nitroglycerin, an explosive. When nitroglycerin is mixed with an absorbent material, forming dynamite, it is less sensitive to shocks than liquid nitroglycerin alone. Nitroglycerin is also used medicinally to dilate the blood vessels of the heart and relieve chest pain.

## 21.11 Aldehydes and Ketones

Goal **18** Identify the structural formulas of the functional groups that distinguish aldehydes and ketones.

Goal **19** Given the name (or structural diagram) of an aldehyde or ketone, write the structural diagram (or name).

Goal **20** Write structural diagrams to show how a specified aldehyde or ketone is prepared from an alcohol.

**Aldehydes** and **ketones** are characterized by the **carbonyl group,**

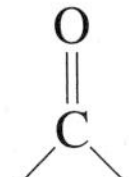

If at least one hydrogen atom is bonded to the carbonyl carbon, the compound is an aldehyde, RCHO; if two alkyl groups are attached, the compound is a ketone, R—CO—R′.

R—C(=O)—H (aldehyde) R—C(=O)—R′ (ketone)

The simplest carbonyl compound is formaldehyde, HCHO, which has two hydrogen atoms bonded to the carbonyl carbon. If a methyl group replaces one of the hydrogens of formaldehyde, the result is acetaldehyde, $CH_3CHO$. Replacement of both formaldehyde hydrogens with methyl groups yields acetone:

H—C(=O)—H (formaldehyde) $CH_3$—C(=O)—H (acetaldehyde) $CH_3$—C(=O)—$CH_3$ (acetone)

Figure 21.25 shows models of formaldehyde and acetone.

Many aldehydes are best known by their common names. The IUPAC nomenclature system for aldehydes employs the name of the parent hydrocarbon, substituting the suffix *-al* for the final *e* to identify the compound as an aldehyde. Thus, the IUPAC name for formaldehyde is methanal, for acetaldehyde, ethanal, and so forth.

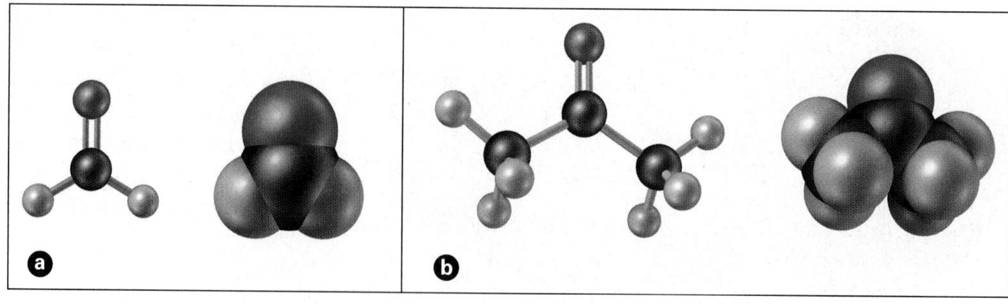

**Figure 21.25** Models of (a) formaldehyde, HCHO, the simplest aldehyde, and (b) acetone, $CH_3OCH_3$, the simplest ketone.

Ketones are named by one of two systems. The first duplicates the method of naming ethers: Identify each alkyl group attached to the carbonyl group, followed by the class name, ketone. Accordingly, methyl ethyl ketone has the structure

$$CH_3-\overset{\overset{\displaystyle O}{\|}}{C}-C_2H_5$$

Under the IUPAC system the number of carbons in the longest chain carrying the carbonyl carbon establishes the hydrocarbon base, which is followed by *-one* to identify the ketone as the class of compound. Methyl ethyl ketone, having four carbons, is called butanone. Two isomers of pentanone would be 2-pentanone and 3-pentanone, the number being used to designate the carbonyl carbon:

$$CH_3-\overset{\overset{\displaystyle O}{\|}}{C}-CH_2CH_2CH_3 \qquad CH_3CH_2-\overset{\overset{\displaystyle O}{\|}}{C}-CH_2CH_3$$

2-pentanone 3-pentanone

Aldehydes and ketones may be prepared by oxidation of alcohols. If the product is to be a ketone, the alcohol must be a *secondary* alcohol, in which the hydroxyl group is bonded to a carbon that is attached to two other carbon atoms:

$$R-\overset{\overset{\displaystyle H}{|}}{\underset{\underset{\displaystyle R'}{|}}{C}}-OH + \tfrac{1}{2}\,O_2 \longrightarrow R-\underset{\underset{\displaystyle R'}{|}}{C}=O + H_2O$$

secondary alcohol ketone

Care must be taken not to overoxidize aldehyde preparations, since aldehydes are easily oxidized to carboxylic acids (see Section 21.12).

Aldehydes and ketones may also be produced by the hydration of alkynes. If the triple bond is on the last carbon in a carbon atom chain, an aldehyde is produced; if the bond is between internal carbons, the result is a ketone. A typical reaction is the commercial preparation of acetaldehyde:

$$H-C\equiv C-H + HOH \longrightarrow \left[\begin{matrix} H & & O-H \\ & C=C & \\ H & & H \end{matrix}\right] \longrightarrow H-\overset{\overset{\displaystyle H}{|}}{\underset{\underset{\displaystyle H}{|}}{C}}-\overset{\displaystyle O}{\overset{\|}{C}}-H$$

(unstable form) acetaldehyde

The double bond of the carbonyl group can engage in addition reactions, just like the double bond in the alkenes. One such reaction is the catalytic hydrogenation of ketones to secondary alcohols, in which the hydroxyl group is bonded to a carbon atom *within* the chain:

$$\underset{\text{ketone}}{R-\overset{\overset{R'}{|}}{C}=O} + H_2 \xrightarrow{\text{catalyst}} \underset{\text{secondary alcohol}}{R-\overset{\overset{R'}{|}}{\underset{\underset{H}{|}}{C}}-O-H}$$

Oxidation reactions occur quite readily with aldehydes, but are resisted by ketones. When an aldehyde is oxidized, the product is a carboxylic acid:

$$\underset{\text{aldehyde}}{R-\overset{\overset{H}{|}}{C}=O} + \tfrac{1}{2}\,O_2 \longrightarrow \underset{\text{carboxylic acid}}{R-C(=O)-OH}$$

Formaldehyde is probably the best-known carbonyl compound. Large quantities are made into polymers such as Bakelite, Formica, and Melmac. Since formaldehyde has been cited as a probable carcinogen, its use in preserving biological specimens has virtually vanished. Acetaldehyde (ethanal) is used in manufacturing organic compounds such as acetic acid and ethyl acetate. Other aldehydes you have probably encountered are benzaldehyde (almond flavor), cinnamaldehyde (cinnamon flavor), and vanillin (vanilla flavor) (Fig. 21.26).

Acetone is the most commercially important ketone, with about 3 billion pounds produced yearly in the United States. It is a solvent used in manufacturing other organic chemicals, drugs, and explosives. Acetone is also found in paint remover and nail polish remover. Methyl ethyl ketone (MEK) is used in the petroleum industry, as a lacquer solvent, and in nail polish remover. Most organic compounds whose names end in *-one* are ketones. You may be familiar with the anti-inflammatory cortisone, or the sex hormones progesterone and testosterone; all are ketones.

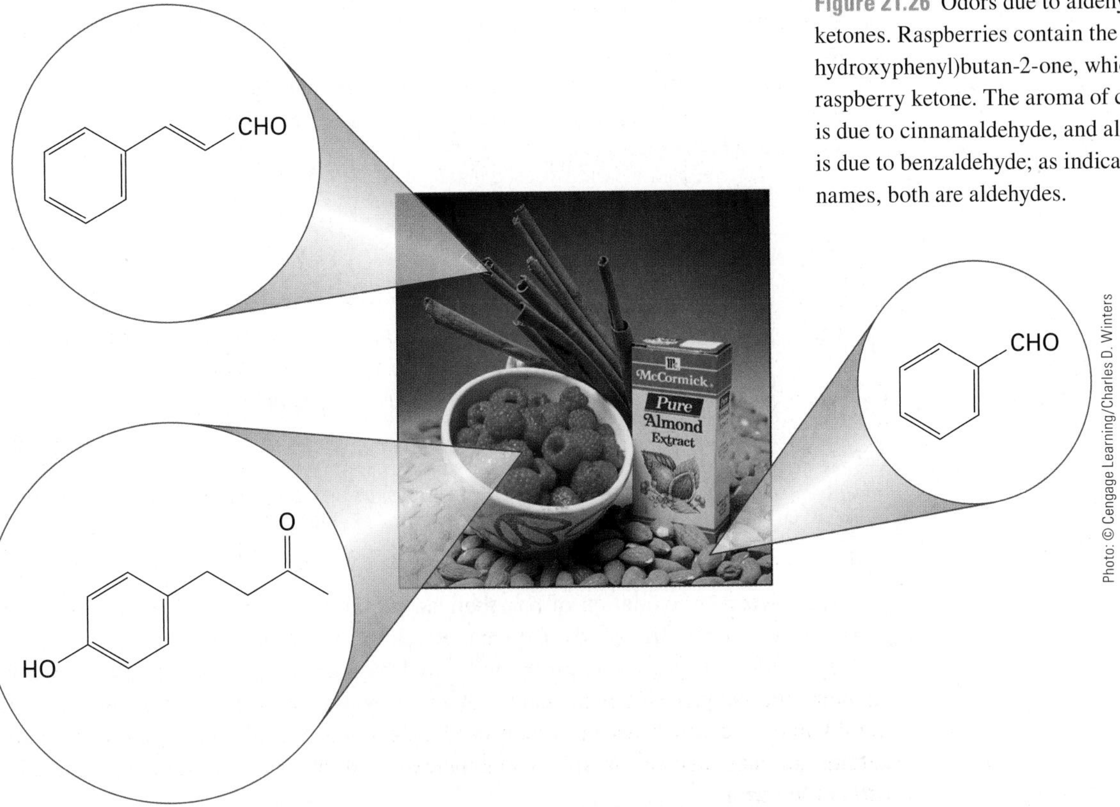


**Figure 21.26** Odors due to aldehydes and ketones. Raspberries contain the ketone 4-(4-hydroxyphenyl)butan-2-one, which is called raspberry ketone. The aroma of cinnamon is due to cinnamaldehyde, and almond scent is due to benzaldehyde; as indicated by their names, both are aldehydes.

Steroid molecules have four rings: three cyclohexane rings and one cyclopentane ring (Section 22.4). Progesterone (*top*) and testosterone (*bottom*) also have ketone functional groups. Both molecules are human sex hormones.

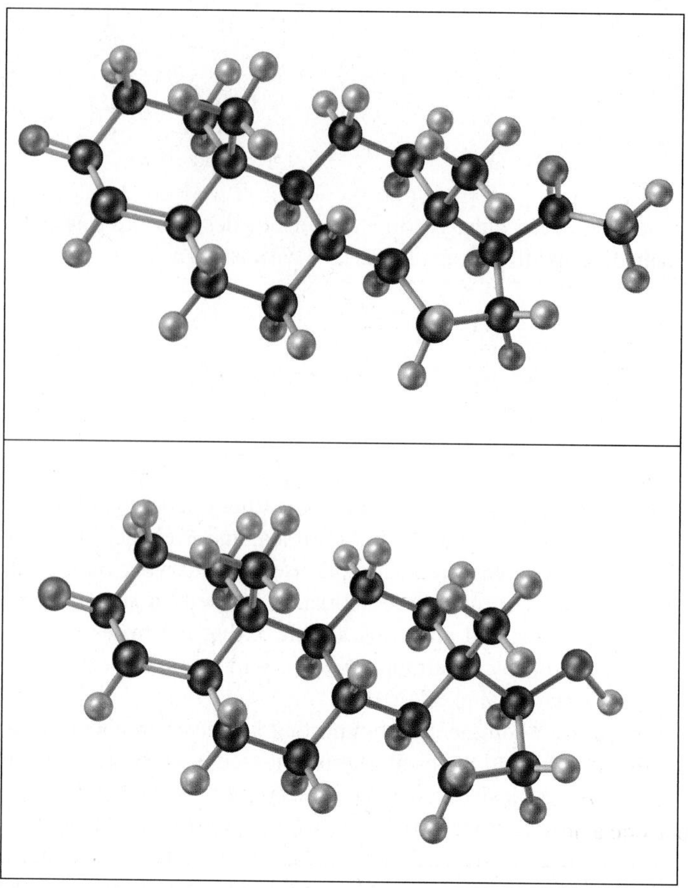

### √ Target Check 21.8

a) Write the name and structural formula of the functional group that identifies an aldehyde or ketone.

b) Write structural diagrams that show the difference between an aldehyde and a ketone.

c) Write structural diagrams for all possible aldehyde or ketone isomers of $C_5H_{10}O$; identify the aldehydes and the ketones.

## 21.12 Carboxylic Acids and Esters

Goal **21** Identify the structural formulas of the functional groups that distinguish carboxylic acids and esters.

Goal **22** Given the name (or structural diagram) of a carboxylic acid or ester, write the structural diagram (or name).

Goal **23** Given the reactants (or products) of an esterification reaction, predict the products (or reactants) of the reaction.

As you have seen, oxidation of an aldehyde produces a **carboxylic acid,** the general formula of which is frequently shown as RCOOH. The functional group, —COOH, is a combination of a carbonyl group and a hydroxyl group, rightly called the **carboxyl group.** You can probably pick out the carboxyl group in the acetic acid model in Figure 21.27(a). In an **ester** the carboxyl carbon may be bonded to a hydrogen atom or an alkyl group, and the carboxyl hydrogen is replaced by another alkyl group, as shown here and in Figure 21.27(b):

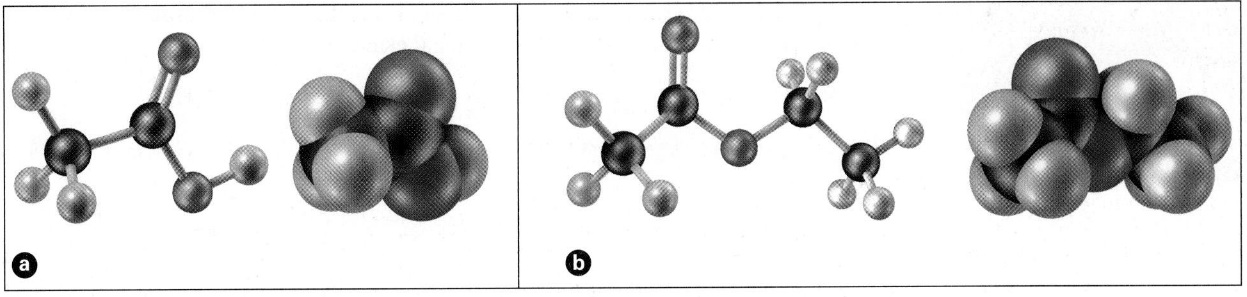

Figure 21.27 Models of (a) acetic acid, $CH_3COOH$, and (b) ethyl acetate, $CH_3COOC_2H_5$.

$$\mathrm{R{-}C({=}O){-}O{-}H} \qquad \mathrm{R{-}C({=}O){-}O{-}R'}$$

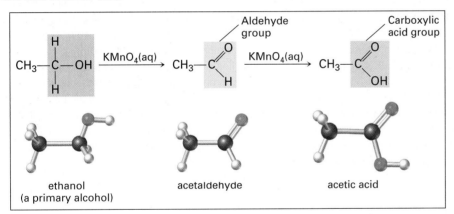

carboxylic acid — ester

The geometry of the carboxyl group results in strong dipole attractions and hydrogen bonding between molecules. As a consequence, boiling points of carboxylic acids tend to be high compared with other compounds of similar molecular mass. Formic acid, HCOOH, for example, boils at 100.5°C. Acids with a relatively small number of carbons are completely miscible in water, but solubility drops off as the aliphatic chain lengthens and the molecule behaves more like a hydrocarbon.

Acetic acid is produced by the stepwise oxidation of ethanol, first to acetaldehyde and then to acetic acid:

$$\mathrm{CH_3{-}CH_2{-}OH} \xrightarrow{\mathrm{KMnO_4(aq)}} \mathrm{CH_3{-}C({=}O){-}H} \xrightarrow{\mathrm{KMnO_4(aq)}} \mathrm{CH_3{-}C({=}O){-}OH}$$

Aldehyde group

Carboxylic acid group

ethanol (a primary alcohol) — acetaldehyde — acetic acid

Potassium permanganate, $KMnO_4$, is a strong oxidizing agent.

Carboxylic acids are weak acids that release a proton from the carboxyl group on ionization.* Acetic acid, for example, ionizes in water as follows:

$$CH_3COOH(aq) \rightleftharpoons CH_3COO^-(aq) + H^+(aq)$$

The ionization takes place but slightly; only about 1% of the acetic acid molecules ionize. The solution consists primarily of molecular $CH_3COOH$. This notwithstanding, acetic acid participates in typical acid reactions such as neutralization

$$CH_3COOH(aq) + OH^-(aq) \rightarrow H_2O(\ell) + CH_3COO^-(aq)$$

and the release of hydrogen on reaction with a metal:

$$2\ CH_3COOH(aq) + Ca(s) \rightarrow 2\ CH_3COO^-(aq) + Ca^{2+}(aq) + H_2(g)$$

Metal acetate salts may be obtained by evaporating the resulting solutions to dryness.

Acetic acid is responsible for the odor and taste of vinegar. Distilled white vinegar (as opposed to cider or malt vinegar, for example) is made from grain alcohol and distilled to remove all compounds but acetic acid and water. The label indicates that this brand has "5% Acidity," indicating that it is a 5% solution of acetic acid.

*In more advanced study of organic reactions, the term *acid* is also used in reference to Lewis acids (Section 17.3). This is why the adjective *carboxylic* is used to identify an organic acid containing the carboxyl group.

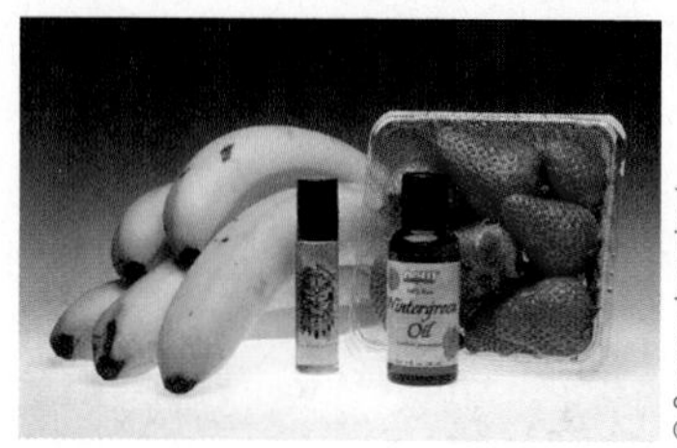


The odors of most fruits and many perfumes are due to molecules that contain the ester functional group.

The reaction between an acid and an alcohol is called **esterification.** The products of the reaction are an ester and water. A typical esterification reaction is

$$\underset{\substack{\text{acetic acid}\\ \textit{acid}}}{H-\overset{\displaystyle H}{\underset{\displaystyle H}{\overset{|}{\underset{|}{C}}}}-\overset{\displaystyle O}{\overset{\|}{C}}-O-H} + \underset{\substack{\text{methanol}\\ \textit{alcohol}}}{H-O-\overset{\displaystyle H}{\underset{\displaystyle H}{\overset{|}{\underset{|}{C}}}}-H} \rightleftharpoons \underset{\substack{\text{methyl acetate}\\ \textit{ester}}}{H-\overset{\displaystyle H}{\underset{\displaystyle H}{\overset{|}{\underset{|}{C}}}}-\overset{\displaystyle O}{\overset{\|}{C}}-O-\overset{\displaystyle H}{\underset{\displaystyle H}{\overset{|}{\underset{|}{C}}}}-H} + HOH$$

Notice how the water molecule is formed: *The acid contributes the entire hydroxyl group*, while *the alcohol furnishes only the hydrogen.*

The names of esters are derived from the parent alcohol and acid. The first term is the alkyl group associated with the alcohol; the second term is the name of the anion derived from the acid. In the preceding example, methanol (methyl alcohol) yields *methyl* as the first term and acetic acid yields *acetate* as the second term.

**P/REVIEW:** A Brønsted–Lowry acid–base reaction involves the transfer of a proton from the acid to the base. An ammonia molecule, with an unshared electron pair on the nitrogen, is the proton receiver—the base—in this reaction. See Section 17.2.

Carboxylic acids engage in typical proton-transfer acid–base-type reactions with ammonia to produce salts. The ammonium salt so produced may then be heated, which causes it to lose a water molecule. The resulting product is called an *amide*. Compared with the original acid, an amide substitutes an $—NH_2$ group for the —OH group of the acid (Section 21.13):

$$\underset{\text{acid}}{R-\overset{\displaystyle O}{\overset{\|}{C}}-O-H} + NH_3 \longrightarrow \underset{\text{salt}}{\left[R-\overset{\displaystyle O}{\overset{\|}{C}}-O\right]^- NH_4^+} \xrightarrow{\text{heat}} \underset{\text{amide}}{R-\overset{\displaystyle O}{\overset{\|}{C}}-NH_2} + H_2O$$

Formic acid and acetic acid are the two most important carboxylic acids. Formic acid is a source of irritation in the bites of ants and other insects or in the scratch of nettles. A liquid with a sharp, irritating odor, formic acid is used in manufacturing esters, salts, and plastics. Acetic acid is present in a concentration of about 5% in vinegar and is responsible for its odor and taste. Acetic acid is among the least expensive organic acids, and is therefore a raw material in many commercial processes that require a carboxylic acid. Sodium acetate is one of several common salts of carboxylic acids. It is used to control the acidity of chemical processes and in preparing soaps and pharmaceutical agents.

The distinctive tangy taste of sourdough bread results from the work of the yeast and lactobacteria that also produce the carbon dioxide that makes the bread rise. Acetic acid and lactic acid, both carboxylic acids, are produced during the fermentation, giving sourdough bread its sour taste.

Ethyl acetate and butyl acetate are two of the relatively few esters produced in large quantity. Both are used as solvents, particularly in manufacturing lacquers. Other esters are used in the plastics industry, and some find application in medicinal fields. Esters are responsible for the odors of most fruits and flowers, leading to their use in the food and perfume industries.

### √ Target Check 21.9

a) Write the name and structural formula of the functional group that defines a carboxylic acid.

b) Describe in words the reactants and products of an esterification reaction.

c) Write structural diagrams for all possible carboxylic acid or ester isomers of $C_4H_8O_2$; identify the carboxylic acid and the esters.

## 21.13 Amines and Amides

Goal **24** Identify the structural formulas of the functional groups that distinguish amines and amides.

Goal **25** Given the name (or structural diagram) of an amine or amide, write the structural diagram (or name).

Goal **26** Given the structural diagram of an amine, or information from which it can be written, classify the amine as primary, secondary, or tertiary.

Goal **27** Given the reactants of a reaction between a carboxylic acid and an amine, predict the products of the reaction.

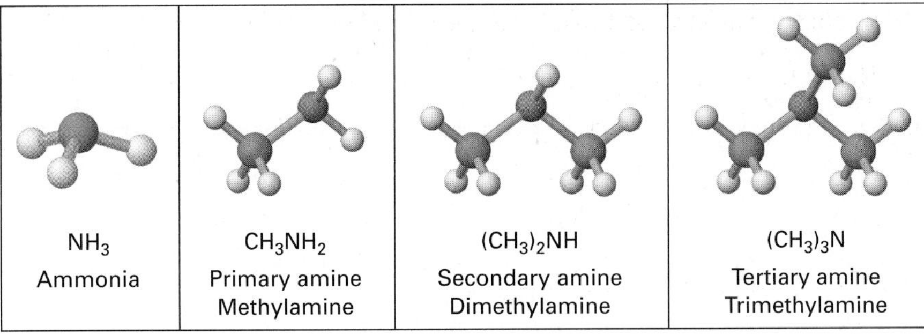


**Figure 21.28** Models of ammonia and the methylamines.

**Amines** are organic derivatives of ammonia, $NH_3$. An amine is formed by replacing one, two, or all three hydrogens in an ammonia molecule with an alkyl group. The number of hydrogens replaced determines whether an amine is primary (one hydrogen replaced), secondary (two), or tertiary (three). IUPAC names are rarely used for amines; they are named in practice by identifying in alphabetical order the alkyl groups that are bonded to the nitrogen atom, using appropriate prefixes if two or three identical groups are present, followed by the suffix *-amine*. Here are some examples:

$$H-\ddot{N}(H)-H \qquad CH_3-\ddot{N}(H)-H \qquad CH_3-\ddot{N}(H)-C_2H_5 \qquad CH_3-\ddot{N}(CH_3)-C_2H_5$$

ammonia; methylamine, primary amine; ethylmethylamine, secondary amine; ethyldimethylamine, tertiary amine

Models of ammonia and the different methylamines are shown in Figure 21.28.

Dimethylamine and trimethylamine are used in making anion-exchange resins. Many chemical products such as dyes, drugs, herbicides, fungicides, soaps, insecticides, and photographic developers are made from amines. The aromatic amine aniline (phenylamine) is used in dye making.

Nicotine, the stimulant in tobacco, is a tertiary amine.

An **amide** is a derivative of a carboxylic acid in which the hydroxyl part of the carboxyl group is replaced by an $—NH_2$, —NHR, or $—NR_2$ group. For example,

$$CH_3-C(=O)-OH \quad \text{becomes} \quad CH_3-C(=O)-NH_2$$

acetic acid → acetamide

by substitution of the $—NH_2$ for the —OH, as shown. An amide is commonly named by replacing the *-ic* or *-oic acid* name of the acid with *amide.*

The amide structure appears in a biochemical system, proteins, as a connecting link between amino acids. The linkage is commonly called a **peptide linkage.** This linkage has the form

$$R-C(=O)-N(H)-C$$

peptide linkage

An **amino acid** is an acid in which an amine group is substituted for a hydrogen atom in the molecule. The amino acids in protein structure have the general formula

$$R-CH(NH_2)-C(=O)-OH$$

in which the amine and carboxyl groups are bonded to the same carbon atom. The peptide linkage is formed when the carboxyl group of one amino acid and the amine group of another combine by removing a water molecule:

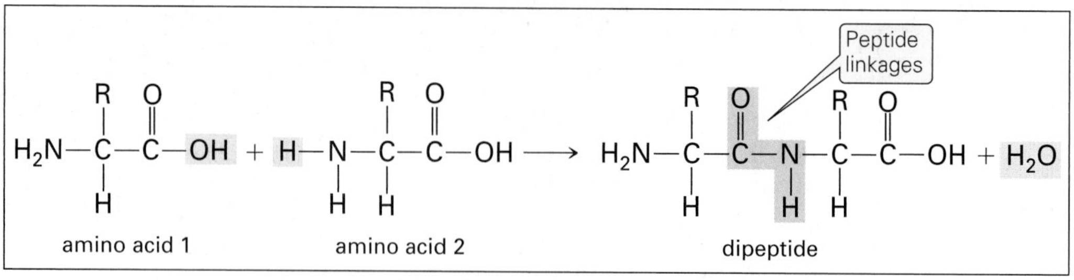

$$\underset{\text{amino acid 1}}{H_2N-\underset{\displaystyle H}{\overset{\displaystyle R}{\overset{|}{\underset{|}{C}}}}-\overset{\displaystyle O}{\overset{\|}{C}}-OH} + \underset{\text{amino acid 2}}{H-\underset{\displaystyle H}{\underset{|}{N}}-\underset{\displaystyle H}{\overset{\displaystyle R}{\overset{|}{\underset{|}{C}}}}-\overset{\displaystyle O}{\overset{\|}{C}}-OH} \longrightarrow \underset{\text{dipeptide}}{H_2N-\underset{\displaystyle H}{\overset{\displaystyle R}{\overset{|}{\underset{|}{C}}}}-\overset{\displaystyle O}{\overset{\|}{C}}-\underset{\displaystyle H}{\underset{|}{N}}-\underset{\displaystyle H}{\overset{\displaystyle R}{\overset{|}{\underset{|}{C}}}}-\overset{\displaystyle O}{\overset{\|}{C}}-OH} + H_2O$$

Peptide linkages

Every living thing has proteins, which are chains of amino acids held together by amide linkages (peptide linkages). There are 20 different amino acids commonly found in proteins. Typical proteins are huge molecules with molar masses ranging from about 34,500 to 50,000,000 g/mol. These proteins perform many functions in a living system. Proteins are discussed in greater detail in Chapter 22.

### √ Target Check 21.10

a) Write the structural formula of the functional group that defines an amine.
b) Write the structural formula of the functional group that defines an amide.

## 21.14 Summary of the Organic Compounds of Carbon, Hydrogen, Oxygen, and Nitrogen

The eight types of organic compounds of carbon, hydrogen, oxygen, and nitrogen you have studied are summarized in Table 21.5.

### Thinking About Your Thinking

#### Classification

**Table 21.5 summarizes a classification scheme for organic compounds based on functional groups, an atom or group of atoms in an organic molecule that is primarily responsible for its chemical properties. Thinking in this way—classifying molecules by functional group—is central to the organization of organic chemistry. Literally millions of different organic molecules are known to exist, and millions of reactions can be used to synthesize or transform these molecules. A body of knowledge so vast must have a classification scheme to give it an organization. Organic functional groups are used by all chemists, everywhere in the world, to organize their knowledge.**

# SECTIONS 21.15–21.16: POLYMERS

The word *plastics* is commonly used to describe a large number of familiar substances, all of which were first developed in the 20th century (Fig. 21.29). These materials are made up of huge molecules having molecular masses that sometimes run into the millions.

One way to make a large molecule is to connect many small chemical units, somewhat as links are joined to form a chain. Each link is called a **monomer,** which, from the Greek, means "having one part." The resulting chain is called a **polymer,** which means "having many parts." The process by which polymers are formed is called **polymerization.** Polymers are classified according to the method by which they are formed, and these two classes guide the organization of the final two sections of this chapter.

**Table 21.5** Summary of Organic Compounds

| Compound Class | General Formula | Functional Group | Names* |
|---|---|---|---|
| Alcohol | R—OH | —OH | Alkyl group + *alcohol:* methyl alcohol<br>Alkane prefix + *-ol:* methanol |
| Ether | R—O—R′ | —O— | Name both alkyl groups + *ether:* ethyl methyl ether<br>Alkyl group + *-oxy-* + alkane: methoxyethane |
| Aldehyde | R—CHO | —C(=O)—H | Common prefix + *-aldehyde:* formaldehyde<br>Alkane prefix + *-al:* methanal |
| Ketone | R—CO—R′ | —C(=O)— | Name both alkyl groups + *ketone:* methyl ethyl ketone; methyl *n*-propyl ketone<br>(Number carbonyl carbon) + alkane prefix + *-one:* butanone; 2-pentanone |
| Acid | R—COOH | —C(=O)—OH | Common name + *acid:* formic acid<br>Alkane prefix + *-oic* + *acid:* methanoic acid |
| Ester | R—CO—OR′ | —C(=O)—OR′ | Alcohol alkyl group + acid anion: methyl acetate<br>Alcohol alkyl group + acid alkane prefix + *-oate:* methyl ethanoate |
| Amine | $RNH_2$<br>$R_2NH$<br>$R_3N$ | —N— (with a third bond below) | Name alkyl group(s) + *-amine:* methylamine<br>*Amino-* + alkane: aminomethane |
| Amide | $R—CONH_2$ | —C(=O)—$NH_2$ | Common acid prefix + *amide:* formamide<br>Alkane prefix + *-amide:* methanamide |

*Common name followed by IUPAC name, with examples of each.

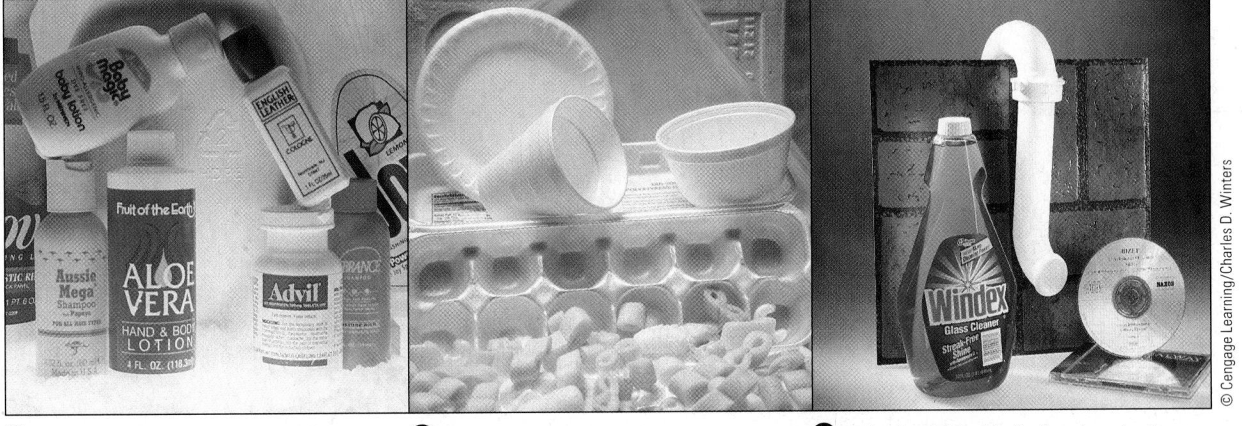




**a** The containers in this photograph are made of polyethylene, the most common plastic.

**b** One form of polystyrene is known as Styrofoam.

**c** Polyvinylchloride is the chemical name from which we derive the abbreviation PVC.

**Figure 21.29** Plastics. Chemists use the word *polymers* to describe what in everyday language we call *plastics.*

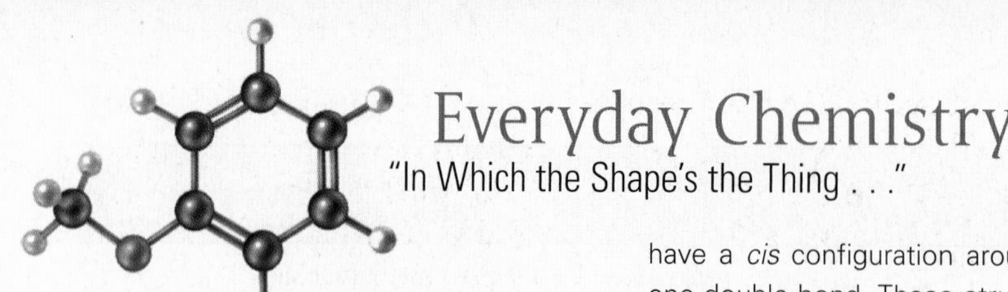

# Everyday Chemistry

## "In Which the Shape's the Thing . . ."

You've probably seen the terms *saturated, monounsaturated,* and *polyunsaturated* used on food labels to describe oils and fats in what we eat. Saturated fats (Fig. 21.30) are cylindrical and tend to pack into masses that are solids at body temperature. Unsaturated fats and oils have a *cis* configuration around at least one double bond. These structures have a bend in them (Fig. 21.31) and do not pack as easily as straight structures. As a result, the unsaturated fats and oils are liquids at body temperature.

Hydrogenation of vegetable oils (seen on food labels as "partially hydrogenated vegetable oils") changes some *cis* fatty acids into saturated fats and also isomerizes some *cis* double bonds into *trans* double bonds. These *trans* double bonds give a cylindrical molecule (Fig. 21.32), like saturated fats. Unfortunately, cylindrical fats, either saturated or *trans* unsaturated, are the fats that lead to blocked arteries. The bend in the fats makes all the difference in the blood.

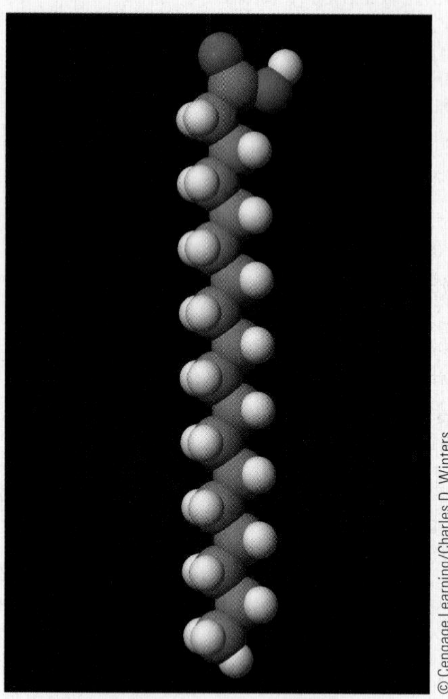



**Figure 21.30** Space-filling model of stearic acid, $CH_3(CH_2)_{16}COOH$. Notice that the molecule is cylindrical and straight.

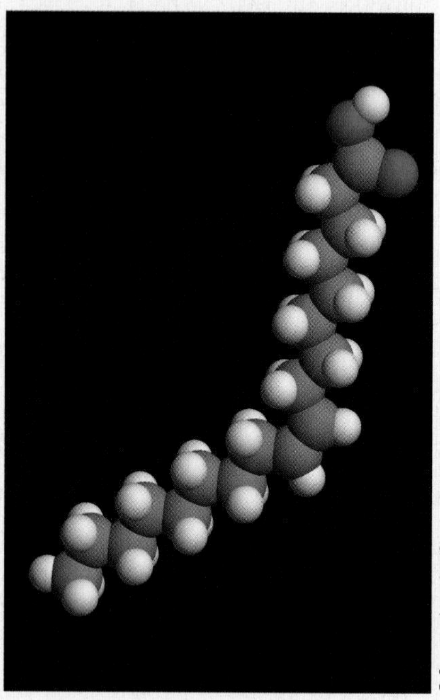



**Figure 21.31** Space-filling model of *cis*-oleic acid, $CH_3(CH_2)_7C{=}C(CH_2)_7COOH$. The two hydrogens at the *cis* double bond are in yellow. The molecule bends at the *cis* double bond.

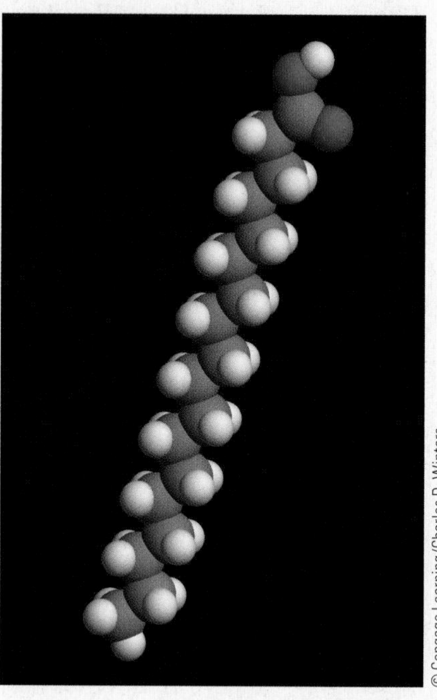



**Figure 21.32** Space-filling model of *trans*-oleic acid. The molecule is straight, similar to a saturated fatty acid.

## 21.15 Chain-Growth Polymers

**Goal** **28** Given the structural diagram or name of an ethylene-like monomer, predict the structural diagram of the product chain-growth polymer; given the structural diagram of a chain-growth polymer, predict the structural diagram and/or the name of the ethylene-like monomer from which it can be formed.

Alkenes form **chain-growth polymers,** which means that the monomers add to a growing polymer chain in sequential addition reactions in a chain-reaction process. The only product is the polymer. The polymerization of an alkene begins when one bond of the double bond between carbon atoms is broken in a molecule. This then generates a new monomer with a broken bond. Each successive carbon atom then has an unshared electron with which to form a bond with a neighboring molecule. The chain continues to build indefinitely in this way.

$$\begin{array}{c}\mathrm{H}\quad\mathrm{H}\\ |\qquad|\\ \mathrm{C}{=}\mathrm{C}\\ |\qquad|\\ \mathrm{H}\quad\mathrm{R}\end{array} + \begin{array}{c}\mathrm{H}\quad\mathrm{H}\\ |\qquad|\\ \mathrm{C}{=}\mathrm{C}\\ |\qquad|\\ \mathrm{H}\quad\mathrm{R}\end{array} + \begin{array}{c}\mathrm{H}\quad\mathrm{H}\\ |\qquad|\\ \mathrm{C}{=}\mathrm{C}\\ |\qquad|\\ \mathrm{H}\quad\mathrm{R}\end{array} \longrightarrow \begin{array}{c}\mathrm{H}\quad\mathrm{H}\\ |\qquad|\\ \cdot\mathrm{C}{-}\mathrm{C}\cdot\\ |\qquad|\\ \mathrm{H}\quad\mathrm{R}\end{array} + \begin{array}{c}\mathrm{H}\quad\mathrm{H}\\ |\qquad|\\ \cdot\mathrm{C}{-}\mathrm{C}\cdot\\ |\qquad|\\ \mathrm{H}\quad\mathrm{R}\end{array} + \begin{array}{c}\mathrm{H}\quad\mathrm{H}\\ |\qquad|\\ \cdot\mathrm{C}{-}\mathrm{C}\cdot\\ |\qquad|\\ \mathrm{H}\quad\mathrm{R}\end{array} \longrightarrow$$

$$\begin{array}{c}\mathrm{H}\quad\mathrm{H}\quad\mathrm{H}\quad\mathrm{H}\quad\mathrm{H}\quad\mathrm{H}\\ |\qquad|\qquad|\qquad|\qquad|\qquad|\\ {-}\mathrm{C}{-}\mathrm{C}{-}\mathrm{C}{-}\mathrm{C}{-}\mathrm{C}{-}\mathrm{C}{-}\\ |\qquad|\qquad|\qquad|\qquad|\qquad|\\ \mathrm{H}\quad\mathrm{R}\quad\mathrm{H}\quad\mathrm{R}\quad\mathrm{H}\quad\mathrm{R}\end{array}$$

Christopher Springmann/Corbisstockmarket.com

**Figure 21.33** Polyethylene film. Sheets are manufactured by squeezing the liquid through a thin opening and inflating it with air.

If R in the monomer is hydrogen, the monomer is ethylene, $C_2H_4$, and the polymer is polyethylene. Ethylene is at the heart of the petrochemical industry. Its annual production in the United States is measured in the billions of pounds. From that mass, billions of pounds of polyethylene are produced, equating to dozens of pounds per person each year (Fig. 21.33). Polyethylene with a molar mass of about 15,000 g/mol is called low-density polyethylene (Fig. 21.34[a]). It is a supple material that folds and bends easily because the intermolecular (induced dipole) forces between the branched carbon chains are weak. These properties make it ideal for sandwich bags and trash bags.

Induced dipole forces increase as branching decreases and molecular size increases. Polymers with long, straight chains, called high-density polyethylene, produce stronger interchain attractions and a corresponding increase in mechanical strength (Fig. 21.34[b]). Plastic milk bottles are made of polyethylene with a molar mass in the 250,000-g/mol range.

When covalent bonds form between carbon chains, a "cross-linked" polymer results and the physical properties change sharply (Fig. 21.34[c]). Cross-linked polyethylene is used for plastic screw caps on soda bottles. This plastic is rigid enough to mold as a solid and has enough mechanical strength to hold the screw thread needed to tighten the cap on the bottle.

If R in the $CH_2{=}CHR$ monomer is a methyl group, $-CH_3$, the monomer is propylene, $CH_2{=}CHCH_3$, and the polymer is polypropylene. We expect the slightly larger R group to increase attractive forces between chains. Polypropylene is used to make plastic bottles, automobile battery cases, fabrics such as Herculon, and indoor-outdoor carpeting.

Halogen-substituted alkenes can also polymerize. These molecules are polar, so the resulting polymer has dipole attractions in addition to induced dipole forces. A familiar example of such a polymer is Saran, the food wrap that clings to itself. Saran is a **copolymer** because it is made from two monomers, $CH_2{=}CHCl$ and $CH_2{=}CCl_2$. The attractive forces between Saran polymer films come from the dipoles caused by the carbon–chlorine bonds. Table 21.6 lists more ethylene-based monomers that give addition polymers.

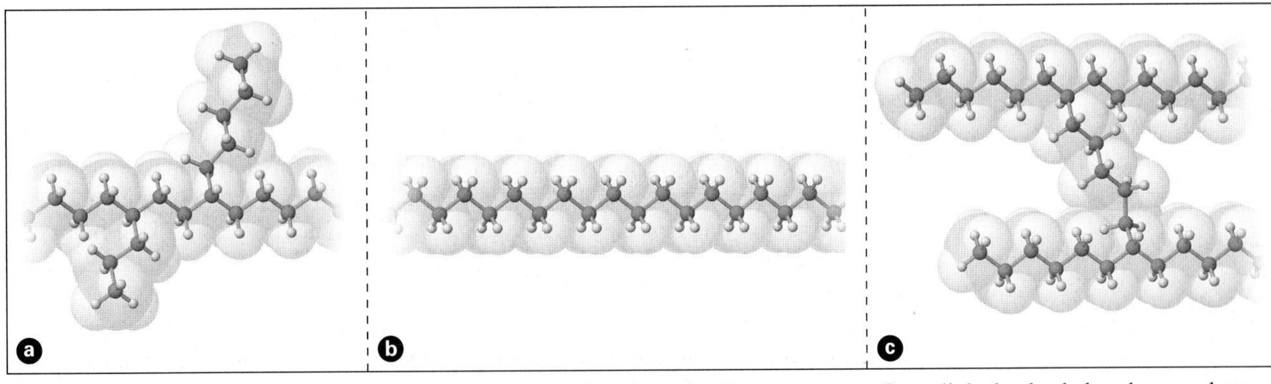


(a) Low-density polyethylene has branched chains.

(b) High-density polyethylene is made of long, straight chains.

(c) Cross-linked polyethylene has covalent bonds between carbon chains.

**Figure 21.34** Molecular structure of forms of polyethylene.

**Table 21.6** Common Polymers Formed from Ethylene-like Monomers

| Monomer Formula | Polymer Formula | Polymer Names | Uses |
|---|---|---|---|
| $CH_2{=}C(C_6H_5){-}H$ Styrene | $-CH_2-CH(C_6H_5)-CH_2-CH(C_6H_5)-$ | Polystyrene Styrofoam | Insulation, packaging |
| $CH_2{=}C(C{\equiv}N){-}H$ Acrylonitrile | $-CH_2-CH(C{\equiv}N)-CH_2-CH(C{\equiv}N)-$ | Polyacrylonitrile Orlon, Acrilan | Fabrics, rugs |
| $CH_2{=}C(O{-}C({=}O){-}CH_3){-}H$ Vinyl acetate | $-CH_2-CH(O{-}C({=}O){-}CH_3)-CH_2-CH(O{-}C({=}O){-}CH_3)-$ | Polyvinylacetate PVA | Chewing gum, paint, glues, safety glass |
| $CH_2{=}C(C({=}O){-}O{-}CH_3){-}CH_3$ Methyl methacrylate | $-CH_2-C(CH_3)(C({=}O){-}O{-}CH_3)-CH_2-C(CH_3)(C({=}O){-}O{-}CH_3)-$ | Lucite Plexiglass | Contact lenses, molded transparent objects |
| $CH_2{=}C(Cl){-}H$ Vinyl chloride | $-CH_2-CH(Cl)-CH_2-CH(Cl)-$ | Polyvinylchloride PVC, Tedlar | Floor tile, pipe |
| $CF_2{=}C(F){-}F$ Tetrafluoroethylene | $-CF_2-CF_2-CF_2-CF_2-$ | Teflon | Coatings, gaskets, bearings |
| $CH_2{=}C(C({=}O){-}O{-}CH_3){-}C{\equiv}N$ Methyl α-cyanoacrylate | $-CH_2-C(C{\equiv}N)(C({=}O){-}O{-}CH_3)-CH_2-C(C{\equiv}N)(C({=}O){-}O{-}CH_3)-$ | Superglue Crazy Glue | Adhesives, battlefield "stitches" |

The attractive forces between non-cross-linked polymer chains are weaker than covalent bonds. As a result, these polymers can be easily recycled by being melted to break *only* the interchain attractive forces, and then the liquid is molded into new shapes. These polymers are called **thermoplastics** and account for over 85% of all plastics sold. To be recycled and reused efficiently, the different thermoplastics must be separated by type, as indicated by the codes in Table 21.7. When processed correctly, the resulting recycled plastic costs about half the price of new plastic. Thermoplastics do not degrade in landfills, but they shouldn't be put there in the first place. Polymers are too valuable to bury.

Polymers having cross-links between carbon chains are generally classified as **thermosets;** these cross-links are covalent bonds, like the bonds making up the polymer chains. As a result, these polymers cannot be easily melted and remolded; there is too much thermal decomposition. Work continues, however, to make a thermoset that can be recycled easily. The only chain-growth polymers classified as thermosets are "superglues."

**Table 21.7** Plastic Container Recycling Codes

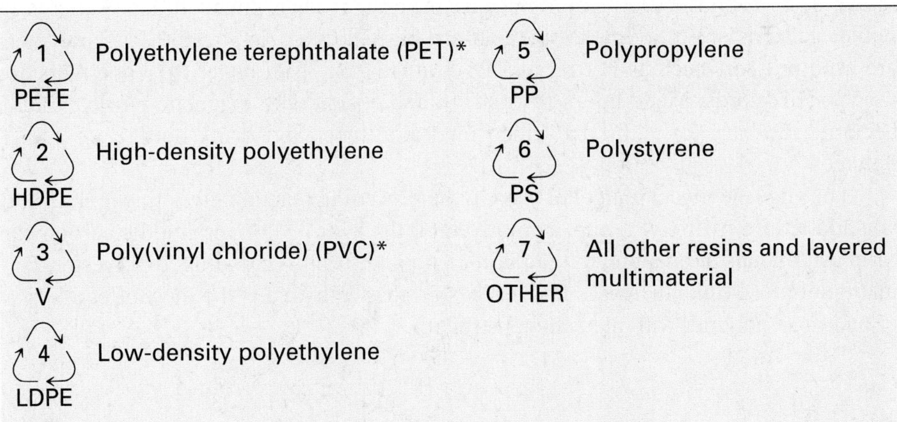

| | | | |
|---|---|---|---|
| 1 PETE | Polyethylene terephthalate (PET)* | 5 PP | Polypropylene |
| 2 HDPE | High-density polyethylene | 6 PS | Polystyrene |
| 3 V | Poly(vinyl chloride) (PVC)* | 7 OTHER | All other resins and layered multimaterial |
| 4 LDPE | Low-density polyethylene | | |

*Bottle codes are different from standard industrial identification to avoid confusion with registered trademarks.

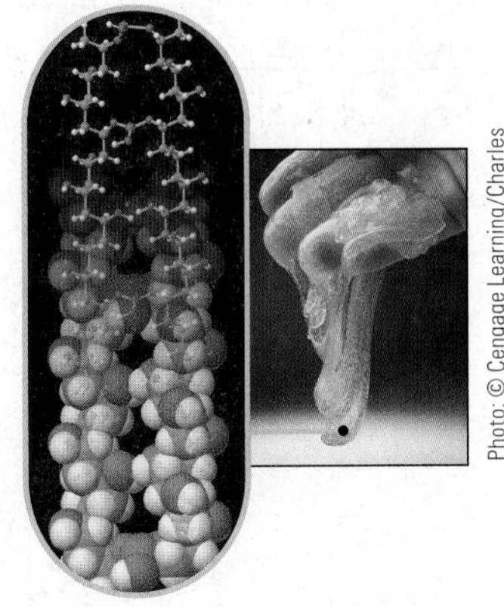

Photo: © Cengage Learning/Charles D. Winters; model, S. M. Young

When mixed with borax, polyvinyl alcohol with a molar mass of about 100,000 g/mol turns into "slime."

### √ Target Check 21.11

*A certain monomer can be written as*

$C_6H_5$ (benzene ring) attached to $CH_2{=}CH$

*Draw a section of the expected chain-growth polymer, which is used as insulation and in packaging. Show at least three monomer units.*

## 21.16 Step-Growth Polymers

**Goal** **29** Given the structural diagrams of a dicarboxylic acid and a dialcohol, predict the structural diagram of the product step-growth polymer; given the structural diagram of a step-growth polymer, predict the structural diagrams of the dicarboxylic acid and dialcohol from which it can be formed.

In Section 21.12 you learned that carboxylic acids react with alcohols to form esters and with ammonia and amines to form amides. In each reaction, a molecule of water is split out in a **condensation reaction.** This is one mechanism by which **step-growth polymers** form. In general, step-growth polymers are produced by the stepwise reaction of monomers with the growing chain, where each bond is formed independently of the others. Polymer chemists use condensation reactions to form polyesters and polyamides. However, to form the polymer chain by repeated condensation reactions, you must use a *di*carboxylic acid (two carboxyl groups), such as terephthalic acid, and a *di*alcohol (two hydroxyl groups), such as ethylene glycol, as shown here.

$$HO{-}\overset{O}{\overset{\|}{C}}{-}C_6H_4{-}\overset{O}{\overset{\|}{C}}{-}OH + H{-}OCH_2CH_2OH \longrightarrow$$

terephthalic acid + ethylene glycol ⟶

$$-\overset{O}{\overset{\|}{C}}{-}C_6H_4{-}\overset{O}{\overset{\|}{C}}{-}OCH_2CH_2{-}O{-} + HOH$$

PET water

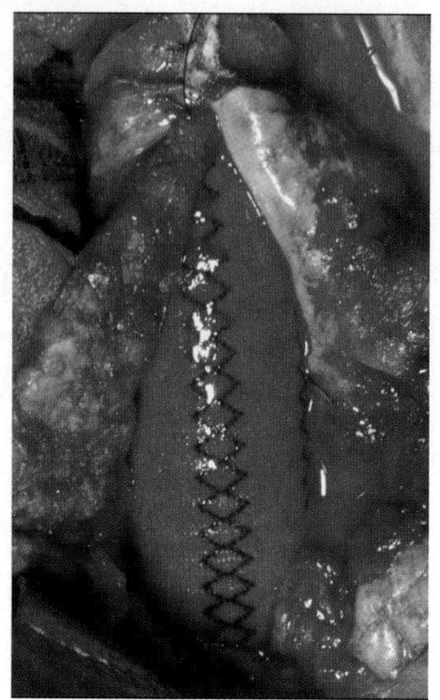
CMSP

**Figure 21.35** Dacron is a PET fiber. It is used in medicine to replace tissues because it is well tolerated by the body. Here, a Dacron patch is used to close a hole between the heart's two upper chambers.

The linear polyester formed above is a polyethylene terephthalate, or PET. It has a molar mass of about 15,000 g/mol. Because the ester group is polar, attractions between polymer chains are of a dipole type and are fairly strong. As a result, PET polymers are used in fibers such as Fortrel and Dacron (Fig. 21.35). Longer PET polymers are used for tire cords. Made into a Mylar film and coated with magnetic particles, PET becomes the base for audio and video recording tape. Soft drink bottles are a PET polymer.

The ethylene glycol dialcohol used in the preceding reaction gives linear chains. If you add a third hydroxyl group, as in glycerol, the hydroxyl on the middle carbon can form ester bonds at an angle to the main chain and form a cross-linked polyester. Formation of cross-links increases molar mass dramatically and is the most effective way of making a polymer with mechanical strength.

$$HOCH_2\overset{OH}{\overset{|}{C}}HCH_2OH + HO-\overset{O}{\overset{\|}{C}}-C_6H_4-\overset{O}{\overset{\|}{C}}-OCH_2\overset{OH}{\overset{|}{C}}HCH_2OH \longrightarrow$$

$$-OCH_2\overset{|}{\overset{O}{\overset{|}{C}}}HCH_2O-\overset{O}{\overset{\|}{C}}-C_6H_4-\overset{O}{\overset{\|}{C}}-OCH_2\overset{|}{\overset{O}{\overset{|}{C}}}HCH_2O- + H_2O$$

cross-linked polyester
(glyptal resin)

Cross-linked polyesters are called *glyptal* resins or *alkyd* resins. These resins have polar ester groups and some free hydroxyl and carboxylic acid groups. As a result, they are water-soluble.

The best-known synthetic polymer is nylon, a polyamide. Nylon was the brainchild of Wallace Carothers, who was hired away from Harvard University in 1928 by the DuPont Company. Carothers was asked to develop a substitute for silk. Silk was known to be a protein, so Carothers's group studied ways of making amide bonds. In 1935, they prepared a product that they named nylon 66, illustrated in Active Figure 21.36. The reactants were a dicarboxylic acid and a diamine: adipic acid and hexamethylenediamine, respectively. Note that each monomer has six carbons, hence the name.

$$\underbrace{HO-\overset{O}{\overset{\|}{C}}CH_2CH_2CH_2CH_2\overset{O}{\overset{\|}{C}}-OH}_{\text{adipic acid}} + \underbrace{H-\overset{H}{\overset{|}{N}}CH_2CH_2CH_2CH_2CH_2CH_2\overset{H}{\overset{|}{N}}-H}_{\text{hexamethylenediamine}} \longrightarrow$$

$$\underbrace{-\overset{O}{\overset{\|}{C}}CH_2CH_2CH_2CH_2\overset{O}{\overset{\|}{C}}-\overset{H}{\overset{|}{N}}CH_2CH_2CH_2CH_2CH_2CH_2\overset{H}{\overset{|}{N}}-}_{\text{nylon 66}} + \underbrace{HOH}_{\text{water}}$$

The hydrogen bonding that can occur between amide linkages produces strong interchain attractive forces, which yield nylon fibers that have good tensile strength (Fig. 21.37). Nylon with molar mass of about 10,000 g/mol can be made into useful fiber. Nylon with molar mass greater than 100,000 g/mol has too much mechanical strength to be used as a fiber. This nylon, however, is mixed with glass fibers and used to make valve covers in automobile engines. Current U.S. production of nylon is measured in the billions of pounds annually.

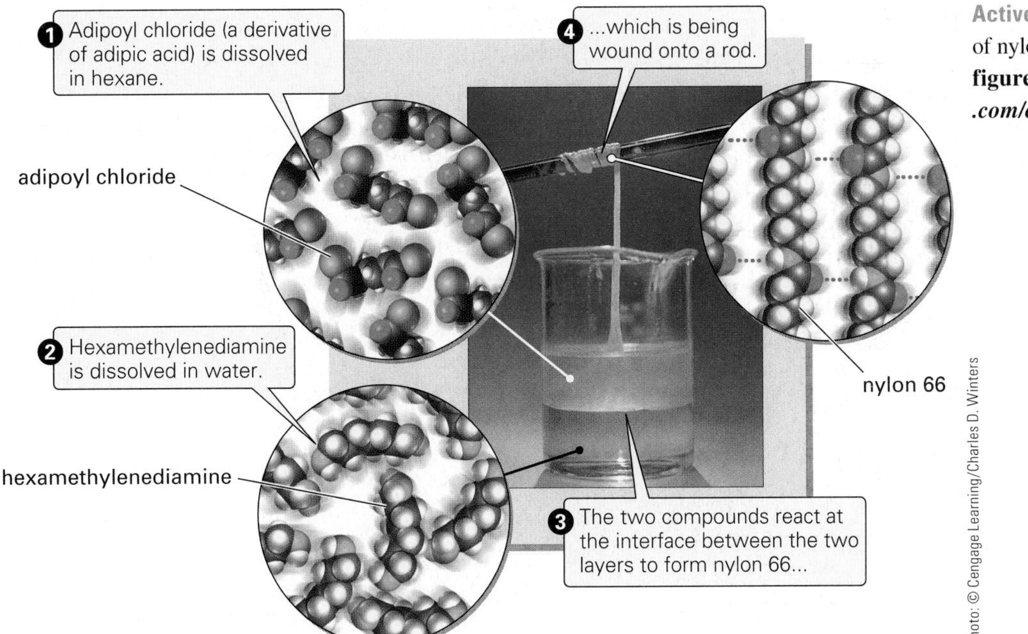
1 Adipoyl chloride (a derivative of adipic acid) is dissolved in hexane.
adipoyl chloride
2 Hexamethylenediamine is dissolved in water.
hexamethylenediamine
3 The two compounds react at the interface between the two layers to form nylon 66...
4 ...which is being wound onto a rod.
nylon 66
Photo: © Cengage Learning/Charles D. Winters

**Active Figure 21.36** Preparation of nylon 66. **Watch this active figure at *http://www.cengage.com/cracolice.***

Hydrogen bonds to adjacent nylon 66 molecules.

$$\cdots NH-CO-(CH_2)_4-CO-NH-(CH_2)_6-NH-CO-(CH_2)_4-CO-NH-(CH_2)_6-NH-CO\cdots$$

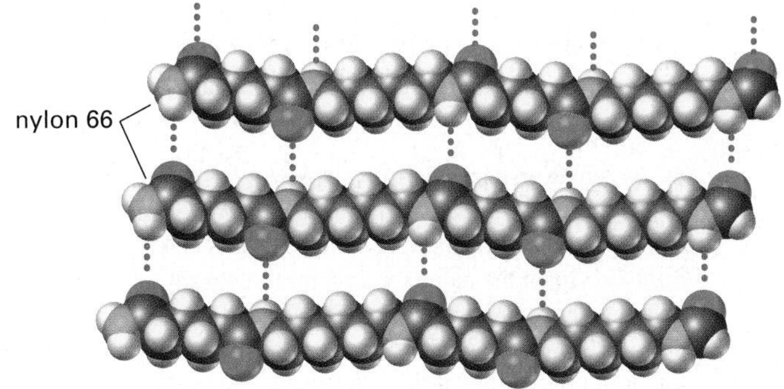
nylon 66

**Figure 21.37** Hydrogen bonding in nylon 66. This property is responsible for the ability of nylon fibers to stretch.

## √ Target Check 21.12

*A portion of the condensation polymer Kodel is given below. Give the structure of the starting materials from which Kodel can be made.*

$$-\overset{O}{\overset{\|}{C}}-C_6H_4-\overset{O}{\overset{\|}{C}}-O-CH_2-C_6H_{10}-CH_2-O-\overset{O}{\overset{\|}{C}}-C_6H_4-\overset{O}{\overset{\|}{C}}-O-CH_2-C_6H_{10}-CH_2-O-$$

# Chapter 21 in Review

Most of the key terms and concepts and many others appear in the Glossary. Use your Glossary regularly.

**21.1 The Nature of Organic Chemistry**

Goal 1 Distinguish between organic and inorganic compounds.

**Key Terms and Concepts: Inorganic chemistry, organic chemistry**

**21.2 The Molecular Structure of Organic Compounds**

Goal 2 Given Lewis diagrams, ball-and-stick models, or space-filling models of two or more organic molecules with the same molecular formula, distinguish between isomers and different orientations of the same molecule.

## HYDROCARBONS

Goal 3 Distinguish between saturated and unsaturated hydrocarbons (or other compounds).
Goal 4 Distinguish between alkanes, alkenes, and alkynes (Sections 21.3 and 21.4).

**Key Terms and Concepts: Hydrocarbons, saturated, unsaturated**

**21.3 Saturated Hydrocarbons: The Alkanes and Cycloalkanes**

Goal 5 Given a formula of a hydrocarbon or information from which it can be written, determine whether the compound can be a normal alkane or a cycloalkane.
Goal 6 Given the name (or structural diagram) of a normal, branched, or halogen-substituted alkane, write the structural diagram (or name).
Goal 7 Given the name (or structural diagram) of a cycloalkane, write the structural diagram (or name).

**Key Terms and Concepts: Alkane, alkyl group, condensed formula, cycloalkane, homologous series, line formula, methyl group, normal alkane, structural diagram, structural formula**

**21.4 Unsaturated Hydrocarbons: The Alkenes and Alkynes**

Goal 8 Given a formula of a hydrocarbon or information from which it can be written, determine whether the compound can be an alkene or an alkyne.
Goal 9 Given the name (or structural diagram) of an alkyne or an alkene, write the structural diagram (or name).
Goal 10 Identify and distinguish between *cis* and *trans* geometric isomers.

**Key Terms and Concepts: Alkene, alkyne, geometric isomers**

**21.5 Aromatic Hydrocarbons**

Goal 11 Distinguish between aliphatic and aromatic hydrocarbons.
Goal 12 Given the name (or structural diagram) of an alkyl- or halogen-substituted benzene compound, write the structural diagram (or name).

**Key Terms and Concepts: Aliphatic, aromatic, benzene ring**

**21.6 Summary of the Hydrocarbons**

**21.7 Sources and Preparation of Hydrocarbons**

**Key Terms and Concepts: Hydrogenation**

**21.8 Chemical Reactions of Hydrocarbons**

Goal 13 Given the reactants in an addition or substitution reaction between (a) an alkane, alkene, alkyne, or benzene and (b) a hydrogen or halogen molecule, predict the products of the reaction.

**Key Terms and Concepts: Addition reaction, halogenation reaction, substitution reaction**

**21.9 Uses of Hydrocarbons**

## ORGANIC COMPOUNDS CONTAINING OXYGEN AND NITROGEN

**Key Terms and Concepts: Functional group**

**21.10 Alcohols and Ethers**

Goal 14 Identify the structural formulas of the functional groups that distinguish alcohols and ethers.
Goal 15 Given the molecular structures of alcohols or ethers, or information from which they may be obtained, predict relative values of boiling points or solubility in water.
Goal 16 Given the name (or structural diagram) of an alcohol or ether, write the structural diagram (or name).
Goal 17 Given the reactants (or products) of a dehydration reaction between two alcohols, predict the products (or reactants) of the reaction.

**Key Terms and Concepts: Alcohol, ether, hydroxyl group**

**21.11 Aldehydes and Ketones**

Goal 18 Identify the structural formulas of the functional groups that distinguish aldehydes and ketones.
Goal 19 Given the name (or structural diagram) of an aldehyde or ketone, write the structural diagram (or name).
Goal 20 Write structural diagrams to show how a specified aldehyde or ketone is prepared from an alcohol.

**Key Terms and Concepts: Aldehyde, carbonyl group, ketone**

**21.12 Carboxylic Acids and Esters**

Goal 21 Identify the structural formulas of the functional groups that distinguish carboxylic acids and esters.
Goal 22 Given the name (or structural diagram) of a carboxylic acid or ester, write the structural diagram (or name).
Goal 23 Given the reactants (or products) of an esterification reaction, predict the products (or reactants) of the reaction.

**Key Terms and Concepts: Carboxyl group, carboxylic acid, ester, esterification**

**21.13 Amines and Amides**

Goal 24 Identify the structural formulas of the functional groups that distinguish amines and amides.
Goal 25 Given the name (or structural diagram) of an amine or amide, write the structural diagram (or name).
Goal 26 Given the structural diagram of an amine, or information from which it can be written, classify the amine as primary, secondary, or tertiary.
Goal 27 Given the reactants of a reaction between a carboxylic acid and an amine, predict the products of the reaction.

**Key Terms and Concepts: Amide, amine, amino acid, peptide linkage**

21.14 Summary of the Organic Compounds of Carbon, Hydrogen, Oxygen, and Nitrogen

## POLYMERS

**Key Terms and Concepts: Monomer, polymer, polymerization**

21.15 Chain-Growth Polymers

Goal 28 Given the structural diagram or name of an ethylene-like monomer, predict the structural diagram of the product chain-growth polymer; given the structural diagram of a chain-growth polymer, predict the structural diagram and/or the name of the ethylene-like monomer from which it can be formed.

**Key Terms and Concepts: Chain-growth polymer, copolymer, thermoplastics, thermosets**

21.16 Step-Growth Polymers

Goal 29 Given the structural diagrams of a dicarboxylic acid and a dialcohol, predict the structural diagram of the product step-growth polymer; given the structural diagram of a step-growth polymer, predict the structural diagrams of the dicarboxylic acid and dialcohol from which it can be formed.

**Key Terms and Concepts: Condensation reaction, step-growth polymer**

# Study Hints and Pitfalls to Avoid

The study of organic chemistry is quite different from the study of most other topics in this book. For example, you may have already noticed that there are no calculations in this chapter. You will have to make some modifications to your "normal" study techniques to accommodate the differences in this chapter.

The most effective way to review material often differs from the most effective way to first learn material. This is especially true with organic chemistry. The best way to first learn the subject matter is in little groups, as we have presented it. However, the best way to review the material is to look at the big picture. Our summary of the hydrocarbons, Table 21.4, and summary of the organic functional groups, Table 21.5, will be particularly useful in your chapter review.

Another useful general approach to studying organic chemistry can be summarized as "memorize, then apply." The general formulas of the functional groups must be memorized, and they must be memorized before you can predict the products of organic reactions. As an example, even though we present both the alcohol functional group and the reactions of alcohols in the alcohols and ethers section, you should review by learning all functional groups, and then learning all reactions for which you are responsible. Many students find it useful to place information to be memorized on flash cards. For example, to memorize functional groups, one flash card would have ester on one side and R—CO—OR′ on the other side.

# Concept-Linking Exercises

*Write a brief description of the relationships among each of the following groups of terms or phrases. Answers to the Concept-Linking Exercises are given after answers to the Target Checks in Appendix III.*

1. Hydrocarbons, saturated, unsaturated
2. Alkane, normal alkane, cycloalkane, homologous series
3. Structural formula, structural diagram, line formula, condensed formula, Lewis diagram
4. Alkene, alkyne, *cis*, *trans*, geometric isomers
5. Aliphatic compounds, aromatic compounds, benzene ring, delocalized electrons
6. Catalytic cracking, fractional distillation, catalytic hydrogenation
7. Addition reaction, substitution reaction, halogenation reaction, hydrogenation reaction
8. Functional group, hydroxyl group, carbonyl group, carboxyl group
9. Amide, peptide linkage, amino acid
10. Monomer, polymer, chain-growth polymer, step-growth polymer

# Small-Group Discussion Questions

*Small-Group Discussion Questions are for group work, either in class or under the guidance of a leader during a discussion section.*

1. Why do the vast majority of known chemical compounds contain carbon?
2. Draw structural diagrams of each isomer of $C_6H_{14}$.
3. There are eight isomers of five-carbon alkyl groups. Draw the structural diagram and state the IUPAC name of each.
4. Which of the following have *cis-trans* isomers? Draw structural diagrams of each, including both *cis-trans* isomers when applicable. (a) $CH_3CH_2CH_2OH$ (b) $C_2H_2$ (c) $CH_3CH{=}CH_2$ (d) $BrCH{=}CHBr$ (e) $BrCH{=}CHCl$ (f) $CH_3CH_2CH{=}CHCH_3$ (g) $(CH_3)_2C{=}C(CH_3)CH_2CH_3$ (h) $(CH_3)_2C{=}CHCH_3$
5. Name each of the molecules below. In parts (e) and (f), [benzene ring bearing $CH_3$] is named toluene.

(a) 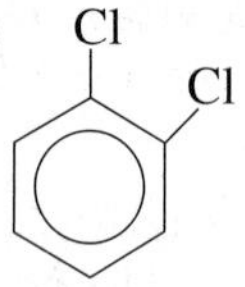 Cl, Cl

(b) 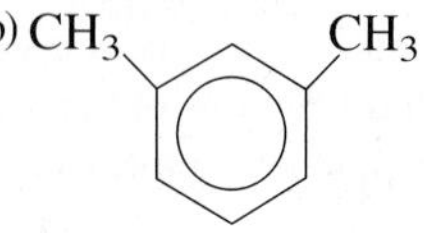 $CH_3$, $CH_3$

(c) 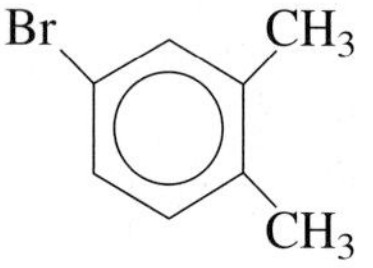 Br, $CH_3$, $CH_3$

(d) 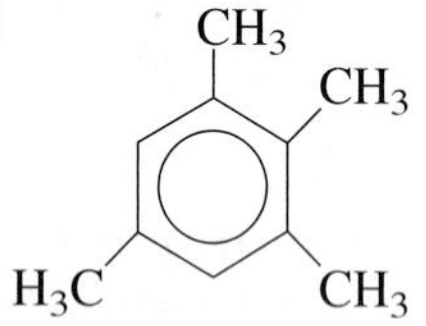 $CH_3$, $CH_3$, $H_3C$, $CH_3$

(e) 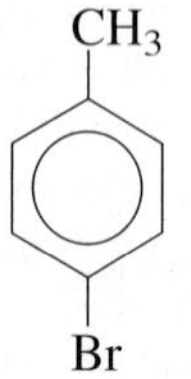 $CH_3$, Br

(f) 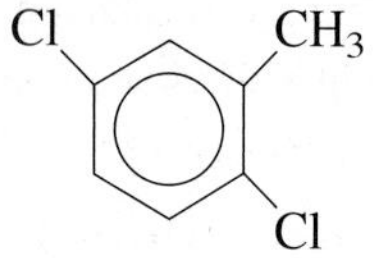 Cl, $CH_3$, Cl

6. Define each of the following terms, and then construct an organizing scheme that allows you to classify molecules into the appropriate categories: hydrocarbon, aliphatic, aromatic, open chain, cyclic, saturated, unsaturated.

7. Predict the products of the following reactions. The structure [cyclohexane ring] represents $C_6H_{12}$, with six carbons bonded to one another. Each carbon has single bonds to two hydrogen atoms.

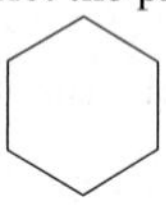

(a) 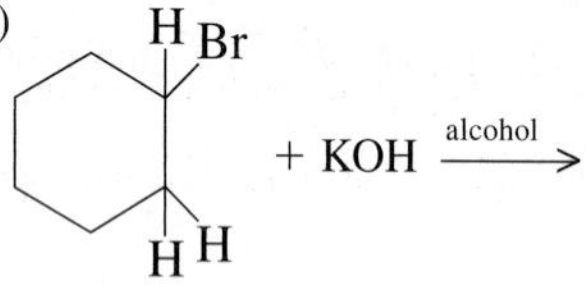 [cyclohexane with H, Br and H, H] + KOH $\xrightarrow{\text{alcohol}}$

(b) 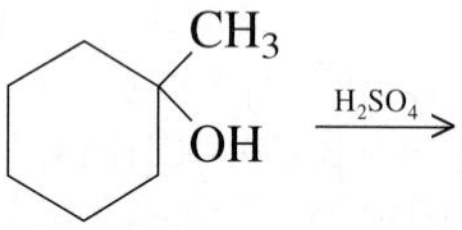 [cyclohexane with $CH_3$, OH] $\xrightarrow{H_2SO_4}$

8. Predict the products of the following reactions.

(a) 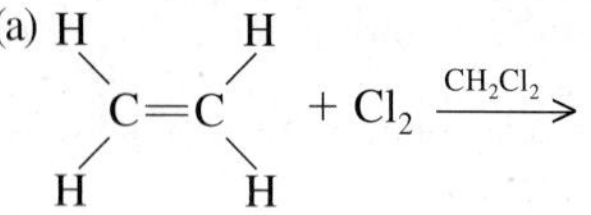 H₂C=CH₂ (H, H / C=C / H, H) + $Cl_2$ $\xrightarrow{CH_2Cl_2}$

(b) $CH_3CH_2CH_2CH{=}CH_2$ + HCl → (1-chloropentane is *not* formed)

(c) 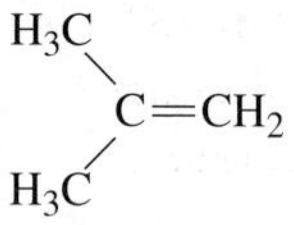 $(H_3C)_2C{=}CH_2$ + HCl → (1-chloro-2-methylpropane is *not* formed)

(d) 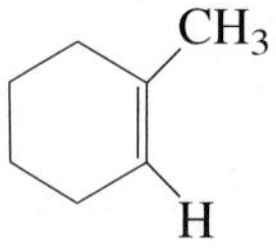 [1-methylcyclohexene: $CH_3$, H] + HBr → (1-bromo-2- methylcyclohexane is *not* formed)

(e) $CH_3CH_2CH{=}CHCH_3$ + HBr → (a mixture of two products is formed)

(f) From the reactions in parts (b) through (e), can you describe the rule that governs the addition of HX to an alkene?

9. Construct a list of things that you can see at this moment that are made of or derived from hydrocarbons.

10. Provide the name for each of the following.

(a) $CH_3OH$ (b) $CH_3CH_2OH$

(c) 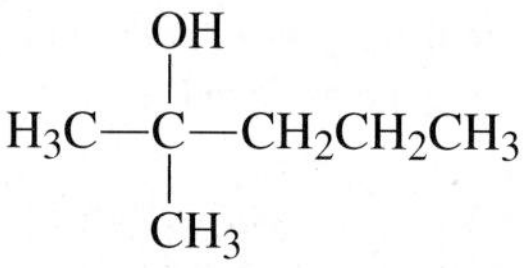 $H_3C{-}C(OH)(CH_3){-}CH_2CH_2CH_3$

(d) 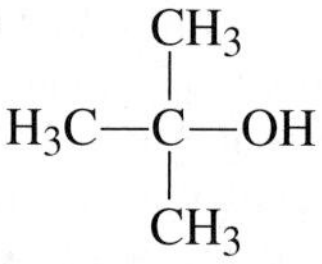 $H_3C{-}C(CH_3)_2{-}OH$

(e) $CH_3OCH_2CH_3$ (f) $CH_3CH_2OCH_2CH_3$

(g) 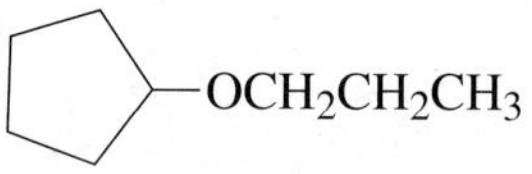 [cyclopentane]$-OCH_2CH_2CH_3$

11. Draw the structural diagrams of: (a) formaldehyde, (b) ethanal, (c) propanal, (d) 2-ethyl-4-methylpentanal, (e) acetone, (f) 3-hexanone, (g) 4-hexen-2-one, (h) 2,4-hexanedione

12. Draw the structural diagrams of the carboxylic acid and alcohol from which each ester is formed.

(a) 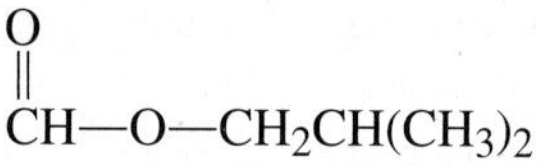 $HC({=}O){-}O{-}CH_2CH(CH_3)_2$

(b) 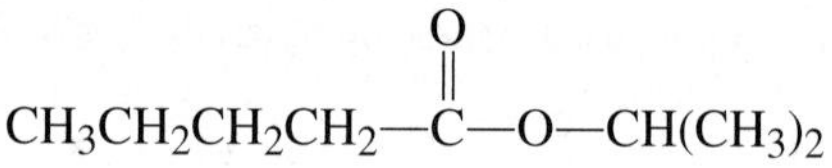 $CH_3CH_2CH_2CH_2{-}C({=}O){-}O{-}CH(CH_3)_2$

(c) [cyclohexane]$-O{-}C({=}O){-}CH_2CH_2CH(CH_3)_2$

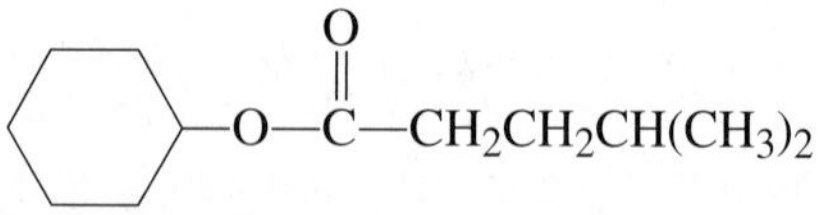

(d) 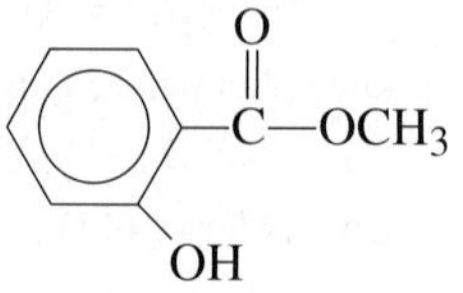 [benzene ring with OH]$-C({=}O){-}OCH_3$

(e) 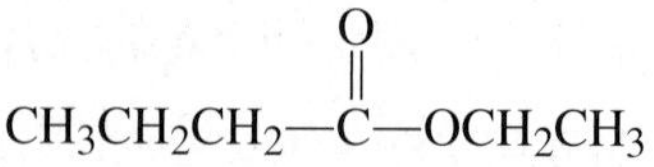 $CH_3CH_2CH_2{-}C({=}O){-}OCH_2CH_3$

13. Draw a structural diagram of each of the following.
    (a) A secondary amine with three carbon atoms.
    (b) A tertiary amine with four carbon atoms.
    (c) An amide with two carbon atoms.
    (d) An amino acid with two carbon atoms.

14. For each compound class that follows, (a) write a general formula, (b) draw the structure of its functional group, (c) draw the structural diagrams of two molecules in the class, and (d) name each of the molecules drawn in part (c): alcohol, ether, aldehyde, ketone, acid, ester, amine, amide.

15. Draw a section of the chain-growth polymer formed from each monomer listed. Include at least four momomer units. (a) ethylene, $CH_2{=}CH_2$, (b) propylene, $CH_2{=}CHCH_3$, (c) vinyl chloride, $CH_2{=}CHCl$, (d) 1,1-dichloroethylene, $CH_2{=}CCl_2$, (e) tetrafluoroethylene, $CF_2{=}CF_2$, (f) acrylonitrile, $CH_2{=}CHCN$, (g) styrene, $CH_2{=}CHC_6H_5$, (h) ethyl acrylate, $CH_2{=}CHCOOCH_2CH_3$

16. Draw structural diagrams of the monomers from which each type of the following step-growth polymers are formed.

(a) Polyamides,
```
  O      O  H       H
  ||     || ..      ..
—C—R—C—N—R′—N—
```

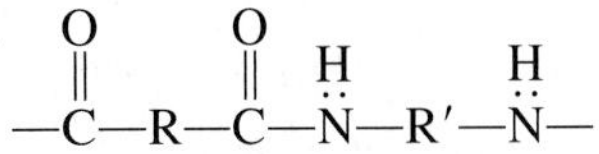

(b) Polyesters,
```
  O      O
  ||     ||
—C—R—C—O—R′—O—
```

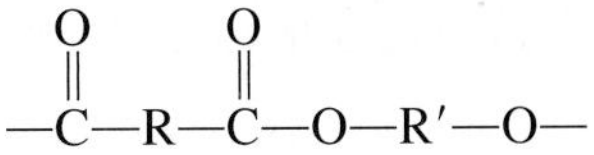

(c) Polycarbonates,
```
      O
      ||
—O—C—O—R—
```

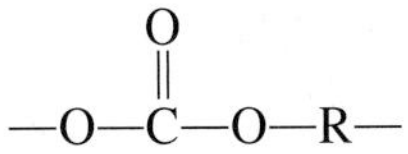

(d) Polyurethanes,
```
  O                 O
  || H       H  ||
—C—N—R—N—C—O—R′—O—
     ..      ..
```

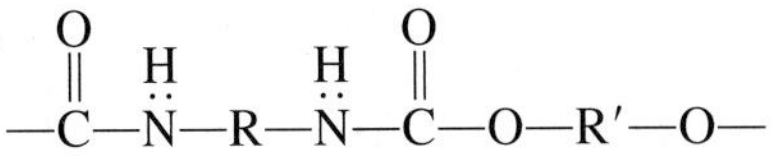

# Questions, Exercises, and Problems

*Blue-numbered questions are answered in Appendix III. ■ denotes problems assignable in OWL. In the first section of Questions, Exercises, and Problems, similar exercises and problems are paired in consecutive odd-even number combinations.*

### Section 21.2: The Molecular Structure of Organic Compounds

1. Would the cyanide ion or the carbonate ion be considered organic? What about the acetate ion?

2. Compare the original and modern definitions of organic chemistry. Why was the definition changed? Which definition includes the other?

3. What is the bond angle around a carbon atom with four single bonds? What word describes this geometry?

4. What is the bond angle around a carbon atom with a double bond and two single bonds? What word describes this molecular geometry?

### Section 21.3: Saturated Hydrocarbons: The Alkanes and Cycloalkanes

5. What is a hydrocarbon? Which, among the following, are hydrocarbons? $CH_3OH$; $C_3H_4$; $C_8H_{10}$; $CH_3CH_2CH_3$

6. Outside of chemistry, what is the meaning of *saturated*? Use structural diagrams to show how the terms *saturated* and *unsaturated* are logically applied to hydrocarbons.

7. Write the molecular formulas of the alkanes having 11 and 21 carbon atoms. How did you arrive at these formulas?

8. Explain why an alkane is an example of a homologous series.

9. What are isomers?

10. Draw structural formulas for all the hydrocarbons with the formula $C_4H_{10}$.

11. Write the molecular formula, line formula, condensed formula, and structural formula of the normal alkane with seven carbon atoms.

12. Write the molecular formula, the line formula, and the condensed formula for the normal alkane having 12 carbon atoms.

13. Write the molecular formula that represents the alkyl groups having two and four carbon atoms.

14. ■ What is the name of the $CH_3CH_2$— group?

15. Write the molecular formula of butane. What is the name of $C_{10}H_{22}$?

16. ■ Give the molecular formula of each of the following compounds: propane, cyclobutane, methane.

17. What is the IUPAC name of the molecule whose carbon skeleton is shown below?
```
        C—C
        |
C—C—C—C—C—C
            |
            C
```

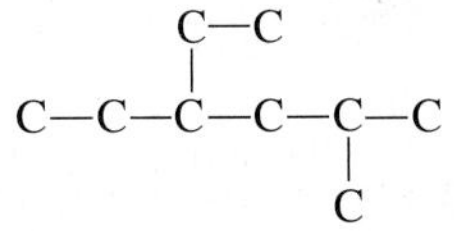

18. ■ What is the molecular formula for 2,2,3-trimethylbutane?

19. Draw the carbon skeleton of 2,3-dimethylpentane.

20. Give the IUPAC name of the molecule whose carbon skeleton is shown below.
```
   C       C
   |       |
C—C—C—C—C
   |
   C
```

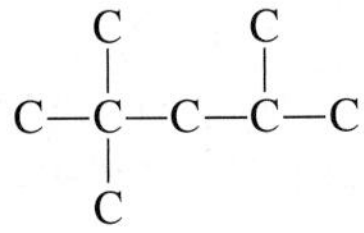

21. Both 1,1,1- and 1,1,2-trichloroethane are used industrially as fat and grease solvents. Draw the structural diagrams of these isomers.

22. ■ Draw structural formulas for all of the compounds that would be named as a bromobutane.

23. Is the general formula of a cycloalkane the same as the general formula of an alkane, $C_nH_{2n+2}$? Draw any structural diagrams to illustrate your answer.

24. How does a cycloalkane differ from a normal alkane?

25. Draw the skeleton diagram of cyclopentane.

26. Name the cycloalkane whose structural diagram resembles a square.

27. Draw a structural diagram of 1-chloro-2-iodocyclopentane.

28. Draw a structural diagram of 1-bromo-3-methylcyclohexane.

*Questions 29–32: Write the names of the compounds whose carbon skeletons are given.*

29.
```
                C
                |
C—C—C—C—C
          |
          C
          |
          C—C
```

30. ■
```
C
|
C—C—C—C—C
          |
          C—C
              |
          C—C—C
```

31.
```
               Cl
               |
Br—C—C—C—C
               |
               Cl
```

32.
```
          Cl            F
          |             |
C—C—C—C—C—C
          |        |
          Cl       F
```

33. Draw the skeleton diagram of 1-chloro-2-iodocyclopentane.

34. Draw the skeleton diagram of 1-bromo-3-methylcyclohexane.

35. Name the molecule whose carbon skeleton is drawn below.

Cl, $CH_2CH_3$ (on cyclohexane ring)

36. Give the IUPAC name of the molecule whose carbon skeleton is shown below.

$CH_2CH_2CH_3$, F (on cyclopentane ring)

## Section 21.4: Unsaturated Hydrocarbons: The Alkenes and Alkynes

37. What is the difference in bonding and in the general molecular formula between an alkene and an alkane with the same number of carbon atoms?

38. ■ Classify each of the following as an alkane, an alkene, or an alkyne and as saturated or unsaturated: $CHCCH_3$, $CH_2CHCH_3$, $CH_3(CH_2)_3CH_3$.

39. Draw the structural formula of trichloroethene, a common dry-cleaning solvent. Why isn't the IUPAC name for this substance 1,1,2-trichloroethene?

40. ■ Give the molecular formula for each of the following compounds: 2-butyne, propyne, 2-butene.

41. Draw the structural formula and explain in words the differences between *cis*-3-heptene and *trans*-3-heptene.

42. ■ Are *cis-trans* isomers possible for $CH_3CH_2CH{=}CHCH_2CH_3$? If they are, write the IUPAC names of the isomers. If *cis-trans* isomers are not possible, write the IUPAC name of the compound.

43. The sex pheromone of the common housefly is *cis*-9-tricosene, where tricosene is the IUPAC name of a 23-carbon alkene. Draw the condensed formula of this molecule, marketed under the name Muscalure.

44. A molecule marketed as Disparlure, the sex pheromone of the gypsy moth, is produced from the molecule shown below. Give the IUPAC name of this molecule. (An 18-carbon alkene is an octadecene.)

```
CH3    H
    \ /
     C              CH3
    / \             /
H3C    (CH2)4   (CH2)9
            \   /
             C=C
            /   \
           H     H
```

45. Give the IUPAC name of the following molecule:

$$CH_3{-}C(CH_3)_2{-}C{\equiv}C{-}CH_2{-}CH_2{-}CH_3$$

```
      CH3
      |
CH3—C—C≡C—CH2—CH2—CH3
      |
      CH3
```

46. Give the IUPAC name of the following molecule:

```
H—C≡C—CH—CH2—CH3
        |
        CH2—CH2—CH3
```

## Section 21.5: Aromatic Hydrocarbons

47. Dimethylbenzenes have the common name xylene. Draw all possible xylene isomers and give their IUPAC names.

48. ■ Draw the structural formulas for all of the compounds that would be named as a bromochlorobenzene.

49. Name the molecule given below.

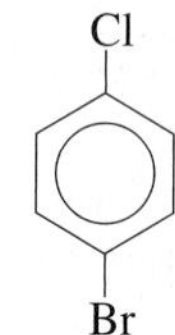


50. ■ Name the molecule given below.

F, Br (on benzene ring)

## Section 21.8: Chemical Reactions of Hydrocarbons

51. Draw skeletal formulas for all the possible dichloro substitution products of propane and give their IUPAC names.

52. ■ Draw skeletal formulas for all the possible dichloro substitution products of butane and give their IUPAC names.

53. Draw skeletal formulas for all the possible dichloro addition products of the normal butenes.

54. Draw skeletal formulas for all the possible dibromo addition products of the pentenes.

55. Write an equation for the hydrogenation of 2-butene. Does the *cis* or *trans* geometry of the butene starting material make a difference in the products obtained?

56. Would you get a different product from the hydrogenation of 1-butene than from the hydrogenation of *trans*-2-butene?

### Section 21.10: Alcohols and Ethers

57. Write the Lewis structures for all possible isomers with the formula $C_4H_{10}O$. Identify them as alcohols or ethers.

58. Show how alcohols and ethers are structurally related to water.

59. Explain why ethers with formula $C_4H_{10}O$ have boiling points between 32°C and 39°C, whereas alcohols with the same formula have boiling points between 82°C and 118°C.

60. Explain why the ether with formula $C_2H_6O$ is very slightly soluble in water, while the alcohol with the same formula is infinitely soluble in water.

61. Write the structural formula for 2-hexanol.

62. ■ Give the IUPAC name of

```
    H   H   H   OH  H
    |   |   |   |   |
H—C—C—C—C—C—H
    |   |   |   |   |
    H   H   H   H   H
```

63. Write the structural formula for butyl ethyl ether.

64. Write the structural formula for ethyl propyl ether.

65. Write a structural equation showing how dipropyl ether might be prepared from an alcohol.

66. How could diisopropyl ether be prepared from an alcohol? The isopropyl group is

```
       CH3
      /
  —CH
      \
       CH3
```

### Section 21.11: Aldehydes and Ketones

67. Write structural formulas for propanal and propanone.

68. ■ What is the molecular formula for dimethyl ketone?

69. Use structural formulas to prepare acetone by oxidation of an alcohol.

70. Use structural formulas to prepare butanal by oxidation of an alcohol.

### Section 21.12: Carboxylic Acids and Esters

71. Write the structural formula for hexanoic acid, a wretched-smelling substance found in goat sweat.

72. ■ What is the molecular formula for formic acid?

73. Write the equation for the reaction between propanoic acid and ethanol, and name the ester formed in this reaction.

74. Write the equation for the reaction between acetic acid and 1-propanol. Name the ester formed in this reaction. Do the same for the reaction between acetic acid and 2-propanol.

### Section 21.13: Amines and Amides

75. Give Lewis diagrams for all amines with the formula $C_2H_7N$ and name them.

76. Give structural formulas for all amines with the formula $C_3H_9N$ and name them.

77. Classify the amines from Question 75 as primary, secondary, or tertiary.

78. Classify the amines from Question 76 as primary, secondary, or tertiary.

79. Write the equation for the reaction between propanoic acid and ammonia. Name the product and give its functional group.

80. Write the equation for the reaction between propanoic acid and diethylamine.

### Section 21.15: Chain-Growth Polymers

81. Draw three repeating units of the chain-growth polymer made from the monomer

```
H   H
|   |
C=C
|   |
H   Br
```

82. Draw three repeating units of the chain-growth polymer made from the monomer

```
H   F
|   |
C=C
|   |
H   F
```

83. The chain-growth polymer shown below is used for ropes, fabrics, and indoor-outdoor carpeting. Give the structure of the monomer from which this polymer was made.

```
[—CH2—CH—CH2—CH—CH2—CH—]
       |        |        |
       CH3      CH3      CH3  n
```

84. The chain-growth polymer shown below is used to thicken motor oil. Give the structure of the monomer from which this polymer was made.

```
[       CH3      CH3      CH3  ]
        |        |        |
 —CH2—C—CH2—C—CH2—C—
        |        |        |
        CH3      CH3      CH3  n
```

85. Draw three repeating units of the chain-growth polymer made from chlorotrifluoroethene.

86. Plastic laboratory ware is usually a chain-growth polymer made from 4-methyl-1-pentene. Draw three repeating units of this polymer.

### Section 21.16: Step-Growth Polymers

87. Lexan is a polycarbonate ester step-growth condensation polymer that is transparent and nearly unbreakable. It is used in "bulletproof" windows (a one-inch-thick Lexan plate will stop a .38-caliber bullet fired from 12 feet), football and motorcycle helmets, and the visors in astronauts' helmets. It is made from the two monomers shown below. Draw two repeating units of this polymer.

```
    O                          CH3
    ||                         |
HO—C—OH      HO—(C6H4)—C—(C6H4)—OH
                               |
                               CH3
```

88. Draw two repeating units of the polyester formed from the two monomers given below.

```
    O           O
    ||          ||
HO—C—CH2—C—OH      HO—CH2—CH2—CH2—OH
```

89. Kevlar is an aramid, a long-chain synthetic polyamide in which at least 85% of the amide linkages are attached directly to two aromatic rings. Because of its great

mechanical strength, Kevlar is used in "bulletproof" clothing and in radial tires. The two monomers used to produce Kevlar are shown below. Draw two repeating units of the Kevlar polymer.

$HO{-}\overset{O}{\overset{\|}{C}}{-}C_6H_4{-}\overset{O}{\overset{\|}{C}}{-}OH \qquad H_2N{-}C_6H_4{-}NH_2$

90. Nomex is a type of nylon called an aramid. It has great heat resistance and is used in "fireproof" clothing worn by firefighters and race car drivers. The two monomers for Nomex are below. Draw two repeating units of the Nomex polymer.

$HO{-}\overset{O}{\overset{\|}{C}}{-}C_6H_4{-}\overset{O}{\overset{\|}{C}}{-}OH \qquad H_2N{-}C_6H_4{-}NH_2$

91. Give the monomers from which the following step-growth condensation polymer can be made.

$\left[\overset{O}{\overset{\|}{C}}{-}C_6H_4{-}\overset{O}{\overset{\|}{C}}{-}O{-}CH_2{-}\underset{CH_3}{\underset{|}{\overset{H}{\overset{|}{C}}}}{-}O\right]_n$

92. Give the monomers from which the following nylon polymer can be made.

$\left[\overset{O}{\overset{\|}{C}}{-}CH_2{-}\overset{O}{\overset{\|}{C}}{-}\overset{H}{\overset{|}{N}}{-}CH_2{-}CH_2{-}\overset{H}{\overset{|}{N}}\right]_n$

93. A leading nylon used in Europe is nylon 6, shown below. Nylon 6 is made by polymerization of a *single, difunctional* reactant. Draw the structural diagram of this reactant.

$\left[\overset{O}{\overset{\|}{C}}{-}CH_2{-}CH_2{-}CH_2{-}CH_2{-}CH_2{-}\overset{H}{\overset{|}{N}}\right]_n$

94. Stanyl is nylon 46, widely used in Europe. Draw the structural diagrams of the Stanyl monomers.

$\left[\overset{O}{\overset{\|}{C}}{-}CH_2{-}CH_2{-}CH_2{-}CH_2{-}\overset{O}{\overset{\|}{C}}{-}\overset{H}{\overset{|}{N}}{-}CH_2{-}CH_2{-}CH_2{-}CH_2{-}\overset{H}{\overset{|}{N}}\right]_n$

### General Questions

95. Distinguish precisely, and in scientific terms, the differences among items in each of the following pairs or groups.
   a) Organic chemistry, inorganic chemistry
   b) Saturated and unsaturated hydrocarbons
   c) Alkanes, alkenes, alkynes
   d) Normal alkane, branched alkane
   e) *Cis* and *trans* isomers
   f) Structural formula, condensed (line) formula, molecular formula
   g) Addition reaction, substitution reaction
   h) Alkane, alkyl group
   i) Monomer, polymer
   j) Aliphatic hydrocarbon, aromatic hydrocarbon
   k) Ortho-, meta-, para-
   l) Alcohol, aldehyde, carboxylic acid
   m) Hydroxyl group, carbonyl group, carboxyl group
   n) Primary, secondary, tertiary alcohols
   o) Alcohol, ether
   p) Aldehyde, ketone
   q) Carboxylic acid, ester
   r) Carboxylic acid, amide
   s) Amine, amide
   t) Primary, secondary, tertiary amine
   u) Chain-growth polymer, step-growth polymer

96. Classify each of the following statements as true or false.
   a) To be classified as organic, a compound must be or have been part of a living organism.
   b) Carbon atoms normally form four bonds in organic compounds.
   c) Only an unsaturated hydrocarbon can engage in an addition reaction.
   d) Members of a homologous series differ by a distinct structural unit.
   e) Alkanes, alkenes, and alkynes are unsaturated hydrocarbons.
   f) Alkyl groups are a class of organic compounds.
   g) Isomers have the same molecular formulas but different structural formulas.
   h) *Cis-trans* isomerism appears among alkenes but not alkynes.
   i) All aliphatic hydrocarbons are unsaturated.
   j) Aromatic hydrocarbons have a ring structure.
   k) An alcohol has one alkyl group bonded to an oxygen atom, and an ether has two.
   l) Carbonyl groups are found in alcohols and aldehydes.
   m) An ester is an aromatic hydrocarbon, made by the reaction of an alcohol with a carboxylic acid.
   n) An amine has one, two, or three alkyl groups substituted for hydrogens in an ammonia molecule.
   o) An amide has the structure of a carboxylic acid, except that $—NH_2$ replaces —OH in the carboxyl group.
   p) A peptide linkage arises when a water molecule forms from a hydrogen from the $—NH_2$ group of one amino acid molecule and an —OH from another amino acid molecule.
   q) Forming cross-links in a polymer makes the polymer more likely to possess low mechanical strength.
   r) Chain-growth polymers are made by a reaction that gives off water as a second product.
   s) Nylon is an example of a step-growth condensation polymer.

97. What is the difference in bonding and in general molecular formula between an alkene and a cycloalkane with the same number of carbon atoms?

98. Draw all isomers of $C_4H_8$.

99. Why is the delocalized structure (I) for benzene more appropriate than the cyclohexatriene structure (II)?

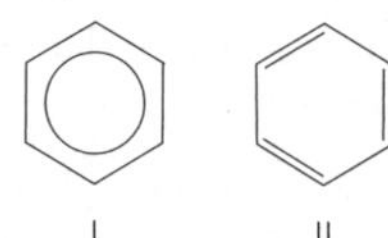

100. Show that the following statement is true: "Every alcohol with two or more carbons is an isomer of at least one ether."

101. Polymers are typically described as having a range of molecular masses, such as "polyvinyl alcohol, molecular mass 31,000–50,000 u." Why do polymers not have unique molecular masses, even though the monomer starting materials do have unique molecular masses?

102. ■ A sample of polystyrene has an average molecular mass of 1,800,000 u. If a single styrene molecule has the formula $C_8H_8$, about how many stryene molecules are in a chain of this polystyrene?

### More Challenging Problems

103. Explain why aldehydes are unstable with regard to oxidation, while ketones are stable. (*Hint:* Look at the number of hydrogen atoms bonded to the carbon that is bonded to the oxygen.)

104. Chemists often use different isotopes such as oxygen-18 to chart the path of organic reactions. If acetic acid reacts with methanol that contains only oxygen-18, show where the oxygen-18 atom exists in the ester product.

105. Can trimethylamine react with a carboxylic acid to form an amide? Explain why or why not using chemical equations.

106. Write three repeating units of the addition polymer that can be made from acetylene. This material, called polyacetylene, conducts electricity because of the alternating single-bond/double-bond pattern in the main chain.

107. Amides are often made by reaction of a carboxylic acid and an amine. This reaction is both acid-catalyzed and reversible. Use these facts to explain why nylon hosiery has a very short wear life in cities where acid rain is common.

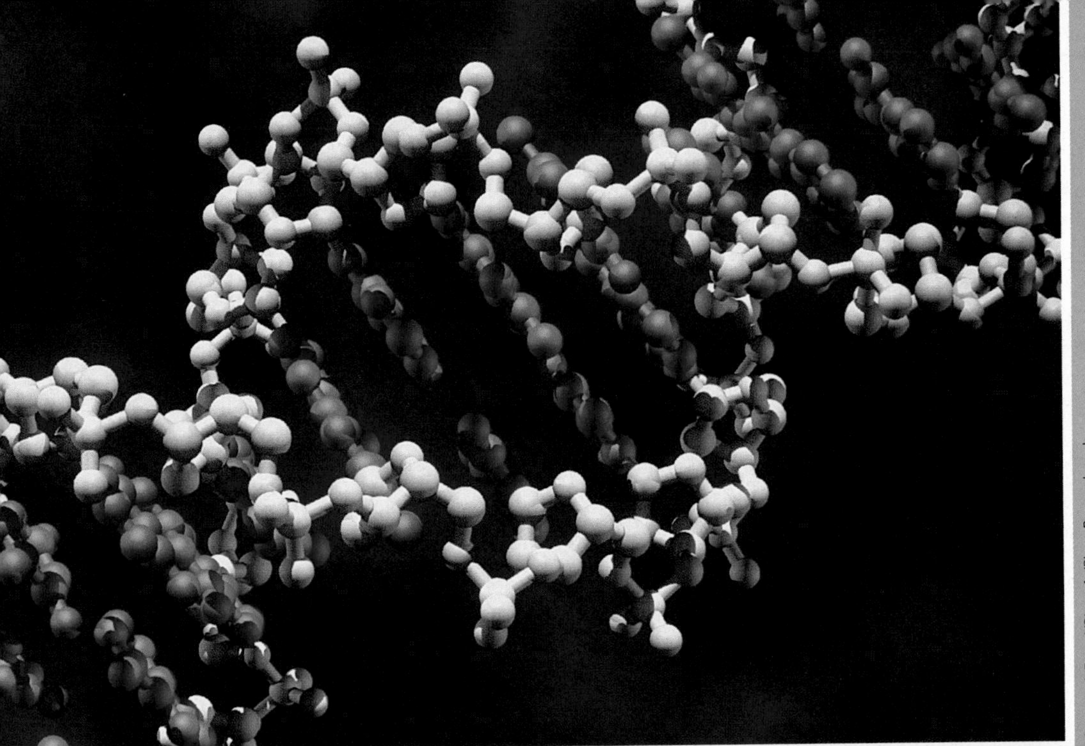

© Kenneth Eward/Biografx/Photo Researchers, Inc.

# 22

CONTENTS



DNA, or deoxyribonucleic acid, is the figurative bridge between biology and chemistry. Scientists who study biological molecules, such as DNA, proteins, and carbohydrates, are known as biochemists. Chemically, DNA consists of two strands that are polymers of nucleic acids (acids prevalently found in the nuclei of cells), held together in a vinelike helix by hydrogen bonds. Biologically, DNA is the storehouse of instructions for the development of all cellular forms of life. Remarkably, the structure of DNA was elucidated only fifty-odd years ago, but the study of biological molecules has now grown to be the most active area of chemical research.

# Biochemistry

This chapter is a brief survey of biochemistry, the chemistry of life. As in Chapter 21, some instructors may prefer the following chapter-wide performance goals rather than the more specific numbered goals within the chapter:

Goal **A** Identify, describe the distinguishing features of, and give an example of a molecule in each of the four major classes of biological molecules: proteins, carbohydrates, lipids, and nucleic acids.

Goal **B** Identify the monomers and describe how the polymeric molecules are assembled in the following: proteins, carbohydrates, nucleic acids.

Goal **C** Describe how an enzyme functions as a biological catalyst.

Goal **D** Describe the process by which protein molecules are constructed from the information encoded in DNA molecules.

If you have studied Chapter 21, especially Sections 21.15 and 21.16, you are ready to tackle this chapter. If you haven't studied Chapter 21 yet, read at least those last two

Online homework for this chapter may be assigned in OWL

sections. Return here with two ideas: (1) Large molecules can be made by joining together many smaller molecules, like links in a chain, and (2) physical and chemical properties of large molecules can be studied and predicted, just as the properties of small molecules are studied and predicted.

**Biochemistry** is the study of life on a molecular level. Although there are no clear dividing lines separating biology and biochemistry or biochemistry and chemistry, biochemists usually focus their research on matter between the cellular and molecular levels (Fig. 22.1). The science of life at a level larger than the single cell is generally the realm of traditional biology, and molecular research on nonbiological molecules and some submolecular research is traditional chemistry. The study of the tiniest bits of matter, those that are smaller than subatomic particles, is classified as modern physics.

**Macromolecules** are polymeric molecules with molar masses starting from about 5000 g/mol and going up from there. In this chapter we introduce the major classes of biological macromolecules. You will see that these classes all feature a pattern of modular assembly, made from surprisingly few different monomers.

## Thinking About Your Thinking
### Classification

**Recall the classification system used by organic chemists described in Chapter 21. Organic molecules are organized according to their functional groups. Biochemists also use functional groups as a classification system, but because biochemical molecules are usually much larger than organic molecules, biochemists also employ a broader classification scheme for the macromolecules they study. The four major classes of biological molecules are (1) proteins (Sections 22.1 and 22.2), (2) carbohydrates (Section 22.3), (3) lipids (Section 22.4), and (4) nucleic acids (Section 22.5). Keep this broad classification system in mind as you study this chapter, and use it to help organize your understanding of biochemistry.**

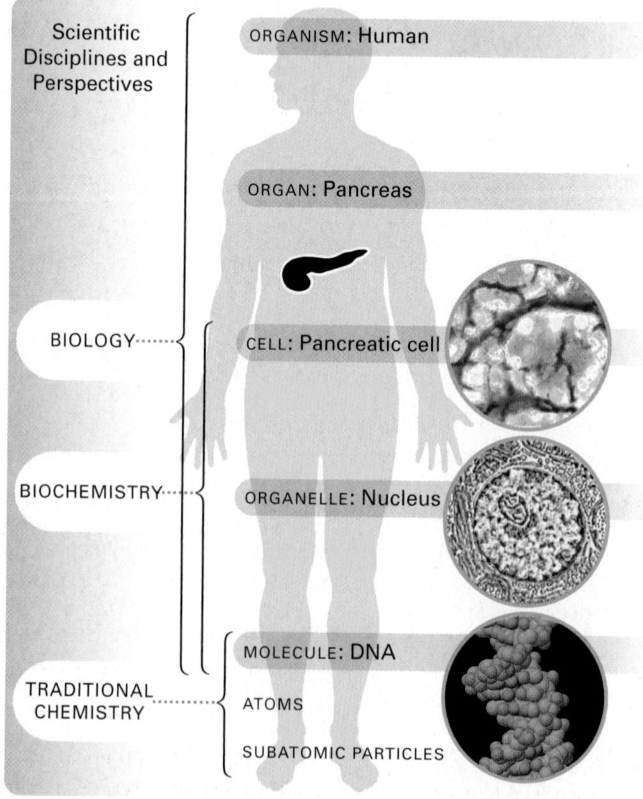


**Figure 22.1** Biology, biochemistry, and traditional chemistry. Biochemists are scientists who study problems of interest to both biologists and chemists. Most biochemists use chemical techniques to study biologically relevant systems, ranging from a single cell to a large macromolecule.

# 22.1 Amino Acids and Proteins

**Goal 1** Given a Lewis diagram of a polypeptide or information from which it may be written, identify the C-terminal amino acid and the N-terminal amino acid.

**Goal 2** Given a table of Lewis diagrams of amino acids and their corresponding three-letter and one-letter abbreviations, draw the Lewis diagram of a polypeptide from its abbreviation.

**Goal 3** Explain the meaning of the terms *primary, secondary, tertiary,* and *quaternary structure* as they apply to proteins.

**Goal 4** Describe how hydrogen bonding results in (a) $\alpha$-helix and (b) $\beta$-pleated-sheet secondary protein structures.

The word *protein* comes from the Greek word *proteios,* meaning "of first importance." Proteins have a wide range of biological functions, and thus they are involved in almost every biochemical process. We will start with the 20 monomers that are the building blocks of proteins, one of nature's polymers. We will then develop protein structure.

An **amino acid** is a molecule that contains both an amine group and a carboxylic acid group. In the amino acid monomers in proteins, the amine group and carboxyl group are attached to the same carbon:

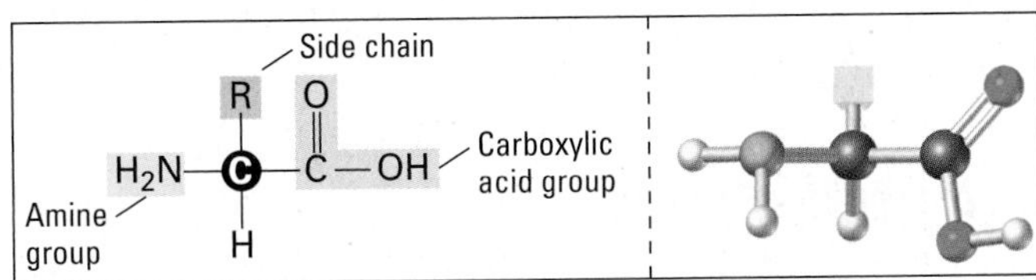


**P/REVIEW:** —$NH_2$ is the amine group in an amino acid and —COOH is the carboxyl group. These structures are introduced in Sections 21.13 and 21.12, respectively.

Twenty amino acids are generally found in proteins in living organisms. Amino acids differ by the identity of the R group, also called a side chain, and can be divided into classes based on the nonpolar or polar nature of the R group or side chain. Table 22.1 shows these amino acids. Note that each amino acid in the table is identified in three ways: by its complete name, by a three-letter abbreviation, and by a single-letter abbreviation.

The symbol R stands for a group of atoms that completes a Lewis diagram.

## Primary Protein Structure

Two amino acids can react with each other in two possible ways. For example, the reaction of glycine and alanine could give either glycylalanine, abbreviated Gly-Ala or G-A,

$$\underset{\text{glycine}}{H_2N{-}CH(H){-}C({=}O){-}OH} + \underset{\text{alanine}}{H{-}N(H){-}CH(CH_3){-}C({=}O){-}OH} \longrightarrow \underset{\text{glycylalanine}}{H_2N{-}CH(H){-}C({=}O){-}N(H){-}CH(CH_3){-}C({=}O){-}OH} + H_2O$$

or alanylglycine, abbreviated Ala-Gly or A-G,

$$\underset{\text{alanine}}{H_2N{-}CH(CH_3){-}C({=}O){-}OH} + \underset{\text{glycine}}{H{-}N(H){-}CH(H){-}C({=}O){-}OH} \longrightarrow \underset{\text{alanylglycine}}{H_2N{-}CH(CH_3){-}C({=}O){-}N(H){-}CH(H){-}C({=}O){-}OH} + H_2O$$

In either case, the amino acid with the free carboxyl group is called the **C-terminal** acid; the amino acid with the free amine group is called the **N-terminal** acid. The

**Table 22.1** The 20 Amino Acids Commonly Found in Proteins[†]

| Amino Acid | ABBREVIATION 3-letter | 1-letter | Structure | Amino Acid | ABBREVIATION 3-letter | 1-letter | Structure |
|---|---|---|---|---|---|---|---|
| | | | ***Nonpolar R Groups*** | | | | |
| Glycine | Gly | G | $H{-}CH(NH_2){-}COOH$ | *Isoleucine | Ile | I | $CH_3{-}CH_2{-}CH(CH_3){-}CH(NH_2){-}COOH$ |
| Alanine | Ala | A | $CH_3{-}CH(NH_2){-}COOH$ | Proline | Pro | P | $H_2C{-}CH_2{-}CH_2{-}N(H){-}CH{-}COOH$ (ring: $H_2C$—$CH_2$, $H_2C$, CH, N–H) |
| *Valine | Val | V | $CH_3{-}CH(CH_3){-}CH(NH_2){-}COOH$ | *Phenylalanine | Phe | F | $C_6H_5{-}CH_2{-}CH(NH_2){-}COOH$ |
| *Leucine | Leu | L | $CH_3{-}CH(CH_3){-}CH_2{-}CH(NH_2){-}COOH$ | *Methionine | Met | M | $CH_3{-}S{-}CH_2CH_2{-}CH(NH_2){-}COOH$ |
| | | | | *Tryptophan | Trp | W | indole(N–H)$-CH_2{-}CH(NH_2){-}COOH$ |
| | | | ***Polar but Neutral R Groups*** | | | | |
| Serine | Ser | S | $HO{-}CH_2{-}CH(NH_2){-}COOH$ | Asparagine | Asn | N | $H_2N{-}C(=O){-}CH_2{-}CH(NH_2){-}COOH$ |
| *Threonine | Thr | T | $CH_3{-}CH(OH){-}CH(NH_2){-}COOH$ | Glutamine | Gln | Q | $H_2N{-}C(=O){-}CH_2CH_2{-}CH(NH_2){-}COOH$ |
| Cysteine | Cys | C | $HS{-}CH_2{-}CH(NH_2){-}COOH$ | Tyrosine | Tyr | Y | $HO{-}C_6H_4{-}CH_2{-}CH(NH_2){-}COOH$ |
| | | | ***Acidic R Groups*** | | | | ***Basic R Groups*** |
| Glutamic acid | Glu | E | $HO{-}C(=O){-}CH_2CH_2{-}CH(NH_2){-}COOH$ | *Lysine | Lys | K | $H_2N{-}CH_2CH_2CH_2CH_2{-}CH(NH_2){-}COOH$ |
| Aspartic acid | Asp | D | $HO{-}C(=O){-}CH_2{-}CH(NH_2){-}COOH$ | ‡Arginine | Arg | R | $H_2N{-}C(=NH){-}NH{-}CH_2CH_2CH_2{-}CH(NH_2){-}COOH$ |
| | | | | ‡Histidine | His | H | imidazole(N, N–H)$-CH_2{-}CH(NH_2){-}COOH$ |

*Essential amino acids that must be part of the human diet. The other amino acids can be synthesized by the body.
[†]The R group in each amino acid is highlighted.
[‡]Growing children also require arginine and histidine in their diet.

term "free" carboxyl or amine group is used to indicate that the functional group occurs at the end of a chain, "free" to form bonds to other amino acids. The bond between the amino acids is called a **peptide linkage,** which is formed when the hydroxyl part of the carboxyl group of an amino acid molecule reacts with a hydrogen of the $-NH_2$ group of another amino acid molecule to form a molecule of water. The two amino acid molecules are then connected through the nitrogen atom.

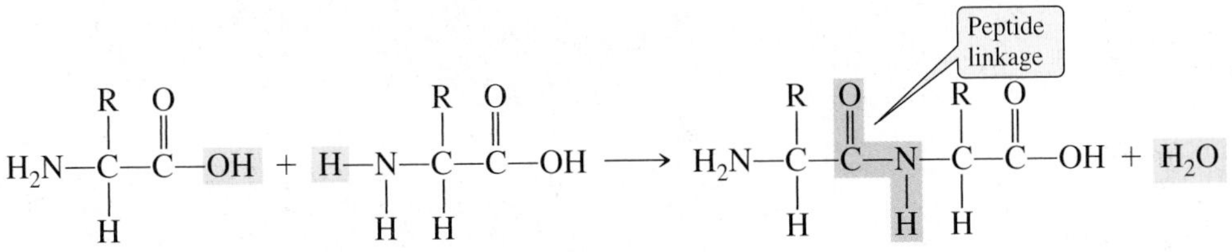

$$\underset{\text{amino acid 1}}{H_2N{-}\overset{R}{\underset{H}{C}}{-}\overset{O}{\overset{\|}{C}}{-}OH} + \underset{\text{amino acid 2}}{H{-}\underset{H}{N}{-}\overset{R}{\underset{H}{C}}{-}\overset{O}{\overset{\|}{C}}{-}OH} \longrightarrow \underset{\text{dipeptide}}{H_2N{-}\overset{R}{\underset{H}{C}}{-}\overset{O}{\overset{\|}{C}}{-}\underset{H}{N}{-}\overset{R}{\underset{H}{C}}{-}\overset{O}{\overset{\|}{C}}{-}OH} + H_2O$$

Peptide linkage

If we add a third amino acid, valine, to the end of our dipeptides above, we could have glycylalanylvaline, Gly-Ala-Val, or G-A-V,

$$NH_2{-}\overset{H}{CH}{-}\overset{O}{\overset{\|}{C}}{-}NH{-}\overset{CH_3}{CH}{-}\overset{O}{\overset{\|}{C}}{-}NH{-}\overset{CH_3-\overset{H}{C}-CH_3}{CH}{-}\overset{O}{\overset{\|}{C}}{-}OH$$

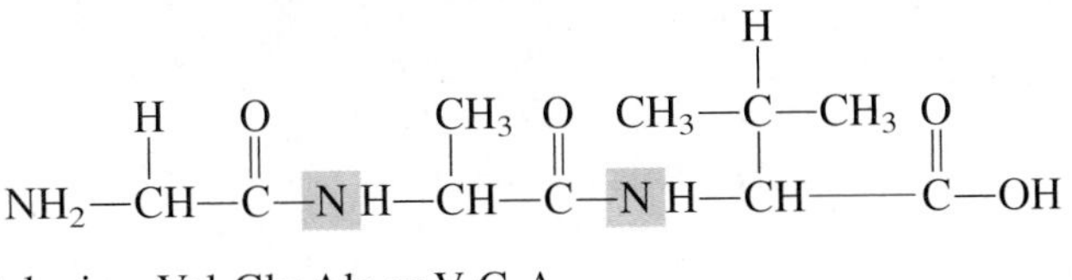

or valylglycylalanine, Val-Gly-Ala or V-G-A,

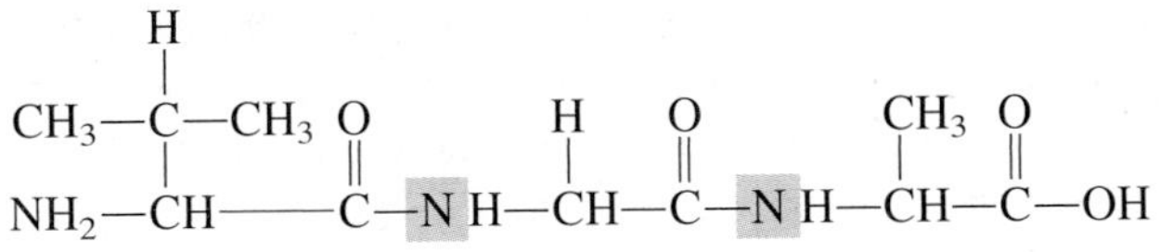

$$NH_2{-}\overset{CH_3-\overset{H}{C}-CH_3}{CH}{-}\overset{O}{\overset{\|}{C}}{-}NH{-}\overset{H}{CH}{-}\overset{O}{\overset{\|}{C}}{-}NH{-}\overset{CH_3}{CH}{-}\overset{O}{\overset{\|}{C}}{-}OH$$

or valylalanylglycine, Val-Ala-Gly or V-A-G,

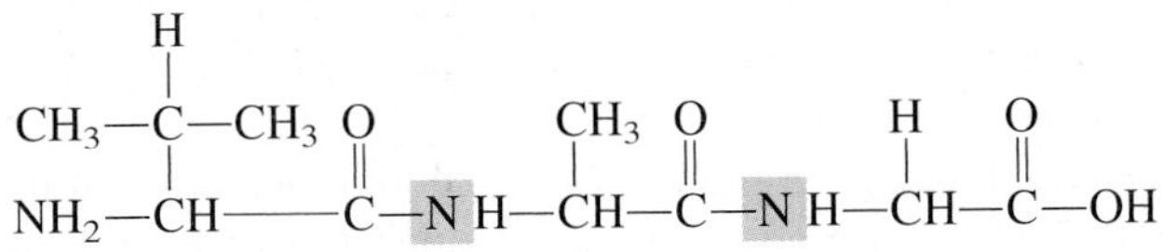

$$NH_2{-}\overset{CH_3-\overset{H}{C}-CH_3}{CH}{-}\overset{O}{\overset{\|}{C}}{-}NH{-}\overset{CH_3}{CH}{-}\overset{O}{\overset{\|}{C}}{-}NH{-}\overset{H}{CH}{-}\overset{O}{\overset{\|}{C}}{-}OH$$

or alanylglycylvaline, Ala-Gly-Val or A-G-V,

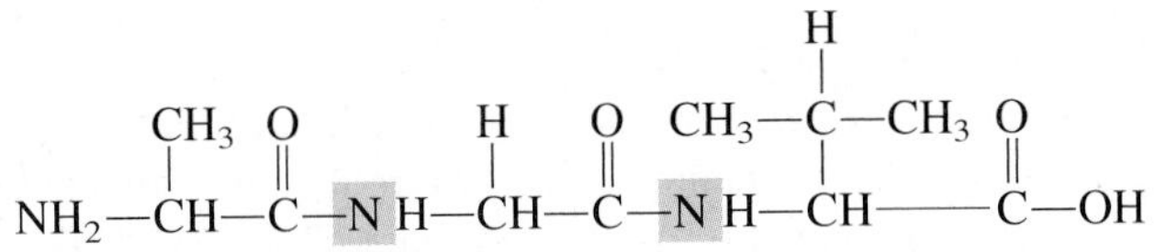

$$NH_2{-}\overset{CH_3}{CH}{-}\overset{O}{\overset{\|}{C}}{-}NH{-}\overset{H}{CH}{-}\overset{O}{\overset{\|}{C}}{-}NH{-}\overset{CH_3-\overset{H}{C}-CH_3}{CH}{-}\overset{O}{\overset{\|}{C}}{-}OH$$

**P/REVIEW:** Lewis diagrams for the peptide linkage reaction appear in Section 21.13. Condensation polymers are formed when monomers combine in this way (Section 21.16). The nitrogen atoms in the peptide linkages in this section are shown in blue.

You can see that the possible number of proteins, from only 20 amino acids, is theoretically infinite. Molecules having fewer than about 50 amino acid residues are called **polypeptides;** a **protein** is a polypeptide chain with more than 50 amino acid residues. The term *residue* refers to the portion of the amino acid monomer that remains after it becomes a part of a polypeptide polymer. When we specify the order of amino acids in a protein, using Lewis structures or the abbreviations shown above, we have specified its **primary structure.**

The dividing line between peptides and proteins is not sharp; they are overlapping terms.

### √ Target Check 22.1

*Using the single-letter abbreviations A, V, and L, as given in Table 22.1, list the tripeptides formed if alanine, valine, and leucine are all mixed together and peptide bonds are allowed to form in all possible combinations.*

## Secondary Protein Structure

Coiled springs are made by winding wire into a spiral. In this process, the primary structure of the metal remains unchanged. However, the **local conformation,** or arrangement in space, of the wire has changed. The spiral, or *helix,* is a regular local conformation that defines the **secondary structure** of the metal atoms in the spring.

Proteins have secondary structures also. The most prevalent secondary structures are the **$\alpha$-helix** and the **$\beta$-pleated sheet** (Fig. 22.2). *Both structures reflect a maximum amount of hydrogen bonding;* these are the most stable conformations possible.

**P/REVIEW:** The electron-sea model of a metallic crystal is introduced in Section 12.8. This model depicts a metal as cations in a definite crystal pattern with relatively freely-moving valence electrons that travel among the ions.

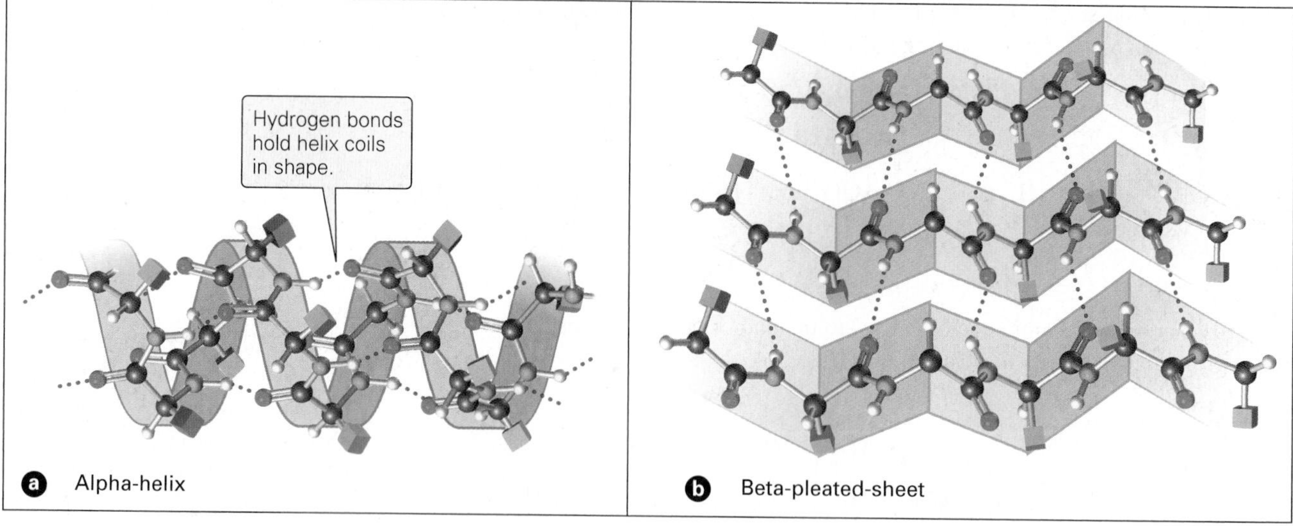
Hydrogen bonds hold helix coils in shape.
a Alpha-helix
b Beta-pleated-sheet

**Figure 22.2** Alpha-helix and beta-pleated-sheet secondary structures. R groups are represented in these ball-and-stick models with green blocks. The $\alpha$-helix structure is a continuous series of loops. The $\beta$-pleated-sheet structure is like a curtain, but with sharp angles, like pleats, instead of rounded curves. In both the $\alpha$-helix and $\beta$-pleated-sheet structures, the helices and sheets are stabilized by hydrogen bonding between N—H and O═C groups.

The $\alpha$-helix is found in fibrous proteins such as wood, hair, and fingernails. These fibers are slightly elastic; stretching a hair, for example, will stretch the hydrogen bonds but will not break the amide bonds.

Look at the two drawings in Figure 22.3. Both represent a small peptide with 11 amino acid residues. The illustration on the left emphasizes the main chain, with carbon atoms shown in gray and the amide nitrogens shown in blue. Note how the chain coils as if it were climbing a right-handed screw thread.

The figure on the right includes all atoms but the R groups, which are represented with single orange spheres. The hydrogen bonds are illustrated with red dots. Note how they are nearly vertical lines connecting an oxygen (red) atom with a hydrogen (white) atom, which is covalently bonded to a nitrogen (blue) atom. Study the figure to see that *every* oxygen and *every* nitrogen atom in the main chain is involved in hydrogen bonding, which governs the secondary structure of the protein. The $\alpha$-helix allows the maximum amount of hydrogen bonding among the amino acids of a single protein chain. Not all amino acids coil to form an $\alpha$-helix; the R groups of the amino acids must be small and nonpolar.

Amino acids such as proline, with large or polar R groups, break the $\alpha$-helix by putting a bend in it.

The tight spiral of the $\alpha$-helix gives a rigid, rodlike protein. The $\alpha$-keratins (found in hair, fingernails, and skin) are examples of this structure. If fibers of an $\alpha$-keratin are steamed and stretched, they almost double their length. The length increase occurs because the intrachain hydrogen bonds that stabilize the $\alpha$-helix are broken, allowing the tightly curled helix to stretch into a more extended, zig-zag conformation called the $\beta$ conformation. When adjacent protein chains are in this conformation, they share *inter*chain hydrogen bonds to form the $\beta$-pleated sheet.

The $\beta$-pleated-sheet secondary structure is formed by hydrogen bonds *between* two adjacent protein chains or when a single polypeptide chain folds back on itself in a repeating fashion. In **silk fibroin,** adjacent protein chains run in opposite directions, called an antiparallel arrangement (Figure 22.4). The pleated sheets can then stack on each other, like the pages of a book. The backbones of these protein chains are already extended. As a result, silk is not elastic. If you try to stretch silk, it will tear.

## Tertiary Protein Structure

Describing the bends in a protein chain defines the **tertiary structure** of the protein. These bends usually fold the protein into a compact, globular shape.

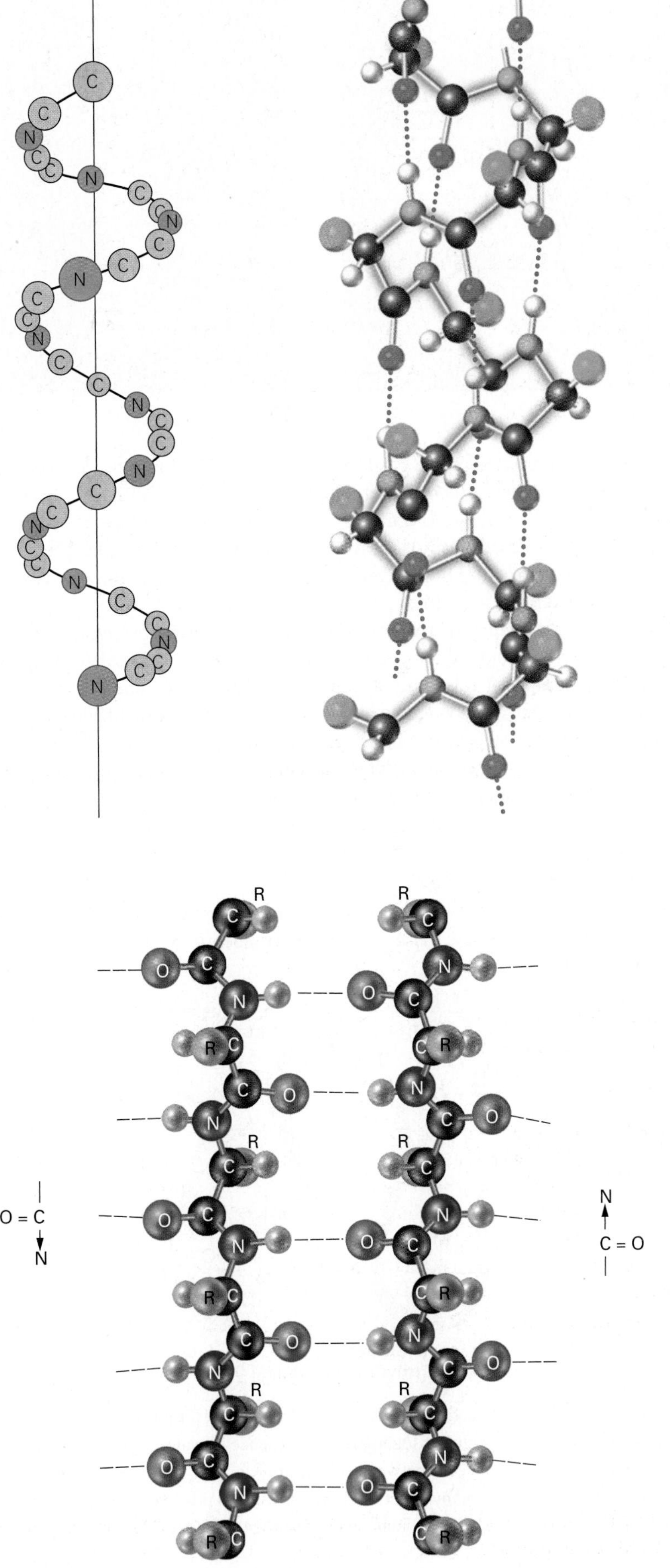


**Figure 22.3** Alpha-helix protein structure.

**Figure 22.4** Beta-pleated-sheet secondary structure. In this depiction of silk, the polar N—H groups in one protein chain form hydrogen bonds with the O═C groups in an adjacent chain. Note how the chains in this depiction run in opposite directions. (The unlabeled small white spheres represent hydrogen.)

Figure 22.5 Macroscopic analogy for secondary and tertiary protein structures. (a) The coiling of this telephone cord is the secondary structure; the looping back is the start of a tertiary structure. (b) The twists and turns of this telephone cord make up a more complicated tertiary structure.

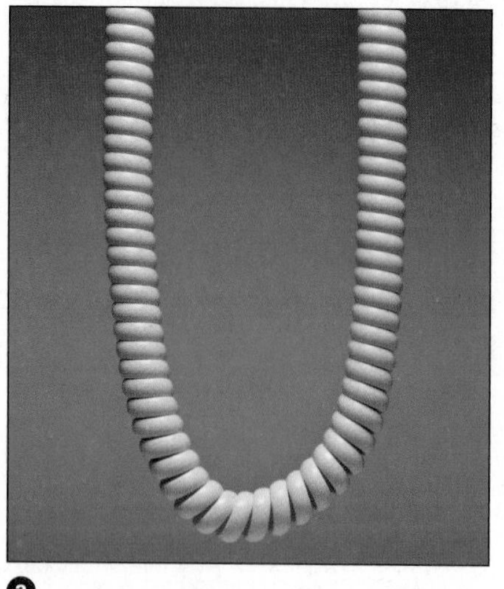

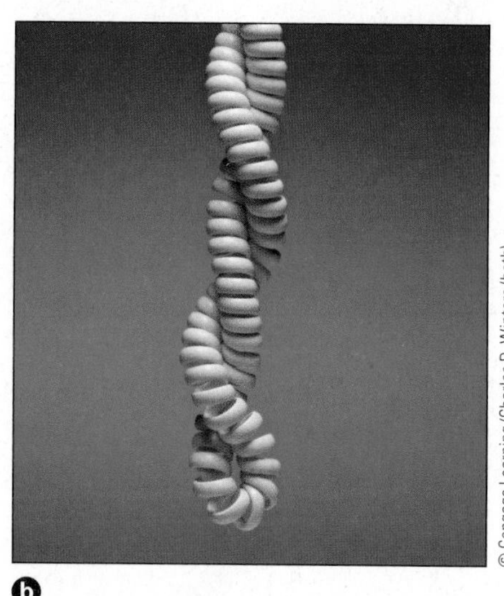



Look at the curly telephone cord of Figure 22.5. The helix of the cord is its secondary structure. When attached to a telephone, the helical cord curls back on itself, beginning a tertiary structure (Figure 22.5[a]). If the curly cord has itself been twisted (Figure 22.5[b]), a description of the bends is needed for a complete account of the three-dimensional structure of the cord. A complete description of the twists and turns of the cord is its tertiary structure.

Permanent waves break and re-form the disulfide linkages in hair. The straight hair is first wrapped on a roller; the disulfide linkages are then broken and allowed to reform as the hair is held curled. The new disulfide linkages give curl to the hair.

The tertiary structure of many proteins is strengthened by interactions between the R groups in the protein chain. Sulfur–sulfur covalent bonds called **disulfide linkages** are one example of this type of interaction. The amino acid cysteine contains a sulfur–hydrogen bond. Two cysteines in a protein can form a disulfide linkage, which stabilizes the protein structure much like a cross-link strengthens the structure of a polymer. Many other noncovalent intermolecular forces contribute to the stability of the tertiary structure of a protein, as shown in Figure 22.6.

## Quaternary Protein Structure

Some proteins are composed of more than one polypeptide chain. The term **quaternary structure** refers to how these chains are arranged in relation to one another. The

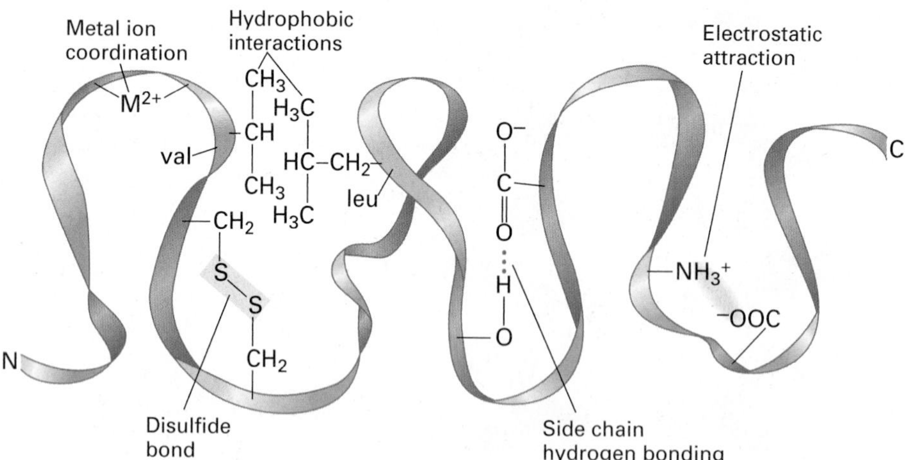


Figure 22.6 Tertiary structure of proteins. Covalent bonds between sulfur atoms are called disulfide bonds. Other weaker noncovalent interactions also serve to stabilize proteins (*left* to *right*): Metal ions, such as $Mg^{2+}$ and $Zn^{2+}$ (symbolized in general as $M^{2+}$), can interact with negatively charged ionized forms of amino acid functional groups. Hydrophobic (*fear of water*) interactions are due to the energetic favorability of nonpolar groups to cluster together in the absence of water. Hydrogen bonding can occur between side-chain R groups. At common biological pH, many functional groups exist as positively charged and negatively charged ions, which are then subject to electrostatic attractions.

Side chains
Backbone
Asparagine (Asn)
Threonine (Thr)
Lysine (Lys)
Primary structure
The sequence of amino acids in a polypeptide chain
Secondary structure
The spatial arrangement of the amino acid sequences into regular patterns such as helices, sheets, and turns
Tertiary structure
The overall three-dimensional shape of a polypeptide chain caused by the folding of various regions
Quaternary structure
The spatial interaction of two or more polypeptide chains in a protein

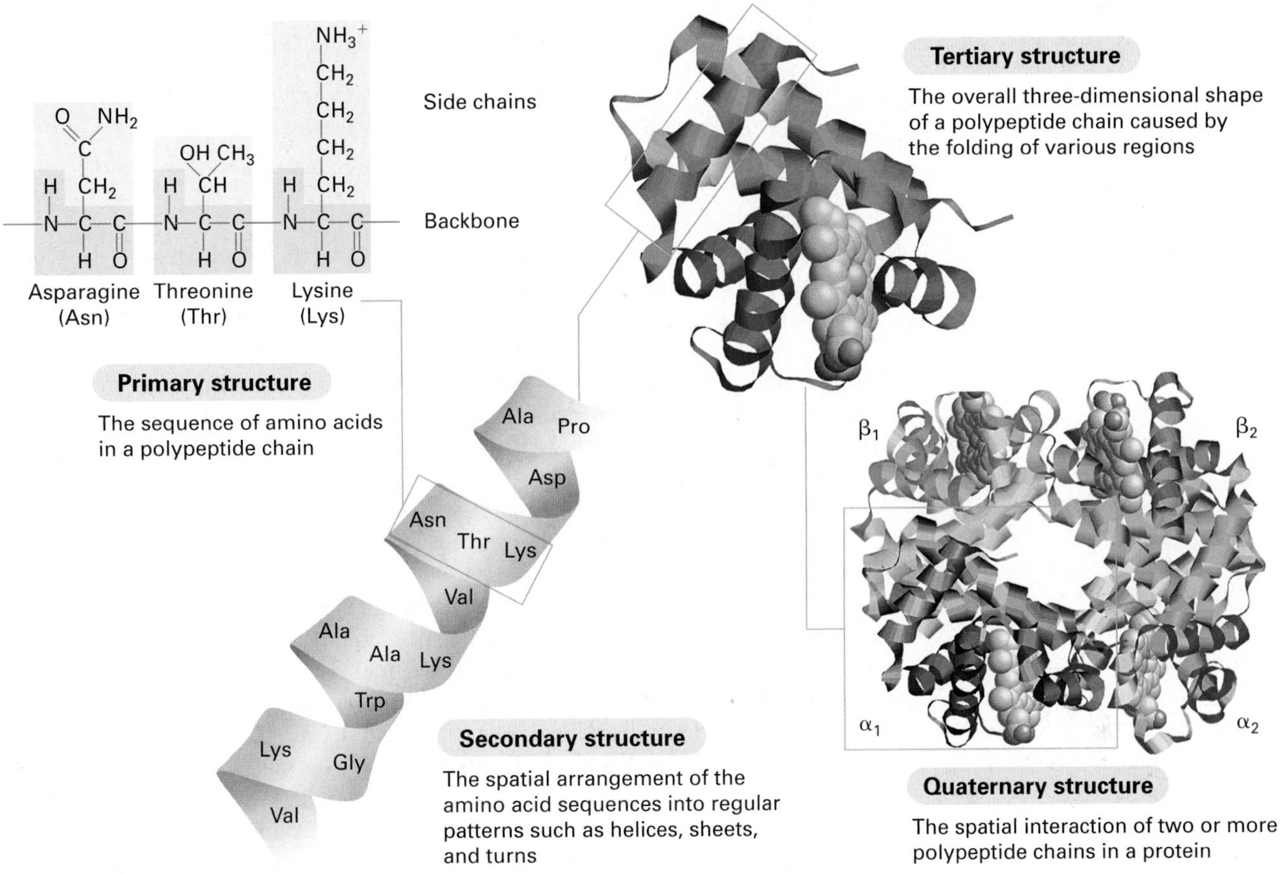

**Figure 22.7** Primary, secondary, tertiary, and quaternary structure of hemoglobin. Four polypeptide chains are shown here, two in blue and two in green, labeled $\alpha_1$, $\alpha_2$, $\beta_1$, and $\beta_2$. The three-dimensional arrangement of those polypeptide chains is the quaternary structure of the protein.

atoms in each individual chain in a protein are held together by covalent bonds, but the chains themselves are attracted to one another only by intermolecular forces. Returning to the telephone-cord analogy, you can think of proteins with quaternary structure as being composed of more than one telephone cord. The first protein for which the complete primary, secondary, tertiary, and quaternary structure was known is hemoglobin (Fig. 22.7). It is composed of four polypeptide chains. Hemoglobin is the molecule found in vertebrate red blood cells that serves to transport oxygen throughout the body. Each human red blood cell contains 300,000,000 hemoglobin molecules!

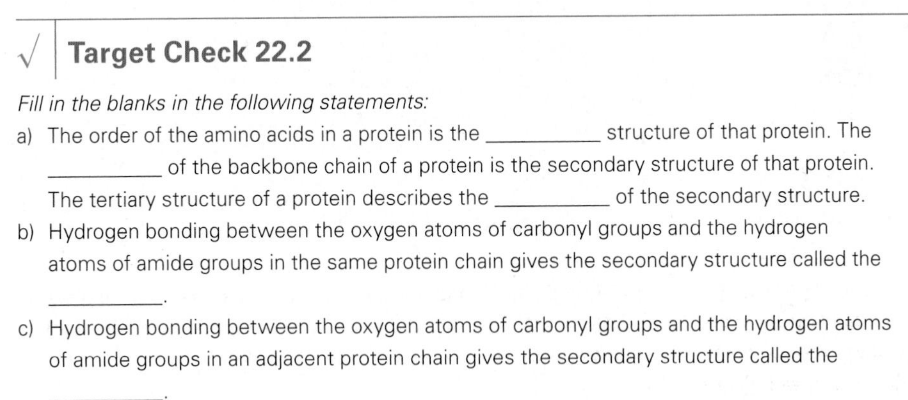

### √ Target Check 22.2

*Fill in the blanks in the following statements:*

a) The order of the amino acids in a protein is the __________ structure of that protein. The __________ of the backbone chain of a protein is the secondary structure of that protein. The tertiary structure of a protein describes the __________ of the secondary structure.

b) Hydrogen bonding between the oxygen atoms of carbonyl groups and the hydrogen atoms of amide groups in the same protein chain gives the secondary structure called the __________.

c) Hydrogen bonding between the oxygen atoms of carbonyl groups and the hydrogen atoms of amide groups in an adjacent protein chain gives the secondary structure called the __________.

## 22.2 Enzymes

**Goal 5** Define the following terms as they apply to enzymes: *substrate, active site, inhibitor.*

**Goal 6** Use the induced fit model to explain enzyme activity.

Recall from Section 9.6 that complete combustion of a carbon–hydrogen–oxygen compound yields carbon dioxide and water:

$$C_6H_{12}O_6(s) + 6\ O_2(g) \rightarrow 6\ CO_2(g) + 6\ H_2O(\ell)$$

This equation also describes the overall process living organisms use to produce the energy needed to live. We eat carbohydrates (the simplest of which is glucose, $C_6H_{12}O_6$) and breathe in oxygen; we breathe out carbon dioxide and water. Sustaining life requires a slow combustion.

We don't catch fire, however, and we can't live at such high temperatures. How do we perform and control the chemical reactions of a combustion process at body temperature?

**P/REVIEW:** In Section 18.4 a catalyst was identified as a substance that speeds the rate of a reaction without being permanently affected.

Each chemical reaction in a living system is controlled by a catalyst called an **enzyme;** the enzyme **substrate** is the reactant the enzyme helps convert to product. Enzymes are proteins that catalyze specific reactions and allow them to occur at body temperature. Scientists have so far isolated well over a thousand enzymes, and there are many more still to be discovered. The catalytic properties of an enzyme exist at the **active site,** a location that binds the enzyme's substrate during the reaction that follows (Fig. 22.8).

The rate of reaction increase caused by an enzyme is amazing. In Chapter 9 we stated that carbonic acid is an unstable substance that decomposes into carbon dioxide and water:

$$H_2CO_3(aq) \rightleftharpoons CO_2(g) + H_2O(\ell)$$

In living systems, formation of $H_2CO_3$ in red blood cells is controlled by the enzyme carbonic anhydrase. *One molecule* of this enzyme can convert 36,000,000 $CO_2$ molecules to $H_2CO_3$ each minute.

Most enzymes are specific; the shape and polarity of the active site allow *only* a specific substrate molecule to bind reversibly to the enzyme. In the **induced fit model,**

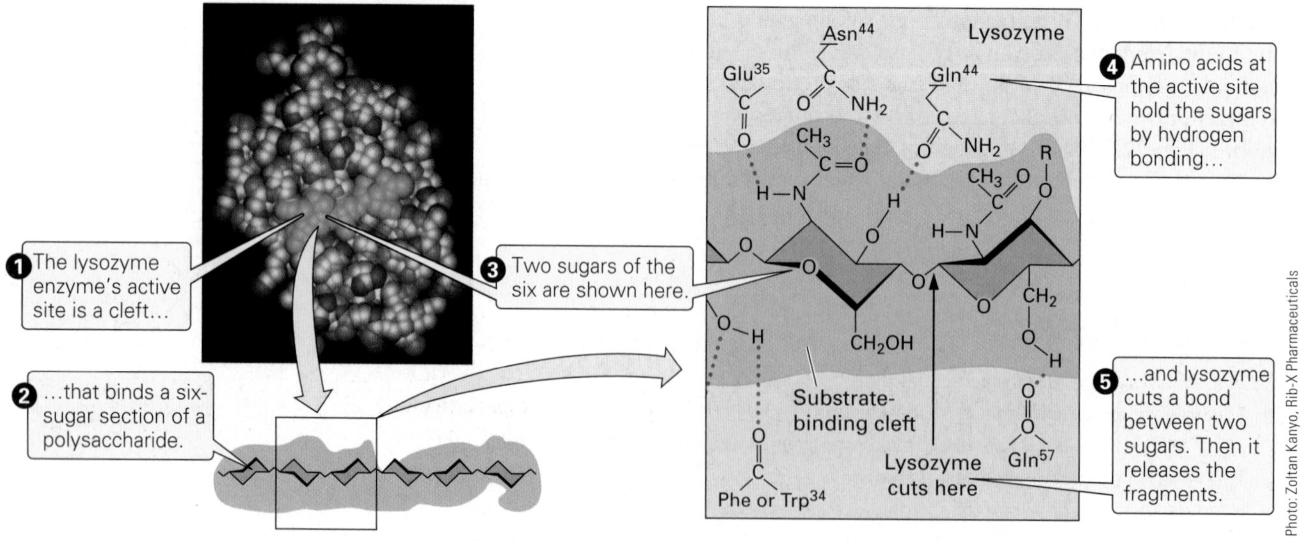


**Figure 22.8** The active site of an enzyme. Lysozyme is nature's bacteria killer. It catalyzes the splitting of polysaccharide molecules (large carbohydrate polymers discussed in Section 22.3) found in the cell walls of bacteria. When the cell walls are sufficiently disintegrated, the bacteria burst. The lysozyme molecule has an indentation that serves as its active site, shown here with a portion of a green polysaccharide molecule within it.

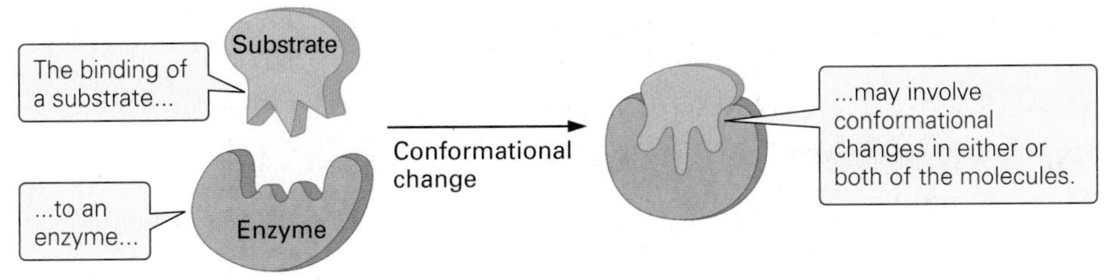


**Figure 22.9** Induced fit model of enzyme activity. The close match between the shape of the substrate and the active site of the enzyme is slightly modified to become an exact match as the two molecules bind. This is called a conformational change, which is a change in the three-dimensional arrangement of the atoms in a molecule.

the shape of the substrate is a close, but not exact, match to the shape of the active site of the enzyme (Fig. 22.9). As the substrate binds to the enzyme, either or both molecules change shape slightly. It is believed that distortion of the shape of the substrate may be a factor in catalyzing the reaction. After the products are released, the enzyme returns to its original shape and is ready to catalyze another reaction.

The preceding model assumes that the substrate and the enzyme bind reversibly. If the substrate remains bound to the enzyme, that enzyme molecule is no longer a catalyst. The irreversibly bound substrate is termed an irreversible **inhibitor.** Unfortunate examples of this inhibition are the effects of nerve gases used as weapons, which inhibit the enzyme acetylcholinesterase, needed for nerve-impulse transmission and muscle contraction. Inhalation or skin absorption of these inhibitors causes difficulty in breathing (you can't move your diaphragm), followed by bronchial constriction, convulsions, and death.

### √ Target Check 22.3

*Identify the true statements, and rewrite the false statements to make them true.*

a) The active site of an enzyme is the portion of the enzyme where the catalytic properties occur.
b) The induced fit model explains why one enzyme helps many reactions to occur faster.
c) An enzyme substrate is the material that is the product of the enzyme-catalyzed reaction.
d) An irreversible enzyme inhibitor cannot be removed from the enzyme's active site.

## 22.3 Carbohydrates

Goal 7 Given a Lewis diagram of a monosaccharide in its open-chain form, determine whether the sugar is an aldose or a ketose.

Goal 8 Distinguish among monosaccharides, disaccharides, and polysaccharides.

Goal 9 Given the Lewis diagrams of two monosaccharides and a description of the bond linking the molecules, draw the Lewis diagram of the resulting disaccharide.

**Carbohydrates** are molecules that were originally thought to be "hydrates of carbon": $(C \cdot H_2O)_n$. We now know that carbohydrates are not "hydrates of carbon"; they are chemically classified as aldehydes or ketones with two or more —OH groups. We shall study carbohydrates in order of increasing size, beginning with simple sugars called monosaccharides, advancing to two-sugar molecules known as disaccharides, and ending the section with complex carbohydrates called polysaccharides (Fig. 22.10).

### Monosaccharides

**Monosaccharides,** or simple sugars, cannot be converted to smaller carbohydrates. In simple sugars, every carbon but one is bonded to a hydroxyl group; the remaining carbon is double-bonded to an oxygen atom, forming a carbonyl group. If the carbonyl is at the end of the carbon chain, the sugar is an **aldose,** derived from the aldehyde

Figure 22.10 Carbohydrate classification.

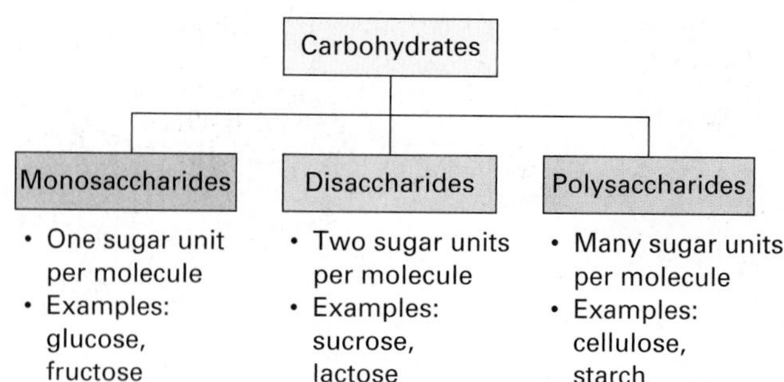


Note how the names of the mono- and disaccharides all end in the suffix *-ose.*

**P/REVIEW:** An aldehyde has a hydrogen atom bonded to the carbon atom of a carbonyl group, and a ketone has two alkyl groups attached to the carbonyl carbon. Both are characterized by the carbonyl group (Section 21.11).

functional group (Section 21.11); if the carbonyl group is within the chain, the sugar is a **ketose,** derived from the ketone functional group. The general ending *-ose* denotes the name of a sugar.

$$\underset{\text{aldehyde}}{R-\overset{O}{\overset{\|}{C}}-H} \qquad \underset{\text{ketone}}{R-\overset{O}{\overset{\|}{C}}-R'}$$

**P/REVIEW:** Chemical equilibria are reversible, dynamic, closed systems in which the forward rate of change is equal to the reverse rate of change. The amounts of substances in an equilibrium are *not* necessarily equal. See Section 18.1.

As you would expect from a molecule with so many polar —OH groups, the monosaccharides are water soluble. They are all white crystalline solids at room temperature and have differing sweetnesses. For example, fructose is about twice as sweet as glucose. There are many simple sugars in your life. We will show each sugar as an open-chain molecule, then in its more common cyclic form. Sugars exist as an equilibrium mixture of open-chain and cyclic forms in water solutions.

The most important simple sugar is an aldohexose called **glucose,** also called dextrose, corn sugar, grape sugar, and blood sugar (Fig. 22.11). The glucose in your bloodstream supplies energy to all of your cells. Because glucose is a simple sugar, you do not need to digest it to obtain energy; in hospitals, intravenous bottles supply glucose to patients who are not able to eat. You will see later that glucose is also the monomer for the plant carbohydrates cellulose and starch.

Diabetes is a metabolic disease characterized by high blood glucose levels.

The equal sign between the two cyclic diagrams indicates different portrayals of the same molecule. The center diagram is called a Haworth projection; the right-hand diagram shows the arrangement of the atoms in three dimensions.

glucose (open chain): H–C(1)=O; H–C(2)–OH; HO–C(3)–H; H–C(4)–OH; H–C(5)–OH; C(6)$H_2OH$

α-glucose (Haworth projection = chair form)

Figure 22.11 A Haworth projection and a ball-and-stick model of the glucose molecule. Compare the diagram with the model to see how the diagram represents the actual molecular structure shown in the model.

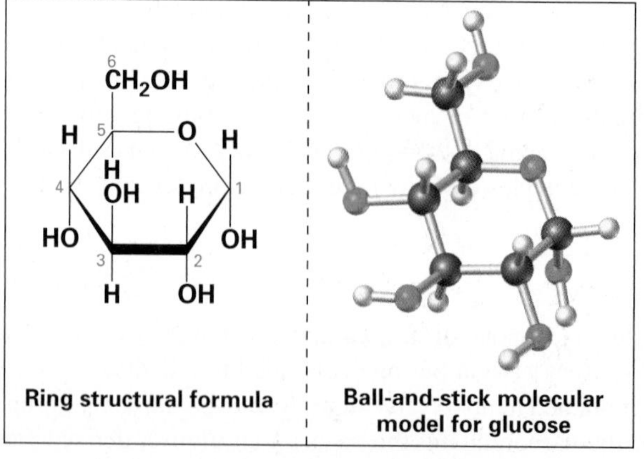

There are two glucose isomers, differing only at carbon-1.

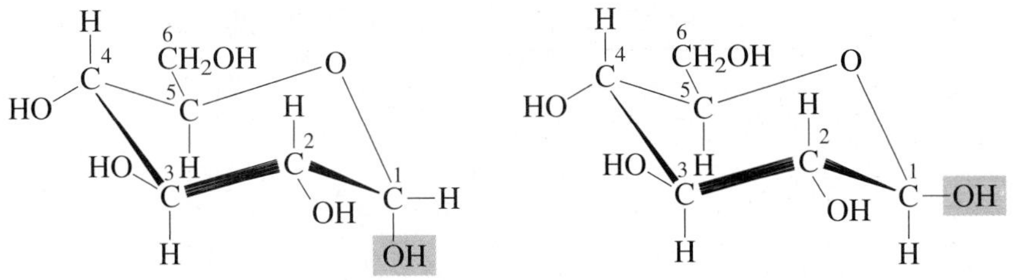

α-glucose — Axial

β-glucose — Equatorial

Carbon-1 is called the anomeric carbon, and these two isomers are often called *anomers.*

Carbon-1 is the only carbon in glucose that is bonded to *two* oxygen atoms. Look at the structures of $\alpha$-glucose and $\beta$-glucose. The bond connecting the OH to carbon-1 in $\alpha$-glucose is nearly vertical (axial); the bond connecting the OH to carbon-1 in $\beta$-glucose is nearly horizontal (equatorial). Later in this chapter we will use the carbon numbering scheme illustrated here to describe bonding between glucose units in polysaccharides.

**Fructose,** also called levulose or fruit sugar, is the sweetest of the sugars. It gives fruits and honey their sweetness and, in the form of high-fructose corn syrup, is widely used in the food and beverage industries. Fructose is also used in many ice creams because it improves the "mouth feel" (texture) of the ice cream by preventing sandiness.

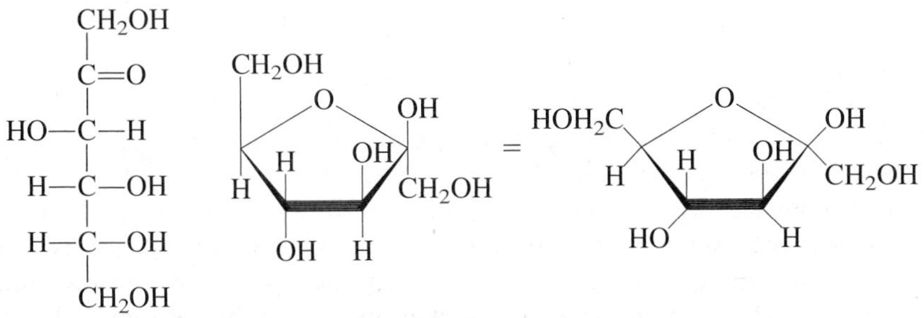

fructose

β-fructose

The five-carbon aldehyde sugar ribose is only slightly sweet but is still an important part of our lives. Ribose is a component of **adenosine triphosphate (ATP),** the molecule involved in transferring chemical energy within cells. Ribose is also a component of ribonucleic acid (RNA), which is directly involved in protein synthesis; deoxyribose is a sugar found in deoxyribonucleic acid (DNA). DNA is the central molecule involved in passing on genetic information in living cells. You will study both RNA and DNA more closely in Section 22.5.

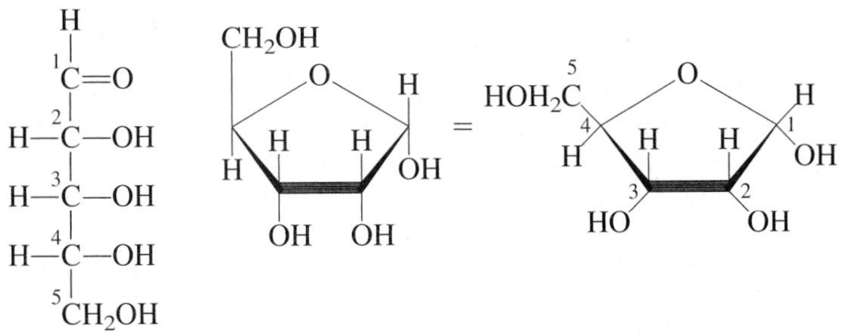

ribose

α-D-ribose

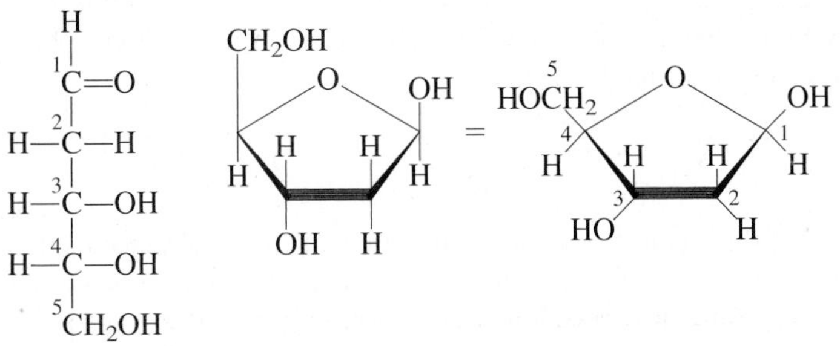

2-deoxyribose

β-D-2-deoxyribose

## Disaccharides

"Putting two simple sugars together" means combining them in a condensation reaction. The oxygen that links the two monomers is shaded in blue. We will use blue to identify atoms that link monomers in other diagrams yet to come.

If you put two simple sugars together, you get a single **disaccharide.** The most well-known disaccharide is **sucrose,** a glucose-fructose combination.

glucose $C_6H_{12}O_6$ + fructose $C_6H_{12}O_6$ ⟶ sucrose $C_{12}H_{22}O_{11}$ + $H_2O$

Carbohydrates are found in many foods such as breads, pastas, and rices. In general, it is believed that the carbohydrates that break down most slowly into simpler sugars are better for your health than those that cause a rapid rise in blood sugar levels.

Other names for sucrose are beet sugar, cane sugar, and table sugar.

Disaccharides and other polysaccharide sugars may be broken up to their simple sugar units. This may be done either in a dilute acid solution or by the proper enzyme. When sucrose is broken up, the product is called invert sugar. Invert sugar is widely used in the food industry because it is sweeter than sucrose and is a creamy product that holds moisture, preventing foods from drying out.

Honey is mostly invert sugar.

The disaccharide **lactose,** also called milk sugar, is found in the milk of mammals. Lactose is a galactose-glucose combination.

lactose

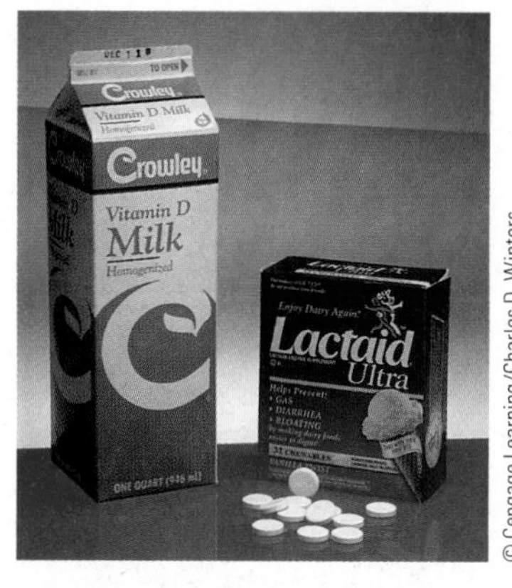



**Figure 22.12** Milk and Lactaid. It is estimated that up to 75% of the world's adult population is lactose (milk sugar) intolerant to some degree. Lactaid contains the enzyme lactase, which catalyzes the reaction of the disaccharide lactose to form the monosaccharides galactose and glucose.

Some adults have lost the ability to synthesize sufficient quantities of the enzyme lactase. As a result, lactose from dairy products is not broken up during digestion, leading to flatulence, diarrhea, and possible dehydration, which can be serious. Reduced-lactose milk and lactase-containing tablets have made life easier for lactose-intolerant adults (Fig. 22.12).

All monosaccharides are converted to glucose during normal metabolic processes. Some infants lack an enzyme needed to change galactose into glucose. This relatively rare genetic disorder is called galactosemia; an infant with galactosemia will suffer mental retardation, cataracts, and liver and kidney disease unless all lactose is quickly removed from the diet. If detected and treated quickly, the symptoms of galactosemia are reversible, and a normal life is possible.

Galactosemia can be detected by prenatal genetic screening.

## Polysaccharides

**Polysaccharides,** or complex carbohydrates, are large molecules, having molar masses between 4000 and 150,000,000 g/mol. You will study three polysaccharides with greatly different biological functions—cellulose, starch, and glycogen. They share a common monomer: glucose.

William E. Weber/Visuals Unlimited

Animals such as cows, deer, and sheep have the ability to utilize cellulose, the primary structural component of plants, as a source of nutrition due to the presence of microorganisms in their digestive tracts that produce the appropriate enzyme. Humans cannot digest cellulose.

**Cellulose** is the major structural component of plants. Cellulose has a molar mass in the 50,000- to 500,000-g/mol range. There is much hydrogen bonding between glucose molecules in the same chain, plus hydrogen bonding between chains. All this hydrogen bonding makes cellulose rigid and insoluble in water. The bond between glucose molecules is defined as $\beta$-1,4. Note the numbering of the atoms and the position of the oxygen atom that connects the two glucose units.

If both bonds from the oxygen atom connecting two sugar molecules are drawn more or less horizontally, the connection is called $\beta$. If one of the bonds is drawn vertically, the connection is called $\alpha$.

cellulose

Starch serves plants as the storehouse of chemical energy, just as fats serve animals. Plant starch comes in two main forms. One is **amylose,** with mainly $\alpha$-1,4 bonds between glucose units and a molar mass between 7000 and 500,000 g/mol. It is also called *soluble starch.*

amylose

The other main form of plant starch is **amylopectin,** which has a molar mass as high as 150,000,000 g/mol (Fig. 22.13). It has the same $\alpha$-1,4 bonds as amylose, but it also has branches designated $\alpha$-1,6. As a result of its high molar mass and its branches, amylopectin is insoluble in water.

amylopectin

**Glycogen,** also called *animal starch* or *liver starch,* is the quick-acting carbohydrate reserve in mammals, including humans. Glycogen has a molar mass between 270,000 and 3,500,000 g/mol and resembles amylopectin. It is found in the liver of mammals and in rested muscle. When muscle activity occurs, glycogen is converted back to glucose 1-phosphate, which is then used for energy.

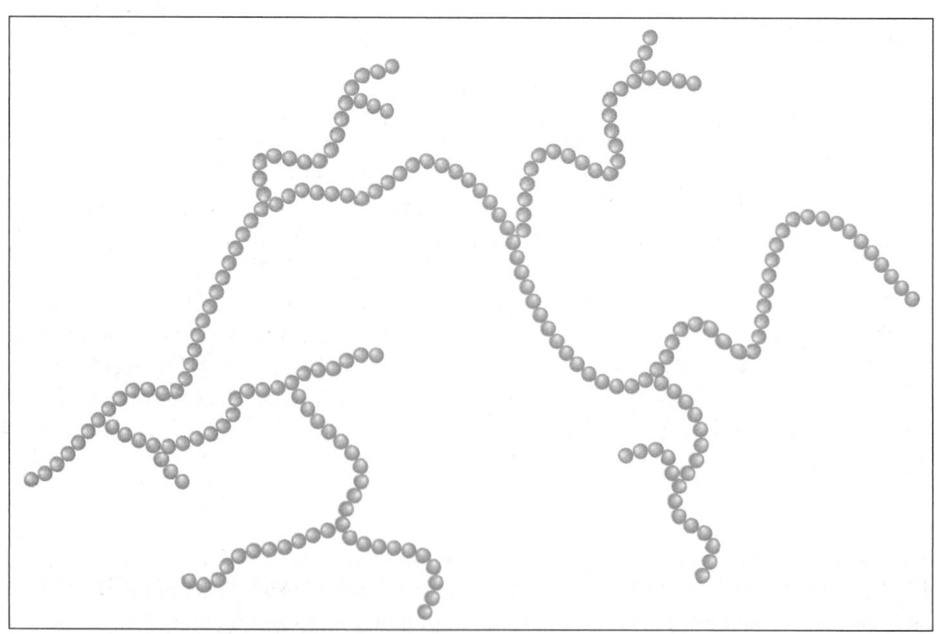

**Figure 22.13** A molecular model of amylopectin, a plant starch. Each sphere represents a glucose monomer. The monomers are linked in linear chains by $\alpha$-1,4 bonds, and branches occur every 24 to 30 glucose units via $\beta$-1,6 bonds. Glycogen, or animal starch, has the same composition and structure except that branching occurs every 8 to 12 glucose units.

✓ **Target Check 22.4**

a) Using the structure for amylopectin, pick out an $\alpha$-1,4 bond. Point out which carbon in this bond is carbon-1 and which is carbon-4.

b) Using the structure for amylopectin, pick out an $\alpha$-1,6 bond. Point out which carbon in this bond is carbon-1 and which is carbon-6.

## 22.4 Lipids

Goal 10 State (a) the defining characteristics and (b) the three major subclassifications of lipids.

Goal 11 Identify the physical property that distinguishes fats from oils.

Goal 12 Identify the structural feature common to all steroid molecules.

**Lipids** are found in living organisms and are insoluble in water but soluble in nonpolar solvents. Lipids may be divided into three classes: (1) fats, oils, and phospholipids, (2) waxes, and (3) lipids (usually) without ester groups, such as steroids.

Figure 22.14 Fats and oils. All are mixtures of triacylglycerols (triglycerides). Cooking oils are liquids at room temperature, and thus they are classified as oil. Butter is solid at room temperature, so it is classified as a fat.



### Fats and Oils

Fats and oils are triesters (the molecule has three ester functional groups) of three long-chain (between 10 and 24 carbons) carboxylic acids. These acids are called **fatty acids.** The alcohol parts of each ester linkage all come from a single molecule of glycerol (also called glycerin), a trihydroxy alcohol.

$$CH_3-CH_2-CH_2-CH_2-CH_2-CH_2-CH_2-CH_2-CH_2-CH_2-CH_2-\overset{\overset{\displaystyle O}{\|}}{C}-O-CH_2$$

$$CH_3-CH_2-CH_2-CH_2-CH_2-CH_2-CH_2-CH_2-CH_2-\overset{\overset{\displaystyle O}{\|}}{C}-O-CH$$

$$CH_3-CH_2-CH_2-CH_2-CH_2-CH_2-CH_2-CH_2-CH_2-CH_2-CH_2-\overset{\overset{\displaystyle O}{\|}}{C}-O-CH_2$$

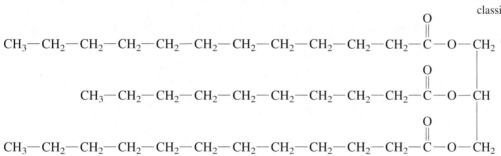

These esters are called triacylglycerols or triglycerides. Fats are triacylglycerols that are solids at room temperature; oils are triacylglycerols that are liquids at room temperature (Fig. 22.14). The composition of the fatty-acid parts of these esters varies with the organism that produced them. Plant oils are usually richer in unsaturated fatty acids than are animal fats.

**P/REVIEW:** An unsaturated molecule has double or triple carbon–carbon bonds (Section 21.4).

Medical evidence correlates a diet high in saturated fats with hardening of the arteries (atherosclerosis) and possible heart attack. Table 22.2 lists the percentage of saturated fatty acids, monounsaturated fatty acids, and polyunsaturated fatty acids in some common dietary fats and oils. This table compares human "depot" (storage) fat with butter and margarines, then with several vegetable oils.

### Phospholipids

**Phospholipids** are a class of lipids that have an alcohol "backbone," two fatty acid residues, and a phosphate group.

polar head

$$
\begin{array}{l}
\hspace{19em} CH_2-O-\overset{\overset{\displaystyle O}{\|}}{P}-O^- \\
\hspace{19em} | \qquad\qquad\ \ | \\
\hspace{19em} | \qquad\qquad\ \ O-CH_2-CH_2-\overset{+}{N}(CH_3)_3 \\
CH_3-(CH_2)_4-CH=CH-CH_2-CH=CH-(CH_2)_7-\overset{\overset{\displaystyle O}{\|}}{C}-O-CH \\
\hspace{19em} | \\
\hspace{7em} CH_3-(CH_2)_{16}-\overset{\overset{\displaystyle O}{\|}}{C}-O-CH_2
\end{array}
$$

two nonpolar tails

A phospholipid has a polar head and two long, nonpolar tails. Phospholipids are found in all animal and vegetable cells, where they are part of cell membranes (Fig. 22.15). The phospholipid shown above is commonly called a **lecithin.** It is used widely in the cosmetic and food industries as an *emulsifier,* a substance that holds two immiscible liquids together as a suspension. You'll find lecithins in many food products such as chocolate, ice cream, and margarine.

## Waxes

Some **waxes** are also esters of fatty acids, but the alcohol portion is a monohydroxyl alcohol with a long carbon chain. Some waxes have a structural function, but most serve as a water-resistant coating on skin, fur, and feathers. The general structure for many waxes is that of an ester:

**Table 22.2** Approximate Fatty Acid Composition of Common Fats and Oils

| Fat or Oil | % Saturated (no double bonds) | % Monounsaturated (1 double bond) | % Polyunsaturated (>1 double bond) |
|---|---|---|---|
| Human | 35 | 55 | 10 |
| Butter | 66 | 31 | 4 |
| Margarine, soft | 18 | 37 | 45 |
| Margarine, stick | 21 | 46 | 33 |
| Coconut | 92 | 6 | 2 |
| Palm kernel | 81 | 18 | 1 |
| Palm | 47 | 43 | 10 |
| Peanut | 18 | 48 | 34 |
| Olive | 14 | 77 | 9 |
| Corn | 13 | 25 | 62 |
| Canola | 6 | 58 | 36 |

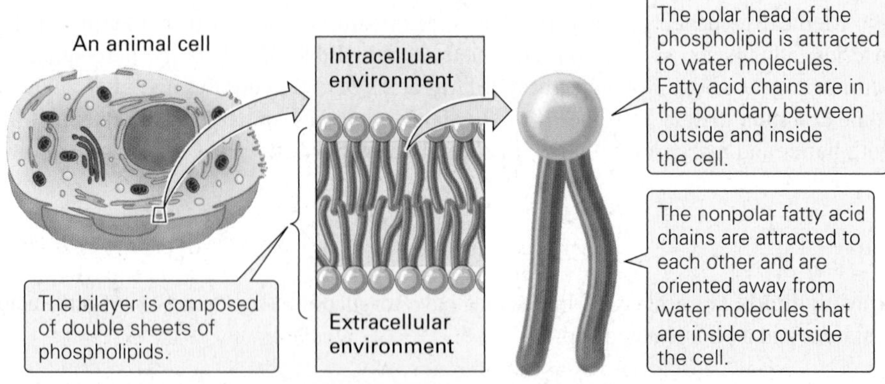
An animal cell
Intracellular environment
Extracellular environment
The bilayer is composed of double sheets of phospholipids.
The polar head of the phospholipid is attracted to water molecules. Fatty acid chains are in the boundary between outside and inside the cell.
The nonpolar fatty acid chains are attracted to each other and are oriented away from water molecules that are inside or outside the cell.

**Figure 22.15** Phospholipids in cell membranes. The polar head is represented by a blue sphere, and the nonpolar tails are shown in red.

$$R-\overset{\overset{\displaystyle O}{\|}}{C}-O-R'$$

where R stands for the alkyl group attached to the carboxylic acid and R′ stands for the alkyl group attached to the hydroxyl oxygen. We will show you three waxes to illustrate how, like fatty acids, waxes can be creamy liquids, soft solids, or hard solids at room temperature, depending on the fatty acid groups and the alcohol groups.

The first is **beeswax,** from bee honeycombs. The major components of beeswax are fatty acid esters of straight-chain alcohols. As with saturated fatty acids, beeswax is a solid at room temperature, melting at 62 to 65°C.

Beeswax is secreted by honeybees and used to build honeycomb cells.

The second wax is **lanolin,** also called wool fat. Lanolin softens the skin and is used in many cosmetics and lotions. Lanolin is a complex mixture of esters of 33 high-molar-mass alcohols and 36 fatty acids. Liquid lanolin has mainly low-molar-mass, branched-chain alcohols and acids (which pack poorly in crystals), while waxy lanolin has higher-molar-mass straight-chain alcohols and acids (which pack better in crystals).

Many plants have waxes that protect their leaves; these coatings serve plants by slowing evaporation of water through the leaves and preventing microbes from entering. The most prized of these waxes is carnauba wax, obtained from Brazilian palm tree leaves. We would expect **carnauba wax,** with $R{=}C_{23}H_{47}-$ to $C_{27}H_{55}-$ and $R'{=}C_{32}H_{65}-$ to $C_{34}H_{69}-$, to be soft, like beeswax. However, carnauba wax melts at 82 to 85°C, the highest melting point of the common waxes. Carnauba wax is used when a hard, shiny surface is desired, such as in floor waxes, and it is cherished by strong-armed car owners who want the hardest possible shine.

Carnauba wax also contains carboxylic acid components having hydroxyl groups. The hydroxyl groups tie together adjacent carbon chains by hydrogen bonding. These weak cross-links make carnauba wax a hard, relatively high-melting wax.

This peanut butter, like all vegetable products, contains no cholesterol. Cholesterol is a product of animal metabolism.

## Steroids

**Steroids,** such as the cholesterol molecule illustrated below, are lipids that all contain the same four-fused-ring system, three six-carbon rings and one five-carbon ring.

HO

Cholesterol is the most common steroid in the body; it is a needed structural molecule found in all normal animal cells. Cholesterol is concentrated in the spinal cord and in the brain. Your brain is about 2% cholesterol by mass.

Although cholesterol is needed to synthesize other steroids, such as those in Figure 22.16, too much cholesterol can lead to blocked blood vessels and high blood pressure, then possibly to heart attack and stroke. We can't live without cholesterol, and we can't live with too much of it, either.

### √ Target Check 22.5

a) Some margarine brands offer three different types of margarine: (1) a thick liquid in a squeeze bottle, (2) a soft solid in a tub, and (3) a harder solid sold in sticks. Describe the relative amounts of saturated and unsaturated fatty acids you would expect to find in these three different types of margarine.
b) Cocoa butter, obtained from chocolate, has (for a vegetable product) a relatively high melting point of 35°C. Does cocoa butter contain many or few saturated fatty acids? Explain.
c) What structural feature is characteristic of a steroid?

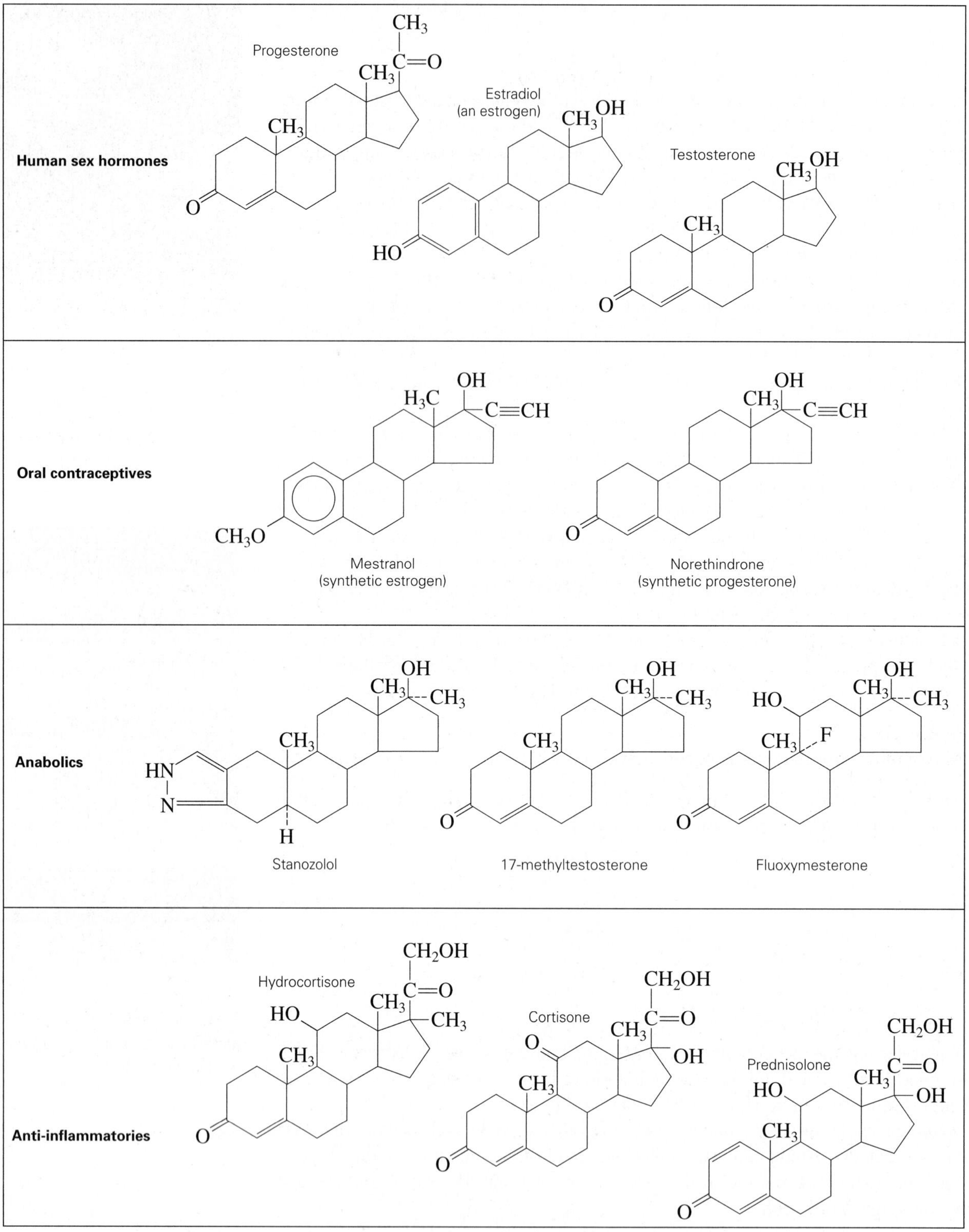

Human sex hormones

Progesterone

Estradiol (an estrogen)

Testosterone

Oral contraceptives

Mestranol (synthetic estrogen)

Norethindrone (synthetic progesterone)

Anabolics

Stanozolol

17-methyltestosterone

Fluoxymesterone

Anti-inflammatories

Hydrocortisone

Cortisone

Prednisolone

**Figure 22.16** Steroids. In mammals, male and female sex hormones are steroids. Oral contraceptives are synthetic steroids. Anabolic steroids increase muscle mass. Adrenocorticoid hormones such as cortisone have amazing anti-inflammatory properties. Hundreds of steroid-based drugs are available by prescription in the United States.

## 22.5 Nucleic Acids

Goal **13** Describe the biological roles of DNA and RNA.

Goal **14** Describe the components of a nucleotide.

Goal **15** Draw Lewis diagrams of adenine, cytosine, guanine, thymine, and uracil.

Goal **16** Determine whether any two DNA or RNA nitrogen bases are complementary.

Goal **17** Describe the process by which a protein molecule is formed.

DOE/Science Source/Photo Researchers, Inc.

James Watson (1928– ) (*left*) and Francis Crick (1916–2004) (*right*). Watson and Crick shared the 1962 Nobel Prize for Physiology and Medicine with Maurice Wilkins for their discovery of the structure of DNA.

There are two types of nucleic acids in all living systems. **Deoxyribonucleic acid (DNA)** stores genetic information and transmits that information to the next generation during cell division. **Ribonucleic acid (RNA)** assists in this process by serving as a "messenger" and as a "switching engine" to transfer the correct amino acid during protein synthesis. DNA molecules have the highest molar masses of any molecules in a living system, up to *several billion* g/mol in higher animals. RNA molecules have molar masses of about 30,000 g/mol.

The nucleic acid monomers are called **nucleotides.** Each nucleotide has three parts: (1) a nitrogen-containing cyclic molecule called a base (the five bases are shown in Fig. 22.17), (2) a sugar, either ribose in RNA or deoxyribose in DNA (Section 22.3), and (3) one or more phosphate groups, attached to the hydroxyl groups of the sugar.*

The primary structure of DNA or RNA is the order of nucleotides joined by covalent bonds between carbon-5 in the sugar of one nucleotide and the phosphate group on carbon-3 of another nucleotide's sugar. The bonds between nucleotides form an alternating pattern: sugar–phosphate–sugar–phosphate (Figure 22.18). The secondary structure of DNA is determined by hydrogen bonds between base pairs on *different* DNA molecules.

Maximum hydrogen bonding occurs between thymine (T) on one DNA strand and adenine (A) on another, and between cytosine (C) on one DNA strand and guanine (G)

| Found only in DNA | Found in both DNA and RNA | Found only in RNA |
|---|---|---|
| thymine | cytosine | uracil |
| | adenine    guanine | |

**Figure 22.17** The nitrogen bases in DNA and RNA.

*If no phosphate groups are attached to the sugar, the two-component base–sugar combination is called a *nucleoside*.

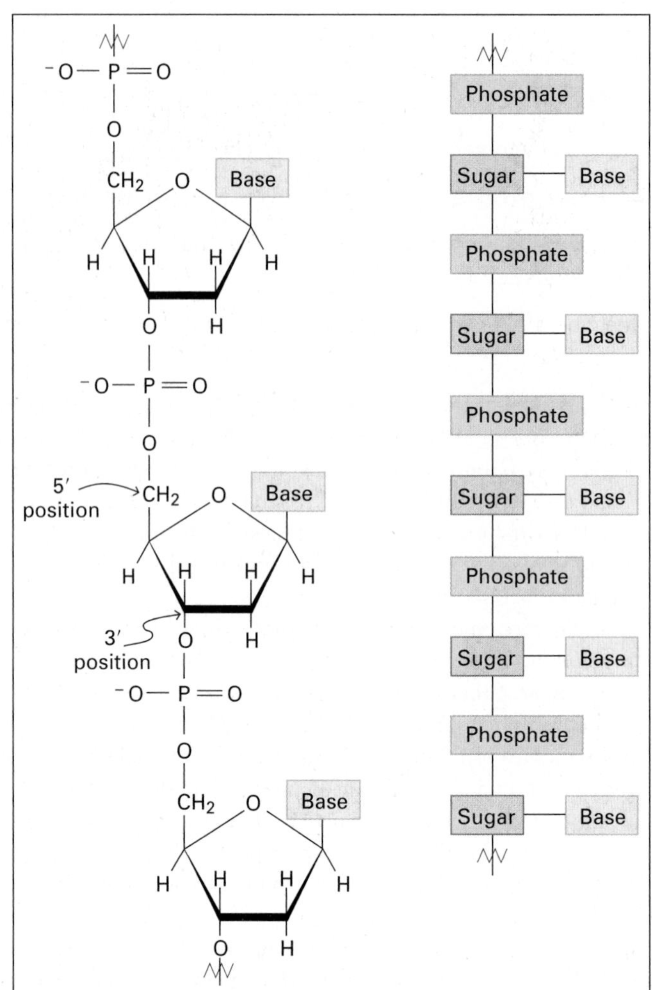


**Figure 22.18** The left drawing shows the sugar–phosphate chain in DNA. The right drawing shows how DNA is a nucleotide (phosphate–sugar–base) polymer.

on another (Fig. 22.19). This gives the famous double helix proposed by James Watson and Francis Crick in 1953 (Fig. 22.20). The two helical DNA strands, with all adenines on one strand hydrogen-bonded to thymines to the other strand, and all guanines on one strand hydrogen-bonded to cytosines on the other strand, are termed *complementary*. In RNA, adenine (A) always pairs with uracil (U), a thymine without a methyl group.

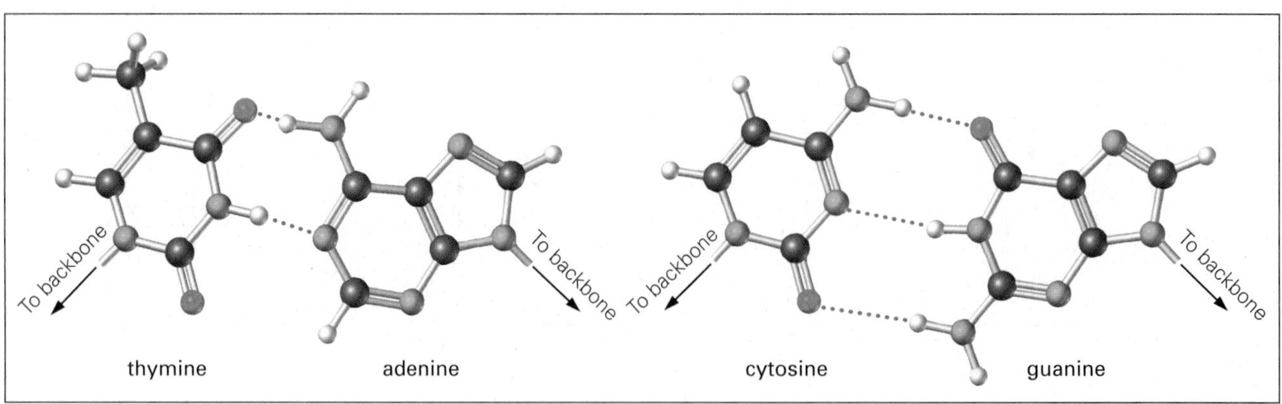


**Figure 22.19** Hydrogen bonding between DNA base pairs. The DNA double helix is held together by thymine-adenine (T-A) and cytosine-guanine (C-G) hydrogen bonds.

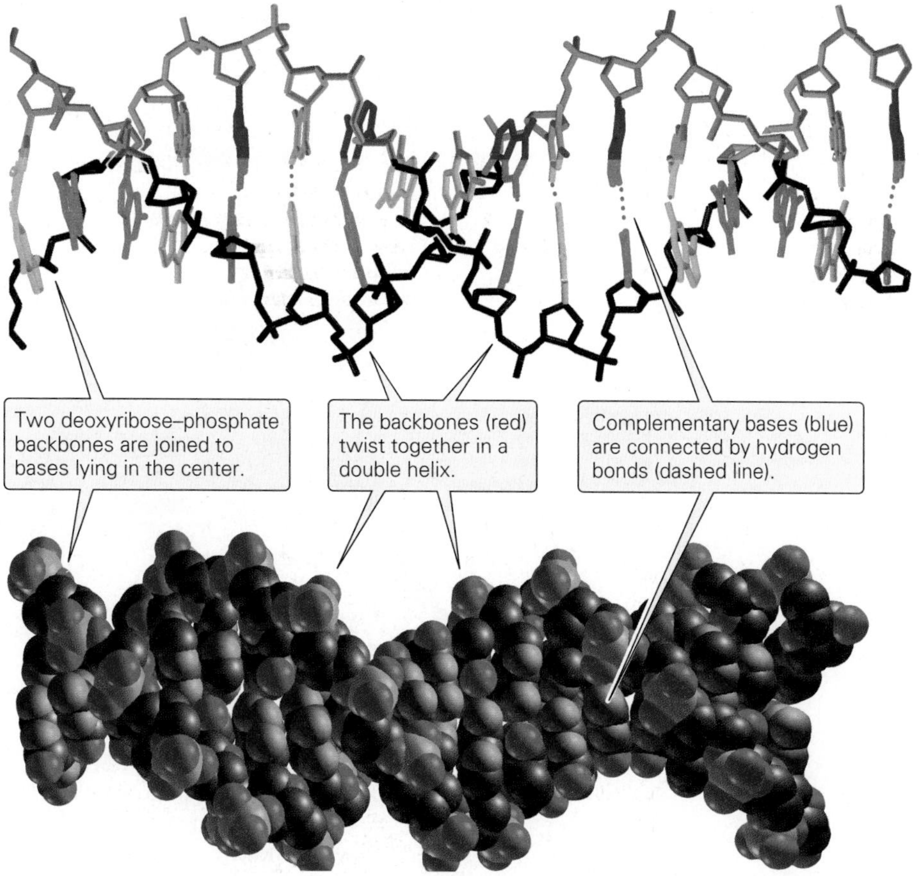

Two deoxyribose–phosphate backbones are joined to bases lying in the center.

The backbones (red) twist together in a double helix.

Complementary bases (blue) are connected by hydrogen bonds (dashed line).

**Figure 22.20** Backbone and space-filling molecular models of DNA.

When cells divide, each daughter cell must contain the complete genetic information of the original cell. To accomplish this, each DNA molecule must duplicate itself. To do so, the hydrogen bonds holding the double helix together are broken, yielding two separate complementary single strands. Each strand of the original DNA molecule then forms new hydrogen bonds to new nucleotide partners, adenine to thymine, guanine to cytosine (Figure 22.21). The sugar–phosphate bonds then form to complete the new strands. The result is two DNA molecules; each new DNA molecule is formed from one strand of the original DNA and one newly constructed complementary strand. This process is called **replication** and occurs during cell division. Replication is the molecular basis of heredity.

Hydrogen bonds are weaker than covalent bonds, and thus DNA can be unwound without damage.

You have seen that proteins, as enzymes, control the chemical reactions in living systems, but what controls the synthesis of these proteins? The DNA molecules in the nucleus of a cell contain plans for making protein molecules. Each segment of DNA that contains information to make a given protein is called a **gene.** When a cell needs to make a specific protein molecule, the appropriate DNA molecule makes a "photocopy" of the protein plan in the form of a complementary **messenger RNA** molecule. This process is called **transcription.**

The messenger RNA leaves the cell nucleus and travels to a **ribosome** (a structure outside the nucleus where protein synthesis occurs) to pick up small molecules of **transfer RNA.** Each transfer RNA molecule carries with it a specific amino acid, and to maximize interchain hydrogen bonding, the transfer RNA bonds to a complementary site on the messenger RNA. As molecules of transfer RNA line up along the messenger RNA, adjacent amino acids carried by the transfer RNA form amide bonds with each other. As these amide bonds form, each amino acid is separated from its transfer RNA, and each transfer RNA is then separated from the larger messenger RNA. When the final transfer RNA is separated from the messenger RNA, a molecule of protein remains. This process is called **translation.**

We have omitted the role of ribosomal RNA for simplicity.

Figure 22.21 DNA replication. As the DNA double helix (*orange*) unwinds, each strand serves as a template on which complementary subunits (*green*) are assembled in the cell. A = adenine, T = thymine, G = guanine, C = cytosine.

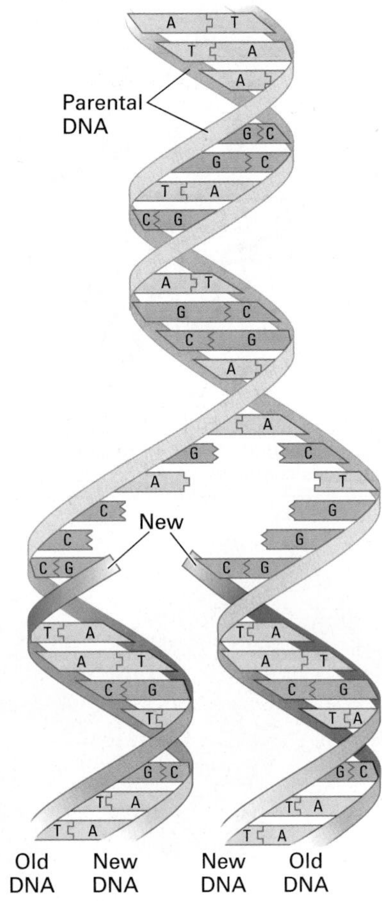


## Thinking About Your Thinking

### Mental Models

**Figures 22.17 through 22.21 are designed to help you form a mental model of the DNA molecule. Each figure provides a different level of detail at which you need to understand the structure and function of DNA. The figures begin with a "close-up" (Fig. 22.17) and then "zoom out" until you see the DNA molecule from a distance in Figure 22.20.**

**Figure 22.17 has you consider the bases in DNA at the level of atoms and the bonds that connect them. Figure 22.18 then shows the sugar–phosphate backbone to which these bases are bonded. The backbone is shown as atoms and bonds in the left, and then the fundamental elements of DNA structure are shown as a schematic drawing on the right. Compare the two representations carefully. Figure 22.19 emphasizes the geometry that allows hydrogen bonding between complementary base pairs.**

**In the top part of Figure 22.20, the sugar–phosphate backbone is in red, and the bases are shown in blue. The bottom part shows you a space-filling model. Here, your perspective is of a segment of the whole molecule. Figure 22.21 is a schematic drawing similar to Figure 22.20 (*top*), but its purpose is to show you how the structure of DNA leads to the mechanism for its duplication.**

**Using different models to represent the same concept at different levels of detail is a common way of thinking in science. These models of DNA are an application of this way of thinking.**

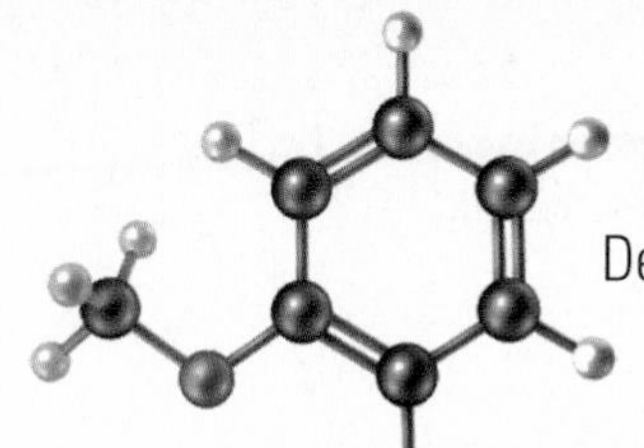

# Everyday Chemistry

## Designer Genes

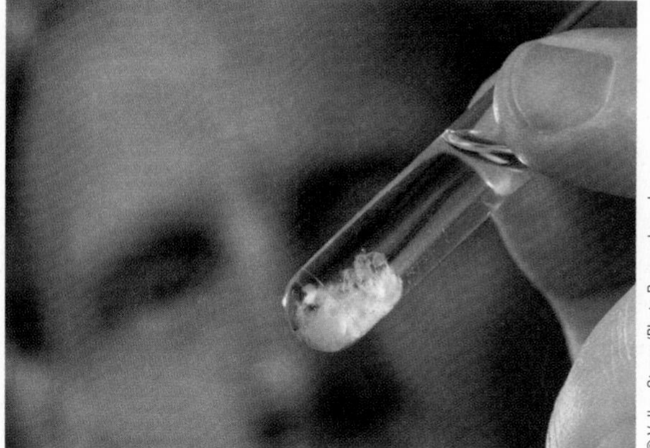

© Volker Steger/Photo Researchers, Inc.

A scientist precipitates solid DNA from solution.

Since the earliest days of human civilization, it has been well known that children tend to resemble their parents. How are these traits passed from generation to generation? Gregor Mendel (1822–1884) investigated this question with his now-famous pea plant experiments. These experiments provided the beginnings of the field of study known as classical genetics. Mendel's "atom of inheritance," now called a gene, was the hypothetical construct he used to explain how physical characteristics were passed through families. Mendel learned that two genes were associated with each trait, one inherited from each parent. Mendel also found that not all genes are created equally. The trait associated with a dominant gene was expressed when paired with a recessive gene.

We now understand genes on the particulate level. A gene is a segment along a DNA molecule (Section 22.5). The information in genes is coded by the four-letter DNA alphabet. A group of three letters specifies a certain amino acid. For example, the DNA sequence (T-A-C) codes for the amino acid tyrosine. Amino acids are assembled to form peptides and eventually proteins (Section 22.1). Each gene codes for one specific protein.

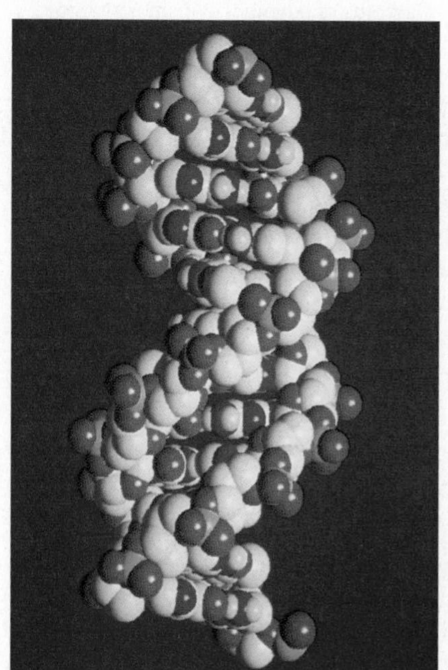

© Cengage Learning/Charles D. Winters

A space-filling molecular model of a segment of DNA.

The genetic code is universal among all life on the earth. All living organisms, including the bacteria that live in your intestines, the plants that produce the air you breathe, your pet cat, and you, share the same genetic code.

The universal genetic code serves as a set of instructions that any cell can read. Take a gene coding for a certain protein out of a human cell and place it in a bacterial cell, and the bacterium faithfully produces that protein. This is the basis of the process known as genetic engineering. Of course, it is not a simple process to move genes from one cell to another, but our knowledge of how to accomplish this continues to grow; it is an active area of research around the world. Genetically engineered products are continually becoming more common. The insulin used by people with diabetes is produced by genetically engineered bacteria. You can probably find genetically engineered products at your local supermarket. Strawberries have been designed to be more resistant to frost, and tomatoes have been engineered to resist rot.

Human society has been practicing a form of genetic engineering for all of recorded history, but it has been called "selective breeding." Cows are bred to produce more milk, dogs are bred to have aesthetically pleasing features, and crops are bred to have better yields. The difference between then and now is that today we are learning to control genes at the particulate level.

Manipulating bacterial cells is not a particularly controversial subject. Changing the genes in human cells is another matter. In 2003, scientists completed the first major stage of a \$3 billion project with the goal of producing a complete map of all human genes. This investigation is known as the Human Genome Project. A genome is a description of all the genes in an organism. The genome sequence has been completed, but the exact number of genes encoded by the human genome is still unknown.

It will be a great day when the gene that codes for a genetic disease such as sickle cell anemia can be identified and altered to prevent the disease. But if we have the knowledge about how to change this gene, we will also have the ability to manipulate other genes. Along with an understanding of the human genome come serious ethical issues. If a genetic predisposition for depression is identified in a fetus, what steps, if any, should parents take? Should they be allowed to have their babies genetically engineered?

Questions such as these about genetic engineering will be issues for society in your lifetime. We encourage you to learn all you can about science, society, ethics, and morals, because such decisions will have a profound effect on the future of the world.

The diagram below is the central tenet of molecular biology, as expressed by Francis Crick in 1958. It still works, and it's a good summary for this section.

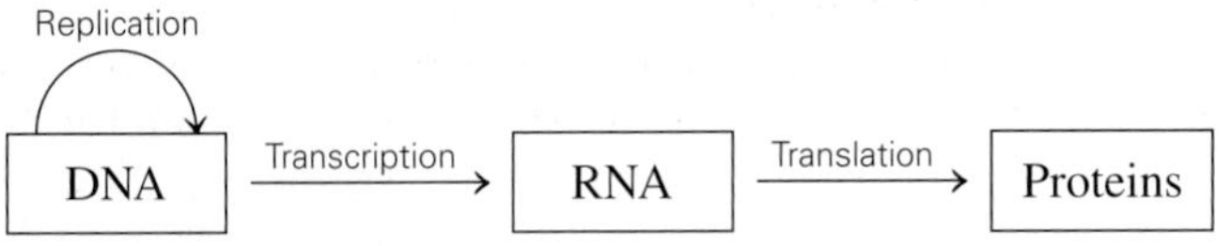


## √ Target Check 22.6

a) Describe the three components of a nucleotide.

b) How does the structure of an RNA nucleotide differ from that of a DNA nucleotide?

# Chapter 22 in Review

Most of the key terms and concepts and many others appear in the Glossary. Use your Glossary regularly.

### 22.1 Amino Acids and Proteins

Goal 1 Given a Lewis diagram of a polypeptide or information from which it may be written, identify the C-terminal amino acid and the N-terminal amino acid.

Goal 2 Given a table of Lewis diagrams of amino acids and their corresponding three-letter and one-letter abbreviations, draw the Lewis diagram of a polypeptide from its abbreviation.

Goal 3 Explain the meaning of the terms *primary, secondary, tertiary,* and *quaternary structure* as they apply to proteins.

Goal 4 Describe how hydrogen bonding results in (a) $\alpha$-helix and (b) $\beta$-pleated-sheet secondary protein structures.

**Key Terms and Concepts: $\alpha$-helix, amino acid, $\beta$-pleated-sheet, biochemistry, C-terminal, disulfide linkages, local conformation, macromolecule, N-terminal, peptide linkage, polypeptide, primary structure, protein, quaternary structure, secondary structure, silk fibroin, tertiary structure**

### 22.2 Enzymes

Goal 5 Define the following terms as they apply to enzymes: *substrate, active site, inhibitor.*

Goal 6 Use the induced fit model to explain enzyme activity.

**Key Terms and Concepts: Active site, enzyme, induced fit model, inhibitor, substrate**

### 22.3 Carbohydrates

Goal 7 Given a Lewis diagram of a monosaccharide in its open-chain form, determine whether the sugar is an aldose or a ketose.

Goal 8 Distinguish among monosaccharides, disaccharides, and polysaccharides.

Goal 9 Given the Lewis diagrams of two monosaccharides and a description of the bond linking the molecules, draw the Lewis diagram of the resulting disaccharide.

**Key Terms and Concepts: Adenosine triphosphate (ATP), aldose, amylopectin, amylose, carbohydrate, cellulose, disaccharide, fructose, glucose, glycogen, ketose, lactose, monosaccharide, polysaccharide, sucrose**

### 22.4 Lipids

Goal 10 State (a) the defining characteristics and (b) the three major subclassifications of lipids.

Goal 11 Identify the physical property that distinguishes fats from oils.

Goal 12 Identify the structural feature common to all steroid molecules.

**Key Terms and Concepts: Beeswax, carnauba wax, fatty acid, lanolin, lecithin, lipid, phospholipid, steroid, wax**

### 22.5 Nucleic Acids

Goal 13 Describe the biological roles of DNA and RNA.

Goal 14 Describe the components of a nucleotide.

Goal 15 Draw Lewis diagrams of adenine, cytosine, guanine, thymine, and uracil.

Goal 16 Determine whether any two DNA or RNA nitrogen bases are complementary.

Goal 17 Describe the process by which a protein molecule is formed.

**Key Terms and Concepts: Deoxyribonucleic acid (DNA), gene, messenger RNA, nucleotide, replication, ribonucleic acid (RNA), ribosome, transcription, transfer RNA, translation**

# Study Hints and Pitfalls to Avoid

This chapter is organized around the four major classes of biological molecules. Recognize what proteins, carbohydrates, and nucleic acids have in common: They are assembled from simple monomer units. Proteins are assembled from amino acids, carbohydrates are assembled from monosaccharides, and nucleic acids are assembled from nucleotides. *Lipid* is a catchall classification that includes fats, oils, phospholipids, waxes, steroids, and some other molecules. Organize your study into these four categories.

Amino acids are the monomers of proteins. Learn how amino acids combine to make peptide chains. Learn the differences among primary, secondary, tertiary, and quaternary protein structures. Enzymes are proteins that catalyze biochemical reactions. The induced fit model will help you understand how an enzyme works.

Monosaccharides are carbohydrates, as are the complex carbohydrates that can be built from them. Take some time to understand the three-dimensional Lewis diagrams used to represent carbohydrates. Learn how to distinguish between an $\alpha$ and a $\beta$ linkage in a polysaccharide.

Nucleotides are the monomers of nucleic acids. C and G (cytosine and guanine) are similar-looking letters; they form a complementary base pair. A and T (adenine and thymine) are letters near the beginning and end of the alphabet; they form a complementary base pair. Uracil (U), near the end of the alphabet, substitutes for thymine, also near the end of the alphabet, in RNA. The summary at the end of Section 22.5 is a good place to start your study of how the information encoded in DNA is used to assemble proteins.

## Concept-Linking Exercises

*Write a brief description of the relationships among each of the following groups of terms or phrases. Answers to the Concept-Linking Exercises are given after answers to the Target Checks in Appendix III.*

1. Primary, secondary, tertiary, quaternary protein structure
2. $\alpha$-helix, $\beta$-pleated sheet, hydrogen bonding
3. Enzyme, enzyme substrate, active site, induced fit model
4. Monosaccharide, disaccharide, polysaccharide, carbohydrate
5. Fat, oil, phospholipid, glycerol, triacylglycerol
6. DNA, RNA, proteins, translation, transcription, replication

## Small-Group Discussion Questions

*Small-Group Discussion Questions are for group work, either in class or under the guidance of a leader during a discussion section.*

1. Does the amino acid glycine have isomers? If so, draw the structure of each. Does alanine have isomers? If so, draw the structure of each isomer. Draw the structure of all isomers (if applicable) of glycine and alanine using wedge-and-dash diagrams. Without drawing the isomers, determine whether or not the other 18 amino acids commonly found in proteins have isomers. Use structural diagrams to illustrate the reaction of alanine and glycine in all possible combinations to form a dipeptide.
2. What is an enzyme and how does it work? How many examples can you identify of products in your home that contain enzymes?
3. Expand the carbohydrate classification scheme illustrated in Figure 22.10 to include as many examples as possible in each category. Draw the structures and/or describe the characteristics of as many of the examples as you can.
4. Define each of the following and draw structural diagrams of two examples from each category: fat, oil, phospholipid, wax, steroid.
5. In their classic 1953 article published in *Nature* magazine that first described the structure of DNA, Watson and Crick stated, "It has not escaped our notice that the specific pairing we have postulated immediately suggests a possible copying mechanism for the genetic material." What pairing, what copying mechanism, and what genetic material were they writing of? Explain in as much detail as you can.

## Questions, Exercises, and Problems

*Blue-numbered questions are answered in Appendix III.* ■ *denotes problems assignable in OWL. In the first section of Questions, Exercises, and Problems, similar exercises and problems are paired in consecutive odd-even number combinations.*

### Section 22.1: Amino Acids and Proteins

1. Draw the Lewis diagram that illustrates the general structure of an amino acid.
2. How are amino acids subdivided by the nature of their R groups?
3. Pick the amino acids from Table 22.1 that have an aromatic R group.
4. How does the R group in proline differ from the other amino acid R groups in Table 22.1?
5. ■ Give in words the name of the tripeptides abbreviated V-T-I and I-V-T. You may use Table 22.1 if you wish. Give the name of the C-terminal amino acid in each tripeptide.
6. Give in words the name of the tetrapeptides abbreviated S-F-G-Y and Y-F-G-S. You may use Table 22.1 if you wish. Give the name of the N-terminal amino acid in each tetrapeptide.
7. Write the Lewis diagram of the tripeptide abbreviated C-G-F. You may use Table 22.1 if you wish.
8. Write the Lewis diagram of the tripeptide abbreviated A-C-F. You may use Table 22.1 if you wish.
9. How does tertiary protein structure differ from quaternary protein structure?
10. How does secondary protein structure differ from tertiary protein structure?
11. Describe in words the hydrogen bonding that occurs in a protein having an $\alpha$-helix secondary structure.
12. Describe in words the hydrogen bonding that occurs in a protein having a $\beta$-pleated sheet secondary structure between adjacent polypeptide chains.

### Section 22.2: Enzymes

13. To what class of biological macromolecules do enzymes belong?
14. Chemists have isolated, purified, studied, and reported on several thousand different enzymes. Why does a living system need so many enzymes?

15. What is an enzyme substrate?

16. What is enzyme specificity?

17. How does an enzyme affect the activation energy (Section 18.3) of a reaction?

18. How does an enzyme differ from an inorganic laboratory catalyst such as $MnO_2$ in the following reaction?

$$2\ KClO_3(s) \xrightarrow{MnO_2,\ heat} 2\ KCl(s) + 3\ O_2(g)$$

19. Many enzymes work best at temperatures near 35°C. Normal body temperature is 37.0°C. What happens to the rates of enzyme-catalyzed reactions when you run a fever?

20. Many enzymes undergo *feedback inhibition*, in which a high concentration of enzyme reaction product slows down or stops enzyme activity. Compare the action of a furnace thermostat to enzyme feedback inhibition.

### Section 22.3: Carbohydrates

21. Give an example of an aldose sugar.

22. Give an example of a ketose sugar.

23. Examine the Lewis structures of $\alpha$-glucose and $\beta$-glucose. What is the difference between these two glucose isomers?

24. What are the three general classes of saccharides? How do they differ?

25. ■ To which saccharide class do the following belong? sucrose, glycogen, fructose.

26. To which saccharide class do the following belong? cellulose, ribose, lactose.

27. Name the simple sugars in lactose.

28. Name the simple sugars in sucrose.

29. Benedict's test is a classic test for some sugars. In this test, an aldehyde is oxidized to a carboxylic acid, and a color change occurs. Which mono- and disaccharide that you have studied would give a positive Benedict's test?

30. A Tollens' test oxidizes aldehydes to carboxylic acids, but does not react with ketones. Which mono- and disaccharides discussed in this chapter would give a negative Tollens' test?

31. Maltose is a disaccharide made from two glucose molecules held together by an $\alpha$-1,4 bond. Draw the structure of maltose.

32. Cellobiose is a plant disaccharide made from two glucose molecules held together by a $\beta$-1,4 bond. Draw the structure of cellobiose.

33. We have enzymes that can digest starch and turn it into energy, but we do not have enzymes that can digest cellulose. Study the structures of starch and cellulose to see how these macromolecules differ. Which bonds can our enzymes break? Which bonds can our enzymes not break?

34. Why does glycogen contain $\alpha$-1,4 bonds rather than $\beta$-1,4 bonds?

### Section 22.4: Lipids

35. What physical property do the three classes of lipids share?

36. What are the three classes of lipids?

37. Are oils usually obtained from animal sources or plant sources?

38. What physical property differentiates fats and oils?

39. Use the letters A, B, and C to stand for three different fatty acids. Draw out all possible triacylglycerols you can make from these acids and one glycerol molecule.

40. Explain why the human body contains only fatty acids with an even number of carbon atoms. (*Hint:* Fatty acids are built from acetic acid molecules.)

41. Use the letters A and B to stand for two different fatty acids in a phospholipid. Draw out all possible phospholipids you can make from these two fatty acids.

42. Using the information provided in the textbook, draw the condensed structure of the nine possible molecules that are classified as beeswax.

### Section 22.5: Nucleic Acids

43. In words, briefly explain the function of DNA and of RNA.

44. Name the two types of nucleic acid polymers present in cells.

45. Draw the Lewis diagram for uracil.

46. Draw the Lewis diagram for thymine.

47. Draw the Lewis diagrams for adenine and thymine.

48. Draw the Lewis diagrams for guanine and cytosine.

49. Draw the Lewis diagram for the sugar ribose. How does this sugar differ from the sugar deoxyribose?

50. Draw the Lewis diagram for the sugar deoxyribose.

51. Draw the nucleoside adenosine, which is an adenine–ribose combination.

52. Draw the nucleoside deoxyadenosine, which is an adenine–deoxyribose combination.

53. Draw the Lewis diagram for the DNA fragment that is *complementary* to the guanine-thymine-adenine DNA fragment.

54. Single-letter codes are often used to stand for the base in a nucleic acid. Draw the Lewis diagram for a DNA fragment having the bases guanine-thymine-adenine, abbreviated G-T-A.

55. ■ Although RNA is single-stranded, the strand sometimes folds back on itself to give a complementary portion. What would be the complementary portion of the RNA fragment having the bases uracil-cytosine-guanine?

56. Draw the Lewis diagram for an RNA fragment having the bases uracil-cytosine-guanine, abbreviated U-C-G.

57. Describe in words the role in protein synthesis of transfer RNA, tRNA.

58. Describe in words the role in protein synthesis of messenger RNA, mRNA.

### General Questions

59. Distinguish precisely, and in scientific terms, the differences between items in each of the following pairs.
    a) Secondary protein structure, tertiary protein structure

b) $\alpha$-helix, $\beta$-pleated sheet

c) Reversible enzyme inhibitors, irreversible enzyme inhibitors

d) $\alpha$-glucose, $\beta$-glucose

e) $\alpha$-1,4 and $\beta$-1,4 linkages

f) Fats, oils

g) Saturated fatty acids, unsaturated fatty acids

h) Triacylglycerols, phospholipids

i) Triacylglycerols, waxes

j) Waxes, steroids

k) Nucleoside, nucleotide

l) Thymine, uracil

m) $\alpha$-helix, double helix

60. Classify each of the following statements as true or false:

a) The most common primary protein structures are the $\alpha$-helix and the $\beta$-pleated sheet.

b) Sulfur–sulfur bonds called disulfide linkages are important in the tertiary protein structure of many proteins.

c) An enzyme substrate is the product of an enzyme-catalyzed reaction.

d) The catalytic properties of an enzyme exist at its active site.

e) Carbohydrates are hydrates of carbon; that is, compounds that consist of carbon and water molecules.

f) Glucose is also called blood sugar.

g) Sucrose is a glucose–fructose combination.

h) Fats and oils are distinguished by their state of matter at room temperature.

i) Steroids have three cyclopentane rings and one cyclohexane ring fused.

j) Nucleic acid monomers are called nucleotides.

k) DNA is a nucleotide (phosphate–sugar–base) polymer.

61. ■ Biochemists use 120 g/mol as the "average molar mass" of a single amino acid in a protein. Hemoglobin is a small protein that transports oxygen from the lungs to the capillaries. The molar mass of hemoglobin is about 64,500 g/mol. Approximately how many amino acids make up hemoglobin?

62. The tobacco mosaic virus is a small virus that has been crystallized in pure form. One complete virus has a molar mass of about 40,000,000 g/mol, of which 38,000,000 g is protein. Use the data in Question 61 to determine the approximate number of amino acids in the proteins of a tobacco mosaic virus.

63. Which of the following biological molecules are polymers: cellulose, proteins, DNA, starch, RNA?

64. What are the monomer units for the polymers listed in Question 63?

65. With what element does protein supply us that carbohydrates or fats or oils do not?

66. What element is found in DNA and RNA, but not in proteins?

#### More Challenging Problems

67. The structure of nylon-66 is given in Section 21.16. Use this structure to explain why nylon fabric is used in waterproof windbreakers, while cotton, a cellulose material, is used where moisture must be absorbed.

68. Rayon is an ester made by treating cotton with acetic acid. Use the cellulose structure given in Section 22.3 to draw a Lewis structure for a short rayon "molecule." (*Hint:* An alcohol-acid esterification reaction occurs randomly.)

69. ■ The base content of a sample of pure mammalian DNA was analyzed to be 21% guanine. In this DNA sample, what is the percentage of cytosine? What is the percentage of adenine? What is the percentage of thymine? Explain.

70. RNA, unlike DNA, is a single-stranded macromolecule. Would you expect the percentage guanine in RNA to equal the percentage cytosine? Why or why not?

# Appendix I

## CHEMICAL CALCULATIONS

A beginning student in chemistry is assumed to have developed calculation skills in earlier mathematics classes. Often chemistry is the first occasion for these skills to be put to the test of practical application. Experience shows that many students who learned these skills, but have not used them regularly, can profit from a review of basic concepts. Others can benefit from a handy reference to the calculation techniques used in chemistry. This section of the Appendix is intended to meet these needs.

## Part A The Hand Calculator

Every chemistry student uses a calculator to solve chemistry problems. A suitable calculator can (1) add, subtract, multiply, and divide; (2) perform these operations in exponential notation; (3) work with logarithms; and (4) raise any base to a power. Calculators that can perform these operations usually have other capabilities, too, such as finding squares and square roots, carrying out trigonometric functions, and offering shortcuts for pi and percentage, enclosures, statistical features, and different levels of storage and recall.

Most calculators operate with one of three logic systems, each with its own order of operations. One is called the Algebraic Operating System (AOS), another is Reverse Polish Notation (RPN), and the third is called Direct Algebraic Logic (DAL). In the examples that follow we will give *general* keyboard sequences for all three systems as they are performed on calculators popular with students. Different brands may vary in details, particularly when some keys are used for more than one function. Some calculators offer an option on the number of digits to be displayed after the decimal point. With or without such an option, the number varies on different calculators. Accordingly, answers in this book may differ slightly from yours. Please consult the instruction book or Web site that accompanies your calculator for specific directions on these or other variations that may appear.

(You may wonder why you should not simply use your instruction book rather than the suggestions that follow. For complete mastery of your calculator, you should do just that. If your present purpose is to learn how to use the calculator for chemistry, these instructions will be much easier. They also include practical suggestions that do not appear in a formal instruction book.)

One precautionary note before we begin: *Never use a calculator as a substitute for thinking.* If a problem is simple and can be solved mentally, do it in your head. You will make fewer mistakes. If you use your calculator, *think* your way through each problem and estimate the answer mentally. Suggestions on approximating answers are given in Part D of this Appendix. If the calculator answer appears reasonable, round it off properly and write it down. Then run through the calculation again to be sure you haven't made a keyboard error. Your calculator is an obedient and faithful servant that will do exactly what you tell it to do, but it is not responsible for the mistakes you make in your instructions.

We just suggested that you round off your answer properly before recording it. The reason for this is that calculator answers to many problems are limited in length only by the display. For example, 273 ÷ 45.6 = 5.9868421 on one calculator. Some calculators can show more numbers, and therefore do. Usually, only the first three or four digits have meaning, and the others should be discarded. Procedures for deciding how many digits to write are given in Section 3.5 on significant figures.

### Reciprocals, Square Roots, Squares, and Logarithms

Finding the reciprocal, square root, square, or logarithm of a number is called a one-number function because only one number must be keyed into the calculator. In this case we are interested in finding $1/x$, $\sqrt{x}$, $x^2$, and log x.

**EXAMPLE:** Find 1/12.34, $\sqrt{12.34}$, $12.34^2$, and log 12.34.

**SOLUTION:**

**AOS or RPN Logic**

| Problem | Press | Display |
|---|---|---|
| 1/12.34 | 12.34 | 12.34 |
| | 1/x | 0.0810373 |
| $\sqrt{12.34}$ | 12.34 | 12.34 |
| | $\sqrt{\ }$ | 3.5128336 |
| $12.34^2$ | 12.34 | 12.34 |
| | $x^2$ | 152.2756 |
| log 12.34 | 12.34 | 12.34 |
| | log | 1.0913152 |

**DAL**

| Problem | Press | Display |
|---|---|---|
| 1/12.34 | 12.34 | 12.34 |
| | $x^{-1}$ | 0.081037277 |
| $\sqrt{12.34}$ | $\sqrt{\ }$ | $\sqrt{\ }$ |
| | 12.34 | $\sqrt{\ }$ 12.34 |
| | = | 3.512833614 |
| $12.34^2$ | 12.34 | 12.34 |
| | $x^2$ | 152.2756 |
| log 12.34 | log | log |
| | 12.34 | log 12.34 |
| | = | 1.09131516 |

## Antilogarithms, $10^x$, and $y^x$

If x is the base-10 logarithm of a number, N, then N = antilog x = $10^x$. Some calculators have a $10^x$ key that makes finding an antilogarithm a one-number function. Other calculators use an inverse function key, sometimes marked INV, to reverse the logarithm function. Again, finding an antilogarithm is a one-number function, but two function keys are used.

**EXAMPLE:** Find the antilogarithm of 3.19.

**SOLUTION:**

**AOS or RPN Logic**

| Press | Display | Press | Display |
|---|---|---|---|
| 3.19 | 3.19 | 3.19 | 3.19 |
| $10^x$ | 1548.8166 | INV | 3.19 |
| | | log | 1548.8166 |

**DAL**

| Press | Display |
|---|---|
| $10^x$ | $10^x$ |
| 3.19 | $10^x$ 3.19 |
| = | 1548.816619 |

The $y^x$ key can be used to raise any base, y, to any power, x. The procedure differs in the three operating systems, as the following example shows.

**EXAMPLE:** Calculate $8.25^{0.413}$

**SOLUTION:**

**AOS or RPN Logic**

| Press | Display | Press | Display |
|---|---|---|---|
| 8.25 | 8.25 | 8.25 | 8.25 |
| $y^x$ | 8.25 | ENTER | 8.25 |
| .413 | 0.413 | .413 | 0.413 |
| = | 2.3905371 | $y^x$ | 2.3905371 |

**DAL**

| Press | Display |
|---|---|
| 8.25 | 8.25 |
| $y^x$ | $y^x$ |
| .413 | $y^x$ 0.413 |
| = | 2.390537059 |

Notice that, even though we always write a decimal fraction less than 1 with a zero before the decimal point, it is not necessary to enter such zeros into a calculator. The zeros are included in the display.

The $y^x$ key can also be used to find an antilogarithm. In that case y = 10 and x is the given logarithm, the exponent to which 10 is to be raised.

## Addition, Subtraction, Multiplication, and Division

For ordinary arithmetic operations the procedures are as follows:

### AOS LOGIC

The procedure for a common arithmetic operation is identical to the arithmetic equation for the same calculation. If you wish to add X to Y, the equation is X + Y = . The calculator procedure is:

1. Key in X.
2. Press the function key (+ for addition).
3. Key in Y.
4. Press =.

The display will show the calculated result.

**EXAMPLE:** Solve 12.34 + 0.0567 = ?

**SOLUTION:**

| Press | Display |
|---|---|
| 12.34 | 12.34 |
| + | 12.34 |
| .0567 | 0.0567 |
| = | 12.3967 |

The numbers in the PRESS column are, in order, Steps 1, 2, 3, and 4 of the procedure. In Step 2, the function key is − for subtraction, × for multiplication, and ÷ for division.

### RPN LOGIC

The procedure for a common arithmetic operation is to key in *both* numbers and then tell the calculator what to do with them. If you wish to add X to Y, you enter X, key in Y, and instruct the calculator to add. The procedure is:

1. Key in X.
2. Press ENTER.
3. Key in Y.
4. Press the function key (+ for addition).

The display will show the calculated result.

**EXAMPLE:** Solve 12.34 + 0.0567 = ?

**SOLUTION:**

| Press | Display |
|---|---|
| 12.34 | 12.34 |
| ENTER | 12.34 |
| .0567 | 0.0567 |
| + | 12.3967 |

The numbers in the PRESS column are, in order, Steps 1, 2, 3, and 4 of the procedure. In Step 4, the function key is − for subtraction, × for multiplication, and ÷ for division.

### DAL

The procedure for a common arithmetic operation is identical to the arithmetic equation for the same calculation. If you wish to add X to Y, the equation is X + Y = . The calculator procedure is:

1. Key in X.
2. Press the function key (+ for addition).
3. Key in Y.
4. Press =.

The display will show the calculated result.

EXAMPLE: Solve 12.34 + 0.0567 = ?

SOLUTION:

| Press | Display |
|---|---|
| 12.34 | 12.34 |
| + | + |
| .0567 | 0.0567 |
| = | 12.3967 |

The numbers in the PRESS column are, in order, Steps 1, 2, 3, and 4 of the procedure. In Step 2, the function key is − for subtraction, × for multiplication, and ÷ for division.

You may wish to confirm the following results on your calculator:

$$12.34 - 0.0567 = 12.2833$$

$$12.34 \times 0.0567 = 0.699678$$

$$12.34 \div 0.0567 = 217.63668$$

## Chain Calculations

A "chain calculation" is a series of two or more operations performed on three or more numbers. To the calculator the sequence is a series of two-number operations in which the first number is always the result of all calculations completed to that point. For example, in X + Y − Z the calculator first finds X + Y = A. The quantity A is already in and displayed by the calculator. All that needs to be done is to subtract Z from it. In the following example, we deliberately begin with a negative number to illustrate the way such a number is introduced to the calculator: −2.45 + 18.7 + 0.309 − 24.6 = ?

| AOS Logic | | RPN Logic | |
|---|---|---|---|
| Press | Display | Press | Display |
| 2.45 | 2.45 | 2.45 | 2.45 |
| +/− | −2.45 | CHS | −2.45 |
| + | −2.45 | ENTER | −2.45 |
| 18.7 | 18.7 | 18.7 | 18.7 |
| + | 16.25 | + | 16.25 |
| .309 | 0.309 | .309 | 0.309 |
| − | 16.559 | + | 16.559 |
| 24.6 | 24.6 | 24.6 | 24.6 |
| = | −8.041 | − | −8.041 |

| DAL | |
|---|---|
| Press | Display |
| +/− | − |
| 2.45 | −2.45 |
| + | + |
| 18.7 | +18.7 |
| + | + |
| .309 | 0.309 |
| − | − |
| 24.6 | −24.6 |
| = | −8.041 |

Notice that it is not necessary to press the = key after each step *in a chain calculation involving only addition and/or subtraction.*

Combinations of multiplication and division are handled the same way. To solve 9.87 × 0.0654 ÷ 3.21:

| AOS Logic | | RPN Logic | |
|---|---|---|---|
| Press | Display | Press | Display |
| 9.87 | 9.87 | 9.87 | 9.87 |
| × | 9.87 | ENTER | 9.87 |
| 0.0654 | 0.0654 | .0654 | 0.0654 |
| × | 0.645498 | × | 0.64549800 |
| 3.21 | 3.21 | 3.21 | 3.21 |
| = | 0.20108972 | = | 0.20108972 |

| DAL | |
|---|---|
| Press | Display |
| 9.87 | 9.87 |
| × | * |
| .0654 | * 0.0654 |
| ÷ | / |
| 3.21 | / 3.21 |
| = | 0.201089719 |

Notice that it is not necessary to press the = key after each step *in a chain calculation involving only multiplication and/or division.*

Combination multiplication/division problems similar to the preceding one usually appear in the form of fractions in which all multipliers are in the numerator and all divisors are in the denominator. Thus, 9.87 × 0.0654 ÷ 3.21 is the same as $\frac{9.87 \times 0.0654}{3.21}$. In chemistry there are often several numerator factors and several denominator factors. A simple calculation you can easily complete in your head brings out some important facts about using calculators for chain calculations. Mentally, right now, calculate $\frac{9 \times 4}{2 \times 6} = ?$

There are several ways to get the answer. The most probable one, if you do it mentally, is to multiply 9 × 4 = 36 in the numerator, and then multiply 2 × 6 = 12 in the denominator. This changes the problem to 36 ÷ 12. Dividing 36 by 12 gives 3 for the answer. This perfectly correct approach is often followed by the beginning calculator user when faced with numbers that cannot be multiplied and divided mentally. It is not the best method, however. It is longer and there is greater probability of error than necessary.

In solving a problem such as $(9 \times 4) \div (2 \times 6)$, you can begin with any number and perform the required operations with other numbers in any order. Logically, you begin with one of the numerator factors. That gives you 12 different calculation sequences that yield the correct answer. They are

| | |
|---|---|
| $9 \times 4 \div 2 \div 6$ | $4 \times 9 \div 2 \div 6$ |
| $9 \times 4 \div 6 \div 2$ | $4 \times 9 \div 6 \div 2$ |
| $9 \div 2 \div 6 \times 4$ | $4 \div 2 \div 6 \times 9$ |
| $9 \div 2 \times 4 \div 6$ | $4 \div 2 \times 9 \div 6$ |
| $9 \div 6 \times 4 \div 2$ | $4 \div 6 \times 9 \div 2$ |
| $9 \div 6 \div 2 \times 4$ | $4 \div 6 \div 2 \times 9$ |

Practice a few of these sequences on your calculator to see how freely you may choose.

There is a common error to avoid in chain calculations. This is to interpret the above problem as $9 \times 4$ divided by $2 \times 6$, which is correct, but then punch it into the calculator as $9 \times 4 \div 2 \times 6$, which is not correct. The calculator interprets these instructions as $9 \times 4 = 36$; $36 \div 2 = 18$; $18 \times 6 = 108$. The last step should be $18 \div 6 = 3$, as in the first setup in the previous list. In chain calculations you must always *divide by each factor* in the denominator.

There are over 100 different sequences by which $\frac{7.83 \times 86.4 \times 291}{445 \times 807 \times 0.302}$ can be calculated. Practice some of them and see if you can duplicate the answer, 1.8152147.

You have seen that in multiplication and division you can take the factors in any order. This is possible in addition and subtraction too, provided that you keep each positive and negative sign with the number that follows it and treat the problem as an algebraic addition of signed numbers. When you mix addition/subtraction with multiplication/division, however, you must obey the rules that govern the order in which arithmetic operations are performed. Briefly, these rules are:

1. Simplify all expressions enclosed in parentheses.
2. Complete all multiplications and divisions.
3. Complete all additions and subtractions.

If your calculator is able to store and recall numbers, it can solve problems with a very complex order of operations. In this book you will find no such problems, but only those that require the simplest application of the first rule. Our comments will be limited to that application, and we will not use the storage capacity of your calculator, as the instruction book would probably recommend.

A typical calculation is

$$6.02 \times (22.1 - 48.6) \times 0.134$$

Recalling that factors in a multiplication problem may be taken in any order, you rearrange the numbers so the enclosed factor appears first:

$$(22.1 - 48.6) \times 6.02 \times 0.134$$

You may then perform the calculation in the order in which the numbers appear.

**AOS or RPN Logic**

| Press | Display | Press | Display |
|---|---|---|---|
| 22.1 | 22.1 | 22.1 | 22.1 |
| − | 22.1 | ENTER | 22.1 |
| 48.6 | 48.6 | 48.6 | 48.6 |
| = | −26.5 | − | −26.5 |
| × | −26.5 | 6.02 | 6.02 |
| 6.02 | 6.02 | × | −159.53 |
| × | −159.53 | .134 | 0.134 |
| .134 | 0.134 | × | −21.37702 |
| = | −21.37702 | | |

**DAL**

| Press | Display |
|---|---|
| 22.1 | 22.1 |
| − | − |
| 48.6 | −48.6 |
| = | −26.5 |
| × | * |
| 6.02 | * 6.02 |
| × | * |
| .134 | * 0.134 |
| = | −21.37702 |

Notice that in a chain calculation involving *both* addition/subtraction *and* multiplication/division *it is necessary to press the = key after each addition or subtraction sequence in the AOS logic system before you proceed to a multiplication/division.*

Sometimes a factor in parentheses appears in the denominator of a fraction, where it is not easily taken as the first factor to enter into the calculator. Again, such problems in this book are relatively simple. They may be solved by working the problem upside down and at the end using the 1/x key to turn it right side up. The process is demonstrated in calculating

$$\frac{13.3}{2.59\,(88.4 - 27.2)}$$

The procedure is to calculate $\frac{2.59\,(88.4 - 27.2)}{13.3}$ and find the reciprocal of the result.

**AOS or RPN Logic**

| Press | Display | Press | Display |
|---|---|---|---|
| 88.4 | 88.4 | 88.4 | 88.4 |
| − | 88.4 | ENTER | 88.4 |
| 27.2 | 27.2 | 27.2 | 27.2 |
| = | 61.2 | − | 61.2 |
| × | 61.2 | 2.59 | 2.59 |
| 2.59 | 2.59 | × | 158.508 |
| ÷ | 158.508 | 13.3 | 13.3 |
| 13.3 | 13.3 | ÷ | 11.9179 |
| = | 11.917895 | 1/x | 0.083907436 |
| 1/x | 0.083907436 | | |

**DAL**

| Press | Display |
|---|---|
| 88.4 | 88.4 |
| − | − |
| 27.2 | − 27.2 |
| = | 61.2 |
| × | * |
| 2.59 | * 2.59 |
| ÷ | / |
| 13.3 | / 13.3 |
| = | 11.91789474 |
| $x^{-1}$ | 0.083907436 |

## Exponential Notation

Modern calculators use exponential notation (Section 3.2) for very large or very small numbers. Ordinarily, if numbers are entered as decimal numbers, the answer appears as a decimal number. If the answer is too large or too small to be displayed, it "overflows" or "underflows" into exponential notation automatically. If you want the answer in exponential notation, even if the calculator can display it as a decimal number, you instruct the machine accordingly. The symbol on the key for this instruction varies with different calculators. EXP, EE, and EEX are common.

A number shown in exponential notation has a space in front of the last two digits, which are at the right side of the display. These last two digits are the exponent. Thus, $4.68 \times 10^{14}$ is displayed as 4.68 14; $2.39 \times 10^{6}$ is 2.39 06. If the exponent is negative, a minus sign is present; $4.68 \times 10^{-14}$ is 4.68 –14.

To key $4.68 \times 10^{14}$ and $4.68 \times 10^{-14}$ into a calculator, proceed as follows.

**AOS or RPN Logic**

| Press | Display | Press | Display |
|---|---|---|---|
| 4.68 | 4.68 | 4.68 | 4.68 |
| EE | 4.68 00 | EEX | 4.68 00 |
| 14 | 4.68 14 | 14 | 4.68 14 |

**DAL**

| Press | Display |
|---|---|
| 4.68 | 4.68 |
| Exp | 4.68 00 |
| 14 | 4.68 14 |

The calculator display now shows $4.68 \times 10^{14.}$ Use the next step only if you wish to change to a negative exponent, $4.68 \times 10^{-14.}$

**AOS Logic or DAL**

| Press | Display |
|---|---|
| +/− | 4.68 −14 |
| Function key: +, −, ×, or ÷ | 4.68 −14 |

**RPN Logic**

| Press | Display |
|---|---|
| CHS | 4.68 −14 |
| ENTER | 4.68 −14 |

The calculator is now ready for the next number to be keyed in, as in earlier examples.

# Part B Arithmetic and Algebra

We present here a brief review of arithmetic and algebra to the point that it is used or assumed in this text. Formal mathematical statement and development are avoided. The only purpose of this section is to refresh your memory in areas where it may be needed.

1) ADDITION. a + b. Example:

$$2 + 3 = 5$$

The result of an addition is a **sum.**

2) SUBTRACTION. a − b. Example:

$$5 - 3 = 2$$

Subtraction may be thought of as the addition of a negative number. In that sense, a − b = a + (–b). Example:

$$5 - 3 = 5 + (-3) = 2$$

The result of a subtraction is a **difference.**

3) MULTIPLICATION. a × b = ab = a · b = a(b) = (a)(b) = b × a = ba = b · a = b(a). The foregoing all mean that **factor** a is to be multiplied by factor b. Reversing the sequence of the factors, a × b = b × a, indicates that factors may be taken in any order when two or more are multiplied together. Examples:

$$2 \times 3 = 2 \cdot 3 = 2(3) = (2)(3) = 3 \times 2 = 3 \cdot 2 = 3(2) = 6$$

The result of a multiplication is a **product.**

**Grouping of Factors** (a)(b)(c) = (ab)(c) = (a)(bc). Factors may be grouped in any way in multiplication. Example:

$$(2)(3)(4) = (2 \times 3)(4) = (2)(3 \times 4) = 24$$

**Multiplication by 1** $n \times 1 = n$. If any number is multiplied by 1, the product is the original number. Examples:

$$6 \times 1 = 6;\ 3.25 \times 1 = 3.25$$

**Multiplication of Fractions**

$$\frac{a}{b} \times \frac{c}{d} \times \frac{e}{f} = \frac{ace}{bdf}$$

If two or more fractions are to be multiplied, the product is equal to the product of the numerators divided by the product of the denominators. Example:

$$4 \times \frac{9}{2} \times \frac{1}{6} = \frac{4}{1} \times \frac{9}{2} \times \frac{1}{6} = \frac{4 \times 9 \times 1}{1 \times 2 \times 6} = \frac{36}{12} = 3$$

4) DIVISION. $a \div b = a/b = \frac{a}{b}$. The foregoing all mean that a is to be divided by b. Example:

$$12 \div 4 = 12/4 = \frac{12}{4} = 3$$

The result of a division is a **quotient.**

**Special Case** If any number is divided by the same or an equal number, the quotient is equal to 1. Examples:

$$\frac{4}{4} = 1;\quad \frac{8-3}{4+1} = \frac{5}{5} = 1;\quad \frac{n}{n} = 1$$

**Division by 1** $\frac{n}{1} = n$. If any number is divided by 1, the quotient is the original number. Examples:

$$\frac{6}{1} = 6;\quad \frac{3.25}{1} = 3.25$$

From this it follows that any number may be expressed as a fraction having 1 as the denominator. Examples:

$$4 = \frac{4}{1};\quad 9.12 = \frac{9.12}{1};\quad m = \frac{m}{1}$$

5) RECIPROCALS. If n is any number, the reciprocal of n is $\frac{1}{n}$; if $\frac{a}{b}$ is any fraction, the reciprocal of $\frac{a}{b}$ is $\frac{b}{a}$. The first part of the foregoing sentence is actually a special case of the second part: If n is any number, it is equal to $\frac{n}{1}$. Its reciprocal is therefore $\frac{1}{n}$.

A reciprocal is sometimes referred to as the **inverse** (more specifically, the multiplicative inverse) of a number. This is because the product of any number multiplied by its reciprocal equals 1. Examples:

$$2 \times \frac{1}{2} = \frac{2}{2} = 1;\quad n \times \frac{1}{n} = \frac{n}{n} = 1$$

$$\frac{4}{3} \times \frac{3}{4} = \frac{12}{12} = 1;\quad \frac{m}{n} \times \frac{n}{m} = \frac{mn}{mn} = 1$$

Division may be regarded as multiplication by a reciprocal:

$$a \div b = \frac{a}{b} = a \times \frac{1}{b}$$

Example: $6 \div 2 = \frac{6}{2} = 6 \times \frac{1}{2} = 3$

$$a \div b/c = \frac{a}{b/c} = a \times \frac{c}{b}$$

Example: $6 \div \frac{2}{3} = \frac{6}{2/3} = 6 \times \frac{3}{2} = 9$

6) SUBSTITUTION. If $d = b + c$, then $a(b + c) = ad$. Any number or expression may be substituted for its equal in any other expression. Example:

$$7 = 3 + 4;\ \text{therefore, } 2(3 + 4) = 2 \times 7$$

7) "CANCELLATION." $\frac{ab}{ca} = \frac{ab}{ac} = \frac{b}{c}$. The process commonly called **cancellation** is actually a combination of grouping of factors (see 3), substitution (see 6) of 1 for a number divided by itself (see 4), and multiplication by 1 (see 3). Note the steps in the following examples:

$$\frac{xy}{yz} = \frac{yx}{yz} = \left(\frac{y}{y}\right)\left(\frac{x}{z}\right) = 1 \times \frac{x}{z} = \frac{x}{z}$$

$$\frac{24}{18} = \frac{6 \times 4}{6 \times 3} = \frac{6}{6} \times \frac{4}{3} = 1 \times \frac{4}{3} = \frac{4}{3}$$

Note that only *factors,* or *multipliers,* can be canceled. There is no cancellation in $\frac{a+b}{a+c}$.

8) ASSOCIATIVE PROPERTIES. An arithmetic operation is associative if the numbers can be grouped in any way.

**Addition** $a + (b - c) = (a + b) - c$. Addition, including subtraction, is associative. Example:

$$3 + 4 - 5 = (3 + 4) - 5 = 3 + (4 - 5) = 2$$

**Multiplication** $(a \times b)/c = a \times (b/c)$. Multiplication, including division (multiplication by an inverse), is associative. (See "Grouping of Factors" under MULTIPLICATION) Example:

$$4 \times 6/2 = (4 \times 6)/2 = 4 \times (6/2) = 12$$

9) EXPONENTIALS. An exponential has the form $B^p$, where B is the **base** and p is the **power** or **exponent**. An exponential indicates the number of times the base is used as a factor in multiplication. For example, $10^3$ means 10 is to be used as a factor 3 times:

$$10^3 = 10 \times 10 \times 10 = 1000$$

A negative exponent tells the number of times a base is used as a divisor. For example, $10^{-3}$ means 10 is used as a divisor 3 times:

$$10^{-3} = \frac{1}{10} \times \frac{1}{10} \times \frac{1}{10} = \frac{1}{10^3}$$

This also shows that an exponential may be moved in either direction between the numerator and denominator by changing the sign of the exponent:

$$\frac{1}{2^3} = 2^{-3}; \quad 3^4 = \frac{1}{3^{-4}}$$

**Multiplication of Exponentials Having the Same Base** $a^m \times a^n = a^{m+n}$. To multiply exponentials, add the exponents. Example:

$$10^3 \times 10^4 = 10^7$$

**Division of Exponentials Having the Same Base** $a^m \div a^n = \frac{a^m}{a^n} = a^{m-n}$. To divide exponentials, subtract the denominator exponent from the numerator exponent. Example:

$$10^7 \div 10^4 = \frac{10^7}{10^4} = (10^7)(10^{-4}) = 10^{7-4} = 10^3$$

**Zero Power** $a^0 = 1$. Any base raised to the zero power equals 1. The fraction $\frac{a^m}{a^m} = 1$ because the numerator is the same as the denominator. By division of exponentials, $\frac{a^m}{a^m} = a^{m-m} = a^0$.

**Raising a Product to a Power** $(ab)^n = a^n \times b^n$. When the product of two or more factors is raised to some power, each factor is raised to that power. Example:

$$(2 \times 5y)^3 = 2^3 \times 5^3 \times y^3 = 8 \times 125 \times y^3 = 1000y^3$$

**Raising a Fraction to a Power** $\left(\frac{a}{b}\right)^n = \frac{a^n}{b^n}$. When a fraction is raised to some power, the numerator and denominator are both raised to that power. Example:

$$\left(\frac{2x}{5}\right)^3 = \frac{2^3x^3}{5^3} = \frac{8x^3}{125} = 0.064x^3$$

**Square Root of Exponentials** $\sqrt{a^{2n}} = a^n$. To find the square root of an exponential, divide the exponent by 2. Example:

$$\sqrt{10^6} = 10^3$$

If the exponent is odd, see below.

**Square Root of a Product** $\sqrt{ab} = \sqrt{a} \times \sqrt{b}$. The square root of the product of two numbers equals the product of the square roots of the numbers. Example:

$$\sqrt{9 \times 10^{-6}} = \sqrt{9} \times \sqrt{10^{-6}} = 3 \times 10^{-3}$$

Using this principle, by adjusting a decimal point, you may take the square root of an exponential having an odd exponent. Example:

$$\sqrt{10^5} = \sqrt{10 \times 10^4} = \sqrt{10} \times \sqrt{10^4} = 3.16 \times 10^2$$

You may use the same technique in taking the square root of a number expressed in exponential notation. Example:

$$\sqrt{1.8 \times 10^{-5}} = \sqrt{18 \times 10^{-6}}$$

$$= \sqrt{18} \times \sqrt{10^{-6}} = 4.2 \times 10^{-3}$$

10) SOLVING AN EQUATION FOR AN UNKNOWN QUANTITY: Most problems in this book can be solved by dimensional analysis methods. There are times, however, when algebra should be used, particularly in relation to the gas laws. Solving an equation for an unknown involves rearranging the equation so that the unknown is the only item on one side and only known quantities are on the other. "Rearranging" an equation may be done in several ways, but the important thing is that *whatever is done to one side of the equation must also be done to the other.* The resulting relationship remains an equality, a true equation. Among the operations that may be performed on both sides of an equation are addition, subtraction, multiplication, division, and raising to a power, which includes taking square root.

In the following examples a, b, and c represent known quantities, and x is the unknown. The object in each case is to solve the equation for x. The steps of the algebraic solution are shown, as well as the operation performed on both sides of the equation. Each example is accompanied by a practice problem that is solved by the same method. You should be able to solve the problem, even if you have not yet reached the point in the book where such a problem is likely to appear. Answers to these practice problems may be found at the end of Appendix I.

(1) $x + a = b$

$x + a - a = b - a$ Subtract a

$x = b - a$ Simplify

PRACTICE: *1) If* $P = p_{O_2} + p_{H_2O}$, *find* $p_{O_2}$, *when* $P =$ *748 torr and* $p_{H_2O} = 24$ *torr.*

(2) $ax = b$

$\frac{\not{a}x}{\not{a}} = \frac{b}{a}$ Divide by a, which is called the **coefficient** of x

$x = \frac{b}{a}$ Simplify

PRACTICE: *2) At a certain temperature,* $PV = k$. *If* $P = 1.23$ *atm and* $k = 1.62$ *L · atm, find V.*

(3) $\frac{x}{a} = b$

$\frac{ax}{a} = ba$ Multiply by a

$x = ba$ Simplify

PRACTICE: *3) In a fixed volume,* $\frac{P}{T} = k$. *Find P if* $k = \frac{2.4 \text{ torr}}{K}$ *and* $T = 300$ *K.*

PRACTICE: *4) For gases at constant volume,* $\frac{P_1}{T_1} = \frac{P_2}{T_2}$. *If* $P_1 = 0.80$ *atm at* $T_1 = 320$ *K, at what value of* $T_2$ *will* $P_2 = 1.00$ *atm?*

*Note:* Procedures (2) and (3) are examples of dividing both sides of the equation by the coefficient of x, which is the same as multiplying both sides by the inverse of the coefficient. This is best seen in a more complex example:

$$\frac{ax}{b} = \frac{c}{d}$$

$$\frac{a}{b}x = \frac{c}{d} \qquad \text{Isolate the coefficient of x}$$

$$\frac{b}{a} \times \frac{a}{b}x = \frac{c}{d} \times \frac{b}{a} \qquad \text{Multiply by the inverse of the coefficient of x}$$

$$x = \frac{cb}{da} \qquad \text{Simplify}$$

(4) $$\frac{b}{ax} = \frac{d}{c}$$

$$\frac{ax}{b} = \frac{c}{d} \qquad \text{Invert both sides of the equation}$$

Proceed as in (3) above.

(5) $$\frac{a}{b + x} = c$$

$$(b + x)\frac{a}{(b + x)} = c(b + x) \qquad \text{Multiply by (b + x)}$$

$$a = c(b + x) \qquad \text{Simplify}$$

$$\frac{a}{c} = \frac{\cancel{c}(b + x)}{\cancel{c}} \qquad \text{Divide by c}$$

$$\frac{a}{c} = b + x \qquad \text{Simplify}$$

$$\frac{a}{c} - b = x \qquad \text{Subtract b}$$

PRACTICE: *5) In how many grams of water must you dissolve 20.0 grams of salt to make a 25% solution? The formula is*

$$\frac{g\ salt}{g\ salt + g\ water} \times 100 = \%;\ or$$

$$\frac{g\ salt}{g\ salt + g\ water} = \frac{\%}{100}$$

## Part C Logarithms

*Sections 17.9 and 17.10 are the only places in this text that use logarithms, and most of what you need to know about logarithms is explained at that point. Comments here are limited to basic information needed to support the text explanations.*

**The common logarithm of a number is the power, or exponent, to which 10 must be raised to be equal to the number.** Expressed mathematically,

$$\text{If } N = 10^x, \text{ then } \log N = \log 10^x = x \qquad \textbf{(AP.1)}$$

The number 100 may be written as the base, 10, raised to the second power: $100 = 10^2$. According to Equation AP.1, 2 is the logarithm of $10^2$, or 100. Similarly, if $1000 = 10^3$, $\log 1000 = \log 10^3 = 3$. And if $0.0001 = 10^{-4}$, $\log 0.0001 = \log 10^{-4} = -4$.

Just as the powers to which 10 can be raised may be either positive or negative, so may logarithms be positive or negative. The changeover occurs at the value 1, which is $10^0$. The logarithm of 1 is therefore 0. It follows that the logarithms of numbers greater than 1 are positive and logarithms of numbers less than 1 are negative.

The powers to which 10 may be raised are not limited to integers. For example, 10 can be raised to the 2.45 power: $10^{2.45}$. The logarithm of $10^{2.45}$ is 2.45. Such a logarithm is made up of two parts. The digit or digits to the left of the decimal are the **characteristic.** The characteristic reflects the size of the number; it is related to the exponent of 10 when the number is expressed in exponential notation. In 2.45, the characteristic is 2. The digits to the right of the decimal make up the **mantissa,** which is the logarithm of the coefficient of the number when written in exponential notation. In 2.45, the mantissa is 0.45.

**The number that corresponds to a given logarithm is its antilogarithm.** In Equation AP.1, the antilogarithm of x is $10^x$, or N. The antilogarithm of 2 is $10^2$, or 100. The antilogarithm of 2.45 is $10^{2.45}$. The value of the antilogarithm of 2.45 can be found on a calculator, as described in Part A of Appendix I: antilog $2.45 = 10^{2.45} = 2.8 \times 10^2$. In this exponential form of the antilogarithm of 2.45 the characteristic, 2, is the exponent of 10, and the mantissa, 0.45, is the logarithm of the coefficient, 2.8. In terms of significant figures, the mantissa matches the coefficient. This is illustrated in Section 17.9.

Because logarithms are exponents, they are governed by the rules of exponents given in Section 9 of Part B of this Appendix. For example, the product of two exponentials to the same base is the base raised to a power equal to the sum of the exponents: $a^m \times a^n = a^{m+n}$. The exponents are added. Similarly, exponents to the base 10 (logarithms) are added to get the logarithm of the product of two numbers: $10^m \times 10^n = 10^{m+n}$. Thus

$$\log ab = \log a + \log b \qquad \textbf{(AP.2)}$$

In a similar fashion, the logarithm of a quotient is the logarithm of the dividend minus the logarithm of the divisor (or the logarithm of the numerator minus the logarithm of the denominator if the expression is written as a fraction):

$$\log a \div b = \log a - \log b \qquad \textbf{(AP.3)}$$

Equation AP.1 is the basis for converting between pH and hydrogen ion concentration in Section 17.9—or between any "p" number and its corresponding value in exponential notation. This is the only application of logarithms in this text. In more advanced chemistry courses you will encounter applications of Equation AP.3 and others that are beyond the scope of this discussion.

Ten is not the only base for logarithms. Many natural phenomena, both chemical and otherwise, involve logarithms to the **base *e*,** which is 2.718. . . . Logarithms to base *e* are known as **natural logarithms.** Their value is 2.303 times greater than a base-10 logarithm. Physical chemistry relationships that appear in base *e* are often converted to base 10 by the 2.303 factor, although modern calculators make it just as easy to work in base *e* as in base 10. The "p" concept, however, uses base 10 by definition.

## Part D Estimating Calculation Results

A large percentage of student calculation errors would never appear on homework or test papers if the student would estimate the answer before accepting the number displayed on a calculator. *Challenge every answer.* Be sure it is reasonable before you write it down.

There is no single "right" way to estimate an answer. As your mathematical skills grow, you will develop techniques that are best for you. You will also find that one method works best on one kind of problem and another method on another problem. The ideas that follow should help you get started in this practice.

In general, estimating a calculated result involves rounding off the given numbers and calculating the answer mentally. For example, if you multiply 325 by 8.36 on your calculator, you will get 2717. To see if this answer is reasonable, you might round off 325 to 300, and 8.36 to 8. The problem then becomes 300 × 8, which you can calculate mentally to 2400. This is reasonably close to your calculator answer. Even so, you should run your calculator again to be sure you haven't made a small error in keying.

If your calculator answer for the preceding problem had been 1254.5, or 38.975598, or 27,170, your estimated 2400 would signal that an error was made. These three numbers represent common calculation errors. The first comes from a mistake that may arise any time a number is transferred from one position (your paper) to another (your calculator). It is called transposition, and appears when two numerals are changed in position. In this case 325 × 3.86 (instead of 8.36) = 1254.5.

The answer 38.875598 comes from pressing the wrong function key: 325 ÷ 8.36 = 38.875598. This answer is so unreasonable—325 × 8.36 = about 38!—that no mental arithmetic should be necessary to tell you it is wrong. But calculators speak to some students with a mystic authority they would never dare to challenge. On one occasion neither student nor teacher could figure out why or how, on a test, the student used a calculator to divide 428 by 0.01, and then wrote down 7290 for the answer, when all he had to do was move the decimal two places!

The answer 27,170 is a decimal error, as might arise from transposing the decimal point and a number, or putting an incorrect number of zeros in numbers like 0.00123 or 123,000. Decimal errors are also apt to appear through an incorrect use of exponential notation, either with or without a calculator.

Exponential notation is a valuable aid in estimating results. For example, in calculating 41,300 × 0.0524, you can regard both numbers as falling between 1 and 10 for a quick calculation of the coefficient: 4 × 5 = 20. Then, thinking of the exponents, changing 41,300 to 4 moves the decimal four places left, so the exponential is $10^4$. Changing 0.0524 to 5 has the decimal moving two places right, so the exponential is $10^{-2}$. Adding exponents gives 4 + (−2) = 2. The estimated answer is 20 × $10^2$, or 2000. On the calculator it comes out to 2164.12.

Another "trick" you can use is to move decimals in such a way that the moves cancel and at the same time the problem is simplified. In 41,300 × 0.0524, the decimal in the first factor can be moved two places left (divide by 100), and in the second factor, two places right (multiply by 100). Dividing and multiplying by 100 is the same as multiplying by 100/100, which is equal to 1. The problem simplifies to 413 × 5.24, which is easily estimated as 400 × 5 = 2000. You can use the same technique to simplify fractions, too. 371,000/6240 can be simplified to 371/6.24 by moving the decimal three places left in both the numerator and denominator, which is the same as multiplying by 1 in the form 0.001/0.001. An estimated 360/6 gives 60 as the approximate answer. The calculator answer is 59.455128. Similarly, 0.000406/0.000839 becomes about 4/8, or 0.5; by calculator, the answer is 0.4839042.

## Answers to Practice Problems in Appendix I

1) $p_{O_2} = P - p_{H_2O} = 748 - 24 = 724 \text{ torr}$

2) $V = \frac{k}{P} = \frac{1.62 \text{ L} \cdot \cancel{\text{atm}}}{1.23 \cancel{\text{atm}}} = 1.32 \text{ L}$

3) $P = kT = \frac{2.4 \text{ torr}}{\cancel{K}} \times 300 \cancel{K} = 720 \text{ torr}$

4) $T_2 = \frac{T_1P_2}{P_1} = \frac{(320\text{K})(1.00 \cancel{\text{atm}})}{0.80 \cancel{\text{atm}}} = 400 \text{ K}$

5) Because of the complexity of this problem, it is easier to substitute the given values into the original equation and then solve for the unknown. The steps in the solution correspond to those immediately preceding the question:

$$\frac{\text{g salt}}{\text{g salt} + \text{g water}} = \frac{\%}{100}$$

$$\frac{20.0}{20.0 + \text{g water}} = \frac{25}{100}$$

$$20.0 = 0.25(20.0 + \text{g water})$$

$$= 5.0 + 0.25(\text{g water})$$

$$20.0 - 5.0 = 0.25(\text{g water})$$

$$\text{g water} = \frac{15.0}{0.25} = 60 \text{ g water}$$

# Appendix II

## THE SI SYSTEM OF UNITS

## Base Units

The International System of Units or *Système International* (SI), which represents an extension of the metric system, was adopted by the 11th General Conference of Weights and Measures in 1960. It is constructed from seven base units, each of which represents a particular physical quantity (Table AP.1).

**Table AP.1 SI Base Units**

| Physical Quantity | Name of Unit | Symbol |
|---|---|---|
| 1) Length | meter | m |
| 2) Mass | kilogram | kg |
| 3) Time | second | s |
| 4) Temperature | kelvin | K |
| 5) Amount of substance | mole | mol |
| 6) Electric current | ampere | A |
| 7) Luminous intensity | candela | cd |

Of the seven units listed in Table AP.1, the first five are particularly useful in introductory and general chemistry. They are defined as follows:

1. The current definition of the *meter* is the distance light travels in a vacuum in 1/299,792,458 second.
2. The *kilogram* represents the mass of the International Prototype Kilogram, a platinum–iridium cylinder kept at the International Bureau of Weights and Measures at Sèvres, France.
3. The *second* is currently defined as the duration of 9,192,631,770 periods of a certain line in the microwave spectrum of cesium-133.
4. The *kelvin* is 1/273.16 of the temperature interval between absolute zero and the triple point of water (0.01°C = 273.16 K).
5. The *mole* is the amount of substance that contains as many entities as there are atoms in exactly 0.012 kg of carbon-12.

### Prefixes Used with SI Units

Decimal fractions and multiples of SI units are designated by using the prefixes listed in Table 3.2. Those that are most commonly used in introductory chemistry are in boldface type.

## Derived Units

In the International System of Units all physical quantities are expressed in the base units listed in Table AP.1 or in combinations of those units. The combinations are called **derived units.** For example, the density of a substance is found by dividing the mass of a sample in kilograms by its volume in cubic meters. The resulting units are kilograms per cubic meter, or $kg/m^3$. Some of the derived units used in chemistry are given in Table AP.2.

**Table AP.2 SI Derived Units**

| Physical Quantity | Name of Unit | Symbol | Definition |
|---|---|---|---|
| Area | square meter | $m^2$ | |
| Volume | cubic meter | $m^3$ | |
| Density | kilograms per cubic meter | $kg/m^3$ | |
| Force | newton | N | $kg \cdot m/s^2$ |
| Pressure | pascal | Pa | $N/m^2$ |
| Energy | joule | J | $N \cdot m$ |

If you have not studied physics, the SI units of force, pressure, and energy are probably new to you. Force is related to acceleration, which has to do with changing the velocity of an object. One **newton** is the force that, when applied for one second, will change the straight-line speed of a 1-kilogram object by 1 meter per second.

A **pascal** is defined as a pressure of one newton acting on an area of one square meter. A pascal is a small unit, so pressures are commonly expressed in kilopascals (kPa), 1000 times larger than the pascal.

A **joule** (pronounced *jool,* as in *pool*) is defined as the work done when a force of one newton acts through a distance of one meter. *Work* and *energy* have the same units. Large amounts of energy are often expressed in kilojoules, 1000 times larger than the joule.

## Some Choices

It is difficult to predict the extent to which SI units will replace traditional metric units in the coming years. This makes it difficult to select and use the units that will be most helpful to the readers of this textbook. Add to that the authors' joy that, after more than 200 years, the United States has finally begun to adopt the metric system, including some units that the SI system would eliminate, and their deep desire to encourage rather than complicate the use of metrics in their native land. With particular apologies to Canadian readers, who are more familiar with SI units than Americans are, we list the areas in which this book does not follow SI recommendations:

1. The SI unit of length is the *metre,* spelled in a way that corresponds to its French pronunciation. In America,

and in this book, it is written *meter,* which matches the English pronunciation.

2. The SI volume unit, *cubic meter,* is huge for most everyday uses. The *cubic decimeter* is 1/1000 as large, and much more practical. When referring to liquids it is customary to replace this six-syllable name with the two-syllable *liter*—or *litre,* for the French spelling. (Which would you rather buy at the grocery store, 2 cubic decimeters of milk, or 2 liters?) In the laboratory the common units are again 1/1000 as large, the *cubic centimeter* for solids and the *milliliter* for liquids.

3. The *millimeter of mercury* has an advantage over the *pascal* or *kilopascal* as a pressure unit because the common laboratory instrument for "measuring" pressure literally measures millimeters of mercury. We lose some of the advantage of this "natural" pressure unit by using its other name, *torr.* Reducing eight syllables to one is worth the sacrifice. For large pressures we continue to use the traditional *atmosphere,* which is 760 torr.

# Appendix III

## ANSWERS

## Chapter 2

### Answers to Target Checks

**1.** The art depicts a model of table salt at the particulate level. The photograph shows salt at the macroscopic level. **2.** Your illustration for the gaseous state should show the particles spaced far apart and distributed throughout the container. The number of particles should be the same for the liquid illustration. Here, the particles will be touching each other, taking the shape of the bottom of the container. See Figure 2.5. **3.** a and d are chemical changes; b and c are physical changes. **4.** Beaker B holds a pure substance because its specific gravity, a physical property, is constant. Beakers A and C hold mixtures because their specific gravities are variable. **5.** b, c, and d are heterogeneous; a is homogeneous. **6.** The distillation apparatus is the better choice. The filtration apparatus will not work to separate the components of the salt water solution because the solution is homogeneous. The filtration apparatus separates solid from liquid in a heterogeneous mixture. **7.** Compounds: a, c, f. Elements: b, d, e. **8.** A compound is a pure substance because it has definite physical and chemical properties. **9.** a and c, repulsion; b, attraction. **10.** (a) Boiling water is endothermic with respect to the water. It must absorb energy in order to boil. (b) The change is an increase in potential energy. **11.** In a scientific context, the term *conserved* means that the quantity of something remains constant before and after a change.

### Answers to Concept-Linking Exercises

*You may have found more relationships or relationships other than the ones given in these answers.*

**1.** Matter is whatever has mass and occupies space. The kinetic molecular theory says matter consists of particles in constant motion. The amount of motion determines the state of matter: A gas has the greatest amount of motion, a solid has the least, and the amount of motion in a liquid is intermediate. **2.** A homogeneous substance has a uniform appearance and composition throughout. It may be pure, consisting of only one substance, or it may be a mixture of two or more substances. A heterogeneous substance has different phases. **3.** An element is a pure substance that cannot be changed into a simpler pure substance. A compound can be changed into simpler pure substances. An atom is the smallest particle of an element. Molecules are tiny particles in which two or more atoms are chemically combined. **4.** A chemical change occurs when one substance disappears and another substance appears. The chemical properties of a substance are the chemical changes that are possible for the substance. A physical change is a change in the form of a substance without a change in its identity. Physical properties can be measured or detected with the five physical senses. **5.** The conservation law states that the total of all matter and energy is conserved in all changes, nuclear and nonnuclear. For nonnuclear changes, the Laws of Conservation of Mass and Energy hold that both mass and energy are conserved independently. **6.** Kinetic energy is associated with particles or other objects in motion. Potential energy is related to the position of a substance in a force field. A change is exothermic if it releases energy to the surroundings, and endothermic if it absorbs energy from the surroundings.

### Answers to Blue-Numbered Questions, Exercises, and Problems

**1.** Macroscopic: c. Microscopic: a, e. Particulate: b, d. **3.** An advantage is that understanding the behavior of particles allows us to predict the macroscopic behavior of samples of matter made from those particles. Chemists can then design particles to exhibit desired macroscopic characteristics, as seen in drug design and synthesis, for example. **5.** Your illustration should resemble the particulate view in Figure 2.5. **7.** A dense gas that is concentrated at the bottom of a container can be poured because its particles can move relative to each other. Chunks of solids, such as sugar crystals, can be poured. **9.** Gases are most easily compressed because of the large spaces between molecules. **11.** Chemical: b, c. Physical: a, d, e. **13.** a, c, d. **15.** Physical—the particles simply change state. **17.** The material is a pure substance, one kind of matter. However, the display is heterogeneous, consisting of two visibly different forms or phases of carbon. **19.** Tap water is a mixture of water and dissolved minerals and gases. Distilled water is pure water. **21.** Pure substances: a, c. Mixtures: b, d. **23.** Pure substances: a, d. Mixtures: b, c. **25.** Examples include glass products, plastic products, aluminum foil, cleaning and grooming solutions, and the air. **27.** b, c, d. **29.** Ice cubes from a home refrigerator are usually heterogeneous, containing trapped air. Homogeneous cubes in liquid water are heterogeneous, having visible solid and liquid phases. **31.** Your sketch should show one type of particle (a pure substance), but in more than one state of matter or molecular form (heterogeneous). **33.** Pick out the ball bearings (no change); use the magnetic property of steel to pick up the ball bearings with a magnet (no change); dissolve the salt in water and filter or pick out the ball bearings (physical change). **35.** The original liquid must be a mixture because the freezing point changed when some of the liquid was removed. The freezing point of a pure substance is the same no matter how much of the substance you have. **37.** Compounds: a, b, c. Elements: d, e, f. **39.** Elements: c, d. Compounds: a, b, e. **41.** Elements: a, b, e. Compounds: c, d. **43.** (a) Elements: 2, 3; compounds: 1, 4, 5. (b) In general, if there are two or more words in the name, the substance is a compound. However, many compounds are known by one-word common names, and one-word names for many carbon-containing compounds are assembled from prefixes, suffixes, and special names for recurring groups. Chloromethane is such a compound. The name of an element is always a single word. **45.** There is no evidence that A is a compound because it has not been broken down into two or more other pure substances by a chemical or physical change. However, only two methods have been tried. A is most likely an element, but the evidence is not conclusive. More tests need to be conducted.

**47.**

| | G, L, S | P, M | Hom, Het | E, C |
|---|---|---|---|---|
| Factory smokestack emissions | All, but mostly G | M | Het | |
| Concrete (in a sidewalk) | S | M | Het | |
| Helium | G | P | Hom | E |
| Hummingbird feeder solution | L | M | Hom | |
| Table salt | S | P | Hom | C |

**49.** Gravitational forces are attractive only; electrostatic forces can be attractive or repulsive. Magnetic forces can be attractive or repulsive also. All three can act simultaneously. **51.** Reactants: $AgNO_3$, NaCl. Products: AgCl, $NaNO_3$. **53.** The reactant Ni is an element; the product $Ni(NO_3)_2$ is a compound. **55.** a, b, c. **57.** Kinetic energy is greatest when the swing moves through its lowest point. Potential energy is at a maximum when the swing is at its highest point. **59.** A gaseous substance has been driven off by the heating process. **61.** The pans will balance without changing the weights. The Law of Conservation of Mass states that mass is neither created nor destroyed in a change. **63.** Examples include electrical energy being converted to mechanical energy (washing machine), light energy (lightbulb), or heat energy (oven). These changes are useful because they are advantageous to you, but they are wasteful because they are not 100% efficient and thus an imperfect use of energy. **66.** True: e, f, i, j, l. False: a, b, c, d, g, h, k. **67.** Nothing. **71.** Mercury, water, ice, carbon. **72.** Yes; nitrogen and oxygen in air are a mixture of two elements. **73.** Yes; nitrogen oxides, for example, occur as at least six different compounds. You also may have thought of carbon monoxide and carbon dioxide. **74.** Rainwater is more pure. Ocean water is a solution of salt and other substances. Ocean water is distilled by evaporation and condensed into rain. **75.** (a) The powder is neither an element nor a compound, both of which have a fixed composition. (b) The contents of the box are homogeneous because the powder has a uniform appearance. (c) The contents must be a mixture of varying composition. **76.** The sources of usable energy now available are limited. If we change them into forms that we cannot use, we risk having an energy shortage in the future. **77.** (a) Neither: The distinction between homogeneous and heterogeneous is not a particulate property. (b) The sample is a mixture because it consists of two different particle types. (c) The particles are compounds because they consist of more than one type of atom. (d) The particles are molecules because they are made up of more than one atom. (e) The sample is a gas because the particles are completely independent of one another. **78.** (a) Reactants: AB, CD. Products: AD, CB. (b) A chemical change is shown. (c) The masses of product particles and the reactant particles are equal. (d) The energy in the container is the same before and after the reaction. **79.** Pure substance. A mixture would have changed boiling temperature during distillation because of a change in composition of the mixture. **80.** The substance is a mixture. If it was a pure substance, its density would not change. **81.** If the objects have opposite charge, there is an attraction between them. An increase in separation will be an increase in potential energy. If they have the same charge (repulsion), the greater distance will be lower in potential energy. **82.** Because each object contains particles with both positive and negative charges, there are both attractive and repulsive forces between the objects. If the net force is one of attraction, the particles move toward each other; if the net force is one of repulsion, the particles separate. When the two forces are balanced, the net force is zero and the particles remain separated at a constant distance. **83.** (a) This is a physical change. (b) The number of particles is not conserved. (c) New particles appear from nowhere. (d) This could be the product of a chemical change.

# Chapter 3

## Answer to Target Check

**1.** a is a GIVEN quantity; b and c are PER expressions.

## Answers to Concept-Linking Exercises

*You may have found more relationships or relationships other than the ones given in these answers.*

**1.** Scientific measurements are made using the metric system of measurement. SI units are included in the metric system. SI is an abbreviation for the French name for the international system of units. The SI system defines seven base units. Examples are kilograms (for mass) and meters (for length). Other quantities are made up of combinations of base units; these are called derived units. Examples are volume and density. **2.** Dimensional analysis is a way to solve problems involving quantities that are proportional to each other. These proportionalities can be stated in PER expressions that are conversion factors by which a GIVEN quantity in one unit is changed to an equivalent WANTED quantity in another unit. The unit PATH lists each unit change in solving the problem. Example: A GIVEN distance can be changed to a WANTED amount of time with speed as the PER expression in miles (or kilometers) per hour. The unit PATH is miles (or kilometers) → hours. **3.** The kilogram is the base unit of mass in the SI system, and the gram is 1/1000 of a kilogram; the smaller unit is more commonly used in the laboratory. Weight is a measure of gravitational attraction that is proportional to mass. A pound is a weight unit. **4.** There is some degree of uncertainty in every physical measurement. In scientific work, measurements are expressed in all digits known accurately plus one digit that is uncertain, which is known as the uncertain digit. Collectively, these digits are significant figures. Significant figures are not applied to exact numbers, which have no uncertainty. **5.** Two quantities are directly proportional if they increase or decrease at the same rate; the ratio of one to the other is constant. Two related variables are inversely proportional to each other if one increases and the other decreases in such a way that their product is a constant. A proportionality is indicated by the operator symbol $\propto$: $A \propto B$. A proportionality may be converted to an equation by inserting a proportionality constant, often symbolized as k: $A = kB$.

## Answers to Blue-Numbered Questions, Exercises, and Problems

**1.** (a) $3.22 \times 10^{-4}$; (b) $6.03 \times 10^{9}$; (c) $6.19 \times 10^{-12}$ **3.** (a) 5,120,000; (b) 0.000000840; (c) 1,920,000,000,000,000,000,000 **5.** (a) $7.29 \times 10^{-3}$; (b) $5.13 \times 10^{-12}$; (c) $2.98 \times 10^{9}$; (d) $4.02 \times 10^{-6}$ **7.** (a) 75.6; (b) 9.41; (c) $3.24 \times 10^{4}$; (d) $1.49 \times 10^{-3}$ **9.** (a) $3.77 \times 10^{-8}$; (b) $2.0 \times 10^{6}$

**11.** (a) $4.65 \times 10^{8}$; (b) $3.0 \times 10^{-7}$ **13.** $406 \text{ mi} \times \frac{1 \text{ hr}}{48 \text{ mi}} = 8.5 \text{ hr}$ **15.** $4.3 \text{ km} \times \frac{1 \text{ hr}}{88 \text{ km}} \times \frac{60 \text{ min}}{\text{hr}} = 2.9 \text{ min}$

**17.** $62 \text{ ft} \times \frac{9 \text{ nails}}{\text{foot}} \times \frac{1 \text{ lb}}{36 \text{ nails}} \times \frac{69 \text{ cents}}{\text{lb}} \times \frac{1 \text{ dollar}}{100 \text{ cents}} = 11 \text{ dollars}$ **19.** $1950 \text{ pesos} \times \frac{1 \text{ dollar}}{218 \text{ pesos}} = 8.94 \text{ dollars}$

**21.** $1 \text{ decade} \times \frac{10 \text{ yr}}{\text{decade}} \times \frac{52 \text{ wk}}{\text{yr}} = 5.2 \times 10^{2} \text{ weeks}$

**23.** Her weight is the same when the elevator is standing still as when it moves at a constant rate. It decreases when the elevator slows; increases when the elevator accelerates. Her mass is constant no matter what the elevator is doing. **25.** The meter. **27.** A kilobuck is $1000. A megabuck is $1,000,000. **29.** 1 mL = 0.001 L. **31.** Megagrams, because a gram is a very small unit when compared to the mass of an automobile. **33.** (a) 0.0574 g; (b) $1.41 \times 10^{3}$ g; (c) $4.54 \times 10^{9}$ mg **35.** (a) $2.17 \times 10^{3}$ cm; (b) 0.517 km; (c) $6.66 \times 10^{4}$ cm **37.** (a) 494 mL; (b) $1.91 \times 10^{3}$ mL; (c) 0.874 L **39.** (a) 711 g; (b) $5.27 \times 10^{5}$ pm; (c) $3.63 \times 10^{5}$ dag **41.** (a) 3; (b) 5; (c) 3; (d) uncertain—2 to 5; (e) 2; (f) 3; (g) 5; (h) 4 **43.** (a) $6.40 \times 10^{-3}$ km; (b) 0.0178 g; (c) $7.90 \times 10^{4}$ m; (d) $4.22 \times 10^{4}$ tons; (e) $6.50 \times 10^{2}$ dollars **45.** $147 \text{ lb} + 67.7 \text{ lb} + 3.6 \times 10^{2} \text{ lb} + 135.43 \text{ lb} = 7.1 \times 10^{2} \text{ lb}$

**47.** 22.93 mL − 19.4 mL = 3.5 mL **49.** $\frac{1}{2}$ mol × $\frac{342.3 \text{ g}}{\text{mol}}$ = 171.2 g; 0.764 mol × $\frac{342.3 \text{ g}}{\text{mol}}$ = 262 g **51.** $\frac{(62.87 \text{ g} - 42.3 \text{ g})}{19 \text{ mL}}$ = 1.1 g/mL

**53.** 0.0715 gal × $\frac{3.785 \text{ L}}{\text{gal}} \times \frac{1000 \text{ cm}^3}{\text{L}}$ = 271 cm$^3$; 2.27 × 10$^4$ mL × $\frac{1 \text{ L}}{1000 \text{ mL}} \times \frac{1 \text{ gal}}{3.785 \text{ L}}$ = 6.00 gal **55.** 515 g × $\frac{1 \text{ lb}}{453.59 \text{ g}}$ = 1.14 lb

**57.** 1.45 × 10$^3$ lb × $\frac{453.59 \text{ g}}{\text{lb}} \times \frac{1 \text{ kg}}{1000 \text{ g}}$ = 658 kg **59.** 115 mg × $\frac{1 \text{ g}}{1000 \text{ mg}}$ = 0.115 g 0.115 g × $\frac{1 \text{ lb}}{453.59 \text{ g}}$ = 2.54 × 10$^{-4}$ lb

**61.** 69.1 kg × $\frac{1000 \text{ g}}{\text{kg}} \times \frac{1 \text{ lb}}{453.59 \text{ g}}$ = 152 lb, a middleweight **63.** 399.9 m × $\frac{100 \text{ cm}}{\text{m}} \times \frac{1 \text{ in.}}{2.54 \text{ cm}} \times \frac{1 \text{ ft}}{12 \text{ in.}}$ (= 1312 ft) × $\frac{1 \text{ yd}}{3 \text{ ft}}$ = 437.3 yd

**65.** 1454 ft × $\frac{12 \text{ in.}}{\text{ft}} \times \frac{2.54 \text{ cm}}{\text{in.}} \times \frac{1 \text{ m}}{100 \text{ cm}}$ = 443.2 m **67.** 29,035 ft × $\frac{12 \text{ in.}}{\text{ft}} \times \frac{2.54 \text{ cm}}{\text{in.}} \times \frac{1 \text{ m}}{100 \text{ cm}} \times \frac{1 \text{ km}}{1000 \text{ m}}$ = 8.8499 km

**69.** 619 gal × $\frac{3.785 \text{ L}}{\text{gal}}$ = 2.34 × 10$^3$ L

**71.**

| Celsius | Fahrenheit | Kelvin |
|---|---|---|
| 69 | 156 | 342 |
| −34 | −29 | 239 |
| −162 | −260 | 111 |
| 2 | 36 | 275 |
| 85 | 185 | 358 |
| −141 | −222 | 132 |

**73.** 37.0°C **75.** 26°C **77.** 136°F **79.** q ∝ m; q = $\Delta H_{fus}$ × m; cal/g; heat energy lost or gained per gram while changing state from liquid to solid or vice versa

**81.** $\Delta H_{fus} = \frac{q}{m} = \frac{7.39 \text{ kcal}}{92 \text{ g}} \times \frac{1000 \text{ cal}}{\text{kcal}}$ = 8.0 × 10$^1$ cal/g **83.** $P \propto \frac{1}{V}$; $P = k' \times \frac{1}{V}$; atm · L **85.** $D \equiv \frac{m}{V} = \frac{166 \text{ g}}{188 \text{ mL}}$ = 0.883 g/mL

**87.** $D \equiv \frac{m}{V} = \frac{18.6 \text{ g}}{15.7 \text{ L}}$ = 1.18 g/L **89.** 471 g × $\frac{1 \text{ cm}^3}{0.736 \text{ g}}$ = 6.40 × 10$^2$ cm$^3$ **91.** 2.0 L × $\frac{0.786 \text{ g}}{\text{mL}} \times \frac{1000 \text{ mL}}{\text{L}}$ = 1.6 × 10$^3$ g

**94.** True: a, b, d, f, h, j, l, m, q, r. False: c, e, g, i, k, n, o, p, s, t. **95.** A 6-foot-tall person is 1.83 m, 18.3 dm, 183 cm, and 1.83 × 10$^3$ mm. The centimeter is thus generally accepted as the preferred unit to express human height without using decimal fractions.

**96.** Sample calculation for a 150-lb person: 150 lb × $\frac{453.59 \text{ g}}{\text{lb}} \times \frac{1 \text{ kg}}{1000 \text{ g}}$ = 68.0 kg = 68,000 g = 68,000,000 mg. Kilograms are best because the number of g and mg are inconveniently large.

**97.** 8.5 in. × $\frac{2.54 \text{ cm}}{1 \text{ in.}}$ = 22 cm; 11 in. × $\frac{2.54 \text{ cm}}{1 \text{ in.}}$ = 28 cm **98.** 126 cans × $\frac{1 \text{ lb}}{21 \text{ cans}} \times \frac{454 \text{ g}}{\text{lb}}$ = (2.7 × 10$^3$ g) × $\frac{1 \text{ cm}^3}{2.7 \text{ g}}$ = 1.0 × 10$^3$ cm$^3$

**99.** ■ 7 oz × $\frac{1 \text{ lb}}{16 \text{ oz}}$ = 0.4 lb; 6.4 lb × $\frac{454 \text{ g}}{\text{lb}}$ = 2.9 × 10$^3$ g = 2.9 kg **101.** ■ 12.0 fl oz × $\frac{1 \text{ qt}}{32 \text{ fl oz}} \times \frac{1 \text{ gal}}{4 \text{ qt}} \times \frac{1 \text{ ft}^3}{7.48 \text{ gal}} \times \frac{64.4 \text{ lb}}{\text{ft}^3} \times \frac{454 \text{ g}}{\text{lb}}$ = 366 g

**103.** 33°F × $\frac{100 \text{ Celsius degrees}}{180 \text{ Fahrenheit degrees}}$ = 18°C **105.** ■ 0.25 cup × $\frac{0.25 \text{ qt}}{\text{cup}} \times \frac{1 \text{ gal}}{4 \text{ qt}} \times \frac{3.785 \text{ L}}{\text{gal}} \times \frac{1000 \text{ mL}}{\text{L}} \times \frac{1 \text{ cm}^3}{1 \text{ mL}} \times \frac{0.86 \text{ g}}{\text{cm}^3}$ = 51 g

**107.** $\frac{\$7.25 \text{ earned}}{\text{hr}} \times \frac{\$(100-23) \text{ take home}}{\$100 \text{ earned}}$ = \$5.5825/hour take-home pay  \$734.26 × $\frac{1 \text{ hr}}{\$5.5825} \times \frac{1 \text{ shift}}{4 \text{ hr}} \times \frac{1 \text{ week}}{5 \text{ shifts}}$ = 6.58 weeks = 7 weeks

# Chapter 4

## Answers to Target Checks

**1.** If $T_2 < T_1$, $V_2 < V_1$, because V ∝ T. The ratio of temperatures must be < 1. **2.** Remember that the Volume–Pressure Law is an inverse proportion: V ∝ 1/P. (a) Because $P_2 < P_1$, $V_2 > V_1$. (b) If $V_2 > V_1$, then $P_2 < P_1$.

## Answers to Concept-Linking Exercises

*You may have found more relationships or relationships other than the ones given in these answers.*

**1.** Substances in the gaseous state of matter can be compressed easily, they have low density, and they can be mixed in all proportions in a fixed volume. All of these properties, to some extent, result from the relatively large open spaces between gas particles. **2.** The kinetic molecular theory describes all matter as made up of tiny particles that are in constant motion. The ideal gas model is the part of the theory that relates to gases. The particle behavior of a gas has particles completely independent of each other, filling the container, and moving in straight lines until colliding or striking the container walls with no loss of energy. **3.** Pressure is the force exerted on a unit area of a surface. The earth is surrounded by gases (the atmosphere) that are drawn to it by gravity, creating atmospheric pressure. A barometer measures pressure directly. Gauge pressure, measured by a mechanical gauge that records zero at atmospheric pressure, measures only the pressure above atmospheric pressure. **4.** Temperature is a measure of the average kinetic energy of particles in a sample. The kelvin is the SI unit for expressing temperature. The kelvin temperature is at absolute zero, 0 K, when no energy can be removed from a system. **5.** For a constant quantity of gas: (a) Charles's Law states that volume is directly proportional to absolute temperature, V ∝ T. (b) Boyle's Law states that volume is inversely proportional to pressure, V ∝ 1/P. (c) These two proportionalities are joined in the Combined Gas Law, which is expressed in the equation $\frac{P_1V_1}{T_1} = \frac{P_2V_2}{T_2}$.

# Answers to Blue-Numbered Questions, Exercises, and Problems

**1.** All gas properties relate in some way to the kinetic character. Specifically, particle motion explains why gases fill their containers. Also, pressure results from the large numbers of particle collisions with the container walls. **3.** Gas molecules move independently of each other in all directions. Therefore, they exert pressure on the container walls, including the top, uniformly in all directions. Liquid and solid pressures are the result of gravity, so they exert a downward pressure on the bottom of the container. Liquid particles can move amongst themselves, so they also exert horizontal pressures against the walls of their container. Solid particles lack this freedom of horizontal movement, so they exert no horizontal pressure on the walls of a container. **5.** Because gas molecules are widely spaced, air is compressible, which makes for a soft and comfortable surface to lie on. The uniform pressure of air in the mattress keeps the person on the mattress off the ground. The low density of a gas allows the entire volume inside the mattress to be filled with a small mass of air, so the mattress can be filled by mouth rather than by a pump. **7.** Gas particles collide with the walls of the container, exerting pressure uniformly in all directions, including the top of the tank and its sides. **9.** Gas particles collide with the walls of the container, exerting pressure uniformly in all directions. **11.** Gas particles are very widely spaced, which leaves room for adding more. **13.** Gas particles are always in motion, "pushing" back the surrounding water. As the bubble rises, there is less liquid pushing on the bubble, so the gas volume increases because the gas particles are pushing against less force. Gas particles are widely spaced, thus gases have lower densities than liquids, so the bubble rises. **15.** Pressure measures force per unit area. Temperature measures the average kinetic energy of the particles in a sample. **17.** Figure 4.9 explains manometer operation.

**19.**

| atm | 1.03 | 0.163 | 1.16 | 0.902 | 0.964 |
|---|---|---|---|---|---|
| psi | 15.2 | 2.40 | 17.1 | 13.3 | 14.2 |
| in. Hg | 30.9 | 4.88 | 34.9 | 27.0 | 28.9 |
| cm Hg | 78.5 | 12.4 | 88.5 | 68.6 | 73.3 |
| mm Hg | 785 | 124 | 885 | 686 | 733 |
| torr | 785 | 124 | 885 | 686 | 733 |
| Pa | $1.05 \times 10^5$ | $1.65 \times 10^4$ | $1.18 \times 10^5$ | $9.14 \times 10^4$ | $9.77 \times 10^4$ |
| kPa | 105 | 16.5 | 118 | 91.4 | 97.7 |
| bar | 1.05 | 0.165 | 1.18 | 0.914 | 0.977 |

**21.** $747 \text{ torr} + 173 \text{ torr} = 920 \text{ torr} = 9.20 \times 10^2 \text{ torr}$ **23.** It is impossible to have a negative kelvin temperature. 0 K, absolute zero, is the lowest temperature that can be approached. The student probably meant to record −18°C = 255 K. **25.** 273 + 31 = 304 K **27.** 273 − 253 = 20 K; 273 − 259 = 14 K **29.** 259 − 273 = −14°C; 299 − 273 = 26°C

**31.** $24.3 \text{ L} \times \frac{(17 + 273) \text{ K}}{(55 + 273) \text{ K}} = 21.5 \text{ L}$ **33.** $(19 + 273) \text{ K} \times \frac{1.34 \text{ L}}{1.26 \text{ L}} = 311 \text{ K}; 311 - 273 = 38°\text{C}$

**35.** Squeezing the bulb reduces gas volume, increasing pressure and forcing some air bubbles out of the pipet. Releasing the bulb increases volume and decreases pressure. Liquid pressure is then greater than gas pressure, so liquid enters the pipet, reducing gas volume until liquid and gas pressures are equal.

**37.** $648 \text{ mL} \times \frac{772 \text{ torr}}{695 \text{ torr}} = 720 \text{ mL} = 7.20 \times 10^2 \text{ mL}$ **39.** $1.22 \text{ atm} \times \frac{7.26 \text{ L}}{3.60 \text{ L}} = 2.46 \text{ atm}$

**41.** $0.717 \text{ L} \times \frac{744 \text{ torr}}{48.6 \text{ atm}} \times \frac{(547 + 273) \text{ K}}{(27 + 273) \text{ K}} \times \frac{1 \text{ atm}}{760 \text{ torr}} = 0.0395 \text{ L}$ **43.** $626 \text{ L} \times \frac{(-58 + 273) \text{ K}}{(25 + 273) \text{ K}} \times \frac{756 \text{ torr}}{0.641 \text{ atm}} \times \frac{1 \text{ atm}}{760 \text{ torr}} = 701 \text{ L}$

**45.** STP conditions have been established so that all data are reported at the same conditions. One bar pressure is easily achieved in a laboratory, but 0°C is not a convenient temperature.

**47.** $28.3 \text{ L} \times \frac{273 \text{ K}}{(23 + 273) \text{ K}} \times \frac{0.985 \text{ bar}}{1 \text{ bar}} = 25.7 \text{ L}$ **49.** $44.5 \text{ mL} \times \frac{(28 + 273) \text{ K}}{273 \text{ K}} \times \frac{1 \text{ bar}}{0.894 \text{ bar}} = 54.9 \text{ mL}$

**51.** $1 \text{ bar} \times \frac{19.6 \text{ L}}{6.85 \text{ L}} \times \frac{(24 + 273) \text{ K}}{273 \text{ K}} = 3.11 \text{ bar}$ **53.** $273 \text{ K} \times \frac{23.7 \text{ L}}{56.2 \text{ L}} \times \frac{2.09 \text{ bar}}{1 \text{ bar}} = 241 \text{ K}; 241 - 273 = -32°\text{C}$

**56.** True: a, d, f, h. False: b, c, e, g. **57.** Dust particles are being pushed around by moving gas particles, which are too small to see.

**59.** ■ $743 \text{ torr} \times \frac{(1.80 + 0.26) \text{ L}}{1.80 \text{ L}} = 850 \text{ torr} = 8.50 \times 10^2 \text{ torr}$ **61.** $3.67 \text{ atm} \times \frac{(7 + 273) \text{ K}}{(25 + 273) \text{ K}} = 3.45 \text{ atm}$

**63.** ■ $(27 + 273) \text{ K} \times \frac{2.00 \text{ atm}}{1.74 \text{ atm}} = 345 \text{ K}; 345 - 273 = 72°\text{C}$

**65.** ■ $738 \text{ torr} \times \frac{1.62 \text{ m}^3}{0.140 \text{ m}^3} \times \frac{(28 + 273) \text{ K}}{(12 + 273) \text{ K}} = 9.02 \times 10^3 \text{ torr}; 9.02 \times 10^3 \text{ torr} \times \frac{14.7 \text{ psi}}{760 \text{ torr}} = 174 \text{ psi absolute}; 174 - 14.7 = 159 \text{ psi gauge}$

**67.** ■ Volume at start: 350 $\text{cm}^3$. Volume at end: 350 − 309 = 41 $\text{cm}^3$. Volume is inversely proportional to pressure.

Compression ratio $= \frac{350 \text{ cm}^3}{41 \text{ cm}^3} = 8.5.$

# Chapter 5

## Answers to Target Checks

**1.** The Law of Multiple Proportions is confirmed. The fixed mass of sulfur is 1.0 g. The ratio of masses of oxygen = 1.0/0.5 = 2/1, which is a ratio of small whole numbers. **2.** True: a, b. (c) The mass of a proton is about 1 u (which is $1.66 \times 10^{-24}$ g). (d) Electrons and protons are electrically charged, but neutrons have no charge (or zero charge). **3.** (a) Atoms are mostly made up of empty space, so they are not hard. They are small and their shape is spherical. True: b, c. **4.** All statements are true. **5.** (a) Four Group 3A/13 elements are metals (Al, Ga, In, Tl). (b) Ten Period-4 elements are transition metals (Sc through Zn).

## Answers to Concept-Linking Exercises

*You may have found more relationships or relationships other than the ones given in these answers.*

**1.** Dalton's atomic theory proposed that matter is composed of indivisible atoms. The nuclear model pictures the atom with a dense nucleus surrounded by electrons. The planetary model has the electrons moving in orbit around the nucleus. Rutherford's scattering experiment indicated that electrons occupy otherwise empty space outside the nucleus. **2.** An atom contains many particles (subatomic particles), the most important of which are electrons, protons, and neutrons. **3.** One atom of carbon-12 is arbitrarily assigned a mass of 12 atomic mass units, u. Thus 1 u is, by definition, exactly 1/12 of the mass of a carbon-12 atom. Both the u and the gram are mass units. 1 u = $1.66 \times 10^{-24}$ g. **4.** All atoms of an element have the same number of protons; this is the atomic number of the element. Atoms of the same element may have different numbers of neutrons and therefore different atomic masses. Atoms of an element with different numbers of neutrons are isotopes. The mass number of an atom is the sum of the number of protons plus the number of neutrons. **5.** Horizontal rows in the periodic table are periods, and vertical columns are groups. **6.** Main group elements are those in the A groups (U.S.) of the periodic table, and transition elements are those in the B groups. In the periodic table, metals are to the left of the stair-step line beginning between atomic numbers 4 and 5 in Period 2 and ending between atomic numbers 84 and 85 in Period 6, and the nonmetals are to the right of this line. Transition metals are the elements in the B groups to the left of the same line, an area that includes all the transition elements. **7.** The term *atomic mass* actually refers to average atomic mass, but the word *average* is omitted. The atomic mass of an element is the average mass of all of its natural isotopes. The atomic mass of an individual isotope of an element is the mass of an atom of that particular isotope. All atoms of a particular isotope have the same mass.

## Answers to Blue-Numbered Questions, Exercises, and Problems

**1.** Yes, see Figure 5.2. **3.** The Law of Definite Composition says that any compound is always made up of elements in the same proportion by mass. Dalton's atomic theory explains this by stating that atoms of different elements combine to form compounds. **5.** The calcium atoms and some of the oxygen atoms are in the calcium oxide; the carbon atoms and some of the oxygen atoms are in the carbon dioxide. **7.** The Law of Multiple Proportions in this case states that the same mass of sulfur combines with masses of fluorine in the ratio of simple whole numbers, 4 to 6. **9.** See Table 5.1.
**11.** Alpha particles and atomic nuclei are positively charged. As an alpha particle approached a nucleus, the repulsion between the positive charges deflected the alpha particle from its path. **13.** The nucleus. **15.** The electrons were thought to travel in circular orbits around the nucleus. **17.** No, the atomic number is the number of protons, and all atoms of an element have the same number of protons. **19.** Mass number is the sum of protons plus neutrons. Isotopes of the same element have different numbers of neutrons, but the same number of protons. The sums must be different. Atoms of different elements *must* have different numbers of protons, and they may have different numbers of neutrons. An atom with one less proton than another may have one more neutron than the other, so their mass numbers would be the same. Example: carbon-14 (6 protons, 8 neutrons) and nitrogen-14 (7 protons, 7 neutrons).

**21.**

| Name of Element | Nuclear Symbol | Atomic Number | Mass Number | Protons | Neutrons | Electrons |
|---|---|---|---|---|---|---|
| Scandium | ${}^{45}_{21}Sc$ | 21 | 45 | 21 | 24 | 21 |
| Germanium | ${}^{76}_{32}Ge$ | 32 | 76 | 32 | 44 | 32 |
| Tin | ${}^{122}_{50}Sn$ | 50 | 122 | 50 | 72 | 50 |
| Chlorine | ${}^{37}_{17}Cl$ | 17 | 37 | 17 | 20 | 17 |
| Sodium | ${}^{23}_{11}Na$ | 11 | 23 | 11 | 12 | 11 |

**23.** The mass of a proton and neutron is close to 1 u, which is $1.66 \times 10^{-24}$ g. It is more convenient to use the u because mass values expressed in u are not tiny fractions that must be expressed in exponential notation. **25.** ■ $6.66 \times 12.0$ u = 79.9 u, Br **27.** The atomic mass of neon is less than the average of the three atomic masses, so the isotope with the lowest mass must be present in the greatest abundance. **29.** $0.604 \times 68.9257$ u + $(1.000 - 0.604) \times 70.9249$ u = 69.7 u, Ga, gallium **31.** $0.5182 \times 106.9041$ u + $0.4818 \times 108.9047$ u = 107.9 u, Ag, silver **33.** $0.5725 \times 120.9038$ u + $0.4275 \times 122.9041$ u = 121.8 u, Sb, antimony **35.** $0.00193 \times 135.907$ u + $0.00250 \times 137.9057$ u + $0.8848 \times 139.9053$ u + $0.1107 \times 141.9090$ u = 140.1 u, Ce, cerium **37.** 18; 21, 39, 57, 89 **39.** (a) Period 4, Group 2A/2; (b) Period 3, Group 4A/14; (c) Period 5, Group 7B/7 **41.** ■ Z = 29, 63.55 u; Z = 55, 132.9 u; Z = 82, 207.2 u **43.** He, 4.003 u; Al, 26.98 u

45.

| Name of Element | Atomic Number | Symbol of Element |
|---|---|---|
| Magnesium | 12 | Mg |
| Oxygen | 8 | O |
| Phosphorus | 15 | P |
| Calcium | 20 | Ca |
| Zinc | 30 | Zn |
| Lithium | 3 | Li |
| Nitrogen | 7 | N· |
| Sulfur | 16 | S |
| Iodine | 53 | I |
| Barium | 56 | Ba |
| Potassium | 19 | K |
| Neon | 10 | Ne |
| Helium | 2 | He |
| Bromine | 35 | Br |
| Nickel | 28 | Ni |
| Tin | 50 | Sn |
| Silicon | 14 | Si |

**48.** True: b*, d, f, j, k, l, q, r, s. False: a, c, e, g, h, i, m, n, o, p. **49.** What was left had to have a positive charge to account for the neutrality of the complete atom. **50.** Sixteen grams of oxygen combines with 46 grams of sodium in sodium oxide, and 32 grams of oxygen combines with 46 grams of sodium in sodium peroxide. The ratio 16/32 reduces to 1/2, a ratio of small, whole numbers. **52.** ■ $0.7215 \times 84.9118\text{ u} + (1 - 0.7215) \times x\text{ u} = 85.4678\text{ u}$; $x = 86.91\text{ u}$ **54.** ■ $y \times 6.10512\text{ u} + (1 - y) \times 7.01600\text{ u} = 6.941\text{ u}$; $y = 0.08234$; 8.234% at 6.10512 u; $1 - 0.08234 = 0.91766$; 91.77% at 7.01600 u **55.** Different e/m ratios for positively charged particles from different elements indicate that, unlike the electron, all positively charged particles are not alike. Either the charge, the mass, or both must vary from element to element. This suggests the presence of at least two particles in varying number ratios. One or both must have a positive charge; others could be electrically neutral. **56.** The planetary model of the atom is similar to the solar system in that electrons orbit the nucleus as planets orbit the sun. Both models are similar in terms of the size of the nucleus/sun being large compared with the electrons/planets and the vast amount of empty space in the atom/solar system. **57.** Chemical properties of isotopes of an element are identical. **58.** ■ 12.09899 u. The difference in masses of the nuclear parts and the sum of the masses of protons and neutrons is what is responsible for nuclear energy in an energy–mass conversion. This is discussed in Chapter 20. **59.** Mass of electron + proton + neutron: 0.000549 u + 1.00728 u + 1.00867 u = 2.01650 u; Electron: (0.000549 u ÷ 2.01650 u) × 100 = 0.0272%; Proton: (1.00728 u ÷ 2.01650 u) × 100 = 49.9519%; Neutron: (1.00867 u ÷ 2.01650 u) × 100 = 50.0208%; The nucleus, containing the protons and neutrons, accounts for 49.9519% + 50.0208% = 99.9727% of the mass of the atom. *Note:* You can obtain the same result by calculating in grams rather than in u. You may have calculated in total mass rather than unit mass, in which case all masses above would be multiplied by 6. Unit masses could not have been used if there were different numbers of electrons, protons, or neutrons.

**60.** (a) $\frac{12.01\text{ u}}{1\text{ C atom}} \times \frac{1.66 \times 10^{-24}\text{ g}}{1\text{ u}} \times \frac{1\text{ C atom}}{1.9 \times 10^{-24}\text{ cm}^3} = 1.0 \times 10^1\text{ g/cm}^3$ (b) In packing carbon atoms into a crystal there are spaces between the atoms. There are no voids in a single atom. (In fact, voids in diamond account for 66% of the total volume, and in graphite, 78%.)
(c) $\frac{1.9 \times 10^{-24}\text{ cm}^3}{(1 \times 10^5)^3} = 2 \times 10^{-39}\text{ cm}^3$ (d) $\frac{12.01\text{ u}}{1\text{ C nucleus}} \times \frac{1.66 \times 10^{-24}\text{ g}}{1\text{ u}} \times \frac{1\text{ C nucleus}}{2 \times 10^{-39}\text{ cm}^3} = 1 \times 10^{16}\text{ g/cm}^3$

(e) $4 \times 10^{-5}\text{ cm}^3 \times \frac{1 \times 10^{16}\text{ g}}{1\text{ cm}^3} \times \frac{1\text{ lb}}{454\text{ g}} \times \frac{1\text{ ton}}{2000\text{ lb}} = 4 \times 10^5\text{ tons}$

# Chapter 6

## Answers to Concept-Linking Exercises

*You may have found more relationships or relationships other than the ones given in these answers.*

**1.** An atom is the smallest unit particle of an element. Two or more atoms can combine chemically to form molecules. A chemical formula represents a formula unit, which may be an actual particle, as with molecular compounds, or an electrically neutral combination of charged particles, as with ionic compounds. **2.** An ion is an atom or chemically bound group of atoms that has an electric charge because of an excess or deficiency of electrons compared with protons. A cation has a positive charge, and an anion has a negative charge. A monatomic ion has only one atom, while a polyatomic ion is made up of two or more atoms. An acid anion is a polyatomic anion that includes an ionizable hydrogen atom. **3.** An acid is a hydrogen-bearing molecular compound that reacts with water to produce a hydrated hydrogen ion and an anion. The hydrated hydrogen ion, $H^+ \cdot (H_2O)_x$, is sometimes represented by the hydronium ion, $H_3O^+$, or simply by the hydrogen ion, $H^+$. **4.** Some solid crystal structures include water molecules. These crystals are called hydrates, and the water molecules in the crystal are water of hydration. A compound whose crystals contain no water molecules is an anhydrous compound.

*Dalton apparently did not make any specific comment about the diameter of an atom, but he did propose that all atoms of an element are identical in every respect. This would include diameters.

# Answers to Blue-Numbered Questions, Exercises, and Problems

**1.** He, Ne, Ar, Kr, Xe, Rn **3.** $F_2$, B, Ni, S (or $S_8$) **5.** Chromium, chlorine, beryllium, iron **7.** Kr, Cu, Mn, $N_2$ **9.** $Cl_2O$, $Br_3O_8$, hydrogen bromide, diphosphorus trioxide **11.** When an atom gains one, two, or three electrons, the particle that results is a monatomic anion. The ion has a negative charge because it has more electrons than protons. **13.** Copper(I) ion, iodide ion, potassium ion, mercury(I) ion, sulfide ion **15.** $Fe^{3+}$, $H^+$, $O^{2-}$, $Al^{3+}$, $Ba^{2+}$ **17.** The formula of an acid usually begins with H. **19.** Monoprotic, 1; diprotic, 2; triprotic, 3
**21.** *The following table contains the names and formulas of all compounds in Table 6.11.*

**Table 6.11** (completed)

| Acid Name | Acid Formula | Ion Name | Ion Formula |
|---|---|---|---|
| Sulfuric | $H_2SO_4$ | Sulfate | $SO_4^{2-}$ |
| Carbonic | $H_2CO_3$ | Carbonate | $CO_3^{2-}$ |
| Chloric | $HClO_3$ | Chlorate | $ClO_3^-$ |
| Hydrofluoric | HF | Fluoride | $F^-$ |
| Bromic | $HBrO_3$ | Bromate | $BrO_3^-$ |
| Sulfurous | $H_2SO_3$ | Sulfite | $SO_3^{2-}$ |
| Arsenic | $H_3AsO_4$ | Arsenate | $AsO_4^{3-}$ |
| Periodic | $HIO_4$ | Periodate | $IO_4^{3-}$ |
| Selenous | $H_2SeO_3$ | Selenite | $SeO_3^{2-}$ |
| Tellurous | $H_2TeO_3$ | Tellurite | $TeO_3^{2-}$ |
| Hypoiodous | HIO | Hypoiodite | $IO^-$ |
| Hypobromous | HBrO | Hypobromite | $BrO^-$ |
| Telluric | $H_2TeO_4$ | Tellurate | $TeO_4^{2-}$ |
| Perbromic | $HBrO_4$ | Perbromate | $BrO_4^-$ |
| Hydrobromic | HBr | Bromide | $Br^-$ |

**23.** An anion or cation that contains an ionizable hydrogen, such as $HSO_4^-$ and $NH_4^+$, can lose the hydrogen and thus behave as an acid. **25.** $HSO_3^-$, $HCO_3^-$ **27.** Hydrogen selenite ion, hydrogen telluride ion **29.** Ammonium ion, cyanide ion **31.** $OH^-$, $Cd^{2+}$

*The following table contains the names and formulas of all compounds in Table 6.13. Some compounds in the table are not actually known.*

**Table 6.13** (completed)

| Ions | Potassium | Calcium | Chromium(III) | Zinc | Silver | Iron(III) | Aluminum | Mercury(I) |
|---|---|---|---|---|---|---|---|---|
| Nitrate | $KNO_3$ | $Ca(NO_3)_2$ | $Cr(NO_3)_3$ | $Zn(NO_3)_2$ | $AgNO_3$ | $Fe(NO_3)_3$ | $Al(NO_3)_3$ | $Hg_2(NO_3)_2$ |
| Sulfate | $K_2SO_4$ | $CaSO_4$ | $Cr_2(SO_4)_3$ | $ZnSO_4$ | $Ag_2SO_4$ | $Fe_2(SO_4)_3$ | $Al_2(SO_4)_3$ | $Hg_2SO_4$ |
| Hypochlorite | KClO | $Ca(ClO)_2$ | $Cr(ClO)_3$ | $Zn(ClO)_2$ | AgClO | $Fe(ClO)_3$ | $Al(ClO)_3$ | $Hg_2(ClO)_2$ |
| Nitride | $K_3N$ | $Ca_3N_2$ | CrN | $Zn_3N_2$ | $Ag_3N$ | FeN | AlN | omit |
| Hydrogen sulfide | KHS | $Ca(HS)_2$ | $Cr(HS)_3$ | $Zn(HS)_2$ | AgHS | $Fe(HS)_3$ | $Al(HS)_3$ | $Hg_2(HS)_2$ |
| Bromite | $KBrO_2$ | $Ca(BrO_2)_2$ | $Cr(BrO_2)_3$ | $Zn(BrO_2)_2$ | Ag $BrO_2$ | $Fe(BrO_2)_3$ | $Al(BrO_2)_3$ | $Hg_2(BrO_2)_2$ |
| Hydrogen phosphate | $K_2HPO_4$ | $CaHPO_4$ | $Cr_2(HPO_4)_3$ | $ZnHPO_4$ | $Ag_2HPO_4$ | $Fe_2(HPO_4)_3$ | $Al_2(HPO_4)_3$ | $Hg_2HPO_4$ |
| Chloride | KCl | $CaCl_2$ | $CrCl_3$ | $ZnCl_2$ | AgCl | $FeCl_3$ | $AlCl_3$ | $Hg_2Cl_2$ |
| Hydrogen carbonate | $KHCO_3$ | $Ca(HCO_3)_2$ | $Cr(HCO_3)_3$ | $Zn(HCO_3)_2$ | $AgHCO_3$ | $Fe(HCO_3)_3$ | $Al(HCO_3)_3$ | $Hg_2(HCO_3)_2$ |
| Acetate | $KC_2H_3O_2$ | $Ca(C_2H_3O_2)_2$ | $Cr(C_2H_3O_2)_3$ | $Zn(C_2H_3O_2)_2$ | $AgC_2H_3O_2$ | $Fe(C_2H_3O_2)_3$ | $Al(C_2H_3O_2)_3$ | $Hg_2(C_2H_3O_2)_2$ |
| Selenite | $K_2SeO_3$ | $CaSeO_3$ | $Cr_2(SeO_3)_3$ | $ZnSeO_3$ | $Ag_2SeO_3$ | $Fe_2(SeO_3)_3$ | $Al_2(SeO_3)_3$ | $Hg_2SeO_3$ |

**33.** $Ca(OH)_2$, $NH_4Br$, $K_2SO_4$ **35.** MgO, $AlPO_4$, $Na_2SO_4$, CaS **37.** $BaSO_3$, $Cr_2O_3$, $KIO_4$, $CaHPO_4$

*The following table contains the formulas and names of all compounds in Table 6.14.*

**Table 6.14** (completed)

| $Na^+$ | $Hg^{2+}$ | $NH_4^+$ |
|---|---|---|
| NaOH, sodium hydroxide | $Hg(OH)_2$, mercury(II) hydroxide | $NH_4OH$, ammonium hydroxide |
| NaBrO, sodium hypobromite | $Hg(BrO)_2$, mercury(II) hypobromite | $NH_4BrO$, ammonium hypobromite |
| $Na_2CO_3$, sodium carbonate | $HgCO_3$, mercury(II) carbonate | $(NH_4)_2CO_3$, ammonium carbonate |
| $NaClO_3$, sodium chlorate | $Hg(ClO_3)_2$, mercury(II) chlorate | $NH_4ClO_3$, ammonium chlorate |
| $NaHSO_4$, sodium hydrogen sulfate | $Hg(HSO_4)_2$, mercury(II) hydrogen sulfate | $NH_4HSO_4$, ammonium hydrogen sulfate |
| NaBr, sodium bromide | $HgBr_2$, mercury(II) bromide | $NH_4Br$, ammonium bromide |
| $Na_3PO_4$, sodium phosphate | $Hg_3(PO_4)_2$, mercury(II) phosphate | $(NH_4)_3PO_4$, ammonium phosphate |
| $NaIO_4$, sodium periodate | $Hg(IO_4)_2$, mercury(II) periodate | $NH_4IO_4$, ammonium periodate |
| $Na_2S$, sodium sulfide | HgS, mercury(II) sulfide | $(NH_4)_2S$, ammonium sulfide |
| $NaMnO_4$, sodium permanganate | $Hg(MnO)_4)_2$,mercury(II) permanganate | $NH_4MnO_4$, ammonium permanganate |
| $Na_2C_2O_4$, sodium oxalate | $HgC_2O_4$, mercury(II) oxalate | $(NH_4)_2C_2O_4$, ammonium oxalate |
| **$Pb^{2+}$** | **$Mg^{2+}$** | **$Ga^{3+}$** |
| $Pb(OH)_2$, lead(II) hydroxide | $Mg(OH)_2$, magnesium hydroxide | $Ga(OH)_3$, gallium hydroxide |
| $Pb(BrO)_2$, lead(II) hypobromite | $Mg(BrO)_2$, magnesium hypobromite | $Ga(BrO)_3$, gallium hypobromite |
| $PbCO_3$, lead(II) carbonate | $MgCO_3$, magnesium carbonate | $Ga_2(CO_3)_3$, gallium carbonate |
| $Pb(ClO_3)_2$, lead(II) chlorate | $Mg(ClO_3)_2$, magnesium chlorate | $Ga(ClO_3)_3$, gallium chlorate |
| $Pb(HSO_4)_2$, lead(II) hydrogen sulfate | $Mg(HSO_4)_2$, magnesium hydrogen sulfate | $Ga(HSO_4)_3$, gallium hydrogen sulfate |
| $PbBr_2$, lead(II) bromide | $MgBr_2$, magnesium bromide | $GaBr_3$, gallium bromide |
| $Pb_3(PO_4)_2$, lead(II) phosphate | $Mg_3(PO_4)_2$, magnesium phosphate | $GaPO_4$, gallium phosphate |
| $Pb(IO_4)_2$, lead(II) periodate | $Mg(IO_4)_2$, magnesium periodate | $Ga(IO_4)_3$, gallium periodate |
| PbS, lead(II) sulfide | MgS, magnesium sulfide | $Ga_2S_3$, gallium sulfide |
| $Pb(MnO_4)_2$, lead(II) permanganate | $Mg(MnO_4)_2$, magnesium permanganate | $Ga(MnO_4)_3$, gallium permanganate |
| $PbC_2O_4$, lead(II) oxalate | $MgC_2O_4$, magnesium oxalate | $Ga_2(C_2O_4)_3$, gallium oxalate |
| **$Fe^{3+}$** | **$Cu^{2+}$** | |
| $Fe(OH)_3$, iron(III) hydroxide | $Cu(OH)_2$, copper(II) hydroxide | |
| $Fe(BrO)_3$, iron(III) hypobromite | $Cu(BrO)_2$, copper(II) hypobromite | |
| $Fe_2(CO_3)_3$, iron(III) carbonate | $CuCO_3$, copper(II) carbonate | |
| $Fe(ClO_3)_3$, iron(III) chlorate | $Cu(ClO_3)_2$, copper(II) chlorate | |
| $Fe(HSO_4)_3$, iron(III) hydrogen sulfate | $Cu(HSO_4)_2$, copper(II) hydrogen sulfate | |
| $FeBr_3$, iron(III) bromide | $CuBr_2$, copper(II) bromide | |
| $FePO_4$, iron(III) phosphate | $Cu_3(PO_4)_2$, copper(II) phosphate | |
| $Fe(IO_4)_3$, iron(III) periodate | $Cu(IO_4)_2$, copper(II) periodate | |
| $Fe_2S_3$, iron(III) sulfide | CuS, copper(II) sulfide | |
| $Fe(MnO_4)_3$, iron(III) permanganate | $Cu(MnO_4)_2$, copper(II) permanganate | |
| $Fe_2(C_2O_4)_3$, iron(III) oxalate | $CuC_2O_4$, copper(II) oxalate | |

**39.** Lithium phosphate, magnesium carbonate, barium nitrate **41.** Potassium fluoride, sodium hydroxide, calcium iodide, aluminum carbonate **43.** Copper(II) sulfate, chromium(III) hydroxide, mercury(I) iodide **45.** Hydrates: $NiSO_4 \cdot 6\ H_2O$, $Na_3PO_4 \cdot 12\ H_2O$; anhydrous compound: KCl **47.** 7; magnesium sulfate heptahydrate **49.** $(NH_4)_3PO_4 \cdot 3\ H_2O$, $K_2S \cdot 5\ H_2O$

*The following list contains the names and formulas of all compounds in Table 6.15.*

$OH^-$, hydroxide ion Sodium bromide, NaBr
PbO, lead(II) oxide Ammonium phosphate, $(NH_4)_3PO_4$
$SO_4^{2-}$, sulfate ion Bromine, $Br_2$
$Cl_2$, chlorine Hydrogen phosphate ion, $HPO_4^{2-}$
$Cl_2O$, dichlorine monoxide Nitrite ion, $NO_2^-$
$Cr(ClO_4)_3$, chromium(III) perchlorate Iron(III) phosphate, $FePO_4$
$H_2PO_4^-$, dihydrogen phosphate ion Nickel sulfide, NiS
$Al_2S_3$, aluminum sulfide Nitrogen, $N_2$
$CuSO_4$, copper(II) sulfate Copper(II) bromide, $CuBr_2$
HF(aq), hydrofluoric acid Oxygen difluoride, $OF_2$
C, carbon Potassium ion, $K^+$

*The following list contains the names and formulas of all compounds in Table 6.16.*

$SeO_4^{2-}$, selenate ion Gallium sulfate, $Ga_2(SO_4)_3$
$HCO_3^-$, hydrogen carbonate ion Perchloric acid, $HClO_4$
Ne, neon Lithium, Li
$N_2O_5$, dinitrogen pentoxide Cobalt(II) chloride hexahydrate, $CoCl_2 \cdot 6\ H_2O$
$HNO_2$, nitrous acid Barium dihydrogen phosphate, $Ba(H_2PO_4)_2$
$CI_4$, carbon tetraiodide Hydrosulfuric acid, $H_2S(aq)$
$BaH_2$, barium hydride Magnesium nitride, $Mg_3N_2$
$CaTeO_3$, calcium tellurite Selenic acid, $H_2SeO_4$
HBrO, hypobromous acid Calcium sulfite, $CaSO_3$
$Fe(NO_3)_2$, iron(II) nitrate Sodium hydride, NaH
$MgSO_4 \cdot 7\ H_2O$, magnesium sulfate heptahydrate Mercury(I) chloride, $Hg_2Cl_2$

**51.** $ClO_4^-$, $BaCO_3$, ammonium iodide, phosphorus trichloride **53.** Hydrogen sulfide ion, beryllium bromide, $Al(NO_3)_3$, $OF_2$ **55.** $Hg_2^{2+}$, $CoCl_2$, silicon dioxide, lithium nitrite **57.** Nitride ion, calcium chlorate, $Fe_2(SO_4)_3$, $PCl_5$ **59.** $SnF_2$, $K_2CrO_4$, lithium hydride, iron(II) carbonate **61.** Nitrous acid, zinc hydrogen sulfate, KCN, CuF **63.** $Mg_3N_2$, $LiBrO_2$, sodium hydrogen sulfite, potassium thiocyanate **65.** Nickel hydrogen carbonate, copper(II) sulfide, $Cr(IO_3)_3$, $K_2HPO_4$ **67.** $SeO_2$, $Mg(NO_2)_2$, iron(II) bromide, silver oxide **69.** Tin(II) oxide, ammonium dichromate, NaH, $H_2C_2O_4$ **71.** $Co_2(SO_4)_3$, $FeI_3$, copper(II) phosphate, manganese(II) hydroxide **73.** Aluminum selenide, magnesium hydrogen phosphate, $KClO_4$, $HBrO_2$ **75.** $Sr(IO_3)_2$, NaClO, rubidium sulfate, diphosphorus pentoxide **77.** Iodine monochloride, silver acetate, $Pb(H_2PO_4)_2$, $GaF_3$ **79.** $MgSO_4$, $Hg(BrO_2)_2$, sodium oxalate, manganese(III) hydroxide

# Chapter 7

## Answers to Concept-Linking Exercises

*You may have found more relationships or relationships other than the ones given in these answers.*

**1.** The terms *atomic mass, molecular mass,* and *formula mass* refer to masses measured on the particulate level. Atomic mass is the mass of an atom, which is usually expressed in atomic mass units (u). Molecular and formula mass refer to the mass of a molecule and a formula unit, respectively. They are also typically expressed in atomic mass units. Molar mass, the mass in grams of one mole of a substance, is a unit that bridges the particulate and macroscopic worlds. Molar mass is numerically equal to atomic, molecular, or formula mass, but it is expressed in grams per mole. **2.** One mole is the amount of any substance that contains the same number of units as the number of atoms in exactly 12 grams of carbon-12. The experimentally determined number of particles in a mole, to three significant figures, is $6.02 \times 10^{23}$, and this number is called Avogadro's number.

## Answers to Blue-Numbered Questions, Exercises, and Problems

**1.** $Al(NO_3)_3$: 1 aluminum atom, 3 nitrogen atoms, 9 oxygen atoms **3.** *Molecular mass* is a term properly applied only to substances that exist as molecules. Sodium nitrate is an ionic compound. **5.** Atomic mass, Ba; molecular mass, $NH_3$ and $Cl_2$ because these are molecular substances; formula mass, all substances, but particularly CaO and $Na_2CO_3$ because neither of the other two terms technically fit these ionic compounds.
**7.** a) 6.941 u Li + 35.45 u Cl = 42.39 u LiCl b) 2(26.98) u Al + 3(12.01 u C) + 9(16.00 u O) = 233.99 u $Al_2(CO_3)_3$ c) 2(14.01 u N) + 8(1.008 u H) + 32.07 u S + 4(16.00 u O) = 132.15 u $(NH_4)_2SO_4$ d) 4(12.01 u C) + 10(1.008 u H) = 58.12 u $C_4H_{10}$ e) 107.9 u Ag + 14.01 u N + 3(16.00 u O) = 169.9 u $AgNO_3$ f) 54.94 u Mn + 2(16.00 u O) = 86.94 u $MnO_2$ g) 3(65.39 u Zn) + 2(30.97 u P) + 8(16.00 u O) = 386.11 u $Zn_3(PO_4)_2$
**9.** The quantity of particles of each is the same; that is, the number of iron atoms and ammonia molecules is the same. **11.** By definition, the mole is the amount of any substance that contains the same number of units as the number of atoms in exactly 12 grams of carbon-12. The definition doesn't say what that number is. Through experiment, we have found that, to three significant figures, there are $6.02 \times 10^{23}$ atoms in 12 grams of carbon-12.

**13.** a) $7.75 \text{ mol } CH_4 \times \frac{6.02 \times 10^{23} \text{ molecules } CH_4}{\text{mol } CH_4} = 4.67 \times 10^{24} \text{ molecules } CH_4$

b) $0.0888 \text{ mol CO} \times \frac{6.02 \times 10^{23} \text{ molecules CO}}{\text{mol CO}} = 5.35 \times 10^{22} \text{ molecules CO}$ c) $57.8 \text{ mol Fe} \times \frac{6.02 \times 10^{23} \text{ atoms Fe}}{\text{mol Fe}} = 3.48 \times 10^{25} \text{ atoms Fe}$

d) $0.81 \text{ mol } MgCl_2 \times \frac{6.02 \times 10^{23} \text{ fu } MgCl_2}{\text{mol } MgCl_2} = 4.88 \times 10^{23} \text{ formula units } MgCl_2$

**15.** a) $2.45 \times 10^{23} \text{ molecules } C_2H_2 \times \frac{1 \text{ mol } C_2H_2}{6.02 \times 10^{23} \text{ molecules } C_2H_2} = 0.407 \text{ mol } C_2H_2$

b) $6.96 \times 10^{24} \text{ atoms Na} \times \frac{1 \text{ mol Na}}{6.02 \times 10^{23} \text{ atoms Na}} = 11.6 \text{ mol Na}$

**17.** Numerically. **19.** a) 3(12.01 g/mol C) + 8(1.008 g/mol H) = 44.09 g/mol $C_3H_8$ b) 6(12.01 g/mol C) + 5(35.45 g/mol Cl) + 16.00 g/mol O + 1.008 g/mol H = 266.32 g/mol $C_6Cl_5OH$ c) 3(58.69 g/mol Ni) + 2(30.97 g/mol P) + 8(16.00 g/mol O) = 366.01 g/mol $Ni_3(PO_4)_2$ d) 65.39 g/mol Zn + 2(14.01 g/mol N) + 6(16.00 g/mol O) = 189.41 g/mol $Zn(NO_3)_2$

**21.** a) $6.79 \text{ g } O_2 \times \frac{1 \text{ mol } O_2}{32.00 \text{ g } O_2} = 0.212 \text{ mol } O_2$ b) $9.05 \text{ g } Mg(NO_3)_2 \times \frac{1 \text{ mol } Mg(NO_3)_2}{148.33 \text{ g } Mg(NO_3)_2} = 0.0610 \text{ mol } Mg(NO_3)_2$

c) $0.770 \text{ g } Al_2O_3 \times \frac{1 \text{ mol } Al_2O_3}{101.96 \text{ g } Al_2O_3} = 0.00755 \text{ mol } Al_2O_3$ d) $659 \text{ g } C_2H_5OH \times \frac{1 \text{ mol } C_2H_5OH}{46.07 \text{ g } C_2H_5OH} = 14.3 \text{ mol } C_2H_5OH$

e) $0.394 \text{ g } (NH_4)_2CO_3 \times \frac{1 \text{ mol } (NH_4)_2CO_3}{96.09 \text{ g } (NH_4)_2CO_3} = 0.00410 \text{ mol } (NH_4)_2CO_3$ f) $34.0 \text{ g } Li_2S \times \frac{1 \text{ mol } Li_2S}{45.95 \text{ g } Li_2S} = 0.740 \text{ mol } Li_2S$

**23.** a) $0.797 \text{ g } KIO_3 \times \frac{1 \text{ mol } KIO_3}{214.0 \text{ g } KIO_3} = 0.00372 \text{ mol } KIO_3$ b) $68.6 \text{ g } BeCl_2 \times \frac{1 \text{ mol } BeCl_2}{79.91 \text{ g } BeCl_2} = 0.858 \text{ mol } BeCl_2$

c) $302 \text{ g } Ni(NO_3)_2 \times \frac{1 \text{ mol } Ni(NO_3)_2}{182.71 \text{ g } Ni(NO_3)_2} = 1.65 \text{ mol } Ni(NO_3)_2$ **25.** a) $0.769 \text{ mol LiCl} \times \frac{42.39 \text{ g LiCl}}{\text{mol LiCl}} = 32.6 \text{ g LiCl}$

b) $57.1 \text{ mol } HC_2H_3O_2 \times \frac{60.05 \text{ g } HC_2H_3O_2}{\text{mol } HC_2H_3O_2} = 3.43 \times 10^3 \text{ g } HC_2H_3O_2$ c) $0.68 \text{ mol Li} \times \frac{6.941 \text{ g Li}}{\text{mol Li}} = 4.7 \text{ g Li}$

d) $0.532 \text{ mol } Fe_2(SO_4)_3 \times \frac{399.91 \text{ g } Fe_2(SO_4)_3}{\text{mol } Fe_2(SO_4)_3} = 213 \text{ g } Fe_2(SO_4)_3$ e) $8.26 \text{ mol } NaC_2H_3O_2 \times \frac{82.03 \text{ g } NaC_2H_3O_2}{\text{mol } NaC_2H_3O_2} = 678 \text{ g } NaC_2H_3O_2$

**27.** a) $0.379 \text{ mol } Li_2SO_4 \times \frac{109.95 \text{ g } Li_2SO_4}{\text{mol } Li_2SO_4} = 41.7 \text{ g } Li_2SO_4$ b) $4.82 \text{ mol } K_2C_2O_4 \times \frac{166.22 \text{ g } K_2C_2O_4}{\text{mol } K_2C_2O_4} = 801 \text{ g } K_2C_2O_4$

c) $0.132 \text{ mol } Pb(NO_3)_2 \times \frac{331.2 \text{ g } Pb(NO_3)_2}{\text{mol } Pb(NO_3)_2} = 43.7 \text{ g } Pb(NO_3)_2$

**29.** a) $29.6\text{ g LiNO}_3 \times \frac{1\text{ mol LiNO}_3}{68.95\text{ g LiNO}_3} \times \frac{6.02 \times 10^{23}\text{ fu LiNO}_3}{\text{mol LiNO}_3} = 2.58 \times 10^{23}\text{ formula units LiNO}_3$

b) $0.151\text{ g Li}_2\text{S} \times \frac{1\text{ mol Li}_2\text{S}}{45.95\text{ g Li}_2\text{S}} \times \frac{6.02 \times 10^{23}\text{ fu Li}_2\text{S}}{\text{mol Li}_2\text{S}} = 1.98 \times 10^{21}\text{ formula units Li}_2\text{S}$

c) $457\text{ g Fe}_2(\text{SO}_4)_3 \times \frac{1\text{ mol Fe}_2(\text{SO}_4)_3}{399.91\text{ g Fe}_2(\text{SO}_4)_3} \times \frac{6.02 \times 10^{23}\text{ fu Fe}_2(\text{SO}_4)_3}{\text{mol Fe}_2(\text{SO}_4)_3} = 6.88 \times 10^{23}\text{ formula units Fe}_2(\text{SO}_4)_3$

**31.** a) $0.0023\text{ g I}_2 \times \frac{1\text{ mol I}_2}{253.8\text{ g I}_2} \times \frac{6.02 \times 10^{23}\text{ molecules I}_2}{\text{mol I}_2} = 5.5 \times 10^{18}\text{ molecules I}_2$

b) $114\text{ g C}_2\text{H}_4(\text{OH})_2 \times \frac{1\text{ mol C}_2\text{H}_4(\text{OH})_2}{62.07\text{ g C}_2\text{H}_4(\text{OH})_2} \times \frac{6.02 \times 10^{23}\text{ molecules C}_2\text{H}_4(\text{OH})_2}{\text{mol C}_2\text{H}_4(\text{OH})_2} = 1.11 \times 10^{24}\text{ molecules C}_2\text{H}_4(\text{OH})_2$

c) $9.81\text{ g Cr}_2(\text{SO}_4)_3 \times \frac{1\text{ mol Cr}_2(\text{SO}_4)_3}{392.21\text{ g Cr}_2(\text{SO}_4)_3} \times \frac{6.02 \times 10^{23}\text{ fu Cr}_2(\text{SO}_4)_3}{\text{mol Cr}_2(\text{SO}_4)_3} = 1.51 \times 10^{22}\text{ formula units Cr}_2(\text{SO}_4)_3$

**33.** a) $4.30 \times 10^{21}\text{ molecules C}_{19}\text{H}_{37}\text{COOH} \times \frac{1\text{ mol C}_{19}\text{H}_{37}\text{COOH}}{6.02 \times 10^{23}\text{ molecules C}_{19}\text{H}_{37}\text{COOH}} \times \frac{310.50\text{ g C}_{19}\text{H}_{37}\text{COOH}}{\text{mol C}_{19}\text{H}_{37}\text{COOH}} = 2.22\text{ g C}_{19}\text{H}_{37}\text{COOH}$

b) $8.67 \times 10^{24}\text{ atoms F} \times \frac{1\text{ mol F}}{6.02 \times 10^{23}\text{ atoms F}} \times \frac{19.00\text{ g F}}{\text{mol F}} = 274\text{ g F}$

c) $7.23 \times 10^{23}\text{ fu NiCl}_2 \times \frac{1\text{ mol NiCl}_2}{6.02 \times 10^{23}\text{ fu NiCl}_2} \times \frac{129.59\text{ g NiCl}_2}{\text{mol NiCl}_2} = 156\text{ g NiCl}_2$

**35.** $1\text{ atom Au} \times \frac{1\text{ mol Au}}{6.02 \times 10^{23}\text{ atoms Au}} \times \frac{197.0\text{ g Au}}{\text{mol Au}} \times \frac{1\text{ troy oz}}{31.1\text{ g}} \times \frac{\$478}{\text{troy oz}} = \$5.03 \times 10^{-21}$

**37.** $0.65\text{ g C}_{12}\text{H}_{22}\text{O}_{11} \times \frac{1\text{ mol C}_{12}\text{H}_{22}\text{O}_{11}}{342.30\text{ g C}_{12}\text{H}_{22}\text{O}_{11}} \times \frac{6.02 \times 10^{23}\text{ C}_{12}\text{H}_{22}\text{O}_{11}\text{ molecules}}{\text{mol C}_{12}\text{H}_{22}\text{O}_{11}} = 1.1 \times 10^{21}\text{ C}_{12}\text{H}_{22}\text{O}_{11}\text{ molecules}$

**39.** a) $4.12 \times 10^{24}\text{ N atoms} \times \frac{1\text{ mol N atoms}}{6.02 \times 10^{23}\text{ atoms}} \times \frac{14.01\text{ g N}}{\text{mol N atoms}} = 95.9\text{ g N}$

b) $4.12 \times 10^{24}\text{ N}_2\text{ molecules} \times \frac{1\text{ mol N}_2\text{ molecules}}{6.02 \times 10^{23}\text{ N}_2\text{ molecules}} \times \frac{28.02\text{ g N}_2}{\text{mol N}_2\text{ molecules}} = 192\text{ g N}_2$

c) $4.12\text{ g N} \times \frac{1\text{ mol N}}{14.01\text{ g N}} \times \frac{6.02 \times 10^{23}\text{ atoms N}}{\text{mol N}} = 1.77 \times 10^{23}\text{ atoms N}$

d) $4.12\text{ g N}_2 \times \frac{1\text{ mol N}_2}{28.02\text{ g N}_2} \times \frac{6.02 \times 10^{23}\text{ N}_2\text{ molecules}}{\text{mol N}_2} = 8.85 \times 10^{22}\text{ N}_2\text{ molecules}$

e) $4.12\text{ g N}_2 \times \frac{1\text{ mol N}_2}{28.02\text{ g N}_2} \times \frac{6.02 \times 10^{23}\text{ N}_2\text{ molecules}}{\text{mol N}_2} \times \frac{2\text{ atoms N}}{\text{molecule N}_2} = 1.77 \times 10^{23}\text{ atoms N}$

**41.** a) $\text{NH}_4\text{NO}_3$ $\frac{2(14.01\text{ g N})}{80.05\text{ g NH}_4\text{NO}_3} \times 100 = 35.00\%\text{ N}$ $\frac{4(1.008\text{ g H})}{80.05\text{ g NH}_4\text{NO}_3} \times 100 = 5.037\%\text{ H}$

$\frac{3(16.00\text{ g O})}{80.05\text{ g NH}_4\text{NO}_3} \times 100 = 59.96\%\text{ O}$ $35.00\% + 5.037\% + 59.96\% = 100.00\%$

b) $\text{Al}_2(\text{SO}_4)_3$ $\frac{2(26.98\text{ g Al})}{342.17\text{ g Al}_2(\text{SO}_4)_3} \times 100 = 15.77\%\text{ Al}$ $\frac{3(32.07\text{ g S})}{342.17\text{ g Al}_2(\text{SO}_4)_3} \times 100 = 28.12\%\text{ S}$ $\frac{12(16.00\text{ g O})}{342.17\text{ g Al}_2(\text{SO}_4)_3} \times 100 = 56.11\%\text{ O}$

$15.77\% + 28.12\% + 56.11\% = 100.00\%$

c) $(\text{NH}_4)_2\text{CO}_3$ $\frac{2(14.01\text{ g N})}{96.09\text{ g (NH}_4)_2\text{CO}_3} \times 100 = 29.16\%\text{ N}$ $\frac{8(1.008\text{ g H})}{96.09\text{ g (NH}_4)_2\text{CO}_3} \times 100 = 8.392\%\text{ H}$ $\frac{12.01\text{ g C}}{96.09\text{ g (NH}_4)_2\text{CO}_3} \times 100 = 12.50\%\text{ C}$

$\frac{3(16.00\text{ g O})}{96.09\text{ g (NH}_4)_2\text{CO}_3} \times 100 = 49.95\%\text{ O}$ $29.16\% + 8.392\% + 12.50\% + 49.95\% = 100.00\%$

d) CaO $\frac{40.08\text{ g Ca}}{56.08\text{ g CaO}} \times 100 = 71.47\%\text{ Ca}$ $\frac{16.00\text{ g O}}{56.08\text{ g CaO}} \times 100 = 28.53\%\text{ O}$ $71.47\% + 28.53\%\text{ O} = 100.00\%$

e) $\text{MnS}_2$ $\frac{54.94\text{ g Mn}}{119.08\text{ g MnS}_2} \times 100 = 46.14\%\text{ Mn}$ $\frac{2(32.07\text{ g S})}{119.08\text{ g MnS}_2} \times 100 = 53.86\%\text{ S}$ $46.14\% + 53.86\% = 100.00\%$

**43.** $454\text{ g LiF} \times \frac{6.941\text{ g Li}}{25.94\text{ g LiF}} = 121\text{ g Li}$ **45.** $57.4\text{ g K}_2\text{SO}_4 \times \frac{2(39.10\text{ g K})}{174.27\text{ g K}_2\text{SO}_4} = 25.8\text{ g K}$ **47.** $146\text{ g Zn} \times \frac{117.43\text{ g Zn(CN)}_2}{65.39\text{ g Zn}} = 262\text{ g Zn(CN)}_2$

**49.** $201\text{ kg Mo} \times \frac{367.1\text{ kg PbMoO}_4}{95.94\text{ kg Mo}} = 769\text{ kg PbMoO}_4$ **51.** $4.17\text{ g Cl} \times \frac{206.98\text{ g Ca(ClO}_3)_2}{2(35.45\text{ g Cl})} = 12.17\text{ g Ca(ClO}_3)_2$

**53.** $C_6H_{10}$ must be a molecular formula because both 6 and 10 are divisible by 2. Its empirical formula is $C_3H_5$. There is no common divisor for 7 and 10, so it can be an empirical formula. An empirical formula can also be a molecular formula.

| | Element | Grams | Moles | Mole Ratio | Formula Ratio | Empirical Formula | Molecular Formula |
|---|---|---|---|---|---|---|---|
| **55.** | C | 52.2 | 4.35 | 2.00 | 2 | | |
| | H | 13.0 | 12.9 | 5.92 | 6 | $C_2H_6O$ | |
| | O | 34.8 | 2.18 | 1.00 | 1 | | |
| **57.** | Fe | 11.89 | 0.213 | 1.00 | 2 | $Fe_2O_3$ | |
| | O | 5.10 | 0.319 | 1.50 | 3 | | |
| **59.** | C | 17.2 | 1.43 | 1.00 | 1 | | |
| | H | 1.44 | 1.43 | 1.00 | 1 | $CHF_3$ | |
| | F | 81.4 | 4.28 | 2.99 | 3 | | |
| **61.** | C | 38.7 | 3.22 | 1.00 | 1 | | $\frac{62.0}{31.0} = 2$ |
| | H | 9.7 | 9.6 | 3.0 | 3 | $CH_3O$ | |
| | O | 51.6 | 3.23 | 1.00 | 1 | | $C_2H_6O_2$ |
| **63.** | Cl | 73.1 | 2.06 | 1.00 | 1 | | $\frac{97}{49} = 2$ |
| | C | 24.8 | 2.06 | 1.00 | 1 | ClCH | |
| | H | 2.1 | 2.1 | 1.0 | 2 | | $Cl_2C_2H_2$ |

**66.** All statements are false.

**67.** Hardly—the mass of $10^{25}$ atoms of copper is 2 pounds: $10^{25}\text{ Cu atoms} \times \frac{1\text{ mol Cu}}{6.02 \times 10^{23}\text{ Cu atoms}} \times \frac{63.55\text{ g Cu}}{\text{mol Cu}} \times \frac{1\text{ lb Cu}}{454\text{ g Cu}} = 2\text{ lb}$

**68.** $85.0\text{ g }P_4 \times \frac{1\text{ mol }P_4}{123.88\text{ g }P_4} \times \frac{6.02 \times 10^{23}\ P_4\text{ molecules}}{\text{mol }P_4} = 4.13 \times 10^{23}\ P_4\text{ molecules}$

$85.0\text{ g }P_4 \times \frac{1\text{ mol }P_4}{123.88\text{ g }P_4} \times \frac{6.02 \times 10^{23}\ P_4\text{ molecules}}{\text{mol }P_4} \times \frac{4\text{ P atoms}}{P_4\text{ molecule}} = 1.65 \times 10^{24}\text{ P atoms}$

**69.** It is a close call. 2 oz is about the capacity of a typical salt shaker. 12 oz is too much for a 10-oz sugar bowl, but only by a little.

$1\text{ mol NaCl} \times \frac{58.44\text{ g NaCl}}{\text{mol NaCl}} \times \frac{1\text{ lb NaCl}}{454\text{ g NaCl}} \times \frac{16\text{ oz NaCl}}{\text{lb NaCl}} = 2\text{ oz NaCl}$

$1\text{ mol }C_{12}H_{22}O_{11} \times \frac{342.30\text{ g }C_{12}H_{22}O_{11}}{\text{mol }C_{12}H_{22}O_{11}} \times \frac{1\text{ lb }C_{12}H_{22}O_{11}}{454\text{ g }C_{12}H_{22}O_{11}} \times \frac{16\text{ oz }C_{12}H_{22}O_{11}}{\text{lb }C_{12}H_{22}O_{11}} = 12\text{ oz }C_{12}H_{22}O_{11}$

**70.** $2.95 \times 10^{22}\text{ air molecules} \times \frac{1\text{ mol air}}{6.02 \times 10^{23}\text{ air molecules}} \times \frac{29\text{ g air}}{\text{mol air}} = 1.42\text{ g air}$

**71.** $5.62 \times 10^{23}\ C_8H_{18}\text{ molecules} \times \frac{1\text{ mol }C_8H_{18}}{6.02 \times 10^{23}\ C_8H_{18}\text{ molecules}} \times \frac{114.22\text{ g }C_8H_{18}}{\text{mol }C_8H_{18}} = 107\text{ g }C_8H_{18}$

**72.** $85.9\text{ g }CO_2 \times \frac{12.01\text{ g C}}{44.01\text{ g }CO_2} = 23.4\text{ g C}$ $\quad 35.5\text{ g }H_2O \times \frac{2(1.008\text{ g H})}{18.02\text{ g }H_2O} = 3.97\text{ g H}$

| Element | Grams | Moles | Mole Ratio | Formula Ratio | Empirical Formula |
|---|---|---|---|---|---|
| C | 23.4 | 1.95 | 1.00 | 1 | $CH_2$ |
| H | 3.97 | 3.94 | 2.02 | 2 | |

**73.**

| | Element | Grams | Moles | Mole Ratio | Formula Ratio | Empirical Formula |
|---|---|---|---|---|---|---|
| a) | Co | 42.4 | 0.719 | 1.00 | 1 | |
| | S | 23.0 | 0.717 | 1.00 | 1 | $CoSO_3$ |
| | O | 34.6 | 2.16 | 3.01 | 3 | |
| b) | $CoSO_3$ | 26.1 | 0.188 | 1.00 | 1 | $CoSO_3 \cdot 5\ H_2O$ |
| | $H_2O$ | 16.9 | 0.938 | 4.99 | 5 | |

43.0 g hydrate − 26.1 g anhydrate = 16.9 g $H_2O$

# Chapter 8

## Answers to Target Checks

**1.** (a) Yes; evidence includes sound, light, smell, and heat emitted. (b) No; physical transformation of liquid water to water vapor. (c) Yes; evidence includes light and heat emitted. (d) No; not a transformation of matter. **2.** (a) False. The equation has seven oxygen atoms on the left and six on the right. (b) False. Never change a chemical formula to balance an equation. $H_2O_2$ is a real compound, but it has nothing to do with this reaction.
**3.** (a) Two hydrogen peroxide molecules decompose to form one oxygen molecule and two water molecules. (b) Two moles of hydrogen peroxide molecules decompose to form one mole of oxygen molecules and two moles of water molecules. (c) Sixty-eight grams ($2 \times 34$) of hydrogen peroxide decompose to form 32 grams of oxygen and 36 grams ($2 \times 18$) grams of water ($68 = 32 + 36$). The two-significant-figure molar masses of the reactant and products are as follows: $H_2O_2$ is 34 g/mol, $O_2$ is 32 g/mol, and $H_2O$ is 18 g/mol. Your amounts may be different from these, but they must be proportional to the ratio of our masses.

## Answers to Equation-Classification Exercise

**3.** Comb **4.** Decomp **5.** SR **6.** DR **7.** SR **8.** Decomp **9.** Comb **10.** DR **11.** DR **12.** SR

## Answers to Equation-Balancing Exercise

**1.** $4\,Na + O_2 \rightarrow 2\,Na_2O$ **2.** $H_2 + Cl_2 \rightarrow 2\,HCl$ **3.** $4\,P + 3\,O_2 \rightarrow 2\,P_2O_3$ **4.** $KClO_4 \rightarrow KCl + 2\,O_2$ **5.** $Sb_2S_3 + 6\,HCl \rightarrow 2\,SbCl_3 + 3\,H_2S$ **6.** $2\,NH_3 + H_2SO_4 \rightarrow (NH_4)_2SO_4$ **7.** $CuO + 2\,HCl \rightarrow CuCl_2 + H_2O$ **8.** $Zn + Pb(NO_3)_2 \rightarrow Zn(NO_3)_2 + Pb$ **9.** $2\,AgNO_3 + H_2S \rightarrow Ag_2S + 2\,HNO_3$ **10.** $2\,Cu + S \rightarrow Cu_2S$ **11.** $2\,Al + 2\,H_3PO_4 \rightarrow 3\,H_2 + 2\,AlPO_4$ **12.** $2\,NaNO_3 \rightarrow 2\,NaNO_2 + O_2$ **13.** $Mg(ClO_3)_2 \rightarrow MgCl_2 + 3\,O_2$ **14.** $2\,H_2O_2 \rightarrow 2\,H_2O + O_2$ **15.** $2\,BaO_2 \rightarrow 2\,BaO + O_2$ **16.** $H_2CO_3 \rightarrow H_2O + CO_2$ **17.** $Pb(NO_3)_2 + 2\,KCl \rightarrow PbCl_2 + 2\,KNO_3$ **18.** $2\,Al + 3\,Cl_2 \rightarrow 2\,AlCl_3$ **19.** $2\,C_6H_{14} + 19\,O_2 \rightarrow 12\,CO_2 + 14\,H_2O$ **20.** $NH_4NO_3 \rightarrow N_2O + 2\,H_2O$ **21.** $3\,H_2 + N_2 \rightarrow 2\,NH_3$ **22.** $2\,Fe + 3\,Cl_2 \rightarrow 2\,FeCl_3$ **23.** $2\,H_2S + 3\,O_2 \rightarrow 2\,H_2O + 2\,SO_2$ **24.** $MgCO_3 + 2\,HCl \rightarrow MgCl_2 + CO_2 + H_2O$ **25.** $2\,P + 3\,I_2 \rightarrow 2\,PI_3$

## Answers to Blue-Numbered Questions, Exercises, and Problems

**1.** a) $CO + H_2O \rightarrow CO_2 + H_2$ (b) Particulate: One molecule of carbon monoxide reacts with one molecule of water to form one molecule of carbon dioxide and one molecule of hydrogen. Molar: One mole of carbon monoxide reacts with one mole of water to form one mole of carbon dioxide and one mole of hydrogen. **3.** $2\,A + B_2 \rightarrow 2\,AB$ **5.** Equations use the lowest whole-number coefficients possible to give the ratio of species in the reaction. $A + B \rightarrow AB$ is the appropriate equation for this reaction. **7.** $X_2 + 3\,Y_2 \rightarrow 2\,XY_3$ **9.** $4\,Li(s) + O_2(g) \rightarrow 2\,Li_2O(s)$ **11.** $4\,B(s) + 3\,O_2(g) \rightarrow 2\,B_2O_3(s)$ **13.** $Ca(s) + Br_2(\ell) \rightarrow CaBr_2(s)$ **15.** $2\,HI(g) \rightarrow H_2(g) + I_2(s)$ **17.** $2\,BaO_2(s) \rightarrow 2\,BaO(s) + O_2(g)$ **19.** $Ca(s) + 2\,HBr(aq) \rightarrow CaBr_2(aq) + H_2(g)$ **21.** $Cl_2(g) + 2\,KI(aq) \rightarrow 2\,KCl(aq) + I_2(s)$ **23.** $CaCl_2(aq) + 2\,KF(aq) \rightarrow CaF_2(s) + 2\,KCl(aq)$ **25.** $2\,NaOH(aq) + MgBr_2(aq) \rightarrow 2\,NaBr(aq) + Mg(OH)_2(s)$ **27.** $H_2SO_4(aq) + Ba(OH)_2(aq) \rightarrow 2\,H_2O(\ell) + BaSO_4(s)$ **29.** $3\,NaOH(s) + H_3PO_4(aq) \rightarrow Na_3PO_4(aq) + 3\,H_2O(\ell)$ **31.** $Pb(NO_3)_2(aq) + 2\,NaI(aq) \rightarrow PbI_2(s) + 2\,NaNO_3(aq)$ **33.** $2\,C_4H_{10}(\ell) + 13\,O_2(g) \rightarrow 8\,CO_2(g) + 10\,H_2O(g)$ **35.** $H_2SO_3(aq) \rightarrow SO_2(aq) + H_2O(\ell)$ **37.** $2\,K(s) + 2\,HOH(\ell) \rightarrow 2\,KOH(aq) + H_2(g)$ **39.** $Zn(s) + 2\,AgClO_3(aq) \rightarrow Zn(ClO_3)_2(aq) + 2\,Ag(s)$ **41.** $(NH_4)_2S(s) + Cu(NO_3)_2(aq) \rightarrow 2\,NH_4NO_3(aq) + CuS(s)$ **43.** $2\,P(s) + 3\,Br_2(\ell) \rightarrow 2\,PBr_3(\ell)$ *or* $P_4(s) + 6\,Br_2(\ell) \rightarrow 4\,PBr_3(\ell)$ **45.** $Ca(OH)_2(s) \rightarrow CaO(s) + H_2O(g)$ **47.** $2\,C_3H_8O_3(\ell) + 7\,O_2(g) \rightarrow 6\,CO_2(g) + 8\,H_2O(g)$ **49.** $2\,Sb(s) + 3\,Cl_2(g) \rightarrow 2\,SbCl_3(s)$ **51.** $2\,KOH(aq) + ZnCl_2(aq) \rightarrow 2\,KCl(aq) + Zn(OH)_2(s)$ **53.** $4\,Al(s) + 3\,C(s) \rightarrow Al_4C_3(s)$ **55.** $3\,Li_2SO_3(aq) + 2\,Na_3PO_4(aq) \rightarrow 2\,Li_3PO_4(s) + 3\,Na_2SO_3(aq)$ **57.** $2\,Cr(s) + 3\,Sn(NO_3)_2(aq) \rightarrow 2\,Cr(NO_3)_3(aq) + 3\,Sn(s)$ **59.** $SO_3(g) + H_2O(\ell) \rightarrow H_2SO_4(aq)$ **61.** $3\,Fe(s) + 4\,H_2O(g) \rightarrow Fe_3O_4(s) + 4\,H_2(g)$ **63.** $Al_4C_3(s) + 12\,H_2O(\ell) \rightarrow 4\,Al(OH)_3(s) + 3\,CH_4(g)$ **65.** $3\,Mg(s) + 2\,NH_3(g) \rightarrow Mg_3N_2(s) + 3\,H_2(g)$ **68.** True: a, b, c, d. False: e. Note: Statement d is true, but it is not "the whole truth." The products of a decomposition reaction may be an element and a compound or two or more compounds. **69.** a) Single replacement $Pb + Cu(NO_3)_2 \rightarrow Pb(NO_3)_2 + Cu$ b) Double replacement $Mg(OH)_2 + 2\,HBr \rightarrow MgBr_2 + 2\,H_2O$ c) $C_5H_{10}O + 7\,O_2 \rightarrow 5\,CO_2 + 5\,H_2O$ d) Double replacement $Na_2CO_3 + CaSO_4 \rightarrow Na_2SO_4 + CaCO_3$ e) Decomposition $2\,LiBr \rightarrow 2\,Li + Br_2$ f) Double replacement $NH_4Cl + AgNO_3 \rightarrow NH_4NO_3 + AgCl$ g) Combination $Ca + Cl_2 \rightarrow CaCl_2$ h) Single replacement $F_2 + 2\,NaI \rightarrow 2\,NaF + I_2$ i) Double replacement $Zn(NO_3)_2 + Ba(OH)_2 \rightarrow Zn(OH)_2 + Ba(NO_3)_2$ j) Single replacement $Cu + NiCl_2 \rightarrow CuCl_2 + Ni$ **70.** Under the circumstances discussed in this book, you may *never* write the formula of a diatomic molecule without a subscript 2 in its formula. **71.** (1) $S + O_2 \rightarrow SO_2$ Combination (2) $2\,SO_2 + O_2 \rightarrow 2\,SO_3$ Combination (3) $SO_3 + H_2O \rightarrow H_2SO_4$ Combination
**72.** $H_2SO_4 + CaCO_3 \rightarrow CaSO_4 + H_2CO_3$; $H_2SO_4 + CaCO_3 \rightarrow CaSO_4 + H_2O + CO_2$ **73.** $4\,Ag + 2\,H_2S + O_2 \rightarrow 2\,Ag_2S + 2\,H_2O$ **74.** $S + 3\,F_2 \rightarrow SF_6$ *or* $S_8 + 24\,F_2 \rightarrow 8\,SF_6$ $S + Cl_2 \rightarrow SCl_2$ *or* $S_8 + 8\,Cl_2 \rightarrow 8\,SCl_2$ $2\,S + Br_2 \rightarrow S_2Br_2$ *or* $S_8 + 4\,Br_2 \rightarrow 4\,S_2Br_2$ **75.** $WO_3 + 3\,H_2 \rightarrow W + 3\,H_2O$; single replacement redox **76.** $Ba(OH)_2(aq) + 2\,HNO_3(aq) \rightarrow Ba(NO_3)_2(aq) + 2\,H_2O(\ell)$ **78.** $HNH_2SO_3(aq) + KOH(aq) \rightarrow H_2O(\ell) + KNH_2SO_3(aq)$

# Chapter 9

## Answer to Target Check

**1.** P is a weak electrolyte because its solution is a poor conductor. This indicates the presence of ions, but in relatively low concentration. G is a strong electrolyte because its solution is a good conductor. The solution contains ions in relatively high concentration. N is a nonelectrolyte because it does not conduct. Its solution contains no ions.

## Answers to Concept-Linking Exercises

*You may have found more relationships or relationships other than the ones given in these answers.*

**1.** An electrolyte is a substance that, when dissolved, yields a solution that conducts electricity. A nonelectrolyte yields a solution that does not conduct electricity. The terms *electrolyte* and *nonelectrolyte* also refer to solutions of dissolved substances. A strong electrolyte is a substance whose solution is a good conductor of electricity. A weak electrolyte conducts electricity, but does so poorly. **2.** A conventional equation shows the complete formulas of the compounds and elements in a chemical reaction. Substances that exist as ions in water solutions are not shown in their ionic form, but rather as the hypothetical unit particle represented by the formula. A total ionic equation replaces the formulas of the dissolved substances with the major species in solution, and it includes spectator ions—those that experience no chemical change. The net ionic equation is written by removing the spectators from the total ionic equation, leaving only the species that undergo a chemical change. **3.** A redox reaction is an electron-transfer reaction. Electrons are transferred from one reacting species to another. An uncombined element appears to replace another element in a compound to yield a single-replacement equation. In ion-combination reactions, the cation from one reactant compound combines with the anion from another. These reactions have double-replacement equations. A precipitation reaction is an ion-combination reaction that yields an insoluble ionic compound. A molecule-formation reaction is an ion-combination reaction that yields a molecular product, usually water or a weak acid. If the product of an ion-combination reaction is water, formed by the reaction of an acid and a hydroxide base, the reaction is a neutralization reaction.

## Answers to Electrolyte-Classifications Exercises

| Formula | Electrolyte Classification | Major Species | Minor Species* |
|---|---|---|---|
| **3.** $C_{12}H_{22}O_{11}$ | Nonelectrolyte | $C_{12}H_{22}O_{11}(aq)$ | none |
| **4.** $HNO_2$ | Weak electrolyte | $HNO_2(aq)$ | $H^+(aq)$, $NO_2^-(aq)$ |
| **5.** HF | Weak electrolyte | HF(aq) | $H^+(aq)$, $F^-(aq)$ |
| **6.** LiF | Strong electrolyte | $Li^+(aq)$, $F^-(aq)$ | none |
| **7.** $HClO_4$ | Strong electrolyte | $H^+(aq)$, $ClO_4^-(aq)$ | $HClO_4(aq)$ |
| **8.** $HCHO_2$ | Weak electrolyte | $HCHO_2(aq)$ | $H^+(aq)$, $CHO_2^-(aq)$ |
| **9.** $NH_4NO_3$ | Strong electrolyte | $NH_4^+(aq)$, $NO_3^-(aq)$ | none |
| **10.** $HC_2H_3O_2$ | Weak electrolyte | $HC_2H_3O_2(aq)$ | $H^+(aq)$, $C_2H_3O_2^-(aq)$ |
| **11.** HCl | Strong electrolyte | $H^+(aq)$, $Cl^-(aq)$ | HCl(aq) |
| **12.** $C_6H_{12}O_6$ | Nonelectrolyte | $C_6H_{12}O_6(aq)$ | none |

*No ionic compound is completely insoluble in water. There are, in fact, very small concentrations of calcium and carbonate ions in solution in the presence of solid calcium carbonate and very small concentrations of aluminum and phosphate ions in solution when that solid is placed in water. Additionally, there are very small concentrations of hydrogen and hydroxide ions present in water or any water solution.

## Answers to Blue-Numbered Questions, Exercises, and Problems

**1.** Solutions of weak electrolytes conduct electricity poorly because only a small quantity of ions forms when the electrolyte dissolves. Solutions of nonelectrolytes do not conduct electricity because essentially no ions exist in the solution. **3.** The movement of ions makes up an electric current in solution. Soluble molecular compounds are generally neutral, and are usually nonelectrolytes. Some molecular compounds can react with water, forming ions as a product, and thus can act as electrolytes. **5.** $NH_4^+(aq)$, $SO_4^{2-}(aq)$; $Mn^{2+}(aq)$, $Cl^-(aq)$ **7.** $Ni^{2+}(aq)$, $SO_4^{2-}(aq)$; $K^+(aq)$, $PO_4^{3-}(aq)$ **9.** $H^+(aq)$, $NO_3^-(aq)$; $H^+(aq)$, $Br^-(aq)$ **11.** $H_2C_4H_4O_4(aq)$, HF(aq) **13.** $3\ Zn^{2+}(aq) + 2\ PO_4^{3-}(aq) \rightarrow Zn_3(PO_4)_2(s)$ **15.** $2\ Fe(s) + 6\ H^+(aq) \rightarrow 3\ H_2(g) + 2\ Fe^{3+}(aq)$ **17.** $Na_2C_2O_4(s) + 2\ H^+(aq) \rightarrow H_2C_2O_4(aq) + 2\ Na^+(aq)$ **19.** $Cu(s) + Li_2SO_4(aq) \rightarrow NR$ **21.** $Ba(s) + 2\ H^+(aq) \rightarrow H_2(g) + Ba^{2+}(aq)$ **23.** $Ni(s) + CaCl_2(aq) \rightarrow NR$ **25.** $C_3H_8(\ell) + 5\ O_2(g) \rightarrow 3\ CO_2(g) + 4\ H_2O(g)$ **27.** $C_2H_5OH(\ell) + 3\ O_2(g) \rightarrow 2\ CO_2(g) + 3\ H_2O(g)$ **29.** $Pb^{2+}(aq) + 2\ I^-(aq) \rightarrow PbI_2(s)$ **31.** $KClO_3(aq) + Mg(NO_2)_2(aq) \rightarrow NR$ **33.** $Ag^+(aq) + Br^-(aq) \rightarrow AgBr(s)$ **35.** $Zn^{2+}(aq) + SO_3^{2-}(aq) \rightarrow ZnSO_3(s)$ **37.** $Pb^{2+}(aq) + CO_3^{2-}(aq) \rightarrow PbCO_3(s)$; $Ca^{2+}(aq) + 2\ OH^-(aq) \rightarrow Ca(OH)_2(s)$ **39.** $H^+(aq) + NO_2^-(aq) \rightarrow HNO_2(aq)$ **41.** $H^+(aq) + C_3H_5O_3^-(aq) \rightarrow HC_3H_5O_3(aq)$ **43.** $OH^-(aq) + HF(aq) \rightarrow H_2O(\ell) + F^-(aq)$ **45.** $2\ H^+(aq) + CO_3^{2-}(aq) \rightarrow H_2O(\ell) + CO_2(aq)$ **47.** $2\ H^+(aq) + SO_3^{2-}(aq) \rightarrow H_2O(\ell) + SO_2(aq)$ **49.** $Ba^{2+}(aq) + SO_3^{2-}(aq) \rightarrow BaSO_3(s)$ **51.** $Cu^{2+}(aq) + 2\ OH^-(aq) \rightarrow Cu(OH)_2(s)$ **53.** $2\ H^+(aq) + MgCO_3(s) \rightarrow Mg^{2+}(aq) + H_2O(\ell) + CO_2(g)$ **55.** $2\ H^+(aq) + Pb(OH)_2(s) \rightarrow Pb^{2+}(aq) + 2\ H_2O(\ell)$ **57.** $2\ CH_3OH(\ell) + 3\ O_2(g) \rightarrow 2\ CO_2(g) + 4\ H_2O(\ell)$ **59.** $HC_7H_5O_2(s) + OH^-(aq) \rightarrow C_7H_5O_2^-(aq) + H_2O(\ell)$ **61.** $Ni(s) + 2\ H^+(aq) \rightarrow Ni^{2+}(aq) + H_2(g)$ **63.** $H^+(aq) + HSO_3^-(aq) \rightarrow H_2O(\ell) + SO_2(aq)$ **65.** $MgSO_4(aq) + NH_4Br(aq) \rightarrow NR$ **67.** $Mg(s) + 2\ H^+(aq) \rightarrow Mg^{2+}(aq) + H_2(g)$ **69.** $Ni(OH)_2(s) + 2\ H^+(aq) \rightarrow Ni^{2+}(aq) + 2\ H_2O(\ell)$ **71.** $F^-(aq) + H^+(aq) \rightarrow HF(aq)$ **73.** $Ag(s) + HCl(aq) \rightarrow NR$ **75.** $2\ Li(s) + 2\ H_2O(\ell) \rightarrow 2\ Li^+(aq) + 2\ OH^-(aq) + H_2(g)$ **77.** $2\ Al(s) + 3\ Cu^{2+}(aq) \rightarrow 2\ Al^{3+}(aq) + 3\ Cu(s)$ **80.** True: a, b, c, h, i, j, k, n. False: d, e, f, g, l, m.

# Chapter 10

## Answer to Target Check

**1.** According to the equation, two particles of A react for every particle of B. There are six particles of A and four particles of B available to react. When three particles of B react, all six particles of A will also react, forming three particles of $A_2B$. There will be one particle of B left.

## Answers to Problem-Classification Exercises

**1.** Mass stoichiometry **2.** Thermochemical stoichiometry **3.** Limiting reactant **4.** Mass stoichiometry **5.** Percent yield **6.** Thermochemical stoichiometry **7.** Limiting reactant **8.** Percent yield

# Answers to Blue-Numbered Questions, Exercises, and Problems

**1.** a) $95.3 \text{ mol } O_2 \times \frac{4 \text{ mol } NH_3}{5 \text{ mol } O_2} = 76.2 \text{ mol } NH_3$ b) $2.89 \text{ mol } NH_3 \times \frac{4 \text{ mol NO}}{4 \text{ mol } NH_3} = 2.89 \text{ mol NO}$ c) $3.35 \text{ mol } H_2O \times \frac{4 \text{ mol NO}}{6 \text{ mol } H_2O} = 2.23 \text{ mol NO}$

**3.** $MgO + H_2O \rightarrow Mg(OH)_2$ $0.884 \text{ mol } Mg(OH)_2 \times \frac{1 \text{ mol MgO}}{1 \text{ mol } Mg(OH)_2} = 0.884 \text{ mol MgO}$ **5.** $2\ SO_2 + O_2 \rightarrow 2\ SO_3$

$3.99 \text{ mol } SO_3 \times \frac{2 \text{ mol } SO_2}{2 \text{ mol } SO_3} = 3.99 \text{ mol } SO_2$ **7.** a) $268 \text{ g } O_2 \times \frac{1 \text{ mol } O_2}{32.00 \text{ g } O_2} \times \frac{4 \text{ mol } NH_3}{5 \text{ mol } O_2} = 6.70 \text{ mol } NH_3$

b) $31.7 \text{ mol } NH_3 \times \frac{6 \text{ mol } H_2O}{4 \text{ mol } NH_3} \times \frac{18.02 \text{ g } H_2O}{\text{mol } H_2O} = 857 \text{ g } H_2O$ c) $404 \text{ g NO} \times \frac{1 \text{ mol NO}}{30.01 \text{ g NO}} \times \frac{4 \text{ mol } NH_3}{4 \text{ mol NO}} \times \frac{17.03 \text{ g } NH_3}{\text{mol } NH_3} = 229 \text{ g } NH_3$

d) $6.41 \text{ g } H_2O \times \frac{1 \text{ mol } H_2O}{18.02 \text{ g } H_2O} \times \frac{4 \text{ mol NO}}{6 \text{ mol } H_2O} \times \frac{30.01 \text{ g NO}}{\text{mol NO}} = 7.12 \text{ g NO}$

**9.** $21.0 \text{ g } C_3H_5(NO_3)_3 \times \frac{1 \text{ mol } C_3H_5(NO_3)_3}{227.10 \text{ g } C_3H_5(NO_3)_3} \times \frac{12 \text{ mol } CO_2}{4 \text{ mol } C_3H_5(NO_3)_3} \times \frac{44.01 \text{ g } CO_2}{\text{mol } CO_2} = 12.2 \text{ g } CO_2$

**11.** $323 \text{ g } C_{17}H_{35}COONa \times \frac{1 \text{ mol } C_{17}H_{35}COONa}{306.45 \text{ g } C_{17}H_{35}COONa} \times \frac{3 \text{ mol NaOH}}{3 \text{ mol } C_{17}H_{35}COONa} \times \frac{40.00 \text{ g NaOH}}{\text{mol NaOH}} = 42.2 \text{ g NaOH}$

**13.** $681 \text{ g } Na_2S_2O_3 \times \frac{1 \text{ mol } Na_2S_2O_3}{158.12 \text{ g } Na_2S_2O_3} \times \frac{1 \text{ mol } Na_2CO_3}{3 \text{ mol } Na_2S_2O_3} \times \frac{105.99 \text{ g } Na_2CO_3}{\text{mol } Na_2CO_3} = 152 \text{ g } Na_2CO_3$

**15.** $616 \text{ mg } NaC_{18}H_{35}O_2 \times \frac{1 \text{ mmol } NaC_{18}H_{35}O_2}{306.45 \text{ mg } NaC_{18}H_{35}O_2} \times \frac{1 \text{ mmol } Ca(C_{18}H_{35}O_2)_2}{2 \text{ mmol } NaC_{18}H_{35}O_2} \times \frac{607.00 \text{ mg } Ca(C_{18}H_{35}O_2)_2}{\text{mmol } Ca(C_{18}H_{35}O_2)_2} = 6.10 \times 10^2 \text{ mg } Ca(C_{18}H_{35}O_2)_2$

**17.** $802 \text{ mg } P_4O_{10} \times \frac{1 \text{ mmol } P_4O_{10}}{283.88 \text{ mg } P_4O_{10}} \times \frac{10 \text{ mmol } Fe_2O_3}{3 \text{ mmol } P_4O_{10}} \times \frac{159.70 \text{ mg } Fe_2O_3}{\text{mmol } Fe_2O_3} = 1.50 \times 10^3 \text{ mg } Fe_2O_3 = 1.50 \text{ g } Fe_2O_3$

**19.** $83.0 \text{ g } NH_4HCO_3 \times \frac{1 \text{ mol } NH_4HCO_3}{79.06 \text{ g } NH_4HCO_3} \times \frac{1 \text{ mol NaCl}}{1 \text{ mol } NH_4HCO_3} \times \frac{58.44 \text{ g NaCl}}{\text{mol NaCl}} = 61.4 \text{ g NaCl}$

**21.** $62.0 \text{ g } NH_3 \times \frac{1 \text{ mol } NH_3}{17.03 \text{ g } NH_3} \times \frac{1 \text{ mol } CaCl_2}{2 \text{ mol } NH_3} \times \frac{110.98 \text{ g } CaCl_2}{\text{mol } CaCl_2} = 202 \text{ g } CaCl_2$ **23.** $H_3PO_4 + 3\ NaOH \rightarrow 3\ H_2O + Na_3PO_4$

$32.6 \text{ g } H_3PO_4 \times \frac{1 \text{ mol } H_3PO_4}{97.99 \text{ g } H_3PO_4} \times \frac{3 \text{ mol NaOH}}{1 \text{ mol } H_3PO_4} \times \frac{40.00 \text{ g NaOH}}{\text{mol NaOH}} = 39.9 \text{ g NaOH}$

**25.** $2\ KOH + Mg(NO_3)_2 \rightarrow 2\ KNO_3 + Mg(OH)_2$ $2.09 \text{ g KOH} \times \frac{1 \text{ mol KOH}}{56.11 \text{ g KOH}} \times \frac{1 \text{ mol } Mg(OH)_2}{2 \text{ mol KOH}} \times \frac{58.33 \text{ g } Mg(OH)_2}{\text{mol } Mg(OH)_2} = 1.09 \text{ g } Mg(OH)_2$

**27.** $2\ NaOH + H_2SO_4 \rightarrow Na_2SO_4 + 2\ H_2O$ $0.521 \text{ g } Na_2SO_4 \times \frac{1 \text{ mol } Na_2SO_4}{142.04 \text{ g } Na_2SO_4} \times \frac{1 \text{ mol } H_2SO_4}{1 \text{ mol } Na_2SO_4} \times \frac{98.08 \text{ g } H_2SO_4}{\text{mol } H_2SO_4} = 0.360 \text{ g } H_2SO_4$

**29.** $Zn + 2\ NH_4Cl \rightarrow ZnCl_2 + 2\ NH_3 + H_2$ $7.05 \text{ g } NH_3 \times \frac{1 \text{ mol } NH_3}{17.03 \text{ g } NH_3} \times \frac{1 \text{ mol Zn}}{2 \text{ mol } NH_3} \times \frac{65.38 \text{ g Zn}}{\text{mol Zn}} = 13.5 \text{ g Zn}$

**31.** $8.18 \text{ g } Na_2S_2O_3 \times \frac{1 \text{ mol } Na_2S_2O_3}{158.12 \text{ g } Na_2S_2O_3} \times \frac{1 \text{ mol NaBr}}{2 \text{ mol } Na_2S_2O_3} \times \frac{102.89 \text{ g NaBr}}{\text{mol NaBr}} = 2.66 \text{ g NaBr (theo)}$ $\frac{2.61 \text{ g NaBr (act)}}{2.66 \text{ g NaBr (theo)}} \times 100 = 98.1\% \text{ yield}$

**33.** $7.25 \text{ g } CaCN_2 \times \frac{1 \text{ mol } CaCN_2}{80.11 \text{ g } CaCN_2} \times \frac{2 \text{ mol } NH_3}{1 \text{ mol } CaCN_2} \times \frac{17.03 \text{ g } NH_3}{\text{mol } NH_3} \times \frac{92.8 \text{ g } NH_3 \text{ (act)}}{100 \text{ g } NH_3 \text{ (theo)}} = 2.86 \text{ g } NH_3 \text{ (act)}$

**35.** $5.00 \times 10^2 \text{ kg } NH_3 \text{ (act)} \times \frac{100 \text{ kg } NH_3 \text{ (theo)}}{88.8 \text{ kg } NH_3 \text{ (act)}} \times \frac{1 \text{ kmol } NH_3}{17.03 \text{ kg } NH_3} \times \frac{3 \text{ kmol } H_2}{2 \text{ kmol } NH_3} \times \frac{2.016 \text{ kg } H_2}{\text{kmol } H_2} = 1.00 \times 10^2 \text{ kg } H_2$

**37.** $7.03 \text{ g } CO_2 \times \frac{1 \text{ mol } CO_2}{44.01 \text{ g } CO_2} \times \frac{1 \text{ mol } C_6H_{12}O_6}{6 \text{ mol } CO_2} \times \frac{180.16 \text{ g } C_6H_{12}O_6}{\text{mol } C_6H_{12}O_6} = 4.80 \text{ g } C_6H_{12}O_6 \text{ (theo)}$ $\frac{3.92 \text{ g } C_6H_{12}O_6 \text{ (act)}}{4.80 \text{ g } C_6H_{12}O_6 \text{ (theo)}} \times 100 = 81.7\% \text{ yield}$

**39.** $62.5 \text{ kg } CH_3COOC_2H_5 \text{ (act)} \times \frac{100 \text{ kg } CH_3COOC_2H_5 \text{ (theo)}}{69.1 \text{ kg } CH_3COOC_2H_5 \text{ (act)}} \times \frac{1 \text{ kmol } CH_3COOC_2H_5}{88.10 \text{ kg } CH_3COOC_2H_5} \times \frac{1 \text{ kmol } CH_3COOH}{1 \text{ kmol } CH_3COOC_2H_5}$

$\times \frac{60.05 \text{ kg } CH_3COOH}{\text{kmol } CH_3COOH} = 61.7 \text{ kg } CH_3COOH$

**41.** a) $N_2(g) + 3\ H_2(g) \rightarrow 2\ NH_3(g)$ b) 4 c) Hydrogen is the limiting reactant. The 12 $H_2$ molecules in the reaction mixture are shown to yield 8 $NH_3$ molecules. The 6 $N_2$ molecules in the reaction mixture can yield 12 $NH_3$ molecules. The excess nitrogen, 2 $N_2$ molecules, is shown as part of the product mixture. **43.** The positive slope occurs in trials where iron is the limiting reactant. Thus the mass of product increases as the mass of the limiting reactant increases. When the mass of iron is greater than 2 g, the mass of product is constant, and the slope of the line is zero. This indicates that bromine is the limiting reactant in these trials.

*Questions 45–49: Both the comparison-of-moles and smaller-amount methods are shown. Differences in answers between the methods are caused by round-offs in calculations.*

| **45. Comparison-of-moles method:** | $BaCl_2$ + | $Na_2CrO_4$ → | $BaCrO_4$ + | 2 NaCl |
|---|---|---|---|---|
| Grams at start | 1.63 | 2.40 | 0 | |
| Molar mass | 208.2 | 161.98 | 253.3 | |
| Moles at start | 0.00783 | 0.0148 | 0 | |
| Moles used (+), produced (−) | −0.00783 | −0.00783 | +0.00783 | |
| Moles at end | 0 | 0.0070 | 0.00783 | |
| Grams at end | 0 | 1.13 | 1.98 | |

**Smaller-amount method:**

$$1.63 \text{ g BaCl}_2 \times \frac{1 \text{ mol BaCl}_2}{208.2 \text{ g BaCl}_2} \times \frac{1 \text{ mol BaCrO}_4}{1 \text{ mol BaCl}_2} \times \frac{253.3 \text{ g BaCrO}_4}{\text{mol BaCrO}_4} = 1.98 \text{ g BaCrO}_4$$

$$2.40 \text{ g Na}_2\text{CrO}_4 \times \frac{1 \text{ mol Na}_2\text{CrO}_4}{161.98 \text{ g Na}_2\text{CrO}_4} \times \frac{1 \text{ mol BaCrO}_4}{1 \text{ mol Na}_2\text{CrO}_4} \times \frac{253.3 \text{ g BaCrO}_4}{\text{mol BaCrO}_4} = 3.75 \text{ g BaCrO}_4$$

$BaCl_2$ is the limiting reactant. The yield is the smaller amount, 1.98 g $BaCrO_4$.

$$1.63 \text{ g BaCl}_2 \times \frac{1 \text{ mol BaCl}_2}{208.2 \text{ g BaCl}_2} \times \frac{1 \text{ mol Na}_2\text{CrO}_4}{1 \text{ mol BaCl}_2} \times \frac{161.98 \text{ g Na}_2\text{CrO}_4}{\text{mol Na}_2\text{CrO}_4} = 1.27 \text{ g Na}_2\text{CrO}_4$$

2.40 g $Na_2CrO_4$ (initial) − 1.27 g $Na_2CrO_4$ (used) = 1.13 g $Na_2CrO_4$ left

**47. Comparison-of-moles method:**

| | $2\ NaIO_3$ + | $5\ NaHSO_3$ → | $I_2$ + others |
|---|---|---|---|
| Kilograms at start | 6.00 | 7.33 | 0 |
| Molar mass | 197.9 | 104.07 | 253.8 |
| Kilomoles at start | 0.0303 | 0.0704 | 0 |
| Kilomoles used (+), produced (−) | −0.0282 | −0.0704 | +0.0141 |
| Kilomoles at end | 0.0021 | 0 | 0.0141 |
| Kilograms at end | 0.42 | 0 | 3.58 |

**Smaller-amount method:**

$$6.00 \text{ kg NaIO}_3 \times \frac{1 \text{ kmol NaIO}_3}{197.9 \text{ kg NaIO}_3} \times \frac{1 \text{ kmol I}_2}{2 \text{ kmol NaIO}_3} \times \frac{253.8 \text{ kg I}_2}{\text{kmol I}_2} = 3.85 \text{ kg I}_2$$

$$7.33 \text{ kg NaHSO}_3 \times \frac{1 \text{ kmol NaHSO}_3}{104.07 \text{ kg NaHSO}_3} \times \frac{1 \text{ kmol I}_2}{5 \text{ kmol NaHSO}_3} \times \frac{253.8 \text{ kg I}_2}{\text{kmol I}_2} = 3.58 \text{ kg I}_2$$

$NaHSO_3$ is the limiting reactant. The yield is the smaller amount, 3.58 kg $I_2$.

$$7.33 \text{ kg NaHSO}_3 \times \frac{1 \text{ kmol NaHSO}_3}{104.07 \text{ kg NaHSO}_3} \times \frac{2 \text{ kmol NaIO}_3}{5 \text{ kmol NaHSO}_3} \times \frac{197.9 \text{ kg NaIO}_3}{\text{kmol NaIO}_3} = 5.58 \text{ kg NaIO}_3$$

6.00 kg $NaIO_3$ (initial) − 5.58 kg $NaIO_3$ (used) = 0.42 kg $NaIO_3$ left

**49. Comparison-of-moles method:**

| | $8\ P_4$ + | $3\ S_8$ → | $8\ P_4S_3$ |
|---|---|---|---|
| Grams at start | 133 | 126 | 0 |
| Molar mass | 123.88 | 256.56 | 220.09 |
| Moles at start | 1.07 | 0.491 | 0 |
| Moles used (+), produced (−) | −1.07 | −0.401 | +1.07 |
| Moles at end | 0 | 0.090 | 1.07 |
| Grams at end | 0 | 23.1 | 235 |

**Smaller-amount method:**

$$133 \text{ g P}_4 \times \frac{1 \text{ mol P}_4}{123.88 \text{ g P}_4} \times \frac{8 \text{ mol P}_4\text{S}_3}{8 \text{ mol P}_4} \times \frac{220.09 \text{ g P}_4\text{S}_3}{\text{mol P}_4\text{S}_3} = 236 \text{ g P}_4\text{S}_3 \quad 126 \text{ g S}_8 \times \frac{1 \text{ mol S}_8}{256.56 \text{ g S}_8} \times \frac{8 \text{ mol P}_4\text{S}_3}{3 \text{ mol S}_8} \times \frac{220.09 \text{ g P}_4\text{S}_3}{\text{mol P}_4\text{S}_3} = 288 \text{ g P}_4\text{S}_3$$

$P_4$ is the limiting reactant. The yield is the smaller amount, 236 g $P_4S_3$. $133 \text{ g P}_4 \times \frac{1 \text{ mol P}_4}{123.88 \text{ g P}_4} \times \frac{3 \text{ mol S}_8}{8 \text{ mol P}_4} \times \frac{256.56 \text{ g S}_8}{\text{mol S}_8} = 103 \text{ g S}_8$

126 g $S_8$ (initial) − 103 g $S_8$ (used) = 23 g $S_8$ left

**51.** a) $0.731 \text{ kcal} \times \frac{4.184 \text{ kJ}}{\text{kcal}} = 3.06 \text{ kJ}$ b) $651 \text{ J} \times \frac{1 \text{ cal}}{4.184 \text{ J}} = 156 \text{ cal}$ c) $6.22 \times 10^3 \text{ J} \times \frac{1 \text{ cal}}{4.184 \text{ J}} \times \frac{1 \text{ kcal}}{1000 \text{ cal}} = 1.49 \text{ kcal}$

**53.** $493 \text{ kJ} \times \frac{1 \text{ kcal}}{4.184 \text{ kJ}} = 118 \text{ kcal} \qquad 118 \text{ kcal} \times \frac{1000 \text{ cal}}{\text{kcal}} = 1.18 \times 10^5 \text{ cal}$ **55.** $\frac{5.8 \times 10^2 \text{ kcal}}{\text{day}} \times \frac{4.184 \text{ kJ}}{\text{kcal}} \times \frac{365 \text{ days}}{\text{year}} = 8.9 \times 10^5 \text{ kJ/year}$

**57.** $2.82 \times 10^3 \text{ kJ} + 6\,CO_2(g) + 6\,H_2O(\ell) \rightarrow C_6H_{12}O_6(s) + 6\,O_2(g)$ $6\,CO_2(g) + 6\,H_2O(\ell) \rightarrow C_6H_{12}O_6(s) + 6\,O_2(g) \qquad \Delta H = +2.82 \times 10^3 \text{ kJ}$

**59.** $286 \text{ kJ} + H_2O(\ell) \rightarrow H_2(g) + \frac{1}{2}\,O_2(g)$ $H_2O(\ell) \rightarrow H_2(g) + \frac{1}{2}\,O_2(g) \qquad \Delta H = +286 \text{ kJ}$ **61.** $C_2H_2(g) + \frac{5}{2}\,O_2(g) \rightarrow 2\,CO_2(g) + H_2O(\ell) + 1.31 \times 10^3 \text{ kJ}$ $C_2H_2(g) + \frac{5}{2}\,O_2(g) \rightarrow 2\,CO_2(g) + H_2O(\ell) \qquad \Delta H = -1.31 \times 10^3 \text{ kJ}$

**63.** $5.80 \text{ kg CaCO}_3 \times \frac{1000 \text{ g CaCO}_3}{\text{kg CaCO}_3} \times \frac{1 \text{ mol CaCO}_3}{100.09 \text{ g CaCO}_3} \times \frac{178 \text{ kJ}}{1 \text{ mol CaCO}_3} = 1.03 \times 10^4 \text{ kJ}$

**65.** $291 \text{ kJ} \times \frac{1 \text{ mol CaO}}{65.3 \text{ kJ}} \times \frac{56.08 \text{ g CaO}}{\text{mol CaO}} = 2.50 \times 10^2 \text{ g CaO}$ **67.** $9.48 \times 10^5 \text{ kJ} \times \frac{2 \text{ mol C}_8\text{H}_{18}}{1.09 \times 10^4 \text{ kJ}} \times \frac{114.22 \text{ g C}_8\text{H}_{18}}{\text{mol C}_8\text{H}_{18}} = 1.99 \times 10^4 \text{ g C}_8\text{H}_{18}$

**70.** True: a, d, e, f. False: b, c. **71.** $35 \text{ g N}_2 \times \frac{1 \text{ mol N}_2}{28.02 \text{ g N}_2} \times \frac{1 \text{ mol Na}_2\text{CO}_3}{1 \text{ mol N}_2} \times \frac{105.99 \text{ g Na}_2\text{CO}_3}{\text{mol Na}_2\text{CO}_3} = 1.3 \times 10^2 \text{ g Na}_2\text{CO}_3$

**72.** $3\,Ca(NO_3)_2 + 2\,Na_3PO_4 \rightarrow Ca_3(PO_4)_2 + 6\,NaNO_3$

$$3.98 \text{ g Na}_3\text{PO}_4 \times \frac{1 \text{ mol Na}_3\text{PO}_4}{163.94 \text{ g Na}_3\text{PO}_4} \times \frac{1 \text{ mol Ca}_3(\text{PO}_4)_2}{2 \text{ mol Na}_3\text{PO}_4} \times \frac{310.18 \text{ g Ca}_3(\text{PO}_4)_2}{\text{mol Ca}_3(\text{PO}_4)_2} = 3.77 \text{ g Ca}_3(\text{PO}_4)_2$$

**73.** $125\text{ g }KO_2 \times \frac{1\text{ mol }KO_2}{71.10\text{ g }KO_2} \times \frac{3\text{ mol }O_2}{4\text{ mol }KO_2} \times \frac{32.00\text{ g }O_2}{\text{mol }O_2} = 42.2\text{ g }O_2$

**74.** a) $6.00\text{ g }H_3C_6H_5O_7 \times \frac{1\text{ mol }H_3C_6H_5O_7}{192.12\text{ g }H_3C_6H_5O_7} \times \frac{3\text{ mol }NaHCO_3}{1\text{ mol }H_3C_6H_5O_7} \times \frac{84.01\text{ g }NaHCO_3}{\text{mol }NaHCO_3} = 7.87\text{ g }NaHCO_3$;

20.0 g $NaHCO_3$ available; $H_3C_6H_5O_7$ is limiting

$6.00\text{ g }H_3C_6H_5O_7 \times \frac{1\text{ mol }H_3C_6H_5O_7}{192.12\text{ g }H_3C_6H_5O_7} \times \frac{3\text{ mol }CO_2}{1\text{ mol }H_3C_6H_5O_7} \times \frac{44.01\text{ g }CO_2}{\text{mol }CO_2} = 4.12\text{ g }CO_2$

b) $20.0\text{ g }NaHCO_3 - 7.87\text{ g }NaHCO_3 = 12.1\text{ g }NaHCO_3$ unreacted

c) When changing the name of an *-ic* acid to an ion, *-ic* changes to *-ate*. Thus citr*ic* acid, $H_3C_6H_5O_7$, becomes citr*ate* ion, $C_6H_5O_7^{3-}$. $Na_3C_6H_5O_7$ is therefore sodium citrate.

**75.** $1.68\text{ g Al} \times \frac{1\text{ mol Al}}{26.98\text{ g Al}} \times \frac{2\text{ mol }Al_2O_3}{4\text{ mol Al}} \times \frac{101.96\text{ g }Al_2O_3}{\text{mol }Al_2O_3} = 3.17\text{ g }Al_2O_3$ $\frac{3.17\text{ g }Al_2O_3}{12.8\text{ g ore}} \times 100 = 24.8\%$ $Al_2O_3$ in the ore

**76.** $1.42\text{ g }KClO_3 \times \frac{1\text{ mol }KClO_3}{122.55\text{ g }KClO_3} \times \frac{89.5\text{ kJ}}{2\text{ mol }KClO_3} = 0.519\text{ kJ}$

**77.** $0.500\text{ ton }Ca(H_2PO_4)_2 \times \frac{2000\text{ lb}}{\text{ton}} \times \frac{1\text{ kg}}{2.20\text{ lb}} \times \frac{1\text{ kmol }Ca(H_2PO_4)_2}{234.05\text{ kg }Ca(H_2PO_4)_2} \times \frac{1\text{ kmol }Ca_3(PO_4)_2}{1\text{ kmol Ca }(H_2PO_4)_2}$

$\times \frac{310.18\text{ kg }Ca_3(PO_4)_2}{\text{kmol }Ca_3(PO_4)_2} \times \frac{100\text{ kg rock}}{79.4\text{ kg }Ca_3(PO_4)_2} = 759\text{ kg rock}$

**79.** $40.1\text{ kg sludge} \times \frac{23.1\text{ kg AgCl}}{100\text{ kg sludge}} \times \frac{1\text{ kmol AgCl}}{143.4\text{ kg AgCl}} \times \frac{4\text{ kmol NaCN}}{2\text{ kmol AgCl}} \times \frac{49.01\text{ kg NaCN}}{\text{kmol NaCN}} = 6.33\text{ kg NaCN}$

**81.** $105\text{ kg }Cl_2\text{ (act)} \times \frac{100\text{ kg }Cl_2\text{ (theo)}}{61\text{ kg }Cl_2\text{ (act)}} \times \frac{1\text{ kmol }Cl_2}{70.90\text{ kg }Cl_2} \times \frac{2\text{ kmol NaCl}}{1\text{ kmol }Cl_2} \times \frac{58.44\text{ kg NaCl}}{\text{kmol NaCl}} \times \frac{100\text{ kg solution}}{9.6\text{ kg NaCl}} = 3.0 \times 10^3\text{ kg solution}$

**83.** $239\text{ mg }Ca(OH)_2 \times \frac{1\text{ mmol }Ca(OH)_2}{74.10\text{ mg }Ca(OH)_2} \times \frac{1\text{ mmol }SnF_2}{1\text{ mmol }Ca(OH)_2} \times \frac{156.7\text{ mg }SnF_2}{1\text{ mmol }SnF_2} = 505\text{ mg }SnF_2$ is needed to treat 239 mg $Ca(OH)_2$.

505 mg $SnF_2$ needed $-$ 305 mg used $= 2.00 \times 10^2$ mg $SnF_2$ additional should be used.

**85.** $454\text{ g Al} \times \frac{1\text{ mol Al}}{26.98\text{ g Al}} \times \frac{1.97 \times 10^3\text{ kJ}}{4\text{ mol Al}} \times \frac{1\text{ kw-hr}}{3.60 \times 10^3\text{ kJ}} = 2.30\text{ kw-hr}$

**87.** $NaCl + AgNO_3 \rightarrow NaNO_3 + AgCl$ $2.056\text{ g AgCl} \times \frac{1\text{ mol AgCl}}{143.4\text{ g AgCl}} \times \frac{1\text{ mol NaCl}}{1\text{ mol AgCl}} \times \frac{58.44\text{ g NaCl}}{\text{mol NaCl}} = 0.8379\text{ g NaCl}$

$1.6240\text{ g mixture} - 0.8379\text{ g NaCl} = 0.7861\text{ g }NaNO_3$ $\frac{0.8379\text{ g NaCl}}{1.6240\text{ g mixture}} \times 100 = 51.59\%\text{ NaCl}$ $\frac{0.7861\text{ g }NaNO_3}{1.6240\text{ g mixture}} \times 100 = 48.41\%\text{ }NaNO_3$

**88.** From the formula, there is 3 mol Cu in 1 mol $Cu_3(PO_4)_2$

$2.637\text{ g }Cu_3(PO_4)_2 \times \frac{1\text{ mol }Cu_3(PO_4)_2}{380.59\text{ g }Cu_3(PO_4)_2} \times \frac{3\text{ mol Cu}}{1\text{ mol }Cu_3(PO_4)_2} \times \frac{63.55\text{ g Cu}}{\text{mol Cu}} = 1.321\text{ g Cu}$ $\frac{1.321\text{ g Cu}}{1.382\text{ g sample}} \times 100 = 95.59\%\text{ Cu}$

**89.** $50.0\text{ mL} \times \frac{1.19\text{ g soln}}{\text{mL}} \times \frac{17.0\text{ g NaOH}}{100\text{ g soln}} \times \frac{1\text{ mol NaOH}}{40.00\text{ g NaOH}} \times \frac{1\text{ mol }Mg(NO_3)_2}{2\text{ mol NaOH}} \times \frac{148.33\text{ g }Mg(NO_3)_2}{\text{mol }Mg(NO_3)_2} = 18.8\text{ g }Mg(NO_3)_2$

**90.** $AgNO_3 + NaCl \rightarrow AgCl + NaNO_3$ $AgNO_3 + NaBr \rightarrow AgBr + NaNO_3$

$0.230\text{ g NaCl} \times \frac{1\text{ mol NaCl}}{58.44\text{ g NaCl}} \times \frac{1\text{ mol }AgNO_3}{1\text{ mol NaCl}} \times \frac{169.9\text{ g }AgNO_3}{\text{mol }AgNO_3} = 0.669\text{ g }AgNO_3$

$0.771\text{ g NaBr} \times \frac{1\text{ mol NaBr}}{102.89\text{ g NaBr}} \times \frac{1\text{ mol }AgNO_3}{1\text{ mol NaBr}} \times \frac{169.9\text{ g }AgNO_3}{\text{mol }AgNO_3} = 1.27\text{ g }AgNO_3$ $0.669\text{ g }AgNO_3 + 1.27\text{ g }AgNO_3 = 1.94\text{ g }AgNO_3$

# Chapter 11

## Answers to Target Checks

**1.** The product of the frequency and wavelength of electromagnetic radiation is a fixed quantity, the speed of light: $\nu \times \lambda = c$. As frequency decreases, wavelength must increase so that their product remains equal to $3.00 \times 10^8$ m/s. **2.** A continuous spectrum contains a rainbow of colors that result from energy at all wavelengths visible to humans. It has no lines or bands. A line spectrum has discrete lines of color that correspond to only some of the wavelengths in the visible spectrum. Continuous spectra result from passing white light, for instance the light generated by a standard lightbulb, through a prism. Line spectra are the product of passing through a prism the light emitted by pure elemental substances in gas discharge tubes. **3.** True: b, c. (a) The speed of automobiles on a highway is *continuous*. (d) The volume of water coming from a faucet is *continuous*. (e) A person's height is *continuous*. (f) Bohr described the orbit of the electron in a *hydrogen* atom. **4.** True: c, d. (a) There is one *s* orbital for each value of *n*. (b) The 3*d* orbitals are at higher energy than the 4*s* orbital. **5.** (a) $C < N < F$. (b) halogen, F; alkali metal, Li; noble gas, Ne; alkaline earth, Be. (c) $O < N < P$. (d) metals, Na, Mg, Al; nonmetals, P, S, Cl, Ar; metalloid, Si.

## Answers to Concept-Linking Exercises

*You may have found more relationships or relationships other than the ones given in these answers.*

**1.** The electromagnetic spectrum refers to the whole range of electromagnetic waves, which includes visible light. In a continuous spectrum of light, the colors blend into each other; there are no separations. In a line spectrum, only separate lines of color appear. **2.** If energy levels change gradually in continuous values of energy, theoretically, between any two values, there is an infinite number of other values. Quantized energy levels have only certain values of energy; "in between" values do not exist. In an one-electron system, the ground state is when the electron is in the lowest quantized energy level, $n = 1$. If the electron is in any level above $n = 1$, it is in an excited state. **3.** The Bohr model of the hydrogen atom pictures the electron in the atom moving in a precisely defined *orbit* around the nucleus. An *orbital* is a region of space around the nucleus in which there is a high probability of finding the electron. An orbital does not define the path of the electron. **4.** The principal energy level ($n$), sublevel ($s, p, d, f$), and electron orbitals (1 $s$ orbital, 3 $p$ orbitals, 5 $d$ orbitals, and 7 $f$ orbitals) represent values that are described mathematically by the quantum mechanical model of the atom. The Pauli exclusion principle limits the occupancy of an orbital to two electrons. **5.** Valence electrons are the electrons in the highest occupied principal energy level when the atom is at ground state. Lewis symbols show the valence electrons as dots placed around the symbol of the element. A filled set of valence electrons, in which the $s$ orbital and all three $p$ orbitals at the highest occupied principal energy level contain two electrons, is referred to as an octet of electrons. **6.** The ionization energy of an element is the amount of energy required to remove one electron from an atom of that element. The second and third ionization energies are the energies required to remove the second and third electrons, respectively, from the atom. **7.** Chemical families are groups of elements that have similar chemical properties. They are located in groups in the periodic table: alkali metals in Group 1A/1; alkaline earths in Group 2A/2; halogens in Group 7A/17; and noble gases in Group 8A/18. Though hydrogen forms compounds that correspond to compounds formed by alkali metals and halogens, other properties exclude hydrogen from either of those families. **8.** Atomic size is affected by the highest occupied principal energy level (the higher the level, the larger the atom) and the nuclear charge (the higher the charge for a given principal energy level, the smaller the atom). **9.** Physically, metals are distinguished from nonmetals by certain properties, such as luster and good electrical conductivity. The character of the elements changes from metal to nonmetal from left to right in any row of the periodic table. The stair-step line in the periodic table separates the metals (left) from the nonmetals (right). The properties of the elements close to the line are neither distinctly metallic nor nonmetallic. These elements are metalloids or semimetals.

## Answers to Blue-Numbered Questions, Exercises, and Problems

**1.** Gamma rays, X-rays, ultraviolet radiation, infrared radiation, microwaves, and radio waves are other parts of the electromagnetic spectrum. X-rays, microwaves, and radio waves are commonly part of everyday vocabulary. **3.** Wavelength, frequency, velocity (speed of light, $c$) **5.** d and e **7.** All colors of light could be emitted because of the infinite number of energy differences possible. White light consists of all visible colors. Infrared and ultraviolet light would also be emitted. **9.** An atom must absorb energy before it can release that energy in the form of light. **11.** An atom with its electron(s) in the ground state cannot emit light because there is no lower energy level to which the electron may fall. Only an atom with electrons in excited states can emit light. **13.** Individual lines result when an electron jumps from $n = 2$ to $n = 1$, $n = 3$ to $n = 2$, $n = 3$ to $n = 1$, etc. **15.** The Bohr model provided an explanation for atomic line spectra in terms of electron energies. It also introduced the idea of quantized electron energy levels in the atom. **17.** In general, energies increase as the principal energy level increases: $n = 1 < n = 2 < n = 3 \ldots < n = 7$. **19.** The total number of sublevels within a given principal energy level is equal to $n$, the principal quantum number. **21.** $s$: 1; $p$: 3; $d$: 5; $f$: 7 **23.** False. Each $p$ sublevel contains three orbitals, each of which can hold two electrons, for a maximum of six electrons in the $p$ sublevel. **25.** True. Each principal energy level has $n$ sublevels. **27.** False. An orbital may hold one or two electrons, or it may be empty. **29.** See Figure 11.10. **31.** 6; 10; 14 **33.** For elements other than hydrogen, the energy of each principal energy level spreads over a range related to the sublevels. **35.** (b) Bohr's quantized electron energy levels appear in quantum theory as principal energy levels. **37.** $n = 2$ has 2 sublevels, $s$ and $p$, so there is no such thing as a $2d$ electron. It is an incorrect symbol. **39.** Sulfur, which is in Period 3, Group 6A/16. **41.** [Ar] substitutes for $1s^22s^22p^63s^23p^6$, and it is used to write shorthand electron configurations for elements with $19 < Z < 35$. **43.** (a) selenium; (b) oxygen; (c) magnesium **45.** (a) nitrogen; (b) aluminum; (c) cobalt **47.** Mg: $1s^22s^22p^63s^2$; Ni: $1s^22s^22p^63s^23p^64s^23d^8$ **49.** Cr: $1s^22s^22p^63s^23p^64s^13d^5$; Se: $1s^22s^22p^63s^23p^64s^23d^{10}4p^4$ **51.** Mg: [Ne]$3s^2$; Ni: [Ar]$4s^23d^8$; Cr: [Ar]$4s^13d^5$; Se: [Ar]$4s^23d^{10}4p^4$ **53.** [Ar]$4s^23d^3$ **55.** Many of the similar chemical properties of elements in the same column of the periodic table are related to the valence electrons. **57.** $3s^23p^1$ or Ȧl: **59.** $ns^2np^4$ **61.** (a) Z = 19, 20, 31, 32. (b) Z = 34, 35, 36. **63.** As you go from left to right across a row in the periodic table, the valence electrons are all in the same principal energy level. As the number of protons in an atom increases, the positive charge in the nucleus increases. This pulls the valence electrons closer to the nucleus, so the atom becomes smaller. **65.** Germanium has $n = 4$ valence electrons; silicon has $n = 3$ valence electrons. When comparing atoms in the same group, the number of occupied energy levels is more influential than nuclear charge in determining size. **67.** Ionization energies of elements in the same chemical family decrease as atomic number increases. **69.** Within a period, the valence electrons are in the same principal energy level. As the number of protons increases across a period, the positive nuclear charge increases, exerting a greater pull on the electrons. Therefore, ionization energy must increase to pull an electron away. **71.** They both have the valence electron configuration $ns^2$. **73.** Halogens **75.** $ns^2np^6$ (Helium is a noble gas with the electron configuration $1s^2$.) **77.** (a) noble gases; (b) alkaline earths **79.** Both chlorine and iodine have seven valence electrons: $ns^2np^5$. **81.** Generally, an element is known as a metal if it can lose one or more electrons and become a positively charged ion. **83.** Most of the elements next to the stair-step line in the periodic table have some properties of both metals and nonmetals. These elements are the metalloids or semimetals. **85.** (a) G, A; (b) R, T **87.** X, Q, R, Z, M, T **89.** Q < X < Z **92.** True: b, e, g, i, k, m, n, p. False: a, c, d, f, h, j, l, o. **93.** The other lines are outside the visible spectrum, in the infrared or ultraviolet regions. **94.** Something that behaves like a wave has properties normally associated with waves, some of which appear in Figure 11.2. **95.** Ba: [Xe]$6s^2$, Tc: [Kr]$5s^24d^5$ **99.** Aluminum atoms have a lower nuclear charge—fewer protons in the nucleus—than chlorine atoms. It is therefore easier to remove a $3p$ electron from an aluminum atom than from a chlorine atom, so the ionization energy of aluminum is lower. **102.** The quantum and Bohr model explanations of atomic spectra are essentially the same. **103.** All species have a single electron. Species with two or more electrons are far more complex. **104.** $Sc^{3+}$ is isoelectronic with an argon atom, $1s^22s^22p^63s^23p^6$. **105.** To form a monatomic ion, carbon would have to lose four electrons. The fourth ionization energy of any atom is very high. **106.** The smaller atoms in Group 5A/15 tend to complete their octets by gaining or sharing electrons, which is a characteristic of nonmetals. Larger atoms in the group tend to lose their highest-energy $s$ electrons and form positively charged ions, a characteristic of metals. **107.** Xenon has a low ionization energy and high reactivity, relative to the other noble gases. This is the same ionization energy trend seen in all groups in the periodic table. **108.** Iron loses two electrons from the $4s$ orbital to form $Fe^{2+}$ and a third from a $3d$ orbital to form $Fe^{3+}$. This is an example of $d$ electrons contributing to the chemical properties of an element. **109.** (a) Ionization energy increases across a period of the periodic table because of increasing nuclear charge and the same principal energy level of the outermost electrons. (b) The breaks in ionization energy trends across Periods 2 and 3 occur just after the $s$ orbital is filled and just after the $p$ orbitals are half-filled.

# Chapter 12

## Answers to Target Checks

**1.** $S^{2-}$ and $Ba^{2+}$. Copper(II), iron(III), and silver ions are all positive ions, which means that they lose electrons to form ions. If they lose electrons, they potentially can be isoelectronic only with the nearest noble-gas atom with a lower atomic number. All have *d* sublevel electrons not present in the noble-gas atom with a lower atomic number. **2.** $K\cdot + \cdot\ddot{F}: \longrightarrow K^+ + [:\ddot{F}:]^- \longrightarrow$ KF crystal **3.** All are true. **4.** **II**, **III**, and **IV** have double bonds. **I** and **II** have triple bonds. All have multiple bonds. **5.** (a) No, it is not possible, under the octet rule, for a single atom to be bonded by double bonds to each of three other atoms. Three double bonds would be 6 electron pairs, placing 12 electrons around the central atom. (b) Four atoms can be bonded by single bonds to the same central atom. At two electrons per bond, there would be eight electrons around the atom, a full octet.

## Answers to Concept-Linking Exercises

*You may have found more relationships or relationships other than the ones given in these answers.*

**1.** Chemical bonds are forces that hold together: (1) atoms in molecules; (2) atoms in polyatomic ions; (3) ions in ionic compounds; (4) atoms in metals. Valence electrons, the outermost electrons, are responsible for these bonding forces. **2.** Ionic bonds are the forces that hold ions in fixed positions in a crystal—a solid with a definite geometric structure. The positively charged ions, cations, and negatively charged ions, anions, arrange themselves in the crystal in a manner that minimizes the potential energy of the crystal. An ionic bond can be thought of as an electron-transfer bond because the atoms that form cations do so by transferring electrons to other atoms to change them into anions. **3.** A covalent bond is formed by the overlap of two atomic orbitals. This produces an electron cloud that is concentrated between the nuclei being bonded. Atoms in molecules are held together by covalent bonds. **4.** Bonding electrons are shared equally in a nonpolar bond; they are shared unequally in a polar bond. Electronegativity is a measure of the ability of an atom to attract bonding electrons toward itself. The distribution of bonding electron charge will be unequal in a polar bond, with the charge density shifted toward the more electronegative atom. **5.** On the particulate level, a metal can be thought of as positive ions surrounded by a sea of freely moving electrons. In this electron-sea model, the electrons that are free to move are the outermost electrons of the metal atoms—the valence electrons. In a metallic bond, a sea of electrons is shared among positively charged metal ions. **6.** An ionic bond forms between oppositely charged ions when the forces of attraction and repulsion balance in the lowest-energy state, forming an ionic crystal. A covalent bond forms between positively charged nuclei when the nuclei share negatively charged electron pairs.

## Answers to Blue-Numbered Questions, Exercises, and Problems

**1.** All monatomic ions of third-period elements that are isoelectronic with a noble-gas atom have the electron configuration of Ne (cations) or Ar (anions): $1s^22s^22p^6$ or $1s^22s^22p^63s^23p^6$. **3.** Any two of $K^+$, $Ca^{2+}$, $Sc^{3+}$. **5.** Any two of $Te^{2-}$, $I^-$, $Cs^+$.

**7.**

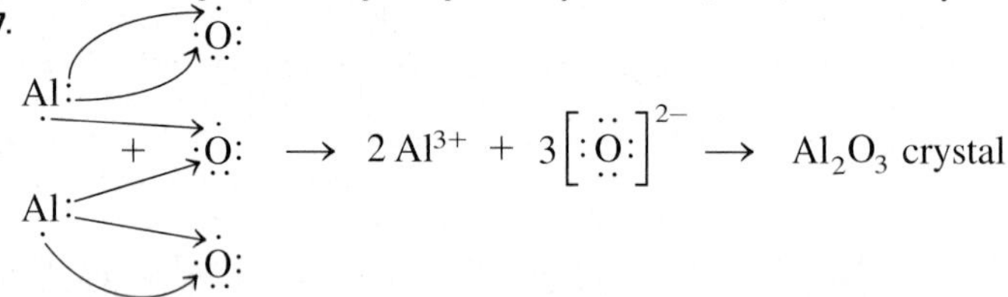

**9.** A potassium atom forms a $K^+$ ion by losing one electron. A chlorine atom can accept one electron to form a $Cl^-$ ion, so it takes one potassium atom to donate the one electron to a single chlorine atom. **11.** Orbital overlap refers to the covalent bond formed when the atomic orbitals of individual atoms extend over one another so that the two nuclei in the bond share electrons.

**13.** $H\cdot + H\cdot + \cdot\ddot{S}\cdot \rightarrow H:\ddot{S}:H$ *or* H—S̈:—H *or* H—S—H

**15.** You should circle the three lone pairs (··) around the chlorine atom. **17.** The energy of a system is reduced when bonds form that reach the noble-gas electron configuration. In a covalent, or electron-sharing, bond, atoms share electrons to achieve the noble-gas configuration. An example is shown in the answer to Question 13 above. The hydrogen atoms obtain the electron configuration of helium and the sulfur atom obtains the electron configuration of argon. **19.** In a nonpolar bond, the charge density of the electron cloud is centered in the region between the bonded atoms. In a polar bond, this charge density is shifted toward the more electronegative atom. **21.** S—O > N—Cl ≈ C—C **23.** S in S—O **25.** Electronegativity values increase from left to right across any row of the table, and they increase from the bottom to top of any column. **27.** An atom bonded by a quadruple bond can conform to the octet rule if it has no unshared electron pairs. **29.** X can be bonded to the maximum of two additional atoms by single bonds. X can be bonded to the minimum of no additional atoms if it has two lone pairs. **31.** The molecule can theoretically have a maximum of an infinite number of atoms: A—B≡C—D . . . . The minimum number is two: : A≡B : **33.** $NO_2$ and NO have an odd number of electrons and thus cannot satisfy the octet rule. **35.** Localized electrons are those that stay near a single atom or pair of atoms; delocalized electrons do not. **37.** The calcium ions in the crystal have a 2+ charge, and there are two electrons for each ion, whereas with potassium metal, the potassium ions have a 1+ charge and there is a 1:1 ion-to-electron ratio. **39.** Alloys are mixtures. They are neither pure substances nor compounds. They lack constant composition and have variable physical properties. **42.** True: a, b, c, d, e, f, h, i. False: g, j, k. **43.** Ions are formed when neutral atoms lose or gain electrons. The electron(s) that is (are) lost by one atom is (are) transferred to another atom. The attraction between the ions is an ionic bond. Covalent bonds are formed when a pair of electrons is shared by the two bonded atoms. Effectively, the electrons belong to both atoms, spending some time near each nucleus. **45.** The K—Cl bond is ionic, formed by "transferring" an electron from a potassium atom to a chlorine atom. The Cl—Cl bond is covalent, formed by two chlorine atoms sharing a pair of electrons. **47.** The $H^+$ ion has no electrons, so it has no electron configuration. **48.** (c) $Mg^{2+}$ is isoelectronic with Ne, not Ar. **49.** 4*p* from bromine and 2*p* from oxygen. **50.** A bond between identical atoms is completely nonpolar. Their attractions for the bonding electrons are equal. **51.** Electronegativities are highest at the upper right corner of the periodic table and lowest at the lower left corner. Therefore, the electronegativity of A is higher than the electronegativity of B. Because X is higher in the table than Y, the electronegativity of X should be larger than that of Y, but because Y is farther to the right, the electronegativity of X should be smaller than Y. Therefore, no prediction can be made for X and Y. **52.** (a) Kr, krypton, Z = 36; (b) The 2−

ion had two electrons added to the neutral atom. The neutral atom is therefore Z = 36 − 2 = 34. Z = 34 is Se, selenium. The ion is $Se^{2-}$, selenide ion. (c) A 1+ ion has one electron removed from the neutral atom, so the neutral atom is Z = 36 + 1 = 37. Z = 37 is Rb, rubidium. The ion is $Rb^{+}$, rubidium ion. **54.** Nonmetal atoms generally have 4, 5, 6, 7, or 8 valence electrons. Fewer electrons have to be added to a nonmetal atom to achieve an octet than would have to be subtracted. For example, a sulfur atom can theoretically achieve an octet by gaining two electrons or losing six. It is simpler for nonmetal atoms to gain electrons to achieve an octet. When two nonmetal atoms combine, the easiest way for both atoms to reach the octet is to share each other's electrons, forming a covalent bond. If the second atom is a metal, however, it has one, two, or possibly three electrons more than an octet. It reaches the octet by giving its electrons to the nonmetal, becoming a positive ion itself, and making the nonmetal atom a negative ion. The two atoms form an ionic bond. **57.** An F—Si bond is more polar than an O—P bond. F has a higher electronegativity than O, and Si has a lower electronegativity than P, based on their relative positions in the periodic table (high at the upper right, low at the lower left. Therefore, the *difference* in electronegativities is largest for F—Si, which makes it the more polar bond. **59.** $AsI_3$ conforms to the octet rule. $AsI_5$ can be formed if each of the lone-pair electrons forms a bonding pair with one electron from an I atom.

```
 ..    ..    ..              .I.     .I.
:I ─── As ─── I:               \     /
 ..          ..          :I ──── As ──── I:
        |                        |
       :I:                      :I:
        ..                       ..
```

(The angles at which these bonds are drawn are related to the spatial arrangements of the bonded atoms. This is considered in Chapter 13. In this chapter, you may limit your interpretations of these diagrams to what atom is bonded to what atom.) **61.** One explanation is that a boron atom is smaller than an aluminum atom. The valence electrons in boron are therefore closer to the nucleus than they are in aluminum. That makes it more difficult to remove the electrons to form an ion.

# Chapter 13

## Answers to Lewis Diagram Recognition Exercises

**1.** Acceptable. Boron is an exception to the octet rule and requires only three bonds.

**2.** Unacceptable. Nitrogen is less electronegative than oxygen, so nitrogen should be the central atom.

```
[ :O─N═O: ]⁻
     |
    :O:
```

**3.** Unacceptable. In a compound with hydrogen, two or more oxygen atoms, and one atom of another nonmetal, hydrogen is bonded to an oxygen atom.

```
:O─Cl─O─H
    |
   :O:
```

**4.** Unacceptable. The middle carbon atom is surrounded by only three electron pairs. Carbon atoms almost always form four bonds and thus usually have no lone pairs.

```
  H  H  H
  |  |  |
H─C═C─C─H
        |
        H
```

**5.** Unacceptable. Carbon does not have a complete octet. H—C≡N: **6.** Acceptable. Beryllium is an exception to the octet rule and requires only two bonds. **7.** Acceptable.

**8.** Unacceptable. A carbon atom must follow the octet rule.

```
   H
   |
H─C─Br:
   |
   H
```

**9.** Unacceptable. Although carbon usually forms four bonds, in this case, it cannot, given the ten valence electrons. This structure accounts for only eight valence electrons. :C≡O: **10.** Unacceptable. Hydrogen is never the central atom in a molecule, and it needs only one bond. H—O—O—H

## Answers to Blue-Numbered Questions, Exercises, and Problems

**1.**

```
H─I:     H─O:     :Cl─N─Cl:
            |          |
            H         :Cl:
```

**3.**

```
O = C = O     :F─S:     [ :O─Br─O: ]⁻
                  |            |
                 :F:          :O:
```

**5.**

```
[ :Br─O: ]⁻

[      :O:      ]⁻      [      :O:      ]⁻
[       |       ]       [       |       ]
[ H─O─P─O─H ]           [ :O─Cl─O: ]
[       |       ]       [       |       ]
[      :O:      ]       [      :O:      ]
```

**7.**

```
    H          H          :Cl:
    |          |            |
:Cl─C─F:   :F─C─F:     :I─C─I:
    |          |            |
   :Cl:       :F:          :I:
```

*Questions 9-17: Where two or more acceptable diagrams are possible, only one representative diagram is shown.*

9. 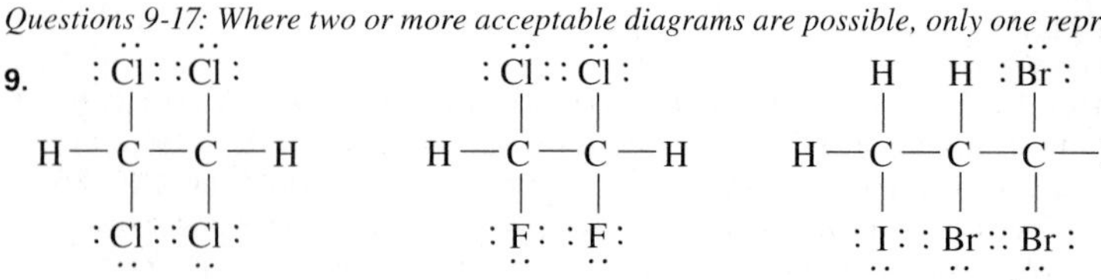

11. 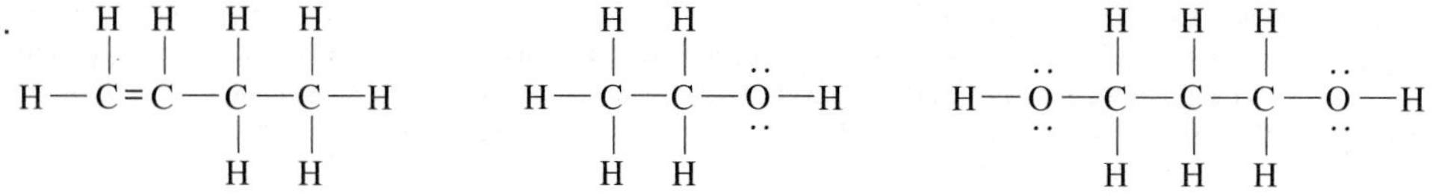

13. 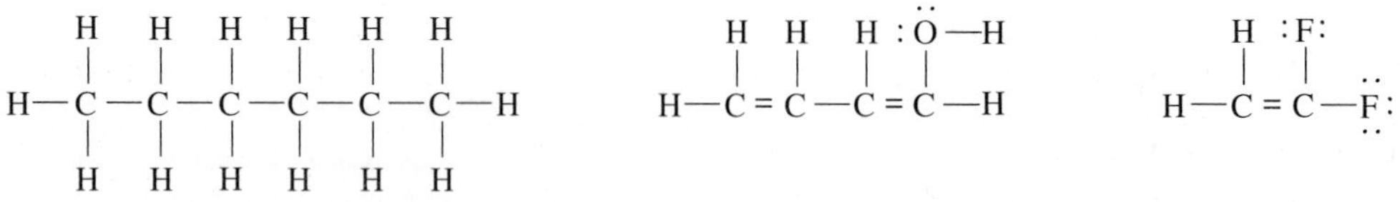

15. 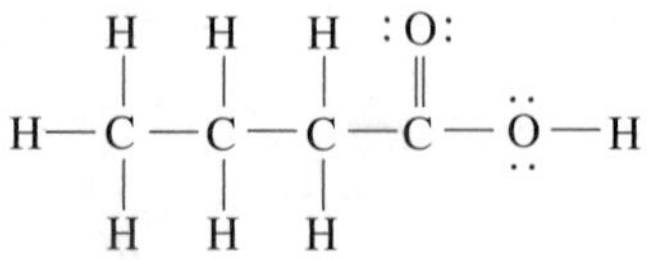

17. 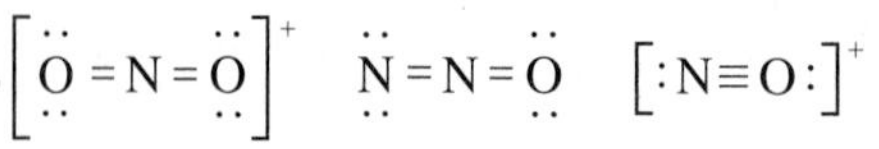

| Substance | Lewis Diagram | Electron Pair Geometry | Molecular Geometry | Wedge-and-Dash Diagram |
|---|---|---|---|---|
| 19. $BCl_3$ | 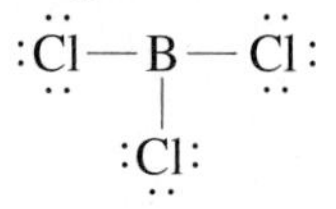 | Trigonal planar | Trigonal planar | 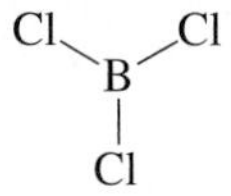 |
| $PH_3$ | 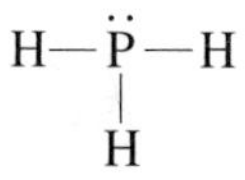 | Tetrahedral | Trigonal pyramidal | 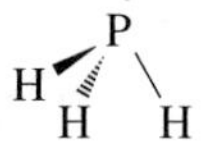 |
| $H_2S$ | 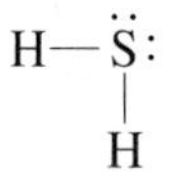 | Tetrahedral | Bent |  |
| 21. $BrO^-$ | 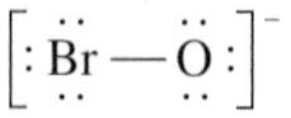 | Linear | Linear | Br—O |
| $ClO_3^-$ | 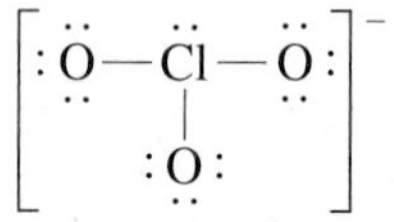 | Tetrahedral | Trigonal pyramidal | 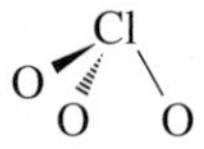 |
| $PO_4^{3-}$ | 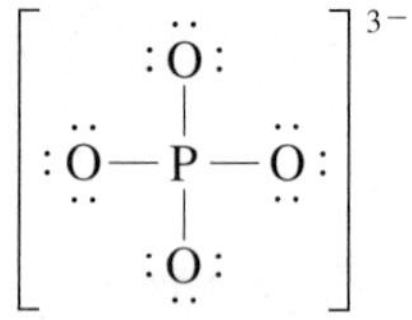 | Tetrahedral | Tetrahedral | 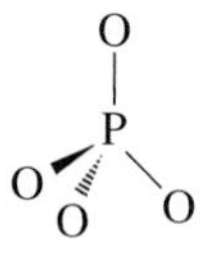 |
| 23. C in $C_3H_7OH$ | 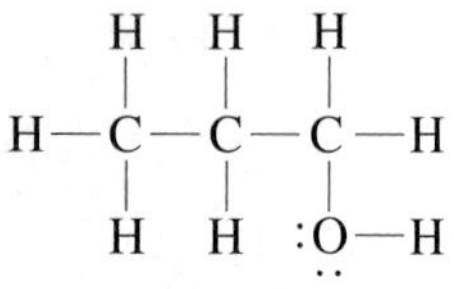 | Tetrahedral | Tetrahedral | |
| 25. N in $C_2H_5NH_2$ | 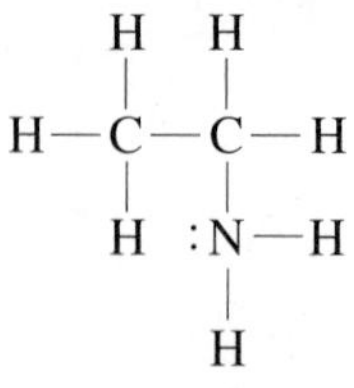 | Tetrahedral | Trigonal pyramidal | |
| 27. C in $C_2H_2$ | H—C≡C—H | Linear | Linear | |

**29.** C in HCN  H—C≡N:  Linear  Linear

**31.** Trigonal pyramidal **33.** Electron-pair and molecular: trigonal planar **35.** Electron-pair: tetrahedral; molecular: trigonal pyramidal **37.** Electron-pair and molecular: tetrahedral **39.** A: 120°, three regions of electron density around C; B: 109°, four regions of electron density around O; C: 109°, four regions of electron density around C; D: 120°, three regions of electron density around C **41.** A: 120°, three regions of electron density around C; B: 120°, three regions of electron density around C; C: 120°, three regions of electron density around C **43.** b, d, and e **45.** $CCl_4$ molecules are nonpolar because the polar bonds are symmetrically arranged, which makes the molecule itself nonpolar. **47.** Both molecules are linear. HF is more polar than HBr. The H end of each is more positive.

**49.** [Lewis structures of $H_2O$ and $CH_3OH$ with bond polarity arrows]

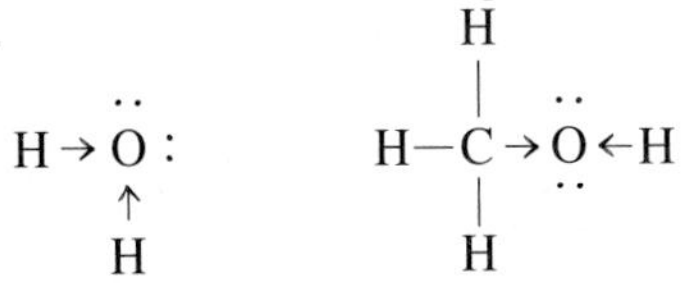

Both molecules are polar because of the bent structure around the oxygen atom and the concentration of negative charge near the highly electronegative oxygen atom. The carbon—oxygen electronegativity difference is slightly less than the hydrogen—oxygen electronegativity difference, so $CH_3OH$ is slightly less polar than $H_2O$.

**51.** a and d **53.** b and c **55.** Organic compounds all contain carbon; inorganic compounds (with some exceptions) do not contain carbon. **57.** $CH_3(CH_2)_6CH_3$, $C_6H_6$, and $C_{18}H_8$ **59.** The structure of a molecule simply describes in two dimensions the relative positions of how its atoms are bonded. The three-dimensional shape of a molecule results from its structure and is the arrangement of the atoms in real three-dimensional space. **61.** A straight line of carbon atoms in a Lewis diagram simply indicates that the atoms are bonded to one another. Single-bonded carbon atoms form tetrahedral bond angles, which form a zigzag chain when in the same plane. **63.** Alcohols, ethers, and water molecules all include an oxygen atom with two lone pairs of electrons that is single-bonded to two other atoms. In an alcohol, one of the bonds to the oxygen atom is to a hydrogen atom and the other is to a carbon atom. In an ether, both bonds to the oxygen atom are to carbon atoms. **65.** Carboxylic acids contain the carboxyl group, —COOH.

**68.** True: a, d, e, g, h*, i, j **69.** [Lewis structure of cyclohexane, $C_6H_{12}$]

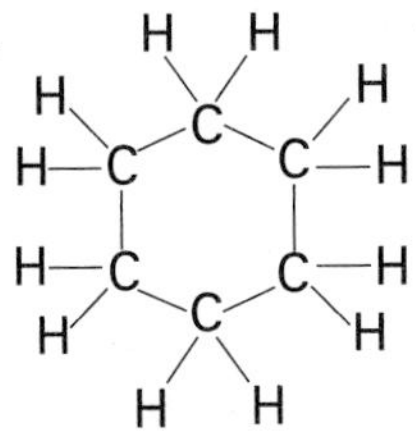

**70.** H—Br:  H—O—Br:  H—O—Br—O:  [Lewis structures of HBr, HOBr, $HBrO_2$, $HBrO_3$, $HBrO_4$]

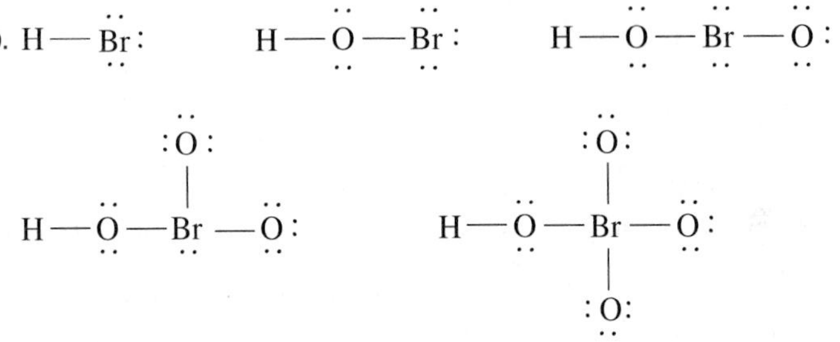

**71.** [Wedge-dash structure of propane]

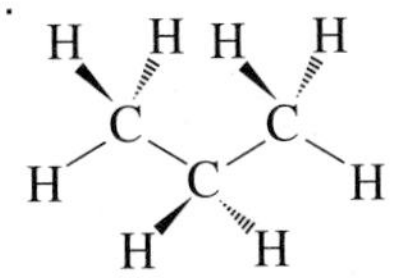

Bond angles between carbon atoms in an alkane are tetrahedral, so the atoms cannot lie in a straight line.

**72.** [Lewis structure of benzene, $C_6H_6$]

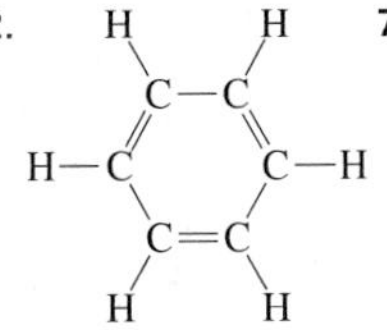

**74.** [Lewis structures]

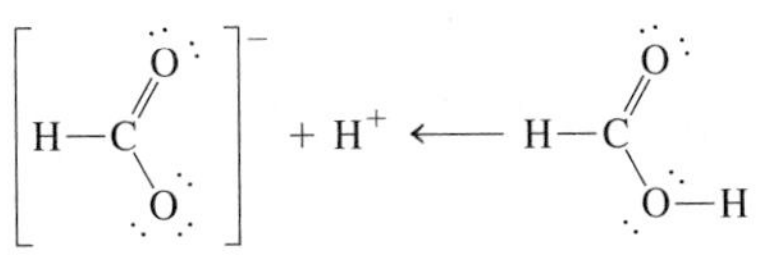

formate ion, $CHO_2^-$ + $H^+$ ⟵ formic acid

**75.** Any two from: [Lewis structures of $C_4H_6$ isomers: H₂C=C=CH—CH₃, H₂C=CH—CH=CH₂, HC≡C—CH₂—CH₃, H₃C—C≡C—CH₃]

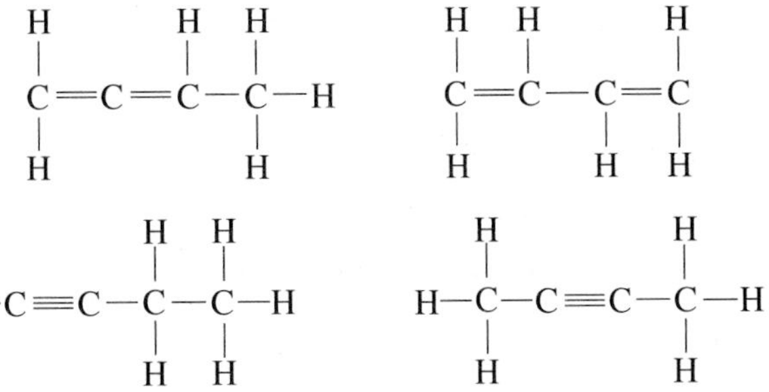

**76.** [Lewis structure of diethyl ether: H₃C—CH₂—O—CH₂—CH₃]

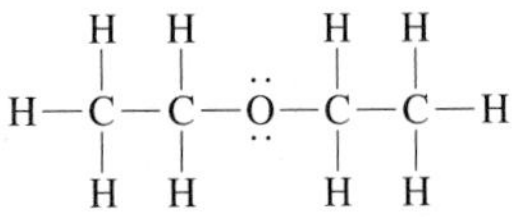

*Statement h is true if central atoms are limited to four electron pairs, as they are in this book.

**77.** All species in the question, including $SeO_4^{2-}$ and $CI_4$, have 32 valence electrons and four atoms to be distributed around a central atom. Consequently, they all have the same tetrahedral shape.

**78.** (a) Trigonal planar with 120° angles around both carbon atoms. (b) Linear. (c) Zigzag carbon chain; all bond angles tetrahedral.

# Chapter 14

## Answers to Target Checks

**1.** In your sketch, the piston should be at the 15 mark on the scale. The 5 two-atom particles that separate form 10 one-atom particles. Five particles remain unseparated, for a total of 15 particles after the reaction. Since $V \propto n$ at constant pressure and temperature, $10 \text{ volume units} \times \frac{15 \text{ particles}}{10 \text{ particles}} = 15$ volume units. **2.** $P = \frac{nRT}{V} = 2.18 \text{ mol} \times \frac{0.0821 \text{ L} \cdot \text{atm}}{\text{mol} \cdot \text{K}} \times 288 \text{ K} \times \frac{1}{5.32 \text{ L}} = 9.69 \text{ atm}$

## Answers to Concept-Linking Exercises

*You may have found more relationships or relationships other than the ones given in these answers.*

**1.** The Law of Combining Volumes states that when gases at the same temperature and pressure react, they do so in ratios of small whole numbers. Avogadro's Law states that the volume occupied by a gas at a fixed temperature and pressure is directly proportional to the amount of gas present, expressed in number of particles (moles). Therefore, the molar ratios that appear in chemical equations are equal to the ratio of volumes of gaseous reactants and products, provided that they are measured at the same temperature and pressure. **2.** The Ideal Gas Law is based on the ideal gas model and the experimentally observed proportionalities among gas pressure, temperature, volume, and amount (number of particles). The constant that combines these proportionalities into the ideal gas equation is called the universal gas constant. **3.** The units in which the variables in the ideal gas equation are expressed are atmospheres or torr (pressure), liters (volume), kelvins (temperature), and moles (amount). **4.** The most common value of the universal gas constant is 0.0821 L · atm/mol · K. If pressure is expressed in torr, the value of the constant is 62.4 L · torr/mol · K. If the ideal gas equation is solved for liters per mole, the result is V/n = RT/P. Substituting standard temperature (273 K) and standard pressure (1 bar) into the right side of the equation gives 22.7 L/mol. This is the molar volume of an ideal gas at STP.

## Answers to Blue-Numbered Questions, Exercises, and Problems

**1.** $n = \frac{PV}{RT}$, so since temperature and pressure are the same for all three samples, n is directly proportional to V. The volume of $1 \times 10^{23}$ hydrogen molecules is the same as the volume of $1 \times 10^{23}$ oxygen molecules and half the volume of $2 \times 10^{23}$ nitrogen molecules.

**3.** $P = \frac{nRT}{V} = 0.0888 \text{ mol} \times \frac{62.4 \text{ L} \cdot \text{torr}}{\text{mol} \cdot \text{K}} \times (36 + 273) \text{ K} \times \frac{1}{5.00 \text{ L}} = 342 \text{ torr}$

**5.** $V = \frac{nRT}{P} = 6.04 \text{ mol} \times \frac{0.0821 \text{ L} \cdot \text{atm}}{\text{mol} \cdot \text{K}} \times (18 + 273) \text{ K} \times \frac{1}{17.2 \text{ atm}} = 8.39 \text{ L}$

**7.** $n = \frac{PV}{RT} = 752 \text{ torr} \times 784 \text{ mL} \times \frac{\text{mol} \cdot \text{K}}{62.4 \text{ L} \cdot \text{torr}} \times \frac{1}{(22 + 273) \text{ K}} \times \frac{1 \text{ L}}{1000 \text{ mL}} = 0.0320 \text{ mol}$

**9.** $T = \frac{PV}{nR} = 756 \text{ torr} \times 15.7 \text{ L} \times \frac{1}{0.810 \text{ mol}} \times \frac{\text{mol} \cdot \text{K}}{62.4 \text{ L} \cdot \text{torr}} = 235 \text{ K}; 235 - 273 = -38°\text{C}$

**11.** $n = \frac{PV}{RT} = 965 \text{ torr} \times 40.0 \text{ L} \times \frac{\text{mol} \cdot \text{K}}{62.4 \text{ L} \cdot \text{torr}} \times \frac{1}{(18 + 273) \text{ K}} = 2.13 \text{ mol}$

**13.** $V = \frac{nRT}{P} = 0.621 \text{ mol} \times \frac{0.0821 \text{ L} \cdot \text{atm}}{\text{mol} \cdot \text{K}} \times (-32 + 273) \text{ K} \times \frac{1}{0.771 \text{ atm}} = 15.9 \text{ L}$

**15.** $MM = \frac{mRT}{PV} = \frac{2.32 \text{ g}}{\text{L}} \times \frac{0.0821 \text{ L} \cdot \text{atm}}{\text{mol} \cdot \text{K}} \times 273 \text{ K} \times \frac{1}{1 \text{ bar}} \times \frac{1.013 \text{ bar}}{\text{atm}} = 52.7 \text{ g/mol}$

**17.** (a) $D = \frac{m}{V} = \frac{(MM)P}{RT} = \frac{29 \text{ g}}{\text{mol}} \times 1 \text{ bar} \times \frac{\text{mol} \cdot \text{K}}{0.0821 \text{ L} \cdot \text{atm}} \times \frac{1}{273 \text{ K}} \times \frac{1 \text{ atm}}{1.013 \text{ bar}} = 1.3 \text{ g/L}$

(b) $D = \frac{m}{V} = \frac{(MM)P}{RT} = \frac{29 \text{ g}}{\text{mol}} \times 751 \text{ torr} \times \frac{\text{mol} \cdot \text{K}}{62.4 \text{ L} \cdot \text{torr}} \times \frac{1}{(20 + 273) \text{ K}} = 1.2 \text{ g/L}$

**19.** $MM = \frac{mRT}{PV} = \frac{m}{V} \times \frac{RT}{P} = \frac{1.61\text{ g}}{L} \times \frac{0.0821\text{ L}\cdot\text{atm}}{\text{mol}\cdot\text{K}} \times (41 + 273)\text{ K} \times \frac{1}{2.61\text{ atm}} = 15.9\text{ g/mol}$

**21.** $MM = \frac{mRT}{PV} = 33.5\text{ g} \times \frac{0.0821\text{ L}\cdot\text{atm}}{\text{mol}\cdot\text{K}} \times (13 + 273)\text{ K} \times \frac{1}{3.25\text{ atm}} \times \frac{1}{8.07\text{ L}} = 30.0\text{ g/mol}$

**23.** $D = \frac{m}{V} = \frac{(MM)P}{RT}$, so when temperature and pressure are constant, density is directly proportional to molar mass. $N_2O_4$ is more dense than $NO_2$.

**25.** $MV = \frac{V}{n} = \frac{RT}{P} = \frac{62.4\text{ L}\cdot\text{torr}}{\text{mol}\cdot\text{K}} \times (30 + 273)\text{ K} \times \frac{1}{1.10\text{ torr}} = 1.72 \times 10^4\text{ L/mol}$ Molar volume is independent of the identity of the gas.

**27.** $MV = \frac{V}{n} = \frac{RT}{P} = \frac{0.0821\text{ L}\cdot\text{atm}}{\text{mol}\cdot\text{K}} \times (21 + 273)\text{ K} \times \frac{1}{0.908\text{ atm}} = 26.6\text{ L/mol}$

**29.** $5.74\text{ g KClO}_3 \times \frac{1\text{ mol KClO}_3}{122.55\text{ g KClO}_3} \times \frac{3\text{ mol O}_2}{2\text{ mol KClO}_3} \times \frac{22.7\text{ L O}_2}{\text{mol O}_2} = 1.59\text{ L O}_2$

**31.** $9.81\text{ L Cl}_2 \times \frac{1\text{ mol Cl}_2}{22.7\text{ L Cl}_2} \times \frac{2\text{ mol KMnO}_4}{5\text{ mol Cl}_2} \times \frac{158.04\text{ g KMnO}_4}{\text{mol KMnO}_4} = 27.3\text{ g KMnO}_4$

*Questions 33 to 40 may be solved by the molar volume method (Section 14.7) or by the ideal gas equation method (Section 14.8). Solution setups are given first for the molar volume method, printed over a tan background. Then, over a green background, the answers are given for the ideal gas equation method. Check your work according to the section you studied.*

**33.** $MV = \frac{V}{n} = \frac{RT}{P} = \frac{62.4\text{ L}\cdot\text{torr}}{\text{mol}\cdot\text{K}} \times (214 + 273)\text{ K} \times \frac{1}{983\text{ torr}} = 30.9\text{ L/mol}$

$598\text{ g FeS}_2 \times \frac{1\text{ mol FeS}_2}{119.99\text{ g FeS}_2} \times \frac{8\text{ mol SO}_2}{4\text{ mol FeS}_2} \times \frac{30.9\text{ L SO}_2}{\text{mol SO}_2} = 308\text{ L SO}_2$

**35.** $2\,H_2O \rightarrow 2\,H_2 + O_2$ $MV = \frac{V}{n} = \frac{RT}{P} = \frac{62.4\text{ L}\cdot\text{torr}}{\text{mol}\cdot\text{K}} \times (28 + 273)\text{ K} \times \frac{1}{728\text{ torr}} = 25.8\text{ L/mol}$

$23.9\text{ L H}_2 \times \frac{1\text{ mol H}_2}{25.8\text{ L H}_2} \times \frac{2\text{ mol H}_2\text{O}}{2\text{ mol H}_2} \times \frac{18.02\text{ g H}_2\text{O}}{\text{mol H}_2\text{O}} = 16.7\text{ g H}_2\text{O}$

**37.** $N_2 + 3\,H_2 \rightarrow 2\,NH_3$ $MV = \frac{V}{n} = \frac{RT}{P} = \frac{0.0821\text{ L}\cdot\text{atm}}{\text{mol}\cdot\text{K}} \times (575 + 273)\text{ K} \times \frac{1}{248\text{ atm}} = 0.281\text{ L/mol}$

$9.16 \times 10^3\text{ g NH}_3 \times \frac{1\text{ mol NH}_3}{17.03\text{ g NH}_3} \times \frac{1\text{ mol N}_2}{2\text{ mol NH}_3} \times \frac{0.281\text{ L N}_2}{\text{mol N}_2} = 75.6\text{ L N}_2$

**39.** $MV = \frac{V}{n} = \frac{RT}{P} = \frac{62.4\text{ L}\cdot\text{torr}}{\text{mol}\cdot\text{K}} \times (447 + 273)\text{ K} \times \frac{1}{896\text{ torr}} = 50.1\text{ L/mol}$

$82.3\text{ L gases} \times \frac{1\text{ mol gases}}{50.1\text{ L}} \times \frac{1\text{ mol NH}_4\text{NO}_3}{3\text{ mol gases}} \times \frac{80.05\text{ g NH}_4\text{NO}_3}{\text{mol NH}_4\text{NO}_3} = 43.8\text{ g NH}_4\text{NO}_3$

**33.** $598\text{ g FeS}_2 \times \frac{1\text{ mol FeS}_2}{119.99\text{ g FeS}_2} \times \frac{8\text{ mol SO}_2}{4\text{ mol FeS}_2} = 9.97\text{ mol SO}_2$

$V = \frac{nRT}{P} = 9.97\text{ mol} \times \frac{62.4\text{ L}\cdot\text{torr}}{\text{mol}\cdot\text{K}} \times (214 + 273)\text{ K} \times \frac{1}{983\text{ torr}} = 308\text{ L SO}_2$

**35.** $2\,H_2O \rightarrow 2\,H_2 + O_2$ $n = \frac{PV}{RT} = 728\text{ torr} \times 23.9\text{ L} \times \frac{\text{mol}\cdot\text{K}}{62.4\text{ L}\cdot\text{torr}} \times \frac{1}{(28 + 273)\text{ K}} = 0.926\text{ mol H}_2$

$0.926\text{ mol H}_2 \times \frac{2\text{ mol H}_2\text{O}}{2\text{ mol H}_2} \times \frac{18.02\text{ g H}_2\text{O}}{\text{mol H}_2\text{O}} = 16.7\text{ g H}_2\text{O}$

**37.** $N_2 + 3\,H_2 \rightarrow 2\,NH_3$ $9.16 \times 10^3\text{ g NH}_3 \times \frac{1\text{ mol NH}_3}{17.03\text{ g NH}_3} \times \frac{1\text{ mol N}_2}{2\text{ mol NH}_3} = 269\text{ mol N}_2$

$V = \frac{nRT}{P} = 269\text{ mol} \times \frac{0.0821\text{ L}\cdot\text{atm}}{\text{mol}\cdot\text{K}} \times (575 + 273)\text{ K} \times \frac{1}{248\text{ atm}} = 75.5\text{ L N}_2$

**39.** $n = \frac{PV}{RT} = 896\text{ torr} \times 82.3\text{ L} \times \frac{\text{mol}\cdot\text{K}}{62.4\text{ L}\cdot\text{torr}} \times \frac{1}{(447 + 273)\text{ K}} = 1.64\text{ mol}$

$1.64\text{ mol gases} \times \frac{1\text{ mol NH}_4\text{NO}_3}{3\text{ mol gases}} \times \frac{80.05\text{ g NH}_4\text{NO}_3}{\text{mol NH}_4\text{NO}_3} = 43.8\text{ g NH}_4\text{NO}_3$

**41.** a) $35.2\text{ L SO}_2 \times \frac{1\text{ L O}_2}{2\text{ L SO}_2} = 17.6\text{ L O}_2$ b) $V_2 = V_1 \times \frac{P_1}{P_2} \times \frac{T_2}{T_1} = 35.2\text{ L} \times \frac{741\text{ torr}}{847\text{ torr}} \times \frac{(17 + 273)\text{ K}}{(26 + 273)\text{ K}} = 29.9\text{ L SO}_2$

$29.9\text{ L SO}_2 \times \frac{1\text{ L O}_2}{2\text{ L SO}_2} = 15.0\text{ L O}_2$

**43.** $V_2 = V_1 \times \frac{P_1}{P_2} \times \frac{T_2}{T_1} = 283\text{ L} \times \frac{2.92\text{ atm}}{0.961\text{ atm}} \times \frac{(31 + 273)\text{ K}}{(21 + 273)\text{ K}} = 889\text{ L}$ $889\text{ L Cl}_2 \times \frac{2\text{ L ClO}_2}{1\text{ L Cl}_2} = 1.78 \times 10^3\text{ L ClO}_2$

**45.** $V_2 = V_1 \times \frac{P_1}{P_2} \times \frac{T_2}{T_1} = 704\text{ mL} \times \frac{159\text{ torr}}{549\text{ torr}} \times \frac{(19 + 273)\text{ K}}{(26 + 273)\text{ K}} = 199\text{ mL}$ $199\text{ mL O}_2 \times \frac{2\text{ L H}_2\text{S}}{3\text{ L O}_2} = 133\text{ mL H}_2\text{S}$

**48.** True: c. False: a, b, d. **49.** $V = \frac{mRT}{(MM)P} = 6.74\text{ g} \times \frac{62.4\text{ L}\cdot\text{torr}}{\text{mol}\cdot\text{K}} \times \frac{(41 + 273)\text{ K}}{733\text{ torr}} \times \frac{1\text{ mol}}{28.05\text{ g}} = 6.42\text{ L}$

**50.** $m = \frac{PV(MM)}{RT} = 775\text{ torr} \times 57.9\text{ L} \times \frac{83.80\text{ g}}{\text{mol}} \times \frac{\text{mol}\cdot\text{K}}{62.4\text{ L}\cdot\text{atm}} \times \frac{1}{(6 + 273)\text{ K}} = 216\text{ g}$

**51.** $D = \frac{m}{V} = \frac{(MM)P}{RT} = \frac{34.09\ g}{mol} \times \frac{0.972\ atm}{287\ K} \times \frac{mol \cdot K}{0.0821\ L \cdot atm} = 1.41\ g/L$

**52.** $MV = \frac{V}{n} = \frac{RT}{P} = \frac{0.0821\ L \cdot atm}{mol \cdot K} \times (17 + 273)\ K \times \frac{1}{0.835\ atm} = 28.5\ L/mol$ At a given temperature and pressure, molar volume is the same for all gases. The identity of the gas, the volume, and the mass are not needed to solve the problem.

**53.** $MM = \frac{mRT}{PV} = 7.60\ g \times \frac{62.4\ L \cdot torr}{mol \cdot K} \times (183 + 273)\ K \times \frac{1}{179\ torr} \times \frac{1}{3.87\ L} = 312\ g/mol$

**54.** $MM = \frac{mRT}{PV} = \frac{1.07\ g}{L} \times \frac{0.0821\ L \cdot atm}{mol \cdot K} \times (18 + 273)\ K \times \frac{1}{0.913\ atm} = 28.0\ g/mol$ The molar mass of nitrogen is twice its atomic mass. Therefore, there must be two atoms in the molecule, $N_2$.

**55.** In algebra, if $x \propto y$, then $x = ky$, where k is a proportionality constant. At a given temperature and pressure, solving Equation 14.6 for density, m/V, yields $D = \frac{m}{V} = \frac{P(MM)}{RT} = \frac{P}{RT} \times MM = k \times MM$ $D = k \times MM$ In this equation, P/RT has the role of a proportionality constant. Hence, molar mass and density are directly proportional to each other at a given temperature and pressure. **56.** Butane, $C_4H_{10}$, has a higher molar mass, 58.12 g/mol, than propane, $C_3H_8$, 44.09 g/mol. At a given temperature and pressure, density is proportional to molar mass. Therefore B, the higher-density gas, must be butane. Its density is $1.37\ g/L \times \frac{58.12\ g/mol}{44.09\ g/mol} = 1.81\ g/L$

**57.** (a) $MM = \frac{mRT}{PV} = 4.08\ g \times \frac{62.4\ L \cdot torr}{mol \cdot K} \times (243 + 273)\ K \times \frac{1}{117\ torr} \times \frac{1}{3.36\ L} = 334\ g/mol$

(b) The difference between the expected molar mass and the molar mass that was found experimentally is 346 − 334 = 12 g/mol. This is the atomic mass of carbon. Perhaps she has one too many carbon atoms in her expected molecule.

# Chapter 15

## Answers to Target Checks

**1.** (a) Gas particles are widely separated compared with liquid particles. (b) A will have the higher surface tension, molar heat of vaporization, boiling point, and viscosity, all of which increase in value with increasing strength of intermolecular forces. B will have the higher vapor pressure, a property that increases in value as the strength of intermolecular attractions decreases. (c) Y should have a higher vapor pressure. If X has a higher molar heat of vaporization than Y, it probably has stronger intermolecular attractions. That should cause X to have a lower vapor pressure. **2.** b and e: true. a: Induced dipole forces are present between all molecules. c: Polar molecules have an unbalanced distribution of electrical charge, but the net charge is zero. d: Intermolecular forces are electrical in character. **3.** (a) tetrahedral, nonpolar, induced dipole; (b) linear, nonpolar, induced dipole; (c) bent, polar, dipole; (d) bent, polar, hydrogen bonding **4.** (a) $CBr_4$ because a $CBr_4$ molecule is larger than a $CCl_4$ molecule and thus has stronger induced dipole forces. (b) $NH_3$ because it has hydrogen bonding and $PH_3$ does not. **5.** b and c: true. a: A liquid–vapor equilibrium is reached when the rate of evaporation is equal to the rate of condensation. **6.** Both true. **7.** Structural particles in a crystalline solid are arranged in a regular geometric order. In an amorphous solid the structural arrangement is irregular. In a polycrystalline solid, microscopic crystals are arranged randomly. **8.** a: true. b: Covalent network solids are usually poor conductors of electricity. c: A solid that melts at 152°C is probably a molecular crystal. d: A soluble molecular crystal is a nonconductor of electricity both as a solid and when dissolved.

## Answers to Concept-Linking Exercises

*You may have found more relationships or relationships other than the ones given in these answers.*

**1.** Dipole forces, induced dipole forces, and hydrogen bonds are all intermolecular attractions—forces that act *between* molecules. Induced dipole forces are also known as dispersion forces, London forces, and London dispersion forces. Covalent bonds are the forces that hold atoms together *within* a molecule. Ionic bonds are the forces that hold ions together *within* a crystal. Covalent and ionic bonds are much stronger than any of the intermolecular forces. **2.** Many physical properties of liquids depend on intermolecular attractions, which can be thought of in terms of the stick-togetherness of particles. Strong intermolecular attractions lead to low vapor pressure, higher molar heat of vaporization, high boiling point, high viscosity, and high surface tension. **3.** The boiling point of a liquid is the temperature at which its vapor pressure is equal to the pressure above its surface. The normal boiling point is the temperature at which the vapor pressure of the liquid is one atmosphere. **4.** The situation that occurs when the condensation rate equals the evaporation rate in a liquid–vapor equilibrium is described as a condition of dynamic equilibrium. The equilibrium is described as dynamic because the particles are continually switching between the liquid and vapor states, although the concentration of particles in the vapor state remains constant. There is higher vapor pressure at higher temperatures because an increase in temperature leads to an increase in the number of particles with sufficient energy to evaporate. **5.** Crystalline solids have an orderly arrangement of particles. Four kinds of crystalline solids are (1) ionic crystals, ions held together by strong ionic bonds, (2) molecular crystals, held together by weak intermolecular forces, (3) covalent network solids, in which atoms are held together by covalent bonds, and (4) metallic crystals, in which positive ions are held together in a sea of electrons. See Figure 12.15 in Section 12.8. **6.** Solids can be classified based on the arrangement of their particles. An amorphous solid is one in which there is no long-range ordering of the particles. In contrast, a crystalline solid has its particles arranged in a repeating, three-dimensional geometric pattern. A polycrystalline solid consists of small crystalline substructures randomly arranged to form the solid. **7.** Heat of vaporization is the amount of energy required to vaporize (liquid → gas) one gram of a substance. Heat of condensation is the amount of heat released when one gram of a substance condenses (gas → liquid). These terms are equal in magnitude but opposite in sign for a given substance. Similarly, heat of fusion refers to the energy required to melt (solid → liquid) one gram of a substance, and heat of solidification is the amount of energy released in freezing (liquid → solid) one gram of a substance. These terms are also equal in magnitude but opposite in sign. Specific heat is the amount of energy required to change the temperature of one gram of a substance by one degree Celsius.

## Answers to Blue-Numbered Questions, Exercises, and Problems

**1.** Each gas fills the volume of the container, $1 \times 10^2$ L. A gas, either pure or a mixture, is made up of tiny particles that are widely separated from each other so that they occupy the whole volume of the container that holds them. **3.** $p_{other} = P_{total} - p_{nitrogen} - p_{oxygen} - p_{argon} = 749$ torr $-$ 584 torr $-$ 144 torr $-$ 19 torr $=$ 2 torr **5.** 733 mm Hg $-$ 16.5 mm Hg $=$ 717 mm Hg **7.** Gas particles are very widely spaced; liquid particles are "touchingly close." When gases are mixed, the particles of one gas can occupy the empty spaces between the particles of the other. When liquids are mixed, particles must be "pushed aside" to make room for others. **9.** Intermolecular attractions are stronger in liquids because of the lack of space between molecules. **11.** The stronger the intermolecular attractions are, the more motion is needed to separate the particles within the liquid, and the higher the boiling point will be. **13.** Energy is required to overcome intermolecular attractions, separate liquid particles from one another, and keep them apart. As more energy is required, the molar heat of vaporization will increase. **15.** The ball bearing will reach the bottom of the water cylinder first. Molecules in syrup have higher intermolecular attractions than those in water, so the syrup is more viscous. Syrup molecules are being pulled apart so the ball bearing can pass through the liquid. This slows the rate of fall. **17.** Intermolecular attractions are stronger in mercury than in water. Mercury therefore has a higher surface tension and clings to itself rather than spreading or penetrating paper. **19.** The wetting agent reduces the surface tension of water, overcoming its intermolecular attractions and allowing it to penetrate the duck's feathers. The duck's buoyancy is reduced, and it sinks. **21.** $NO_2$ has the highest boiling point, which suggests strong intermolecular attractions. It should also have the highest molar heat of vaporization. **23.** Only $N_2O$ is a liquid at $-90$°C, so it alone has a measurable equilibrium vapor pressure as that term is used in this unit. NO is a gas, and its vapor pressure is its gas pressure. $NO_2$, a solid at $-90$°C, probably has a very small vapor pressure. **25.** Other things being equal, dipole forces are stronger than induced dipole forces. Induced dipole forces are likely to be larger than dipole forces when the molecules are very large. **27.** $NH(CH_3)_2$, hydrogen bonding; $CH_2F_2$, dipole; $C_3H_8$, induced dipole. **29.** Dipole forces are attractive forces between polar molecules; hydrogen bonds are stronger dipole-like forces between polar molecules in which hydrogen is bonded to a highly electronegative element, usually nitrogen, oxygen, or fluorine. **31.** $NH_3$ has the higher boiling point because it has relatively strong hydrogen bonds between its molecules, versus relatively weak induced dipoles between $CH_4$ molecules. **33.** Argon has the higher boiling point because its atoms are larger than those of neon and thus the strength of the induced dipole forces in Ar are greater than in Ne. **35.** The hydrogen atom consists of a single proton and a single electron. When the bonding electron pair is shifted away from the atom, the proton is responsible for the strength of the hydrogen bond. See the discussion on hydrogen bonds in Section 15.3. **37.** (a) Induced dipoles; (b) Hydrogen bonding **39.** Induced dipole forces act among all molecules. In (e), induced dipole forces are the major intermolecular force. In addition, dipole forces are present in (a) through (d) and the major intermolecular force in (b). Molecules (a), (c), and (d) have hydrogen bonding as the major intermolecular force. **41.** $CS_2$ should have the higher melting and boiling points because its molecules are larger than otherwise similar $CO_2$ molecules so the induced dipole forces are stronger in $CS_2$. **43.** $CH_4$ should have the higher vapor pressure as a liquid at a given temperature because only weak induced dipoles are present. Relatively stronger dipole forces are present in $CH_3F$. **45.** Rates of change in opposite directions are equal. See the discussion in Section 15.4. **47.** At higher temperatures, a greater percentage of molecules in the liquid state have enough energy to vaporize into the gaseous state. **49.** The vapor pressure of the compound exceeds the external pressure, so it is a gas. A compound boils, or changes from a liquid to a gas, when its vapor pressure equals the external pressure. **51.** The water is delivered at very high pressure, pressure greater than the vapor pressure at the temperature of delivery. **53.** Low-boiling liquids and a low heat of vaporization are both characteristic of relatively weak intermolecular attractions. A liquid with weak attractions would therefore exhibit both properties. **55.** N should have both the lower boiling point and lower molar heat of vaporization. **57.** Ice is a crystalline solid. Ice crystals have a definite geometric order and ice melts at a definite, constant temperature. **59.** A: Network solid. B: Ionic solid.

**61.** $\frac{44.8 \text{ kJ}}{61.2 \text{ g}} = 0.732$ kJ/g **63.** $227 \text{ g} \times \frac{4.2 \text{ kJ}}{\text{g}} = 9.5 \times 10^2$ kJ **65.** $79.4 \text{ kJ} \times \frac{1 \text{ g}}{0.880 \text{ kJ}} = 90.2$ g

**67.** $23.8 \text{ g } C_3H_6O \times \frac{1 \text{ mol } C_3H_6O}{58.08 \text{ g } C_3H_6O} \times \frac{32.0 \text{ kJ}}{\text{mol } C_3H_6O} = 13.1$ kJ **69.** $3.30 \text{ kg} \times \frac{23 \text{ kJ}}{\text{kg}} = 76$ kJ **71.** $\frac{2.51 \text{ kJ}}{36.9 \text{ g}} \times \frac{1000 \text{ J}}{\text{kJ}} = 68.0$ J/g

**73.** $4.45 \text{ kJ} \times \frac{1000 \text{ J}}{\text{kJ}} \times \frac{1 \text{ g}}{112 \text{ J}} = 39.7$ g **75.** A. From $q = m \times c \times \Delta T$, when q and m are equal for two objects, c is inversely proportional to $\Delta T$.

**77.** $q = 467 \text{ g} \times \frac{0.39 \text{ J}}{\text{g} \cdot °\text{C}} \times (31 - 68)°\text{C} = -6.7 \times 10^3$ J **79.** $q = 2.30 \text{ kg} \times \frac{1000 \text{ g}}{\text{kg}} \times \frac{0.13 \text{ J}}{\text{g} \cdot °\text{C}} \times (22 - 88)°\text{C} \times \frac{1 \text{ kJ}}{1000 \text{ J}} = -2.0 \times 10^1$ kJ

**81.** $\Delta T = \frac{q}{m \times c} = 1.47 \text{ kJ} \times \frac{1000 \text{ J}}{\text{kJ}} \times \frac{1}{144 \text{ g}} \times \frac{\text{g} \cdot °\text{C}}{0.38 \text{ J}} = 27°\text{C}$ $33 - 27 = 6$°C **83.** J (boiling point), K (freezing point) **85.** H, I **87.** C

**89.** Gas condenses at boiling point, J; liquid cools from boiling point, J, to freezing point K. **91.** P – O

**93.** $q \text{ (heat solid)} = 127 \text{ g} \times \frac{2.06 \text{ J}}{\text{g} \cdot °\text{C}} \times [0 - (-11)]°\text{C} \times \frac{1 \text{ kJ}}{1000 \text{ J}} = 2.9 \text{ kJ}$ $q \text{ (melt solid)} = 127 \text{ g} \times \frac{333 \text{ J}}{\text{g}} \times \frac{1 \text{ kJ}}{1000 \text{ J}} = 42.3 \text{ kJ}$

$q \text{ (heat liquid)} = 127 \text{ g} \times \frac{4.18 \text{ J}}{\text{g} \cdot °\text{C}} \times (21 - 0)°\text{C} \times \frac{1 \text{ kJ}}{1000 \text{ J}} = 11 \text{ kJ}$ Total $q = 2.9 \text{ kJ} + 42.3 \text{ kJ} + 11 \text{ kJ} = 56 \text{ kJ}$

**95.** $q \text{ (cool liquid)} = 689 \text{ g} \times \frac{0.51 \text{ J}}{\text{g} \cdot °\text{C}} \times (420 - 552)°\text{C} \times \frac{1 \text{ kJ}}{1000 \text{ J}} = -46 \text{ kJ}$ $q \text{ (freeze liquid)} = 689 \text{ g} \times \frac{-112 \text{ J}}{\text{g}} \times \frac{1 \text{ kJ}}{1000 \text{ J}} = -77.2 \text{ kJ}$

$q \text{ (cool solid)} = 689 \text{ g} \times \frac{0.39 \text{ J}}{\text{g} \cdot °\text{C}} \times (21 - 420)°\text{C} \times \frac{1 \text{ kJ}}{1000 \text{ J}} = -1.1 \times 10^2 \text{ kJ}$ Total $q = -46 \text{ kJ} + (-77.2 \text{ kJ}) + (-1.1 \times 10^2 \text{ kJ}) = -2.3 \times 10^2 \text{ kJ}$

**97.** $q \text{ (heat solid)} = 941 \text{ kg} \times \frac{1000 \text{ g}}{\text{kg}} \times \frac{0.27 \text{ J}}{\text{g} \cdot °\text{C}} \times (264 - 26)°\text{C} \times \frac{1 \text{ kJ}}{1000 \text{ J}} = 6.0 \times 10^4 \text{ kJ}$

$q \text{ (melt solid)} = 941 \text{ kg} \times \frac{1000 \text{ g}}{\text{kg}} \times \frac{29 \text{ J}}{\text{g}} \times \frac{1 \text{ kJ}}{1000 \text{ J}} = 2.7 \times 10^4 \text{ kJ}$

$q \text{ (heat liquid)} = 941 \text{ kg} \times \frac{1000 \text{ g}}{\text{kg}} \times \frac{0.21 \text{ J}}{\text{g} \cdot °\text{C}} \times (339 - 264)°\text{C} \times \frac{1 \text{ kJ}}{1000 \text{ J}} = 1.5 \times 10^4 \text{ kJ}$ Total $q = 6.0 \times 10^4 \text{ kJ} + 2.7 \times 10^4 \text{ kJ} + 1.5 \times 10^4 \text{ kJ}$

$= 10.2 \times 10^4 \text{ kJ} = 1.02 \times 10^5 \text{ kJ}$

**100.** True: a, c, f, h, i, r. False: b, d, e, g, j, k, l, m, n, o, p, q. **101.** Both molecules have induced dipole and dipole forces. $CH_3OH$ has hydrogen bonding and $CH_3F$ does not. The molecules are about the same size. It is reasonable to predict stronger intermolecular forces in $CH_3OH$ and therefore a higher boiling point, and $CH_3F$ will have the higher equilibrium vapor pressure. **102.** Large molecules having strong induced dipole forces may have stronger intermolecular attractive forces than small molecules with hydrogen bonding and therefore exhibit greater viscosity. **103.** Reducing volume increases vapor concentration, which causes a temporary increase in the rate of condensation until equilibrium is once again established. Evaporation rate, which depends only on temperature, is not affected. **104.** The final ether vapor pressure in Containers A and C is the greatest—the equilibrium vapor pressure. Container B has the lowest vapor pressure, having all evaporated before reaching equilibrium. **105.** (757 + 22.4) torr = 779 torr **106.** (b) Evaporation at constant rate begins immediately when liquid is introduced. At that time condensation rate is zero. Net rate of increase in vapor concentration is at a maximum, so rate of vapor pressure increase is a maximum at start. Later condensation rate is more than zero but less than evaporation rate. Net rate of increase in vapor concentration is less than initially, so rate of vapor pressure increase is less than initially. At equilibrium, evaporation and condensation rates are equal. Vapor concentration and therefore vapor pressure remain constant. **107.** All of the liquid evaporated before the vapor concentration was high enough to yield a condensation rate equal to the evaporation rate. At lower-than-equilibrium vapor concentration, the vapor pressure is lower than the equilibrium vapor pressure. More liquid must be introduced to the flask until some excess remains and the pressure stabilizes in order to measure equilibrium vapor pressure. **108.** The liquid can be boiled by reducing the pressure and thus the boiling temperature.

**109.** $72.0 \text{ g} \times \frac{4.18 \text{ J}}{\text{g} \cdot {}^\circ\text{C}} \times (25.5 - 19.2){}^\circ\text{C} = -[141 \text{ g} \times c \times (25.5 - 89.0){}^\circ\text{C}] \qquad c = 0.21 \text{ J/g} \cdot {}^\circ\text{C}$

**110.** $q_{Al} = 54.1 \text{ g} \times \frac{0.90 \text{ J}}{\text{g} \cdot {}^\circ\text{C}} \times (-9 - 17){}^\circ\text{C} = -1.3 \times 10^3 \text{ J} = -1.3 \text{ kJ}$ $q \text{ (cool water)} = 408 \text{ g} \times \frac{4.18 \text{ J}}{\text{g} \cdot {}^\circ\text{C}} \times (0 - 17){}^\circ\text{C} = -2.9 \times 10^4 \text{ J} = -29 \text{ kJ}$

$q \text{ (freeze water)} = 408 \text{ g} \times -\frac{333 \text{ J}}{\text{g}} = -1.36 \times 10^5 \text{ J} = -136 \text{ kJ}$ $q \text{ (cool ice)} = 408 \text{ g} \times \frac{2.06 \text{ J}}{\text{g} \cdot {}^\circ\text{C}} \times (-9 - 0){}^\circ\text{C} = -8 \times 10^3 \text{ J} = -8 \text{ kJ}$

Total $q = -1.3 \text{ kJ} + (-29 \text{ kJ}) + (-136 \text{ kJ}) + (-8 \text{ kJ}) = -174 \text{ kJ}$

**111.** Dissolve the compounds and check for electrical conductivity. The ionic potassium sulfate solute will conduct, whereas the molecular sugar solute will not. **112.** Without a regular and uniform structure in an amorphous solid, some intermolecular forces are stronger than others. The weak forces are more easily overcome than the stronger forces, and thus a lower temperature is needed for melting in some parts of the solid than in other parts. **113.** As temperature drops, the equilibrium vapor pressure drops below the atmospheric vapor pressure. The air becomes first saturated, then supersaturated, and condensation (dew) begins to form. **114.** Energy lost by lemonade = Energy gained by ice; Let M = Mass of ice

$175 \text{ g} \times \frac{4.18 \text{ J}}{\text{g} \cdot {}^\circ\text{C}} \times (23 - 5){}^\circ\text{C} = \text{M g} \times \frac{2.06 \text{ J}}{\text{g} \cdot {}^\circ\text{C}} \times 8{}^\circ\text{C} + \text{M g} \times \frac{333 \text{ J}}{\text{g}} + \text{M g} \times \frac{4.18 \text{ J}}{\text{g} \cdot {}^\circ\text{C}} \times 5{}^\circ\text{C} \quad \text{M} = 36 \text{ g}$

# Chapter 16

## Answers to Target Checks

**1.** True: a and b. (c) A solution is always made up of two or more pure substances. (d) The different parts of a solution are too small to be detected visually. **2.** (a) The solutes are A and B; water is the solvent. (b) Degree of saturation cannot be estimated without knowing the solubility of the compound. (c) The question has no meaning because the term *dilute* compares solutions of the same solute, not different solutes. **3.** (a) Equal to. (b) Less than. (c) More than. The steadily increasing crystallization rate "subtracts from" the constant dissolving rate, reducing the net rate as time goes on. The net rate eventually reaches zero at equilibrium. **4.** Main criterion: Do substances have similar intermolecular attractions? If yes, they are probably soluble. Look for similarities in polarity, hydrogen bonding capability, and size, each of which contributes to similar intermolecular forces.

## Answers to Concept-Linking Exercises

*You may have found more relationships or relationships other than the ones given in these answers.*

**1.** A solution is a homogeneous mixture, made up of two or more pure substances. Two solutions of the same substance may have different percentage compositions. **2.** Solubility is a measure of how much solute will dissolve in a given amount of solvent at a given temperature. A solution holding that amount of solute is saturated; less than that amount, unsaturated; and more than that amount, supersaturated. **3.** Solubility is a measure of how much solute will dissolve in a given amount of solvent at a given temperature. In general, substances with similar intermolecular forces will dissolve in one another. The greater the partial pressure of a solute gas over a liquid solvent is, the greater its solubility will be. The solubilities of most solids increase with increasing temperature. The solubilities of gases in liquids are generally lower at higher temperatures. **4.** A primary standard may be weighed accurately, dissolved, and diluted to an accurately determined volume, yielding a solution whose concentration is known with a high degree of accuracy. It may be used to standardize—find the concentration of—a second solution by titration, using an indicator to signal when the two reactants are present in precisely the molar quantities by which they react. Concentration is expressed in molarity, moles of solute per liter of solution, mol/L.
**5.** Normality is the number of equivalents per liter of solution. For acids, one equivalent is the quantity that yields one mole of hydrogen ions in solution. For bases, one equivalent is the quantity that reacts with one mole of hydrogen ions. Equivalent mass is the number of grams per equivalent.
**6.** The depression of the freezing point and elevation of the boiling point of a solvent are proportional to the solute particle concentration of a solution expressed in molality, moles of solute particles per kilogram of solvent. The molal freezing-point depression constant and molal boiling-point elevation constant are proportionality constants for those proportionalities. These are colligative properties because they depend on solute particle concentration without regard to the identity of the solute particles.

## Answers to Blue-Numbered Questions, Exercises, and Problems

**1.** True. Provided that there is no chemical reaction, gases will always combine to form a homogeneous mixture, which is, by definition, a solution.
**3.** If particles were visible, there would be distinctly different phases and the mixture would not be homogeneous and therefore not a solution.
**5.** a) Salt is solute; water is solvent. When a solid is dissolved in a liquid, the solid is the solute and the liquid is the solvent. b) Copper is solute; silver is solvent. c) Oxygen is solute; nitrogen is solvent. For (b) and (c), the solute is the substance present in the smaller amount. **7.** A saturated solution is at its solubility limit, so it is concentrated. A concentrated solution is not necessarily saturated. A concentrated solution contains a relatively large amount of solute; a saturated solution cannot hold more solute. **9.** a) All salt added will be solid in the beaker. b) Some or all salt added will dissolve; little or no salt will be solid in the beaker. c) All added salt and some previously dissolved salt will be solid in the beaker. **11.** Temperature. Usually, the higher the temperature, the higher the solubility. **13.** Acetic acid is soluble in water because it is dispersed uniformly throughout the solution. It is also miscible, a term usually used to express the solubility of liquids in each other. **15.** Cations are attracted to the negative portion of the water molecule; anions are attracted to the positive portion. **17.** As dissolved solute particles move through the solution, they come into contact with each other and with undissolved solute and return to the solid state. **19.** When dissolving begins, the crystallization rate is zero. Dissolving rate remains constant, and as the

solution becomes more concentrated, crystallization rate increases. Concentration of the solute in solution continues to increase until the crystallization rate equals the dissolving rate. **21.** Never. **23.** A supersaturated solution is unstable, and a physical disturbance such as stirring will start crystallization. **25.** Finely dividing a solid offers more surface area per unit of mass. Stirring or agitating the solution prevents concentration buildup at the solute surface, which minimizes crystallization rate. All physical processes speed up at higher temperatures because particle movement is more rapid.
**27.** (a) formic acid; (c) methylamine. Both compounds exhibit hydrogen bonding in water. Hexane and tetrafluoromethane are nonpolar and only exhibit induced dipole forces. **29.** Glycerine exhibits hydrogen bonding, as does water, so they are miscible. Hexane is nonpolar. **31.** Carbon dioxide, $CO_2$.

**33.** $\frac{2.32 \text{ g solute}}{(2.32 + 81.0) \text{ g solution}} \times 100 = 2.78\%$

**35.** $415 \text{ g solution} \times \frac{58.0 \text{ g } NH_4NO_3}{100 \text{ g solution}} = 241 \text{ g } NH_4NO_3$ $415 \text{ g solution} - 241 \text{ g } NH_4NO_3 = 174 \text{ g } H_2O = 174 \text{ mL } H_2O$

**37.** $2.41 \text{ g KI} \times \frac{1 \text{ mol KI}}{166.0 \text{ g KI}} = 0.0145 \text{ mol KI}$ $\frac{0.0145 \text{ mol KI}}{50.0 \text{ mL}} \times \frac{1000 \text{ mL}}{\text{L}} = 0.290 \text{ M KI}$

**39.** $18.0 \text{ g } NiCl_2 \times \frac{1 \text{ mol } NiCl_2}{129.59 \text{ g } NiCl_2} = 0.139 \text{ mol } NiCl_2$ $30.0 \text{ g } NiCl_2 \cdot 6\,H_2O \times \frac{1 \text{ mol } NiCl_2 \cdot 6\,H_2O}{237.69 \text{ g } NiCl_2 \cdot 6\,H_2O} = 0.126 \text{ mol } NiCl_2 \cdot 6\,H_2O$

The anhydrous compound has more moles and thus has the higher concentration: $\frac{0.139 \text{ mol } NiCl_2}{90.0 \text{ mL}} \times \frac{1000 \text{ mL}}{\text{L}} = 1.54 \text{ M } NiCl_2$

**41.** $2.50 \times 10^2 \text{ mL} \times \frac{1 \text{ L}}{1000 \text{ mL}} \times \frac{0.058 \text{ mol } AgNO_3}{\text{L}} \times \frac{169.9 \text{ g } AgNO_3}{\text{mol } AgNO_3} = 2.5 \text{ g } AgNO_3$

**43.** $2.50 \text{ L} \times \frac{1.40 \text{ mol KOH}}{\text{L}} \times \frac{56.11 \text{ g KOH}}{\text{mol KOH}} = 196 \text{ g KOH}$ **45.** $5.19 \text{ mol } H_2SO_4 \times \frac{1 \text{ L}}{18 \text{ mol } H_2SO_4} = 0.29 \text{ L}$

**47.** $8.33 \text{ g NaCl} \times \frac{1 \text{ mol NaCl}}{58.44 \text{ g NaCl}} \times \frac{1 \text{ L}}{0.132 \text{ mol NaCl}} = 1.08 \text{ L}$ **49.** $55.7 \text{ mL} \times \frac{1 \text{ L}}{1000 \text{ mL}} \times \frac{0.204 \text{ mol } AgNO_3}{\text{L}} = 0.0114 \text{ mol } AgNO_3$

**51.** $25.0 \text{ mL} \times \frac{1 \text{ L}}{1000 \text{ mL}} \times \frac{0.0841 \text{ mol } KMnO_4}{\text{L}} = 0.00210 \text{ mol } KMnO_4$

**53.** $\frac{101.11 \text{ g } KNO_3}{\text{mol } KNO_3} \times \frac{3.30 \text{ mol } KNO_3}{\text{L}} \times \frac{1 \text{ L}}{1000 \text{ mL}} \times \frac{1 \text{ mL}}{1.15 \text{ g soln}} \times 100 = 29.0\% \; KNO_3$

**55.** $44.9 \text{ g } C_{10}H_8 \times \frac{1 \text{ mol } C_{10}H_8}{128.16 \text{ g } C_{10}H_8} = 0.350 \text{ mol } C_{10}H_8$ $\frac{0.350 \text{ mol } C_{10}H_8}{175 \text{ g}} \times \frac{1000 \text{ g}}{\text{kg}} = 2.00 \text{ m}$

**57.** $4.00 \times 10^2 \text{ g ethanol} \times \frac{1 \text{ kg ethanol}}{1000 \text{ g ethanol}} \times \frac{4.70 \text{ mol } (CH_3CH_2)_2NH}{\text{kg ethanol}} \times \frac{73.14 \text{ g } (CH_3CH_2)_2NH}{\text{mol } (CH_3CH_2)_2NH} = 138 \text{ g } (CH_3CH_2)_2NH$

**59.** $97.7 \text{ mg NaCl} \times \frac{1 \text{ g NaCl}}{1000 \text{ mg NaCl}} \times \frac{1 \text{ mol NaCl}}{58.44 \text{ mol NaCl}} \times \frac{1 \text{ kg } H_2O}{2.80 \times 10^{-3} \text{ mol NaCl}} \times \frac{1000 \text{ g } H_2O}{\text{kg } H_2O} \times \frac{1 \text{ mL } H_2O}{1 \text{ g } H_2O} = 597 \text{ mL } H_2O$

**61.** Equivalent mass is the mass of a substance that reacts with one mole of hydrogen or hydroxide ions. LiOH has one mole of $OH^-$ ions, so equivalent mass = molar mass. $H_2SO_4$ can release one or two moles of $H^+$ ions, so equivalent mass = molar mass or $\frac{1}{2}$ of molar mass. **63.** 1 eq/mol $HNO_2$; 1 eq/mol $H_2SeO_4$ **65.** 2 eq/mol $Cu(OH)_2$; 3 eq/mol $Fe(OH)_3$ **67.** 47.02 g $HNO_2$/eq; 144.98 g $H_2SeO_4$/eq **69.** 48.78 g $Cu(OH)_2$/eq; 35.62 g $Fe(OH)_3$/eq

**71.** $2.25 \text{ g KOH} \times \frac{1 \text{ eq KOH}}{56.11 \text{ g KOH}} = 0.0401 \text{ eq KOH}$ $\frac{0.0401 \text{ eq KOH}}{2.50 \times 10^2 \text{ mL}} \times \frac{1000 \text{ mL}}{\text{L}} = 0.160 \text{ N KOH}$

**73.** $7.50 \times 10^2 \text{ mL} \times \frac{1 \text{ L}}{1000 \text{ mL}} \times \frac{0.200 \text{ eq } NaHSO_4}{\text{L}} \times \frac{120.07 \text{ g } NaHSO_4}{\text{eq}} = 18.0 \text{ g } NaHSO_4$

**75.** $6.69 \text{ g } H_2C_2O_4 \times \frac{1 \text{ eq } H_2C_2O_4}{90.04 \text{ g } H_2C_2O_4} = 0.0743 \text{ eq } H_2C_2O_4$ $\frac{0.0743 \text{ eq } H_2C_2O_4}{2.00 \times 10^2 \text{ mL}} \times \frac{1000 \text{ mL}}{\text{L}} = 0.372 \text{ N } H_2C_2O_4$

**77.** (a) 0.965 M NaOH; (b) 0.119 M $H_3PO_4$ **79.** $73.1 \text{ mL} \times \frac{1 \text{ L}}{1000 \text{ mL}} \times \frac{0.834 \text{ eq NaOH}}{\text{L}} = 0.0610 \text{ eq NaOH}$

**81.** $0.788 \text{ eq } KMnO_4 \times \frac{1 \text{ L}}{0.492 \text{ eq } KMnO_4} = 1.60 \text{ L}$ **83.** $M_d = \frac{17 \text{ M} \times 45.0 \text{ mL}}{1.5 \text{ L}} \times \frac{1 \text{ L}}{1000 \text{ mL}} = 0.51 \text{ M } HC_2H_3O_2$

**85.** $V_c = \frac{0.69 \text{ M} \times 7.50 \times 10^2 \text{ mL}}{16 \text{ M}} = 32 \text{ mL } HNO_3$ **87.** $V_c = \frac{2.9 \text{ eq/L} \times 3.0 \text{ L}}{18 \text{ mol/L}} \times \frac{1 \text{ mol}}{2 \text{ eq}} \times \frac{1000 \text{ mL}}{\text{L}} = 2.4 \times 10^2 \text{ mL } H_2SO_4$

**89.** $N_d = \frac{15 \text{ mol/L} \times 15.0 \text{ mL}}{2.50 \times 10^2 \text{ mL}} \times \frac{2 \text{ eq}}{\text{mol}} = 1.8 \text{ N } H_3PO_4$ **91.** $MgCl_2 + 2\,NaOH \rightarrow Mg(OH)_2 + 2\,NaCl$

$25.0 \text{ mL} \times \frac{1 \text{ L}}{1000 \text{ mL}} \times \frac{0.398 \text{ mol } MgCl_2}{\text{L}} \times \frac{1 \text{ mol } Mg(OH)_2}{1 \text{ mol } MgCl_2} \times \frac{58.33 \text{ g } Mg(OH)_2}{\text{mol } Mg(OH)_2} = 0.580 \text{ g } Mg(OH)_2$

**93.** $2\,Na_3PO_4 + 3\,Ca(NO_3)_2 \rightarrow 6\,NaNO_3 + Ca_3(PO_4)_2$

$100.0 \text{ mL} \times \frac{1 \text{ L}}{1000 \text{ mL}} \times \frac{0.130 \text{ mol } Ca(NO_3)_2}{\text{L}} \times \frac{1 \text{ mol } Ca_3(PO_4)_2}{3 \text{ mol } Ca(NO_3)_2} \times \frac{310.18 \text{ g } Ca_3(PO_4)_2}{\text{mol } Ca_3(PO_4)_2} = 1.34 \text{ g } Ca_3(PO_4)_2$

**95.** $2.00 \text{ L} \times \frac{273 \text{ K}}{295 \text{ K}} \times \frac{789 \text{ torr}}{1 \text{ bar}} \times \frac{1.013 \text{ bar}}{760 \text{ torr}} \times \frac{1 \text{ mol } H_2}{22.7 \text{ L}} = 0.0857 \text{ mol } H_2$ *or* $n = \frac{PV}{RT} = 789 \text{ torr} \times 2.00 \text{ L} \times \frac{\text{mol} \cdot \text{K}}{62.4 \text{ L} \cdot \text{torr}} \times \frac{1}{295 \text{ K}} = 0.0857 \text{ mol } H_2$

$0.0857 \text{ mol } H_2 \times \frac{6 \text{ mol NaOH}}{3 \text{ mol } H_2} \times \frac{1 \text{ L}}{1.50 \text{ mol NaOH}} \times \frac{1000 \text{ mL}}{\text{L}} = 114 \text{ mL}$

**97.** $NH_2SO_3H + NaOH \rightarrow NH_2SO_3Na + H_2O$ $8.74 \text{ g } NH_2SO_3H \times \frac{1 \text{ mol } NH_2SO_3H}{97.10 \text{ g } NH_2SO_3H} \times \frac{1 \text{ mol NaOH}}{1 \text{ mol } NH_2SO_3H} \times \frac{1 \text{ L}}{0.842 \text{ mol NaOH}} \times \frac{1000 \text{ mL}}{\text{L}} = 107 \text{ mL}$

**99.** $5.038 \text{ g } HC_7H_5O_2 \times \frac{1 \text{ mol } HC_7H_5O_2}{122.12 \text{ g } HC_7H_5O_2} \times \frac{1 \text{ mol } Na_2CO_3}{2 \text{ mol } HC_7H_5O_2} = 0.02063 \text{ mol } Na_2CO_3$ $\frac{0.02063 \text{ mol } Na_2CO_3}{51.89 \text{ mL}} \times \frac{1000 \text{ mL}}{\text{L}} = 0.3976 \text{ M } Na_2CO_3$

**101.** $H_2C_4H_2O_4 + 2\ KOH \rightarrow K_2C_4H_2O_4 + 2\ H_2O$ $1.45\ g\ H_2C_4H_2O_4 \times \frac{1\ mol\ H_2C_4H_2O_4}{116.07\ g\ H_2C_4H_2O_4} \times \frac{2\ mol\ KOH}{1\ mol\ H_2C_4H_2O_4} = 0.0250\ mol\ KOH$

$\frac{0.0250\ mol\ KOH}{50.0\ mL} \times \frac{1000\ mL}{L} = 0.500\ M\ KOH$

**103.** $37.80\ mL \times \frac{1\ L}{1000\ mL} \times \frac{0.4052\ mol\ NaHCO_3}{L} \times \frac{1\ mol\ H_2SO_4}{2\ mol\ NaHCO_3} = 7.658 \times 10^{-3}\ mol\ H_2SO_4$

$\frac{7.658 \times 10^{-3}\ mol\ H_2SO_4}{20.00\ mL} \times \frac{1000\ mL}{L} = 0.3829\ M\ H_2SO_4$

**105.** At 2 eq/mol, 0.3976 M $Na_2CO_3$ = 0.7952 N $Na_2CO_3$ **107.** At 1 eq/mol, 0.500 M KOH = 0.500 N KOH

**109.** $N_2 = \frac{0.405\ N \times 39.8\ mL}{25.0\ mL} = 0.645\ N\ Na_2CO_3$ **111.** $N_2 = \frac{0.402\ N \times 42.2\ mL}{50.0\ mL} = 0.339\ N\ H_2C_4H_4O_6$

**113.** $N_2 = \frac{0.208\ N \times 16.3\ mL}{20.0\ mL} = 0.170\ N\ H_3PO_4$ The normality of the acid depends on how many hydrogens react.

**115.** $\frac{1.21\ g}{30.7\ mL \times 0.170\ eq/L} \times \frac{1000\ mL}{L} = 232\ g/eq$

**117.** Partial pressure is a colligative property because it depends on the number of particles and is independent of their identity (for an ideal gas).

**119.** $27.2\ g\ C_6H_5NH_2 \times \frac{1\ mol\ C_6H_5NH_2}{93.13\ g\ C_6H_5NH_2} = 0.292\ mol\ C_6H_5NH_2$ $\frac{0.292\ mol\ C_6H_5NH_2}{1.20 \times 10^2\ g\ H_2O} \times \frac{1000\ g\ H_2O}{kg\ H_2O} = 2.43\ m$

$\Delta T_b = \frac{0.52°C}{m} \times 2.43\ m = 1.3°C$ $T_b = 101.3°C$ $\Delta T_f = \frac{1.86°C}{m} \times 2.43\ m = 4.52°C$ $T_f = -4.52°C$

**121.** $2.12\ g\ C_{10}H_8 \times \frac{1\ mol\ C_{10}H_8}{128.16\ g\ C_{10}H_8} = 0.0165\ mol\ C_{10}H_8$ $\frac{0.0165\ mol\ C_{10}H_8}{32.0\ g\ C_6H_6} \times \frac{1000\ g\ C_6H_6}{kg\ C_6H_6} = 0.516\ m$

$\Delta T_f = \frac{5.10°C}{m} \times 0.516\ m = 2.63°C$ $T_f = 5.50°C - 2.63°C = 2.87°C$

**123.** $\Delta T_f = 16.6°C - 14.1°C = 2.5°C$ $m = 2.5°C \times \frac{1\ m}{3.90°C} = 0.64\ m$

**125.** $m = (100.28 - 100.00)°C \times \frac{1\ m}{0.52°C} = 0.54\ m$ $6.00 \times 10^2\ g\ H_2O \times \frac{1\ kg\ H_2O}{1000\ g\ H_2O} \times \frac{0.54\ mol\ solute}{kg\ H_2O} = 0.32\ mol\ solute$

$\frac{16.1\ g}{0.32\ mol} = 5.0 \times 10^1\ g/mol$

**127.** $m = 9.6°C \times \frac{1\ m}{3.56°C} = 2.7\ m$ $90.0\ g\ phenol \times \frac{1\ kg\ phenol}{1000\ g\ phenol} \times \frac{2.7\ mol\ solute}{kg\ phenol} = 0.24\ mol\ solute$ $\frac{12.4\ g}{0.24\ mol} = 52\ g/mol$

**129.** $11.4\ g\ C_2H_5OH \times \frac{1\ mol\ C_2H_5OH}{46.07\ g\ C_2H_5OH} = 0.247\ mol\ C_2H_5OH$ $\frac{0.247\ mol\ C_2H_5OH}{2.00 \times 10^2\ g\ solvent} \times \frac{1000\ g\ solvent}{kg\ solvent} = 1.24\ m\ C_2H_5OH$

$K_f = \frac{28.7°C - 22.5°C}{1.24\ m} = 5.0°C/m$

**132.** True: a, d, h, j, m. False: b, c, e, f, g, i, k, l. **133.** The bubbles are dissolved air (mostly nitrogen and oxygen) that becomes less soluble at higher temperatures. **134.** It raises the boiling point. **135.** No **136.** Distillation is one method. It is used to separate petroleum into its components, which includes many products such as gases, gasoline, kerosene, fuel oil, lubricating oil, and asphalt. **138.** Attractions between solute particles and attractions between solvent particles. **139.** Dissolve more than 1.02 g of silver acetate at a temperature greater than 20°C; then cool the solution without disturbance. Crystallization does not occur at the solubility limit because solute particles are not properly organized for crystal formation. **140.** Finely powdered pure sugar has more surface area than granular sugar and therefore dissolves more quickly. Both sweeten coffee equally.

**141.** $\frac{18.0\ g\ HCl}{100\ g\ soln} \times \frac{1\ mol\ HCl}{36.46\ g\ HCl} \times \frac{1.09\ g\ soln}{mL\ soln} \times \frac{1000\ mL\ soln}{L\ soln} = 5.38\ M\ HCl$

**142.** $1.00 \times 10^2\ mL\ B \times \frac{0.879\ g\ B}{mL\ B} \times \frac{1\ kg\ B}{1000\ g\ B} \times \frac{0.254\ mol\ C_4H_8O}{kg\ B} \times \frac{72.10\ g\ C_4H_8O}{mol\ C_4H_8O} = 1.61\ g\ C_4H_8O$

**143.** $2.50 \times 10^2\ mL \times \frac{1\ L}{1000\ mL} \times \frac{0.500\ eq\ H_2C_2O_4}{L} \times \frac{126.07\ g\ H_2C_2O_4 \cdot 2\ H_2O}{2\ eq\ H_2C_2O_4} = 7.88\ g\ H_2C_2O_4 \cdot 2\ H_2O$

**145.** $2\ Fe(NO_3)_3 \rightarrow 2\ Fe(OH)_3 \rightarrow Fe_2O_3$ $35.0\ mL \times \frac{1\ L}{1000\ mL} \times \frac{0.516\ mol\ Fe(NO_3)_3}{L} \times \frac{1\ mol\ Fe_2O_3}{2\ mol\ Fe(NO_3)_3} \times \frac{159.70\ g\ Fe_2O_3}{mol\ Fe_2O_3} = 1.44\ g\ Fe_2O_3$

**146.** This limiting reactant problem is solved by the methods described in Sections 10.4 and 10.6.

$2\ NaOH + CuSO_4 \rightarrow Cu(OH)_2 + Na_2SO_4$ $25.0\ mL \times \frac{1\ L}{1000\ mL} \times \frac{0.350\ mol\ NaOH}{L} \times \frac{1\ mol\ Cu(OH)_2}{2\ mol\ NaOH} \times \frac{97.57\ g\ Cu(OH)_2}{mol\ Cu(OH)_2} = 0.427\ g\ Cu(OH)_2$

$45.0\ mL \times \frac{1\ L}{1000\ mL} \times \frac{0.125\ mol\ CuSO_4}{L} \times \frac{1\ mol\ Cu(OH)_2}{1\ mol\ CuSO_4} \times \frac{97.57\ g\ Cu(OH)_2}{mol\ Cu(OH)_2} = 0.549\ g\ Cu(OH)_2$ 0.427 g of $Cu(OH)_2$ will precipitate.

**148.** $16.80\ mL \times \frac{1\ L}{1000\ mL} \times \frac{0.629\ mol\ AgNO_3}{L} \times \frac{1\ mol\ Cl^-}{1\ mol\ AgNO_3} = 0.0106\ mol\ Cl^-$ $\frac{0.0106\ mol\ Cl^-}{25.00\ mL} \times \frac{1000\ mL}{L} = 0.424\ M\ Cl^-$

**150.** $Na_2CO_3 + 2\ HCl \rightarrow 2\ NaCl + H_2O + CO_2$

$41.24\ mL \times \frac{1\ L}{1000\ mL} \times \frac{0.244\ mol\ HCl}{L} \times \frac{1\ mol\ Na_2CO_3}{2\ mol\ HCl} \times \frac{105.99\ g\ Na_2CO_3}{mol\ Na_2CO_3} \times \frac{1000\ mg\ Na_2CO_3}{g\ Na_2CO_3} = 533\ mg\ Na_2CO_3$

$\frac{533\ mg\ Na_2CO_3}{694\ mg\ sample} \times 100 = 76.8\%\ Na_2CO_3$

**151.** $19.58\ mL \times \frac{0.201\ mmol\ NaOH}{mL\ NaOH} \times \frac{1\ mmol\ NaH_2PO_4}{1\ mmol\ NaOH} \times \frac{119.98\ mg\ NaH_2PO_4}{mmol\ NaH_2PO_4} = 472\ mg\ NaH_2PO_4$

$\frac{472 \text{ mg}}{599 \text{ mg}} \times 100 = 78.8\%\ NaH_2PO_4$ $100 - 78.8 = 21.2\%\ Na_2HPO_4$

**152.** a) $2\ KI + Pb(NO_3)_2 \rightarrow PbI_2 + 2\ KNO_3$ $60.0 \text{ mL} \times \frac{1 \text{ L}}{1000 \text{ mL}} \times \frac{0.322 \text{ mol KI}}{\text{L}} \times \frac{1 \text{ mol } PbI_2}{2 \text{ mol KI}} \times \frac{461.0 \text{ g } PbI_2}{\text{mol } PbI_2} = 4.45 \text{ g } PbI_2$

$20.0 \text{ mL} \times \frac{1 \text{ L}}{1000 \text{ mL}} \times \frac{0.530 \text{ mol } Pb(NO_3)_2}{\text{L}} \times \frac{1 \text{ mol } PbI_2}{1 \text{ mol } Pb(NO_3)_2} \times \frac{461.0 \text{ g } PbI_2}{\text{mol } PbI_2} = 4.89 \text{ g } PbI_2$ $4.45 \text{ g } PbI_2$ will precipitate

| b) | **2 KI** | + | **$Pb(NO_3)_2$** | → | **$PbI_2$** | + | **2 $KNO_3$** |
|---|---|---|---|---|---|---|---|
| Volume at start, mL | 60.0 | | 20.0 | | | | |
| Volume at start, L | 0.0600 | | 0.0200 | | | | |
| Molarity, mol/L | 0.322 | | 0.530 | | | | |
| Moles at start | 0.0193 | | 0.0106 | | | | |
| Moles used (−), produced (+) | −0.0193 | | −0.00965 | | +0.00965 | | +0.0193 |
| Moles at end | 0 | | 0.0010 | | 0.00965 | | 0.0193 |

Total volume = 0.0600 L + 0.0200 L = 0.0800 L $\frac{0.0193 \text{ mol } KNO_3}{0.0800 \text{ L}} \times \frac{1 \text{ mol } K^+}{1 \text{ mol } KNO_3} = 0.241 \text{ M } K^+$

c) $\frac{0.0010 \text{ mol } Pb(NO_3)_2}{0.0800 \text{ L}} \times \frac{1 \text{ mol } Pb^{2+}}{1 \text{ mol } Pb(NO_3)_2} = 0.013 \text{ M } Pb^{2+}$

**153.** A small sample of air is a homogeneous mixture and is therefore a solution. The atmosphere is a very tall sample that becomes less dense at higher elevations. The atmosphere is therefore not homogeneous; consequently, it is not a solution. **154.** The density of a solution must be known in order to convert concentrations based on mass only (percentage, molality) to those based in volume (molarity, normality).

# Chapter 17

## Answers to Target Checks

**1.** An acid produces an $H^+$ ion and a base yields an $OH^-$ ion. **2.** Brønsted–Lowry (BL) and Arrhenius (AR) acids both yield protons; they are the same. AR bases all have hydroxide ions to receive protons; BL bases are anything that can receive protons. All AR bases are BL bases, but not all BL bases are AR bases. **3.** The term *proton-transfer reaction* describes what happens in a Brønsted–Lowry acid–base reaction. **4.** Water can be a Lewis base because it has unshared electron pairs. It cannot be a Lewis acid because it has no vacant orbital to receive an electron pair from a Lewis base.

## Answers to Concept-Linking Exercises

*You may have found more relationships or relationships other than the ones given in these answers.*

**1.** Arrhenius and Brønsted–Lowry acids are both associated with the hydrogen ion, or proton. An Arrhenius acid is a substance that, when added to water, increases the hydrogen ion concentration. A Brønsted–Lowry acid is a substance that has a proton that can be removed by a base. A Lewis acid is any species that has a vacant valence orbital and can accept an electron pair, which includes the hydrogen ion. **2.** The solution of an Arrhenius base contains hydroxide ions. A Brønsted–Lowry base removes protons in an acid–base reaction. A Lewis base has a pair of unshared electrons that can form a bond with a substance that has an empty valence orbital. **3.** The Brønsted–Lowry acid–base theory identifies an acid–base reaction as a proton-transfer reaction in which a proton is transferred from a proton source, the acid, to the proton remover, the base. A substance that can be a proton source or a proton remover is amphoteric. **4.** A reversible reaction is one in which the products react and re-form the reactants, as the equation is written. Reading the equation from left to right is the forward direction; reading from right to left is the reverse direction. The direction in which the products are the species present in higher concentration is the favored direction. **5.** According to the Lewis acid–base theory, a base is any species that has an unshared electron pair, an electron pair donor, that may form a bond with a species having a vacant valence orbital, which is an acid or an electron-pair acceptor. **6.** An acid is a substance that can release a proton in a reaction. The species that remains is the conjugate base of the original acid. Together they constitute a conjugate acid–base pair. **7.** A strong acid releases protons easily in a proton-transfer reaction and readily engages in an acid–base reaction. A weak acid holds its protons and tends not to engage in proton-transfer reactions. **8.** The water equilibrium is the reaction in which water dissociates into hydrogen and hydroxide ions: $H_2O(\ell) \rightleftharpoons H^+(aq) + OH^-(aq)$. The product of the ion concentrations in an aqueous solution equals $10^{-14}$ at 25°C and is called the water constant, $K_w$: $K_w = [H^+][OH^-] = 10^{-14}$. **9.** Hydrogen-ion concentration, $[H^+]$, is often expressed in terms of pH, the negative of the logarithm of the concentration: $pH = -\log [H^+]$. Similarly, pOH is the negative of the logarithm of the hydroxide ion concentration: $pOH = -\log [OH^-]$. **10.** The logarithm of a number, N, is the exponent, x, to which a base (10 in a decimal system) must be raised to be equal to the original number: $N = 10^x$. An antilogarithm, M, is the number produced when the base is raised to the power of a given logarithm, y: $10^y = M$. **11.** A number, N, written in exponential notation has the form $C \times 10^x$ in which C is the coefficient and $10^x$ is the exponential. If the logarithm of N is written in decimal form, the number to the left of the decimal is called the characteristic and what follows the decimal is the mantissa. The characteristic of the logarithm expresses the exponent in the exponential of the number, and the mantissa is the logarithm of the coefficient.

## Answers to Blue-Numbered Questions, Exercises, and Problems

**1.** The classical properties of acids and bases are listed in the introduction to the chapter. As an example of how a property relates to the ion associated with it, an acid–base neutralization is $H^+ + OH^- \rightarrow H_2O$. **3.** An Arrhenius base is a source of $OH^-$ ions, whereas a Brønsted–Lowry base is a proton remover. The two are in agreement, as the $OH^-$ ion is an excellent proton remover. Other substances, however, can also remove protons, so there are other bases according to the Brønsted–Lowry concept. **5.** In the reaction shown below, $AlCl_3$, a Lewis acid, accepts an electron pair from $Cl^-$, a Lewis base, in a Lewis acid–Lewis base neutralization reaction.

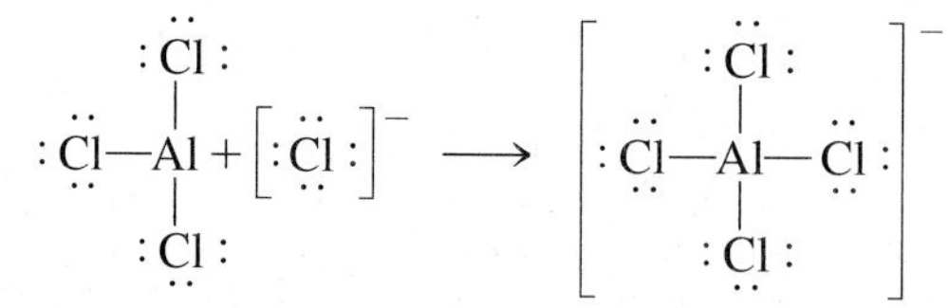

$$AlCl_3 + [:\ddot{Cl}:]^- \longrightarrow [AlCl_4]^-$$

**7.** $BF_3$ is a Lewis acid because the empty valence orbital of the boron atom accepts an electron pair from the oxygen atom in $C_2H_5OC_2H_5$, a Lewis base because it donates the electron pair. **9.** $F^-$; $HPO_4^{2-}$; $HNO_2$; $H_3PO_4$ **11.** Acids: $HSO_4^-$ (forward) and $HC_2O_4^-$ (reverse); bases: $C_2O_4^{2-}$ (forward) and $SO_4^{2-}$ (reverse) **13.** $HSO_4^-$ and $SO_4^{2-}$; $HC_2O_4^-$ and $C_2O_4^{2-}$ **15.** $HNO_2$ and $NO_2^-$; $HC_3H_5O_2$ and $C_3H_5O_2^-$ **17.** $NH_4^+$ and $NH_3$; $H_2PO_4^-$ and $HPO_4^{2-}$ **19.** A strong base has a strong attraction for protons, while a weak base has little attraction for protons. Stronger bases are at the bottom of the right column in Table 17.1 and weaker bases are at the top. **21.** $H_2O < HClO < HC_2O_4^- < H_2SO_3$ **23.** $CN^- > ClO^- > HSO_3^- > H_2O > Cl^-$ **25.** $HC_3H_5O_2 + PO_4^{3-} \rightleftharpoons C_3H_5O_2^- + HPO_4^{2-}$ Forward **27.** $HSO_4^- + CO_3^{2-} \rightleftharpoons SO_4^{2-} + HCO_3^-$ Forward **29.** $H_2CO_3 + NO_3^- \rightleftharpoons HCO_3^- + HNO_3$ Reverse **31.** $NO_2^- + H_3O^+ \rightleftharpoons HNO_2 + H_2O$ Forward **33.** $HSO_4^- + HC_2O_4^- \rightleftharpoons H_2SO_4 + C_2O_4^{2-}$ Reverse $HSO_4^- + HC_2O_4^- \rightleftharpoons SO_4^{2-} + H_2C_2O_4$ Reverse **35.** A strong acid releases protons readily; a strong reducing agent releases electrons readily. A strong base attracts protons strongly; a strong oxidizing agent attracts electrons strongly. **37.** The very small value for $K_w$ indicates that water ionizes to a very small extent. **39.** An acidic solution has a higher $H^+$ concentration than $OH^-$ concentration. The solution is therefore acidic: $10^{-5} > 10^{-9}$. **41.** $10^{-12}$ M **43.** (a) neutral; (b) weakly basic; (c) strongly basic **45.** Basic. Water solutions with pH = 7 ($[H^+] = [OH^-] = 10^{-7}$) are neutral, pH > 7 ($[H^+] < 10^{-7}$ and $[OH^-] > 10^{-7}$) are basic, and those with pH < 7 ($[H^+] > 10^{-7}$ and $[OH^-] < 10^{-7}$) are acidic.

| pH | pOH | $[H^+]$ | $[OH^-]$ | |
|---|---|---|---|---|
| **47.** 5 | 9 | $10^{-5}$ | $10^{-9}$ | weakly acidic |
| **49.** 13 | 1 | $10^{-13}$ | $10^{-1}$ | strongly basic |
| **51.** 10 | 4 | $10^{-10}$ | $10^{-4}$ | strongly basic |
| **53.** 9 | 5 | $10^{-9}$ | $10^{-5}$ | weakly basic |
| **55.** 4.40 | 9.60 | $4.0 \times 10^{-5}$ | $2.5 \times 10^{-10}$ | |
| **57.** 4.06 | 9.94 | $8.7 \times 10^{-5}$ | $1.1 \times 10^{-10}$ | |
| **59.** 0.55 | 13.45 | $2.8 \times 10^{-1}$ | $3.5 \times 10^{-14}$ | |
| **61.** 6.60 | 7.40 | $2.5 \times 10^{-7}$ | $4.0 \times 10^{-8}$ | |

**64.** True: a, b, c, e, f, g, h. False: d, i, j, k, l **65.** Yes, ammonia has a proton, so its proton theoretically can be removed by hydroxide ion: $OH^- + NH_3 \rightarrow H_2O + NH_2^-$. Additionally, hydroxide ion has a proton and ammonia has an unshared electron pair, so the opposite reaction is also theoretically possible: $OH^- + NH_3 \rightarrow O^{2-} + NH_4^+$. **66.** An amphoteric substance can be an acid by losing a proton, or a base by gaining a proton. $HX^-$ is a general formula of an amphoteric substance: $HX^- \rightarrow H^+ + X^{2-}$ (acid reaction); $HX^- + H^+ \rightarrow H_2X$ (base reaction). Examples of $HX^-$ without carbon include $HSO_4^-$ and $H_2PO_4^-$. **67.** pCl = 7.126 **68.** No. A Brønsted–Lowry acid is a proton source. A proton is a hydrogen ion. If there are no hydrogen atoms in a substance, it cannot donate a hydrogen ion to another species. **69.** Chemists generally prefer the relative simplicity of the pH scale. Most people find that working with numbers such as pH = 4 is much more convenient than the equivalent hydrogen ion concentration $1 \times 10^{-4}$ M or 0.0001 M. **70.** $[Br^-]$ = antilog (−7.2) = $6 \times 10^{-8}$ M **71.** When a proton is removed from an $H_3X$ species, a single positive charge is being pulled away from a particle with a single minus charge, $H_2X^-$. When a proton is removed from an $H_2X^-$ species, a single positive charge is being pulled away from a particle with a double minus charge, $HX^{2-}$. The loss of the second proton is energetically more difficult, so $H_2X^-$ is a weaker acid than $H_3X$. **72.** There can be no proton transfer without a proton—an $H^+$ ion. **73.** Carbonate ion is a proton acceptor: $H^+ + CO_3^{2-} \rightarrow HCO_3^-$. **74.** $SO_3 + H_2O \rightarrow H_2SO_4$, sulfuric acid **75.** $CaO + H_2O \rightarrow Ca(OH)_2$, calcium hydroxide **76.** The nitrogen dioxide can react with water to form nitrous and nitric acid: $2\,NO_2(g) + H_2O(\ell) \rightarrow HNO_2(aq) + HNO_3(aq)$. It is also possible for the nitrogen dioxide to combine with oxygen and water in the atmosphere to form nitric acid: $4\,NO_2(g) + 2\,H_2O(\ell) + O_2(g) \rightarrow 4\,HNO_3(aq)$.

# Chapter 18

## Answers to Target Checks

**1.** An equilibrium between a solution and excess undissolved solute can be contained in an open beaker if there is no significant evaporation of solvent (see Section 16.3). **2.** Your drawings may have shown "glancing" collisions, collisions that have insufficient kinetic energy, and/or collisions with improper orientations, all of which do not produce a chemical change. **3.** A barrier is something that prevents or limits some event. Activation energy limits a reaction to the fraction of the intermolecular collisions that have enough kinetic energy and proper orientation. **4.** (a) As temperature drops, the fraction of collisions with enough kinetic energy to meet the activation energy requirement drops significantly, which reduces reaction rate. Collision frequency also drops, which reduces reaction rate slightly. (b) A catalyst increases reaction rate by providing a reaction path with a lower activation energy. (c) At higher concentrations collisions are more frequent, so reaction rate increases. **5.** (a) The forward reaction is favored. (b) Appreciable quantities of all species are present at equilibrium.

## Answers to Concept-Linking Exercises

*You may have found more relationships or relationships other than the ones given in these answers.*

**1.** A reversible reaction reaches equilibrium if the forward reaction rate is equal to the reverse reaction rate. **2.** The collision theory of chemical reactions declares that reactant particles must collide in order to react. An energy-reaction coordinate graph traces the energy of the system before, during, and after collision. The activated complex is the temporary high-energy combination of reacting particles during a collision. Activation energy is the increase in energy as reactant particles change to the activated complex. **3.** Temperature, a catalyst, and reactant concentrations are three variables that determine the rate of a chemical reaction. **4.** Le Chatelier's Principle says that if an equilibrium system is disturbed in a way that makes the rates of forward and reverse reactions unequal, the equilibrium adjusts itself to counteract the disturbance partially until a new equilibrium is established. Temperature and concentration are two such disturbances. A change in gas volume also disturbs an equilibrium if the number of gas molecules on opposite sides of the equation is unequal. **5.** An equilibrium constant (symbol K) is a ratio of the product of the concentrations of each species on the right-hand side of an equilibrium equation, each raised to a power equal to its coefficient in the equilibrium equation, divided by the comparable product of the concentrations of the species on the left-hand side of the equilibrium equation, each raised to a power equal to its coefficient in the equilibrium equation. $K_{sp}$ identifies an equilibrium constant for a solubility equilibrium; *sp* means "solubility product." $K_a$ identifies an acid equilibrium; *a* is for "acid."

## Answers to Blue-Numbered Questions, Exercises, and Problems

**1.** In a dynamic equilibrium, opposing changes continue to occur at equal rates. An equilibrium in which nothing is changing—a book resting on a table, for example—is called a static equilibrium. **3.** Both systems can reach equilibrium. At equilibrium, the salt dissolves at the same rate at which it crystallizes. Whether the container is open or closed is of no importance. **5.** The system is not an equilibrium because energy must be supplied constantly to keep it in operation. Also, the water is circulating, not moving reversibly in two opposing directions. **7.** There will be no reaction if the orientation of the colliding particles is unfavorable. **9.** $\Delta E = b - c$; $E_a = a - c$

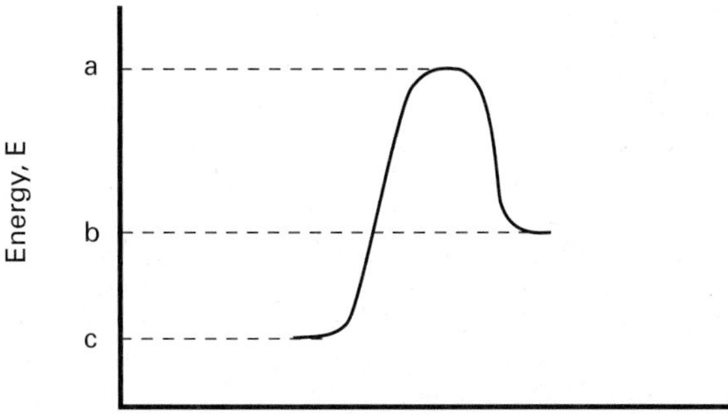


**11.** $E_a$ (forward) $= a - b$; $E_a$ (reverse) $= a - c$. Both activation energies are positive. Point a is the highest on the curve; $a > b$ and $a > c$. **13.** An activated complex is an unstable intermediate species formed during a collision of two reacting particles. The properties of an activated complex cannot be described because the complex decomposes almost as soon as it forms. **15.** At a higher temperature, a larger fraction of the molecules have enough kinetic energy to engage in a reaction-producing collision, so reaction rates are higher. Also, collisions are more frequent. At low temperature, a smaller fraction of the collisions produce reactions and there are fewer collisions, so the reaction rate is slower. **17.** The equilibria are identical. Equilibrium will be reached more quickly in the system with the catalyst. **19.** An increase in the concentration of A will increase the reaction rate, while a decrease in the concentration of B will decrease the reaction rate. The net effect depends on the size of the two changes. **21.** The reverse reaction rate reaches its maximum at equilibrium. **23.** See Figure 18.9. **25.** If $O_2$ concentration is decreased, equilibrium will shift in the reverse direction, the direction in which more $O_2$ will be produced. **27.** If $O_2$ concentration is increased, equilibrium will shift in the forward direction, the direction in which more $O_2$ will be consumed. **29.** Removal of $NH_3$ will shift the equilibrium forward, the direction in which some additional $NH_3$ will be produced. **31.** Increasing volume decreases the pressure. The equilibrium shifts to the left (5 moles of gases on the left versus 3 moles of gases on the right) to cause a pressure increase. **33.** It will not shift because there will be no change in the total number of molecules of gases. **35.** If heat is removed, the equilibrium will shift in the direction that produces heat, the forward direction. **37.** Cool the system. Removing heat causes the reaction to shift in the direction that produces heat, the forward direction. That increases the $SO_3$ yield. **39.** $Ca(OH)_2(s) \rightleftharpoons Ca^{2+}(aq) + 2\ OH^-(aq)$ is the equilibrium equation. $H^+$ ions from the acid combine with $OH^-$ ions to form water molecules. This reduces the $OH^-$ ion concentration and causes a forward shift in the equilibrium. The process continues until all the $Ca(OH)_2$ is dissolved.

**41.** $K = \dfrac{[CO_2][H_2]}{[CO][H_2O]}$ **43.** $K = \dfrac{[CO][H_2]}{[H_2O]}$ **45.** $K = [Zn^{2+}]^3[PO_4^{3-}]^2$ **47.** $K = \dfrac{[H_3O^+][NO_2^-]}{[HNO_2]}$ **49.** $K = \dfrac{[Cu^{2+}][NH_3]^4}{[Cu(NH_3)_4^{2+}]}$

**51.** The equilibrium constant expression for the given equation is $K = \dfrac{[NO_2]^2}{[NO]^2[O_2]}$. If the equation is written in reverse, the equilibrium constant expression is inverted. If different sets of coefficients are used, both the expression and its numerical value change. For example:

$$NO + \tfrac{1}{2}\,O_2 \rightleftharpoons NO_2 \qquad K_1 = \frac{[NO_2]}{[NO][O_2]^{1/2}} \qquad 4\ NO + 2\ O_2 \rightleftharpoons 4\ NO_2 \qquad K_2 = \frac{[NO_2]^4}{[NO]^4[O_2]^2}$$

The equilibrium constants are not equal: $K_2 = K^2 = K_1^{\,4}$. **53.** The equilibrium will be favored in the forward direction. If K is very large, at least one factor in the denominator must be very small, indicating that at least one reactant has been almost completely consumed. **55.** The equilibrium will be favored in the reverse direction. A very small equilibrium constant results when the concentration of one species on the right-hand side of the equation is very small. **57.** (a) The equilibrium is favored in the forward direction. $H_2SO_3$ is one of the acids that is unstable, decomposing to $H_2O$ and $SO_2$, as indicated. K is large for this equilibrium. (b) The equilibrium is favored in the forward direction. $HC_2H_3O_2$ is a weak acid. Nearly all of the ions will combine to form the molecule. **59.** $[Co^{2+}] = 3.7 \times 10^{-6}$; $[OH^-] = 2 \times (3.7 \times 10^{-6}) = 7.4 \times 10^{-6}$; $K_{sp} = [Co^{2+}]\,[OH^-]^2 = (3.7 \times 10^{-6})(7.4 \times 10^{-6})^2 = 2.0 \times 10^{-16}$

**61.** $$\frac{8.7\ \text{mg}\ Ag_2CO_3}{250\ \text{mL}} \times \frac{1\ \text{g}\ Ag_2CO_3}{1000\ \text{mg}\ Ag_2CO_3} \times \frac{1\ \text{mol}\ Ag_2CO_3}{275.8\ \text{g}\ Ag_2CO_3} \times \frac{1000\ \text{mL}}{\text{L}} = 1.3 \times 10^{-4}\ \text{M}\ Ag_2CO_3$$

$[Ag_2CO_3] = [CO_3^{2-}] = 1.3 \times 10^{-4}$ M $[Ag^+] = 2 \times (1.3 \times 10^{-4}\ \text{M}) = 2.6 \times 10^{-4}$ M $K_{sp} = [Ag^+]^2\,[CO_3^{2-}] = (2.6 \times 10^{-4})^2(1.3 \times 10^{-4}) = 8.8 \times 10^{-12}$

**63.** $K_{sp} = [Ag^+]\,[IO_3^-]$ $[Ag^+] = [IO_3^-] = s$ $s^2 = 2.0 \times 10^{-8}$ $s = 1.4 \times 10^{-4}$ M

$$\frac{1.4 \times 10^{-4}\ \text{mol}\ AgIO_3}{\text{L}} \times \frac{282.8\ \text{g}\ AgIO_3}{\text{mol}\ AgIO_3} \times 0.100\ \text{L} = 4.0 \times 10^{-3}\ \text{g/100 mL}\ (0.100\ \text{L} = 100\ \text{mL})$$

**65.** $K_{sp} = [Mn^{2+}][OH^-]^2$ $[Mn^{2+}] = s$ $[OH^-] = 2s$ $(s)(2s)^2 = 4s^3 = 1.0 \times 10^{-13}$ $s = 2.9 \times 10^{-5}$ M

**67.** $K_{sp} = [Ca^{2+}][C_2O_4^{2-}] = [Ca^{2+}](0.22) = 2.4 \times 10^{-9}$ $[Ca^{2+}] = 1.1 \times 10^{-8}$ M

$2.50 \times 10^2 \text{ mL} \times \frac{1.1 \times 10^{-8} \text{ mol } CaC_2O_4}{1000 \text{ mL}} \times \frac{128.10 \text{ g } CaC_2O_4}{\text{mol } CaC_2O_4} = 3.5 \times 10^{-7} \text{ g } CaC_2O_4$

**69.** $[H^+] = 10^{-2.12} = 7.6 \times 10^{-3}$ M $K_a = \frac{[H^+][C_4H_5O_3^-]}{[HC_4H_5O_3]} = \frac{(7.6 \times 10^{-3})^2}{0.22} = 2.6 \times 10^{-4}$ $\frac{7.6 \times 10^{-3}}{0.22} \times 100 = 3.5\%$ ionized

**71.** $K_a = \frac{[H^+][C_2H_3O_2^-]}{[HC_2H_3O_2]} = \frac{[H^+]^2}{[HC_2H_3O_2]}$ $[H^+] = \sqrt{(1.8 \times 10^{-5})(0.35)} = 2.5 \times 10^{-3}$ M pH = 2.60

**73.** $24.0 \text{ g } NaC_2H_3O_2 \times \frac{1 \text{ mol } NaC_2H_3O_2}{82.03 \text{ g } NaC_2H_3O_2} = 0.293 \text{ mol } NaC_2H_3O_2$ $5.00 \times 10^2 \text{ mL} \times \frac{1 \text{ L}}{1000 \text{ mL}} = 0.500 \text{ L}$

$\frac{0.293 \text{ mol } NaC_2H_3O_2}{0.500 \text{ L}} = 0.586 \text{ M } NaC_2H_3O_2 = 0.586 \text{ M } C_2H_3O_2^-$ $[H^+] = 1.8 \times 10^{-5} \times \frac{0.12}{0.586} = 3.7 \times 10^{-6}$ pH = 5.43

**75.** $\frac{[HC_2H_3O_2]}{[C_2H_3O_2^-]} = \frac{10^{-4.25}}{1.8 \times 10^{-5}} = 3.1$ **77.** $[CO]$ at start $= \frac{0.351}{3.00} = 0.117$ $[Cl_2]$ at start $= \frac{1.340}{3.00} = 0.447$ $[Cl_2]$ at end $= \frac{1.050}{3.00} = 0.350$

| | $CO(g)$ + | $Cl_2(g)$ ⇌ | $COCl_2$ |
|---|---|---|---|
| Initial | 0.117 | 0.447 | 0 |
| Reacting | −0.097 | −0.097 | +0.097 |
| Equilibrium | 0.020 | 0.350 | 0.097 |

$K = \frac{0.097}{(0.020)(0.350)} = 14$

**80.** True: b, d, f, g, i, j, l, m, n. False: a, c, e, h, k, o. **81.** Kinetic energies are greater at Time 1 because at Time 2 some of that energy has been converted to potential energy in the activated complex. **83.** a) High pressure to force reaction to the smaller number of gaseous product molecules. b) High temperature, at which all reaction rates are faster. **84.** A manufacturer cannot use an equilibrium, which is a closed system from which no product can be removed. **85.** The higher temperature is used to speed the reaction rate to an acceptable level. Lower pressure is dictated by limits of mechanical design and safety. **86.** (a) $Ca(OH)_2\,(s) \rightleftharpoons Ca^{2+}\,(aq) + 2\,OH^-\,(aq)$. (b) (1) Adding a strong base or soluble calcium compound would increase $[OH^-]$ and $[Ca^{2+}]$, respectively, causing a shift in the reverse direction and reducing the solubility of $Ca(OH)_2$. (2) Adding an acid to reduce $[OH^-]$ by forming water, adding a cation that will reduce $[OH^-]$ by precipitation, or adding an anion whose calcium salt is less soluble than $Ca(OH)_2$ would cause a shift in the forward direction, increasing the solubility of $Ca(OH)_2$. (c) (1) Any anion that reacts with calcium ion to form an ionic compound that is less soluble than $Ca(OH)_2$ will cause a forward shift, increasing $[OH^-]$. (2) An acid that will form water with $OH^-$ or a cation that will precipitate $OH^-$ will reduce $[OH^-]$. **87.** Add $NO_2$: R—I—I—D—I. Reduce temperature: F—D—D—I—I. Add $N_2$: None. Remove $NH_3$: R—D—I—D—D. Add a catalyst: None. **88.** The "truth" of the statement depends on how you interpret it. A system *at equilibrium* is neither endothermic nor exothermic. The system is closed; heat energy neither enters nor leaves. The thermochemical *equation* for the equilibrium is endothermic in one direction and exothermic in the other direction. Both directions describe the same equilibrium. **89.** Temperature affects reaction rates in forward and reverse directions differently. Therefore, the value of an equilibrium constant depends on temperature. If you change the temperature, the value of K changes. **90.** $2\,SO_2(g) + O_2(g) \rightleftharpoons 2\,SO_3(g)$ and $2\,SO_3(g) \rightleftharpoons 2\,SO_2(g) + O_2(g)$ both express the equilibrium described. The K expression for one equation is the reciprocal of the other (write the equilibrium constant expression for each equation to see this, if necessary). If any fraction is greater than 1, its reciprocal is less than 1. Put another way, if the numerator is greater than the denominator, the fraction is greater than 1. In the reciprocal, the numerator is less than the denominator, so the fraction is less than 1. **91.** The system can reach equilibrium only when it is closed, that is, when no water is running in the house; it cannot be reached in an open system while hard water enters the softener and soft water leaves. $[Ca^{2+}]$ is relatively high in the hard water that enters the softener. By Le Chatelier's Principle, the reaction is favored in the forward direction in which $Ca^{2+}$ ions in the water are replaced by $Na^+$ ions. This also means the $Na^+$ ions in the resin are replaced by $Ca^{2+}$ ions from the water. Eventually the $Na^+$ ions are used up and must be replenished. This is done by running water with a high sodium ion concentration through the softener. This forces the reaction in the reverse direction, again according to Le Chatelier's Principle. $Na^+$ ions replace the $Ca^{2+}$ ions on the resin, and the $Ca^{2+}$ ions are flushed down the drain.

*Answer to the "trick question" in Study Hints:* The reverse reaction rate also increases. All reaction rates are higher at higher temperatures. But the increases in the forward and reverse rates are not the same. That's what destroys the equilibrium—temporarily.

# Chapter 19

## Answers to Target Checks

**1.** All common batteries, including those used to power portable MP3 players, are voltaic (galvanic) cells because they cause an electric current to flow in an external circuit, rather than simply carrying current from an external source, which is the role of an electrolytic cell. **2.** (a) 0, (b) −2, (c) the sum of the oxidation numbers of atoms in a molecule equals zero, (d) equal to ion charge, (e) +1, (f) ion charge equal to sum of oxidation number of elements in the ion. **3.** A strong oxidizing agent is strong because it has a strong attraction for electrons. It is able to take them from many other species, thereby oxidizing them. Once it has gained an electron, it does not give it up readily, which it must do to function as a reducing agent. It therefore reduces few other species; it is a weak reducing agent.

## Answers to Concept-Linking Exercises

*You may have found more relationships or relationships other than the ones given in these answers.*

**1.** An electric current passing through a solution—the electrolyte—is electrolysis. Current enters and leaves the electrolyte through electrodes immersed in the solution. If the current flows spontaneously, the system is called a voltaic or galvanic cell. If current must come from an outside source, the system is an electrolytic cell. **2.** Oxidation is a loss of electrons. An oxidation half-reaction equation describes oxidation and has electrons as a product of the reaction. Reduction is a gain of electrons. A reduction half-reaction equation describes reduction and has electrons as one of the reactants. **3.** An oxidation number is a number assigned to an element by arbitrary rules to account for electrons transferred in redox reactions. An increase in oxidation number identifies oxidation of the element, and a decrease identifies reduction. **4.** An oxidizing agent, also known as an oxidizer, oxidizes another species by taking electrons from it. A reducing agent, or reducer, reduces another species by giving electrons to it. **5.** A strong oxidizing agent has a strong attraction for electrons and is able to draw electrons from many species; a weak oxidizing agent attracts electrons weakly and takes them from few other species. A strong reducing agent has a weak hold on electrons and gives them up easily to many species; a weak reducing agent holds its electrons tightly and gives them up to few other species.

## Answers to Blue-Numbered Questions, Exercises, and Problems

**1.** See the discussion in Section 19.1. **3.** Oxidation: a, c, d. Reduction: b. **5.** Reduction **7.** Oxidation

**9.**
$$2\ Cr \rightarrow 2\ Cr^{3+} + 6\ e^-$$
$$3\ Cl_2 + 6\ e^- \rightarrow 6\ Cl^-$$
$$\overline{3\ Cl_2 + 2\ Cr \rightarrow 2\ Cr^{3+} + 6\ Cl^-}$$

**11.** The second equation is the oxidation half-reaction equation. Overall: $2\ NiOOH + 2\ H_2O + Cd \rightarrow 2\ Ni(OH)_2 + Cd(OH)_2$
**13.** Examples include any item that runs on batteries, such as watches and calculators. **15.** Yes. A galvanic cell causes current to flow through an external circuit by electrochemical action. An electrolytic cell is a cell through which a current driven by an external source passes. **17.** +3, −2, +4, +6 **19.** +3, +5, +6, +5 **21.** (a) Bromine reduced from 0 to −1; (b) Lead oxidized from +2 to +4 **23.** (a) Iodine reduced from +7 to −1; (b) Oxygen reduced from 0 to −2 **25.** (a) Nitrogen oxidized from +4 to +5; (b) Chromium oxidized from +3 to +6 **27.** Chlorine is the oxidizing agent, and the bromide ion is the reducing agent. **29.** $PbO_2$ is the oxidizing agent, oxidizing Pb. Pb is the reducing agent, reducing the lead in $PbO_2$. **31.** From Table 19.2, Zn is a stronger reducer than $Fe^{2+}$. A strong reducer releases electrons to an oxidizer more readily than a weak reducer releases them.
**33.** $Na^+ < Fe^{2+} < Cu^{2+} < Br_2$ **35.** $Br_2 + 2\ I^- \rightleftharpoons 2\ Br^- + I_2$; forward reaction favored **37.** $2\ H^+ + 2\ Br^- \rightleftharpoons H_2 + Br_2$; reverse reaction favored
**39.** $2\ NO + 4\ H_2O + 3\ Fe^{2+} \rightleftharpoons 2\ NO_3^- + 8\ H^+ + 3\ Fe$; reverse reaction favored **41.** A strong acid releases protons readily; a strong reducer releases electrons readily. A strong base attracts protons strongly; a strong oxidizer attracts electrons strongly.

**43.**
$$S_2O_3^{2-} + 5\ H_2O \rightarrow 2\ SO_4^{2-} + 10\ H^+ + 8\ e^-$$
$$4\ Cl_2 + 8\ e^- \rightarrow 8\ Cl^-$$
$$\overline{S_2O_3^{2-} + 5\ H_2O + 4\ Cl_2 \rightarrow 2\ SO_4^{2-} + 10\ H^+ + 8\ Cl^-}$$

**45.**
$$4\ NO_3^- + 8\ H^+ + 4\ e^- \rightarrow 4\ NO_2 + 4\ H_2O$$
$$Sn + 3\ H_2O \rightarrow H_2SnO_3 + 4\ H^+ + 4\ e^-$$
$$\overline{4\ NO_3^- + 4\ H^+ + Sn \rightarrow 4\ NO_2 + H_2O + H_2SnO_3}$$

**47.**
$$2\ MnO_4^- + 16\ H^+ + 10\ e^- \rightarrow 2\ Mn^{2+} + 8\ H_2O$$
$$5\ C_2O_4^{2-} \rightarrow 10\ CO_2 + 10\ e^-$$
$$\overline{2\ MnO_4^- + 16\ H^+ + 5\ C_2O_4^{2-} \rightarrow 2\ Mn^{2+} + 8\ H_2O + 10\ CO_2}$$

**49.**
$$Cr_2O_7^{2-} + 8\ H^+ + 6\ e^- \rightarrow Cr_2O_3 + 4\ H_2O$$
$$2\ NH_4^+ \rightarrow N_2 + 8\ H^+ + 6\ e^-$$
$$\overline{Cr_2O_7^{2-} + 2\ NH_4^+ \rightarrow Cr_2O_3 + 4\ H_2O + N_2}$$

**51.**
$$4\ NO_3^- + 16\ H^+ + 12\ e^- \rightarrow 4\ NO + 8\ H_2O$$
$$3\ As_2O_3 + 15\ H_2O \rightarrow 6\ AsO_4^{3-} + 30\ H^+ + 12\ e^-$$
$$\overline{3\ As_2O_3 + 7\ H_2O + 4\ NO_3^- \rightarrow 6\ AsO_4^{3-} + 14\ H^+ + 4\ NO}$$

**54.** True: a, c, g. False: b, d, e, f. **55.** This "property of an acid" is more correctly described as the property of an acid (hydrogen ion) acting as an oxidizing agent. The $H^+$ ion reacts only with metals whose ions are weaker oxidizing agents, located below hydrogen in Table 19.2. **56.** All reactants were in acidic solutions. Water is available in large amounts in any aqueous solution, as is $H^+$ in an acidic solution. **57.** In the electrolytic cell, the force that moves the charges through the circuit is *outside* the cell. In the voltaic cell, the cell itself is the *source* of the force that moves the charged particles.
**58.** The statement is (c) sometimes correct. In a simple element ↔ monatomic ion redox reaction the statement is correct. The *element* oxidized or reduced can always be identified by a change in oxidation number. The oxidizing or reducing *agent,* however, is a *species* that contains the element being oxidized or reduced, and it may be an element, a monatomic ion, a polyatomic ion, such as $MnO_4^-$, or a compound. **59.** Oxidation occurs at the anode: $2\ Cl^- \rightarrow Cl_2 + 2\ e^-$. Reduction occurs at the cathode: $Na^+ + e^- \rightarrow Na$. **60.** Your (1) and (2) numbers may be the reverse of ours. (a) (1) $Cu \rightarrow Cu^{2+} + 2\ e^-$. (2) $Cu^{2+} + 2\ e^- \rightarrow Cu$. (b) (1) occurs at the "+" electrode and (2) at the "−" electrode. (c) "+" is the anode, where oxidation occurs. "−" is the cathode, where reduction occurs. (d) Charge is carried through the electrolyte by $Cu^{2+}$ ions moving from anode to cathode. ($H^+$ and $SO_4^{2-}$ ions also move, but without an identifiable "flow" of charge that is responsible for electrolysis in the solution.) (e) The bubbles are hydrogen. They come from the only ion that can be "deposited" as a gas, $H^+$. This occurs at the cathode, the same electrode at which copper is deposited. Any electrons that are used to reduce $H^+$ ions instead of copper ions cause the mass of copper deposited to be less than the mass of copper dissolved. (What happens to the concentration of $Cu^{2+}$ ion in the solution?* [The answer to this question appears at the bottom of the page.]) In one sentence: Hydrogen gas, instead of copper, is reduced (deposited) at the cathode according to the half-reaction $2\ H^+(aq) + 2\ e^- \rightarrow H_2(g)$. **61.** Zinc is used to prevent harmful galvanic action (corrosion) in the equipment. The conditions for galvanic action are present: two metals in contact with an electrolyte (seawater) and in metal-to-metal contact with each other. If one of the metals is going to corrode, it will be the strongest reducing agent of the group. Zinc is a stronger reducing agent than iron or copper. When it goes, it is simply and cheaply replaced. **62.** The moisture on your tongue becomes an electrolyte for the passage of electric current from one metal to the other. The tingle you "taste" is caused by that current.

# Chapter 20

## Answers to Target Checks

**1.** a) $\alpha$: $^4_2He$, 2+; $\beta$: $^{\ 0}_{-1}e$, 1− b) Gamma > Beta > Alpha **2.** Geiger and scintillation counters "count" individual radioactive emissions by measuring electric current (Geiger) or light intensity (scintillation) produced by the radiation. These are both proportional to the intensity of radiation and are interpreted in that way. This intensity can be expressed as so many counts per unit of time. **3.** Don't smoke! **4.** (a) Nuclear bombardment reactions produce isotopes that do not exist in nature. (b) To be accelerated, a particle must have an electrical charge. **5.** Isotopes produced in a fission reaction have smaller atomic numbers and smaller mass numbers than the starting isotope. **6.** An isotope produced by a fusion reaction has a larger atomic number and mass number than the starting isotopes.

---

*The concentration of copper increases over time.

## Number of Radioactive ${}^{40}_{19}K$ Atoms in a 150-Pound Person

$$225 \text{ g K} \times \frac{6.02 \times 10^{23} \text{ K atoms}}{39.10 \text{ g K}} \times \frac{0.0118\ {}^{40}_{19}\text{K atoms}}{100 \text{ K atoms}} = 4.09 \times 10^{20}\ {}^{40}_{19}\text{K atoms}$$

## Answers to Concept-Linking Exercises

*You may have found more relationships or relationships other than the ones given in these answers.*

**1.** Alpha and beta particles and gamma rays are products of radioactive decay. Alpha particles have a positive charge, beta particles have a negative charge, and gamma radiation is uncharged. The penetrating power of these emissions increases in the order: alpha < beta < gamma. **2.** A cloud chamber, Geiger counter, and scintillation counter are all instruments used to detect radioactivity. A cloud chamber is a container filled with air and a supersaturated vapor. When radiation passes through the container, some air molecules ionize and vapor condenses on them, leaving a visible vapor trail. Radiation detection in a Geiger counter occurs in a tube filled with a gas. When radiation passes through the tube, gas molecules ionize and produce a current that can be quantitatively measured by a detector. A scintillation counter uses a transparent solid to detect radiation. When radiation passes through the solid, the particles absorb energy and then release it in the form of flashes of light, which can be quantitatively counted by a detector.
**3.** Radiocarbon dating is used to determine the age of fossils. It is based on measuring the ratio of carbon-14 to carbon-12 in the fossil. When an organism is alive, the ${}^{14}C$:${}^{12}C$ ratio in the organism is the same as in the environment. Carbon-14 is radioactive and slowly decays, with a half-life of $5.73 \times 10^3$ years. It is continually replenished in a living organism, however, so the ${}^{14}C$:${}^{12}C$ ratio remains constant as long as the organism is alive. Carbon-12 is a stable isotope, so it does not decay. When the organism dies, radioactive ${}^{14}C$ continues to decay but it is no longer replenished. The age of a fossil can be determined by comparing the ${}^{14}C$:${}^{12}C$ ratio in the fossil to that in the environment. **4.** A natural radioactive decay series describes the fate of natural unstable isotopes, from the starting radioactive isotope to the final stable isotope. Uranium-238 is an example of a natural radioactive isotope. It decays to lead-206 through a series of eight alpha emissions and six beta emissions. **5.** Nuclear bombardment is the process by which two particles are made to collide with sufficient kinetic energy to form new particles. A particle accelerator is an instrument used to carry out bombardment reactions. The transuranium elements—those with higher atomic numbers than that of uranium, Z = 92—can be produced via nuclear bombardment in a particle accelerator. **6.** Nuclear fission is a nuclear reaction in which a large nucleus splits into two smaller nuclei. A nuclear power plant uses the energy released in a fission reaction to generate electricity. Nuclear fusion is a nuclear reaction in which two small nuclei combine to form a larger nucleus. Solar energy results from fusion reactions. **7.** A chain reaction has one of its own reactants as a product, allowing the original reaction to continue. The minimum quantity of matter necessary for a chain reaction to continue is its critical mass. A breeder reactor is a nuclear reactor in which fissionable fuel is produced from nonfissionable isotopes.

## Answers to Blue-Numbered Questions, Exercises, and Problems

**1.** *Nuclide* is a general term that refers to the nucleus of any atom. An isotope is a specific kind of atom of an element that has a specific nuclear composition. **3.** It refers to the spontaneous decomposition of the nucleus. **5.** Alpha particles are helium nuclei with a mass number of 4 and a 2+ charge. Beta particles are electrons with a mass number of 0 and a 1− charge. Gamma rays are high-energy electromagnetic radiation having no mass and no charge. Penetration power increases in the order alpha < beta < gamma. **7.** The collision of a radioactive emission with an atom or molecule may rearrange the electrons in the target, possibly ionizing it and causing a potentially harmful chemical change. **9.** A Geiger counter is a device for detecting and measuring radiation. Figure 20.4 describes how it works. **11.** A Geiger counter "clicks" when a radioactive emission is detected; a scintillation counter counts pulses of light generated by radiation. Both devices can measure radiation as well as detect it. The Geiger counter actually measures electric current; the scintillation counter measures light pulses. Both devices express their measurements as counts per unit time. **13.** A gamma camera is immobile while it takes a picture, creating essentially a two-dimensional image of an object. The scanner moves as it takes many pictures. Computer enhancement and combination of many pictures allow three-dimensional-like images to be constructed. **15.** Figure 20.8 shows that radon is the greatest source of background radiation, accounting for 55% of typical exposure. Internal sources and medical X-rays each contribute 11%. **17.** Two half-lives have passed. $1/4 = (1/2)^2$, where the exponent is the number of half-lives. **19.** $(1/2)^7 = 1/128$ **21.** $R = 12.9 \text{ g} \times (1/2)^3 = 1.6 \text{ g}$ **23.** From 2006 to 2062 is 56 years, or two half-lives. Hence, $R = 654 \text{ g} \times (1/2)^2 = 1.6 \times 10^2 \text{ g}$. From 2006 to 2082 is 76 years, or 76/28 = 2.7 half-lives. Hence, $R = 654 \text{ g} \times (1/2)^{2.7} = 1.0 \times 10^2 \text{ g}$. **25.** (a) 12 min/(3.1 min/half-life) = 3.9 half-lives. From the graph, R/S = 0.067. $0.067 \times 84.6 \text{ g} = 5.7 \text{ g}$ remain. By the equation, $R = 84.6 \times (1/2)^{3.9} = 5.7 \text{ g}$ remain. (b) R/S = 3.48 g/84.6 g = 0.0411. From the graph, this is 4.6 half-lives. 4.6 half-lives × 3.1 min/half-life = 14 minutes. **27.** R/S = 1.05 g/9.53 g = 0.110. From the graph, this is 3.25 half-lives. 83.2 hr/3.25 half-lives = 25.6 hr/half-life = $t_{1/2}$. **29.** R/S = 0.55. From the graph, this is 0.86 half-life. 0.86 half-life × $5.73 \times 10^3$ yr/half-life = $4.9 \times 10^3$ years. **31.** The original nucleus disintegrates into a ${}^{4}_{2}He$ nucleus and another nucleus. The final nuclide has two fewer protons and two fewer neutrons than the original. This is a transmutation, a change from one element to another, because the number of protons changes. **33.** ${}^{228}_{89}Ac \rightarrow {}^{0}_{-1}e + {}^{228}_{90}Th$ ${}^{212}_{83}Bi \rightarrow {}^{0}_{-1}e + {}^{212}_{84}Po$ **35.** ${}^{216}_{84}Po \rightarrow {}^{4}_{2}He + {}^{212}_{82}Pb$ ${}^{234}_{92}U \rightarrow {}^{4}_{2}He + {}^{230}_{90}Th$
**37.** "Nuclear chemical properties of lead" is meaningless for two reasons. First, the chemical properties of all isotopes of lead are the same. Second, the nuclear properties of lead isotopes are specific for each individual isotope. **39.** The count will remain at 5000/minute. The radioactivity of an element is independent of the form of the element, whether it is a pure element or in a compound. **41.** Radioactivity is spontaneous, while nuclear bombardment reactions are produced by projecting a nuclear particle into another nuclear particle. **43.** Electrons, protons, positrons, and alpha particles can be accelerated in particle accelerators. The particle must have an electrical charge to be accelerated. **45.** Natural versus artificial radioactivity is not a function of atomic number. Isotopes of many elements are naturally radioactive, and artificially radioactive isotopes of many elements can be created.
**47.** All elements in the lanthanide series have atomic numbers less than 92, so none is a transuranium element. The elements in the actinide series with atomic numbers greater than 92 are transuranium elements. **49.** (a) ${}^{44}_{21}Sc$, (b) ${}^{257}_{103}Lr$, (c) ${}^{1}_{1}H$ **51.** Both fission and fusion reactions result in a release of energy. In fission reactions, a larger nucleus splits into smaller nuclei. In a fusion reaction, two small nuclei combine to form a larger nucleus.
**53.** A chain reaction is a reaction that has as a product one of its own reactants; that product becomes a reactant, thereby allowing the original reaction to continue. For a chain reaction to continue, there must be enough fissionable material to react with the neutrons given off. **55.** No: The products of a fusion reaction are not one of the reactants. **57.** Nuclear fusion is more promising than fission as an energy source because it produces more energy per given amount of fuel. Fusion fuel is more abundant, and fusion reactions generate no hazardous radioactive waste. Fusion's major drawback is the extremely high temperature needed to initiate the process. **60.** True: a, d, f, g, j, k. False: b, c, e, h, i, m. The answer to l is left to you. **61.** Natural fissionable isotopes are rare. Plutonium-234 is produced from the most abundant uranium isotope, uranium-238. **62.** Presumably, it takes an infinite time for all of a sample of radioactive matter to decay.

**63.** $$1 \text{ lb} \times \frac{454 \text{ g}}{\text{lb}} \times \frac{1 \text{ mol U}}{235 \text{ g U}} \times \frac{2.0 \times 10^{10} \text{ kJ}}{1 \text{ mol U}} \times \frac{1 \text{ ton coal}}{2.5 \times 10^{7} \text{ kJ}} = 1.5 \times 10^{3} \text{ tons coal}$$

**64.** Loss of mass of a specific isotope is difficult to measure because its decay product is another isotope that typically is mixed with the reactant. Rate of decay is best measured by Geiger counters or other devices described in Section 20.3 because their measurements are independent of the mass of the sample or the compound in which the radionuclide is found. **65.** (a) A > C > B. At the end of one half-life, half of the original A would remain, leaving the other half to be divided between B and C. Half of B disintegrates in one day, so more than half of what was produced in days 1 through 5 has passed along to C. (b) C > A > B. At the end of two half-lives, A is down to 1/4 of the starting amount. Most of the 3/4 that disintegrated has passed through B to C. **66.** R/S = 22.7 units/29.0 units = 0.783. From Figure 20.9, this corresponds to 0.35 half-life. The half-life of carbon-14 is $5.73 \times 10^3$ years. 0.35 half-lives $\times$ $5.73 \times 10^3$ years/half-life = $2.0 \times 10^3$ years. Radiocarbon dating indicates that the cloth is about 2000 years old, which places it at the beginning of the Christian era. **67.** Emission of a beta particle would change a calcium atom into a scandium atom: ${}^{47}_{20}Ca \rightarrow {}^{0}_{-1}e + {}^{47}_{21}Sc$.
**68.** US is $\frac{238}{(238 + 32)} \times 100 = 88\%$ uranium by mass. $US_2$ is 79% uranium by mass. Thus, for samples of equal mass, there are more uranium atoms in US than in $US_2$. Only the radioactive element, uranium, contributes to radioactivity, so US will exhibit the greater amount of radioactivity. The radioactivity of 0.5 mole of US will be the same as that of 0.5 mole of $US_2$ because both samples contain the same number of uranium atoms. **69.** The reaction described by Equation 20.7 is more apt to be a chain reaction. Three neutrons per uranium atom are produced in the process in Equation 20.7, two neutrons per uranium atom are produced in the process in Question 54. The greater the number of neutrons, the more likely it is that there will be a chain reaction.

# Chapter 21

## Answers to Target Checks

**1.** (a) $CH_4$ and $CH_3NH_2$ are organic; NaCl and CO are inorganic. CO is inorganic because organic compounds must have both carbon and hydrogen. (b) Some everyday definitions of the term *organic* include (1) food grown with "natural" fertilizers, (2) derived from living organisms, (3) organized, and (4) essential. Note that none of these everyday definitions is based on consideration of the particulate-level composition of matter.

**2.**

```
   Br H  Cl          H  Cl Br          H  H  Cl
   |  |  |           |  |  |           |  |  |
 H—C—C—C—H        H—C—C—C—H        H—C—C—C—H
   |  |  |           |  |  |           |  |  |
   H  H  H           H  H  H           H  H  Br
```

**3.** (a) $C_7H_{16}$ and $C_9H_{20}$ are normal alkanes. $C_5H_{10}$ and $C_{11}H_{22}$ could be cycloalkanes. (b) $—C_5H_{11}$

(c)

```
   H  H  F  I  H  H  H  H  H  H
   |  |  |  |  |  |  |  |  |  |
 H—C—C—C—C—C—C—C—C—C—C—H
   |  |  |  |  |  |  |  |  |  |
   H  H  F  H  H  H  H  H  H  H
```

(d) $CH_3CH_2$ and $CH_2CH_3$ on a cyclohexane ring

**4.** (a) $C_8H_{16}$ is the only molecule that can be a straight-chain alkene. $C_4H_6$ and $C_7H_{12}$ could be cycloalkenes. (b)

```
F       H
 \     /
  C=C
 /     \
H       F
```

(c) $C_4H_6$ and $C_7H_{12}$ can be alkynes. (d) The isomeric pentynes are 1-pentyne and 2-pentyne. The structural formula for 2-pentyne is

```
   H         H  H
   |         |  |
 H—C—C≡C—C—C—H
   |         |  |
   H         H  H
```

**5.**

F, F, F on a benzene ring; Cl, Cl on a benzene ring

1,3,5-trifluorobenzene  *p*-dichlorobenzene

**6.** (a)

```
   H  H  H  H
   |  |  |  |
 H—C—C—C—C—H
   |  |  |  |
   H  H  Br Br
```

(b)

```
   H  H  H  H  H  H
   |  |  |  |  |  |
 H—C—C—C—C—C—C—H
   |  |  |  |  |  |
   H  H  H  H  H  Cl
```

+ HCl You may have correctly substituted the chlorine atom for any of the hydrogen atoms on the reactant molecule or indicated that the product is a mixture of the possible substitution products.

(c) Cl on a benzene ring

**7.** (a) Hydroxyl group, —OH (b) —O—, where both bonds from oxygen are to carbon atoms (c) Alcohols: $CH_3CH_2CH_2—OH$; $CH_3CHCH_3$ (with OH on the central C)

Ether: $CH_3—O—CH_2CH_3$

**8.** (a) Carbonyl group:

```
\
 C=O
/
```

(b) Aldehyde:

```
R
 \
  C=O
 /
H
```

Ketone:

```
R
 \
  C=O
 /
R'
```

(c)

$CH_3CH_2CH_2CH_2{-}\overset{O}{\overset{\|}{C}}{-}H$ aldehyde $\quad$ $CH_3CH_2{-}\underset{}{\overset{CH_3}{\overset{|}{CH}}}{-}\overset{O}{\overset{\|}{C}}{-}H$ aldehyde

$CH_3CH_2CH_2{-}\overset{O}{\overset{\|}{C}}{-}CH_3$ ketone $\quad$ $CH_3{-}\overset{CH_3}{\overset{|}{CH}}{-}CH_2{-}\overset{O}{\overset{\|}{C}}{-}H$ aldehyde

$CH_3CH_2{-}\overset{O}{\overset{\|}{C}}{-}CH_2{-}CH_3$ ketone $\quad$ $CH_3{-}\underset{\underset{CH_3}{|}}{\overset{CH_3}{\overset{|}{C}}}{-}\overset{O}{\overset{\|}{C}}{-}H$ aldehyde

**9.** (a) Carboxyl group: $-\overset{O}{\overset{/\!/}{C}}{-}O{-}H$ (b) acid + alcohol → ester + water

(c)

$CH_3CH_2CH_2{-}\overset{O}{\overset{\|}{C}}{-}OH$ acid $\quad$ $CH_3CH_2{-}\overset{O}{\overset{\|}{C}}{-}O{-}CH_3$ ester $\quad$ $H{-}\overset{O}{\overset{\|}{C}}{-}O{-}CH_2CH_2CH_3$ ester

$CH_3{-}\overset{CH_3}{\overset{|}{CH}}{-}\overset{O}{\overset{\|}{C}}{-}OH$ acid $\quad$ $CH_3{-}\overset{O}{\overset{\|}{C}}{-}O{-}CH_2CH_3$ ester $\quad$ $H{-}\overset{O}{\overset{\|}{C}}{-}O{-}\overset{CH_3}{\overset{|}{CH}}{-}CH_3$ ester

**10.** (a) $R_1{-}\underset{\underset{R_3}{|}}{N}{-}R_2$ One or two of the Rs may be H. (b) $-\overset{O}{\overset{/\!/}{C}}{-}NH_2$ **11.** $-CH_2{-}\underset{H}{\overset{C_6H_5}{\overset{|}{C}}}{-}CH_2{-}\underset{H}{\overset{C_6H_5}{\overset{|}{C}}}{-}CH_2{-}\underset{H}{\overset{C_6H_5}{\overset{|}{C}}}{-}$

**12.** $HO{-}\overset{O}{\overset{\|}{C}}{-}C_6H_4{-}\overset{O}{\overset{\|}{C}}{-}OH + HO{-}CH_2{-}C_6H_{10}{-}CH_2{-}OH$

## Answers to Concept-Linking Exercises

*You may have found more relationships or relationships other than the ones given in these answers.*

**1.** Hydrocarbons are binary compounds of carbon and hydrogen. A hydrocarbon is saturated when each carbon is single-bonded to four other atoms. An unsaturated hydrocarbon has two or more carbon atoms double- or triple-bonded to other carbon atoms. **2.** Each carbon atom in an alkane forms a single bond to four other atoms. There are no multiple bonds. A normal alkane is one in which all carbon atoms are in a continuous chain; there are no branches off the main chain. A homologous series is a series of compounds in which each member differs from the one before it by a $—CH_2—$ unit. The alkane series—$CH_4$, $CH_3CH_3$, $CH_3CH_2CH_3$, and so on—is an example of a homologous series. A cycloalkane is a molecule composed of a ring of carbon atoms in which all carbon atoms are saturated. **3.** A Lewis diagram shows how atoms are arranged and bonded in a molecule or ion. It is also known as a structural formula or a structural diagram. A line formula is a "shorthand" Lewis diagram in which bonds between a chain carbon and atoms or other branches from the chain are not shown precisely. When the $CH_2$ units in a line formula are grouped together, the result is a condensed formula. **4.** An alkene is an aliphatic hydrocarbon in which two carbon atoms are double-bonded to each other. Alkene isomers that exist because of the double-bond geometry are called *cis-trans,* or geometric, isomers. If identical groups of atoms are on the same side of the double bond, they are *cis* to one another. Identical groups on opposite sides of the double bond are *trans*. An alkyne is an aliphatic hydrocarbon in which two carbon atoms are triple-bonded to each other. **5.** Aromatic compounds are those that contain the benzene ring. These compounds have delocalized bonding electrons, that is, bonding electrons that belong to the molecule as a whole rather than to a specific bond in the benzene ring. Non-aromatic hydrocarbons are aliphatic hydrocarbons. **6.** Catalytic cracking and fractional distillation are processes used in petroleum refining. Catalytic cracking is the process in which long-chain hydrocarbons are split into shorter molecules. Hydrocarbons are separated into fractions that boil at different temperatures in the process of fractional distillation. Alkanes are prepared by catalytic hydrogenation of an alkene, where the alkene reacts with hydrogen in the presence of a catalyst to form the corresponding alkane. **7.** An addition reaction adds atoms across a double or triple bond. Alkanes undergo substitution reactions where a hydrogen atom is replaced by another atom. Halogenation reactions are addition or substitution reactions with a halogen. Hydrogenation is an addition reaction with hydrogen. **8.** A functional group is an atom or group of atoms that establishes the identity of a class of compounds and determines its chemical properties. The hydroxyl group, —OH, identifies alcohols. The carbonyl group, $\overset{O}{\overset{\|}{C}}$, characterizes both aldehydes and ketones. The carboxyl group, $-\overset{O}{\overset{/\!/}{C}}{-}O{-}H$, distinguishes carboxylic acids.

**9.** An amide is a carboxylic acid derivative in which the hydroxyl part of the carboxyl group is replaced with an $NH_2$, NHR, or $NR_2$ group. An amino acid is an acid in which an amine group is substituted for a hydrogen atom in the molecule. The amide functional group appears in proteins as a peptide linkage between amino acids. **10.** A polymer is a many-part chemical compound formed by bonding two or more one-part chemical species known as monomers. A chain-growth polymer results from combining alkene monomers in such a way that one bond in the double bond is broken, leaving each monomer with an unshared electron that is used to form a bond with the neighboring molecule. A step-growth polymer results from the reaction of dicarboxylic acid molecules with dialcohol molecules. The —OH group from the acid combines with the —H of the hydroxyl group from the alcohol to form a water molecule, and the remainder of the molecules bond to form the polymer.

# Answers to Blue-Numbered Questions, Exercises, and Problems

**1.** The cyanide ion, $CN^-$, and the carbonate ion $CO_3^{2-}$, are not organic because they do not contain hydrogen (they contain carbon, but not *both* carbon and hydrogen). The acetate ion, $C_2H_3O_2^-$, is organic. **3.** 109.5°; tetrahedral **5.** A hydrocarbon is a compound made up of carbon and hydrogen atoms. $C_3H_4$, $C_8H_{10}$, and $CH_3CH_2CH_3$ are hydrocarbons. **7.** $C_{11}H_{24}$; $C_{21}H_{44}$. Alkanes have the general formula $C_nH_{2n+2}$. **9.** Isomers are compounds having the same molecular formula but different structural formulas. **11.** $C_7H_{16}$ is the molecular formula; $CH_3CH_2CH_2CH_2CH_2CH_2CH_3$ is the line formula; $CH_3(CH_2)_5CH_3$ is the condensed formula. The structural formula is

```
    H  H  H  H  H  H  H
    |  |  |  |  |  |  |
 H—C—C—C—C—C—C—C—H
    |  |  |  |  |  |  |
    H  H  H  H  H  H  H
```

**13.** $C_2H_5$— $C_4H_9$—

**15.** $C_4H_{10}$; decane **17.** 2-methyl-4-ethylhexane **19.**

```
   C  C
   |  |
C—C—C—C—C
```

**21.**

```
   Cl H               H  Cl
   |  |               |  |
Cl—C—C—H          Cl—C—C—H
   |  |               |  |
   Cl H   (1,1,1)     Cl H   (1,1,2)
```

**23.** The general formula of a cycloalkane is $C_nH_{2n}$. Cyclobutane, $C_4H_8$, is an example:

```
   H  H
   |  |
H—C—C—H
   |  |
H—C—C—H
   |  |
   H  H
```

**25.** (cyclopentane ring)

**27.**

```
      Cl  H
   H   \ /   I
    \   C   /
 H—C       C—H
     \     /
   H—C—C—H
      |   |
      H   H
```

**29.** 4-ethylheptane **31.** 1-bromo-3,3-dichlorobutane **33.** (cyclopentane ring with Cl and I on adjacent carbons) **35.** 1-chloro-2-ethylcyclohexane

**37.** An alkene has one or more double bonds; an alkane has only single bonds. The general formula for an alkane is $C_nH_{2n+2}$, for an alkene, $C_nH_{2n}$.

**39.**

```
Cl      Cl
  \    /
   C=C
  /    \
Cl      H
```

This molecule can have only two additional atoms attached to each of its two carbons, so the first two chlorines go on one carbon atom. The third chlorine must go on the other carbon atom, thus, 1,1,2 is the only possible arrangement. It does not need to be specified.

**41.**

```
   H          H                H          CH2CH2CH3
     \      /                    \      /
      C=C                         C=C
     /      \                    /      \
CH3CH2   CH2CH2CH3          CH3CH2       H
```

Think of a line drawn between the two lines that depict a double bond. In *cis*-3-heptene, the hydrogen atoms attached to the double bonded carbons are on the same side of that line. In *trans*-3-heptene, the hydrogen atoms attached to the double bonded carbons are on opposite sides of that line.

**43.**

```
CH3(CH2)6CH2        CH2(CH2)11CH3
            \      /
             C=C
            /      \
           H        H
```

**45.** 2,2-dimethyl-3-heptyne

**47.**

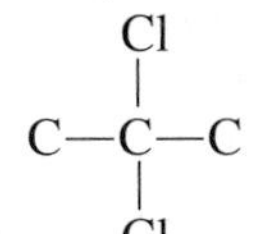

1,2-dimethylbenzene 1,3-dimethylbenzene 1,4-dimethylbenzene

**49.** 1-bromo-4-chlorobenzene. Because both substituents are halogen atoms, the lower number is given to the first halogen in the alphabet.

**51.**

```
Cl            Cl Cl                Cl          Cl       Cl
|             |  |                 |           |        |
C—C—C         C—C—C             C—C—C         C—C—C
|                                  |
Cl                                 Cl
```

1,1-dichloropropane 1,2-dichloropropane 2,2-dichloropropane 1,3-dichloropropane

**53.**

```
   Cl Cl         Cl Cl
   |  |          |  |
C—C—C—C         C—C—C—C
```

55. $H_3C{-}CH{=}CH{-}CH_3 + H_2 \xrightarrow{\text{catalyst}} H_3C{-}CH_2{-}CH_2{-}CH_3$

Because there is only one straight-chain butane molecule, it makes no difference if the starting material is *cis*- or *trans*-2-butene.

57. $CH_3{-}CH_2{-}CH_2{-}CH_2{-}OH$ and $CH_3{-}CH_2{-}CH(OH){-}CH_3$ both alcohols

$CH_3{-}CH(CH_3){-}CH_2{-}OH$ and $CH_3{-}C(CH_3)(OH){-}CH_3$ both alcohols

$CH_3{-}CH_2{-}O{-}CH_2{-}CH_3$ and $CH_3{-}CH_2{-}CH_2{-}O{-}CH_3$ both ethers

59. Although ethers have two carbon–oxygen bonds that are polar, the dipoles of these two bonds almost cancel each other by geometry. The forces of attraction in an ether are then weak dipole-dipole. In an alcohol, the hydroxyl proton on one alcohol molecule can hydrogen bond with the lone pair electrons of the oxygen atom on another alcohol molecule. The higher the forces of attraction, the higher the boiling point.

61. $CH_3{-}CH(OH){-}CH_2{-}CH_2{-}CH_2{-}CH_3$

63. $CH_3{-}CH_2{-}CH_2{-}CH_2{-}O{-}CH_2{-}CH_3$

65. $CH_3{-}CH_2{-}CH_2{-}\mathbf{OH} + \mathbf{H}{-}O{-}CH_2{-}CH_2{-}CH_3 \longrightarrow CH_3{-}CH_2{-}CH_2{-}O{-}CH_2{-}CH_2{-}CH_3 + \mathbf{HOH}$

67. $CH_3{-}CH_2{-}CHO$ and $CH_3{-}CO{-}CH_3$

69. $CH_3{-}CH(OH){-}CH_3 + \tfrac{1}{2}O_2 \longrightarrow CH_3{-}CO{-}CH_3 + H_2O$

71. $CH_3{-}CH_2{-}CH_2{-}CH_2{-}CH_2{-}COOH$

73. $CH_3{-}CH_2{-}CO{-}\mathbf{OH} + \mathbf{H}{-}O{-}CH_2{-}CH_3 \longrightarrow CH_3{-}CH_2{-}CO{-}O{-}CH_2{-}CH_3 + \mathbf{HOH}$

The organic reaction product is ethyl propanoate, an ester.

75. $CH_3{-}\ddot{N}H{-}CH_3$ dimethylamine

$H_2\ddot{N}{-}CH_2{-}CH_3$ ethylamine

77. Dimethylamine is a secondary amine; ethylamine is a primary amine.

**79.** $H-CH_2-CH_2-C(=O)-OH + H-\ddot{N}H-H \longrightarrow H-CH_2-CH_2-C(=O)-\ddot{N}H-H + HOH$

The organic reaction product is propanamide, an amide.

**81.** $\left[-CH_2-CHBr-CH_2-CHBr-CH_2-CHBr-\right]$

**83.** $H_2C=CH(CH_3)$

**85.** $\left[-CClF-CF_2-CClF-CF_2-CClF-CF_2-\right]$

**87.** $\left[-C(=O)-O-C_6H_4-C(CH_3)_2-C_6H_4-O-C(=O)-O-C_6H_4-C(CH_3)_2-C_6H_4-O-\right]$

**89.** $\left[-C(=O)-C_6H_4-C(=O)-NH-C_6H_4-NH-C(=O)-C_6H_4-C(=O)-NH-C_6H_4-NH-\right]$

**91.** $HO-C(=O)-C_6H_4-C(=O)-OH$ $\qquad$ $HO-CH_2-CH(CH_3)-OH$

**93.** $HO-C(=O)-CH_2-CH_2-CH_2-CH_2-CH_2-NH_2$

**96.** True: b, c, d, g, h, j, k, n, o, p, s. False: a, e, f, i, l, m, q, r. **97.** An alkene and a cycloalkane with the same number of carbon atoms have the same molecular formula, $C_nH_{2n}$. The alkene has a double bond between two carbon atoms, and there is no closed loop of carbon atoms. The cycloalkane has only single bonds, and the carbon atoms are assembled in a closed ring. **99.** Experimentally, there is only one type of carbon–carbon bond in benzene, not two. Benzene also does *not* undergo the addition reactions typical of alkenes. The delocalized structure reminds us that benzene is different from both alkanes and alkenes. **101.** The number of monomers in different polymers varies, so they have no definite molecular mass.

**102.** $\frac{1.8 \times 10^6 \text{ u}}{\text{polymer}} \times \frac{\text{molecule}}{104 \text{ u}} = 1.7 \times 10^4$ molecules/polymer

**103.** For an aldehyde to be oxidized to a carboxylic acid, an oxygen atom must be inserted between the carbonyl carbon and the hydrogen bonded to it. If there is no hydrogen atom bonded to the carbonyl atom, as in a ketone, there can be no oxidation.

**104.** The esterification equation is

$$CH_3-C(=O)-OH + H\overset{*}{O}-CH_3 \longrightarrow CH_3-C(=O)-\overset{*}{O}-CH_3 + HOH$$

The asterisk on the oxygen atom in methanol identifies the oxygen-18 atom. It is the presence of the radioactive oxygen in the ester product, rather than in the water product, that shows that the water molecule is made up from a hydroxyl from the acid and only a hydrogen atom from the alcohol.

**105.** Look at the equation describing formation of a peptide linkage:

$$H_2N-CHR-C(=O)OH + H-NH-CHR'-C(=O)OH \longrightarrow H_2N-CHR-C(=O)-NH-CHR'-C(=O)OH + HOH$$

[OH from the first molecule and H from the second are circled; the C(=O)–NH bond is marked "peptide link"]

This condensation reaction requires loss of a water molecule, one hydrogen of which must come from the amine. A tertiary amine has no hydrogens to lose. Trimethylamine (or any tertiary amine) *cannot* form an amide.

**106.** $\left[CH=CH-CH=CH-CH=CH\right]_n$

**107.** Because the acid-catalyzed condensation reaction in which amides are made is reversible, the amides can be decomposed in the presence of an acid catalyst and water. Acid rain gives precisely that combination.

# Chapter 22

## Answers to Target Checks

**1.** A-V-L, A-L-V, V-A-L, V-L-A, L-V-A, L-A-V **2.** (a) primary, local conformation, overall three-dimensional shape; (b) $\alpha$-helix; (c) $\beta$-pleated sheet **3.** True: a and d. (b) The induced fit model explains why one enzyme helps *a single reaction* to occur faster. (c) An enzyme substrate is *the reactant* in an enzyme-catalyzed reaction. **4.** (a) First find in the amylopectin structure a shaded oxygen atom in the main chain. One bond from this oxygen atom is drawn vertically and the second bond goes slightly up and to the right. The carbon atom attached to the vertical bond is carbon-1 (carbon-1 is the only carbon atom in glucose that is bonded to two oxygen atoms). The carbon atom attached to the up-and-right bond is carbon-4. (b) Find the shaded oxygen atom that leads off the main chain to the branch. One bond from this oxygen atom is drawn vertically and the second bond goes down and to the left. The carbon atom attached to the vertical bond is carbon-1. The carbon atom attached to the down-and-left bond is carbon-6 (shown as $—CH_2$). **5.** (a) The squeezable liquid margarine is the lowest of the three in saturated (higher-melting-point) fatty acids and highest in unsaturated (lower-melting-point) fatty acids. The soft solid has more saturated fatty acids than the liquid and fewer unsaturated fatty acids than the liquid. The stick margarine is highest in saturated fatty acids and lowest in unsaturated fatty acids. It most resembles butter, also sold in sticks. (b) Judging by its relatively high melting point, cocoa butter contains many saturated fatty acids. (c) Steroids are characterized by four fused rings, as shown here:

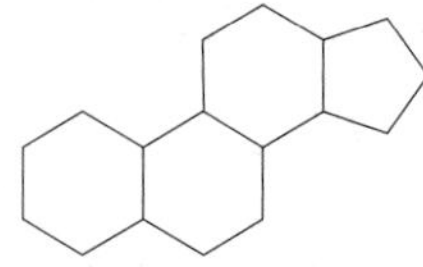

**6.** a) The components of a nucleotide are (1) a nitrogen-containing base, (2) a sugar, and (3) phosphate groups. b) RNA uses the sugar ribose and the bases adenine, cytosine, guanine, and uracil. DNA uses the sugar deoxyribose and the bases adenine, cytosine, guanine, and thymine.

## Answers to Concept-Linking Exercises

*You may have found more relationships or relationships other than the ones given in these answers.*

**1.** Primary protein structure is the linear amino acid residue sequence. Secondary structure is the regular local conformation maintained by hydrogen bonding. Tertiary structure describes the three-dimensional arrangement of the fully folded polypeptide chain. Quaternary structure describes how two or more polypeptide chains combine to make a protein. **2.** The two most common secondary protein structures are the $\alpha$-helix and the $\beta$-pleated sheet. Both structures reflect a maximum amount of hydrogen bonding. The $\alpha$-helix secondary structure is formed by hydrogen bonds within a chain; the $\beta$-pleated-sheet secondary structure is formed by hydrogen bonds between adjacent chains in different molecules or when a single chain folds back on itself. **3.** Enzymes are biochemical catalysts; they speed up the rate of a biochemical reaction. The enzyme substrate is the reactant in the enzyme-catalyzed reaction. The active site is that part of the enzyme to which the substrate binds during the reaction. The induced fit model is used to explain enzyme activity. The shapes of the substrate and enzyme are modified as the molecules bind and induce a fit. **4.** Monosaccharides, disaccharides, and polysaccharides are carbohydrates: aldehydes or ketones with two or more —OH groups. A monosaccharide is the simplest carbohydrate, one that cannot be broken down to simpler carbohydrates. A disaccharide is a chemical combination of two monosaccharides; a polysaccharide is a combination of many monosaccharides. **5.** Fats and oils have a glycerol "backbone" residue bonded to two fatty acid residues, forming a type of molecule known as a triacylglycerol. If a macroscopic sample of a triacylglycerol is solid at room temperature, it is classified as a fat; if a sample is liquid at room temperature, it is classified as an oil. A phospholipid is similar to a fat or oil, but its components are an alcohol backbone, fatty acid residues, and a phosphate group. **6.** Replication is the process by which two new DNA molecules form by building a new complementary strand on each strand of the original molecule. Transcription is the process by which a segment of a DNA molecule is copied in the form of a complementary messenger RNA molecule. The messenger RNA molecule then bonds to complementary transfer RNA molecules, each with a specific amino acid. This process is called translation. The amino acids then bond to form a protein.

## Answers to Blue-Numbered Questions, Exercises, and Problems

**1.**

```
            :O:
             ‖
   ..             ..
H—N—C—C—O—H
     |    |       ..
    H   H
```

**3.** Phenylalanine, tyrosine, and tryptophan

**5.** V-T-I is valylthreonylisoleucine. I-V-T is isoleucylvalylthreonine. The C terminal acid in V-T-I is isoleucine, in I-V-T, threonine.

**7.**

```
                                   (benzene ring)
                                        |
      SH                                |
      |                                 |
      CH2  O        H    O        CH2  O
      |    ‖        |    ‖        |    ‖
NH2—CH—C—NH—CH—C—NH—CH—C—OH
```

**9.** Tertiary protein structure describes the overall three-dimensional shape of a polypeptide chain caused by the folding of various regions. Quaternary protein structure describes how multiple polypeptide chains are arranged in relation to one another. **11.** The $\alpha$-helix secondary structure involves hydrogen bonding between the hydrogen attached to the peptide link nitrogen and a peptide link oxygen of an amino acid farther down the *same* protein chain. **13.** Enzymes are usually proteins. **15.** An enzyme substrate is a reactant that the enzyme helps change to product in the enzyme-catalyzed reaction. **17.** They lower the activation energy of a reaction. **19.** When you run a fever, enzyme-catalyzed reactions run faster than at normal body temperature. This may help the body fight off illness more quickly. **21.** Examples of aldose sugars are glucose, ribose, and deoxyribose. Aldose usually refers only to monosaccharides. **23.** Find carbon-1 (the only carbon bonded to two oxygen atoms) in both structures. In $\alpha$-glucose, the OH attached to carbon-1 is vertical, either pointed down or up. In $\beta$-glucose, the OH attached to carbon-1 is horizontal, or nearly so. **25.** Sucrose is a disaccharide, glycogen is a polysaccharide, and fructose is a monosaccharide. **27.** Galactose and glucose. **29.** Glucose, ribose, deoxyribose, and lactose would give a positive Benedict's test because all these sugars can have an aldehyde group in the open-chain form.

**31.**

**33.** Starch has $\alpha$-1,4 bonds; cellulose has $\beta$-1,4 bonds. Our enzymes can break the $\alpha$-1,4 bonds, but not the $\beta$-1,4 bonds. **35.** Immiscible in water. **37.** Plant sources. **39.**

| | | | |
|---|---|---|---|
| A—O—$CH_2$ | A—O—$CH_2$ | B—O—$CH_2$ | B—O—$CH_2$ |
| B—O—CH | C—O—CH | A—O—CH | C—O—CH |
| C—O—$CH_2$ | B—O—$CH_2$ | C—O—$CH_2$ | A—O—$CH_2$ |

**41.**

$CH_2$—O—P(=O)(—$O^-$)—O—$X^+$
A—O—CH
B—O—$CH_2$

$CH_2$—O—P(=O)(—$O^-$)—O—$X^+$
B—O—CH
A—O—$CH_2$

**43.** DNA is the storehouse of genetic information in all life forms. Messenger RNA carries instructions for protein synthesis from DNA to ribosomes. Transfer RNA delivers specific individual amino acids to the ribosome. **45.** See Figure 22.17. **47.** See Figure 22.17. **49.** See the Lewis diagram in Section 22.3. Ribose has a hydroxyl group at carbon 2; deoxyribose does not.

**51.**

**53.** C–A–T:

**55.** A–G–C. **57.** Transfer RNA picks up an amino acid molecule and carries it to a protein being synthesized by a ribosome. **60.** True: b, d, f, g, h, j, k. False: a, c, e, i.

**61.** $\frac{64{,}500 \text{ g}}{\text{mole hemoglobin}} \times \frac{1 \text{ mole amino acid}}{120 \text{ g}} = 538 \frac{\text{amino acids}}{\text{hemoglobin}}$

**63.** All are polymers. **64.** Glucose is the monomer in cellulose; amino acids, in proteins; nucleotides (adenine, cytosine, guanine, thymine), in DNA; glucose, in starch, and nucleotides (adenine, cytosine, guanine, uracil), in RNA. **65.** Nitrogen in found in all proteins; if you picked sulfur (from cysteine or methionine), that is also true. **66.** Phosphorus. **67.** The coil of the $\alpha$-helix and the strong hydrogen bonding within the protein chains keep water from soaking into the nylon; there is no further hydrogen bonding to be made by the water to the nylon. In cotton, however, there is little hydrogen bonding between the cellulose chains, so water molecules can form hydrogen bonds with (and therefore soak into) the hydroxyl groups on the sugars that make up cellulose.

**68.**

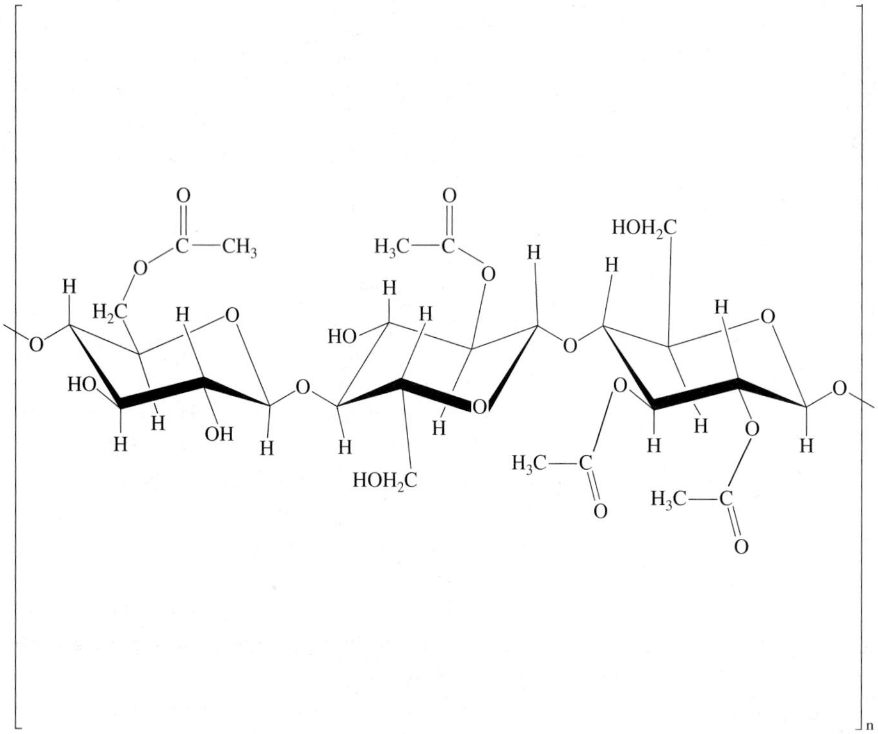

**69.** Because guanine and cytosine are complementary base pairs in DNA, there must also be 21% guanine. If guanine + cytosine = 42%, then adenine and thymine must equal 58%. Adenine is then 29%, as is thymine. **70.** Because RNA is single-stranded, there is no complementary RNA strand. As a result, the percentage of guanine has no relationship to the percentage of cytosine.

# Glossary*

$\propto$ is (directly) proportional to.

$\equiv$ is exactly equal to; is defined as.

$\Sigma$ the sum of all values of.

**$6.02 \times 10^{23}$** the number of units in one mole.

**absolute temperature** *see Kelvin temperature scale.*

**absolute zero** the lowest temperature. At this temperature, 0 K, equal to $-273.15$°C or $-459.67$°F, molecular motion is at a minimum.

**acid** a substance that yields hydrogen (hydronium) ions in aqueous solution (Arrhenius definition); a substance from which a proton can be removed via a chemical reaction (Brønsted–Lowry definition); a substance that forms covalent bonds by accepting a pair of electrons (Lewis definition).

**acid anion** a negatively charged ion that has a proton that can be removed by reaction with a base.

**acid (equilibrium) constant ($K_a$)** *see equilibrium constant.*

**acidic solution** an aqueous solution in which the hydrogen-ion concentration is greater than the hydroxide-ion concentration; a solution in which the pH is less than 7.

**actinides** elements 90 (Th) through 103 (Lr).

**activated complex** an intermediate molecular species presumed to be formed during the interaction (collision) of reacting molecules in a chemical change.

**activation energy** the energy barrier that must be overcome to start a chemical reaction.

**active site** the location on an enzyme that binds the substrate.

**activity series** a list of metal elements and hydrogen in order of reactivity in a single-replacement oxidation–reduction reaction.

**addition polymer** a polymer formed by monomers binding to each other without forming any other product.

**addition reaction** the reaction of an organic compound with another compound to form a single product.

**adenosine triphosphate (ATP)** the molecule involved in transferring chemical energy within cells.

**alcohol** an organic compound consisting of an alkyl group and at least one hydroxyl group, having the general formula ROH.

**aldehyde** a compound consisting of a carbonyl group bonded to a hydrogen on one side and a hydrogen, alkyl, or aryl group on the other, having the general formula RCHO.

**aldose** any of a class of monosaccharides containing an aldehyde group.

**aliphatic hydrocarbon** an alkane, alkene, or alkyne.

**alkali metal** a metal from Group 1A/1 of the periodic table.

**alkaline** basic; having a pH greater than 7.

*The abbreviation q.v. stands for the Latin *quod vide,* literally meaning "which see." It tells you that a term in a definition is also another entry in the glossary.

**alkaline earth metal** a metal from Group 2A/2 of the periodic table.

**alkane** a saturated hydrocarbon containing only single bonds, in which each carbon atom is bonded to four other atoms.

**alkene** an unsaturated hydrocarbon containing a double bond, in which each carbon atom that is double-bonded is bonded to a maximum of three atoms.

**alkyl group** an alkane hydrocarbon group lacking one hydrogen atom, having the general formula $C_nH_{2n+1}$, and frequently symbolized by the letter R.

**alkyne** an unsaturated hydrocarbon containing a triple bond, in which each carbon atom that is triple-bonded is bonded to a total of two atoms.

**alloy** a solid mixture of two or more elements that has macroscopic metallic properties.

**alpha helix** a coiled secondary protein structure that is stabilized by hydrogen bonds.

**alpha ($\alpha$) particle** the nucleus of a helium atom, often emitted in nuclear disintegration.

**amide** a derivative of a carboxylic acid in which the hydroxyl group is replaced by an —$NH_2$ group, having the general formula $RCONH_2$.

**amine** an ammonia derivative in which one or more hydrogens are replaced by an alkyl group.

**amino acid** a carboxylic acid containing both an amine group and a variable alkyl group R. Amino acids of the general form $RCH(NH_2)COOH$ can form amide bonds with each other to form peptides and proteins.

**amorphous** without definite structure or shape.

**amphiprotic, amphoteric** pertaining to a substance that can act as an acid or a base.

**amylopectin** a highly branched, high molar mass polysaccharide that is one of two main forms of plant starch.

**amylose** a mostly unbranched, soluble polysaccharide that is one of two main forms of plant starch.

**angstrom** a length unit equal to $10^{-10}$ m.

**angular (or bent)** (molecular geometry) arrangement of three electron pairs and two bonded atoms surrounding a central atom in a molecule or ion to form a 120° angle between the two bonded atoms and the central atom.

**anhydride (anhydrous)** a substance that is without water or from which water has been removed.

**anion** a negatively charged ion.

**anode** the electrode at which oxidation occurs in an electrochemical cell.

**antilogarithm** the number whose logarithm is a given number.

**aqueous** pertaining to water.

**aromatic hydrocarbon** a hydrocarbon containing a benzene ring.

**Arrhenius acid–base theory** *see acid and base.*

**artifical radioactivity** *see induced radioactivity.*

**atmosphere** (pressure unit) a unit of pressure based on atmospheric pressure at sea level and capable of supporting a mercury column 760 mm high.

**atom** the smallest particle of an element that can combine with atoms of other elements to form chemical compounds.

**atomic mass** the average mass of the atoms of an element compared with an atom of carbon-12 at exactly 12 atomic mass units. Also called *atomic weight.*

**atomic mass unit (u)** a unit of mass that is exactly $^1/_{12}$ of the mass of an atom of carbon-12.

**atomic number (Z)** the number of protons in an atom of an element.

**atomic weight** *see atomic mass.*

**average kinetic energy** for an ideal gas with all particles at equal mass, the energy due to particle motion, which is directly proportional to absolute temperature.

**Avogadro's Law** the volume of a gas at constant temperature and pressure is proportional to the number of particles.

**Avogadro's number** the number of carbon atoms in exactly 12 grams of carbon-12; the number of units in 1 mole ($6.02 \times 10^{23}$).

**balanced equation** an equation describing a chemical reaction that has the same number of atoms of each element and the same total charge for both reactants and products.

**ball-and-stick model** a three-dimensional representation of a molecule that uses balls to represent atoms and sticks to represent electron pairs.

**bar** (pressure unit) 100 kilopascals.

**barometer** a laboratory device for measuring atmospheric pressure.

**base** a substance that yields hydroxide ions in aqueous solution (Arrhenius definition); a substance that removes protons in chemical reaction (Brønsted–Lowry definition); a substance that forms covalent bonds by donating a pair of electrons (Lewis definition); mathematically, any positive real number not equal to one that has the form b in the function $kb^e$.

**base unit** one of seven units used to express the seven base quantities that make up the fundamental foundation of the International System of Units.

**basic solution** an aqueous solution in which the hydroxide-ion concentration is greater than the hydrogen-ion concentration; a solution in which the pH is greater than 7.

**beeswax** a wax that is a mixture of several substances including esters, acids, and hydrocarbons with fatty acid esters of straight-chain alcohols as the major components.

**bent (or angular)** (molecular geometry) arrangement of four electron pairs and two bonded atoms surrounding a central atom in a molecule or ion to form a 109.5° angle between the two bonded atoms and the central atom.

**benzene ring** a planar six-carbon structural unit connected by delocalized electrons that is found in aromatic hydrocarbons, including the benzene molecule.

**beta ($\beta$) particle** a high-energy electron, often emitted in nuclear disintegration.

**beta-pleated sheet** a pleated-sheet secondary protein structure stabilized by hydrogen bonds.

**binary compound** a compound consisting of two elements.

**biochemistry** the study of life on a molecular level.

**boiling point** the temperature at which vapor pressure becomes equal to the pressure above a liquid; the temperature at which vapor bubbles form spontaneously any place within a liquid.

**boiling-point elevation** the difference between the boiling point of a solution and the boiling point of the pure solvent.

**bombardment (nuclear)** the striking of a target nucleus by an atomic particle, causing a nuclear change.

**bond** *see chemical bond.*

**bond angle** the angle formed by the bonds between two atoms that are bonded to a common central atom.

**bonding electrons** the electrons transferred or shared in forming chemical bonds; valence electrons.

**Boyle's Law** the pressure of a fixed quantity of a gas at constant temperature is inversely proportional to volume, $P \propto (1/V)$.

**breeder reactor** a nuclear reactor designed to create new fissionable nuclear fuel from nonfissionable isotopes.

**Brønsted–Lowry acid–base theory** *see acid and base.*

**buffer** a solution that resists a change in pH.

**buret** a glass tube of uniform width calibrated to accurately measure volume of liquid delivered through an adjustable-flow stopcock at the bottom of the tube.

**calorie** a unit of heat energy equal to 4.184 joules.

**Calorie (food)** a unit of heat energy equal to 4184 joules.

**calorimeter** a laboratory device for measuring heat flow.

**carbohydrate** a class of organic compounds consisting mainly of polyhydroxy aldehydes or ketones. Carbohydrates form the supporting tissue of plants and serve as food for animals, including people.

**carbonyl group** an organic functional group, C=O, characteristic of aldehydes and ketones.

**carboxyl group** an organic functional group, —COOH, characteristic of carboxylic acids.

**carboxylic acid** an organic acid containing the carboxyl group, having the general formula RCOOH.

**carnauba wax** a wax that is a mixture of several substances including esters of fatty acids, fatty alcohols, acids, and hydrocarbons; its hardness is due to carboxylic acid components with hydroxyl groups that tie together adjacent chains by hydrogen bonding.

**catalyst** a substance that increases the rate of a chemical reaction by lowering activation energy. The catalyst is either a non-participant in the reaction, or it is regenerated. *See also* inhibitor.

**cathode** the negative electrode in a cathode ray tube; the electrode at which reduction occurs in an electrochemical cell.

**cation** a positively charged ion.

**cell, electrolytic** a cell in which electrolysis occurs as a result of an externally applied electrical potential.

**cell, galvanic** *see cell, voltaic.*

**cellulose** a polysaccharide that serves as the major structural component of plants.

**cell, voltaic** a cell in which an electrical potential is developed by a spontaneous chemical change. Also called a galvanic cell.

**Celsius temperature scale** a system of temperature measurement based on assignment of 0°C to the freezing point of water and 100°C to the boiling point of water, with 100 equally divided degrees between the reference points (historical); a system of measurement based on assignment of −273.15°C to absolute zero and 273.16 kelvin = 0.01°C (modern).

**chain-growth polymer** a polymer formed as monomers add to a growing polymer chain in sequential addition reactions in a chain-reaction process.

**chain reaction** a reaction that has as a product one of its own reactants; that product becomes a reactant, thereby allowing the original reaction to continue.

**charge cloud** *see electron cloud.*

**charge density** the amount of electric charge per unit volume.

**Charles's Law** the volume of a fixed quantity of a gas at constant pressure is directly proportional to absolute temperature, $V \propto T$.

**chemical bond** a general term that sometimes includes all of the electrostatic attractions among atoms, molecules, and ions, but more often refers to covalent and ionic bonds. *See covalent bond, ionic bond.*

**chemical change** a change in which one or more substances disappear and one or more new substances form.

**chemical equation** a symbolic representation of chemical change, with the formulas of the beginning substances to the left of an arrow that points to the formulas of the substances formed.

**chemical family** a group of elements having similar chemical properties because of similar valence electron configuration, appearing in the same column of the periodic table.

**chemical formula** *see formula, chemical.*

**chemical nomenclature** a system of names used in chemistry.

**chemical properties** the types of chemical change a substance is able to experience.

**chemical reaction** *see chemical change.*

**closed system** a sample of matter that is isolated so that it cannot exchange matter with the rest of the universe.

**cloud chamber** a device in which condensation tracks form behind radioactive emissions as they travel through a supersaturated vapor.

**coefficient** in a chemical equation, numbers used to equalize the number of atoms before and after the chemical change; mathematically, the quantity c when a number is written in exponential notation in the form $c \times 10^e$.

**colligative properties** physical properties of mixtures that depend on the concentration of particles, irrespective of their identity.

**collision theory of chemical reactions** a particulate-level model of chemical reactivity that features sufficient kinetic energy and proper orientation as necessary conditions for a reaction-producing collision.

**colloid** a nonsettling dispersion of aggregated ions or molecules intermediate in size between the particles in a true solution and those in a suspension.

**combination (synthesis) reaction** a reaction in which two or more substances combine to form a single product.

**Combined Gas Law** the volume of a fixed quantity of a gas is proportional to temperature and inversely proportional to pressure, $V \propto (T/P)$.

**combustion** the process of burning.

**common ion effect** the percentage ionization of a weak acid or low-solubility ionic compound is decreased when an ion in common with the acid or ionic compound is introduced to the solution.

**compound** a pure substance that can be broken down into two or more other pure substances by a chemical change.

**concentrated** adjective for a solution with a relatively large amount of solute per given quantity of solvent or solution.

**condensation** the act of condensing.

**condensation polymer** a polymer formed through condensation reactions.

**condensation reaction** a chemical change in which two molecules or functional groups react to form a larger molecule and a small molecule, such as water.

**condense** to change from a vapor to a liquid or solid.

**condensed structural formula** a symbolic representation of an organic compound that expands the fundamental formula to suggest the arrangement of atoms while still fitting on a standard line of text.

**conductor** a substance that readily conveys electricity.

**conjugate acid–base pair** a Brønsted–Lowry acid and the base derived from it when it loses a proton, or a Brønsted–Lowry base and the acid developed from it when it removes a proton.

**continuous spectrum** the band of colors that results from electromagnetic radiation emissions over a range of wavelengths.

**conventional equation** a chemical equation written with conventional formulas for soluble aqueous ionic compounds and strong acids.

**conversion factor** the relationship between different units of measurement that express the same quantity.

**coordinate covalent bond** a bond in which both bonding electrons are furnished by only one of the bonded atoms.

**copolymer** a polymer formed from two or more different monomers.

**coulomb** a unit of electrical charge.

**covalent bond** the chemical bond between two atoms that share a pair of electrons.

**covalent network solid** a crystalline solid made of atoms that are connected by a network of covalent bonds, essentially forming one large molecule.

**critical mass** the minimum quantity of fissionable material needed to sustain a nuclear chain reaction.

**crystalline solid** a solid in which the ions and/or molecules are arranged in a definite geometric pattern.

**C-terminal** the end of a polypeptide or protein chain terminated by an amino acid with a free carboxyl group.

**cubic centimeter ($cm^3$)** a unit of volume equal to the volume of a cube with a 1 cm length, width, and height.

**cycloalkane** an alkane (q.v.) that has one or more rings of carbon atoms.

**Dalton's atomic theory** a model of matter based on the idea that each element is made of particles called atoms.

**Dalton's Law of Partial Pressures** the total pressure exerted by a mixture of gases is the sum of the partial pressures of the gases in the mixture.

**decompose** to change chemically into simpler substances.

**decomposition reaction** a reaction in which a single compound breaks down into simpler substances.

**defining equation** an equation used to establish the defintion of a unit or a property of a substance.

**delocalized electrons** electrons in a molecule that are not restricted to remaining near a single atom or between two atoms in a covalent bond.

**density** the mass of a substance per unit volume.

**deoxyribonucleic acid (DNA)** a large nucleotide (q.v.) polymer found in the cell nucleus. DNA contains genetic information and controls protein synthesis.

**derived unit** a unit of measurement derived from the seven base units in the International System of Units.

**diatomic** having two atoms.

**dilute** adjective for a solution with a relatively small amount of solute per given quantity of solvent or solution; verb meaning to reduce the concentration of a solution by adding solvent.

**dimensional analysis** a problem-solving method that emphasizes algebraic cancellation of units in calculation setups.

**dipole** a polar molecule.

**dipole forces** a type of intermolecular attractive force of intermediate relative strength that occurs between the positive pole of one polar molecule and the negative pole of another.

**diprotic acid** an acid capable of yielding two protons per molecule in complete ionization.

**directly proportional** two quantites that have a constant ratio; expressed as $y \propto x$ and $y = kx$, where k is a nonzero constant.

**disaccharide** a sugar composed of two monosaccharides; two sugar units per molecule.

**discrete** discontinuous; individually distinct.

**dispersion forces** weak electrical attractions between molecules, temporarily produced by the shifting of internal electrons.

**dissolve** to pass into solution.

**distillation** the process of separating components of a mixture by boiling off and condensing the more volatile component.

**distilled water** water that has been purified by distillation.

**disulfide linkages** sulfur-sulfur covalent bonds that strengthen the tertiary structure of many proteins.

**double bond** a covalent chemical bond formed by the sharing of two pairs of electrons between two bonded atoms.

**double-replacement equation (reaction)** a chemical equation with the form $AX + BY \rightarrow AY + BX$, where the reactants are two compounds, and the ions appear to exchange partners.

**dynamic equilibrium** a state in which opposing changes occur at equal rates, resulting in zero net change over a period of time.

**electrode** a conductor by which electric charge enters or leaves an electrolyte.

**electrolysis** the passage of electric charge through an electrolyte.

**electrolyte** a substance that, when dissolved, yields a solution that conducts electricity; a solution or other medium that conducts electricity by ionic movement.

**electrolytic cell** *see cell, electrolytic.*

**electromagnetic radiation** energy in the form of electric and magnetic waves, including gamma rays, X-rays, ultraviolet, visible, and infrared light, microwaves, and radio waves.

**electron** subatomic particle carrying a unit negative charge and having a mass of $9.1 \times 10^{-28}$ gram, or $^{1}/_{1837}$ of the mass of a hydrogen nucleus, found outside the nucleus of the atom.

**electron (charge) cloud** the region of space around or between atomic nuclei that is occupied by electrons.

**electron configuration** the orbital arrangement of electrons in ions or atoms.

**electron-dot diagram (structure)** *see Lewis diagram.*

**electron-pair angle** the angle formed by any two electron pairs in a molecule or ion and the central atom between them.

**electron-pair geometry** the arrangement of electron pairs around a central atom in a molecule or ion.

**electronegativity** a scale of the relative ability of an atom of one element to attract the electron pair that forms a single covalent bond with an atom of another element.

**electron orbit** the circular or elliptical path supposedly followed by an electron around an atomic nucleus, according to the Bohr theory of the atom.

**electron orbital** a mathematically described region within an atom in which there is a high probability that an electron will be found.

**electron-pair geometry** a description of the distribution of bonding and unshared electron pairs around a bonded atom.

**electron-pair repulsion** the principle that electron-pair geometry is the result of repulsion between electron pairs around a bonded atom, causing them to be as far apart as possible.

**electron-sea model** a particulate-level model of a metallic crystal that features a definite crystal pattern of positive ions with valence electrons moving relatively freely among the ions.

**electron-transfer reaction** a chemical change in which one or more electrons are transferred from one substance, the reducing agent, to another, the oxidizing agent.

**electrostatic force** the force of attraction or repulsion between electrically charged objects.

**element** a pure substance that cannot be decomposed into other pure substances by ordinary chemical means.

**elemental symbol** *see symbol (chemical).*

**empirical formula** a formula that represents the lowest integral ratio of atoms of the elements in a compound.

**endothermic** a change that absorbs energy from the surroundings, having a positive $\Delta H$, an increase in enthalpy.

**energy** the ability to do work.

**enthalpy** the heat content of a chemical system.

**enthalpy of reaction** *see heat of reaction.*

**enzyme** a protein molecule that catalyzes chemical reactions.

**equilibrium** *see dynamic equilibrium.*

**equilibrium constant** with reference to an equilibrium equation, the ratio in which the numerator is the product of concentrations of the species on the right-hand side of the equation, each raised to a power corresponding to its coefficient in the equation, and the denominator is the corresponding product of the species on the left side of the equation; symbol: K, $K_c$, or $K_{eq}$.

**equilibrium vapor pressure** *see vapor pressure.*

**equivalent** the quantity of an acid (or base) that yields or reacts with one mole of $H^+$ (or $OH^-$) in a chemical reaction; the quantity of a substance that gains or loses one mole of electrons in a redox reaction.

**equivalent mass** the mass in grams per equivalent.

**ester** an organic compound formed by the reaction between a carboxylic acid and an alcohol, having the general formula R—CO—OR′.

**esterification** the reaction between a carboxylic acid and an alcohol, yielding an ester and water.

**ether** an organic compound in which two alkyl groups are bonded to the same oxygen, having the general formula R—O—R′.

**exact number** a number with no uncertainty, such as a counting number or a number established by definition.

**excess reactant** the reactant(s) in a chemical reaction that remain when the reaction is complete.

**excited state** the state of an atom in which one or more electrons have absorbed energy—becoming "excited"—to raise them to energy levels above ground state.

**exothermic reaction** a reaction that gives off energy to its surroundings.

**exponent** a number, as e in $10^e$, denoting the power to which another number (the base) is to be raised.

**exponential** a number, called the base, raised to some power, called the exponent.

**exponential notation** a method of writing numbers in the form: a.bcd $\times$ $10^e$.

**exponential notation, standard** a method of writing numbers in the form: a.bcd $\times$ $10^e$, where a.bcd is a number equal to or greater than 1 and less than 10.

**Fahrenheit temperature scale** a system of temperature measurement based on assignment of 32°F to the freezing point of water and 212°F to the boiling point of water, with 180 equally divided degrees between the reference points (historical); a system of measurement based on assignment of −273.15°C to absolute zero, 273.16 kelvin = 0.01°C, and $T_{°F} - 32 = 1.8T_{°C}$ (modern).

**family** *see chemical family.*

**fat** an ester formed from glycerol and three fatty acids. Fats are solids at room temperature. Also called *triacylglycerols* or *triglycerides.*

**fatty acid** a long-chain carboxylic acid, typically having between 10 and 24 carbon atoms.

**filtration** the process of passing a liquid (or gas) through a filter to separate components of a mixture based on relative particle sizes.

**fission** a nuclear reaction in which a large nucleus splits into two smaller nuclei.

**force field** a region of space in which a force is effectively operative.

**formula, chemical** a combination of chemical symbols and subscript numbers that represents the elements in a pure substance and the ratio in which the atoms of the different elements appear.

**formula mass (weight)** the mass in u of one formula unit of a substance; the molar mass of formula units of a substance.

**formula unit** a real (molecular) or hypothetical (ionic) unit particle represented by a chemical formula.

**fractional distillation** the separation of a mixture into fractions whose components boil over a given temperature range.

**freezing-point depression** the difference between the freezing point of a solution and the freezing point of the pure solvent.

**fructose** a ketone monosaccharide with the formula $C_6H_{12}O_6$ that is the sugar found in fruits and honey and widely used in industry in the form of high-fructose corn syrup as a sweetner.

**functional group** an atom or a group of atoms that establishes the identity of a class of compounds and determines its chemical properties.

**fusion** the process of melting; also, a nuclear reaction in which two small nuclei combine to form a larger nucleus.

**galvanic cell** *see cell, voltaic.*

**gamma ($\gamma$) ray** a high-energy electromagnetic emission in radioactive disintegration.

**gas** the state of matter characterized by particles that are independent and have a macroscopic-level variable shape and volume.

**gauge pressure** the pressure above atmospheric pressure.

**Gay-Lussac's Law** the pressure exerted by a fixed quantity of gas at constant volume is directly proportional to absolute temperature, $P \propto T$.

**Geiger counter** an electrical device for detecting and measuring the intensity of radioactive emission.

**gene** typically, a sequence of nucleotides in a segment of DNA that contains the information needed to produce a protein.

**geometric isomers** two compounds having the same molecular formulas but different geometric configurations around a structurally rigid bond.

**given quantity (*GIVEN*)** in a dimensional analysis setup, the quantity given in the problem statement that will need to be converted to an equivalent amount of another unit.

**glucose** an aldehyde monosaccharide with the formula $C_6H_{12}O_6$ that is the sugar used as an energy source by most living organisms.

**glycogen** a polysaccharide that is the quick-acting carbohydrate reserve in mammals; animal starch.

**gram (g)** 1/1000 the mass of a kilogram (q.v.).

**ground state** the state of an atom in which all electrons occupy the lowest possible energy levels.

**group (periodic table)** the elements making up a vertical column in the periodic table.

**half-life ($t_{1/2}$)** the time required for the disintegration of one-half of the radioactive atoms in a sample.

**half-reaction** the oxidation or reduction half of an oxidation–reduction reaction.

**halide ion** $F^-$, $Cl^-$, $Br^-$, or $I^-$.

**halogen** the name of the chemical family consisting of fluorine, chlorine, bromine, iodine, and astatine; any member of the halogen family.

**halogenation reaction** the reaction of an organic compound with a halogen.

**heat of fusion (solidification)** the heat flow when one gram of a substance changes between a solid and a liquid at constant pressure and temperature. *See also molar heat of fusion (solidification).*

**heat of reaction** the change of enthalpy in a chemical reaction.

**heat of vaporization (condensation)** the heat flow when one gram of a substance changes between a liquid and a vapor at constant pressure and temperature. *See also molar heat of vaporization (condensation).*

**heterogeneous** having a nonuniform composition, usually with visibly different parts or phases.

**homogeneous** having a uniform appearance and uniform properties throughout.

**homologous series** a series of compounds in which each member differs from the one next to it by the same structural unit.

**hydrate** a crystalline solid that contains water of hydration.

**hydrated hydrogen ion** *see hydronium ion.*

**hydrated ion** an ion in solution surrounded by water molecules.

**hydrocarbon** an organic compound consisting of carbon and hydrogen.

**hydrogenation** addition of hydrogen to a double or triple bond to produce a saturated product.

**hydrogen bond** an intermolecular bond (attraction) between a hydrogen atom in one molecule and a highly electronegative atom (fluorine, oxygen, or nitrogen) of another polar molecule; the polar molecule may be of the same substance containing the hydrogen, or of a different substance.

**hydronium ion** a hydrated hydrogen ion, $H_3O^+$.

**hydroxyl group** an organic functional group, —OH, characteristic of alcohols.

**ideal gas** a hypothetical gas that behaves according to the ideal gas model over all ranges of temperature and pressure.

**ideal gas equation** the equation $PV = nRT$ that relates quantitatively the pressure, volume, quantity, and temperature of an ideal gas.

**ideal gas model** a representation of a gas as identical, volumeless particles that move in straight lines, undergo collisions with no total loss of energy, and do not attract or repel other particles.

**immiscible** insoluble (usually used only in reference to liquids).

**indicator** a substance that changes from one color to another, used to signal the end of a titration.

**induced dipole forces** a type of intermolecular attractive force of varying relative strength, depending on molecular size, that occurs between nonpolar molecules because of shifting electron clouds within the molecules.

**induced fit model** a model used to explain enzyme function that features a close match between the shape of the substrate and the active site of the enzyme that becomes slightly modified to an exact fit as the two molecules bind.

**induced radioactivity** decay of a radionuclide that was formed from a previously stable nucleus that artifically was made radioactive.

**inhibitor** a substance added to a chemical reaction to retard its rate; sometimes called a *negative catalyst*; a molecule that decreases enzyme activity.

**inorganic chemistry** the chemistry of all chemical compounds except the hydrocarbons and most of their derivatives.

**International System of Units** (abbreviated SI from the French *Le Système International d'Unités*) a subset of metric units that is used to express physical quantities in terms of seven base units and in combinations of those units.

**inversely proportional** two quantites that have a constant product; expressed as $y \propto 1/x$ and $xy = k$, where k is a nonzero constant.

**ion** an atom or group of covalently bonded atoms that is electrically charged because of an excess or deficiency of electrons.

**ion-combination reaction** when two solutions are combined, the formation of a precipitate or molecular compound by a cation from one solution and an anion from the second solution.

**ionic bond** the chemical bond arising from the attraction forces between oppositely charged ions in an ionic compound.

**ionic compound** a compound in which ions are held by ionic bonds.

**ionic crystal** a cryatalline solid composed of oppositely charged ions held together by strong electrostatic forces.

**ionic equation** a chemical equation in which dissociated compounds are shown in ionic form.

**ionizable hydrogen** a hydrogen atom in a molecule or ion that can be removed by reaction with a base.

**ionization** the formation of an ion from a molecule or atom.

**ionization energy** the energy required to remove an electron from an atom or ion.

**isoelectronic** having the same electron configuration.

**isomers** two compounds having the same molecular formulas but different structural formulas and different physical and chemical properties.

**isotopes** two or more atoms of the same element that have different atomic masses because of different numbers of neutrons.

**IUPAC** International Union of Pure and Applied Chemistry.

**joule** the SI energy unit, defined as a force of one newton applied over a distance of one meter; 1 joule = 0.239 calorie.

**K** the symbol for the kelvin, the absolute temperature unit; the symbol for an equilibrium constant. $K_a$ is the constant for the ionization of a weak acid; $K_{sp}$ is the constant for the equilibrium between a slightly soluble ionic compound and a saturated solution of its ions; $K_w$ is the constant for the ionization of water.

**Kelvin temperature scale** an absolute temperature scale with 0 K at absolute zero, or −273.15°C, and the magnitude of the kelvin unit as 1/273.16 of the difference between absolute zero and the triple point of water, 273.16 K.

**ketone** a compound consisting of a carbonyl group bonded on each side to an alkyl group, having the general formula R—CO—R′.

**ketose** any of a class of monosaccharides containing a ketone group.

**kilogram (kg)** a unit of mass equal to the mass of the International Prototype Kilogram.

**kinetic energy** energy of motion; translational kinetic energy is equal to $^1/_2 \times$ mass $\times$ (velocity)$^2$.

**kinetic molecular theory** the general theory that all matter consists of particles in constant motion, with different degrees of freedom distinguishing among solids, liquids, and gases.

**kinetic theory of gases** the portion of the kinetic molecular theory that describes gases and from which the model of an ideal gas is developed.

**lactose** a disaccharide made from galactose and glucose with the formula $C_{12}H_{22}O_{11}$ that is also called milk sugar because it sweetens the milk of mammals.

**lanolin** a wax that is a complex mixture of esters of 33 high-molar-mass alcohols and 36 fatty acids; also called *wool fat.*

**lanthanides** elements 58 (Ce) through 71 (Lu).

**Law of Combining Volumes** when gases at the same temperature and pressure react, the reacting and product volumes are in a ratio of small whole numbers.

**Law of Conservation of Energy** in a nonnuclear change, energy is conserved.

**Law of Conservation of Mass** in a nonnuclear change, mass is conserved.

**Law of Constant Composition** *see Law of Definite Composition.*

**Law of Definite Composition** any compound is always made up of elements in the same proportion by mass.

**Law of Multiple Proportions** when two elements combine to form more than one compound, the different weights of one element that combine with the same weight of the other element are in a simple ratio of whole numbers.

**Le Chatelier's Principle** if an equilibrium system is subjected to a change, processes occur that tend to counteract partially the initial change, thereby bringing the system to a new position of equilibrium.

**lecithin** any of a group of phospholipids that have an alcohol backbone, two fatty acid residues, a phosphate group, and a quaternary saturated amine in the head group.

**Lewis acid–base theory** *see acid and base.*

**Lewis diagram, structure, or symbol** a diagram representing the valence electrons and covalent bonds in an atomic or molecular species.

**limiting reactant** the reactant first totally consumed in a reaction, thereby determining the maximum yield possible.

**linear** arrangement of two electron pairs surrounding a central atom in a molecule or ion to form a 180° angle between the pairs and the central atom (electron-pair geometry); arrangement of two electron pairs and two bonded atoms surrounding a central atom in a molecule or ion to form a 180° angle between the bonded atoms and the central atom (molecular geometry).

**line formula** a condensed formula reprsenting an organic compound that includes each $CH_2$ unit in an alkane chain.

**line spectrum** the spectral lines that appear when light emitted from a sample is analyzed in a spectroscope.

**lipid** any of a group of compounds that are found in living organisms, insoluble in water, and soluble in nonpolar solvents.

**liquid** the state of matter characterized by particles in contact with one another that move freely among themselves and macroscopic-level variable shape and constant volume.

**liter (L)** 0.001 $m^3$ (exactly).

**local conformation** the arrangement in space of atoms in a molecule, with consideration restricted to a segment of a larger molecule.

**logarithm** the power to which a base must be raised to produce a given number.

**London (dispersion) forces** *see induced dipole forces.*

**lone pair** a pair of valence electrons in a molecule that are not used for bonding.

**macromolecular crystal** a crystal made up of a large but indefinite number of atoms covalently bonded to each other to form a huge molecule. Also called a *network solid.*

**macromolecule** a polymeric molecule with molar mass $\geq$ 5000 g/mol.

**macroscopic** consideration of matter on a scale observable by the human eye; it is measurable with conventional apparatus.

**main group element** an element from one of the A Groups (IUPAC Groups 1–2 and 13–18) of the periodic table.

**major species** in an acid solution, the species present in greatest abundance.

**manometer** a laboratory device for measuring gas pressure.

**mass** a property reflecting the quantity of matter in a sample.

**mass number (A)** the total number of protons plus neutrons in the nucleus of an atom.

**mass spectrometer** a laboratory device in which a flow of gaseous ions may be analyzed in regard to their charge and/or mass.

**materials science** the field of scientific study that investigates the particulate-level structure of materials and their macroscopic properties.

**matter** that which occupies space and has mass.

**messenger RNA** an RNA molecule transcribed from a DNA template that carries genetic information to the site of protein synthesis.

**metal** a substance that possesses metallic properties, such as luster, ductility, malleability, and good conductivity of heat and electricity; an element that loses electrons to form monatomic cations.

**metallic bond** forces of attraction between delocalized electrons in a metallic crystal and the metal ions.

**metallic character** exhibition of the physical and chemical properties of metals.

**metallic crystal** a crystalline solid made of an orderly, repeating pattern of positive ions through which delocalized valence electrons move relatively freely.

**metalloid (semimetal)** an element that has both metallic and nonmetallic properties.

**meter (m)** the length of the path traveled by light in a vacuum in 1/299,792,458 second.

**methyl group** the one-carbon alkyl structural group $—CH_3$.

**metric system** a system of measurement used by most of the world that is based on a small number of basic units and a standard set of prefixes that represent multiples of 10.

**microscopic** consideration of matter on a scale not observable by the human eye but observable with a classic microscope.

**milliliter (mL)** 0.001 liter.

**millimeter of mercury (mm Hg)** (pressure unit) the pressure exerted at the base of a column of mercury 1 mm high.

**minor species** in an acid solution, the species present in lesser abundance.

**miscible** soluble (usually used only in reference to liquids).

**mixture** a sample of matter containing two or more pure substances.

**model** a representation of something else.

**molal boiling-point elevation constant** the ratio of boiling point elevation to molality for a dilute ideal solution.

**molal freezing-point depression constant** the ratio of freezing point depression to molality for a dilute ideal solution.

**molality** solution concentration expressed in moles of solute per kilogram of solvent.

**molar heat of fusion (solidification)** the heat flow when one mole of a substance changes between a solid and a liquid at constant temperature and pressure.

**molar heat of vaporization (condensation)** the heat flow when one mole of a substance changes between a liquid and a vapor at constant temperature and pressure.

**molarity** solution concentration expressed in moles of solute per liter of solution.

**molar mass (weight)** the mass of one mole of any substance.

**molar volume** the volume occupied by one mole, usually of a gas.

**mole** the quantity of any species that contains the same number of units as the number of atoms in exactly 12 grams of carbon-12.

**molecular compound** a compound whose fundamental particles are molecules rather than ions.

**molecular crystal** a molecular solid in which the molecules are arranged according to a definite geometric pattern.

**molecular formula** a description of the composition of a molecule that lists each element in the molecule by chemical symbol and the number of atoms of each element, if more than one, with a subscript after the symbol.

**molecular geometry** a description of the shape of a molecule.

**molecular mass (weight)** the number that expresses the average mass of the molecules of a compound compared to the mass of an atom of carbon-12 at a value of exactly 12; the average mass of the molecules of a compound expressed in atomic mass units.

**molecule** the smallest unit particle of a pure substance that can exist independently and possess the identity of the substance.

**monatomic** having only one atom.

**monomer** the individual chemical structural unit from which a polymer may be developed.

**monoprotic acid** an acid capable of yielding one proton per molecule in complete ionization.

**monosaccharide** the simplest sugars, which cannot be converted to smaller carbohydrates; one sugar unit per molecule.

**multiple bond** a covalent chemical bond formed by the sharing of two or more pairs of electrons between two bonded atoms.

**negative catalyst** *see inhibitor.*

**net ionic equation** an ionic equation from which all spectators have been removed.

**network solid** a crystal made up of a large, indefinite number of atoms covalently bonded to each other to form a huge molecule. Also called a *macromolecular crystal.*

**neutralization** the reaction between an acid and a base to form a salt and water; any reaction between an acid and a base.

**neutron** an electrically neutral subatomic particle having a mass of $1.7 \times 10^{-24}$ gram, approximately equal to the mass of a proton, or 1 atomic mass unit, found in the nucleus of the atom.

**newton** the SI unit of force, equal to $kg \cdot m^2/s^2$.

**noble gas** the name of the chemical family of relatively unreactive elemental gases appearing in Group 8A/18 of the periodic table.

**nomenclature** a system of names used in a particular science.

**nonconductor** a substance that does not readily convey electricity.

**nonelectrolyte** a substance that, when dissolved, yields a solution that is a nonconductor of electricity; a solution or other fluid that does not conduct electricity by ionic movement.

**nonmetal** a substance that possesses nonmetallic properties, such as being dull and brittle, and a poor conductivity of heat and electricty; an element that forms covalent bonds or gains electrons to form monatomic anions.

**nonpolar** pertaining to a bond or molecule having a symmetrical distribution of electric charge.

**normal alkane** a straight-chain alkane.

**normal boiling point** the temperature at which a substance boils in an open vessel at one atmosphere pressure.

**normality** solution concentration in equivalents per liter.

**N-terminal** the end of a polypeptide or protein chain terminated by an amino acid with a free amine group.

**nuclear bombardment** direction of high-energy particles such as protons, neutrons, and alpha particles at a larger nucleus.

**nuclear charge** the electrical charge due to the protons in the nucleus of an atom.

**nuclear model of the atom** a model of the atom that features a relatively tiny, dense nucleus that contains most of the mass of the atom and all of its positive charge.

**nuclear symbol** a symbol for an isotope of an element in the form $^{\text{mass number}}_{\text{atomic number}}\text{Sy}$, where Sy is the chemical symbol of the element.

**nucleotide** compound consisting of a nitrogen-containing base, a sugar, and one or more phosphate groups. Nucleotides are the monomers for the polymers DNA and RNA. DNA contains the sugar deoxyribose; RNA contains the sugar ribose.

**nucleus** the extremely dense central portion of the atom that contains the neutrons and protons that constitute nearly all the mass of the atom and all of the positive charge.

**nuclide** an atomic nucleus, typically identified by its atomic number and mass number.

**octet rule** the general rule that atoms tend to form stable bonds by sharing or transferring electrons until the atom is surrounded by a total of eight electrons.

**oil** an ester formed from glycerol and three fatty acids. Oils are liquids at room temperature. Also called *triacylglycerols* or *triglycerides.*

**orbit** *see electron orbit.*

**orbital** *see electron orbital.*

**organic chemistry** the chemistry of carbon compounds other than CO, $CO_2$, $CN^-$, and $CO_3^{2-}$.

**overlap (orbital)** the merging of two atomic orbitals and their associated electrons to create a covalent bond.

**oxidation** chemical reaction with oxygen; a chemical change in which the oxidation number (state) of an element is increased; also, the loss of electrons in a redox reaction.

**oxidation number** a number assigned to each element in a compound, ion, or elemental species by an arbitrary set of rules. Its two main functions are to organize and simplify the study of oxidation–reduction reactions and to serve as a base for one branch of chemical nomenclature.

**oxidation–reduction reaction** a chemical change in which electrons are transferred from one species to another.

**oxidation state** *see oxidation number.*

**oxidizer, oxidizing agent** the substance that takes electrons from another species, thereby oxidizing it.

**oxyacid** an acid that contains oxygen.

**oxyanion** an anion that contains oxygen.

**partial pressure** the pressure one component of a mixture of gases would exert if it alone occupied the same volume as the mixture at the same temperature.

**particle accelerator** a device that uses electrical fields to increase the kinetic energy of charged particles that bombard nuclei.

**particulate** consideration of matter on a scale too small to be observable with the human eye or a conventional microscope; it must be modeled.

**pascal** (pressure unit) one newton per square meter.

**Pauli exclusion principle** the principle that says, in effect, that no more than two electrons can occupy the same orbital.

**peptide** an amino acid polymer typically containing 50 or fewer amino acids. Also called *polypeptide.*

**peptide linkage** a covalent bond formed by the reaction of the carboxyl group of one amino acid and the amine group of another, yielding a peptide and water.

**percent** the amount of one part of a mixture per 100 total parts in the mixture.

**percentage composition** the percentage by mass of each element in a compound.

**percentage concentration by mass** grams of solute per 100 grams of solution.

**percent yield** the actual yield of a chemical reaction expressed as a percentage of theoretical yield.

**per expression (*PER*)** in a dimensional analysis setup, two quantities that are directly proportional to one another.

**period (periodic table)** a horizontal row of the periodic table.

**periodic table** a table of chemical elements arranged in order of increasing atomic number with elements with similar chemical properties arranged in vertical columns.

**pH** a way of expressing hydrogen-ion concentration; the negative of the logarithm of the hydrogen-ion concentration.

**phase** a visibly distinct part of a heterogeneous sample of matter.

**phospholipid** a class of lipids that have an alcohol backbone, two fatty acid residues, and a phosphate group, with a polar head and a long, nonpolar tail.

**phosphorescence** emission of light by a substance after absorbing energy from a source of electromagnetic radiation; emission may continue for some time after exposure to the energy source stops.

**photon** a massless quantum particle that carries the electromagnetic force.

**physical change** a change in the physical form of a substance without changing its chemical identity.

**physical properties** properties of a substance that can be observed and measured without changing the substance chemically.

**planetary model of the atom** a model of the atom that features electrons orbiting the nucleus as planets orbit the sun.

**pOH** a way of expressing hydroxide-ion concentration; the negative logarithm of the hydroxide-ion concentration.

**polar** pertaining to a bond or molecule having an asymmetrical distribution of electric charge.

**polyatomic** pertaining to a species consisting of more than one atom; usually said of polyatomic ions.

**polycrystalline** a solid made of small crystals arranged in a random or a directed order.

**polymer** a chemical compound formed by bonding two or more monomers. Polymers with molar mass $>$ 5000 g/mol are also called *macromolecules.*

**polymerization** the reaction in which monomers combine to form polymers.

**polypeptide** an amino acid polymer typically containing fewer than about 50 amino acid residues.

**polyprotic acid** an acid capable of yielding more than one proton per molecule on complete ionization.

**polysaccharide** complex carbohydrates made up of many monosaccharides; many sugar units per molecule.

**positron** a subatomic particle with a charge of +1 and the same mass as an electron.

**potential energy** energy possessed by a body by virtue of its position in an attractive and/or repulsive force field.

**precipitate** a solid that forms when two solutions are mixed.

**pressure** force per unit area.

**primary standard** a soluble solid of reasonable cost used in a titration that is very stable and pure, preferably with a high molar mass, that can be weighed accurately.

**primary structure** the amino acid sequence in a protein.

**principal energy level(s)** the main energy levels within the electron arrangement in an atom. They are quantized by a set of integers beginning at $n = 1$ for the lowest level, $n = 2$ for the next, and so forth; also called the *principal quantum number.*

**principal quantum number (*n*)** *see principal energy levels.*

**product** a substance formed as a result of a chemical change.

**proportionality constant** the nonzero constant k in the equation that expresses the relationship between two variables x and y, y = kx.

**proportional reasoning** the thinking skill that involves recognizing relationships of the type $y \propto x$, $y \propto 1/x$, $y = mx$, and $y = m(1/x)$, and comparison of proportions.

**protein** an amino acid polymer typically containing more than 50 amino acids.

**proton** a subatomic particle carrying a unit positive charge and having a mass of $1.7 \times 10^{-24}$ gram, almost the same as the mass of a neutron, found in the nucleus of the atom.

**proton-transfer reaction** a chemical change in which a proton is transferred from one substance, the acid, to another, the base.

**pure substance** a sample consisting of only one kind of matter, either compound or element.

**quantization of energy** the existence of certain discrete electron energy levels within an atom such that electrons may have any one of these energies but no energy between two such levels.

**quantum jump (leap)** movement of an electron within an atom from one energy level to another without measurable passage through energy levels in between.

**quantum mechanical model of the atom** an atomic concept that recognizes four quantum numbers by which electron energy levels may be described.

**quaternary structure** the arrangement of polypeptide chains in a protein composed of more than one chain.

**R** a symbol used to designate any alkyl group; the ideal gas constant, having a value of 0.0821 L · atm/mol · K.

**radioactive decay series** a description of a series of nuclear reactions, starting from a radioacitve isotope and ending with a stable isotope.

**radioactivity** the spontaneous emission of rays and/or particles from an atomic nucleus.

**radiocarbon dating** a method of determining the age of the remains of plants and animals that is based on comparing the carbon-14-to-carbon-12 ratio in the atmosphere to the same ratio in the object to be dated.

**radionuclide** a radioactive atomic nucleus.

**reactant** a substance that will be destroyed in a chemical change.

**reactivity** the relative capacity of a species to undergo chemical change.

**redox** a term coined from REDuction–OXidation to refer to oxidation–reduction reactions.

**reducer, reducing agent** the substance that loses electrons to another species, thereby reducing it.

**reduction** a chemical change in which the oxidation number (state) of an element is reduced; also, the gain of electrons in a redox reaction.

**rem** roentgen equivalent in man; a unit of measure of radiation dose that accounts for both the absorbed dose and a quality factor that depends on the type of radiation.

**representative element** *see main group element.*

**resonance structures** two or more equivalent Lewis structures for a molecule or ion that are created by changing only the positions of the electrons, where the actual species is an average of the resonance structures.

**reversible reaction** a chemical reaction in which the products may react to re-form the original reactants.

**ribonucleic acid (RNA)** a nucleotide (q.v.) polymer found in the nucleus and other parts of the cell.

**ribosome** structure found in the cell outside the nucleus where protein synthesis occurs based on the coding of the messenger RNA.

**round off** to express as a number with fewer digits.

**Rutherford scattering experiments** experiments conducted by Ernest Rutherford and collaborators that provided evidence for the nuclear model of the atom.

**salt** the product of a neutralization reaction other than water; an ionic compound that does not contain the hydrogen ion, $H^+$, the oxide ion, $O^{2-}$, or the hydroxide ion, $OH^-$.

**saturated hydrocarbon** a hydrocarbon that contains only single bonds, in which each carbon atom is bonded to four other atoms.

**saturated solution** a solution of such concentration that it is or would be in a state of equilibrium with excess solute present.

**scientific notation** *see exponential notation.*

**scintillation counter** an instrument used to measure radioactivity that is based on a transparent solid that phosphoresces after absorbing energy from radiation.

**secondary structure** the regular local conformational structures $\alpha$-helicies and $\beta$-pleated sheets in a protein that reflect a maximum amount of hydrogen bonding.

**semimetal** *see metalloid.*

**significant figures** the digits in a measurement that are known to be accurate plus one uncertain digit.

**silk fibroin** a type of protein consisting of layers of antiparallel $\beta$-pleated sheets.

**single bond** a covalent chemical bond formed by the sharing of one pair of electrons between two bonded atoms.

**single-replacement equation (reaction)** a chemical equation with the form $A + BX \rightarrow AX + B$, where the reactants are an element and a compound, and the element appears to replace one of the ions in the compound.

**SI unit** a unit associated with the International System of Units.

**solid** the state of matter characterized by particles vibrating in fixed positions and macroscopic-level constant shape and constant volume.

**solubility** the quantity of solute that will dissolve in a given quantity of solvent or in a given quantity of solution, at a specified temperature, to establish an equilibrium between the solution and excess solute; frequently expressed in grams of solute per 100 grams of solvent.

**solubility product constant** *under K, see* $K_{sp}$.

**soluble** pertaining to a substance that will dissolve in a suitable solvent.

**solute** the substance dissolved in the solvent, sometimes not clearly distinguishable from the solvent (see below), but usually the lesser of the two.

**solution** a homogeneous mixture of two or more substances of molecular or ionic particle size, the concentration of which may be varied, usually within certain limits.

**solution inventory** a precise identification of the chemical species present in a solution, in contrast with the solute from which they may have come; that is, sodium ions and chloride ions, rather than sodium chloride.

**solvent** the medium in which the solute is dissolved; *see solute*.

**space-filling model** a three-dimensional representation of a molecule representing its outer boundaries using spheres of different color to represent different atoms.

**specific gravity** the ratio of the density of a substance to the density of some standard, usually water at 4°C.

**specific heat** the quantity of heat required to raise the temperature of one gram of a substance one degree Celsius.

**spectator (ion)** a species present at the scene of a reaction but not participating in it.

**spectrometer** a laboratory instrument used to analyze spectra.

**spectrum** (plural **spectra**) the result of dispersing a beam of light into its component colors; also the result of dispersing a beam of gaseous ions into its component particles, distinguished by mass and electric charge.

**speed of light (c)** the speed of all electromagnetic radiation in open space: 299,792,458 m/s.

**spontaneous** pertaining to a change that appears to take place by itself, without outside influence.

**stable** pertaining to that which does not change spontaneously.

**standardize** determination of the concentration of a solution to be used in a titration by titrating it against a primary standard.

**standard temperature and pressure (STP)** arbitrarily defined conditions of temperature (0°C) and pressure (1 bar) (IUPAC definition) at which gas volumes and quantities are frequently measured and/or compared.

**state of matter** one of four principal conditions in which matter typically exists: gas, liquid, solid, and plasma.

**state symbol** symbols used in chemical equations to indicate the physical state of the species: (s) for solid, ($\ell$) for liquid, (g) for gas, and (aq) for aqueous.

**static electricity** an accumulation of stationary electric charge.

**step-growth polymer** a polymer produced by the stepwise reaction of monomers with the growing chain, where each bond is formed independently of the others.

**steroid** any of a group of lipids that have 17 carbon atoms arranged in three cyclohexane rings and one cyclopentane ring fused together within the larger molecule.

**stoichiometry** the quantitative relationships among the substances involved in a chemical reaction, established by the equation for the reaction.

**STP** abbreviation for standard temperature and pressure (see above).

**strong acid** an acid that ionizes almost completely in aqueous solution; an acid that loses its protons readily.

**strong base** a soluble compound that produces $OH^-$ when dissolved in $H_2O$; a base that has a strong attraction for protons.

**strong electrolyte** a substance that, when dissolved, yields a solution that is a good conductor of electricity because of nearly complete ionization or dissociation.

**strong oxidizer (oxidizing agent)** an oxidizer that has a strong attraction for electrons.

**strong reducer (reducing agent)** a reducer that releases electrons readily.

**structural formula (diagram)** a Lewis diagram (q.v.) without unshared electron pairs.

**subatomic particle** a particle smaller than an atom; in chemistry, electron, neutron, and proton.

**sublevels** the levels into which the principal energy levels are divided according to the quantum mechanical model of the atom; usually specified *s*, *p*, *d*, and *f*.

**substitution reaction** a reaction in which a hydrogen atom in an alkane is replaced by an atom of another element.

**substrate** the reactant molecule that an enzyme helps convert to products.

**sucrose** a disaccharide made from glucose and fructose with the formula $C_{12}H_{22}O_{11}$ that is commonly used as table sugar.

**supersaturated** pertaining to a state of solution concentration that is greater than the equilibrium concentration (solubility) at a given temperature and/or pressure.

**surface tension** a property of the surface of a liquid that minimizes the area of the surface and causes the surface to act like an elastic membrane.

**suspension** a mixture that gradually separates by settling.

**symbol (chemical)** a one- or two-letter abbreviation used to represent in writing the name of a chemical element.

**terminal atom** (Lewis diagram) an atom that is bonded to only one other atom.

**tertiary structure** the three-dimensional structure of a protein.

**tetrahedral** related to a tetrahedron; usually used in reference to the orientation of four covalent bonds radiating from a central atom toward the vertices of a tetrahedron, or to the 109.5° angle formed by any two corners of the tetrahedron and the central atom as its vertex; arrangement of four electron pairs surrounding a central atom in a molecule or ion to form a 109.5° angle between any two pairs and the central atom (electron-pair geometry); arrangement of four electron pairs and four bonded atoms surrounding a central atom in a molecule or ion to form a 109.5° angle between any two bonded atoms and the central atom (molecular geometry).

**tetrahedron** a regular four-sided solid having congruent equilateral triangles as its four faces.

**thermal** pertaining to heat.

**thermochemical equation** a chemical equation that includes an energy term, or for which $\Delta H$ is indicated.

**thermochemical stoichiometry** stoichiometry expanded to include the energy involved in a chemical reaction, as defined by the thermochemical equation.

**thermoplastic** a polymer that can be easily melted and molded into new shapes because it has weak attractive forces between non-cross-linked polymer chains.

**thermoset** a polymer with a relatively rigid structure because of cross-links between carbon chains.

**titration** the controlled and measured addition of one solution into another.

**torr** a unit of pressure essentially equal to the pressure unit millimeter of mercury (mm Hg).

**total ionic equation** a chemical equation written with ion formulas for soluble aqueous ionic compounds and strong acids.

**total ionization** the removal of all ionizable hydrogen from an acid molecule or ion.

**transcription** synthesis of RNA where the genetic information encoded in DNA is copied to a complementary messenger RNA molecule.

**transfer RNA** a small RNA molecule that carries with it a specific amino acid that is transferred to a polypeptide chain as the transfer RNA molecule bonds to a complementary site on messenger RNA in a ribosome.

**transition element, transition metal** an element from one of the B groups (IUPAC Groups 3–12) of the periodic table.

**translation** the process that occurs in a ribosome where the genetic information in messenger RNA is decoded to produce a protein molecule.

**transmutation** conversion of an atom from one element to another by means of a nuclear change.

**transuranium elements** elements whose atomic numbers are greater than 92.

**trigonal (triangular) planar** arrangement of three electron pairs surrounding a central atom in a molecule or ion to form a 120° angle between any two pairs and the central atom (electron-pair geometry); arrangement of three electron pairs and three bonded atoms surrounding a central atom in a molecule or ion to form a 120° angle between any two bonded atoms and the central atom (molecular geometry).

**trigonal (triangular) pyramidal** arrangement of four electron pairs and three bonded atoms surrounding a central atom in a molecule or ion to form a 109.5° angle between any two bonded atoms and the central atom (molecular geometry).

**triple bond** a covalent chemical bond formed by the sharing of three pairs of electrons between two bonded atoms.

**triprotic acid** an acid capable of yielding three protons in complete ionization.

**uncertain digit** the digit in a measured quantity that cannot be accurately measured; the last digit written when expressing a measured quantity.

**uncertainty (in measurement)** that which is not accurately measurable.

**unit path (*PATH*)** in a dimensional analysis setup, the sequence of units to be followed to convert between the given quantity (*GIVEN*) and the wanted quantity (*WANTED*).

**universal gas constant (R)** the proportionality constant equal to the product of pressure and volume divided by the product of number of particles and absolute temperature for an ideal gas; 0.0821 L · atm/mol · K = 62.4 L · torr/mol · K = 8.314 J/mol · K.

**unsaturated hydrocarbon** a hydrocarbon that contains one or more multiple bonds.

**unsaturated solution** a solution with a concentration that is less than the solubility limit.

**valence electrons** the highest-energy *s* and *p* electrons in an atom, which determine the bonding characteristics of an element.

**valence shell electron-pair repulsion theory (VSEPR)** a model used to predict and explain molecular geometry that is based on mutual repulsion among electron pairs (or multiple bonds) in the valence shell of a central atom in a molcule or ion and the resulting tendency of the electron pairs to minimize potential energy by distributing themselves as far away from each other as possible.

**van der Waals forces** a general term for all kinds of weak intermolecular attractions.

**vapor** a gas.

**vaporize, vaporization** changing from a solid or liquid to a gas.

**vapor pressure** the pressure or partial pressure exerted by a vapor that is in contact with its liquid phase. The term often refers to the pressure or partial pressure of a vapor that is in equilibrium with its liquid state at a given temperature.

**viscosity** the resistance of a liquid to flow.

**volatile** that which vaporizes easily.

**voltaic cell** *see cell, voltaic.*

**wanted quantity (*WANTED*)** in a dimensional analysis setup, the quantity asked for in the problem statement that will result from an equivalent amount of another unit.

**water (equilibrium) constant ($K_w$)** $K_w = [H^+][OH^-] = 1.0 \times 10^{-14}$ at 25°C.

**water of crystallization, water of hydration** water molecules that are included as structural parts of crystals formed from aqueous solutions.

**wave equation** an equation that describes the propagation of waves.

**wave-particle duality** all matter and energy possess both wave-like and particle-like properties.

**wax** various substances that are mixtures that contain esters of fatty acids and are solid at room temperature, malleable, and insoluble in water.

**weak acid** an acid that ionizes only slightly in aqueous solution; an acid that does not donate protons readily.

**weak base** a base that dissociates only slightly in aqueous solution; a base that has a weak attraction for protons.

**weak electrolyte** a substance that, when dissolved, yields a solution that is a poor conductor of electricity because of limited ionization or dissociation.

**weak oxidizer (oxidizing agent)** an oxidizer that has a weak attraction for electrons.

**weak reducer (reducing agent)** a reducer that does not release electrons readily.

**weight** a measure of the force of gravitational attraction.

**yield** the amount of product from a chemical reaction.

**Z** atomic number.

# Index

Page numbers in *italics* indicate figures and illustrations; page numbers followed by "n" indicate footnotes; page numbers followed by "t" indicate tables.

# Chapter in Review 2

| | |
|---|---|
| GOAL 1 **Identify and explain the differences among observations of matter at the macroscopic, microscopic, and particulate levels.** | **Matter**, anything that has mass and takes up space; can be studied at three levels: macroscopic (seen with human eye), microscopic (seen with light microscope), and particulate (cannot be directly seen). |
| GOAL 2 **Define the term *model* as it is used in chemistry to represent pieces of matter too small to see.** | Chemists use symbols and **models** to represent particulate matter: chemical formula, Lewis diagram, ball-and-stick model, space-filling model. |
| GOAL 3 **Identify and explain the differences among gases, liquids, and solids in terms of (a) visible properties, (b) distance between particles, and (c) particle movement.** | **Kinetic molecular theory:** Particles of matter always moving. *Kinetic* means motion. **Gas:** variable shape and volume, independent particle movement. **Liquid:** variable shape, constant volume, independent particle movement beneath the surface. **Solid:** constant shape and volume, particles vibrate in fixed positions. |
| GOAL 4 **Distinguish between physical and chemical properties at both the particulate level and the macroscopic level.** | Matter is described by its properties, which may be physical or chemical. Changes in matter may also be physical or chemical. **Physical property:** can be observed and measured without a chemical change. |
| GOAL 5 **Distinguish between physical and chemical changes at both the particulate level and the macroscopic level.** | **Physical change:** change in form with no change in identity. **Chemical change:** identity of original substance changes to something new. |
| GOAL 6 **Distinguish between a pure substance and a mixture at both the macroscopic level and the particulate level.** | **Pure substance:** one chemical with distinct set of physical and chemical properties; cannot be separated by physical changes. **Mixture:** two or more pure substances; properties vary, depending on relative amounts of pure substances; components can be separated by physical changes. |
| GOAL 7 **Distinguish between homogeneous and heterogeneous matter.** | **Homogeneous:** same appearance, composition, properties throughout. **Solution:** a homogenous mixture. **Heterogeneous:** different phases visible, variable properties in different parts of sample. |
| GOAL 8 **Describe how distillation and filtration rely on physical changes and properties to separate components of mixtures.** | Separations usually based on different physical properties of components. **Distillation** separates based on volatilities of components. Water boiled (physical change) to separate it from mixture. **Filtration** separates based on particle sizes. A porous medium is used to separate mixture components based on size (a physical property). |
| GOAL 9 **Distinguish between elements and compounds.** | **Element:** pure substance, cannot be decomposed chemically into other pure substances. **Compound:** pure substance that can be decomposed chemically into other pure substances. |
| GOAL 10 **Distinguish between elemental symbols and the formulas of chemical compounds.** | **Elemental symbol:** capital letter, sometimes followed by small letter. **Formula of compound:** elemental symbols of elements in compound. Subscripts show number of atoms of each element. |
| GOAL 11 **Distinguish between atoms and molecules.** | **Atom:** the smallest unit particle of an element. **Molecule:** the smallest unit particle of a pure substance that can exist independently and retain the identity of that substance. |
| GOAL 12 **Match electrostatic forces of attraction and repulsion with combinations of positive and negative charge.** | Objects with the same charge repel. Objects with opposite charges attract. |
| GOAL 13 **Distinguish between reactants and products in a chemical equation.** | **Equation:** reactants $\rightarrow$ products |
| GOAL 14 **Distinguish between exothermic and endothermic changes.** | **Exothermic:** energy released; **endothermic:** energy absorbed. |

| | |
|---|---|
| GOAL 15 **Distinguish between potential energy and kinetic energy.** | **Potential energy:** energy because of position in force field. Tendency toward reducing energy to the smallest amount possible is a driving force for chemical reactions. **Kinetic energy:** energy of motion. |
| GOAL 16 **State the meaning of, or draw conclusions based on, the Law of Conservation of Mass.** | **The Conservation Law:** Mass can be changed to energy. Total mass + energy is constant. **The Law of Conservation of Mass:** Mass conserved in chemical change; neither created nor destroyed. |
| GOAL 17 **State the meaning of, or draw conclusions based on, the Law of Conservation of Energy.** | **The Law of Conservation of Energy:** Energy is conserved in ordinary change; neither created nor destroyed. |

## Chapter 2 Test Yourself

1) Copper, Cu, has the following properties: (i) does not react with neon; (ii) density of 8.92 grams per cubic centimeter. Identify the *correct* statement among the following:
   a. Both (i) and (ii) are physical properties
   b. (i) is a physical property; (ii) is a chemical property
   c. (i) is a chemical property; (ii) is a physical property
   d. Both (i) and (ii) are chemical properties
2) All of the following are physical changes *except:*
   a. defrosting a frozen pizza
   b. putting extra garlic on the pizza
   c. slicing a cooked pizza
   d. digesting the pizza
3) Which state of matter expands to fill the container that holds it?
   a. solid b. liquid c. gas
4) Particle movement is most restricted in which state?
   a. solid b. liquid c. gas
5) Among the following, identify the *incorrect* classification:
   a. A flawless diamond is homogeneous
   b. Raisin bread is heterogeneous
   c. Filtered spring water is homogeneous
   d. A bubbly carbonated beverage is homogeneous
6) Identify the *incorrect* statement among the following:
   a. Water is a pure substance, but salt water is a mixture
   b. Air is a mixture, but oxygen is a pure substance
   c. Salt is a pure substance; sugar is a pure substance; but salt and sugar together are a mixture
   d. Nitrogen and hydrogen are both pure substances, but ammonia, the compound formed by them, is a mixture
7) All of the following are compounds *except:*
   a. carbon dioxide b. sodium chloride c. hydrogen d. water
8) In the chemical equation given below, identify the reactants and the products, then state which are elements and which are compounds:

   $H_2 + I_2 \rightarrow 2\ HI$ (hydrogen and iodine react to form hydrogen iodide)
9) In ordinary chemical and physical changes, the mass of the reactants is _____ the mass of the products.
   a. always less than b. always more than c. usually less than d. the same as
10) The electrostatic force between a positively charged object and a negatively charged object is one of:
    a. attraction b. repulsion c. neither attraction nor repulsion
11) As you make a cup of coffee, you first heat the water to brew the coffee, then let the mixture cool so you can drink it. The physical changes undergone by the water are:
    a. endothermic then exothermic
    b. endothermic then endothermic
    c. exothermic then endothermic
    d. exothermic then exothermic
12) As you catch a ball that was thrown very high, the potential energy of the ball just before the catch is _____; the kinetic energy of the ball just before the catch is _____.
    a. high; high b. high; low c. low; high d. low; low
13) In a nonnuclear chemical or physical change, the _____ energy is always conserved, although its _____ sometimes changes.
    a. form of; total b. total; form c. kinetic; potential energy d. potential; kinetic energy

**Answers to Chapter 2 Test Yourself**

1) c 2) d 3) c 4) a 5) d 6) d 7) c

8) The reactants are hydrogen, $H_2$, and iodine, $I_2$; the product is hydrogen iodide, HI. The $H_2$ and the $I_2$ are both elements; the HI is a compound.

9) d 10) a 11) a 12) c 13) b

# Chapter in Review 3A

| | |
|---|---|
| GOAL 1 **Write in exponential notation a number given in ordinary decimal form; write in ordinary decimal form a number given in exponential notation.** | Any decimal number can be written in **exponential notation**. Exponential notation expresses a number as a coefficient C (between 1 and 9.99 . . .) multiplied by 10 raised to the e power, in general, $C \times 10^e$. When e is larger than 0, $10^e$ is larger than 1; when e is smaller than 0, $10^e$ is smaller than 1. (Remember that $10^0 = 1$.) |
| GOAL 2 **Using a calculator, add, subtract, multiply, and divide numbers expressed in exponential notation.** | To add, subtract, multiply, or divide numbers in exponential notation, following the instructions that are appropriate for your calculator. |
| GOAL 3 **In a problem, identify given and wanted quantities that are related by a *PER* expression. Set up and solve the problem by dimensional analysis.** | (1) Identify and write down the *GIVEN* quantity, including units; (2) Identify and write down the units of the *WANTED* quantity; (3) Write the *PER/PATH*; (4) Write the calculation setup, including units; (5) Calculate the answer; (6) Check the answer to be sure both the number and the units make sense. |
| GOAL 4 **Distinguish between mass and weight.** | The **mass** of an object *does not change* in different gravitational fields; the **weight** of that object *does change.* |
| GOAL 5 **Identify the metric units of mass, length, and volume.** | The SI metric unit of mass is the **kilogram, kg**. The metric unit of length is the **meter, m.** The SI unit of volume is the **cubic meter, $m^3$.** |
| GOAL 6 **State and write with appropriate metric prefixes the relationship between any metric unit and its corresponding kilounit, centiunit, and milliunit.** | The important metric prefixes for this course are *kilo-* (1000), *centi-* (0.01) and *milli-* (0.001). |
| GOAL 7 **Using Table 3.1, state and write with appropriate metric prefixes the relationship between any metric unit and other larger and smaller metric units.** | Because the metric system is decimal based, conversions between larger and smaller metric units involve simply moving the decimal point. |
| GOAL 8 **Given a mass, length, or volume expressed in metric units, kilounits, centiunits, or milliunits, express that quantity in the other three units.** | Use dimensional analysis to convert metric units. |
| GOAL 9 **State the number of significant figures in a given quantity.** | The number of **significant figures** in a measurement is the number of digits that are known accurately plus one digit that is uncertain. |
| GOAL 10 **Round off given numbers to a specified number of significant figures.** | To round off a number to the proper number of significant figures, leave the uncertain digit unchanged if the digit to its right is less than 5. Increase the uncertain digit by one if the digit to its right is 5 or greater. |
| GOAL 11 **Add or subtract given quantities and express the result in the proper number of significant figures.** | In addition and subtraction, round off a sum or difference to the first column that has an uncertain digit. |
| GOAL 12 **Multiply or divide given measurements and express the result in the proper number of significant figures.** | In multiplication and division, round off a product or quotient to the same number of significant figures as the smallest number of significant figures in any factor. |
| GOAL 13 **Given a metric–USCS conversion table and a quantity expressed in any unit in Table 3.2, express that quantity in corresponding units in the other system.** | Use dimensional analysis to convert metric units tao USCS units and vice versa. |
| GOAL 14 **Given a temperature in either Celsius or Fahrenheit degrees, convert it to the other scale.** | The Fahrenheit and Celsius temperature scales are related by the equation $\mathbf{T_{°F} - 32 = 1.8\, T_{°C}}$. |

GOAL 15 **Given a temperature in Celsius degrees or kelvins, convert it to the other scale.**

The Kelvin and Celsius temperature scales are related by the equation $\mathbf{T_K = T_{°C} + 273}$.

GOAL 16 **Write a mathematical expression indicating that one quantity is directly proportional to another quantity.**

If y is directly proportional to x, it is expressed mathematically as $y \propto x$.

GOAL 17 **Use a proportionality constant to convert a proportionality to an equation.**

To convert the proportionality $y \propto x$ into an equation, insert a proportionality constant, k: $y = kx$.

GOAL 18 **Given the values of two quantities that are directly proportional to each other, calculate the proportionality constant, including its units.**

If $y = kx$, then $k = y \div x$.

GOAL 19 **Write the defining equation for a proportionality constant and identify units in which it might be expressed.**

Units are set by definition and the defining equation. For example, for density, the units are mass units over volume units. Examples: $kg/m^3$, $g/cm^3$, g/mL, g/L.

GOAL 20 **Given two of the following for a sample of a pure substance, calculate the third: mass, volume, and density.**

The defining equation for density is: $\textbf{Density} \equiv \frac{\textbf{mass}}{\textbf{volume}}$. Because the defining equation for density is a *PER* expression, density problems may be solved by either dimensional analysis or algebra.

# Chapter in Review 3B

## Chapter 3 Test Yourself

1) Where decimal numbers are given, write exponential numbers; where exponential numbers are given, write their decimal equivalents:
(a) 413,400 (b) $6.91 \times 10^{7}$ (c) 0.00103 (d) $1.47 \times 10^{-4}$

2) Perform the following operations; leave 3 digits in your answers:
(a) $4.1 \times 10^{-6} + 1.59 \times 10^{-5} =$ (b) $6.7 \times 10^{3} + 2.61 \times 10^{4} =$
(c) $7.14 \times 10^{3} - 3.9 \times 10^{2} =$ (d) $8.34 \times 10^{-1} - 3.6 \times 10^{-2} =$

3) Perform the following operations; leave 3 digits in your answers:
(a) $(1.16 \times 10^{-3})(6.32 \times 10^{-11}) =$ (b) $(4.62 \times 10^{-6})(2.17 \times 10^{8}) =$
(c) $\dfrac{(9.76 \times 10^{7})(8.17 \times 10^{3})}{(1.23 \times 10^{1})} =$ (d) $\dfrac{-4.39 \times 10^{4}}{(107)(7.11 \times 10^{1})} =$

4) The Lagrange points are points in space between the earth and moon; at these points the gravity of the earth exactly cancels the gravity of the moon. What would be the mass and the weight of a 70 kg person at one of these points? (453.59 g = 1 lb)

5) How many significant figures are in the measured quantity 0.099 gram?

6) Round off 2.6034 kilometers to three significant figures.

7) Express the following sum to the correct number of significant figures. All numbers represent measured quantities.
$16.08 + 0.043 + 121.80 + 7.99463 =$

8) Express the following to the correct number of significant figures. All numbers are from measurements.
$2.193 \times \dfrac{5.876}{4.88} \times \dfrac{0.065}{64.06} =$

*Questions 9–14: A correct setup, beginning with the given quantity, is required for a correct answer.*

9) How many dollars can you earn in a part-time job in three months if your hourly wage is \$6.45, you average 13.6 hours per week, and there are 4.33 weeks in a month?

10) If 2.54 cm ≡ 1 inch, how many centimeters are in 45.0 inches?

11) How many millimeters are in 40.1 meters?

12) Convert 9.45 kilometers to meters.

13) State the number of quarts in 3440 $cm^3$, if 1 liter = 1.06 quart.

14) How many ounces are in 439 centigrams?

15) A temperature of 14°F is what temperature in °C?

16) A temperature of 71°C is what temperature in °F?

17) A temperature of 312°C is what temperature in kelvins?

18) A 7.6 $cm^3$ piece of metal has a mass of 65.588 g. Calculate its density.

19) Find the mass of 58.8 mL of a solution having a density of 1.16 g/mL.

**Answers to Chapter 3 Test Yourself**

1) (a) $4.13400 \times 10^{5}$ (b) 69,100,000 (c) $1.03 \times 10^{-3}$ (d) 0.000147 2) (a) $2.00 \times 10^{-5}$ (b) $3.28 \times 10^{4}$ (c) $6.75 \times 10^{3}$ (d) $7.98 \times 10^{-1}$ 3) (a) $7.33 \times 10^{-14}$ (b) $1.00 \times 10^{3}$ (c) $6.48 \times 10^{10}$ (d) −5.77 4) At the Lagrange points, as well as anywhere else in the universe, a 70 kg person has a mass of 70 kg. When there is no gravity, the person has no weight. 5) The measured quantity 0.099 gram has two significant figures. 6) To three significant figures, 2.6034 kilometers rounds to 2.60 kilometers. 7) 145.92 8) 0.0027 9) 1139 dollars 10) 114 cm 11) $4.01 \times 10^{4}$ mm 12) $9.45 \times 10^{3}$ m 13) 3.65 qt 14) 0.155 oz 15) −10°C 16) 160°F 17) 585 K 18) 8.6 $g/cm^3$ 19) 68.2 g

# Chapter in Review 4

GOAL 1 **Describe five macroscopic characteristics of gases.**

Five **macroscopic characteristics of gases** are

1. Gases may be compressed.
2. Gases expand to fill their containers uniformly.
3. All gases have low density.
4. Gases may be mixed.
5. A confined gas exerts constant pressure on the walls of its container uniformly in all directions.

GOAL 2 **Explain or predict physical phenomena relating to gases in terms of the ideal gas model.**

The five main features of the **ideal gas model** are

1. Gases consist of particles moving at any given instant in straight lines.
2. Molecules collide with each other and with the container walls without loss of total kinetic energy.
3. Gas molecules are very widely spaced.
4. The actual volume of molecules is negligible compared to the space they occupy.
5. Gas molecules behave as independent particles; attractive forces between them are negligible.

GOAL 3 **Given a gas pressure in atmospheres, torr, millimeters (or centimeters) of mercury, inches of mercury, pascals, kilopascals, bars, or pounds per square inch, express that pressure in each of the other units.**

Common pressure units and their relationships to one another are

1 atm = 760 mm Hg = 760 torr = $1.013 \times 10^5$ Pa = 101.3 kPa = 1.013 bar = 29.92 in. Hg = 14.69 psi

GOAL 4 **Define *pressure* and interpret statements in which the term *pressure* is used.**

**Pressure is defined as force per unit area.** Pressure is usually measured with a manometer or its mechanical equivalent.

GOAL 5 **Given a temperature in degrees Celsius, convert it to kelvins, and vice versa.**

The relationship between temperature measured in kelvins and degrees Celsius is $\mathbf{T_K = T_{°C} + 273}$.

GOAL 6 **Describe the relationship between the volume and temperature of a fixed quantity of an ideal gas at constant pressure, and express that relationship as a proportionality, an equation, and a graph.**

**Charles's Law** states that the volume of a fixed quantity of gas at constant pressure is directly proportional to absolute temperature, $V \propto T$ and $V = kT$. The plot of volume versus temperature is a straight line that passes through the origin.

GOAL 7 **Given the initial volume (or temperature) and the initial and final temperatures (or volumes) of a fixed quantity of gas at constant pressure, calculate the final volume (or temperature).**

For a given sample of gas at constant pressure, $\frac{V_1}{T_1} = \frac{V_2}{T_2}$.

GOAL 8 **Describe the relationship between the volume and pressure of a fixed quantity of an ideal gas at constant temperature, and express that relationship as a proportionality, an equation, and a graph.**

**Boyle's Law** states that for a fixed quantity of gas at constant temperature, pressure is inversely proportional to volume, $P \propto 1/V$ and $P = k(1/V)$ or $PV = k$. The plot of pressure versus the inverse of volume is a straight line that passes through the origin.

| | |
|---|---|
| GOAL 9 **Given the initial volume (or pressure) and initial and final pressures (or volumes) of a fixed quantity of gas at constant temperature, calculate the final volume (or pressure).** | For a given sample of gas at constant temperature, $P_1V_1 = P_2V_2$. |
| GOAL 10 **For a fixed quantity of a confined gas, given the initial volume, pressure, and temperature and the final values of any two variables, calculate the final value of the third variable.** | The **Combined Gas Law** states that for a fixed quantity of gas, $\frac{P_1V_1}{T_1} = \frac{P_2V_2}{T_2}$. Given the initial (or final) values of all three variables and the final (or initial) values of two, the unknown value is calculated with the Combined Gas Law. |
| GOAL 11 **State the values associated with standard pressure and temperature (STP) for gases.** | Standard temperature and pressure, **STP,** are defined as **273 K (0°C)** and **1 bar** pressure. |

## Chapter 4 Test Yourself

1) Pick the statement about the ideal gas model that is incorrect:
   a) The volume of the particles, or molecules, is negligible compared with the volume occupied by the gas.
   b) There are large attractive forces between molecules in an ideal gas.
   c) Intermolecular collisions occur without loss of kinetic energy.
   d) Gas molecules are in constant motion.
   e) Gas molecules are independent of each other.
2) A pressure of 0.836 atmosphere is equal to
   a) 836 torr
   b) 63.5 cm Hg
   c) 732 mm Hg
   d) 0.846 kPa
   e) none of these
3) A gas occupies 0.610 L at 0.103 atm pressure. What volume will the gas occupy at 1.62 atm pressure, if the temperature is held constant?
4) A gas occupies 2.14 L at 40°C; what volume does this gas occupy if the temperature is lowered to 20°C, with pressure remaining constant?
5) Initially a gas occupies 4.80 L at 744 torr and 32°C. What volume will it fill at 811 torr and 64°C?
6) A gas occupies 1.24 L at STP. Find the volume it will occupy at 21°C and 1.21 bar.

**Answers to Chapter 4 Test Yourself**

1) b 2) b 3) 0.0388 L 4) 2.00 L 5) 4.87 L 6) 1.10 L

# Chapter in Review 5A

GOAL 1 **Identify the main features of Dalton's atomic theory.**

The main features of **Dalton's atomic theory** are:

1) Each element is made up of tiny, individual particles called atoms.
2) Atoms are indivisible; they cannot be created or destroyed.
3) All atoms of each element are identical in every respect.
4) Atoms of one element are different from atoms of any other element.
5) Atoms of one element may combine with atoms of other elements, usually in the ratio of small, whole numbers, to form chemical compounds.

GOAL 2 **State the meaning of, or draw conclusions based on, the Law of Multiple Proportions.**

The **Law of Multiple Proportions** states that when two elements combine to form more than one compound, the different weights of one element that combine with the same weight of the other element are in a simple ratio of whole numbers.

GOAL 3 **Identify the three major subatomic particles by symbol, charge, and approximate atomic mass, expressed in atomic mass units.**

| Subatomic Particle | Symbol | Fundamental Charge | Mass (u) |
|---|---|---|---|
| Electron | $e^-$ | 1− | 0 |
| Proton | p or $p^+$ | 1+ | 1 |
| Neutron | n or $n^0$ | 0 | 1 |

GOAL 4 **Describe and/or interpret the Rutherford scattering experiments and the nuclear model of the atom.**

The **Rutherford scattering experiments** were designed so that positively charged alpha particles were directed at thin metal foils. Most particles passed through the foils, but some were deflected at large angles.

The main features of the **nuclear model of the atom** are:

1) Every atom contains an extremely small, extremely dense nucleus.
2) All of the positive charge and nearly all of the mass of an atom are concentrated in the nucleus.
3) The nucleus is surrounded by a much larger volume of nearly empty space that makes up the rest of the atom.
4) The space outside the nucleus is very thinly populated by electrons, the total charge of which exactly balances the positive charge of the nucleus.

GOAL 5 **Explain what isotopes of an element are and how they differ from each other.**

Atoms of the same element that have different masses are called **isotopes.** The mass differences between atoms of an element are caused by different numbers of neutrons.

GOAL 6 **For an isotope of any element whose chemical symbol is known, given one of the following, state the other two: (a) nuclear symbol, (b) number of protons and neutrons, (c) atomic number and mass number.**

An isotope can be represented by a **nuclear symbol** that has the form ${}^{A}_{Z}Sy$, where A is the **mass number** of the element, Z is the **atomic number** of the element, and Sy is the chemical symbol of the element. The mass number is the sum of the number of protons plus the number of neutrons. The atomic number is the number of protons.

GOAL 7 **Identify the features of Dalton's atomic theory that are no longer considered valid, and explain why.**

Two features of Dalton's atomic theory are no longer considered valid:

1) *Atoms are indivisible.* The existence of ions, charged particles that form when atoms gain or lose electrons, introduced in Chapter 6, invalidates this postulate.
2) *Atoms of an element are identical.* The existence of isotopes invalidates this postulate.

GOAL 8 **Define and use the atomic mass unit (u).**

1 **atomic mass unit (u)** $\equiv \frac{1}{12}$ the mass of one carbon-12 atom

GOAL 9 **Given the relative abundances of the natural isotopes of an element and the atomic mass of each isotope, calculate the atomic mass of the element.**

The average mass of all atoms of an element as they occur in nature is called the **atomic mass.** The atomic mass may be calculated from the masses of the natural isotopes of that element and the percentage abundance of each isotope.

GOAL 10 **Distinguish between groups and periods in the periodic table and identify them by number.**

The periodic table arranges the elements into seven **periods** (horizontal rows) and eighteen **groups** (vertical columns) in order of atomic numbers and periodic recurrence of physical and chemical properties.

GOAL 11 **Given the atomic number of an element, use a periodic table to find the symbol and atomic mass of that element, and identify the period and group in which it is found.**

Each box in the periodic table gives the atomic number (Z), the elemental symbol, and the atomic mass of an element.

GOAL 12 **Given an elemental symbol or information from which it can be identified, classify the element as either a main group or transition element and either a metal or nonmetal.**

Elements in the A groups (1, 2, and 13 to 18) of the periodic table are called **main group elements.** Elements in the B groups (3 to 12) are called **transition elements.** The stair-step line that begins between atomic numbers 4 and 5 in Period 2 and ends between 84 and 85 in Period 6 separates the **metals** on the left from the **nonmetals** on the right.

GOAL 13 **Given the name of the symbol of an element in Figure 5.9, write the other.**

The names and symbols of 35 common elements are to be learned, using the periodic table as a memory aid: H, He, Li, Be, B, C, N, O, F, Ne, Na, Mg, Al, Si, P, S, Cl, Ar, K, Ca, Cr, Mn, Fe, Co, Ni, Cu, Zn, Br, Kr, Ag, Sn, I, Ba, Hg, Pb.

# Chapter in Review 5B

## hapter 5 Test Yourself

*structions:* You may refer to a "clean" periodic table.

One of the postulates of Dalton's Atomic Theory says that atoms can neither be created nor destroyed. This postulate supports the _____.
(a) Law of Definite Composition (b) Law of Conservation of Mass
(c) existence of isotopes (d) nuclear model of Lord Rutherford

Which subatomic particle, proton, neutron, or electron, is the lightest?
(a) proton, (b) neutron, (c) electron

If a very small amount of neon gas is placed in a previously empty glass tube, which is then sealed, the neon glows when electricity is passed through the tube. While glowing, neon particles flow toward *both* the negative and positive ends of the tube. This particle flow shows that
(a) neon atoms are indivisible (b) neon is a gas made up of atoms
(c) neon atoms are electrically neutral (d) neon atoms contain positively and negatively charged parts

Rutherford's scattering experiments showed that the electrons
(a) are outside the nucleus in almost vacant space (b) are packed tightly together in the nucleus
(c) are packed tightly together outside of the nucleus (d) occupy very little space in the nucleus

Two atoms are identified by the symbols $^{22}Ne$ and $^{23}Ne$. Which of the following statements about these atoms is false?
(a) The atoms contain the same number of protons; (b) The masses of the atoms are different; (c) $^{22}Ne$ has fewer neutrons than $^{23}Ne$; (d) The number of electrons equals the number of protons in $^{22}Ne$, but not in $^{23}Ne$

Select the correct statement about $^{43}_{22}Ti$:
(a) The nucleus contains 22 protons and 26 neutrons, mass number = 48; (b) The nucleus contains 22 protons and 21 neutrons, atomic number = 22; (c) The nucleus contains 21 protons and 22 neutrons, mass number = 43; (d) The nucleus contains 22 protons and 43 neutrons, atomic number = 22

An atomic mass unit (u) is
(a) the mass of an atom in grams (b) $\frac{1}{12}$ the mass of an atom of carbon-12
(c) the mass of an atom of carbon-12 (d) the mass of an atom compared with the mass of a carbon-12 atom

The mass numbers of two natural isotopes of an imaginary element and their relative abundances are, respectively, 94.0 u—82.4% and 99.0 u—18.6%. What is the average atomic mass of this imaginary element?
(a) 94.0 u, (b) 95.9 u, (c) 96.5 u, (d) 98.6 u

Select the correct placement in the periodic table for nitrogen, N.
(a) Group 5A/15, Period 1, (b) Group 2, Period 5A/15, (c) Group 7, Period 2A/2, (d) Group 5A/15, Period 2

0) Select the correct statement about an element for which the atomic number is 16.
(a) The element is sulfur, S, for which Z = 16, Group 6A/16, Period 3; (b) The element is sulfur, S, for which A = 16, Group 6A/16, Period 3; (c) The element is oxygen, O, for which Z = 16, Group 6A/16, Period 2; (d) The element is oxygen, O, for which A = 16, Group 6A/16, Period 2

1) Write the name of the element for each symbol given, and the chemical symbol for each element given. Br, Mg, Pb, Fe, Ag, K, Hg, sodium, nickel, phosphorus, calcium, silicon, fluorine, manganese

**Answers to Chapter 5 Test Yourself**

1) b 2) c 3) d 4) a 5) d 6) b 7) b 8) b 9) d 10) a
11) Br, bromine; Mg, magnesium; Pb, lead; Fe, iron; Ag, silver; K, potassium; Hg, mercury; sodium, Na; nickel, Ni; phosphorus, P; calcium, Ca; silicon, Si; fluorine, F; manganese, Mn

# Chapter in Review 6A

**GOAL 1 Given a name or formula of an element in Figure 5.8, write the other.**

The names and formulas of the 35 elements in Figure 5.8 should already be in memory. The **seven diatomic elements** $H_2$, $N_2$, $O_2$, $F_2$, $Cl_2$, $Br_2$, and $I_2$ must be learned. Remember: **H**orses **N**eed **O**ats **F**or **Cl**ear **Br**own **I**'s.

**GOAL 2 Given the name or formula of a binary molecular compound, write the other.**

Two nonmetals or a nonmetal and a metalloid form chemical bonds with each other to form **binary molecular compounds.** The name of a binary molecular compound is the name of the first element followed by the name of the second element, modified with an *-ide* suffix. Prefixes are used to indicate the number of atoms of each element in the molecule.

Memorize the **number prefixes** used in chemical names: *mono-* = 1, *di-* = 2, *tri-* = 3, *tetra-* = 4, *penta-* = 5, *hexa-* = 6, *hepta-* = 7, *octa-* = 8, *nona-* = 9, *deca-* = 10.

**GOAL 3 Given the name or the formula of water, write the other; given the name or the formula of ammonia, write the other.**

Two common binary molecular compounds with nonsystematic names are **water, $H_2O$, and ammonia, $NH_3$.**

**GOAL 4 Given the name or formula of an ion in Figure 6.3, write the other.**

Ions are charged particles. A **cation** has a positive charge; an **anion** has a negative charge. The name of a monatomic cation is the name of the element, followed by the word *ion.* The name of a monatomic anion is the name of the element, changed to end in *-ide,* followed by the word *ion.* The formula of a monatomic ion is the symbol of the element followed by its electrical charge, written in superscript. The charge of ions formed from main-group elements corresponds to the group number: for metals, 1A/1, 1+; 2A/2, 2+; 3A/3, 3+; and for nonmetals, 5A/15, 3−; 6A/16, 2−; 7A/17, 1−. Some transition elements commonly form more than one ion. For these ions, its oxidation state is added to the elemental name. The oxidation state is written in parentheses immediately after the name. For example, the formula of the iron(II) ion is $Fe^{2+}$. Three common transition elements normally form only one ion. The charges on these ions must be memorized: nickel ion, $Ni^{2+}$; zinc ion, $Zn^{2}$; silver ion, $Ag^+$.

**GOAL 5 Given the name (or formula) of an acid or ion in Table 6.5, write its formula (or name).**

An **acid** ionizes in water to give $H^+$ and an anion. A general equation used to describe acid ionization is $HX \rightarrow H^+ +X X^-$. Five ***-ic* acids** must be memorized: carbonic acid, $H_2CO_3$; nitric acid, $HNO_3$; phosphoric acid, $H_3PO_4$; sulfuric acid, $H_2SO_4$; chloric acid, $HClO_3$. A system is used to name acids with different numbers of oxygens than the related *-ic* acids. Learn the system.

| Number of oxygen atoms compared with *-ic* acid | Acid Prefix and/or Suffix | | Anion Prefix and/or Suffix | |
|---|---|---|---|---|
| One more | *per-* | *-ic* | *per-* | *-ate* |
| Same | | *-ic* | | *-ate* |
| One fewer | | *-ous* | | *-ite* |
| Two fewer | *hypo-* | *-ous* | *hypo-* | *-ite* |
| No oxygen | *hydro-* | *-ic* | | *-ide* |

Remember: **Ick**! **I ate** a poison**ous** b**ite**!

**GOAL 6 Given the name (or formula) of an ion formed by the step-by-step ionization of a polyprotic acid from a Group 4A/14, 5A/15, or 6A/16 element, write its formula (or name).**

**Acid anions** are named in the same way as oxyanions, with the term hydrogen or dihydrogen added to indicate the number of hydrogen ions bonded to the oxyanion.

**GOAL 7 Given the name (or formula) of the ammonium ion or hydroxide ion, write the corresponding formula (or name).**

The ammonium ion is $NH_4^+$; the hydroxide ion is $OH^-$.

GOAL 8 **Given the name of any ionic compound made up of ions included in Goals 4 through 7, or other ions whose formulas are given, write the formula of that compound.**

To write the **formulas for ionic compounds,** write the formula for the cation, then the anion. Use subscripts to indicate the number of each ion needed so that the total charge of the compound is zero.

GOAL 9 **Given the formula of an ionic compound made up of identifiable ions, write the name of the compound.**

To **name ionic compounds,** name the cation, then the anion.

GOAL 10 **Given the formula of a hydrate, state the number of water molecules associated with each formula unit of the anhydrous compound.**

**Hydrates** are ionic compounds that exist with a definite number of water molecules in their crystal structure. Waters of hydration are indicated by the "·" symbol before the number of water molecules.

GOAL 11 **Given the name (or formula) of a hydrate, write its formula (or name). (This goal is limited to hydrates of ionic compounds for which a name and formula can be written based on the rules of nomenclature presented in this book.)**

Prefixes are used to indicate the number of water molecules in a formula unit of a hydrate. These are the same as the prefixes used in naming binary molecular compounds.

**A BRIEF SUMMARY OF THE NOMENCLATURE SYSTEM BY EXAMPLE**

| Substance | Example Name | Example Formula |
|---|---|---|
| Element | Hydrogen<br>Helium | $H_2$<br>He |
| Compound made of 2 nonmetals | Diphosphorus pentoxide<br>Carbon tetrachloride | $P_2O_5$<br>$CCl_4$ |
| Acid | Sulfuric acid<br>Phosphoric acid | $H_2SO_4$<br>$H_3PO_4$ |
| Monatomic cation | Sodium ion<br>Aluminum ion | $Na^+$<br>$Al^{3+}$ |
| Monatomic anion | Chloride ion<br>Sulfide ion | $Cl^-$<br>$S^{2-}$ |
| Polyatomic anion: total ionization | Nitrate ion<br>Carbonate ion | $NO_3^-$<br>$CO_3^{2-}$ |
| Polyatomic anion: step-by-step ionization | Hydrogen carbonate ion<br>Dihydrogen phosphate ion | $HCO_3^-$<br>$H_2PO_4^-$ |
| Ionic Compound | Mercury(II) bromite<br>Chromium(III) iodate | $Hg(BrO_2)_2$<br>$Cr(IO_3)_3$ |
| Hydrate | Copper(II) sulfate pentahydrate<br>Sodium carbonate decahydrate | $CuSO_4 \cdot 5\ H_2O$<br>$Na_2CO_3 \cdot 10\ H_2O$ |

# Chapter in Review 6B

## Chapter 6 Test Yourself

*structions:* For each name given, write the formula; for each formula given, write the name. You may use a "clean" eriodic table. Te is the symbol for tellurium, Z = 52.

| | | | |
|---|---|---|---|
| omine | $H_2$ | nitrogen | $O_2$ |
| lorine | $F_2$ | magnesium oxide | $CaI_2$ |
| dium fluoride | $Ba^{2+}$ | aluminum nitride | $Na_2S$ |
| lcium phosphide | $Cl^-$ | potassium bromide | $Li_2O$ |
| arium sulfide | $AlCl_3$ | lithium fluoride | $Mg_3P_2$ |
| anganese(III) ion | $Cu^{2+}$ | phosphorus tribromide | NaClO |
| erchloric acid | $S_2F_2$ | ammonium phosphate | $H_2S$ |
| on(II) nitrate | $HIO_4$ | potassium bromite | $Na_2TeO_3$ |
| dium hydrogen carbonate | $Mg(H_2PO_4)_2$ | sulfuric acid | $PbCl_2$ |
| ydrofluoric acid | $K_2CO_3$ | potassium iodide | $Si_2F_6$ |
| xygen difluoride | CuI | magnesium sulfate heptahydrate | $BaCl_2 \cdot 2\ H_2O$ |

**Answers to Chapter 6 Test Yourself**

| | | | |
|---|---|---|---|
| $Br_2$ | hydrogen | $N_2$ | oxygen |
| $Cl_2$ | fluorine | MgO | calcium iodide |
| NaF | barium ion | AlN | sodium sulfide |
| $Ca_3P_2$ | chloride ion | KBr | lithium oxide |
| BaS | aluminum chloride | LiF | magnesium phosphide |
| $Mn^{3+}$ | copper(II) ion | $PBr_3$ | sodium hypochlorite |
| $HClO_4$ | disulfur difluoride | $(NH_4)_3PO_4$ | hydrogen sulfide or hydrosulfuric acid |
| $Fe(NO_3)_2$ | periodic acid | $KBrO_2$ | sodium tellurite |
| $NaHCO_3$ | magnesium dihydrogen phosphate | $H_2SO_4$ | lead(II) chloride |
| HF | potassium carbonate | KI | disilicon hexafluoride |
| $OF_2$ | copper(I) iodide | $MgSO_4 \cdot 7\ H_2O$ | barium chloride dihydrate |

# Chapter in Review 7

**GOAL 1 Given the formula of a chemical compound (or a name from which the formula may be written), state the number of atoms of each element in the formula unit.**

A **chemical formula** tells how many atoms of each element are present in the formula unit of a substance. The number of each atom is given by a subscript following the symbol of that atom or a group of atoms. If the number is one, it is omitted in the formula.

**GOAL 2 Distinguish among atomic mass, molecular mass, and formula mass.**

**Atomic mass** is the average mass of all atoms of an element as they occur in nature. It is measured relative to the assignment of a mass of 12 u to an atom of carbon-12. **Molecular (or formula) mass** is the average mass of molecules (or formula units) compared with the mass of an atom of carbon-12, which is 12 atomic mass units.

**GOAL 3 Calculate the formula (molecular) mass of any compound whose formula is given (or known).**

The formula mass of a compound is equal to the sum of all of the atomic masses in the formula unit: **Formula mass = $\Sigma$ atomic masses in the formula unit.**

**GOAL 4 Define the term *mole*. Identify the number of objects that corresponds to one mole.**

One **mole** of anything contains the same number of objects as the number of atoms in exactly 12 grams of carbon-12. This experimentally determined value is **Avogadro's number, $N_A$, $6.02 \times 10^{23}$.**

**GOAL 5 Given the number of moles (or units) in any sample, calculate the number of units (or moles) in the sample.**

Use dimensional analysis to **convert between moles and number of units:**

$$\text{\# of moles} \times \frac{6.02 \times 10^{23}\ \text{units}}{\text{mol}} = \text{\# of units}$$

$$\text{\# of units} \times \frac{1\ \text{mol}}{6.02 \times 10^{23}\ \text{units}} = \text{\# of moles}$$

**GOAL 6 Define *molar mass* or interpret statements in which the term molar mass is used.**

**Molar mass** is the mass in grams of one mole of a substance.

**GOAL 7 Calculate the molar mass of any substance whose chemical formula is given (or known).**

The molar mass of any substance in grams per mole is numerically equal to the atomic, molecular, or formula mass of that substance in atomic mass units.

**GOAL 8 Given any one of the following for a substance whose formula is given (or known), calculate the other two: (a) mass, (b) number of moles, (c) number of formula units, molecules, or atoms.**

*Molar mass* is the connecting link between the macroscopic world, in which we measure quantities in grams, and the particulate world, in which we count the number of units, usually grouped in moles. Using $N_A$ for Avogadro's number and MM for molar mass,

$$\text{units} \xrightarrow{N_A} \text{mol} \xrightarrow{MM} \text{g} \quad \textit{or} \quad \text{g} \xrightarrow{MM} \text{mol} \xrightarrow{N_A} \text{units}$$

**Changing from formula units to mass or vice versa is a two-step dimensional analysis conversion.**

**GOAL 9 Calculate the percentage composition of any compound whose formula is given (or known).**

The **percentage composition** of a compound is the percentage by mass of each element in the compound. **Percent** is the amount of one part of a mixture per 100 total parts in the mixture. To calculate the percentage of each element,

$$\%\ \text{Element} = \frac{\text{total molar mass of element in compound}}{\text{Molar mass of compound}} \times 100$$

If you calculate the percentage composition of a compound correctly, the sum of all percents must be 100%.

**GOAL 10 Given the mass of a sample of any compound whose formula is given (or known), calculate the mass of any element in the sample; or, given the mass of any element in the sample, calculate the mass of the sample or the mass of any other element in the sample.**

To find the amount of any element in a known amount of compound, use percentage as a conversion factor, grams of the element PER 100 grams of the compound.

**GOAL 11 Distinguish between an empirical formula and a molecular formula.**

An **empirical formula** gives the simplest whole-number ratio of atoms of the elements in a compound. Empirical formulas are calculated from percentage composition data. They are also found from the mass of each element in a sample of a compound. Empirical formulas may or may not be the actual molecular formulas of compounds. The molar mass of the compound is needed to determine molecular formulas from simplest formulas.

**GOAL 12 Given data from which the mass of each element in a sample of a compound can be determined, find the empirical formula of the compound.**

To **find an empirical formula:**

1) Find the masses of different elements in a sample of the compound.
2) Convert the masses into moles of atoms of the different elements.
3) Determine the ratio of moles of atoms.
4) Express the moles of atoms as the smallest possible ratio of integers.
5) Write the empirical formula, using the number for each atom in the integer ratio as the subscript in the formula.

**GOAL 13 Given the molar mass and empirical formula of a compound, or information from which they can be found, determine the molecular formula of the compound.**

To **find the molecular formula of a compound:**

1) Determine the empirical formula of the compound.
2) Calculate the molar mass of the empirical formula unit.
3) Determine the molar mass of the compound (this will be given in problems in this textbook).
4) Divide the molar mass of the compound by the molar mass of the empirical formula unit to get n, the number of empirical formula units per molecule.
5) Write the molecular formula.

# Chapter 7 Test Yourself

Instructions: You may use a "clean" periodic table.

1) How many nitrogen atoms and how many oxygen atoms are in one molecule of dinitrogen pentoxide?
2) Write the formula of the compound composed of one phosphorus atom and five fluorine atoms. Name this compound.
3) Identify the incorrect statement among the following:
   (a) A mole is that quantity of a substance that contains $6.02 \times 10^{23}$ particles of that substance.
   (b) One mole of any substance contains the same number of particles as one mole of any other substance.
   (c) One mole of any substance contains the same number of particles as the number of atoms in exactly 12 grams of carbon-12.
   (d) A mole is a quantity of a substance that has a mass of exactly 12 grams.
4) Identify the correct statement among the following:
   (a) The molar mass of atoms of an element is numerically equal to its atomic mass.
   (b) The molar mass is the mass in u of one mole of any substance.
   (c) Molar mass and atomic mass are always equal, both numerically and in units.
   (d) Molar masses of compounds are always equal to atomic masses of those compounds.
5) Calculate the molar mass of potassium chloride.
6) What is the mass of $1.06 \times 10^{24}$ molecules of carbon monoxide?
7) What mass of sodium nitrate contains 1.62 g of nitrogen?
8) How many moles of potassium sulfate are in 37.5 grams?
9) Calculate the percentage composition of barium hydroxide.
10) From the following, pick those that are empirical formulas. $C_{10}H_6$; $C_2H_6O_2$; $CH_3$; $C_2H_2$; $C_{30}H_{50}O$
11) A compound has the percentage composition 92.25% C and 7.75% H. Calculate the empirical formula for this compound. The molar mass of this compound is 78 g/mol. Calculate the true molecular formula of this compound.

**Answers to Chapter 7 Test Yourself**

1) Dinitrogen pentoxide, $N_2O_5$: 2 nitrogen atoms and 5 oxygen atoms
2) $PF_5$, phosphorus pentafluoride 3) d 4) a
5) KCl: 39.10 + 35.45 = 74.55 g/mol 6) 49.3 g CO
7) 9.83 g $NaNO_3$ 8) 0.215 mol $K_2SO_4$
9) % Ba = 80.14%, % O = 18.68%, % H = 1.177% H
10) Both $CH_3$ and $C_{30}H_{50}O$ could be empirical formulas; the others are not empirical formulas because they are not in "lowest terms."
11) Empirical formula: CH; Molecular formula: $C_6H_6$

# Chapter in Review 8

**The overarching goals for this chapter are:**

1) Learn the mechanics of writing an equation.
2) Learn how to identify four different kinds of reactions.
3) Learn how to predict the products of each kind of reaction and write the formulas of those products.
4) Given potential reactants, write the equations for the probable reactants.

GOAL 1 **Describe five types of evidence detectable by human senses that usually indicate a chemical change.**

**Five types of evidence indicate the possibility of a chemical change:**

1) A color change
2) The formation of a solid
3) The formation of a gas
4) The absorption or release of heat energy
5) The emission of light energy

GOAL 2 **Distinguish between an unbalanced and a balanced chemical equation, and explain why a chemical equation needs to be balanced.**

A **chemical equation** is a shorthand description of a chemical reaction. A **balanced** chemical equation reflects the Law of Conservation of Mass. The subscripts in a chemical equation may never be changed simply to balance the equation. Changing a subscript in the formula of a substance changes the chemical identity of that substance (Law of Definite Composition).

GOAL 3 **Given an unbalanced chemical equation, balance it by inspection.**

A formal approach to **balancing a chemical equation** is:

1) Place a "1" in front of the formula with the largest number of atoms (the starting formula). If two formulas have the same number of atoms, select the one with the greater number of elements.
2) Insert coefficients that balance the elements that appear in compounds. Use fractional coefficients, if necessary. Choosing elements in the following order is usually easiest:
   a) Elements in the starting formula that are in only one other compound
   b) All other elements from the starting formula
   c) All other elements in compounds
3) Place coefficients in front of formulas of uncombined elements that balance those elements. Use fractional coefficients, if necessary.
4) Clear fractions, if any, by multiplying all coefficients by the lowest common denominator. Remove any "1" coefficients that remain.
5) Check to be sure the final equation is balanced.

GOAL 4 **Given a balanced chemical equation or information from which it can be written, describe its meaning on the particulate, molar, and macroscopic levels.**

The **coefficients** in a balanced chemical equation have two common interpretations:

1) **Particulate-level interpretation**: The coefficients represent the number of atoms, molecules, or formula units of each species.
2) **Molar interpretation:** The coefficients represent the number of moles of each species.

To interpret coefficients at the macroscopic level, you need to combine the molar interpretation with the molar mass of each species.

GOAL 5 **Write the equation for the reaction in which a compound is formed by the combination of two or more simpler substances.**

A reaction in which two or more substances combine to form a single product is a **combination reaction** or synthesis reaction.
Reactants: Any combination of elements and/or compounds
Product: One compound
Equation type: $A + X \rightarrow AX$

GOAL 6 **Given a compound that is decomposed into simpler substances, either compounds or elements, write the equation for the reaction.**

A **decomposition reaction** occurs when a single compound breaks down into simpler substances.
Reactant: One compound
Products: Any combination of elements and compounds
Equation type: $AX \rightarrow A + X$

| | |
|---|---|
| GOAL 7 **Given the reactants of a single-replacement reaction, write the equation for the reaction.** | In a **single-replacement reaction,** it looks as if one element is replacing another in a compound. If the reactant element is a metal, it replaces the metal or hydrogen in the compound. If the element is a nonmetal, it replaces the nonmetal in a compound.<br>Reactants: Element (A) plus a solution of either an acid or an ionic compound (BX)<br>Products: An ionic compound (usually in solution) (AX) plus an element (B)<br>Equation type: $A + BX \rightarrow AX + B$ |
| GOAL 8 **Given the reactants in a double-replacement precipitation or neutralization reaction, write the equation.** | In a **double-replacement reaction,** it looks as if the ions of the two reactants change partners.<br>Reactants: Solutions of two compounds, each with positive and negative ions (AX + BY)<br>Products: Two new compounds (AY + BX), which may be a solid, water, an acid, or an aqueous ionic compound<br>Equation type: $AX + BY \rightarrow AY + BX$<br><br>When one or more products of a double-replacement reaction is a solid that is formed from reactants in solution, the reaction is **a precipitation reaction.** When one reactant is an **acid** and the other is a **base,** the double-replacement reaction yields water as one product, and the reaction is a **neutralization reaction.** |

# Chapter 8 Test Yourself

Instructions: You may use a "clean" periodic table.

1) List the five types of evidence detectable by human senses that usually indicate a chemical change.
2) Classify each of the following statements as true or false:
   (a) Chemical formulas that contain the same elements should be adjusted so that they have the same number of atoms when balancing equations. Example: In the unbalanced equation $H_2O_2 \rightarrow H_2O + O_2$, change $H_2O_2$ to $H_2O$.
   (b) Compound formulas should be adjusted to account for diatomic elements when balancing equations. Example: In the unbalanced equation $NaCl \rightarrow Na + Cl_2$, change NaCl to $NaCl_2$.
   (c) Coefficients should be placed inside of formulas so that they apply only to part of that formula when balancing equations. Example: In the unbalanced equation $Pb(NO_3)_2 + NaCl \rightarrow PbCl_2 + NaNO_3$, change NaCl to Na2Cl.
   (d) Formulas of species that are not in the reaction description should be added to either the product side or the reactant side, as necessary, when balancing equations. Example: In the reaction description "nitrogen reacts with hydrogen to form ammonia," add $O_2$ to the reactant side and $H_2O$ to the product side.
3) Potassium reacts with bromine to form potassium bromide. If 10 moles of bromine react, how many moles of potassium are needed and how many moles of potassium bromide will form?

*In each of the remaining questions, a description of a chemical reaction is provided. Classify the reaction type of each as combination, decomposition, single replacement, or double replacement, and then write and balance the chemical equation. You do not need to include state symbols.*

4) Solid calcium oxide is formed from its elements.
5) Solid barium oxide and liquid water result from the decomposition of solid barium hydroxide.
6) Hydrogen gas and a solution of lithium hydroxide are the products of the reaction between solid lithium metal and liquid water.
7) Potassium hydroxide solution reacts with a solution of copper(II) nitrate. Copper(II) hydroxide is a solid product; the other product is aqueous.
8) Aqueous solutions of hydrobromic acid and sodium hydroxide react.
9) Lead reacts with a solution of copper(II) nitrate.
10) When chloric acid is poured over solid calcium carbonate, carbon dioxide bubbles off, leaving water and aqueous calcium chlorate as the other products. (You do not have to classify this reaction.)

**Answers to Chapter 8 Test Yourself**

1) Color change, formation of a solid, formation of gas, absorption or release of heat energy, emission of light energy.
2) All are false.
3) $2\ K + Br_2 \rightarrow 2\ KBr$; 20 moles of K and 20 moles of KBr.
4) Combination: $2\ Ca(s) + O_2(g) \rightarrow 2\ CaO(s)$
5) Decomposition: $Ba(OH)_2(s) \rightarrow BaO(s) + H_2O(\ell)$
6) Single replacement: $2\ Li(s) + 2\ H_2O(\ell) \rightarrow H_2(g) + 2\ LiOH(aq)$
7) Double replacement: $2\ KOH(aq) + Cu(NO_3)_2(aq) \rightarrow Cu(OH)_2(s) + 2\ KNO_3(aq)$
8) Double replacement: $HBr(aq) + NaOH(aq) \rightarrow H_2O(\ell) + NaBr(aq)$
9) Single replacement: $Pb(s) + Cu(NO_3)_2(aq) \rightarrow Pb(NO_3)_2(aq) + Cu(s)$
10) $2\ HClO_3(aq) + CaCO_3(s) \rightarrow CO_2(g) + H_2O(\ell) + Ca(ClO_3)_2(aq)$

# Chapter in Review 9

GOAL 1 **Distinguish among strong electrolytes, weak electrolytes, and nonelectrolytes.**

A **solution** is a homogeneous mixture. A solute may be classified as a strong **electrolyte,** a weak electrolyte, or a nonelectrolyte according to the ability of its water solution to conduct electricity. If a solution conducts electricity, ions must be present as solute particles.

GOAL 2 **Given the formula of an ionic compound (or its name), write the formulas of the ions present when it is dissolved in water.**

When an ionic compound dissolves in water, its solution consists of water molecules and ions surrounded by water molecules. The ions are identified by separating the compound into its ions.

GOAL 3 **Explain why the solution of an acid may be a good conductor or a poor conductor of electricity.**

An **acid** is a hydrogen-bearing molecule or ion that releases a hydrogen ion in water solution. Acids are classified as strong or weak, depending on the extent to which the original compound ionizes when dissolved in water. When a **strong acid** dissolves, it dissociates into ions. The **major species** present in the solution are ions, and the **minor species** present are un-ionized molecules. When a **weak acid** dissolves, it does not dissociate into ions to a large extent. The solution inventory is mainly un-ionized molecules.

GOAL 4 **Given the formula of a soluble acid (or its name), write the major and minor species present when it is dissolved in water.**

There are **seven common strong acids.** Their names and formulas must be memorized: nitric acid, $HNO_3$; sulfuric acid, $H_2SO_4$; hydrochloric acid, HCl; hydrobromic acid, HBr; hydroiodic acid, HI; chloric acid, $HClO_3$; perchloric acid, $HClO_4$. If an acid is not one of the seven strong acids, it is a weak acid. Ions are the major species in the solutions of two kinds of substances: (1) all soluble ionic compounds and (2) the seven strong acids. Neutral molecules are the major species in solutions of everything else, primarily (1) weak acids, (2) weak bases, and (3) water.

GOAL 5 **Distinguish among conventional, total ionic, and net ionic equations.**

A **conventional equation** shows the formulas of the reactants written on the left side of an arrow and the formulas of the products on the right side of the arrow. Strong acids and ionic compounds designated (aq) in a conventional equation are rewritten in a **total ionic equation** with the formulas of the major species in solution. A total ionic equation is made into a **net ionic equation** by removing the spectators, those species that are on both sides of the total ionic equation.

GOAL 6 **Given two substances that may engage in a single-replacement redox reaction and an activity series by which the reaction may be predicted, write the conventional, total ionic, and net ionic equations for the reaction that will occur, if any.**

A **single-replacement redox reaction** has the form $A + BX \rightarrow AX + B$. The compounds BX and AX are usually aqueous and the elements A and B are usually solid metals or $H_2(g)$. Prediction of a single-replacement redox reaction is made by referring to an activity series. A more active element will replace the dissolved ions of any less active element.

GOAL 7 **Write the equation for the complete oxidation or burning of any compound containing only carbon and hydrogen or only carbon, hydrogen, and oxygen.**

The general unbalanced equation for a **complete oxidation (burning)** of a compound that consists of only carbon and hydrogen or only carbon, hydrogen, and oxygen is $C_xH_yO_z + O_2(g) \rightarrow CO_2(g) + H_2O(\ell)$. As a rule, these equations are most easily balanced if you take the elements carbon, hydrogen, and oxygen in that order.

GOAL 8 **Predict whether a precipitate will form when known solutions are combined; if a precipitate forms, write the net ionic equation. (Reference to a solubility table or a solubility guidelines list may or may not be allowed.)**

An **ion-combination reaction** occurs when the cation (positively charged ion) from one reactant combines with the anion (negatively charged ion) from another to form a product compound. The conventional equation is a **double-replacement** type in which the ions appear to change partners: $AX + BY \rightarrow AY + BX$. When a product in this type of reaction is an insoluble ionic compound, the solid is called a **precipitate** and the reaction is a **precipitation reaction.**

GOAL 9 **Given the product of a precipitation reaction, write the net ionic equation.**

Your instructor will have you memorize **solubility guidelines** or allow you to have access to the guidelines or a solubility table. Use the method suggested by your instructor to predict precipitation reactions.

**GOAL 10 Given reactants for a double-replacement reaction that yield a molecular product, write the conventional, total ionic, and net ionic equation.**

The reaction of an acid often leads to an ion combination that yields a **molecular product** instead of a precipitate. Except for the difference in the product, the equations are written in exactly the same way. Just as you had to recognize an insoluble product and not break it up in total ionic equations, you must recognize a molecular product and not break it into ions. Water and weak acids are the two kinds of molecular products you will find. **Neutralization** reactions are the most common molecular-product reactions, HX + MOH → HOH + MX. There are two points by which you can identify a molecular-product reaction: (1) one reactant is an acid, usually strong, and (2) one product is water or a weak acid.

**GOAL 11 Given reactants that form $H_2CO_3$, $H_2SO_3$, or "$NH_4OH$" by ion combination, write the net ionic equation for the reaction.**

Three ion combinations yield molecular products that are not the products you would expect:

$2\,H^+(aq) + CO_3^{2-}(aq) \rightarrow H_2CO_3(aq) \rightarrow CO_2(g) + H_2O(\ell)$

$2\,H^+(aq) + SO_3^{2-}(aq) \rightarrow H_2SO_3(aq) \rightarrow SO_2(aq) + H_2O(\ell)$

$NH_4^+(aq) + OH^-(aq) \rightarrow \text{"}NH_4OH\text{"} \rightarrow NH_3(aq) + H_2O(\ell)$

When ion combinations yield **unstable** substances, the right side of the net ionic equation has the formulas of the stable decomposition products.

# **Chapter 9** Test Yourself

*Instructions: You may use a "clean" periodic table. You may use a solubility table or a list of solubility rules if your instructor allows it.*

1) A soluble salt is a _____ electrolyte.
   (a) weak
   (b) non-
   (c) strong
2) Write the formulas of the major species present in solutions of the following compounds:
   (a) calcium chloride
   (b) hydroiodic acid
   (c) acetic acid, $HC_2H_3O_2$
   (d) aluminum nitrate

*Questions 3–8: Write the conventional, total ionic, and net ionic equation for each reaction that occurs, if any. Write NR if no reaction occurs.*

3) Solutions of sodium sulfate and potassium carbonate are mixed.
4) Hydrochloric acid is added to ammonium carbonate solution.
5) Sodium formate solution, $NaCHO_2(aq)$, is added to hydrobromic acid.
6) Lead(II) nitrate and nickel sulfate solutions are mixed.
7) Solid calcium hydroxide is treated with nitric acid.
8) Hydroiodic acid is poured on nickel metal (hydrogen gas is below nickel metal in the activity series).

**Answers to Chapter 9 Test Yourself**

1) c
2) a) $CaCl_2$: $Ca^{2+}(aq) + 2\,Cl^-(aq)$; b) HI: $H^+(aq) + I^-(aq)$; c) $HC_2H_3O_2$: $HC_2H_3O_2(aq)$; d) $Al(NO_3)_3$: $Al^{3+}(aq) + 3\,NO_3^-(aq)$
3) $2\,Na^+(aq) + SO_4^{2-}(aq) + 2\,K^+(aq) + CO_3^{2-}(aq) \rightarrow$ NR
4) $2\,HCl(aq) + (NH_4)_2CO_3(aq) \rightarrow 2\,NH_4Cl(aq) + \text{"}H_2CO_3\text{"}$;
   $2\,HCl(aq) + (NH_4)_2CO_3(aq) \rightarrow 2\,NH_4Cl(aq) + H_2O(\ell) + CO_2(g)$; $2\,H^+(aq) + 2\,Cl^-(aq) + 2\,NH_4^+(aq) + CO_3^{2-}(aq) \rightarrow$
   $2\,NH_4^+(aq) + 2\,Cl^-(aq) + H_2O(\ell) + CO_2(g)$; $2\,H^+(aq) + CO_3^{2-}(aq) \rightarrow H_2O(\ell) + CO_2(g)$
5) $NaCHO_2(aq) + HBr(aq) \rightarrow NaBr(aq) + HCHO_2(aq)$; $Na^+(aq) + CHO_2^-(aq) + H^+(aq) + Br^-(aq) \rightarrow$
   $Na^+(aq) + Br^-(aq) + HCHO_2(aq)$; $CHO_2^-(aq) + H^+(aq) \rightarrow HCHO_2(aq)$
6) $Pb(NO_3)_2(aq) + NiSO_4(aq) \rightarrow PbSO_4(s) + Ni(NO_3)_2(aq)$; $Pb^{2+}(aq) + 2\,NO_3^-(aq) + Ni^{2+}(aq) + SO_4^{2-}(aq) \rightarrow$
   $PbSO_4(s) + Ni^{2+}(aq) + 2\,NO_3^-(aq)$; $Pb^{2+}(aq) + SO_4^{2-}(aq) \rightarrow PbSO_4(s)$
7) $Ca(OH)_2(s) + 2\,HNO_3(aq) \rightarrow Ca(NO_3)_2(aq) + 2\,HOH(\ell)$;
   $Ca(OH)_2(s) + 2\,H^+(aq) + 2\,NO_3^-(aq) \rightarrow Ca^{2+}(aq) + 2\,NO_3^-(aq) + 2\,H_2O(\ell)$; $Ca(OH)_2(s) + 2\,H^+(aq) \rightarrow Ca^{2+}(aq) + 2\,H_2O(\ell)$
8) $2\,HI(aq) + Ni(s) \rightarrow NiI_2(aq) + H_2(g)$; $2\,H^+(aq) + 2\,I^-(aq) + Ni(s) \rightarrow Ni^{2+}(aq) + 2\,I^-(aq) + H_2(g)$; $2\,H^+(aq) + Ni(s) \rightarrow$
   $Ni^{2+}(aq) + H_2(g)$

# Chapter in Review 10

**GOAL 1 Given a chemical equation, or a reaction for which the equation is known, and the number of moles of one species in the reaction, calculate the number of moles of any other species.**

The **coefficients** in a chemical equation express the **mole relationships** between the different substances in the reaction. The coefficients may be used in a dimensional analysis conversion from moles of one substance to moles of another.

**GOAL 2 Given a chemical equation, or a reaction for which the equation can be written, and the number of grams or moles of one species in the reaction, find the number of grams or moles of any other species.**

The **mass-to-mass stoichiometry path** is:

1) Change the mass of the given species to moles

2) Change the moles of the given species to moles of the wanted species

3) Change the moles of the wanted species to mass

This three-step method is at the heart of almost all stoichiometry problems.

**GOAL 3 Given two of the following, or information from which two of the following may be determined, calculate the third: theoretical yield, actual yield, percent yield.**

The efficiency of a reaction is stated in **percent yield,** in which the **actual yield** is expressed as a percent of the **theoretical yield** calculated by stoichiometry:

$$\%\text{ yield} = \frac{\text{actual yield}}{\text{theoretical yield}} \times 100$$

You need to be able to solve three types of percent yield problems:

1) Given: Actual and theoretical yields — Wanted: Percent yield

Solve by: $\%\text{ yield} = \frac{\text{actual yield}}{\text{theoretical yield}} \times 100$

2) Given: Reactant quantity and percent yield — Wanted: Product quantity

Solve by: Dimensional analysis

3) Given: Product quantity and percent yield — Wanted: Reactant quantity

Solve by: Dimensional analysis

**GOAL 4 Identify and describe or explain limiting reactants and excess reactants.**

A **limiting reactant** is the reactant totally consumed in a reaction, thereby determining the maximum yield possible. An **excess reactant** is a reactant that has a quantity that is in excess of the amount needed to completely react with the limiting reactant.

**GOAL 5 Given a chemical equation, or information from which it may be determined, and initial quantities of two or more reactants, (a) identify the limiting reactant, (b) calculate the theoretical yield of a specified product, assuming complete use of the limiting reactant, and (c) calculate the quantity of the reactant initially in excess that remains unreacted.**

The **comparison-of-moles method** for solving limiting reactant problems is:

1) Convert the number of grams of each reactant to moles

2) Identify the limiting reactant

3) Calculate the number of moles of each species that reacts or is produced

4) Calculate the number of moles of each species that remains after the reaction

5) Change the number of moles of each species to grams

The **smaller-amount method** for solving limiting reactant problems is:

1) Calculate the amount of product that can be formed by the initial amount of each reactant
   a) The reactant that yields the smaller amount of product is the limiting reactant
   b) The smaller amount of product is the amount that will be formed when all of the limiting reactant is used up

2) Calculate the amount of excess reactant that is used by the total amount of limiting reactant

3) Subtract from the amount of excess reactant present initially the amount that is used by all of the limiting reactant; the difference is the amount of excess reactant that is left

**GOAL 6 Given energy in one of the following units, calculate the other three: joules, kilojoules, calories, and kilocalories.**

The SI unit of energy is the **joule,** a force of one newton (kg · m/s$^2$) applied for a distance of one meter. Another energy unit used by chemists is the **calorie,** which is equal to 4.184 joules.

The **joule (J)** and the **calorie (cal)** are small energy units, so units 1000 times larger, the **kilojoule (kJ)** and the **kilocalorie (kcal)** are often used. The food energy **Calorie** is the thermochemical kilocalorie.

**GOAL 7 Given a chemical equation, or information from which it may be written, and the heat (enthalpy) of reaction, write the thermochemical equation either (a) with ΔH to the right of the conventional equation or (b) as a reactant or product.**

The amount of heat given off or absorbed in a chemical reaction is called the **enthalpy of reaction,** symbolized by ΔH.

The ΔH of a reaction may be included in the chemical equation as a reactant or a product, or it may be written next to the equation. The equation is then called a **thermochemical equation.**

For an endothermic reaction, ΔH is positive; heat is a reactant in the thermochemical equation. For an exothermic reaction, ΔH is negative; heat is a product in the thermochemical equation.

**GOAL 8 Given a thermochemical equation, or information from which it may be written, calculate the amount of energy released or added for a given amount of reactant or product; alternately, calculate the mass of reactant required to produce a given amount of energy.**

Using the ΔH of a thermochemical equation and the coefficient of any substance in that equation, you can convert in either direction from moles of that substance to amount of heat absorbed or produced.

**Summary of stoichiometry pattern:**

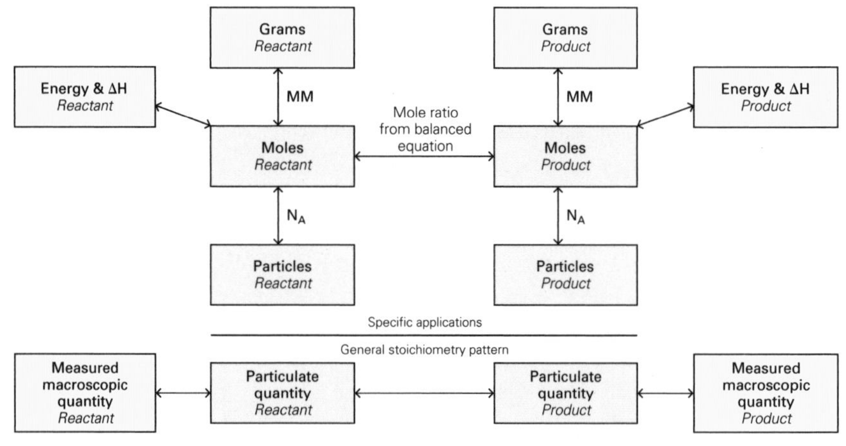


# Chapter 10 Test Yourself

*Instructions:* You may use a "clean" periodic table.

Questions 1–4 and 9 refer to the equation: $2\ C_5H_{10} + 15\ O_2 \rightarrow 10\ CO_2 + 10\ H_2O$

1) How many moles of carbon dioxide result from the reaction of 3 moles of $C_5H_{10}$?
2) What mass of oxygen must react to form 12.0 moles of carbon dioxide?
3) If the reaction consumes 4.12 grams of oxygen, how many grams of water are also produced?
4) The reaction of 6.81 grams of $C_5H_{10}$ yields 19.2 grams of carbon dioxide. Find the percentage yield.
5) Calculate the grams of iron(III) oxide produced if 15.2 grams of iron and 7.41 grams of oxygen react to form the compound until one of them is completely consumed.
6) Identify the substance that is in excess in Question 5, and calculate the mass of that substance that remains unreacted.
7) Express 127 joules as kilocalories.
8) Write in two different forms the thermochemical equation showing the decomposition of one mole of steam into hydrogen gas and oxygen gas if ΔH for this reaction is +286 kJ.
9) ΔH for the reaction in Questions 1–4 is −6.08 kJ. How many grams of $C_5H_{10}$ must be consumed to release 2.22 kJ?

**Answers to Chapter 10 Test Yourself**

1) 15 mol $CO_2$
2) 576 g $O_2$
3) 1.55 g $H_2O$
4) 89.7% yield (21.4 g CO theo)
5) 21.7 g $Fe_2O_3$
6) 0.88 g $O_2$ unreacted
7) 0.0304 kcal
8) $H_2O(g) \rightarrow H_2(g) + (1/2)\ O_2(g)$ $\quad \Delta H = +286$ kJ
$286\ kJ + H_2O(g) \rightarrow H_2(g) + (1/2)\ O_2(g)$
9) 51.2 g $C_5H_{10}$

# Chapter in Review 11A

| | |
|---|---|
| GOAL 1 **Define and describe electromagnetic radiation.** | **Electromagnetic radiation** is a form of energy that has wave-like properties that include gamma rays, X-rays, ultraviolet radiation, visible light, infrared radiation, microwaves, and radio waves. |
| GOAL 2 **Distinguish between continuous and line spectra.** | A **continuous spectrum** is a spectrum that is distributed over a continuous range of wavelengths. A **line spectrum** has discrete lines. |
| GOAL 3 **Describe the Bohr model of the hydrogen atom.**<br>GOAL 4 **Explain the meaning of quantized energy levels in an atom and show how these levels relate to the discrete lines in the spectrum of that atom.** | The **Bohr model of the hydrogen atom** restricts the electron to certain **quantized energy levels.** The electron can have certain definite energies, but never may it have an energy between the quantized values. The spectrum of the atoms of an element is the result of energy released as electrons in an excited state drop to a lower energy level. |
| GOAL 5 **Distinguish between ground state and excited state.** | The lowest quantized energy level of the atom is called the **ground state.** All energy levels above the ground state are called **excited states.** |
| GOAL 6 **Identify the principal energy levels in an atom and state the energy trend among them.** | The **quantum mechanical model** of the atom identifies **principal energy levels** and **sublevels** within each principal energy level. In general, energies increase as the principal quantum numbers increase: $n = 1 < n = 2 < n = 3 \ldots < n = 7$. |
| GOAL 7 **For each principal energy level, state the number of sublevels, identify them, and state the energy trend among them.** | The total number of sublevels within a given principal energy level is equal to *n*, the principal quantum number. For any given value of *n*, energy increases through the sublevels in the order $\mathbf{s} < \mathbf{p} < \mathbf{d} < \mathbf{f}$. |
| GOAL 8 **Sketch the shapes of *s* and *p* orbitals.** | Figure 11.11 illustrates the **shapes** of *s* and *p* orbitals. |
| GOAL 9 **State the number of orbitals in each sublevel.** | There is one orbital for every *s* sublevel. All *p* sublevels have three orbitals, all *d* sublevels have five, and all *f* sublevels have seven. This **1–3–5–7 sequence of odd numbers** continues through higher sublevels. |
| GOAL 10 **State the restrictions on the electron population of an orbital.** | An **orbital** may be unoccupied, occupied by one electron, or occupied by two electrons. |
| GOAL 11 **Use a periodic table to list electron sublevels in order of increasing energy.** | The periodic table is a guide to the order of **increasing energy of sublevels:** $1s < 2s < 2p < 3s < 3p < 4s < 3d < 4p$ and so on. |
| GOAL 12 **Referring only to a periodic table, write the ground-state electron configuration of an atom of any element up to atomic number 36.** | Writing **electron configurations** requires the ability to list sublevels in order of increasing energy, knowledge of the maximum number of electrons in each sublevel, and the ability to use the periodic table to establish the number of electrons in the highest occupied energy sublevel of the atom. |
| GOAL 13 **Using *n* for the highest occupied energy level, write the configuration of the valence electrons of any main group element.** | Atoms in the same group of the periodic table (same column) have the same highest occupied sublevel electron configurations, or the same $ns^xnp^y$ **valence electron** configurations. The highest occupied principal energy level value (*n*) increases as you go down a group. |
| GOAL 14 **Write the Lewis (electron-dot) symbol for an atom of any main group element.** | Valence electrons are depicted by **Lewis symbols,** which are also called electron-dot symbols. |
| GOAL 15 **Predict how and explain why atomic size varies with position in the periodic table.** | The **sizes of atoms** in the periodic table increase as you move down a group, but decrease as you move left to right across a period. This is explained by the highest occupied energy level and nuclear charge. |
| GOAL 16 **Predict how and explain why first ionization energy varies with position in the periodic table.** | **First ionization energy,** the energy required to remove an electron from a neutral atom, decreases as you move down a group and increases as you move across a period. |

| | |
|---|---|
| GOAL 17 **Explain, from the standpoint of electron configuration, why certain groups of elements make up chemical families.** | Groupsv of elements in the periodic table exhibit similar behavior and are called **chemical families.** The chemical properties of the elements in a family are similar because they have the same valence electron configuration. |
| GOAL 18 **Identify in the periodic table the following chemical families: alkali metals, alkaline earths, halogens, noble gases.** | Alkali metals, Group 1A/1; alkaline earths, Group 2A/2; halogens, Group 7A/17; noble gases, Group 8A/18. |
| GOAL 19 **Identify metals and nonmetals in the periodic table.** | The elements in the periodic table are classified as **metals or nonmetals,** based on their chemical behavior. Metals are to the left of the stair-step line, nonmetals are to the right, and seven **metalloids** hug the line. |
| GOAL 20 **Predict how and explain why metallic character varies with position in the periodic table.** | **Metallic character** increases from right to left across any row of the periodic table and from top to bottom in any group. A metal can lose one or more electrons and become a positively charged ion. As the energy required to remove an electron decreases, the metallic character of an element increases. |

# Chapter in Review 11B

## Chapter 11 Test Yourself

*Instructions:* You may refer to a "clean" periodic table.

1) The Bohr model of the atom provides all of the following except (a) an explanation of the line spectra of the elements, (b) a description of electron behavior in many-electron atoms, (c) evidence that electron energies are quantized, (d) calculation of energies of known lines in the hydrogen spectrum, (e) the radius of the electron orbit in a hydrogen atom.

2) Identify the false statement about electron energies: (a) Electron energies are quantized in excited states, but not quantized in the ground state, (b) Light spectra of the elements are experimental evidence of the quantization of electron energies, (c) Lines in the spectrum of an element are produced by electrons dropping from a high energy level to a lower energy level, (d) Energy must be absorbed to raise an electron from ground state to an excited state, (e) Electrons cannot possess an energy between two quantized energy levels.

3) Identify the incorrect statement among the following: (a) Except for $n = 1$, each principal energy level has three $p$ orbitals, (b) There are three sublevels when $n = 3$, (c) Electrons in the $5d$ orbitals have higher energies than electrons in the $5p$ orbitals, (d) The $n = 4$ sublevels are at higher overall energies than the corresponding $n = 5$ sublevels, (e) A $3s$ orbital is at higher energy than a $1s$ orbital.

4) Identify the incorrect statement about electron orbitals: (a) An orbital may be occupied by no more than two electrons, (b) All energy sublevels contain the same number of orbitals, (c) For a given atom, the $3p$ orbitals are larger than the $2p$ orbitals, but smaller than the $4p$ orbitals, (d) At a given $d$ sublevel, the maximum number of $d$ electrons is 10, (e) The orbital sketched to the right is a $p$ orbital.

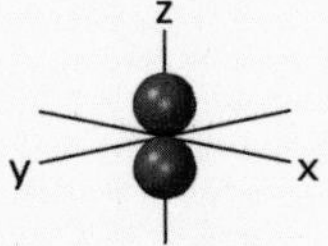


5) Using the symbolism $ns^x np^y$, where x and y are whole numbers, write the general electron configuration that is responsible for the family properties of the alkali metals.

6) Write the Lewis symbol for oxygen, and then for any alkaline earth element.

7) Write the electron configuration for an atom of chlorine.

8) Write the electron configuration for an atom of titanium, Z = 22.

9) The halogens chlorine and bromine form the insoluble silver chloride, AgCl, and silver bromide, AgBr, upon reaction with silver. These halogens exhibit these reactions with silver because: (a) elemental silver has its outermost electrons in the $n = 5$ quantum level, (b) chlorine and bromine are in the same period of the periodic table, (c) silver is a reactive metal, (d) chlorine and bromine have similar electron configurations in their highest energy occupied orbitals, (e) silver and these two halogens all have similar electron configurations in their highest energy-occupied orbitals.

10) List the symbols of the following elements in order of increasing first ionization energy: helium, lithium, boron, oxygen, neon.

11) Which trio of atomic numbers is arranged in order of increasing atomic size? (a) 54–18–2, (b) 52–33–14, (c) 8–9–17, (d) 17–35–34, (e) 12–19–20

12) Consider atoms with atomic numbers 7, 11, 16, 20, 43, and 53, and then (a) write the atomic numbers of the metals listed and (b) write the atomic numbers of the nonmetals listed.

**Answers to Chapter 11 Test Yourself**

1) b 2) a 3) d 4) b

5) $ns^1$. The alkali metals share an $ns^1$ highest occupied energy level electron configuration. In this description, $n$ is the principal quantum number of the highest occupied energy level when the atom is in its ground state. For lithium, $n = 2$; for sodium, $n = 3$; for potassium, $n = 4$, and so forth. The $s$ refers to the $s$ sublevel in the highest occupied energy level. The $p$, when needed, refers to the $p$ sublevel in the same principal energy level. Superscript numbers identify the number of electrons in those sublevels. The $ns^1$ indicates that the highest occupied energy level of a ground state alkali metal is occupied by one electron in the $s$ orbital.

6) The Lewis symbol for oxygen is the letter O surrounded by six dots, two pairs on each of two sides, and single dots on the other two sides. The Lewis symbol for an alkaline earth element is **X** : X can be Be, Mg, Ca, Sr, Ba, or Ra.

7) Cl: $1s^2 2s^2 2p^6 3s^2 3p^5$ or $[Ne]3s^2 3p^5$ 8) Ti: $1s^2 2s^2 2p^6 3s^2 3p^6 4s^2 3d^2$ or $[Ar]4s^2 3d^2$

9) d 10) Li < B < O < Ne < He 11) d

12) (a) The metals are Z = 11 (Na), 20 (Ca), and 43 (Tc); (b) The nonmetals are Z = 7 (N), 16 (S), and 53 (I)

# Chapter in Review 12

GOAL 1 **Define and distinguish between cations and anions.**

A **cation** is a positively charged ion; an **anion** is a negatively charged ion.

GOAL 2 **Identify the monatomic ions that are isoelectronic with a given noble-gas atom and write the electron configuration of those ions.**

The formation of monatomic ions that are **isoelectronic** with neon atoms illustrates the pattern that is duplicated for other noble gases. A neon atom has ten electrons, including a full octet of valence electrons. Its electron configuration is $1s^2 2s^2 2p^6$. Nitrogen, oxygen, and fluorine atoms form anions by gaining enough electrons to reach the same configuration. Sodium, magnesium, and aluminum atoms form cations by losing valence electrons to reach the same configuration.

GOAL 3 **Use Lewis symbols to illustrate how an ionic bond can form between monatomic cations from Groups 1A, 2A, and 3A (1, 2, 13) and anions from Groups 5A, 6A, and 7A (15–17) of the periodic table.**

Ionic bonds between atoms form when atoms of a metal lose one, two, or three electrons to form a cation that is isoelectronic with a noble gas and atoms of a nonmetal gain one, two, or three electrons to form an anion that is also isoelectronic with a noble gas. An **ionic bond,** also called an electron-transfer bond, is formed because of the electrostatic attraction between oppositely charged ions.

GOAL 4 **Describe, use, or explain each of the following with respect to forming a covalent bond: electron cloud, charge cloud, or charge density; valence electrons; half-filled electron orbital; filled electron orbital; electron sharing; orbital overlap; octet rule or rule of eight.**

Two atoms in a molecule are held together by a **covalent bond** when they share one or more pairs of electrons. The **electron cloud** or **charge cloud** formed by the two bonding electrons is concentrated in the region between the two nuclei. The bonding electrons count as **valence electrons** for each bonded atom. Covalent bonds form by the **overlap** of half-filled electron orbitals. The stability of a noble-gas electron configuration—the **octet rule** or **rule of eight**—is a result of the minimization of energy associated with that configuration.

GOAL 5 **Use Lewis symbols to show how covalent bonds are formed between two nonmetal atoms.**

Covalent bonds form between two nonmetals, both of which have atoms that are one, two, or even three electrons short of a noble gas electron configuration. This is accomplished through **covalent bonding,** in which electrons are **shared.**

GOAL 6 **Distinguish between bonding electron pairs and lone pairs.**

**Bonding electron pairs** are represented in a Lewis diagram as two dots or a straight line drawn between atoms. Both formats represent the covalent bond that hold the atoms together. Unshared pairs are also shown in Lewis diagrams. These are also called **lone pairs.**

GOAL 7 **Distinguish between polar and nonpolar covalent bonds.**

In a **nonpolar covalent bond,** the bonding electrons are shared equally by the bonded atoms. In a **polar covalent bond,** the nucleus of one atom attracts the shared electrons more strongly than the other.

GOAL 8 **Predict which end of a polar bond between identified atoms is positive and which end is negative. (You may refer to a periodic table.)**

The relative ability of atoms of an element to attract electron pairs in covalent bonds is expressed by the **electronegativity** of the element. The **polarity** of a bond is estimated by the difference in electronegativities of the bonded atoms. The atom with the higher electronegativity is the negative end of the bond. The atom with the lower electronegativity is the positive end of the bond.

GOAL 9 **Rank bonds in order of increasing or decreasing polarity based on periodic trends in electronegativity values or actual values, if given. (You may refer to a periodic table.)**

There is a periodic trend in the electronegativity of elements. In general, electronegativity increases from left to right across a period and decreases down a group. The greater the difference in electronegativity between two bonded elements, the more polar the bond will be.

GOAL 10 **Distinguish among single, double, and triple bonds, and identify these bonds in a Lewis diagram.**

The sharing of one pair of electrons by two bonded atoms is called a **single bond.** When two atoms are bonded by two pairs of electrons, it is a **double bond.** When two atoms are bonded by three pairs of electrons, the bond is called a **triple bond.** All four electrons in a double bond and all six electrons in a triple bond are counted as valence electrons for the bonded atoms. **Multiple bonds** is a general term that includes double and triple bonds.

GOAL 11 **Describe how metallic bonding differs from ionic and covalent bonding.**

**Metallic bonding** occurs because of attractive forces between negatively charged valence electrons moving among positively charged metal ions. In covalent bonds, the bonding electrons are localized between two specific atoms. Electrons in a metallic bond are **delocalized** because the bonding electrons do not stay near any single atom or pair of atoms. The nature of the metallic bond explains many of the properties of metals, such as electrical conductivity and the ability to bend and be stretched into thin wires.

Portable Content Card for Cracolice's *Introductory Chemistry: An Active Learning Approach*

GOAL 12 **Sketch a particulate-levvel illustration of the electron-sea model of metallic bonding.**

The **electron-sea model** of a metallic crystal is characterized by monatomic ions in a crystal pattern with the highest-energy valence electrons free to move among the ions. The positively charged metal ions are held in fixed positions in the crystal because of their attraction to the negatively charged valence electrons that move among the ions.

# Chapter 12 Test Yourself

Instructions: You may refer to a "clean" periodic table.

1) Which of the following is not an anion?
   a) $F^-$ b) $NO_3^-$ c) $HPO_4^{2-}$ d) $K^+$ e) $N^{3-}$
2) Which of the following is not isoelectronic with a noble gas?
   a) $Na^+$ b) $Cl^-$ c) $S^{2-}$ d) $Al^{3+}$ e) $Zn^{2+}$
3) A bond formed by the transfer of an electron from one atom to another is called a(n) _______ bond.
   a) ionic b) covalent c) metallic d) nonpolar e) Lewis
4) Identify the incorrect statement among the following:
   a) The bond between hydrogen atoms in H—H is a covalent bond.
   b) When two bonding electrons are shared by two atoms, they count as valence electrons for only the least electronegative atom.
   c) The $Cl_2$ bond is formed by the overlap of half-filled 3*p* orbitals of two chlorine atoms.
   d) The stability of a noble-gas electron configuration contributes to the formation of a covalent bond.
   e) When a covalent bond forms, the charge density is concentrated in the region between the two nuclei.
5) How many bonding electron pairs (____) and how many lone pairs (____) are in a nitrogen molecule? : N≡N :
6) Identify each of the following bonds as either primarily ionic or primarily covalent: a) Na—F b) Ba—Br c) C—H d) N—O
7) Identify the incorrect statement among the following:
   a) The bond between two atoms of the same element is probably less polar than the bond between two atoms of different elements.
   b) The distribution of electronic charge in a polar bond is not symmetrical.
   c) A bonding electron pair does not tend to be closer to either of the bonded atoms if the bond is polar.
   d) One end of a polar bond is said to be the negative pole.
   e) You can estimate the polarity of a bond by calculating the difference between the electronegativity values for the bonded elements.
8) Which of the following best describes the C—O bond?
   a) Nonpolar
   b) Polar covalent with C as the + end of the bond
   c) Polar covalent with O as the + end of the bond
   d) Ionic
   e) Metallic
9) Rank the following bonds in order of decreasing polarity:

   C—C C—H C—N C—O
10) Identify the incorrect statement among the following:
   a) An alloy is a solid mixture of two or more elements that has macroscopic metallic properties.
   b) The fact that metals are good electrical conductors can be explained theoretically by the nature of the metallic bond.
   c) The electrons in a metallic bond are localized.
   d) A metal effectively consists of metal ions surrounded by an electron sea.
   e) Metals are found to the left of the stair-step line in the periodic table.

**Answers to Chapter 12 Test Yourself**

1) d 2) e 3) a 4) b 5) 3 bonding electron pairs and 2 lone pairs
6) a and b are primarily ionic; c and d are primarily covalent
7) c 8) b 9) C—O > C—N > C—H > C—C 10) c

# Chapter in Review 13A

GOAL 1 **Draw the Lewis diagram for any molecule or polyatomic ion made up of main group elements.**

The procedure for **drawing a Lewis diagram** is

1. Count the total number of valence electrons. Adjust for charge on ions.
2. Place the least electronegative atom(s) in the center of the molecule.
3. Draw a tentative diagram. Join atoms by single bonds. Add unshared pairs to complete the octet around all atoms except hydrogen.
4. Calculate the number of valence electrons in your tentative diagram and compare it with the actual number of valence electrons. If the tentative diagram has too many electrons, remove a lone pair from the central atom and from a terminal atom, and replace them with an additional bonding pair between those atoms. If the tentative diagram still has too many electrons, repeat the process.
5. Check the Lewis diagram. Hydrogen atoms must have only one bond, and all other atoms should have a total of four electron pairs.

GOAL 2 **Describe the electron-pair geometry when a central atom is surrounded by two, three, or four electron pairs.**

**Electron-pair geometry** describes the arrangement of two, three, or four pairs of electrons, either shared or unshared, around a central atom.

| Electron Pairs | Geometry | Electron-Pair Angles |
|---|---|---|
| 2 | Linear | 180° |
| 3 | Trigonal planar | 120° |
| 4 | Tetrahedral | 109.5° |

GOAL 3 **Given or having derived the Lewis diagram of a molecule or polyatomic ion in which a second-period central atom is surrounded by two, three, or four pairs of electrons, predict and sketch the molecular geometry around that atom.**

**Molecular geometry** describes the arrangement of two, three, or four atoms around a central atom to which they are all bonded.

| Electron Pairs | Bonded Atoms | Molecular Geometry | Bond Angle |
|---|---|---|---|
| 2 | 2 | Linear | 180° |
| 3 | 3 | Trigonal planar | 120° |
| 3 | 2 | Angular | 120° |
| 4 | 4 | Tetrahedral | 109.5° |
| 4 | 3 | Trigonal pyramidal | 109.5° |
| 4 | 2 | Bent | 109.5° |

GOAL 4 **Draw a wedge-and-dash diagram of any molecule for which a Lewis diagram can be drawn.**

The procedure for **drawing a wedge-and-dash diagram** is

1. When two atoms are in the same plane as the page, they are connected with a solid line of uniform width.
2. When an atom is behind the plane of the page, it is connected to the central atom by a line that is dashed. The width of the dashed line increases as it moves away from the central atom.
3. When an atom is in front of the plane of the page, it is connected to the central atom by a line that is wedge-shaped. The width of the wedge-shaped line increases as it moves away from the central atom.

GOAL 5 **For a molecule with more than one central atom and/or multiple bonds, draw the Lewis diagram and predict and sketch the molecular geometry around each central atom, and draw a wedge-and-dash diagram of the molecule.**

It is the **number of regions of electron density** that surround a central atom that determines the electron-pair geometry around that atom. A region of electron density can be a single, double, or triple bond, or a lone pair. No matter the number of pairs of bonding electrons between two atoms, each region of electron density is distributed as far away from other regions of electron density as possible, as predicted by VSEPR theory.

GOAL 6 **Given or having determined the Lewis diagram of a molecule, predict whether the molecule is polar or nonpolar.**

**Molecular polarity** depends on both bond polarity and molecular geometry. A **polar molecule** is one in which there is an asymmetrical distribution of charge, resulting in positive and negative poles. A **nonpolar molecule** has either nonpolar bonds or polar bonds that cancel, resulting in no overall regions of positive and negative charge. If the central atom of a molecule has no lone pairs and all atoms

GOAL 7 **Distinguish between organic compounds and inorganic compounds.**

GOAL 8 **Distinguish between hydrocarbons and other organic compounds.**

GOAL 9 **On the basis of structure and the geometry of the identifying group, distinguish among alcohols, ethers, and carboxylic acids.**

bonded to it are identical, the molecule is nonpolar. If these conditions are not met, the molecule is polar.

The majority of all chemical compounds that have been characterized are classified as **organic compounds**—compvounds based upon the carbon atom.

**Hydrocarbons** are made of carbon and hydrogen. The **alkanes** are a hydrocarbon family with all single bonds.

If a hydrogen atom in a hydrocarbon, $CH_4$, for example, is replaced by a hydroxyl group, —OH, the resulting molecule is an **alcohol,** $CH_3$—OH. An alcohol may be thought of as a water molecule (H—OH) in which one hydrogen atom is replaced by a hydrocarbon group ($CH_3$—OH).

An **ether** may be thought of as a water molecule (H—O—H) in which both hydrogen atoms are replaced by hydrocarbon groups ($CH_3$—O—$CH_3$). Alcohols and ethers with the same number of carbon atoms are isomers.

If a hydrogen atom in a hydrocarbon is replaced by a carboxyl group, —COOH, the resulting molecule is a **carboxylic acid.** The most common carboxylic acid is acetic acid, written as $HC_2H_3O_2$ or $CH_3COOH$.

# Chapter in Review 13B

## hapter 13 Test Yourself

*structions:* Draw Lewis diagrams for each substance in Questions 1–8. You may refer to a "clean" periodic table.

$OF_2$ 2) $NH_3$ 3) $SeO_4^{2-}$ (Se, Z = 34) 4) $NO_3^-$ 5) $H_3PO_4$ 6) $C_3H_4$ 7) $C_3H_7F$ 8) $CH_3CH_2COOH$

*r Questions 9–12, draw wedge-and-dash diagrams and predict the electron-pair and molecular geometry of the substances om Questions 1–4.*

$OF_2$ 10) $NH_3$ 11) $SeO_4^{2-}$ 12) $NO_3^-$

*r Questions 13–14, draw Lewis diagrams and wedge-and-dash diagrams for the substance indicated, and state whether it is npolar or polar. If polar, indicate the electropositive and electronegative regions on the sketch.*

) $BHF_2$ 14) $BeBr_2$

) Write the Lewis structures of all the organic compounds having the formula $C_2H_6O$. Identify the alcohols and ethers.

---

**Answers to Chapter 13 Test Yourself**

1) 2) 3) 4) 2− 5)

6) or 7) or 8)

9) EP Geo: tetrahedral; Molecular Geo: bent

10) EP Geo: tetrahedral; Molecular Geo: trigonal pyramidal

11) EP Geo: tetrahedral; Molecular Geo: tetrahedral

12) EP Geo: trigonal planar; Molecular Geo: trigonal planar

13) δ+ δ−

14) nonpolar

15) alcohol ether

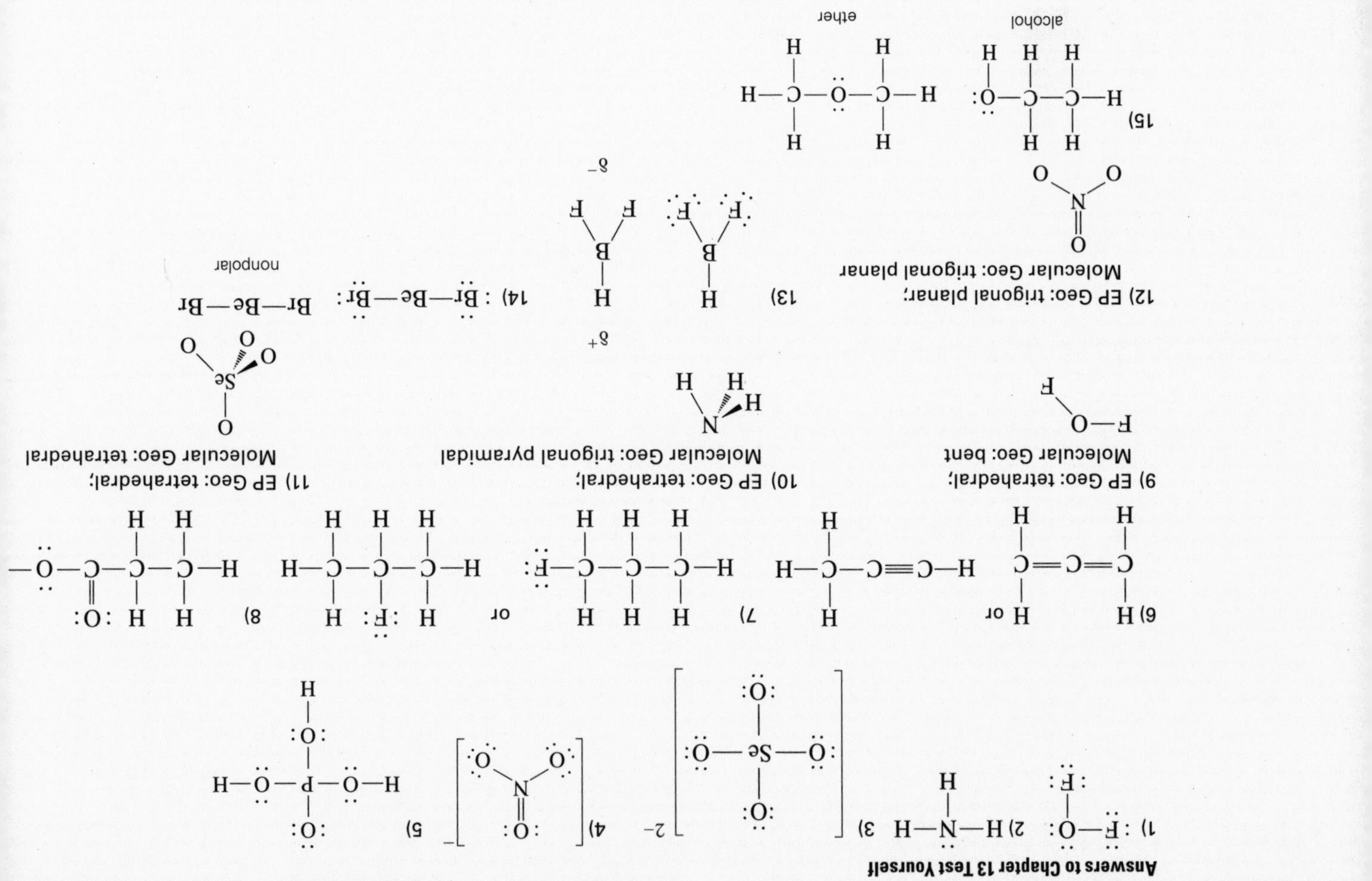

# Chapter in Review 14

Important ideas to review from Chapter 4:

**Charles's Law** states that at constant pressure, the volume of a fixed quantity of a gas is directly proportional to the absolute temperature, $V \propto T$.

**Boyle's Law** states that at constant temperature, the volume of a fixed quantity of a gas is inversely proportional to its pressure, $V \propto 1/P$.

Charles's and Boyle's Laws can be coupled as the **Combined Gas Law:**

$$\frac{P_1V_1}{T_1} = \frac{P_2V_2}{T_2}.$$

GOAL 1 **If pressure and temperature are constant, state how volume and amount of gas are related and explain phenomena or make predictions based on that relationship.**

**Avogadro's Law** states that equal volumes of two gases at the same temperature and pressure contain the same number of molecules, $V \propto n$.

GOAL 2 **Explain how the ideal gas equation can be constructed by combining Charles's, Boyle's, and Avogadro's Laws, and explain how the ideal gas equation can be used to derive each of the three two-variable laws.**

Since $V \propto T$, $V \propto 1/P$, and $V \propto n$, it follows that $V \propto T \times (1/P) \times n$. Inserting a proportionality constant R, $V = RT(1/P)n$, or, rearranging, $\mathbf{PV = nRT}$. When pressure and amount are constant, $V = kT$, which is Charles's Law. When temperature and amount are constant, $PV = k$, which is Boyle's Law. When pressure and temperature are constant, $V = kn$, which is Avogadro's Law.

GOAL 3 **Given values for all except one of the variables in the ideal gas equation, calculate the value of the remaining variable.**

The **ideal gas equation** is $PV = nRT$. Two values of the **universal gas constant, R,** are 0.0821 L · atm/mol · K and 62.4 L · torr/mol · K. Given all the values in the ideal gas equation except one, the remaining value may be calculated. Substituting m/MM for its equivalent, n, in the ideal gas equation gives $PV = \frac{m}{MM}RT$.

GOAL 4 **Calculate the density of a known gas at any specified temperature and pressure.**

Solving the $PV = (m/MM)RT$ form of the ideal gas equation for m/V, which is **density,** yields $D = \frac{m}{V} = \frac{(MM)P}{RT}$.

GOAL 5 **Given the density of a pure gas at a specified temperature and pressure, or information from which it may be found, calculate the molar mass of that gas.**

The density of a gas is directly proportional to its molar mass, $\mathbf{D \propto MM}$. Either molar mass or density can be calculated from the other using the ideal gas equation.

GOAL 6 **Calculate the molar volume of any gas at any given temperature and pressure.**

**Molar volume is the volume occupied by one mole of a gas, V/n.** The molar volume of any ideal gas at STP (0°C and 1 bar) is 22.7 L/mol. This quantity is useful in calculations involving moles, mass, volume, density, and molar mass of a gas measured at STP.

GOAL 7 **Given the molar volume of a gas at any specified temperature or pressure, or information from which the molar volume may be determined, and either the number of moles in or the volume of a sample of that gas, calculate the other quantity.**

Molar volume is $MV = \frac{V}{n} = \frac{RT}{P}$. Once molar volume is determined, it can be used to convert between the macroscopic volume of a gas and the particulate-level number of particles, grouped in moles.

**GOAL 8 Given a chemical equation, or a reaction for which the equation can be written, and the mass or number of moles of one species in the reaction, or the STP volume of a gaseous species, find the mass or number of moles of another species, or the STP volume of another gaseous species.**

22.7 liters per mole is a dimensional analysis conversion factor that can be used to convert between the volume of a gas at STP and the number of particles of that gas, counted in moles. 22.7 L/mol can be used *only* for ideal gases at STP. If your stoichiometry skills are rusty, review Section 10.2.

**GOAL 9 Given a chemical equation, or a reaction for which the equation can be written, and the mass or number of moles of one species in the reaction, or the volume of any gaseous species at a given temperature and pressure, find the mass or number of moles of any other species, or the volume of any other gaseous species at a given temperature and pressure.**

A **gas stoichiometry** problem at non-STP conditions can be solved by finding the molar volume of the gas and then following the stoichiometry path (molar volume method presented in Section 14.7) or by applying the ideal gas equation and then following the stoichiometry path or by following the stoichiometry path and then applying the ideal gas equation (ideal gas equation method presented in Section 14.8).

**GOAL 10 Given a chemical equation, or a reaction for which the equation can be written, and the volume of any gaseous species at a given temperature and pressure, find the volume of any other gaseous species at a given temperature and pressure.**

The ratio of volumes of gases in a reaction is the same as the ratio of moles, provided that the gas volumes are measured at the same temperature and pressure. Thus the coefficients in a balanced chemical equation can be used to convert between volumes, as long as the volumes are at the same temperature and pressure.

# Chapter 14 Test Yourself

1) One of two identical containers holds oxygen, and the other container holds chlorine. Both gases exert a pressure of 1.19 atm at 21°C. Which statement is incorrect?
   a) The number of molecules of oxygen is the same as the number of molecules of chlorine.
   b) The mass of oxygen is equal to the mass of chlorine.
   c) The number of moles of oxygen is equal to the number of moles of chlorine.
   d) The number of oxygen atoms is equal to the number of chlorine atoms.
2) There is 0.028 mol of an ideal gas in a 0.377 L container at 293 torr. What is the temperature of the gas (°C)?
3) What is the density (g/L) of ammonia at STP?
4) Find the molar mass of a gas if 0.460 L, measured at 819 torr and 22°C, has a mass of 0.369 gram.
5) What is the molar volume of fluorine gas at −17°C and 1.03 atm?
6) The molar volume of hydrogen bromide gas at 14°C and 772 torr is 23.2 L/mol. How many moles of gas are in a 1.25 L vessel at these conditions?
7) Carbon dioxide can be removed from a closed-container breathing apparatus by reaction with potassium superoxide: $4\ KO_2(s) + 2\ CO_2(g) \rightarrow 2\ K_2CO_3(s) + 3\ O_2(g)$. Calculate the mass of potassium superoxide needed to remove an STP volume of 10.0 L of carbon dioxide.
8) Calculate the mass (in grams) of zinc that must react to produce 148 mL of hydrogen gas at 767 torr and 24°C by the reaction $Zn(s) + 2\ HCl(aq) \rightarrow H_2(g) + ZnCl_2(aq)$.
9) What volume of oxygen, measured at 0.891 atm and 18°C is needed to burn completely 4.18 L of butane measured at 1.34 atm and 38°C? The gas-phase reaction is $2\ C_4H_{10}(g) + 13\ O_2(g) \rightarrow 8\ CO_2(g) + 10\ H_2O(g)$.

**Answers to Chapter 14 Test Yourself**

1) b 2) $-2.10 \times 10^2$ °C 3) 0.770 g/L 4) 18.0 g/mol 5) 20.4 L/mol 6) 0.0539 mol 7) 62.6 g $KO_2$ 8) 0.400 g Zn 9) 38.2 L $O_2$

# Chapter in Review 15A

GOAL 1 **Given the partial pressure of each component in a mixture of gases, find the total pressure.**

The **partial pressure** of a gas in a gaseous mixture is the pressure that gas alone would exert in the same volume at the same temperature.

GOAL 2 **Given the total pressure of a gaseous mixture and the partial pressures of all components except one, or information from which those partial pressures can be obtained, find the partial pressure of the remaining component.**

The total pressure of a gas mixture is the sum of the partial pressures of all gases in that mixture, $P = p_1 + p_2 + p_3 + \cdots$. This is **Dalton's Law of Partial Pressures.**

GOAL 3 **Explain the differences between the physical behavior of liquids and gases in terms of the relative differences among particles and the effect of those distances on intermolecular forces.**

Important **properties of liquids** (and comparisons with gases) include:

Liquid particles are "touchingly close." Gases can be compressed; liquids cannot.

The attractions among liquid particles are sufficiently strong to hold them together. Gases expand to fill their containers; liquids do not.

A given number of liquid particles occupies a much smaller volume than the same number of particles occupies as a gas. Gases have low densities; liquids have relatively high densities.

There is no space between particles of a liquid, so combining liquids must increase volume. Gases may be mixed in a fixed volume; liquids cannot.

GOAL 4 **For two liquids, given comparative values of physical properties that depend on intermolecular attractions, predict the relative strengths of those attractions; or, given a comparison of the strengths of the intermolecular attractions, predict the relative values of physical properties that the attractions cause.**

**Properties of liquids are related to intermolecular attractions:**

**Vapor pressure** is the partial pressure of a vapor in equilibrium with its liquid state at a given temperature. Liquids with strong intermolecular attractions have lower vapor pressures than liquids with weak intermolecular attractions.

**Molar heat of vaporization** is the energy required to change one mole of a liquid to a gas, while at constant temperature and pressure. Liquids with strong intermolecular attractions have higher molar heats of vaporization than liquids with weak intermolecular attractions.

**Boiling point** is the temperature at which vapor pressure becomes equal to the pressure above a liquid. Liquids with strong intermolecular attractions have higher boiling points than liquids with weak intermolecular attractions.

**Viscosity** is the ability of a liquid to flow. Liquids with strong intermolecular attractions have higher viscosities than liquids with weak intermolecular attractions.

**Surface tension** is the energy required to break through the surface of a liquid or to disrupt a drop of liquid and spread the material into a film. Liquids with strong intermolecular attractions have stronger surface tension than liquids with weak intermolecular attractions.

GOAL 5 **Identify and describe or explain induced dipole forces, dipole forces, and hydrogen bonds.**

**Induced dipole forces** are comparatively weak intermolecular attractions between nonpolar molecules. They are the result of temporary dipoles caused by shifting electron density in molecules. Induced dipole forces vary directly with surface area and may be large if the molecules are large.

**Dipole forces** give an electrostatic attraction between polar molecules.

Exceptionally strong dipole-like forces called **hydrogen bonds** arise between molecules that have hydrogen atoms bonded to a highly electronegative atom. This atom, usually nitrogen, oxygen, or fluorine, must have at least one unshared electron pair.

GOAL 6 **Given the structure of a molecule, or information from which it may be determined, identify the significant intermolecular forces present.**

In general, a liquid with nonpolar molecules will have only **induced dipole forces** acting among the molecules. The larger the molecules, the greater the strength of the attractive forces. Liquids with polar molecules have **dipole forces** acting among the molecules. The more polar the molecules, the stronger the forces. Liquids with molecules that have a hydrogen atom bonded to an atom that is small and highly electronegative and has at least one unshared pair of electrons have **hydrogen bonds** acting among the molecules.

GOAL 7 **Given the molecular structures of two substances, or information from which they may be obtained, compare or predict relative values of physical properties that are related to them.**

All other things being equal, intermolecular attractive forces increase in the order: induced dipoles $<$ dipole forces $<$ hydrogen bonds. In large molecules, however, induced dipole forces can be the most important forces acting to determine values of physical properties.

GOAL 8 **Describe or explain the equilibrium between a liquid and its own vapor and the process by which it is reached.**

**Equilibrium** is defined as the condition in which the rates of opposing changes are equal. In a liquid-vapor equilibrium, the rate of evaporation is equal to the rate of condensation. Such an equilibrium is achieved in a closed flask by starting with the movement of molecules is one direction, from liquid to vapor. The condensation rate is initially zero. As the number of molecules in the vapor state increases, the condensation rate increases. Simultaneously, the evaporation rate stays constant. As long as the rate of evaporation is greater than the rate of condensation, the vapor concentration will rise. Eventually, the evaporation and condensation rates become equal, and equilibrium is achieved.

GOAL 9 **Describe the relationship between vapor pressure and temperature for a liquid–vapor system in equilibrium; explain this relationship in terms of the kinetic molecular theory.**

The partial pressure exerted by a vapor in equilibrium with a liquid is the **equilibrium vapor pressure** at the existing temperature.

Equilibrium vapor pressure increases with increasing temperature. At higher temperatures, a larger fraction of the liquid sample has enough energy to evaporate.

GOAL 10 **Describe the process of boiling and the relationships among boiling point, vapor pressure, and surrounding pressure.**

The **boiling point** of a liquid is the temperature at which the vapor pressure of that liquid is equal to or slightly greater than the surrounding pressure. Normal boiling point is the boiling point at one atmosphere of pressure.

GOAL 11 **Describe the typical relative density relationship between the solid and liquid phase of a substance, and explain why water is an exception to this trend.**

**Water,** a molecule necessary for life, breaks almost all the rules for predicting physical properties of liquids due to its extremely strong hydrogen bonding, which leads to exceptionally strong intermolecular attractions among its molecules.

Some of the **unusual properties of water** include an anomalously high boiling point, high surface tension, high heat of vaporization, low vapor pressure, high viscosity, an exceptional ability as a solvent, and its unusual state (liquid) at common temperatures and pressures. The fact that solid water (ice) floats on liquid water is also unusual.

GOAL 12 **Distinguish among amorphous, polycrystalline, and crystalline solids.**

Solids can be classified based upon particle arrangement. A **crystalline solid** has its particles arranged in a repeating pattern. An **amorphous solid** has no long-range order among its particles. A **polycrystalline solid** has groups of particles arranged in a pattern, but these groups are randomly arranged.

GOAL 13 **Distinguish among the following types of crystalline solids: ionic, molecular, covalent network, and metallic.**

Crystalline solids can be classified based upon the forces that hold the particles together:

**Ionic crystals** are composed of oppositely charged ions that are held together by strong ionic bonds. Ionic crystals typically have a high melting temperature, they are frequently water soluble, and they have very low electrical conductivities.

**Molecular crystals** are made of small, discrete molecules held together by relatively weak intermolecular forces. Molecular crystals are typically soft, have a low melting temperature, and are generally insoluble in water. They are usually nonconductors.

**Covalent network solids** are composed of atoms that are covalently bonded to each other to form a single, indefinite-size network. Covalent network solids are almost always insoluble in any common solvent, poor conductors of electricity, and they have high melting points.

**Metallic crystals** are made of a lattice of positive ions through which valence electrons move freely. The freely moving electrons make metals excellent conductors of electricity. Metallic crystals are insoluble in common solvents, malleable, and ductile (can be pressed into thin sheets and drawn into wire), and they have a wide range of melting points.

# Chapter in Review 15B

| | |
|---|---|
| OAL 14 **Given two of the following, lculate the third: (a) mass of a pure bstance changing between the liquid d vapor (gaseous) states, (b) heat of porization, (c) energy change.** | **Heat of vaporization** (or condensation) is the energy transferred when one gram of a substance changes between the liquid and gaseous states. |
| OAL 15 **Given two of the following, lculate the third: (a) mass of a pure bstance changing between the solid d liquid states, (b) heat of fusion, energy change.** | **Heat of fusion** (or solidification) is the energy transferred when one gram of a substance changes between the liquid and solid states. |
| OAL 16 **Given three of the following quantities, calculate the fourth: energy change, (b) mass of a pure bstance, (c) specific heat of the subance, (d) temperature change, or inil and final temperatures.** | **Specific heat** is the amount of energy needed to change the temperature of one gram of a substance one degree Celsius.<br>Heat transfer questions involve either three or four factors. You can solve them using either dimensional analysis or algebra. |
| OAL 17 **Sketch, interpret, or ideny regions in a graph of temperature rsus energy for a pure substance er a temperature range from below e melting point to above the boiling int.** | For a pure substance, a graph of temperature versus energy added to or removed from that substance expresses temperature change in the sloped regions and change of state in the horizontal regions. Heat flow from one point to another on the graph is the sum of the heat flows for each step in the process. |
| OAL 18 **Given (a) the mass of a pure bstance, (b) $\Delta H_{vap}$ and/or $\Delta H_{fus}$ of the bstance, and (c) the average specific at of the substance in the solid, liqd, and/or vapor state, calculate the tal heat flow in going from one state d temperature to another state and mperature.** | |

# Chapter in Review 15C

## hapter 15 Test Yourself

*structions:* You may use a "clean" periodic table.

Calculate the total pressure in a gaseous mixture with partial pressures of 0.48 atm He, 0.23 atm Ne, and 0.88 atm Ar.

The primary reason intermolecular forces are stronger in liquids than in gases is that
a) liquids are cooler than gases. b) molecules are closer to each other in liquids.
c) liquids weigh more than gases. d) the liquid state is between the gaseous state and the solid state.

The bulb of a medicine dropper is depressed and the tip is immersed into liquid A. When the bulb is released the dropper fills quickly. After cleaning the dropper the identical procedure is followed with liquid B, but the filling is slower, more sluggish. From these observations
a) it is reasonable to predict that intermolecular attractions are stronger in A than in B.
b) it is reasonable to predict that intermolecular attractions are about the same in A and in B.
c) it is reasonable to predict that intermolecular attractions are stronger in B than in A.
d) no reasonable prediction can be made about the relative strengths of intermolecular attractions in A and B.

Select from the following the statement about intermolecular forces that is incorrect.
a) Attractions between molecules without hydrogen bonding are generally weaker than attractions between hydrogen bonded molecules of about the same size.
b) Induced dipole forces exist only between nonpolar molecules.
c) Dipole forces exist only between polar molecules.
d) Hydrogen bonding would not be evident in molecules in which all the hydrogen atoms were covalently bonded to selenium ($Z = 34$) and no other element.

What type(s) of intermolecular forces would you expect in $CHF_3(\ell)$?
a) Induced dipole only b) Dipole and induced dipole c) Induced dipole and hydrogen bonding d) Dipole only

Based on the generalizations developed in the chapter, which of the following is listed in order of increasing boiling point?
a) $OF_2 < Cl_2O < Br_2O$ b) $Cl_2O < OF_2 < Br_2O$ c) $Cl_2O < Br_2O < OF_2$ d) $Br_2O < OF_2 < Cl_2O$

Closed system A consists of liquid $CCl_4$ in equilibrium with its own vapor at 20°C. System B is identical to system A, except that the equilibrium temperature is 40°C. Identify the incorrect statement among the following:
a) The vapor pressure of A is less than that of B. b) The evaporation rate of A is less than that of B.
c) The condensation rate of A is less than that of B. d) The vapor concentration of A is greater than that of B.

Acetone, a highly volatile liquid, is placed in a container, and the container is sealed. The liquid disappears. The vapor pressure of the acetone in the container
a) is lower than the equilibrium vapor pressure. b) is higher than the equilibrium vapor pressure.
c) is the same as the equilibrium vapor pressure. d) cannot be compared to the equilibrium vapor pressure without additional information.

The normal boiling points of the two major components of air, nitrogen and oxygen, are −196°C and −183°C, respectively. Liquid air at −200°C is in a closed cylinder with gaseous helium (boiling point −269°C) at 760 torr above the surface of the liquid. If the pressure of helium is slowly released,
a) nitrogen will begin to boil before oxygen. b) nitrogen and oxygen will begin to boil at the same time.
c) oxygen will begin to boil before nitrogen. d) you cannot predict which liquid will boil first.

0) Silly Putty is a slightly elastic solid that bounces like a rubber ball when dropped. This elasticity suggests that Silly Putty is a(n) ______ solid.
a) molecular crystalline b) ionic crystalline c) amorphous d) network crystalline

1) Germane melts at −165°C, is insoluble in water, and does not conduct electricity when melted. Germane most likely forms a(n) ______ crystal.
a) ionic b) metallic c) network d) molecular

2) Calculate the heat of vaporization in kJ/g of sodium metal if 10.0 kJ of energy is needed to boil 2.35 grams of the metal already at the boiling point.

3) How much heat, in kJ, is needed to melt 123 grams of Ni(s), already at its melting point, if the heat of fusion for Ni(s) is $3.10 \times 10^2$ J/g?

14) The graph below presents a temperature versus energy plot for a pure substance. Identify the region(s) on the graph (i) where only liquid exists and (ii) the heat of vaporization.

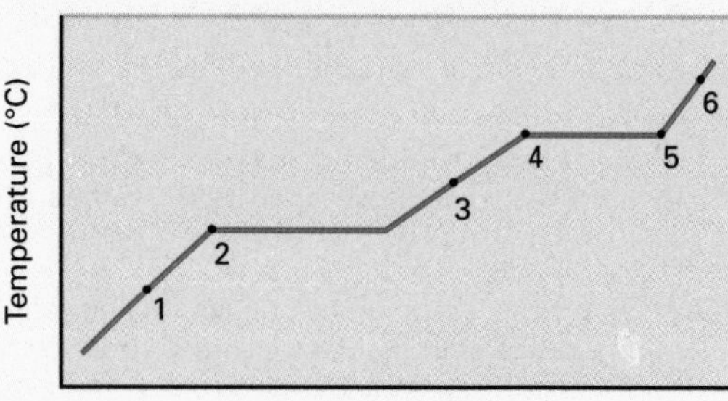


15) Calculate the heat absorbed by 3636 grams of lubricating oil in a car engine as it warms from a garage temperature of 16°C to an operating temperature of 110°C. The specific heat of the oil is 3.6 J/g · °C.

16) A 65.0 gram sample loses 1.92 kJ when cooling from 114.6°C to 31.2°C. Calculate the specific heat of the sample.

17) What is the total heat flow if 44 grams of $H_2O(\ell)$, initially at 82°C is cooled to $H_2O(s)$ at −23°C? The specific heat of $H_2O(\ell)$ is 4.18 J/g · °C; the specific heat of $H_2O(s)$ is 2.1 J/g · °C. The heat of fusion of water is 335 J/g, and the freezing point of water is 0.0°C.

**Answers to Chapter 15 Test Yourself**

1) 0.48 atm + 0.23 atm + 0.88 atm = 1.59 atm  2) b  3) c  4) b  5) b  6) a  7) d  8) a  9) a  10) c  11) d  12) 4.26 kJ
13) 38.1 kJ
14) (i) Liquid only exists in the sloped region that includes 3. (ii) Heat of vaporization is the region between 4 and 5.
15) $1.2 \times 10^3$ kJ  16) 0.354 J/g · °C  17) −32 kJ

# Chapter in Review 16A

GOAL 1 **Define the term *solution,* and, given a description of a substance, determine if it is a solution.**

**A solution is a homogeneous mixture.** Solutions have variable physical properties, which are determined by the composition of the mixture.

GOAL 2 **Distinguish among terms in the following groups: *solute* and *solvent; concentrated* and *dilute; solubility, saturated, unsaturated,* and *supersaturated; miscible* and *immiscible.***

The major component of a solution is called the **solvent;** the minor components are called the **solutes.** A **concentrated** solution has a relative large quantity of solute per quantity of solvent; a **dilute** solution has a relatively small quantity of the same solute per quantity of solvent. A solution whose concentration is at the solubility limit for a given temperature is a **saturated** solution. If the concentration is less than the solubility limit, the solution is **unsaturated.** A **supersaturated** solution has a concentration greater than the normal solubility limit. Liquids are **miscible** if they dissolve in each other in all proportions. Liquids that are insoluble in each other are **immiscible.**

GOAL 3 **Describe the formation of a saturated solution from the time excess solid solute is first placed into a liquid solvent.**

When a soluble ionic solid solute is placed in water, water molecules surround the ions, helping them move away from their positions in the crystal. In solution, the ions are **hydrated,** or surrounded by water molecules. The **dissolving rate** is constant throughout the process. When dissolving has just begun, the **crystallization rate** is zero, and it will increase until it equals the dissolving rate and **equilibrium** is established.

GOAL 4 **Identify and explain the factors that determine the time required to dissolve a given amount of solute or to reach equilibrium.**

A **finely divided solid** dissolves more rapidly because of greater surface area. **Stirring** or **agitating** a solution makes it dissolve more rapidly because of prevention of buildup of concentration at the solute surface. **Higher temperature** causes a solution to dissolve more rapidly because of faster particle movement.

GOAL 5 **Given the structural formulas of two molecular substances, or other information from which the strength of their intermolecular forces may be estimated, predict if they will dissolve appreciably in each other, and state the criteria on which your prediction is based.**

Substances with **similar intermolecular forces** will usually dissolve in one another.

GOAL 6 **Predict how and explain why the solubility of a gas in a liquid is affected by a change in the partial pressure of that gas over the liquid.**

In most dilute solutions, **the solubility of a gas is directly proportional to the partial pressure of the gas over the surface of the liquid.**

GOAL 7 **Given mass of solute and of solvent or solution, calculate percentage concentration.**

$$\text{\% by mass} = \frac{\text{g solute}}{\text{g solution}} \times 100$$

GOAL 8 **Given mass of solution and percentage concentration, calculate mass of solute and solvent.**

$$\text{\% by mass} = \frac{\text{g solute}}{\text{g solute} + \text{g solvent}} \times 100$$

GOAL 9 **Given two of the following, calculate the third: moles of solute (or data from which it may be found), volume of solution, molarity.**

$$\text{Molarity} = \text{M} \equiv \frac{\text{moles solute}}{\text{liter solution}} = \frac{\text{mol}}{\text{L}}$$

GOAL 13 **Given any three of the following, calculate the fourth: (a) volume of concentrated solution, (b) molarity of concentrated solution, (c) volume of dilute solution, (d) molarity of dilute solution.**

The key relationship for calculations involving dilution of concventrated solutions is $\mathbf{M_c \times V_c = M_d \times V_d}$, in which M is molarity, V is volume, c is concentrated, and d is diluted.

| | |
|---|---|
| GOAL 14 **Given the quantity of any species participating in a chemical reaction from which the equation can be written, find the quantity of any other species, either quantity being measured in (a) grams, (b) volume of solution at specified molarity, or (c) (if gases have been studied) volume of gas at given temperature and pressure.** | For all **stoichiometry** problems, a macroscopic measurable quantity (mass, energy, volume of a gas at known pressure and temperature, or volume of a solution of known concentration) is changed to the number of particles, grouped in moles. The mole ratio from the balanced chemical equation is then used to change to the number of particles of another species involved in the chemical change. Finally, the number of moles is changed to a macroscopic measurable quantity. For solution stoichiometry, volume of solution and concentration, typically molarity, can be used to convert to number of moles. |
| GOAL 15 **Given the volume of a solution that reacts with a known mass of a primary standard and the equation for the reaction, calculate the molarity of the solution.** | **Titration** is the controlled and measured addition of one solution into another. Titration problems are solution stoichiometry problems. |
| GOAL 16 **Given the volumes of two solutions that react with each other in a titration, the molarity of one solution, and the equation for the reaction or information from which it can be written, calculate the molarity of the second solution.** | A **buret** is a device that measures delivered volumes precisely. An **indicator** is a substance that exhibits different color in solution at different solution acidities. A solution is **standardized** when its concentration is determined by reaction with a substance that can be weighed accurately, which is called a **primary standard.** |

## Chapter 16 Test Yourself

*Instructions:* You may use a "clean" periodic table.

1) A solution that can dissolve more of a substance than it currently contains is said to be
   a) supersaturated. b) saturated. c) unsaturated. d) immiscible. e) miscible.
2) When undissolved solute is present in an unsaturated solution, the rate of dissolving is _____ the rate of crystallization.
   a) greater than b) equal to c) less than
3) The time required between adding excess solute to a solvent and reaching equilibrium with a saturated solution can be reduced by all of the following except
   a) raising the temperature. b) stirring the mixture.
   c) illuminating the solution. d) reducing solute particle size.
4) Considering the structural formulas of $CH_3OH$ and $CH_3CH_2OH$, it is logical to expect these liquids are
   a) miscible because their intermolecular attractions are similar.
   b) immiscible because one molecule is highly polar and the other nonpolar.
   c) immiscible because one molecule has hydrogen bonding and the other does not.
   d) immiscible because their molecular masses and sizes are so different.
5) A cylinder contains liquid water and nitrogen, oxygen, and carbon dioxide gases. Only carbon dioxide dissolves appreciably in the water, and there is an equilibrium between the dissolved and undissolved carbon dioxide. More nitrogen gas is forced into this cylinder, at constant temperature. The solubility of the carbon dioxide in the water
   a) increases greatly. b) increases slightly. c) remains unchanged. d) decreases slightly.
6) How would you make 312 grams of a 0.90% sodium chloride solution?
7) How many mL of 0.415 M $H_2SO_4$ do you need to have 0.716 mole $H_2SO_4$?
8) A 62.5 mL sample of 12.0 M $HNO_3$ is diluted to a final concentration of 0.812 M. What is the final volume of this solution?

*Questions 9–12 refer to the equation* $Na_2CO_3(aq) + 2\ HCl(aq) \rightarrow CO_2(g) + 2\ NaCl(aq) + H_2O(\ell)$

9) What volume of $CO_2(g)$, measured at 756 torr and 23°C, is obtained from the reaction of 24.16 mL of 0.0872 M $Na_2CO_3$ with excess HCl?
10) What volume of 0.123 M HCl is needed to react with 14.27 mL of 0.102 M $Na_2CO_3$?
11) What is the molarity of the HCl if 19.46 mL react with 0.317 grams of sodium carbonate?
12) What is the molarity of the HCl if 22.4 mL react with 11.7 mL of 0.113 M $Na_2CO_3$?

**Answers to Chapter 16 Test Yourself**

1) c 2) a 3) c 4) a 5) c 6) 2.8 g NaCl + 309 g $H_2O$ 7) $1.73 \times 10^3$ mL 8) 924 mL 9) 0.0515 L $CO_2$ 10) 23.7 mL HCl
11) 0.307 M HCl 12) 0.118 M HCl

# Chapter in Review 16B

This Portable Content card covers optional Sections 16.7 on Molality, 16.8 on Normality, 16.13 on Titration Using Normality, and 16.14 on Colligative Properties of Solutions.

GOAL 10 **Given two of the following, calculate the third: moles of solute (or data from which it may be found), mass of solvent, molality.**

**Molality** is moles of solute in one kilogram of solvent. The symbol for molality is m (note that molality is lowercase m and molarity is uppercase M).

$$m \equiv \frac{\text{mol solute}}{\text{kg solvent}}$$

Molality is used in situations where temperature independence is important. Neither the number of solute particles nor the mass of the solvent varies with temperature. In contrast, molarity is temperature dependent because the volume of a solution varies with temperature.

GOAL 11 **Given an equation for a neutralization reaction, state the number of equivalents of acid or base per mole and calculate the equivalent mass of the acid or base.**

One **equivalent** of an acid is defined as that amount of acid that yields one mole of hydrogen ion in a specific reaction. One equivalent of a base is that amount of base that reacts with one equivalent of an acid.

The **equivalent mass** of a substance is the number of grams of the substance per equivalent. To calculate equivalent mass, divide molar mass by equivalents per mole:

$$\frac{\text{g/mol}}{\text{eq/mol}} = \frac{\text{g}}{\text{mol}} \times \frac{\text{mol}}{\text{eq}} = \frac{\text{g}}{\text{eq}}, \text{ equivalent mass}$$

GOAL 12 **Given two of the following, calculate the third: equivalents of acid or base (or data from which they can be found), volume of solution, normality.**

**Normality** is the number of equivalents of solute in one liter of solution. Units of normality are equivalents per liter, or eq/L. The symbol for normality is N.

$$N \equiv \frac{\text{equivalents solute}}{\text{liter solution}} = \frac{\text{eq}}{\text{L}}$$

GOAL 17 **Given the volume of a solution that reacts with a known mass of a primary standard and the equation for the reaction, calculate the normality of the solution.**

In a reaction, **the number of equivalents of acid and base that react with each other are equal.** This idea of equal numbers of equivalents is the basis of the normality system.

There are two ways to calculate the number of equivalents in a sample of a substance:

1) If you know the mass of the substance and its equivalent mass, use the equivalent mass as a conversion factor to get equivalents.

2) If the sample is a solution and you know its volume and normality, multiply one by the other: $V \times N = \text{eq}$.

GOAL 18 **Given the volumes of two solutions that react with each other in a titration and the normality of one solution, calculate the normality of the second solution.**

For an acid-base titration, $\mathbf{V_{acid} \times N_{acid} = V_{base} \times N_{base}}$.

The properties of a solution that depend only on the number of solute particles present, without regard to their identity, are called **colligative properties.**

GOAL 19 **Given (a) the molality of a solution, or data from which it may be found, (b) the normal freezing or boiling point of the solvent, and (c) the freezing- or boiling-point constant, find the freezing or boiling point of the solution.**

**Freezing point depression and boiling point elevation** are colligative properties that are directly proportional to the molal concentration of any solute. These proportionality constants are called the molal freezing point constant and the molal boiling point constant, respectively. The values of these constants depend only on the chemical identity of the solvent in the solution.

**GOAL 20 Given the freezing-point depression or boiling-point elevation and the molality of a solution, or data from which they may be found, calculate the molal freezing-point constant or molal boiling-point constant.**

For freezing point depression, $\Delta T_f = K_f m$.

For boiling point elevation, $\Delta T_b = K_b m$.

**GOAL 21 Given (a) the mass of solute and solvent in a solution, (b) the freezing-point depression or boiling-point elevation, or data from which they may be found, and (c) the molal freezing/boiling-point constant of the solvent, find the approximate molar mass of the solute.**

Freezing point depression experiments may be used to determine molar mass. To calculate the molar mass of a solute from freezing-point depression or boiling-point elevation data,

1) Calculate molality from $m = \Delta T_f/K_f$ or $m = \Delta T_b/K_b$. Express as mol solute/kg solvent.

2) Using molality as a conversion factor between moles of solute and kilograms of solvent, find the number of moles of solute.

3) Use the defining equation for molar mass, MM = g/mol, to calculate the molar mass of the solute.

# Chapter 16 Test Yourself: Optional Sections

*Instructions:* You may use a "clean" periodic table.

**Molality and Colligative Properties Test Yourself**

1) Adding 1.60 g acetic acid, $HC_2H_3O_2$, to 21.47 g of solvent raises the boiling point of a solution 3.14°C. What is $K_b$ for this solvent?
2) If 5.22 grams of naphthalene, $C_{10}H_8$, are dissolved in 35.81 grams of cyclohexane, $C_6H_6$, what is the molality of the resulting solution?
3) If pure cyclohexane freezes at 6.5°C and the molal freezing point constant $K_f$ is 20.2°C/m for cyclohexane, what is the freezing point of the solution made in Question 2?
4) A solution is made by dissolving 4.18 g of an unknown solid in 19.89 g solvent. The freezing point falls by 4.31°C. If $K_f$ for the solvent is 5.48°C/m, what is the molar mass of the unknown solid?

**Normality Test Yourself**

*All questions refer to the equation* $Na_2CO_3(aq) + 2\,HCl(aq) \rightarrow CO_2(g) + 2\,NaCl(aq) + H_2O(\ell)$.

1) What is the equivalent mass of sodium carbonate?
2) A volume of 17.9 mL 0.119 N $Na_2CO_3$ contains how many equivalents of base?
3) What is the normality of the hydrochloric acid if 21.42 mL react with 0.288 grams of sodium carbonate?
4) What is the normality of the sodium carbonate if 11.3 mL react with 25.1 mL of 0.0995 N HCl?

**Answers to Chapter 16 Test Yourself: Optional Sections**

**Molality and Colligative Properties**

1) 2.53°C/m 2) 1.14 m $C_{10}H_8$ 3) −16.5°C 4) 268 g/mol

**Normality**

1) 52.995 g $Na_2CO_3$/eq 2) 0.00213 eq $Na_2CO_3$ 3) 0.253 N HCl 4) 0.221 N $Na_2CO_3$

# Chapter in Review 17

GOAL 1 **(Optional) Distinguish between an acid and a base according to the Arrhenius theory of acids and bases.**

The hydrogen ion acid–hydroxide ion base model of acids and bases is based on chemical properties and is known as the **Arrhenius theory** of acids and bases.

GOAL 2 **Given the equation for a Brønsted–Lowry acid–base reaction, explain how or why it can be so classified.**

According to the **Brønsted–Lowry theory,** an acid–base reaction involves a transfer of a proton from one substance, the acid, to another, the base. An acid is a proton source; a base is a proton remover. When writing an equation, there must be a Brønsted–Lowry base whenever there is a Brønsted–Lowry acid.

GOAL 3 **Given the formula of a Brønsted–Lowry acid and the formula of a Brønsted–Lowry base, write the net ionic equation for the reaction between them.**

Acid–base reactions are reversible; they reach a state of equilibrium according to the general equation $HA + B \rightleftharpoons A^- + AB^+$.

GOAL 4 **(Optional) Distinguish between a Lewis acid and a Lewis base. Given the structural formula of a molecule or ion, state if it can be a Lewis acid, a Lewis base, or both, and explain why.**

According to the **Lewis theory** of acids and bases, an acid is an electron-pair acceptor, and a base is an electron-pair donor.

GOAL 5 **(Optional) Given the structural equation for a Lewis acid–base reaction, explain how or why it can be so classified.**

Structurally, the most common feature of **Lewis acids** in this course is that they tend to be positive ions or species with a central atom with less that a full octet of valence electrons. **Lewis bases** tend to be negative ions or species with central atoms with one or more unshared electron pairs.

GOAL 6 **Define and identify conjugate acid–base pairs.**

Two substances whose formulas differ only by a proton (a hydrogen ion) are a **conjugate acid–base pair.**

GOAL 7 **Given the formula of an acid or a base, write the formula of its conjugate base or acid.**

If an acid has the general form HA, its conjugate base has the general form $A^-$. If a base has the general form B, its conjugate acid has the general form $HB^+$.

GOAL 8 **Given a table of the relative strengths of acids and bases, arrange a group of acids or a group of bases in order of increasing or decreasing strength.**

The relative strengths of Brønsted–Lowry acids and bases may be decided from their positions in a table of relative strengths of acids and bases.

GOAL 9 **Given the formulas of a potential Brønsted–Lowry acid and a Brønsted–Lowry base, write the equation for the possible proton-transfer reaction between them.**

**Proton–transfer reactions** involve two conjugate acid–base pairs.

GOAL 10 **Given a table of the relative strengths of acids and bases and information from which a proton-transfer reaction equation between two species in the table may be written, write the equation and predict the direction in which the reaction will be favored.**

Strong acids release protons readily; weak acids do not. Strong bases remove protons readily; weak bases do not. When a proton-transfer reaction reaches equilibrium, the proton transfer yields the weaker conjugate acid and the weaker conjugate base.

GOAL 11 **(If Chapter 19 has been studied) Compare and contrast acid–base reactions with redox reactions.**

Acid–base reactions resemble redox reactions in a number of ways, including the transfer of a subatomic particle, specialized names, relative strengths of species, and equilibrium.

**GOAL 12 Given the hydrogen or hydroxide ion concentration of water or a water solution, calculate the other value.**

Water itself is both a weak acid and a weak base. $K_w$ = **[H⁺][OH⁻] = 1.0 × 10⁻¹⁴** at 25°C. A neutral water solution has pH = 7.00 at 25°C. An acidic water solution has pH less than 7.00 at 25°C; a basic water solution has pH greater than 7.00 when measured at 25°C.

**GOAL 13 Given any one of the following, calculate the remaining three: hydrogen or hydroxide ion concentration expressed as 10 raised to an integral power or its decimal equivalent, pH, and pOH expressed as an integer.**

Because concentrations of $H^+(aq)$ or $OH^-(aq)$ are usually small, but can vary over wide ranges, they are usually expressed in exponential notation and as logarithms. **pH = −log[H⁺]** and **pOH = −log[OH⁻]**. From these equations, you can obtain **[H⁺] = 10⁻ᵖᴴ** and **[OH⁻] = 10⁻ᵖᴼᴴ**. In water solutions at 25°C, pH + pOH = 14.00.

**GOAL 14 (Optional) Given any one of the following, calculate the remaining three: hydrogen ion concentration, hydroxide ion concentration, pH, and pOH.**

The new skills needed to complete this optional section are not chemical, but mathematical. The ideas concerning pH, pOH, [H⁺], and [OH⁻] are the same as in Section 17.9; only the numbers have been changed.

# Chapter 17 Test Yourself

*Instructions:* You may refer to Table 17.1.

1) Which of these properties is not related to the $H^+(aq)$ ion?
   a) turns litmus indicator red b) feels slippery on the skin. c) tastes sour d) reacts with and neutralizes a base
2) According to the Brønsted–Lowry acid–base theory, an acid is a(n)
   a) electron source. b) electron remover. c) proton source. d) proton remover.
3) According to the Lewis acid–base theory, a base in a(n)
   a) electron source. b) electron remover. c) proton source. d) proton remover.
4) Identify the Lewis acid among the following:
   a) $NH_3$ b) $Cl^-$ c) $O_2$ d) $K^+$
5) For the reaction $HX^- + HY^- \rightleftharpoons H_2X + Y^{2-}$, pick the conjugate acid–base pair among the following:
   a) $HX^-$, $HY^-$ b) $HX$, $Y^{2-}$ c) $HY^-$, $H_2X$ d) $Y^{2-}$, $HY^-$ e) $HX^-$, $Y^{2-}$
6) The stronger of two Brønsted–Lowry bases tends to
   a) lose protons more easily. b) lose protons less easily.
   c) remove protons more readily. d) remove protons less readily.
7) Using Table 17.1, identify the strongest acid among the following:
   a) HF b) $SO_3^{2-}$ c) $NH_4^+$ d) $H_2PO_4^-$
8) At 25°C, the numeric value of $K_w$ is
   a) $1.0 \times 10^{-7}$. b) $1.0 \times 10^{-14}$. c) $1.0 \times 10^{14}$. d) 14.00.
9) A solution with a pH of 7.6 would be classified as a) strongly acidic. b) about neutral. c) strongly basic.
10) In a solution in which $[OH^-] = 10^{-5}$, pH and pOH are, respectively,
    a) $10^{-14}$ and $10^{-9}$. b) 5 and 9. c) 5 and 14. d) 9 and 5. e) 14 and 5.
11) Which of the following solutions is most basic?
    a) $[OH^-] = 10^{-5}$ b) pH = 7 c) pOH = 9 d) $[H^+] = 10^{-2}$

*Questions 12 and 13:* For each pair of reactants below, complete the equation for a single proton transfer reaction; state if the reaction would be favored in the forward or reverse direction. You may use Table 17.1.

12) $HCN(aq) + SO_3^{2-}(aq) \rightleftharpoons$ 13) $HSO_3^-(aq) + HPO_4^{2-}(aq) \rightleftharpoons$

14) A solution has a pH of 2.87. Calculate its pOH, [H⁺], and [OH⁻].

**Answers to Chapter 17 Test Yourself**

1) b 2) c 3) a 4) d 5) d 6) c 7) a 8) b 9) b 10) d 11) a

12) $HCN(aq) + SO_3^{2-}(aq) \rightleftharpoons CN^-(aq) + HSO_3^-(aq)$ Reverse

13) $HSO_3^-(aq) + HPO_4^{2-}(aq) \rightleftharpoons SO_3^{2-}(aq) + H_2PO_4^-(aq)$ Reverse;
$HSO_3^-(aq) + HPO_4^{2-}(aq) \rightleftharpoons H_2SO_3(aq) + PO_4^{3-}(aq)$ Reverse

14) pH = 2.87; pOH = 14.00 − pH = 14.00 − 2.87 = 11.13;
$[H^+] = 10^{-pH} = 10^{-2.87} = 1.3 \times 10^{-3}$; $[OH^-] = 10^{-pOH} = 10^{-11.13} = 7.4 \times 10^{-12}$

# Chapter in Review 18A

GOAL 1 **Identify a chemical equilibrium by the conditions it satisfies.**

Four conditions characterize every **equilibrium**:

1) The change is reversible and can be represented by an equation with a double arrow.
2) The equilibrium system is "closed"—closed in the sense that no substance can enter or leave the immediate vicinity of the equilibrium.
3) The equilibrium is **dynamic.**
4) The things that are equal in an equilibrium are the forward rate of change (from left to right in the equation) and the reverse rate of change (from right to left).

GOAL 2 **Distinguish between reaction-producing molecular collisions and molecular collisions that do not yield reactions.**

All chemical reactions start with **molecular collisions,** but not all molecular collisions give a chemical reaction. The particles must have (1) enough kinetic energy and (2) the proper orientation.

GOAL 3 **Sketch and/or interpret an energy-reaction coordinate graph. Identify (a) the activated complex region, (b) activation energy, and (c) $\Delta E$ for the reaction.**

An **energy-reaction coordinate graph** shows potential energies of reactants, activated complex, and products in a reaction, and the activation energy as the reaction proceeds in either direction, as well as $\Delta E$ for the reaction.

GOAL 4 **State and explain the relationship between reaction rate and temperature.**

Reaction rates are higher at higher temperatures because a larger fraction of the sample has enough kinetic energy to participate in reaction-producing collisions. The energy of collision must be enough to overcome the mutual repulsion of the valence electrons of the reacting particles.

GOAL 5 **Using an energy-reaction coordinate graph, explain how a catalyst affects reaction rate.**

A **catalyst** increases reaction rate by providing an alternative reaction path with a lower activation energy.

GOAL 6 **Identify and explain the relationship between reactant concentration and reaction rate.**

Collision rates are higher at higher concentrations, so reaction rates are higher at higher reactant concentrations.

GOAL 7 **Trace and explain the changes in concentrations of reactants and products that lead to chemical equilibrium.**

For any reversible reaction in a closed system, whenever the opposing reactions are occurring at different rates, the faster reaction will gradually become slower, and the slower reaction will become faster. Finally, the reaction rates become equal, and equilibrium is established.

GOAL 8 **Given the equation for a chemical equilibrium, predict the direction in which the equilibrium will shift because of a change in the concentration of one species.**

According to **Le Chatelier's Principle,** if you do anything to alter the rates of reaction in an equilibrium, the equilibrium responds in a way that counteracts partially the initial change until a new equilibrium is reached. When the concentration of a species in an equilibrium reaction is changed, a shift will occur in the direction that tries to return the substance disturbed to its original condition.

GOAL 9 **Given the equation for a chemical equilibrium involving one or more gases, predict the direction in which the equilibrium will shift because of a change in the volume of the system.**

If a gaseous equilibrium is compressed, the increased pressure will be partially relieved by a shift in the direction of fewer gaseous molecules; if the system is expanded, the reduced pressure will be partially restored by a shift in the direction of more gaseous molecules.

GOAL 10 **Given a thermochemical equation for a chemical equilibrium, or information from which it can be written, predict the direction in which the equilibrium will shift because of a change in temperature.**

A **thermochemical equation** is one that includes a change in energy. Including the energy term in a thermochemical equation, rather than showing $\Delta H$ separately, makes it easier to predict the Le Chatelier effect of a change in temperature. In the equation, think of energy as you would a substance being added or removed. An increase in temperature is interpreted as the "addition of heat," and a lowering of temperature is the "removal of heat."

GOAL 11 **Given any chemical equilibrium equation, or information from which it can be written, write the equilibrium constant expression.**

An equilibrium system can be described by an **equilibrium constant, K.** The constant K is a concentration ratio; the form of the ratio depends on how the equilibrium equation is written. For the general reaction a A + b B $\rightleftharpoons$ c C + d D, $K = \frac{[C]^c [D]^d}{[A]^a [B]^b}$. When writing an equilibrium constant expression, use only the concentrations of gases, (g), or dissolved substances, (aq). Do not include solids, (s), or liquids, ($\ell$).

GOAL 12 **Given an equilibrium equation and the value of the equilibrium constant, identify the direction in which the equilibrium is favored.**

If an equilibrium constant is very large (>100), the forward reaction is favored; if the constant is very small (<100), the reverse reaction is favored. If the constant is neither large nor small, appreciable quantities of all species are present at equilibrium.

# Chapter 18 Test Yourself

*Instructions:* You may refer to a "clean" periodic table.

1) Which is the incorrect statement about equilibrium systems?
   a) Forward and reverse reaction rates are the same at equilibrium.
   b) The amounts of products and reactants are the same when the system reaches equilibrium.
   c) No substance can enter or leave an equilibrium system.
   d) Both chemical and physical equilibria exist.
2) Which is the correct statement about collision theory?
   a) Most collisions are effective collisions.
   b) Particles with low kinetic energy are most likely to have effective collisions.
   c) Particles with high kinetic energy always have effective collisions.
   d) Most collisions do not lead to a chemical reaction.
3) Consider the energy-reaction graph shown on the right. Which statement concerning the reaction depicted by this graph is correct?
   a) The products are of lower energy than the reactants.
   b) The $\Delta E$ for the forward reaction is less than zero.
   c) The products are of higher energy than the reactants.
   d) The activation energy is the energy at point c minus the energy at point b.

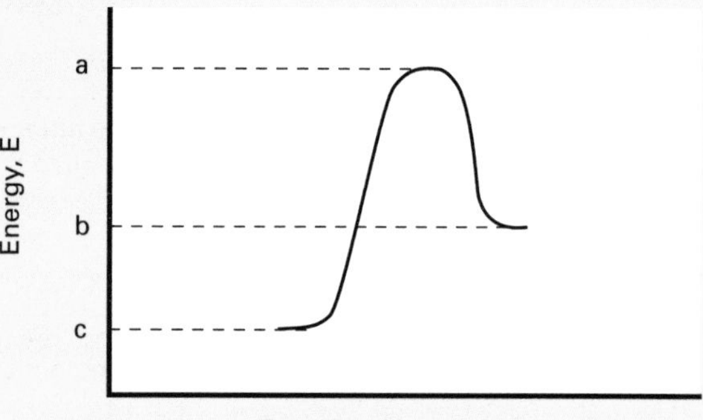


4) All of the following increase reaction rate except
   a) lowering the temperature. b) adding a catalyst. c) raising concentrations. d) lowering the activation energy.

*Questions 5–7 refer to* $2\ CO(g) + 2\ NO(g) \rightleftharpoons 2\ CO_2(g) + N_2(g) + 747\ kJ$

5) Adding some NO(g) causes the equilibrium to
   a) shift to the left. b) shift to the right. c) remain unchanged.
6) Increasing the container volume causes the equilibrium to
   a) shift to the left. b) shift to the right. c) remain unchanged.
7) Lowering the temperature causes the equilibrium to
   a) shift to the left. b) shift to the right. c) remain unchanged.
8) The reaction $A(g) + B(g) \rightleftharpoons AB(g)$ has an equilibrium constant of $2.1 \times 10^7$. This reaction is favored
   a) very strongly in the reverse direction. b) slightly in the reverse direction.
   c) very strongly in the forward direction. d) slightly in the forward direction.

*Questions 9–11: Write the equilibrium constant expressions for the following reactions.*

9) $2\ CO(g) + O_2(g) \rightleftharpoons 2\ CO_2(g)$ 10) $PbBr_2(s) \rightleftharpoons Pb^{2+}(aq) + 2\ Br^-(aq)$ 11) $NH_3(aq) + H_2O(\ell) \rightleftharpoons NH_4^+(aq) + OH^-(aq)$

**Answers to Chapter 18 Test Yourself**

1) b 2) d 3) c 4) a 5) b 6) a 7) b 8) c

9) $K = \frac{[CO_2]^2}{[CO]^2[O_2]}$ 10) $K = [Pb^{2+}][Br^-]^2$ 11) $K = \frac{[NH_4^+][OH^-]}{[NH_3]}$

# Chapter in Review 18B

This Portable Content card covers optional Section 18.9 on Equilibrium Calculations.

**GOAL 13 Given the solubility product constant or the solubility of a slightly solute compound (or data from which the solubility can be found), calculate the other value.**

No ionic compound is completely insoluble. The equilibrium constant for a low-solubility compound is the **solubility product constant, $K_{sp}$**. The solubility product constant for any ionic compound has the form

$$A_xB_y(s) \rightleftharpoons x\,A^{y+}(aq) + y\,B^{x-}(aq) \qquad K_{sp} = [A^{y+}]^x\,[B^{x-}]^y$$

The first two steps in solving any solubility product constant problem are the same:

1) Write the equilibrium equation for the reaction.
2) Write the solubility product constant expression.

To find $K_{sp}$ from solubility:

3) Determine the molar concentrations of the ions in solution.
4) Substitute into the $K_{sp}$ expression and solve.

To find solubility from $K_{sp}$:

3) Assign a variable to represent one of the ionic species in the equilibrium.
4) Determine the concentration of the other ionic species in terms of the same variable.
5) Substitute the concentrations from Steps 3 and 4 into the $K_{sp}$ expression, equate to the $K_{sp}$ value, and solve.

**GOAL 14 Given the solubility product constant of a slightly soluble compound and the concentration of a solution having a common ion, calculate the solubility of the slightly soluble compound in the solution.**

The solubility of a low solubility substance is reduced when a common ion—one already present in the solution—is introduced from another source. This is called the **common ion effect.**

**GOAL 15 Given the formula of a weak acid, HA, write the equilibrium equation for its ionization and the expression for its acid constant, $K_a$.**

If HA is the formula of a weak acid, its ionization equation and equilibrium constant expression are

$$HA(aq) \rightleftharpoons H^+(aq) + A^-(aq) \qquad K_a = \frac{[H^+][A^-]}{[HA]}$$

The equilibrium constant is the **acid constant, $K_a$**. The undissociated molecule is the major species in the solution, and the $H^+$ ion and the conjugate base of the acid, $A^-$, are the minor species. Weak acids ionize only slightly when dissolved in water. The ionization of a weak acid is usually so small that it is negligible compared with the initial concentration of the acid.

**GOAL 16 Given any two of the following three values for a weak acid, HA, calculate the third: (a) the initial concentration of the acid; (b) the pH of the solution, the percentage dissociation of the acid, or $[H^+]$ or $[A^-]$ at equilibrium; (c) $K_a$ for the acid.**

**Percentage ionization,** as all other percentage concepts, is the ratio of the part to the whole, multiplied by 100:

$$\%\ \text{ionization} = \frac{\text{amount of solute ionized}}{\text{total solute present}} \times 100$$

To determine $K_a$ from pH or percent ionization of an acid with a given concentration, determine the values of $[H^+]$ and $[A^-]$ with $[H^+] = 10^{-pH}$, substitute the values into the $K_a$ expression, and solve. To determine percent ionization and pH from known molarity and $K_a$, if the ionization of the acid is the only source of $H^+$ and $A^-$ ions, start with the acid constant equation, multiply both sides by [HA], and substitute $[H^+]$ for its equal $[A^-]$:

$$K_a\,[HA] = [H^+][A^-] = [H^+]^2 \qquad [H^+] = \sqrt{K_a\,[HA]}$$

**GOAL 17 For a weak acid, HA, given $K_a$, $[H^+]$, and $[A^-]$, or information from which they may be obtained, calculate the pH of the buffer produced.**

A **buffer** is a solution that resists changes in pH because it contains relatively high concentrations of both a weak acid and a weak base. The acid is able to consume any $OH^-$ that may be added, and the base can react with $H^+$, both without significant change in either [HA] or $[A^-]$. To find the pH of a buffer solution, solve the acid constant expression for $[H^+]$:

$$[H^+] = K_a \times \frac{[HA]}{[A^-]}$$

**GOAL 18 Given $K_a$ for a weak acid, HA, determine the ratio between [HA] and $[A^-]$ that will produce a buffer of specified pH.**

A buffer can be tailor-made for any pH simply by adjusting the $[HA]/[A^-]$ ratio to the proper value. Solving the acid constant expression for that ratio gives

$$\frac{[HA]}{[A^-]} = \frac{[H^+]}{K_a}$$

**GOAL 19 Given equilibrium concentrations of species in a gas-phase equilibrium, or information from which they can be found, and the equation for the equilibrium, calculate the equilibrium constant.**

When an equilibrium involves only gases, the changes in the starting concentrations are not negligible. It often helps to trace these changes by assembling them into a table. The columns are headed by the species in the equilibrium just as they appear in the reaction equation. The three lines give the initial concentration of each substance, the change in concentration as the system reaches equilibrium, and the equilibrium concentration:

| | a A | + | b B | $\rightleftharpoons$ | c C | + | d D |
|---|---|---|---|---|---|---|---|
| mol/L at start | | | | | | | |
| mol/L change, + or − | | | | | | | |
| mol/L at equilibrium | | | | | | | |

# Chapter 18 Test Yourself: Optional Section 18.9

*Instructions:* You may use a "clean" periodic table.

1) Calculate the solubility of silver sulfate, in moles per liter. $K_{sp} = 1.2 \times 10^{-5}$.
2) What is the pH of a 1.2 M $HCHO_2$ solution ($K_a = 2.0 \times 10^{-4}$)?
3) A solution contains 6.10 g $HCHO_2$ ($K_a = 2.0 \times 10^{-4}$) and 43.1 g $NaCHO_2$. At what pH is this solution buffered?
4) At 325°C, the system $N_2(g) + 3\ H_2(g) \rightleftharpoons 2\ NH_3(g)$ reaches equilibrium with $[N_2] = 0.057$ M, $[H_2] = 0.17$ M, $[NH_3] = 0.042$ M. Find K for the system at this temperature.

**Answers to Chapter 18 Test Yourself: Optional Section 18.9**

1) $Ag_2SO_4(s) \rightleftharpoons 2\ Ag^+(aq) + SO_4^{2-}(aq) \qquad K_{sp} = [Ag^+]^2\,[SO_4^{2-}] = 1.2 \times 10^{-5}$

Let s = solubility of $Ag_2SO_4(s)$ in moles per liter; therefore,

$[Ag^+] = 2s$ and $[SO_4^{2-}] = s \qquad K_{sp} = [2s]^2\,[s] = 1.2 \times 10^{-5} = 4s^3 \qquad s = 0.014$ M

2) *GIVEN:* $K_a = 2.0 \times 10^{-4}$; 1.2 M $HCHO_2$ *WANTED:* pH

*EQUATION:* $[H^+] = \sqrt{K_a\,[HCHO_2]} = \sqrt{(2.0 \times 10^{-4})(1.2)} = 0.015$ M

*EQUATION:* $pH = -\log\,[H^+] = -\log\,(0.015) = 1.82$

3) $6.10\text{ g } HCHO_2 \times \frac{1\text{ mol } HCHO_2}{46.03\text{ g } HCHO_2} = 0.133\text{ mol } HCHO_2$

$43.1\text{ g } NaCHO_2 \times \frac{1\text{ mol } NaCHO_2}{68.01\text{ g } NaCHO_2} = 0.634\text{ mol } NaCHO_2 = 0.634\text{ mol } CHO_2^-$

Let V = solution volume in L; therefore $[HCHO_2] = 0.133/V$ and $[CHO_2^-] = 0.634/V$

*EQUATION:* $[H^+] = K_a \times \frac{[HCHO_2]}{[CHO_2^-]} = 2.0 \times 10^{-4} \times \frac{0.133/V}{0.634/V} = 4.2 \times 10^{-5}$

*EQUATION:* $pH = -\log\,[H^+] = -\log\,(4.2 \times 10^{-5}) = 4.38$

4) $K = \frac{[NH_3]^2}{[N_2][H_2]^3} = \frac{(0.042)^2}{(0.057)(0.17)^3} = 6.3$

# Chapter in Review 19A

GOAL 1 **Describe and explain oxidation and reduction in terms of electron transfer.**

**Oxidation** is a loss of electrons or an increase in oxidation number. **Reduction** is a gain of electrons or a reduction in oxidation number.

GOAL 2 **Given an oxidation half-reaction equation and a reduction half-reaction equation, combine them to form a net ionic equation for an oxidation-reduction reaction.**

Oxidation-reduction (redox) reactions can be divided into an **oxidation half-reaction** equation and a **reduction half-reaction** equation. Addition of these equations gives a balanced equation for a redox reaction.

GOAL 3 **Distinguish among electrolytic cells, voltaic cells, and galvanic cells.**

A **voltaic cell** is a cell in which an electrical potential is developed by a spontaneous chemical change. It is also called a **galvanic cell.** An **electrolytic cell** is a cell in which electrolysis occurs as a result of an externally applied electrical potential.

GOAL 4 **Describe and identify the parts of an electrolytic or voltaic (galvanic) cell and explain how it operates.**

An electrolytic cell is made up of a container holding an ionic solution called an **electrolyte** and two **electrodes.** Ions in the electrolyte move to oppositely charged electrodes, where chemical reactions occur. In a voltaic cell, chemical changes occur at the electrodes, causing electricity to flow in an outside circuit. The solutions in each half-cell are connected by a **salt bridge.**

GOAL 5 **Given the formula of an element, molecule, or ion, assign an oxidation number to each element in the formula.**

**Oxidation numbers** are assigned as follows: an element is 0; a monatomic ion is the charge on the ion; combined oxygen is −2 (peroxides, −1; superoxides, −1/2; $OF_2$, +2); combined hydrogen is −1 ($H^-$, −1); the sum of oxidation numbers of all atoms in a polyatomic species is equal to its charge.

GOAL 6 **Describe and explain oxidation and reduction in terms of change in oxidation numbers.**

Oxidation is an increase in oxidation number. Reduction is a decrease in oxidation number.

GOAL 7 **Given a redox equation, identify the oxidizing agent, the reducing agent, the element oxidized, and the element reduced.**

The element that loses electrons is **oxidized;** the species from which electrons are removed is the **reducing agent.** The element that gains electrons is **reduced;** the species that removes electrons is an **oxidizing agent.**

GOAL 8 **Distinguish between strong and weak oxidizing agents.**

A **strong oxidizing agent** has a strong attraction for electrons. A **weak oxidizing agent** attracts electrons only slightly. A **strong reducing agent** releases electrons readily. A **weak reducing agent** holds on to its electrons.

GOAL 9 **Given a table of the relative strengths of oxidizing and reducing agents, arrange a group of oxidizing agents or a group of reducing agents in order of increasing or decreasing strength.**

Table 19.2 in the textbook list oxidizing agents in order of decreasing strength on the left-hand side of the half-reaction equation and lists reducing agents in order of increasing strength on the right-hand side.

GOAL 10 **Given a table of the relative strengths of oxidizing and reducing agents, and information from which an electron-transfer reaction equation between two species in the table may be written, write the equation and predict the direction in which the reaction will be favored.**

When a reversible reaction equation is read from left to right, the **forward reaction,** or the reaction in the **forward direction,** is described; from right to left, the change is the **reverse reaction,** or in the **reverse direction.** A strong oxidizing agent takes electrons from a strong reducing agent to produce weaker reducing and oxidizing agents, respectively. The reaction is said to be **favored** in the direction pointing to the weaker oxidizing and reducing agents. Use Table 19.2 to identify the stronger and weaker oxidizing and reducing agents.

| | |
|---|---|
| GOAL 11 **Compare and contrast redox reactions with acid–base reactions.** | An acid–base reaction is a transfer of protons; a redox reaction is a transfer of electrons. An acid is a proton source; a reducing agent is an electron source. A base is a proton remover; an oxidizing agent is an electron remover. Certain species can behave as an acid in one reaction and a base in another; certain species can act as an oxidizing agent in one reaction and a reducing agent in another. Acids and bases are classified as strong or weak based on their ability to release and attract protons; oxidizing agents and reducing agents are classified as strong or weak based on their ability to release and attract electrons. The favored direction of an acid–base equilibrium can be predicted from acid–base strength; the favored direction of a redox equilibrium can be predicted from oxidizing agent–reducing agent strength. |
| GOAL 12 **(Optional) Given the before and after formulas of species containing elements that are oxidized and reduced in an acidic solution, write the oxidation and reducing half-reaction equations and the net ionic equation for the reaction.** | To **write a redox equation for a half-reaction in acidic solution,** (1) write a skeleton equation that includes the element oxidized or reduced, (2) balance the element oxidized or reduced, (3) balance elements other than H or O, if any, (4) balance oxygen by adding water molecules, (5) balance hydrogen by adding $H^+$, (6) balance the charge by adding electrons to the more positive side, and (7) recheck the equation to be sure it is balanced in both atoms and charge. |

# Chapter in Review 19B

## hapter 19 Test Yourself

*uestions 1 and 2: Use the following half-reactions:* $Ca(s) \rightarrow Ca^{2+}(aq) + 2\,e^-$; $Al^{3+}(aq) + 3\,e^- \rightarrow Al(s)$

) Identify the reduction half-reaction equation and the oxidation half-reaction equation. Explain your reasoning.

) Rewrite the half-reaction equations in such form that they may be added to produce a balanced redox equation. Write the balanced redox equation.

*uestions 3–5: Use the following redox equation:* $2\,CrO_4^{2-}(aq) + 16\,H^+(aq) + 3\,Cu(s) \rightarrow$ $Cr^{3+}(aq) + 3\,Cu^{2+}(aq) + 8\,H_2O(\ell)$.

) The element that loses electrons is ______________. Its oxidation number changes from _____ to _____.

) The element reduced is ______________. Its oxidation number changes from _____ to _____.

) The oxidizing agent is ______________. The reducing agent is ______________.

) If element Q is very likely to form $Q^-$ ions in a redox reaction, element Q is a ____________________ agent.
a) strong reducing b) strong oxidizing c) weak reducing d) weak oxidizing

*uestions 7-8: Use this hypothetical table of relative strengths oxiding agents and reducing agents. In this table, the stron-st oxidizing agent is W.*

$W \rightarrow W^-$
$X^+ \rightarrow X$
$Y^+ \rightarrow Y$
$Z \rightarrow Z^-$

) Identify the stronger oxidizing agent between $Y^+$ and $X^+$; identify the stronger reducing agent between $W^-$ and Y.

) Determine whether or not each of the following redox reactions will proceed in the forward direction.
a) $Z + X \rightleftharpoons Z^- + X^+$ b) $X^+ + Y \rightleftharpoons X + Y^+$ c) $Z^- + W \rightleftharpoons Z + W^-$
d) $Y + Z \rightleftharpoons Y^+ + Z^-$ e) $X^+ + W^- \rightleftharpoons X + W$

*uestion 9 is based on optional Section 19.8.*

) Consider the redox reaction $ClO_3^-(aq) + I^-(aq) \rightarrow I_2(aq) + Cl^-(aq)$
a) Write the oxidation half-reaction equation.
b) Write the reduction half-reaction equation.
c) Add the half-reaction equations to produce a balanced redox equation.

**Answers to Chapter 19 Test Yourself**

1) The reduction half-reaction equation is $Al^{3+}(aq) + 3\,e^- \rightarrow Al(s)$. The reactant gains electrons. The oxidation half-reaction equation is $Ca(s) \rightarrow Ca^{2+}(aq) + 2\,e^-$. The reactant loses electrons.

2)
$$\begin{array}{ll} 3\,Ca(s) & \rightarrow 3\,Ca^{2+}(aq) + 6\,e^- \\ 2\,Al^{3+}(aq) + 6\,e^- & \rightarrow 2\,Al(s) \\ \hline 3\,Ca(s) + 2\,Al^{3+}(aq) & \rightarrow 3\,Ca^{2+}(aq) + 2\,Al(s) \end{array}$$

3) The element that loses electrons is copper. Its oxidation number changes from 0 to +2.

4) The element reduced is chromium. Its oxidation number changes from +6 to +3.

5) The oxidizing agent is $CrO_4^{2-}$. The reducing agent is Cu.

6) b 7) Stronger oxidizer: $X^+$; Stronger reducer: Y 8) b and c will proceed in the forward direction.

9)
$$\begin{array}{lll} \text{a)} & 3\,[2\,I^-(aq) & \rightarrow I_2(aq) + 2\,e^-] \\ \text{b)} & 6\,e^- + ClO_3^-(aq) + 6\,H^+(aq) & \rightarrow Cl^-(aq) + 3\,H_2O(\ell) \\ \hline \text{c)} & 6\,I^-(aq) + ClO_3^-(aq) + 6\,H^+(aq) & \rightarrow 3\,I_2(aq) + Cl^-(aq) + 3\,H_2O(\ell) \end{array}$$

# Chapter in Review 20

| | |
|---|---|
| GOAL 1 **Define radioactivity.** | **Radioactivity** is the spontaneous emission of particles or electromagnetic radiation resulting from the decay, or breaking up, of an atomic nucleus. |
| GOAL 2 **Name, identify from a description, or describe three types of radioactive emissions.** | Three types of natural radioactivity, **alpha ($\alpha$)**, **beta ($\beta$)**, and **gamma ($\gamma$) rays**, can be formed when a nucleus decays. These emissions differ in their masses, charges, and ability to penetrate matter. |
| GOAL 3 **Identify the function of a Geiger counter and describe how it operates.** | The hand piece of a **Geiger counter** has a thin window through which ionizing radiation passes, entering a tube filled with argon gas. The gas is momentarily ionized, and the ions complete a circuit between two electrodes. The signal is amplified to produce clicking from a speaker and register on a meter. The frequency of the clicks indicates the radiation intensity. |
| GOAL 4 **Explain how exposure to radiation may harm or help living systems.** | The harmful effects of radiation on living systems come from its ability to break chemical bonds and thereby destroy healthy tissue. The key to radiation therapy is selective radiation of unhealthy tissue, which can eliminate it or reduce it. |
| GOAL 5 **Describe or illustrate what is meant by the half-life of a radioactive substance.** | The rate at which a radioactive substance decays is measured by its **half-life,** the time it takes for one half the radioactive atoms in a sample to decay. |
| GOAL 6 **Given the starting quantity of a radioactive substance, Figure 20.9, and two of the following, calculate the third: half-life, elapsed time, quantity of isotope remaining.** | If S is the starting quantity of a radioactive substance and R is the amount that remains after n half-lives, then $\mathbf{R = S \times (1/2)^n}$. Figure 20.9 is a half-life decay curve for a radioactive substance, a plot of fraction of substance remaining versus half-lives. |
| GOAL 7 **Describe a natural radioactive decay series.** | Products of radioactive decay may be other nuclei that undergo further decay, forming a **natural radioactive decay series.** |
| GOAL 8 **Given the identity of a radioactive isotope and the particle it emits, write a nuclear equation for the emission.** | An equation for radioactive decay is balanced for nuclear charge (number of protons) and nuclear mass (number of protons and neutrons). |
| GOAL 9 **List or identify four ways in which nuclear reactions differ from ordinary chemical reactions.** | (1) Chemical properties are the same for all isotopes of an element. Nuclear properties of the isotopes of an element are quite different. (2) Radioactivity is independent of the state of chemical combination of the radioactive isotope. (3) Nuclear reactions can result in the formation of different elements. In chemical reactions, atoms keep their identities. (4) The amount of energy per quantity of reactant for nuclear changes is much larger than for chemical changes. |
| GOAL 10 **Define or identify nuclear bombardment reactions.** | A **nuclear bombardment** reaction occurs when one nuclide is bombarded with another. |
| GOAL 11 **Distinguish natural radioactivity from induced radioactivity produced by bombardment reactions.** | **Induced or artificial radioactivity** is generated when a stable nuclide is made radioactive by combination with another nuclide. |
| GOAL 12 **Define or identify transuranium elements.** | All elements having atomic numbers greater than 92 are called the **transuranium elements.** |
| GOAL 13 **Identify and describe uses for synthetic radionuclides.** | **Synthetic radionuclides** have many uses, including ionizing air in a smoke detector and killing bacteria, molds, and yeasts to preserve food, among other industrial and scientific applications. |
| GOAL 14 **Define or identify a nuclear fission reaction.** | A nucleus that splits into lighter nuclei undergoes nuclear **fission.** |
| GOAL 15 **Define or identify a chain reaction.** | A **chain reaction** occurs when a product of one reaction is a reactant in the next step of the reaction pathway. |

| | |
|---|---|
| GOAL 16 **Describe how a nuclear power plant differs from a fossil-fueled power plant.** | The turbine, generator, and condenser in a nuclear power plant are similar to those found in any fuel-burning power plant. The nuclear fission reaction has three main components: the fuel elements, control rods, and moderator. |
| GOAL 17 **Define or identify a nuclear fusion reaction.** | Two light nuclei that are joined to form a heavier nucleus undergo nuclear **fusion.** |

## Chapter 20 Test Yourself

*Instructions:* You may use a "clean" periodic table.

1) Which natural radioactive emission is neither attracted nor repelled by either a positive or a negative charge?
   a) alpha ray b) beta ray c) gamma ray d) both a and b
2) Ionizing radiation has an effect on body tissue because
   a) the tissue gets a positive charge b) the tissue gets a negative charge.
   c) physical changes occur. d) chemical changes occur.
3) Geiger counter tubes operate when
   a) the gas in the tube is ionized. b) the gas in the tube fluoresces.
   c) supersaturated vapor condenses on the ions in the tube. d) a clicking sound is heard.
4) What fraction of a radioisotope is left after 3 half-lives have passed?
   a) 1/27 b) 1/9 c) 1/8 d) 1/6 e) 1/3
5) In 24 hours, the mass of a radioisotope changed from 3.7 mg to 0.20 mg. Determine the half-life of this isotope. You may use Figure 20.9.
6) Balance the nuclear equation ${}^{221}_{87}Fr \rightarrow$ ________________ $+ {}^{4}_{2}He$
7) The three natural radioactive decay sequences all end with a ______ isotope as the final product.
   a) ${}_{92}U$ b) ${}_{90}Th$ c) ${}_{82}Pb$ d) ${}_{91}Pa$
8) Which of the following pairs have the same chemical properties?
   a) ${}^{239}_{94}Pu^{4+}$ and ${}^{238}_{94}Pu$ b) ${}^{239}_{94}Pu$ and ${}^{238}_{94}Pu^{4+}$ c) ${}^{239}_{94}Pu^{4+}$ and ${}^{238}_{94}Pu^{4+}$ d) all have the same chemical properties
9) Induced radioactivity comes from
   a) stable products of bombardment reactions. b) background radiation.
   c) radioactive products of bombardment reactions. d) cosmic radiation.
10) Which is not a transuranium element?
    a) ${}^{237}_{93}Np$ b) ${}^{247}_{97}Bk$ c) ${}^{244}_{94}Pu$ d) ${}^{237}_{91}Pa$

*The following are the answer choices for Questions 11–14:*
a) ${}^{235}_{92}U + {}^{1}_{0}n \rightarrow {}^{144}_{54}Xe + {}^{90}_{38}Sr + 2\,{}^{1}_{0}n$ b) ${}^{96}_{42}Mo + {}^{2}_{1}H \rightarrow {}^{97}_{43}Tc + {}^{1}_{0}n$ c) ${}^{3}_{2}He + {}^{3}_{2}He \rightarrow {}^{4}_{2}He + 2\,{}^{1}_{1}H$ d) ${}^{20}_{8}O \rightarrow {}^{20}_{9}F + {}^{0}_{-1}e$

11) Which reaction above is a nuclear bombardment reaction?
12) Which reaction above is a nuclear fission reaction?
13) Which reaction above could be used in a nuclear chain reaction?
14) Which reaction above is a nuclear fusion reaction?

**Answers to Chapter 20 Test Yourself**

1) c 2) d 3) a 4) c 5) R/S = 0.054; From Figure 20.9, fraction remaining = 0.054 corresponds to 4.2 half-lives; $\frac{24\text{ hr}}{4.2\text{ half-lives}} = 5.7\text{ hr/half-life}$
6) ${}^{217}_{85}At$ 7) c 8) c 9) c 10) d 11) b 12) a 13) a 14) c

# Chapter in Review 21A

GOAL 1 **Distinguish between organic and inorganic compounds.**

**Organic compounds** are those that contain carbon atoms. **Inorganic compounds** contain elements other than carbon. Although carbonates, cyanides, and oxides of carbon contain carbon, they are traditionally classified as inorganic.

GOAL 2 **Given Lewis diagrams, ball-and-stick models, or space-filling models of two or more organic molecules with the same molecular formula, distinguish between isomers and different orientations of the same molecule.**

Compounds with the same molecular formula but different molecular structures are called **isomers.**

GOAL 3 **Distinguish between saturated and unsaturated hydrocarbons (or other compounds).**

**Hydrocarbons** are made of carbon and hydrogen. A saturated hydrocarbon has only single bonds. An unsaturated hydrocarbon has one or more double or triple bonds between carbon atoms.

GOAL 4 **Distinguish between alkanes, alkenes, and alkynes.**

The **alkanes** are a hydrocarbon family where each carbon atom forms single bonds to four other atoms and there are no multiple bonds. The **alkenes** are hydrocarbons with at least one double bond; the **alkynes** are hydrocarbons with at least one triple bond.

GOAL 5 **Given a formula of a hydrocarbon or information from which it can be written, determine whether the compound can be a normal alkane or a cycloalkane.**

A hydrocarbon with all carbon atoms in the molecule in a continuous chain is a **normal alkane**. The normal and branched alkanes have the general formula $C_nH_{2n+2}$, where n is the number of carbon atoms in the compound. **Cycloalkanes** have all carbon–carbon single bonds, with at least some of the carbon atoms forming a ring.

GOAL 6 **Given the name (or structural diagram) of a normal, branched, or halogen-substituted alkane, write the structural diagram (or name).**

An alkane is named by using a prefix to indicate the number of carbon atoms in the longest chain: meth- = 1, eth- = 2, prop- = 3, but- = 4, pent- = 5, hex- = 6, hept- = 7, oct- = 8, non- = 9, dec- = 10. Numbers tell to which carbon atom(s) different functional groups are attached. The suffix *-ane* denotes an alkane. Removing an H atom from an alkane gives an **alkyl** functional group. A general symbol for an alkyl *r*esidue or *r*adical is R.

GOAL 7 **Given the name (or structural diagram) of a cycloalkane, write the structural diagram (or name).**

Cycloalkanes are named according to the number of carbon atoms in the ring with the prefix *cyclo-*.

GOAL 8 **Given a formula of a hydrocarbon or information from which it can be written, determine whether the compound can be an alkene or an alkyne.**

The **alkenes** are hydrocarbons with at least one double bond; the **alkynes** are hydrocarbons with at least one triple bond.

GOAL 9 **Given the name (or structural diagram) of an alkyne or an alkene, write the structural diagram (or name).**

The suffix *-ene* denotes an alkene. The suffix *-yne* denotes an alkyne. In naming unsaturated compounds, the position of the double or triple bond is specified by number.

GOAL 10 **Identify and distinguish between *cis* and *trans* geometric isomers.**

Double bonds can give **geometric isomers**, also called *cis-trans* isomers. Two alkyl groups can be on the same side (*cis*) or on opposite sides (*trans*) of the double bond.

GOAL 11 **Distinguish between aliphatic and aromatic hydrocarbons.**

Any hydrocarbon that does not contain a benzene ring is an **aliphatic** hydrocarbon. A hydrocarbon with one or more benzene rings, which may be substituted, is an **aromatic** hydrocarbon.

GOAL 12 **Given the name (or structural diagram) of an alkyl- or halogen-substituted benzene compound, write the structural diagram (or name).**

The carbon on a benzene ring to which a functional group is bonded is identified by number. *Ortho-*, *meta-*, and *para-* prefixes are also used for disubstituted rings.

GOAL 13 **Given the reactants in an addition or substitution reaction between (a) an alkane, alkene, alkyne, or benzene and (b) a hydrogen or halogen molecule, predict the products of the reaction.**

An alkene or alkyne is capable of reacting via an **addition reaction,** where atoms of an element (or compound) are added to the unsaturated hydrocarbon. A reaction in which a hydrogen atom in an alkane is replaced by an atom of another element is a **substitution reaction**.

GOAL 14 **Identify the structural formulas of the functional groups that distinguish alcohols and ethers.**

The general formula of an **alcohol** is R—OH. The general formula for an ether is R—O—R′.

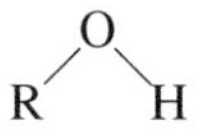

R–O–H (alcohol) R–O–R′ (ether)

alcohol 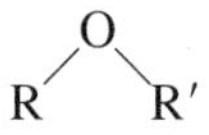 ether

GOAL 15 **Given the molecular structures of alcohols or ethers, or information from which they may be obtained, predict relative values of boiling points or solubility in water.**

In general, the boiling points of alcohols increase as the number of carbon atoms increases. The boiling point of an alcohol is always much higher than that of the alkane with the same number of carbon atoms. The 1-to-3-carbon alcohols are completely soluble in water, and the solubility drops off as the alkyl chain lengthens. The boiling points of ethers are lower than the corresponding alcohols. The solubility of ethers in water is about the same as the solubility of the isomeric alcohols.

GOAL 16 **Given the name (or structural diagram) of an alcohol or ether, write the structural diagram (or name).**

The suffix *-ol* denotes an alcohol. The word ether, preceded by the names of the attached alkyl groups is the name of the ether.

GOAL 17 **Given the reactants (or products) of a dehydration reaction between two alcohols, predict the products (or reactants) of the reaction.**

Ethers can be prepared by dehydrating alcohols. The —H from the hydroxyl group of one molecule reacts with the —OH hydroxyl group of another molecule, producing a R—O—R′ ether and a water molecule.

GOAL 18 **Identify the structural formulas of the functional groups that distinguish aldehydes and ketones.**

**Aldehydes** and **ketones** have a carbon atom double bonded to an oxygen atom. In an aldehyde, the carbon is at the end of the chain; in a ketone, the carbon is inside the chain.

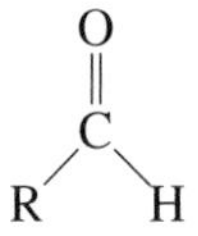

R–C(=O)–H R–C(=O)–R′

aldehyde 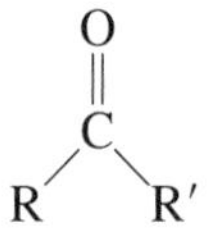 ketone

GOAL 19 **Given the name (or structural diagram) of an aldehyde or ketone, write the structural diagram (or name).**

The IUPAC system uses the suffix *-al* for aldehydes and the suffix *-one* for ketones. Ketones may also be named like ethers, but using the word ketone.

GOAL 20 **Write structural diagrams to show how a specified aldehyde or ketone is prepared from an alcohol.**

Aldehydes and ketones are prepared by oxidation of alcohols or hydrations of alkynes. Aldehydes are themselves easily oxidized; ketones resist oxidation. Aldehydes and ketones may be reduced to alcohols.

GOAL 21 **Identify the structural formulas of the functional groups that distinguish carboxylic acids and esters.**

The general formula of a **carboxylic acid** is RCOOH. The functional group, —COOH, is a combination of a carbonyl group and a hydroxyl group called a carbonyl group. In an **ester**, the carboxyl hydrogen is replaced by another alkyl group, RCOOR′.

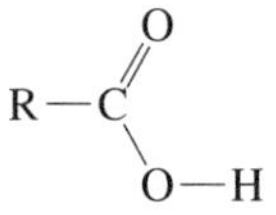

R—C(=O)—O—H R—C(=O)—O—R′

carboxylic acid 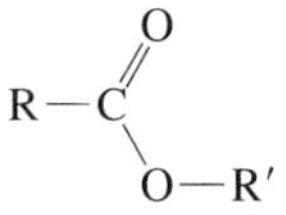 ester

GOAL 22 **Given the name (or structural diagram) of a carboxylic acid or ester, write the structural diagram (or name).**

IUPAC uses the suffix *-oic* and the word *acid* to denote carboxylic acids. Esters have two word names. The first word is the alkyl group from the alcohol and the second is the anion derived from the acid. (Remember *-ic* → *-ate* in anions of acids.)

GOAL 23 **Given the reactants (or products) of an esterification reaction, predict the products (or reactants) of the reaction.**

The reaction between an acid and an alcohol is called **esterification**. The products of the reaction are an ester and water. The acid contributes the entire hydroxyl group, while the alcohol furnishes only the hydrogen.

GOAL 24 **Identify the structural formulas of the functional groups that distinguish amines and amides.**

**Amines** are organic derivatives of ammonia, $NH_3$. An amine is formed by replacing one, two, or three hydrogens in an ammonia molecule with an alkyl group. An **amide** is a derivative of a carboxylic acid in which the hydroxyl part of the carboxyl group is replaced by an $NH_2$ group.

# Chapter in Review 21B

**OAL 25 Given the name (or structural agram) of an amine or amide, write e structural diagram (or name).**

The IUPAC system names amines like ethers, but using the word *amine.* Name an amide by replacing the *-oic acid* suffix with the word *amide.*

**OAL 26 Given the structural diagram an amine, or information from which can be written, classify the amine as imary, secondary, or tertiary.**

The number of ammonia hydrogens replaced determines whether an amine is primary (one hydrogen replaced), secondary (two), or tertiary (three).

**OAL 27 Given the reactants of a reacn between a carboxylic acid and amine, predict the products of the action.**

A carboxylic acid and an amine react by removing a water molecule:

$$\begin{array}{c} O \\ \| \\ R-C-OH \end{array} + \begin{array}{c} H-N-R' \\ | \\ H \end{array} \longrightarrow \begin{array}{c} O \\ \| \\ R-C-N-R' \\ \quad\;\; | \\ \quad\;\; H \end{array} + H_2O$$

**OAL 28 Given the structural diagram name of an ethylene-like monomer, edict the structural diagram of the oduct chain-growth polymer; given e structural diagram of an chainowth polymer, predict the strucral diagram and/or the name of the hylene-like monomer from which it n be formed.**

Small molecules called **monomers** join together to form **polymers. Chain-growth polymers** are formed by repeated addition reaction of an alkene monomer to give an alkane-like polymer chain. For example, polyethylene is formed from ethylene monomers:

$$\begin{array}{c} H\;\;H \\ |\;\;\;| \\ C{=}C \\ |\;\;\;| \\ H\;\;H \end{array} + \begin{array}{c} H\;\;H \\ |\;\;\;| \\ C{=}C \\ |\;\;\;| \\ H\;\;H \end{array} + \begin{array}{c} H\;\;H \\ |\;\;\;| \\ C{=}C \\ |\;\;\;| \\ H\;\;H \end{array} \longrightarrow \begin{array}{c} \;\;H\;\;\;H\;\;\;H\;\;\;H\;\;\;H\;\;\;H \\ \;\;|\;\;\;\;|\;\;\;\;|\;\;\;\;|\;\;\;\;|\;\;\;\;| \\ -C-C-C-C-C-C- \\ \;\;|\;\;\;\;|\;\;\;\;|\;\;\;\;|\;\;\;\;|\;\;\;\;| \\ \;\;H\;\;\;H\;\;\;H\;\;\;H\;\;\;H\;\;\;H \end{array}$$

**OAL 29 Given the structural diagrams a dicarboxylic acid and a dialcohol, edict the structural diagram of the oduct step-growth polymer; given e structural diagram of a stepowth polymer, predict the structural agrams of the dicarboxylic acid and alcohol from which it can be formed.**

One type of **step-growth polymer** is formed by repeated condensation reactions to give a polymer chain with repeated ester or amide functional groups. For example, nylon 66 is formed from a 6-carbon dicarboxylic acid and a 6-carbon diamine:

$$\begin{array}{c} O \qquad\qquad\quad\;\; O \\ \| \qquad\qquad\quad\;\; \| \\ HO-C-(CH_2)_4-C-OH \end{array} + \begin{array}{c} H \qquad\qquad\qquad H \\ | \qquad\qquad\qquad | \\ H-N-(CH_2)_6-N-H \end{array} \longrightarrow$$

$$\begin{array}{c} O \qquad\qquad\quad\; O\;\;H \qquad\qquad\quad H \\ \| \qquad\qquad\quad\; \|\;\;\;| \qquad\qquad\quad | \\ -C-(CH_2)_4-C-N-(CH_2)_6-N- \end{array} + H_2O$$

## Chapter 21 Test Yourself

*Instructions:* You may refer to a "clean" periodic table.

1) The general formula $C_nH_{2n}$ is the formula for the hydrocarbons called
   a) aliphatics. b) alkanes. c) alkynes. d) alkenes.

2) The radical $C_4H_9$— depicts a _____ group. a) hexyl b) pentyl c) butyl d) propyl

3) The molecule below has the IUPAC name

$$\begin{array}{c} \quad CH_3\;H \\ \quad\;\; |\quad\;\; | \\ H-C-C-CH_3 \\ \quad\;\; |\quad\;\; | \\ \quad CH_3\;CH_3 \end{array}$$

   a) 1,1,2-trimethylpropane. b) 2-methylbutane. c) 1,2-dimethylbutane. d) 2,3-dimethylbutane.

4) Draw the carbon skeleton of 1,1-diethyl-3-methylcyclohexane.

5) Name the compound below:

$$\begin{array}{ccc} H_3CH_2C & & H \\ & \diagdown\;\diagup & \\ & C{=}C & \\ & \diagup\;\diagdown & \\ H & & CH_2CH_3 \end{array}$$

   a) 1-ethyl-1-propene b) *trans*-3-hexene c) *cis*-2-hexene d) *cis*-3-hexene

6) The catalyzed reduction of an alkyne with one mole of hydrogen gas gives a(n) _____ hydrocarbon.
   a) alkane   b) alkene   c) cycloalkane   d) aromatic

7) Alkenes usually give _____ reactions; aromatic hydrocarbons like benzene usually give _____ reactions.
   a) addition; addition   b) substitution; addition   c) addition; substitution   d) substitution; substitution

8) The aromatic hydrocarbon commonly called mesitylene is actually 1,3,5-trimethylbenzene. Draw the Lewis diagram for mesitylene.

9) Draw and give IUPAC names for all the straight chain (no branches) alcohols having the formula $C_6H_{14}O$.

10) Primary alcohols can be oxidized to _____; secondary alcohols can be oxidized to _____.
   a) aldehydes; carboxylic acids   b) ketones; aldehydes
   c) ketones; carboxylic acids   d) aldehydes; ketones

11) What is the name of the compound below?

$$H_3CH_2CH_2C-\overset{\displaystyle O}{\overset{\|}{C}}-CH_2CH_3$$

   a) 3-hexanone   b) butyl ethyl ketone   c) 3-hexanol   d) 4-hexanone

12) The following compound is called pineapple oil in the flavor and fragrance industry. What is its "official" name?

$$H_3CH_2CH_2C-\overset{\displaystyle O}{\overset{\|}{C}}-O-CH_2CH_3$$

   a) butyl ethanoate   b) butyl acetate   c) butyl ethanone   d) ethyl butanoate

13) The condensation reaction between a carboxylic acid and an amine gives a(n)
   a) ester.   b) anhydride.   c) amide.   d) carboxylic acid amine.

14) The addition polymer KEL-F is formed from chlorotrifluoroethene. Draw three repeating units of Kel-F.

15) The addition polymer shown below is commonly called povidone and is used in eye drops as a lubricant. Give the monomer from which povidone can be made.

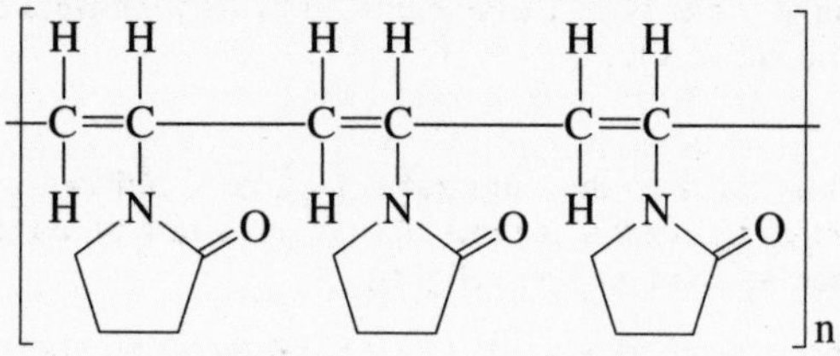

H H   H H   H H
C=C   C=C   C=C
H N   H N   H N
O   O   O
n

**Answers to Chapter 21 Test Yourself**

1) d   2) c   3) d   4) $CH_3CH_2$ $CH_2CH_3$ $CH_3$   5) b   6) b   7) c   8) $CH_3$ $H_3C$ $CH_3$

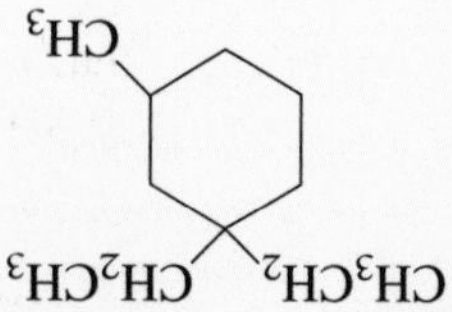

9) OH OH OH

$H_2C-CH_2CH_2CH_2CH_2CH_2$   $H_3C-C(H)-CH_2CH_2CH_2CH_3$   $H_3CH_2C-C(H)-CH_2CH_2CH_3$

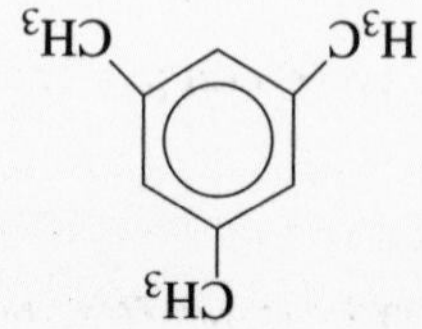

1-hexanol   2-hexanol   3-hexanol

10) d   11) a   12) d   13) c

14) Cl F Cl F Cl F
C C C C C C
F F F F F F
n

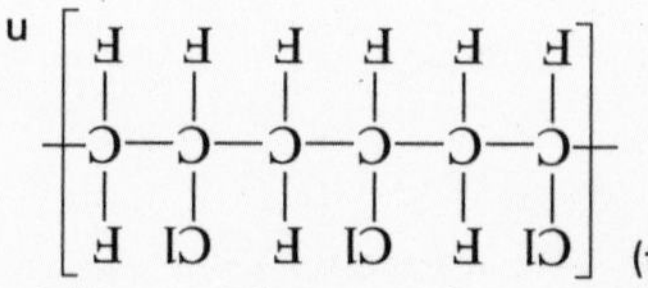

15) H H
C=C
H N
O

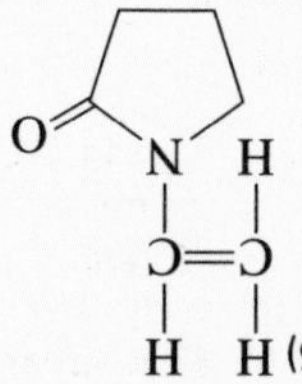

# Chapter in Review 22A

**GOAL 1 Given a Lewis diagram of a polypeptide or information from which it may be written, identify the C-terminal acid and the N-terminal acid.**

The amino acid with the free carboxyl group at the end of a polypeptide chain is called the **C-terminal acid;** the amino acid with the free amine group is called the **N-terminal acid**.

**GOAL 2 Given a table of Lewis diagrams of amino acids and their corresponding three-letter and one-letter abbreviations, draw the Lewis diagram of the polypeptide from its abbreviation.**

The bond between amino acids is called a **peptide linkage,** which is formed when the hydroxyl part of the carboxyl group of an amino acid molecule reacts with a hydrogen of the $—NH_2$ group of another amino acid molecule to form a molecule of water.

**GOAL 3 Explain the meaning of the terms *primary, secondary, tertiary,* and *quaternary structure* as they apply to proteins.**

The **primary structure** of a protein is its sequence of amino acids. The **secondary structure** of a protein is the local spatial layout of the amino acid backbones. The **tertiary structure** of a protein describes the spatial arrangement of the entire polypeptide chain. The **quaternary structure** of a protein describes how its polypeptide chains are arranged to form the protein molecule.

**GOAL 4 Describe how hydrogen bonding results in (a) $\alpha$-helix and (b) $\beta$-pleated-sheet secondary protein structures.**

Secondary structures allow for maximum hydrogen bonding and the greatest stability. The **$\alpha$-helix** structure is a continuous series of loops. The **$\beta$-pleated-sheet** structure is like a curtain, but with sharp angles, like pleats.

**GOAL 5 Define the following terms as they apply to enzymes: *substrate, active site, inhibitor.***

An enzyme's **substrate** binds at the enzyme's **active site.** Enzyme **inhibitors** compete with the substrate for the active site.

**GOAL 6 Use the induced fit model to explain enzyme activity.**

In the **induced fit model,** the shape of the substrate is a close, but not exact, match to the shape of the active site of the enzyme. As the substrate binds to the enzyme, either or both molecules change shape slightly.

**GOAL 7 Given a Lewis diagram of a monosaccharide in its open-chain form, determine whether the sugar is an aldose or a ketose.**

If the carbonyl group of a monosaccharide is at the end of the carbon chain, the sugar is an **aldose;** if the carbonyl group is within the chain, the sugar is a **ketose.**

**GOAL 8 Distinguish among monosaccharides, disaccharides, and polysaccharides.**

**Monosaccharides** are simple sugars that cannot be converted to smaller carbohydrates. **Disaccharides** are formed from two simple sugars. **Polysaccharides** are formed from many sugars.

**GOAL 9 Given the Lewis diagrams of two monosaccharides and a description of the bond linking the molecules, draw the Lewis diagram of the resulting disaccharide.**

Two monosaccharides are combined by a reaction between two —OH groups, resulting in a —O— bond between the monomers and a water molecule.

**GOAL 10 State (a) the defining characteristics and (b) the three major subclassifications of lipids.**

**Lipids** are found in living organisms and are insoluble in water but soluble in nonpolar solvents. The three major classes are: (1) fats, oils, and phospholipids, (2) waxes, and (3) lipids (usually) without ester groups such as steroids.

**GOAL 11 Identify the physical property that distinguishes fats from oils.**

**Fats** are triacylglycerols that are solids at room temperature; oils are triacylglycerols that are liquids at room temperature.

**GOAL 12 Identify the structural feature common to all steroid molecules.**

**Steroids** have three cyclohexane rings and one cyclopentane ring fused together.

**GOAL 13 Describe the biological roles of DNA and RNA.**

**DNA** stores and transmits genetic information. **RNA** serves as a messenger and a switching engine.

**GOAL 14 Describe the components of a nucleotide.**

Each **nucleotide** is composed of a cyclic nitrogen compound called a base, a sugar, and a phosphate group.

| | |
|---|---|
| GOAL 15 **Draw Lewis diagrams of adenine, cytosine, guanine, thymine, and uracil.** | There are five nucleic acid nitrogen bases: thymine (T), cytosine (C), adenine (A), guanine (G), and uracil (U). |
| GOAL 16 **Determine whether any two DNA or RNA nitrogen bases are complementary.** | In DNA, A and T form a complementary pair; C and G form the other complementary pair. |
| GOAL 17 **Describe the process by which a protein molecule is formed.** | In **translation,** messenger RNA is decoded and individual amino acids are brought by different transfer RNA molecules to be assembled into proteins. |

# Chapter in Review 22B

## Chapter 22 Test Yourself

*Questions 1 and 2: You may refer to Table 23.1. Consider the dipeptide Gly-Ala.*

1) What is the C-terminal acid?

2) Draw the structures of the two individual amino acids, then show how they react to form a peptide bond.

3) Some proteins are composed of more than one polypeptide chain. The term _____ structure refers to how these chains are arranged in relation to one another.
a) primary  b) secondary  c) tertiary  d) quaternary  e) ternary

4) Hydrogen bonding in an $\alpha$-helix protein secondary structure is best described as:
a) at a minimum, and occurring between amino acids in the same chain
b) at a maximum, and occurring between amino acids in the same chain
c) at a minimum, and occurring between amino acids in adjacent chains
d) at a maximum, and occurring between amino acids in adjacent chains

5) Enzymes are: a) carbohydrates  b) fats  c) lipids  d) proteins  e) DNA

6) The mechanism by which an enzyme catalyzes the reaction of a substrate is analogous to:
a) a lock and key  b) peas in a pod  c) a door and latch
d) bread and wheat  e) a bicycle wheel and spokes

7) When looking at a Lewis diagram of a monosaccharide in open-chain form, a ketose is distinguished by having:
a) the carbonyl at the end of the chain
b) the carbonyl within the chain
c) all the hydroxyl groups on one side of the carbon chain
d) hydroxyl groups on both sides of the carbon chain
e) five carbons rather than six

8) Lactose is a galactose–glucose combination. Which of the following statements is true?
a) Lactose is a monosaccharide; galactose and glucose are disaccharides
b) Lactose and galactose are monosaccharides; glucose is a disaccharide
c) Lactose, galactose, and glucose are all monosaccharides
d) Lactose is a disaccharide; galactose and glucose are monosaccharides
e) Lactose is a disaccharide; galactose and glucose are polysaccharides

9) Consider the bonds from the oxygen atom connecting two sugar molecules. If one is drawn vertically, the connection is called (i); if both are drawn horizontally, the connection is called (ii).
a) (i) a; (ii) b  b) (i) b; (ii) a  c) (i) a; (ii) g  d) (i) g; (ii) a  e) (i) b; (ii) g

10) Which of the following is not classified as a lipid?
a) fats  b) oils  c) waxes  d) steroids  e) polysaccharides

11) Fats are triacylglycerols that are (i) at room temperature; oils are triacylglycerols that are (ii) at room temperature.
a) (i) esterified; (ii) acids  b) (i) acids; (ii) esterified  c) (i) saturated; (ii) unsaturated
d) (i) unsaturated; (ii) saturated  e) (i) solids; (ii) liquids

12) Which of the following best describes the class of lipids structurally characterized by three cyclohexane rings and one cyclopentane ring?
a) oils  b) steroids  c) waxes  d) phospholipids  e) fats

13) What type of nucleic acid is responsible for serving as a messenger and as a switching engine to transfer the correct amino acid during protein synthesis?
a) acetic acid  b) fatty acid  c) carboxylic acid  d) ribonucleic acid  e) deoxyribonucleic acid

14) Which of the following is not a component of a nucleotide?
a) nitrogen base  b) phosphate  c) protein  d) ribose or deoxyribose  e) all are nucleotide components

15) The Lewis diagram of what molecule is shown below?
a) adenine b) cytosine c) guanine d) thymine e) uracil

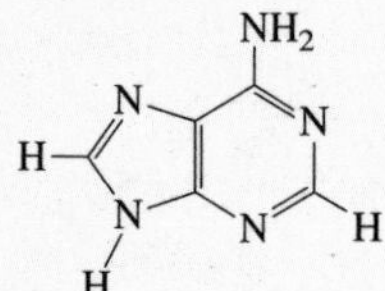

16) Which of the following DNA bases form a complementary pair?
a) A—A b) A—C c) A—G d) A—T e) none of the given pairs is complementary

17) What term best describes the final step in protein synthesis?
a) duplication b) transcription c) translation d) replication e) interpretation

---

**Answers to Chapter 22 Test Yourself**

1) alanine

2)

$$H_2N{-}\underset{H}{\overset{R}{C}}{-}\overset{O}{\overset{\|}{C}}{-}OH + H{-}N{-}\underset{H}{\overset{R'}{C}}{-}\overset{O}{\overset{\|}{C}}{-}OH \longrightarrow H_2N{-}\underset{H}{\overset{R}{C}}{-}\overset{O}{\overset{\|}{C}}{-}\overset{H}{N}{-}\overset{R'}{C}{-}\overset{O}{\overset{\|}{C}}{-}OH + H_2O$$

3) d 4) b 5) d 6) a 7) b 8) d 9) a 10) e 11) e 12) b 13) d 14) c 15) a 16) d 17) c